Recognition Receptors in Biosensors

For full color versions of select images,
please visit http://www.springer.com/978-1-4419-0918-3

Mohammed Zourob
Editor

Recognition Receptors in Biosensors

Springer

Editor
Dr. Mohammed Zourob
GDG ENVIRONNEMENT LTÉE
430, rue Saint-Laurent, 2e étage
Trois-Rivières (Quebec),
Canada G8T 6H3
mohammed.zourob@gdg.ca

Professor at:
Institut national de la recherche scientifique
Center- Énergie, Matériaux et Télécommunications
1650, boul. Lionel-Boulet
Varennes (Québec)
Canada J3X 1S2
zourob@emt.inrs.ca

ISBN: 978-1-4419-0918-3 e-ISBN: 978-1-4419-0919-0
DOI 10.1007/978-1-4419-0919-0
Springer New York Dordrecht Heidelberg London

Library of Congress Control Number: 2009930469

Printed on acid-free paper

Springer is part of Springer Science+Business Media (www.springer.com)

Preface

Recognition receptors play a key role in the successful implementation of chemical and biosensors. Molecular recognition refers to non-covalent specific binding between molecules, one of which is typically a macromolecule or a molecular assembly, and the other is the target molecule (ligand or analyte). Biomolecular recognition is typically driven by many weak interactions such as hydrogen bonding, metal coordination, hydrophobic forces, van der Waals forces, pi-pi interactions and electrostatic interaction (due to permanent charges, dipoles, and quadrupoles) the polarization of charge distributions by the interaction partner leading to induction and dispersion forces, and Pauli-exclusion-principle-derived inter-atomic repulsion, and a strong, "attractive" force arising largely from the entropy of the solvent and termed the hydrophobic effect. In recent years, there has been much progress in understanding the forces that drive the formation of such complexes, and how these forces are relate to the physical properties of the interacting molecules and their environment allows rational design of molecules and materials that interact in specific and desired ways.

This book presents a significant and up-to-date review of the various recognition elements, their immobilization, characterization techniques by a panel of distinguished scientists. This work is a comprehensive approach to the recognition receptors area presenting a thorough knowledge of the subject and an effective integration of these receptors on sensor surfaces in order to appropriately convey the state-of the-art fundamentals and applications of the most innovative approaches.

This book is comprized of 21 chapters written by 32 researchers who are actively working in USA, Canada, France, Switzerland, Ireland, Germany, Spain, Italy and the United Kingdom. The authors were requested to adopt a pedagogical tone in order to accommodate the needs of novice researchers such as graduate students and post-doctoral scholars as well as of established researchers seeking new avenues. This has resulted in duplication of some material, which we have chosen to retain, because we know that many readers will pick only a specific chapter to read at a certain time.

We have divided this book into four sections. The first part comprises three chapters, which are devoted to introduce biomolecular recognition, surface sensitization and recognition receptors immobilization, and an overview of the analytical tools, which can be used for surface characterization. The second part (chapters 4–11) describe the different natural recognition receptors (enzymes, antibodies, cells, tissues, plants tissue, peptides, Carbohydrates, Nucelic acid and bacteriophages) used in biosensors. It covers the theory behind each technique and delivers a detailed state-of-the-art review for all the new technologies used. The third part (chapters 12–20) covers the recognition receptors which can be prepared and engineered by human to mimic natural molecules and used up to date in biosensors and microarrays arena. It describes in detail the use of engineered antibodies, genetically engineered proteins, genetically engineered cells, Photosynthetic proteins, oligonucleotides, aptamers, phage display, molecularly imprinted polymers and biomemetic as recognition elements. The fourth part contains one Chapter which covers briefly the kinetics and thermodynamics of association/dissociation of analytes to the recognition receptors.

This book is intended to be a primary source both on fundamental and practical information of where the recognition receptors area is now and where it is headed in the future. We anticipate that the book will be helpful to academics, practitioners and professionals working in various fields; to name a few biologist, biotechnologists, biochemists, analytical chemists, biomedical, physical, microsystems engineering, nanotechnology, veterinary science, medicine, food QA, bioterrorism and security. As well as allied health, healthcare and surveillance. Since the fundamentals were also reviewed, we believe that the book will appeal to advanced undergraduate and graduate students who study in areas related to chemical and biosensors.

We gratefully acknowledge all authors for their participation and contributions, which made this book a reality. We give many thanks to Olivier Laczka and Joseph Piliero for the book cover design. Last, but not least, I thank my family for their patience and enthusiastic support of this project.

May 2009 Mohammed Zourob

Foreword

Recognition Receptors in Biosensors takes us back to first principles to make sure we really understand the science behind molecular recognition. The first part, in particular, begins at the atomic level and develops a framework for selection and integration of biomolecules into sensors. The chapters in the next two parts expound on the various candidates for natural and synthetic receptors. This book is not light reading: it delves deeply into the details of recognition molecules, with extended descriptions of structure, function, and requirements for operation in sensor devices. For the serious student entering the field of biosensor science and technology, it will provide a sound foundation on which to build. The book is didactic and comprehensive. The chapters are selected and organized in such a way as to provide a valuable reference and a transparent learning tool.

Researchers expanding into receptor-based recognition science would benefit from a careful exploration of relevant chapters early in their endeavors. I certainly would have benefitted from the chapter by Chevolot et al. when I was developing carbohydrate-based arrays for pathogen detection. The chapter provides an excellent summary of basic carbohydrate chemistry, carbohydrate oligomer structures responsible for recognition, methods for immobilization, and experience with oligosaccharides in different types of biosensors. Even the impact of the linkage chemistry and oligosaccharide density on the recognition function are explored.

While there are also similar chapters exploring the use of "popular receptors" such as antibodies, oligonucleotides, phages, and cells, I was particularly pleased to see a chapter on the use of plants for environmental monitoring. The biochemistry of plants is not as widely understood as that of bacteria, viruses, and animals, and this lack of understanding has retarded the integration of plants into the armament of widely used recognition elements. Yet plants can be self-replicating, self-repairing, self-sustaining, and capable of automated signal amplification. The chapter by Basu and Kovalchuk describes biochemical pathways that can be harnessed for sensor functions, methods for introducing new recognition capabilities or signal generation into plants, and applications for sensing environmental toxicants.

Considering the potential use of plants for low-cost, energy-free, wide-spectrum monitoring, this is a chapter we should all read.

The section of the book focused on synthetic receptors is an impressive compendium of how combinatorial methods can expand the recognition capabilities of all types of molecules including chemical as well as biological constructs. The list of references in each chapter is an additional resource, and the quality of the figures increases clarity and reading pleasure. The advantages of each approach are well presented by the authors, although in some cases the reader needs to figure out the limitations for himself; having the chapters side by side makes it easier to see which approaches are most appropriate for which applications.

Washington, DC Frances S. Ligler

Contents

Contributors

Victor Acha Sam and Ann Barshop Institute for Longevity and Aging Studies, University of Texas Health Science Center, San Antonio, TX 78245, USA

Maher Albitar Department of Hematology/Oncology, Quest Diagnostics-Nichols Institute, 33608 Ortega Highway, San Juan Capistrano, CA, USA, maher.x.Albitar@questdiagnostics.com

Thomas Andrews Sam and Ann Barshop Institute for Longevity and Aging Studies, University of Texas Health Science Center, San Antonio, TX 78245, USA

E. Baldrich Institut de Microelectrònica de Barcelona – Centro Nacional de Microelectrónica (IMB-CNM), CSIC, Campus Universidad Autónoma de Barcelona, 08193 – Bellaterra, Barcelona, Spain, Eva.baldrich@imb-cnm.csic.es

Saikat Kumar Basu Department of Biological Sciences, Hepler Hall, University of Lethbridge, 4401 University Drive, Lethbridge, AB, Canada T1K 3M4

Loïc J. Blum Institut de Chimie et Biochimie Moléculaires et Supramoléculaires, Laboratoire de Génie Enzymatique, Membranes Biomimétiques et Assemblages Biomoléculaires, université Lyon 1/CNRS, 43 boulevard du 11 novembre 1918, Villeurbanne, 69622, France, Loic.Blum@univ-lyon1

Yves Chevalier Laboratoire d'Automatique et de Génie des Procédés, Université de Lyon, UMR CNRS 5007, 43 Bd 11 Novembre 1918, 69622 Villeurbanne Cedex, France

Yann Chevolot Institut des Nanotechnologies de Lyon UMR CNRS 5270, Université de Lyon, Equipe Chimie et Nanotechnologie, Ecole Centrale de Lyon, 36 Avenue Guy de Collongue, 69134 Ecully Cedex, France, yann.chevolot@ec-lyon.fr

Louise O'Connor Molecular Diagnostics Research Group, National Centre for Biomedical Engineering Science, National University of Ireland, Galway, Ireland, louise.oconnor@nuigalway.ie

Jonathan D. Dattelbaum Department of Chemistry, University of Richmond, Richmond, VA 23173, USA, jdattelb@richmond.edu

Sylvia Daunert Department of Chemistry, University of Kentucky, Lexington, KY 40506-0055, USA, daunert@uky.edu

Amber C. Donahue Department of Hematology/Oncology, Quest Diagnostics-Nichols Institute, Ortega Highway, San Juan Capistrano, CA, USA

Vincent Dugas Laboratoire des Sciences Analytiques, Université de Lyon, UMR CNRS 5180, 43 Bd 11 Novembre 1918, 69622 Villeurbanne Cedex, France, vincent.dugas@univ_lyon1.fr

Abdelhamid Elaissari Laboratoire d'Automatique et de Génie des Procédés, Université de Lyon, UMR CNRS 5007, 43 Bd 11 Novembre 1918, 69622 Villeurbanne Cedex, France

Ifejesu Eni-Olorunda Chemical Engineering Department, University of Mississippi, University, MS 38677-1848, USA

Indraneel Ghosh Department of Chemistry, University of Arizona, Tucson, AZ 85721, USA, ghosh@u.arizona.edu

Maria Teresa Giardi Institute of Crystallography, National Council of Research, Department of Molecular Design and of Agrofood, Area of Research of Rome, Via Salaria Km 29.300, 00016 Monterotondo scalo, Rome, Italy, mariateresa.giardi@mlib.ic.cnr.it

Barry Glynn Molecular Diagnostics Research Group, National Centre for Biomedical Engineering Science, National University of Ireland, Galway, Ireland, barry.glynn@nuigalway.ie

Keith E. Herold Fischell Department of Bioengineering, University of Maryland, College Park, MD 20742, USA, herold@umd.edu

Neal A.E. Hopkins Detection Department, Defence Science and Technology Laboratory, Porton Down, Salisbury, UK SP4 0JQ, DSTL/PUB33855, NAHOPKINS@mail.dstl.gov.uk

Peter J. Hornsby Sam and Ann Barshop Institute for Longevity and Aging Studies, University of Texas Health Science Center, San Antonio, TX 78245, USA, hornsby@uthscsa.edu

Qin Huang Sam and Ann Barshop Institute for Longevity and Aging Studies, University of Texas Health Science Center, San Antonio, TX 78245, USA

Kalju Kahn Department of Chemistry and Biochemistry, University of California, Santa Barbara, CA 93106, USA, kalju@chem.ucsb.edu

Igor Kovalchuk Department of Biological Sciences, Hepler Hall, University of Lethbridge, 4401 University Drive, Lethbridge, AB, Canada T1K 3M4, igor.kovalchuk@uleth.ca

Anjali Kumari Struss Department of Chemistry, University of Kentucky, Lexington, KY 40506-0055, USA

Emmanuelle Laurenceau Institut des Nanotechnologies de Lyon UMR CNRS 5270, Université de Lyon, Equipe Chimie et Nanotechnologie, Ecole Centrale de Lyon, 36 Avenue Guy de Collongue, 69134 Ecully Cedex, France

Béatrice D. Leca-Bouvier Institut de Chimie et Biochimie Moléculaires et Supramoléculaires, Laboratoire de Génie Enzymatique, Membranes Biomimétiques et Assemblages Biomoléculaires, université Lyon 1/CNRS, 43 boulevard du 11 novembre 1918, Villeurbanne, 69622, France, leca@univ-lyon1

Soojin Lim Department of Chemistry, Portland State University, Portland, OR 97201, USA

Hans Jörg Mathieu Materials Science Institute, École Polytechnique Fédérale de Lausanne (EPFL), CH-1015 Lausanne, Switzerland, Hansjoerg.Mathieu@EPFL.ch

Scott C. Meyer Department of Chemistry, University of Arizona, Tucson, AZ 85721, USA

François Morvan Institut des Biomolécules Max Mousseron, UMR 5247 CNRS - Université Montpellier 1 - Université Montpellier 2, DACAN, CC 1704, Place E. Bataillon, 34095 Montpellier Cedex 5, France

George M. Murray Department of Materials Science and Engineering, Center for Laser Applications, University of Tennessee Space Institute, 411 B. H. Goethert Parkway MS 24, Tullahoma, TN 37388, USA, gmurray@utsi.edu

Patrizia Pasini Department of Chemistry, University of Kentucky, Lexington, KY 40506-0055, USA

Kevin W. Plaxco Department of Chemistry and Biochemistry, University of California, Santa Barbara, CA 93106, USA

Avraham Rasooly FDA Center for Devices and Radiological Health, Silver Spring, MD 20903, USA and the NIH-National Cancer Institute, Rockville, MD 20852, USA, rasoolya@mail.nih.gov

Steven Ripp The Center for Environmental Biotechnology, The University of Tennessee, 676 Dabney Hall, Knoxville, Tennessee 37996, USA, saripp@utk.edu

Ajit Sadana Chemical Engineering Department, University of Mississippi, University, MS 38677–1848, USA, cmsadana@oblemiss.edu

Dhiraj K. Sardar Department of Physics and Astronomy, University of Texas at San Antonio, San Antonio, TX 78249, USA

Hans-Jörg Schneider FR 8.1 Organische Chemie, Universität des Saarlandes, D 66041 Saarbrücken, Germany, Chemie, h-j.schneider@mx.uni-saarland.de

Eliane Souteyrand Institut des Nanotechnologies de Lyon UMR CNRS 5270, Université de Lyon, Equipe Chimie et Nanotechnologie, Ecole Centrale de Lyon, 36 Avenue Guy de Collongue, 69134 Ecully Cedex, France

Glen E. Southard Department of Chemistry, MIPSolutions, Inc., 421 Wakara Way #203, Salt Lake City, UT 84108, USA

Robert M. Strongin Department of Chemistry, Portland State University, Portland, OR 97201, USA

Ibtisam E. Tothill Cranfield University, Cranfield, Bedfordshire, MK43 0AL, UK, i.tothill@cranfield.ac.uk

Jean-Jacques Vasseur Institut des Biomolécules Max Mousseron, UMR 5247 CNRS – Université Montpellier 1 – Université Montpellier 2, DACAN, CC1704, Place E. Bataillon, 34095 Montpellier Cedex 5, France

Sébastien Vidal Institut de Chimie et Biochimie Moléculaires et Supramoléculaires (ICBMS), Laboratoire de Chimie Organique 2 – Glycochimie, UMR 5246 CNRS, Bâtiment 308 – CPE Lyon, 43 Boulevard du 11 Novembre 1918, F-69622 Villeurbanne, France

Part I
Sensor Surface Chemistry and Receptor Immobilization

Chapter 1
Principles of Biomolecular Recognition

Kalju Kahn and Kevin W. Plaxco

Abstract Biomolecular recognition, the process by which biomolecules recognize and bind to their molecular targets, typically highly specific, high affinity and reversible, and is generalizable to an effectively unlimited range of aqueous analytes. Consequently, it has been exploited in a wide range of diagnostic and synthetic technologies. Biomolecular recognition is typically driven by many weak interactions working in concert. The most important of these interactions include (i) the electrostatic interaction due to permanent charges, dipoles, and quadrupoles, (ii) the polarization of charge distributions by the interaction partner leading to induction and dispersion forces, (iii) Pauli-exclusion principle-derived inter-atomic repulsion, and (iv) a strong, "attractive" force arising largely from the entropy of the solvent and termed the hydrophobic effect. Because the aqueous environment significantly reduces the impact of electrostatic and induction interactions, the hydrophobic effect is often the dominant force stabilizing the formation of correct biomolecule–target complexes. The other effects are nevertheless important in defining the specificity of the macromolecule toward its target by destabilizing binding events in which a less-than-ideal network of interactions between two partners would be established.

Keywords Molecular recognition · Binding thermodynamics · Electrostatic interaction · London dispersion · Hydrophobic effect

K. Kahn (✉)
Department of Chemistry and Biochemistry, University of California, Santa Barbara, CA, 93106, USA
e-mail: kalju@chem.ucsb.edu

M. Zourob (ed.), *Recognition Receptors in Biosensors*,
DOI 10.1007/978-1-4419-0919-0_1, © Springer Science+Business Media, LLC 2010

Abbreviations

CCSD	Coupled cluster singles and doubles
DNA	Deoxyribonucleic acid
MP2	Møller–Plesset perturbation theory of the second order
NMR	Nuclear magnetic resonance
RNA	Ribonucleic acid
TIP4P	Transferable intermolecular potential–4 point

Symbols

ΔG	Free energy change
ΔU	Internal energy change
ΔS	Entropy change
ΔH	Enthalpy change
ΔV	Volume change
R	Universal gas constant
T	Absolute temperature
F	Force
q_i	Charge on particle i
ε_0	Dielectric permittivity of vacuum
r	Distance between the centers of two objects
μ	Dipole moment
k	Boltzmann constant
Θ	Quadrupole moment
α	Polarizability
I_i	First ionization potential of atom or molecule i
σ	Size parameter in the Lennard-Jones potential
ε	Softness parameter in the Lennard Jones potential; dielectric constant
h	Planck constant
ν	Vibrational frequency; effective volume of the molecule
σ	Rotational symmetry number
λ_B	Bjerrum length
N_A	Avogadro constant
n_i	Refractive index of solute i.

1.1 What Is Molecular Recognition?

Molecules in solution collide billions of times a second. In most cases the "complexes" formed by these collisions are weak, short-lived, and nonspecific. But when the surface features of one molecule are complementary to those of its partner – that is, when the attractive forces generated by the interactions of the features outweigh the repulsive forces and outweigh the entropic costs of bringing them together – stronger, long-lived, and specific interactions can be established. These specific complexes have proven to have significant biological and technological significance. Molecular recognition refers to the process of such specific binding between molecules, one of which is typically a macromolecule or molecular assembly, and the other is the target molecule to which it binds, be it a small molecule or another macromolecule (ligand, analyte). Understanding the forces that drive the formation of such complexes and how those forces are related to the physical properties of the interacting molecules and their environment allows rational design of molecules and materials that interact in specific, desired ways.

The importance of molecular recognition in different branches of science is reflected in a multitude of terms that describe the process and the players. For example, the molecular recognition of cholesterol by cyclodextrin is described as the formation of "host-guest complex" by an organic chemist while the molecular recognition of cholesterol by the enzyme cholesterol oxidase would be used as an example of "substrate binding" by a biochemist. In the traditional terminology, with roots in the study of biosignaling, the macromolecular component of an interaction is called a *receptor*, and a small molecule that upon binding to its receptor elicits a normal biological response is called a *ligand*. Physiologists often make a distinction between a natural ligand and agonists, which are molecules that bind to the receptor and elicit a response similar to the natural ligand. Furthermore, physiologists would call a molecule that binds to the receptor without eliciting the biological response, but blocks the binding of a natural ligand, an *antagonist*. Most biochemists call a small molecule that binds to a macromolecule a ligand for this macromolecule without regard for its physiological consequence. They would use the term "receptor" narrowly to designate proteins that send signals in response to ligand binding. Instead, functionally descriptive terms such as "storage protein," "transporter," "permease," "immunoglobulin," "promoter region" and many others are used to describe macromolecules or the parts of macromolecules that bind other molecules. In the language of an immunologist, membrane-bound receptors are called receptors, but soluble receptors become antibodies, and anything that binds to antibodies are called antigens. In analytical chemistry, the interaction partner that is designed to bind the analyte is often called a molecular sensor or biosensor; molecules that also interact with the biosensor may give false positive readings or act by preventing the analyte binding, leading to false negatives. Because of the interdisciplinary nature of research in molecular recognition, it is important to be familiar with terminology used in other fields. For example, if one were designing a biosensor for the detection of a sugar, the considerable body of work that has been performed to elucidate the mechanisms of carbohydrate-modifying enzymes may be of interest.

Characteristic features	Commonly used terminology for "small" interaction partner		
	Traditional physiology	Classical enzymology	Design of molecular sensors
Molecule that the macromolecular partner was designed for	Ligand	Substrate	Analyte, target
Molecule that binds and causes a response similar to natural ligand	Agonists	Alternative substrate	False positive
Molecule that binds but prevents binding of the natural ligand	Antagonists	Inhibitor	False negative
Molecule that binds and enhances binding of the natural ligand	Potentiator	Activator	Co-target

Molecular recognition typically arises due to the contributions of many weak, reversible interactions. Fortunately, while our understanding of these interactions is limited enough to render the ab initio (from basic principles) design of molecular recognition systems a cutting-edge challenge (Jiang et al. 2008), the physics underlying them are straight-forward enough to achieve at least semi-quantitative description of the key players. Specifically, only half a dozen key forces dominate molecular recognition in biological systems. We describe them here in detail.

Before we launch into our description of the various energetic contributions to molecular recognition, we must note an important issue. Entropy, and not just enthalpy (in effect, internal energy), matters. This is especially important considering that most biorecognition occurs in water which, because of its high molar density (55.5 mol/L – meaning there are a lot of water molecules to move around and "order"), contributes very significant entropic effects to any recognition event occurring within it. More generally, the environment in which the recognition event is taking place is always critical because it is the *relative energy and entropy of the bound state with respect to the unbound state* that determines the affinity, not the absolute free energy of the complex. We will return to this important issue again and again in our discussion.

1.2 General Principles of Interaction Thermodynamics

From the thermodynamic viewpoint, molecular recognition occurs because the free energy of the receptor–ligand complex is lower than the free energy of the unbound receptor plus the free ligand. Thus, one approach toward understanding molecular recognition is to ask which factors contribute to the free energies of the ligand, receptor, and the complex. To do so, it is conceptually helpful to decompose the free energy into the enthalpic and entropic components and discuss these separately. The rationale for such a decomposition is that some aspects of molecular recognition, such as direct electrostatic attraction between positively charged ligand and the negatively charged binding pocket, mainly affect the enthalpic term. Others, such as the restriction of possible motions in the ligand upon binding, are largely described by the entropy. However, sometimes such partitioning also leads to undesirable complications, such as the often-cited enthalpy-entropy compensation. Here, we will first outline the basic principles of the thermodynamic treatment of interacting systems, then discuss the origin and characteristics of various forces that attract or repel two bodies.

1.2.1 Free Energy, Enthalpy, and Entropy in Interacting Systems

We will be concerned with the Gibbs free energy, which is applicable to processes that occur at a constant pressure and temperature. Most molecular recognition events fall into this category (because most of biology and, indeed, most of chemistry, occurs at a fixed pressure of 1 atm) but the reader should be aware that processes occurring in closed microfluidic devices with fixed volumes could lead to changes in pressure, and the Helmholtz free energy is more appropriate in such cases.

The Gibbs free energy of binding can be written as:

$$\Delta G_{\text{bind}} = \Delta H_{\text{bind}} - T\Delta S_{\text{bind}} \tag{1.1}$$

where ΔH_{bind} and ΔS_{bind} are the enthalpy of binding and entropy of binding, respectively. In general, both the enthalpy and entropy depend on external factors, such as pressure, temperature, or strength of the external electric and magnetic fields. The enthalpy change can be further decomposed into the energy and the work term:

$$\Delta H_{\text{bind}} = \Delta E_{\text{bind}} + \Delta PV_{\text{bind}} \tag{1.2}$$

In the absence of external electric and magnetic fields, the energy of a molecule reduces to its internal energy: $\Delta E_{\text{bind}} = \Delta U_{\text{bind}}$. For processes that occur at constant pressure in the absence of external fields – again, quite common conditions for biosensor applications – the enthalpy of binding becomes:

$$\Delta H_{\text{bind}} = \Delta U_{\text{bind}} + P\Delta V_{\text{bind}} \tag{1.3}$$

The work term $P\Delta V_{\text{bind}}$is significant for processes that lead to the change of the system's volume. For example, it should not be ignored when describing the association of two molecules in the gas phase. However, because water is largely incompressible, the volume change, and thus the contributions of work to the free energy, is typically insignificant in biorecognition. For example, a typical change in the partial molar volume upon a protein–ligand binding is less than 2% (Chalikian and Filfil 2003). For our purposes, the enthalpy associated with a biorecognition event can be considered interchangeable with the internal energy change that occurs upon binding. Thus, for the purpose of the following discussion, we ignore the small volume term and write:

$$\Delta G_{\text{bind}} \approx \Delta U_{\text{bind}} - T\Delta S_{\text{bind}} \tag{1.4}$$

The thermodynamic quantities ΔG_{bind}, ΔU_{bind}, and ΔS_{bind}, only depend on the nature of the free and bound states and not on the path or rate by which the binding was achieved. In other words, these quantities are state functions (meaning that, for example, the free energy difference between state A and state B is independent of the path by which one transits between the two), and we can write:

$$\Delta G_{\text{bind}} = G_{\text{complex}} - G_{\text{receptor}} - G_{\text{ligand}} \tag{1.5}$$

The free energies of the complex, receptor, and ligand, are complex functions of the internal structure and dynamics of these molecules. Furthermore, in solution the free energies G_{complex}, G_{receptor}, and G_{ligand} implicitly include contributions from the solvent. The change of solvation is often a major driving force in association, a topic that is discussed at length later.

We also note in passing that the binding free energy and the binding enthalpy are directly observable properties. The standard free energy can be obtained rigorously from the experimental ligand–receptor equilibrium binding constant K_{bind} via:

$$\Delta G^0_{\text{bind}} = -RT \ln K_{\text{bind}} \tag{1.6}$$

The standard binding enthalpy can be calculated from the temperature-dependence of the equilibrium association constant (i.e., by measuring the equilibrium constant at various temperatures and fitting the resulting observations) via the van't Hoff equation:

$$\frac{\mathrm{d} \ln K_{\text{bind}}}{\mathrm{d}T} = \frac{\Delta H^0_{\text{bind}}}{RT^2} \tag{1.7}$$

or measured directly via isothermal calorimetry (Leavitt and Freire 2001).

1.2.2 Interaction Energy in the Association of Two Semi-Rigid Molecules in the Gas Phase

We will now lay down a path toward understanding how the state functions, such as internal energy and entropy, change upon complex formation. We will start with a hypothetical case in which the receptor and ligand are in the gas phase. While such gas-phase complexes will never be seen in biology, which has been optimized during billions of years of evolution for optimal performance in an aqueous environment, they reveal important connections between molecular structure, dynamics, and thermodynamic state functions.

Let us consider the internal energy of a molecule. It has a potential energy component that arises from a multitude of interactions between parts of the molecule, and a kinetic energy component that arises from the vibration of atoms around their equilibrium positions, and the rotation of groups about single bonds. The potential energy is uniquely determined by the structure of the molecule. The kinetic energy components depend on the temperature and on derivatives of potential energy with respect to structural changes. In principle, once we know the structure of the molecule, we can calculate how the energy of each chemical bond or each long-range interaction between charged groups contributes to the potential energy. Similarly, once we know how the potential energy changes when the structure changes, we can calculate the kinetic energy of vibrations and internal rotations.

Before continuing with the analysis of internal energy, we will need to make another simplifying assumption: we will assume for a moment that the free receptor, free ligand, and the complex are all semi-rigid structures. They are semi-rigid in the sense that individual atoms can still vibrate around their equilibrium positions, and small groups, such as the methyl group can rotate, but the overall molecule is characterized by one particular conformation. The binding of a ligand is possible if the conformation of the free ligand matches the shape of the pre-existing binding pocket in the receptor. Such a simplified model for binding was first considered by the Nobel-prize laureate Emil Fisher who wrote in 1894: "To use a metaphor, I would say that the enzyme and substrate must fit together like lock and key in order to exert a chemical effect on each other" (Fischer 1894; Cramer 1995). While we now know that the assumption of full rigidity is generally too restrictive, the lock-and-key model allows us to introduce the concept of interaction energies as seen below. The assumption of a rigid ligand and a rigid receptor are also used in many computational docking programs that allow the rapid identification of small molecules that fit to the pre-existing pockets in macromolecules.

What can we say about the internal energy changes when a rigid ligand binds to a rigid receptor? Recalling that internal energy is a state function, we can write:

$$\Delta U_{\text{bind}} = U_{\text{complex}} - U_{\text{receptor}} - U_{\text{ligand}} \tag{1.8}$$

The internal energy of the complex that is formed between a ligand and its receptor can be generally expressed as:

$$U_{\text{complex}} = U_{\text{receptor in complex}} + U_{\text{ligand in complex}} + U_{\text{interaction}} \tag{1.9}$$

Because our receptor is semi-rigid, its internal energy is expected to change little upon the formation of the complex ($U_{\text{receptor}} \approx U_{\text{receptor in complex}}$). Similarly, if the ligand does not undergo a significant conformational change upon binding, its internal energy does not change appreciably upon binding ($U_{\text{ligand}} \approx U_{\text{ligand in complex}}$). Substituting (1.9) into (1.8), we arrive to the conclusion that upon the binding of a rigid ligand to a rigid receptor in the gas phase, the internal energy difference between the complex and the free molecules is closely equal to the interaction energy between the two components in the complex:

$$\Delta U_{\mathrm{bind}} \approx U_{\mathrm{interaction}} \tag{1.10}$$

What, then, are the interactions between noncovalently bound molecular species that drive binding? At the very fundamental level, the interaction energy arises from the interaction of electrons and nuclei in one molecule with electrons and nuclei in other molecule. Unfortunately, considerable mathematical complexity arises when one tries to describe such interactions using quantum mechanics. As a result of this complexity, we are usually unable to make rational design decisions based on the analysis of the quantum mechanics on the interacting electrons and nuclei. In some cases, quantum mechanical calculations can provide valuable guidance, but in most cases the size of the system is simply too large for meaningful quantum mechanical calculations. Fortunately, however, the interaction problem can be considerably simplified by dividing the total interaction energy into several individual contributions that are relatively easy to understand (Stone 2008). One usually considers the following contributions:

1. First-order electrostatic interactions involving permanent charges and multipoles
2. Second-order electrostatic energy involving induction due to permanent polarization of charges and multipoles
3. London dispersion attraction between temporary induced dipoles
4. Steric repulsion between electron clouds due to the Pauli exclusion principle that prohibits two electrons with the same spin occupying the same space
5. Charge transfer between electron-rich and electron-deficient structures

Notice that the noncovalent interactions described above do not include sharing a pair of electrons, which is a hallmark of a covalent bond. However, some of the interactions that play a role in chemical and biological recognition have a slight covalent character. For example, while the attraction between the carbonyl and amide moieties in proteins or nucleic acids arises mainly from the electrostatic interaction between the C=O dipole and the H–N dipole, it also involves some sharing of the lone pair electrons on oxygen with the partially vacant antibonding orbital on amide hydrogen. For typical interactions, the covalent character may be in the order of a few percent (Grzesiek et al. 2004).

The usefulness of partitioning the interaction energy into the contributions listed above lies in our ability to correlate molecular structures with the strength of the interaction. We will now describe each of the five major interactions in detail.

1.3 Interaction Energies in the Gas Phase

The theory of interaction of molecules in the gas phase is generally well understood and a large number of computer programs are available that can calculate the interaction energy between two molecules in the gas phase with reasonable accuracy. Here, we discuss the nature of the forces felt by molecules in the gas-phase in quantitative detail in order to, later, understand how these same forces define biological recognition in the more relevant, aqueous environment.

Most of the forces felt in the gas phase are, ultimately, electrostatic in nature. In order to gain conceptual understanding, the first-order electrostatic interactions between complex charge distributions of real molecules is typically approximated by expanding the charge distribution into terms corresponding to the permanent charge, permanent dipole, permanent quadrupole, and permanent octupole. Typically, the first nonvanishing local multipole dominates the interaction: we do not need to worry about the quadrupole and octupole moments when describing interaction of two strong dipoles. The interaction energy formulas can be readily obtained based on the Coulomb law and geometric considerations of multipoles. The interaction energy formulas given later in this chapter are derived assuming that the interaction distance is large when compared to the size of the dipole or quadrupole. This assumption is not entirely valid for typical intermolecular interactions, however, the comparison with rigorous calculations indicates that these formulas describe interactions between real molecules reasonably well. The main advantage of multipole expansion is that the features of the lowest multipoles – charge, dipole, and quadrupole – can be readily understood intuitively based on the chemical structures. Readers interested in a more rigorous mathematical treatment of molecules in multipole expansion should consult a book "The Theory of Intermolecular Forces" by Anthony Stone (Stone 1996). The second order induction effects are harder to characterize, and many molecular modeling programs that calculate the interaction energy based on molecular mechanics models treat polarization only approximately. The London dispersion and steric repulsion are fundamentally quantum mechanical phenomena and now advanced mathematical methods are needed to derive the interaction energy formulas. In practice, these short-range forces are described together via simple empirical potentials. Charge transfer is a quantum mechanical phenomenon and its contribution is usually estimated via quantum chemical calculations.

1.3.1 First-Order Electrostatic Interactions Involving Permanent Charges and Multipoles

Electrostatic interactions are common in biological recognition because receptor proteins and nucleic acids invariably contain multiple charged groups, as often do their ligands. Moreover, despite the large dielectric constant of water (charges attract or repel almost 80 times less effectively in water than they do in vacuo), electrostatic interactions achieve chemically relevant energies across much longer distances than any of the other nonbonding forces and thus electrostatics are typically the strongest nonbonded interactions at intermediate to longer distances (i.e., greater than a few ångstroms).

Electrostatic interaction between charges and multipoles can be further divided according to the nature of multipole moments in two molecules:

1. Electrostatic interaction between two permanent charges (e.g., between Na^+ and Cl^-)
2. Electrostatic interaction between a permanently charged ion and a dipole (e.g., between Mg^{2+} and the strong dipole set up in water by the significant difference in electronegativity between oxygen and hydrogen)
3. Electrostatic interaction between a permanently charged ion and a quadrupole (e.g., between Li^+ and the charges distributed around the net-neutral benzene ring)
4. Electrostatic interaction between two permanent dipoles (e.g., between two molecules of dimethyl ether)
5. Electrostatic interaction between a permanent dipole and a permanent quadrupole (e.g., between the NH group of amide and the net-neutral benzene ring)
6. Electrostatics interaction between two permanent quadrupoles (e.g., between two benzene molecules)

1.3.1.1 Two Point Charges

Electrostatic interaction between ionic species is probably the easiest to understand. Like charges repel, and opposite charges are attracted with a force (in vacuum) given by Coulomb's law:

$$F_{\mathrm{el}}(r) = \frac{1}{4\pi\varepsilon_0}\frac{q_1 \cdot q_2}{r^2} \tag{1.11}$$

The force between two ions is a conservative force, meaning that the work done by moving one ion relative to another does not depend on the path of movement; the force is a function of the distance between them such that:

$$F_{\mathrm{el}}(r) = -\frac{\mathrm{d}E_{\mathrm{el}}(r)}{\mathrm{d}r}, \tag{1.12}$$

where $E_{\mathrm{el}}(r)$is a distance dependent potential energy function. The potential energy of an interaction can be generally calculated as the work needed to bring two interacting particles from infinite distance to the distance r:

$$E_{\mathrm{el}}(r) = -\int_{\infty}^{r} \frac{1}{4\pi\varepsilon_0}\frac{q_1 \cdot q_2}{r^2} + E_{\mathrm{el}}(\infty) \tag{1.13}$$

The interaction energy of two infinitely separated particles is zero by definition and after integration, the electrostatic energy between two point charges is given by:

$$E_{\mathrm{el}}(r) = \frac{1}{4\pi\varepsilon_0}\frac{q_1 \cdot q_2}{r} \tag{1.14}$$

A convenient expression in the more familiar kJ/mol units is obtained by substituting the value $\varepsilon_0 = 8.854 \times 10^{-12}$ C^2/J*m for the vacuum permittivity, expressing the charge in multiples of electron's charge, and multiplying the result with the Avogadro's constant. The result in terms of charge values and the distance (in ångstroms) between point charges is:

$$E_{\mathrm{el}}(r) = 1389.35 \frac{Z_1 \cdot Z_2}{r} \text{ kJ/mol} \tag{1.15}$$

The interaction energy is negative (favoring ion pairing) at all distances in the case of two oppositely charged ions, and positive (favoring ion separation) in case of two like charges. The interaction is the strongest of the noncovalent interactions at medium and long range: a cation and an anion separated by 4 Å in vacuum are bound about as strongly as two covalently connected carbon atoms in a typical organic molecule. The interaction is also long range, decaying as the inverse first power in distance. At what distance does such an interaction become negligibly small? The interaction does not lead to an association if the interaction energy is smaller than the average translational kinetic energy due to random thermal motions. The average kinetic energy of a particle in classical mechanics is $\frac{3}{2}RT$(3.7 kJ/mol at room temperature); the electrostatic interaction between two univalent ions in the gas phase reaches this threshold at an amazing 374 Å, which is ten times larger than a typical biological macromolecule. Of course, as noted above, the dielectric constant of water is nearly 80, and thus this same cutoff under biologically relevant, aqueous conditions is less than 5 Å, which is only a few times the diameter of a water molecule.

1.3.1.2 Point Charge and Dipole

The interaction between a point charge and a fixed dipole arising from two point charges can be also easily understood when one considers that such a dipole can be represented by two oppositely charged point charges. When the negative end of the dipole points toward a cation, the interaction is attractive. When the positive end of the dipole points toward a cation, the ion and dipole repel. When the dipole is perpendicular to the normal of the cation's surface, the interaction energy vanishes. The charge–dipole interaction energy is given by:

$$E_{\mathrm{charge-dipole}}(r, \theta) = -\frac{1}{4\pi\varepsilon_0} \frac{q \cdot \mu \cdot \cos\theta}{r^2} \tag{1.16}$$

A convenient expression in the more familiar kJ/mol units is obtained by substituting the value $\varepsilon_0 = 8.854 \times 10^{-12}$ C^2/J*m for the vacuum permittivity, expressing the charge in multiples of electron's charge, expressing the dipole moment in debye units, and multiplying the result with the Avogadro's constant.

The expression in terms of charge values, dipole moment (in debye), and distance (in ångstroms) between a point charge and distant dipole is:

$$E_{\text{charge-dipole}}(r,\theta) = -289.3\frac{Z \cdot \mu(D) \cdot \cos\theta}{r^2} \quad \text{kJ/mol} \tag{1.17}$$

The interaction between an ion and a fixed dipole is weaker than the interaction between two ions: it takes about 67 kJ/mol to break the interaction between a Mg^{2+} ion and a water molecule that are 4 Å apart in the gas phase. Furthermore, the strength of ion-dipole interaction falls off more rapidly ($1/r^2$) than the strength of ion–ion interaction ($1/r$). For example, the attraction energy between Mg^{2+} and a water molecule becomes smaller than the average kinetic energy of particles beyond 17 Å.

In the gas phase, the dipole far away from the charge will rotate and the angle θ will vary. In the case of a freely rotating dipole, rotationally averaged interaction energies should be considered:

$$\begin{aligned} E_{\text{charge-rotating dipole}}(r) &= KT - \left(\frac{1}{4\pi\varepsilon_0}\right)\frac{q \cdot \mu}{r^2}\coth\left[\frac{q \cdot \mu}{4\pi\varepsilon_0\ KTr^2}\right] \\ &\approx -\left(\frac{1}{4\pi\varepsilon_0}\right)^2\frac{q^2 \cdot \mu^2}{3kT}\frac{1}{r^4} \end{aligned} \tag{1.18}$$

The rotation of the dipole decreases its attraction to the nearby charge and shortens the distance at which the rotating dipole feels a significant pull due to the nearby charge. Calculations based on the rotating dipole model predict that a water molecule is not significantly attracted to the Mg^{2+} ion in the gas phase when their separation is larger than 13 Å.

Ion–dipole interactions are ubiquitous in polar solvents as every charged group is typically solvated by several polar solvent molecules. In many cases, the solvated molecules form a hydration shell with a discrete structure, for example Mg^{2+} in water is surrounded by six water molecules in octahedral geometry. In other cases, the solvent shell is more dynamic and rapidly fluctuates between different arrangements of polar solvent molecules around the cation. The solvation shell structure is also determined by the requirement of favorable solvent-solvent interactions and thus the size of the ion plays a major role in determining the solvent shell structure: Ions that are too small or too large do not support regular arrangements of solvent molecules around them.

1.3.1.3 Point Charge and Quadrupole

Highly symmetric linear or planar molecules such as CO_2 or benzene do not exhibit any net dipole moment because, while their constituent bonds may be polarized, these dipoles cancel due to molecular symmetry. However, such molecules have higher moments such as the quadrupole moment (Williams 1993). In a significantly oversimplified picture, the quadrupole moment of CO_2 can be related to the partial negative charge on two oxygen atoms, and to a compensating positive charge on the carbon atom. Specifically, the quadrupole moment of CO_2 (or similar linear quadrupoles) can be calculated as:

$$\Theta = -4q_{\mathrm{O}} \cdot r_{\mathrm{CO}}{}^{2} \tag{1.19}$$

where q_{O} is the partial charge on each oxygen atom, and r_{CO} is the carbon–oxygen distance in the molecule. Similarly, the quadrupole moment of aromatic rings can be related to the excess negative charge density at the region shared by the π electrons, and to the excess positive charge in the plane of the ring. Even an uncharged molecule will interact with other charges if its quadrupole moment is not zero. The strength of this interaction is related to the magnitude of the quadrupole moment and the orientation of the molecule with respect to the point charge. A full mathematical description of the quadrupole charge interaction is beyond the scope of this chapter, but a conceptual understanding can be gained by considering the interaction of a linear quadrupole with a point charge. The interaction energy between a linear quadrupole with a quadrupole moment Θ, and a point charge q is given by:

$$E_{\mathrm{charge-quadrupole}}(r,\theta) = \frac{1}{4\pi\varepsilon_0}\frac{q \cdot \Theta \cdot (1 + 3\cos(2\theta))}{4r^3} \tag{1.20}$$

The expression in terms of the charge value, quadrupole moment (in debye–ångstrom), and distance (in ångstroms) between a point charge and the center of a distant linear quadrupole is:

$$E_{\mathrm{charge-quadrupole}}(r,\theta) = 72.3\,\frac{Z \cdot \Theta(D \cdot \mathrm{\AA}) \cdot (1 + 3\cos(2\theta))}{r^3}\quad \mathrm{kJ/mol} \tag{1.21}$$

In this model, an attraction is predicted when the test charge lies along the axis of the linear quadrupole: a cation near the oxygen of CO_2 along the line of molecular axis is attracted to the nonpolar carbon dioxide! The charge–quadrupole interaction energy of a monovalent cation ($Z = +1$) and CO_2 ($\Theta = -4.5$ debye Å or -15×10^{-40} C m^2) in this alignment at 4 Å separation between the carbon and the cation is 20 kJ/mol. The attraction along this line is expected to fall of as $1/r^3$ and, in case of CO_2-like linear quadrupole and monovalent ion, becomes weaker than the thermal energy at a distance of 7 Å. A weak repulsion is predicted when the cation lies on the line normal to the axis of CO_2. The interaction vanishes when the angle between the axis of the linear dipole and the line connecting the dipole to the point charge is about 54.74° or a multiple of this value.

The interaction of point charges with quadrupoles contributes significantly to what is known in chemistry as the cation–π interaction (Ma and Dougherty 1997; Zacharias and Dougherty 2002). The π–bonds in the aromatic ring of benzene set up a substantial quadrupole moment ($\Theta \approx -29 \times 10^{-40}$ C m^2) with a negative charge density concentrated above and below the molecular plane near the center of the molecule. As a result, benzene and the benzene-like amino acids phenylalanine, tyrosine, and tryptophan interact strongly with cations both in the gas phase and solution (Dougherty 1996). For example, in the gas phase a potassium ion binds more strongly to benzene ($\Delta H^0 = -80$ kJ/mol) than it does to a water molecule ($\Delta H^0 = -75$ kJ/mol) (Sunner et al. 1981). We note, however, that second order induction–polarization (see below) effects appear to be at least as important as these first-order electrostatic attractions in stabilizing interactions between cations and aromatic moieties (Tsuzuki et al. 2001). In cases of strong induction, the interaction between a cation and aromatic system can be powerful. For example, the strength of Mg^{2+}–benzene interaction has been estimated to be comparable to covalent bonds in the gas phase. Given their strength, it is not surprising that cation–π interactions contribute to a variety of biologically important recognition events. For example, cation–π interactions between the positive quaternary amine on the substrate and aromatic amino acids in the protein are responsible for the binding of acetylcholine to acetylcholine esterase and, more generally, appear to be a common theme in the binding of neurotransmitters acetylcholine, γ-aminobutyric acid, and serotonin to their respective membrane receptors in the nerve cells (Ngola and Dougherty 1996).

1.3.1.4 Two Dipoles; Hydrogen Bonding

All but the most symmetric molecules have a nonzero dipole moment, and thus dipole–dipole interactions are common throughout biorecognition. While not nearly as strong as interactions involving charged species, the large number of dipole–dipole interactions plays a crucial role in determining structures of proteins, nucleic acids, and carbohydrates.

Dipole–dipole interactions are strongly orientation-dependent. The general expression for the interaction energy of two dipoles μ_1 and μ_2 that are separated by a distance r which is much greater than the length of the dipoles is:

$$E_{\text{dipole-dipole}}(r, \theta_1, \theta_2, \varphi) = -\frac{1}{4\pi\varepsilon_0}\frac{\mu_1 \cdot \mu_2}{r^3}[2\cos\theta_1\cos\theta_2 - \sin\theta_1\sin\theta_2\cos\phi] \tag{1.22}$$

The angles θ_1 and θ_2 define the orientation of the two dipoles with respect to an axis line that joins the centers of the dipole; the angle ϕ measures the degree of rotation of one dipole with respect to another along the axis line. When the dipoles are parallel ($\theta_1 = \theta_2$; μ_1 and μ_2 have the same sign) and lie in the same plane ($\phi = 0$), the above expression simplifies to:

$$E_{\text{dipole-dipole}}(r, \theta) = -\frac{1}{4\pi\varepsilon_0}\frac{\mu_1 \cdot \mu_2}{2r^3}[1 + 3\cos(2\theta)] \tag{1.23}$$

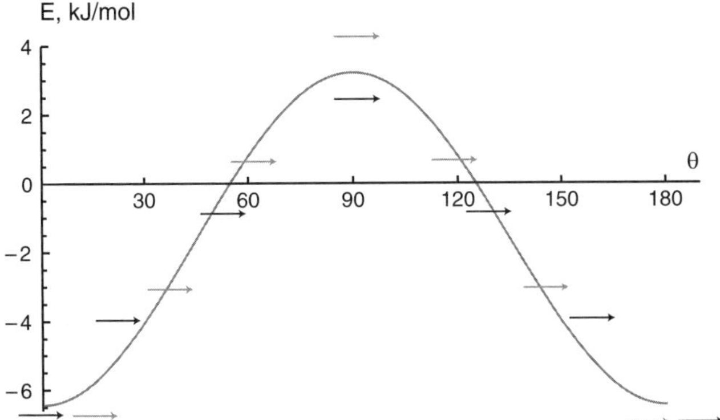

Fig. 1.1 Interaction profile of two parallel dipoles ($\mu_1 = \mu_2 = 1.85$ Debye) in the gas phase at center-to-center separation of 4 Å as a function of the angle between the dipole axis and the line connecting centers of the two dipoles

An analysis of this expression (Fig. 1.1) reveals that the attraction is strongest when the two parallel dipoles are aligned in a collinear manner head-to-tail. When the parallel dipoles are exactly on top of each other ($\theta = 90°$ on Fig. 1.1) they repel maximally, and when the angle θ reaches 57.74° the interaction energy is zero. (On a distantly related note, this so-called magic angle is particularly important for the suppression of dipolar coupling in solid state NMR.) When the dipoles are antiparallel ($\theta_1 = \theta_2$; μ_1 has a sign opposite to μ_2), the graph is inverted with respect to the x-axis. The two antiparallel dipoles repel in head-to-head in-line alignment but attract when aligned on top of each other. It would appear that the head-to-tail attraction of parallel dipoles is stronger than the on-top attraction of antiparallel dipoles but this is true only if the distance between the centers of dipoles is held constant. Dipolar molecules and groups are often ellipsoidal which permits a closer approach of the two dipoles in the on-top alignment. Because of the $1/r^3$ distance dependence, the alignment in which the centers of the two on-top dipoles are at 3 Å distance is more favorable than the alignment in which the centers of two head-to-tail dipoles are at 4 Å distance.

The interaction strength between static dipoles falls off as the third power of the distance between the dipoles. Thus, the dipole–dipole interaction is a shorter-range than the charge–charge or the charge–dipole interaction. While offering significant stabilization at short intermolecular distances, the energy of the dipole–dipole interaction between two unpolarized dipoles with $\mu = 1.85$ becomes less than the thermal energy at the separation larger than 4.8 Å for the optimal head-to-tail alignment (although, again, this distance will be significantly smaller in water). A somewhat different treatment is needed when the two dipoles can rotate with respect to each other. For rotating dipoles, angle-averaged interaction energy (Keesom energy) can be obtained by averaging the interaction energy over the Boltzmann-weighted ensemble of all possible orientations. The approximate interaction energy of two rotating dipoles in the weak interaction limit ($\mu_1 \cdot \mu_2 / 4\pi\varepsilon_0 r^3 < kT$) is:

$$E_{\text{Keesom}}(r) = -\left(\frac{1}{4\pi\varepsilon_0}\right)^2 \frac{2\mu_1{}^2 \cdot \mu_2{}^2}{3kT} \frac{1}{r^6} \tag{1.24}$$

The Keesom energy shows $1/r^6$ distance dependence and contributes significantly to van der Waals energy of small molecules in the gas phase (Magnasco et al. 2006). It is likely to play a minor role in biomolecular interactions where the interacting dipoles are larger and have only limited mobility.

The best-known interaction involving two dipoles is the hydrogen bond in which one of the participants (donor) contains a partially positively charged hydrogen atom and the other participant (acceptor) provides a pair of nonbonding electrons (Buckingham et al. 2008). In order for the hydrogen to have the requisitely large partial positive charge, it must be covalently bonded to a strongly electronegative element, such as fluorine, oxygen, or nitrogen. In the language of chemists, a strong hydrogen bond is formed when the donor is sufficiently acidic and the acceptor is sufficiently basic (Lii and Allinger 2008). As with any interaction between real molecules, the accurate treatment of hydrogen bonding requires the consideration of induction-polarization, weak London dispersion forces, and as in case of any noncovalent interaction, of short-range steric repulsion.

There has been some debate about the covalent character of the hydrogen bond. The idea, first proposed by Linus Pauling, found some experimental support in Compton scattering and NMR couplings through hydrogen-bonds in late 1990 (Isaacs et al. 1999). However, it is now understood that the observed anisotropy in the Compton scattering profile can be explained by a repulsive superposition of molecular orbitals (Romero et al. 2001) and that the observation of NMR couplings through hydrogen bonds, while consistent with covalent character, does not constitute a proof of covalent nature of hydrogen bonding (Grzesiek et al. 2004). The dominant role of electrostatic interactions is also supported by the observation that classical (i.e., non-quantum-mechanical), nonpolarizable models of water developed for computer simulations, such as the TIP4P model of water (Jorgensen et al. 1983), do a remarkably good job describing the liquid properties and phase diagram of water despite ignoring the potentially covalent nature of the hydrogen bond (Guillot 2002; Aragones et al. 2007).

The orientation dependence described above should be considered in the rational design of ligands that would bind via hydrogen bonding. Not only is it important that the hydrogen bond donors and acceptors be in proximity, but the angle between the hydrogen bond donor axis and the hydrogen bond acceptor axis should not be too large. As discussed above, the interaction of polarizable dipoles is usually the strongest in the case of in-line alignment. If the hydrogen bond donor and acceptor are perpendicular ($\theta = 45°$), the net attraction between the two moieties is only a quarter of attraction that can occur in the in-line alignment. Figure 1.2 illustrates some hydrogen bonding geometries.

Hydrogen bonds play a significant role in defining the native three-dimensional structures of nucleic acids, proteins and folded (packed) polysaccharides. For example, cellulose and chitin owe their tough, rigid structures to extensive

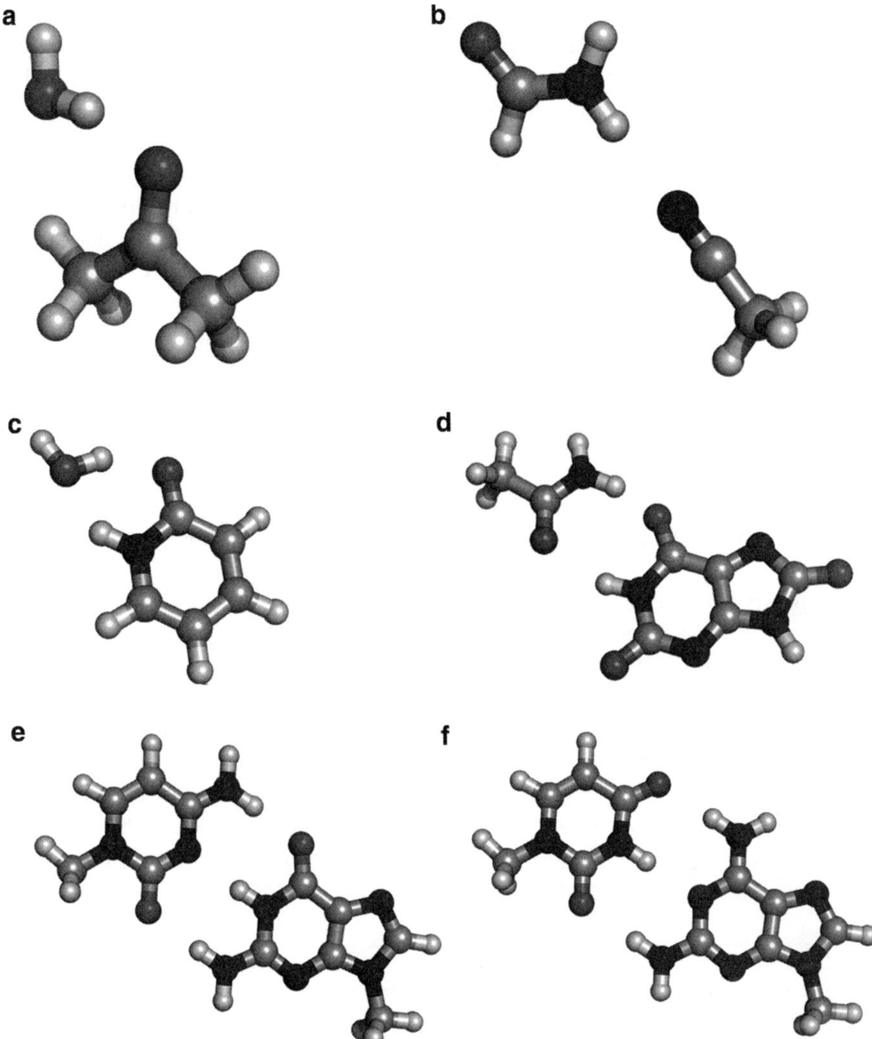

Fig. 1.2 Complexes in which hydrogen bonding plays a major stabilizing role. (**a**) Acetone–water. The distance between the carbonyl oxygen and the water hydrogen in this classical hydrogen-bonding complex is 1.9 Å, and the oxygen–hydrogen–oxygen angle is 163°. Notice that an additional weak attraction between the methyl hydrogen and the water oxygen is possible in this configuration. The interaction energy between acetone and water in the gas phase is 25 kJ/mol according to our CCSD/aug-cc-pVTZ calculations using MP2/aug-cc-pVTZ optimized geometries. (**b**) Acetonitrile–formamide. The distance between the nitrile nitrogen and formamide hydrogen in this atypical hydrogen-bonded complex is 2.1 Å. The interaction energy between acetonitrile and formamide in the gas phase is 21 kJ/mol according to our CCSD/aug-cc-pVTZ calculations using MP2/aug-cc-pVTZ optimized geometries. (**c**) Pyridone–water. The distance between the carbonyl oxygen and the water hydrogen in this bidentate hydrogen bonding complex is 1.8 Å, and the distance between the amide hydrogen and the water oxygen is 1.9 Å. The interaction energy between pyridone and water in the gas phase is about 50 kJ/mol according to

two-dimensional networks of hydrogen bonds between covalently linked carbohydrate chains. Biological macromolecules typically contain multiple hydrogen bond acceptor and hydrogen bond donor groups in close proximity. For example, the peptide bond contains a strong donor (the amide proton) and a strong acceptor (the carbonyl oxygen). This has two important consequences. First, ligand binding specificity is improved because for the optimal binding, all available hydrogen-bond interactions must be made, and only a few ligands have a structure that permits that. Second, hydrogen bonding between some biological macromolecules and their ligands is cooperative. The cooperativity can be understood by considering a simple model system consisting of three dipoles, e.g., the noncyclic water trimer. The polarization of the first water by the second increases its dipole moment, and makes the interaction between the first and the third water molecule more favorable. Such cooperativity has a significant effect on properties of biological macromolecules and contributes to changes in the receptor structure upon ligand binding.

1.3.1.5 Dipole and Quadrupole; π-Facial Hydrogen Bonding

Analogously with the charge–quadrupole interaction between a cation and benzene, the partially positive charged atoms in dipolar molecules can also favorably interact with the negatively charged regions of a quadrupole. Such interactions are, however much weaker than the cation–π interactions. For example, the interaction energy between a dipolar water molecule and quadrupolar benzene in the gas phase complex is “only” about 16 kJ/mol. Comparison with this value with the water–water interaction energy in a gas-phase water dimer reveals that the water–benzene interaction is about 25% weaker than water–water attraction (Feller 1999). Given the abundance of dipolar groups and aromatic residues in biological systems, it is not surprising that a number of dipole–quadrupole interactions have been observed

Fig. 1.2 (continued) our MP2/aug-cc-pVTZ calculations. Notice that the hydrogen bond to the carbonyl oxygen is shorter, and apparently stronger than a similar hydrogen bond in the acetone–water complex because of the cooperative nature of hydrogen bonding. (**d**) Urate dianion–acetamide (as a model for glutamine side chain). Bidentate interactions such as this are commonly used in the biological recognition of amide-containing ligands. Computational and experimental results suggest that the distance between the carbonyl of urate and hydrogen of acetamide is significantly shorter than the distance between the amide nitrogen of urate and the carbonyl oxygen of acetamide. This suggests the presence of a repulsive secondary interaction of the carbonyl of acetamide with the carbonyls of the negatively charged urate. E) 1-Methylcytosine–9-methylguanine (as a model for the GC base pair in double-stranded DNA). The interaction energy (neglecting London dispersion) of these two molecules in the gas phase is nearly 100 kJ/mol based on our HF/6-31+G(d,p) calculations. This is an example of a fairly strong association that determines the structure and function of biological macromolecules such as DNA, transfer RNA, and ribosomal RNA. F) 1-Methyluracil–2,6-diamino-9-methylpurine. The interaction energy (neglecting London dispersion) of these two molecules in the gas phase is about 50 kJ/mol based on our HF/6-31 + G (d,p) calculations. Notice that despite having three hydrogen bonds similar to the GC base pair, this complex is significantly weaker due to unfavorable secondary interactions of the amide hydrogen in the pyrimidine with the two nearby hydrogens from amino groups in the purine

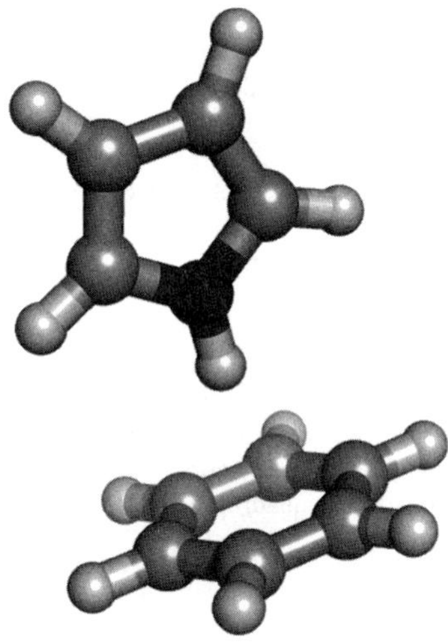

Fig. 1.3 π-facial hydrogen bonding between dipolar pyrrole and quadrupolar benzene. Our quantum chemical calculations suggest that the electrostatic interaction energy in this complex is about 6 kJ/mol when the pyrrole nitrogen is 4 ångstroms from the center of the benzene ring. The total interaction energy, including London dispersion, is 15 kJ/mol at the MP2/aug-cc-pVTZ level

in proteins (Burley and Petsko 1986). Most commonly, amide or hydroxyl hydrogen points toward the π-face of the aromatic ring, hence the interaction is also called the π-facial hydrogen bonding. A typical example of the π-facial hydrogen bonding is the interaction of pyrrole with benzene (Fig. 1.3).

1.3.1.6 Two Quadrupoles

Symmetric molecules typically lack a net dipole moment but often possess a quadrupole moment (Williams 1993). In such cases, weak quadrupole–quadrupole interactions are possible. Because of the weakness of the quadrupole–quadrupole interaction and its complex angular dependence, the prediction of optimal alignments of two interacting molecules usually requires detailed quantum mechanical calculations. Figure 1.4 illustrates the electrostatic potential of the parallel-slipped carbon dioxide dimer.

The formation of such weakly bound (binding energy about 4 kJ/mol) dimers leads to deviations from the ideal gas law at lower temperatures. Other well-characterized systems (Fig. 1.5) in which an electrostatic interaction between quadrupoles plays a significant stabilizing role are the T-shaped benzene dimer, the face-to-face stacking of benzene and hexafluorobenzene, and the planar arrangement of 1,4-benzoquinones in the dimer (Plokhotnichenko et al. 1999). Interaction energies between monomers in such systems are typically in the order of 10–20 kJ/mol in the gas phase. However, these are net interaction energies due to the combined effect of quadrupole–quadrupole, induction, and dispersion energies, which will be discussed later.

Fig. 1.4 Slipped-parallel CO_2 dimer rendered along with its multipole-derived electrostatic potential iso-surfaces. The *lighter gray* indicates a region of positive electrostatic potential; the darker gray a region of negative potential. Our ab initio calculations show that the total binding energy of CO_2 dimer is about 5 kJ/mol of which about 0.4 kJ/mol is due to the electrostatic quadrupole–quadrupole interaction

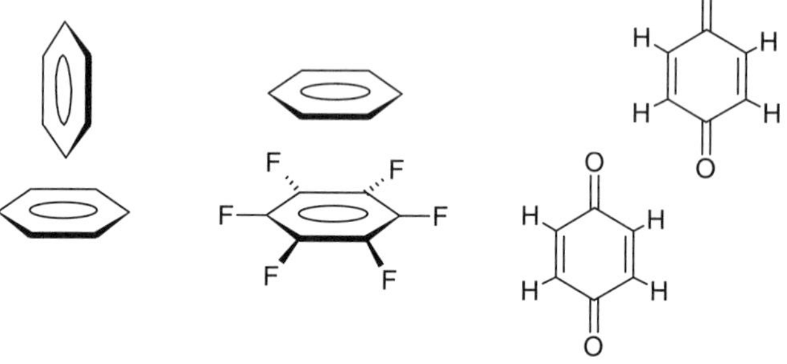

Fig. 1.5 Quadrupole–quadrupole interactions contribute to the formation of weak molecular complexes between molecules that are usually considered non-polar because their permanent dipole moment is zero

1.3.2 Second-Order Induction–Polarization Energy

Ions and polar molecules create electric fields in the space surrounding them. This electric field can affect the distribution of a charge around any nearby molecules. Specifically, in response to an applied electric field the electrons in a molecular orbital will alter their distribution in a manner that mirrors the applied electric field. This increase in the polarization (charge separation) of the orbital always occurs in the direction that *increases* the dipole moments of favorably interacting molecules. This effect thus leads to a favorable electrostatic interaction. The change of the dipole moment can be substantial. For example, the net dipole moment of water increases from 1.86 debye for an isolated molecule in vacuo to 2.95 ± 0.2 debye in the liquid (Gubskaya and Kusalik 2002). To account for polarization, many computational models employ dipole moments that are enhanced relative to those observed in the gas phase. For example, the popular TIP4P model employs $\mu = 2.18$ debye (Guillot 2002).

The induced dipole moment is proportional to the field strength and the proportionality constant is called the electronic polarizability. The electronic polarizability is related to the volume available for the displaced electrons and thus large atoms are more polarizable than small atoms. One can derive the expressions for the second order ion–induced dipole and permanent dipole–induced dipole interactions analogously to the first order electrostatic ion–dipole and dipole–dipole interactions. The interaction between a charge and the nonrotating, induced dipole is given as:

$$E_{\text{charge-induced dipole}}(r) = -\frac{1}{(4\pi\varepsilon_0)^2}\frac{q^2 \cdot \alpha}{2r^4} \tag{1.25}$$

The induction interaction shows stronger distance dependence than the first order electrostatic energy because the inducing field decreases as the distance increases. Also, as some of the field energy is spent on the polarization of the charge distribution, the induction interaction is weaker than the corresponding interaction between an ion and a permanent dipole of the same magnitude. The interaction energy expression in terms of the charge value (Z), polarizability volume (α in 10^{-24} cm^3), and distance (in ångstroms) between an ion and an induced dipole is:

$$E_{\text{charge-induced dipole}}(r) = -694.7\frac{Z^2 \cdot \alpha \cdot (10^{-24}\text{cm}^3)}{r^4} \quad \text{kJ/mol} \tag{1.26}$$

The induction energy by a permanent charge is large at close distances. For example, because of this effect the gas-phase attraction between a Na^+ ion (radius 1.02 Å) and even the small (i.e., not very polarizable) methane molecule (effective radius 2.07 Å, $\alpha = 2.60 \times 10^{-24}$ cm^3) is -20 kJ/mol at 3 Å. The induction energy depends on the *square* of the ion charge and is thus even more significant for divalent cations or anions such as sulfate or phosphate. For example, Mg^{2+} (ionic radius 0.72 Å) induces polarization in benzene (effective radius 3.6 Å, $\alpha = 10.40 \times 10^{-24}$ cm^3) that stabilizes the complex by 86 kJ/mol when the center of Mg ion is 4.3 Å from the center of benzene. The effect, however, drops off relatively rapidly with distance; the attraction between sodium and methane drops to below the thermal energy at just 4.7 Å.

The induction energy due to the interaction of a polar molecule with a molecule that lacks a permanent dipole moment is given as:

$$E_{\text{dipole-induced dipole}}(r,\theta) = -\frac{1}{(4\pi\varepsilon_0)^2}\frac{\mu^2 \cdot \alpha}{2\,r^6}(1 + 3\cos^2\theta) \tag{1.27}$$

where μ is the permanent dipole moment of the polar partner, α is the electronic polarizability of the nonpolar partner, r is the separation between the centers of the molecules, and θ defines the orientation of the polar molecule with respect to the line joining the centers of polar and nonpolar molecule. However, this formula can be also used to estimate the *additional* energy gain due to *additional* polarization of already polar molecule by a nearby polar molecule. The interaction energy expression in terms of the dipole moment (in debye), polarizability volumes (in 10^{-24} cm^3), and distance (in ångstroms) between an optimally aligned permanent dipole and an induced dipole in the gas phase reads:

$$E_{\text{dipole-induced dipole}}(r) = -120.4\frac{\mu(D)^2 \cdot \alpha \cdot (10^{-24}\text{cm}^3)}{r^6} \quad \text{kJ/mol} \qquad (1.28)$$

The strength of dipole–induced dipole interaction energy varies as the inverse sixth power of distance and this interaction is important only at very short distances. For weak dipoles ($\mu < 1$ debye) and slightly polarizable ($\alpha < 3.5$) molecules, the induction energy does not exceed thermal energy, and is not sufficient to bind or strongly orient two molecules. The induction energy becomes significant for stronger dipoles. For example, induction contribution is about one quarter of the total hydrogen bond energy in hydrogen fluoride ($\mu = 1.9$ debye, $\alpha = 0.83 \times 10^{-24}$ cm^3) dimer and about one third of the total hydrogen bond energy in water dimer ($\mu = 2.1$ for water monomer in the dimer)(Gregory et al. 1997). As seen with these examples, the induced increase of the dipole moment by nearby dipoles can be large, and the induction energy contribution to intermolecular interactions may be significant in nonpolar environments.

The electronic polarizability in molecules is strongly anisotropic and contributes to the observed geometry of noncovalent complexes. For example, it is easier to polarize the carbonyl group along the C–O axis than the axis perpendicular to it. Thus, the dipolar interactions that involve carbonyl groups or similar polarizable moieties are typically more stable in the head-to-tail arrangement of dipoles. Thus, the linearity of hydrogen bonds to carbonyl groups arises from the combination of a favorable permanent dipole–dipole interaction when the dipoles are aligned head-to-tail and from the maximum polarization contribution in this alignment.

1.3.3 London Dispersion

Above we have seen that a nearby charge (or even higher order moment, such as a dipole) will induce a counteracting dipole in any nearby polarizable electron cloud. It turns out, however, that even species lacking a charge (or even any higher moments) can induce the formation of a dipole in an effect now termed London dispersion after the physicist who first explained it. Like all induced dipoles, the London dispersion force is always attractive. The effect is, at its heart, purely quantum mechanical but can be understood semi-classically by considering two interacting spherical atoms (London 1937). Electrons are in constant (in this classical picture) motion around the nucleus and at any given instant the charge distribution of atom is slightly asymmetric. This asymmetry means that an atom has an instantaneous fluctuating dipole moment that can induce a fluctuating dipole in a nearby atom. The net effect of these correlated fluctuations is that on average, the interaction between fluctuating mutually inducing dipoles is an attractive interaction. The London dispersion interaction is usually given by the formula:

$$E_{\text{disp}}(r) = -\left(\frac{1}{4\pi\varepsilon_0}\right)^2 \left(\frac{I_1 I_2}{I_1 + I_2}\right) \frac{3}{2} \frac{\alpha_1 \cdot \alpha_2}{r^6} \tag{1.29}$$

Here α_1 and α_2 are the polarizabilities of the two atom or molecules and I_1 and I_2 are the first ionization potentials. The expression for homodimers ($\alpha_1 = \alpha_2$; $I_1 = I_2$) in terms of polarizability volumes (in 10^{24} cm^3) as a function of distance in ångstroms reads:

$$E_{\text{disp}}(r) = -\frac{3\,I\alpha^2}{4\,r^6} \tag{1.30}$$

The strength of London dispersion is a strong function of the electronic structure of atoms and molecules because electronic polarizabilities vary significantly between different compounds. Generally, however, polarizability goes with the number of electrons in an atom, as inner shell electrons (close to the nucleus) are in tight orbitals that are recalcitrant to these perturbations. A good illustrative example is the increase of the boiling points of noble gasses, which only attract one another (and thus condense) via London dispersion. The helium atom, which contains only two 1s electrons, is quite difficult to polarize ($\alpha = 0.2 \times 10^{-24}$ cm^3, $I = 2372$ kJ/mol) and helium has the lowest boiling point, at 4.2 K, of any atom or molecule. As the number of outer shell electrons (and their distance from the nucleus) increases, however, the boiling points of the noble gasses increases steadily reaching 212 K for radon ($\alpha = 4.9 \times 10^{-24}$ cm^3, $I = 1037$ kJ/mol). The steady increase of boiling points in the noble gas series is illustrated on Fig. 1.6.

The London dispersion energy in noble gas dimers remains smaller than the thermal energy because the van der Waals radius of atoms also increases. As a result, these atoms behave as simple monatomic gasses. London dispersion between highly polarizable small molecules such as carbon dioxide can exceed the average thermal energy and such gasses show significant deviation from the ideal gas law at lower temperatures.

Macromolecular receptors and their ligands are largely made of carbon, oxygen, nitrogen, and hydrogen. Because the London dispersion effect is dependent on the number of outer shell electrons, sp^3 hybridized N, O and C atoms are similarly polarizable, and thus, for example, the London dispersion between two water molecules and between water and methane are quite similar. In this respect, London dispersion can be thought of as a universal attraction for which the optimal separation between the two interacting moieties is more important than the matching one

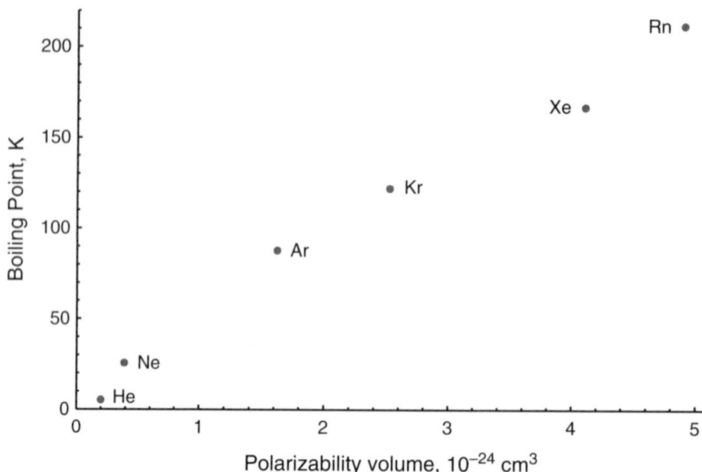

Fig. 1.6 The London-dispersion force is the only attractive interaction between noble gas atoms, and thus the boiling points of the noble gasses are defined solely by the magnitude of this force. Since London dispersion forces are themselves dependent on the polarizability of the atoms involved, the boiling points increase with increasing atomic number as the number of electrons in the atom – and thus its polarizability – increases

Fig. 1.7 Structure of the thyroid hormone triiodothyronine. Notice that thyroid hormones contain several highly polarizable groups and atoms

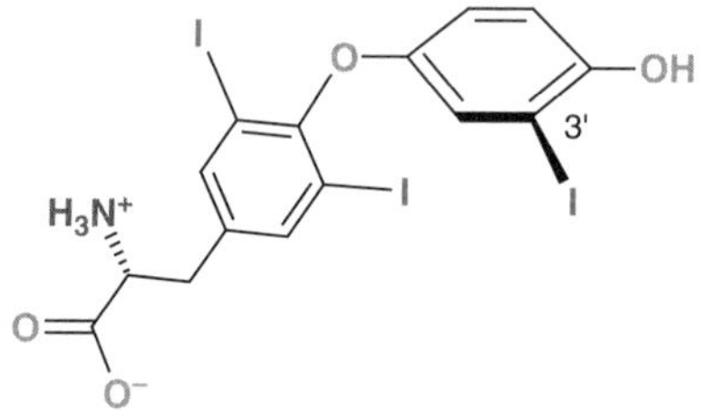

type of atom with another. However, atoms such as sulfur, chlorine, and bromine, are significantly more polarizable and are attracted strongly to other atoms. In general, molecules that are easily polarizable either because they contain aromatic rings or large atoms could derive a significant fraction of their binding affinity from the London dispersion force. For example, it has been estimated that the iodine atom in the position 3′ contributes about 15 kJ/mol toward the net binding affinity (52 kJ/mol) of the hormone thyroxin for its receptor (Fig. 1.7) (Bolger and Jorgensen 1980). Analysis of a crystal structure of the complex between triiodothyronine and the human thyroid hormone receptor β reveals that the 3′ iodine atom interacts with several rather polarizable moieties in the binding site, including the π-bonds of phenylalanine and the sulfur of methionine (Sandler et al. 2004). Of note, mutation of the relevant methionine (to threonine) in the human thyroid hormone receptor leads to a clinically significant resistance to the thyroid hormone (Bayer et al. 2004).

The inverse sixth power distance dependence of London dispersion interaction implies that this interaction is important only at very short distances. "Very short," however, is dependent on the size of interacting atoms. In hydrogen fluoride dimer, the dispersion contribution becomes smaller than the thermal energy at about 3 Å separation between the two molecules (in the gas phase). In the complex between triiodothyronine and the human thyroid hormone receptor a strong interaction with methionine and phenylalanine occurs over 4–5 Å. Using the experimental polarizabilities and ionization potentials of methyl iodide and dimethyl sulfide, the London dispersion between these two model compounds in the gas phase can be estimated to be about 10 kJ/mol at 4 Å separation.

1.3.4 Steric Repulsion (Pauli Exclusion) and Modeling of van der Waals Forces

London dispersion and other induced-dipole interactions are the main attractive forces between molecules in the gas phase. They are thus responsible for an easily observable deviation that real gasses make from ideal gas behavior: whereas, the volume of an ideal gas is simply inversely proportional to the pressure, the volume of a real gas does not expand quite as much when the pressure is reduced. Likewise, real gasses deviate from ideal gas behavior at low temperatures; whereas, an ideal gas would continue to contract *ad infinitum* as the temperature is reduced, real gasses contract only until the temperature reaches their boiling point, below which little further contraction is observed. In the mid nineteenth century Johannes Diderik van der Waals set out to define an equation of state for gasses that would capture these deviations. In order to deal with them, he added two empirical parameters to the ideal gas law to obtain the relationship:

$$\left(P + \frac{a'}{v^2}\right)(v - b') = kT \tag{1.31}$$

where, v is the molar volume divided by Avogadro's constant ($v = V_m/N_A$). The first empirical parameter, a', corrects for the less than expected expansion at high temperatures. The second, b', corrects for the lack of contraction observed below the boiling point; it is the volume of the rigid molecule.

Van der Waals relationship was empirical; the mechanistic origins of these deviations were not understood at the time. We now know their origins and have lumped the nonbonding forces giving rise to them into a single nonbonding force called the attractive van der Waals force. Thus, the parameter a' has its origin in attractive forces arising from induced dipoles – the London dispersion force. But the van der Waals equation also contains parameter b', which arises because two occupied molecular orbitals cannot penetrate each other. Here we discuss the origins of this repulsive interaction.

The Pauli exclusion principle states that no two electrons can have the same quantum numbers. Because of this, filled orbitals (orbitals containing two electrons of opposite spin) repel one another. This quantum mechanical effect profoundly shapes our universe; it is why two objects cannot be in the same place at the same time. At the molecular level, this same effect is responsible for the specific shape and volume of a molecule. It is thus thanks to steric repulsion that the receptors have well-defined surface cavities into which ligands fit. The steep distance dependence of Pauli repulsion dictates that two atoms can approach only to a certain well-defined distance and thus we can talk about specific inter-atomic distances between the atoms in the receptor–ligand complex.

There is no simple classical formula to rigorously describe this quantum mechanical repulsive force. However, it can be shown that the repulsion decays exponentially with the distance between interacting atoms and is important only at very short distances – typically just a few ångstroms. The repulsion becomes overwhelmingly strong when two atoms approach beyond a certain limit – termed their van der Waals radii – and is negligibly weak at distances about twice the sum of their van der Waals radii. (We can talk about specific inter-atomic distances between the atoms in the receptor–ligand complex because the repulsive force between atoms is such a steep function of distance.) Various functional forms have been used to describe the net effect of steric repulsion and attractive forces arising from induced dipoles. In practical calculation of van der Waals interactions in liquids and biological molecules, the empirical Lennard–Jones potential is typically employed:

$$E_{\mathrm{LJ}}(r) = 4\varepsilon\left[\left(\frac{\sigma}{r}\right)^{12} - \left(\frac{\sigma}{r}\right)^{6}\right] \tag{1.32}$$

The first term in the Lennard–Jones potential describes the repulsion arising from the Pauli exclusion principle, and the second term describes the induced-dipole attraction. The parameter σ is related to the minimum-energy interaction distance and is usually evaluated as a sum of atomic van der Waals radii. The parameter ε (softness constant) is related to the depth of the interaction well and is usually calculated as a geometric mean of atomic softness parameters. The inverse twelfth-power distance dependence only approximates the true Pauli repulsion, which is a complex, quantum mechanical interaction dependent on the precise shape and density of the interacting orbitals. But the approximation is good enough for most calculations and has the benefit of supporting a rapid evaluation of interaction energy in computers (we need to only square $(\sigma/r)^6$ instead of evaluating the exponent of distance for every pair of interactions.

Practical calculations of London dispersion and steric repulsion energy in biomolecular systems are complicated by the fact that most molecular mechanics force fields calculate the interaction energy by summing pair-wise contributions from all interacting atoms. The parameters depend on "atomic polarizabilities", but they are not the polarizabilities of free atoms because the molecular environment strongly affects polarizabilities, especially in the case of extended π-electron systems. In practice, empirical fitting is used to obtain a set of σ and ε values that reproduces the liquid properties or the gas second virial coefficient.

1.3.5 Charge Transfer

Interactions that we have discussed so far arise because of the electrostatic force between static or fluctuating charge distributions. By including induction, we allow these charge distributions to change within a molecule in response to charges around the molecule. But so far the electrons that create the charge distribution around the nuclei have remained on molecular orbitals of that molecule. In chemistry, we know that a strong interaction – a chemical bond – can be made when one molecule shares its pair of electrons with another molecule. We will not discuss chemical bonding in this review because such interactions are typically irreversible under the given set of conditions. However, there is an interaction, called the charge transfer, which does not involve a full sharing of a specific electron pair but still contributes to the attraction between two molecules. In this case, an electron-rich region from one molecule can donate some electron density to the electron-deficient region in another molecule without forming a covalent bond. The interaction appears relatively weak in the absence of other stabilizing forces, but the charge transfer mechanism allows the selective capture of electron-deficient ligands into electron-donating cavities (Ko et al. 2007).

Charge transfer interactions are common in inorganic chemistry and typically involve a metal cation with vacant d-orbitals, and an electron-rich moiety such as carbonyl or amine. A complex formation by a charge transfer mechanism is also observed between electron-rich aromatic hydrocarbons (e.g., anline) and electron-deficient aromatic structures, such as *p*-dinitrobenzene (Hunter et al. 2001). The theoretical description of charge-transfer interactions is challenging because of the need to use quantum mechanical methods. In qualitative terms, the complex can be thought of as a superposition of a system in which there is no charge transfer and a system in which there is a full transfer of one electron from the donor to the acceptor. In the ground electronic state, the system with no charge transfer usually dominates, but the electron can be promoted from a highest occupied molecular orbital on the donor to the excited state orbital that is localized on the acceptor moiety. Such an electronic excitation from the donor orbital to an acceptor orbital is promoted by light and the appearance of strong color when two colorless molecules interact is usually a telltale sign of charge transfer interaction. Biologically relevant systems that are stabilized by charge transfer include chlorophyll and heme, in which Mg^{2+} and Fe^{2+}, respectively, are bound in a pophyrin ring. Charge transfer interactions have been employed in the construction of various supramolecular assemblies that function as catalysts, drug delivery or sequestering systems, or specific sensors for analytes. One of the most interesting artificial supramolecular assemblies utilizes charge transfer to harvest sunlight and drive an autonomous nanomotor (Balzani et al. 2006). An interesting review of structures that are stabilized by charge transfer interactions has been recently published (Ko et al. 2007).

1.4 Thermodynamics of Association in the Gas Phase

At this point, a critical reader might be wondering if there are any free molecules left in the air if one considers the all the possible attractive forces that we have described between molecules. The combination of quadrupole–quadrupole and London dispersion forces should lead to the dimerization, if not the precipitation of carbon dioxide. The reason why carbon dioxide and strongly interacting dipolar molecules such as water and ammonia mainly exist as monomers in the gas phase is mainly due to the entropic effects that we have neglected so far. In this part, we will outline the basic thermodynamics of gas phase association, and briefly comment on the applicability of these ideas for association in solutions.

1.4.1 Thermodynamic Contributions from Nuclear Motions

So far, the energies of interaction were calculated assuming that the nuclei of interacting particles are placed in a specific, fixed configuration. In reality, nuclei of atoms and molecules move in space, and these movements are a source of additional thermodynamic contributions to interaction free energy. Consider the simplest association reaction in which two identical atoms form a dimer. Both of the atoms and the dimer have a certain translational energy and entropy while the dimer also contributes due to its rotational and vibrational degrees of freedom. More degrees of freedom generally mean higher entropy, and thus more favorable free energy. Upon dimerization of spherical atoms, we lose three translational degree of freedom but gain two rotational (diatomic molecule has two rotational degrees of freedom) and one vibrational degree of freedom. The energy and entropy contributions for atoms and molecules in the ideal gas are readily calculated via statistical thermodynamics assuming that the vibrations are harmonic and the molecule rotates as a rigid body. The general equations for energies, entropies, and heat capacities for polyatomic molecules are given in standard statistical mechanics textbooks (McQuarrie 2000). Here, we reproduce formulas relevant for the description of association of two spherical atoms into a diatomic molecule:

$$E_{\mathrm{tr}} = \frac{3}{2}RT \tag{1.33}$$

$$E_{\mathrm{rot}} = RT \quad \text{(linear molecule)} \tag{1.34}$$

$$E_{\text{vib}} = R\left(\frac{\Theta_v}{2} + \frac{\Theta_v}{\exp\left[\frac{\Theta_v}{T}\right] - 1}\right) \quad \text{where vibrational temperature } \Theta_v = \frac{h\nu}{k} \tag{1.35}$$

$$S_{\text{tr}} = \frac{5}{2}R + R\ln\left[\frac{V}{N\Lambda^3}\right] \quad \text{where thermal wavelength } \Lambda = \sqrt{\frac{h^2}{2\pi m\,kT}} \tag{1.36}$$

$$S_{\text{rot}} = R + R\ln\left[\frac{T}{\sigma\Theta_r}\right] \quad \text{where rotational temperature } \Theta_r = \frac{h^2}{8\pi^2 kI} \tag{1.37}$$

$$S_{\text{vib}} = R\frac{\frac{\Theta_v}{T}}{\exp\left[\frac{\Theta_v}{T}\right] - 1} - R\ln\left[1 - \exp\left[-\frac{\Theta_v}{T}\right]\right] \tag{1.38}$$

In these equations, ν is the harmonic vibrational frequency (in hertz), σ is the symmetry number (2 for a homonuclear diatomic, 1 for heteronuclear diatomic), m is the mass of the atom or molecule (in kg) and I is the moment of inertia ($\mu{\cdot}r^2$) of the diatomic molecule. To calculate the molar translational entropy in the gas phase, V is taken as the molar volume of the gas and N becomes the Avogadro's constant. In practice, thermodynamic corrections are usually calculated using computational chemistry software that evaluates the molecular geometry and harmonic vibrational frequencies based on quantum mechanical or molecular mechanical models.

An analysis of each of these contributions for the formation of a dimer from identical monomer gives rise to the following observations:

1. Translational energy contribution (–3.7 kJ/mol at room temperature) always favors binding. If the work term is considered as a part of translational enthalpy, then the translational enthalpy correction (–6.2 kJ/mol) also favors binding.
2. Rotational energy contribution favors dissociation in case of atom–atom association but offers a small preferential stabilization to the complex in the case of molecule–molecule association. In general, this contribution does not exceed –3.7 kJ/mol. Together, the maximum contribution of translational and rotational modes to association enthalpy in the gas phase is $4RT$ (9.9 kJ/mol).
3. Vibrational energy contribution generally favors dissociation because the complex has an additional vibrational energy. The vibrational energy term is made up of two parts: the zero-point energy, which is temperature independent, and the thermal energy. In the case of weak complexes, the magnitude of vibrational energy contribution is moderate (3–5 kJ/mol per vibration) because the vibrational frequency ν is in the order of a several tens to few hundreds of cm^{-1}. In the limit of $\nu \rightarrow 0$, zero-point energy vanishes, and the thermal contribution to vibrational energy reduces to RT. In the case of nonlinear polyatomics, six vibrational degrees of freedom are gained, and the minimal vibrational energy contribution is $6RT$ (14.9 kJ/mol at room temperature).

4. Translational entropy always strongly favors dissociation because of loss of three translational degrees of freedom upon complex formation. The magnitude of the destabilization increases logarithmically with the molecular weight of the monomer. For two small molecules, the loss of translational entropy at room temperature costs about 50 kJ/mol. In the case of dimerization of two macromolecules, the translational entropy term may destabilize the complex by over 70 kJ/mol at 25°C. It should be kept in mind that such enormous destabilization is a specific feature of gas phase association reactions.
5. Rotational entropy contribution depends on the symmetry and geometry of interacting molecules. It usually favors dissociation because the number of rotational degrees of freedom is decreased upon complex formation. However, in case of association of small symmetric atoms or molecules, this term may favor complex formation (symmetric monomers have no, or low rotational entropy). On the other hand, when two asymmetric large monomers give a dimer with a twofold symmetry, the dimer may experience an entropic destabilization of about 20 kJ/mol at room temperature.
6. Vibrational entropy contribution is difficult to predict because it depends strongly on the strength of the complex. If we assume that the intrinsic vibrational frequencies of the two monomers are not altered upon formation of the complex, then the vibrational entropy favors complex formation (creation of additional degrees of freedom). This assumption is reasonable in the case of weak complexes and such complexes may enjoy a sizable stabilization (Fig. 1.8). Strong complexes are characterized by high vibrational frequencies

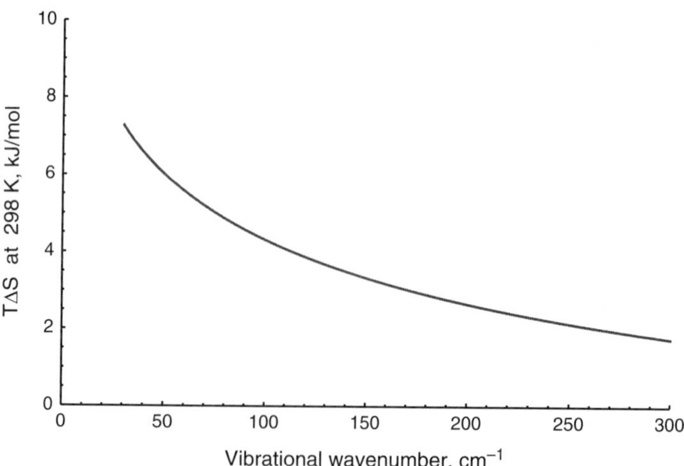

Fig. 1.8 Vibrational entropy stabilizes complexes, but the extent to which it does depends on the vibrational frequencies involved. This figure also provides an example of enthalpy–entropy compensation: strong specific interactions between two molecules lead to lowering of the enthalpy but because stronger non-covalent bonds tend to have higher vibrational frequencies, the entropic gain is smaller in such complexes

for the binding modes; in this case the entropic stabilization is more modest. The association of nonlinear polyatomic molecules leads to six new vibrational modes, and the entropic stabilization of the complex may be as high as 30 kJ/mol.

In summary, the dominant contributions from translational and rotational entropy strongly favor the dissociation of molecular complexes in the gas phase. This effect is partially offset by translational end rotational energy terms. In complexes where the new intermolecular vibrational modes are soft, vibrational entropy further stabilizes the complex but not enough to overcome the unfavorable translational and rotational entropy. As a rule, molecular complexes are strongly destabilized in the gas phase.

1.4.2 Conformational Entropy

Most molecules of interest in the area of biorecognition have additional internal degrees of freedom. Some of these degrees, such as the internal rotation around symmetric methyl groups, modify the free energy of the bound and unbound state by approximately equal degrees. But internal rotations can also give rise to a number of unique conformers for each ligand. In general, the conformers differ in their energy due to intramolecular interactions of the same types discussed earlier. The conformer found in the binding site is usually not the one with the lowest energy in the free state and such "bioactive conformers" may be present in minute quantities because the Boltzmann distribution law dictates that the probability of a conformer decreases exponentially with its relative energy. Even if the bioactive conformer is isoenergetic with all the other conformers, its probability of existence is reduced because the same molecule can equally well adopt a multitude of alternative conformations. Thus, there is a significant entropy penalty when a flexible ligand is forced to adopt a single, specific conformation in the receptor's binding pocket (Chang et al. 2007). The conformational entropy of a molecule that can exist in multiple conformations is given by:

$$S_{\mathrm{conf}} = -R\sum_i p_i \ln[p_i] \quad \text{with} \quad p_i = \frac{\exp\left[-\frac{E_i}{kT}\right]}{\sum_i \exp\left[-\frac{E_i}{kT}\right]} \tag{1.39}$$

Quantitative analysis of this entropy loss, however, is challenging because of our limited ability to accurately calculate conformational energies in larger molecules, especially in the presence of solvent. In the extreme case of N isoenergetic conformers, the conformational entropy simplifies considerably to:

$$S_{\mathrm{conf}}(\mathrm{iso}) = R\ln[N] \tag{1.40}$$

For isoenergetic conformers, the conformational entropy contributions to the free energy exceeds the thermal energy when $N > 4$. Due to the logarithmic dependence, the conformational energy entropy contribution for typical organic molecules is expected to be less than 15 kJ/mol. Chemical methods for reducing the available conformational space of a ligand, such as cyclization, provide a powerful strategy for enhancing binding energetics. Classic studies by Bruice, for example, have demonstrated that significant rate accelerations can be achieved for intramolecular reactions by limiting the number of possible conformations the molecule can adopt (Bruice and Pandit 1960). Further experimental data clearly illustrates that the rate acceleration in such model systems correlates well with the activation entropy (Bruice and Bradbury 1968). The conformational restriction can be readily achieved via the introduction of covalent bonds between otherwise flexible parts of the molecule. For example, in medicinal chemistry one often turns to rigid analogs via chain-ring transformations while attempting to optimize the binding affinity of a lead compound.

1.5 Interaction Energies in the Aqueous Environment

So far we have been discussing electrostatic effects in molecular recognition effectively in vacuo. Biomolecular recognition, however, does not take place in a vacuum, but rather invariably occurs in an aqueous solution, which has profound effects on the extent to which nonbonded interactions enhance binding. Above, in describing electrostatic interactions, we touched on this issue when noting that the high dielectric constant of water reduces electrostatic effects by a factor of almost 80 in aqueous solution. In the following lines, we describe in more detail the ways in which water modulates, and even mediates, molecular recognition.

1.5.1 Effects of Water on Electrostatics

Achieving a quantitative description of the effect of water on intermolecular interactions is challenging. If we treat water as a homogeneous dielectric, the interaction of two point charges is described by a modification of Coulomb's law:

$$E_{\mathrm{el}}(r) = \frac{1}{4\pi\varepsilon_0}\frac{q_1 \cdot q_2}{\varepsilon r} \tag{1.41}$$

The static dielectric constant ε of liquid water is high and slightly temperature dependent (87.9 at 0°C, 55.3 at 100°C). This high dielectric arises due to water's strong dipole moment and exceptionally high molar density (the water concentration in the neat liquid is 55.5 M, which is the highest of any molecular liquid and thus there are more dipoles per cubic ångstrom). Therefore, water effectively shields (weakens) charge–charge interactions. For example, the electrostatic attraction between a cation and anion in a dilute solution would become weaker than the energy of thermal motions at a distance of 4.7 Å at 25°C (ε = 78.4). To achieve 99% association at room temperature by electrostatic attraction alone, the centers of cation and anion must approach to within 1.55 Å. However, the van der Waals repulsion is so significant at such a short distance that ion pairs (except ones involving protons) are not stable in dilute aqueous solution. For example, the experimentally measured thermodynamic dissociation constant for the Li^+F^- ion pair, which, because it is made up of two of the smallest ions, is one of the most stable, is only 0.41 M (Manohar and Atkinson 1993). Ion association becomes more prominent as temperature increases, largely due to the decrease in the dielectric constant.

The static dielectric constant of aqueous solution in the presence of small diffusible ions is higher than that of pure water. This salt-dependent screening of electrostatic interactions is called the Debye–Hückel screening, and the distance dependence of the effective dielectric constant is approximately given by:

$$\varepsilon(r, I) = \varepsilon(\mathrm{H_2O}) \cdot \exp\left(\sqrt{8\pi\lambda_{\mathrm{B}}N_{\mathrm{A}}I} \cdot r\right) \tag{1.42}$$

where I is the ionic strength of the solution, and λ_{B} is the Bjerrum length of the medium, which for water is about 7 Å. The rapid increase of the effective dielectric of electrolyte solutions with the ionic strength and distance means that under physiological conditions ($I \approx 0.15$ M), the attraction of two oppositely charged monovalent ions is insignificant unless they are in a direct contact pair.

The ion–dipole interactions in water are attenuated even more significantly. When a dipole is separated from an ion by solvent, the dipole is free to rotate. In this case, the square of the static dielectric constant enters into the formula of the rotationally averaged interaction energy. The obvious consequence is that the attraction between a dipolar molecule and a point charge in water is very weak. In a similar manner, the attraction of two solvent-separated dipoles is negligible in aqueous milieu. In summary, while the electrostatic forces are enormous in the vacuum, they only play a minor role in the stability of intermolecular complexes between small molecules.

1.5.2 Effect of Water on Induction and van der Waals Forces

The attenuation of London dispersion force by solvent is also a complex topic. At a very approximate level, we can understand the effects of aqueous solvation on van der Waals forces by noting that the polarizabilities of C, O and N are all quite similar – they are neighbors on the periodic table and thus have similar numbers and arrangements of electrons. Because of this, the London dispersion forces associated with the interactions of, for example, two carbon atoms (in a ligand–receptor complex) will likely be quite similar to the forces felt due to the interaction between the carbon and an oxygen from the water solvent (which is a liquid and, again, of very high molar density, and thus it can "nestle" in quite close to any solute, satisfying the attraction). For this reason, van der Waals attractions are generally well satisfied for solvated ligands (and solvated receptor binding sites) and, at best, equally favorable in the receptor–ligand complex, rendering the effect energetically neutral for the formation of most complexes in aqueous solution.

Keep in mind, though, that the preceding description, however helpful, really is a vast simplification. In order to understand this effect in more realistic detail, first recall that the dispersion interaction is mediated by electromagnetic fields that oscillate at frequencies characteristic to the interacting atoms or molecules. In terms of quantum electrodynamics, the dispersion interaction occurs because radiation emitted by the first molecule is reflected back off the second and returns. If the solvent has identical polarizability and characteristic frequency spectrum as the two solutes, the second solute would not reflect the electromagnetic field back any more than the solvent and would be essentially invisible (McLachlan 1965). As a result of this, two solutes that are surrounded by the solvent feel an effective force that also depends on the solute–solvent and solvent–solvent attractions. An approximate expression for this effective interaction energy in terms molecular properties of the solvent (0) and solutes (1 and 2) has been given by Hamaker: (Hamaker 1937)

$$E_{\mathrm{disp(Hamaker)}} = -\left(\frac{1}{4\pi\varepsilon_0}\right)^2 \frac{3}{2r^6} \times \left\{ \frac{I_1 I_2}{I_1 + I_2}\alpha_1\alpha_2 - \frac{I_o I_1}{I_0 + I_1}\alpha_0\alpha_1 - \frac{I_o I_2}{I_0 + I_2}\alpha_0\alpha_2 + \frac{I_0^2}{2I_0}\alpha_0^2 \right\} \quad (1.43)$$

The Hamaker model predicts attenuated attractive dispersion interaction between two identical molecules regardless of the nature of the solvent. This model, however, does not fully describe how the solvent modulates the pair-wise molecule–molecule dispersion interactions. First, the characteristic frequencies of the electromagnetic waves that mediate the London dispersion are in the ultraviolet and optical spectral region; water attenuates such electromagnetic waves with a dielectric constant of ~3. Second, solvent exerts many-body effects on the two solutes, modifying their response. Detailed microscopic description of such many-body effects has been provided by Sinanoğlu for small molecules and Vilker for macromolecules (Kestner and Sinanoğlu 1965). Alternatively, the complicated many-body effects are captured via the macroscopic approach in which the solutes and the solvent are described by their static dielectric constants (epsilon) and high-frequency refractive indexes. A concise treatment of the latter model has been presented by Israelachvili (Israelachvili 1992). In terms of refractive indexes of solutes 1 and 2, and the refractive index of solvent 3, the interaction energy can be expressed as (Erbil 2006):

$$E_{\mathrm{disp}}(r) = -\frac{\sqrt{3}h\nu a_1 a_2}{2r^6} \times \frac{(n_1^2 - n_3^2)(n_2^2 - n_3^2)}{\sqrt{n_1^2 + 2n_3^2}\sqrt{n_2^2 + 2n_3^2}\left[\sqrt{n_1^2 + 2n_3^2} + \sqrt{n_2^2 + 2n_3^2}\right]} \frac{1}{r^6} \tag{1.44}$$

This equation has interesting consequences. First, two molecules will interact via London dispersion through water only if their refractive indexes differ from that of water. Second, London dispersion between identical molecules ($n_1 = n_2$) is always attractive unless the refractive indexes of the solvent and the solute are identical. This result aligns well with the everyday observation *like dissolves like*. But repulsive London dispersion arises if one solute has a larger refractive index than solvent and the other has smaller refractive index than solvent. The theory predicts that water ($n = 1.33$) and benzene ($n = 1.501$) repel when dissolved in ethanol ($n = 1.361$). Indeed, ethanol solution that contains small amounts of water and benzene boils as an azeotrope 13°C below the boiling point of pure ethanol.

1.5.3 Effect of Water on Thermodynamic Contributions from Nuclear Motions

Equations (1.33)–(1.38) are strictly valid for molecules in the gas phase and efforts to adapt these to describe an association of molecules in solution have been only partially successful. It appears well accepted that the ideal gas formulas are appropriate for the estimation of translational and rotational enthalpy terms, which together are expected to favor the binding process by 3*RT* (7.4 kJ/mol at room temperature). The estimation of vibrational enthalpy and entropy changes during binding is technically more involved because of the need to calculate the change of vibrational frequencies between the free and the bound molecules (Tidor and Karplus 1994). The biggest issue lies with our ability to calculate the entropic contributions from translational and rotational modes (Finkelstein and Janin 1989). For example, the volume term in the Sackur–Tetrode equation (1.36) is not clearly defined in the case of dissolved particles: does the solute have the whole volume of the container available, or is it limited to a region defined by the solvent cage? Similarly, it is not clear how much does the solvent affect the rotational freedom of dissolved molecules. How much translational and rotational freedom does the bound ligand have in the binding pocket? Moreover, shall one include the bound water molecules to the mass and moment of inertia that enter the translational and rotational entropy formulas?

If one believes that the gas phase formulas for calculating entropy are appropriate for solution phase interactions, one absolutely needs to invoke additional sources of (favorable) entropy change. This additional favorable entropy is needed to reconcile the large, unfavorable entropy typically calculated (using these formulas) for molecular associations in solution and the small, sometimes favorable entropy of association actually observed experimentally. For example, the release of bound ions and water molecules upon complexation (whether due to the loss of ion hydration, or the "hydrophobic effect," which will be discussed in detail later) is expected to lead to a significant increase in the overall entropy of the system. Unfortunately, we are not much better equipped to calculate the entropy due to such processes because of our limited understanding of the structure of aqueous solution. If one believes that the gas phase formulas are not suitable for the description of translational and rotational degrees of freedom in solution, one can rely on experimental data, calculate all easy-to-calculate entropic terms (i.e., change of vibrational energy in the receptor and ligand), and assign the leftover to a total translational and rotational entropy change in the system (Tamura and Privalov 1997; Yu et al. 1999; Siebert and Amzel 2003). Such works have led to an estimate that the use of ideal gas law entropies overestimates the solution entropies by an order of magnitude (Tamura and Privalov 1997; Yu et al. 1999). In many cases, the translational and rotational entropy contributions at room temperature seem to roughly cancel out the small enthalpic term (Yu et al. 1999; Yu et al. 2001).

As is clear from the above discussion, our understanding of entropic effects in molecular recognition is hampered by our limited understanding of the role that solvent plays in molecular associations in water. So far, we have considered the solvent merely as a homogeneous medium that screens electrostatic and van der Waals interactions. In fact, however, one needs to just consider a case of simple ion association to see that the role that water plays in biomolecular recognition goes well beyond these effects. Recall that the well-founded dielectric model predicts that the distance to which two ions can approach determines the strength of the interaction: two small ions (e.g., Li^+ and F^-) should associate moderately, a small cation (Li^+) and a large anion (I^-) would bind weakly, and two large oppositely charged ions (e.g., Cs^+ and I^-) would not associate significantly because the sum of their ionic radii approaches the distance at which thermal energy becomes comparable with the electrostatic attraction. However, anomalous x-ray diffraction experiments show that Cs^+ and I^- associate more extensively than Na^+ and I^- in water (Ramos et al. 2005). These results suggest that the specific structures of hydration shells around ions play a significant role in determining the interaction between solutes in aqueous solutions. It turns out that the specific hydration shells also exist around nonpolar, "hydrophobic" moieties that do not interact directly with water molecules, and the change of these hydration shells upon molecular interactions greatly affects binding thermodynamics.

1.5.4 Hydrophobic Effect

Given the ability of water to form hydration shells around polar as well as nonpolar moieties, it is not surprising that molecular recognition strongly depends on the properties of liquid water. This unusual driving force – called the hydrophobic effect – unlike all of the other interactions we have described, is not driven by interactions between "hydrophobic" atoms (atoms that do not participate in favorable electrostatic interactions with water) directly, but rather by the unfavorable thermodynamics of solvating these atoms, which indirectly stabilizes their association because the de-solvated, associated state lowers the free energy of the solvent.

In order to fully appreciate the origin of the hydrophobic effect, a brief review of water structure is in order. The two hydrogen bond donors and two hydrogen bond acceptors in a water molecule are nicely arrayed at near tetrahedral angles (~109° apart), allowing water to form a diamond lattice in which every donor and acceptor can be satisfied. This "full satisfaction" only occurs in the solid (the expansion of water to form the diamond lattice is why, unlike the vast majority of substances, water expands upon freezing), but hydrogen bonding remains largely satisfied also in liquid water. For example, even at 90°C there remain, on average, two to three hydrogen bonds per water molecule (Yoshii et al. 2001). Consider, then, what would happen if one were to place an inert (e.g., non-hydrogen-bonding) solute into such partially ordered liquid? Some of the hydrogen bond donors and acceptors at the interface would find themselves missing a partner, at significant energetic cost. Breaking a hydrogen bond in water costs in the order of 10–20 kJ/mol in enthalpy and, given the extremely high molar density of water, the number of effected hydrogen bonds is generally quite high. It is too high, in fact. At room temperature and below, the waters at the interface are thought to perform complex molecular "yoga" so that they can continue to hydrogen bond to other water molecules despite the presence of the noninteracting solute. This, however, comes at a cost. The reduced freedom of the waters at the interface (which can now only hydrogen bond in certain, specific directions) leads to a significant entropic cost, corresponding to a free energy cost of about 0.06–0.1 kJ/mol for each and every square ångstrom of inert surface on the solute (Eisenberg and McLachlan 1986; Olsson et al. 2008). At higher temperatures, above physiological temperatures, this entropic cost becomes too great and some hydrogen bonds become broken. The enthalpic cost of breaking these, however, leads to a similar free energy cost per square ångstrom, and thus the hydrophobic effect is largely temperature independent (Southall et al. 2002). A free energy of 0.06–0.1 kJ/mol $Å^2$ is enormous: the accessible surface area of the side chain of valine, a medium-sized amino acid is estimated at 157 $Å^2$ (Frommel 1984), and thus costs 10–16 kJ/mol to solvate. For this reason, substances that do not hydrogen bond and are not charged readily form intra- or inter-molecular aggregates that serve to reduce the total surface area exposed to water and thus reduce this free energy burden.

1.5.5 Interactions of Dissolved Ligands with Macromolecules in Solution

We can now turn our attention to the process of central interests: the association of a ligand with its receptor. From the above discussion, however, one can surmise that

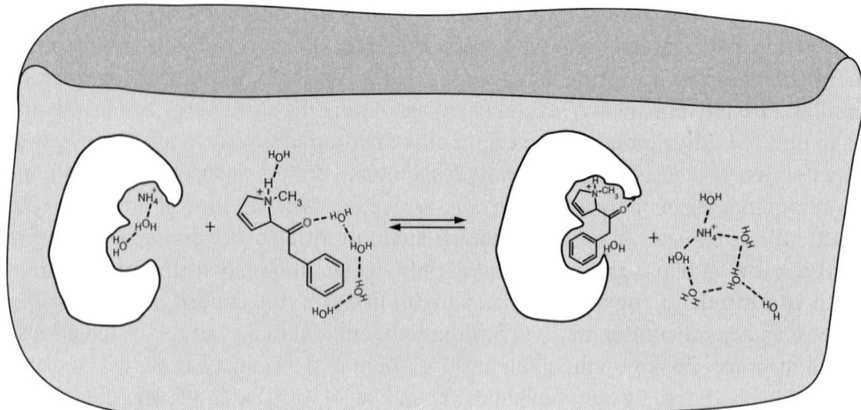

Fig. 1.9 The energetics of water-solute and water-water interactions both contribute significantly to the binding of solvated flexible ligand to a solvated macromolecule. The figure illustrates that one has to consider specific hydration and change in the solvent structure in addition to screening the electrostatic and oscillating electromagnetic fields

the relevant process is not the association of a naked receptor and ligand but instead involves water-attenuated interactions between the receptor and the ligand, as well as changes in conformation and solvation of the two free molecules (Fig. 1.9).

Let us analyze the possible role of hydrogen bonding once more in such biorecognition. Consider for a moment the case of a ligand that contains one hydrogen bond acceptor (the carbonyl in the Fig. 1.9). In its free, unbound state the ligand will be bathed in a sea of hydrogen bond donors; water, as we noted above, has an enormous molar density (again: the highest of *any* molecular substance) and thus the ligand will find itself facing an amazing 110 M concentration of hydrogen bond donors (two per water molecule). Under these conditions its hydrogen bond acceptor will be fully satisfied. Since, at best, all of the hydrogen-bonds will also be satisfied in the bound complex, hydrogen-bonds tend to be energetically nearly neutral for molecular recognition events in aqueous solution. Similarly, the strong dipole moment of water very effectively solvates ions, and thus ion–ion interactions in ligand–receptor complexes likewise add little to the free energy of complex formation. A possible exception might be an ion–ion interaction in a very nonpolar environment deep within the protein, where the effective dielectric is much lower than in water.

The discussion above does not mean that hydrogen bonds, ion pairs, or cation–π interactions are unimportant in biorecognition. In fact, while they play little role in stabilizing complexes relative to the free energy of the free ligand and receptor, they serve a crucial role in determining *which* ligand will bind to which receptor. That is, they tend to destabilize incorrect associations more than they stabilize (relative, again, to the free state) correct associations. Indeed, given the ~10–20 kJ/mol cost of breaking a hydrogen bond, the failure to match up even a single hydrogen bond in a complex can reduce binding affinity by several thousand fold. Analysis of group contributions to desolvation enthalpies shows that the energetic cost of desolvating polar moieties varies significantly with their structure (Cabani et al. 1981). In one extreme, an amide group, which is capable of cooperative hydrogen bonding with several donor and acceptor moieties, has desolvation enthalpy of 60 kJ/mol. Desolvation of un-conjugated hydroxyl and amino groups requires 30–40 kJ/mol while desolvation of nitro and cyano groups takes about 20 kJ/mol. Consider a scenario where a group in the ligand can only accept a single hydrogen bond in the complex. In this scenario, substitution of a hydroxyl or amino group with the cyano group is expected to lead to a stronger binding not because the cyano group is a better hydrogen bond acceptor (acetonitrile has indeed a larger dipole moment than methanol or methyl amine) (Gadhi et al. 1995), but because desolvation of the cyano group is much more affordable than desolvation of the hydroxyl or amino groups. Such an effect has been observed recently in a series of thrombin inhibitors: the cyano analogs bind about ten times stronger than basic amino analogs. In absolute terms, the cyano group contributes at least 10 kJ/mol to the binding via a accepting a single hydrogen bond from the backbone amide while the aliphatic amino group appears to contribute only about 5 kJ/mol via interaction with the active site aspartate (Lee et al. 1997).

The contributions of both the van der Waals attractive and repulsive terms to stabilizing receptor–ligand interactions are, like those of hydrogen bonding, also rather limited when the recognition takes place in aqueous solution. Recall that the London dispersion attractions are quite similar between C, O and N, and thus the dispersion forces of the free ligand and free receptor are well satisfied by water. Likewise, because water is a liquid, it can easily move to relax steric repulsions. Thus both attraction and repulsion tend to be energetically neutral with respect to the formation of most ligand–receptor complexes. Bad steric clashes or large voids, however, typically destabilize incorrect complexes, and thus, as was the case with hydrogen bonds, in aqueous solution (i.e., in most all of biology) evolution tends to utilize van der Waals forces to enforce specificity rather than affinity. Thus, in total, we see that for a good many biological interactions the hydrophobic effect is a primary driving force stabilizing the complex, and van der Waals and electrostatic interactions are there primarily to destabilize incorrect complexes, thus ensuring the specificity of the interaction.

1.6 A Synthesis

Molecular recognition, the specific, noncovalent interaction between two or more molecules, involves a range of forces including simple electrostatics, the special cases of hydrogen bonding, and purely quantum mechanical effects such as the London dispersion and steric repulsion. When the recognition occurs in solution – particularly in water, with its high degree of order, enormous molar density and exceptional hydrogen bonding potential – solvent effects play a major role by attenuating the interaction between polar groups. Finally, the entropies of the free ligand and receptor, the ligand–receptor complex and – again, especially in water – the solvent also play important, sometimes dominant roles in defining the thermodynamics of association.

From the thermodynamics viewpoint one can achieve high-affinity molecular recognition via optimization of a number of entropic and enthalpic contributions. Because of the complexity of calculating the entropic contributions to molecular recognition, practical ligand design often relies on empirical rules, which in many cases are system specific (Freire 2005). One general rule is that more hydrophobic ligands generally bind with more favorable entropy due to the hydrophobic effect. The hydrophobic surface area is a good predictor of this contribution. Another general rule that we arrived to when discussing gas phase thermodynamics but also applies in solutions is that restricting the conformation of the ligand into the shape that is complementary with the receptor surface is entropically favorable. Yet another fairly general rule is the 55% rule, which states that optimal binding affinity is achieved when about 55% of the available cavity is occupied by the ligand (Mecozzi and Rebek 1998). Higher occupancy leads to severe restriction of motions of the ligand in the bound complex and drives down entropy. With lower occupancy, the strongly distance-dependent stabilizing interactions are becoming too weak.

Optimization of enthalpic components is more challenging because the polar and London dispersion interactions that contribute to interaction energy between the ligand and the receptor also occur between these molecules and the solvent. The key here is that a polar group in the receptor should be matched with a complementary polar group in the ligand, and every polar group in the ligand shall interact favorably with a polar group in the protein. A significant enthalpic cost is accrued when the polar groups that are desolvated cannot meet their favorable interaction partners in the complex. Because of the unique polarity (and, in case of interactions involving dipoles and quadrupoles, strong orientation dependence) of electrostatic interactions, only a few molecular structures allow for the correct positioning of polar groups to perfectly match the receptor. Thus, the number of ligands that bind to a particular receptor is rather limited, and finding these ligands in the vast chemical space is challenging. By the same argument, the number of receptor proteins that can meet all the polar interactions in a molecule of a typical drug-like complexity is so small that in most cases only closely homologous proteins are able to bind the ligand. Thus, even though electrostatics, including the special case of hydrogen bonding, and the van der Waals attraction, are greatly ameliorated in water, they modulate the specificity of receptor–ligand interactions. A molecule with too few polar interaction sites may bind to proteins with high affinity, but it would bind to many nonpolar sites indiscriminately.

Taking this all in, we come to a view that seems most consistent with all of the available evidence and with a rational analysis of various contributions to biomolecular recognition: the hydrophobic effect is the driving force behind a binding of a typical ligand to the receptor and, in general, electrostatics (including hydrogen bonds) and van der Waals interactions are largely relegated to the role of ensuring specificity by destabilizing incorrect associations.

1.7 Concluding Remarks

Biomolecular recognition, the process by which biomolecules recognize and bind to their molecular targets, is characterized by high specificity, high affinity, reversibility, and rapid binding kinetics. These attributes arise as a natural consequence of an underlying principle: biomolecular recognition is typically driven by many weak interactions working in concert. This leads to high specificity: it is unlikely that any molecule except the specific target will support a sufficient number of favorable interactions to produce high-affinity binding. It follows that molecular recognition is also generalizable: these individually weak interactions can be re-arranged to produce new binding specificities allowing one to recognize effectively unlimited range of aqueous analytes. Thus, molecular recognition is a cornerstone for a wide range of diagnostic and synthetic technologies. The interactions underlying biomolecular recognition arise from the electrostatic interaction of permanent charge distributions, induction and London dispersion due to polarization of charge distributions, Pauli-exclusion-principle-derived repulsion, and a strong, "attractive" force arising largely from the entropy of the solvent and termed the hydrophobic effect. Indeed, because biomolecular recognition usually takes place in water, which tends to reduce the impact of electrostatic, induction, and London dispersion interactions, this hydrophobic effect is often the dominant force stabilizing the correct biomolecule–target complex. The other nonbonding forces are then relegated to ensure the specificity of the interaction by destabilizing incorrect binding events.

References

Aragones JL, Noya EG, Abascal JLF, Vega C (2007) Properties of ices at 0 K: A test of water models. J Chem Phys 127:154518

Balzani V, Clemente-Leon M, Credi A, Ferrer B, Venturi M, Flood AH, Stoddart JF (2006) Autonomous artificial nanomotor powered by sunlight. Proc Natl Acad Sci USA 103:1178–1183

Bayer Y, Fasshauer M, Paschke R (2004) The novel missense mutation methionine 442 threonine in the thyroid hormone receptor beta causes thyroid hormone resistance: a case report. Exp Clin Endocrinol Diabetes 112:95–97

Bolger MB, Jorgensen EJ (1980) Molecular interactions between thyroid hormone analogs and the rat liver nuclear receptor. Partitioning of equilibrium binding free energy changes into substituent group interactions. J Biol Chem 255:10271–10278

Bruice TC, Bradbury WC (1968) The gem effect. IV. Activation parameters accompanying the increased steric requirements of 3, 3′-substituents in the solvolysis of mono(p-bromophenyl) glutarates. J Am Chem Soc 90:3808–3812

Bruice TC, Pandit UK (1960) The effect of geminal substitution, ring size, and rotamer distribution on the intramolecular nucleophilic catalysis of the hydrolysis of monophenyl esters of dibasic acids and the solvolysis of the intermediate anhydrides. J Am Chem Soc 82:5858–5865

Buckingham AD, Del Bene JE, McDowell SAC (2008) The hydrogen bond. Chem Phys Lett 463:1–10

Burley SK, Petsko GA (1986) Amino–aromatic interactions in proteins. FEBS Lett 203:139–143

Cabani S, Gianni P, Mollica V, Lepori L (1981) Group contributions to the thermodynamic properties of nonionic organic solutes in dilute aqueous solution. J Solution Chem 10:563–595
Chalikian TV, Filfil R (2003) How large are the volume changes accompanying protein transitions and binding? Biophys Chem 104:489–499
Chang CA, Chen W, Gilson MK (2007) Ligand configurational entropy and protein binding. Proc Natl Acad Sci USA 104:1534–1539
Cramer F (1995) Biochemical correctness: Emil Fischer's lock and key hypothesis, a hundred years after – an essay. Pharm Acta Helv 69:193–203
Dougherty DA (1996) Cation–π interactions in chemistry and biology: a new view of benzene, Phe, Tyr, and Trp. Science 271:163–168
Eisenberg D, McLachlan AD (1986) Solvation energy in protein folding and binding. Nature 319:199–203
Erbil HY (2006) Surface chemistry of solid and liquid interfaces. Wiley-Blackwell, New York
Feller D (1999) Strength of the benzene–water hydrogen bond. J Phys Chem A 103:7558–7561
Finkelstein AV, Janin J (1989) The price of lost freedom: entropy of bimolecular complex formation. Protein Eng 3:1–3
Fischer E (1894) Einfluss der Configuration auf die Wirkung der Enzyme. Ber 27:2985–2993
Freire E (2005) A thermodynamic guide to affinity optimization of drug candidates. Protein Rev 3:291–307
Frommel C (1984) The apolar surface area of amino acids and its empirical correlation with hydrophobic free energy. J Theor Biol 111:247–260
Gadhi J, Lahrouni A, Legrand J, Demaison J (1995) Dipole moment of CH_3CN. J Chim Phys Phys – Chim Biol 92:1984–1992
Gregory JK, Clary DC, Liu K, Brown MG, Saykally RJ (1997) The water dipole moment in water clusters. Science 275:814–817
Grzesiek S, Cordier F, Jaravine V, Barfield M (2004) Insights into biomolecular hydrogen bonds from hydrogen bond scalar couplings. Prog Nucl Magn Reson Spectrosc 45:275–300
Gubskaya AV, Kusalik PG (2002) The total molecular dipole moment for liquid water. J Chem Phys 117:5290–5302
Guillot B (2002) A reappraisal of what we have learnt during three decades of computer simulations on water. J Mol Liq 101:219–260
Hamaker HC (1937) The London–van der Waals attraction between spherical particles. Physica 4:1058–1072
Hunter CA, Lawson KR, Perkins J, Urch CJ (2001) Aromatic interactions. J Chem Soc Perkin Trans 2:651–669
Isaacs ED, Shukla A, Platzman PM, Hamann DR, Barbiellini B, Tulk CA (1999) Covalency of the hydrogen bond in ice: A direct X-ray measurement. Phys Rev Lett 82:600–603
Israelachvili JN (1992) Intermolecular and surface forces, 2nd edn. Academic Press, New York
Jiang L, Althoff EA, Clemente FR, Doyle L, Röthlisberger D, Zanghellini A, Gallaher JL, Betker JL, Tanaka F, Barbas CFI, Hilvert D, Houk KN, Stoddard BL, Baker D (2008) De novo computational design of retro-aldol enzymes. Science 319:1387–1391
Jorgensen WL, Chandrasekhar J, Madura JD, Impey RW, Klein ML (1983) Comparison of simple potential functions for simulating liquid water. J Chem Phys 79:926–935
Kestner NR, Sinanoğlu O (1965) Intermolecular forces in dense media. Discuss Faraday Soc 40:266–267
Ko YH, Kim E, Hwang I, Kim K (2007) Supramolecular assemblies built with host-stabilized charge-transfer interactions. Chem Commun 13:1305–1315
Leavitt S, Freire E (2001) Direct measurement of protein binding energetics by isothermal titration calorimetry. Curr Opin Struct Biol 11:560–566
Lee S-L, Alexander R, Smallwood A, Trievel R, Mersinger L, Weber PC, Kettner C (1997) New inhibitors of thrombin and other trypsin-like proteases: Hydrogen bonding of an aromatic cyano group with a backbone amide of the P1 binding site replaces binding of a basic side chain. Biochemistry 36:13180–13186

Lii J-H, Allinger NL (2008) The important role of lone-pairs in force field (MM4) calculations on hydrogen bonding in alcohols. J Phys Chem A 112:11903–11913
London F (1937) The general theory of molecular forces. Trans Faraday Soc 33:8–26
Ma JC, Dougherty DA (1997) The Cation–π Interaction. Chem Rev 97:1303–1324
Magnasco V, Battezzati M, Rapallo A, Costa C (2006) Keesom coefficients in gases. Chem Phys Lett 428:231–235
Manohar S, Atkinson G (1993) The effect of high pressure on the ion pair equilibrium constant of alkali metal fluorides: a spectrophotometric study. J Solution Chem 22:859–872
McLachlan AD (1965) Effect of the medium on dispersion forces in liquids. Discuss Faraday Soc 40:239–245
McQuarrie DA (2000) Statistical Mechanics, 2nd edn. University Science Books, Sausalito, CA
Mecozzi S, Rebek J Jr (1998) The 55% solution: a formula for molecular recognition in the liquid state. Chem Eur J 4:1016–1022
Ngola SM, Dougherty DA (1996) Evidence for the importance of polarizability in biomimetic catalysis Involving cyclophane receptors. J Org Chem 61:4355–4360
Olsson TSG, Williams MA, Pitt WR, Ladbury JE (2008) The thermodynamics of protein-ligand interaction and solvation: insights for ligand design. J Mol Biol 384:1002–1017
Plokhotnichenko AM, Radchenko ED, Stepanian SG, Adamowicz L (1999) *p*-Quinone dimers: H-bonding vs. stacked interaction. Matrix-isolation infrared and ab initio study. J Phys Chem A 103:11052–11059
Ramos S, Neilson GW, Barnes AC, Buchanan P (2005) An anomalous X-ray diffraction study of the hydration structures of Cs^+ and I^- in concentrated solutions. J Chem Phys 123:214501–214510
Romero AH, Silvestrelli PL, Parrinello M (2001) Compton scattering and the character of the hydrogen bond in ice Ih. J Chem Phys 115:115–123
Sandler B, Webb P, Apriletti JW, Huber BR, Togashi M, Lima STC, Juric S, Nilsson S, Wagner R, Fletterick RJ, Baxter JD (2004) Thyroxine–thyroid hormone receptor interactions. J Biol Chem 279:55801–55808
Siebert X, Amzel LM (2003) Loss of translational entropy in molecular associations. Proteins: Struct Funct Bioinf 54:104–115
Southall NT, Dill KA, Haymet ADJ (2002) A view of the hydrophobic effect. J Phys Chem B 106:521–533
Stone AJ (1996) The theory of intermolecular forces. Oxford University Press, New York
Stone AJ (2008) Intermolecular potentials. Science 321:787–789
Sunner J, Nishizawa K, Kebarle P (1981) Ion–solvent molecule interactions in the gas phase. The potassium ion and benzene. J Phys Chem 85:1814–1820
Tamura A, Privalov PL (1997) The entropy cost of protein association. J Mol Biol 273:1048–1060
Tidor B, Karplus M (1994) The contribution of vibrational entropy to molecular association. The dimerization of insulin. J Mol Biol 238:405–414
Tsuzuki S, Yoshida M, Uchimaru T, Mikami M (2001) The origin of the cation/π interaction: the significant importance of the induction in Li^+ and Na^+ complexes. J Phys Chem A 105: 769–773
Williams JH (1993) The molecular electric quadrupole moment and solid-state architecture. Acc Chem Res 26:593–598
Yoshii N, Miura S, Okazaki S (2001) A molecular dynamics study of dielectric constant of water from ambient to sub- and supercritical conditions using a fluctuating-charge potential model. Chem Phys Lett 345:195–200
Yu YB, Lavigne P, Kay CM, Hodges RS, Privalov PL (1999) Contribution of translational and rotational entropy to the unfolding of a dimeric coiled-coil. J Phys Chem B 103:2270–2278
Yu YB, Privalov PL, Hodges RS (2001) Contribution of translational and rotational motions to molecular association in aqueous solution. Biophys J 81:1632–1642
Zacharias N, Dougherty DA (2002) Cation–π interactions in ligand recognition and catalysis. Trends Pharmacol Sci 23:281–287

Chapter 2
Surface Sensitization Techniques and Recognition Receptors Immobilization on Biosensors and Microarrays

Vincent Dugas, Abdelhamid Elaissari, and Yves Chevalier

Abstract The quality of a biosensing system relies on the interfacial properties where bioactive species are immobilized. The design of the surface includes both the immobilization of the bioreceptor itself and the overall chemical preparation of the transducer surface. Hence, the sensitivity and specificity of such devices are directly related to the accessibility and activity of the immobilized molecules. The inertness of the surface that limits the nonspecific adsorption sets the background noise of the sensor. The specifications of the biosensor (signal-to-noise ratio) depend largely on the surface chemistry and preparation process of the biointerface. Lastly, a robust interface improves the stability and the reliability of biosensors. This chapter reports in detail the main surface coupling strategies spanning from random immobilization of native biospecies to uniform and oriented immobilization of site-specific modified biomolecules. The immobilization of receptors on various shapes of solid support is then introduced. Detection systems sensitive to surface phenomena require immobilization as very thin layers (two-dimensional biofunctionalization), whereas other detection systems accept thicker layers (three-dimensional biofunctionalization) such as porous materials of high specific area that lead to large increase of signal detection. This didactical overview introduces each step of the biofunctionalization with respect to the diversity of biological molecules, their accessibility and resistance to nonspecific adsorption at interfaces.

Keywords Functionalization · Biofunctionalization · Surface chemical modification · Native biomolecules · Modified biomolecules · Staudinger ligation · Click-chemistry · Native chemical ligation · Expressed protein ligation · Silanization · Self-assembled monolayer · Entrapment · Nanoparticles · Sol–gel process · Adsorption · Chemisorption · Silica · Silicon · Gold layer · Streptavidin · Biotin ·

V. Dugas (✉)
Laboratoire des Sciences Analytiques, UMR5180, Bâtiment CPE, 43 Boulevard 11 Novembre 1918, 69622 Villeurbanne Cedex, France
e-mail: vincent.dugas@univ-lyon1.fr

M. Zourob (ed.), *Recognition Receptors in Biosensors*,
DOI 10.1007/978-1-4419-0919-0_2, © Springer Science+Business Media, LLC 2010

Protein · DNA · Carbohydrate · Enzyme · Ligand capture · Protein capture · Site-directed immobilization · Site-specific immobilization

Abbreviations

μ-TAS	Micro-total analysis system
AAPS	N-(2-aminoethyl)-3-aminopropyltrimethoxysilane
ALD	Atomic layer deposition
APTS	Aminopropyltriethoxysilane
Asp	Aspartic acid
BSA	Bovin serum albumin
DIOS	Desorption/ionization on silicon
DMP	Dimethyl pimelimidate
DMS	Dimethyl suberimidate
DMSO	Dimethylsulfoxide
DNA	Desoxyribonucleic acid
DTT	Dithiothreitol
EDTA	Ethylenediamine tetraacetic
ELISA	Enzyme-Linked immunosorbent assay
ENFET	Enzymatic field-effect transistor
EPL	Express protein ligation
ET	Electron transfer
Glu	Glutamic acid
GOD	Glucose oxidase
GPTS	3-glycidoxypropyltriethoxysilane
IDA	Iminodiacetic acid
IPL	Intein-mediated protein ligation
ISE	Ion-selective electrodes
ISFET	Ion-selective field-effect transistor
ITO	Indium Titanium oxide
Lys	Lysine
M_2C_2H	4-(N-maleimidomethyl)cyclohexan-1-carboxylhydrazide
MALDI	Matrix-assisted laser desorption/ionization
MESNA	2-Mercaptoethansulfonate
MPAA	(4-carboxymehtyl)thiophenol
mRNA	Messenger ribonucleic acid
NCL	Native Chemical Ligation
NHS	N-hydroxysuccinimide
NTA	Nitrolotriacetic acid
ODN	Oligodesoxyribonucleotides
PAMAM	Poly(amino)amine
PCP	Peptide carrier protein
PCR	Polymerase chain reaction

PDITC	Phenylenediisothiocyanate
PDMS	Polydimethylsiloxane
PEG	Poly(ethylene glycol)
PNA	Peptide nucleic acid
SAM	Self-assembled monolayer
SIAB	Succinimidyl 4-(N-iodoacetyl)aminobenzoate
SMCC	Succinimidyl-4-(N-maleimidomethyl)cyclohexane-1-carboxylate
SMPB	Succinimidyl-4-(N-maleimidophenyl)butyrate
SPDP	N-succinimidyl-3(2-pyridyldithio)propionate
s-SIAB	Sulfosuccinimidyl 4-(N-iodoacetyl)aminobenzoate
TCEP	Tris(2-carboxyethyl)phosphine
TEOS	Tetraethoxysilane
TMOS	Tetramethoxysilane

2.1 Introduction

Sensors and further biosensors can be straightforwardly depicted as devices that convert a physical or biological event into measurable (electrical) signal. Biosensors exhibit two elementary parts; the sensitive part where specific biological events take place and a transducing system that mediates the biological events into quantifiable signal. The transducer work relies on various physical principles: mechanical (e.g. weight or topographic measurement), optical (e.g. fluorescence emission, absorbance of surface evanescent waves) or electrochemical (e.g. potentiometric or conductimetric measurements). Common feature of detection techniques is the biosensitive layer that confines the biological event in the very close vicinity of the transducer. A target biomolecule is detected when a specific interaction/recognition takes place in the sensitive layer where a bioreceptor has been attached (tethered probe, ligand or substrate). The specific interaction gives rise to a "chemical signal" at the surface that the transducer is sensitive to. Instances of chemical signals are variation of local concentration of ionic species or pH in the sensitive layer, formation of absorbing or fluorescent complex species, electron transfer to a conducting surface, variation of refractive index of the sensitive layer, etc. The performance of a biosensor comes from (1) its ability to immobilize receptors while maintaining their natural activity, (2) the bioavailability (accessibility) of the receptors to targets in solution, (3) a low nonspecific adsorption to the solid support. These specifications govern the specificity and sensitivity of such devices and can be tailored by an appropriate choice of the solid–liquid interface where the bioreceptors are immobilized. The chemical preparation of the surface is a key parameter; the physico–chemical properties of the interface play an important role in achieving optimal recognition of the target and limiting the nonspecific adsorption. Lastly, the stability of the sensitive part of a biosensor depends on the sustainability of the surface functionalization. Immobilization of bioreceptors through robust and stable covalent bonds is a good means to gain stability. The optimum elaboration and use of biosensors calls for sure background knowledge about the surface preparation and immobilization process of biospecies.

This presentation gives a detailed overview of the individual steps of surface modification (functionalization) and immobilization of bioreceptors (biofunctionalization) onto solid supports. In the first section, the adsorption of biomolecules to solid–liquid interface is addressed. The simple utilization of adsorption and its main drawbacks are presented. A definite improvement in terms of reliability, selectivity, and sustainability brings in covalent immobilization as an alternative to adsorption. The main coupling strategies allowing covalent immobilization of native biomolecules are reviewed, together with the design of the surface properties aimed at improving the accessibility of analytes in solution by oriented immobilization of modified biomolecules. The last section introduces the chemical modification of a wide variety of solid supports used (flat solid supports, porous materials, polymer coatings and nanoparticles) as biosensing elements.

2.2 Adsorption, Chemical Grafting, and Entrapment

2.2.1 From Adsorption to Grafting, a Historical Perspective

The development of biological analysis techniques based on the detection of solid-supported biomolecular interactions began in the 1970's. Thus, ELISA (Engvall and Perlman 1971), Southern blot (Southern 1975) and Northern blot (Burnette 1981) techniques have been introduced and still remain widely used. The principle is simple and relies on the specific recognition between a molecule in solution and a partner molecule immobilized onto a solid support. Immobilization makes the detection easier and more sensitive because the molecules are concentrated on the surface and the recognition events to be detected are precisely localized. The molecular biologists familiar to the blotting techniques define the "unknown" molecule that is to be identified as the "target" and the well-characterized molecule that recognizes the target as the "probe." In the original blotting techniques, the target is immobilized on the support and probe is in solution. Immobilization of the molecules from the analyte solution proceeds through adsorption. Therefore, the surface where the recognition takes place is chosen such that most molecules of interest readily adsorb. Adsorption is not specific and there is no need for such techniques.

These standard blotting techniques consist of assaying a panel of probe molecules in solution by exposing them to the target immobilized on a solid support. This is conducted in parallel experiments; the number of individual tests corresponds to the number of available probe molecules. The reverse approach is much more suitable for large scale analysis. Modern blotting techniques consist in attaching known probes to a surface and contacting them to the target solution as a single test. Several probes are attached to the same support as localized spots. A large number of different probes can be analyzed during a single test depending on the density of spots, therefore on the area of support in contact with the solution and on the spotting technology.

This method, where known chemical species are immobilized on a surface, takes after the so-called modified electrodes that electrochemists were developing simultaneously during the same period (Guilbault and Montalvo 1969, 1970). Thus, electrodes are made specific by immobilization of chemical species at their surface. The chemical reaction with the analyte at the electrode surface gives a chemical event (red-ox reaction, pH change, etc.) that is detected. The transduction is therefore limited to electrochemical detection.

In the 1980s, the blotting techniques evolved to multispots formats (multispot or dot) called arrays. These progresses are largely due to better control of the chemistry and physical chemistry of the support surface, synthesis of modified biomolecules, and manufacturing technologies of spot arrays. Early tests by blotting techniques were conducted on a small number of individual spots, giving "low-density arrays" or "macroarrays." The biological solutions were simply deposited as spots on glass slides or sheets of polymer materials (nitrocellulose or Nylon membranes for instance). More sensitive detection was achieved by using porous membranes of large available surface area that allowed larger loading capacities. The traditional process aimed at the immobilization of biomolecules is quite rough; it consists in depositing a drop of solution and letting it dry on the surface. Therefore, the biomolecules are presumably adsorbed.

There are several drawbacks related to the immobilization by means of adsorption only. Adsorption depends on the interaction of the biomolecule and the surface, so that the amount of adsorbed molecules may vary from spot to spot. Release to the solution is possible when the solid support is immersed in the analyte solution; this causes a loss of the signal and possible cross contamination of the spots. The reversal of the roles of probes and target (now probes are immobilized) allows the immobilization process to work again. Adsorption was used at the early beginning because it was very simple and could be readily implemented in the biology research laboratory. Since the probes are known molecules that are isolated and purified before their deposition by spotting, optimized chemistry can be done. Modified probe molecules are used to improve their immobilization. A decisive progress is made by replacing the adsorption by chemical grafting. Chemical grafting firstly avoids release to solution, which is the main drawback of adsorption. Other definite improvements are several. Chemical grafting allows reducing the background noise coming from nonspecific adsorption because optimized surface chemistry can be implemented for that purpose. Thorough cleaning of the surface (repeated rinsing) is possible in order to remove the nongrafted molecules and contaminants. Lastly, the device may be reusable if a correct regeneration process is devised. Possible reutilization is often a valuable benefit, even for disposable devices, because one wishes to repeat the analysis under the same condition in order to calculate a statistical mean value, the standard error of the mean, and to assess the variability of the sample. Therefore, the tendency replaces adsorption by chemical grafting.

Technological advances have made it possible to carry out chemical reactions at the micrometer scale. The full benefits of microtechnology or nanotechnology can be achieved if the chemical and physico–chemical properties of the solid surface are improved in parallel. The adaptation of immobilization technique to miniaturization is made while preserving the bioactivity and increasing the accessibility of target molecules in solution. For instance, the use of elaborated devices for the localized synthesis of oligonucleotides onto a solid surface at the micrometer scale allows the manufacture of DNA microarrays with several hundred thousand spots per square centimeter (Gershon 2002). Besides technological improvements, new methods are intimately related to progresses in the field of molecular biology and synthesis and/or modification of biomolecules (e.g. protein engineering, fusion proteins, tagging).

Definite improvements have involved new grafting methods to the surfaces but their implementation in the everyday practice of the users took much time. Although early attempts to oriented covalent linkage (one anchoring point preserving of the recognition sites) of DNA date back to 1964 (Gilham 1964), most biosensors or affinity tests up until the middle 1990s relied on simple adsorption or random grafting of the bioprobes (reaction through the DNA exocyclic amines (Millan et al. 1994) or phosphates (Sun et al. 1998)). Progresses in regio-specific grafting reactions emerged when the optimization of rough strategies reached their limits. Uncontrolled immobilization is only suitable for biological tests where the quantity of available biological products is not limited or for detection of a single species such as in ELISA tests.

Efforts have been focused on DNA analysis before the year 2000 which lead to the advent of DNA chips. Proteins chips came few years later (MacBeath and Schreiber 2000). Although the technology is currently able to achieve over 400,000 parallel measurements in a single test (DNA chip), the complexity of biological molecules leads to many artifacts that make the exploitation of these "high-density" microarrays troublesome. A tendency was to increase the amount of information (the density of spots) in order to collect redundant information that could be used to eliminate false responses. Bioinformatics appeared and became an independent discipline that the end-users could not control. Because of such difficulties, a current trend in proteins microarrays is lowering the number of parallel measurements to improve the robustness of the results.

Therefore, the increase of chemical signal allowed more sensitive analyses and miniaturization of the devices, giving "high-density arrays" (microarrays). This straight idea is to be tempered because the loss of selectivity is a limitation. Increasing the signal by improving the biochemical recognition sensitivity (higher density and bioavailability of attached probe molecules) often leads to the concomitant loss of selectivity. On the contrary, decreasing the background noise by preventing false recognition events improves the signal-to-noise ratio and the selectivity. Controlling the physical chemistry of surface preparation and grafting process is the way that helps improving the detection sensitivity and selectivity.

Lastly, the technologies derived from blotting techniques may fail in some specific cases because the confinement of the biochemical recognition at a transducer surface is troublesome. An example is the recognition by carbohydrates and lectins that is very sensitive to the local environment. This requires looking for quite new strategy of analysis. As example, new strategies make use of molecular recognition in solution. Methods based on Biobarcode (Lehmann 2002; Nam et al. 2003; Winssinger et al. 2004; Diaz-Mochon et al. 2006; Bornhop et al. 2007; Chevolot et al. 2007) and nanoparticles are presented in the last part of this review.

Finally, new biosensor paradigms deal with miniaturization and integration of different features ranging from the preparation of the biological sample to the detection (micro-Total Analysis System (μ-TAS) or lab-on-a-chip). The adaptation of the surface chemistry is once again sought to tailor surfaces with different functionalities depending on the compartment system.

2.2.2 *Adsorption and Grafting: Physical Chemistry and Thermodynamics*

2.2.2.1 Surface Interactions or Chemical Grafting?

There is a confusion regarding the precise nature of adsorption, physisorption, chemisorption and chemical grafting because the words do not have the same acceptance in different scientific communities. It is difficult to eliminate the confusion coming from contradictory definitions. Explanatory definitions are given here in order to replace semantic disputes by scientific understanding. A more detailed discussion of the same type can be found in (Rouquerol et al. 1999).

Adsorption from solution takes place if the molecules interact with the surface by means of attractive interactions. Therefore, part of the molecules goes to the surface and a residual part remains in solution. A dynamic equilibrium is established where molecules in solution adsorb at the surface, while adsorbed molecules leave the surface. The system is equilibrated when the rate of adsorption and desorption are equal. The surface concentration and residual concentration are related, the larger the concentration in solution, the larger the adsorbed amount. The surface concentration is termed surface excess; it is expressed in mol/m^2. From a thermodynamic point of view, the standard chemical potential of the molecule at the surface is lower than that in solution, so that the molecules are in a more favorable environment at the surface. The interactions may be of various physico-chemical origins and the standard free energy of adsorption may have enthalpic and entropic contributions of various relative magnitudes.

Grafted molecules are attached to the surface by means of a chemical bond (or several). Grafted molecules do not leave the surface, even if repulsive interactions are operated.

The difference between the adsorbed and grafted is really large and has important implications regarding sensor application. Adsorbed molecules are in equilibrium with a residual concentration in solution. Molecules of the same nature are present in solution and should not be removed because it would shift the adsorption equilibrium. On the contrary, grafted molecules are irreversibly attached, so that the solution may be free of residual grafts in solution. Residual concentration can be safely removed from the solution by rinsing the surface.

Therefore, there are two major differences between adsorbed and grafted molecules:

- There is a residual concentration of molecules in solution in case of adsorption and such molecules are able to do the chemical recognition in bulk solution. Such unwanted molecules can be eliminated in case of grafting.
- Adsorbed molecules can be desorbed according to the reversible adsorption equilibrium, whereas grafted are irreversibly attached. In particular, rinsing out a surface-bearing adsorbed molecules causes washing them off.

The above view is oversimplified however. The difference between adsorption and grafting is a matter of binding energy. Adsorption involves interactions between molecules with interaction energies of the same magnitude as thermal energy. Therefore, Brownian motion can cause adsorption and desorption. The energy of a chemical bond is orders of magnitude larger, so that it is considered irreversible. But there are molecular interactions that are in between the two. The hydrogen bond can easily be broken. Polymers that bind to surface sites by several anchoring points cannot be desorbed by rinsing the surface with the pure solvent; adsorption looks irreversible, actually it is very slow (Fleer et al. 1993).The very strong biotin–avidin complex behaves like a chemical bond. A chemical bond can also be cleaved. For example, an ester bond hydrolyses when it is immersed in water and the hydrolyzed grafts can react again with the free surface sites. This process that looks like an adsorption–desorption process is so slow, however, that hydrolysis is called a chemical degradation. There is actually no fundamental difference between surface hydrolysis and desorption, this is just a matter of energy of activation that sets the release rate. The same indefinite difference is found between chemical reaction and complex formation in bulk solution.

The difference between physisorption and chemisorption is also a matter of interaction energy and this adds to the general confusion. Physisorption is adsorption by means of weak (physical) interactions while chemisorption involves stronger (chemical) interactions. Some authors consider chemisorption as irreversible attachment, therefore chemical grafting. The crossover from physical to chemical has never been precisely given in terms of energy (it depends on who is speaking), so that this is just a semantic discussion.

2.2.2.2 What are the Different Forces Acting on Molecules at the Vicinity of Surface and What About the Specificity?

The interactions that are responsible for adsorption are the same as in solution. Adsorption of an invited molecule to a host surface site can be viewed as a complex formation where one of the partners is attached to the surface. The interactions are the same in this case. The presence of adsorption sites is not necessary however; attraction to the surface is enough. As example, adsorption of a hydrophobic molecule (surfactant) from aqueous solution to the oil–water interface does not involve localized adsorption sites. Generally speaking, specific adsorption often requires well-defined adsorption sites.

The interactions that are most relevant in the manufacturing and functioning of chemical sensors are the following:

- Hydrophobic interaction comes from the poor solvation of molecules or surface by water. Hydrophobic molecules are pushed one to each other by water solvent. This interaction is not specific since every nonpolar molecule or material is hydrophobic.
- Electrostatic interaction is either attractive or repulsive according to the respective signs of the electrically charged partners. This is also highly nonspecific since it only depends on the electrical charge and the ionic strength of the medium. The precise nature of the ionic species does not matter. Electrostatic interaction is strong and long-ranged.
- Polar interactions such as dipole–dipole interactions are electrostatic in origin but are not called as such. They are weaker and short-ranged, they can be either attractive or repulsive according to the orientations of the dipoles or multipoles.
- Hydrogen bond is an acid-base interaction that involves acidic and basic sites. It is oriented but not specific. The presence of both acidic and basic sites is enough for bonding.
- Complex formation involves sharing external electron orbitals of partners. It is highly oriented and closely depends of the chemical nature of the partner, so that it is specific.

Therefore, most interactions are not specific. Selectivity, especially chemical recognition comes from the combination of several localized interactions. Even the hydrophobic interaction can become specific when it is localized. As an example, proteins bear well-localized hydrophobic domains that can favorably interact with a patterned hydrophobic/hydrophilic surface if the separation between hydrophobic areas fit the distance between the hydrophobic domains of the protein. A single hydrogen bond, although oriented, is not specific; high selectivity is often brought about by combination of several oriented hydrogen bonds between the partners as for example between the bases of nucleic acids or in the biotin–avidin complex.

2.2.3 *Kinetic Aspects of Adsorption, Desorption, and Grafting*

- The adsorption rate is related to the concentration in solution. High concentrations are preferred and often compulsory, even if the fraction that remains finally adsorbed is small. In case of scarcely available materials, slow adsorption from dilute solution is a problem. Drying the solution on the surface helps the adsorption of the dilute probe to be adsorbed but also forces the unwanted materials to the surface.

- Very slow desorption allows working quite safely in sensor applications. Adsorbed polymers desorb very slowly because of the multiple anchoring to the surface (Fleer et al. 1993). It is often considered irreversible. Therefore, polymer deposits are quite satisfactory and bear further surface treatments. Immobilization by means of entrapment of large species (proteins) in gels relies on the same idea. This is not a permanent attachment but tight entrapment which slows down enough the release with respect to the time scale of the utilization.
- Grafting is irreversible but involves a chemical reaction that is very slow if the probe is too dilute. Therefore immobilization by grafting materials of poor availability may be long. Long reaction time leaves opportunities for side-reactions, chemical degradations, etc. Adsorption causes a local increase of concentration at the surface that accelerates the grafting reaction. It is therefore often wise to perform the grafting reaction under conditions of strong adsorption and go back to weak adsorption conditions for recovering the selective biochemical recognition and low background signal.

2.2.4 Nonspecific Adsorption as a Source of Background Signal

In sensor applications, the target molecule should bind to the surface by specific interactions only in order to obtain a selective detection. Nonspecific interactions cause adsorption of unwanted molecules that induce an interfering signal. This is an important origin of background signal (noise). There are so many nonspecific interactions listed above that reducing the background signal is a difficult task.

The difficulties encountered with early devices relying on adsorption come from the uncontrolled nonspecific adsorption. Thus, sensitizing a surface by adsorption of a selective probe to the surface requires an interaction that remains active after the functionalization. A high adsorbed density aimed at obtaining a high detection level requires a strong interaction of the probe with the surface and consequently a high interaction of everything with the surface, leading to a high background signal. Typically adsorption of a negatively charged probe protein can be achieved to a positively charged surface; but every anionic species present in the analyte will bind to the surface by nonspecific interactions and the consequence will be a disastrous selectivity. Therefore, functionalization by adsorption does not allow to reach high sensitivity.

Grafting allows much more freedom because the physico-chemical characteristics of the surface can be changed after grafting the probe without wondering about probe release to the solution. Repulsion of grafted molecules out from the surface does not cause the loss of the attached probes and often appears advantageous because the recognition site is pushed far from the surface, increasing the bioavailability.

2.3 Classification of the Main Immobilization Pathways

2.3.1 Specificities of Biomolecules for Chemical Coupling

Most of the molecules implied in biomolecular or cellular recognition mechanisms are water-soluble (nucleic acids, proteins, carbohydrates, etc.). Exceptions to this rule are integral membrane proteins that are permanently attached to biological membranes. Therefore, the coupling of biological molecules onto solid support should preferably proceed in aqueous solvent.

In addition to the solubility constraints, the preservation of the activity of immobilized molecules requires keeping nondenaturizing conditions. For example, temperature, salt and pH conditions are the important factors for proteins stability; the immobilization process should be adapted to such conditions. For instance, proteins are spotted to arrays surface in aqueous buffered solution containing large amount of low volatility moistening agent like glycerol to prevent evaporation of the droplets (MacBeath and Schreiber 2000; Lee and Kim 2002). The probes to be immobilized in their native (active) form are mainly coupled to the surface via reactive functional groups such as amino, hydroxyl, thiol, carboxyl and aldehyde groups. Covalent immobilization must avoid using the functional groups that are involved in the functional conformation of biomolecules (base stacking, protein folding) or/and in the recognition mechanisms.

Lastly, biological molecules are considered as chemical reagents during the immobilization process. Therefore, their concentration should preferably be that of conventional chemical reactions of organic chemistry, that is approximately 10 wt%. The concentration of reagents sets the reaction rate; it should not be calculated from the very small amount required for covering the surface with a dense layer. This is a difficult issue because biological molecules such as DNA, proteins or carbohydrates are available in small quantities owing to their expenses. So, the coupling step of biological molecules onto solid support is often done under unfavorable conditions with diluted aqueous solution in the micromolar range (Dugas et al. 2004).

Main strategies of immobilization of biomolecules rely on (1) physical adsorption onto solid support; (2) entrapment into polymeric networks (gels); (3) chemical grafting by means of covalent binding onto reactive groups of native molecules; (4) chemical grafting onto modified biomolecules (linker, polymerization); (5) biochemical approaches to conjugate modified biomolecules (e.g. tag expressed protein).

Obviously, immobilization chemistry must be specific, fast and provide stable chemical bonds. Large scale, high-throughput manufacturing of arrays also requires that the chemistry is simple and reproducible (Podyminogin et al. 2001).

Robustness is also important because the chemistry is to be implemented in technology or biology laboratories that are neither equipped nor skilled for undertaking sophisticated chemistry under strict conditions.

2.3.2 *Strategy of Chemical Grafting*

Biomolecules cannot be coupled directly to the surface. Therefore, a chemical modification of the surface is performed before the biomolecule immobilization step itself. The surface modification often involves several steps, at least two. The most difficult step of the grafting process is the binding of the biomolecule. Indeed it has to accommodate the different constraints presented above: aqueous solvent, low concentration, low temperature, etc. The strategy consists in choosing the grafting reaction of the biomolecule first and adapting the surface functionalization to this choice. Once the chemical functionality of the surface is chosen, a process is devised for attaching this reactive function to the surface. The surface chemistry of the transducer is the next constraint to be managed. Indeed, the transducer in not chosen for its ability to be grafted but for its sensitivity to the chemical signal. Typical surfaces are inorganic materials such as silica, glass, gold, etc. There is not so much choice regarding reactions and reagents for grafting organic molecules to such inorganic surface. The involved chemistry is often quite vigorous, that is, not compatible with the functionality to be attached. For example, silanes cannot contain carboxylic acid or hydroxyl group; primary amino group is compatible with ethoxysilane but reacts strongly with chlorosilane. Because of this limitation, a coupling agent is often used in a second step after the first derivatization step performed directly on the transducer surface. A coupling agent is a difunctional molecule that reacts by one end to functionalized surface and leaves its second group for further reaction with the biomolecules. Coupling agents (Fig. 2.1) are

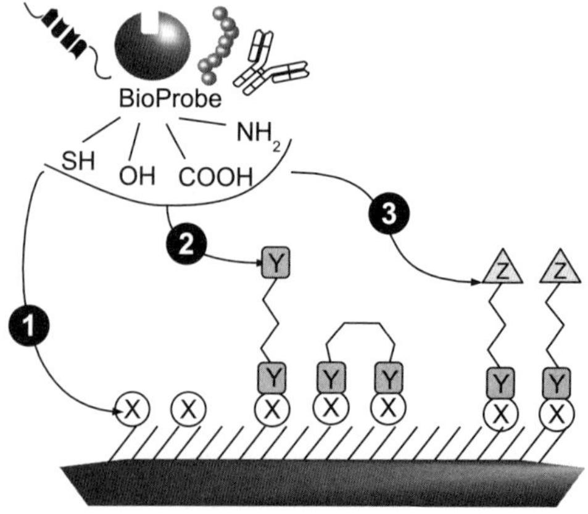

Fig. 2.1 Mains strategies of chemical grafting; (*1*) direct reaction to active surface, (*2*) activation of the surface with homodifunctional linker, and (*3*) activation of the surface with heterodifunctional linker

either homodifunctional if the two reactive functions are identical (e.g. phenylenediisothiocyanate) or heterodifunctional if they are different. The coupling agent is used for functionalization of the surface and/or for the chemical modification of the biomolecules as well.

The reactions of biomolecules to functionalized surfaces will be presented first, followed by a second part introducing the surface functionalization in several steps.

2.3.3 *Chemical Grafting of Native Biomolecules and Associated Surface Biofunctionalization*

2.3.3.1 Immobilization Through Amino Groups of Biomolecules

The most widely used route to conjugate biomolecules to solid supports makes use of reactions involving primary amino groups ($-NH_2$). The utility of amines stems from their high nucleophilic character and the existence of a wide variety of amine-base coupling chemistry suitable for use under aqueous conditions. The lysine residues of proteins have amino groups in ε-position. Figure 2.2 summarizes the main coupling routes of amines. Many reactions groups are acylation reactions leading to C–N bonds.

The unprotonated amino group is the nucleophilic (thus reactive) form of amines that exists in basic medium. The reactivity of amines in water depends on the pH according to the equilibrium between the protonated (acidic) and unprotonated (basic) forms given by the pK_a value. Increasing pH shifts equilibrium toward the unprotonated form and makes the reaction kinetics faster. A side-reaction occurring at high pH is the hydrolysis of acylation reagents that is observed for long reaction

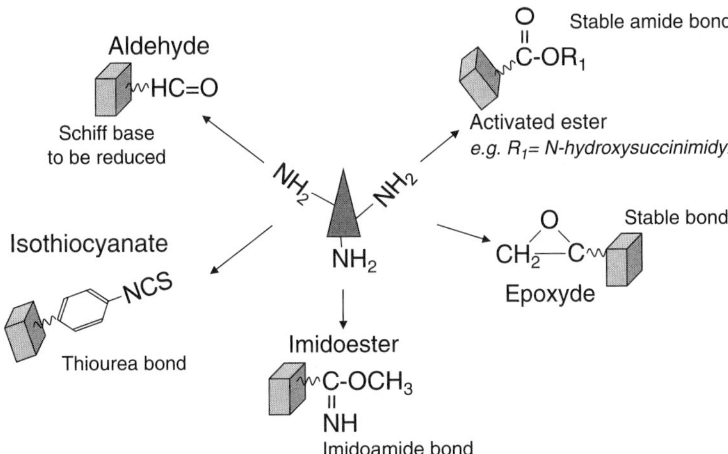

Fig. 2.2 Mains strategies of chemical grafting via amine groups

times in water. Anhydrous polar organic solvents are best suited if the biomolecules can resist. The competition between the acylation of amines and hydrolysis depends on the relative concentrations of amine and water. Hence, high concentration of amine is preferred.

2.3.3.1.1 Functionalization Using Epoxide

Surface immobilization of biomolecules onto solid support has been extensively studied for peptide synthesis (Anderson et al. 1964) and chromatographic separation since the 1960's (Sundberg and Porath 1974). Two efficient and suitable chemical routes (epoxy and activated ester) have emerged. They are currently used to immobilize molecules through their amine groups. As example, affinity chromatography columns are prepared by cross-linking reactive glycidyl groups to agarose (Sundberg and Porath 1974). Addition of biomolecules to epoxide ring involves nucleophilic primary or secondary amine, sulfhydryl groups or, less commonly hydroxyl groups (Wheatley and Schmidt 1999). The reaction rate of amine addition is optimal at pH = 11 (Sundberg and Porath 1974) because the hydrolysis reaction rate does not increase much in the pH range 7–11 (Wheatley and Schmidt 1999). The epoxy-activated phases have been criticized as being unreactive when compared to the other activated phases because of the low reactivity of amino groups at neutral pH (Ernst-Cabrera and Wilchek 1988). Milder conditions must be used in case of pH-sensitive molecules or solid support (Sundberg and Porath 1974). In particular silica or glass that are soluble in alkaline medium. Consequently, the use of strong alkaline solutions during immobilization of DNA to glass slides leads to inconsistent results because of the degradation of the underlying silane layer (Pirrung et al. 2000). Such activated supports give excellent results when favorable conditions are met with (Kusnezow et al. 2003; Oh et al. 2007).

Chemo-selective covalent coupling of hydrazide (R-$NHNH_2$) to epoxide-coated surface has been reported. The hydrazide group reacts more rapidly than thiol and amine functional groups (Lee and Shin 2005).

Epoxide functionalization of organic or inorganic supports is quite easy when compared to some other functional groups. Silica or glasses are derivatized by silane coupling agent (3-glycidoxypropyltriethoxysilane) or coated with epoxy-resins. Surface alcohol or amine groups of polymer substrates can be cross-linked by difunctional monomers (e.g. 1,3-butanediol diglycidyl ether) or by photografting glycidyl methacrylate monomers (Eckert et al. 2000).

2.3.3.1.2 Functionalization Using Activated Ester

Esters of N-hydroxysuccinimide (NHS-ester) are conveniently prepared from reaction of carboxylic acid and NHS under mild conditions in the presence of carbodiimide coupling agent (Cuatrecasas and Parikh 1972; Staros et al. 1986).

NHS-ester as other activated esters (e.g. the sulfonated ester derivative of N-hydroxysulfosuccinimide that has higher water solubility), N-hydroxybenzotriazole, p-nitrophenol, tetrafluorophenol allow acylation of primary amines in high yield. Aminolysis of these activated esters is notably much faster than hydrolysis, enabling reaction to be achieved in aqueous medium (Lee et al. 2003; Salmain and Jaouen 2003). Unlike other nucleophilic reactive groups, NHS-ester has low activity with secondary amines, alcohols, phenols (including tyrosine) and histidine. NHS-esters can be stored for long periods in cold and dry conditions. NHS-ester functional groups are thus versatile; they are widely used in peptide synthesis, preparation of biomolecules conjugates, and in surface immobilization (ligand affinity chromatography; biosensors and microarrays).

2.3.3.1.3 Functionalization Using Aldehyde

Aldehyde reacts reversibly with amines to form Schiff's base (imine group) (equilibrium 1 in Fig. 2.3). In order to get sustainable immobilization, imine groups must be reduced into stable secondary amines by borohydrides treatment, for example (reaction 2 in Fig. 2.3). It is worth noticing that this reaction deactivates residual unreacted aldehyde groups into alcohol groups that do not cause nonspecific adsorption. Aldehydes are moisture-sensitive and are easily oxidized into carboxylic acids (Zammatteo et al. 2000).

Functionalization of solid support by aldehyde reactive groups is classically performed by oxidation of alcohol function (Zammatteo et al. 2000) or reduction of alkenes (Hevesi et al. 2002) and can be also obtained by modification of aminated surface by glutaraldehyde (Yao 1983; Yershov et al. 1996).

2.3.3.1.4 Imidoester Functionalization

Difunctional imidoester (also called imidoether) reagents, dimethyl pimelimidate (DMP) or dimethyl suberimidate (DMS) (Fig. 2.16a) for example, react

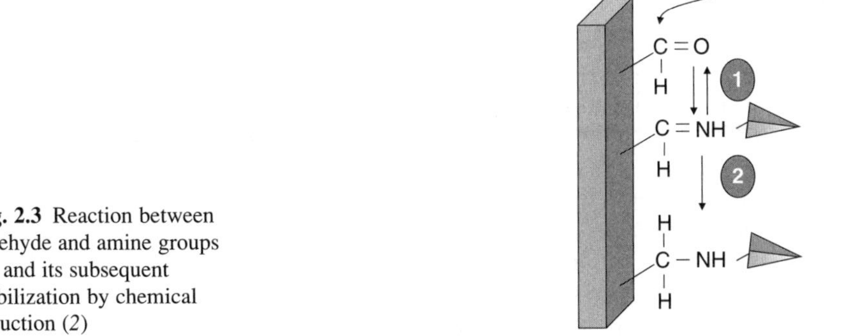

Fig. 2.3 Reaction between aldehyde and amine groups (*1*) and its subsequent stabilization by chemical reduction (*2*)

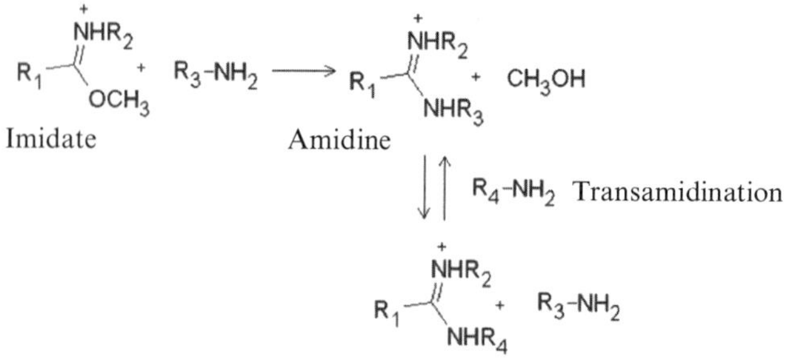

$R_1C(=\overset{+}{N}HR_2)OCH_3 + R_3\text{-}NH_2 \longrightarrow R_1C(=\overset{+}{N}HR_2)NHR_3 + CH_3OH$

Imidate Amidine

$R_4\text{-}NH_2$ Transamidination

$R_1C(=\overset{+}{N}HR_2)NHR_4 + R_3\text{-}NH_2$

Scheme 2.1 Proposed mechanism of amidination and transamidination reaction (Adapted from Sundaram 1979)

rapidly with primary amines to form amidine bonds. Although imidoester homodifunctional cross-linkers have been used in biochemistry to study protein structure, only few examples concern the immobilization of ligands to solid support (Morris et al. 1975; Beier and Hoheisel 1999). Hydrophobic activated supports seem stable against wet storage (Sundaram 1979) when compared to soluble homologues which are extremely sensitive to moisture (half life of a few tenths of minute).

The reactive species are the neutral amino form of the primary amine and the protonated form of the imidoester (called "imidate") that both exist in different pH domains. Therefore, an optimum pH is determined between pH 7 and 10 (Sundaram 1979).

The amidine bond formed retains a net charge at neutral pH. It is important to note that amidine undergoes a subsequent transamidination in the presence of another nucleophile (Scheme 2.1). Thus immobilization through amidine bond may be reversible (Sundaram 1979).

2.3.3.1.5 Isothiocyanate Functionalization

Isocyanate (R-NCO) group is highly reactive but also amenable to deterioration during storage. Isothiocyanate (R-NCS) group is a less reactive alternative to isocyanate. It is quite stable in water and reacts with nucleophiles to yield thiocarbamate (reaction with thiol groups) or thiourea bonds (reaction with amines). Reaction of isothiocyanate on amine groups is well-known in biochemistry and serves to N-terminal peptide/protein sequencing. Despite this well-known reaction, immobilization of biomolecules to isothiocyanate functional groups proceeds mostly through reaction with thiol groups.

The preferred route to activate support with isothiocyanate functional groups implements modification of aminated support with phenylenediisothiocyanate cross-linker (PDITC) (Guo et al. 1994; Thiel et al. 1997; Beier and Hoheisel 1999). The thiourea bond is stable for use in microarray applications (Guo et al. 1994; Lindroos et al. 2001). It has been reported however that the antibody conjugates prepared with fluorescent isothiocyanates showed degradation over time (Banks and Paquette 1995).

2.3.3.2 Immobilization Through Sulfhydryl Groups of Biomolecules

The chemical immobilization of complex recognition molecules like proteins onto the active layer of a biosensor is quite challenging. Specific concerns are: (1) attach these molecules by one anchoring point in order to obtain a repeatable, uniform and oriented layer; (2) avoid alteration of the active sites or denaturation that leads to loss of activity. The regio-specificity of immobilization using either amine or carboxylic acid moieties is difficult to control because there is large number of these hydrophilic residues (Lys, Glu, Asp) on the periphery of proteins where binding takes place. This leads to heterogeneity of the attachment and random orientation of the protein on the surface. Unlike amine or carboxylic groups, proteins possess a limited number of cysteine residues at their periphery. Covalent coupling strategies using thiol groups allow site-specific reactions. The sulfur atom of cysteine belongs to form a sulfhydryl (or thiol) group. The low likelihood of cysteine's presence or accessibility in protein restricts the use of thiol groups to immobilization of native molecules. This turns out advantageous in order to introduce solvent accessible reactive groups on protein surface. Cysteine residues involved in disulfide bonds can be chemically or enzymatically cleaved (Brogan et al. 2003) before a subsequent reaction with activated surfaces. A nondenaturizing method of cleavage of internal disulfide bridges of proteins by UV-light has recently been reported (Neves-Petersen et al. 2006). Proteins that do not bear cysteine residues can be genetically modified to engineered site-specific thiol groups distal from the active sites of the molecule (Firestone et al. 1996; Yeung and Leckband 1997; Backer et al. 2006).

Thiol groups allow very specific reactions because they are very reactive toward various chemical functions that are quite stable in water. The simplest approach is the reaction of thiol group onto supported thiol groups by formation of disulfide linkage. But disulfide bonds are unstable under reducing conditions. Alternatives involve formation of stable thioether bonds by Michael addition on maleimide groups or reaction with haloacetamide groups.

Thiols are prone to self-oxidation. The reactions of oxygen with thiols in aqueous solution give disulfides quite easily (pH 7–9). Radiolytic oxidation is also possible (Bagiyan et al. 2003). In order to regenerate the free reactive form, disulfide dimers are cleaved immediately before the coupling reaction (Chassignol and Thuong 1998). Dithiothreitol (DTT), β-mercaptoethanol or tris(2-carboxyethyl)phosphine (TCEP) are common disulfide-reducing agents. An additional step of elimination of the reducing agent from the solution before coupling the biomolecule to the support is required.

Phosphorothioates do not form dimers (Chassignol and Thuong 1998) and function as nucleophiles without prior reduction (Zhao et al. 2001). The elegant reaction of phosphorothioates modified oligonucleotides to bromoacetamide activated support (Pirrung et al. 2000) avoids the oxidation side-reaction of thiols into disulfides.

2.3.3.2.1 Sulfhydryl Functionalization

The use of disulfide bond for immobilization of biomolecules is of particular interest for oriented immobilization of antigens and other proteins. A chemical reduction step induces cleavage of the disulfide bridge in the Hinge region of antibodies, yielding two fragments formed of a heavy and light chain and a free thiol group. The disulfide bridge is located distal from the antigen-binding site. Immobilization through this precise anchoring point allows oriented immobilization of antigen while maintaining its availability to antibody recognition. Reversible formation of disulfide linkage with thiol-functionalized surface have been investigated on glass slides modified with mercaptosilane coupling agent (Neves-Petersen et al. 2006).

N-succinimidyl-3(2-pyridyldithio)propionate (SPDP or sulfo-SPDP, Fig. 2.16) reacts with surface amino groups by the NHS moiety to form stable amide bonds (Gad et al. 2001; Hertadi et al. 2003). The 2-pyridyl disulfide group at the other end reacts with sulfhydryl residues borne by biomolecules to form a disulfide linkage (Ohtsuka et al. 2004). Interestingly, 2-pyridyl disulfide can be selectively reduced to thiol, even in the presence of other disulfide groups. It is worth to note that SPDP heterocross-linkers are also used to add thiol functionalities to nonthiolated proteins (Carlsson et al. 1978) for their subsequent specific immobilization.

2.3.3.2.2 Maleimide Functionalization

The double bond of maleimide undergoes a Michael addition reaction with sulfhydryl groups to form stable thioether bonds. It is also reactive toward unprotonated primary amines and hydroxyde anions (Shen et al. 2004). Interestingly, reaction of maleimide and sulfhydryl groups is specific in the pH range 6.5–7.5. The reaction of maleimide with sulfhydryl at pH 7 proceeds at a rate 1,000 times faster than the reaction with amines.

As depicted in Fig. 2.16, maleimide surface functionalization is achieved by coupling hetero-difunctional cross-linkers to surface terminal amino groups. During and after derivatization of surface amine groups by difunctional cross-linkers, maleimide groups are exposed to reaction with amino groups. Once grafted, the close proximity of the maleimide groups and residual amino groups remaining on the solid support may lead to a fast decrease of surface activity. A capping reaction of the amine groups with a large excess of highly reactive reagents such as isothiocyanate or succinimidyl ester groups is to be made in order to avoid the deactivation by residual amines. The main maleimide heterodifunctional cross-linkers aimed at immobilization of biomolecules (Chrisey et al. 1996; Strother et al., 2000a, b; Shen et al. 2004) are the succinimidyl-4-(N-maleimidomethyl) cyclohexane-1-carboxylate (SMCC) and succinimidyl-4-(N-maleimidophenyl) butyrate (SMPB) or its sulfonated analogue sulfo-SMPB. Maleimide-activated solid supports cannot be stored for long time because of base-catalyzed addition of water to the imide bond and ring-opening the maleimide by hydrolysis (Shen et al. 2004; Xiao et al. 1997).

2.3.3.2.3 Haloacetyl Derivatives Functionalization

Condensation of thiol with haloacetyl moieties (bromoacetyl, iodoacetyl) creates stable thioether bond. A comparative study (Chrisey et al. 1996) of the iodoacetamide and maleimide reactive groups for oligonucleotide grafting indicated a strong dependence of immobilization performance with the solvent used. Functionalization of aminated supports with haloacetyl functional cross-linkers is currently done by succinimidyl-4-(N-iodoacetyl)aminobenzoate (SIAB) or its sulfonated analogues sulfosuccinimidyl-4-(N-iodoacetyl)aminobenzoate (s-SIAB). Direct functionalization of glass slides with bromoacetyl functional groups has been implemented by modification of clean glass with haloalkylsilanes (Firestone et al. 1996; Pirrung et al. 2000; Zhao et al. 2001). The very reactive iodoacetamido-functionality is sensitive to light and hydrolysis, so that the bromoacetamide group may be preferred.

2.3.4 Immobilization of Modified Biomolecules

The modification of biomolecules in perspective to their specific immobilization was considered for various purposes. These modifications are aimed at implementing a specific immobilization process such as electropolymerization (pyrole-derivatized biomolecules), or determining a site-specific anchoring point (protein tag, chemo-selective reactive site and/or site-specific reactive groups) for a uniform and oriented immobilization.

2.3.4.1 Covalent and Selective Immobilization of Biomolecules

The sensitivity and selectivity of biosensors depend on the quality of the interface and the activity of the tethered biomolecules. This activity is intimately correlated to the surface configuration of the immobilization, that is, the active site must remain accessible (Fig. 2.4).

Straightforward immobilization techniques rely on either adsorption, or direct covalent attachment of biomolecules to chemically activated surfaces (MacBeath and Schreiber 2000). The adsorption of proteins to interfaces causes possible denaturation and/or limits the availability of active sites. Adsorption is often associated to losses of activity and poor selectivity. Because adsorption is reversible,

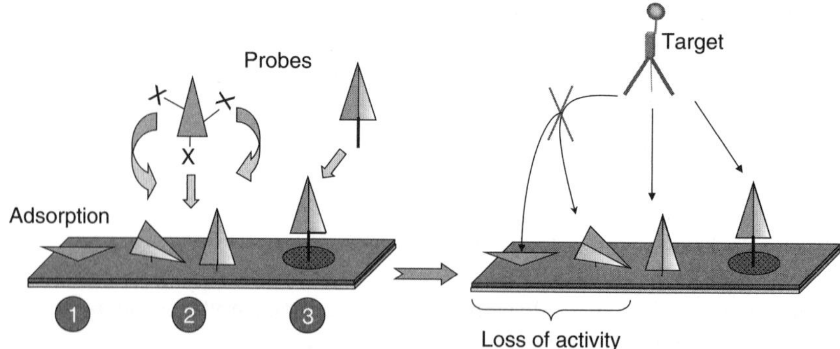

Fig. 2.4 Schematic representation of the different way of bioprobes immobilization. A probe can be adsorbed to the surface (*1*), immobilized by random reactions (*2*) and site-specifically immobilized to the surface. The adsorption and random immobilization results in loss of activity compared to the uniform orientation of site-specifically immobilized bioprobes

the surface does not withstand long-time reaction or washes aimed at enhancing sensitivity and selectivity. Unlike DNA, there is a large diversity of proteins of various size, physico-chemical properties, solubility or biological activities. Reactive amino or thiol groups of large protein molecules are scattered throughout the biomolecules. Hence, grafted proteins are randomly immobilized under various orientations. Protein immobilization after site-directed modification of proteins allow selective attachment of proteins to a solid support in a uniform orientation (Hodneland et al. 2002). Modifications are made by recombinant peptide tags, fusion proteins, or posttranslational modifications. Proper orientation of the immobilized proteins compared to random immobilization improve the biological activity (homogeneous-binding activity, kinetics of reactions) and thus measured signals (Firestone et al. 1996; Vijayendran and Leckband 2001; Zhu et al. 2001).

The diversity of biomolecules led to various specific strategies. The specific immobilization of DNA strands, antibodies, proteins and small molecules by means of oriented attachment is addressed in this overview.

2.3.4.2 DNA Probes

Nucleic acid stands are linear polymeric molecules formed by a chain of nucleotides borne by a sugar-phosphate backbone. The four nucleotides (Adenine, Guanine, Cytosine, Tyrosine) of the DNA sequence display internal amino-groups that are involved in the complementary base-pairing hybridization. These internal amino-groups must stay free and fully functional after immobilization.

Two main strategies are used to elaborate DNA microarrays. Prefabricated DNA strands such as long c-DNA fragments or synthetic oligonucleotide probes can be individually addressed on a solid support by micro-deposition techniques. The use of contact or noncontact deposition techniques allows manufacturing high-density

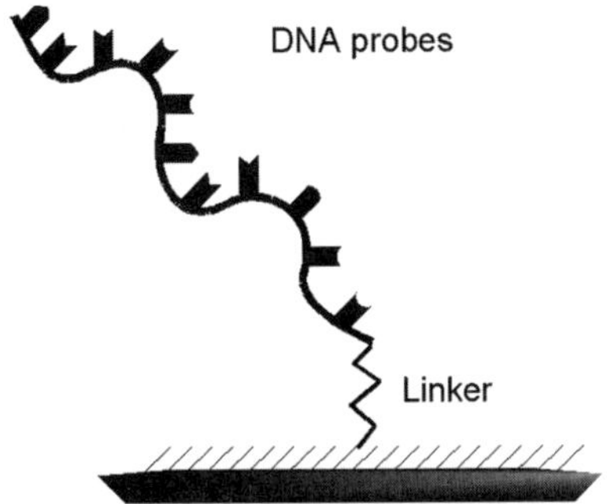

Fig. 2.5 Schematic representation of oriented DNA probe immobilization

DNA microarrays of up to 50,000 distinct spots on a glass slide. Another approach takes advantage of the progresses in DNA synthesis to fabricate each sequence *in situ* on the glass slide. The phosphoramidite chemistry of the standard DNA synthesis on cartridge has been modified and combined to spatially addressable techniques in order to synthesize in parallel up to 500,000 oligoprobes on one square centimeter(Fodor et al. 1991; Maskos and Southern 1992a, b). The DNA synthesis is beyond the scope of our discussion and we will focus on the immobilization of prefabricated DNA strands at the biosensor surface.

DNA strands can be uniformly and selectively immobilized by means of a reaction at one of their 5′ or 3′ end chain (Fig. 2.5). DNA molecules are conveniently end-modified by a reactive moiety called linker. A wide variety of linkers are available on synthetic oligonucleotides. Short DNA strands (oligonucleotides) modified by amino-, thiol-, hydrazide, phophorothioates or biotin-linkers are currently used for DNA immobilization. Similar modifications can be introduced into long DNA strands during PCR amplification using 5′ end-modified oligonucleotide primers (Saiki et al. 1989; Guo et al. 1994; Pirrung et al. 2000; Raddatz et al. 2002; Han et al. 2005).

Key parameter is to prevent interaction between the nucleobases and the surface, it has been reported that DNA molecule become totally inaccessible for hybridization when only 3% of its bases are involved in the covalent linkage (Bunemann 1982).

End modification of DNA allows not only introducing a site-specific group for covalent and oriented attachment, but allows inserting a spacer link between the nucleic acid probes and the surface. The spacer is intended to improve the mobility of the immobilized probes and thus their accessibility by the complementary strands, and move away the DNA probes from the surface to limit the adsorption and steric effects of the surface. Large improvements of hybridization yields have been reported by optimization of the spacer molecule and surface coverage (Guo et al. 1994; Herne and Tarlov 1997; Shchepinov et al. 1997). Indeed, accessibility is hampered by steric and electrostatic hindrances for too tightly packed surface-bound DNA strands.

After immobilization onto reactive surface, unreacted groups on the surface must be deactivated. In order to limit the adsorption of DNA molecules, deactivation involves hydrophilic molecules (e.g. alcohols, poly(ethylene oxide)). The widespread manufacturing of DNA microarray with DNA on glass slides involves amino-modified DNA probes. DNA probes are selectively tethered by their amino terminus because the terminal primary amino groups are much more reactive than the internal amino-groups of the DNA bases (Beier and Hoheisel 1999; Dugas et al. 2004).

Beside these standard covalent immobilization techniques some punctual works report on original DNA immobilization techniques such as silanized nucleic acids (Dolan et al. 2001), benzaldehyde-modified oligoprobes (Podyminogin et al. 2001). The direct and oriented immobilization of unmodified oligonucleotides onto zirconium-functionalized solid support takes advantage of the stronger interaction between the terminal phosphate group, $ROPO_3^{2-}$ compared to the phosphodiester groups of the backbone (Bujoli et al. 2005). This allows specifically tethered oligoprobes by means of oriented configuration. The incorporation of short segments of guanine oligomers as spacer, leads to a twofold increase in hybridization signal.

2.3.4.3 Antibodies

Owing to their high specificity, antibodies are widely used for purification/concentration of specific molecules in biological fluid by immunoaffinity chromatography or for diagnostic purposes in immunoassays. Whatever the nature of the attached molecular probe, uniform layers of well-oriented molecules are obtained by site-directed reaction, keeping the binding site accessible. Three specific reaction sites leading to three main strategies (Lu et al. 1996) of oriented immobilization are used for antibodies (Fig. 2.6).

The "biochemical approach" takes advantage of the selective binding of the Fc fragment of antibodies to specific receptors (protein A, protein G). This approach involves a first biofunctionalization of the solid support by Fc receptors (Vijayendran and Leckband 2001; Grubor et al. 2004; Briand et al. 2006) and subsequent binding of the antibody by its Fc fragment.

The two others approaches involve chemical immobilization of antibodies to solid support:

- The reduction of the disulfide bonds linking the heavy and light chain of the Fab fragment in the Hinge region is made by chemical, photochemical or enzymatic treatments. This creates site-specific thiol groups far from the binding sites (Brogan et al. 2003). Straightforward immobilization of antibody fragments containing thiol site-specific group can be realized directly onto gold surface (Brogan et al. 2003). However, higher immobilization efficiencies are obtained on thiol reactive surfaces.

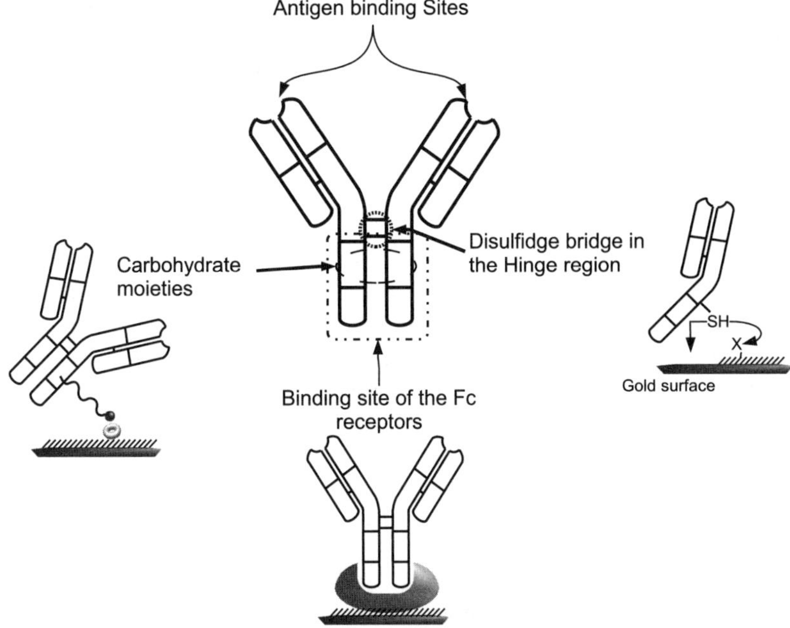

Fig. 2.6 Schematic representation of antibody molecule with the three sites dedicated for oriented immobilization

– A reactive site-specific group is created on the carbohydrate moieties located at the CH_2 domain of the Fc fragment. The carbohydrate is modified by specific ring-opening. For example, sodium periodate ($NaIO_4$) oxidizes vicinal diols of the carbohydrate into aldehyde reactive groups. Enzymatic oxidation (Solomon et al. 1990) with neuraminidase and galactose oxidase have also been reported for avoiding possible damages of the antigen-binding site during chemical oxidation. Direct grafting of antibodies is obtained by reaction of the aldehyde groups onto amino or hydrazide functionalized solid supports (Arenkov et al. 2000). Indirect grafting using heterodifunctional cross-linkers (4-(N-maleimidomethyl)cyclohexan-1-carboxylhydrazide (M_2C_2H), or Biotin-LC-hydrazide) also gave excellent results (Vijayendran and Leckband 2001). The hydrazide end group of the cross-linker reacts on the aldehyde carbohydrate to form a stable hydrazone bond.

2.3.4.4 Proteins

The production of proteins relies on the expression of genetically modified (cloned) host cells (e.g. *Escherichia Coli*, Yeast, phage). Uniform and specific attachment of proteins onto solid surface involves the incorporation of an accessible site-specific group onto the proteins. The processes for the immobilization the site-specific reactive groups must be compatible with the functional groups of proteins and proceed under mild conditions in aqueous solution. Preferred modifications of native proteins are either genetic recombination of peptide tags or biochemical grafting of ligands (e.g. biotin).

2.3.4.4.1 Genetic Fusion of Peptide Tags

Fusion of site-specific peptide reactive moiety to native proteins is obtained with high selectivity by genetic recombination. Modifications are mostly done at the N-terminal or C-terminal extremity of native proteins in order to limit potential interference on the biological activity upon fusion of the reported group. Because oriented immobilization is intended to improve the biological activity, it must be ascertained that the genetically modified biomolecules display identical biological properties (folding, accessibility of the active site and the site-specific group for attachment, binding affinity) as the unmodified molecules or that cell expression is not underexpressed or truncated (Kindermann et al. 2003; Backer et al. 2006).

Various reported groups have been investigated for site-specific immobilization of proteins to functionalized solid supports. Straightforward approaches implement the simple incorporation of one amino acid (e.g. cysteine residue) or oligopeptide (e.g. polyhistidine tail). The thiol group of cysteine can be used for the thiol specific binding reactions to the surface as described in the previous section. Immobilization by means of a selective reaction to a thiol group limits the reaction of other naturally occurring nucleophilic groups of the protein (amino groups of lysines). Polyhistidine tails are selectively immobilized onto metal-chelate surface.

Other approaches involve species of larger size such as enzymatic domains or protein splicing elements. Proteogenic sequences have been used to covalently attach fused proteins to the solid surface or to introduce chemo-selective reactive sites or affinity tags in a subsequent posttranslational modifications step.

2.3.4.4.2 Metal-Binding Peptide Ligands

One-third of all proteins are metallo-proteins, and many of the reactions that are most critical to life are catalyzed by metallo-enzymes (Rosenzweig and Dooley 2006). Metal cations bound to specific peptide sequence or protein folding are active sites of many biological processes (oxygenic photosynthesis, nitrous fixation, replication and transcription). Selective adsorption of peptides and proteins comprising neighboring histidine residues was achieved to Ni^{2+} and Cu^{2+} ions bound to the surface (Giedroc et al. 1986; Hochuli et al. 1987). Metal ions form octahedral coordination complex with six ligands coming from the functional surface or the protein. Ligands such as iminodiacetic acid (IDA) (Fig. 2.7) or nitrolotriacetic acid (NTA) (Hochuli et al. 1987; Schmid et al. 1997; Cha et al. 2004) are subsequently covalently bound to solid surface and loaded with the metal cation. The tridentate IDA and quadridentate NTA chelating group form respectively three and four

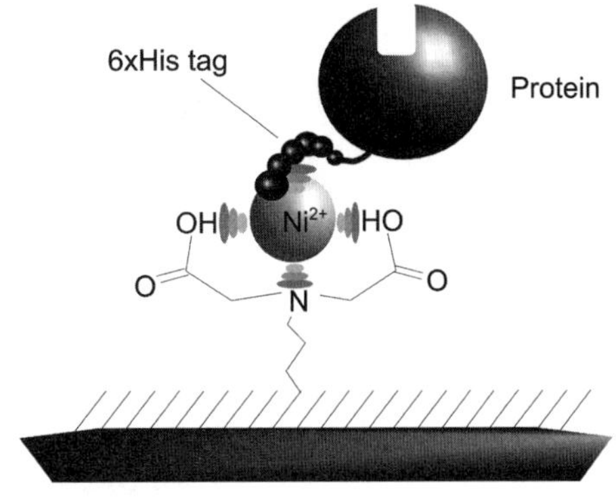

Fig. 2.7 Schematic representation of the metal-binding immobilization. The surface is chemically modified by iminodiacetic acid (IDA) moiety. The biomolecule is tag with a polyhistidine tail represented by the six black balls

coordination complexes. The remaining coordination sites filled by water molecules are readily displaced by stronger ligands (histidine) of the protein (Hochuli et al. 1987). Proteins genetically modified with polyhistidine tails (at least six histidines) at their N- or C-terminus were prepared for their oriented grafting by a chain end. Poly-His tag generally does not interfere with the structure or function of proteins (Cha et al. 2004). Uniform and oriented immobilization of 5,800 fusion proteins onto nickel-coated glass slides gave a protein microarray of higher sensitivity when compared to the protein immobilized through primary amines reactions on aldehyde coated slide (random orientation) (Zhu et al. 2001).

Limitations with respect to the immobilization through interaction between His-tagged proteins and metal complex attached to the surface come from the reversibility of the interaction (Schmid et al. 1997). The His-tag can be removed from the metal-bearing surface by exposure to reducing conditions, EDTA, detergents, large excess of ligands (imidazole).

2.3.4.4.3 Binding of Biochemical Affinity Ligands

Some very specific, stable and efficient biochemical interactions have been used to tether proteins to solid support by means of selective and oriented bond. Among the different partners investigated, the streptavidin–biotin system is the most widely used and known system to immobilize biomolecules onto solid support. Other recombinant affinity tag systems based on leucine zippers (Zhang et al. 2005) or split-intein systems (Kwon et al. 2006) efficiently attach recombinant proteins to support via uniform and oriented way.

2.3.4.4.3.a Streptavidin–Biotin System

Oriented immobilization of fused proteins can be made using affinity ligands of high selectivity. The very specific interaction of biotin and streptavidin is characterized by the strongest known noncovalent interaction ($K_a = 10^{15}\ M^{-1}$).

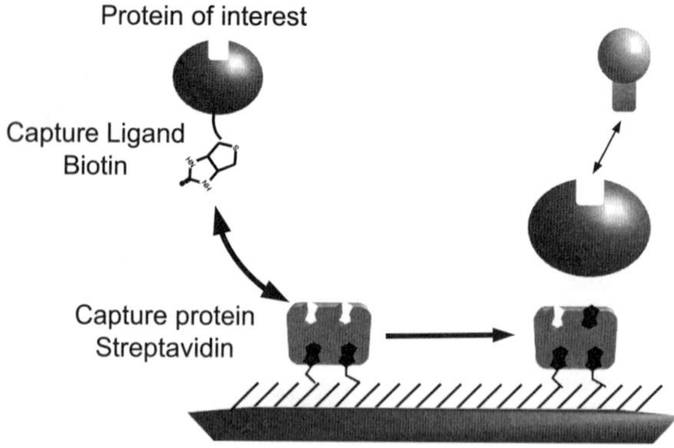

Fig. 2.8 Schematic representation of oriented immobilization of protein by affinity through the streptavidin/biotin interaction

Implementation of such affinity ligands involves the posttranslational modification of the recombinant proteins with a heterodifunctional cross-linker agent containing biotin prior to immobilization. The site-specific protein biotinylation have been developed through very specific reactions relying on peculiar biochemical activities (Lue et al. 2004a, b; Tan et al. 2004a, b; Yin et al. 2004). One technique takes advantage of the protein-splicing activity (inteins-mediated biotinylation) (Lue et al. 2004a, b; Tan et al. 2004a, b); another technique derived from the inhibition of protein synthesis by aminonucleoside antibiotic products (puromycin analogs); a last technique is based on non-ribosomal peptide synthetase activity peptide carrier protein tag (PCP tag) (Yin et al. 2004). PCP tags present several advantages when compared to the intein-mediated approaches. Firstly, the PCP tag made of a small protein domain (80 aminoacids) (Yin et al. 2004) is compatible with a wide range of proteins, making the PCP tag technique quite universal. Secondly, the high efficiency of the chemical ligation allows complete labeling reactions within 30 min requiring only micromolar concentration of reagents. The efficiency of intein-mediated biotinylation is highly dependent upon the intein fusions, so that a specific adaptation of the general scheme is required to each protein of interest.

Biotinylated proteins are immobilized onto avidin or streptavidin coated solid support (Fig. 2.8). Avidin (or streptavidin) is a tetrameric glycoprotein with four identical subunits. Avidin-coated supports were currently obtained by reaction of the natural amino groups of proteins to amine reactive surface. This biofunctionalization leads to randomly immobilized streptavidin. But, the specific interaction between biotinylated proteins and streptavidin leads to oriented immobilization of the molecule of interest. A possible trouble might be the control of the immobilization density and orientation. Oriented immobilization of the streptavidin has been performed by immobilization of the streptavidin on biotin functionalized solid support (Yeung and Leckband 1997; Vijayendran and Leckband 2001).

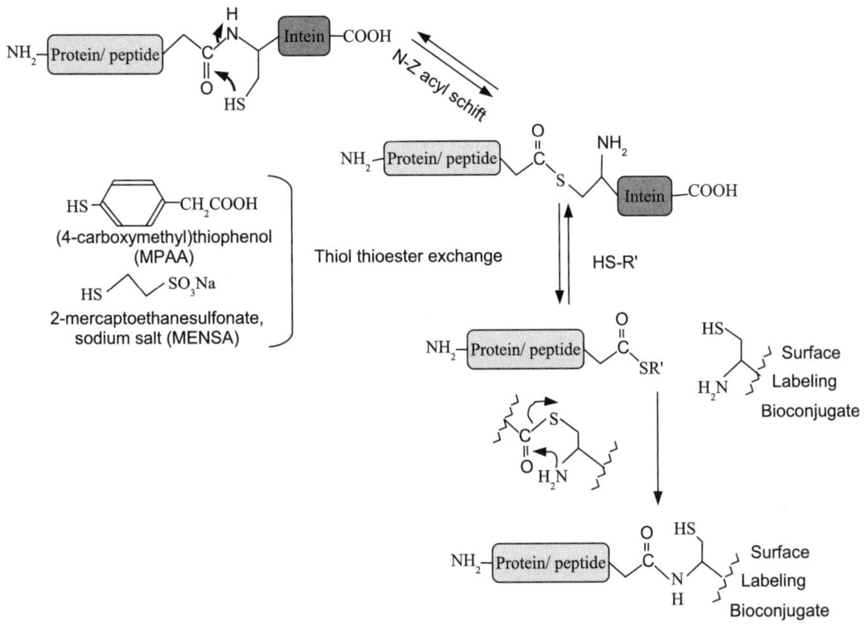

Fig. 2.9 Schematic representation of intein rearrangement and Native Chemical Ligation

2.3.4.4.3.b Other Affinity Tag Systems

On the same footing as for the well-known biotin–avidin pair, various affinity systems can be designed. Selective attachment of recombinant proteins through an artificial polypeptide scaffold system based on heterodimeric association of separate leucine zipper pair (Zhang et al. 2005). Recombinant proteins incorporate a basic component of the leucine system as affinity tag while an acidic component is attached to the surface. The heterodimeric association comes from efficient coiled–coil interaction (affinity constant of 10^{-15} M). The small size of the zipper tag (43 aminoacids) and the high specificity of the polypeptide scaffold are intended to be implemented in direct immobilization of various recombinant proteins from crude cell lysates without time-consuming purification steps.

Kwon et al. (2006) have developed a traceless capture ligand approach based on the use of protein trans-splicing mechanism. Intein fragments are known to promote protein splicing activity. These authors use a variant in which the intein self-processing domain is split into two fragments. One fragment is incorporate to the protein while the other one is attached to the surface. The specific interaction of these protein and ligand capture domains leads to a functional protein-splicing domain that, in parallel, splices the split intein out in the solution and attach by mean of covalent grafting the recombinant protein to the surface. This approach allows removing the affinity capture ligand after immobilization but suffers from long reaction time (up to 16 h) like the other intein-mediated approaches mentioned previously.

2.3.4.4.4 Enzymatic- and Chemo-Selective Ligation Reactions

Site-specific mutagenesis is used to introduce a unique cysteine residue (Firestone et al. 1996; Yeung and Leckband 1997) at the end of a long fusion tag of 15 aminoacids (N- or C-terminus) (Backer et al. 2006). The site-specific free sulfhydryl group can react directly on appropriately derivatized surface or conjugate with thiol-directed bifunctional cross-linker agent. Because thiol groups tend spontaneously to form disulfide bridges, the use of free sulfhydryl group involves a reduction step (with DTT for example) prior to the immobilization reaction.

2.3.4.4.4.a Chemo-Selective Ligation Reactions

Native chemical ligation. Native Chemical Ligation (NCL) involves the chemoselective reaction between a peptide-α-thioester and a cysteine-peptide to yield a native peptide bond at the ligation site. The carboxy-terminal α-thioester (COSR) group is generated with an intein vector (Fig. 2.10). Inteins are protein domains that catalyze protein splicing. More than 200 intein sequences (containing from 134 to 1,650 aminoacids) have been identified with a wide diversity of specific activity. Inteins can be one element or split domain that upon reconstitution allow protein splicing of the regions flanking by autocleavage (Perler and Adam 2000). Specifically engineered intein expression systems allow to perform single splice-junction cleavage (Fig. 2.9). Considering N-terminal intein expressed protein, the thiol group of cysteine side chain is involved in an S–N

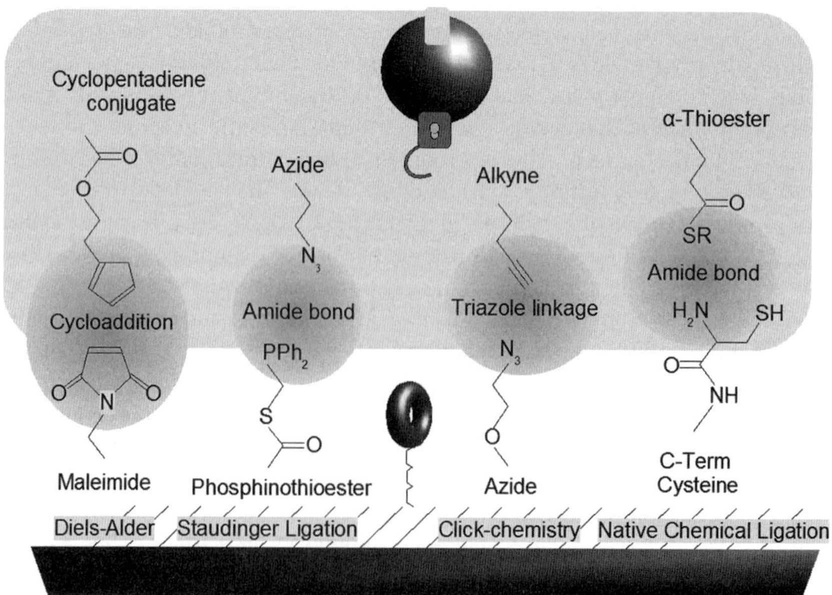

Fig. 2.10 Examples of immobilization of proteins by chemo-selective ligation

acyl shift to form a thioester bond. Unfortunately, the alkyl thioester residues generated at the C-terminal amino acid show extremely low rate of ligation. Trans-thioesterification with exogeneous thiols (e.g. MESNA or MPAA) have been reported to promote the *in situ* formation of more active thioester (Johnson and Kent 2006).

Peptide or protein thioester such as mercaptoethanesulfonic acid (MESNA) thioester reacts in neutral aqueous medium with an N-terminal cysteine residue, yielding a native peptide bond (amide bond) at the ligation site.

Intein-mediated protein ligation (IPL) or Express protein ligation (EPL) has been investigated for chemo-selective immobilization of proteins and peptides onto cysteine functionalized solid support (Camarero et al. 2004; Helms et al. 2007). Direct ligation to cysteine-modified biosensors surface required relatively high proteins concentration (100 μM) (Camarero et al. 2004). Whereas, NCL is extensively used for the preparation of proteins by synthesis or semi-synthesis as well as site-specific bioconjugation, few reports deal with direct immobilization onto solid support for biosensors or microarrays manufacturing.

In addition to direct immobilization, IPL/EPL has expanded the scope of chemo-selective ligations of proteins and peptides to solid support by incorporation of some unnatural functionality. The ligation process generates site-specific reactive groups that drive in a subsequent immobilization step onto appropriately functionalized solid support. EPL/IPL have mediated the incorporation of chemical (C-terminal azide, alkyne or cyclopentadiene) or biochemical reactive (e.g. biotin moiety) groups. Protein farnesyl transferase like EPL/IPL has been used to incorporate chemical reactive groups by modification of cysteine residue located at the C-terminus (Gauchet et al. 2006). The reactivity of these site-specific chemical reactive groups is orthogonal to that of biomolecules and allows specific chemical coupling in a second step. Most of the ligation reactions implemented by these two-step ligation strategies belong to the Staudinger ligation, click-chemistry or Diels–Alder ligation (Fig. 2.10). These chemo-selective reactions proceed under mild conditions and in aqueous solution, preferably in the absence of any potentially denaturing cosolvent (Dantas de Araújo et al. 2006).

Chemo-selective ligation. The Staudinger reaction allows conversion of an azide group into amine by phosphine or phosphite reducing agents. In the Staudinger ligation, an electrophilic trap (methyl ester or sulfhydryl group) undergoes an intramolecular rearrangement that leads to the ligation of the azide substituent by formation of a stable amide linkage. Staudinger ligation involves azido peptides or azido proteins and surface functionalized by phosphine groups. Oxidation of phosphine groups is a possible problem that can be limited by a suitable functionalized building block (Watzke et al. 2006). Azido-modified proteins are immobilized in 4h reaction at minimum concentration of 50 μM. For example, Soellner et al. (2003) have reported a traceless version of the Staudinger ligation using the fast and complete reaction of azido peptide or proteins to diphenylphosphinomethanethiol-modified surface (Fig. 2.10). Immobilization of azido peptide (5 nM in DMF/H_2O 50:50) reached 67% yield within 1 min. The use of wet organic solvent for Staudinger ligation decrease the proteins nonspecific adsorption (Gauchet et al. 2006).

Chemo-selective covalent immobilization by click-chemistry involves reaction between azide-derivatized solid support and alkyne-derivatized proteins or peptides, yielding the chemically robust triazole linkage. Like NCL and Staudinger ligation, "click-chemistry" is fully compatible with the functional groups found in proteins. Maximum surface density for proteins immobilization by "click-chemistry" was achieved within 2 h for protein concentration of 20 μM (Gauchet et al. 2006).

The Diels–Alder reaction involves the condensation of a diene (a molecule with two conjugated double bonds) and a dienophile (an alkene) into a cyclic product. The Diels–Alder reaction (Yousaf and Mrksich 1999; Dantas de Araújo et al. 2006) has been investigated for immobilization of biomolecules modified by a site-specific diene (cyclopentadiene, hexadienyl ester) onto solid surface bearing the corresponding dienophile counterpart (respectively, 1,4-benzohydroquinone or maleimide functionalized surface). As a side-reaction that imparts selectivity, surface-bound maleimide reacts with the free thiol functions of the cysteine residues of proteins; a preprotection step by treatment with Ellmann's reagent avoids this undesired reaction (Dantas de Araújo et al. 2006). Reaction is amenable in aqueous solution of dilute proteins (8 μM).

2.3.4.4.4.b Enzyme-Based Immobilization Reaction

Others methods of selective and covalent immobilization of proteins to surface are based on covalent bonding catalyzed by enzymes. Figure 2.11 depicts the reconnaissance of specific partners (capture protein and ligand) leading to selective biochemical immobilization. In such strategies, fusion proteins comprise a capture protein (enzyme) and the protein to be immobilized at the surface. The capture

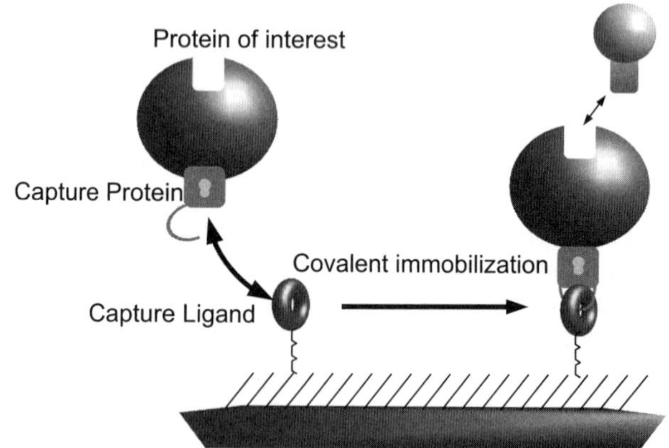

Fig. 2.11 Schematic representation of site-specific chemical or enzymatic immobilization of protein. The reactive group on the surface is named capture ligand while its counterpart link to the protein is the capture protein group

protein is a protein with a specific enzyme activity leading to the formation of a covalent link with the ligand. For example, serine esterase (cutinase) (Hodneland et al. 2002) or DNA repair protein (hAGT) (Kindermann et al. 2003) have been successfully fused to proteins and used to selectively graft them to, respectively, phosphonate or O^6-benzylguanine ligands attached to the solid surface. The highly specific interaction between both partners (e.g. hAGT fusion proteins) allows immobilization of recombinant proteins without time-consuming purification step (Kindermann et al. 2003). The covalent nature of binding ensures sustained immobilization of proteins. Fusion should not alter the properties of both partners and the enzyme is preferably small in order to minimize steric effects during the immobilization.

All these immobilization methods have been studied in the frame of proteomic applications. Large scale protein productions were required for protein microarrays manufacturing. The production and use of site-specific biotin labeling of proteins should not be a bottleneck with respect to these high-throughput technologies. Direct labeling of proteins from cell lysates and its subsequent use for microarray manufacturing without time-consuming and expensive protein purification have been successfully investigated (Lue et al., 2004a, b; Tan et al., 2004a, b; Yin et al. 2004). Due to potential problems that may arise during protein expression in a host cell and in order to simplify the process of protein microarray preparation, intein-mediated approaches (protein biotinylation or self-spliced split intein) in a cell-free protein synthesis have been also proposed (Lue et al. 2004a, b; Tan et al. 2004a, b; Kwon et al. 2006).

2.3.4.5 Small Molecules

An original approach for the generation of protein microarrays combines an original mRNA-peptide fusion synthesis and an addressable immobilization via hybridization to surface-bound DNA capture probes. The mRNA-peptide fusion process relies on the covalent linkage between the translated peptide and its own encoding mRNA (Lin and Cornish 2002; Weng et al. 2002). The DNA chip technology currently enables to produce microarrays with hundreds to thousands of different sequences (up to 60 bases length). The mRNA fusion method is amenable to the generation of large library size of fusion proteins (10^{15} distinct sequences). Limits will come from the selectivity of the hybridization. In addition, although there is no practical limit on the size of the mRNA use, limitations due to proper protein folding could be encountered in the large peptide sizes.

In this method, the fused peptides are addressed to the solid surface and are subsequently used to investigate interaction with specific ligands (targets) in solution. Several groups (Winssinger et al. 2004; Diaz-Mochon et al. 2006; Chevolot et al. 2007) have followed a similar approach, that is, by addressing DNA or PNA tagged biomolecules to DNA microarray. However, in these studies, they take advantage of this addressability to perform the binding reaction step directly in solution prior to addressing each compound onto the microarray. Emphasis is about

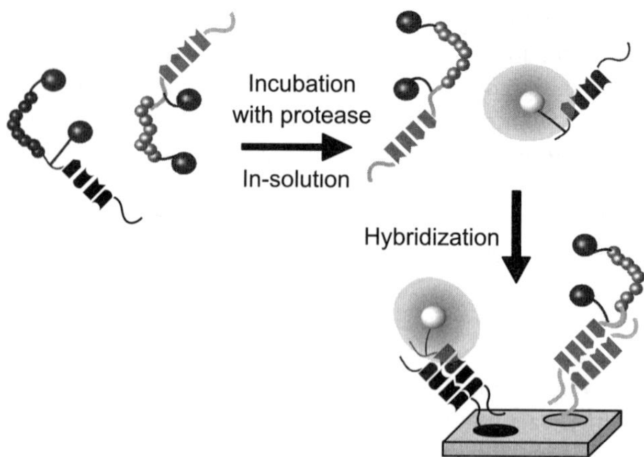

Fig. 2.12 Schematic representation of the specific immobilization of DNA encoded peptides

the higher activity in solution when compared to reaction where one partner is immobilized at solid surface leading to excellent sensibility. The production of chimerical biomolecules coupled to library of nucleic acid tags is a hurdle task.

DNA-based carbohydrate biochips were successfully realized by immobilization of oligonucleotide glycoconjugate molecules that present a nucleic acid sequence allowing hybridization to DNA chips (Chevolot et al. 2007). For example the screening of a library of protease substrates for global analysis of protease cleavage specificity was achieved (Fig. 2.12) by combining in-solution enzymatic reaction of fluorogenic protease substrate and binding the library to specific locations on DNA chips (Winssinger et al. 2004; Diaz-Mochon et al. 2006; Diaz-Mochon et al. 2007). It is worth noticing that a peptide nucleic acid (PNA) tag was bound to the substrate instead of an oligonucleotide because it was not cleaved by the enzymatic reactions. PNA is a nucleic acid analog in which the sugar phosphate backbone of DNA has been substituted by a synthetic uncharged peptide backbone of N-(2-aminoethyl) glycine units. PNA is chemically stable and resistant to enzymatic hydrolysis.

2.4 Surface Functionalization

Biosensors make use of a large panel of detection techniques ranging from electrochemistry to optics. The type and properties of the solid support do not matter so much for some detection techniques such as fluorescence. The support is said *passive*. Its properties may contribute to the sensitivity of the transduction of the chemical signal however. On the contrary, *active* supports or surface layers are absolutely required for transduction of the signal. For example, electronic detection requires conductive materials (vitreous carbon, gold surface, etc.). In addition, analytical properties such as sensibility, detection limit, specificity and lifetime depend on the bulk and surface properties of the solid support and the sensitizing layer (functionalized surface). Sensibility and detection limit can be improved by (1) an appropriate design of the biomolecular coupling (i.e., uniform and oriented covalent coupling), (2) increasing the loading capacity of active molecule (surface coverage) while avoiding steric or electronic hindrance, and (3) limiting the adsorption of biomolecules that leads to denaturation of active species and/or to background signal in case of nonspecific adsorption.

The aim of the surface functionalization is to place detection biomolecules close to the transducer and in an environment similar to that of their natural medium where biological interactions take place. Thus, the functional solid support can be viewed as a mixed interface. The interface is composed of specific sites for immobilization overhanging an inert underlayer (Fig. 2.13).

The type and surface coverage of reactive sites are tuned according to the immobilization strategy, the nature of biomolecules (size, charge) and specificities of analysis (e.g., concentration range to be detected, ionic strength). Signal intensities increase with the number of receptors immobilized per unit surface, until an optimum. Beyond optimum surface coverage, steric and/or electrostatic hindrance lead to lower accessibility and loss of detection signal (Herne and Tarlov 1997; Houseman and Mrksich 2002).

Biological molecules (DNA, proteins) are electrically charged. The size and shape of these biomolecules are governed by intramolecular electrostatic forces that

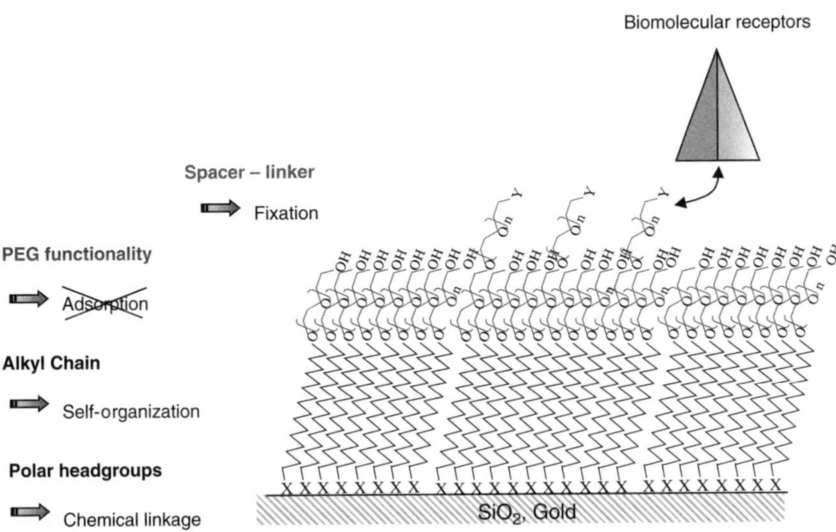

Fig. 2.13 Schematic representation of ideal surface modification for immobilization of biomolecules. Mineral support (silica, gold) are modified by organic layer. Amphiphile molecules with a polar head group and long hydrophobic tails arrange in dense layer. The inorganic polar head groups (silane or thiol) ensure a stable and robust anchorage to the support. The outer surface is terminated by hydrophilic PEG functionality to minimize nonspecific adsorption. Longer linkers bring site-specific groups for the immobilization of molecular receptors

can be modified by varying the ionic strength. In addition to the molecular surface area that controls steric hindrance, electrostatic forces also play an important role in immobilization and in biomolecular interactions. For example, Herne and Tarlov (1997) have reported the formation of densely packed single-strand DNA (ssDNA) monolayer by increasing the ionic strength conditions; but hybridization was prevented because of steric and electrostatic factors. Careful optimization of the DNA coverage and surface properties allowed improving the hybridization signal. Similarly, immobilization of small molecules (ligands) as dense monolayer (high surface coverage) limits their accessibility due to steric hindrance by larger biomolecules (Houseman and Mrksich 2002).

Improved accessibility of biomolecules is achieved by attaching them to the surface through a linker that moves them away from the solid interface. The linker must be sufficiently long to eliminate unwanted steric interference from the support and preferably hydrophilic enough to be swollen in aqueous solution (Guo et al. 1994; Hodneland et al. 2002).

The interface underneath the immobilized receptors is either the bare solid support or a layer of organic molecules that confers specific properties in relation to the biomolecular assay. This sublayer or surrounding matrix plays an important role in the functionalization and on the analytical properties of biosensors. The physico-chemical properties of the sublayer are tailored to limit the biomolecular adsorption. It may also be designed to protect the solid surface or the grafted surface from degradation.

Extensive literature deals with the adsorption of biomolecules at solid–liquid interface (Ostuni et al. 2001). Because adsorption is governed by dispersive interactions (Lifshitz–van der Waals), hydrophilic surfaces tend to resist the adsorption of biomolecules. But the hydrophilic character is not the sole relevant physico-chemical characteristic. Otsuni et al. (2001) reported that surfaces that resist protein adsorption present four common properties: they are polar, hydrogen-bond acceptors, not hydrogen-bond donors, electrically neutral. In biosensor applications, grafted poly(ethylene glycol) (PEG) is widely used to limit the adsorption of biomolecules to solid surface. Such surfaces grafted with PEG are often known as "PEGylated."

The variety of detection techniques and the complexity of parameters to be adjusted have lead to extensive literature about materials and their functionalization. It is described in three parts devoted to: (1) 2D surface functionalization; (2) 3D membranes and porous layers; (3) nanoparticles and nanostructures.

2.4.1 2D Immobilization: Grafting of Monolayers

Very few materials allow direct immobilization of biological molecules onto their bare native surface. Controlled immobilization that meets specificities of biosensor applications requires particular surface chemistry in order to maintain biological activity and minimize nonspecific adsorption. The formation of thin organic films on mineral surface is widely used to promote the immobilization of biomolecules to solid surface. In addition to bringing adequate specifications to the solid/liquid interface, the organic film must fulfill several properties: reproducibility, homogeneity, thermal stability and chemical stability with respect to both environmental conditions (UV, biofouling, hydrolysis) and chemical reactions performed during the immobilization/functionalization process.

At the molecular scale, surface modification involves organic heterodifunctional molecules. One extremity is intended to react with the solid support while the other extremity brings appropriate functional end-groups for site-specific immobilization of bioreceptors. Various organic molecules and chemical reactions are involved depending on the nature of the solid support. For example, hydroxyl-terminated surfaces of inorganic oxides like silica, glass or titanium dioxide can be derivatized by silane coupling agents, gold surfaces are modified by adsorption of alkanethiolate reagents, hydride-terminated silicon surface undergo chemical modification by either hydrosilylation or coupling reaction with organometal compounds.

2.4.1.1 Alkylsilane Functionalization

The general formula of alkylsilane $R_nSiX_{(4-n)}$ shows a dual behavior, organic and inorganic. R is organic moiety and X is a hydroxyl or a reactive hydrolysable group (–Cl, –OMe, –OEt, etc.) that reacts onto hydroxyl-terminated supports. Reaction of silane molecules to surface silanol groups of silica leads to the formation of quite stable siloxane bonds (Si–O–Si). The siloxane bond is thermally stable and is relatively chemically stable. The siloxane bridge can be cleaved by strong alkaline solution and fluorhydric acid.

The functionality of silane is the number of reactive groups present at the silicon atom. Monofunctional silanes ($n = 3$) having only one reactive group at the silicon extremity bind to silica by means of a single bond. Multifunctional silanes ($n = 1$ or 2) bear several (2 or 3) reactive groups the silicon atom and can bind to the surface by several bonds. Binding by several bonds is stronger but the reaction is more difficult to be controlled in a reproducible manner.

An appropriate choice of the functional "tail group" (R) introduces the specific surface functionalization regarding the subsequent immobilization of bioreceptors. Some functional group cannot be used directly however because the silyl head-group (alkoxy-, chloro-, dimethylamino-silane) is reactive toward several organic functions (carboxylic acid, alcohol, aldehyde, etc.) (Beier and Hoheisel 1999; Zammatteo et al. 2000; Dugas et al. 2004; Wang et al. 2005) or because the functional extremity can interfere with the film formation at the surface (e.g., amino groups) (Martin et al. 2005). Incompatible tail groups are either protected during grafting or chemically synthesized on the solid support after silane grafting. For example, a silane bearing a *tert*-butyl ester group has been grafted because activated acids and their acid precursors have not been chemically compatible with silanes; it has been subsequently converted into the corresponding acid and activated ester in a third step (Dugas et al. 2004). Aldehyde functional groups can be

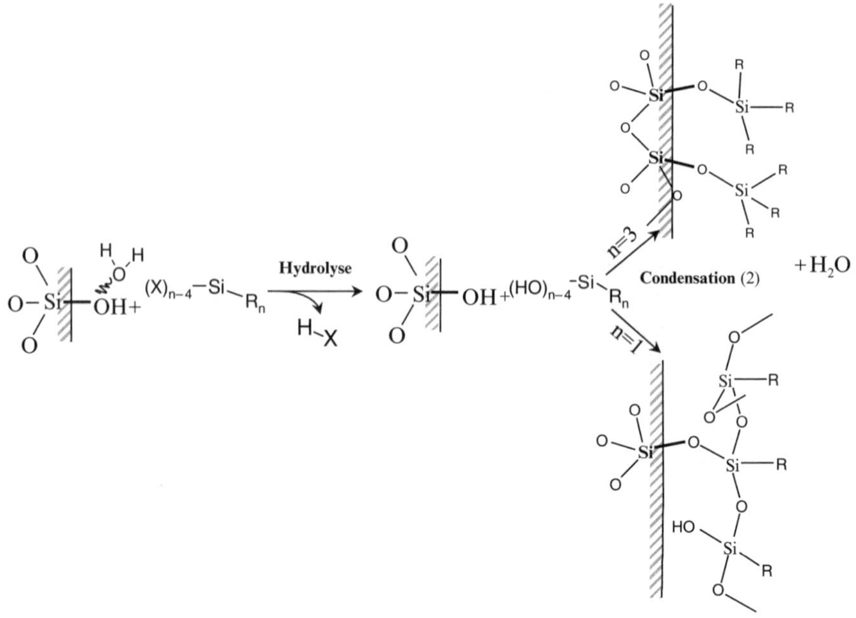

Scheme 2.2

obtained by reduction of alkenes (Hevesi et al. 2002). Several approaches have been proposed for grafting maleimido-terminated silane layer (Wang et al. 2005). As previously mentioned, many research applications involve surface modification by amino or epoxy groups followed by postgrafting modification by heterodifunctional linkers.

2.4.1.1.1 Reaction Mechanism

Two reaction mechanisms have been proposed for silane grafting (Boksanyi et al. 1976; Tripp and Hair 1995; Clark and Macquarrie 1998):

- The first mechanism "hydrolysis and condensation" is a two-step reaction. In the first step, substituents of the silicon atom are hydrolyzed to form silanols. The alkylsilanols establish hydrogen bonds with the free hydroxyl groups on the surface or with adsorbed water molecules on the surface. The covalent bonding is carried out by condensation and release of water molecule (Scheme 2.2). Water has a catalytic effect; its presence and concentration should be carefully controlled in order to ensure reproducibility.
- The second proposed mechanism corresponds to a direct condensation reaction between the silicon headgroups and the surface hydroxyls during a thermal treatment (Scheme 2.3).

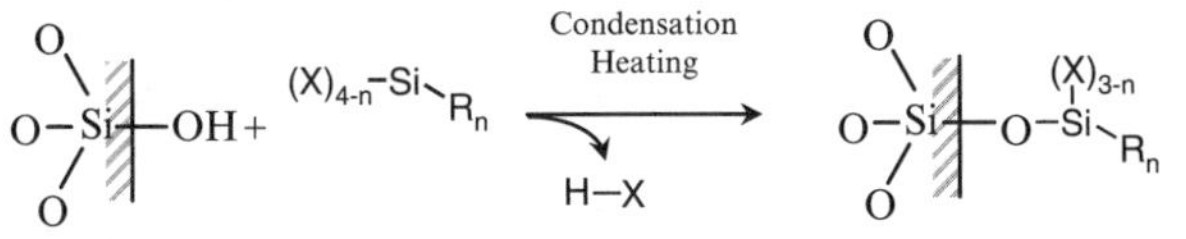

$$(O)_3Si{-}OH + (X)_{4-n}{-}Si{-}R_n \xrightarrow[\text{H}-\text{X}]{\text{Condensation, Heating}} (O)_3Si{-}O{-}Si(X)_{3-n}{-}R_n \qquad (3)$$

Scheme 2.3

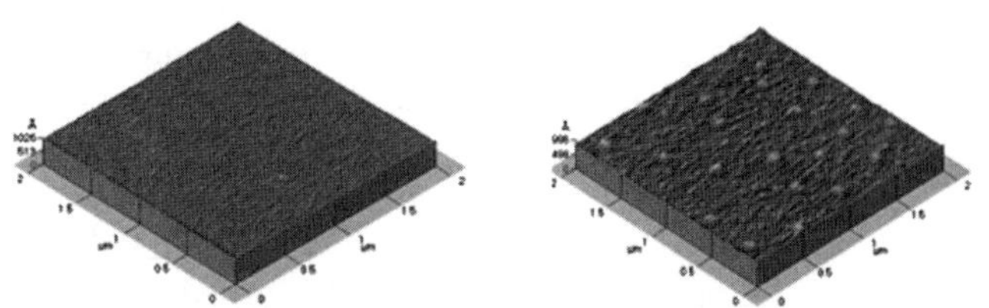

Fig. 2.14 AFM images of bare Si/SiO_2 wafers (right) and after 12 h of silanization with GPTS

The silane grafting mechanisms with surface oxide have not been completely characterized due to the extreme difficulty of following the chemical modifications that occur at the extreme surface of solid support. It is worth noting that no direct analysis (*in situ*) has characterized separately the hydrolysis step prior to the surface condensation. But some studies report low or no chemical grafting of silane molecules under strictly anhydrous conditions. But surprisingly, trimethylsilanol molecules are less "reactive" than their chlorinated homologues (trichlorosilane) (Azzopardi and Arribart 1994).

Water "catalyzes" the silane grafting reaction by causing the hydrolysis of silane leaving groups. Multifunctional silanes are hydrolyzed in solution into silanol intermediates that may undergo intermolecular polycondensation in solution. The small polymeric species can then react with the surface and lead to the formation an inhomogeneous layer.

AFM images in Fig. 2.14 are 3D representations of flat silicon dioxide surface after cleaning (left) and after silanization with the trifunctional glycidoxypropyltrimethoxysilane (GPTS) (right). The surface modified by GPTS exhibits nodules of 0.3 μm diameter and up to 150 Å height containing polycondensed silane.

There is vast literature dealing with *ex situ* characterization of alkylsilane derivatized support. The chemical grafting has been proven by infrared (Blitz et al. 1988; Angst and Simmons 1991; Azzopardi and Arribart 1994; Tripp and Hair 1995) and NMR spectroscopy (Sindorf and Maciel 1981, 1983; Kinney et al. 1993) on silica powders (high specific area). Nevertheless, some *in situ* studies performed on silicon dioxide allowed a more detailed description of the grafting mechanisms of various silanes in organic solutions (Azzopardi and Arribart 1994).

The choice of the type of silane is of importance. The silanization reaction is largely related to relevant experimental conditions. The nature of the silane and its concentration (Boksanyi et al. 1976; Szabo et al. 1984; Azzopardi and Arribart 1994; Tripp and Hair 1995), the nature of the solvent and its water content

(Tripp and Hair 1992; McGovern et al. 1994; Navarre et al. 2001), temperature (Boksanyi et al. 1976) and time (Gobet and Kovats 1984; Azzopardi and Arribart 1994) are the most important parameters.

The silane grafting reaction is a condensation of silane head groups to the free hydroxy groups of the surface. Thus the solid surface must be adequately prepared prior to the silanization reaction in order to create a constant density of such hydroxy groups (silanols) and improve the density of silane grafting and repeatability of the reaction (McGovern et al. 1994; Cras et al. 1999; Fadeev and McCarthy 1999). Surface cleaning removes contamination that may hamper the reaction to the surface hydroxyls. Surface cleaning steps under oxidizing conditions (e.g. "Piranha" solution ($H_2O_2 + H_2SO_4$), UV-light + ozone, oxygen-plasma treatment) and treatment in hot water are also intended to increase the hydroxyl surface density (e.g. on glass slides). After cleaning, the strongly hydrophilic hydroxylated surface is covered by several layers of hydration water molecules. It is therefore highly recommended to dry the freshly cleaned surfaces in order to get reproducible results.

The formation of monolayers at the surface of oxide surface proceeds through different ways depending that the silane is mono or multifunctional.

Trifunctional silanes such as trimethoxy or trichlorosilane react with surface hydroxyls by means of their three reactive groups, but this reaction is most often incomplete because it is not possible to attach the same silicon atom to the three oxygen's present at fixed positions at the surface, keeping the bonds lengths and angles close to reasonable values. Furthermore, the possible hydrolysis of silane taking place in case of incomplete drying of the silica surface leads to a concomitant polycondensation of the hydrolyzed silanes and condensation with the surface silanols; an ill-defined polycondensate of organosilane grows from the surface and a thick and rough final silane layer results (Tripp and Hair 1995; Yoshida et al. 2001). The formation of monolayers with trifunctional silane thus requires precise control of the experimental conditions (humidity, solvent, temperature).

Such troubles do not occur with monofunctional silanes that bind to the surface by means of a single bond. Even in the case of hydrolysis of the silane next to the surface, the polycondensation of hydrolyzed monomers lead to dimers that cannot react with the solid support. The grafting is directly related to the density of silanol groups on the surface. But monofunctional silanes have a lower reactivity than multifunctional ones (Evans and White 1968). This has to be compensated by the choice of a highly reactive leaving group such as dimethylamino (Szabo et al. 1984).

2.4.1.1.2 Alkoxysilane Film Properties

Three different silanization protocols can be described from the literature. Short or volatile silanes can be involved in gas phase silanization reaction. Clean and dry supports are exposed to vapor of silane molecules. This technique allows accurate control of experimental conditions in order to avoid hydrolysis of silane in solution. A variant of this technique called Atomic layer deposition (ALD) allowed the preparation of homogeneous aminosilylated silica surfaces at low temperatures ($\leq$150°C) in a reproducible way by a solvent-free procedure (Ek et al. 2004).

Grafting of nonvolatile silanes can be carried out in two steps called "impregnation" and "condensation" (Chevalier et al. 2000; Dugas et al. 2004). A solution of the silane in anhydrous volatile solvent (e.g. pentane) is deposited on the surface and the solvent is evaporated under reduced pressure. This first "impregnation" step leaves a thin layer of pure silane on the surface. The grafting of silanes ("condensation") is performed by heating at elevated temperature (>140°C) under dry atmosphere. Highly reproducible grafting densities are obtained by this technique with monofunctional silane (Dugas et al. 2004).

More common silanization protocols involve reaction in solution. Trialkoxysilanes are widely used for immobilization purposes. Aminopropyltriethoxysilane (APTES) and glycidoxypropyltrimethoxysilane (GPTS) are the most widely used silanes for chemical surface modification. Grafting reactions are currently carried out from aqueous alcohol solution (95% ethanol, 5% water) under slightly acidic conditions (pH = 4.5–5.5). This simple route for surface modification often leads to irreproducible results (Yoshida et al. 2001). As example, the analysis by ellipsometry of DNA layers immobilized by the intermediate of the trifunctional glycidoxypropyltrimethoxysilane (GPTS) has revealed the nonuniform grafting density and layer thickness that is larger than expected on the basis of the length of the oligonucleotides (Gray et al. 1997). In addition, the terminal functional groups can also interfere with the chemical grafting and reduce the uniformity of the monolayer. In particular, the amino group of APTES adsorbs by hydrogen bonding to silanols groups on the oxide surface (Horr and Arora 1997; Sieval et al. 2001) (Fig. 2.15).

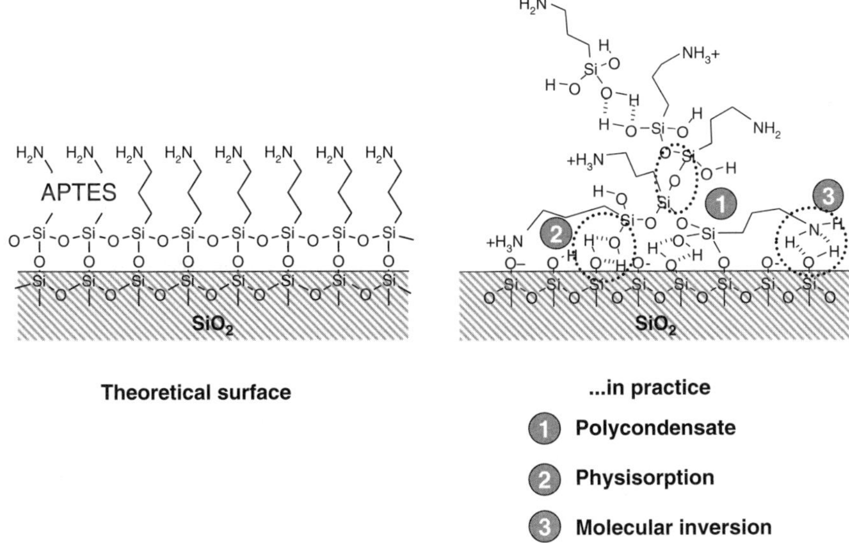

Fig. 2.15 Schematic representation of intended layer and effective layers formed with APTES

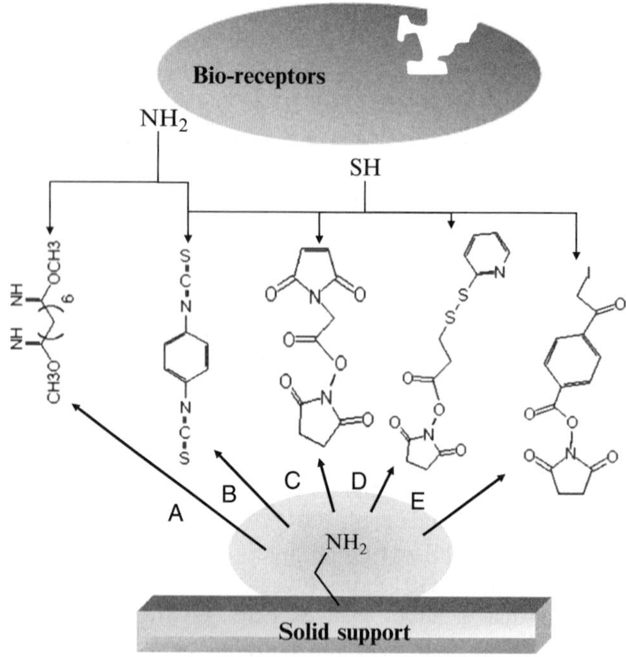

Fig. 2.16 Schematic representation of the versatile derivatization of aminated support. (**A**) dimethylsuberimidiate; (**B**) 1,4-phenylene diisothiocyanate (PDITC); (**C**) N-succinimidyl-3-maleimidopropionate (SMP); (**D**) N-succinimidyl-3-(2-pyridyldithio) propionate; (**E**) N-succinimidyl (4-iodoacetyl) aminobenzoate (SIAB)

Once a functional silane has been grafted, the functional group can be modified by reaction with a coupling agent. Amine functional groups are extensively used for surface functionalization because they are quite versatile. γ-Aminopropyltriethoxysilane (APTES) or N-(2-aminoethyl)-3-aminopropyltrimethoxysilane (AAPS) coupling agents are widely used to modify hydroxylated supports (silica's, glasses). Aminated supports are then treated with homo- or hetero-difunctional reagents (or cross-linkers) allowing to create a large panel of active surfaces toward nucleophilic groups (Fig. 2.16). Water-soluble cross-linkers (sulfonate analogues) give consistently higher coupling yields than their insoluble counterparts (Adessi et al. 2000).

As described previously, surface functionalization with epoxy, activated esters or aldehyde functions can also be prepared by reaction of homodifunctional molecules (Yershov et al. 1996; Volle et al. 2003).

The formation of well-controlled films from the short-chain functional silanes can be achieved by precise control of the experimental conditions (silane concentration, reaction time and water content) (Hu et al. 2001). Larger density can be reached by combining chemical grafting and self-organization of the grafts as follows. The elaboration of dense, homogeneous and stable monolayers can be prepared from long-chain trifunctional silanes by self-assembly. A self-assembled monolayer, SAM, is formed when surfactant molecules adsorb as a dense and organized monomolecular layer on a surface. Dispersive interaction (van der Waals) along the long enough hydrophobic tails is the driving force of the self-organization. For example octadecyltrichlorosilane shows self-assembling behavior. The terminal functional group at the extremity of the alkyl chain must be protected in order to avoid interference during the film formation (Martin et al. 2005). In addition, this terminal "tail group" must be sterically small enough to limit interference during monolayer organization driven by alkyl chain interactions.

SAMs formation is realized under careful control of the solvent composition, moisture and temperature conditions. Trichlorosilanes compared to trialkoxysilanes are highly moisture sensitive.

SAMs of pure alkylsilane (R is nonfunctional alkyl) self organize quite easily. Terminal functional groups at the extremity of the alkyl chain tend to destabilize the monolayer. For example, long-chain of octadecyltrichlorosilane ($CH_3(CH_2)_{17}SiCl_3$) form dense monolayers while longer chains (21 CH_2) are required to obtain well-ordered monolayers with a hydrophilic "tail group" such as poly(ethylene glycol) or protected amino groups (Navarre et al. 2001; Martin et al. 2005).

True dense SAMs provide stable and well-oriented layers. The insulating polymeric surface coating helps to protect the Si–O–Si bonds between the coupling agents and the surface. SAMs can withstand harsher conditions than less ordered film tethered to the surface. It can serve to form a dielectric barrier for electronic detection devices. However the benefits coming from the SAM properties (dense packing, well-ordered monolayer) hamper the reactivity of the surface groups used for covalent coupling of bioreceptors or biomolecular interaction (Fryxell et al. 1996). Incorporation of the desired functional groups in an inert background monolayer (deposition of a mixed film) avoids the crowding effect of the functional substituents at the surface. Mixed films can be prepared either in one shot (Martin et al. 2005; Hoffmann and Tovar 2006) or in a two-step process (Harder et al. 1997).

2.4.1.1.3 Silane Coupling Applied to Various Materials

Silanes are used to modify hydrated surface materials ranging from glass slides to aluminum oxide materials:

- Glass is widely used in microarray techniques owing to its chemical stability, transparency, low-fluorescence background and flatness. Surface modification involves silane grafting.
- Miniaturization of biosensors has lead to use materials coming from microelectronics. This trend calls upon semiconductor materials like crystalline silicon wafers. Thermal silicon dioxide (SiO_2) or silicon nitride (Si_3N_4) are among the materials that are commonly used as insulating or passivating layer for electronic detection systems (e.g., field-effect transistors, GenFET, Genosensors). These materials are also implemented to fabricate miniaturized cantilevers that are used to detect biomolecular interaction by deflection of the cantilever.

Indium Titanium Oxide (ITO) (Moore et al. 2006) is a conducting material deposited as thin films in electrochemical detection system.

- Specific optical detection techniques rely on the use of high refractive index layer such as titanium oxide (TiO_2) (Trummer et al. 2001; Dubruel et al. 2006) that can be chemically derivatized by alkoxysilanes.
- Integrated system (Lab on chip or μTAS) involves hybrid materials composed of glass, silicon, silicon dioxide and/or often elastomer materials. Polydimethylsiloxane (PDMS) elastomers are widespread in such miniaturized systems. The surface of PDMS is modified by silane molecules in order to limit biomolecule adsorption and to promote bioreceptors immobilization.
- Porous materials like porous silicon (Bessueille et al. 2005) or anodized aluminum oxide (Lee et al. 2007) present hydroxyl that can be turned out to chemical surface modification.

2.4.1.2 Hydrosilylation of Silicon Support

Siloxane bridges that ensure anchoring of the alkoxysilane to hydroxylated materials like glass are alkaline sensitive chemical bonds. Silicon-based materials coming from the microelectronic industry are used to implement miniaturized system that can be thermally oxidized into a silica layer at their surface for undertaking the classical silanization reaction on silica surface. Alternatively, they can be used without surface oxidation. Raw bare silicon exhibits silicon–hydride surface groups (Si–H) that can be derivatized by difunctional organic molecules to form stable silicon–carbon bonds, potentially allowing the incorporation of a wide range of chemical groups. Silicon surfaces coated with a self-assembled alkyl monolayer are very stable. For example, derivatized porous silicon withstands to harsh alkaline conditions ($pH > 10$) (Buriak et al. 1999).

Hydrogen terminated silicon surfaces are prepared just before modification because rapid oxidation in air leads to a thin silicon dioxide passivating layer. Silicon surfaces are prepared by a short hydrogen fluoride treatment just before the subsequent chemical modification. Different grafting routes were assessed, using reaction with alkenes, alkynes (Buriak and Allen 1998), peroxides (Linford et al. 1995) with organometallic agents (Tao and Maciel 2000). The well-known chemistry in solution encompasses thermal- (Sieval et al. 2001), radical-, Lewis acid- (Boukherroub et al. 1999) or metal complexes catalysis (Zazzera et al. 1997) and UV-light mediated (Boukherroub and Wayner 1999) hydrosilylation reactions.

Surface modification by long-chain alkenes or alkyne have been characterized by IR spectroscopy, ellipsometry, X-ray reflectometry to form dense and organized monolayers. The covalent Si–C linkages are very stable (UV-light, fluorhydric acid, alkaline conditions). The efficiency of molecular grafting largely depends on the steric hindrance of the molecules (Buriak et al. 1999). Insertion of alkenes or alkynes into surface Si–H groups yields alkyl or alkenyl termination, respectively. Both chemistries are remarkably tolerant to a wide range of chemical functionalities (Buriak et al. 1999). But as for the silane grafting reaction, the terminal functional groups of the ω-alkenes have to be protected to prevent direct reaction with the Si surface which results in disordered monolayers (Strother et al. 2000a, b; Sieval et al. 2001). The formation of mixed monolayers with amino-terminated function has been reported with control over the surface density of the amino-groups (Sieval et al. 2001).

Silicon devices are now utilized for microelectromechanical systems (MEMS), microfluidics and micro-total analytical systems (μ-TAS), biosensors and other bioanalytical applications. The chemical modification of silicon surfaces have been applied to systems of various shapes; flat silicon substrates (Strother et al. 2000a, b), nanowires (Streifer et al. 2005), nanoparticles (Nelles et al. 2007) and porous layers (Mengistu et al. 2005). In addition, the great stability of the Si–C bonds when compared to the siloxane bridge allows to use the biosensors under harsher conditions or to reuse the biosensors without or with only minor losses of activity (Strother et al. 2000a, b).

2.4.1.3 Organo-sulfur Precursors for Surface Functionalization

Intensive literature deals with the formation, properties and applications of self-assembled monolayers of alkylthiolates on metals. The high affinity of thiols (R–SH), disulfides (R–S–S–R) and sulfides (R–S–R) for metals (gold, silver, copper, palladium, platinum, and mercury) drive the formation of well-defined organic surfaces with useful chemical functionalities displayed at the exposed interfaces for biosensor applications. Gold is usually the surface material of choice due to its chemical inertness and the well-defined structure of the obtained film (densely packed and ordered array of long chain molecules). Gold does not oxidize at room temperature, does not react with atmospheric oxygen and is biocompatible (compared to silver that oxidizes readily in air and is toxic to cells (Dowling et al. 2001)).

Self-assembled monolayers of organosulfur precursors spontaneously form on gold upon immersion of gold surfaces in a solution of precursors. Film formation and properties of SAMs on gold are out of the scope of this report and are exposed in details elsewhere (Ulman 1996; Love et al. 2005).

Flat gold surfaces are commonly prepared by physical vapor deposition (PVD) techniques, electrodeposition or electroless deposition on various supported materials (silicon wafers, glass, mica and plastic substrates). The strong adsorption of alkylthiols onto gold surface is able to displace miscellaneous impurities or contaminant already present at the surface (Love et al. 2005; Luderer and Walschus 2005). Cleaned surfaces are thus not absolutely essential to produce alkylthiolate SAM. Kinetics of the SAM formation are however affected because desorption of contaminants may be slow. Therefore, preparation of SAMs on gold is easy and reproducible. Common procedures for the cleaning of gold surfaces are chemical treatment with highly oxidizing solutions (e.g. "piranha solution"), electrochemical cleaning in sulfuric acid solution (Luderer and Walschus 2005; Lao et al. 2007) and thermal treatment, annealing in a flame for instance (Briand et al. 2006).

The formation of SAMs on gold surface with organosulfur compounds follows two distinct kinetics. The first very fast step corresponds to the adsorption of molecules from solution to the surface and the second slow step where monolayer organizes into a well-order state. Adsorption takes from millisecond or minutes in millimolar solutions to 1 h in micromolar solutions (Ulman 1996; Love et al. 2005). The second step is the reorganization of the initially adsorbed monolayer into a crystalline configuration. The 2D crystallization on the surface is slow and may require hours to several days until monolayer reaches tightly packed SAM with minimum defects (Ulman 1996; Luderer and Walschus 2005).

Film formed from alkylthiols or alkyl disulfides lead to identical film properties. Upon the adsorption process these organosulfur compounds form alkanethiolate species ($R–S^-$):

$$RSH + Au \rightarrow RS - Au + e^- + \frac{1}{2} H_2$$

$$RS - SR + Au \rightarrow RS - Au + e^-$$

The most common protocol for preparing SAMs on gold and other materials is immersion of clean gold surface into dilute (1–2 mM) solution of ethanolic alkylthiol (or disulfide) for 12–18 h at room temperature. Film properties (reproducibility, physico-chemicals behavior of the organic interface, SAM defects, density) depend on the nature of the sulfur compound (head groups, length of the alkyl tail, nature of the terminal group for heterodifunctional molecules), and on reaction factors (immersion time, concentration of adsorbate, temperature, solution composition) (Love et al. 2005).

For biosensor application purpose, mixed SAMs are easily prepared. Mixed monolayers allow to control the density of the reactive species (diluting), to increase its accessibility and to bring chemical inertness to the underlying surface.

2.4.1.3.1 Direct Biofunctionalization of Gold Surfaces

Straightforward immobilization of native proteins, peptide (via cysteine groups) or thiols modified DNA strands have been reported on gold surface (Steel et al. 1998; Brogan et al. 2003; Ohtsuka et al. 2004). Tightly packed monolayers were obtained with oligonucleotide but immobilization is less efficient with other molecules like oligopeptides (Ohtsuka et al. 2004). Misorientation is possible however when the other functional groups (amines) bind to the gold surface. Herne and Tarlov (1997) reported that thiol-derivatized DNA molecules mostly interact with gold surface through the sulfur atom of the thiol group. Small extent of adsorption by nucleobases of oligoprobes took place and resisted even hard washings conditions. But interestingly, displacement experiments, where the gold surface functionalized with HS-DNA was contacted with high concentrations of small thiolated molecule

(6-mercapto-1-hexanol), have resulted in nearly complete displacement of adsorbed molecules.

Supramolecular assembly of hindered thiolated molecules has been reported to form compact and homogeneous monolayers on gold electrodes. Collagen-like peptides exhibit rod-like conformation and have been used to prepare SAMs onto gold electrode (Yemini et al. 2006). Enzymes were covalently immobilized through the amino terminal group of the peptides. The collagen-like peptides confers biocompatibility properties to the sensors surface while ensuring a stable attachment of enzymes to the biosensors.

2.4.1.3.2 Functionalization of Gold Surface for Covalent Coupling

Various thiol precursors have been used to tailor the surface functionalities of gold surface. Bioreceptors are covalently immobilized through amine or thiols reactive groups (see Sect. 2.2). A large panel of difunctional thiol molecules is available for straightforward SAMs formation on gold surfaces. As for the alkylsilane mixed monolayers, different synthetic routes allow the formation of mixed monolayers (Love et al. 2005). The simplest one consists in coadsorption of mixtures of alkane thiol precursors.

In order to increase the accessibility and limit the influence of the underlying surface, the site-specific groups which serve to immobilize bioreceptors (–COOH (Briand et al. 2006), ligand protein (Hodneland et al. 2002)) are placed at the extremity of a spacer. These ω-substituted alkanethiols are mixed with shorter alkanethiols terminated by hydroxyl or oligo(ethylene glycol) groups.

In order to move away bioreceptors from the organic interface, immobilization strategies have been initiated through a "sandwich" configuration. A first receptor is immobilized (streptavidin (Grubor et al. 2004), protein G (Briand et al. 2006), S-layer protein (Pleschberger et al. 2004) and DNA (Lao et al. 2007)) and serves as a spacer to specifically immobilize the biological probe through oriented configuration. In addition to increase accessibility, these strategies allow to immobilize bioprobes under oriented and selective configurations.

Direct detection, label-free, measurements call upon "active" surface layers involved in the detection system. For example, electrodes are the sensitive elements of electrochemical detection techniques. Gold surfaces are also used to propagate surface plasmon waves in the conventional surface plasmon resonance (SPR) technique. Surface chemistry will be tailored in order to match the detection technique specifications. Achievement of dense and compact hydrophobic monolayers with low defects, allows formation of insulating layers of low dielectric permittivity, therefore small Faradaic background at the electrode biointerface regarding their electrical properties. The electrode/solution interface in potentiostatic measurements is described equivalent to several capacitors in series (Berggren et al. 2001). The electrical equivalent circuit is different for amperometric and voltametric detection techniques that rely on the detection of electroactive species (red-ox, electron transfer proteins). The hurdle consists in establishing an electrical connection between the receptor and the signal transducers (electrode surface). Direct electron transfer (ET) through the organic interface represents the alternative approach to the use and incorporation of "electrochemical" mediators (Zhang and Li 2004). One strategy to the direct electron transfer is DNA-mediated protein ET (Lao et al. 2007). In this case, the interface helps to efficiently drive electrons to the conductive electrode.

2.4.1.4 Reactivity Issues

Slow kinetics of grafting reaction is the main concern of bioimmobilization. Slow reaction rates come from the very low concentrations of species to be grafted and detrimental close vicinity to the surface. The slow reaction rate cannot be dealt with using extremely long reaction time because the reactive surface is open to competitive side-reactions during the same time. Therefore highly reactive species are selected, so that the reaction rate of solid-supported molecules is largely controlled by diffusion. Of course, the concentration gradients that drive diffusion are low when the mean concentration itself is low. Diffusion can be supplemented by stirring and mixing the solution at the solid surface. Stirring inside miniaturized devices is a technological challenge. A common practice in array manufacturing using spotters is letting the deposited drops dry. Drying the drops improves immobilization rate because the reagents are concentrated *in situ*; but no reaction can take place when the drops are dry to completion because the reagents are no longer mobile. Therefore, drying is useful but it has to be controlled in a chamber with residual humidity.

Additional aspects related to the underlying surface also interfere onto reactivity. The close proximity of the outermost functional groups and the hard surface modifies the mobility behavior (and reactivity) when compared to reaction in solution. It is well-known that the rates of interfacial reactions are slower than in solution. There are obvious causes coming from geometrical reasons. One of the reagents is motionless because it is attached to the surface; therefore mutual diffusion coefficient is roughly reduced by a factor of 2. Secondly, diffusion is restricted to half a space, which causes a supplementary reduction of rate by a factor of 2. The actual losses of reaction rates with respect to solution kinetics are much larger than the rough geometrical factor of 4 however. There are several additional phenomena that are significantly operating. Words often found in the literature are "steric hindrance", "reduced mobility", "lack of bioavailability". This is difficult to go beyond the words as long as fundamental studies have definitely reached clear-cut conclusions.

The parameters affecting surface reactions are of two distinct types: one type pertaining to the accessibility to the reactive sites; the second type deals with the reaction itself that requires the correct mutual orientation of the reagents and the desolvation of the reaction centers. So-called steric effects have various origins. The underlying material contributes since buried and/or misoriented reactive groups are not reactive. The high surface concentration of reactive functional sites affects the grafting kinetics at high conversion; lateral steric hindrance and crowding effects take place when the biomolecules of large size are immobilized up to high grafting densities.

The problem of accessibility and misorientation depends on the attack trajectory dictated by the reaction mechanism. This is the case for the SN_2 reaction where the terminal "tail group" attached at the surface undergoes "backside attack" by incoming nucleophilic reagents and the leaving group has to diffuse off the reaction site (otherwise the back-reaction takes place). The hindrance by the underlying layer dictates the reaction rate (Fryxell et al. 1996). Grubor et al. (2004) have compared the reactivity of two NHS-ester ω-functionalized thiols of different chain lengths (C_3 and C_{11}) adsorbed onto gold surface. SAMs made of short thiol precursor (3 $-CH_2-$) are of lower packing density than longer alkyl-chain thiol precursor (11 $-CH_2-$); the film is less ordered and less rigid. The hydrolysis and acylation reaction rates of the NHS end group at the top side of the SAM surface are strongly influenced by the presence of the SAM and the structure of the SAM appears of definite importance. The rate of base-catalyzed succinimidyl ester hydrolysis at the surface is about 1,000 times slower than in solution; the rate of hydrolysis and amidation is significantly lower for long-chain when compared to short-chain SAMs. Similar steric effects on SN_2 reaction on planar SAMs have been reported by several authors (Hostetler et al. 1998; Templeton et al. 1998; Dordi et al. 2003).

Moving the reactive groups away from the surface via hydrophilic spacer arms can improve the solubilisation of the receptors and minimize the background influences of the underlying interface.

In addition to the hindrance arising from the position of the reactive site relative to the surface, lateral steric effects or crowding effects take place (Chechik et al. 2000; Houseman and Mrksich 2002). This arises when bulky reactants are involved when compared to the supported reactive sites and their loading surface densities. Crowding at the surface may even contribute for small molecules in case of unfavorable interactions. Vicinal dipolar moieties such as carbonyl group of ester or carboxylic acid functionalized SAMs can establish hydrogen-bonds and/or undergo condensation reactions (e.g. condensation of carboxylic acids into anhydride (Fryxell et al. 1996)). Hydrogen bonding between polar functional tail groups (e.g. ester groups), lead to molecular segregation when these molecules are coadsorbed with "diluent" molecules to form mixed monolayer (Fryxell et al. 1996; Rao et al. 1999).

Positive contribution of the highly ordered and oriented reactive end-groups present at the surface of SAMs has also been reported. The apparent increase in reaction rate is attributed to the favorable orientation of the reactive groups (Chechik et al. 2000).

Lastly, electrostatic interactions, either attraction or repulsion, often contribute to the reaction rates when the surface is electrostatically charged because it bears ionic groups. Electrostatic interactions are long-ranged and depend on the ionic strength. This is clear that electrostatic interactions do not change the reactivity itself but attract to the charged surface ionic species of opposite charge or repel ions having charge of the same sign. The surface charge depends on the pH when the ionic groups are either acidic or basic. The variation of the surface charge with respect to pH does not obey simple mass action law because electrostatic interactions attract or repel H^+ and OH^- ions. Such effect is often described in terms of apparent pK_a. The intrinsic acid-base properties of the surface groups are not changed however; electrostatic interactions only just increase or decrease the local concentration of ions in the close vicinity of the charged surface. Interactions between the biomolecule and the support are also under the influence of electrostatic forces (Yeung and Leckband 1997).

Electrostatically driven adsorption is a major origin of nonspecific adsorption that causes background noise in the detection signal. Careful control of surface charge is quite an important issue because nonionic hydrophilic layers are generally preferred. However, one can take advantage of adsorption, and particularly electrostatic adsorption, for improving or getting a better control over the chemical grafting. Thus, it has been noticed that aminated surfaces were highly reactive because their charge was positive in moderate pH conditions. Indeed, adsorption helps the reaction from very dilute solutions because it creates a local concentration of reagents in close vicinity of the surface. Let us point out that most biomolecules are anionic and strongly adsorb to cationic surfaces. Regarding aminated surfaces, the cationic form, $-NH_3{}^+$, of the primary amino group causes adsorption whereas the reactive species is the neutral amine form, $-NH_2$. A compromise is to be found between favorable adsorption and the surface density of reactive species; therefore there is an optimum pH for the grafting reaction. Strong adsorption helps grafting but also causes nonspecific adsorption during the utilization for the detection of negatively charged biospecies. Therefore, the presence of residual amino groups at the surface is detrimental. A "capping" step consisting in the deactivation of amino groups by reaction of strong electrophilic reagent after grafting is advisable. Capping yields either a neutral derivative, preferably hydrophilic, in order to prevent hydrophobic adsorption and biofouling, or an anionic derivative that creates electrostatic repulsions for anionic species (the reaction of succinic anhydride makes the conversion of primary amines into anionic monosuccinimidyl amides).

The overall issues affecting interfacial reactions may account for the apparent variability of results when different strategies are compared. The chemical reactivity for a given immobilization reaction is related to the chemical reaction itself and the interfacial properties of the underlying support. The influence of the support is consequential because the reaction conditions are quite different of those usually used in chemical synthesis; in particular the concentrations of reagents are very low.

2.4.2 3D Immobilization: Thick Layers, Entrapment Methods

The performances of biosensor devices are expressed in terms of detection specifications such as sensitivity (lowest detectable quantity), concentration range of

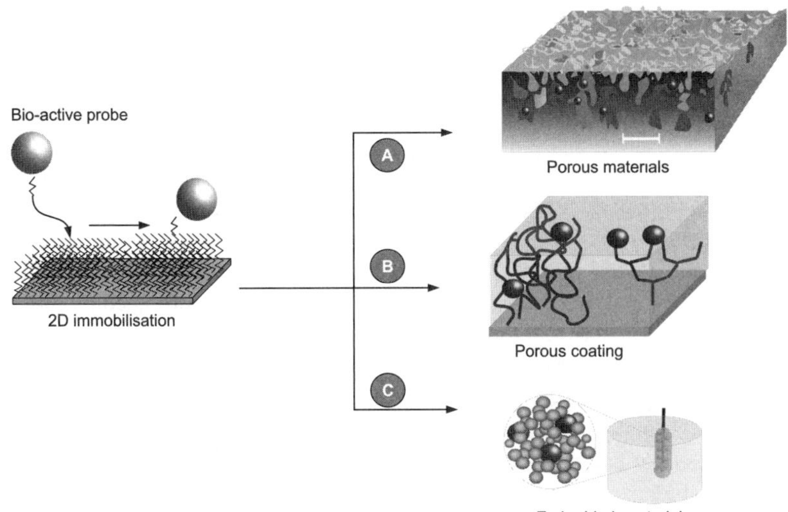

Fig. 2.17 Schematic representation of the three different approaches used to elaborate high specific area support for immobilization. (**A**) Immobilization of bioprobes onto porous materials, (**B**) immobilization of bioprobes onto polymeric layers deposited or synthesized onto flat support, and (**C**) entrapment of bioprobes during porous materials elaboration (sol–gel process)

linear response, and stability properties. Stability of the sensitive layer is mainly ensured by covalent grafting between reactive biospecies and the surface.

For a given appropriate surface preparation, signal intensity for most of biosensor devices depends on the surface density of immobilized species. The loading capacity of flat support (2D) is limited by the external surface area. A way to increase the sensitivity is increasing the specific area of the surface by elaboration of thick immobilization layers named 3D supports (Beattie et al. 1995; Fidanza et al. 2001). Different approaches have been envisaged to take advantage of 3D layers in the field of biosensors (Fig. 2.17). A straightforward approach is based on the biofunctionalization of porous materials with high specific area. A second common approach involves the formation or deposition of thick polymeric matrices onto flat supports. The active species are immobilized (chemically attached or simply entrapped) inside a gel made of, cross-linked polymer materials. A last approach to 3D biofilms is the direct entrapment or encapsulation of biocomponents inside porous materials during the synthesis (sol–gel strategy).

The activity on such layers depends on, (1) the mass transfer resistance (diffusion of analytes to receptors inside the 3D network); (2) the influence of the surrounding surfaces onto the biomolecules adsorption and denaturation (high surface area makes the biointerface more sensitive to the physico-chemical properties); (3) stability and/or leaching of the immobilized or embedded biocomponents into the liquid.

2.4.2.1 Porous Materials

Biologists are accustomed to use thick polymeric supports since long times (e.g. nitrocellulose membranes). The biomolecules are immobilized by deposition onto the membranes. Their huge binding capacity allows for large signal amplification. Widespread at macro-scale, the advent of glass slides coated with this kind of membrane has opened the way to membrane-based microarray technology. These porous materials suffer from nonspecific binding (Kusnezow et al. 2003) and release of the adsorbed probes. In order to improve the stability while maintaining a high binding capacity, the covalent immobilization of biomolecules has been investigated into various inorganic porous matrices.

Porous layers have been elaborated by means of several methods. Electrochemical etching is used to prepare porous silicon (p-Si), nanocrystalline silicon support (nc-Si) (Zhu et al. 2007) or porous aluminum oxide (Lee et al. 2007). Deposition of colloidal silica particles onto flat glass supports and subsequent thermal sintering yields either porous silica (Fidanza et al. 2001) or a porous glass containing 3D network of ordered micro channels, also called micro channel glass (Benoit et al. 2001).

The immobilization of bioreceptors makes use of the same strategies according to surface chemistry of the support as described above. Freshly prepared p-Si surface exhibits silicon–hydride termination. Oxidation by thermal or chemical treatment yields hydroxylated silicon dioxide layer which can be functionalized through alkylsilane derivatization. Functionalization with silanes also applies to porous silica, aluminum oxide and micro channel glass.

The covalent immobilization of biomolecules into these inorganic porous layers improves the stability of the biointerface with respect to washings and repeated utilizations. For instance, oligoprobes immobilized onto p-Si or nc-Si sustained up to 25 cycles of hybridization/stripping (Bessueille et al. 2005; Zhu et al. 2007). In addition, porous glass or p-Si bring about large increases of detection signal (up to tenfold) when compared to flat glass support (Fidanza et al. 2001; Bessueille et al. 2005). The greater loading capacities obtained on such porous layer have allowed to extend the linear dynamic detection range (Zhu et al. 2007) and increase the sensitivity (Cheek et al. 2001).

As a drawback, slow diffusion of molecules inside the 3D matrix causes an increase of response time related to the time required to reach equilibrium (Fidanza et al. 2001). A compromise has to be found to improve the diffusion of biomolecules while maintaining as great a specific surface as possible. Parameters to be optimized are the density and specific area of porous material, thickness of the deposit. In case of anodic etching of silicon substrate, the porosity and thickness of the porous layer can be adjusted by suitable choice of the initial silicon support (doping type and concentration of charge carriers, crystal planes orientation) (Ronnebeck et al. 1999) and/or by optimizing the etching conditions (Janshoff et al. 1998).

Diffusion inside 3D matrix can be forced by making the analyte flow through the porous layer. The most advanced solution to the diffusion resistance problem encountered in 3D matrices is to flow the analyte solution through free-standing porous membrane. As an example, a "flow-thru" device (Cheek et al. 2001) consists in a 0.5 mm thick micro channel glass support made of a parallel and ordered stacking of 5 μm diameter micro channels oriented perpendicular to the surface plane. The liquid is allowed to flow through each micro channel assimilated to an individual micro reactor. Kinetics of chemical recognition is accelerated because the mass transport distance is greatly reduced when compared to flat surfaces. In addition, the increase of available surface area gives a 96-fold increase in signal intensity when compared to flat glass.

In addition to the high specific area and loading capacity, these materials have attracting specific properties that can be used for detection purposes in biosensor applications. For instance, the electronic or optical properties of porous silicon make it suitable as a transducer in biological sensors (Bayliss et al. 1995; Thust et al. 1996; Janshoff et al. 1998; Schoning et al. 2000; Archer et al. 2004; Lie et al. 2004). Several photonic porous silicon–based structures have been investigated as label-free detection systems. These systems take advantage of optical interferometric methods (Bragg mirrors, rugate filters or optical microcavity) for very sensitive detection of small molecules, DNA and proteins interactions (Tinsley-Bown et al. 2005; Koh et al. 2007; Rendina et al. 2007). Another porous silicon–based method for detection of label-free biomolecules relies upon surface-enhanced laser desorption/ionization mass spectrometry (Wei et al. 1999). Compared to the widespread Matrix-assisted laser desorption/ionization (MALDI) technique, desorption/ionization on silicon (DIOS) is a matrix-free desorption technique that offers high sensitivity, no matrix interference (Zhouxin Shen et al. 2004). This technique is amenable for small molecules including peptides, glycolipids or carbohydrates (Wei et al. 1999), DNA, proteins (Mengistu et al. 2005), and enzyme activity analysis (Zhouxin Shen et al. 2004). It is worth to note the outstanding use of porous silicon layers for label-free detection of biomolecules. This opens the frame to new developments in optically active materials for biosensor applications.

New trend for high-throughput screening of biomolecular interactions implement a "bar-code strategy." Indeed, shortcomings of flat microarray substrates are related to the extremely slow diffusion process of analytes throughout the surface and more generally in confined domains of small volume. Improvement of reaction rates by mixing small volumes of dilute solutions is a technological hurdle (Edman et al. 1997; Adey et al. 2002; Hui Liu et al. 2003; Toegl et al. 2003; McQuain et al. 2004).

One way to overcome the diffusion limitation of flat surface immobilization is to remote the biorecognition step in solution. Thousands of different probes that are attached on the labeled particles are dispersed in solution of analyte; the recognition event and the label are detected simultaneously. To meet the detection requirements, encoded micrometer-sized porous-silicon particles have been used to screen fluorescence-tagged proteins (Cunin et al. 2002).

2.4.2.2 Organic Polymer Coatings and Gels

A widely used technology makes use of polymer coatings. Macromolecules are either grafted to the surface, or cross-linked for avoiding their leakage into solution.

Water-insoluble polymers are often used in chemical sensors sensitization. This method is derived from the usual technology of the Ion-Selective Electrodes membranes. There may be some applications of water-insoluble polymers to biosensors. A well-known example which may be considered as biosensor is the potassium selective electrode and the ISFET sensors derived from it where the cyclic peptide complexing agent valinomycin is entrapped in a PVC membrane. But most cases require aqueous environment in the surroundings of the biochemical recognition site. Selected polymers are hydrophilic; the polymer layer is swollen by water in order to allow the diffusion of biospecies to the recognition site inside the aqueous environment of the polymer layer. Therefore, grafted water-soluble polymers or hydrogels are used as immobilization matrices on the sensor surface. Polymeric layers can be obtained by grafting prefabricated polymers (brushes, dendrimers) or synthesizing the polymer layer *in situ* on the surface by chemically initiated polymerization, photopolymerization, or electropolymerization (Korri-Youssoufi et al. 1997). The 3D architecture can be made of macromolecular chains tethered to the solid surface by one chain end. The attached macromolecules look like long spacers bearing one or several reactive group per chain (brushes (Pirri et al. 2006; Schlapak et al. 2006)). More complex architecture can be found with branched polymers and starburst structures (dendrimers). Star polymers and dendrimers are branched polymers of well-defined architecture of the polymer backbone that should be considered on the same footing as branched polymers. Such a precise structure probably does not bring about a significant difference with respect to the less ordered architecture of randomly branched polymers. It may not be worth implementing the complex chemistry required for the synthesis of dendrimers for immobilization purposes. A possible advantage of the (over)controlled chemistry of dendrimer is the possibility of attaching the reactive groups at the periphery of the dendrimer macromolecules. Hydrogels made of hydrophilic cross-linked polymers are another class. The classification into two different classes as grafted macromolecules and hydrogels, is difficult and arbitrary however since the differences between the two may appear quite slight. The wide variety of polymeric structures allows depositing layers of thicknesses ranging from few 10 nanometers to few micrometers, giving rise to different surface behavior and analytical properties. The deposition, grafting or synthesis of polymer layers have been realized onto chemically modified surfaces that ensure strong adhesion, chemical grafting or surface induced polymerization.

2.4.2.2.1 Polymer Coatings, Brushes, and Dendrimers

These coatings lead to relatively thin layers (<100 nm) when compared to inorganic porous materials or gels that can easily reach up to few micrometers thickness (Kato et al. 2003). Various polymer architectures as brush, grafted branched polymers, 3D dendritic polymers give technological answers to the several queries of the analyst (Fig. 2.18). "Polymer brush" refers to linear macromolecules adsorbed or grafted by one of their chain end. This term was introduced by Alexander and de

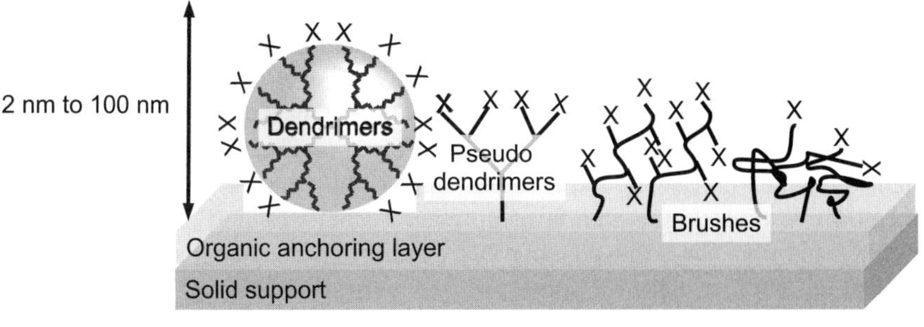

Fig. 2.18 Schematic representation of various surfaces modified with thin polymeric layers

Gennes for describing the dense regime of tethered chains where the density is such that the chains are constrained to stretch perpendicular to the surface (De Gennes 1987). The meaning of the word "brush" is currently extended to end-tethered polymer chains, whatever the grafting density.

There are two main strategies for chemical grafting of polymers: grafting reactive polymers to surface groups ("grafting to") or initiating the living polymerization from surface bound initiators ("grafting from") (Zhao and Brittain 2000). Grafting from process is currently developing fast thanks to the recent discovery of new living polymerization reactions that do not require the extreme dryness conditions of ionic polymerization reactions (Advincula 2006). In particular, controlled radical polymerization allows the *in situ* synthesis of vinyl polymers (acrylates, methacrylates, vinyl ethers) (Bergbreiter and Kippenberger 2006; Buchmeiser 2006; Tsujii et al. 2006). Photo initiated grafting is also promising (Dyer 2006; Matsuda 2006).

The primary aim of polymer coating is to place the bioprobes away from the solid interface in order to limit hindrance phenomena and make the probes available to the targets in solution. The chemical grafting of PEG polymers onto flat glass surface (appropriately functionalized) brings on very beneficial properties regarding the biological interactions. The hydrophilic PEG polymer chain is well-known for its resistance to protein adsorption. The PEG polymer acts as a hydrophilic noncharged spacer improving the solubilisation of the biospecies, avoiding the steric hindrance and limiting surface effects such as adsorption onto hydrophobic patches of the surface that may lead to denaturation or loss of activity (Shchepinov et al. 1997; Wang et al. 2002). Each PEG molecule is tethered by one end and leaves only one reactive group toward biomolecules immobilization. In case of DNA immobilization and hybridization, the PEGylated surface yielded fourfold increase in signal hybridization compared to 2D silanized support; another benefit was lowering the nonspecific adsorption by a factor of 13 (Schlapak et al. 2006). Additional advantage concerns the macroscopic improvement of the surface homogeneity since the soft the PEG layer of 3.5 nm thickness masks inhomogeneities (roughness) of the underlying silane anchoring layer.

Polymer brushes with multiple reactive groups increase the density of recognition sites. As example Piri et al. (2006) reported a "grafting from" approach where each macromolecule of the brush contained several reactive groups (~100) giving rise to larger signal compared to flat glass slides functionalized with the same reactive group (epoxy-functional silane). Despite the direct synthesis of polymer chain onto the solid support that would allow dense layers of parallel polymer chains, the surface concentration of polymer chains was fairly low (0.24 pmol/ cm^2). In addition, the corresponding DNA grafting density (0.3×10^{13} molecule/ cm^2) stood in the range of most grafting densities reported for 2D surface functionalization (Dugas et al. 2004). By comparison, the surface preparation by immobilization of prefabricated dendrimers (poly(amino)amine, G4 PAMAM) onto silanized glass slides gave DNA grafting density of 9×10^{13} DNA strands/ cm^2 (150 fmol/mm^2).

Various approaches to elaborate polymer linkers (Beier and Hoheisel 1999) or chemically grafted functional polymers including dendrimers (Benters et al. 2001; Le Berre et al. 2003) have been reported. While the binding capacity of such nanometric structures are no necessarily higher the excellent accessibility of probes grafted on such dendrimer coatings lead to higher immobilization and hybridization efficiencies than those of 2D functionalized glass slides (Le Berre et al. 2003).

Linear polymer brushes and branched polymer coatings allow showing up the bioprobes in a configuration that resembles that of free solution (Pirri et al. 2006) and provide an excellent accessibility of the target. In addition, such hydrophilic polymer coatings show lower non-specific adsorption and hence low background signal and improved detection sensitivity (10–100-fold higher than 2D functionalized glass slide) (Le Berre et al. 2003). It is worth to note that no trouble regarding resistance to diffusion was reported, probably due to the relatively small thicknesses of the layers.

Lastly, polymer coatings allow improving the uniformity of the functionalized layer when compared to silane functionalized layers for example (Schlapak et al. 2006). The spots of microarrays obtained with polymer coating are uniform and homogenous (Benters et al. 2001; Le Berre et al. 2003). This result is particularly important since the ability to extract meaningful information from microarray experiments depends considerably on the spot quality (Diehl et al. 2001; Rickman et al. 2003; Dugas et al. 2005; Derwinska et al. 2007).

2.4.2.2.2 Gels, Hydrogels

The entrapment inside a thick layer of hydrogel is the first immobilization technique that was used for proteins (Guilbault and Montalvo 1969, 1970). This is a historical technology that still finds wide acceptance because of its simplicity. Thus, an aqueous solution of protein and water-soluble polymer is mixed with a cross-linking agent and drying at the surface of the sensor. The surface is ready to be used after it has been washed with water. As typical example, proteins are immobilized

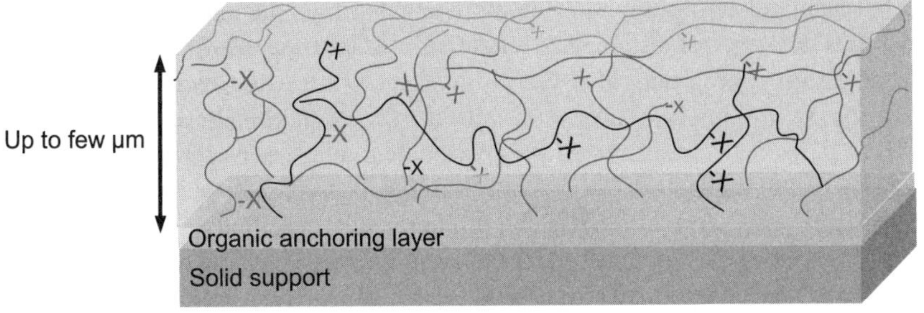

Fig. 2.19 Schematic representation of hydrogel coating

in a gel of bovine serum albumin cross-linked with glutaraldehyde. Bioprobes are either entrapped inside the gel but may also have reacted with the cross-linking agent, so that they are chemically attached to the polymer network constituting the gel. Of course there is only poor control over the chemical groups that have reacted and the orientation of the active centers.

An important benefit of the hydrogel technology lies in the immobilization of fragile biomolecules that should be kept in aqueous media for remaining active. One drawback inherent to spotting nanolitre droplets of protein solution on glass slides for microarray fabrication is the easy evaporation of water. Even in 100% humidity atmosphere, preventing drying very small drops is difficult. Preservation of protein stability and functionality in hydrated state involve the conservation of homogeneous water environment. Moisturizing additives are often added to the buffered immobilization solution: glycerol (MacBeath and Schreiber 2000), carbohydrates (trehalose, sucrose) (Kusnezow et al. 2003), or low molar mass poly (ethylene oxide) (Lee and Kim 2002). The use of highly hydrated polymers (hydrogels) provides a suitable liquid-like 3D environment for keeping the activity and allowing biological interactions. In addition hydrogels allow high loading capacities. Hydrogels (Fig. 2.19) are cross-linked hydrophilic polymers with interstitial spaces that contain as much as 90–99% w/w water (Hynd et al. 2007; Ulijn et al. 2007). Hydrogel matrices used in the field of biosensors are mostly based on polyacrylamide, polysaccharide (e.g. agarose) or substituted polysaccharide (e.g. carboxymethylated dextran) that are cross-linked. These hydrated coatings provide hydrophilic surfaces that prevent nonspecific adsorption of proteins. The manufacturing of hydrogel coatings can be realized by grafting prefabricated polymers (Zhou et al. 2004), UV-light or chemically initiated polymerization onto chemically derivatized solid support (Guschin et al. 1997; Wang et al. 2002). The coating can be realized collectively by casting the gel over the whole area of the support, or localized by photolithography (Guschin et al. 1997) or by dispensing gel drops onto the solid support (Rubina et al. 2004). The thickness of hydrogel films ranges from 200 nm (Zhou et al. 2004) to 30 µm (Arenkov et al. 2000; Angenendt et al. 2002).

The chemical immobilization of biomolecules into precoated 3D hydrophilic films uses a range of established chemistry. Carboxymethylated dextran films present carboxylate groups that readily react onto aminated biomolecules via an activation step with N-hydroxysuccinimide for example (Löfås and Johnsson 1990). Agarose gels are activated by a periodate oxidation ($NaIO_4$) step that leaves aldehyde reactive groups (Wang et al. 2002; Wei et al. 2004). Like agarose gel, prefabricated polyacrylamide gels can be chemically modified in order to immobilize biomolecules (Arenkov et al. 2000). Modifications involve incorporation of aldehyde groups or hydrazide groups that react respectively onto aminated biomolecules (proteins, DNA) or aldehyde groups that can be site-specifically introduced onto the polysaccharide portion of antibodies. Proteins can be entrapped by cogelation with Bovine Serum Albumin (BSA) using glutaraldehyde as cross-linking agent (Wan et al. 1999). Glutaraldehyde is a short homodifunctional cross-linker reagent widespread for protein conjugating. The aldehyde functionality reacts readily with amino groups and creates a network of cross-linked proteins. Coupling is not oriented however; proteins inside the thick layer display nonhomogeneous binding activity.

Taking advantage of the flexible synthesis of polyacrylamide gels by changing or modifying the monomers composition of the starting solution, it was possible to tailor the porosity of the gel matrices in order to improve access of macromolecules inside the gel (Arenkov et al. 2000). Copolymerization with acrylic-labeled biomolecules (Brueggemeier et al. 2005) or biomolecules bearing nucleophilic groups (–SH, –NH_2) (Dyukova et al. 2005) is a straightforward means to immobilize biomolecules with high efficiency (Rubina et al. 2004).

Gel coating–based microarrays are produced by spotting dilute solution of bioprobes onto the gel layer leading to uneven distribution of probes and leaving most material immobilized on the top layer of the gel. By contrast, copolymerization allows manufacturing gel drop with homogeneous distribution of probes within the gel and with high immobilization efficiency. For example, up to 87% of oligonucleotides modified with methacrylamide groups present in the starting solution have been immobilized (Dyukova et al. 2005).

Control of the porosity and loading capacity allows the immobilization of well-spaced (unhindered) biomolecules in aqueous environment and improves the diffusion of analyte through the gel matrix.

3D hydrophilic gel films are intended to provide a stable, low-fluorescence background, low nonspecific adsorption and high loading capacity for immobilization (Arenkov et al. 2000; Dyukova et al. 2005). The most valuable improvement of hydrogel films is holding biomolecules inside a solution-like environment far away from detrimental interfacial effects. This is crucial to maintain fragile biomolecules in active from upon immobilization and utilization conditions.

For example, Wang et al. (2002) have compared the efficiency of peculiar molecular beacons oligoprobes immobilized onto solid support regarding the discrimination of single nucleotide mismatches. Molecular beacons are hairpin-shaped oligoprobes obtained through a self-hybridization of complementary sequences present on a DNA strand. A bound fluorophore has its fluorescence quenched in the hairpin structure; fluorescence is recovered when the hybridization of complementary DNA strands in solution opens the beacon. Immobilization of molecular beacons onto aldehyde-activated glass slides (2D chemistry) gives high fluorescence background and relatively low increase in fluorescence upon hybridization because the hydrophobic and electrostatic interactions between molecular beacons and the underlying surface destabilize the stem loop. Molecular beacons immobilized inside agarose film provide improved quenching efficiency and excellent discrimination of single-nucleotide polymorphism when compared to flat supports (Wang et al. 2002).

The loading capacity of immobilized probes per unit area is two to three orders of magnitude higher than that for flat support. Combination of the high loading capacity and the low nonspecific adsorption on hydrogel films leads to higher sensitivity (Angenendt et al. 2002; Brueggemeier et al. 2005). The large linear range of detection of hydrogel microchips allows quantitative assessment of antibody–lectin interactions (Dyukova et al. 2005), or enzymatic activities (e.g., inhibition of tyrosine kinase activity) in complex biological medium (Brueggemeier et al. 2005).

Diffusion inside thick porous membranes is slow however, which leads to quite long response times (Wang et al. 2002; Zubtsov et al. 2006). It would be wide to increase the area of gel spots to accelerate the diffusion (Rubina et al. 2004); of course, this is done at the expense of the spot density of the arrays. Interestingly, stirring the target solution on top of the surface not only accelerates the mass transport of targets to the hydrogel surface but also contributes to the penetration of targets inside the gel. Mechanical stirring (by flowing with peristaltic pump) leads to approximately fivefold acceleration in reaction–diffusion kinetics for hydrogel-based microchips (Zubtsov et al. 2006).

Hydrogel arrays have been investigated to immobilize and analyze oligonucleotides, proteins, antibodies, glycans and living cells (Proudnikov et al. 1998; Dyukova et al. 2005). Hydrogel microchips are compatible with various detection techniques. Colorimetric techniques (chemiluminescence) were implemented to monitor enzymatic activities (Brueggemeier et al. 2005). Investigation of protein–protein, enzymatic activity or enzyme–inhibitor interactions have been performed on hydrogel protein arrays by MALDI-MS (Gavin et al. 2005). Surface plasmon resonance techniques were used with relatively thin hydrogel films (200 nm) (Löfås and Johnsson 1990; Zhou et al. 2004). Adaptation of the detection to thick layers may be necessary in some instances. Thus, hydrogel coated slides of microarrays may show rather high variability of spot geometry and homogeneity (Afanassiev et al. 2000; Angenendt et al. 2002) which makes confocal detection of fluorescence worse than a robust classical detection (Derwinska et al. 2007).

Finally, bioresponsive hydrogels materials open the way to original label-free detection modes. Bioactive materials undergo macroscopic changes of the hydrogel network (swelling/collapse or solution-to-gel transition) in response to selective biological stimuli ((Ulijn et al. 2007) and references within).

2.4.2.2.3 Structured Coatings

The internal structure of thick coatings allows the control of the diffusion inside. This is of primary importance for indirect transduction modes where the signal does not come from the biorecognition itself but from secondary-products of the bioreaction. For example, the historical urea-sensitive enzyme electrode (Guilbault and Montalvo 1969, 1970) detects the local rise of pH when urea is enzymatically decomposed into ammonia. Electrochemical devices often require the accumulation of mobile species in the vicinity of the transducer surface: conductometric sensors detect variations of local concentration of ionic species; amperometric sensors often require electron transfer by mean of a red-ox mediator.

2.4.2.2.3.a Functional Multilayer Coatings

It has been recognized that the direct transfer of the ion-selective electrodes (ISE) membrane technology to chemical sensors has not been satisfactory. ISE membranes are self-supported polymer films where ion-sensitive chemical species are incorporated. Such species are mobile and partition between polymer membrane and solution. Coatings on solid surface do not require being self-supported and their thickness and internal structure may be optimized with respect to diffusion issues. An additional issue is the adhesion of the coating to the surface which can be ensured either by chemical grafting the materials or by a grafted sub layer (silane layer). The hydrophilic character of the membrane is essential for limiting nonspecific adsorption and fouling. Therefore, quite complicated multilayer coatings appeared for the manufacture of ion-sensitive field-effect transistor (ISFET), giving a definite improvement of the signal and reduction of background. The following example helps understanding the issues and benefits.

An interesting two-layers coating was proposed for the urea-sensitive enzymatic field-effect transistor (ENFET) and conductometric sensors made of interdigitated electrodes on a solid support (Shulga et al. 1993; Soldatkin et al. 1993, 1994; Jdanova et al. 1996). Thus, the hydrolysis of urea by urease yields ammonia that increases the pH at the surface of the FET device. Glucose oxidase catalyzes conversion of glucose into gluconolactone that yields gluconic acid upon hydrolysis; therefore, the local pH decreases in the presence of glucose. Urease is immobilized at the surface of the FET inside a classical gel of BSA cross-linked with glutaraldehyde. The detection signal comes from the ionization of the surface groups of the silica or silicon nitride insulator, which persists as long as the acidic or basic species stay close to the surface. The buffer in the analyte solution cancels out the signal. A definite increase of the signal is obtained when the diffusion of the acidic/basic species is blocked by an over layer deposited on top of the cross-linked BSA layer aimed at the immobilization of the enzyme. The external layer made of a polyelectrolyte (Nafion, polyvinylpyridine) opposes ion transport but does not slow down too much the diffusion of the neutral analyte (glucose, urea) to the internal layer.

2.4.2.2.3.b Structured Organic–Inorganic Particle Coatings

A gel of inorganic particles can be used as an immobilization matrix in the same way as the polymer gels described above. Organic particles could be used as well. The inorganic particles functionalized with organic grafted or mixed with organic biomolecules form a hybrid organic–inorganic material that is deposited as a thin film onto the solid support. The hybrid material stays at the surface of the support if the inorganic and organic compounds are stuck together as a water-insoluble and nondispersible material. Both organic and inorganic materials are held together by means of physical forces. This process is a simple and mild variant of the sol–gel method for the preparation of porous silica matrix described in the next section. The process is mild since it does not involve chemical reaction in the presence of the biomolecules.

Deposition of thin films of inorganic particles, mainly clays, has been introduced by Kotov (Kotov et al. 1997) and implemented further for the manufacture of "clay electrodes" (Mousty 2004). Basically, clay particles (laponite) and enzymes are precipitated on the support (Poyard et al. 1996). The clay gel consists in a concentrated oriented phase of clay swollen by water; macroscopic orientation (nematic phase) is achieved as thin films at the surface of electrodes (Chevalier et al. 2002). The laponite particles in the gel are held together in concentrated enough electrolyte medium; clay particles delaminate because of electrostatic repulsions in case of low ionic strength. Cationic species (surfactants, polymers) are also added possibly for inducing the sustained coagulation of the particles (Besombes et al. 1997; Coche-Guérente et al. 1998; Coche-Guérente et al. 1999). Fast coagulation conditions leave a loose network of particles that immobilizes enzymes and keeps the possible diffusion of analyte to the biomolecules. Tighter immobilization in a dense gel can be achieved by a slower coagulation rate. As for any immobilization process in a gel, optimization of the properties is a compromise between the sustainable immobilization of biospecies and the necessary diffusion of the analytes to the recognition site.

2.4.2.2.3.c Layer-by-Layer Alternate Polyelectrolyte Coatings

Electrostatic adsorption of electrically charged biospecies is quite easy to carry out but suffers the limitations of adsorption: poor control of the orientation of biospecies and massive nonspecific adsorption that leads to high background in sensor applications. In spite of these drawbacks, electrostatic adsorption can be utilized as a ready deposition process of thick polymer coatings. One simple way consists in heterocoagulation of negatively and positively charged species that precipitate onto the surface. This is identical to the well-known "complex coacervation" process used in the field of microencapsulation; it leads to high loadings of each charged partners inside a gel swollen by water (Gibson et al. 1996).

Better control of the thickness of the polymer coating is achieved by the layer-by-layer deposition of oppositely charged polyelectrolytes that leads to alternate polyelectrolyte layers. Such materials can be favorably used in many coating applications (Jaber and Schlenoff 2006), including sensors applications (Caruso et al. 1997; Chluba et al. 2001; He et al. 2002; Ferreyra et al. 2003; Ferreyra et al. 2004; Rodriguez and Rivas 2004; Coche-Guérente et al. 2005; Miscoria et al. 2006). The structure of the deposit is a well-ordered alternate stack of both polyelectrolyte layers in favorable case. Actually, the structure is most often quite disordered but the thickness of the deposit can be nicely controlled by the number of elementary deposition steps; the surface charge of the full layer is that of the last deposited polyelectrolyte (Ferreyra et al. 2003; Ferreyra et al. 2006).

Alternate polyelectrolyte-nanoparticle multilayers are fabricated in the same way. Various particles such as gold nanoparticles (Santos et al. 2005), CdS semiconductor nanoparticles (Kotov et al. 1995), organic polymer particles (Kong et al. 1994)andclay (laponite as in the previous section but deposited layer-by-layer) (van Duffel et al. 1999).

2.4.2.3 Entrapment in Inorganic Gels

Bioactive species can be embedded within the inorganic nanostructure made of oxide materials using the sol–gel process. This technology involves mixture of alkoxysilane precursors or ionic salts (e.g., sodium silicate) in solution with functional biocomponents. *In situ* synthesis of a porous inorganic support is a straightforward and convenient way to entrap (immobilize) functional biocomponents.

2.4.2.3.1 Silica Sol–Gel Mechanism

In general, the sol–gel reaction of metal oxides precursors such as tetramethoxysilane (TMOS) or tetraethoxysilane (TEOS) proceeds at room temperature through three steps:

(1) "activation" of the metal oxides precursors by acid or base-catalyzed hydrolysis (Fig. 2.20).

It is worth noticing that by-products of the hydrolysis of alkoxysilanes are organic molecules (methanol or ethanol from methoxy or ethoxysilane precursors respectively).

(2) Polycondensation of the activated precursors or "monomers" to oligomers leading to formation of transparent and stable nanometric dispersions of silica particles (the "sol") (Fig. 2.21). The solid content of the dispersions of nanoparticles ranges from 4 to 20 wt% (Böttcher 2000).

The size of the sol particles and cross-linking within the particles depend upon the pH and solution concentration and composition. The low viscosity of sols make

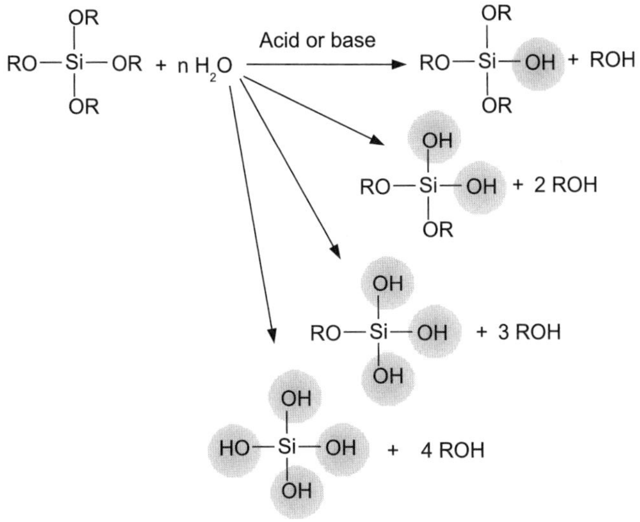

$$Si(OR)_4 + n\,H_2O \xrightarrow{\text{Acid or base}} (RO)_3Si\text{—}OH + ROH$$

$$\rightarrow (RO)_2Si(OH)_2 + 2\,ROH$$

$$\rightarrow RO\text{—}Si(OH)_3 + 3\,ROH$$

$$\rightarrow Si(OH)_4 + 4\,ROH$$

Fig. 2.20 Products of the hydrolysis of the alkoxysilane precursors in solution

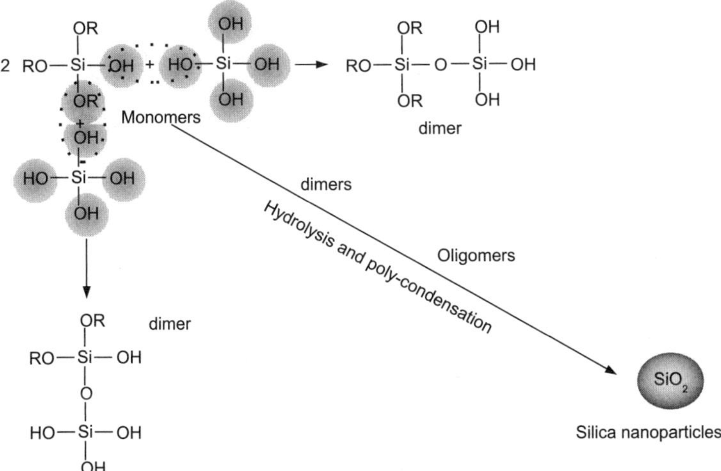

Fig. 2.21 Condensation and polycondensation of alkoxysilane and hydrolyzed alkoxysilane and formation of the silica nanoparticles in solution (sol)

them compatible with various deposition techniques (spin coating, spray drying, mold casting) in order to produce thin film or bulk materials.

(3) Formation of a three dimensional silica network by further polymerization and cross-linking (chemical sintering) of inorganic particles. This solidification of the sol leads to a solid network holding the liquid (the "gel") (Fig. 2.22).

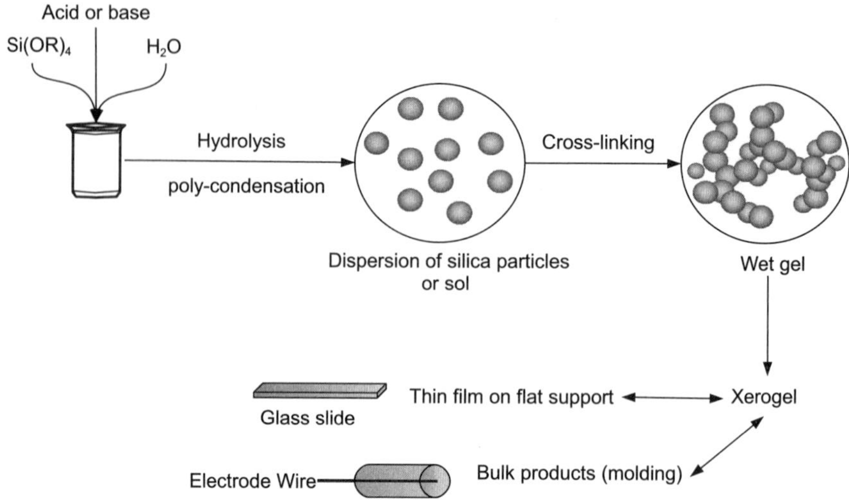

Fig. 2.22 Schematic representation of the sol–gel process and formation of xerogels

Subsequent drying of the gel affords xerogels. The physical properties of the 3D matrix depend on the size of particles and the extent of cross-linking occurring before gelation. The solvent evaporation after deposition of the wet gel as thin films or after mold casting leads to the formation of thin films of xerogels (0.1–2 μm) or bulk xerogels that replicate the shape of the mold. For example, sol–gel mixture can be cast into test-tubes with copper electrode wire; the resulting electrode will support silica sol–gel with embedded biomolecules (Thenmozhi and Narayanan 2007).

2.4.2.3.2 Bioactive Sol–Gel Silica Layers

Immobilization of a wide range of viable biomolecules ranging from enzymes to whole cells was made in silica gels (Avnir et al. 1994; Böttcher 2000). The silica sol–gel process is a simple and efficient fabrication process of inorganic porous materials. The preparation of bioactive sol–gel materials is realized through simple addition of the bioactive components either before or after hydrolysis of the precursors to the sol–gel mixture (Fig. 2.23). It is carried out at room temperature and is well-suited to the direct encapsulation of temperature sensitive biomolecules. Owing to the good mechanical, thermal and photochemical stability of the silica matrix, entrapped biomolecules are protected from external harsh conditions encountered in some applications (e.g. environmental analyses).

Organic solvents may cause proteins denaturation or affect the viability of cell systems. They are either introduced intentionally for solubilizing the precursors or they are by-products of the sol–gel reaction. Aqueous route for sol–gel preparation involving sodium silicate precursors or evaporation of the alcohol by-products before addition of proteins, avoid the unfavorable effects caused by the organic

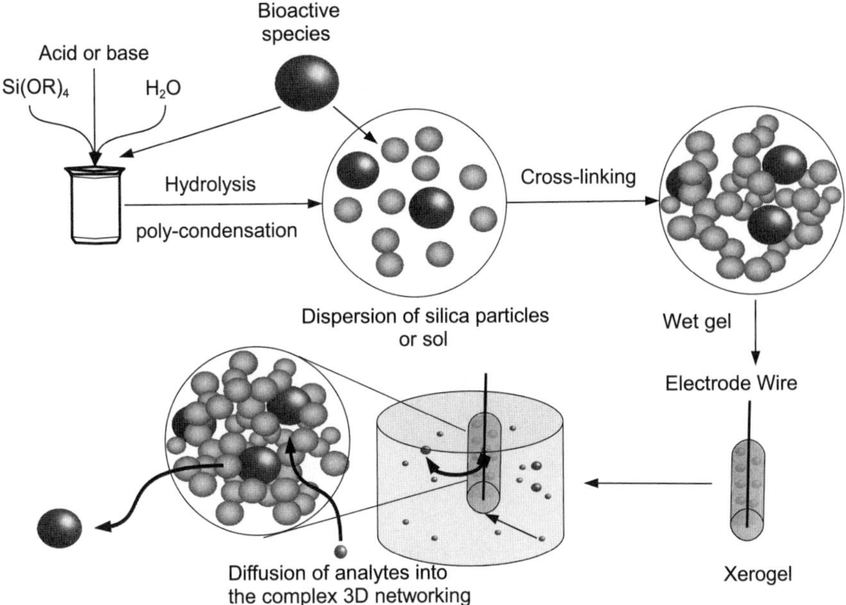

Fig. 2.23 Schematic representation of the sol–gel process and bioreceptors embedding

molecules (Carturan et al. 2004; Lin et al. 2007). Acidic or basic environments also are severe limitations with regards to cell immobilization (Carturan et al. 2004). Various stabilizing agents have been proposed to improve the stability of biomolecules during the sol–gel reaction; loss of activity was often observed however (Besanger and Brennan 2003). Utilization of functional silanes that release such stabilizing agents upon hydrolysis is an elegant alternative (Besanger and Brennan 2003). Functional silanes also allow the preparation of hybrid sol–gel materials with the optimum hydrophilic/hydrophobic balance that provides a suitable medium for enzymatic activity (e.g. lipases that possess hydrophobic domains) (Thenmozhi and Narayanan 2007). Similarly, membrane proteins can be immobilized (stabilized) in the phospholipid bilayer of liposomes prior to silica sol–gel entrapment (Besanger and Brennan 2003).

2.4.2.3.3 Activity and Stability of Biomolecules Doped Sol–Gel Silica

Performances of sol–gel biofilms with regards to biosensor applications depend on the activity of the bioactive receptors within the inorganic matrix and the storage stability. The catalytic activity of bioreceptors entrapped inside a 3D network is related to (1) the diffusion of analytes inside the 3D network where receptors are immobilized, (2) the extent of denaturation of receptors during the sol–gel process and (3) the leaching or release of the immobilized molecules during use.

2.4.2.3.3.a Leaching

Leaching occurs in the early use of biosensors. After an initial burst release, the remaining biomolecules may be strongly immobilized, allowing long-term stability if the matrix does not get physically damaged (Böttcher 2000). Leaching affects the reproducibility and sensitivity of biosensors.

Process parameters leading to increase the pore sizes (soluble additives, pH) increase the leaching of biomolecules (Lin et al. 2007). For instance, glycerol or glycerated silane precursors used for stabilizing proteins and improving silica–protein interactions, tend to increase the pore sizes and thus the leaching (Lin et al. 2007). Favorable electrostatic interactions between silica and proteins can be used to limit leaching (Rezwan et al. 2005).

Of course leaching can be prevented by covalent immobilization of the entrapped biomolecules (Thenmozhi and Narayanan 2007).

2.4.2.3.3.b Diffusion

Surface reactions are often diffusion-controlled. Diffusion in porous materials is mainly function of the porosity and the viscosity of solution. Porosity of porous materials is defined as the opens pores' space in a material according to IUPAC recommendations (Sing et al. 1985). Pores are classified into three categories: (1) macropores with widths exceeding 50 nm, (2) mesopores of widths between 2 and 50 nm and (3) micropores below 2 nm. Immobilization of biomolecules takes place in the mesoporosity. Porosity of the sol–gel materials can be controlled by the pH or the water/silane ratio of the sol–gel mixture. The experimental conditions are adjusted to limit leaching while optimizing the diffusion of analytes within the silica sol–gel matrix.

2.4.2.3.3.c Intrinsic Biological Activity

Intrinsic biological activity is referred to the activity of the immobilized biocomponents when compared to the nonimmobilized ones. Proteins can undergo denaturation. Cell systems may disrupt upon sol–gel reaction or be poisoned by experimental conditions. As consequences, activity of such bioactive components is altered.

Proteins activity measured by the Michaelis constant K_m varies strongly according to the protein type. Glucose oxidase (GOD) entrapped in sol–gel layers exhibits K_m values similar to free GOD in solution, while lipase activity is only 3–29% of the free enzyme (Böttcher 2000).

While some bacteria and yeast (e.g., *Saccharomyces cerevisiae*) survive the experimental conditions of sol–gel reaction, more global approaches allowing entrapment of functional microorganisms are challenging. Various solutions are protection of cell by alginate microencapsulation prior to sol–gel process (Heichal-Segal et al. 1995), retarded addition of cells just before the sol–gel transition, and gas-phase synthesis of sol–gel prior to its deposition on top of the cells (Carturan et al. 2004).

2.4.2.3.4 In Conclusion

Biocompatible silica sol–gel processes are the convenient way to prepare bioactive films. By controlling the processing parameters and the sol–gel solution a wide panel of biological species can be embedded ranging from nucleic acid to cell system while maintaining their viability. A one-step immobilization method allows the formation of films or bulk products of various shapes (e.g., electrodes, thin film). Biosensors with excellent long-term stability (up to 120 days) are prepared with biomolecules immobilized inside mechanically and photochemically stable inorganic matrix (Thenmozhi and Narayanan 2007).

The silica sol–gel can be transparent in case of optical detection systems (Yamanaka et al. 1992; Lin et al. 2007) or enclose electronic mediators such as ferrocene, or graphite powders to improve response and reproducibility of electrochemical biosensors (Gun et al. 1994; Kunzelmann and Bottcher 1997; Li et al. 1997; Thenmozhi and Narayanan 2007).

2.4.3 Immobilization onto Colloidal Particles

Since many years, the colloids (latexes, magnetic particles, silica particles, fluorescent particles, etc.) offer multiple potentialities of applications, in particular in the fields of pharmaceutical, medical and biological sciences (Arshady 1993; Elaissari et al. 2003a, b, c). The major interest of these dispersed materials (simple or composite) is related to their use as support of biomolecules in the biomedical diagnosis. The first applications of colloidal particles in biomedical diagnostic tests are based on the agglutination process in which latex particles sensitized by antibodies or antigens are used (Fig. 2.24). The biological molecules immobilized,

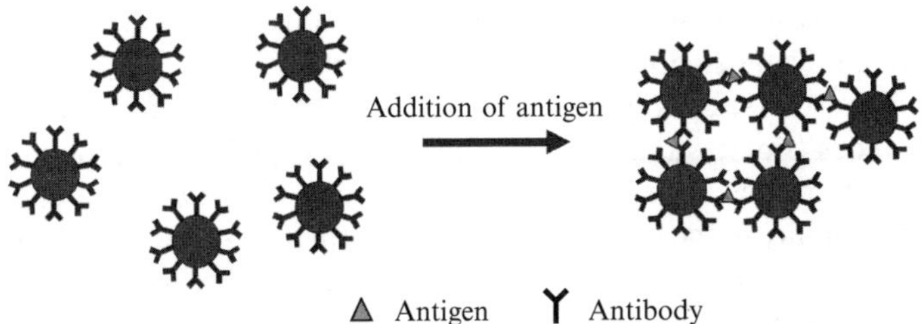

Fig. 2.24 Antibody-coated particles agglutinated by the specific antigen molecules (Stoll et al. 1993; Ouali et al. 1995)

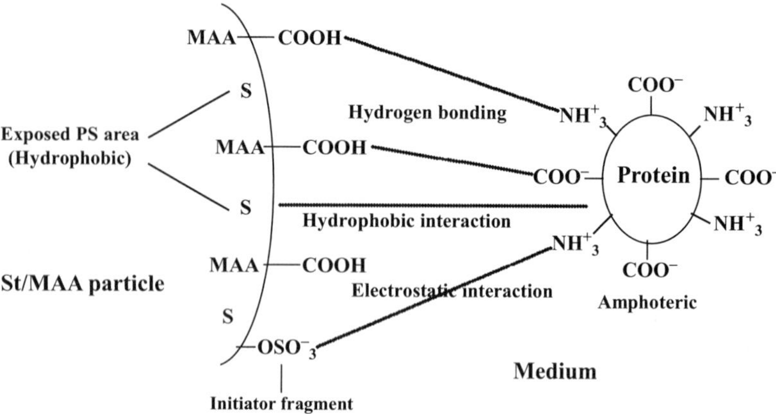

Fig. 2.25 Schematic representation of the possible interaction forces between proteins and polystyrene latex particles bearing sulfate and carboxylic groups. (St/MAA for styrene Methacrylic Acid particles). The interaction forces involving the adsorption process are classified into four interactions, that is, (*1*) hydrophobic interaction, (*2*) electrostatic interaction, (*3*) hydrogen bonding, and (*4*) Van der Waals interaction. The hydrophobic interaction has a major role in protein adsorption phenomena, especially the adsorption of proteins onto the low-charge particle surface

on latex particles can be antigens or antibodies. The immobilization of these molecules is mainly via covalent grafting or physical adsorption.

The specificity and the sensitivity efficiency of such traditional diagnostic are directly related to the surface particles properties of the latex particles and to the accessibility of the receptor. The interactions between receptor and reactive particles are strongly dependent on the colloidal and surface properties of the dispersion, and on the physical-chemistry properties of the receptor (Fig. 2.25). The immobilization (adsorption and/or covalent grafting) of receptor is generally studied by taking into account the influence of physical-chemistry parameters such as; pH, salinity, buffer nature, temperature, surface nature and the presence of competitive adsorbing agent. The immobilized receptors onto colloidal particles are characterized with respect to their conformation and biological activity. In this direction, various polymer-based colloids have been prepared for *in vivo* and *in vitro* biomedical applications (Elaissari et al. 2003a, b, c).

To reach those first conventional applications, the receptors can be antibodies, oligonucleotides (single-stranded DNA fragments), peptides and proteins. The use of particles bearing well appropriate reactive groups and surface properties may confer more stability to the elaborated conjugates (particle–receptor). In addition to the stability aspect, the reactivity of the receptor can be enhanced when the conformation was adequately designed. The needed reactive particles are elaborated using many heterophase processes (emulsion, dispersion, precipitation, self-assembly, physical processes) (Elaissari et al. 2003a, b, c).

Nowadays, colloidal particles are of great interest not only in conventional diagnostic, but also in microsystems based on microfluidics in which the sample preparation and the detection of the targeted disease label. Among, these colloids, only the magnetic particles contribute to the realization of the systems of diagnoses or analytical automated. In these novel technological applications, the specifications list of the desired particles is heavy to answer and still a challenging research area.

With the development of nano-biotechnologies, sensitive colloidal particles have received increasing attention due to their exhaustive panel of specifications:

(1) The colloidal particle offer high specific surface (6–60 m^2/g) compared to flattened surfaces such as silica wafer. Such high surface in dispersed media enhances the kinetic of the capture of the target. Indeed, the diffusion phenomena are limited in disperse media compared to flat support.
(2) To their possible guidance when they have magnetic property and to their possible intrinsic properties (i.e. fluorescent, conducting stimuli-responsive, etc.). In fact, colloidal particles bearing well appropriate receptor are used as solid support of biomolecules in various biological applications such as in immunoassay and DNA diagnostic, cell detection, and also in drug delivery area.

The combination of Microsystems technology and biosensor is the key point answering the above-mentioned criteria. In this direction, the use of nanocolloids has been recently explored.

The main objective of biomedical diagnostic research is to develop microsystems making it possible to analyze complex mediums such as the biological fluids and matrixes. Indeed, because of volumes of very reduced samples, of analyses in a significant number and requirements in terms of speed and low cost, an extreme miniaturization of the analytical systems is made necessary today (Fig. 2.26).

In brief, the formation of the new chemical bond, linking the macro-biomolecule to the particles, requires well defined water chemistry, a maximization of receptor–particle interactions in order to bring both counterparts within a shorter distance than the length of a chemical bond (otherwise no reaction takes place) and control of the colloidal stability of the particles and particle/receptor during and after the functionalization process.

To perform the covalent grafting of biomolecules onto colloidal particles is more complicated when compared to the use of flat solid surface. In fact, the general problem to solve is mainly related to colloidal stability of the particles during the covalent grafting process. In addition, the activation step of reactive groups of both polymer based particles and biomolecules (or receptors) should be performed in water phase. In fact, in order to avoid the alteration of the activity of proteic-based biomolecules, the immobilization onto particles is incontestably performed in water-based process. Whereas, in the case of single-stranded DNA fragments (i.e. oligonucleotides) or peptides, the use of polar and miscible solvents with water during the activation and chemical grafting is tolerated in some conditions.

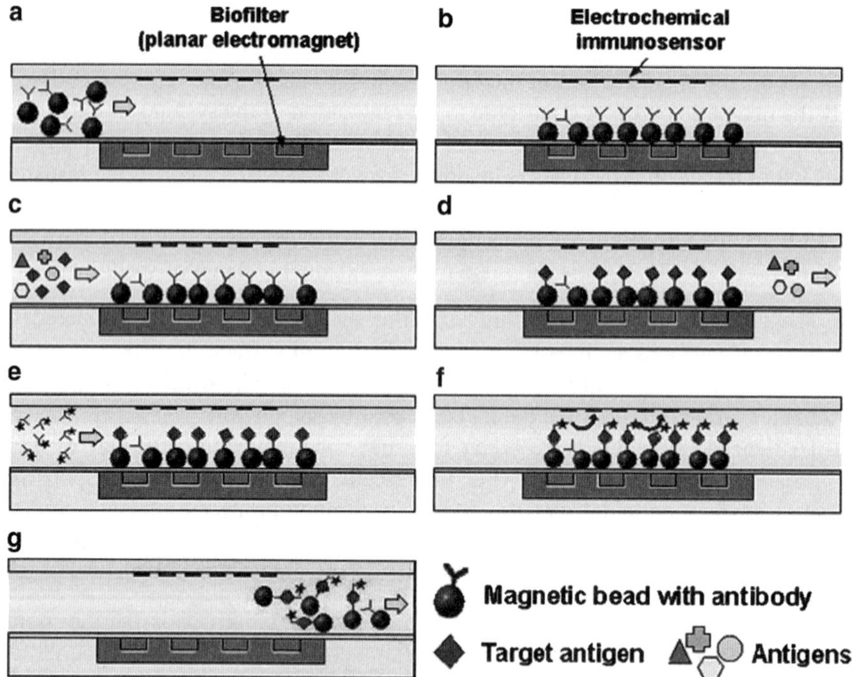

Fig. 2.26 Principle of the microchip assay developed by Choi et al. **a** Introduction of the magnetic beads into the microchannel. **b** Immobilization of the beads on the surface of the biofilter due to magnetic field. **c** Introduction of the sample. **d** Interaction between the bead and the target analyte. **e** Introduction of the secondary labelled antibody. **f** Electrochemical detection after the introduction of the substrate. **g** Washing (With permission from Choi et al. (2002).)

In the following parts, some basic chemical grafting processes are presented in order to provide the readers the well-established approaches.

2.4.3.1 Chemical Grafting of Biomolecules onto Activated Groups of the Particles

The covalent binding of macromolecules (or biomolecules) bearing primary amine group has been widely performed onto carboxylic containing particles. In this case, the activation of the carboxylic groups is first realized before adding the macromolecules at suitable pH, salinity and temperature. This covalent coupling method has been used for the immobilization of proteins, antibodies, modified peptides and then extended to expensive biomolecules such as oligonucleotides bearing amino-link spacer arm at its 5′ position (Fig. 2.27). This approach is due to easy automated synthesis of amino-link oligonucleotides.

The general problem of the immobilization process of receptor onto surface charged particles (i.e. carboxylic containing particles) is related to the colloidal

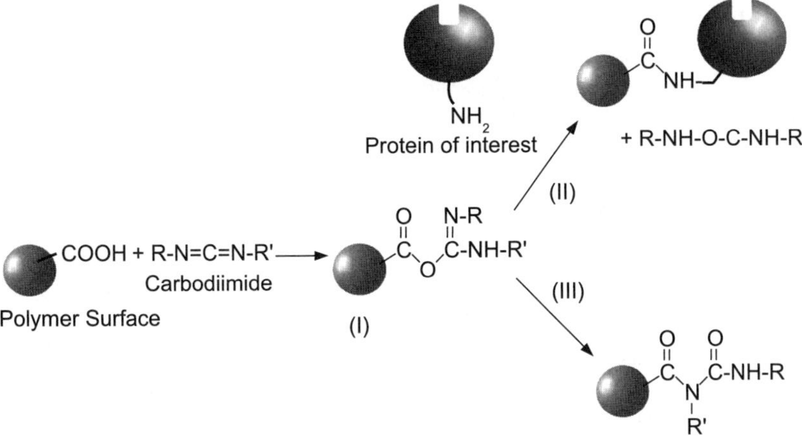

Fig. 2.27 Activation of carboxylic containing particles using charged activating agent 1-Ethyl-3-(3-dimethylaminopropyl) carbodiimide hydrochloride (EDAC)

stability of the particles during the activation step. Indeed, the activation step (using water-soluble activating agent) of the particles leads to low charged surfaces and in some cases to amphoteric surface character. Consequently, the colloidal stability of the activated dispersion is highly sensitive to the salt concentration used during the receptor immobilization process. To avoid the loss of the colloidal stability of the particles during the activation step, high concentrations of noncharged surfactant are used. Other than the use of surfactant, it may dramatically reduce the covalent grafting yield of the receptor due to the screening effect of the surface-active groups. To prevent those phenomena, the amount and the nature of the used surfactant should be well selected or to use activated receptors onto active stable colloidal particles.

2.4.3.2 Chemical Grafting of Activated Biomolecules onto Reactive Groups of the Particles

In order to avoid the above-mentioned problems related to the colloidal stability and to the screening effect induced by the use surfactant, the activation of biomolecules followed by the grafting onto reactive particles was generally favored. In this case, biomolecules are activated using water-soluble homobifunctional activating agent. Such approach needs the purification step so as to isolate the activated biomolecules before their chemical grafting process. In this case, only small surfactant amount is needed and in some cases, surfactant free process is used. The general problem of such methodology is related to (1) the possible denaturation

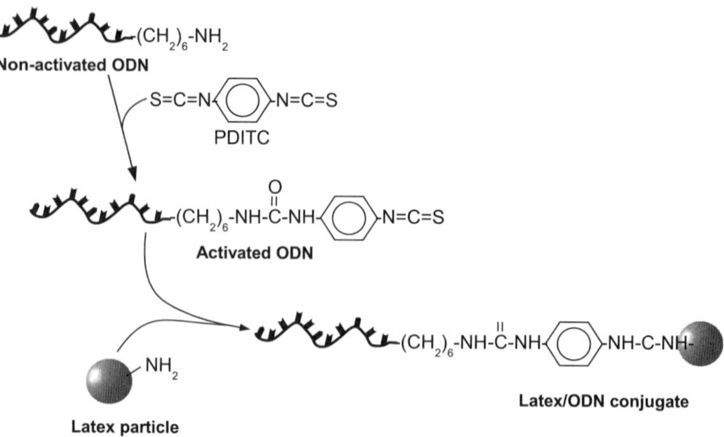

Fig. 2.28 Covalent immobilization of ODNs onto aminated-latex particles. Step 1: activation reaction of the ODNs by PDITC. Step 2 : coupling of the activated ODNs onto aminated-latex particles (Ganachaud et al. 2000)

or activity reduction of the activated biomolecules, and to (2) the separation process of activated and nonactivated biomolecules.

For illustration of such approach, the immobilization of oligonucleotides onto amino-containing latex particles is presented. The immobilization of oligonucleotides (ODNs) onto colloidal particles can be performed through various methods. To perform an irreversible immobilization of ODNs bearing amino-containing spacer arm onto particles surfaces, a chemical grafting has to be considered onto carboxylic, amine and aldehyde containing particles. The covalent grafting of single-stranded DNA fragment bearing amino-link spacer arm at its 5′ position (oligonucleotide) onto amino-containing particles (i.e. polystyrene latexes) is presented and discussed since such immobilization it is not well-known.

This chemical immobilization process is performed in various steps as depicted in (Fig. 2.28):

(1) The activation of the 5′ amino group of the oligonucleotide molecules using homobifunctional reagent such as phenylenediisothiocyanate (PDITC) (Ganachaud et al. 2001). The use of high amount of PDITC may lead to ODN–ODN conjugates.
(2) The chemical reaction between the activated oligonucleotide and the amine group of the particles. In order to avoid the aggregation phenomena, the reaction is performed at $pH > 8$ and in the presence of noncharged surfactant. High pH is needs in order to have amine group rather ammonium and the surfactant is preserve the colloidal stability of the particles.
(3) The ODN/particle conjugates are washed using classical surfactant free and moderate salinity buffers (10 mM borate buffer, pH 10.7, 0.2 M NaCl, 0.3 wt% Triton X-405). To some extend, the final particles are hairy like particles.

The pertinent result that can be deduced from such particular system is related to the surface charge density. In fact, the surface reactive group density play non-negligible role in the adsorption process and consequently in the residual grafted amount of oligonucleotides after washing step. The grafted amount was found to decrease with increasing the incubation pHs. The observed behavior revealed that the adsorption is the key parameter that controls the final immobilization (via chemical grafting) process of activated ODN. In addition, the residual grafted amount of ODN increased with increasing the initial activated oligonucleotide molecules concentration until reaching a plateau conditions. The reached plateau values correspond to possible surface saturation, which is principally limited by steric hindrance. Whereas, the chain length and the microstructure of oligonucleotide are reported to be totally marginal in nature (Ganachaud et al. 2000; Elaissari et al. 2003a, b, c; Ganachaud et al. 2003).

In order to enhance the accessibility of chemically grafted biomolecules or receptors onto particles surface, the chemical binding methodology has been oriented to the use of selective chemistry or to the use of polymer-like spacer arm as coupling agent.

2.4.3.3 Chemical Grafting of Functional Biomolecules onto Active Particles

The only chemical grafting, which can be directly performed without any addition of surface-active agent is based on the use of high reactive moieties, such as covalent grafting of amine onto aldehyde compound (Fig. 2.3). Such approach has been used for instance, for performing the chemical grafting of amine containing oligonucleotides onto latex particles bearing aldehyde function (Charreyre et al. 1999). In order to maintain the colloidal stability of such reactive aldehyde containing latex particles charged initiators (e.g. potassium persulfate) are generally used in the polymerization process. Consequently, the final particles contain both charged initiator fragments and aldehyde groups. It is interesting to notice that the aldehyde function can be easily deduced from the periodate oxidation of saccharide compound (i.e. lipomaltonamide) (Charreyre et al. 1999; Beattie et al. 1995).

2.4.3.4 Receptor Covalent Grafting Mediated by Reactive Polymer

In some specific biomedical applications, the direct chemical grafting of receptors onto particles surface, leads to low stable sensitive particles and to drastic reduction of the sensitivity. This sensitivity problem is mainly related to the interfacial conformation of the immobilized receptor and to the loss in the reactivity induced by the mobility reduction. In order to favor the interfacial mobility of the immobilized molecules, three approaches have been used: (1) the chemical modification of the used receptor (i.e. single-stranded DNA fragment bearing amino-link spacer

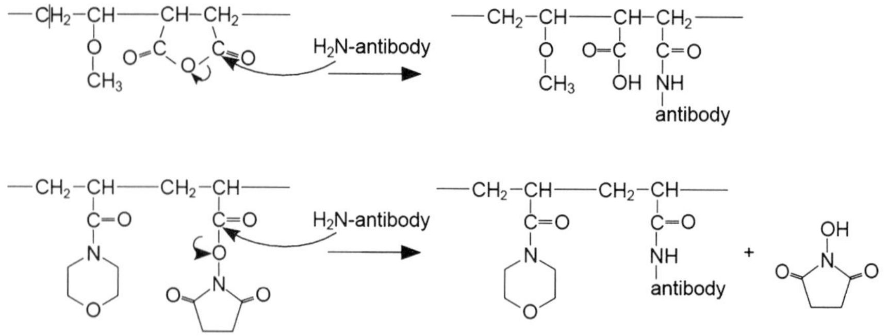

Fig. 2.29 Chemical grafting of antibody onto reactive anhydride or active-ester-based copolymers

arm), (2) the use of reactive hairy like particles and (3) the use of reactive polymers as macro-coupling agent rather than classical small molecules as above discussed.

The use of reactive polymer was found to be exiting methodology leading to interesting results. This approach has been used for the covalent binding of single-stranded DNA fragment bearing amino-link spacer arm (Erout et al. 1996a, 1996b) and antibodies (Rossi et al. 2004). The used reactive polymers are active-ester or anhydride based copolymers (Yang et al. 1990; Erout et al. 1996a, b) (Fig. 2.29).

Experimentally, the suitable biomolecule was first grafted on the reactive polymer using DMSO–water mixture in order to enhance the chemical grafting yield by reducing the hydrolysis kinetic (Erout et al. 1996a, b; Ladaviére et al. 2000). The polymer–biomolecule conjugates were extracted using HPLC technique. Before the total hydrolysis of the reactive groups of the used polymer, the chemical grafting of the conjugates was performed in the presence of surfactant free amino-containing latex particles (Rossi et al. 2004; Veyret et al. 2005).

The total hydrolysis of the residual reactive esters (or anhydride) leads to hairy like and hydrated sensitive particles (Rossi et al. 2004).

2.4.3.5 Immobilization of Cis-diol-Containing Receptor onto Particles Bearing Boronic Acid

Various works took advantage of this specific and strong specific complexation reaction by designing several synthetic phenyl boronic acid–containing materials (polymers and particles) (Camli et al. 2002; Uguzdogan et al. 2002; Elmas et al. 2004) (Elmas et al. 2002) mainly for biomedical application purposes. In fact, the specific recognition of cis-diol function by the boronic acid derivatives was used for controlling the reversible immobilization of proteins, enzymes or any biomolecules bearing glucose compounds. Indeed, this specific interaction is sensitive to the pH of the medium. The boronic acid exhibits a trigonal structure at low pH and tetragonal one at basic pH (pH $>$ pKa $=$ 8.8). Thus, the specific complexation reaction is favored at basic pH, because of the stability of boronate form. The shift to acidic medium leads to the decomplexation and thereafter to the release of the

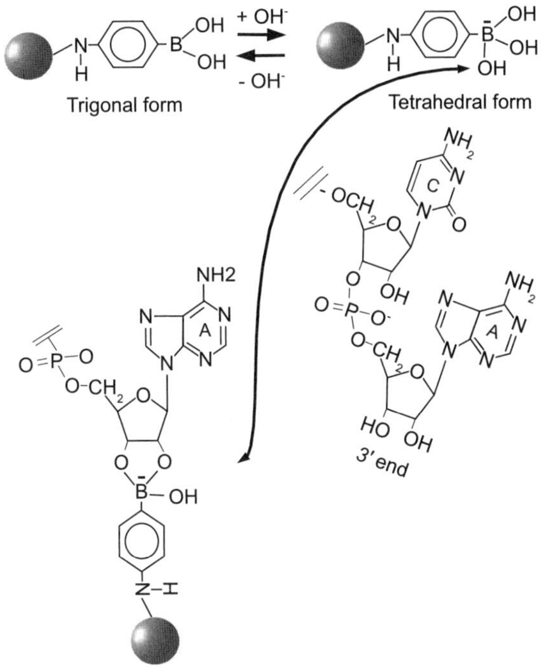

Fig. 2.30 Immobilization of RNA bearing cis-diol onto phenyl boronic acid containing support

cis-diol-containing molecules. It should be noted that the yield of the complexation reaction between the boronic acid and the polyols compound depends mainly on the availability of the cis-diol functions on the selected receptor. The use of such approach for RNA molecules immobilization onto latex particles bearing boronic acid groups was performed using surfactant free condition (Camli et al. 2002) (Fig. 2.30).

2.4.3.6 Dipol–Dipol Interaction

Biomolecules can be tethered onto polymer particles either by covalent coupling or physical adsorption. Standard chemical immobilization techniques or more specific adsorption do not allow addressing the final activity issue, as these methods have no control on the orientation of biomolecules on the support.

One alternative way to avoid such a disadvantage is to perform the immobilization process via regio–selective interactions, which can take place at a well-defined site of the biomolecule, not involved in the molecular recognition properties. In this direction, dipole–dipole interactions can be considered a promising approach because some biomolecules bear polar groups at a specific position, which could interact with polarizable groups containing particles surface (Fig. 2.31). So far, cyano groups have been recognized to specifically interact with sugar moieties

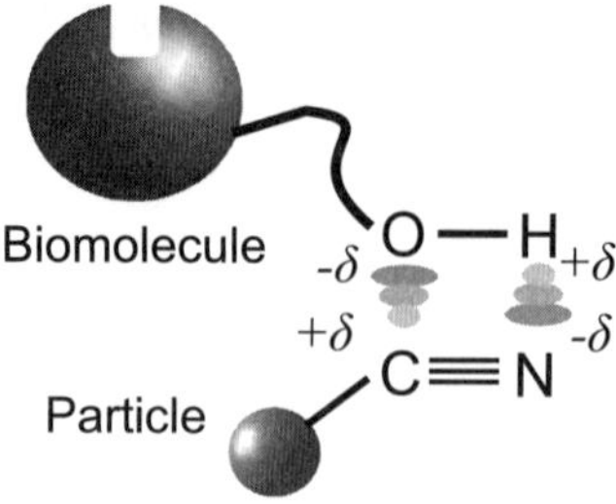

Fig. 2.31 Dipol–dipol cyano-sugar moieties interaction illustration

located in some biomolecules (e.g. antibodies). As a strongly polar group, cyano-sugar moieties interaction can be considered strong adsorption in terms of energy level. Such specific interaction is principally sensitive to the pH of the medium.

2.5 Concluding Remarks

Biofunctionalization is first a matter of defining an appropriate strategy according to the type of target molecules and the transducer technology. Thus, the detection mode (in solution of at the surface) is to be selected early; the type of functionalization as a thin 2D layer or a thick porous membrane. The criteria for this preliminary choice are the required sensitivity and the possible want for quantitative analysis.

Most coupling strategies for immobilizing native biomolecules make use of reactions with their amino and thiol groups. The chemical reaction can be selective by an appropriate choice of reactive group and experimental conditions (e.g. pH). However, it is a difficult task to keep the biological activity homogeneous over a population of biomolecules owing to the random nature of the coupling. The advent of miniaturized and highly parallelized reliable tests (microarrays for example) requires a good accessibility of the target in solution. The surface properties and the linkage between the solid support and the bioreceptors must be tailored to reach these analytical properties. Homogeneous immobilization is attained by introduction of a site-specific reactive group distal from the bioactive sites by modification of the biomolecules.

The wide variety of transducing techniques has lead to implement the immobilization step on a large panel of solid support and under different shapes (ranging from thin monolayer to thick porous membranes). Despite this wide diversity, homogeneous and uniform microenvironments at the solid/liquid interface are the most relevant parameters that govern the quality of a biointerface. The steric effects (receptors binding sites near the surface, tightly package of neighboring proteins) are among others parameters to be taken into consideration. The manufacturing of well-controlled 2D nanostructures (vertical and horizontal configuration) can be obtained by combining the SAM technology with selective site-directed strategy of biomolecules immobilization.

Thick interfaces have been considered in order to increase the surface loading of biomolecules and thus the detection signal. The 3D-embedding matrix can be simply considered passive membranes with high specific surface, or as active membranes with protective properties toward immobilized biomolecules or conferring specific properties for the detection. But it is often considered that porous materials limit the diffusion of target molecules.

As presented in this chapter, there are several strategies of biomolecule attachment to solid supports. The complex nature of biology, the ever growing demand of highly sensitive device asks for the development of new specific strategy to each new application.

References

Adessi C, Matton G, Ayala G, Turcatti G, Mermod J-J, Mayer P, Kawashima E (2000) Solid phase DNA amplification: characterisation of primer attachment and amplification mechanisms. Nucleic Acids Res 28:e87

Adey NB, Lei M, Howard MT, Jensen JD, Mayo DA, Butel DL, Coffin SC, Moyer TC, Slade DE, Spute MK, Hancock AM, Eisenhoffer GT, Dalley BK, McNeely MR (2002) Gains in sensitivity with a device that MIXES microarray hybridization solution in a 25-μm-thick chamber. Anal Chem 74:6413–6417

Advincula R (2006) Polymer brushes by anionic and cationic surface-initiated polymerization (SIP). Surface-initiated polymerization I. In: Jordan R (ed) Surface-initiated polymerization I. Springer Berlin, Heidelberg, pp 107–136

Afanassiev V, Hanemann V, Wolfl S (2000) Preparation of DNA and protein micro arrays on glass slides coated with an agarose film. Nucleic Acids Res 28:e66

Anderson GW, Zimmerman JE, Callahan FM (1964) The use of esters of N-hydroxysuccinimide in peptide synthesis. J Am Chem Soc 86:1839–1842

Angenendt P, Glokler J, Murphy D, Lehrach H, Cahill DJ (2002) Toward optimized antibody microarrays: a comparison of current microarray support materials. Anal Biochem 309: 253–260

Angst DL, Simmons GW (1991) Moisture absorption characteristics of organosiloxane self-assembled monolayers. Langmuir 7:2236–2242

Archer M, Christophersen M, Fauchet PM (2004) Macroporous silicon electrical sensor for DNA hybridization detection. Biomed Microdevices 6:203–211

Arenkov P, Kukhtin A, Gemmell A, Voloshchuk S, Chupeeva V, Mirzabekov A (2000) Protein microchips: use for immunoassay and enzymatic reactions. Anal Biochem 278:123–131

Arshady R (1993) Microspheres for biomedical applications: preparation of reactive and labelled microspheres. Biomaterials 14:5–15

Avnir D, Braun S, Lev O, Ottolenghi M (1994) Enzymes and other proteins entrapped in sol-gel materials. Chem Mater 6:1605–1614

Azzopardi MJ, Arribart H (1994) In situ FTIR study of the formation of an organosilane layer at the silica/solution interface. J Adhesion 46:103–115

Backer MV, Patel V, Jehning BT, Claffey KP, Backer JM (2006) Surface immobilization of active vascular endothelial growth factor via a cysteine-containing tag. Biomaterials 27: 5452–5458

Bagiyan GA, Koroleva IK, Soroka NV, Ufimtsev AV (2003) Oxidation of thiol compounds by molecular oxygen in aqueous solutions. Russ Chem Bull 52:1135–1141

Banks PR, Paquette DM (1995) Comparison of three common amine reactive fluorescent probes used for conjugation to biomolecules by capillary zone electrophoresis. Bioconjug Chem 6:447–458

Bayliss SC, Hutt DA, Zhang Q, Harris P, Phillips NJ, Smith A (1995) Structural study of porous silicon. Thin Solid Films 255:128–131

Beattie KL, Beattie WG, Meng L, Turner SL, Coral-Vazquez R, Smith DD, McIntyre PM, Dao DD (1995) Advances in genosensor research. Clin Chem 41:700–706

Beier M, Hoheisel JD (1999) Versatile derivatisation of solid support media for covalent bonding on DNA-microchips. Nucleic Acids Res 27:1970–1977

Benoit V, Steel A, Torres M, Yu Y-Y, Yang H, Cooper J (2001) Evaluation of three-dimensional microchannel glass biochips for multiplexed nucleic acid fluorescence hybridization assays. Anal Chem 73:2412–2420

Benters R, Niemeyer CM, Wöhrle D (2001) Dendrimer-activated solid supports for nucleic acid and protein microarrays. ChemBioChem 2:686–694

Bergbreiter D, Kippenberger A (2006) Hyperbranched surface graft polymerizations. Springer Berlin, Heidelberg

Berggren C, Bjarnason B, Johansson G (2001) Capacitive biosensors. Electroanalysis 13:173–180

Besanger TR, Brennan JD (2003) Ion sensing and inhibition studies using the transmembrane ion channel peptide gramicidin a entrapped in sol-gel-derived silica. Anal Chem 75:1094–1101

Besombes J-L, Cosnier S, Labbé P (1997) Improvement of poly(amphiphilic pyrrole) enzyme electrodes via the incorporation of synthetic laponite-clay-nanoparticles. Talanta 44:2209–2215

Bessueille F, Dugas V, Vikulov V, Cloarec J-P, Souteyrand É, Martin J-R (2005) Assessment of porous silicon substrate for well-characterised sensitive DNA chip implement. Biosens Bioelectron 21:908–916

Blitz JP, Murthy RSS, Leyden DE (1988) Studies of silylation of Cab-O-Sil with methoxymethylsilanes by diffuse reflectance FTIR spectroscopy. J Colloid Interface Sci 121:63–69

Boksányi L, Liardon O, Kováts ESz (1976) Chemically modified silicon dioxide surfaces reaction of n-alkyldimethylsilanols and n-oxaalkyldimethylsilanols with the hydrated surface of silicon dioxide - the question of the limiting surface concentration. Adv Colloid Interface Sci 6:95–137

Bornhop DJ, Latham JC, Kussrow A, Markov DA, Jones RD, Sorensen HS (2007) Free-solution, label-free molecular interactions studied by back-scattering interferometry. Science 317: 1732–1736

Böttcher H (2000) Bioactive sol-gel coatings. J Prakt Chem 342:427–436

Boukherroub R, Morin S, Bensebaa F, Wayner DDM (1999) New synthetic routes to alkyl monolayers on the Si(111) surface. Langmuir 15:3831–3835

Boukherroub R, Wayner DDM (1999) Controlled functionalization and multistep chemical manipulation of covalently modified Si(111) surfaces. J Am Chem Soc 121:11513–11515

Briand E, Salmain M, Herry J-M, Perrot H, Compere C, Pradier C-M (2006) Building of an immunosensor: how can the composition and structure of the thiol attachment layer affect the immunosensor efficiency? Biosens Bioelectron 22:440–448

Brogan KL, Wolfe KN, Jones PA, Schoenfisch MH (2003) Direct oriented immobilization of F (ab') antibody fragments on gold. Anal Chim Acta 496:73–80

Brueggemeier SB, Wu D, Kron SJ, Palecek SP (2005) Protein-acrylamide copolymer hydrogels for array-based detection of tyrosine kinase activity from cell lysates. Biomacromolecules 6:2765–2775

Buchmeiser M (2006) Metathesis polymerization to and from surfaces. In: Jordan R (ed) Surface-initiated polymerization I. Springer Berlin, Heidelberg, pp 137–171

Bujoli B, Lane SM, Nonglaton G, Pipelier M, Léger J, Talham DR, Tellier C (2005) Metal phosphonates applied to biotechnologies: a novel approach to oligonucleotide microarrays. Chemistry Eur. J 11:1980–1988

Bunemann H (1982) Immobilization of denatured DNA to macroporous supports: II. Steric and kinetic parameters of heterogeneous hybridization reactions. Nucleic Acids Res 10:7181–7196

Buriak JM, Allen MJ (1998) Lewis acid mediated functionalization of porous silicon with substituted alkenes and alkynes. J Am Chem Soc 120:1339–1340

Buriak JM, Stewart MP, Geders TW, Allen MJ, Choi HC, Smith J, Raftery D, Canham LT (1999) Lewis acid mediated hydrosilylation on porous silicon surfaces. J Am Chem Soc 121: 11491–11502

Burnette WN (1981) "Western blotting": electrophoretic transfer of proteins from sodium dodecyl sulfate–polyacrylamide gels to unmodified nitrocellulose and radiographic detection with antibody and radioiodinated protein A. Anal Biochem 112:195–203

Camarero JA, Kwon Y, Coleman MA (2004) Chemoselective attachment of biologically active proteins to surfaces by expressed protein ligation and its application for "protein chip" fabrication. J Am Chem Soc 126:14730–14731

Camli ST, Senel S, Tuncel A (2002) Nucleotide isolation by boronic acid functionalized hydrophilic supports. Colloids Surf A Physicochem Eng Asp 207:127–137

Carlsson J, Drevin H, Axèn R (1978) Protein thiolation and reversible protein-protein conjugation. N-succinimidyl 3-(2-pyridyldithio)propionate, a new heterobifunctional reagent. Biochem J 173:723–737

Carturan G, Toso RD, Boninsegna S, Monte RD (2004) Encapsulation of functional cells by sol-gel silica: actual progress and perspectives for cell therapy. J Mater Chem 14: 2087–2098

Caruso F, Niikura K, Furlong DN, Okahata Y (1997) 2. Assembly of alternating polyelectrolyte and protein multilayer films for immunosensing. Langmuir 13:3427–3433

Cha T, Guo A, Jun Y, Pei D, Zhu X-Y (2004) Immobilization of oriented protein molecules on poly(ethylene glycol)-coated Si(111). Proteomics 4:1965–1976

Charreyre MT, Revilla J, Elaissari A, Pichot C, Gallot B (1999) Surface functionalization of polystyrene nanoparticles with liposaccharide monomers: preparation, characterization and applications. J Biocat Compat Polym 14:64–90

Chassignol M, Thuong NT (1998) Phosphodisulfide bond: a new linker for the oligonucleotide conjugation. Tetrahedron Lett 39:8271–8274

Chechik V, Crooks RM, Stirling CJM (2000) Reactions and reactivity in self-assembled monolayers. Adv Mater 12:1161–1171

Cheek BJ, Steel AB, Torres MP, Yu Y-Y, Yang H (2001) Chemiluminescence detection for hybridization assays on the flow-thru chip, a three-dimensional microchannel biochip. Anal Chem 73:5777–5783

Chevalier Y, Coche-Guérente L, Labbé P (2002) Small angle neutron scattering studies and kinetic analysis of laponite-enzyme hydrogels in view of biosensors construction. Mater Sci Eng C 21:81–89

Chevalier Y, Elbhiri Z, Chovelon J-M, Jaffrezic-Renault N (2000) Chemically grafted field-effect transistors for the recognition of ionic species in aqueous solutions. Prog Colloid Polym Sci 115:243–248

Chevolot Y, Bouillon C, Vidal S, Morvan F, Meyer A, Cloarec J-P, Jochum A, Praly J-P, Vasseur J-J, Souteyrand É (2007) DNA-based carbohydrate biochips: a platform for surface glyco-engineering. Angew Chem Int Ed Engl 46:2389–2402

Chluba J, Voegel J-C, Decher G, Erbacher P, Schaaf P, Ogier J (2001) Peptide hormone covalently bound to polyelectrolytes and embedded into multilayer architectures conserving full biological activity. Biomacromolecules 2:800–805

Choi V, Oh KW, Thomas J, Heineman W, Halsall HB, Nevin JH, Helmicki A, Henderson HT, Ahn CH (2002) Integrated microfluidic biochemical detection system for protein analysis with magnetic bead-based sampling capabilities. Lab on a Chip 2:27–30

Chrisey L, Lee G, O'Ferrall C (1996) Covalent attachment of synthetic DNA to self-assembled monolayer films. Nucleic Acids Res 24:3031–3039

Clark JH, Macquarrie DJ (1998) Catalysis of liquid phase organic reactions using chemically modified mesoporous inorganic solids. Chem Commun 8:853–860

Coche-Guérente L, Desbrières J, Fatisson J, Labbé P, Rodriguez MC, Rivas G (2005) Physicochemical characterization of the layer-by-layer self-assembly of polyphenol oxidase and chitosan on glassy carbon electrode. Electrochim Acta 50:2865–2877

Coche-Guérente L, Desprez V, Labbé P (1998) Characterization of organosilasesquioxane-intercalated-laponite-clay modified electrodes and (bio)electrochemical applications. J Electroanal Chem 458:73–86

Coche-Guérente L, Desprez V, Labbé P, Therias S (1999) Amplification of amperometric biosensor responses by electrochemical substrate recycling. Part II. Experimental study of the catechol-polyphenol oxidase system immobilized in a laponite clay matrix. J Electroanal Chem 470:61–69

Cras JJ, Rowe-Taitt CA, Nivens DA, Ligler FS (1999) Comparison of chemical cleaning methods of glass in preparation for silanization. Biosens Bioelectron 14:683–688

Cuatrecasas P, Parikh I (1972) Adsorbents for affinity chromatography. Use of N-hydroxysuccinimide esters of agarose. Biochemistry 11:2291–2299

Cunin F, Schmedake TA, Link JR, Li YY, Koh J, Bhatia SN, Sailor MJ (2002) Biomolecular screening with encoded porous-silicon photonic crystals. Nat Mater 1:39–41

Dantas de Araújo A, Palomo JM, Cramer J, Köhn M, Schröder H, Wacker R, Niemeyer C, Alexandrov K, Waldmann H (2006) Diels-Alder ligation and surface immobilization of proteins. Angew Chem Int Ed Engl 45:296–301

De Gennes PG (1987) Polymers at an interface; a simplified view. Adv Colloid Interface Sci 27:189–209

Derwinska K, Sauer U, Preininger C (2007) Reproducibility of hydrogel slides in on-chip immunoassays with respect to scanning mode, spot circularity, and data filtering. Anal Biochem 370:38–46

Diaz-Mochon JJ, Bialy L, Bradley M (2006) Dual colour, microarray-based, analysis of 10,000 protease substrates. Chem Commun 38:3984–3986

Diaz-Mochon JJ, Tourniaire G, Bradley M (2007) Microarray platforms for enzymatic and cell-based assays. Chem Soc Rev 36:449–457

Diehl F, Grahlmann S, Beier M, Hoheisel JD (2001) Manufacturing DNA microarrays of high spot homogeneity and reduced background signal. Nucleic Acids Res 29:e38

Dolan PL, Wu Y, Ista LK, Metzenberg RL, Nelson MA, Lopez GP (2001) Robust and efficient synthetic method for forming DNA microarrays. Nucleic Acids Res 29:e107

Dordi B, Schonherr H, Vancso GJ (2003) Reactivity in the confinement of self-assembled monolayers: chain length effects on the hydrolysis of N-hydroxysuccinimide ester disulfides on gold. Langmuir 19:5780–5786

Dowling DP, Donnelly K, McConnell ML, Eloy R, Arnaud MN (2001) Deposition of anti-bacterial silver coatings on polymeric substrates. Thin Solid Films 398–399:602–606

Dubruel P, Vanderleyden E, Bergada M, De Paepe I, Chen H, Kuypers S, Luyten J, Schrooten J, Van Hoorebeke L, Schacht E (2006) Comparative study of silanisation reactions for the biofunctionalisation of Ti-surfaces. Surf Sci 600:2562–2571

Dugas V, Broutin J, Souteyrand É (2005) Droplet evaporation study applied to DNA chip manufacturing. Langmuir 21:9130–9136

Dugas V, Depret G, Chevalier Y, Nesme X, Souteyrand É (2004) Immobilization of single-stranded DNA fragments to solid surfaces and their repeatable specific hybridization: covalent binding or adsorption? Sensors Actors B 101:112–121

Dyer D (2006) Photoinitiated synthesis of grafted polymers. In: Jordan R (ed) Surface-initiated polymerization I. Springer Berlin, Heidelberg, pp 47–65

Dyukova VI, Dementieva EI, Zubtsov DA, Galanina OE, Bovin NV, Rubina AY (2005) Hydrogel glycan microarrays. Anal Biochem 347:94–105

Eckert AW, Grobe D, Rothe U (2000) Surface-modification of polystyrene-microtitre plates via grafting of glycidylmethacrylate and coating of poly-glycidylmethacrylate. Biomaterials 21:441–447

Edman C, Raymond D, Wu D, Tu E, Sosnowski R, Butler W, Nerenberg M, Heller M (1997) Electric field directed nucleic acid hybridization on microchips. Nucleic Acids Res 25:4907–4914

Ek S, Iiskola EI, Niinisto L (2004) Atomic layer deposition of amino-functionalized silica surfaces using N-(2-aminoethyl)-3-aminopropyltrimethoxysilane as a silylating agent. J Phys Chem B 108:9650–9655

Elaissari A, Veyret R, Mandrand B, Chatterjee J (2003a) In: Elaissari A (eds) Colloidal biomolecules, biomaterials, and biomedical applications. Marcel Deker, New York, pp 1–26

Elaissari A, Ganachaud F, Pichot C (2003b) Biorelevant latexes and microgels for the interaction with nucleic acids. Top Curr Chem 227:169–193

Elaissari A, Sauzedde F, Montagne F, Pichot C (2003c) In: Elaissari A (ed) Colloids polymers, synthesis and characterization, Surfactant science series. Marcel Deker, New York, pp 285–318

Elmas B, Onur MA, Senel S, Tuncel A (2002) Temperature controlled RNA isolation by N-isopropylacrylamide-vinylphenyl boronic acid copolymer latex. Colloid Polym Sci 280: 1137–1146

Elmas B, Onur MA, Senel S, Tuncel A (2004) Thermosensitive N-isopropylacrylamide-vinylphenyl boronic acid copolymer latex particles for nucleotide isolation. Colloids and Surf A 232:253–259

Engvall E, Perlman P (1971) Enzyme-linked immunosorbent assay (ELISA). Quantitative assay of immunoglobulin G. Immunochemistry 8:871–874

Ernst-Cabrera K, Wilchek M (1988) High-performance affinity chromatography. Trends Analyt Chem 7:58–63

Erout M-N, Elaissari A, Pichot C, Llauro M-F (1996a) Radical-initiated copolymers of N-vinyl pyrrolidone and N-acryloxy succinimide: kinetic and microstructure studies. Polymer 37:1157–1165

Erout M-N, Troesch A, Pichot C, Cros P (1996b) Preparation of conjugates between oligonucleotides and N-vinylpyrrolidone/N-acryloxysuccinimide copolymers and applications in nucleic acid assays to improve sensitivity. Bioconjug Chem 7:568–575

Evans B, White TE (1968) Adsorption and reaction of methylchlorosilanes at an "Aerosil" surface. J Catalysis 11:336–341

Fadeev AY, McCarthy TJ (1999) Trialkylsilane monolayers covalently attached to silicon surfaces: wettability studies indicating that molecular topography contributes to contact angle hysteresis. Langmuir 15:3759–3766

Ferreyra N, Coche-Guérente L, Fatisson J, Teijelo ML, Labbé P (2003) Layer-by-layer self-assembled multilayers of redox polyelectrolytes and gold nanoparticles. Chem Commun:2056–2057

Ferreyra N, Coche-Guérente L, Labbé P (2004) Construction of layer-by-layer self-assemblies of glucose oxidase and cationic polyelectrolyte onto glassy carbon electrodes and electrochemical study of the redox-mediated enzymatic activity. Electrochim Acta 49:477–484

Ferreyra NF, Forzani ES, Teijelo ML, Coche-Guérente L, Labbé P (2006) Unraveling the spatial distribution of immunoglobulins, enzymes, and polyelectrolytes within layer-by-layer self-assembled multilayers. Ellipsometric studies. Langmuir 22:8931–8938

Fidanza J, Glazer M, Mutnick D, McGall G, Frank C (2001) High capacity substrates as a platform for a DNA probe array genotyping assay. Nucleosides Nucleotides Nucleic Acids 20:533–538

Firestone MA, Shank ML, Sligar SG, Bohn PW (1996) Film architecture in biomolecular assemblies. Effect of linker on the orientation of genetically engineered surface-bound proteins. J Am Chem Soc 118:9033–9041

Fleer GJ, Cohen Stuart MA, Scheutjens JMHM, Cosgrove T, Vincent B (1993) Polymers at interfaces. Chapman & Hill, London

Fodor SPA, Read JL, Pirrung MC, Stryer L, Lu AT, Solas D (1991) Light-directed, spatially addressable parallel chemical synthesis. Science 251:767–773

Fryxell GE, Rieke PC, Wood LL, Engelhard MH, Williford RE, Graff GL, Campbell AA, Wiacek RJ, Lee L, Halverson A (1996) Nucleophilic displacements in mixed self-assembled monolayers. Langmuir 12:5064–5075

Gad M, Machida M, Mizutani W, Ishikawa M (2001) Method for orienting DNA molecules on mica surfaces in one direction for atomic force microscopy imaging. J Biomol Struct Dyn 19:471–477

Ganachaud F, Elaissari A, Pichot C (2000) Covalent grafting of polythymidylic acid onto amine-containing polystyrene latex particles. J Biomater Sci Polym Ed 11:931–945

Ganachaud F, Laayoun A, Chaix C, Delair T, Pichot C, Elaissari A (2001) Oligodeoxyribonucleotide activation with 2, 4-phenylenediisothiocyanate and their covalent grafting onto amine-functionalized latex microspheres. J Dispersion Sci Technol 22:473–484

Ganachaud F, Pichot C, Elaissari A (2003) In: Elaissari A (ed) Colloidal Biomolecules, Biomaterials and Biomedical Applications, Surfactant science Series. Marcel Deker, New York, pp 253–286

Gauchet C, Labadie GR, Poulter CD (2006) Regio- and chemoselective covalent immobilization of proteins through unnatural amino acids. J Am Chem Soc 128:9274–9275

Gavin IM, Kukhtin A, Glesne D, Schabacker D, Chandler DP (2005) Analysis of protein interaction and function with a 3-dimensional MALDI-MS protein array. Biotechniques 39:99–107

Gershon D (2002) Microarray technology: an array of opportunities. Nature 416:885–891

Gibson TD, Pierce BLJ, Hulbert JN, Gillespie S (1996) Improvements in the stability characteristics of biosensors using protein-polyelectrolyte complexes. Sens Actuators B 33:13–18

Giedroc DP, Keating KM, Martin CT, Williams KR, Coleman JE (1986) Zinc metalloproteins involved in replication and transcription. J Inorg Biochem 28:155–169

Gilham PT (1964) The synthesis of polynucleotide-celluloses and their use in the fractionation of polynucleotides. J Am Chem Soc 86:4982–4985

Gobet J, Kováts ESz (1984) Preparation of hydrated silicon dioxide for reproducible chemisorption experiments. Adsorp Sci Technol 1:77–92

Gray DE, Case-Green SC, Fell TS, Dobson PJ, Southern EM (1997) Ellipsometric and interferometric characterization of DNA probes immobilized on a combinatorial array. Langmuir 13:2833–2842

Grubor NM, Shinar R, Jankowiak R, Porter MD, Small GJ (2004) Novel biosensor chip for simultaneous detection of DNA-carcinogen adducts with low-temperature fluorescence. Biosens Bioelectron 19:547–556

Guilbault GG, Montalvo JG (1969) Urea-specific enzyme electrode. J Am Chem Soc 91: 2164–2165

Guilbault GG, Montalvo JG (1970) Enzyme electrode for the substrate urea. J Am Chem Soc 92:2533–2538

Gun G, Tsionsky M, Lev O (1994) Voltammetric studies of composite ceramic carbon working electrodes. Anal Chim Acta 294:261–270

Guo Z, Guilfoyle RA, Thiel AJ, Wang R, Smith LM (1994) Direct fluorescence analysis of genetic polymorphisms by hybridization with oligonucleotide arrays on glass supports. Nucleic Acids Res 22:5456–5465

Guschin D, Yershov G, Zaslavsky A, Gemmell A, Shick V, Proudnikov D, Arenkov P, Mirzabekov A (1997) Manual manufacturing of oligonucleotide, DNA, and protein microchips. Anal Biochem 250:203–211

Han J, Lee H, Nguyen NY, Beaucage SL, Puri RK (2005) Novel multiple 5′-amino-modified primer for DNA microarrays. Genomics 86:252–258

Harder P, Bierbaum K, Woell C, Grunze M, Heid S, Effenberger F (1997) Induced orientational order in long alkyl chain aminosilane molecules by preadsorbed octadecyltrichlorosilane on hydroxylated Si(100). Langmuir 13:445–454

He PL, Hu NF, Zhou G (2002) Assembly of electroactive layer-by-layer films of hemoglobin and polycationic poly(diallyldimethylammonium). Biomacromolecules 3:139–146

Heichal-Segal O, Rappoport S, Braun S (1995) Immobilization in alginate-silicate sol-gel matrix protects [beta]-glucosidase against thermal and chemical denaturation. Biotechnology 13:798–800

Helms B, van Baal I, Merkx M, Meijer EW (2007) Site-specific protein and peptide immobilization on a biosensor surface by pulsed native chemical ligation. ChemBioChem 8:1790–1794

Herne TM, Tarlov MJ (1997) Characterization of DNA probes immobilized on gold surfaces. J Am Chem Soc 119:8916–8920

Hertadi R, Gruswitz F, Silver L, Koide A, Koide S, Arakawa H, Ikai A (2003) Unfolding mechanics of multiple OspA substructures investigated with single molecule force spectroscopy. J Mol Biol 333:993–1002

Hevesi L, Jeanmart L, Remacle J (2002) Method for obtaining a surface activation of a solid support for building biochip microarrays. P/N 20020076709 United States

Hochuli E, Dobeli H, Schacher A (1987) New metal chelate adsorbent selective for proteins and peptides containing neighbouring histidine residues. J Chromatogr 411:177–184

Hodneland CD, Lee Y-S, Min D-H, Mrksich M (2002) Supramolecular chemistry and self-assembly special feature: Selective immobilization of proteins to self-assembled monolayers presenting active site-directed capture ligands. Proc Natl Acad Sci USA 99:5048–5052

Hoffmann C, Tovar GEM (2006) Mixed self-assembled monolayers (SAMs) consisting of methoxy-tri(ethylene glycol)-terminated and alkyl-terminated dimethylchlorosilanes control the non-specific adsorption of proteins at oxidic surfaces. J Colloid Interface Sci 295:427–435

Horr TJ, Arora PS (1997) Determination of the acid-base properties for 3-amino, 3-chloro and 3-mercaptopropyltrimethoxysilane coatings on silica surfaces by XPS. Colloids Surf A 126:113–121

Hostetler MJ, Wingate JE, Zhong C-J, Harris JE, Vachet RW, Clark MR, Londono JD, Green SJ, Stokes JJ, Wignall GD, Glish GL, Porter MD, Evans ND, Murray RW (1998) Alkanethiolate gold cluster molecules with core diameters from 1.5 to 5.2 nm: core and monolayer properties as a function of core size. Langmuir 14:17–30

Houseman BT, Mrksich M (2002) Carbohydrate arrays for the evaluation of protein binding and enzymatic modification. Chem Biol 9:443–454

Hu M, Noda S, Okubo T, Yamaguchi Y, Komiyama H (2001) Structure and morphology of self-assembled 3-mercaptopropyltrimethoxysilane layers on silicon oxide. Appl Surf Sci 181: 307–316

Hui Liu R, Lenigk R, Druyor-Sanchez R, Yang J, Grodzinski P (2003) Hybridization enhancement using cavitation microstreaming. Anal Chem 75:1911–1917

Hynd MR, Frampton JP, Burnham M-R, Martin DL, Dowell-Mesfin NM, Turner JN, Shain W (2007) Functionalized hydrogel surfaces for the patterning of multiple biomolecules. J Biomed Mater Res 81A:347–354

Jaber JA, Schlenoff JB (2006) Recent developments in the properties and applications of polyelectrolyte multilayers. Curr Opin Colloid Interface Sci 11:324–329

Janshoff A, Dancil K-PS, Steinem C, Greiner DP, Lin VS-Y, Gurtner C, Motesharei K, Sailor MJ, Ghadiri MR (1998) Macroporous p-type silicon fabry-perot layers. Fabrication, characterization, and applications in biosensing. J Am Chem Soc 120:12108–12116

Jdanova AS, Poyard S, Soldatkin AP, Jaffrezic-Renault N, Martelet C (1996) Conductometric urea sensor. Use of additional membranes for the improvement of its analytical characteristics. Anal Chim Acta 321:35–40

Johnson ECB, Kent SBH (2006) Insights into the mechanism and catalysis of the native chemical ligation reaction. J Am Chem Soc 128:6640–6646

Kato K, Uchida E, Kang ET, Uyama Y, Ikada Y (2003) Polymer surface with graft chains. Prog Polym Sci 28:209–259

Kindermann M, George N, Johnsson N, Johnsson K (2003) Covalent and selective immobilization of fusion proteins. J Am Chem Soc 125:7810–7811

Kinney DR, Chuang IS, Maciel GE (1993) Water and the silica surface as studied by variable-temperature high-resolution proton NMR. J Am Chem Soc 115:6786–6794

Koh Y, Jin Kim S, Park J, Park C, Cho S, Woo H-G, Ko YC, Sohn H (2007) Detection of avidin based on rugate-structured porous silicon interferometer. Bull Korean Chem Soc 28: 2083–2088

Kong W, Wang LP, Gao ML, Zhou H, Zhang X, Li W, Shen JC (1994) Immobilized bilayer glucose-isomerase in porous trimethylamine polystyrene-based on molecular deposition. J Chem Soc Chem Commun:1297–1298

Korri-Youssoufi H, Garnier F, Srivastava P, Godillot P, Yassar A (1997) Toward bioelectronics: specific DNA recognition based on an oligonucleotide-functionalized polypyrrole. J Am Chem Soc 119:7388–7389

Kotov NA, Dékány I, Fendler JH (1995) Layer-by-layer self-assembly of polyelectrolyte-semiconductor nanoparticle composite films. J Phys Chem 99:13065–13069

Kotov NA, Haraszti T, Turi L, Zavala G, Geer RE, Dekány I, Fendler JH (1997) Mechanism of and defect formation in the self-assembly of polymeric polycation-montmorillonite ultrathin films. J Am Chem Soc 119:6821–6832

Kunzelmann U, Bottcher H (1997) Biosensor properties of glucose oxidase immobilized within SiO_2 gels. Sens Actuators B Chem 39:222–228

Kusnezow W, Jacob A, Walijew A, Diehl F, Hoheisel JD (2003) Antibody microarrays: an evaluation of production parameters. Proteomics 3:254–264

Kwon Y, Coleman MA, Camarero JA (2006) Selective immobilization of proteins onto solid supports through split-intein-mediated protein trans-splicing. Angew Chem Int Ed Engl 45:1726–1729

Ladaviére C, Lorenzo C, Elaissari A, Mandrand B, Delair T (2000) Electrostatically driven immobilization of peptides onto (maleic anhydride-alt-methyl vinyl ether) copolymers in aqueous media. Bioconjug Chem 11:146–152

Lao R, Wang L, Wan Y, Zhang J, Song S, Zhang Z, Fan C, He L (2007) Interactions between cytochrome c and DNA strands self-assembled at Gold electrode. Int J Mol Sci 8:136–144

Le Berre V, Trévisiol E, Dagkessamanskaia A, Sokol S, Caminade A-M, Majoral J-P, Meunier B, Francois J (2003) Dendrimeric coating of glass slides for sensitive DNA microarrays analysis. Nucleic Acids Res 31:e88

Lee C-S, Kim B-G (2002) Improvement of protein stability in protein microarrays. Biotechnol Lett 24:839–844

Lee J-C, An JY, Kim B-W (2007) Application of anodized aluminium oxide as a biochip substrate for a Fabry–Perot interferometer. J Chem Technol Biotechnol 82:1045–1052

Lee JW, Louie YQ, Walsh DP, Chang Y-T (2003) Nitrophenol resins for facile amide and sulfonamide library synthesis. J Comb Chem 5:330–335

Lee MR, Shin I (2005) Fabrication of chemical microarrays by efficient immobilization of hydrazide-linked substances on epoxide-coated glass surfaces. Angew Chem Int Ed Engl 44:2881–2884

Lehmann V (2002) Biosensors: barcoded molecules. Nat Mater 1:12–13

Li J, Chia LS, Goh NK, Tan SN, Ge H (1997) Mediated amperometric glucose sensor modified by the sol-gel method. Sens Actuators B Chem 40:135–141

Lie LH, Patole SN, Pike AR, Ryder LC, Connolly BA, Ward AD, Tuite EM, Houlton A, Horrocks BR (2004) Immobilisation and synthesis of DNA on Si(111), nanocrystalline porous silicon and silicon nanoparticles. Faraday Discuss 125:235–249

Lin H, Cornish VW (2002) Screening and selection methods for large-scale analysis of protein function. Angew Chem Int Ed Engl 41:4402–4425

Lin T-Y, Wu C-H, Brennan JD (2007) Entrapment of horseradish peroxidase in sugar-modified silica monoliths: toward the development of a biocatalytic sensor. Biosens Bioelectron 22:1861–1867

Lindroos K, Liljedahl U, Raitio M, Syvanen A-C (2001) Minisequencing on oligonucleotide microarrays: comparison of immobilisation chemistries. Nucleic Acids Res 29:e69

Linford MR, Fenter P, Eisenberger PM, Chidsey CED (1995) Alkyl monolayers on silicon prepared from 1-alkenes and hydrogen-terminated silicon. J Am Chem Soc 117:3145–3155

Löfås S, Johnsson B (1990) A novel hydrogel matrix on gold surfaces in surface plasmon resonance sensors for fast and efficient covalent immobilization of ligands. J Chem Soc Chem Commun:1526–1528

Love JC, Estroff LA, Kriebel JK, Nuzzo RG, Whitesides GM (2005) Self-assembled monolayers of thiolates on metals as a form of nanotechnology. Chem Rev 105:1103–1170

Lu B, Smyth MR, O'Kennedy R (1996) Oriented immobilization of antibodies and its applications in immunoassays and immunosensors. Analyst 121:29R–32R

Luderer F, Walschus U (2005) Immobilization of oligonucleotides for biochemical sensing by self-assembled monolayers: thiol-organic bonding on gold and silanization on silica surfaces. In: Wittmann C (ed) Immobilisation of DNA on chips I. Springer Berlin, Heidelberg, pp 37–56

Lue RY, Chen GY, Zhu Q, Lesaicherre ML, Yao SQ (2004a) Site-specific immobilization of biotinylated proteins for protein microarray analysis. Methods Mol Biol 264:85–100

Lue RYP, Chen GYJ, Hu Y, Zhu Q, Yao SQ (2004b) Versatile protein biotinylation strategies for potential high-throughput proteomics. J Am Chem Soc 126:1055–1062

MacBeath G, Schreiber SL (2000) Printing proteins as microarrays for high-throughput function determination. Science 289:1760–1763

Martin P, Marsaudon S, Thomas L, Desbat B, Aimé J-P, Bennetau B (2005) Liquid mechanical behavior of mixed monolayers of amino and alkyl silanes by atomic force microscopy. Langmuir 21:6934–6943

Maskos U, Southern EM (1992a) Oligonucleotide hybridizations on glass supports - a novel linker for oligonucleotide synthesis and hybridization properties of oligonucleotides synthesized *in situ*. Nucl Acids Res 20:1679–1684

Maskos U, Southern EM (1992b) Parallel analysis of oligodeoxyribonucleotide (oligonucleotide) interactions. I. Analysis of factors influencing oligonucleotide duplex formation. Nucleic Acids Res 20:1675–1678

Matsuda T (2006) Photoiniferter-driven precision surface graft microarchitectures for biomedical applications. In: Jordan R (ed) Surface-initiated polymerization I. Springer Berlin, Heidelberg, pp 67–106

McGovern ME, Kallury KMR, Thompson M (1994) Role of solvent on the silanization of glass with octadecyltrichlorosilane. Langmuir 10:3607–3614

McQuain MK, Seale K, Peek J, Fisher TS, Levy S, Stremler MA, Haselton FR (2004) Chaotic mixer improves microarray hybridization. Anal Biochem 325:215–226

Mengistu TZ, DeSouza L, Morin S (2005) Functionalized porous silicon surfaces as MALDI-MS substrates for protein identification studies. Chem Commun 45:5659–5661

Millan KM, Saraullo A, Mikkelsen SR (1994) Voltammetric DNA biosensor for cystic fibrosis based on a modified carbon paste electrode. Anal Chem 66:2943–2948

Miscoria SA, Desbrières J, Barrera GD, Labbé P, Rivas GA (2006) Glucose biosensor based on the layer-by-layer self-assembling of glucose oxidase and chitosan derivatives on a thiolated gold surface. Anal Chim Acta 578:137–144

Moore E, O'Connell D, Galvin P (2006) Surface characterisation of indium-tin oxide thin electrode films for use as a conducting substrate in DNA sensor development. Thin Solid Films 515:2612–2617

Morris DL, Campbell J, Hornby WE (1975) The preparation of nylon-tube-supported hexokinase and glucose 6-phosphate dehydrogenase and the use of the co-immobilized enzymes in the automated determination of glucose. Biochem J 147:593–603

Mousty C (2004) Sensors and biosensors based on clay-modified electrodes - new trends. Appl Clay Sci 27:159–177

Nam J-M, Thaxton CS, Mirkin CA (2003) Nanoparticle-based bio-bar codes for the ultrasensitive detection of proteins. Science 301:1884–1886

Navarre S, Choplin F, Bousbaa J, Bennetau B, Nony L, Aimé J-P (2001) Structural characterization of self-assembled monolayers of organosilanes chemically bonded onto silica wafers by dynamical force microscopy. Langmuir 17:4844–4850

Nelles J, Sendor D, Ebbers A, Petrat F, Wiggers H, Schulz C, Simon U (2007) Functionalization of silicon nanoparticles via hydrosilylation with 1-alkenes. Colloid Polym Sci 285: 729–736

Neves-Petersen MT, Snabe T, Klitgaard S, Duroux M, Petersen SB (2006) Photonic activation of disulfide bridges achieves oriented protein immobilization on biosensor surfaces. Protein Sci 15:343–351

Oh Y-H, Hong M-Y, Jin Z, Lee T, Han M-K, Park S, Kim H-S (2007) Chip-based analysis of SUMO (small ubiquitin-like modifier) conjugation to a target protein. Biosens Bioelectron 22:1260–1267

Ohtsuka K, Kajiki R, Waki M, Nojima T, Takenaka S (2004) Immobilization of sunflower trypsin inhibitor (SFTI-1) peptide onto a gold surface and analysis of its interaction with trypsin. Analyst 129:888–889

Ostuni E, Chapman RG, Holmlin RE, Takayama S, Whitesides GM (2001) A survey of structure-property relationships of surfaces that resist the adsorption of protein. Langmuir 17: 5605–5620

Ouali L, Stoll S, Pefferkorn E, Elaissari A, Lanet V, Pichot C, Mandrand B (1995) Coagulation of antibody-sensitized latexes in the presence of antigen. Polym Adv Tech 6:541–546

Perler FB, Adam E (2000) Protein splicing and its applications. Curr Opin Biotechnol 11:377–383

Pirri G, Chiari M, Damin F, Meo A (2006) Microarray glass slides coated with block copolymer brushes obtained by reversible addition chain-transfer polymerization. Anal Chem 78: 3118–3124

Pirrung MC, Davis JD, Odenbaugh AL (2000) Novel reagents and procedures for immobilization of DNA on glass microchips for primer extension. Langmuir 16:2185–2191

Pleschberger M, Saerens D, Weigert S, Sleytr UB, Muyldermans S, Sara M, Egelseer EM (2004) An S-layer heavy chain camel antibody fusion protein for generation of a nanopatterned sensing layer to detect the prostate-specific antigen by surface plasmon resonance technology. Bioconjugate Chem 15:664–671

Podyminogin MA, Lukhtanov EA, Reed MW (2001) Attachment of benzaldehyde-modified oligodeoxynucleotide probes to semicarbazide-coated glass. Nucleic Acids Res 29:5090–5098

Poyard S, Jaffrezic-Renault N, Martelet C, Cosnier S, Labbé P, Besombes J-L (1996) A new method for the controlled immobilization of enzyme in inorganic gels (laponite) for amperometric glucose biosensing. Sens Actuators B 33:44–49

Proudnikov D, Timofeev E, Mirzabekov A (1998) Immobilization of DNA in Polyacrylamide Gel for the Manufacture of DNA and DNA-Oligonucleotide Microchips. Anal Biochem 259: 34–41

Raddatz S, Mueller-Ibeler J, Kluge J, Wass L, Burdinski G, Havens JR, Onofrey TJ, Wang D, Schweitzer M (2002) Hydrazide oligonucleotides: new chemical modification for chip array attachment and conjugation. Nucleic Acids Res 30:4793–4802

Rao J, Yan L, Xu B, Whitesides GM (1999) Using surface plasmon resonance to study the binding of vancomycin and its dimer to self-assembled monolayers presenting D-ALA-D-Ala. J Am Chem Soc 121:2629–2630

Rendina I, Rea I, Rotiroti L, De Stefano L (2007) Porous silicon-based optical biosensors and biochips. Physica E 38:188–192

Revilla J, Elaissari A, Pichot C, Gallot B (1995) Surface functionalization of polystyrene latex-particles with a liposaccharide monomer. Polym Adv Technol 6:455–464

Rezwan K, Meier LP, Gauckler LJ (2005) Lysozyme and bovine serum albumin adsorption on uncoated silica and AlOOH-coated silica particles: the influence of positively and negatively charged oxide surface coatings. Biomaterials 26:4351–4357

Rickman DS, Herbert CJ, Aggerbeck LP (2003) Optimizing spotting solutions for increased reproducibility of cDNA microarrays. Nucl Acids Res 31:e109

Rodriguez MC, Rivas GA (2004) Assembly of glucose oxidase and different polyelectrolytes by means of electrostatic layer-by-layer adsorption on thiolated gold surface. Electroanalysis 16:1717–1722

Ronnebeck S, Carstensen J, Ottow S, Foll H (1999) Crystal orientation dependence of macropore growth in n-type silicon. Electrochem Solid-State Lett 2:126–128

Rosenzweig AC, Dooley DM (2006) Bioinorganic chemistry. Curr Opin Chem Biol 10:89–90

Rossi S, Lorenzo-Ferreira C, Battistoni J, Elaissari A, Pichot C, Delair T (2004) Polymer mediated peptide immobilization onto amino-containing N-isopropylacrylamide-styrene core-shell particles. Colloid Polym Sci 282:215–222

Rouquerol F, Rouquerol J, Sing K (1999) Solids, adsorption by powders and porous. Academic Press, New York

Rubina AY, Pan'kov SV, Dementieva EI, Pen'kov DN, Butygin AV, Vasiliskov VA, Chudinov AV, Mikheikin AL, Mikhailovich VM, Mirzabekov AD (2004) Hydrogel drop microchips with immobilized DNA: properties and methods for large-scale production. Anal Biochem 325:92–106

Saiki RK, Walsh PS, Levenson CH, Erlich HA (1989) Genetic analysis of amplified DNA with immobilized sequence-specific oligonucleotide probes. Proc Natl Acad Sci USA 86: 6230–6234

Salmain M, Jaouen G (2003) Side-chain selective and covalent labelling of proteins with transition organometallic complexes. Perspectives in biology. C R Chimie 6:249–258

Santos HA, Chirea M, Garcia-Morales V, Silva F, Manzanares JA, Kontturi K (2005) Electrochemical study of interfacial composite nanostructures: polyelectrolyte/gold nanoparticle multilayers assembled on phospholipid/dextran sulfate monolayers at a liquid-liquid interface. J Phys Chem B 109:20105–20114

Schlapak R, Pammer P, Armitage D, Zhu R, Hinterdorfer P, Vaupel M, Fruhwirth T, Howorka S (2006) Glass surfaces grafted with high-density poly(ethylene glycol) as substrates for DNA oligonucleotide microarrays. Langmuir 22:277–285

Schmid EL, Keller TA, Dienes Z, Vogel H (1997) Reversible oriented surface immobilization of functional proteins on oxide surfaces. Anal Chem 69:1979–1985

Schoning MJ, Kurowski A, Thust M, Kordos P, Schultze JW, Luth H (2000) Capacitive microsensors for biochemical sensing based on porous silicon technology. Sens Actuators B Chem 64:59–64

Shchepinov M, Case-Green S, Southern EM (1997) Steric factors influencing hybridisation of nucleic acids to oligonucleotide arrays. Nucleic Acids Res 25:1155–1161

Shen G, Anand MFG, Levicky R (2004) X-ray photoelectron spectroscopy and infrared spectroscopy study of maleimide-activated supports for immobilization of oligodeoxyribonucleotides. Nucl Acids Res 32:5973–5980

Shulga AA, Strikha VI, Soldatkin AP, Elskaya AV, Maupas H, Martelet C, Cléchet P (1993) Removing the influence of buffer concentration on the response of enzyme field-effect transistors by using additional membranes. Anal Chim Acta 278:233–236

Sieval AB, Linke R, Heij G, Meijer G, Zuilhof H, Sudholter EJR (2001) Amino-terminated organic monolayers on hydrogen-terminated silicon surfaces. Langmuir 17:7554–7559

Sindorf DW, Maciel GE (1981) Silicon-29 CP/MAS NMR studies of methylchlorosilane reactions on silica gel. J Am Chem Soc 103:4263–4265

Sindorf DW, Maciel GE (1983) Silicon-29 NMR study of dehydrated/rehydrated silica gel using cross polarization and magic-angle spinning. J Am Chem Soc 105:1487–1493

Sing KSW, Everett DH, Haul RAW, Moscou L, Pierotti RA, Rouquerol J, Siemieniewska T (1985) Reporting physisorption data for gas/solid systems with special reference to the determination of surface area and porosity (Recommendations 1984). Pure Appl Chem 57:603–619

Soellner MB, Dickson KA, Nilsson BL, Raines RT (2003) Site-specific protein immobilization by Staudinger ligation. J Am Chem Soc 125:11790–11791

Soldatkin AP, Elskaya AV, Shulga AA, Jdanova AS, Dzyadevich SV, Jaffrezic-Renault N, Martelet C, Clechet P (1994) Glucose-sensitive conductometric biosensor with additional nafion membrane - reduction of influence of buffer capacity on the sensor response and extension of its dynamic-range. Anal Chim Acta 288:197–203

Soldatkin AP, Elskaya AV, Shulga AA, Netchiporouk LI, Hendji AMN, Jaffrezic-Renault N, Martelet C (1993) Glucose-sensitive field-effect transistor with additional nafion membrane - reduction of influence of buffer capacity on the sensor response and extension of its dynamic-range. Anal Chim Acta 283:695–701

Solomon B, Koppel R, Schwartz F, Fleminger G (1990) Enzymic oxidation of monoclonal antibodies by soluble and immobilized bifunctional enzyme complexes. J Chromatogr A 510:321–329

Southern EM (1975) Detection of specific sequences among DNA fragments separated by gel electrophoresis. J Mol Biol 98:503–517

Staros JV, Wright RW, Swingle DM (1986) Enhancement by N-hydroxysulfosuccinimide of water-soluble carbodiimide-mediated coupling reactions. Anal Biochem 156:220–222

Steel AB, Herne TM, Tarlov MJ (1998) Electrochemical quantitation of DNA immobilized on gold. Anal Chem 70:4670–4677

Stoll S, Lanet V, Pefferkorn É (1993) Kinetics and modes of destabilization of antibody-coated polystyrene latices in the presence of antigen: reactivity of the system IgG-IgM. J Colloid Interface Sci 157:302–311

Streifer JA, Kim H, Nichols BM, Hamers RJ (2005) Covalent functionalization and biomolecular recognition properties of DNA-modified silicon nanowires. Nanotechnology 16: 1868–1873

Strother T, Cai W, Zhao X, Hamers RJ, Smith LM (2000a) Synthesis and characterization of DNA-modified silicon (111) surfaces. J Am Chem Soc 122:1205–1209

Strother T, Hamers RJ, Smith LM (2000b) Covalent attachment of oligodeoxyribonucleotides to amine-modified Si (001) surfaces. Nucleic Acids Res 28:3535–3541

Sun X, He P, Liu S, Ye J, Fang Y (1998) Immobilization of single-stranded deoxyribonucleic acid on gold electrode with self-assembled aminoethanethiol monolayer for DNA electrochemical sensor applications. Talanta 47:487–495

Sundaram PV (1979) The reversible immobilization of proteins on nylon activated through the formation of a substituted imidoester, and its unusual properties. Biochem J 183:445–451

Sundberg L, Porath J (1974) Preparation of adsorbents for biospecific affinity chromatography. Attachment of group-containing ligands to insoluble polymers by means of bifunctional oxiranes. J Chromatogr 90:87–98

Szabo K, Ngoc LH, Schneider P, Zeltner P, Kováts ESz (1984) Monofunctional (dimethylamino) silane as silylating agent. Helv Chim Acta 67:2128–2142

Tan L-P, Chen GYJ, Yao SQ (2004a) Expanding the scope of site-specific protein biotinylation strategies using small molecules. Bioorg Med Chem Lett 14:5735–5738

Tan LP, Lue RY, Chen GY, Yao SQ (2004b) Improving the intein-mediated, site-specific protein biotinylation strategies both in vitro and in vivo. Bioorg Med Chem Lett 14:6067–6070

Tao T, Maciel GE (2000) Reactivities of silicas with organometallic methylating agents. J Am Chem Soc 122:3118–3126

Templeton AC, Hostetler MJ, Kraft CT, Murray RW (1998) Reactivity of monolayer-protected gold cluster molecules: steric effects. J Am Chem Soc 120:1906–1911

Thenmozhi K, Narayanan SS (2007) Surface renewable sol-gel composite electrode derived from 3-aminopropyl trimethoxy silane with covalently immobilized thionin. Biosens Bioelectron 23:606–612

Thiel AJ, Frutos AG, Jordan CE, Corn RM, Smith LM (1997) In situ surface plasmon resonance imaging detection of DNA hybridization to oligonucleotide arrays on gold surfaces. Anal Chem 69:4948–4956

Thust M, Schoning MJ, Frohnhoff S, Arens-Fischer R, Kordos P, Luth H (1996) Porous silicon as a substrate material for potentiometric biosensors. Meas Sci Technol 7:26–29

Tinsley-Bown A, Smith RG, Hayward S, Anderson MH, Koker L, Green A, Torrens R, Wilkinson A-S, Perkins EA, Squirrell DJ, Nicklin S, Hutchinson A, Simons AJ, Cox TI (2005) Immunoassays in a porous silicon interferometric biosensor combined with sensitive signal processing. Physica Status Solidi A Appl Res 202:1347–1356

Toegl A, Kirchner R, Gauer C, Wixforth A (2003) Enhancing results of microarray hybridizations through microagitation. J Biomol Tech 14:197–204

Tripp CP, Hair ML (1992) Reaction of alkylchlorosilanes with silica at the solid/gas and solid/liquid interface. Langmuir 8:1961–1967

Tripp CP, Hair ML (1995) Reaction of methylsilanols with hydrated silica surfaces: the hydrolysis of trichloro-, dichloro-, and monochloromethylsilanes and the effects of curing. Langmuir 11:149–155

Trummer N, Adnyi N, Varadi M, Szendrö I (2001) Modification of the surface of integrated optical wave-guide sensors for immunosensor applications. Fresenius J Anal Chem 371: 21–24

Tsujii Y, Ohno K, Yamamoto S, Goto A, Fukuda T (2006) Structure and properties of high-density polymer brushes prepared by surface-initiated living radical polymerization. In: Jordan R (ed) Surface-initiated polymerization I. Springer Berlin, Heidelberg, pp 1–45

Uguzdogan E, Denkbas EB, Tuncel A (2002) RNA-sensitive N-isopropylacrylamide/vinylphenylboronic acid random copolymer. Macromol Biosci 2:214–222

Ulijn RV, Bibi N, Jayawarna V, Thornton PD, Todd SJ, Mart RJ, Smith AM, Gough JE (2007) Bioresponsive hydrogels. Mater Today 10:40–48

Ulman A (1996) Formation and structure of self-assembled monolayers. Chem Rev 96: 1533–1554

van Duffel B, Schoonheydt RA, Grim CPM, De Schryver FC (1999) Multilayered clay films: atomic force microscopy study and modeling. Langmuir 15:7520–7529

Veyret R, Delair T, Pichot C, Elaissari A (2005) Functionalized magnetic emulsion for genomic applications. Curr Org Chem 9:1099–1106

Vijayendran RA, Leckband DE (2001) A quantitative assessment of heterogeneity for surface-immobilized proteins. Anal Chem 73:471–480

Volle JN, Chambon G, Sayah A, Reymond C, Fasel N, Gijs MA (2003) Enhanced sensitivity detection of protein immobilization by fluorescent interference on oxidized silicon. Biosens Bioelectron 19:457–464

Wan K, Chovelon J-M, Jaffrezic-Renault N, Soldatkin AP (1999) Sensitive detection of pesticide using ENFET with enzymes immobilized by cross-linking and entrapment method. Sens Actuators B Chem 58:399–408

Wang H, Li J, Liu H, Liu Q, Mei Q, Wang Y, Zhu J, He N, Lu Z (2002) Label-free hybridization detection of a single nucleotide mismatch by immobilization of molecular beacons on an agarose film. Nucleic Acids Res 30:e61

Wang Y, Cai J, Rauscher H, Behm RJ, Goedel WA (2005) Maleimido-terminated self-assembled monolayers. Chemistry 11:3968–3978

Watzke A, Köhn M, Gutierrez-Rodriguez M, Wacker R, Schröder H, Breinbauer R, Kuhlmann J, Alexandrov K, Niemeyer CM, Goody RS, Waldmann H (2006) Site-selective protein immobilization by Staudinger ligation. Angew Chem Int Ed Engl 45:1408–1412

Wei J, Buriak JM, Siuzdak G (1999) Desorption-ionization mass spectrometry on porous silicon. Nature 399:243–246

Wei Y, Ning G, Hai-Qian Z, Jian-Guo W, Yi-Hong W, Wesche K-D (2004) Microarray preparation based on oxidation of agarose-gel and subsequent enzyme immunoassay. Sens Actuators B Chem 98:83–91

Weng S, Gu K, Hammond PW, Lohse P, Rise C, Wagner RW, Wright MC, Kuimelis RG (2002) Generating addressable protein microarrays with PROfusion covalent mRNA-protein fusion technology. Proteomics 2:48–57

Wheatley JB, Schmidt DE Jr (1999) Salt-induced immobilization of affinity ligands onto epoxide-activated supports. J Chromatogr A 849:1–12

Winssinger N, Damoiseaux R, Tully DC, Geierstanger BH, Burdick K, Harris JL (2004) PNA-encoded protease substrate microarrays. Chem Biol 11:1351–1360

Xiao SJ, Textor M, Spencer ND, Wieland M, Keller B, Sigrist H (1997) Immobilization of the cell-adhesive peptide Arg-Gly-Asp-Cys (RGDC) on titanium surfaces by covalent chemical attachment. J Mater Sci Mater Med 8:867–872

Yamanaka SA, Nishida F, Ellerby LM, Nishida CR, Dunn B, Valentine JS, Zink JI (1992) Enzymatic activity of glucose oxidase encapsulated in transparent glass by the sol-gel method. Chem Mater 4:495–497

Yang HJ, Cole CA, Monji N, Hoffman AS (1990) Preparation of a thermally phase-separating copolymer, poly(N-isopropylacrylamide-co-N-acryloxysuccinimide), with a controlled number of active esters per polymer-chain. J Polym Sci A 28:219–226

Yao T (1983) A chemically-modified enzyme membrane electrode as an amperometric glucose sensor. Anal Chim Acta 148:27–33

Yemini M, Xu P, Kaplan DL, Rishpon J (2006) Collagen-like peptide as a matrix for enzyme immobilization in electrochemical biosensors. Electroanalysis 18:2049–2054

Yershov G, Barsky V, Belgovskiy A, Kirillov E, Kreindlin E, Ivanov I, Parinov S, Guschin D, Drobishev A, Dubiley S, Mirzabekov A (1996) DNA analysis and diagnostics on oligonucleotide microchips. Proc Natl Acad Sci USA 93:4913–4918

Yeung C, Leckband D (1997) Molecular level characterization of microenvironmental influences on the properties of immobilized proteins. Langmuir 13:6746–6754

Yin J, Liu F, Li X, Walsh CT (2004) Labeling proteins with small molecules by site-specific posttranslational modification. J Am Chem Soc 126:7754–7755

Yoshida W, Castro RP, Jou J-D, Cohen Y (2001) Multilayer alkoxysilane silylation of oxide surfaces. Langmuir 17:5882–5888

Yousaf MN, Mrksich M (1999) Diels-Alder reaction for the selective immobilization of protein to electroactive self-assembled monolayers. J Am Chem Soc 121:4286–4287

Zammatteo N, Jeanmart L, Hamels S, Courtois S, Louette P, Hevesi L, Remacle J (2000) Comparison between different strategies of covalent attachment of DNA to glass surfaces to build DNA microarrays. Anal Biochem 280:143–150

Zazzera LA, Evans JF, Deruelle M, Tirrell M, Kessel CR, Mckeown P (1997) Bonding organic molecules to hydrogen-terminated silicon wafers. J Electrochem Soc 144:2184–2189

Zhang K, Diehl MR, Tirrell DA (2005) Artificial polypeptide scaffold for protein immobilization. J Am Chem Soc 127:10136–10137

Zhang W, Li G (2004) Third-generation biosensors based on the direct electron transfer of proteins. Anal Sci 20:603–609

Zhao B, Brittain WJ (2000) Polymer brushes: surface-immobilized macromolecules. Prog Polym Sci 25:677–710

Zhao X, Nampalli S, Serino AJ, Kumar S (2001) Immobilization of oligodeoxyribonucleotides with multiple anchors to microchips. Nucl Acids Res 29:955–959

Zhou Y, Andersson O, Lindberg P, Liedberg B (2004) Protein microarrays on carboxymethylated dextran hydrogels: immobilization, characterization and application. Microchim Acta 147: 21–30

Zhouxin Shen EPG, Gamez A, Apon JV, Fokin V, Greig M, Ventura M, Crowell JE, Blixt O, Paulson JC, Stevens RC, Finn MG, Siuzdak G (2004) A mass spectrometry plate reader: monitoring enzyme activity and inhibition with a desorption/ionization on silicon (DIOS) platform. ChemBioChem 5:921–927

Zhu H, Bilgin M, Bangham R, Hall D, Casamayor A, Bertone P, Lan N, Jansen R, Bidlingmaier S, Houfek T, Mitchell T, Miller P, Dean RA, Gerstein M, Snyder M (2001) Global analysis of protein activities using proteome chips. Science 293:2101–2105

Zhu ZQ, Zhu B, Zhang J, Zhu JZ, Fan CH (2007) Nanocrystalline silicon-based oligonucleotide chips. Biosens Bioelectron 22:2351–2355

Zubtsov DA, Ivanov SM, Rubina AY, Dementieva EI, Chechetkin VR, Zasedatelev AS (2006) Effect of mixing on reaction-diffusion kinetics for protein hydrogel-based microchips. J Biotechnol 122:16–27

Chapter 3
Analytical Tools for Biosensor Surface Chemical Characterization

Hans Jörg Mathieu

Abstract Development and improvement of surface-sensitive biosensors is made possible by the use of analytical tools. Spectroscopies available today are known for their extreme sensitivity and quantitation of functional chemical groups within the first nanometers of the surface. The first part of this chapter highlights Time-of-Flight Secondary Ion Mass Spectrometry (ToF-SIMS) based on the detection of ionic masses with sensitivity limits down to the femtomol limit and X-ray Photoelectron Spectroscopy (XPS) using emitted photoelectrons from the sample to determine the binding states of the electrons within their chemical environment. Quantitation of the detected molecular groups is possible within the first 5–10 monolayers of a device is introduced. Performance of the two spectroscopies is only available under ultra-high vacuum conditions, e.g., no direct in-situ analytical measurements are possible. In the second part, examples from the author's own laboratory are presented. The development of cell-based biosensor prototypes is shown making use of the strengths of the available analytical spectroscopies. The second example presents the application of glyco-engineering to surface sensitive devices. The third example illustrates the beneficial use of XPS to determine the different binding states of functional groups containing carbon, oxygen and/or nitrogen links during the preparation and development of immunosensors.

Keywords Biosensors · Time-of-Flight Secondary Ion Mass Spectrometry · X-ray Photoelectron Spectroscopy · Peptide adsorption · Imaging · Glycol-engineering · Cell guidance · MAD · MAD-Gal · Functional molecular groups · Lactose aryl diazirine · Neural Red uptake · Immunosensors · ELISA · Fluorescence reader · Allyldextran · Avidin · Streptavidin · Neutravidin · Glutaraldehyde · Polystyrene · Silicon wafer · Diamond substrate · Evanescent field

---

H.J. Mathieu
Materials Science Institute, École Polytechnique Fédérale de Lausanne (EPFL), CH-1015, Lausanne, Switzerland
e-mail: Hansjoerg.Mathieu@EPFL.ch

M. Zourob (ed.), *Recognition Receptors in Biosensors*,
DOI 10.1007/978-1-4419-0919-0_3, © Springer Science+Business Media, LLC 2010

Abbreviations

APC	Allophycocyanin
CDPGYIGSR	Olligopeptide
ELISA	Enzyme-linked immuno sorbent assay
ESCA	Electron spectroscopy for chemical analysis
Gal	Galactose
IgG	Immunoglobulin G
MAD	Maleimide
NR	Neural red
PS	Polystyrene
RF	Radio frequency
ROI	Region of interest
SIMS	Secondary ion mass spectrometry
TIRF	Total internal reflectance fluorescence
ToF-SIMS	Time-of-flight secondary ion mass spectrometry
XPS	X-ray photoelectron spectroscopy

3.1 Introduction

Biosensors are diagnostic tools used for the rapid detection of metabolites, drugs, hormones, antibodies, and antigens. Such a biosensor is a sensor coated with a biologically sensitive material. A wide variety of receptor molecules can be used including enzyme, antibody, affinity ligands, etc., that interact with a biomolecule present in the sample. The receptors immobilize molecules by adsorption, covalent cross-linking or entrapment. Control of biomolecular architecture on surfaces in conjunction with the retention of biological activities of biomolecules offers a wide range of applications in biosensing, cell guidance and molecular electronics including extension of semiconductor microlithography to surface bioengineering with appropriate molecular tools enabling their immobilization (Nicolini 1995; Biosensors and Bioelectronics 2006; Blitz and Gunk'ko 2006).

3.2 Surface Chemical Analysis

In this chapter, we present the analytical tools for the control of the chemistry on solid surfaces followed by examples of functionalized surfaces, such as silicon, diamond and polymers for the detection of the presence of functional groups of biomolecules. Highly surface sensitive spectroscopic methods such as Time-of-Flight Secondary Ion Mass Spectrometry (ToF-SIMS) and X-ray Photoelectron Spectroscopy (XPS) are introduced. Surface sensitivity is found to exhibit a sub-monolayer molecular sensitivity limit, a few nanometers down to the femtomol level combined with a lateral resolution in the sub-micrometer (ToF-SIMS) or micrometer range (XPS). As a draw-back one has to accept that analytical measurements are performed under ultra-high vacuum conditions (10^{-9} mbar range or below). The principle of surface analysis is illustrated in Fig. 3.1. A primary beam of ions in case of Tof-SIMS or X-rays in XPS is directed toward the sample surface, which emits a signal from a depth of 2–10 nm. Interaction of these primary particles within the first few monolayers leads to an emission of molecular fragments and/or electrons. Secondary ions (ToF-SIMS) or photo-electrons are detected and counted by the appropriate analyzer, which determines either the mass (ToF-SIMS) or the electron binding energy (XPS) of the emitted species. Electron energy of elements Li–U or ion mass spectra of all elements and isotopes (ToF-SIMS) are obtained. Maps of the lateral distribution of functional molecular groups or selected masses are available as well (Mathieu et al. 2003a, b). Depth profiles may be acquired by varying either the electron take-off angle (XPS) or by changing the primary particle beam energy.

3.2.1 Secondary Ion Mass Spectrometry (SIMS)

Static Time of Flight SIMS (ToF-SIMS) owes its interest to high surface sensitivity, an information depth limited to the top-surface and molecular imaging capabilities (Vickerman and Gilmore 2009). Selected examples given here review recent applications like characterization of engineered heterogeneous bioactive surfaces and biosensors with molecular imaging of cells and quantification of biomolecules in real biological samples. These examples illustrate advantages and possible limitations of ToF-SIMS in this field of research. Over the past years ToF-SIMS has demonstrated high efficiency in providing characterization of the elemental (including H and isotopes) and molecular composition of the top surface (<2 nm). Combining low surface damage and high mass range with high surface sensitivity

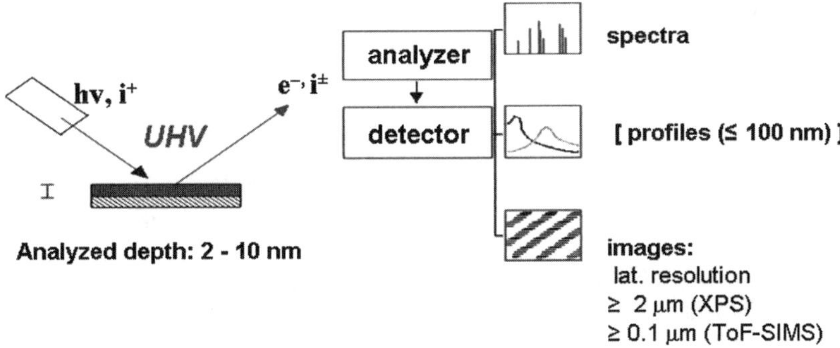

Fig. 3.1 Principle of surface analysis (in–out)

(10^7–10^{11} atoms/cm^2) and high mass resolution ($m/\Delta m > 5{,}000$ for $^{18}Si^+$), ToF-SIMS provides results no other methods can provide in spite of the lack of easy quantification of SIMS data. Moreover, high lateral resolution (<0.1 μm) ToF-SIMS images provide helpful information on the chemical mapping of the surface.

Basic concepts and examples of the SIMS technique are described in detail in various seven books and review papers and are not detailed here (see Mathieu et al. 2003b; Briggs and Seah 1992; Benninghoven et al. 1987; Benninghoven 1994; Vickerman et al. 1996; Bertrand and Weng 1996; Léonard and Mathieu 1999). Briefly, the basic concept of secondary ion mass spectrometry (SIMS) consists of bombarding the sample surface using a 10–30 keV energy primary ion beam. Primary ions penetrate the surface and transfer energy and momentum to surface atoms further engaged in collisions with neighboring atoms, creating a perturbed surface region not limited to the primary ion impact site (compare Fig. 3.2.). Less than 5% of the sputtered particles are ionized (Sigmund 1981). The emitted sputtered ionized species are separated as a function of the ratio mass over electric charge, *m*/*z*. The positively and negatively charged species are detected in two separate acquisitions.

In studies with ToF-SIMS under *static* conditions, the total area impacted by primary ions is limited to a small part of the surface area in order to minimize the surface modification inherent to the technique. For spectra acquisition, an ion dose well below 10^{13} ions/cm^2, preferably in the range of 10^{11}–10^{12} ions/cm^2 considering a monolayer surface density of 10^{15} particles/cm^2 is applied. Applications for which a high current density is used, e.g., depth profiling, experiments are referenced as *dynamic* SIMS. In that case – not shown here – the surface may be etched rapidly during analysis to monitor changes of elemental composition (Odom 1994; Schueler 1992).

Despite the lack in complete understanding of the fundamentals of the technique, static SIMS has been successfully applied in various fields from semiconductors to polymers (Benninghoven et al. 1997; Gillen et al. 1998). The sensitivity is limited by the number of secondary ions that actually reach the detector. The ToF-analyzer used in static SIMS has improved mass transmission, independent of the ion mass, mass resolution, mass range and sensitivity.

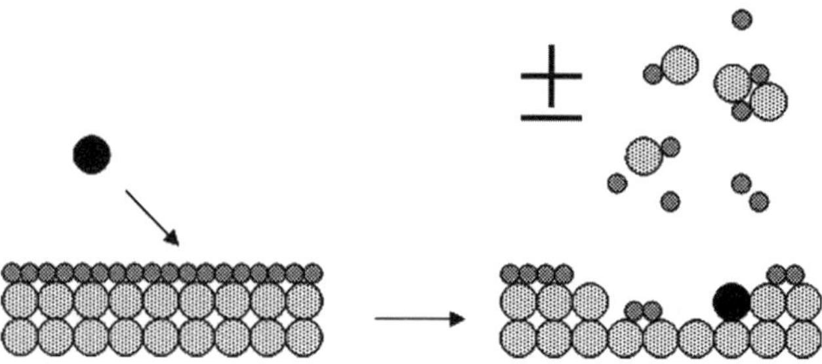

Fig. 3.2 Principle of secondary ion emission

In the case of ToF-SIMS, the sample surface is bombarded using very short (ns range) pulsed ion beams (compare Fig. 3.3.). Between two consecutive pulses (repetition rate in the 5–10 kHz range), secondary ions are extracted and are electrostatically accelerated into a field free drift region with a kinetic energy E_{kin} of

$$zV_o = \frac{m}{2}v^2 \tag{3.1}$$

where z is the ion charge, V_0 the accelerating potential between sample (S) and immersion lens (IL), m the ion mass, and v the ion velocity. Lighter ions will have higher velocities and hence will reach the detector at the end of the drift region earlier than the heavier masses (Briggs and Seah 1990). As illustrated in Fig. 3.3. the flight time t of the ion, m/z can be deduced from the following relation:

$$t = \frac{L_0}{v} = \frac{L}{\sqrt{2zV_0}}\sqrt{m} \tag{3.2}$$

where L_0 is the effective length of the spectrometer between the sample S and the detector D. In the case of insulating materials, charge neutralization is achieved using a pulsed low energy electron beam. The ToF-SIMS instruments include imaging capabilities with liquid metal ion sources (Ga^+, Bi^+) with a beam diameter in the range of 200 nm.

ToF-SIMS is more sensitive (limit $<10^9$ atoms/cm^2) than X-ray Photoelectron Spectroscopy (XPS) (10^{12} atoms/cm^2) (Mathieu 2001; Briggs and Seah 1990). Under primary bombardment with a focused ion beam, the solid surface emits secondary particles. The bombardment with primary ions such as Ga^+, Bi^+ or molecular ion sources like Au_3^+ (Vickerman and Gilmore 2009) or C_{60} (Justes et al. 1997) provokes the emission of neutral, positively or negatively charged fragments and clusters. Only a fraction of the first monolayer of the surface layer is perturbed depending on the flux of the primary ion beam, which is kept well below 10^{12} particles/cm^2. This number corresponds to 1‰ of the number of particles of the first monolayer. Once this limit is crossed, degradation of the surface takes

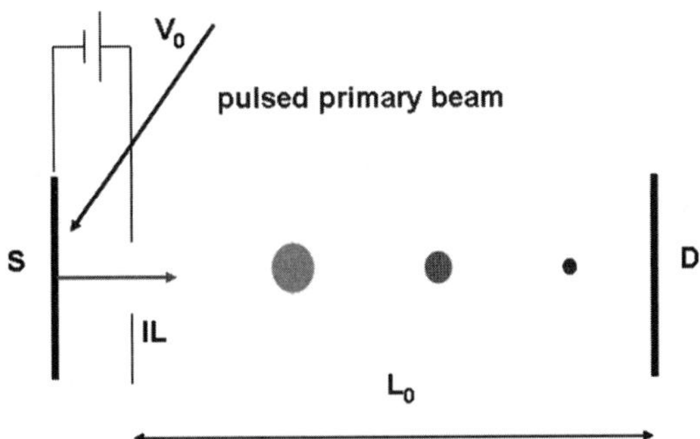

Fig. 3.3 ToF-SIMS analyzer principle: S sample, V_{acc} acceleration voltage, I_p pulsed ion beam, L_0 flight length, IL immersion lens, and D detector

place. During further bombardment one speaks of the *dynamic mode*, which is when the top-surface will be destroyed, leading to a depth-profile-type of information (Mathieu et al. 2003b; Vickerman and Gilmore 2009).

The intensity $I_A^{\pm}$ for emitted positively or negatively charged secondary ions measured for a given target is described by the following equation (Mathieu et al. 2003b; Vickerman and Gilmore 2009).

$$I_A^{\pm} = I_p Y_{tot} \gamma_{A(M)}^{\pm} f_A c_A \eta_S \quad (3.3)$$

with

$$Y_M^{\pm} = Y_{tot} \gamma_{A(M)}^{\pm} \quad (3.4)$$

Here I_p is the primary ion current, Y_{tot} the total neutral sputter yield of species A in the matrix M, $\gamma_{A(M)}^{\pm}$ the ionization probability of species A in the matrix M, f_A the isotopic abundance of element A, c_A the atomic concentration and η_S an instrumental constant concerning the analyzer collection efficiency (transmission). $\gamma_A^{\pm}$ may vary over several orders of magnitude. All elemental fragments (H–U) of the molecules and their isotopes are detectable. Due to its high sensitivity, surface contamination may influence the ionization probability, $\gamma_{A(M)}^{\pm}$, strongly. Furthermore, $\gamma_A^{\pm}$ is very dependent upon matrix effects that influence strongly the number of emitted ions compared to emitted neutrals. This ion/neutral ratio is typically 10^{-4} or smaller for polymers and biomaterials (Vickerman and Gilmore 2009). The influence of the matrix on the ionization, the $\gamma_{A(M)}^{\pm}$ factor in (3.3), represents the severest limitation of this technique for quantitative analysis of both inorganic and organic materials.

To illustrate the importance of the mass resolution for the discrimination of surface impurities, from the sample species, Fig. 3.4 shows a spectrum acquired with a ToF spectrometer, a relative mass resolution of $m/\Delta m = 6{,}900$. It shows a typical fragment of a molecule fragment $C_4H_2NO_2^-$ of maleimide at 96.009 amu. Due to this mass resolution one can clearly distinguish between the maleimide the sulfate molecule SO_4^- at 95.952 amu.

3.2.2 X-Ray Photoelectron Spectroscopy

X-Ray Photoelectron Spectroscopy (XPS), originally called Electron Spectroscopy for Chemical Analysis (ESCA) by the nobel-prize winner Kai M. Siegbahn uses as primary energy source a monochromatic X-Ray radiation that causes a photo-ionization of the atoms of the sample surface under examination (see Fig. 3.5). The incoming X-ray with an energy hv provokes the emission of a photoelectron e^-_{photo}. θ_i and θ are the angles of incidence and take-off with respect to the surface normal, respectively. Each emitted photo-electrons has a very distinctive binding energy depending on the emitting molecule. Analysis of the binding energy

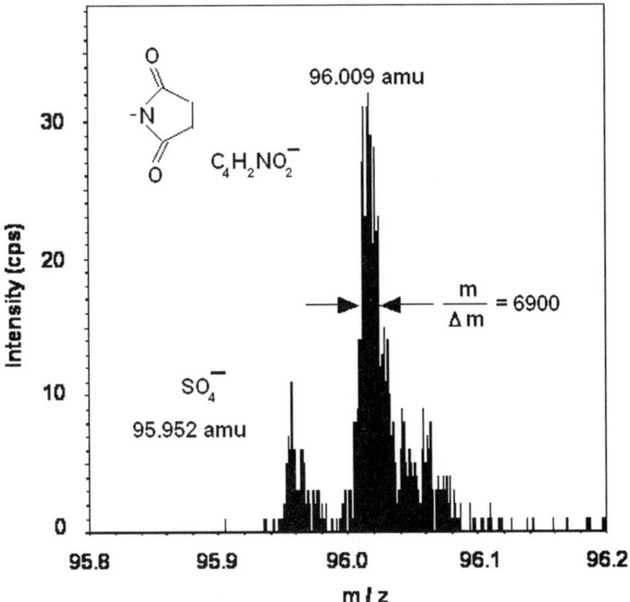

Fig. 3.4 Time-of-Flight mass spectrum of negative ions of $C_4H_2NO_2^-$ at 96.009 amu (Maleimide) which is clearly separated from SO_4^- at 95.952 amu using a mass resolution of $m/\Delta m = 6{,}900$

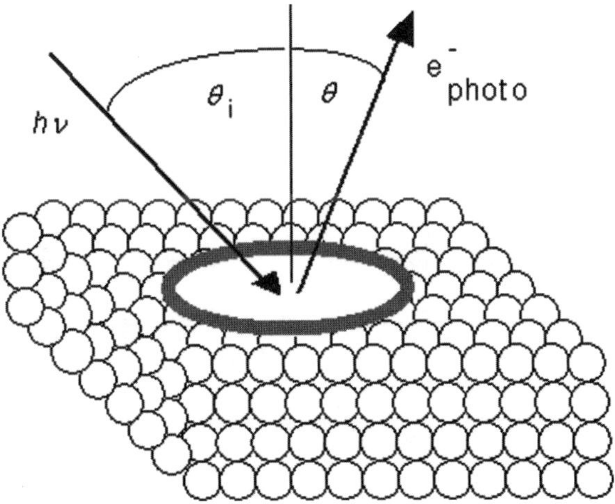

Fig. 3.5 X-ray Photoelectron Spectroscopy (XPS) principle: A primary X-ray beam with energy $h\nu$ is directed towards the sample surface under an angle i with respect to surface normal. After interaction of the X-ray with the electrons of the surface atoms photo-electrons are emitted under a take-off angle θ with respect to surface normal

provides identification of the elements and their state of oxidation. By scanning the surface photoelectron images may be obtained, which allows determining the surface composition within 3–10 nm.

XPS provides qualitative and quantitative information of each atom and molecule present at the sample surface (Mathieu et al. 2003b; Vickerman and Gilmore 2009; Mathieu 2001; Briggs and Seah 1990). All elements at a concentration greater than 0.1% of an atomic layer (10^{13} atoms/cm^2), except for the elements H and He are detectable. XPS-analysis uses monochromatic X-Rays (Al$K_{\alpha 1}$ with source energy of 1486.6 eV or non-monochromatic Mg$K_{\alpha 12}$ with energy of 1256.6 eV). Monochromatic X-rays are used for optimum chemical state sensitivity (energy resolution), lower background radiation, and lower radiation damage caused by X-rays, and stray electron beam irradiation during XPS analysis, but may lead to more severe charging problems during sample charge-up.

Emission of photo-electrons with a well-defined binding energy, E_B, during X-ray irradiation is described in Fig. 3.6. During the excitation process photo-ionization

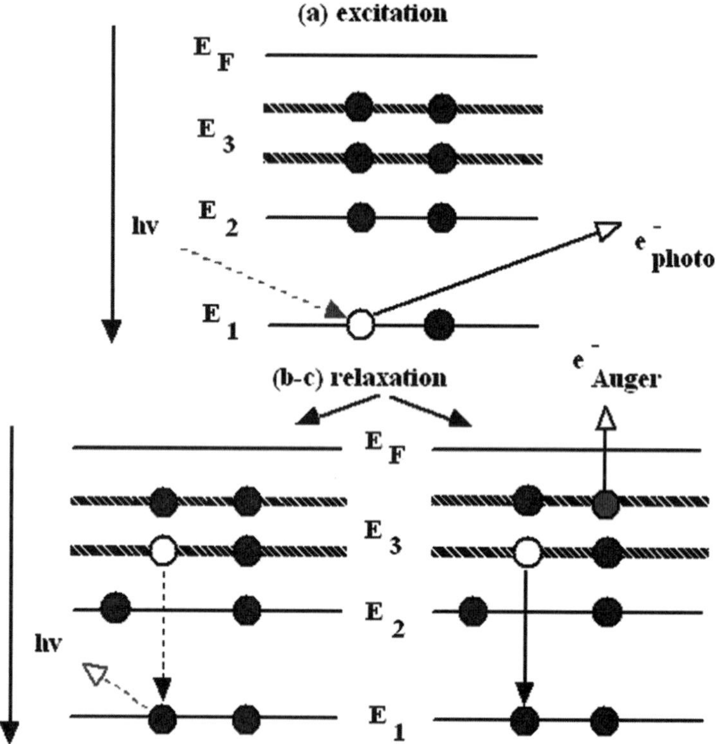

Fig. 3.6 XPS processes: during X-ray bombardment with an energy of *hv* a photo-electron is liberated with a binding energy corresponding to the energy level with binding energy E1 (**a**). During the relaxation process (**b** or **c**) either fluorescence (emission of secondary X-ray) or an Auger electron is emitted with a kinetic energy corresponding to the difference of binding energy of the levels three levels involved

takes place (a), which is followed by a relaxation process via the emission of a secondary X-ray fluorescence (b) or an Auger-electron (c). During relaxation, the vacant electron site of the atom is occupied by an electron from a higher energy level, which results in a fluorescent energy emission $h\nu$. In step (c), after filling of the vacant site, the energy difference of the two energy levels involved is transferred to a third electron with a specific elemental dependant kinetic energy, the so-called Auger electron. This relaxation process is the preferred process for light elements like oxygen, carbon or nitrogen.

Application of the law of conservation of energy during the photoemission allows determining the principal equation for XPS (Mathieu et al. 2003b; Vickerman and Gilmore 2009; Briggs and Seah 1990):

$$E_{\mathrm{B}} = h\nu - E_{\mathrm{kin}} - \Phi_{\mathrm{A}} \tag{3.5}$$

with E_{B}: binding energy

E_{kin}: kinetic energy of the photoelectron measured by the spectrometer proportional to the Fermi energy level

Φ_{A}: work function of analyzer A defined by

$$\Phi_{\mathrm{A}} = h\nu - E_{\mathrm{B}} - E_{\mathrm{kin}} \tag{3.6}$$

with Φ_{A}, a constant, determined by calibration of the analyzer (ISO International Standard 15472 2001).

A concentric hemispheric analyzer (CHA) is usually used, which is illustrated in Fig. 3.7. During analysis the substrate is irradiated by the X-rays with an energy $h\nu$. It is the lens I which determines the accepted area for analysis in conjunction with the diaphragm. Take-off angle $\theta_{\mathrm{s}} = 90° - \theta$ is complimentary to the take-off angle θ. Lens II retards the energy of the photo-electrons to pass energy E_{pass}, a value selected by the operator. E_{pass} determines the energy resolution of the analyzer. The photo-electrons are filtered between the two concentric plates of the analyzer applying an electric potential to focus the electrons at the detector. The intensity of the electrons is a quantitative measure sent to the computer for further data treatment.

The measured XPS intensity, I_{A}, is used for quantification. The relation between I_{A} and the atomic concentration c_{A} of an element or chemical component A at depth z is (Mathieu et al. 2003b; Vickerman and Gilmore 2009; Briggs and Seah 1990):

$$I_{\mathrm{A}} = kI_{\mathrm{s}}S_{\mathrm{A}} \int_0^{\Lambda} c_{\mathrm{A}}(z) \exp\left(-\frac{z}{\lambda \cos \theta}\right) \mathrm{d}z \tag{3.7}$$

with k as an instrument variable, I_{s} the primary electron beam current, S_{A} the elemental sensitivity and λ the inelastic mean free path of the photoelectron trajectory multiplied by $\cos\theta$ where θ is the take-off angle of the emitted electron

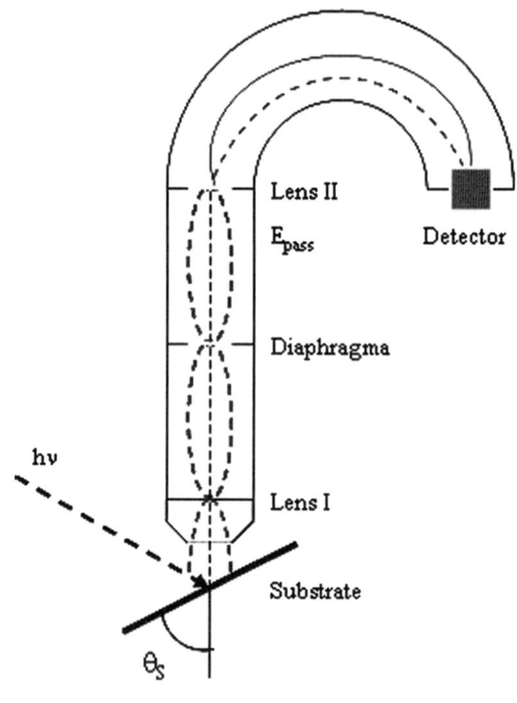

Fig. 3.7 Concentric hemispherical analyzer (CHA) used for XPS analysis: During irradiation of the substrate with $h\nu$ and emission under an angle θ_s (with respect to the surface) $= 90° - \theta$ lens I and the diaphragm determine the area accepted by the analyzer. The potential of lens II is incremented by finite steps to retard the energy of the electrons to a value E_{pass} selected by the operator. It is the pass energy that determines the energy resolution of the measured spectra. Energy filtering takes place within the hemispherical part by applying an appropriate potential before the electrons reach the detector. The number of electrons reaching the analyzer is counted and converted into a signal that is displayed on the system computer

with respect to the surface normal. Due to the exponential term in (3.7) a reasonable upper limit for integration is the escape depth Λ

$$\Lambda = 3\lambda \cos\theta \tag{3.8}$$

Typical values for Λ for polymers are between 3 and 10 nm indicating the shallow information depth of XPS (Mathieu et al. 2003b; ISO International Standard 15472 2001; Powell and Jablonski 1999). Quantification to determine the atomic concentration of element A, c_A, is performed by the application of the simple formula

$$c_A = \frac{I_A/S_A}{\sum_{i=1}^{n} I_i/S_i} \tag{3.9}$$

applying elemental sensitivity factors S_i ($i = 1, 2, ...n$) found in the literature (Mathieu 2001; Briggs and Seah 1990) by summing over the number of elements, i, taken into account. The sum of the molar fractions taken into account is one. As for the polymers accuracy of a few percent this is typically obtained by use of (3.9).

From (3.8), one can see that measurements at different take-off angles allow probing sample composition at different depths. Thus such angle resolved XPS (ARXPS) is an elegant way of obtaining a depth profile in a non-destructive manner and to detect surface contaminations at less than 2 nm.

Analyzing polymers and biosensors with organic molecules, charging effects and possible degradation of samples have to be taken into account. Emitted photoelectrons carry a negative charge and may lead to a positive charge build-up. This effect can be compensated by supplying the sample surface with low energy electrons for charge neutralization. The reader is referred to for complementary information (Frydman et al. 1997; Buchwalter and Czornyj 1990; Chaney and Barth 1987; Storp 1985; Coullerez et al. 1999; Beamson and Briggs 1992 Clark and Brennan 1986).

Imaging is possible with a lateral resolution limit of a few microns for the state-of-the-art spectrometers. As for ToF-SIMS, all measurements are carried out under ultra high vacuum conditions (UHV). A fast entry lock is usually available for a transfer of a sample within minutes from atmospheric pressure to 10^{-8} mbar. A comparison of the two spectroscopic methods described above is given in Table 3.1 indicating the important benchmarks of either one.

3.3 Examples

In the second part of this chapter practical examples from our laboratory are presented in which the two surface spectroscopic methods were applied for the development of prototype biosensors.

3.4 Cell-Based Biosensor Prototypes

The development of silicon-based devices incorporating bioactive molecules is a subject of longstanding interest in the fields of biosensors and bioelectronics (Kleinfeld et al. 1988; Fodor et al. 1991). The use of semiconductor fabrication technology particularly facilitated the proliferation of this trend by providing, inter alia, capabilities to realize patterned deposition of biomolecules in miniature device

Table 3.1 Comparison of XPS and ToF-SIMS performances

Technique	Information	Λ	Surface sensitivity
XPS	• Atomic concentration • Functional groups	3–10 nm	10^{12}–10^{13} at/cm^2
ToF-SIMS	• Specific molecular fragments • Chemical structure	<2 nm	10^{7}–10^{11} at/cm^2

configurations and other microsystems (Fodor et al. 1991; Connolly 1994). In such applications, the outermost surface forms the primary interface through which the device can associate with or interrogate its surrounding environment. Therefore, the uppermost layers govern the surface phenomena that may ensue upon device exposure to various envisaged applications (Flounders et al. 1997; Hagenhoff 1995). In the event that the application ultimately requires an association of the synthetic system with a biological element, it is necessary that its surface is appropriately tailored to accommodate the foreign species. This may be achieved by a variety of surface modification techniques that alter device surface properties such that the desired surface/foreign-species interaction is attained (Connolly 1994; Benninghoven et al. 1993).

If living cells are to contact the surface, even the smallest quantities of contamination material could pose an appreciable cytotoxic hazard (Göpel 1995). Therefore, a careful evaluation of the surface composition and properties of the device is critical towards successful development of such systems. The current review presents the results towards the realization of a neural cell-based biochip sensing system, whose surface has been engineered to possess two key features: the first feature is a surface strategy for the biopatterning or spatial control of tissue deposition and cell adhesion at the micron level. Thin films of amorphous Teflon (Teflon AF®) were employed for this task. The second feature is to optimize the cell/electronics interface, i.e., promote and enhance coupling between the cell and the micro-electrode surface. The primary candidates were the laminin-derived synthetic oligopeptides that have been previously demonstrated to enhance cell attachment and outgrowth via highly specific receptor-ligand interactions (Black 1992; Graf et al. 1987). Here, we deal with the characterization and evaluation of a biochip surface engineered to exhibit the dual surface property of inhibiting cell attachment everywhere else, while inducing it on the micro-electrode regions (Makohliso et al. 1999). More specifically, we constructed a biochip surface consisting of an array of gold micro-electrodes isolated from each other by a thin film of amorphous Teflon®. As a final step, cell adhesion promoting molecules were grafted onto gold micro-electrode surfaces. The synthetic nanopeptide CDPGYIGSR, whose sequence resembles a cell-attachment promoting subdomain of laminin, was utilized, and it was crucial that this biomolecule binds only to the gold electrodes and nowhere else. The intention was to immobilize this biomolecule directly onto gold via the thiol side group of its terminal cysteine (C) residue (Makohliso et al. 1999). This immobilization strategy is an advantageous simplification over other oligopeptide surface grafting protocols (Ranieri et al. 1995; Massia and Hubbell 1991), as it requires no surface pre-treatment and coupling agents. The biochip fabrication process consists of several steps that included the use of polymeric resins for substrate photolithographic patterning, and fluorosilane compounds for improving fluoropolymer film adhesion.

Some of the materials used for biochip construction (e.g., gold) are prone to surface contamination (Bain et al. 1989; Troughton et al. 1988) and some have been implicated in effecting cytotoxicity (e.g., aluminum) (Mueller and Bruinink 1994; Xie et al. 1996). It is, therefore, important to carefully assess the resulting composite surface. ToF-SIMS proved to be the ideal tool for probing the surface chemical details of this system. Its high mass resolution capabilities allow to distinguish between similar molecular structures (e.g., isotopes and isobar ions) (Benninghoven et al. 1993; Mantus et al. 1993; Brummel et al. 1994; Bertrand and Weng 1996) thus making it particularly suitable for analyzing the complex molecular environment encountered in the present application. The ToF-SIMS imaging feature provides a capability of mapping the distribution of surface species with micron lateral resolution (Brummel et al. 1994; Bertrand and Weng 1996).

The fabrication process is summarized in Fig. 3.8 (Makohliso et al. 1999). Using a positive contrast photoresist, the gold electrode array pattern (including line contacts) was photolithographically imprinted onto a thermally oxidized silicon [100] wafer. The next step involved deposition of a thin layer of amorphous Teflon®, and photolithographically imprint a pattern such that only the electrode surfaces are exposed.

The geometric arrangement of the biochip surface structures consists of a square gold patch (4 mm^2) that could serve as the reference electrode in biochip electrical measurements, and is henceforth termed the ground pad. An array of 16 micro-electrodes is located at the center. The dimensions of individual micro-electrodes fabricated ranged from squares of 20 to 60 μm in length, with an inter-separation distance of either 100 or 400 μm. The typical thickness of the amorphous Teflon® layer was about 0.4 to 0.5 μm, meaning that the micro-electrodes were receded into grooves of about 0.5 μm in depth. The final step was to deposit the oligopeptides on the gold surfaces of the biochip. This was accomplished as described above, by

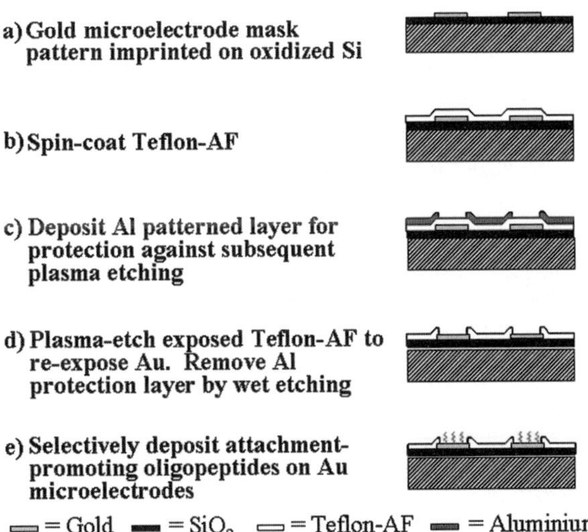

Fig. 3.8 Schematic representation of the biochip micro-fabrication process. Features not drawn to scale, (Reprinted with permission from Makohliso et al. (1999), copyright 2008 American Chemical Society)

covering the biochip surfaces with an oligopeptide solution, and allowing the reaction to progress overnight. The surface would then be removed from the reaction well, thoroughly rinsed with distilled, de-ionized water and the blown dry with argon. Surface analysis was carried out immediately thereafter.

ToF-SIMS spectra were obtained by applying a biased ±3 kV with respect to the grounded extraction electrode, for positive and negative mode analysis respectively using a Ga^+ source operated at 15 keV and giving a DC ion beam current of 850 pA pulsed at a frequency of 5 kHz. The lateral distribution of surface ion species on the biochips was obtained in the imaging mode. The ion images are obtained from the (x, y) raster position of the primary ion beam over the sample surface. The region-of-interest (ROI) analysis facility allows one to selectively obtain spectral data from user-defined areas of interest of the full image. We carried out the ROI analysis in "full raster" mode; i.e., the full raster images were acquired, but only the spectra from the predefined regions are displayed.

The chemical formula and molecular weight of the oligopeptide reference sample CDPGYIGSR-NH_2 is $C_{40}H_{64}O_{13}N_{13}S$ and 966 amu, respectively. Its spectrum of the whole peptide is shown in Fig. 3.9.

The ground pads of the devices just before and after peptide depositions were compared. Distribution maps are shown in Fig. 3.10 illustrating typical positive ion ToF-SIMS images of a micro-electrode: $^{197}Au^+$-ion distribution before and after peptide modification, respectively (a–b), and (c) the peptide-derived ion after modification ($C_4H_8N^+$).

Spectra of the micro-electrode pads are shown in Fig. 3.11 in the region-of-interest (ROI) for three areas: (a) electrode pad borders with the $C_4H_8H^+$ peptide-derived ion, (b) fragment of Teflon AF®, and (c) the border. This analysis revealed peptide presence only on the micro-electrodes (as intended), but not on the Teflon AF® or elsewhere, and correspondingly the intensity of the gold peak had decreased on areas where the peptide could be observed, in agreement with earlier results on the ground pads.

It was shown that the objective of this study to investigate the surface chemical properties of a micro structured and multi-molecular biochip surface, comprising

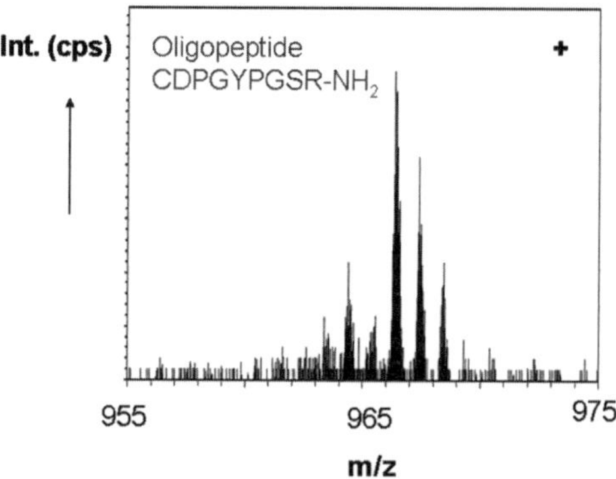

Fig. 3.9 Positive ion spectrum of Oligopeptide CDPFYPGSR-NH_2. (Reprinted with permission from Makohliso et al. (1999), copyright 2008, American Chemical Society)

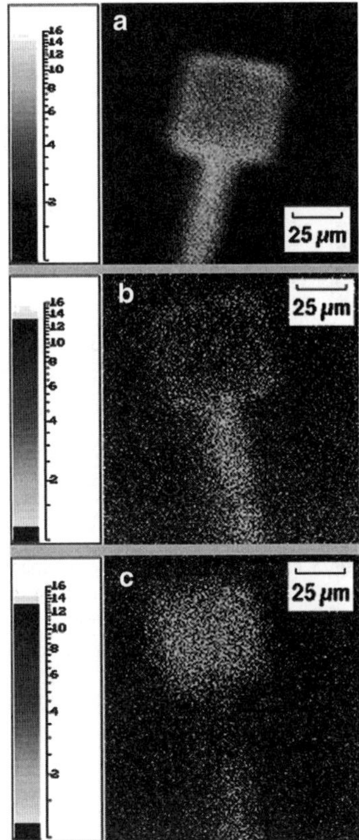

Fig. 3.10 Positive ToF-SIMS imaging of a defective microelectrode. The gold ion (196.600–197.400 amu) is imaged prior to surface modification with peptide (**a**), and after peptide modification (**b**), (**c**) distribution of a peptide-derived ion ($C_4H_8N^+$; 70.016–70.144 amu). (Reprinted with permission from Makohliso et al. (1999), copyright 2008, American Chemical Society)

an array of gold microelectrodes incorporating a cell-attachment promoting oligopeptide (H-CDPGYIGSRNH_2), and a thin film of amorphous Teflon® in between the electrodes and everywhere else, was attained. It was demonstrated that the multistep biochip micro fabrication process introduced negligible or no surface contamination that could hinder surface incorporation of the oligopeptide or effect cytotoxicity. The protocol employed for immobilizing the biomolecule was highly specific to gold surfaces, and did not lead to inadvertent adsorption of the peptide on the amorphous Teflon AF®. Altogether, these results demonstrate a role for a surface engineering perspective in the development of biochip and biosensor technology. The use of DNA molecules and their subunits for purposes of realizing novel supramolecular materials and nanoelectronic applications has been gaining attention lately (Braun et al. 1998; Deming 1997; Bethell and Schiffrin 1996). Within that context, the future possibility of utilizing the oligopeptide template and the electrically addressable micro-electrode sites is envisaged as a foundation for engineering novel self assembling or supramolecular biochip and biosensor technology.

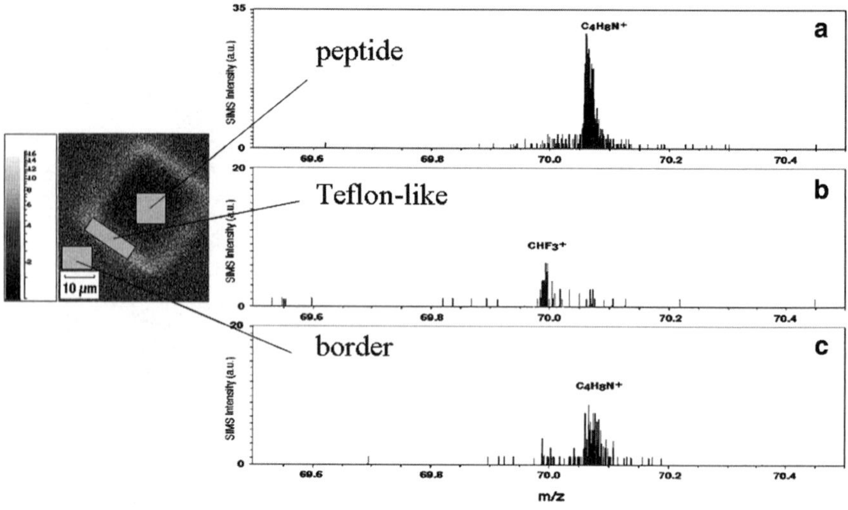

Fig. 3.11 Positive ion ToF-SIMS spectra of biochip surface obtained from the microelectrode and its surrounding regions, via the "region-of-interest" (ROI) analysis. (**a**) spectrum of the central region of interest (ROI) the micro-electrode pad (ROI #1). (**b**) spectrum of the amorphous teflon region (ROI #2). spectrum of the region at the border of the gold micro-electrode and amorphous teflon (ROI #3). Reprinted with permission from Makohliso et al. (1999), copyright 2008, American Chemical Society

3.4.1 Glyco-Engineering

Control of biomolecular architecture on surfaces in conjunction with the retention of biological activities of biomolecules offers a wide range of applications in biosensing, cell guidance and molecular electronics (Nicolini 1995). It will be demonstrated that semiconductor microlithography can be extended to surface bioengineering with appropriate molecular tools enabling light addressable biomolecule immobilization. The aim of the results addressed here (Sigrist et al. 1995; Clémence et al. 1995) is to provide experimental evidence for the controlled binding of the bifunctional MAD reagent shown in Fig. 3.12 to material surfaces by light-dependent reactions. MAD carries a diazirine function that will be lost during light activation (350 nm), leading to carbene-mediated binding to solid supports, and a maleimide function for thermochemical reactions with thiolated reagents (Collioud et al. 1993; Gao et al. 1997).

This example presented here applies ToF-SIMS and XPS to characterize the grafting of the reagent *N*-(*m*-(3-(trifluoromethyl) diazirine-3-yl) phenyl)-4-maleimidobutyramide (MAD) to various substrates: silicon, silicon nitride and diamond. Here I present only the results of the grafting of MAD onto diamond samples. This reagent is stable and not degraded during the 350 nm illumination. Indeed, biological tests showed that MAD activation by light does not impair biological

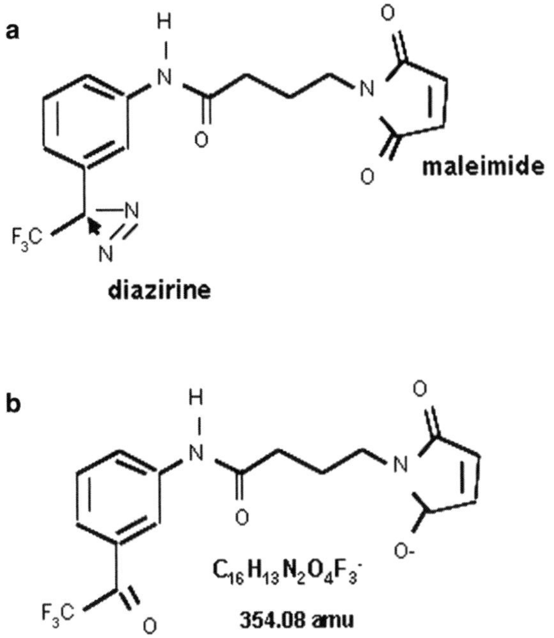

Fig. 3.12 Chemical structure of (**a**) MAD and (**b**) fragment after illumination

activities (Léonard et al. 1998a; Chevolot et al. 1999). Photo-activation leads to carbene-mediated grafting to solid supports. XPS determines atomic constituents and chemical shifts, ToF-SIMS identifies molecular peaks and characteristic fragments of the photo-immobilized molecule. Interpretation of surface analysis data suggests that diamond is the substrate with the highest MAD grafting efficiency and that the formation of C–O bonds upon diazirine photo-activation is thus involved (Léonard et al. 1998a, b; Chevolot et al. 1999).

Surface grafting is studied using XPS to providing both qualitative and quantitative analysis of a surface (Briggs and Seah 1990; Beamson and Briggs 1992). The photo-bonding of the molecule is characterized by comparing the influence of molecule deposition, intensive washing and grafting by illumination. The carbene-mediated grafting is versatile and leads to binding to different substrates (Gao et al. 1997). Here, only the grafting of the basic reagent (MAD) to diamond is represented (Léonard et al. 1998a, b; Chevolot et al. 1999).

Pristine samples were washed for 5 min ultrasonically in hexane and then in ethanol. Then they were dried for 2 h at room temperature under a vacuum. A MAD solution (0.25 mM in ethanol, 10 μl) was deposited as a droplet released by a syringe. Samples were dried for 2 h at room temperature under a vacuum. For photo-bonding, the samples were irradiated for 20 min using the Stratalinker 350 nm light source with an irradiance of 0.95 mW cm^{-2} and washed again (Gao et al. 1997). The investigated samples were referenced as follows: (a) after MAD deposition, (b) sample a illuminated, (c) (sample a not illuminated and then washed) and (d) (sample b washed). The resulting ToF-SIMS negative ion spectra in the mass range 340–400 amu are illustrated in Fig. 3.13 to demonstrate the efficiency of deposition and illumination. Inspection reveals that a peak at *m*/*z* 366.09 is only observed for the sample on which the molecules are deposited (corresponding to $(M + H)^-$ with $M = 366.09$ amu) while a peak at *m*/*z* 354.08 is detected for illuminated samples (before and after washing). Moreover, both peaks are not detected when no photo-illumination was performed before washing (c). The peak at *m*/*z* 354.07 confers with a fragment $[M - 2N + O]^-$ of 354.07 *m*/*z* and indicates the reaction with oxygen-containing sites suggesting the chemical interaction of the photo-generated carbene with the surface-bonded oxygen (compare Fig. 3.12a, b (Léonard and Mathieu 1999; Léonard et al. 1998a). The peak at 350.98

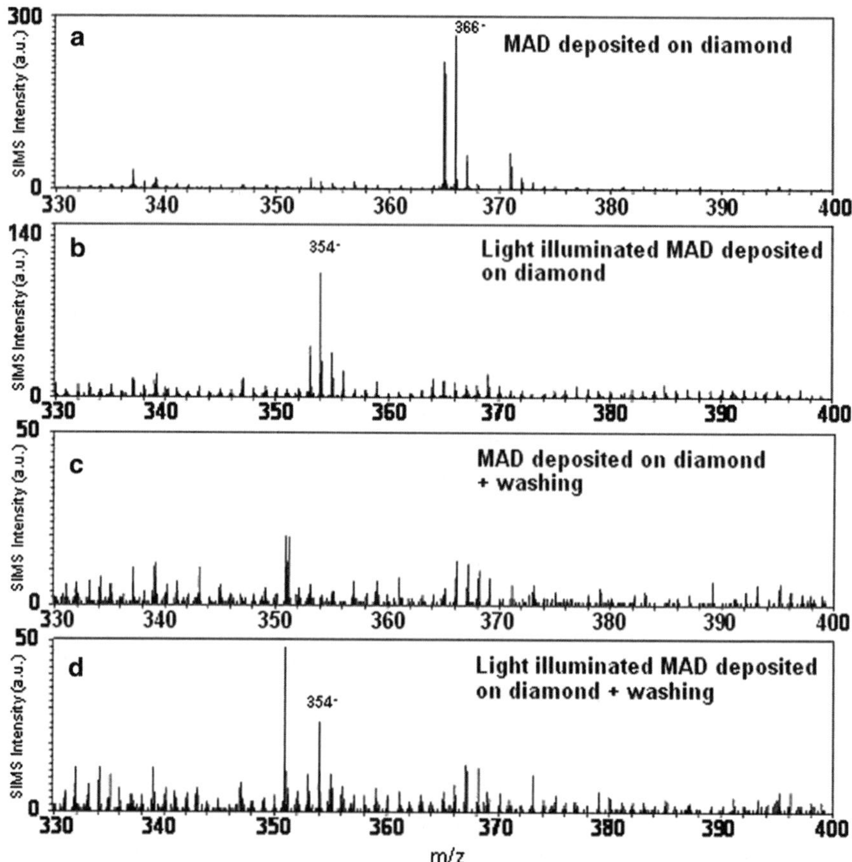

Fig. 3.13 Negative-mode spectra in the mass range 330–400 amu for each step of the process of MAD photo-immobilization on diamond: (**a**) after MAD deposition; (**b**) sample a illuminated; (**c**) sample a not illuminated and then washed; (**d**) sample (a) illuminated and washed (Reproduced with permission from Léonard et al. (1998a), copyright 2008 John wiley & sons Ltd)

m/*z* is a contamination peak after washing. Comparison of the spectra for the various steps of the process gave indirect proof by showing that the molecules are mostly removed by washing when they are not photo-illuminated before this final washing. Differences are observed in the extent of grafting.

The carbene-mediated grafting of the photo-activatable reagent MAD to diamond (Léonard et al. 1998a) was taking place when comparing the various steps of the photo-immobilization of these molecules. It appears that when the molecules are only deposited on the substrates, they are almost completely removed by washing. If illumination is performed before this washing, they remain at the surface to an extent that depended on the substrate, diamond being the substrate for which the highest grafting efficiency was observed. Complementary to this indirect proof of the grafting of the molecule only under irradiation, it has been possible to identify high-mass molecular ToF-SIMS peaks, indicating that the reaction occurred on oxygen-containing sites. However, it is probable that other reaction sites are involved due to the difference in grafting efficiency from one substrate to another.

In a second step of the glyco-engineering example evidence is provided for the controlled binding of the photo-activatable reagent MAD-Gal shown in Fig. 3.14. (Léonard et al. 2001) to diamond again, by light-dependent reactions. *N*-[*m*-(3-(Trifluoromethyl)diazirine-3-yl)phenyl]-4-(-3-thio(-1-D-galactopyrannosyl)-maleimidyl)butyramide (MAD-Gal) is obtained by derivatization of MAD with a thiogalactose, 1-thio-β-D-galactopyrannose. Reagent characterization described elsewhere confirms the structure of MAD-Gal (Léonard et al. 1998b). During photo-immobilization of MAD-Gal the diazirine functions are lost and the carbenes react with surface atoms. Biological tests in previous studies showed that the light activation of MAD does not impair biological activities (Collioud et al. 1993; Gao et al. 1997).

Table 3.2 displays the atomic concentrations of different binding states of MAD-Gal on diamond determined by XPS (Léonard et al. 1998b). It illustrates the possibility to determine the presence of functional groups at the various steps of the process. The small analyzed area of 0.12 mm^2, allowed acquiring spectra on each sample at a 45° take-off angle.

Figure 3.15 shows nitrogen N1s XPS data revealing in (a) before illumination two contributions at 400.3 and 402.3 eV corresponding respectively to the amide and diazirine functions. In (b) only the amide functionality is observed. Curve

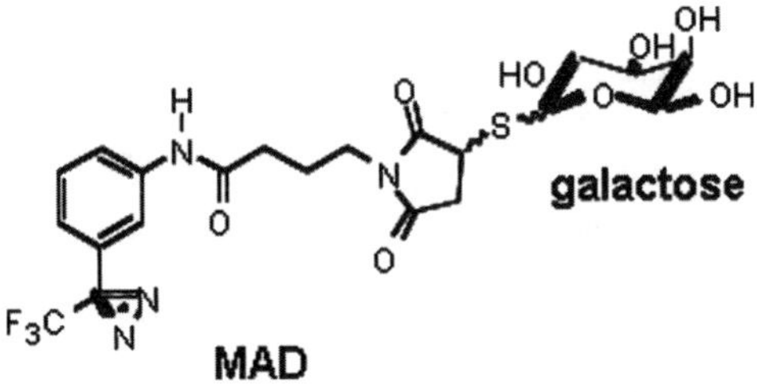

Fig. 3.14 MAD-Gal Chemical structure of MAD-Gal (Reprinted with permission from Léonard et al. (1998a), copyright 2008 John wiley & sons Ltd)

Table 3.2 Concentrations determined by XPS (at%) for the various contributions of the C 1s line for all the steps of the MAD-Gal photo-immobilization process on diamond (average of three areas) applying sensitivity factors for (C1s) 0.296, (O1s) 0.711, (N1s) 0.477, (F1s) 1.000, and (S2p) 0.570, (Léonard et al. 1998b)

C1s	Pristine A	A + washed B	MAD-Gal C	+Light D	C + washed E	D + washed F
C–H	86.6 ± 0.2	84.5 ± 0.8	40.1 ± 5.8	40.4 ± 12.2	84.0 ± 1.3	82.6 ± 0.8
C–C	11.8 ± 0.2	13.1 ± 0.6	27.3 ± 2.3	28.4 ± 4.9	13.7 ± 1.4	12.6 ± 0.4
C–O	1.6 ± 0.1	1.9 ± 0.3	20.9 ± 3.6	19.2 ± 5.6	1.9 ± 0.1	3.8 ± 0.7
C=O	0	0.6 ± 0.01	8.3 ± 0.5	8.9 ± 1.9	0.4 ± 0.01	1.1 ± 0.3
C–F	0	0	3.4 ± 0.6	3.2 ± 0.7	0	0

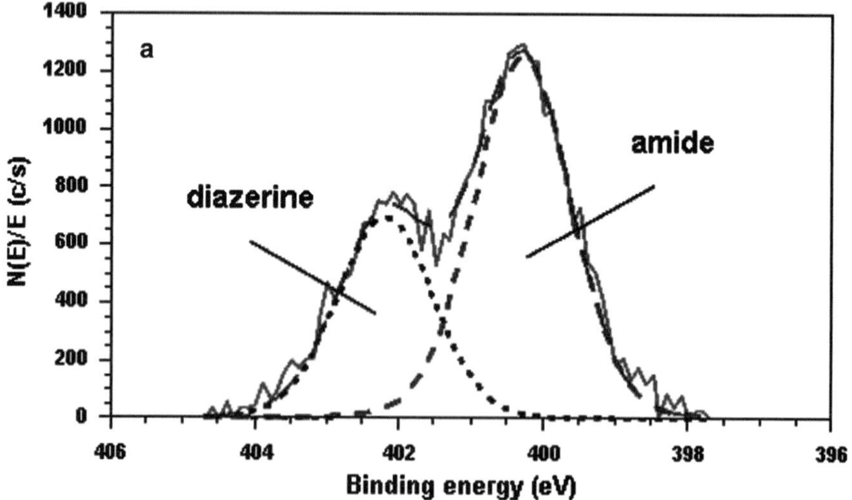

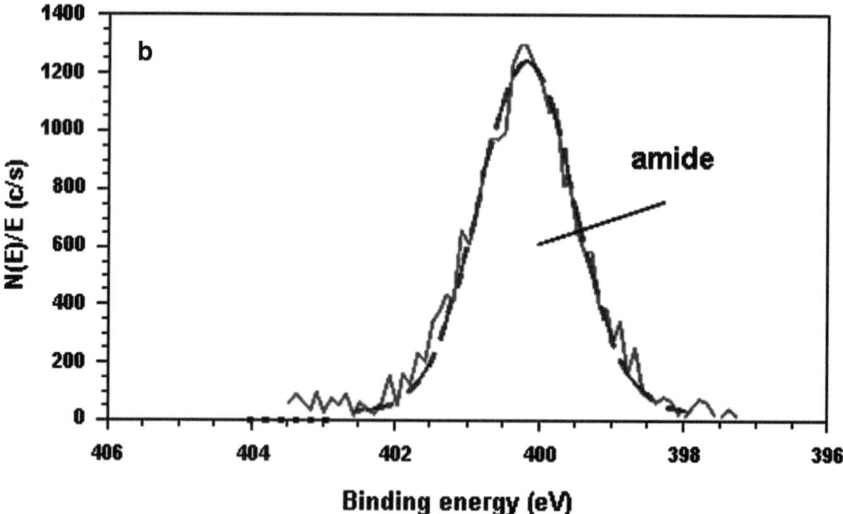

Fig. 3.15 XPS N1s data on diamond of (**a**) MAD-Gal deposited - sample C of Table 3.2; (**b**) MAD-Gal deposited and illuminated – sample D of Table 3.2. (Reprinted with permission from Léonard et al. (1998b), copyright 2008 John wiley & sons Ltd)

fitting results of carbon C1s are illustrated in Fig. 3.16 for sample C, with five contributions at 284.2, 285, 286.24, 288.22 and 292.55 eV corresponding to C–C (diamond), C–H (hydrocarbon), C–N/C–O, N–C=O/O–C–S and most importantly by a peak at 292.55 eV corresponding to CF (compare Table 3.2). The difference in relative intensity (%) of the diamond C–C peak at 284.2 eV illustrates also the presence of MAD-Gal at the surface of samples C, D and F. Quantitative analysis of and at the surface was calculated for the different elements at different preparation steps using (3.9) given in Table 3.3 (Léonard et al. 1998b).

Surface glycoengineering is of crucial importance because glycosylated molecules are involved in cell recognition, in species discrimination, in blood coagulation cascade, and in the asialoglycoprotein system (Léonard et al. 2001; Lee and Lee 1995; Monsigny 1995; Petrak 1994). The principle of surface glycoengineering is based on photochemistry through derivatization of carbohydrates with a diazirines as the photo-activatable group.

When dealing with surface immobilized biomolecules, an important issue is related to orientation of bioactive groups to allow optimal "key-lock" interactions with wanted receptors. All surface analysis measurements are performed in UHV

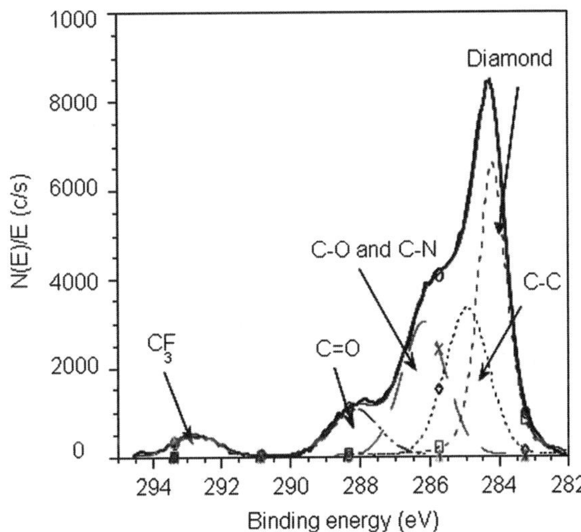

Fig. 3.16 XPS C1s data of MAD-Gal on diamond (sample E of Table 3.2). (Reprinted with permission from Léonard et al. (1998b), copyright 2008 John wiley & sons Ltd)

Table 3.3 MAD photo-immobilization (concentrations in at%) on diamond (Léonard et al. 1998b)

XPS	Pristine A	Washed B	MAD-Gal C	C + light D	+Washed C + washed	+Washed D + washed
C	94.0 ± 0.1	94.0 ± 0.6	69.9 ± 3.2	69.9 ± 4.1	92.3 ± 1.4	88.7 ± 2.0
O	6.1 ± 0.1	6.0 ± 0.6	15.4 ± 1.0	17.5 ± 1.8	7.5 ± 1.1	9.7 ± 1.7
N	0	0	5.7 ± 1.3	4.2 ± 0.6	0.1 ± 0.2	0.7 ± 0.1
F	0	0	78 ± 0.8	7.5 ± 1.5	0.1 ± 0.1	0.8 ± 0.2
S	0	0	1.2 ± 0.1	1.0 ± 0.3	0	0.1 ± 0.1

(hydrophobic) environment leading to possible surface reorganization. The analyzed surface conformation is then different from that observed in a hydrophilic environment such as biological fluids. The characterization of biosurfaces using biological tests specific to the biofunctions immobilized at the surface is the best way to verify surface availability (including orientation) of these biomolecules. Polystyrene being a typical material for dishes in biological testing, the photo-immobilization of MAD-Gal was exploited on polystyrene dishes and not on diamond for biological testing. Biological availability of the galactose residues was probed with Allo A lectin which recognizes specifically β-galactose and on the cell level with primary rat hepatocytes which possess at their surface a specific receptor for galactose. Photo-immobilization of MAD-Gal on polystyrene dishes was ascertained by ToF-SIMS and XPS (Léonard et al. 2001). Allo A lectin and rat hepatocytes tests indicated specific interactions with galactose residues of MAD-Gal modified surfaces (Chevolot et al. 1999).

The next step is the application of the strategy to more complex carbohydrates. I illustrate here the use of XPS and ToF-SIMS for the surface immobilization of the synthesized photo-activatable reagent shown in Fig. 3.17 5-carboxamidopentyl *N*-[*m*-[3-(trifluoromethyl)diazirin-3-yl] phenyl β-D-galactopyranosyl]-(1→4)-1-thio-β-D-glucopyranoside) (Lactose aryl diazirine). Reagent synthesis and chemical characterization described elsewhere (Chevolot et al. 2001) confirmed the structure of the Lactose aryl diazirine. The expected photo-immobilization of such a compound is expected to be similar to that of MAD-Gal. The diazirine functions are lost and the carbenes react with surface atoms. Again, biological tests showed that the 350 nm light irradiation of the photo-reagents studied here does not degrade molecules nor impair biological activities of surface immobilized residues (Collioud et al. 1993; Léonard et al. 1998a; Gao et al. 1997). Concerning the positive ion spectra, no significant difference was observed in the mass range 0–100 amu for each step of the process of lactose aryl diazirine and MAD-Gal photo-immobilization on diamond.

The comparison of the high mass range of the positive ion spectra was more specific to distinguish between galactose and lactose (Léonard et al. 2001). As

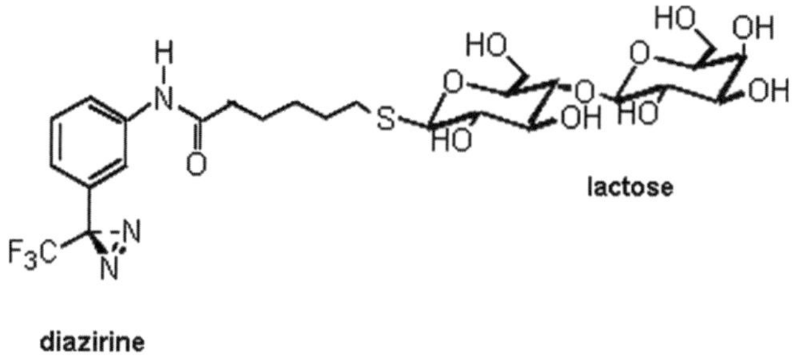

Fig. 3.17 Chemical structure of lactose aryl diazirine (Reprinted with permission from Léonard et al. (2001), copyright 2008 John wiley & sons Ltd)

illustrated in Fig. 3.18, a high mass peak at 557 *m/z* was observed in the case of MAD-Gal for sample C (molecules deposited on the surface) and another one at 573 *m/z* for sample D (after light irradiation). The difference corresponds to an oxygen atom and this is consistent with the above mentioned possible reaction of the photo-reactive part with a surface O atom. Inspection reveals that corresponding peaks were observed for the immobilized and light irradiated lactose aryl diazirine molecules at 650 and 666 *m/z*, respectively. The mass difference between the two molecules is 93 amu; all these peaks are characteristic fragments of the molecules. It is assumed that peaks at 557 and 650 *m/z* correspond to $(M - 2N + Na)^{+}$ and those appearing at 573 and 666 *m/z* to $(M - 2\,N + O + Na)^{+}$.

Plasma membranes display arrays of receptors that interact with specific ligands. Among these interactions, carbohydrate-protein interactions play a key role in a number of regulatory processes in cells (Lee and Lee 1995; Monsigny 1995; Petrak

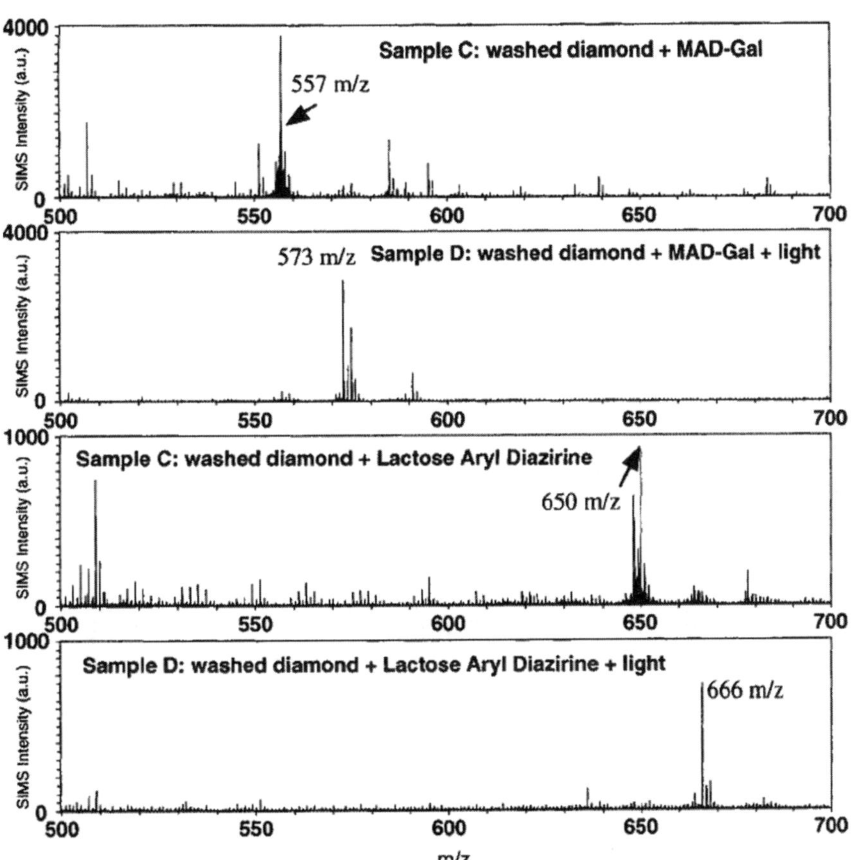

Fig. 3.18 Positive ion spectra of diamond in the range 500–700 amu for MAD-Gal steps C and D (illuminated), and lactose aryl diazerine (C) washed, and (D) washed plus light (Reprinted with permission from Léonard et al. (2001), copyright 2008 John wiley & sons Ltd)

1994; Chevolot et al. 2001; Gao et al. 1997; Varki 1993). Thus, oriented surface immobilization of carbohydrates allows one to mimic the specific interactions that occur naturally at cell surfaces and can create a matrix on which specific cellular functions of cultured cells might be investigated.

Finally we show here, how ToF-SIMS and XPS are applied for semi-quantitative analysis and that they can be correlated: Fig. 3.19a shows that ToF-SIMS F-normalized intensity and XPS fluorine atomic percentage increased as a function of the concentration of MAD-Gal solution used for surface immobilization on polystyrene. The intensity of F^- and F atomic % are related to the surface density of photo-bonded molecules, and thus MAD-Gal density increased with the

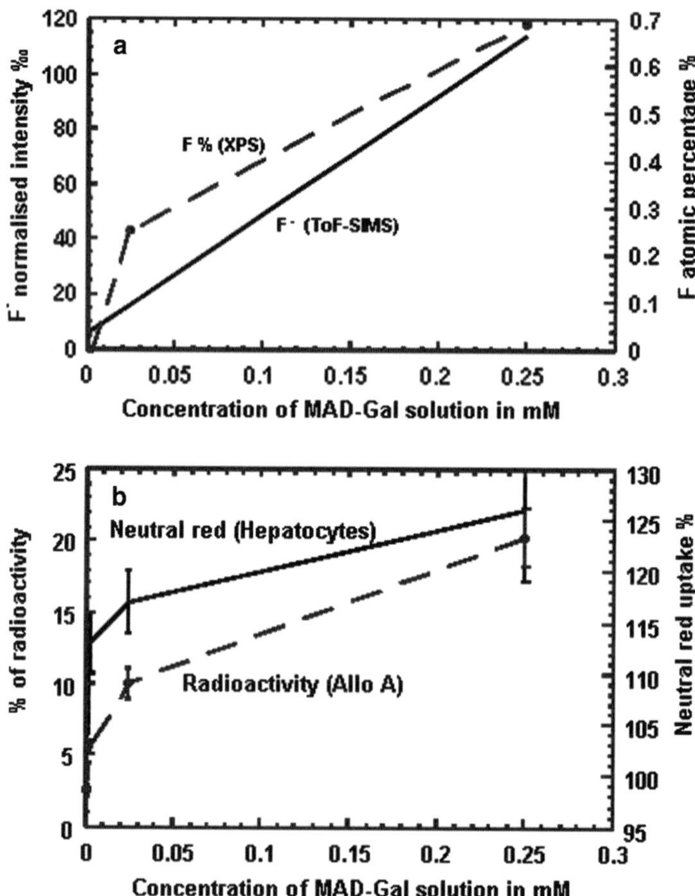

Fig. 3.19 Normalized intensity values (absolute intensity of F^-/(total intensity – H-intensity)) and fluorine atomic percentages are displayed as a function of the MAD-Gal concentration used for immobilization. The intensity of surface characteristic signatures increases with increasing density of MAD-Gal. (Reprinted with permission from Blitz and Gunk'ko (2006), copyright Springer)

concentration of applied MAD-Gal solution. This observation is confirmed by previous results obtained with photo-bonded 4-(2,5-dioxo-2,5-dihydropyrrol-1-yl)-*N*-[3-(3-trifluoromethyl-3*H*-diazirin-3-yl) phenyl] butyramide (MAD) on silicon nitride (Blitz and Gunk'ko 2006; Varki 1993).

On the cellular level, functionality of the galactose and lactose residues was tested with primary rat hepatocytes. The MAD-Gal coated surface altered hepatocellular function. Neural Red (NR) uptake (Fig. 3.19b) increases for hepatocytes cultured on MAD-Gal coated PS surfaces and the increase correlates with the density of galactose molecules on the surface. MTT formation and NR uptake increased up to a concentration of 2.5 mM galactose and showed no saturation effect (Chevolot 1999; Chevolot et al. 2001). This is in agreement with Fig. 3.20. where the NR uptake is related to the lysosomal activity. NR uptake decreases with decreasing galactose residues C0–C2). NR uptake is similar to that of surface A (plain PS). Hepatocytes possess higher NR uptake on asialofetuin coated surfaces than on surface A, but lower than on MAD-Gal grafted PS.

Table 3.4 shows XPS and ToF-SIMS analysis of MAD-Gal grafted poly-Styrene (PS). Atomic percentages of the elements were observed with XPS. Zero stands for absent or below detection limit. Sample A corresponds to the plain PS, sample B corresponds to the surface on which MAD-Gal was deposited and washed without

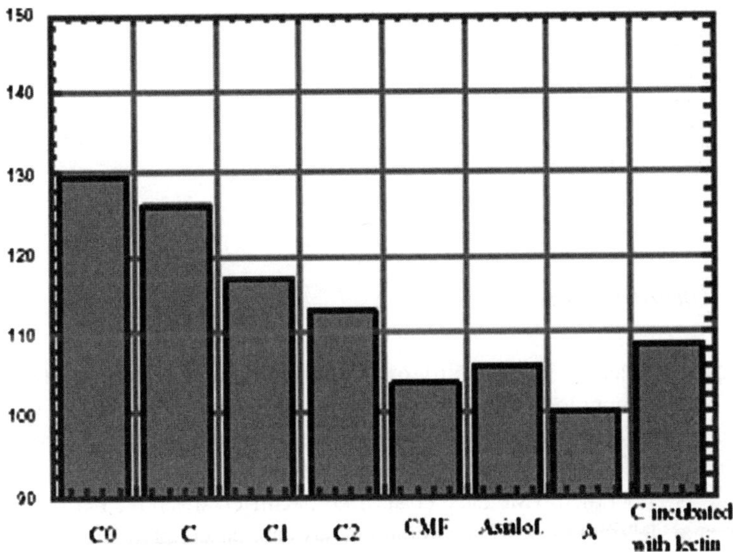

Fig. 3.20 Neutral Red (NR) uptake is related to the lysosomal activity. Galactose residues (C0, C, C1, C2 corresponding to 2.5, 0.25, 0.025 and 0.0025 mM). The NR uptake is similar to that of surface A (plain polystyrene) when the surface was incubated with a lectin (RCA) that protects the galactose residues. Hepatocytes possess higher NR uptake on asialofetuin coated surfaces than on surface A, but lower than on MAD-Gal grafted PS. (Reprinted with permission from Chevolot et al. (2001), copyright 2008 Elsevier)

Table 3.4 XPS and ToF-SIMS analysis of MAD-Gal grafted on poly-Styrene (PS). Atomic percentages (%) of the elements were observed with XPS: (A) virgin polystyrene, (B) after MAD-Gal deposition and washing without light activation, (C) MAD-Gal with light activation and washed. Fluorine and nitrogen are signatures of the diazirine and are only observed on sample C. ToF-SIMS normalized intensity of selected ions characteristic of MAD-Gal molecule (F^- and CF_3^-). Normalized intensity is the ratio of the absolute intensity of the selected ion over the total corrected intensity which is the total intensity minus H-intensity (Chevolot et al. 2001)

XPS (at%)	Sample A	Sample B	Sample C
N	0	0	0.28 ± 0.26
F	0	0	0.45 ± 0.16
S	0	0	0
C	93.7 ± 0.6	94.1 ± 1.0	95.5 ± 1.2
O	6.1 ± 0.6	5.9 ± 1.0	3.8 ± 0.8
ToF-SIMS			
Corrected total intensity (counts) $\times 10^4$	110.4 ± 1.8	106.4 ± 12.9	132.2 ± 15.7
F^-	0.9 ± 0.13	26 ± 4	205 ± 53
CF_3^-	0.008 ± 0.003	0.2 ± 0.08	1.2 ± 0.5

light activation and sample C is the surface on which MAD-Gal was deposited, light activated and washed. MAD-Gal is expected to be observed only on sample C. Fluorine and nitrogen are signatures of the molecule and are only observed on sample C.

In conclusion, it was shown that Diazirine-containing photo-reagents containing mono- and disaccharides were successfully immobilized on polystyrene. Biological activity of the modified polystyrene was demonstrated with Allo A lectin, primary rat hepatocytes and a-2,6-sialyltransferase. Different responses of lectin Allo A and hepatocytes to the galactose and lactose modified surfaces indicate that the biological activity depends not only on the carbohydrate but also on structure of the spacer within the biomolecule.

3.4.2 Immunosensors

Immunoassays identify and quantify organic compounds that combine the principles of both immunology and chemistry. They use labeled antibodies (radioisotopes, enzymes, etc.) that have been developed to bind with a target compound. High binding specificity and detection sensitivity are of great interest as far as this technique is concerned. (Metzger et al. 1999; Kim et al. 2001; Pei et al. 2001; Pearson et al. 1998). Fluorescence immunoassays are based on monitoring the light emitted from a fluorescent label, which is chemically conjugated with antigen or antibody while excited by light of a suitable wavelength. This non-isotopic detection system offers equal or superior sensitivity and has mostly replaced the classical radioisotopic labels in routine use. A biosensor that integrates antibody fragments is called an immunosensor (Peterman et al. 1988). For non-isotopic antibodies labeled immunoassays, the antibody-antigen complexes can be directly monitored without a prior separation step between the bound and the free tracer (Luppa et al. 2001; Ratner et al. 1987; Ratner et al. 1996). The Enzyme-Linked Immuno Sorbent Assay (ELISA) is a very common technique for antigen measurements. It is a heterogeneous, solid phase assay (Catt and Tregear 1967; Diamandis and Christopoulos 1996).

Once the receptor is adsorbed on the polymer, it binds to its ligand. There are two available techniques to perform an ELISA test. These are the competitive technique and the sandwich technique. The sandwich type technique, illustrated schematically in Fig. 3.21, uses the following procedure: the solid phase is coated with an excess of anti-analyte antibody and an excess of a second enzyme-conjugated anti-analyte antibody is added. Then, in a second step, the analyte is pipeted inside the well. During the incubation, antibodies bind the analytes specifically. If there are changes due to the presence of the enzyme conjugate bound to the immune complex, a positive test or color change will occur. Typical solid-phase techniques such as ELISA are both specific and sensitive, but they take time, particularly because of the various washing, adding steps and the incubation time. Also they produce waste during the process that one must get rid of. Thus, the design of immunosensors suitable for real time monitoring and with lower detection limits than classical assays is of great interest in the field of immunoassays. Each avidin molecule binds with 4 biotins with high affinity and selectivity. The first antibody close to the bottom of the well is a Biotin-conjugated Goat Mouse IgG. It binds to the chip surface via biotin/avidin conjugation. The second antibody is a crosslinked APC Goat Mouse IgG – it is labeled with APC (Allophycocyanin), which is a

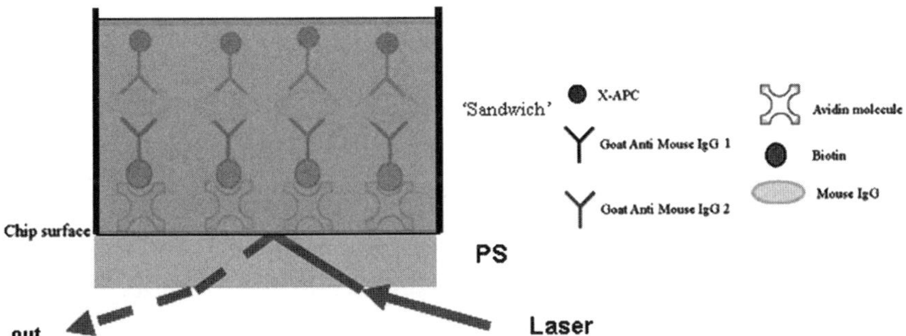

Fig. 3.21 Fluorescent test (schematically): Each avidin molecule binds with 4 biotins with high affinity and selectivity. The first antibody – the black "Y" is a Biotin-conjugated Goat Mouse IgG. It binds to the chip surface via biotin/avidin conjugation. The second antibody – the green "Y" is a Crosslinked APC Goat Mouse IgG – it is labeled with APC (Allophycocyanin), which is a fluorescence dye. It is optimally excited by helium neon laser at 635 nm with emission of light at 650 nm. Mouse IgG (the yellow oval mark) is used as analyte in the test. After the two antibodies bind with Mouse IgG at specific sites, the whole conjugate system will become fluorescence sensitive because of the existing of APC – not to scale. The increase of fluorescence counts with time is directly proportional to the amount of fluorophore in the solution above the surface, if they are capable to bind with a specific ligand immobilized on the surface where the fluorescence exiting evanesence field of photons is generated

fluorescence dye. It is optimally excited by helium neon laser at 635 nm (compare Fig. 3.22). Mouse IgG is used as analyte in the test. After the two antibodies bind with Mouse IgG at specific sites, the whole conjugate system will become fluorescence sensitive because of APC. The number of reflected fluorescence photons is counted. The increase of fluorescence counts with time is directly proportional to the amount of fluorophore in the solution above the surface, if they are capable to bind with a specific ligand immobilized on the surface where the fluorescence exiting evanescence field of photons is generated. Such a sensor with Polystyrene as the base material has been proposed, developed and patented by Diamed ADS (Quapil and Schawaller 2002; Schawaller and Quapil 2003). The laser light is totally reflected from the bottom of the well of the mirror-like PS device as shown in Figs. 3.21 and 3.22.

Before describing the chemical modification of the PS surface, I recall briefly the evanescent field theory: the measurement of light absorbed or emitted by either reactants or products of a biological system is popular in immunosensor design. This is done by optical immunosensors. They can be designed to respond to a wide range of radiations. Many optical principles can be used to design optical immunosensors. The principle interest here is of total internal reflectance spectroscopy (TIRF) (Harrick 1967). This technique involves an evanescent wave and requires fluorescent labeled molecules. For optical immunosensors based on the emission of an evanescent wave, light entering the device is directed towards the sensing surface and then reflected back again while the evanescent field excites the labeled molecules immobilized on the reversed side of the sensing surface.

Evanescence occurs when a light beam propagating through a medium of refractive index n_1 meets the interface with a medium having lower refractive index n_2; undergoing a total reflection when its incidence angle θ is higher than a critical angle θ_c defined by

$$\theta_c = \sin^{-1}\left(\frac{n_2}{n_1}\right) \tag{3.10}$$

where θ is the angle between the incident beam and the axis normal to the plane of the interface. Despite being totally reflected the incident beams establishes an

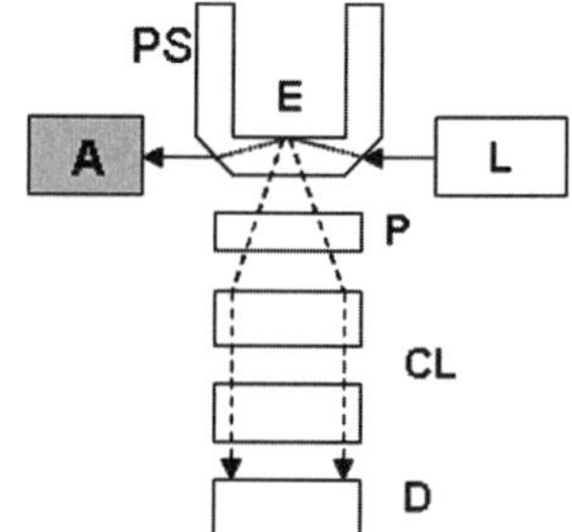

Fig. 3.22 Principle of the Polystyrene reader: Laser (L), PS well, evanescent field E, light absorber (C), polarizing filter (P), collecting lenses (CL) and photon counter (D). The result of such an experiment is a plot of the number of fluorescence photons per second as a function of time, in this system we study the period of 100s to 700s

electromagnetic field through the second media inside the well. This is the "evanescent wave." It penetrates a small distance of the order of a wavelength of the incident light into the solution, where it propagates parallel to the plane of the interface. The evanescent wave provokes fluorescence from molecules in the solution close to the interface.

The evanescent electric field intensity $I(z)$ decays exponentially with perpendicular distance z from the interface. Therefore, molecules, which are more than a wavelength or so away from the interface are not excited.

$$I(z) = I_0 e(-z/d) \tag{3.11}$$

with d the penetration depth for angles of incidence $\theta > \theta_c$.

The fluorescence reader proposed in the patents (Quapil and Schawaller 2002) is based on the evanescent field method described. Figure 3.22 shows the principle arrangement of the reader: PS represents the polystyrene well and (E) is its bottom surface, where the evanescence occurs. A polarized Laser light is transmitted (L) to the well. The light wave undergoes total reflection at the bottom surface of the well (E) and is reflected towards an absorber (A). The mean height of the evanescent field wave is produced to a maximum of about 200 nm. The evanescent field passes through the solution containing the fluorophores. Excitation takes place which leads to a fluorescence radiation. The fluorescence radiation passes through a polarization filter (P) and a set of collecting lenses and an interferences filter (CL) before reaching the photon counter (D). The fluorescence signal proportional to the amount of fluorophores on the PS surface is monitored by a laptop computer. Typical data acquisition takes place during several minutes.

The polystyrene (PS) biosensing chip serves as a medical diagnostic tool. PS (γ-PS) was proposed in the work of Gao et al. (2004). γ-PS was functionalized by covalent bonding with allyldextran, which is dextran substituted with allyl group asindicated in Scheme 3.1, a three step process: (1) γ-activation, (2) covalent coating of allyldextran molecules on PS surface initiated by free radicals in the PS resulting from γ-irradiation; (3) functionalisation of dextran molecules by

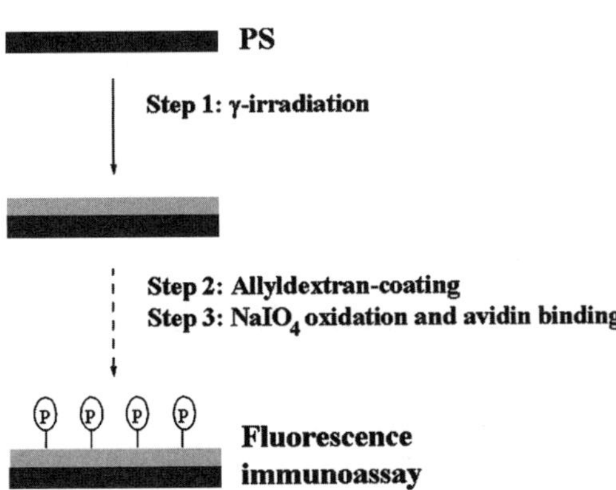

Scheme 3.1 Three step surface modification procedure (Gao et al. 2004)

periodate oxidation followed by protein-binding. The dextran molecule shown in Fig. 3.23 is a polymer formed by glycosidic bonds linking the carbon atoms number 1 and 6 of consecutive α-D-glucopyranose rings. The surface covered with a layer of dextran matrix provides a hydrophilic environment suitable for biospecific interactions and enhances the immobilization capacity of biomolecules.

Allyldextran coating is achieved by adding allyldextran solution to the γ-radiation activated PS chip and incubating it over-night. The result of the coating analyzed by XPS is shown in Fig. 3.24 at two different take-off angles at 0° and 70° with respect to the surface normal to detect differences in the homogeneity of the coating with depth. Analysis at 0° measured with respect to the surface provides an information depth of 8–10 nm while analysis at 70° provides information from less than 3 nm. For the coated allyldextran layer and part of PS substrate underneath the presence of C and O, their binding state, as well as their atomic concentrations are identified.

Inspection of the functional groups of the C1s peak in Fig. 3.24 reveals four contributions at 285.0, 285.5, 286.8, and 288.2 eV corresponding to C–C (aromatic carbons on PS), C–H (aliphatic carbons from main chains of PS and covalently bonded allyldextran), C–O and C–OH of the dextran-ring, and O–C–O (connected to dextran-ring), respectively. It also shows a "shake up" peak of aromatic ring at 292.0 eV. Comparison of the data collected at the two angles of emission provides supporting evidence for a thin coating owing to the strong decrease in C–aromatic belonging to PS and the great increase in dextran-related C components such as

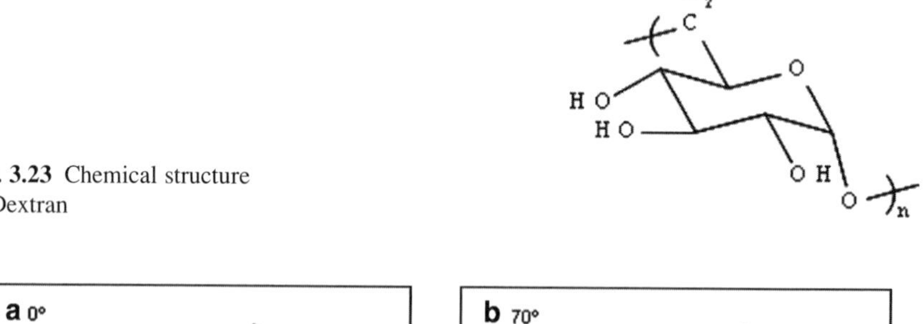

Fig. 3.23 Chemical structure of Dextran

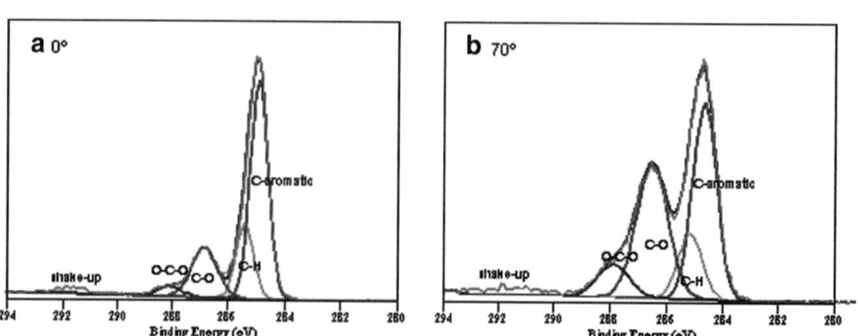

Fig. 3.24 High resolution XPS C1s peaks of γ-PS coated with 100 μg/ml Allyldextran (**a**) at take-off angle $\theta = 0°$ and (**b**) 70° (Reprinted from Gao et al. (2004), copyright 2008 John wiley & sons Ltd)

C–O, C–OH, and O–C–O. The thickness of the allyldextran coating was calculated to be approx. 4 ± 1 nm with a coverage of 13% for allyldextran compared to dextran at 16% (Gao et al. 2004).

The allyldextran coated surfaces are successfully activated by oxidation using $NaIO_4$. Figure 3.25 shows the high resolution C1s spectra of (a) periodate-oxidized PS-dextran with a high extent of oxidative conversion. This result gives information of both coated oxidized allyldextran layer and part of PS substrate underneath. C=O component at 288.4 eV is observed to be 7.2 at% proving that the ring of cyclic ether of dextran moiety opened and changed to the corresponding aldehyde. Components C–O and C–OH on the dextran-ring at 286.8 eV and O–C–O of dextran moiety at 288.2 eV decrease when Streptavidin or Neutravidin are subsequently bound to the iodate activated dextran surface (Fig. 3.25b, c) (Gao et al. 2004). The polyaldehyde obtained from oxidation of dextran polymer is conjugated with Streptavidin and Neutravidin. The reaction proceeds via Schiff base formation followed by reductive amination by using sodium cyanoborohydride to create stable secondary amine linkages (Idage and Badrinarayanan 1998). Surfaces conjugated with Streptavidin or Neutravidin indicated the presence of 8.0 at% nitrogen at 398 eV. XPS analysis shows that there are about 12 at% N in pure Streptavidin and Neutravidin. Therefore we estimate that the coating of Streptavidin and

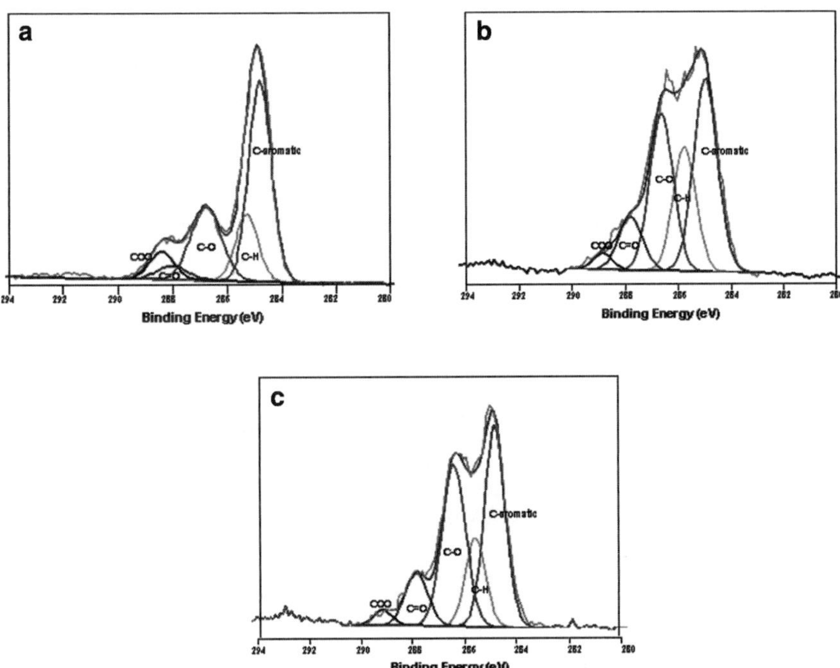

Fig. 3.25 XPS high resolution C1s spectra of PS at 70° take-off angle (**a**) after $NaIO_4$ oxidation, (**b**) after Streptavidin conjugation and (**c**) after Neutravidin conjugation (Reprinted from Gao et al. (2004), copyright 2008 John wiley & sons Ltd)

Neutravidin on PS surface is about 67%. C-aromatic from PS substrate, C–O and C–OH from dextran moieties and C–N from Streptavidin or Neutravidin molecules are observed. In addition, such surfaces turned out to be very hydrophilic: their contact-angles are too low (<5°) to be measurable.

This hydrophilic surface with biotin binding functionality is applied for the measurement of biological samples such as serum or other biological fluids because they reduce non-specific binding while presenting a high biotin binding activity to generate an increased signal intensity of the biosensor reported elsewhere. The covalent surface modification and immobilization methods were shown to be successful for enhanced biological coupling. γ-Activated polystyrene surface treated with allyldextran, sodium periodate and Streptavidin and Neutravidin are highly hydrophilic with low non-specific adsorption properties, and give a highly promising result for binding with biotin derivative biomolecules. Optical fluorescence measurements on these modified surfaces are commercialized elsewhere (Quapil and Schawaller 2002; Schawaller and Quapil 2003).

Finally, we will compare the γ-radiated sample to an NH_3 RF plasma activated surface (Gao et al. 2005). We focus on optimizing a method of surface modification of the PS biochip without the use of γ-radiation. The surface modification process is shown in Scheme 3.2. In step 1 a NH_3 RF plasma is used to graft amino groups onto PS surfaces; in step 2–3 the PS surface carrying primary amino groups is chemically activated with glutaraldehyde (OHC–CH_2–CH_2–CH_2-CHO) before coupling Neutravidin covalently, respectively. The fluorescence immunoassay tests on the Avidin modified PS surface were performed on the basis of Neutravidin-Biotin binding. Aldehydes react with primary and secondary amines to form Schiff bases, which can be converted to an alkylamine linkage by using reducing reagents. Glutaraldehyde is used for protein conjugation. It reacts with amine groups to create cross-links by one of several routes. Under reducing conditions, the aldehydes on both ends of glutaraldehyde will couple with amines to form secondary amine linkages (Kim et al. 2001). Using a pH 6 a successful reaction between

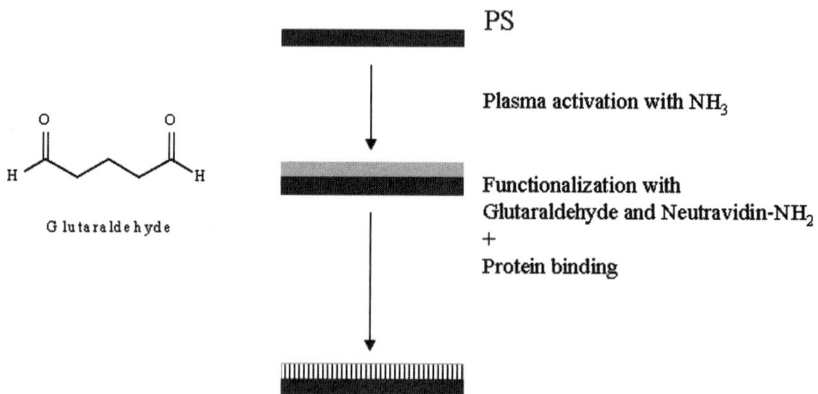

Scheme 3.2 NH3 Plasma modification process (Gao et al. 2005)

glutaraldehyde molecules having two oxygen atoms and five carbon atoms and amino groups on PS surfaces was observed. As shown in Fig. 3.26 by XPS, a significant increase in C=O and C-aliphatic peaks and an increase of the C–N peak after Neutravidin coating are observed. When comparing the γ-activated and plasma modified PS surfaces of Figs. 3.24 and 3.25, one notices the same functional groups at the PS surfaces: One observes, however, different activities in particular, the relative C–N concentration in Fig. 3.24b is almost absent after the plasma modification. Such a hydrophilic surface with biotin binding functionality is applied for the measurement of biological samples such as serum or other biological fluids because they reduce non-specific binding while presenting a high biotin binding activity to generate an increased signal intensity of the biosensor.

Contact angle analysis (Table 3.5) provides information on the surface hydrophobicity/hydrophilicity and molecular mobility at the air–solid–water interface. The contact angle, α, is defined by (3.12)

$$\cos\alpha = \frac{\sigma_{\mathrm{SV}} - \sigma_{\mathrm{SL}}}{\sigma_{\mathrm{LV}}} \tag{3.12}$$

With σ the respective surface energies between solid (S), liquid (L) and vacuums (V) are indicated by their indices. When water is the liquid, hydrophilic surfaces are

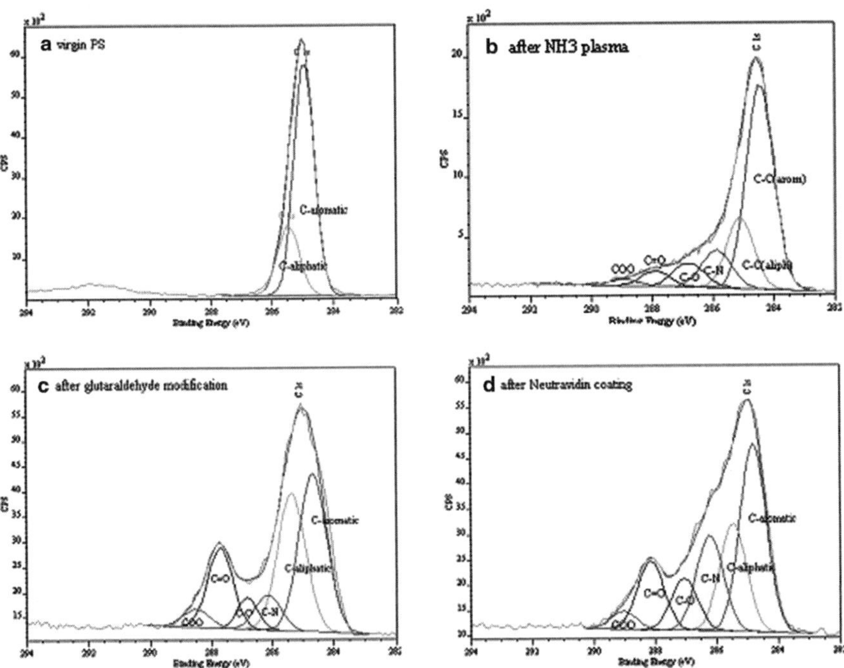

Fig. 3.26 High energy resolution XPS spectra of C1s (**a**) virgin PS, (**b**) after NH_3 plasma, after (**c**) after Glutaraldehyde modification and (**d**) Neutravidin coating of the NH_3 RF plasma activated PS surface

Table 3.5 Contact angle measurements on PS (Gao et al. 2005). Data are results of at least 10 measurements on different samples with a sessile drop method

Sample	Contact-angle (degree)
PS	83 ± 2
γ-PS	75 ± 2
γ-PS coated with 100 μg/ml allyldextran	46 ± 3

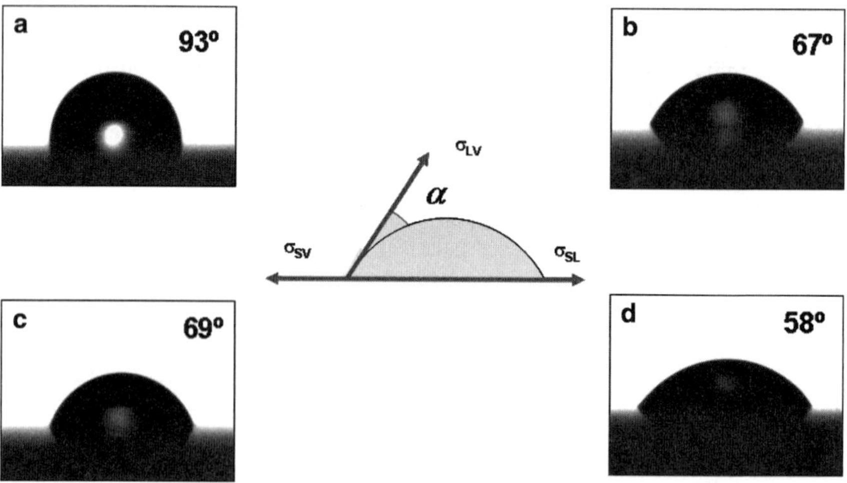

Fig. 3.27 Contact angle measurements on PS (**a**) virgin, (**b**) treated with NH_3 plasma, (**c**) with Glutaraldehyde and (**d**) coated with Glutaradehyde and Neutravidin

characterized by a small contact angle. Virgin PS is a hydrophobic surface with a contact angle of $83 \pm 2°$. γ-irradiation introduces some polar functional groups on the PS surface and makes the surface more hydrophilic ($75 \pm 2°$). After the treatment with Allyldextran, the contact angle remains and decreases to a hydrophilic value of $46 \pm 3°$.

In Fig. 3.27 contact angles after a NH_3 plasma are presented. The contact angle decreases from (a) $93 \pm 2°$ for virgin PS to (b) $67 \pm 7°$ after NH_3 plasma treatment. In (c) application of Glutaradehyde leads to almost no change, e.g., $69 \pm 4°$, before reaching (d) $58 \pm 7°$ after coating with Neutravidin. Comparison to Table 3.5 shows that the PS surface becomes equally more hydrophilic with the applied surface plasma-modification in agreement with the XPS results of Figs. 3.24–3.26 for the γ-treated PS surfaces. This indicates that contact angle measurements can be used as a fast and an easy way to measure surface energy changes despite the lack of chemical state identification of XPS. PS surfaces modified either by γ-irradiation or NH_3-plasma give comparable results with similar sensor activities.

3.5 Concluding Remarks

After an introduction into the first principles of ion and electron spectroscopies, e.g., Time-of Flight Secondary Ion Mass Spectrometry (ToF-SIMS) and X-ray Photoelectron Spectroscopy (XPS) to chemically analyze the top monolayers of biosensitive device within the sub-micrometer range the following practical examples were presented:

1. The surface chemical properties of a micro structured and multi-molecular biochip surface, comprising an array of gold microelectrodes incorporating a cell-attachment promoting oligopeptide (H-CDPGYIGSRNH_2), and a thin film of amorphous Teflon® in between the electrodes and everywhere else, were identified. It was demonstrated that the multistep biochip micro fabrication process introduced negligible or no surface contamination that could hinder surface incorporation of the oligopeptide or effect cytotoxicity.
2. Diazirine-containing photo-reagents containing mono- and disaccharides were successfully immobilized on polystyrene. Biological activity of the modified polystyrene was demonstrated with Allo A lectin, primary rat hepatocytes and a-2,6-sialyltransferase. Different responses of lectin Allo A and hepatocytes to the galactose and lactose modified surfaces indicate that the biological activity depends not only on the carbohydrate but also on structure of the spacer within the biomolecule.
3. The covalent surface modification and immobilization methods were shown to be successful for enhanced biological coupling. γ-Activated polystyrene surface treated with allyldextran, sodium periodate and Streptavidin and Neutravidin are highly hydrophilic with low non-specific adsorption properties, and give a highly promising result for binding with biotin derivative biomolecules. Comparing the γ-activated and plasma modified PS surfaces for the development of an immunosensor one notices the same functional groups at the PS surfaces. Such a hydrophilic surface with biotin binding functionality is applied for the measurement of biological samples such as serum or other biological fluids because they reduce non-specific binding while presenting a high biotin binding activity to generate an increased signal intensity of the biosensor.

References

Bain CD, Troughton EB, Tao YT, Evall J, Whitesides GM, Nuzzo NG (1989) Formation of monolayers by the coadsorption of thiols on gold: variation in the length of the head group. J Am Chem Soc 111:321–335

Benninghoven A, Rüdenauer FG, Werner HW (1987) Secondary ion mass spectrometry: basic concepts, instrumental aspects, applications and trends. Wiley, New York

Benninghoven A, Hagenhoff B, Niehuis E (1993) Surface MS: probing real-world samples. Anal Chem 65:630A–640A

Benninghoven A (1994) Surface Analysis by Secondary Ion Mass Spectrometry (SIMS). Surf Sci 246:299–300

Benninghoven A, Hagenhoff B, Werner HW (1997) Secondary Ion Mass Spectrometry (Proceedings SIMS X). Wiley, Chichester

Beamson G, Briggs D (1992) High resolution XPS of organic polymers. The Scienta ESCA300 database. Wiley, Chichester

Bethell D, Schiffrin DJ (1996) Supramolecular chemistry – nanotechnology and nucleotides. Nature 382:581

Bertrand P, Weng LT (1996) Time-of-Flight Secondary Ion Mass Spectrometry (Tof-SIMS). Mikrochimica Acta Suppl 13:167–182

Biosensors and Bioelectronics (2006) Selected Papers from the Ninth World Congress on Biosensors. Toronto: Elsevier, vol. 22, issues 9–10

Black J (1992) Biological performance of materials, 2nd edn, ch. 16. Marcel Dekker, New York

Blitz JP, Gunk'ko VM (eds) (2006) Surface chemistry in biomedical and environmental science, nato science series, series II: mathematics, physics and chemistry, vol 228. Springer, Dordrecht

Braun E, Eichen Y, Sivan U, Ben-Yoseph G (1998) DNA-templated assembly and electrode attachment of a conducting silver wire. Nature 391:775–778

Briggs D, Seah MP (1990) Practical surface analysis. Volume 1 – Auger and photoelectron spectroscopy, 2nd edn. Wiley, Chichester

Briggs D, Seah MP (1992) Practical surface analysis, Vol 2: Ion and neutral spectroscopy. Wiley, Chichester

Brummel CL, Lee INW, Zhou Y, Benkovic SJ, Winograd N (1994) A mass spectrometric solution to the address problem of combinatorial libraries. Science 264:399–402

Buchwalter LP, Czornyj G (1990) Poly(methyl methacrylate) degradation during X-ray photoelectron spectroscopy analysis. J Vac Sci Tech A8(2):781–784

Catt K, Tregear GW (1967) Solid-phase radioimmunoassay in antibody-coated tubes. Science 158 (3808):1570–1572

Chaney R, Barth G (1987) An ESCA study on the X-ray induced changes in polymeric materials. Fresen J Anal Chem 326:143

Chevolot Y, Bucher O, Léonard D, Mathieu HJ, Sigrist H (1999) Synthesis and characterization of a photoactivatable glycoaryl-diazirine for surface glycoengineering. Bioconjug Chem 10:169–175

Chevolot Y (1999) Surface photoimmobilisation of aryl diazirine containing carbohydrates-tools toward surface glycoengineering. Ph.D. Thesis EPFL Lausanne: no. 1995

Chevolot Y, Martins J, Milosevic N, Léonard D, Zeng S, Malissard M, Berger EG, Maier P, Mathieu HJ, Crout DHG, Sigrist H (2001) Immobilization on polystyrene of diazirine derivates of mono- and disaccarides: biological activities of modified surfaces. Bioorg Med Chem 9:2943–2953

Clark DT, Brennan WJ (1986) An ESCA investigation of low energy electron beam interactions with polymers. I: PTFE. J Electron Spectrosc Rel Phenom 41:399–410

Clémence JF, Ranieri JP, Aebischer P, Sigrist H (1995) Photoimmobilization of a bioactive laminin fragment and pattern-guided selective neuronal cell attachment. Bioconjug Chem 6:411–417

Collioud A, Clémence JF, Sänger M, Sigrist H (1993) Oriented and covalent immobilization of target molecules to solid supports: synthesis and application of a light-activatable and thiol-reactive cross-linking reagent. Bioconjug Chem 4:528–536

Connolly P (1994) Bioelectronic interfacing: micro- and nanofabrication techniques for generating predetermined molecular arrays. Trends Biotechnol 12:123–127

Coullerez G, Chevolot Y, Léonard D, Xanthopoulos N, Mathieu HJ (1999) Degradation of polymers (PVC, PTFE, M-F) during X-ray photoelectron spectroscopy (XPS-ESCA) analysis. J Surf Anal 5:235–239

Deming TJ (1997) Facile synthesis of block copolypeptides of defined architecture. Nature 390:386–389

Diamandis EP, Christopoulos TK (1996) Fluorescence immunoassays. Academic Press, San Diego

Flounders AW, Brandon DL, Bates AH (1997) Patterning of immobilized antibody layers via photolithography and oxygen plasma exposure. Biosens Bioelectron 12:447–456

Fodor SP, Read JL, Pirrung MC, Stryer L, Lu AT, Solas D (1991) Light-directed, spatially addressable parallel chemical synthesis. Science 251:767–773

Frydman E, Cohen H, Maoz R, Sagiv J (1997) Monolayer damage in XPS measurements as evaluated by independent methods. Langmuir 13:5089–5106

Gao H, Luginbühl R, Sigrist H (1997) Bioengineering of silicon nitride. Sens Actuators B 38–39:38–41

Gao X, Schawaller M, Mathieu HJ (2004) Surface functionalization for biomedical applications. Surf Interface Anal 36:1507–1512

Gao X, Schawaller M, Mathieu HJ (2005) Diagnosis of the plasma used for modification of polystyrene. Unpublished data

Gillen G, Lareau R, Bennett J, Stevie F (1998) Secondary Ion Mass Spectrometry (Proceedings SIMS XI). Wiley, Chichester

Göpel W (1995) Heiduschka P interface analysis in biosensor design. Biosens Bioelectron 10(9–10):853–883

Graf J, Iwamoto Y, Sasaki M, Martin GR, Kleinman HK, Robey FA, Yamada Y (1987) Identification of an amino acid sequence in laminin mediating cell attachment, chemotaxis, and receptor binding. Cell 48:989–996

Hagenhoff B (1995) Surface mass spectrometry – application to biosensor characterization. Biosens Bioelectron 10:885–894

Harrick N (1967) Internal reflection spectroscopy. Interscience, New York

Idage SB, Badrinarayanan S (1998) Surface modification of polystyrene using Nitrogen plasma. An X-ray photoelectron spectroscopy study. Langmuir 14:2780–2785

ISO International Standard 15472 (2001) Surface chemical analysis – X-ray photoelectron spectrometers – calibration of energy scales. ISO, Geneva

Justes DR, Harris RD, Stipdonk MJV, Schweikert AE (1997) A comparison of Cs and C_{60} primary projectiles for the characterization of GaAs and Si Surfaces. Proceedings SIMS XI. Wiley, Orlando, New York, pp 581–584

Kleinfeld D, Kahler KH, Hockberger PE (1988) Controlled outgrowth of dissociated neurons on patterned substrates. J Neurosci 8:4098–4120

Kim SY, Jo Y-A, Choi J, Choi MJ (2001) Characterization of s-triazine antibodies and comparison of enzyme immunoassay and biotin-avidin enzyme immunoassay for the determination of s-triazine. Microchem Journal 68:163–172

Lee YC, Lee RT (1995) Carbohydrate–protein interactions: basis of glycobiology. Acc Chem Res 28:321–327

Léonard D, Chevolot Y, Bucher O, Sigrist H, Mathieu HJ (1998a) TOF-SIMS and XPS study of photoactivatable reagents designed for surface glycoengineering. Part 1: *N*-[*m*-(3-(trifluoromethyl) diazirine-3-yl)phenyl]-4-maleimido-butyramide (MAD) on silicon, silicon nidride and diamond. Surf Interface Anal 26:783–792

Léonard D, Chevolot Y, Bucher O, Haenni W, Sigrist H, Mathieu HJ (1998b) TOF-SIMS and XPS study of photoactivatable reagents designed for surface glycoengineering. Part 2: *N*-[*m*-(3-(trifluoromethyl)diazirine-3-yl)phenyl]-4-(-3-thio(-1-D-galactopyrannosyl)-maleimidyl)butyramide (MAD-Gal) on diamond. Surf Interface Anal 26:793–799

Léonard D, Mathieu HJ (1999) Characterisation of biomaterials using ToF-SIMS. Fresenius J Anal Chem 365:3–11

Léonard D, Chevolot Y, Heger F, Martins J, Crout HG, Sigrist H, Mathieu HJ (2001) ToF-SIMS and XPS study of photoactivable reagents designed for surface glycoengineering. Part III: 5-Carboxamidopentyl-*N*-[*m*-[3-(trifluoromethyl)diazirin-3-yl]phenyl-β-D-glactopyranolsyl]-(1⇒4)-1-thio-β-D-glucopyranoside (lactose aryl diazirine) on diamond. Surf Interface Anal 31:457–466

Luppa PB, Sokoll LJ, Chan DW (2001) Immunosensors-principles and applications to clinical chemistry. Clin Chim Acta 314:1–26

Makohliso SM, Leonard D, Giovangrandi L, Mathieu HJ, Illegems M, Aebischer P (1999) Surface characterization of a biochip prototype for cell-based biosensor applications. Langmuir 15:2940–2946

Mantus DS, Ratner BD, Carlson BA, Moulder JF (1993) Static secondary ion mass spectrometry of. adsorbed proteins. Anal Chem 65:1431–1438

Massia SP, Hubbell JA (1991) An RGD spacing of 440 nm is sufficient for integrin alpha V beta 3-mediated fibroblast spreading and 140 nm for focal contact and stress fiber formation. J Cell Biol 114:1089–1100

Mathieu HJ (2001) Elemental analysis by AES, XPS and SIMS. In: Alfassi ZB (ed) Non-destructive elemental analysis. Blackwell Science, Oxford, pp 201–231

Mathieu HJ, Chevolot Y, Ruiz-Taylor L, Léonard D (2003) Engineering and characterization of polymer surfaces for biomedical applications. Advances in polymer Science, vol 162. Springer, Berlin Heidelberg, pp 1–31

Mathieu HJ, Bergmann E, Gras R (2003b) Analyse et Technologie des Surfaces. Presses Polytechniques et Universitaires Romandes, Lausanne

Metzger SW, Natesan M, Yanavich C, Schneider J, Lee GU (1999) Development and characterization of surface chemistries for microfabricated biosensors. J Vac Sci Technol A 17:2623–2628

Monsigny M (1995) Glycotechnologies: soft biotechnologies. Biofutur 142:27–32

Mueller JP, Bruinink A (1994) Neurotoxic effects of aluminium on embryonic chick brain cultures. Acta Neuropathol 88:359–366

Nicolini C (1995) From neural chip and engineered biomolecules to bioelectronic devices: an overview. Biosens Bioelectron 10:105–127

Odom RW (1994) Characterisation of biomaterials using ToF-SIMS. Appl Spectr Rev 29:67–70

Pearson JE, Kane JW, Petraki-Kallioti I, Gill A, Vadgama P (1998) Surface plasmon resonance: a study of the effect of biotinylation on the selection of antibodies for use in immunoassays. J Immunol Methods 221:87–94

Pei R, Cheng Z, Wang E, Yang X (2001) Amplification of antigen–antibody interactions based on biotin labeled protein–streptavidin network complex using impedance spectroscopy. Biosens Bioelectron 16:355–361

Peterman JH, Tarcha PJ, Chu VP, Butler JE (1988) The immunochemistry of sandwich ELISA. IV, The antigen capture capacity of antibody covalently attached to bromoacetyl polytyrene. J Immunol Methods 111:271–275

Petrak K (1994) Theme issue on glycolipids in drug delivery. Adv Drug Deliv Rev 13(3):211

Quapil G, Schawaller M (2002) Apparatus for performing immunoassays. Patents: WO0114859A1 and EP 1079226A1

Powell CJ, Jablonski A (1999) Evaluation of calculated and measured electron attenuation lengths. J Phys Chem Ref Data 28:19–62

Ranieri JP, Bellamkonda R, Bekos EJ, Vargo TG, Gardella JA, Aebischer P (1995) Neuronal cell attachment to fluorinated ethylene propylene films with covalently immobilized laminin oligopeptides YIGSR and IKVAV. II. J Biomed Mater Res 29:779–785

Ratner BD, Johnston AB, Lenk TJ (1987) Biomaterial surfaces. J Biomed Materials Res 21(1 suppl A):59–89

Ratner BD, Hoffman AS, Schoen FJ, Lemons JE (1996) Biomaterials science – an introduction to materials in medicine. Academic Press, New York

Schawaller M, Quapil G (2003) A cuvette for a reader device for assaying substances using the evanescence field method. Patent: EP20020013189

Schueler BW (1992) Imaging by time-of-flight secondary ion mass spectrometry. Microsc Microanal Microstruct 3:119–139

Sigmund P (1981) Sputtering by particle bombardment I. Physical sputtering of single-element solids. Springer, Berlin, pp 9–72

Sigrist H, Collioud A, Clémence J, Gao H, Luginbühl R, Sänger M, Sundarababu G (1995) Surface immobilisation of biomolecules by light. Opt Eng 34:2339–2348

Storp S (1985) Radiation damage during surface analysis. Spectrochim Acta B40:745–756

Troughton EB, Bain CD, Whitesides GM, Nuzzo RG, Allara DL, Porter MD (1988) Monolayer films prepared by the spontaneous self-assembly of symmetrical and unsymmetrical dialkyl sulfides from solution onto Gold substrates: structure, properties, and reactivity of constituent functional groups. Langmuir 4:365–385

Varki A (1993) Biological roles of oligosaccharides – all of the theories are correct. Glycobiol 3:97–130

Vickerman JC, Briggs D, Henderson DA (1996) The Wiley Static SIMS Library. Wiley, Chichester

Vickerman JC, Gilmore IS (2009) Surface analysis – the principal techniques, 2nd edn. Wiley, Chichester

Xie CX, Mattson MP, Lovell MA, Yokel RA (1996) Intraneuronal aluminum potentiates iron-induced oxidative stress in cultured rat hippocampal neurons. Brain Res 743:271–277

Part II
Natural Recognition Receptors

Chapter 4
Enzyme for Biosensing Applications

Béatrice D. Leca-Bouvier and Loïc J. Blum

Abstract Enzymes are very efficient biocatalysts, which have the ability to specifically recognize their substrates and to catalyze their transformation. These unique properties make the enzymes powerful tools to develop analytical devices. Enzyme-based biosensors associate intimately a biocatalyst-containing sensing layer with a transducer. The transformations catalyzed by an enzyme come with the variations of some physicochemical parameters. The role of the transducer is to convert those physicochemical signals into a measurable electrical signal. In biosensors, enzymes are generally immobilized on or close to the transducer. Depending on the chemical and physical characteristics of the enzyme support, different immobilization techniques can be implemented. The transduction mode will be adapted to the physicochemical parameter that is monitored. The variation of the concentration of a substrate or a product in the course of an enzymatic reaction can be detected with the help of a physical or chemical sensor, which then acts as a transducer. Electrochemical biosensors have been then developed involving oxidoreduction reactions. Optical biosensors are based either on fluorescence, absorbance, and bioluminescence or chemiluminescence measurements. Enzymatic reactions are usually associated with a high enthalpy change, which results in a temperature variation that can be recorded using a thermistor. Gravimetric biosensors are based on a mass variation induced by an enzymatic reaction.

Due to their proteic nature, enzymes are often fragile and this instability results in a decrease in the enzyme activity and consequently in a decrease in the biosensor performances. Although, fortunately, not all enzymes are concerned, this can limit the development of enzyme-based biosensors. The first biosensors described were the size of a pH electrode but now the progress in the transducer technology makes the fabrication of miniaturized systems possible and this allows the

L.J. Blum (✉)
Institut de Chimie et Biochimie Moléculaires et Supramoléculaires, Laboratoire de Génie Enzymatique, Membranes Biomimétiques et Assemblages Biomoléculaires, Université Lyon 1/CNRS, 43 boulevard du 11 novembre 1918, Villeurbanne, 69622, France
e-mail: Loic.Blum@univ-lyon1

M. Zourob (ed.), *Recognition Receptors in Biosensors*,
DOI 10.1007/978-1-4419-0919-0_4, © Springer Science+Business Media, LLC 2010

development of small-sized multi-biosensors and the integration of miniaturized biosensors in lab-on-a-chip-type devices.

Keywords Enzyme · Biosensor · Biocatalysis · Electrochemical detection · Amperometric detection · Potentiometric detection · ENFET · LAPS · Conductometric detection · Impedimetric detection · QCM · SAW · Microcantilever · Calorimetric detection · Optical · SPR · Opt(r)ode · Oxidoreductase · Luciferase · GOD · Chemiluminescence · Electrochemiluminescence · Bioluminescence · Luminol · FAD · NAD · PQQ · Cofactor · Nanoparticles · Carbon nanotubes

Abbreviations

AChE	Acetylcholinesterase
ADP	Adenosine 5′-diphosphate
AMP	Adenosine 5′-monophosphate
ATP	Adenosine 5′-triphosphate
BSA	Bovine serum albumin
BuChE	Butyrylcholinesterase
CL	Chemiluminescence
CNT	Carbon nanotube
DAMAB	*N,N*-Didecylaminomethylbenzene
DH	Dehydrogenase
ECL	Electrochemiluminescence
EDC	Ethyl-3-[1-dimethylaminopropyl]carbodiimide
ENFET	Enzyme field-effect transistor
EuTC	Europium (III) tetracycline
FAD	Flavin adenine dinucleotide
FET	Field-effect transistor
FMN	Flavin mononucleotide
GFOR	Glucose–fructose oxidoreductase
GOD	Glucose oxidase
HRP	Horseradish peroxidase
IDA	Interdigitated array
ISE	Ion-selective electrode
ISFET	Ion-sensitive field-effect transistor
ITO	Indium tin oxide
LAPS	Light-addressable potentiometric sensor
MWCNT	Multiwall carbon nanotube
NAD	Nicotinamide adenine dinucleotide
NBD-PE	Nitrobenzoxadiazole dipalmitoylphosphatidylethanolamine
NHS	*N*-Hydroxysuccinimide

NP	Nanoparticle
NTA	Nitriloacetic acid
OD	Oxidase
OPH	Organophosphorus hydrolase
PDDA	Poly (diallyldimethylammonium chloride)
PEG	Poly (ethylene glycol)
PQQ	Pyrroloquinoline quinone
PVA-SbQ	Poly (vinyl alcohol) bearing styrylpyridinium groups
QCM	Quartz crystal microbalance
Ru-bipy	Ruthenium (II) trisbipyridyl
Ru-dpp	Ruthenium (II) tris(4,7-diphenyl-1,10-phenanthroline)
Ru-phen	Ruthenium (II) tris(1,10-phenanthroline)
SAM	Self-assembled monolayer
SAW	Surface acoustic wave
SPR	Surface plasmon resonance
SWCNT	Singlewall carbon nanotube
TTF-TCNQ	Tetrathiafulvalene-tetracyanoquinodimethane

4.1 Introduction

Biosensors based on enzymes as recognition elements represent the most extensively studied ones. The high specificity of enzyme–substrate interactions and the usually high turnover rates of biocatalysts are at the origin of sensitive and specific enzyme-based biosensor devices. Enzymes were historically the first molecular recognition elements included in biosensors and continue to be the basis for a significant number of publications reported for biosensors in general. The first biosensor, known as "enzyme electrode," was presented by Clark and Lyons (1962). The enzyme glucose oxidase (GOD) was coupled with an amperometric oxygen electrode, thus allowing glucose concentration to be measured. Since this period, enzyme electrodes for the detection of a variety of other substances have been described by associating the relevant immobilized enzymes to different kinds of transducers: electrochemical, optical, acoustic or calorimetric.

4.2 Biocatalysis and Enzyme Specificity

Enzymes are biological catalysts, which allow the chemical reactions essential to the life and multiplication of cells to occur at high rates and with a specificity that prevents the formation of undesirable products. Except a particular group of RNA, enzymes are proteins. Their molecular weights range from about 10,000 to over

1,000,000. Enzymes can be monomeric or oligomeric. In the latter case, they generally have an even number of subunits.

The catalytic activity of enzymes depends on the integrity of the conformation of the native protein. The primary, secondary, tertiary, and if the case arises, quaternary structures of these macromolecular entities are essential for the expression of the catalytic activity.

To be fully active, some enzymes require the presence of a small organic or inorganic molecule called a cofactor. When this is the case, the active enzyme including the cofactor is termed holoenzyme whereas the sole protein part is called apoenzyme or apoprotein. The cofactor is either a metal ion (e.g. Fe^{2+}, Mg^{2+}, Zn^{2+}, Mn^{2+}) or a more complex organic molecule, then called a coenzyme. (e.g. NAD^{+}, FAD). A prosthetic group is a cofactor tightly bound to the enzyme.

The compound transformed in the course of an enzyme-catalyzed reaction is called the substrate.

A major feature of enzymes in comparison with other catalysts is specificity. This characteristic arises from the complementary structures of the substrate and its binding site on the enzyme that is the particular region of the enzyme where enzyme/substrate interactions occur. Not only enzymatic reactions are substrate-specific but they are also product-specific, contrary to uncatalyzed reactions or reactions catalyzed by chemical catalysts which often generate by-products. It is often said that the enzyme specificity is dual because an enzyme is specific for the substrate it acts on and also specific for the type of reaction catalyzed. For example, glucose oxidase (GOD) catalyzes the oxidation of glucose by molecular oxygen with such a specificity that this enzyme acts only on β-D-glucose, the α-anomer being not oxidized. In addition, GOD catalyzes only an oxido-reduction reaction and cannot act on glucose for another type of reaction.

More than a century ago, only a few enzymes were known and they were generally identified by adding the suffix -ase to the name of the substrate transformed. The names of some enzymes were also composed by adding this suffix to a word describing their activity. For example, urease is the enzyme that catalyzes the hydrolysis of urea and glucose DH (dehydrogenase) catalyzes the removal of hydrogen from glucose. However, as the number of discovered enzymes increased (more than 3,000 enzymes were listed in 1992), some confusion arose because the same enzyme might have several names or the same name was used for different enzymes. Thus, it appeared necessary to adopt systematic rules to name and classify enzymes.

4.2.1 Classification of Enzymes

The International Union of Biochemistry and Molecular Biology (formerly the International Union of Biochemistry) has established a system of enzyme nomenclature based on the catalyzed reaction. All enzymes are divided into six major classes. Each class is subdivided into several subclasses. Subclasses are subdivided into sub-subclasses. The type of reaction catalyzed together with the name(s) of the substrate(s) provides a basis for naming individual enzymes. Each enzyme is assigned a code number, prefixed by EC (for Enzyme Commission), consisting of four digits separated by dots, a systematic name and a recommended name. The first digit indicates to which of the six classes the enzyme belongs; the second digit indicates the subclass; the third digit gives the sub-subclass and the fourth digit is the serial number of the enzyme in its sub-subclass. So, the second and third digits describe the kind of reaction catalyzed but there is no general rule because the meanings of these digits are defined separately for each class.

The six major classes are: (1) oxidoreductases, (2) transferases, (3) hydrolases, (4) lyases, (5) isomerases, (6) ligases.

A web version of the enzyme nomenclature including the principles of classification, an historical introduction and the full list of enzymes can be found at http://www.chem.qmul.ac.uk/iubmb/enzyme/. Some examples for each class of enzymes are given below.

4.2.1.1 Oxidoreductases

Oxidoreductases catalyze oxidoreduction reactions. The substrate that is oxidized is considered as a hydrogen donor. The second digit indicates the hydrogen or electron donor (1, a –CHOH– group; 2, aldehyde or ketone; 3, –CH–CH– group, and so on). The third digit refers to the hydrogen or electron acceptor.

The systematic name is donor:acceptor oxidoreductase. The common name is dehydrogenase or reductase. Oxidase is only used if O_2 is the acceptor. Some examples of oxidoreductases are:

Code number	Common name	Systematic name	Reaction
EC 1.1.1.1	Alcohol dehydrogenase	Alcohol: NAD^+oxidoreductase	An alcohol + NAD^+ = an aldehyde or ketone + NADH + H^+
EC 1.1.3.4	Glucose oxidase	β-D-Glucose:oxygen 1-oxidoreductase	β-D-Glucose + O_2 = D-glucono-1,5-lactone + H_2O_2
EC 1.6.2.5	NADPH-cytochrome-$c_2$reductase	NADPH:ferricytochrome-$c_2$oxidoreductase	NADPH + 2ferricytochromec_2 = $NADP^+$ + H^+ + 2ferrocytochromec_2

4.2.1.2 Transferases

Transferases catalyze a group transfer (e.g. methyl, acyl, glycosyl, amino, phosphate) from a donor to an acceptor. Subclasses (second digit) are defined by the type of group transferred. For example, aspartate transaminase (EC 2.6.1.1, L-aspartate:2-oxoglutarate aminotransferase) catalyzes the transfer of the α-amino group of L-aspartate on 2-oxoglutarate to form oxaloacetate (i.e. deaminated aspartate) and L-glutamate (i.e. aminated 2-oxoglutarate):

$$\text{L-Aspartate} + \text{2-oxoglutarate} \Leftrightarrow \text{oxaloacetate} + \text{L-glutamate}$$

4.2.1.3 Hydrolases

Hydrolases catalyze hydrolytic cleavage of C–O, C–N, O–P and C–S bonds. These enzymes can be considered as a special group of transferases since the reaction catalyzed is the transfer of a specific group on water. For this class, the second digit indicates the type of bond hydrolyzed and the third one normally specifies the nature of the substrate. Alkaline phosphatase (EC 3.1.3.1, phosphate-monoester phosphohydrolase (alkaline optimum)) catalyzes the hydrolysis of phosphomonoester into an alcohol and phosphate:

$$\text{A phosphate monoester} + H_2O \Leftrightarrow \text{an alcohol} + \text{phosphate}$$

This enzyme has a rather broad specificity and can act on a variety of substrates including synthetic molecules and nucleotides.

4.2.1.4 Lyases

Lyases are enzymes catalyzing the nonhydrolytic cleavage of C–C, C–O, C–N, and other bonds by elimination, leaving double bonds or rings, or conversely the addition of groups to double bonds. The second digit in the code number indicates the bond broken whereas the third digit gives further information on the group eliminated. As an example, pyruvate decarboxylase (EC 4.1.1.1, 2-oxo-acid carboxy-lyase (aldehyde-forming)) catalyzes the decarboxylation of a keto-acid (e.g. pyruvate) to an aldehyde and carbon dioxide:

$$\text{A 2 - oxo acid} \rightarrow \text{an aldehyde} + CO_2$$

4.2.1.5 Isomerases

Isomerases catalyze isomerization reactions and are classified according to the type of isomerism. Thus, the second digit indicates the type of isomerization reaction and the third digit the type of substrate.

Mutarotase (EC 5.1.3.3, aldose 1-epimerase) catalyzes the mutarotation of α-D-glucose into β-D-glucose.

4.2.1.6 Ligases

Ligases catalyze the formation of new bonds between two molecules coupled with the breakdown of a nucleoside triphosphate, most often ATP. The second and the third digits of the code number indicate the type of new bonds synthesized. Pyruvate carboxylase (EC 6.4.1.1, pyruvate:carbon-dioxide ligase (ADP-forming)) catalyzes the carboxylation of pyruvate in the presence of ATP and bicarbonate to form oxaloacetate:

$$\text{Pyruvate} + \text{ATP} + \text{HCO}_3^- \rightarrow \text{ADP} + \text{phosphate} + \text{oxaloacetate}$$

4.3 Enzyme Immobilization

Enzyme immobilization on the transducer surface is of critical importance. Several criteria must be taken into consideration when selecting a method of immobilization: the enzyme has to be stable during the reaction, the reagents forming cross-linking bonds should not reach the enzyme's active center which must be protected, the washing of unbound enzyme must not be detrimental for the immobilized enzyme, the mechanical properties of the carrier should be considered. Techniques for immobilization can be broadly classified into several categories, namely adsorption, entrapment, cross-linking and covalent binding. Combination of one or more of these techniques can also be envisioned.

4.3.1 Confinement Within a Semipermeable Membrane

The first biosensor was based on GOD immobilization on an oxygen electrode via a semipermeable dialysis membrane (Clark and Lyons 1962). Dialysis (ultrafiltration) membranes have been used in some biosensor configurations (Lee et al. 1994a; Lojou and Bianco 2004). Enzyme is imprisoned between the electrode and a dialysis membrane having a cut-off inferior to the size of the enzyme, while the substrate is able to freely diffuse through the dialysis membrane. Such immobilization method allows small amounts of enzyme to be used and the integrity of the enzyme to be preserved.

4.3.2 Adsorption

Physical adsorption is the easiest and the less denaturating method for enzyme immobilization. The procedure consists of simple deposition of enzyme onto the electrode material and attachment through weak bonds such as Van der Waals and electrostatic interactions. The main advantages are simplicity, cheapness and the possibility of subsequent enzyme (sensor) regeneration. However, biosensors with adsorbed enzymes suffer from poor operational and storage stability, so that adsorption followed by cross-linking has also been used for the immobilization of enzymes. A layer-by-layer technique based on alternate layers of oppositely charged polyelectrolyte and enzyme has been reported, with cholesterol OD (oxidase) and poly (diallyldimethylammonium chloride) (PDDA) acting as a cationic polyelectrolyte on a multiwall carbon nanotube (MWCNT)-modified electrode (Guo et al. 2004). Penicillinase has also been immobilized by adsorption on an ISFET to design a penicillin-sensitive ENFET (Poghossian et al. 2001a). Recently, acetylcholinesterase (AChE) and choline OD have been successfully adsorbed onto the surface of carbon nanotube-based screen-printed electrodes (Lin et al. 2004; Joshi et al. 2005a).

4.3.3 Affinity Interactions

Adsorption can also involve stronger bonds, even if they are not as strong as the covalent ones. Immobilization of enzymes can be achieved by affinity interactions between functional groups such as lectins, (strept)avidin, or metal chelates on an activated electrode surface and an affinity tag such as carbohydrate residues, biotin, histidine or cysteine which is present or genetically engineered at a specific location in the protein sequence and which does not affect the activity nor the folding of the protein. It provides a basis for controlled and oriented immobilization of the enzyme on different supports (Bucur et al. 2004; Andreescu and Marty 2006). The strong affinity link between a Ni complexed nitriloacetic acid (NTA) chelate and the hexahistidine residue has been used to immobilize a $(His)_6$-AChE by affinity linkage, the tag being genetically engineered in the protein sequence. HRP modified with histidine has also been immobilized on imidodiacetic acid – Sepharose beads to design an optical biosensor based on luminol chemiluminescence (CL) for hydrogen peroxide detection (Tsafack et al. 2000b). Immobilization on imidodiacetic acid-Sepharose beads was shown to enhance the sensing layer stability and to enable the immobilization of a larger amount of enzyme. The mannose residue naturally present in AChE has also been used to create an affinity binding with concanavalin A (a lectin which has multiple sites with high affinity for carbohydrates) deposited on the surface of the electrode. Based on the affinity complex between GOD (glycoprotein) and concanavalin A, a cross-linked GOD/peroxidase monolayer has also been assembled on the lectin layer and associated with a fluoride-sensitive FET to detect glucose (Köneke et al. 1996). The strong affinity bond between (strept)avidin–biotin has also been used, for example, in an impedimetric biosensor for hydrogen peroxide detection, with streptavidin-conjugated HRP attached on a mixed SAM formed by 1,2-dipalmitoyl-sn-glycero-3-phosphoethanolamino-*N*-(biotinyl) and 1,6-mercaptohexadecanoic acid (Esseghaier et al. 2008).

4.3.4 Entrapment

Physical immobilization methods such as entrapment in photopolymerized, electropolymerized or sol–gel matrices have also been widely used. These methods are easy to perform and allow a simple one-step fabrication of the electrodes in which enzyme, mediators and additives can be simultaneously deposited in the same layer. They do not require a modification of the biological component. Biosensors based on physically entrapped enzymes are often characterized by an increased operational and storage stability. The activity of the enzyme is preserved during the immobilization process. However, limitations such as leaching of biocomponents and possible diffusion barriers can restrict the performances of the biosensors. Poly (vinyl alcohol) bearing styrylpyridinium groups (PVA-SbQ) has been used to efficiently entrap enzymes such as choline OD (Leca et al. 1995), alcohol DH (Leca and Marty 1997a) aldehyde DH or AChE (Noguer et al. 1999) to give amperometric biosensors for choline, ethanol or pesticide detection, respectively. Highly sensitive optical biosensors have also been designed using this photopolymerizable immobilization matrix (Leca and Blum 2000; Tsafack et al. 2000a; Godoy et al. 2005). Numerous enzymes have also been entrapped into electropolymerized films (Cosnier 2003; Wang and Musameh 2005). For instance, an amperometric biosensor based on tyrosinase entrapment in electrogenerated polythiophene film has been reported (Védrine et al. 2003). Electrodeposition of a conducting polymer such as polypyrrole or polyaniline in the presence of the enzyme enables control of the polymer/enzyme layer thickness, which is extremely simple and rapid but requires a significant amount of enzyme. Polypyrrole has also been used for electrical wiring of enzymes and carbon nanotubes (CNTs) to the underlying electrode, the CNTs being coentrapped with GOD in the polymer network (Wang and Musameh 2005). Enzymes have also been immobilized in hydrogels or sol–gel materials (Salimi et al. 2004; Joshi et al. 2005b). Sol–gel is well-suited for fabrication of optical sensor membranes because this material does not absorb in the near UV and visible rays (Wolfbeis et al. 2000). Sol–gel immobilization enables the activity of biocomponents to be retained over a long time. A broad range of enzymes has been successfully immobilized on CNT-incorporated redox hydrogels to yield sensitive biosensors (Joshi et al. 2005b). L-Amino acid OD has been coimmobilized with MWCNTs in a sol–gel matrix (Gavalas et al. 2004). Urease has also been immobilized by the sol–gel method to design a conductometric biosensor for urea determination (Lee et al. 2000a). Apart from hydrogels and sol–gels, other composite electrodes have been reported, for example by coimmobilizing GOD and MWCNTs in a Nafion matrix (Tsai et al. 2005).

4.3.5 Cross-linking

Enzymes can also be cross-linked with glutaraldehyde and bovine serum albumin (BSA) to form a polymer network located on a support in optical biosensors (Rhines and Arnold 1989; Schaffar and Wolfbeif 1990; Zhang et al. 1997). Several conductometric devices based on the immobilization of enzymes in a gel obtained by coreticulation with glutaraldehyde in the presence of BSA have also been reported (Watson et al. 1987; Bilitewski et al. 1992; Shul'ga et al. 1994). A conductometric biosensor for nitrate determination has also been developed using nitrate reductase immobilized on a thin-film electrode by cross-linking with BSA in the presence of glutaraldehyde, Nafion and methyl viologen (Wang et al. 2006). The glutaraldehyde cross-linking procedures are attractive due to their simplicity and the strong chemical bonding achieved. The main drawback is the possibility of activity losses due to the distortion of the active enzyme conformation and the chemical alterations of the active site during cross-linking (Willner and Katz 2000).

4.3.6 Covalent Binding

Covalent binding is an extensively used technique for the immobilization of enzymes. Biocatalysts can be covalently linked to the support through functional groups they contain and that are not essential for their catalytic activity. Many procedures have been reported to covalently link an enzyme to a support (Willner and Katz 2000). The enzyme can be coupled with surface amino residues by using glutaraldehyde. The amino functional groups of lysine residues in enzymes can be linked to surface aldehyde functional groups to yield imine bonds or to surface carboxy groups to generate amide bonds. Aromatic amines linked to surfaces enable the generation of the surface-bound diazonium salt, which reacts with tyrosine or histidine residues to link the enzymes to the solid support through azo groups. Covalent bonding is also possible through the carbohydrate moiety when a glycoprotein is concerned. In general, functional aldehyde group can be introduced in a glycoprotein by oxidizing the carbohydrate moiety by periodate oxidation without significantly affecting the enzyme activity. The procedure provides increased stability of the enzyme but requires a high amount of bioreagent and is generally poorly reproducible. Optical biosensors have often been reported where enzymes have been covalently immobilized onto preactivated polyamide or poly (vinylidenedifluoride) membrane supports such as Immunodyne (Gautier et al. 1990b; Blum et al. 1993; Hlavay and Guilbault 1994; Spohn et al. 1995), Biodyne (Xie et al. 1992b; Wolfbeis and Li 1993) or Immobilon (Hlavay et al. 1994; Mascini 1995). Cholesterol OD has been immobilized in polyaniline on an ITO electrode, by covalent immobilization via ethyl-3-[1-dimethylaminopropyl]carbodiimide (EDC)/*N*-hydroxysuccinimide (NHS), with an optical detection by spectrophotometry at 500 nm using *o*-dianisidine (Dhand et al. 2007). GOD has also been covalently immobilized on semiconducting SWCNTs via a linker molecule in an ENFET format (Chen et al. 2001; Besteman et al. 2003). An impedimetric biosensor for glucose detection has also been reported. The amine groups of GOD were coupled to the acidic head groups of gold-mercaptopropionic acid self-assembled monolayers through the formation of imide using carbodiimide and NHS.

4.4 Enzyme-Based Biosensors: Different Transduction Modes

4.4.1 Electrochemical Detection

The first enzyme electrode developed by Clark and Lyons (1962) was a first-generation electrochemical biosensor. Measurements were based on the monitoring of the oxygen consumed by the enzyme-catalyzed reaction. Updike and Hicks (Updike and Hicks 1967) further developed this principle by using a differential two oxygen electrode system in order to correct for the oxygen background variation in samples. Guilbault and Lubrano (1973) described an enzyme electrode for the measurement of blood glucose based on amperometric monitoring of the hydrogen peroxide product. Biosensors for glucose detection are still the most widely used, although many improvements (generations) have been added since the 1960s. Electrochemical detection techniques use predominantly enzymes as biorecognition elements. They play a prevailing role in the design and development of biosensors. As portable, low-cost and simple-to-operate analytical devices, electrochemical biosensors present considerable advantages over conventional analytical instruments including optical, piezoelectric and thermal biosensors. Other advantages are their robustness, easy miniaturization and ability to be used in turbid media with optically absorbing and fluorescing compounds (Grieshaber et al. 2008). According to the official nomenclature recommended by IUPAC (Thevenot et al. 2001), electrochemical transducers include amperometric, potentiometric, conductometric, impedimetric and semiconductor field-effect principles.

4.4.1.1 Amperometric Biosensors

A wide range of amperometric enzyme electrodes, differing in enzyme nature, electrode design or material, immobilization approach, or membrane composition, has been described. Three generations of amperometric enzyme electrodes have evolved during the past 40 years. Mostly, oxygen and hydrogen peroxide based systems are classified as "first-generation" biosensors whereas mediated systems are "second-generation" biosensors. More recently described, "third-generation" biosensors based on direct electron transfer between an enzyme and an electrode have also been developed. As underlined by Schuhmann's group (Habermüller et al. 2000), "the border between mediated and direct electron transfer is often not easy to draw."

First-generation biosensors. Since the first biosensor (Clark and Lyons 1962) based on GOD fixed on top of a Clark oxygen electrode by means of a semipermeable membrane, many biosensors based on oxidases, detecting either oxygen consumption or the enzymatically produced hydrogen peroxide have been developed. First-generation glucose biosensors rely on the use of GOD and involve reduction of the flavin group (FAD) in the enzyme by reaction with glucose to give the reduced form of the enzyme ($FADH_2$) followed by its reoxidation by molecular oxygen to regenerate the oxidized form of the enzyme (FAD). The drawback of oxygen based sensors is the dependence on the oxygen concentration which is difficult to maintain at a constant level. A differential two oxygen electrode system can be used to correct this effect (Updike and Hicks 1967). To decrease the interference of coexisting electroactive compounds during amperometric (anodic) measurement of hydrogen peroxide, one useful avenue is to employ a permselective coating that minimizes the access of these constituents to the electrode surface. Different polymers, multilayers, and mixed layers with transport properties based on charge, size, or polarity have thus been used for blocking coexisting electroactive compounds, especially electropolymerized films such as poly(phenylenediamine), polyphenol, overoxidized polypyrrole (Sasso et al. 1990; Palmisano et al. 1993, 2000), size-exclusion cellulose acetate films (Sternberg et al. 1988; Vaidya and Wilkins 1994) or the negatively charged ones (sulfonated) Nafion or Kodak AQ ionomers (Moussy et al. 1994).

Second-generation biosensors. Artificial redox mediators have been used instead of natural cosubstrates to design second-generation biosensors. The electron transfer mediator shuttles the electrons between the redox center of the enzyme and the electrodes. Electron transfer kinetics must be fast to compete with the regeneration of the enzyme via reduction of oxygen. Using these electron-carrying mediators, measurements become largely independent of oxygen partial pressure and can be carried out at lower potentials without interference from coexisting electroactive species. The mediator should also possess good electrochemical properties (such as a low pH-independent redox potential). Most commonly used mediators are ferrocene and its derivatives, ferrocyanide, methylene blue, benzoquinone or *N*-methyl phenazine. A reasonable use of NAD^+-dependent enzymes (DH) that need a high potential to oxidize NADH can also be envisioned (Leca and Marty 1997b). In the simplest configuration, artificial redox mediators such as ferrocene derivatives, quinones or Os-complexes are added to the sample and used as freely-diffusing electron transfer shuttles (Schlapfer et al. 1974). Adsorption of low-soluble redox mediators on the electrode surface followed by the immobilization of the enzyme in a second layer has also been reported (Cass et al. 1984). This process can suffer from sample contamination and unsatisfactory long-term stability due to the possible mediator leakage. The mediator can also be immobilized by entrapment in carbon paste (Kaku et al. 1994), hydrogel (Lange and Chambers 1985) or in conducting polymers (Kajiya et al. 1991; Seker and Becerik 2004). The mediator may be electrochemically codeposited with enzyme, as described for example with GOD entrapped in a polypyrrole film with hydroquinone sulfonate (Kajiya et al. 1991) or with *p*-benzoquinone (Seker and Becerik 2004). Electrodeposition of HRP in polypyrrole has also been reported on the surface of ferrocenecarboxylic acid mediated sol–gel derived composite carbon electrode (Tian et al. 2001). Slow mediator leakage can occur from the electrode thus limiting long-term stability. Such a drawback is not present with covalently linked mediators, where electron transfer does not occur via a shuttle mechanism but rather via a sequence of hopping reactions between redox mediators (relays) that are covalently attached to an enzyme or to a matrix. Two approaches have been described to design the so-called "reagentless" biosensors: "electroenzymes" and "wired enzymes." On one hand, the protein itself has been modified with covalently bound redox mediators (Degani and Heller 1987; Tsai and Cass 1995; Bartlett et al. 1997). Chemical modification of GOD with electron-relay groups (covalent attachment of ferrocene or TTF) represents an attractive route for facilitating the electron transfer between the GOD redox center and the electrode surface (Degani and Heller 1987; Bartlett et al. 1997). A printed carbon electrode has also been modified with adsorbed ferrocene-modified HRP. The electrode has been shown to be suitable for the determination of hydrogen peroxide in the range from 10^{-6} to $5\,10^{-5}$M (Tsai and Cass 1995). On the other hand, "wired" enzyme electrodes have been obtained by using redox hydrogels or redox conducting polymers. The redox species have been covalently bound to the backbone of the immobilization matrix, leading to redox polymers suitable to "wire" enzymes with the electrode surface. "Redox hydrogels" mainly consist of poly (vinyl pyridine) (Heller 1990; Pishko et al. 1990; Csoeregi et al. 1995; Iyer et al. 2003), poly(vinyl imidazole) (Ohara et al. 1994) or poly(allyl amine) (Calvo et al. 2001; Fei et al. 2003) backbone with covalently bound osmium-complexes. Such redox polymers allow a high mediator loading to be achieved and a good sensitivity to be obtained, coupled with a very high selectivity (e.g. negligible interferences from ascorbic or uric acids at +0.2 V vs. SCE) (Ohara et al. 1994). Conducting polymers (such as polypyrrole) can be modified with redox mediators, either prior to the electropolymerization process by using mediator-modified monomers (Schuhmann et al. 1991) or after the formation of the polymer film (Hiller et al. 1996). Integration of monomers with covalently-bound ferrocene units (Foulds and Lowe 1988) or osmium complexes (Schuhmann et al. 1993) into polypyrrole has been described. The reconstitution process has also been developed as a versatile method to electrically contact redox enzymes with electrodes and as a generic approach to develop amperometric biosensors (Riklin et al. 1995). It has been recently reviewed by Willner's group (Zayats et al. 2008). The principle is based on the reconstitution of the apo-enzyme on a relay-cofactor-functionalized electrode. Different relay units (molecular redox-active relays, molecular redox-active "shuttles," redox-active polymers such as polyaniline, gold nanoparticles (NPs), CNTs) have been used to electrically communicate FAD-containing enzymes (flavoenzymes) such as GOD (Willner et al. 1996; Xiao et al. 2003; Patolsky et al. 2004), cholesterol OD (Vidal et al. 2004) or pyrroloquinoline quinone (PQQ)-containing enzymes such as PQQ-dependent glucose DH (Zayats et al. 2005) with electrodes. Very sensitive amperometric biosensors have been obtained, with high selectivity. Oxygen-insensitive amperometric electrodes for glucose detection which are also insensitive to common glucose sensing interferants such as ascorbic acid or uric acid have been reported (Willner et al. 1996; Xiao et al. 2003). The PQQ electron relay units have been shown to mediate electron transfer between the FAD sites and the electrode (Willner et al. 1996; Vidal et al. 2004). Gold nanoparticles (NPs) have also been shown to act as "electrical nanoplugs" (relay units) for the electrical wiring of FAD (Xiao et al. 2003) or PQQ (Zayats et al. 2005) redox-active centers of GOD or PQQ-dependent glucose DH, respectively. Carbon nanotubes (CNT) can also provide a favorable surface orientation and act as an electrical connector between the enzyme redox center and the electrode surface. Efficient direct electrical connection with GOD has been reported by Willner's group (Patolsky et al. 2004) and by Gooding's group (Liu et al. 2005) in connection to aligned SWCNT arrays.

Third-generation biosensors are based on direct electron transfer between the enzyme and the electrode. Some examples are given in Table 4.1, showing performances in terms of linear range of detection. The absence of a mediator is the main advantage, leading to a very high selectivity (owing to the very low operating potential, close to that of the redox potential of the enzyme). A third-generation biosensor for lactose detection has been reported, based on a newly discovered cellobiose DH, immobilized by simple physical adsorption on a solid spectrographic graphite electrode, leading to a linear range from 10^{-6} to 10^{-4}M of lactose (Table 4.1, Stoica et al. 2006). However, efficient direct electron transfer at conventional electrodes has been reported only for few redox enzymes. Special materials have been used for modification of electrode surface to achieve a direct electron transfer between enzymes and electrodes, among which are organic conducting salts, conducting polymers, DNA, nanoparticles (NPs) or carbon nanotubes (CNTs). Even if there is no convincing evidence for direct electron transfer, TTF-TCNQ is reported to have interesting electrochemical properties as electrode material for biosensors of the third generation. It has been used with different enzymes such as GOD (Khan et al. 1996; Palmisano et al. 2002; Pandey et al. 2003; Llopis et al. 2005; Cano et al. 2008), xanthine OD (Korell and Spichiger 1993) or peroxidase (Korell and Spichiger 1994). Several configurations have been described for the incorporation of TTF-TCNQ as part of these biosensors, i.e. grown at the surface of a shapable electroconductive film (Khan et al. 1996), through the nonconducting (overoxidized) polypyrrole film (Palmisano et al. 2002), forming a paste electrode (Pandey et al. 2003), integrated in an epoxy-graphite biocomposite (Llopis et al. 2005) or in a recently reported PVC/TTF-TCNQ composite containing only the TTF-TCNQ salt as the conducting phase (Cano et al. 2008). A detection limit of 8.5 10^{-6}M of glucose was obtained in batch (Cano et al. 2008), a value similar to that obtained for other recent glucose biosensors that use gold NP-modified electrodes (Mena et al. 2005). TTF-TCNQ/MWCNTs nanocomposite films have also been recently proposed for the first time to immobilize HRP and to design a performant third-generation biosensor for H_2O_2 detection (Cao et al. 2008). A linear range from 5 10^{-6} to 1 10^{-3}M of H_2O_2 was reported. The TTF-TCNQ/MWCNTs film seems to be a good HRP immobilization matrix to achieve direct electron transfer between the enzyme and the gold electrode. The adsorption of MWCNTs on the electrode surface improved the growth of tree-like TTF-TCNQ crystals for direct electron communication between the active center of HRP and the electrode. Mediatorless third-generation glucose biosensors based on the GOD/polypyrrole system have also been suggested (Yabuki et al. 1989; Koopal et al. 1991). However, the relatively high anodic potential of this

Table 4.1 Examples of enzyme-based third-generation amperometric biosensors

Enzyme/detected substance	Modified electrode	Potential	Linear range (M)	References
HRP/H_2O_2	Functionalized nanocrystalline diamond	+0.05 V vs. Ag/AgCl	10^{-4}–4.5×10^{-2}	Rubio-Retama et al. (2006)
	Silver NPs/cysteamine/gold electrode	−0.2 V vs. Ag/AgCl	3.3×10^{-6}–9.4×10^{-3}	Ren et al. (2005)
	Gold NPs/silica sol–gel/gold electrode	−0.1 V vs. SCE	1.6×10^{-6}–3.2×10^{-3}	Di et al. (2005)
		−0.25 V vs. Ag/AgCl	5×10^{-6}–10^{-2}	Jia et al. (2002)
	DNA films/gold electrode	−0.2 V vs. Ag/AgCl	10^{-5}–9.7×10^{-3}	Song et al. (2006)
	MWCNTs/gold electrode	+0.2 V vs. SCE	5×10^{-7}–10^{-5}	Xu et al. (2005b)
GOD/glucose	Growing TTF-TCNQ tree-like crystals	+0.25 V vs. Ag/AgCl	Up to 4×10^{-2}	Khan et al. (1996)
			Up to 8×10^{-3}	Palmisano et al. (2002)
	Oxidized boron-doped diamond electrode	+0.35 V vs. SCE	6.7×10^{-5}–2×10^{-3}	Wu and Qu (2006)
	TTF-TCNQ – sol–gel in graphite paste electrode	+0.3 V vs. Ag/AgCl	Up to 8×10^{-2}	Pandey et al. (2003)
	Colloidal gold modified carbon paste electrode	−0.5 V vs. SCE	4×10^{-5}–2.8×10^{-4}	Liu and Ju (2003)
	PVC/TTF-TCNQ composite electrode	+0.2 V vs. Ag/AgCl	10^{-4}–8×10^{-3}	Cano et al. (2008)
Superoxide dismutase/O_2^-	SAMs on gold electrode	+0.3 V or−0.2 V vs. Ag/AgCl	1.3×10^{-8}–1.3×10^{-7}	Tian et al. (2002)
	SAMs with gold NPs on carbon fiber microelectrode	+0.25 V or−0.15 V vs. Ag/AgCl	1.3×10^{-8}–10^{-7}	Tian et al. (2005)
	Gold NPs/silica sol–gel	−0.15 V vs. SCE	5×10^{-8}–4×10^{-7}	Di et al. (2007)
Tyrosinase/bisphenol-A	Oxidized boron-doped diamond electrode	−0.3 V vs. Ag/AgCl	10^{-6}–10^{-4}	Notsu et al. (2002)
Putrescine OD/putrescine	MWCNTs dispersed in PDDA/glassy carbon electrode	−0.45 V vs. Ag/AgCl	5×10^{-6}–10^{-4}	Rochette et al. (2005)
Cellobiose DH/lactose	Adsorption/spectrographic graphite electrode	+0.3 V vs. Ag/AgCl	10^{-6}–10^{-4}	Stoica et al. (2006)

system compared to the redox potential of FAD/$FADH_2$ suggests the possibility of electron transfer mediated by oligomeric pyrroles present on the surface. Carbon nanotubes have also been used to develop composite biosensors, such as MWCNTs/chitosan (Qian and Yang 2006; Zhou et al. 2007) or polyaniline/MWCNTs (Luo et al. 2006). HRP has been immobilized by electrostatic adsorption on a gold electrode modified with DNA (Song et al. 2006) or silver NPs (Ren et al. 2005) which were adsorbed on self-assembled monolayers of cysteamine. Hydrogen peroxide detection was performed at −0.2 V vs. Ag/AgCl and gave a linear range extending from 10^{-5} to $9.7\ 10^{-3}$M with DNA (Table 4.1, Song et al. 2006) or from $3.3\ 10^{-6}$ to $9.4\ 10^{-3}$M with silver NPs (Table 4.1, Ren et al. 2005). DNA not only provided a biocompatible microenvironment for the enzyme, but also acted as a conductive polymer allowing direct electron transfer between HRP and the electrode surface and greatly amplifying the surface coverage of HRP molecules (Song et al. 2006). Superoxide dismutase-based third-generation biosensors for superoxide anion detection have also been reported (Table 4.1, Tian et al. 2002, 2005; Di et al. 2007). Electron transfer of superoxide dismutase could be efficiently facilitated by self-assembled monolayer (SAM) of thiols and disulfides confined on gold electrodes (Tian et al. 2002) and was adapted to implantable and miniature third-generation carbon fiber microelectrode biosensors using cysteine modified gold NPs for enzyme immobilization (Tian et al. 2005). GOD and HRP have been immobilized onto TiO_2-based nanostructured silicon surfaces that could be easily miniaturized at a low cost (Viticoli et al. 2006). Linear responses extended from 5 10^{-6} to $5.5\ 10^{-4}$M and from 10^{-6} to $2\ 10^{-4}$M for glucose and hydrogen peroxide, respectively, with a response time of a few seconds (7 and 6 s, for glucose and hydrogen peroxide, respectively). MWCNTs dispersed in PDDA have been used to design an efficient biosensor for putrescine detection to be rapid and selective, allowing direct electron transfer between putrescine OD and the glassy carbon electrode (Table 4.1, Rochette et al. 2005). Boron-doped polycrystalline diamond electrodes have also been used for direct electron transfer HRP-based biosensors (Tatsuma et al. 2000). HRP has also been covalently immobilized on the surface of functionalized nanocrystalline diamond thin films, thus leading to a third-generation biosensor for hydrogen detection with a linear range extending from 10^{-4} to $4.5\ 10^{-2}$M at +0.05 V vs. Ag/AgCl (Table 4.1, Rubio-Retama et al. 2006). A tyrosinase-modified boron-doped diamond electrode has been described for determining phenols (Table 4.1, Notsu et al. 2002). Oxidized boron-doped diamond electrodes also indicated recently some promise for mediator-free glucose detection based on direct electron transfer. In this case, GOD was immobilized on the carboxylated electrode surface by being cross-linked with glutaraldehyde in the presence of BSA and oxygen had no effect on the electron transfer. The response time was less than 5 s and the linear range extended from $6.7\ 10^{-5}$ to 2 10^{-3}M of glucose (Table 4.1, Wu and Qu 2006). Many improvements have been performed, but it seems that progress still remains to be made in the charge transport between FAD redox center and electrodes (Wang 2008).

4.4.1.2 Potentiometric Biosensors

The most common potentiometric devices are pH electrodes. A glass membrane or another membrane electrode is used for measuring the membrane potential resulting from the difference in the concentrations of H^+ or other positive ions across the membrane. Three types of ion-selective electrodes (ISEs) are used in biosensors: glass electrodes for cations (e.g. normal pH electrodes) (selectivity of the membrane is determined by the composition of the glass), glass pH electrodes coated with a gas-permeable membrane selective for CO_2, NH_3 or H_2S (change in pH of a sensing solution between the membrane and the electrode), solid-state electrodes where the glass membrane is replaced by a thin membrane of a specific ion conductor made from a mixture of silver sulfide and a silver halide (e.g. iodide electrode responding to I^- or cyanide ions). The use of ion-selective membranes can thus make these transducers sensitive to various ions (e.g. H^+, F^-, I^-, Cl^-) in addition to gases such as CO_2 and NH_3. A potentiometric biosensor for detection of tributyrin and urea has been reported (Basu et al. 2005). Change of pH was measured, by production of butyric acid and ammonia, respectively. The range of detection varied from $5\ 10^{-3}$ to $1.5\ 10^{-2}$M for tributyrin and from $1.5\ 10^{-2}$ to $6\ 10^{-2}$M for urea. Potentiometric biosensors can be fabricated using large numbers of ions and are thus compatible with many enzyme-catalyzed reactions: H^+ cation produced by GOD, penicillinase, urease, lipase, etc.; $NH_4{}^+$ cation produced by L-AAOD, asparaginase, urease, etc.; I^- anion consumed by peroxidase; CN^- anion produced by β-glucosidase (Merkoçi et al. 1999). A renewable potentiometric biosensor for determining Hg^{2+} has been reported using urease inhibition (Yang et al. 2006). Gold nanoparticles (NPs) were chemically adsorbed on the PVC-NH_2 matrix membrane pH electrode surface containing *N,N*-didecylaminomethylbenzene (DAMAB) as a neutral carrier and urease was then immobilized on the gold NPs. The linear range was $9\ 10^{-8}$–$2\ 10^{-6}$M for Hg^{2+} detection. The fabrication of a polyaniline-based biosensor for choline determination has also been reported (Langer et al. 2004). Choline OD was immobilized inside nanostructured polyaniline layers of a controlled porosity and micrometer or nanometer thickness. A linear range extending up to $3.5\ 10^{-2}$M was reported. Polypyrrole sensitivity to NH_3 has also been used to produce potentiometric biosensors (Pandey and Mishra 1988). The interaction of NH_3 with polypyrrole was used to detect urea with urease immobilized in an electrodeposited polypyrrole layer. A low sensitivity was obtained, with strong interference in the presence of potassium and sodium. An electrode for creatinine detection has also been developed by coimmobilization of creatininase, creatinase and sarcosine OD in a polypyrrole matrix (Yamato et al. 1995). Disposable potentiometric strips have recently been fabricated exclusively by means of screen-printing technology (Tymecki et al. 2005). Urease was incorporated in a graphite-ruthenium dioxide paste used for the printing of pH-sensitive films. Potentiometric detection of millimolar concentrations of urea and heavy metal ions was reported.

4.4.1.3 ENFETs, Capacitive Electrolyte–Insulator–Semiconductor and LAPS Biosensors

Silicon-type (semiconductor-based) field-effect biosensors comprise three types of devices that have been coupled with enzyme-catalyzed reactions: enzyme-field-effect transistors (ENFETs, enzyme FETs) and two capacitive field-effect biosensors which are capacitive electrolyte–insulator–semiconductor biosensors and light-addressable potentiometric sensors (LAPS).

Biosensors based on ion-selective field-effect transistors (ISFETs), considered earlier as a category of potentiometric sensors, are now, according to the last IUPAC technical report on electrochemical biosensors (Thevenot et al. 2001), separated into the different classes of electrochemical sensors. The concept to design an ENFET, i.e. to combine enzymes with an ISFET was realized in 1979 and 1980 with biosensors for urea (Danielsson et al. 1979) or penicillin (Caras and Janata 1980) detection, respectively. Later, a multitude of ISFETs were developed with the application of a wide variety of enzymes (Dzyadevych et al. 2006 and references therein). These robust systems with small size and weight are attractive because of their miniaturization and their compatibility with advanced microfabrication technology. Most reported enzyme-modified FETs are built-up of pH-sensitive ISFETs, where hydrogen ions are produced or consumed by the enzymatic reaction. Many ENFETs differing in their sensor design or gate material, enzyme-membrane composition or immobilization method have been reported for detection of glucose (Yin et al. 2001; Park et al. 2002; Luo et al. 2004a, b), urea (Rebriiev and Starodub 2004), penicillin (Poghossian et al. 2001a, b; Poghossian and Schöning 2004;), lactate (Zayats et al. 2000), cyanide (Volotovsky and Kim 1998), organophosphorus pesticides (Abdelmalek et al. 2001; Simonian et al. 2004) or glycoalkaloids (Soldatkin et al. 2005). BuChE-immobilized Langmuir–Blodgett films have been reported to realize ENFET for detection of trichlorfon, an organophosphorus pesticide (Abdelmalek et al. 2001). Enzyme/stearylamine mixed Langmuir–Blodgett films were immobilized onto the gate of the ISFET (Si/SiO_2) and treated by glutaraldehyde vapor to improve the stability. A concentration of 0.1 ppm could clearly be detected using a 30 min incubation period. A highly sensitive and long lifetime penicillin-sensitive ENFET has also been developed by adsorption of penicillinase on a Ta_2O_5-gate ISFET (Poghossian et al. 2001a). A low detection limit of $5\ 10^{-6}$M was obtained. A pH-sensitive organic field-effect transistor based on GOD immobilized on a Ta_2O_5 has also been reported, with a linear response observed between 10^{-5} and 10^{-2}M (Bartic et al. 2003). Recent activities in nanobiotechnology rely on the use of enzyme-nanoparticle hybrids as functional components of nanoscale enzyme-based FETs (Xu et al. 2005a). GOD has also been covalently immobilized to the nanotube sidewall of a single wall carbon nanotube (SWCNT) transistor and was found to induce a clear change in the transistors' conductance (Besteman et al. 2003). Direct real-time detection of glucose was achieved by using GOD modified-SWCNTs in a FET device through the monitoring of the change in the conductance upon the binding of glucose to the enzyme. On the other hand, a conducting polymer nanojunction sensor has also been used for the detection of glucose (Forzani et al. 2004). GOD was electrochemically entrapped into polyaniline. Upon glucose exposure, the hydrogen peroxide that was produced then oxidized polyaniline, thus triggering an increase in the polyaniline conductivity which constituted the analytical signal. The small size of the nanojunction sensor led to very fast responses (<200 ms).

Enzyme-modified capacitive field-effect biosensors have also been designed for the detection of various analytes (Poghossian and Schöning 2007 and references therein). Nevertheless, in spite of their very simple structure, these biosensors are less investigated than ENFETs.

Enzyme-modified LAPS have also been realized for the detection of penicillin (Poghossian et al. 2001b), urea and butyrylcholine (Mourzina et al. 2004). Enzyme-based LAPS for the determination of urea and butyrylcholine have been elaborated with photocurable polymeric enzyme membranes of urease and BuChE, respectively (Mourzina et al. 2004). LAPS is a silicon-based detector that takes advantage of the photovoltaic effect to selectively determine the point of measurement. By scanning with a focused light source, it is possible to determine the spatially resolved surface potential distribution along the interface of the sample and substrate surfaces. The attractive feature of LAPS is their addressability but the main disadvantage is the necessity of a light pointer and the light sensitivity. Like ENFETs and capacitive field-effect biosensors, these miniaturized devices are suitable for multisensory applications.

4.4.1.4 Conductometric and Impedimetric Biosensors

Rather insignificant attention has been paid to conductometric biosensors in comparison with other transducers. Their advantages rely on the fact that they do not need the use of a reference electrode, they are insensitive to light and they can be miniaturized and integrated easily by using a cheap thin-film standard technology. The main disadvantage is that the ionic species produced have to significantly change the total ionic strength to obtain reliable measurement. Conductometric technique involves the measurement of changes in conductance due to the migration of ions. Conductometric electrodes are mainly based on interdigitated structures. They have been used mainly for elaborating biosensors for urea determination (Watson et al. 1987; Bilitewski et al. 1992; Lee et al. 2000b; Castillo-Ortega et al. 2002; Ilangovan et al. 2006). The first reported conductometric biosensor exhibited a detection range of 10^{-4}–10^{-2}M urea (Watson et al. 1987). Conducting polymers have been used to develop glucose, urea, lipids and hemoglobin (Contractor et al. 1994) biosensing devices. An enzymatic biosensor based on pH-responsive polypyrrole and a penicillinase membrane has been proposed (Nishizawa et al. 1992). The enzyme reaction acidified the polypyrrole film and a change of conductivity was registered. The same principle has also been used with polyaniline (Mabeck and Malliaras 2006 and reference therein) and GOD (Hoa et al. 1992) or HRP

(Bartlett et al. 1998). A biosensor for urea detection based on sol–gel-immobilized urease on a microfabricated interdigitated array (IDA) gold electrode has been reported (Lee et al. 2000b). It can be used in urine sample by employing a differential measurement format based on an active biosensor with sol–gel immobilized urease and a reference sensor with BSA. A sol–gel-immobilized urease-based conductometric biosensor has been developed on a thick-film interdigitated electrode (Ilangovan et al. 2006). The linear range extended from 10^{-3} to 1.5 10^{-2} M of urea. The same electrode has been used for detection of heavy metals (cadmium, copper, lead) concentrations ranging from 10^{-4} to 10^{-2}M. Conductometric planar electrodes and pH-sensitive FETs have been compared and demonstrated similar performances (short response time, high operational stability), the former systems being technologically preferable since they are cheaper and easier to manufacture, therefore promising for practical use. Conductometric biosensors are still a novel trend in the field of biosensors, with promising future development.

Few studies on enzyme-based impedimetric biosensors have been reported due to the inherent time-consuming characteristics (recording of the full impedance spectrum within a broad region of frequencies) of such techniques. They are seldom used to detect substrates or products of an enzymatic reaction. They are preferably used for the characterization of enzyme-based biosensors (Guan et al. 2004). Based on the HRP-catalyzed formation of an insoluble, non conductive (insulating) product in the presence of H_2O_2, impedimetric analysis of hydrogen peroxide has been performed and extended to glucose (Patolsky et al. 1999) or acetylcholine (Alfonta et al. 2000) detection by coupling GOD (bienzyme system) or choline OD and AChE (trienzyme system), respectively. It can be noticed that such a principle based on H_2O_2-induced HRP-catalyzed precipitation of an insoluble product has also been used to design microgravimetric QCM biosensors. Recently, a biosensor for glucose detection has also been reported, based on faradaic impedance transducers and a soluble mediator, *p*-benzoquinone (Shervedani et al. 2006). GOD was immobilized covalently on gold mercaptopropionic acid self-assembled monolayers to determine glucose using electrochemical impedance spectroscopy. Values of the charge transfer resistances gradually decreased upon addition of glucose. The linear range for glucose detection extended from 5 10^{-3} to 1.4 10^{-1}M. As underlined by the authors, this method is still time consuming. Urease has also been covalently attached to silanised polysilicon interdigitated electrode surfaces using glutaraldehyde (de la Rica et al. 2006). The hydrolysis of urea caused an increase of the conductivity of the solution thus allowing impedance measurements of urea concentrations in the range from 10^{-5} to 10^{-4}M. Very recently, impedance spectroscopy has also been performed on immobilized streptavidin-HRP layer for biosensing purposes (Esseghaier et al. 2008). Streptavidin-conjugated HRP was attached on a mixed autoassembled monolayer through biotin–streptavidin interaction. The charged transfer resistance decreased upon enzymatic reaction with hydrogen peroxide and in the presence of ferrocyanide, thus enabling H_2O_2 detection in the range from 10^{-5} to 10^{-1}M.

4.4.2 Gravimetric Detection: Quartz Crystal Microbalance (QCM), Surface Acoustic Wave (SAW) Devices, Microcantilevers

Piezoelectric systems are considered to be very sensitive transducers for biosensors. Basic principle lies in the binding of molecular species to the surface of the crystal resulting in a mass change, which eventually leads to a change in the frequency of oscillation of the crystal. Few piezoelectric quartz crystal microbalance (QCM) biosensors based on enzymes have been reported in the literature. Biocatalytic reactions that stimulate the precipitation of an insoluble substrate have generally been used. The coimmobilization of HRP, choline OD and AChE has been reported on Au-quartz crystals, thus providing a frequency change of ca. 80 Hz in the presence of 10^{-5}M acetylcholine (Alfonta et al. 2000).The HRP-biocatalyzed oxidation of 3,3′,5,5′-tetramethylbenzidine in the presence of the generated hydrogen peroxide led to an insoluble product which precipitated onto the electrode surface, thus leading to a mass increase that could be measured by a microgravimetric QCM method. The same tri-enzyme system has been used with 3.3′-diaminobenzidine that is oxidized in an insoluble precipitate (Karousos et al. 2002). Carbaryl and dichlorvos could be determined down to 1 ppm. Both three-enzyme systems are characterized by response times in the order of 30–45 min. Similar techniques where the detection is based on the enzyme-induced precipitate formation have also been reported in the literature. AChE has been immobilized by simple adsorption on the gold surface of quartz crystals. The enzymatic conversion of 3-indolyl acetate leading to the production of a blue precipitate (indigo pigment) has been used to detect paraoxon (detection limit of 5 10^{-8}M) and carbaryl (detection limit of 10^{-7}M) (Abad et al. 1998). Determination of hydrogen peroxide and glucose has also been reported by QCM-based enzyme biosensors, the oxidation of 4-chloro-1-naphtol resulting in an insoluble product which could then be detected (Patolsky et al. 1999). In these systems, enzymes (HRP or HRP/GOD) were covalently immobilized onto an Au-quartz crystal. AT-cut quartz crystals have been used for fabrication of biosensors to measure cholesterol in real time up to 3 10^{-4}M (Martin et al. 2003). The quartz crystal technique has been shown to offer advantages over conventional electrochemical and optical biosensing strategies as by virtue of its mass/viscoelasticity sensing mode, it is not affected by electrochemical and optical interferants.

SAW biosensors have mainly been reported under immunosensor formats (Länge et al. 2008). The first SAW enzyme sensors had been developed in 1992/1993. They were based on the pH change associated with an enzyme–substrate reaction. GOD has been used for the detection of glucose (Inoue et al. 1992). Urease has also been used for the detection of urea on a SAW device (Kondoh et al. 1993). Another SAW urease-based biosensor has been reported (Liu et al. 1995). The detection limit was reported to be 0.5 μg/mL of urea. The results obtained using the SAW technique were consistent with the results obtained by a conventional colorimetric technique. Cholinesterase has also been coupled to the sensor surface to detect fenitrothion, an organophosphorus pesticide (Kondoh et al. 1994).

Microcantilever sensors are micromechanical devices based on the shifting of resonance frequency (*dynamic* deflection) and changes of surface forces (surface stress) (*static* deflection) due to molecular adsorption onto the cantilever. Adsorption onto the device composed of two chemically different surfaces produces a differential stress between the two surfaces and induces bending (Goeders et al. 2008). One main advantage of acoustic techniques is the cost of the apparatus that can be rather low. Limitation of this transduction method involves format and calibration requirements. Few enzymatic microcantilevers have been described in the literature. The first microcantilever-based enzymatic assay involved GOD immobilization onto a gold-coated silicon cantilever with glutaraldehyde following the coating of the gold surface with poly-L-lysine (Subramanian et al. 2002). Quantifiable deflection of the cantilever was observed in the presence of glucose from $5\ 10^{-3}$ to $4\ 10^{-2}$M. Cantilever deflection upon exposition to glucose resulted from surface-induced stresses. GOD has also been electrostatically immobilized within an alternately charged polyelectrolyte multilayer structure that comprised of poly (sodium 4-styrenesulfonate) and polyethyleneimine (Yan et al. 2004, 2005). The multilayer approach provided improved performance. Millimomar glucose concentrations were detected. A GOD-based cantilever biosensor enabling millimolar glucose concentration detection has also been reported, based on cross-linking of the enzyme with BSA chemisorbed onto the surface (Pei et al. 2004). The authors attributed the deflection mechanism of the cantilever to changes in the local chemical environment of the coating layer, but the mechanism behind the surface-induced stress observed for this type of biosensor remains unknown.

4.4.3 Calorimetric Detection

"Thermal biosensors" based on enzymatic reactions measure enthalpy changes. The thermal measurement scheme provides unique opportunities for detecting a wide range of substances and serves as an important complement to other biosensor detection schemes. The most sensitive and practically useful calorimetric sensor arrangements involve continuous flow operation (namely, flow injection analysis), although several attempts have been made to create calorimetric enzyme probes simply by coating the temperature transducer with an enzyme layer (true "biosensor"!). However, these measurements, which are carried out differentially using an inactive reference probe, suffer from a poor sensitivity. Typically, the so-called "thermal biosensors" employ a flow injection analysis scheme using an immobilized enzyme reactor, together with a differential temperature measurement scheme. The configuration usually involves a pair of thermal transducers, such as thermistors or thermopiles, positioned across the enzyme column. Even if such systems are commonly described in the literature as biosensors, it must be underlined that the so-called "thermal biosensors" should not be considered as such because they are not true biosensors that intimately associate the enzymatic sensing layer and the transducer. "Calorimetric biosensors" have been reviewed (Ramanathan and Danielsson 2001; Danielsson et al. 2005). The enthalpy changes for enzymatic catalysis is around −10 to −200 kJ/mol (−28 kJ/mol for hexokinase with glucose, −49 kJ/mol for uricase with urate, −80 kJ/mol for GOD with glucose, −100 kJ/mol for catalase with H_2O_2, −225 kJ/mol for NADH DH with NADH) (Danielsson et al. 2005), which is adequate for determination of the substrate concentrations at clinically interesting levels for a range of metabolites. Oxidases generally offer higher sensitivity than dehydrogenases because of higher heat of reaction ($-\Delta H = 75$–100 kJ/mol) and have no extra cofactor requirement. Thermometric detection is especially advantageous when multiple reactions are involved, since the sensitivity of the assay is determined by the sum of all the reaction enthalpies. Thus, it is advantageous to coimmobilize an oxidase with catalase which consumes the H_2O_2 produced in the OD-catalyzed reaction. The high protonation enthalpy of a buffer, like Tris, can also be used to enhance the total enthalpy of proton-producing reactions. Increases in detection limit can also be obtained in substrate or coenzyme recycling enzyme systems, where the net enthalpy change of each turn of the cycle adds to the overall enthalpy change. With conventional "thermal biosensors" (enzyme thermistors), determinations of substances are in the 10^{-5}–10^{-1} M range for most enzymatic reactions. Thermal biosensors are constructed in a wide range of formats and sizes using a number of fabrication technologies. Research in the area of calorimetric biosensors nowadays focuses on miniaturization and microfabrication of calorimetric devices to increase sample throughput, decrease reagent volumes, and to eventually integrate the device in fully automated lab systems.

4.4.4 *Optical Detection*

An excellent review has recently been written on optical biosensors (Borisov and Wolfbeis 2008). Advantages of optical techniques involve the speed and reproducibility of the measurements. The main drawback of optical measurements is the high cost of the apparatus. Moreover, these instruments are generally larger and not practical for on-site measurements. In general the optical transducers of most common enzyme biosensors are based on optical techniques such as absorbance, reflectance, fluorescence (the most often applied), bio-/chemi-luminescence or SPR. Some enzymatic optical biosensors based on SPR detection are presented, as well as the more numerous ones based on opt(r)odes.

4.4.4.1 Surface Plasmon Resonance

Few enzyme-based SPR-based optical biosensors have been described in the literature, as SPR biosensors mainly rely on immunosensors or affinity sensors.

The surface plasmon resonance (SPR) technique has been used in combination with covalently immobilized cholesterol OD onto the 4-fluoro-3-nitro-azidobenzene modified self-assembled monolayer (SAM) of octadecanethiol (Arya et al. 2006) and poly (3-hexylthiophene) (Arya et al. 2007b). The optical devices showed linearity in the 50–500 mg/dL range of cholesterol in both cases. The same authors have also shown that aminosilane SAM modified ITO surface can be used for the covalent immobilization of cholesterol OD for the fabrication of optical cholesterol biosensors (Arya et al. 2007a), with a detection limit of 25 mg/dL and a stability of ca. 10 weeks. Covalently immobilized cholesterol OD onto the 11-amino-1-undecanethiol SAM can also be used for the fabrication of SPR based cholesterol biosensors (Solanki et al. 2007). SPR has also been used for hydrogen peroxide detection, HRP being immobilized in a polyaniline film (Kang et al. 2001). Such system is based on the fact that SPR is sensitive to the transformation between the oxidized and reduced states of polyaniline by the enzymatic reaction. After each measurement, it can also be reused by electrochemically reducing the polyaniline film.

4.4.4.2 Optrodes

Opt(r)odes refer to fiber-optic biosensors. Two main types of optical biosensors can be distinguished, based on direct or indirect detection. Direct optical biosensors are those based on changes of intrinsic optical properties either of the enzyme itself or of the species that are generated or consumed (chromogenic or fluorogenic substrates or any cosubstrates) during the catalytic reaction. Indirect optical biosensors are represented by chemical sensors that make use of various indicator dyes. These chemical opt(r)odes rely on measurements of absorbance, reflectance or fluorescence whereas direct optical biosensors also rely on bio- and chemi-luminescence reactions.

Indirect detection (chemical opt(r)odes). The indicator layer consists in an immobilized indicator dye that is often sensitive to oxygen or pH. Such a layer is not needed in optical biosensors based on bio-/chemi-luminescence. The concentration of an analyte can be related to the amount of oxygen consumed, hydrogen peroxide produced or variation in pH due to the immobilized enzyme-catalyzed reaction. Table 4.2 shows some examples of enzyme-based biosensors using chemical opt(r)odes.

Oxygen opt(r)odes. Oxygen acts as a dynamic quencher of the luminescence of some indicator dyes such as pyrene butyric acid that was first associated to GOD for glucose detection or to other oxidases for xanthine, lactate or cholesterol detection (Opitz and Lubbers 1988). Other oxygen probes have been used, such as pyrene, decacyclene (Trettnak and Wolfbeis 1989a; Dremel et al. 1992) and their derivatives (Schaffar and Wolfbeif 1990; Dremel et al. 1992) or the widely used ruthenium(II) complexes with ligands such as bipyridyl (Ru-bipy), 1,10-phenanthroline (Ru-phen) and 4,7-diphenyl-1,10-phenanthroline (Ru-dpp) (Li and Walt 1995; Rosenzweig and Kopelman 1996; Wolfbeis et al. 2000; Wu and Choi 2004; Zhou et al. 2005). A fiber optic microbiosensor for glucose detection incorporating a

Table 4.2 Examples of enzyme-based optical biosensors using chemical opt(r)odes

Chemical opt(r)ode	Indicator reagent	Enzyme/detected substance	Analytical range (M)	References
Oxygen	Decacyclene	Lactate monooxygenase/lactate	3×10^{-4}–6×10^{-3}	Trettnak and Wolfbeis (1989a)
	Decacyclene	Lactate OD/lactate	2×10^{-5}–5×10^{-4}	Dremel et al. (1992)
		GOD/glucose	5×10^{-5}–10^{-3}	
	Decacyclene	GOD/glucose	10^{-5}–2×10^{-3}	Schaffar and Wolfbeif (1990)
	Pt-octaethylporphine	GOD/glucose	3×10^{-5}–1.2×10^{-3}	Papkovsky (1993),Papkovsky et al. (1993)
	Ru-phen	GOD/glucose	7×10^{-4}–10^{-2}	Rosenzweig and Kopelman (1996)
	Ru complex	GOD/glucose	5×10^{-5}–2×10^{-4}	Wu and Choi (2004)
	Ru-dpp	GOD/glucose	10^{-4}–8×10^{-4}	Zhou et al. (2005)
	Ru-dpp	Cholesterol OD/cholesterol	7×10^{-5}–1.8×10^{-2}	Wu and Choi (2003)
	Ru(II) complex	Alcohol OD/ethanol	5×10^{-4}–9×10^{-3}	Mitsubayashi et al. (2003)
pH	Carboxy SNARF-1	OPH/paraoxon	8×10^{-7}–1.5×10^{-5}	Russell et al. (1999)
	Carboxynaphthofluorescein	OPH/diisopropyl phosphorofluoridate	2×10^{-6}–4×10^{-4}	Viveros et al. (2006)
	FITC	AChE/acetylcholine	5×10^{-4}–2×10^{-2}	Doong and Tsai (2001)
	Prussian Blue	Urease/urea	2×10^{-3}–1.2×10^{-2}	Koncki et al. (2001)
	Acryloylfluorescein	Penicillinase/penicillin	2.5×10^{-4}–10^{-2}	Healey and Walt (1995)
	FITC	Penicillinase/penicillin	10^{-4}–2.5×10^{-2}	Xie et al. (1992b)
	Phenol red	Penicillinase	5×10^{-4}–8×10^{-3}	Busch et al. (1993b)
	Phenol red	Urease/urea	4×10^{-5}–2.5×10^{-4}	Gauglitz and Reichert (1992)
	HPTS	GOD/glucose	10^{-4}–2×10^{-3}	Trettnak et al. (1989)
NH_3	HPTS	Urease/urea	10^{-4}–5×10^{-3}	Xie et al. (1991)
	Brilliant yellow	Urease/urea	10^{-4}–10^{-2}	Mascini (1995)
	Bromothymol blue	Urease/urea	2.5×10^{-4}–8×10^{-3}	Xie et al. (1990)
	Octadecyl dichlorofluorescein	Urease/urea	10^{-4}–10^{-2}	Chen and Wang (2000)
	Nile Blue	Urease/urea	10^{-5}–10^{-3}	Wolfbeis and Li (1993)
	Carboxyfluorescein	Urease/urea	5×10^{-5}–2.5×10^{-3}	Rhines and Arnold (1989)
	Carboxyfluorescein	Glutamate OD/glutamate	10^{-6}–1.2×10^{-5}	Kar and Arnold (1992)
H_2O_2	Prussian blue	GOD/glucose	5×10^{-5}–2×10^{-3}	Lenarczuk et al. (2001),Koncki et al. (2001)
	EuTc	GOD/glucose	10^{-4}–5×10^{-3}	Wolfbeis et al. (2003)

ruthenium oxygen probe and GOD and needing small sample volumes has been described (Rosenzweig and Kopelman 1996). Due to its small size and the absence of membrane support, the response time of the sensor was only 2 s. A linear dynamic range between $7\,10^{-4}$ and 10^{-2}M was observed (Table 4.2). Platinum (II) and palladium(II) porphyrins represent another group of viable luminescent oxygen indicators. A phosphorescent platinum (II) complex with octaethylporphyrin has been used as an oxygen probe in a glucose biosensor (Papkovsky 1993). Glucose could be detected between $3\,10^{-5}$ and $1.2\,10^{-3}$M (Table 4.2). Biosensors using an oxidase-type enzyme and an oxygen probe have been designed such as those for lactate (Cattaneo et al. 1992), ethanol and methanol (Mitsubayashi et al. 2003), cholesterol (Wu and Choi 2003) or sulfite (Xie et al. 1994) detection. The use of two sensors, one sensitive to oxygen only and the other sensitive to both oxygen and glucose, can be a way to avoid the effect of varying oxygen supply (Wolfbeis et al. 2000). A fiber-optic flow-through biosensor based on an oxygen probe for online monitoring of glucose has been reported, using a reference oxygen sensor to compensate for interferences and showing high selectivity (Pasic et al. 2006, 2007). The linear range of the sensor depended basically on the thickness of outer membrane which acted as a diffusion barrier for the glucose molecules. By varying the thickness of the outer membrane, the glucose sensitivity could be adjusted and the desired dynamic range could be obtained, enabling detection up to 20 mM of glucose.

pH opt(r)ode. The amount of protons consumed or produced during an enzyme reaction can also be used for optical detection. pH opt(r)odes have been used in biosensors for penicillin and urea detection for example, by measuring decrease in pH upon penicillinase hydrolysis or increase in pH upon urease reaction. Fluorescein-derived indicators are usually contained in a hydrogel (often a polyacrylamide gel) and are used almost exclusively in order to monitor pH changes in a neutral media. Biosensors based on this principle suffer from the fact that pH changes depend on the buffer capacity of the sample medium, which often is unknown and can hardly be compensated for. Biosensors based on pH transducers make use of absorption-based pH indicators that have been coupled with penicillinase (Goldfinch and Lowe 1984; Busch et al. 1993b) or urease (Gauglitz and Reichert 1992). OPH hydrolysis of pesticides such as paraoxon is accompanied by a release of protons, which makes possible determination of organophosphorus pesticides using pH transducers (Russell et al. 1999; Viveros et al. 2006). Prussian Blue has also been found to exhibit a pH-dependent intrinsic absorption. It has been chemically incorporated along with urease into polypyrrole films (Koncki et al. 2001). Urea could be detected in the range from $2\,10^{-3}$ to $1.2\,10^{-2}$M (Table 4.2). An interesting approach has been described, with a pH-insensitive fluorophore linked to cadaverine whose amine moiety is responsible for the pH-dependent swelling of the cross-linked acrylamide-based hydrogel placed at the end of an optical fiber (McCurley 1994). The decrease in pH with penicillinase has also been monitored by changes of emission intensity of pH indicators (Xie et al. 1992b; Healey and Walt 1995). Detection of penicillin was in the range 10^{-4}–10^{-2}M (Table 4.2). Biosensors based on fluorescent pH indicators have also been reported for acetylcholine (detection limit of $5\,10^{-4}$M) or organophosphorus and carbamate pesticides (detection limit of 27 ppb paraoxon) detection (Doong and Tsai 2001).

Ammonia opt(r)ode. Urease-catalyzed hydrolysis of urea leads to formation of ammonium ions. Many urea biosensors are based on the determination of ammonia gas produced during hydrolysis of urea (Rhines and Arnold 1989; Xie et al. 1991; Mascini 1995; Chen and Wang 2000). In this type of ammonia transducers, a pH indicator is contained in a buffer solution positioned behind a gas-permeable membrane, in which the enzyme is immobilized. Gaseous ammonia diffuses through the membrane and dissolves in the buffer. This results in an increase of pH of the internal solution and in deprotonation of the indicator. Changes in absorbance or fluorescence intensity of the indicator are related to the ammonia concentration present in the external solution and, thus, to the level of urea. Urea detection was reported in a ca. two decade range beginning from 10^{-4} or 10^{-5}M (Table 4.2). Ammonia transducer has also been used in an optical biosensor for glutamate detection (Kar and Arnold 1992). Glutamate OD catalyzes the oxidative deamination of glutamate with concomitant production of ammonia and leads to glutamate detection from 10^{-6} to $1.2\,10^{-5}$M (Table 4.2).

Hydrogen peroxide opt(r)ode. The amount of hydrogen peroxide produced during the enzymatic reaction can also be related to the concentration of an oxidase-substrate compound. Optical detection of glucose with a GOD-modified Prussian Blue film has been reported (Lenarczuk et al. 2001). The colored form of Prussian Blue was formed from the colorless one (Prussian White) upon oxidation by hydrogen peroxide. Glucose could be detected from $5\,10^{-5}$ to $2\,10^{-3}$M (Table 4.2). An europium probe serving as a hydrogen peroxide indicator has also recently been coupled with an iridium–trispyridine complex as an oxygen indicator for glucose detection (Borisov and Wolfbeis 2008). The luminescent europium-(III) tetracycline complex (EuTc) probe is excitable by visible light and responds to H_2O_2 at neutral pH by ca. 15-fold increase in luminescence intensity. The probe immobilized into a hydrogel was successfully used for sensing of glucose (Wolfbeis et al. 2003). Detection of glucose concentrations in the 10^{-4}–$5\,10^{-3}$M range was obtained (Table 4.2).

Other opt(r)ode. It was recently discovered that the europium tetracycline (EuTc) complex can act as a luminescent probe for nucleoside phosphates including AMP, ADP, and ATP. The probe can be excited with the 405 nm laser diode and is nonspecific, but the response to the various nucleoside phosphates is different. It has been applied to the determination of the activity of soluble kinases (Schäferling and Wolfbeis 2007).

Direct optical biosensors based on the fact that enzymes are intrinsically optically detectable have been described. Few biosensors are based on the intrinsic optical properties of an enzyme and exploit optical changes (absorbance, fluorescence) of the biomolecule after its interaction with the target analyte. They suffer from a low sensitivity. For example, enzymes using FAD as a coenzyme can undergo intrinsic spectral changes upon catalysis. The intrinsic fluorescence of GOD, which contains the fluorophore flavine adenine dinucleotide (FAD) has been exploited (Trettnak and Wolfbeis 1989b). The optical properties of the enzyme are slightly different for the oxidized (FAD) and the reduced form ($FADH_2$) that is produced during reaction with glucose. A narrow analytical range ($5\ 10^{-4}$–$8\ 10^{-4}$M) was obtained. Similarly, the intrinsic absorption of GOD has been used to obtain a biosensor with a wider analytical range for glucose (10^{-3}–10^{-2}M) (Chudobova et al. 1996). The intrinsic fluorescence of GOD has also been used for the determination of glucose in serum (in the range from $5\ 10^{-4}$ to $2\ 10^{-2}$ M) by exploiting the UV fluorescence of the protein part of GOD (Sierra et al. 1997). The intrinsic fluorescence of GOD immobilized on photopolymerized polyacrylamide gave a linear range between $1.7\ 10^{-3}$ and $1.1\ 10^{-2}$M of glucose (de Marcos et al. 2006). An increase in fluorescence intensity has been found with fluorescein-labeled GOD in the presence of glucose, thus giving rise to a linear response from $5.5\ 10^{-4}$ to $5.5\ 10^{-3}$M of glucose (de Marcos et al. 1999). Intrinsic absorption properties of HRP have also been used, as HRP undergoes spectral changes upon H_2O_2 binding (Sanz et al. 2007). Both GOD and HRP were thus entrapped in a polyacrylamide gel matrix and a linear response was observed between $1.5\ 10^{-6}$ and $3\ 10^{-4}$M of glucose. The intrinsic optical properties of enzymes have also been used for determination of nitrate (absorbance of nitrate reductase) (Aylott et al. 1997) and nitrite ions (absorbance of cytochrome *cd*1 nitrite reductase) (Ferretti et al. 2000).

Direct optical biosensors based on optically detectable species that are generated or consumed during the reaction have also been reported. Such sensors usually consist of a membrane that contains the immobilized enzyme. Chromogenic or fluorogenic substrates and any cosubstrates are added to the sample into which the sensor is placed. One type of direct optical biosensor is based on hydrolysis of a substrate catalyzed by a hydrolase-type enzyme. The principle was first demonstrated by immobilizing alkaline phosphatase to catalyze hydrolysis of *p*-nitrophenyl phosphate, which results in the formation of yellow *p*-nitrophenolate (Arnold 1985). A linear dependence of the change in absorbance vs. the concentration of the substrate from $2.5\ 10^{-5}$ to $3.5\ 10^{-4}$M was observed. A synthetic yellow substrate can be converted into a blue product by AChE and inhibition of the reaction by pesticides can be monitored spectroscopically (Trettnak et al. 1993). Indoxyl acetate has also been used as a substrate for AChE (Navas Diaz and Ramos Peinado 1997). The fluorescence intensity of indoxyl was related to the concentration of the inhibiting organophosphorus pesticide (detected from 0.1 to 58 mg/L). Enzymatic hydrolysis of *o*-nitrophenyl acetate has also been used for determination of organophosphates (limit of detection of 2 ppm, Choi et al. 2001). Cholesterol OD has also been covalently immobilized on polyaniline on an ITO electrode, the optical detection being performed by spectrophotometry at 500 nm using *o*-dianisidine (Dhand et al. 2007). Linearity was observed from 25 to 400 mg/dL of cholesterol. Hydrolysis of some organophosphorus pesticides by OPH can also produce detectable chromophoric products (Mulchandani et al. 1999). The $2\ 10^{-6}$M lower detection limit based on the hydrolysis of organophosphates by the enzyme OPH was still tenfold higher than the one based on AChE inhibition. An ultrasensitive biosensor for the pesticide parathion has been designed, using an array of colloidal polymer particles (which diffract light in the visible spectral region) introduced into a polyacrylamide-based hydrogel (Walker and Asher 2005). AChE was covalently attached to the hydrogel backbone where it irreversibly bound parathion, which in turn resulted in the formation of a charged product. This induced swelling of the hydrogel network and resulted in a shift of the wavelength of the diffracted light that was proportional to the concentration of the analyte. The sensor was able to detect parathion in the fM to pM concentration range. The reduced form of nicotinamide adenine dinucleotide (NADH) can also be monitored via its characteristic absorption at λ_{max} 350 nm and emission at λ_{max} 450 nm. Flow-injection DH-based biosensors have been described, with NAD^+ added in the solution to detect glucose (detection limit of 6 10^{-4}M, Narayanaswamy and Sevilla 1988), lactate (2 10^{-4}–10^{-3}M detection range, Li et al. 2002a), pyruvate (detection limit of 8.4 10^{-6} M, Zhang et al. 1997), glutamate (detection limit of 2 10^{-7}M, Cordek et al. 1999) or alcohols (10^{-3}–10^{-1}M ethanol detection range, Busch et al. 1993a) using glucose DH, lactate DH, glutamate DH, and alcohol DH, respectively. Enzymatic oxidation of lactate by NAD^+ results in the formation of pyruvate and NADH, so that the reaction can be used not only for determination of lactate but also in the reverse direction for pyruvate (Zhang et al. 1997), NADH being supplied instead of NAD^+. The dual-enzyme (lactate OD and lactate DH) biosensor for pyruvate was clearly superior to the single-enzyme case and offered high sensitivity and good detection limits (Zhang et al. 1997). A poly (ethylene glycol) molecular weight-enlarged NAD^+ (PEG-NAD^+) has also been used instead of NAD^+ to design a self-contained biosensor (Lee et al. 1994b). A pair of dehydrogenase-type enzymes (for substrate detection and for regeneration of the coenzyme) and PEG-NAD^+ were enclosed in the sensing compartment between the ultrafiltration membrane and the fiber-optic tip. L-Phe was detected from 6 10^{-4} to 6 10^{-3}M using L-Phe DH and L-Ala was detected from 4.5 10^{-4} to 4.5 10^{-3}M using L-Ala DH. The unique enzyme glucose–fructose oxidoreductase (GFOR) is able to dehydrogenate glucose to gluconolactone and to simultaneously reduce fructose to sorbitol. Both the intrinsic absorbance and fluorescence of NADH can be measured to enable optical detection of both substrates. In the GFOR-based biosensor (Lee et al. 1994a) the enzyme was cross-linked with glutaraldehyde and placed between an optical fiber and a dialysis membrane. Glucose was sensed via the increase in fluorescence due to formation of NADH, and fructose was sensed via the decrease in fluorescence due to consumption of NADH. The system could be regenerated by passing fructose or glucose solutions, respectively, over it. Glucose and fructose could be detected from 5.5 10^{-5} to 5.5 10^{-2}M and from 2.8 10^{-4} to 3.3 10^{-1}M, respectively. ADP has also been determined via the fluorescence of NADH that is formed in the sequence of reactions catalyzed by the enzymes pyruvate kinase, hexokinase, and glucose-6-phosphate DH (Schubert 1993). ADP could be determined at concentrations as low as 10^{-7}M.

Direct optical biosensors also refer to those based on bioluminescence. Apart from measurements of its intrinsic absorbance or fluorescence, NADH can be detected with much higher sensitivity via reactions catalyzed by bacterial enzymes and resulting in blue-green bioluminescence (Blum 1997a). There are two enzyme systems, which have been mainly used to design optical biosensors based on bioluminescence reactions: the firefly luciferase, which is specific for ATP, and the bacterial oxidoreductase/luciferase enzyme system, which is specific for NAD (P) H. Provided that ATP and NAD(P)H are the limiting substrates in the above reactions, then the measured light intensities are proportional to the substrate concentrations.

NAD (P)H can be measured according to the sequence reaction catalyzed by the bacterial enzymes, NAD(P)H:FMN oxidoreductase and bacterial luciferase, in the presence of flavin mononucleotide (FMN) and a long-chained aldehyde (R-CHO), e.g. decanal:

$$\mathrm{NADH} + \mathrm{H}^+ + \mathrm{FMN} \overset{\text{oxidoreductase}}{\longrightarrow} \mathrm{NAD}^+ + \mathrm{FMNH}_2$$

$$\mathrm{FMNH}_2 + \mathrm{R} - \mathrm{CHO} + \mathrm{O}_2 \overset{\text{bacterial luciferase}}{\longrightarrow} \mathrm{FMN} + \mathrm{R} - COOH + \mathrm{H}_2\mathrm{O} + h\nu(\lambda_{\max} = 490\,\mathrm{nm})$$

Highly sensitive optical biosensors exhibiting wide dynamic ranges (e.g. 10^{-9}–10^{-6}M or $5\,10^{-9}$–$5\,10^{-7}$M for NADH (Gautier et al. 1990a, 1991)) have been obtained. Biosensors for NADH use enzymes immobilized onto a polymer support (a preactivated polyamide membrane), while the long-chained aldehyde needs to be continuously supplied to the solution. The cofactor FMN has been either added in the solution (Gautier et al. 1990a) or noncovalently entrapped in a matrix of poly (vinyl alcohol), which allowed its controlled release in the vicinity of the immobilized enzymes (self-contained biosensor, which do not require the addition of coreactants, Gautier et al. 1991). The successful coentrapment of both coreactants (FMN and *N*-decyl aldehyde) in a poly (vinyl alcohol) matrix was reported, demonstrating that a reagentless bioluminescent sensor for NADH was possible. By coimmobilizing some auxiliary NAD (P)-dependent DH-type enzymes with the bacterial luminescence enzyme system, biosensors for ethanol, sorbitol, and oxaloacetate have also been produced (Gautier et al. 1990b). For ethanol and sorbitol biosensors, the appearance of NADH was the monitored response, whereas sorbitol determination relied on the consumption of NADH. Oxaloacetate, ethanol and sorbitol were detected from $3\,10^{-9}$ to $2\,10^{-6}$M, from $4\,10^{-7}$ to $7\,10^{-5}$M and from $2\,10^{-8}$ to 10^{-5}M, by using malate DH, alcohol DH or sorbitol DH, respectively.

The luciferin/luciferase bioluminescent system from the firefly was adapted to the determination of adenosine triphosphate (ATP). Firefly luciferase catalyzes the production of light, in the presence of ATP, Mg^{2+}, molecular oxygen, and luciferin:

$$\mathrm{ATP} + \mathrm{luciferin} + \mathrm{O}_2 \overset{\text{luciferase, Mg}^{2+}}{\longrightarrow} \mathrm{AMP} + \mathrm{oxyluciferin} + \mathrm{PPi} + \mathrm{CO}_2 + h\nu(\lambda_{\max} = 560\,\mathrm{nm})$$

As in the case of bioluminescent determination of NADH, biosensors for ATP are extremely sensitive (limits of detection lower than 1 pmol). Luciferin could be added to the sample solution (Gautier et al. 1990a; Blum et al. 1993) or incorporated into acrylic (Eudragit) microspheres entrapped in a film of poly(vinyl alcohol) (Michel et al. 1998). Such a controlled-release system allowed the determination of ATP via firefly luciferase entrapped in a collagen membrane, as well as that of adenosine monophosphate (AMP) and adenosine diphosphate (ADP) by entrapping two additional enzymes, adenylate kinase and creatine kinase, responsible for conversion of AMP and ADP into ATP. The sensitivity for ATP was significantly lower and the limit of detection was significantly higher (10 pmol) than for sensors using luciferin in solution.

Direct optical biosensors based on chemiluminescence have also been reported (Blum 1997b). Apart from optical measurements with chemical opt(r) ode (indicator dyes), H_2O_2 can be detected with much higher sensitivity via chemiluminescence reactions. Most chemiluminescent biosensors are based on the oxidation of luminol (5-amino-2,3-dihydrophthalazine-1,4-dione) catalyzed by horseradish peroxidase (HRP):

$$\text{Luminol} + 2H_2O_2 + OH^- \xrightarrow{\text{HRP}} 3\text{ - aminophthalate} + N_2 + 3H_2O + h\nu(\lambda_{max} = 430\,\text{nm})$$

Many of these devices show satisfactory detection limits and ranges, benefiting from the high sensitivity of the light measurement systems such as photomultiplier tubes. Bioanalytical applications of luminol chemiluminescence (Marquette and Blum 2006), as well as luminol electrochemiluminescence (Marquette and Blum 2006, 2008), have been recently reviewed by our group. Apart from peroxidase-catalyzed oxidation of luminol, its electrochemical oxidation is usually considered as the second most efficient way of triggering the CL reaction. Transition metal cations (Co^{2+}, Cu^{2+}, Fe^{2+}...) and their complexed forms (ferrocene, ferricyanide...) can also be used as catalysts with mitigated performances. Detection of hydrogen peroxide via *chemi-* and *electrochemiluminescence* of luminol has been used in several biosensor devices. The enhanced chemiluminescence (CL) reaction of the luminol–H_2O_2–HRP system with HRP immobilized in a sol–gel matrix has been used to develop a biosensor for *p*-iodophenol, *p*-coumaric acid, 2-naphthol and hydrogen peroxide (Ramos et al. 2001). The detection limits obtained for *p*-iodophenol, *p*-coumaric acid and 2-naphthol were $8.3\,10^{-7}$M, $1.5\,10^{-8}$ and $4.8\,10^{-8}$M, respectively. Direct enzyme immobilization onto the end of the optical fiber allowed the construction of a remote enhanced CL biosensor which has been applied to H_2O_2 detection. Hydrogen peroxide could be detected from $1.7\,10^{-5}$ to $1.2\,10^{-4}$ M (detection limit of $1.7\,10^{-5}$M). Peroxidase-catalyzed CL of luminol has been used for lactate (10^{-10}–$5\,10^{-6}$mol, in a compartmentalized system, Blum et al. 1995), ethanol ($3\,10^{-6}$–$7.5\,10^{-4}$M, Xie et al. 1992a), lysine ($5\,10^{-6}$–10^{-2}M, Preuschoff et al. 1994) sulfite (10^{-6}–10^{-4}M, Hlavay and Guilbault 1994) xanthine ($3\,10^{-6}$–$3\,10^{-4}$M, Hlavay et al. 1994) or hypoxanthine ($5\,10^{-7}$–10^{-3}M, Spohn et al. 1995) detection. An optical fiber bienzyme sensor based on the luminol chemiluminescent reaction has also been developed and demonstrated to be sensitive to glucose (Li et al. 2002b). GOD and HRP were coimmobilized in a sol–gel film derived from tetraethyl orthosilicate. The calibration plots for glucose were established by the optical fiber glucose sensor fabricated by attaching the bienzyme silica gel onto the glass window of the fiber bundle. The linear range was $2\,10^{-4}$–$2\,10^{-3}$M and the detection limit was approximately $1.2\,10^{-4}$M. Fiber optic biosensors based on the electrochemiluminescence of luminol have also been developed for glucose, lactate (Marquette and Blum 1999; Marquette et al. 2001) or choline (10^{-11}–$3\,10^{-8}$mol, Tsafack et al. 2000a, b; Marquette et al. 2001) flow injection analysis. A glassy carbon electrode was polarized at +425 mV vs. a platinum pseudoreference electrode. After optimization of the reaction conditions and physicochemical parameters influencing the sensor response, the measurement of hydrogen peroxide could be performed in the range $1.5\,10^{-12}$–$3\,10^{-8}$mol. GOD or lactate OD were immobilized on polyamide and collagen membranes. With collagen as the enzymatic support, the detection limits for glucose and lactate were $6\,10^{-11}$ and $3\,10^{-11}$mol, respectively, whereas with the enzymatic polyamide membranes, the corresponding values were $1.5\,10^{-10}$ and $6\,10^{-11}$mol. Finally, using a glassy carbon electrode in a FIA system and under optimal conditions, the detection limits were 30 pmol, 60 pmol, 0.6 nmol and 10 pmol for lactate, glucose, cholesterol and choline, respectively (Marquette and Blum 1999; Marquette et al. 2001). The electrochemiluminescence of luminol can also be triggered using screen-printed electrodes instead of glassy carbon macroelectrodes. The miniaturization of electrochemiluminescence-based biosensors has been realized using screen-printed electrodes, with choline OD as a model H_2O_2-generating oxidase (Leca and Blum 2000; Leca et al. 2001; Marquette et al. 2001). Screen-printed electrodes appear to be powerful devices that can be successfully coupled with an optical detection based on luminol electrochemiluminescence. They offer a possibility to conceive miniaturized devices, since electrodes of several tens μm size can be reproducibly achieved by screen-printing. This represents a step towards miniaturization of electrochemiluminescence-based sensors, thus giving performant disposable or reusable biosensors (Leca et al. 2001; Godoy et al. 2005). An original immobilization method for luminol has also been reported on these carbon-based screen-printed electrodes, thus leading to interesting reagentless disposable devices for detection of H_2O_2 or any oxidase-substrate compounds (Sassolas et al. 2009). In general, analysis with immobilized enzymes using optical fiber CL detection offers several advantages, such as high sensitivity, simple instrumentation (no light source needed) and miniaturized analytical systems.

4.5 Concluding Remarks

Since the first enzyme electrode has been reported, numerous biosensors have been described, most of them based on enzymatic systems. The specificity of the enzymatic recognition is both the strength and the weakness of enzyme-based biosensors. The stability of the bioreceptor can limit the development of true operational biosensors. One way to circumvent this problem is to develop single-use sensing layers. The miniaturization and the development of multiparametric devices allow this kind of approach without increasing drastically the cost price. However, in some cases, a repeated or continuous use is required and the enzyme operational stability becomes a real critical point. Different tracks have been explored, sometimes with a real success. They are briefly mentioned below.

It is often claimed that the immobilization of enzymes improves their stability; however, this is intrinsically true only in some cases. Indeed, the apparent stability observed after immobilization is generally due to diffusional limitations induced by mass transport phenomena, which mask the decrease of the enzyme activity. Of course, this particular property can be exploited to artificially enhance the stability of the bioreceptor, but it is obtained only with a high enzyme loading in or on the support and obviously, this procedure consumes a great amount of enzymes.

The use of some additives, generally water-binding agents such as polyols or sugars (sucrose, mannitol, sorbitol...), may preserve the activity of some enzymes. However, a high concentration is required for efficiency and often, additives can be used only on storage and not on operation.

Chemical modifications can sometimes increase enzyme stability for example by producing succinimide derivatives. However, no systematic approach can be developed and very often the chemical modifications are ineffective.

Thanks to recombinant DNA technology, recombinant enzymes may have improved properties in terms of stability and catalytic efficiency. Protein engineering allows producing mutant enzymes having greater thermostability than the parent proteins.

Some thermostable enzymes can also be isolated from thermophilic microorganisms which have an optimal growth temperature above 60°C. For extreme thermophiles the optimal growth temperature is at 90°C, even up to 110°C for certain archaebacteria. Some enzymes derived from those microorganisms are highly thermostable biocatalysts. Unfortunately, only a few of them have been put on the market and are of poor interest for biosensor development. The experimental problems encountered in the cultivation of high-temperature microorganisms still limit the progress in this way. However, the possibility of cloning enzymes from thermophilic organisms into mesophiles would facilitate the exploitation of high temperature microorganisms in biotechnology.

Robustness, sensitivity and accuracy depend mainly on the bioreceptor and on the associated instrumentation that is the sampling and transducing systems as well as the signal processing unit. Reliable mechanical and electronic components are now widely available so that the associated instrumentation is not a limiting factor.

References

Abad JM, Pariente F, Hernandez L, Abruna HD, Lorenzo E (1998) Determination of organophosphorus and carbamate pesticides using a piezoelectric biosensor. Anal Chem 70 (14):2848–2855

Abdelmalek F, Shadaram M, Boushriha H (2001) Ellipsometry measurements and impedance spectroscopy on Langmuir–Blodgett membranes on Si/SiO_2 for ion sensitive sensor. Sens Actuators B Chem 72(3):208–213

Alfonta L, Katz E, Willner I (2000) Sensing of acetylcholine by a tricomponent-enzyme layered electrode using faradaic impedance spectroscopy, cyclic voltammetry, and microgravimetric quartz crystal microbalance transduction methods. Anal Chem 72(5):927–935

Andreescu S, Marty J-L (2006) Twenty years research in cholinesterase biosensors: from basic research to practical applications. Biomol Eng 23(1):1–15

Arnold MA (1985) Enzyme-based fiber optic sensor. Anal Chem 57(2):565–566

Arya SK, Solanki PR, Singh RP, Pandey MK, Datta M, Malhotra BD (2006) Application of octadecanethiol self-assembled monolayer to cholesterol biosensor based on surface plasmon resonance technique. Talanta 69(4):918–926

Arya SK, Prusty AK, Singh SP, Solanki PR, Pandey MK, Datta M, Malhotra BD (2007a) Cholesterol biosensor based on *N*-(2-aminoethyl)-3-aminopropyl-trimethoxysilane self-assembled monolayer. Anal Biochem 363(2):210–218

Arya SK, Solanki PR, Singh SP, Kaneto K, Pandey MK, Datta M, Malhotra BD (2007b) Poly-(3-hexylthiophene) self-assembled monolayer based cholesterol biosensor using surface plasmon resonance technique. Biosens Bioelectron 22(11):2516–2524

Aylott JW, Richardson DJ, Russell DA (1997) Optical biosensing of nitrate ions using a sol–gel immobilized nitrate reductase. Analyst 122:77–80

Bartic C, Campitelli A, Borghs S (2003) Field-effect detection of chemical species with hybrid organic/inorganic transistors. Appl Phys Lett 82(3):475–477

Bartlett PN, Booth S, Caruana DJ, Kilburn JD, Santamaria C (1997) Modification of glucose oxidase by the covalent attachment of a tetrathiafulvalene derivative. Anal Chem 69(4):734–742

Bartlett PN, Birkin PR, Wang JH, Palmisano F, De Benedetto G (1998) An enzyme switch employing direct electrochemical communication between horseradish peroxidase and a poly (aniline) film. Anal Chem 70(17):3685–3694

Basu I, Subramanian RV, Mathew A, Kayastha AM, Chadha A, Bhattacharya E (2005) Solid state potentiometric sensor for the estimation of tributyrin and urea. Sens Actuators B Chem 107(1):418–423

Besteman K, Lee JO, Wiertz FGM, Heering HA, Dekker C (2003) Enzyme-coated carbon nanotubes as single-molecule biosensors. Nano Lett 3(6):727–730

Bilitewski U, Drewes W, Schmid RD (1992) Thick film biosensors for urea. Sens Actuators B Chem 7(1–3):321–326

Blum LJ (1997a) Bioluminescence-based fiberoptic sensors. In: Bio- and chemi-luminescent sensors. World Scientific, Singapore, pp 113–141

Blum LJ (1997b) Chemiluminescence-based fiberoptic sensors. In: Bio- and chemi-luminescent sensors. World Scientific, Singapore, pp 143–179

Blum LJ, Gautier SM, Coulet PR (1993) Design of bioluminescence-based fiber optic sensors for flow-injection analysis. J Biotech 31(3):357–368

Blum LJ, Gautier SM, Berger A, Michel PE, Coulet PR (1995) Multicomponent organized bioactive layers for fiber-optic luminescent sensors. Sens Actuators B Chem 29(1–3):1–9

Borisov SM, Wolfbeis OS (2008) Optical biosensors. Chem Rev 108(2):423–461

Bucur B, Andreescu S, Marty J-L (2004) Affinity methods to immobilize acetylcholinesterases for manufacturing biosensors. Anal Lett 37(8):1571–1588

Busch M, Gutberlet F, Hobel W, Polster J, Schmidt HL, Schwenk M (1993a) The application of optodes in FIA-based fermentation process control using the software package FIACRE. Sens Actuators B Chem 11(1–3):407–412

Busch M, Hobel W, Polster J (1993b) Software FIACRE: bioprocess monitoring on the basis of flow injection analysis using simultaneously a urea optode and a glucose luminescence sensor. J Biotech 31(3):327–343

Calvo EJ, Etchenique R, Pietrasanta L, Wolosiuk A, Danilowicz C (2001) Layer-by-layer self-assembly of glucose oxidase and Os(Bpy)2ClPyCH2NH-poly(allylamine) bioelectrode. Anal Chem 73(6):1161–1168

Cano M, Luis Ávila J, Mayén M, Mena ML, Pingarrón J, Rodríguez-Amaro R (2008) A new, third generation, PVC/TTF-TCNQ composite amperometric biosensor for glucose determination. J Electroanal Chem 615(1):69–74

Cao Z, Jiang X, Xie Q, Yao S (2008) A third-generation hydrogen peroxide biosensor based on horseradish peroxidase immobilized in a tetrathiafulvalene-tetracyanoquinodimethane/multiwalled carbon nanotubes film. Biosens Bioelectron 24(2):222–227

Caras S, Janata J (1980) Field effect transistor sensitive to penicillin. Anal Chem 52(12):1935–1937

Cass AEG, Davis G, Francis GD, Hill HAO, Aston WJ, Higgins IJ, Plotkin EV, Scott LDL, Turner APF (1984) Ferrocene-mediated enzyme electrode for amperometric determination of glucose. Anal Chem 56(4):667–671

Castillo-Ortega MM, Rodriguez DE, Encinas JC, Plascencia M, Mendez-Velarde FA, Olayo R (2002) Conductometric uric acid and urea biosensor prepared from electroconductive polyaniline-poly(*n*-butyl methacrylate) composites. Sens Actuators B Chem 85(1–2):19–25

Cattaneo MV, Male KB, Luong JHT (1992) A chemiluminescence fiber-optic biosensor system for the determination of glutamine in mammalian cell cultures. Biosens Bioelectron 7(8):569–574

Chen H, Wang E (2000) Optical urea biosensor based on ammonium ion selective membrane. Anal Lett 33(6):997–1011

Chen RJ, Zhang Y, Wang D, Dai H (2001) Noncovalent sidewall functionalization of single-walled carbon nanotubes for protein immobilization. J Am Chem Soc 123(16):3838–3839

Choi J-W, Kim Y-K, Lee I-H, Min J, Lee WH (2001) Optical organophosphorus biosensor consisting of acetylcholinesterase/viologen hetero Langmuir–Blodgett film. Biosens Bioelectron 16(9–12):937–943

Chudobova I, Vrbova E, Kodicek M, Janovcova J, Kas J (1996) Fibre optic biosensor for the determination of D-glucose based on absorption changes of immobilized glucose oxidase. Anal Chim Acta 319(1–2):103–110

Clark JLC, Lyons C (1962) Electrode systems for continuous monitoring in cardiovascular surgery. Ann N Y Acad Sci 102:29–45

Contractor AQ, Sureshkumar TN, Narayanan R, Sukeerthi S, Lal R, Srinivasa RS (1994) Conducting polymer-based biosensors. Electrochim Acta 39(8–9):1321–1324

Cordek J, Wang X, Tan W (1999) Direct immobilization of glutamate dehydrogenase on optical fiber probes for ultrasensitive glutamate detection. Anal Chem 71(8):1529–1533

Cosnier S (2003) Biosensors based on electropolymerized films: new trends. Anal Bioanal Chem 377(3):507–520

Csoeregi E, Schmidtke DW, Heller A (1995) Design and optimization of a selective subcutaneously implantable glucose electrode based on "wired" glucose oxidase. Anal Chem 67 (7):1240–1244

Danielsson B, Lundström I, Mosbach K, Stiblert L (1979) On a new enzyme transducer combination: the enzyme transistor. Anal Lett 12(11):1189–1199

Danielsson B, Paul W, Alan T, Colin P (2005) SENSORS: calorimetric/enthalpimetric. In: Encyclopedia of analytical science. Elsevier, Oxford, pp. 237–245

de la Rica R, Fernández-Sánchez C, Baldi A (2006) Polysilicon interdigitated electrodes as impedimetric sensors. Electrochem Commun 8(8):1239–1244

de Marcos S, Galindo J, Sierra JF, Galban J, Castillo JR (1999) An optical glucose biosensor based on derived glucose oxidase immobilised onto a sol–gel matrix. Sens Actuators B Chem 57(1–3):227–232

de Marcos S, Sanz V, Andreu Y, Galbán J (2006) Comparative study of polymeric supports as the base of immobilisation of chemically modified enzymes. Microchim Acta 153(3):163–170

Degani Y, Heller A (1987) Direct electrical communication between chemically modified enzymes and metal electrodes. I. Electron transfer from glucose oxidase to metal electrodes via electron relays, bound covalently to the enzyme. J Phys Chem 91(6):1285–1289

Dhand C, Singh SP, Arya SK, Datta M, Malhotra BD (2007) Cholesterol biosensor based on electrophoretically deposited conducting polymer film derived from nano-structured polyaniline colloidal suspension. Anal Chim Acta 602(2):244–251

Di J, Shen C, Peng S, Tu Y, Li S (2005) A one-step method to construct a third-generation biosensor based on horseradish peroxidase and gold nanoparticles embedded in silica sol–gel network on gold modified electrode. Anal Chim Acta 553(1–2):196–200

Di J, Peng S, Shen C, Gao Y, Tu Y (2007) One-step method embedding superoxide dismutase and gold nanoparticles in silica sol–gel network in the presence of cysteine for construction of third-generation biosensor. Biosens Bioelectron 23(1):88–94

Doong R-A, Tsai H-C (2001) Immobilization and characterization of sol–gel-encapsulated acetylcholinesterase fiber-optic biosensor. Anal Chim Acta 434(2):239–246

Dremel BAA, Li SY, Schmid RD (1992) On-line determination of glucose and lactate concentrations in animal cell culture based on fibre optic detection of oxygen in flow-injection analysis. Biosens Bioelectron 7(2):133–139

Dzyadevych SV, Soldatkin AP, El'skaya AV, Martelet C, Jaffrezic-Renault N (2006) Enzyme biosensors based on ion-selective field-effect transistors. Anal Chim Acta 568(1–2):248–258

Esseghaier C, Bergaoui Y, Tlili A, Helali S, Abdelghani A (2008) Impedance spectroscopy on immobilized streptavidin horseradish peroxidase layer for biosensing. Sens Actuators B Chem 134:112–116

Fei J, Wu Y, Ji X, Wang J, Hu S, Gao Z (2003) An amperometric biosensor for glucose based on electrodeposited redox polymer/glucose oxidase film on a gold electrode. Anal Sci 19(9):1259–1263

Ferretti S, Russell DA, Sapsford KE, Lee S-K, MacCraith BD, Oliva AG, Vidal M, Richardson DJ (2000) Optical biosensing of nitrite ions using cytochrome cd_1 nitrite reductase encapsulated in a sol–gel matrix. Analyst 125:1993–1999

Forzani ES, Zhang H, Nagahara LA, Amlani I, Tsui R, Tao N (2004) A conducting polymer nanojunction sensor for glucose detection. Nano Lett 4(9):1785–1788

Foulds NC, Lowe CR (1988) Immobilization of glucose oxidase in ferrocene-modified pyrrole polymers. Anal Chem 60(22):2473–2478

Gauglitz G, Reichert M (1992) Spectral investigation and optimization of pH and urea sensors. Sens Actuators B Chem 6(1–3):83–86

Gautier SM, Blum LJ, Coulet PR (1990a) Alternate determination of ATP and NADH with a single bioluminescence-based fiber-optic sensor. Sens Actuators B Chem 1(1–6):580–584

Gautier SM, Blum LJ, Coulet PR (1990b) Fibre-optic biosensor based on luminescence and immobilized enzymes: microdetermination of sorbitol, ethanol and oxaloacetate. J Biolumin Chemilumin 5(1):57–63

Gautier SM, Blum LJ, Coulet PR (1991) Cofactor-containing bioluminescent fibre-optic sensor: new developments with poly(vinyl alcohol) matrices. Anal Chim Acta 255(2):253–258

Gavalas VG, Law SA, Christopher Ball J, Andrews R, Bachas LG (2004) Carbon nanotube aqueous sol–gel composites: enzyme-friendly platforms for the development of stable biosensors. Anal Biochem 329(2):247–252

Godoy S, Leca-Bouvier B, Boullanger P, Blum LJ, Girard-Egrot AP (2005) Electrochemiluminescent detection of acetylcholine using acetylcholinesterase immobilized in a biomimetic Langmuir–Blodgett nanostructure. Sens Actuators B Chem 107(1):82–87

Goeders KM, Colton JS, Bottomley LA (2008) Microcantilevers: sensing chemical interactions via mechanical motion. Chem Rev 108(2):522–542

Goldfinch MJ, Lowe CR (1984) Solid-phase optoelectronic sensors for biochemical analysis. Anal Biochem 138(2):430–436

Grieshaber D, MacKenzie R, Voros J, Reimhult E (2008) Electrochemical biosensors – sensor principles and architectures. Sensors 8:1400–1458

Guan J-G, Miao Y-Q, Zhang Q-J (2004) Impedimetric biosensors. J Biosci Bioeng 97(4):219–226

Guilbault GG, Lubrano GJ (1973) An enzyme electrode for the amperometric determination of glucose. Anal Chim Acta 64(3):439–455

Guo M, Chen J, Li J, Nie L, Yao S (2004) Carbon nanotubes-based amperometric cholesterol biosensor fabricated through layer-by-layer technique. Electroanalysis 16(23):1992–1998

Habermüller K, Mosbach M, Schuhmann W (2000) Electron-transfer mechanisms in amperometric biosensors. Fresenius J Anal Chem 366(6):560–568

Healey BG, Walt DR (1995) Improved fiber-optic chemical sensor for penicillin. Anal Chem 67(24):4471–4476

Heller A (1990) Electrical wiring of redox enzymes. Acc Chem Res 23(5):128–134

Hiller M, Kranz C, Huber J, Bäuerle P, Schuhmann W (1996) Amperometric biosensors produced by immobilization of redox enzymes at polythiophene-modified electrode surfaces. Adv Mater 8(3):219–222

Hlavay J, Guilbault GG (1994) Determination of sulphite by use of a fiber-optic biosensor based on a chemiluminescent reaction. Anal Chim Acta 299(1):91–96

Hlavay J, Haemmerli SD, Guilbault GG (1994) Fibre-optic biosensor for hypoxanthine and xanthine based on a chemiluminescence reaction. Biosens Bioelectron 9(3):189–195

Hoa DT, Kumar TNS, Punekar NS, Srinivasa RS, Lal R, Contractor AQ (1992) A biosensor based on conducting polymers. Anal Chem 64(21):2645–2646

Ilangovan R, Daniel D, Krastanov A, Zachariah C, Elizabeth R (2006) Enzyme based biosensor for heavy metal ions determination. Biotechnol Biotechnol Equip 20:184–189

Inoue Y, Kato Y, Sato K (1992) Surface acoustic wave method for in situ determination of the amounts of enzyme–substrate complex formed on immobilized glucose oxidase during catalytic reaction. J Chem Soc Faraday Trans 88:449–454

Iyer R, Pavlov V, Katakis I, Bachas LG (2003) Amperometric sensing at high temperature with a "wired" thermostable glucose-6-phosphate dehydrogenase from *Aquifex aeolicus*. Anal Chem 75(15):3898–3901

Jia J, Wang B, Wu A, Cheng G, Li Z, Dong S (2002) A method to construct a third-generation horseradish peroxidase biosensor: self-assembling gold nanoparticles to three-dimensional sol–gel network. Anal Chem 74(9):2217–2223

Joshi KA, Tang J, Haddon R, Wang J, Chen W, Mulchandani A (2005a) A disposable biosensor for organophosphorus nerve agents based on carbon nanotubes modified thick film strip electrode. Electroanalysis 17(1):54–58

Joshi PP, Merchant SA, Wang Y, Schmidtke DW (2005b) Amperometric biosensors based on redox polymer-carbon nanotube-enzyme composites. Anal Chem 77(10):3183–3188

Kajiya Y, Sugai H, Iwakura C, Yoneyama H (1991) Glucose sensitivity of polypyrrole films containing immobilized glucose oxidase and hydroquinonesulfonate ions. Anal Chem 63(1):49–54

Kaku T, Karan HI, Okamoto Y (1994) Amperometric glucose sensors based on immobilized glucose oxidase–polyquinone system. Anal Chem 66(8):1231–1235

Kang X, Cheng G, Dong S (2001) A novel electrochemical SPR biosensor. Electrochem Commun 3(9):489–493

Kar S, Arnold MA (1992) Fiber-optic ammonia sensor for measuring synaptic glutamate and extracellular ammonia. Anal Chem 64(20):2438–2443

Karousos NG, Aouabdi S, Way AS, Reddy SM (2002) Quartz crystal microbalance determination of organophosphorus and carbamate pesticides. Anal Chim Acta 469(2):189–196

Khan GF, Ohwa M, Wernet W (1996) Design of a stable charge transfer complex electrode for a third-generation amperometric glucose sensor. Anal Chem 68(17):2939–2945

Koncki R, Lenarczuk T, Radomska A, Glab S (2001) Optical biosensors based on Prussian blue films. Analyst 126:1080–1085

Kondoh J, Matsui Y, Shiokawa S (1993) New biosensor using shear horizontal surface acoustic wave device. Jpn J Appl Phys 32:2376–2379

Kondoh J, Matsui Y, Shiokawa S, Wlodarski WB (1994) Enzyme-immobilized SH-SAW biosensor. Sens Actuators B Chem 20(2–3):199–203

Köneke R, Menzel C, Ulber R, Schügerl K, Scheper T, Saleemuddin M (1996) Reversible coupling of glucoenzymes on fluoride-sensitive FET biosensors based on lectin-glucoprotein binding. Biosens Bioelectron 11(12):1229–1236

Koopal CGJ, de Ruiter B, Nolte RJM (1991) Amperometric biosensor based on direct communication between glucose oxidase and a conducting polymer inside the pores of a filtration membrane. J Chem Soc Chem Commun 1691–1692

Korell U, Spichiger UE (1993) Membraneless immobilization of xanthine oxidase on organic conducting salt/silicone oil electrodes. Electroanalysis 5(9–10):869–876

Korell U, Spichiger UE (1994) Novel membranes amperometric peroxide biosensor based on a tetrathiafulvalene-*p*-tetracyanoquinodimethane electrode. Anal Chem 66(4):510–515

Lange MA, Chambers JQ (1985) Amperometric determination of glucose with a ferrocene-mediated glucose oxidase/polyacrylamide gel electrode. Anal Chim Acta 175:89–97

Länge K, Rapp B, Rapp M (2008) Surface acoustic wave biosensors: a review. Anal Bioanal Chem 391(5):1509–1519

Langer JJ, Filipiak M, Kecinska J, Jasnowska J, Wlodarczak J, Buladowski B (2004) Polyaniline biosensor for choline determination. Surf Sci 573(1):140–145

Leca B, Blum LJ (2000) Luminol electrochemiluminescence with screen-printed electrodes for low-cost disposable oxidase-based optical sensors. Analyst 125(5):789–791

Leca B, Marty JL (1997a) Reagentless ethanol sensor based on a NAD-dependent dehydrogenase. Biosens Bioelectron 12(11):1083–1088

Leca B, Marty JL (1997b) Reusable ethanol sensor based on a NAD(+)-dependent dehydrogenase without coenzyme addition. Anal Chim Acta 340(1–3):143–148

Leca B, Morelis RM, Coulet PR (1995) Design of a choline sensor via direct coating of the transducer by photopolymerization of the sensing layer. Sens Actuators B Chem 27(1–3):436–439

Leca BD, Verdier AM, Blum LJ (2001) Screen-printed electrodes as disposable or reusable optical devices for luminol electrochemiluminescence. Sens Actuators B Chem 74(1–3):190–193

Lee S-J, Saleemuddin M, Scheper T, Loos H, Sahm H (1994a) A fluorometric fiber-optic biosensor for dual analysis of glucose and fructose using glucose–fructose-oxidoreductase isolated from *Zymomonas mobilis*. J Biotech 36(1):39–44

Lee SJ, Scheper T, Buckmann AF (1994b) Application of a flow injection fibre optic biosensor for the analysis of different amino acids. Biosens Bioelectron 9(1):29–32

Lee W-Y, Kim S-R, Kim T-H, Lee KS, Shin M-C, Park J-K (2000a) Sol–gel-derived thick-film conductometric biosensor for urea determination in serum. Anal Chim Acta 404(2):195–203

Lee W-Y, Lee KS, Kim T-H, Shin M-C, Park J-K (2000b) Microfabricated conductometric urea biosensor based on sol–gel immobilized urease. Electroanalysis 12(1):78–82

Lenarczuk T, Wencel D, Glab S, Koncki R (2001) Prussian blue-based optical glucose biosensor in flow-injection analysis. Anal Chim Acta 447(1–2):23–32

Li L, Walt DR (1995) Dual-analyte fiber-optic sensor for the simultaneous and continuous measurement of glucose and oxygen. Anal Chem 67(20):3746–3752

Li C-I, Lin Y-H, Shih C-L, Tsaur J-P, Chau L-K (2002a) Sol–gel encapsulation of lactate dehydrogenase for optical sensing of L-lactate. Biosens Bioelectron 17(4):323–330

Li YX, Zhu LD, Zhu GY, Zhao CA (2002b) A chemiluminescence optical fiber glucose biosensor based on co-immobilizing glucose oxidase and horseradish peroxidase in a sol–gel film. Chem Res Chin Univ 18(1):12–15

Lin Y, Lu F, Wang J (2004) Disposable carbon nanotube modified screen-printed biosensor for amperometric detection of organophosphorus pesticides and nerve agents. Electroanalysis 16(1–2):145–149

Liu S, Ju H (2003) Reagentless glucose biosensor based on direct electron transfer of glucose oxidase immobilized on colloidal gold modified carbon paste electrode. Biosens Bioelectron 19(3):177–183

Liu D, Ge K, Chen K, Nie L, Yao S (1995) Clinical analysis of urea in human blood by coupling a surface acoustic wave sensor with urease extracted from pumpkin seeds. Anal Chim Acta 307(1):61–69

Liu J, Chou A, Rahmat W, Paddon-Row MN, Gooding JJ (2005) Achieving direct electrical connection to glucose oxidase using aligned single walled carbon nanotube arrays. Electroanalysis 17(1):38–46

Llopis X, Merkoçi A, del Valle M, Alegret S (2005) Integration of a glucose biosensor based on an epoxy-graphite-TTF·TCNQ-GOD biocomposite into a FIA system. Sens Actuators B Chem 107(2):742–748

Lojou É, Bianco P (2004) Membrane electrodes for protein and enzyme electrochemistry. Electroanalysis 16(13–14):1113–1121

Luo X-L, Xu J-J, Zhao W, Chen H-Y (2004a) Glucose biosensor based on ENFET doped with SiO_2 nanoparticles. Sens Actuators B Chem 97(2–3):249–255

Luo X-L, Xu J-J, Zhao W, Chen H-Y (2004b) A novel glucose ENFET based on the special reactivity of MnO_2 nanoparticles. Biosens Bioelectron 19(10):1295–1300

Luo X, Killard AJ, Morrin A, Smyth MR (2006) Enhancement of a conducting polymer-based biosensor using carbon nanotube-doped polyaniline. Anal Chim Acta 575(1):39–44

Mabeck J, Malliaras G (2006) Chemical and biological sensors based on organic thin-film transistors. Anal Bioanal Chem 384(2):343–353

Marquette CA, Blum LJ (1999) Luminol electrochemiluminescence-based fibre optic biosensors for flow injection analysis of glucose and lactate in natural samples. Anal Chim Acta 381(1):1–10

Marquette C, Blum L (2006) Applications of the luminol chemiluminescent reaction in analytical chemistry. Anal Bioanal Chem 385(3):546–554

Marquette C, Blum L (2008) Electro-chemiluminescent biosensing. Anal Bioanal Chem 390 (1):155–168

Marquette CA, Leca BD, Blum LJ (2001) Electrogenerated chemiluminescence of luminol for oxidase-based fibre-optic biosensors. Luminescence 16(2):159–165

Martin SP, Lamb DJ, Lynch JM, Reddy SM (2003) Enzyme-based determination of cholesterol using the quartz crystal acoustic wave sensor. Anal Chim Acta 487(1):91–100

Mascini M (1995) Enzyme-based optical-fibre biosensors. Sens Actuators B Chem 29(1–3):121–125

McCurley MF (1994) An optical biosensor using a fluorescent, swelling sensing element. Biosens Bioelectron 9(7):527–533

Mena ML, Yáñez-Sedeño P, Pingarrón JM (2005) A comparison of different strategies for the construction of amperometric enzyme biosensors using gold nanoparticle-modified electrodes. Anal Biochem 336(1):20–27

Merkoçi A, Braga S, Fàbregas E, Alegret S (1999) A potentiometric biosensor for D-amygdalin based on a consolidated biocomposite membrane. Anal Chim Acta 391(1):65–72

Michel PE, Gautier-Sauvigne SM, Blum LJ (1998) Luciferin incorporation in the structure of acrylic microspheres with subsequent confinement in a polymeric film: a new method to develop a controlled release-based biosensor for ATP, ADP and AMP. Talanta 47(1):169–181

Mitsubayashi K, Kon T, Hashimoto Y (2003) Optical bio-sniffer for ethanol vapor using an oxygen-sensitive optical fiber. Biosens Bioelectron 19(3):193–198

Mourzina IG, Yoshinobu T, Ermolenko YE, Vlasov YG, Schöning MJ, Iwasaki H (2004) Immobilization of urease and cholinesterase on the surface of semiconductor transducer for the development of light-addressable potentiometric sensors. Microchim Acta 144(1):41–50

Moussy F, Jakeway S, Harrison DJ, Rajotte RV (1994) In vitro and in vivo performance and lifetime of perfluorinated ionomer-coated glucose sensors after high-temperature curing. Anal Chem 66(22):3882–3888

Mulchandani A, Pan S, Chen W (1999) Fiber-optic enzyme biosensor for direct determination of organophosphate nerve agents. Biotechnol Prog 15(1):130–134

Narayanaswamy R, Sevilla F (1988) An optical fibre probe for the determination of glucose based on immobilized glucose dehydrogenase. Anal Lett 21(7):1165–1175

Navas Diaz A, Ramos Peinado MC (1997) Sol–gel cholinesterase biosensor for organophosphorus pesticide fluorimetric analysis. Sens Actuators B Chem 39(1–3):426–431

Nishizawa M, Matsue T, Uchida I (1992) Penicillin sensor based on a microarray electrode coated with pH-responsive polypyrrole. Anal Chem 64(21):2642–2644

Noguer T, Leca B, Jeanty G, Marty JL (1999) Biosensors based on enzyme inhibition: detection of organophosphorus and carbamate insecticides and dithiocarbamate fungicides. Field Anal Chem Technol 3(3):171–178

Notsu H, Tatsuma T, Fujishima A (2002) Tyrosinase-modified boron-doped diamond electrodes for the determination of phenol derivatives. J Electroanal Chem 523(1–2):86–92

Ohara TJ, Rajagopalan R, Heller A (1994) "Wired" enzyme electrodes for amperometric determination of glucose or lactate in the presence of interfering substances. Anal Chem 66 (15):2451–2457

Opitz N, Lubbers DW (1988) Electrochromic dyes, enzyme reactions and hormone–protein interactions in fluorescence optic sensor (optode) technology. Talanta 35(2):123–127

Palmisano F, Centonze D, Guerrieri A, Zambonin PG (1993) An interference-free biosensor based on glucose oxidase electrochemically immobilized in a non-conducting poly(pyrrole) film for continuous subcutaneous monitoring of glucose through microdialysis sampling. Biosens Bioelectron 8(9–10):393–399

Palmisano F, Rizzi R, Centonze D, Zambonin PG (2000) Simultaneous monitoring of glucose and lactate by an interference and cross-talk free dual electrode amperometric biosensor based on electropolymerized thin films. Biosens Bioelectron 15(9–10):531–539

Palmisano F, Zambonin PG, Centonze D, Quinto M (2002) A disposable, reagentless, third-generation glucose biosensor based on overoxidized poly(pyrrole)/tetrathiafulvalene-tetracyanoquinodimethane composite. Anal Chem 74(23):5913–5918

Pandey PC, Mishra AP (1988) Conducting polymer-coated enzyme microsensor for urea. Analyst 113:329–331

Pandey PC, Upadhyay S, Sharma S (2003) TTF-TCNQ functionalized ormosil based electrocatalytic biosensor: a comparative study on bioelectrocatalysis. Electroanalysis 15(13):1115–1119

Papkovsky DB (1993) Luminescent porphyrins as probes for optical (bio)sensors. Sens Actuators B Chem 11(1–3):293–300

Papkovsky DB, Olah J, Kurochkin IN (1993) Fibre-optic lifetime-based enzyme biosensor. Sens Actuators B Chem 11(1–3):525

Park K-Y, Choi S-B, Lee M, Sohn B-K, Choi S-Y (2002) ISFET glucose sensor system with fast recovery characteristics by employing electrolysis. Sens Actuators B Chem 83(1–3):90–97

Pasic A, Koehler H, Schaupp L, Pieber T, Klimant I (2006) Fiber-optic flow-through sensor for online monitoring of glucose. Anal Bioanal Chem 386(5):1293–1302

Pasic A, Koehler H, Klimant I, Schaupp L (2007) Miniaturized fiber-optic hybrid sensor for continuous glucose monitoring in subcutaneous tissue. Sens Actuators B Chem 122(1):60–68

Patolsky F, Zayats M, Katz E, Willner I (1999) Precipitation of an insoluble product on enzyme monolayer electrodes for biosensor applications: characterization by faradaic impedance spectroscopy, cyclic voltammetry, and microgravimetric quartz crystal microbalance analyses. Anal Chem 71(15):3171–3180

Patolsky F, Weizmann Y, Willner I (2004) Long-range electrical contacting of redox enzymes by SWCNT connectors. Angew Chem Int Ed Engl 43(16):2113–2117

Pei J, Tian F, Thundat T (2004) Glucose biosensor based on the microcantilever. Anal Chem 76(2):292–297

Pishko MV, Katakis I, Lindquist S-E, Ye L, Heller BAGA (1990) Direct electrical communication between graphite electrodes and surface adsorbed glucose oxidase/redox polymer complexes. Angew Chem Int Ed Engl 29(1):82–84

Poghossian A, Schöning MJ (2004) Detecting both physical and (bio-)chemical parameters by means of ISFET devices. Electroanalysis 16(22):1863–1872

Poghossian A, Schöning MJ (2007) Chemical and biological field-effect sensors for liquids – status report. In: Marks RS, Cullen DC, Karube I, Lowe CR, Weetall HH (eds) Handbook of biosensors and biochips. Wiley, Chichester, pp 395–411

Poghossian A, Schoning MJ, Schroth P, Simonis A, Luth H (2001a) An ISFET-based penicillin sensor with high sensitivity, low detection limit and long lifetime. Sens Actuators B Chem 76(1–3):519–526

Poghossian A, Yoshinobu T, Simonis A, Ecken H, Luth H, Schoning MJ (2001b) Penicillin detection by means of field-effect based sensors: EnFET, capacitive EIS sensor or LAPS? Sens Actuators B Chem 78(1–3):237–242

Preuschoff F, Spohn U, Janasek D, Weber E (1994) Photodiode-based chemiluminometric biosensors for hydrogen peroxide and L-lysine. Biosens Bioelectron 9(8):543–549

Qian L, Yang X (2006) Composite film of carbon nanotubes and chitosan for preparation of amperometric hydrogen peroxide biosensor. Talanta 68(3):721–727

Ramanathan K, Danielsson B (2001) Principles and applications of thermal biosensors. Biosens Bioelectron 16(6):417–423

Ramos MC, Torijas MC, Diaz AN (2001) Enhanced chemiluminescence biosensor for the determination of phenolic compounds and hydrogen peroxide. Sens Actuators B Chem 73–75(1):71

Rebriiev AV, Starodub NF (2004) Enzymatic biosensor based on the ISFET and photopolymeric membrane for the determination of urea. Electroanalysis 16(22):1891–1895

Ren C, Song Y, Li Z, Zhu G (2005) Hydrogen peroxide sensor based on horseradish peroxidase immobilized on a silver nanoparticles/cysteamine/gold electrode. Anal Bioanal Chem 381(6):1179–1185

Rhines TD, Arnold MA (1989) Fiber-optic biosensor for urea based on sensing of ammonia gas. Anal Chim Acta 227:387–396

Riklin A, Katz E, Wiliner I, Stocker A, Buckmann AF (1995) Improving enzyme–electrode contacts by redox modification of cofactors. Nature 376(6542):672–675

Rochette JF, Sacher E, Meunier M, Luong JHT (2005) A mediatorless biosensor for putrescine using multiwalled carbon nanotubes. Anal Biochem 336(2):305–311

Rosenzweig Z, Kopelman R (1996) Analytical properties and sensor size effects of a micrometer-sized optical fiber glucose biosensor. Anal Chem 68(8):1408–1413

Rubio-Retama J, Hernando J, Lopez-Ruiz B, Hartl A, Steinmuller D, Stutzmann M, Lopez-Cabarcos E, AntonioGarrido J (2006) Synthetic nanocrystalline diamond as a third-generation biosensor support. Langmuir 22(13):5837–5842

Russell RJ, Pishko MV, Simonian AL, Wild JR (1999) Poly(ethylene glycol) hydrogel-encapsulated fluorophore–enzyme conjugates for direct detection of organophosphorus neurotoxins. Anal Chem 71(21):4909–4912

Salimi A, Compton RG, Hallaj R (2004) Glucose biosensor prepared by glucose oxidase encapsulated sol–gel and carbon-nanotube-modified basal plane pyrolytic graphite electrode. Anal Biochem 333(1):49–56

Sanz V, de Marcos S, Galban J (2007) A reagentless optical biosensor based on the intrinsic absorption properties of peroxidase. Biosens Bioelectron 22(6):956–964

Sasso SV, Pierce RJ, Walla R, Yacynych AM (1990) Electropolymerized 1,2-diaminobenzene as a means to prevent interferences and fouling and to stabilize immobilized enzyme in electrochemical biosensors. Anal Chem 62(11):1111–1117

Sassolas A, Blum LJ, Leca-Bouvier BD (2008) Electrogeneration of polyluminol and chemiluminescence for new disposable reagentless optical sensors. Anal Bioanal Chem 390(3):865–871

Sassolas A, Blum LJ, Leca-Bouvier B (2009) Polymeric luminol on pre-treated screen-printed electrodes for the design of performant reagentless (bio)sensors. Sens Actuators B Chem 139:214–221

Schäferling M, Wolfbeis OS (2007) Europium tetracycline as a luminescent probe for nucleoside phosphates and its application to the determination of kinase activity. Chemistry 13 (15):4342–4349

Schaffar BPH, Wolfbeif OS (1990) A fast responding fibre optic glucose biosensor based on an oxygen optrode. Biosens Bioelectron 5(2):137–148

Schlapfer P, Mindt W, Racine PH (1974) Electrochemical measurement of glucose using various electron acceptors. Clin Chim Acta 57(3):283–289

Schubert F (1993) A fiber-optic enzyme sensor for the determination of adenosine diphosphate using internal analyte recycling. Sens Actuators B Chem 11(1–3):531–535

Schuhmann W, Lammert R, Hämmerle M, Schmidt H-L (1991) Electrocatalytic properties of polypyrrole in amperometric electrodes. Biosens Bioelectron 6(8):689–697

Schuhmann W, Kranz C, Huber J, Wohlschläger H (1993) Conducting polymer-based amperometric enzyme electrodes. Towards the development of miniaturized reagentless biosensors. Synth Met 61(1–2):31–35

Seker S, Becerik I (2004) A neural network model in the calibration of glucose sensor based on the immobilization of glucose oxidase into polypyrrole matrix. Electroanalysis 16 (18):1542–1549

Shervedani RK, Mehrjardi AH, Zamiri N (2006) A novel method for glucose determination based on electrochemical impedance spectroscopy using glucose oxidase self-assembled biosensor. Bioelectrochemistry 69(2):201–208

Shul'ga AA, Soldatkin AP, El'skaya AV, Dzyadevich SV, Patskovsky SV, Strikha VI (1994) Thin-film conductometric biosensors for glucose and urea determination. Biosens Bioelectron 9(3):217–223

Sierra JF, Galban J, Castillo JR (1997) Determination of glucose in blood based on the intrinsic fluorescence of glucose oxidase. Anal Chem 69(8):1471–1476

Simonian AL, Flounders AW, Wild JR (2004) FET-based biosensors for the direct detection of organophosphate neurotoxins. Electroanalysis 16(22):1896–1906

Solanki PR, Arya SK, Nishimura Y, Iwamoto M, Malhotra BD (2007) Cholesterol biosensor based on amino-undecanethiol self-assembled monolayer using surface plasmon resonance technique. Langmuir 23(13):7398–7403

Soldatkin AP, Arkhypova VN, Dzyadevych SV, El'skaya AV, Gravoueille J-M, Jaffrezic-Renault N, Martelet C (2005) Analysis of the potato glycoalkaloids by using of enzyme biosensor based on pH-ISFETs. Talanta 66(1):28–33

Song Y, Wang L, Ren C, Zhu G, Li Z (2006) A novel hydrogen peroxide sensor based on horseradish peroxidase immobilized in DNA films on a gold electrode. Sens Actuators B Chem 114(2):1001–1006

Spohn U, Preuschoff F, Blankenstein G, Janasek D, Kula MR, Hacker A (1995) Chemiluminometric enzyme sensors for flow-injection analysis. Anal Chim Acta 303(1):109–120

Sternberg R, Bindra DS, Wilson GS, Thevenot DR (1988) Covalent enzyme coupling on cellulose acetate membranes for glucose sensor development. Anal Chem 60(24):2781–2786

Stoica L, Ludwig R, Haltrich D, Gorton L (2006) Third-generation biosensor for lactose based on newly discovered cellobiose dehydrogenase. Anal Chem 78(2):393–398

Subramanian A, Oden PI, Kennel SJ, Jacobson KB, Warmack RJ, Thundat T, Doktycz MJ (2002) Glucose biosensing using an enzyme-coated microcantilever. Appl Phys Lett 81(2):385–387

Tatsuma T, Mori H, Fujishima A (2000) Electron transfer from diamond electrodes to heme peptide and peroxidase. Anal Chem 72(13):2919–2924

Thevenot DR, Toth K, Durst RA, Wilson GS (2001) Electrochemical biosensors: recommended definitions and classification. Biosens Bioelectron 16(1–2):121–131

Tian F, Xu B, Zhu L, Zhu G (2001) Hydrogen peroxide biosensor with enzyme entrapped within electrodeposited polypyrrole based on mediated sol–gel derived composite carbon electrode. Anal Chim Acta 443(1):9–16

Tian Y, Mao L, Okajima T, Ohsaka T (2002) Superoxide dismutase-based third-generation biosensor for superoxide anion. Anal Chem 74(10):2428–2434

Tian Y, Mao L, Okajima T, Ohsaka T (2005) A carbon fiber microelectrode-based third-generation biosensor for superoxide anion. Biosens Bioelectron 21(4):557–564

Trettnak W, Wolfbeis OS (1989a) A fiber optic lactate biosensor with an oxygen optrode as the transducer. Anal Lett 22(9):2191–2197

Trettnak W, Wolfbeis OS (1989b) Fully reversible fibre-optic glucose biosensor based on the intrinsic fluorescence of glucose oxidase. Anal Chim Acta 221:195–203

Trettnak W, Leiner MJP, Wolfbeis OS (1989) Fibre-optic glucose sensor with a pH optrode as the transducer. Biosensors 4(1):15–26

Trettnak W, Reininger F, Zinterl E, Wolfbeis OS (1993) Fiber-optic remote detection of pesticides and related inhibitors of the enzyme acetylcholine esterase. Sens Actuators B Chem 11(1–3):87–93

Tsafack VC, Marquette CA, Leca B, Blum LJ (2000a) An electrochemiluminescence-based fibre optic biosensor for choline flow injection analysis. Analyst 125:151–155

Tsafack VC, Marquette CA, Pizzolato F, Blum LJ (2000b) Chemiluminescent choline biosensor using histidine-modified peroxidase immobilised on metal-chelate substituted beads and choline oxidase immobilised on anion-exchanger beads co-entrapped in a photocrosslinkable polymer. Biosens Bioelectron 15(3–4):125–133

Tsai W-C, Cass AEG (1995) Ferrocene-modified horseradish peroxidase enzyme electrodes. A kinetic study on reactions with hydrogen peroxide and linoleic hydroperoxide. Analyst 120:2249–2254

Tsai YC, Li SC, Chen JM (2005) Cast thin film biosensor design based on a Nafion backbone, a multiwalled carbon nanotube conduit, and a glucose oxidase function. Langmuir 21 (8):3653–3658

Tymecki L, Zwierkowska E, Koncki R (2005) Strip bioelectrochemical cell for potentiometric measurements fabricated by screen-printing. Anal Chim Acta 538(1–2):251–256

Updike SJ, Hicks GP (1967) The enzyme electrode. Nature 214:986–988

Vaidya R, Wilkins E (1994) Effect of interference on amperometric glucose biosensors with cellulose acetate membranes. Electroanalysis 6(8):677–682

Védrine C, Fabiano S, Tran-Minh C (2003) Amperometric tyrosinase based biosensor using an electrogenerated polythiophene film as an entrapment support. Talanta 59(3):535–544

Vidal J-C, Espuelas J, Castillo J-R (2004) Amperometric cholesterol biosensor based on in situ reconstituted cholesterol oxidase on an immobilized monolayer of flavin adenine dinucleotide cofactor. Anal Biochem 333(1):88–98

Viticoli M, Curulli A, Cusma A, Kaciulis S, Nunziante S, Pandolfi L, Valentini F, Padeletti G (2006) Third-generation biosensors based on TiO_2 nanostructured films. Mater Sci Eng C 26 (5–7):947–951

Viveros L, Paliwal S, McCrae D, Wild J, Simonian A (2006) A fluorescence-based biosensor for the detection of organophosphate pesticides and chemical warfare agents. Sens Actuators B Chem 115(1):150–157

Volotovsky V, Kim N (1998) Cyanide determination by an ISFET-based peroxidase biosensor. Biosens Bioelectron 13(9):1029–1033

Walker JP, Asher SA (2005) Acetylcholinesterase-based organophosphate nerve agent sensing photonic crystal. Anal Chem 77(6):1596–1600

Wang J (2008) Electrochemical glucose biosensors. Chem Rev 108(2):814–825

Wang J, Musameh M (2005) Carbon-nanotubes doped polypyrrole glucose biosensor. Anal Chim Acta 539(1–2):209–213

Wang X, Dzyadevych SV, Chovelon J-M, Renault NJ, Chen L, Xia S, Zhao J (2006) Development of a conductometric nitrate biosensor based on methyl viologen/Nafion composite film. Electrochem Commun 8(2):201–205

Watson LD, Maynard P, Cullen DC, Sethi RS, Brettle J, Lowe CR (1987) A microelectronic conductimetric biosensor. Biosensors 3:101–115

Willner I, Katz E (2000) Integration of layered redox proteins and conductive supports for bioelectronic applications. Angew Chem Int Ed Engl 39(7):1180–1218

Willner I, Heleg-Shabtai V, Blonder R, Katz E, Tao G, Buckmann AF, Heller A (1996) Electrical wiring of glucose oxidase by reconstitution of FAD-modified monolayers assembled onto Au-electrodes. J Am Chem Soc 118(42):10321–10322

Wolfbeis OS, Li H (1993) Fluorescence optical urea biosensor with an ammonium optrode as transducer. Biosens Bioelectron 8(3–4):161–166

Wolfbeis OS, Oehme I, Papkovskaya N, Klimant I (2000) Sol–gel based glucose biosensors employing optical oxygen transducers, and a method for compensating for variable oxygen background. Biosens Bioelectron 15(1–2):69–76

Wolfbeis OS, Schäferling M, Dürkop A (2003) Reversible optical sensor membrane for hydrogen peroxide using an immobilized fluorescent probe, and its application to a glucose biosensor. Microchim Acta 143(4):221–227

Wu XJ, Choi MMF (2003) Hydrogel network entrapping cholesterol oxidase and octadecylsilica for optical biosensing in hydrophobic organic or aqueous micelle solvents. Anal Chem 75(16):4019–4027

Wu XJ, Choi MMF (2004) An optical glucose biosensor based on entrapped-glucose oxidase in silicate xerogel hybridised with hydroxyethyl carboxymethyl cellulose. Anal Chim Acta 514(2):219–226

Wu J, Qu Y (2006) Mediator-free amperometric determination of glucose based on direct electron transfer between glucose oxidase and an oxidized boron-doped diamond electrode. Anal Bioanal Chem 385(7):1330–1335

Xiao Y, Patolsky F, Katz E, Hainfeld JF, Willner I (2003) "Plugging into enzymes": nanowiring of redox enzymes by a gold nanoparticle. Science 299(5614):1877–1881

Xie X, Suleiman AA, Guilbault GG (1990) A urea fiber optic biosensor based on absorption measurement. Anal Lett 23(12):2143–2153

Xie X, Suleiman AA, Guilbault GG (1991) Determination of urea in serum by a fiber-optic fluorescence biosensor. Talanta 38(10):1197–1200

Xie X, Suleiman AA, Guilbault GG, Yang Z, Z-a S (1992a) Flow-injection determination of ethanol by fiber-optic chemiluminescence measurement. Anal Chim Acta 266(2):325–329

Xie X, Suleiman AA, Guilbault GG (1992b) A fluorescence-based fiber optic biosensor for the flow-injection analysis of penicillin. Biotechnol Bioeng 39(11):1147–1150

Xie X, Shakhsher Z, Suleiman AA, Guilbault GG, Yang Z, Z-a S (1994) A fiber optic biosensor for sulfite analysis in food. Talanta 41(2):317–321

Xu J-J, Zhao W, Luo X-L, Chen H-Y (2005a) A sensitive biosensor for lactate based on layer-by-layer assembling MnO_2 nanoparticles and lactate oxidase on ion-sensitive field-effect transistors. Chem Commun 792–794

Xu JJ, Wang G, Zhang Q, Xia XH, Chen HY (2005b) Third generation horseradish peroxidase biosensor based on self-assembling carbon nanotubes to gold electrode surface. Chin Chem Lett 16(4):523–526

Yabuki S-I, Shinohara H, Aizawa M (1989) Electro-conductive enzyme membrane. J Chem Soc Chem Commun 945–946

Yamato H, Ohwa M, Wernet W (1995) A polypyrrole/three-enzyme electrode for creatinine detection. Anal Chem 67(17):2776–2780

Yan X, Ji H-F, Lvov Y (2004) Modification of microcantilevers using layer-by-layer nanoassembly film for glucose measurement. Chem Phys Lett 396(1–3):34–37

Yan X, Xu XK, Ji HF (2005) Glucose oxidase multilayer modified microcantilevers for glucose measurement. Anal Chem 77(19):6197–6204

Yang Y, Wang Z, Yang M, Guo M, Wu Z, Shen G, Yu R (2006) Inhibitive determination of mercury ion using a renewable urea biosensor based on self-assembled gold nanoparticles. Sens Actuators B Chem 114(1):1–8

Yin L-T, Chou J-C, Chung W-Y, Sun T-P, Hsiung K-P, Hsiung S-K (2001) Glucose ENFET doped with MnO_2 powder. Sens Actuators B Chem 76(1–3):187–192

Zayats M, Kharitonov AB, Katz E, Buckmann AF, Willner I (2000) An integrated NAD^+-dependent enzyme-functionalized field-effect transistor (ENFET) system: development of a lactate biosensor. Biosens Bioelectron 15(11–12):671–680

Zayats M, Katz E, Baron R, Willner I (2005) Reconstitution of apo-glucose dehydrogenase on pyrroloquinoline quinone-functionalized Au nanoparticles yields an electrically contacted biocatalyst. J Am Chem Soc 127(35):12400–12406

Zayats M, Willner B, Willner I (2008) Design of amperometric biosensors and biofuel cells by the reconstitution of electrically contacted enzyme electrodes. Electroanalysis 20(6):583–601

Zhang W, Chang H, Rechnitz GA (1997) Dual-enzyme fiber optic biosensor for pyruvate. Anal Chim Acta 350(1–2):59–65

Zhou Z, Qiao L, Zhang P, Xiao D, Choi M (2005) An optical glucose biosensor based on glucose oxidase immobilized on a swim bladder membrane. Anal Bioanal Chem 383(4):673–679

Zhou Q, Xie Q, Fu Y, Su Z, Jia X, Yao S (2007) Electrodeposition of carbon nanotubes-chitosan-glucose oxidase biosensing composite films triggered by reduction of *p*-benzoquinone or H_2O_2. J Phys Chem B 111(38):11276–11284

Chapter 5
Antibodies in Biosensing

Amber C. Donahue and Maher Albitar

Abstract Biosensing, the detection of biological phenomena with accuracy and precision, is a rapidly growing and increasingly divergent field. The requirement for a stable, adaptable, and highly specific recognition receptor is of paramount importance to the design of a good biosensing assay. Among the receptors currently used in these applications, the antibody is perhaps the only one designed by evolution to be a natural biosensor. A greater understanding of the genetics of antibodies has given researchers the ability to manipulate their structure and to take advantage of the vast range of possible specificities. The platforms that rely on antibodies as recognition receptors, including enzyme-linked immunosorbent assay (ELISA), flow cytometry, radioimmunoassay (RIA), immunohistochemistry and immunocytochemistry, and antibody arrays, have evolved significantly to include new techniques and greater multiplexing opportunities. Here we give an overview of these techniques, and the ways in which the development of these biosensing assays are revolutionizing basic research, diagnostics, and therapeutics.

Keywords Horseradish peroxidase · Alkaline phosphatase · Streptavidin · Immunoprecipitation · Chromatin immunoprecipitation · Radioimmunoassay · Enzyme immunoassay · Enzyme-linked immunosorbent assay · Sandwich ELISA · ELISpot · Dot plot · Luminex · ChIP-on-chip · Multispot ELISA

M. Albitar (✉)
Department of Hematology/Oncology, Quest Diagnostics-Nichols Institute, 33608 Ortega Highway, San Juan Capistrano, CA92675, USA
e-mail: maher.x.Albitar@questdiagnostics.com

M. Zourob (ed.), *Recognition Receptors in Biosensors*,
DOI 10.1007/978-1-4419-0919-0_5,

Abbreviations

ELISA	Enzyme-linked immunosorbent assay
RIA	Radioimmunoassay
DNA	Deoxyribonucleic acid
Ig	Immunoglobulin
HRP	Horseradish peroxidase
ALP	Alkaline phosphatase
IP	Immunoprecipitation
SDS	Sodium dodecyl sulfate
PAGE	Polyacrylamide gel electrophoresis
ChIP	Chromatin immunoprecipitation
PVDF	Polyvinylidene fluoride
PCR	Polymerase chain reaction
CML	Chronic myeloid leukemia
WT	Wildtype
ECL	Electrochemiluminescent
ELISpot	Enzyme-linked immuno-sorbent spot assay
AEC	3-amino-9-ethylcarbazole
BCIP	5-bromo-4-chloro-3-indolyl phosphate
NBT	Nitro blue tetrazolium chloride
IHC	Immunohistochemistry
ICC	Immunocytochemistry
CBA	Cytometric bead array
MFI	Mean fluorescence intensity
RBCs	Red blood cells

5.1 Introduction

The detection and measurement of biological phenomena are at the heart of unraveling their complexity. Scientists have become increasingly aware of the importance of understanding the role of each component of a biological system, both individually and with respect to one another. Be it the interaction of the different cells that form a human organ, or the protein signaling cascade that transmits a signal from the cell surface to the nucleus to stimulate gene transcription in a single cell, the characteristics and interactions of these complex systems are best understood and most informative when considered in the context of their pathway or organ (Aggarwal and Lee 2003). The need to examine these systems as a whole has driven the creation of platforms such as DNA microarrays, and the expansion and integration of the fields of genomics, proteomics, transcriptomics, and metabolomics (Tan et al. 2008). These platforms, which take multiplex assays to new heights, provide a wealth of information about large numbers of analytes simultaneously and are proving increasingly valuable to basic researchers and clinicians alike.

Biosensing technologies have become ever more sophisticated and streamlined over time. One such technology is the antibody recognition system, whose past and current uses have inspired much of the current thinking leading to the use of antibodies as a basis for large-scale multiplex biosensing platforms (Hudelist et al. 2005; Haab 2006; Ling et al. 2007). Designed by evolution to bind targets with high specificity and high affinity, antibodies have been and will remain an invaluable tool in the discovery and measurement of biological phenomena. While most of the work is currently ex vivo, the use of antibodies in vivo has already shown significant progress and benefits. Antibodies are currently used for biosensing of specific targets in the body, in order to deliver radioactive isotopes or cytotoxic drugs (reviewed in (Ricart and Tolcher 2007)). Antibodies as biosensors have also been used for the imaging of tumors, or for visualizing a specific biological process such as tumor shrinkage or rate of growth (van Dongen et al. 2007; Sofou and Sgouros 2008; Tanaka and Fukase 2008; Weber et al. 2008). These types of applications for antibodies will likely become more common as the engineering of antibodies becomes more sophisticated, increasing the potential of using antibodies in vivo for the targeting of specific lesions or tumors, or even for the neutralization of specific biological processes.

5.2 Antibodies: An Overview

5.2.1 Immunoglobulin Expression In Vivo

Antibodies, or soluble forms of immunoglobulin (Ig), are uniquely suited to the requirements of biosensing, with a vast array of possible specificities and a structure that is one of the more stable among mammalian proteins. Produced by B lymphocytes, immunoglobulin is initially expressed as a transmembrane protein and functions as the antigen receptor for the cell. Through the variation introduced during the transcription process, the immunoglobulin expressed by each B cell is different, allowing each cell to recognize a different motif or epitope. Following recognition of this epitope, the specificity of the antibodies secreted by the activated B cells is sharpened further by additional mutations (Janeway et al. 2004). In this way, the immune response is carefully honed to be highly specific, and for decades researchers have exploited this talent of the immune system. However, the ex vivo generation of antibodies is becoming the standard for the purposes of research, diagnostics, and therapy.

5.2.2 Antibody Formats

The versatility of antibodies is further enhanced by the large number of formats available. Researchers and clinicians can choose monoclonal or polyclonal antibodies from a large number of species. The antibodies can be used as intact molecules or as one of several types of fragments, and can be chemically conjugated to many types of reporter molecules. These possibilities make antibodies amenable to an even wider range of assay platforms.

5.2.2.1 Monoclonal vs. Polyclonal Antibodies

The polyclonal antibody preparations used by researchers consist of a mixture of immunoglobulins with multiple specificities, all directed against the epitope of interest. Most polyclonal antibodies are generated by the injection of a peptide or purified full-length recombinant protein into a rabbit, although polyclonal antibodies can be derived from many species. This range of species contributes greatly to the multiplexing flexibility of antibodies. The injected peptide, or immunogen, is carefully chosen to mimic a specific and preferably unique region of interest in a target molecule. The injected animal's immune system will mount an antibody response against the foreign peptide or recombinant protein, generating a number of soluble antibodies able to bind to the immunogen. These polyclonal antibodies can then be isolated from the animal to yield a polyclonal antiserum, or be further purified via affinity chromatography (Cooper and Paterson 2008).

Monoclonal antibodies, on the other hand, are generally preferred because of the greater confidence that a monoclonal antibody is specific and has a high affinity for the target. The first monoclonal antibodies, developed by Kohler and Milstein in the mid-1970s, relied on an immune response elicited in the animal against an injected immunogen, much like polyclonal antibodies (Kohler and Milstein 1975). In the case of monoclonal antibodies, however, multiple antibody-producing daughter B cells are isolated from the spleen of the injected animal after several days. The antibody-producing cells are fused with myeloma cells to create a number of hybridomas. These hybridomas are capable of proliferating in culture indefinitely, and should produce relatively large amounts of the antibodies expressed by the original daughter B cells. Single hybridomas are separated and expanded in culture to create monoclonal populations, and the antibodies are screened for specificity and affinity (Mechetner 2007; Zhang et al. in press).

Several technologies have since been developed for more rapid, cost-effective, and technologically simpler generation of monoclonal antibodies. Recombinant techniques have made possible chimeric or humanized antibodies, which combine the DNA expressing the binding site of a mouse monoclonal antibody with human antibody DNA (Donzeau and Knappik 2007). Bacterial expression of antibodies has also come to the fore in recent years. These developments make possible the selection of desirable antibody specificities by phage display, with the displayed antibody fragments generated from the plasma cells of human donors or from the spleen of an immunized animal. More and more frequently, however, these phage libraries are generated by genetic engineering (discussed in greater detail in (Donzeau and Knappik 2007)). The highly specific high-affinity monoclonal antibodies required for therapies, diagnostics, and basic research are created using these methods.

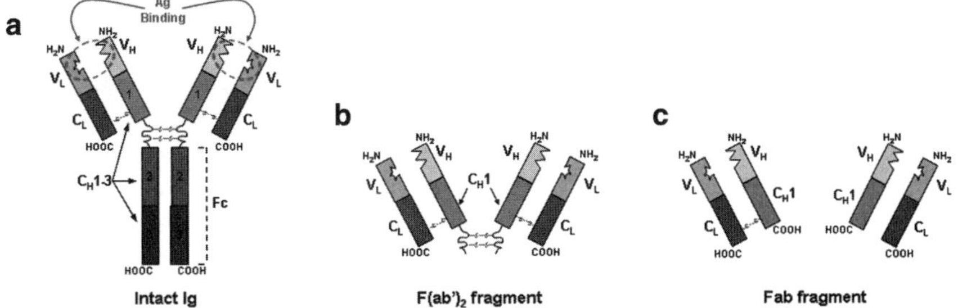

Fig. 5.1 Intact immunoglobulin and common antibody fragments. (**a**) Schematic representation of an intact immunoglobulin molecule. Each heavy chain consists of three constant domains (C_H1–3), and the variable domain (V_H). C_H1 and C_H2 are linked by the flexible hinge region, which forms two disulfide bonds with the hinge region of the complementary heavy chain. Each light chain consists of one constant domain (C_L) and one variable domain (V_L), and is associated with the heavy chain through a disulfide bond proximal to the carboxy termini of the two chains (COOH). The antigen-binding regions of the molecule (Ag Binding) are found at the amino termini of the V_H/V_L pairs (NH_2) and are circled. The Fc portion of the molecule, consisting of C_H2–3, is indicated. Domain labels are constant throughout the figure. (**b**) The $F(ab')_2$ antibody fragment. Enzymatic digestion of intact immunoglobulin with pepsin results in the cleavage of the molecule at the hinge region, maintaining the disulfide bonds and yielding the $F(ab')_2$ fragment. (**c**) Papain cleaves the hinge region of intact immunoglobulin just above the disulfide bonds, generating two Fab fragments. Fab fragments can also be created through genetic manipulation. The heavy and light chains can associate noncovalently (*right*), or may maintain a disulfide bond near the carboxy termini (*left*)

5.2.2.2 Antibody Engineering

This topic is discussed in greater detail elsewhere in this text. Here we will discuss the engineered antibodies most often used in biosensing. Briefly, antibodies can be used in a number of ways, ranging from the intact molecule to a number of different fragments thereof (Fig. 5.1). Fab fragments, which are composed of the entire light chain and the variable and first constant region of the heavy chain, can be used in a number of assay platforms. These fragments can form stable H/L heterodimers without being covalently linked, or can be joined by a C-terminal disulfide bond (Fig. 5.1b) (Donzeau and Knappik 2007). A similar fragment, the $F(ab')_2$ fragment, also retains this disulfide bond which covalently links the two chains of the Fab fragment (Fig. 5.1c). Smaller fragments and multivalent engineered antibodies will likely enjoy increasing use in biosensing assays in coming years.

5.2.2.3 Antibody Conjugation

In order to visualize and quantify the binding of an antibody to its target, the antibody can be chemically linked to a number of different reporter molecules (Haugland 2001). Some reporters, such as those used in flow cytometry, are laser-activated fluorescent molecules called fluorophores or fluorochromes (see

Sect. 5.3.5). Other reporters are enzymatic and rely on chemical reactions. In this case, antibodies are conjugated to an enzyme such as horseradish peroxidase (HRP) or alkaline phosphatase (ALP). When incubated with chromogenic substrates, as is often the case in a classic enzyme-linked immunosorbent assay (ELISA), the products generated by the enzyme are intensely colored, and are measured with a spectrophotometer. Alternatively, these enzyme-conjugated antibodies can be incubated with a chemiluminescent substrate. The product of these reactions gives off light, which is captured and measured by multiple types of documentation systems.

For greater flexibility in multiplexing, antibodies are also often conjugated to biotin (Hirsch and Haugland 2005). The interaction of biotin with streptavidin is highly specific, and one of the most stable noncovalent bonds known. Fluorophores and enzymes like HRP and ALP can be linked to streptavidin, allowing the researcher to use a particular biotinylated antibody in multiple assay platforms, or to vary the detection within one assay platform, for instance by pairing the biotinylated antibody with different streptavidin-conjugated fluorophores as in flow cytometry (Haugland and Bhalgat 2008).

5.2.3 Anti-Antibodies System

For a number of assay formats, it is sometimes necessary to use a pair of antibodies for detection (see Fig. 5.6 for a schematic illustration). The first antibody which is specific for the target is known as the primary antibody. In cases where the primary antibody itself is not conjugated to a reporter molecule, a reporter-conjugated secondary antibody can be used. Secondary antibodies are generally anti-species antibodies, or antibodies directed against the immunoglobulin of another species. For example, many secondary antibodies used for detecting mouse monoclonal primary antibodies are made in goat. Mouse immunoglobulin is used as the immunogen, and the result is a polyclonal goat anti-mouse immunoglobulin antibody preparation that can be conjugated to a reporter molecule and used to detect the primary antibody wherever it is bound to the target. In general, however, simpler assays in which the primary antibody is directly conjugated to a reporter are preferred, both to minimize complexity and possible cross reactions, and to minimize cost.

5.3 Antibodies as Biosensors: Various Technologies

The most fundamental requirements for a good biosensing assay are that it be specific and sensitive. To that end, each biosensing assay needs a good recognition receptor. These receptors must therefore be highly specific, stable, and versatile if they are to be successful in the rapidly evolving techniques now available. In the antibody, evolution has designed a nearly perfect recognition receptor, and scientists have capitalized on its strengths and found ways to improve its weaknesses until the array of antibody-based biosensing applications is nearly overwhelming.

5.3.1 Techniques Utilizing Immunoprecipitation

Antibodies have been used for a number of years as a tool to capture and concentrate targets through the technique of immunoprecipitation (IP) (Bonifacino et al. 2001). IP involves the incubation of the lysate or fluid with antibody specific for the target (Fig. 5.2). Following capture of the target protein by the specific antibody, the antibody is bound by bacterial Protein A, Protein G, or a combination of both, which has been coated onto beads. These beads can then be spun out of solution and the supernatant removed or transferred to another tube. The protein of interest, which has been greatly concentrated through binding to the beads and should be the only protein remaining, can then be visualized via immunoblotting. In some cases, the protein of interest is present at such low concentrations that an extremely large number of cells or a large volume of bodily fluid is required for detection, presenting difficulties for the detection assay, which can be circumvented through concentration of the target with IP. In other instances, the process of IP is used more to minimize background, either by isolating the target from all other proteins in the biological sample for analysis, or by specifically removing an unwanted protein known to interfere with detection of the intended target.

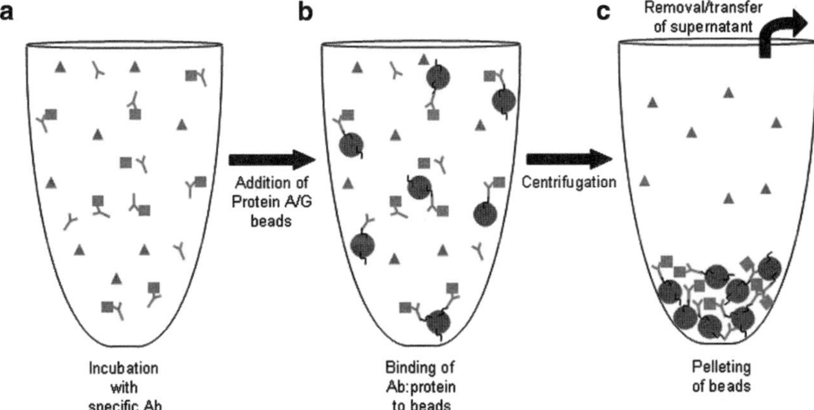

Fig. 5.2 Immunoprecipitation. (**a**) Cell lysate or other biological sample is incubated with specific antibody (Ab), which binds to the target (squares) in solution. (**b**) Microbeads coated with bacterial Protein A, Protein G, or a combination of both are added to the solution. The antibodies, whether bound to target protein or free, will be bound by the bacterial proteins coating the bead. (**c**) Following centrifugation, the beads and their cargo of antibody and target protein will form a pellet at the bottom of the tube. The supernatant, now depleted of the target protein, can be transferred to another tube or discarded. These schematic representations of antibodies and their targets will be used for all subsequent figures

5.3.1.1 Immunoblotting

Immunoblotting (IB) or Western blotting does not always require the use of IP, but is of interest here as a biosensing assay in its own right, as it also makes use of antibodies as recognition receptors (Gallagher et al. 2004). The proteins in a sample, which have been denatured by sodium dodecyl sulfate (SDS) and separated by mass via polyacrylamide gel electrophoresis (SDS-PAGE), are transferred to a membrane for detection (Fig. 5.3). The membrane is incubated with specific primary antibody, followed by enzyme-conjugated secondary anti-species antibody. A chemiluminescent substrate is then added, generating a band of light at the site of the target protein which was traditionally detected by exposure to autoradiography film. Camera-based documentation systems are also becoming increasingly popular for the detection and quantitation of IB bands. Despite being an older technique, IB provides an opportunity to physically view the interactions

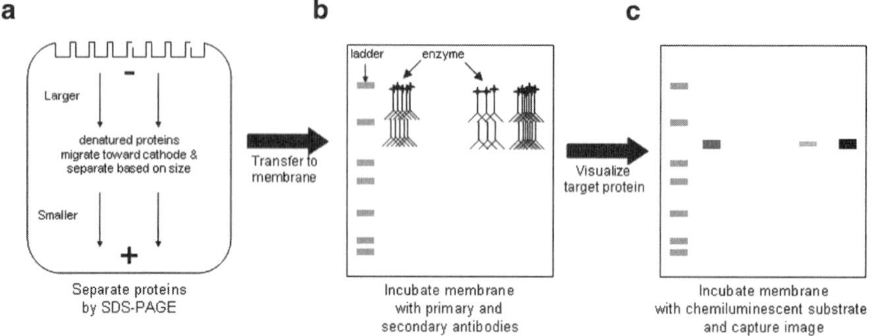

Fig. 5.3 Immunoblotting. (**a**) Samples are denatured in lysis buffer, loaded onto a polyacrylamide gel, and separated by electrophoresis (PAGE). The presence of sodium dodecyl sulfate (SDS) in the buffer masks the native charges of the proteins and lends an overall negative charge, allowing the proteins to migrate toward the cathode according to size, with smaller proteins traveling farther through the matrix than large proteins (SDS-PAGE). Proteins can also be analyzed by their native conformations under non-denaturing conditions in the absence of SDS (not shown). (**b**) Separated proteins are transferred to a nitrocellulose or polyvinylidene fluoride (PVDF) membrane via the application of electrical current. The membrane is then probed with primary antibody specific for the target protein or residue, followed by an enzyme-conjugated secondary anti-species antibody (more detail on secondary antibodies and reporters is given in Fig. 5.6). Note that a greater amount of target will provide more binding sites for primary antibody and thus a greater number of reporter-conjugated secondary antibodies, as shown. A molecular weight standard containing multiple proteins of known molecular weights is usually included in each experiment (ladder), to provide an estimation of the distribution of the sample proteins. The proteins in these ladders are often dyed, sometimes with multiple colors, to allow visualization on the membrane. (**c**) The target is visualized by incubating the membrane with the chemiluminescent substrate of the reporter enzyme, which emits light. The signal is captured by exposure to autoradiography film, or by a camera-based gel-documentation system. The quantity of target can then be extrapolated from signal intensity and/or band size, although this measure is not truly quantitative, but relative to the other samples in that experiment only

of an antibody with the proteins in a sample, and thus remains a gold standard for the analysis of an antibody's specificity and off-target interactions.

5.3.1.2 Chromatin Immunoprecipitation Assays

Although the antibody does not directly provide the readout in the chromatin immunoprecipitation (ChIP) assay, a specific antibody is nonetheless crucial to the detection of the correct protein–DNA interaction (Weinmann and Farnham 2002; Wells and Farnham 2002). Many types of proteins, including but not limited to transcription factors, have the ability to bind DNA. By chemically crosslinking the proteins bound to DNA, lysing the cells, sonicating the sample to create small chromatin fragments, and then immunoprecipitating the protein of interest with a specific antibody, the DNA sequence targeted by that protein can be isolated from the rest of the genome (Fig. 5.4). After reversal of the crosslink and proteinase K degradation to remove the protein, the DNA fragments are amplified via polymerase chain reaction (PCR), using primers designed for predicted binding sequences. The target can also be analyzed by DNA dot blot, and in some instances where a putative binding site is not known, the DNA released from the protein can be subcloned and sequenced to determine the site(s). The ChIP assay makes possible a greater understanding of transcriptional regulation, and can also be multiplexed in the ChIP-on-chip assays described in Sect. 5.4.1.

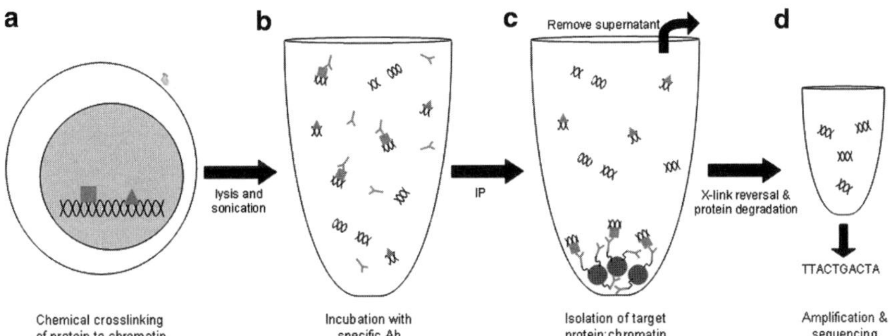

Fig. 5.4 Chromatin immunoprecipitation (ChIP). (**a**) Proteins associated with chromatin in living cells are chemically crosslinked (X-link) to the DNA to preserve the association. The cells are then lysed and the sample is sonicated to fragment the chromatin. (**b**) Antibodies specific for the target protein bind the protein–chromatin complexes, allowing immunoprecipitation (IP) of the complexes with Protein A/G coated microbeads as described in Fig. 5.2. (**c**) The target protein–chromatin complexes are isolated by centrifugation and pelleting of the IP beads, and the supernatant is transferred to another tube or discarded. The chemical crosslink is reversed and/or the target protein is eliminated by proteinase K degradation, leaving only the DNA sequence(s) bound by the protein. (**d**) The DNA is amplified by PCR and sequenced, revealing the consensus DNA binding sequence(s) of the protein of interest

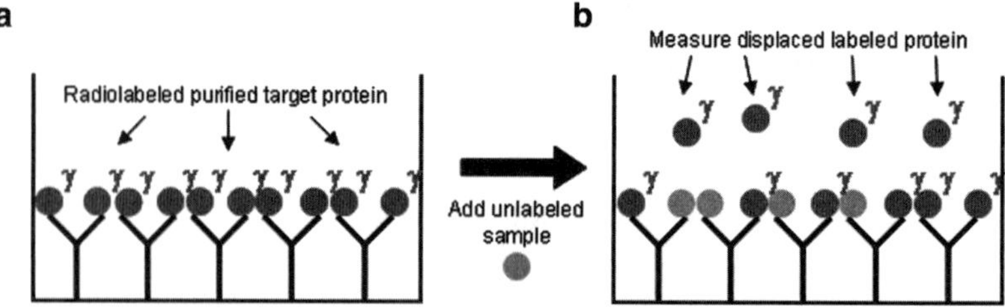

Fig. 5.5 Radioimmunoassay. (**a**) Purified target protein is radiolabeled, often with the gamma isotope of iodine (γ), and incubated with immobilized specific antibody. Sample containing unlabeled target protein is then added to the well. (**b**) The unlabeled target protein competes with the purified radiolabeled protein for binding to the antibodies, displacing some of the radiolabeled protein when present at high enough concentrations. The unbound protein is removed from the well, and the radioactivity of the displaced radiolabeled protein is measured to give an indirect measure of the amount of unlabeled target protein present in the sample

5.3.2 *Radioimmunoassays*

One of the first highly sensitive methods for measuring the levels of proteins such as hormones in the blood was radioimmunoassay (RIA) (e.g. (Zamrazilova et al. 2007). RIA is a classic competition assay in which a known quantity of purified target protein is radiolabeled, generally with a gamma radioisotope of iodine. This "hot" protein is mixed with an immobilized specific antibody, and then with the biological sample containing the unlabeled "cold" target protein (Fig. 5.5). The cold protein will compete with the radiolabeled protein for binding to the antibody, and the amount of radioactivity that has been displaced can be measured, giving an indirect measure of the amount of the target protein present in the sample. Although technology has moved away from assays requiring radioactivity, the development of the RIA allowed some of the first sensitive and specific measures of important hormones such as insulin in human blood (Yalow and Berson 1960). The technique has largely been replaced by enzymatic immunoassays.

5.3.3 *Enzymatic Immunoassays*

Enzyme immunoassays (EIAs) are the classic antibody-based biosensing applications and a mainstay in immunology research and in diagnostics of human disease (Porstmann and Kiessig 1992). The most common of these, the ELISA (Hornbeck 2001), has been used to detect relevant biological information in cell lysates and in nearly all bodily fluids, including whole blood, bone marrow aspirates, serum, plasma, cerebrospinal fluid, urine, and more. The assay is generally performed in microtiter plates of 96 or more wells, and the often identical treatment of each well and the availability of robotic systems and automated plate washers make the ELISA an excellent platform for high-throughput screening assays. Further, the assay design is amenable to several reporter types, giving researchers great flexibility with respect to the desired type of signal readout.

ELISA designs can vary greatly (Fig. 5.6) (Hornbeck 2001). The simplest ELISA assays require only a single reporter-conjugated primary antibody, which

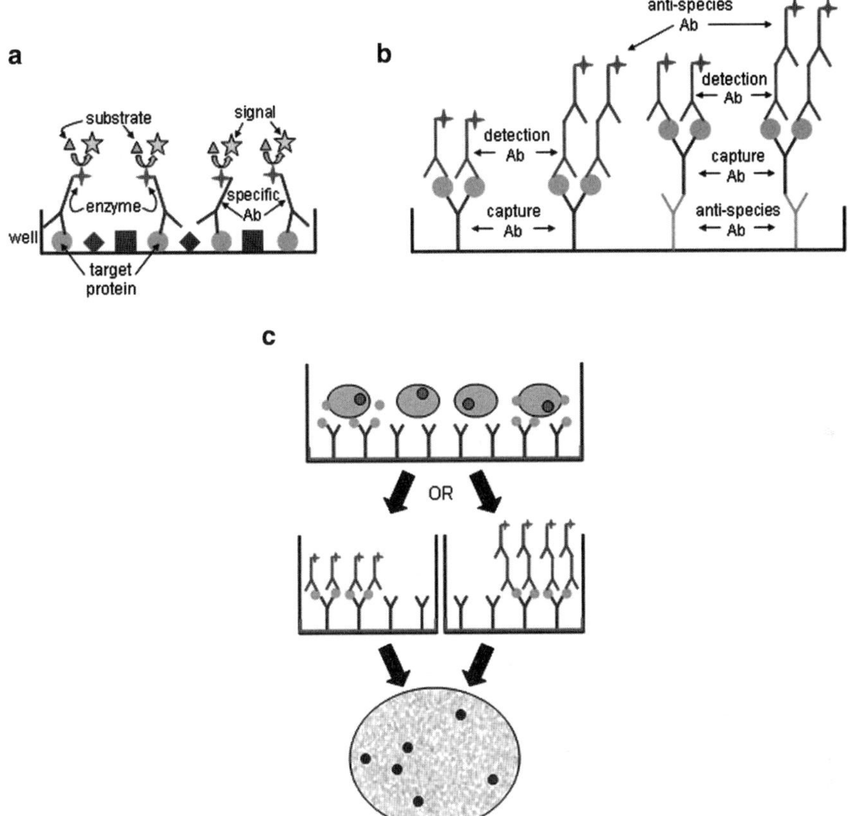

Fig. 5.6 ELISA and ELISpot. (**a**) The simplest ELISA consists of proteins adsorbed to the surface of a well, and incubated with specific enzyme-conjugated antibodies. After binding of the antibodies to the target protein, the well is washed, and the colorimetric or chemiluminescent substrate is added. The reporter enzyme acts on the substrate, generating signal in the form of color or light, respectively. (**b**) The sandwich ELISA and its possible variations. The specific capture antibody can be directly coated onto the surface of the well, or be bound itself by an anti-species antibody. After capture of the target protein, the target is bound by the detection antibody, which can be conjugated to a reporter itself, or bound by a reporter-conjugated secondary anti-species antibody. Each of these permutations is represented. (**c**) The ELISpot assay consists of specific antibodies adsorbed to the surface of a membrane-backed well. Cells are stimulated to secrete the target protein, which is captured by the antibodies in the immediate vicinity of the activated cell (*top*). After removal of the cells, the target protein is bound by the detection antibody, which can be conjugated itself (*center left*) or detected by an enzyme-conjugated secondary antibody (*center right*). The end result of the ELISpot is a blue/black precipitate on the membrane, each spot corresponding to a single target-producing cell (*bottom*)

detects target protein adsorbed to the surface of the well (Fig. 5.6a). The more common "sandwich" ELISAs can utilize from two to four antibodies (Fig. 5.6b). Researchers often prefer this format for the greater level of specificity afforded by using two target-specific antibodies to first capture and then detect the target protein. The capture antibody is bound to the plate, either by direct adsorption or through interaction with a corresponding anti-species antibody that is bound to the plate instead. Target protein is bound by the capture antibody, irrelevant proteins are washed away, and the target is then bound by the detection antibody. This detection antibody can be conjugated to a reporter, or can be detected itself by a secondary reporter-conjugated anti-species antibody. The major caveat of sandwich ELISAs is the requirement that, if using a secondary anti-species antibody for detection, the capture and detection antibodies must be derived from different species to prevent binding of the secondary antibody to both.

The versatility of the sandwich ELISA makes possible the detection of specialized protein motifs, including different isoforms created by alternative splicing (e.g. (Bulanova et al. 2007), and post-translational modifications such as phosphorylation, acetylation, glycosylation, methylation, ubiquitination, S-nitrosation, and even cleavage of the protein (Takada et al. 1995; Kocinsky et al. 2005; Bulanova et al. 2007; Spoettl et al. 2007; Liang et al. 2008; Sumbayev and Yasinska 2008). These modifications can reveal a great deal about the signaling occurring in cells, such as the activation status of specific pathways or the turnover rate of important proteins. The target protein is bound by the capture antibody, unbound nonspecific protein is washed away, and the post-translational change in the target, if present, can then be revealed by a detection antibody specific for the modification of interest. The strategy can also be reversed, for example, to determine whether a protein of interest is among those captured by a primary antibody directed against total phosphotyrosine. Also possible is the detection of fusion proteins such as BCR-Abl, which is one of the causative agents in chronic myeloid leukemia (CML). For example, an anti-BCR capture antibody should immobilize both the wildtype (WT) and fused version of BCR, while the anti-Abl detection antibody would reveal only those captured proteins that also express Abl. This versatility of sandwich ELISAs is augmented by the increasing availability of multiplexing technologies, as well as the use of electrochemiluminescent (ECL) substrates and advanced detection systems (see Sect. 5.4.4).

A variation on the ELISA platform which allows the detection and enumeration of single cells secreting specific proteins is known as the enzyme-linked immunosorbent spot assay (ELISpot) (Fig. 5.2c) (Czerkinsky et al. 1983). Membrane-backed wells are coated with monoclonal or polyclonal antibodies specific for the secreted protein of interest, including such targets as cytokines and antigen-specific immunoglobulin (Klinman and Nutman 2001; Lycke and Coico 2001). Cells are then added to the wells and stimulated to produce the target proteins. As the proteins of interest are secreted by a cell, they are captured in the immediate vicinity of that cell by the immobilized antibodies, effectively marking the location of the active cell. When the incubation period has elapsed, the wells are washed to remove the cells and any debris, and the wells are incubated with a second specific antibody, making the ELISpot assay a sandwich assay. The detection antibody is generally directly conjugated to biotin, or bound by a biotin-conjugated secondary antibody. The wells are then incubated with streptavidin-conjugated HRP followed by 3-amino-9-ethylcarbazole (AEC), or streptavidin-ALP followed by 5-bromo-4-chloro-3-indolyl phosphate and nitro blue tetrazolium chloride (BCIP/NBT). These substrates yield colored precipitates on the membrane, providing the blue/black spot for quantitation. ELISpot protocols have also begun to make use of fluorescent reporters as well.

The ELISpot assay provides valuable information in that it allows quantification of the number of cells responding to the stimulus. The assay is very valuable for the detection of rare cell populations, down to about 1 in 100,000 cells, which is made possible by the immediate capture of the secreted protein in the vicinity of the activated cell before it is diluted, degraded, or bound by other cells. The assay also requires fewer cells per well than other assays designed to gather similar information, which is a boon for researchers or diagnosticians assaying precious sample types. A growing number of secreted proteins are being assayed in this manner, and advances in detection systems allow the counting of multiple reporter types through the use of light filters, while software developments have made possible the automated counting of ELISpot plates, giving the platform greater throughput as well.

5.3.4 Immunocytochemical and Immunohistochemical Assays

Pathologists and researchers use the related techniques of immunohistochemistry (IHC) and immunocytochemistry (ICC) to identify specific cells or to examine the subcellular localization of important proteins such as tumor markers, as well as markers of proliferation and apoptosis (Fig. 5.7) (Hofman 2002; Polak and Van Noorden 2002). Suspension cells or cells taken from a smear (ICC), or intact tissue sections (IHC), are incubated with specific primary antibodies which, as in ELISAs, can be directly conjugated to a reporter or detected with a reporter-conjugated secondary anti-species antibody. These techniques make use of both enzymatic and fluorescent reporters. Researchers often also include other standard dyes or antibodies that label specific cellular structures, such as the nucleus. The samples are examined using advanced microscopy and computer-based image analysis systems.

The value of IHC/ICC lies in the ability to determine both whether a protein is expressed in the cells being examined and the subcellular location of the protein. Some proteins, such as transcription factors, are regulated wholly or in part by localization. For example, many transcription factors are sequestered in the cytoplasm and only translocate to the nucleus upon activation. In some malignancies, genetic lesions can include mutations which lead to improper localization of these and other important proteins. Antibodies directed against these proteins, as well as antibodies specific for post-translational modifications, different isoforms, and

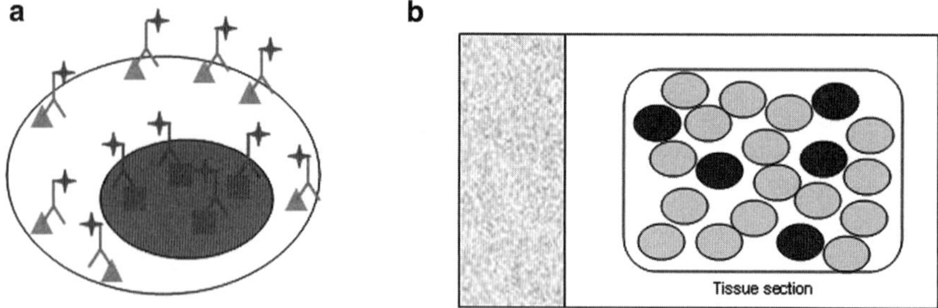

Fig. 5.7 Immunocytochemistry and immunohistochemistry. (**a**) Simplified schematic of ICC, depicting a single cell probed for two specific proteins. One protein is found to be localized to the cytoplasm, proximal to the plasma membrane, while the other protein is localized to the nucleus. This nuclear localization is confirmed by a co-stain which identifies the nucleus. (**b**) Simplified schematic of IHC, depicting a slide-mounted tissue section. Only a few cells in the tissue section express the protein for which the sample has been stained. IHC and ICC can make use of both colored stains and fluorescent markers, and often require microscopes with multiple excitation and/or emission filters (not shown)

mutant proteins, are extremely valuable as diagnostic and prognostic tools for pathologists when used as recognition receptors in IHC/ICC.

5.3.5 Flow Cytometry

The considerable versatility of antibodies is perhaps best demonstrated in flow cytometry (Sharrow 2002). Antibodies can be conjugated to a wide array of reporter fluorophores, which absorb the energy of laser light and then emit the energy at a different specific wavelength. Multiple lasers and an intricate series of filters and computer-controlled time delays allow the most advanced cytometers to record up to 11 parameters simultaneously for each cell that passes through the instrument (Fig. 5.8). The further adaptability provided by biotin-conjugated antibodies and streptavidin-conjugated fluorophores makes possible a staggering number of analyte combinations that can be examined for a given cell.

Initially, flow cytometry was a cell-based platform designed to measure the expression levels of proteins expressed on the cell surface. The ability to examine multiple surface markers simultaneously on intact cells allows researchers to discover and characterize the numerous and intricate subsets of cells in the body. Moreover, this platform allows diagnosticians to find and enumerate populations of abnormal cells, most notably for circulating hematological malignancies of the peripheral blood and bone marrow. In recent years, however, flow cytometry has expanded to include the analysis of intracellular and soluble proteins, as well as cellular DNA content (Figs. 5.9 and 5.10).

Although ELISAs and immunoblotting are powerful techniques for the elucidation of events occurring inside the cell, both actually provide what could be

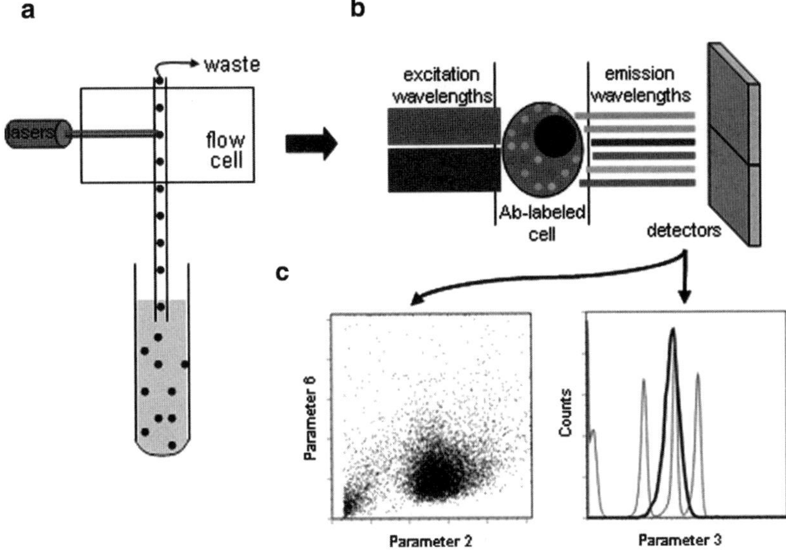

Fig. 5.8 Basic principles of flow cytometry. (**a**) Cells that have been incubated with fluorophore-conjugated antibodies are drawn from the sample tube into the machine, where they pass the beams(s) of laser light in single file and continue into a waste receptacle. (**b**) As the cells pass the interrogation point, the bound fluorophores are excited by the laser light. The excited fluorophores then emit light at slightly different wavelengths, which is captured by detectors after passing through a complex system of optics (not shown). (**c**) Software manipulation of the recorded light signals results in data that can be analyzed in many ways and combinations. Each target assayed, or parameter, can be analyzed in tandem with any other in a dot plot (*left*; see Fig. 5.9 for more details), or analyzed singly in the form of histograms and then compared to the histograms of other samples (*right*)

considered an average measure. Unless highly purified, cell preparations are a heterogeneous mix of different cell types, which may express differential levels of proteins of interest, or respond differently to stimuli. When cells are lysed, more abundant cell types mask the response of less numerous subsets. It is also possible that even cells within highly purified populations may express slightly different levels of a given protein, or respond heterogeneously to a specific stimulus. For these reasons, the ability to combine lineage-defining surface staining with intracellular staining of fixed and permeabilized cells is an invaluable tool, allowing the analysis of intracellular events side by side in mixed populations of cells (Fig. 5.9) (Darzynkiewicz and Huang 2004; June and Moore 2004; Donahue and Fruman 2007; Donahue et al. 2007; Foster et al. 2007; Schulz et al. 2007). This intracellular staining has made possible the analysis of expression and post-translational modification of proteins, and even the elucidation of signal transduction networks in single cells (Irish et al. 2006).

To make possible the analysis of mixed samples, for example two different patients or a single patient pre and post treatment, Krutzik and Nolan also developed a cell-based "barcoding" method. Each sample is first incubated with a

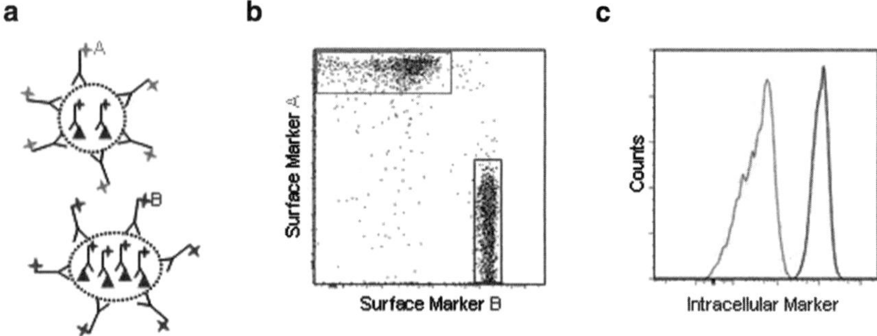

Fig. 5.9 Surface and intracellular staining of permeabilized cells. (**a**) Mixed cell populations are labeled with antibodies specific for surface markers that identify subsets such as different lineages, different activation states, and others. Two different cell subsets are indicated here by shape, and by binding to two different surface marker antibodies, represented here by reporters A and B, which will be seen by the cytometer as different parameters. The cells are then permeabilized to allow passage of antibodies across the membrane, represented by the *dashed line* surrounding the cell. Permeabilized cells are incubated with antibodies specific for the intracellular target reporter C, which will be seen by the cytometer as a third parameter that is the same for all cells). (**b**) After sample acquisition by the flow cytometer, the different cell subsets are differentiated by their expression of the surface markers for which they were stained. Comparison of two parameters is generally done with a dot plot, in which each dot represents a single cell. Surface marker A (y-axis) is present at high levels on the round cell, while surface marker B (x-axis) is absent, indicating that these cells will fall in the *top left corner* of the dot plot. Conversely, the oval cell shows high levels of marker B and low levels of marker A, placing them in the *lower right corner* of the dot plot. These expression patterns create two distinct populations in the dot plot. Gates can then be drawn around the populations (*rectangles*), telling the software to consider only those cells falling within the gate in downstream analyses. (**c**) The cells within each gate are analyzed for levels of the intracellular protein. Levels are suggested by the intensity of the staining for the third parameter (*intracellular marker C*). The diagram in (**a**) depicts the oval cell as having a higher level of the target intracellular protein, and this is reflected by the surface marker B-positive (Marker B+) histogram falling farther to the right on the scale than the Marker A+ histogram, indicating a higher intensity of staining in the marker B-positive cells than in the marker A-positive cells

different fluorescent signature that emits light at a discrete wavelength (Krutzik and Nolan 2006). The samples are then mixed together for standard surface and intracellular antibody staining, and the samples can be differentiated during acquisition and analysis by the barcode signature. The ability to enumerate cell types and observe intracellular events on a cell-by-cell basis is of enormous importance, and is already providing excellent insights to diagnosticians and researchers alike.

5.3.6 *Bead-Based Assays*

The ability to detect soluble proteins in cell culture supernatant and body fluids was also historically the province of ELISAs and immunoblots. However, recent advances in flow cytometry include microbead technologies that are similar to the

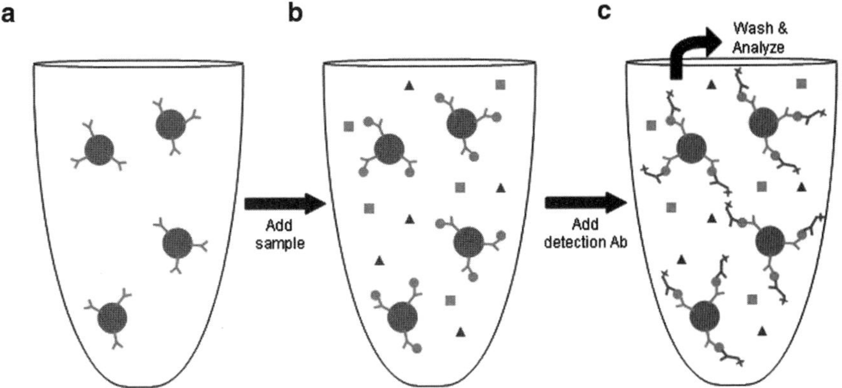

Fig. 5.10 Bead-based flow cytometry assays. (**a**) Capture antibodies are coated on microspheres. (**b**) The beads are incubated with proteins in solution (e.g. lysate, cell culture supernatant, or plasma), and bind only the target protein. (**c**) The target protein is bound by fluorophore-conjugated detection antibody, the sample is washed to remove unbound detection antibody, and analyzed by flow cytometry

sandwich ELISA in design, and allow detection of soluble protein, including intracellular proteins released into the blood by dying leukemia cells (Kellar and Douglass 2003; Jilani et al. 2008a, b). Capture antibody is coated onto beads rather than a microtiter plate, the beads are incubated with the body fluid or supernatant, and the captured protein is then bound by the specific detection antibody. As with a sandwich ELISA, bead assays can make use of two to four antibodies, although fewer antibodies are generally preferred (Fig. 5.10). In addition to the simple detection of whole protein in solution, bead-based assays can also be used to detect post-translational modifications and chromosomal translocations. For example, Fig. 5.11 depicts the detection of the fusion protein BCR-Abl via bead-based assay, which is similar in principle to the sandwich ELISA example described in Sect. 5.3.3 and has already been used to detect the mutation status of CML patients (Jilani et al. 2008a). Despite these similarities to the sandwich ELISA, the multiplexing opportunities for bead-based assays are slightly different, however, and include the cytometric bead array and Luminex technologies.

Although the most advanced cytometers can measure up to 11 parameters, it is often difficult to calibrate the instrument with large numbers of parameters in use, as the emission spectra for the different reporters overlap to varying degrees. As an alternative to multiple fluorophores for the measurement of soluble proteins, one multiplexing option for flow cytometry is to assay several analytes using the same fluorophore. In this assay, known as a cytometric bead array (CBA), each capture antibody is conjugated to beads of a different size (Fig. 5.12a) (Morgan et al. 2004). One parameter measured by flow cytometers is the size of the particle flowing past the detector, meaning that multiple bead sizes will be recognized as discrete populations, allowing researchers to easily distinguish between the analytes. Like

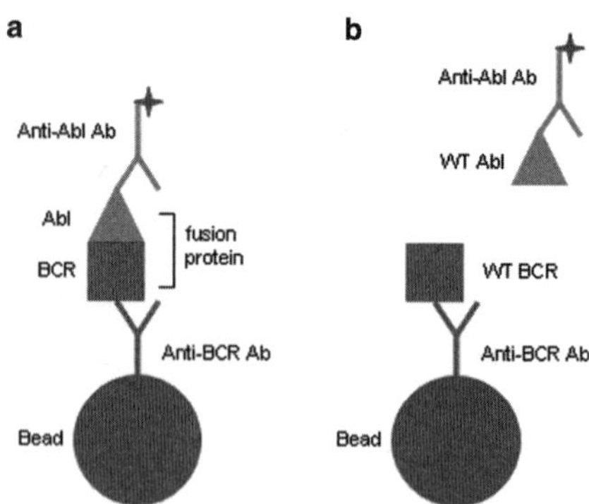

Fig. 5.11 Detecting chromosomal translocations via bead-based flow cytometry. Microbeads are coated with an anti-BCR capture antibody. Both WT BCR and BCR-Abl fusion protein present in the sample will be captured by the beads. A fluorophore-conjugated anti-Abl detection antibody capable of binding both WT Abl and BCR-Abl is then added, and the sample is analyzed by flow cytometry. Note that, in principle, this assay can be set up in the reverse orientation, with anti-Abl antibody on the bead and anti-BCR as the detection antibody. (**a**) The only scenario in which signal will be detected by the cytometer is one in which BCR is bound by the bead, and the fused Abl is bound by the detection antibody. (**b**) Signal will not be detected by the instrument following the binding of either WT protein. The Abl:anti-Abl antibody complex will remain in the supernatant when the beads are pelleted and will be washed away. WT BCR bound to beads will be present in the sample analyzed by the cytometer, but will not be detected by the instrument due to the lack of fluorophore-conjugated detection antibody binding

multispot ELISAs (Sect. 5.4.4), this approach allows the measurement of several analytes side by side in the same sample.

In an improvement over the ELISA, however, recent advances in flow cytometry now also allow quantitation of the number of antibody molecules bound to a bead. A tube containing four groups of beads with increasing levels of bound reporter fluorophore is acquired with each experiment, creating a standard curve relating the known number of fluorophore molecules bound to the standard beads to the mean fluorescence intensity (MFI) of the fluorophore recorded by the instrument. The number of reporter-conjugated secondary antibodies bound to the bead in an experimental sample can then be extrapolated from the MFI of the sample using this curve. This quantitation of antibodies bound can also be applied to the antibodies bound to proteins on the surface of intact cells as well as intracellular proteins in fixed and permeabilized cells (Pannu et al. 2001).

Using a unique combination of the principles of flow cytometry and microbead assays, Luminex technology has created a methodology purportedly capable of analyzing up to 100 analytes simultaneously in a single well (see Luminex Corporation for examples of this technology). Polystyrene microspheres are color-coded using titrations of red and infrared dyes, creating a different color signature for each analyte (Fig. 5.12b). These beads are then coated with capture antibodies and combined in a single well, where they are incubated with cell lysate or biological fluids, followed by a detection antibody in another sandwich-style assay. The beads

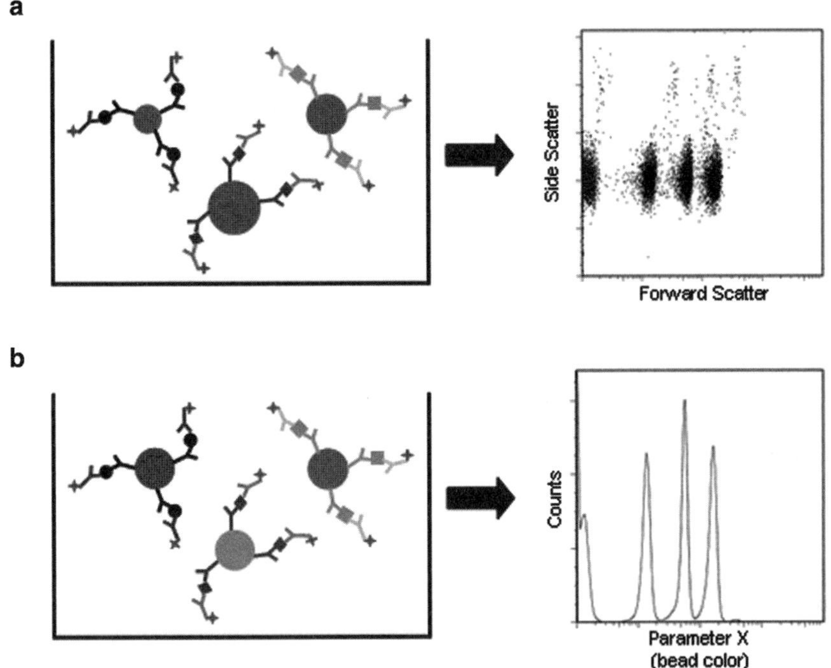

Fig. 5.12 Cytometric bead array and Luminex technologies. (**a**) The CBA platform consists of the antibodies specific for each target being conjugated to beads of a different size. The beads are incubated with the sample at the same time, allowing capture of the target proteins. The beads are then incubated with detection antibodies for each target, all conjugated to the same fluorophore (*left*). When analyzed, the different bead sizes are recognized by the cytometer via the forward and side scatter parameters, and are identifiable as discrete populations that can be analyzed separately via gating (*right*). (**b**) Luminex technology makes use of beads of the same size which have been impregnated with dyes of slightly different wavelengths. Each set of beads is coated with a different capture antibody, incubated with sample to capture target protein, and detected with a fluorophore-conjugated detection antibody (*left*). The cytometer-based analysis instrument detects the slight variations in the color of the bead (Parameter X), creating discrete populations based on bead color which can be gated (*right*). The reporter fluorophore intensities within each population can then be analyzed, yielding information about the concentration of each target analyte

are then analyzed in a manner based on the principles of flow cytometry, in which the internal dyes are excited by a red laser to reveal the color code of the bead, thus identifying what target should be captured by the antibody coated onto that bead (Gu et al. 2008). Any reporter that is bound through recognition of the target protein by the detection antibody is excited by a green laser and recorded as well, allowing measurement of the protein captured. This technology holds the potential to provide nearly as much information about a sample as some forms of antibody microarrays or multispot ELISAs (discussed in Sect. 5.4).

5.3.7 *Additional Techniques*

Other biosensing assays that make use of antibodies as recognition receptors include some classic diagnostic and typing tests. The hemagglutination assay, for example, has been used for decades to determine blood type within the ABO system. Antibodies directed against the A antigen are added to an aliquot of red blood cells (RBCs), and the sample is inspected for the process of agglutination, or clumping of the cells. Agglutination of the sample indicates that the blood type is either A or AB, while a lack of clumping means the RBCs are type B or type O. A second aliquot is incubated with antibodies against the B antigen, and agglutination in this case suggests B or AB blood, depending on the result of the A-antigen test. A lack of agglutination with antibodies against either the A or B antigens indicates type O blood. Although simple, the hemagglutination blood-typing test is a common example of a widely used antibody-based biosensing assay.

The related techniques of immunoelectrophoresis and immunodiffusion represent another dated but powerful example of biosensing via antibody. The proteins in a mixed sample are first separated within a matrix, usually an agar gel, by electrophoresis. An antibody or antiserum is then added to a trough in the gel, and the proteins and antibodies are allowed to diffuse outward through the gel in the process of immunodiffusion. Precipitates are formed where they meet only if recognition and binding occur, creating precipitin arcs which can be visualized and indicate the presence of the target. Immunodiffusion can also be performed without immunoelectrophoresis, by placing the sample to be tested in the center of a ring of possible interactants (Fig. 5.13). Immunoelectrophoresis and immunodiffusion have been used to determine the number of antigens recognized by patient serum in a mixed sample of antigens, such as those derived from a paraiste, for example (Chargui et al. 2007). They are also used in the study of protein variants in

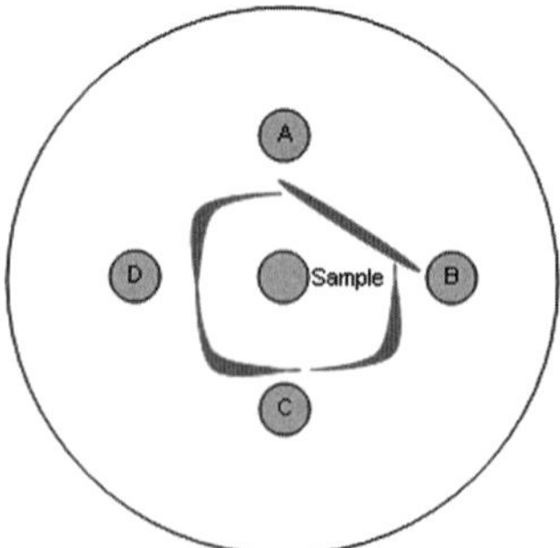

Fig. 5.13 Immunodiffusion. An agar plate is made with multiple wells surrounding the central sample well. The geometry of the wells is crucial to the interpretation of the results. The sample (e.g. serum or other bodily fluid) is placed in the center and allowed to diffuse outward radially. The substances whose interactions are being tested are placed in the outer wells, and also diffuse through the agar. The presence and nature of the interaction of the sample with these substances is demonstrated by the shape of the resulting precipitin arc

disorders such as von Willebrand disease (Michiels et al. 2006). Most often, however, they were used to detect the relative levels of antibody isotypes in the blood of patients, checking for abnormally high or low levels as compared to a normal sample assayed simultaneously. These assays have been largely replaced by immunoblotting in recent years, and are not particularly amenable to multiplexing.

5.4 Multiplexing Methodologies

Researchers are constantly developing new methods for multiplexing the various platforms introduced above. The ability to examine many targets simultaneously provides a more detailed picture of a given system than do analytes measured singly. Although the process of multiplexing often creates new hurdles, such as minimizing background signal and analyzing the vast amount of data generated, finding highly specific recognition receptors will always represent the major challenge. In this section, we discuss some of the multiplexing and high-throughput solutions that have been developed for the antibody-based platforms we have described.

5.4.1 ChIP-on-Chip Assays

By pairing the specificity of protein–DNA interactions provided by ChIP assays with the extensive probe sets offered by DNA microarrays, ChIP-on-chip assays allow researchers to examine the binding of specific proteins to DNA across the whole genome (Negre et al. 2006). After performing the standard protocol of ChIP assays, any DNA sequences bound by the protein of interest are amplified by PCR with random primers and labeled with a fluorescent dye. The labeled DNA is then hybridized to the microarray and analyzed for binding to the DNA sequences spotted on the array. Some ChIP-on-chip array systems can support multiple reporter dyes, allowing the comparison of different conditions or of the DNA sequences bound by different proteins. The data generated by these assays provide insight into the regulatory processes of DNA transcription, replication, and repair, as well as other cellular processes including apoptosis and cell cycle entry. It is also possible to generate maps of gene regulatory networks in order to predict and measure the efficacy of therapeutics on disease states in patients (Wei et al. 2006). Though they measure different aspects of cellular dynamics, the wealth of data provided by ChIP-on-chip assays is reminiscent of that generated by antibody microarrays.

5.4.2 Antibody Microarrays

Like its older cousin the DNA microarray, the antibody microarray allows the detection of a very large number of analytes in a mixed sample (Haab 2005;

Borrebaeck and Wingren 2007). In many ways, most antibody microarray formats are also similar to the ELISA, albeit on a larger scale (Fig. 5.14). Although antibody arrays are valuable for basic research applications, they are thought to be nearly invaluable as diagnostic and prognostic tools for cancer. A wealth of information can be collected from a small volume of sample, and more importantly, patterns within this information can be characterized and relationships between proteins can be analyzed. There are multiple designs in use for antibody microarrays, but the first consideration in choosing a protocol is determining whether it will be antibody or protein bound to the array.

The earliest antibody arrays mimicked DNA arrays, with the monoclonal antibody probes spotted onto the surface of the array, followed by incubation with labeled proteins for detection (Fig. 5.14a) (Borrebaeck and Wingren 2007). In some cases, the proteins are directly labeled with reporters, and in others, with biotin or

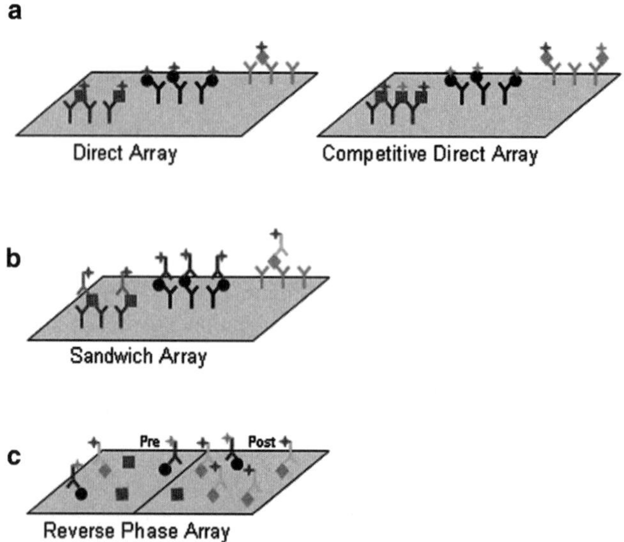

Fig. 5.14 Antibody array formats. (**a**) Direct antibody arrays involve the spotting of specific antibodies onto a surface. The array is then incubated with reporter-labeled proteins (*left*). The identity of a target protein that binds to the array is determined by matching the location of the signal to the known layout of the antibodies. In a competitive direct array, the proteins in two separate samples are labeled with distinct reporters, and incubated with the array simultaneously (*right*). The target proteins will compete for binding to the antibodies on the array, and the relative signal intensities will indicate which sample contained greater quantities of each protein assayed. (**b**) The sandwich antibody array is highly similar to the sandwich ELISA depicted in Fig. 5.6b, simply with a large number of capture antibody specificities combined into a single assay and requiring one small volume of analyte. (**c**) The reverse-phase array consists of the proteins in a sample being adsorbed to the array surface, followed by detection with reporter-conjugated antibodies as in Fig. 5.6a. Although the number of targets that can be analyzed simultaneously is limited here, the value of the reverse-phase array is that it allows multiple samples to be analyzed side by side. The example represented here is pre- and post therapy, and the changes in protein expression resulting from the treatment are clear

digoxigenin for indirect detection. Researchers can also directly compare two samples by labeling each with a different reporter and incubating them together in a competition assay (Fig. 5.14a). This design is often referred to as a direct antibody array, and is the most amenable to truly large numbers of analytes, being limited primarily by the availability of specific antibodies and space. Most commercially available arrays offer targets numbering in the hundreds. Major concerns with direct arrays are limited sensitivity and specificity, background levels, and the possibility that direct conjugation of the protein with a reporter or indirect label may somehow mask the epitope, interfering with the ability of the antibody to recognize the protein.

To address these concerns, other versions of the antibody microarray include both capture and detection antibodies, truly becoming sandwich ELISAs on a grand scale (Fig. 5.14b) (Jaras et al. 2007). Using two specific antibodies to detect proteins of interest greatly increases the specificity and sensitivity of the assay, and reduces background problems. Moreover, this method does not require the direct labeling of proteins, removing the potential interference of the label with the antibody–target interaction. One limiting factor with the sandwich antibody microarray, however, is the sometimes inadequate availability of good matched antibody pairs. Another caveat is that concern about cross-reactivity among detection antibodies limits the number of targets in sandwich microarrays as compared to direct arrays. Despite the smaller number of possible targets allowed by the sandwich method, highly customized arrays are becoming extremely valuable for diagnostic, prognostic, and research purposes. For example, arrays are designed to study particular groups of targets known to be breast cancer markers, or for screening the effects of drug candidates on their target cells.

In contrast to the designs described above, the reverse-phase antibody microarray begins with the immobilization of a complex mixture of sample proteins on the array surface rather than antibodies (Fig. 5.14c) (Jaras et al. 2007). The protein spots are then probed with specific antibodies. Reverse-phase arrays allow multiple samples to be spotted on a single array, providing side-by-side analysis. The method also makes analysis of insoluble proteins easier. However, nonspecific interactions are a concern for this assay design as well, and the number of antibodies that can be used on a given array is limited due to restricted reporter multiplexing options. Despite these limitations, reverse-phase antibody microarrays also provide a wealth of useful information for researchers and clinicians.

5.4.3 Multiplexing IHC/ICC

Advances have been made in automation and higher throughput solutions for IHC/ICC in recent years. Tissue arrays, which allow multiple patients' samples to be placed on a single slide, greatly increase the speed and uniformity of preparations. In addition, new automation systems and sophisticated software reduce the amount of time spent screening slides and thus improve throughput.

5.4.4 Multispot ELISAs

The ability to detect a number of targets side by side in a single aliquot of sample can provide an abundance of important information. New ELISA technologies are making this possible via methods such as multispot wells, in which several primary antibodies are carefully coated onto the plate in discrete areas, or spots (Fig. 5.15a). Companies are offering up to 100 spots per well, creating a profusion of information approaching that of antibody microarrays (see MesoScale Discoveries for examples of the technologies described). Further, the ability to conjugate antibodies to chemical linkers is paving the way for even greater flexibility, allowing researchers to mix and match the capture antibodies they choose to assay together (Fig. 5.15b). These linkers are used with specialized plates in which each spot in a given well has been coated with a molecule designed to partner with only one type of linker. Each capture antibody is conjugated to a different linker, and the antibodies can then be incubated in the plate simultaneously. Each antibody will bind only to the spot corresponding to its linker, and the entire well can then be incubated with sample, followed by a mixture of detection antibodies. These multiplexing ELISA platforms are generally accompanied by camera-based detection systems.

The complexity of the data generated by these sorts of multiplexing assays requires that the data be recorded and quantitated by software specially designed for that purpose. Although colorimetric substrates are a possibility, greater ease and utility is afforded by the use of chemiluminescent substrates whose emitted light can be easily recorded and measured. In more advanced ELISA platforms, the light-generating enzymatic reaction is triggered by a computer-controlled electrical

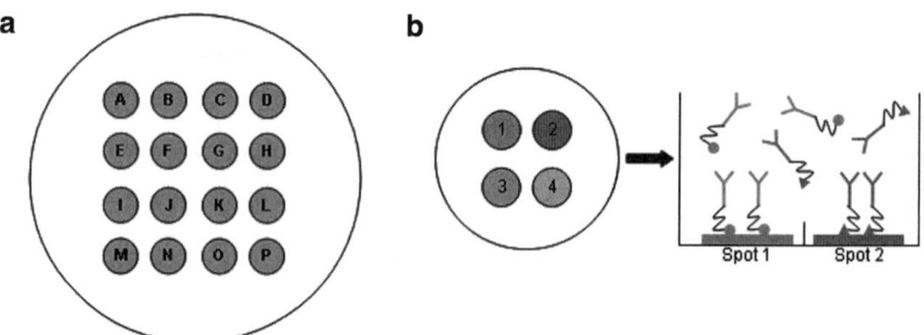

Fig. 5.15 Multispot ELISAs. (**a**) A schematic representation of a 16-spot multispot ELISA well. Each spot, or letter, corresponds to a different capture antibody that is carefully applied to the plate in one discrete area, usually by robot. A single sample can then be incubated in the well, and 16 different sandwich ELISAs are performed simultaneously on one small volume of analyte. (**b**) Chemical linkers can create multispot assays without robotic spotting of the capture antibodies, allowing mixing and matching of desired analytes. Each capture antibody is conjugated to one of several chemical linkers, and incubated simultaneously in the well. Each linker binds only to its corresponding spot, isolating each capture antibody in one specific region of the plate. Multiple sandwich ELISAs can then be performed as in (**a**)

current, making it a true ECL system. These systems require specialized plates, in which each well contains a small electrode. Computer-controlled current is applied to the wells to stimulate the enzymatic reaction, usually parsing the plate into sectors that are read sequentially. These multiplexing platforms and computer-controlled reporting techniques represent the most recent advances in ELISA technology, capitalizing on the specificity, affinity, and adaptability of antibodies for use as recognition receptors to detect soluble proteins of interest.

5.5 Concluding Remarks

The term biosensing takes in a very large array of different types of assays, each of which depends heavily on the recognition receptor used. A good recognition receptor should be highly specific, stable, and versatile, and few possibilities fit the bill as well as antibodies. Antibodies can be chemically conjugated to many different reporter types, and they can be coated onto many different surfaces. They can be raised against nearly any biological material, from ssDNA to protein. The genetics of antibodies are well understood and easily manipulated to produce an array of specificities for screening. For these reasons, antibodies have long been an invaluable tool in research and diagnostics biosensing applications. Advances in the platforms that make use of antibodies as recognition receptors have created extremely powerful technologies which have already begun to revolutionize research, drug discovery, and diagnostics. Multiplexing opportunities provide the ability to discover and analyze the components of entire signal transduction pathways in a single sample, screen large numbers of analytes for their utility as biomarkers in cancer, and greatly increase the turnaround time of diagnostic and prognostic testing. Further, as biosensing assays increase in sensitivity and specificity, earlier detection of low-level biomarkers will become possible, as well as more comprehensive monitoring of disease before and after therapy. In future, we will be limited primarily by the quality of the antibodies we are able to create.

References

Aggarwal K, Lee KH (2003) Functional genomics and proteomics as a foundation for systems biology. Brief Funct Genomic Proteomic 2:175–184

Bonifacino JS, Dell'Angelica, EC, Springer TA (2001) Immunoprecipitation. Curr Protoc Immunol. Chapter 8:Unit 8.3

Borrebaeck CA, Wingren C (2007) High-throughput proteomics using antibody microarrays: an update. Expert Rev Mol Diagn 7:673–686

Bulanova E, Budagian V, Duitman E, Orinska Z, Krause H, Ruckert R, Reiling N, Bulfone-Paus S (2007) Soluble Interleukin IL-15Ralpha is generated by alternative splicing or proteolytic cleavage and forms functional complexes with IL-15. J Biol Chem 282:13167–13179

Chargui N, Haouas N, Gorcii M, Akrout Messaidi F, Zribi M, Babba H (2007) Increase of canine leishmaniasis in a previously low-endemicity area in Tunisia. Parasite 14:247–251

Cooper HM, Paterson Y (2008) Production of polyclonal antisera. Curr Protoc Mol Biol. Chapter 11:Unit 11.12

Czerkinsky CC, Nilsson LA, Nygren H, Ouchterlony O, Tarkowski A (1983) A solid-phase enzyme-linked immunospot (ELISPOT) assay for enumeration of specific antibody-secreting cells. J Immunol Methods 65:109–121

Darzynkiewicz Z, Huang X (2004) Analysis of cellular DNA content by flow cytometry. Curr Protoc Immunol Chapter 5:Unit 5.7

Donahue AC, Fruman DA (2007) Distinct signaling mechanisms activate the target of rapamycin in response to different B-cell stimuli. Eur J Immunol 37:2923–2936

Donahue AC, Kharas MG, Fruman DA (2007) Measuring phosphorylated Akt and other phosphoinositide 3-kinase-regulated phosphoproteins in primary lymphocytes. Methods Enzymol 434:131–154

Donzeau M, Knappik A (2007) Recombinant monoclonal antibodies. In: Albitar M (ed) Monoclonal antibodies: methods and protocols. Humana Press, Totowa, pp 15–31

Foster B, Prussin C, Liu F, Whitmire JK, Whitton JL (2007) Detection of intracellular cytokines by flow cytometry. Curr Protoc Immunol Chapter 6:Unit 6.24

Gallagher S, Winston SE, Fuller SA, Hurrell JG (2004) Immunoblotting and immunodetection. Curr Protoc Mol Biol. Chapter 10:Unit 10.18

Gu AD, Xie YB, Mo HY, Jia WH, Li MY, Li M, Chen LZ, Feng QS, Liu Q, Qian CN, Zeng YX (2008) Antibodies against Epstein-Barr virus gp78 antigen: a novel marker for serological diagnosis of nasopharyngeal carcinoma detected by xMAP technology. J Gen Virol 89: 1152–1158

Haab BB (2005) Antibody arrays in cancer research. Mol Cell Proteomics 4:377–383

Haab BB (2006) Applications of antibody array platforms. Curr Opin Biotechnol 17:415–421

Haugland RP (2001) Antibody conjugates for cell biology. Curr Protoc Cell Biol. Chapter 16: Unit 16.5

Haugland RP, Bhalgat MK (2008) Preparation of avidin conjugates. Methods Mol Biol 418:1–12

Hirsch JD, Haugland RP (2005) Conjugation of antibodies to biotin. Methods Mol Biol 295: 135–154

Hofman F (2002) Immunohistochemistry. Curr Protoc Immunol Chapter 21:Unit 21.24

Hornbeck P (2001) Enzyme-linked immunosorbent assays. Curr Protoc Immunol Chapter 2: Unit 2.1

Hudelist G, Singer CF, Kubista E, Czerwenka K (2005) Use of high-throughput arrays for profiling differentially expressed proteins in normal and malignant tissues. Anticancer Drugs 16:683–689

Irish JM, Czerwinski DK, Nolan GP, Levy R (2006) Kinetics of B cell receptor signaling in human B cell subsets mapped by phosphospecific flow cytometry. J Immunol 177:1581–1589

Janeway CA, Travers PJ, Walport MJ, Shlomchick MJ (2004) Immunobiology: the immune system in health and disease. Garland Science, New York

Jaras K, Ressine A, Nilsson E, Malm J, Marko-Varga G, Lilja H, Laurell T (2007) Reverse-phase versus sandwich antibody microarray, technical comparison from a clinical perspective. Anal Chem 79:5817–5825

Jilani I, Kantarjian H, Faraji H, Gorre M, Cortes J, Ottmann O, Bhalla K, O'Brien S, Giles F, Albitar M (2008a) An immunological method for the detection of BCR-ABL fusion protein and monitoring its activation. Leuk Res 32:936–943

Jilani I, Kantarjian H, Gorre M, Cortes J, Ottmann O, Bhalla K, Giles FJ, Albitar M (2008b) Phosphorylation levels of BCR-ABL, CrkL, AKT and STAT5 in imatinib-resistant chronic myeloid leukemia cells implicate alternative pathway usage as a survival strategy. Leuk Res 32:643–649

June CH, Moore JS (2004) Measurement of intracellular ions by flow cytometry. Curr Protoc Immunol Chapter 5:Unit 5.5

Kellar KL, Douglass JP (2003) Multiplexed microsphere-based flow cytometric immunoassays for human cytokines. J Immunol Methods 279:277–285

Klinman DM, Nutman TB (2001) ELISPOT assay to detect cytokine-secreting murine and human cells. Curr Protoc Immunol Chapter 6:Unit 6.19

Kocinsky HS, Girardi AC, Biemesderfer D, Nguyen T, Mentone S, Orlowski J, Aronson PS (2005) Use of phospho-specific antibodies to determine the phosphorylation of endogenous Na+/H+ exchanger NHE3 at PKA consensus sites. Am J Physiol Renal Physiol 289:F249–F258

Kohler G, Milstein C (1975) Continuous cultures of fused cells secreting antibody of predefined specificity. Nature 256:495–497

Krutzik PO, Nolan GP (2006) Fluorescent cell barcoding in flow cytometry allows high-throughput drug screening and signaling profiling. Nat Methods 3:361–368

Liang Z, Wong RP, Li LH, Jiang H, Xiao H, Li G (2008) Development of pan-specific antibody against trimethyllysine for protein research. Proteome Sci 6:2

Ling MM, Ricks C, Lea P (2007) Multiplexing molecular diagnostics and immunoassays using emerging microarray technologies. Expert Rev Mol Diagn 7:87–98

Lycke NY, Coico R (2001) Measurement of immunoglobulin synthesis using the ELISPOT assay. Curr Protoc Immunol Chapter 7:Unit 7.14

Mechetner E (2007) Development and characterization of mouse hybridomas. In: Albitar M (ed) Monoclonal antibodies: methods and protocols. Humana Press, Totowa, pp 1–13

Michiels JJ, Berneman Z, Gadisseur A, van der Planken M, Schroyens W, van de Velde A, van Vliet H (2006) Classification and characterization of hereditary types 2A, 2B, 2C, 2D, 2E, 2M, 2N, and 2U (unclassifiable) von Willebrand disease. Clin Appl Thromb Hemost 12:397–420

Morgan E, Varro R, Sepulveda H, Ember JA, Apgar J, Wilson J, Lowe L, Chen R, Shivraj L, Agadir A, Campos R, Ernst D, Gaur A (2004) Cytometric bead array: a multiplexed assay platform with applications in various areas of biology. Clin Immunol 110:252–266

Negre N, Lavrov S, Hennetin J, Bellis M, Cavalli G (2006) Mapping the distribution of chromatin proteins by ChIP on chip. Methods Enzymol 410:316–341

Pannu KK, Joe ET, Iyer SB (2001) Performance evaluation of QuantiBRITE phycoerythrin beads. Cytometry 45:250–258

Polak JM, Van Noorden S (2002) Introduction to Immunocytochemistry. BIOS Scientific Publishers Ltd., Oxford

Porstmann T, Kiessig ST (1992) Enzyme immunoassay techniques. An overview. J Immunol Methods 150:5–21

Ricart AD, Tolcher AW (2007) Technology insight: cytotoxic drug immunoconjugates for cancer therapy. Nat Clin Pract Oncol 4:245–255

Schulz KR, Danna EA, Krutzik PO, Nolan GP (2007) Single-cell phospho-protein analysis by flow cytometry. Curr Protoc Immunol Chapter 8:Unit 8.17

Sharrow SO (2002) Overview of flow cytometry. Curr Protoc Immunol Chapter 5:Unit 5.1

Sofou S, Sgouros G (2008) Antibody-targeted liposomes in cancer therapy and imaging. Expert Opin Drug Deliv 5:189–204

Spoettl T, Hausmann M, Klebl F, Dirmeier A, Klump B, Hoffmann J, Herfarth H, Timmer A, Rogler G (2007) Serum soluble TNF receptor I and II levels correlate with disease activity in IBD patients. Inflamm Bowel Dis 13:727–732

Sumbayev VV, Yasinska IM (2008) Protein S-nitrosation in signal transduction: assays for specific qualitative and quantitative analysis. Methods Enzymol 440:209–219

Takada K, Nasu H, Hibi N, Tsukada Y, Ohkawa K, Fujimuro M, Sawada H, Yokosawa H (1995) Immunoassay for the quantification of intracellular multi-ubiquitin chains. Eur J Biochem 233:42–47

Tan K, Tegner J, Ravasi T (2008) Integrated approaches to uncovering transcription regulatory networks in mammalian cells. Genomics 91:219–231

Tanaka K, Fukase K (2008) PET (positron emission tomography) imaging of biomolecules using metal-DOTA complexes: a new collaborative challenge by chemists, biologists, and physicians for future diagnostics and exploration of in vivo dynamics. Org Biomol Chem 6:815–828 van Dongen GA, Visser GW, Lub-de Hooge MN, de Vries EG, Perk LR (2007) Immuno-PET: a navigator in monoclonal antibody development and applications. Oncologist 12:1379–1389

Weber WA, Czernin J, Phelps ME, Herschman HR (2008) Technology Insight: novel imaging of molecular targets is an emerging area crucial to the development of targeted drugs. Nat Clin Pract Oncol 5:44–54

Wei CL, Wu Q, Vega VB, Chiu KP, Ng P, Zhang T, Shahab A, Yong HC, Fu Y, Weng Z, Liu J, Zhao XD, Chew JL, Lee YL, Kuznetsov VA, Sung WK, Miller LD, Lim B, Liu ET, Yu Q, Ng HH, Ruan Y (2006) A global map of p53 transcription-factor binding sites in the human genome. Cell 124:207–219

Weinmann AS, Farnham PJ (2002) Identification of unknown target genes of human transcription factors using chromatin immunoprecipitation. Methods 26:37–47

Wells J, Farnham PJ (2002) Characterizing transcription factor binding sites using formaldehyde crosslinking and immunoprecipitation. Methods 26:48–56

Yalow RS, Berson SA (1960) Immunoassay of endogenous plasma insulin in man. J Clin Invest 39:1157–1175

Zamrazilova L, Kazihnitkova H, Lapcik O, Hill M, Hampl R (2007) A novel radioimmunoassay of 16alpha-hydroxy-dehydroepiandrosterone and its physiological levels. J Steroid Biochem Mol Biol 104:130–135

Zhang ZJ, Albitar M. Monoclonal antibodies. In: Walker JM, Rapley, R (eds) Molecular biomethods handbook. Humana Press, Totowa, pp 545–559

Chapter 6
Peptides as Molecular Receptors

Ibtisam E. Tothill

Abstract The use of specific and sensitive sensing layers for molecular diagnosis and biosensor developments is crucial for producing a successful device. The sensing layers are required to be stable and robust for sample analysis (serum, urine, water, soil extracts and foods), storage and application in field conditions. Therefore, the technology is advancing to replace nature molecules with synthetic materials that are more stable and robust for sensing purposes. A range of novel approaches have recently been used based on synthetic chemistry and computational methodologies to complement natural affinity systems with synthetic ligands. Peptides have emerged as one of the promising approaches to synthetic biomimics. The characteristic properties of synthetic peptides can render them as potential alternatives to antibodies and natural receptors for biosensor application. Different methodologies are used today to design and discover sensitive and selective peptides for specific analytes. These include computational chemistry, combinatorial chemistry, phage display technology and molecular imprinting. This chapter introduces the concept of using peptides as sensing materials and cover methods of their design, selection, synthesis and use as receptors in sensors and diagnostics applications. Problems and challenges facing this technology are also discussed.

Keywords Peptides · Molecular receptors · Biosensors · Synthetic receptors

Abbreviations

DNA	Deoxyribonucleic acid
SPE	Solid phase extraction columns
HPLC	High-performance liquid chromatography

I.E. Tothill
Cranfield University, Cranfield, Bedfordshire, MK43 0AL, UK
e-mail: i.tothill@cranfield.ac.uk

M. Zourob (ed.), *Recognition Receptors in Biosensors*,
DOI 10.1007/978-1-4419-0919-0_6, © Springer Science+Business Media, LLC 2010

GC	Gas chromatography
CAMD	Computer-aided molecular design
MIPs	Molecularly Imprinted Polymers
PEGA	Polyethylene glycol acrylamide
HTS	High-throughput screening
DCL	Dynamic combinatorial library
NMQ	N-methyl quinuclidinium iodide
Ach	Acetylcholine iodide
ab MCR	Aqueous-based Multi-Component Reaction
NMR	Nuclear magnetic resonance
QSAR	Quantitative structure-activity relationships
SPR	Surface plasmon resonance
QCM	Quartz crystal microbalance

6.1 Introduction

The specific and selective binding capabilities inherent in biomolecules have been exploited in biosensor devices to produce either affinity or catalytic based biosensors. Most diagnostics systems and biomedical products are also protein based and protein–protein or protein–peptide recognition is usually implemented. In natural systems, reactions and catalysis are conducted through a complex evolutionary mechanism based on recognising and distinguishing molecular entities and therefore molecular recognition is the base for most of the biological processes. Reactions such as ligand-receptor binding, antibody–antigen interaction and substrate–enzyme reactions have therefore been employed in the development of many diagnostics and analytical methods. Recognition elements (sensing layers/receptors), such as nucleic acids (single-stranded DNA to bind its complementary sequence), protein lipids and their derivatives, enzymes, antibodies, cell receptors, lectins (plant proteins used to bind carbohydrates) are all used to date in methods such as enzyme assays, immunoassays, biosensors and various affinity interaction and separation techniques (Tothill 2001). Because of their specificity and catalytic properties, enzymes have found widespread use as the sensing element in biosensors with antibodies (polyclonal and monoclonal antibodies) following as the next most common receptors. Commercially available enzymes and antibodies for the detection and analysis of different analytes exist but in some cases are very expensive and scarce or lack the required stability which hinders the commercialisation of biosensors. Antibody synthesis against small molecules is usually more difficult since small molecules require to be conjugated (linked) to a large protein before being injected into the animal to elicit sufficient antibody production. This process can therefore be labour-intensive and expensive especially for analytes such as toxins, pesticides, phenols etc. Due to the toxicity of many environmental pollutants, they are also more likely to cause an acute toxic response in the boosted animal while implementing antibody production procedures. The polyclonal antibodies produced will also partly recognise the protein conjugate that it was raised against and therefore, may require affinity purification to increase the antibody's specificity. Obtaining molecules that specifically recognise and bind small molecules is therefore cumbersome and difficult and in some cases are not possible. However, biological molecules have shown advantages over chemically synthesised affinity ligands in that they exhibit higher specificity and sensitivity in many applications. Many conventional analytical methods such as solid phase extraction columns (SPE), high-performance liquid chromatography (HPLC) and, gas chromatography (GC) techniques also require new separation materials that demonstrate higher specificity and efficiency for more specialised application. Therefore, the use of novel methods to produce new materials for phase separation systems is also required to fill the gap in separation chemistry. A range of alternative strategies and approaches are now being used to develop different type of recognition materials for affinity interactions which include biological as well as biomimics (Toko 2001; Tothill et al. 2001).

Studying and understanding biological systems and how natural evolution was able to create high affinity molecules such as enzymes and antibodies and implementing this knowledge by using molecular engineering combined with innovative bioinformatics solutions, we should be able to develop materials with similar properties by molecular design. The principle of bio-recognition and selective interaction of molecules has been the subject of a wide range of scientific research and has been found to be based on the induced fit principle rather than just the key and lock hypothesis. Therefore, to mimic such natural systems the developments of molecular entities that conferee similar molecular flexibility is essential to synthesis materials with high specificity. The use of new technologies such as recombinant antibodies, antibody fragments and molecular engineered antibodies have now created new range of molecules that can be used for sensors and diagnostics (Tothill 2003). However, more recently novel approaches based on synthetic chemistry and computational methodologies to complement natural affinity systems have been developed.

To date, the research into the development of new affinity materials (receptors) have moved into several new dimensions with state-of-the art approaches are being used to discover affinity receptors for analytes separation and detection. Synthetic peptides are one of the new promising approaches and the design and synthesis of new peptides is expanding rapidly.

6.2 Peptides as Receptor Molecules

Peptides have emerged as one of the most interdisciplinary science and are being applied into a range of technologies including drug discovery, biomimetics for diagnostics applications and cosmetics. Attention to artificial receptors or biomimics has increased in recent years in order to solve the most critical problems related to the application of biological molecules for sensing applications. These include; low stability in high or low pH and high temperature; poor performance in organic solvents; high cost of production for some enzymes and antibodies and also the lack of specific biological receptors for certain small pollutants and toxins (Nakamura et al. 2001, 2005). As biological compounds often offer the most exquisite selectivity and sensitivity, efforts have been put in place to make them more stable by applying different stabilising procedures. This has helped in increasing the shelf life and durability of some sensor devices applying biological materials as receptor molecules (Tothill and Turner 1998). However, some biosensor applications require stability far beyond what can be exhibited by biological molecules and hence, research into the use of biomimics was introduced to fill the gap in this area of application. The search for possible solutions has taken a divers direction; the use of molecules produced by extremophiles; the use of genetic manipulation to improve bimolecular properties; the use of artificial analogues to substitute natural biomolecules etc. To date, the filed of artificial receptors (biomimics) comprise a vast array of compounds and molecules including peptides. Synthetic peptides have emerged as one of the most promising alternative molecules to replace the antibodies and proteins in sensors and diagnostics techniques.

A range of biochemical processes are conducted using peptides such as signal transduction (e.g. neuropeptides), metabolism (e.g. hormones), cell growth (e.g. Ras-protein) and immune defence (e.g. MHC-receptors) (Schmuck 2001). Molecular recognition by peptides is based on non-covalent interactions, such as hydrogen-bonding, salt-bridges, hydrophobic interaction and van der Waals interactions. Peptides are of special interest for a wide range of industrial, pharmaceutical and medical purposes (Eguchi and Kahn 2002; Lam and Renil 2002; Sundaram et al. 2002; Sadler and Tam 2002; Sawyer and Chorev 2003). They are compounds made from ten or less amino acids linked together with covalent peptide bonds, with polypeptides having longer sequences (Fig. 6.1). Over 500 amino acids exist in nature with only 20 observed in all species (Berg et al. 2006). Therefore, a vast number of peptides can be synthesised and used for different application (Welling et al. 1990). However, most of the synthesised peptides use the 20 essential amino acids. More recently non-essential amino acids have also been used, but cost and availability may constrain their application. The hydrophilicity/hydrophobicity properties of the peptides are very important for their affinity as well as predicting

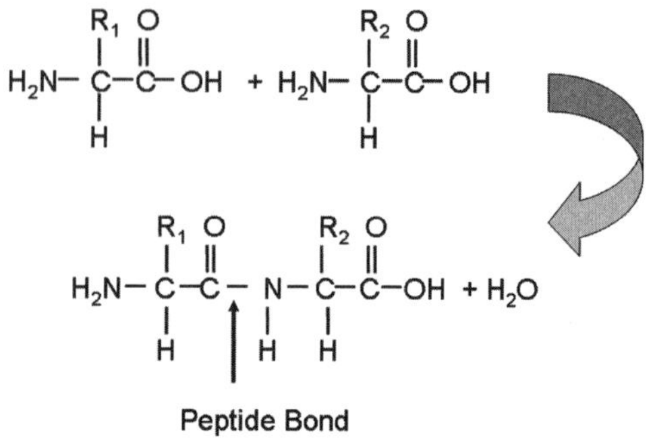

$$H_2N-C(R_1)(H)-C(=O)-OH + H_2N-C(R_2)(H)-C(=O)-OH \rightarrow H_2N-C(R_1)(H)-C(=O)-N(H)-C(R_2)(H)-C(=O)-OH + H_2O$$

Peptide Bond

Fig. 6.1 Peptide bond formation between two amino acids

if the selected peptide structures can be chemically synthesised. This has also played a major role in the selection process of amino acids used for peptide construction and synthesis. The sensitivity and specificity of the peptide also depend on the analyte target that is synthesised to detect and this includes other characteristic that may be molecularly engineered in the peptide. In the pharmaceutical Industry, there has been great interest in using unnatural amino acids (α and β amino acids) to synthesis drugs. Ma (2003), reported that the α unnatural amino acids have been used more than the β amino acids in order to improve the availability and dynamics and reduce conformational flexibility.

As the biologically active sites of the antibody molecules are polypeptide in nature, it is possible to recreate the affinity of the antibody by using a peptide sequence. A range of methods have been used to design peptides. Recently, interest has been undertaken in the use of computer-aided molecular design (CAMD) methods for the design of novel molecules such as peptides possessing desired characteristics (Patel et al. 1998). With the current methods of producing polyclonal antibodies, variations in the quality and concentration of the antibodies occur between batches (Tozzi et al. 2003a, b) which make it difficult to standardise test methods using these antibodies. However, since peptides are synthesised using a standard instrumentation, the quality of the produced peptides and their rapid production method, their reproducibility make them more attractive to implement in diagnostics methods than antibodies. Peptides should also have an advantage over Molecularly Imprinted Polymers (MIPs) as sensing receptors due to their flexibility in an induced fit conformation.

To date there are several approaches used to design and develop peptides and these include:

1. Combinatorial chemistry
2. Phage display technology
3. Rational design approach using computational methods
4. Molecular imprinting

6.3 Combinatorial Chemistry

The combinatorial library technique (Combinatorial chemistry) is a fast developing area of receptor discovery and has been widely used by the drug industry to both discover and optimise lead compounds as new drugs. This technique has also been used to synthesis libraries of compounds for the screening of novel affinity ligands that can be used as materials in affinity chromatography or new receptors in sensors and biosensors. Combinatorial libraries (illustrated in Fig. 6.2, Tothill et al. 2001), consist of a large array of diverse molecular entities, generated by the systematic and repetitive covalent connexion of a set of different "building blocks" of varying structures. The number of possible compounds (n) generated in a single library can be found using the relationship:

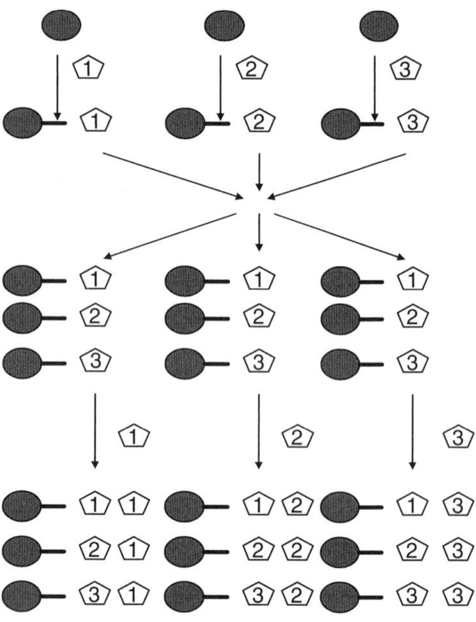

Fig. 6.2 Schematic of split and mix combinatorial synthesis (Tothill et al. 2001)

$$n = b^x$$

where b is the number of building blocks for each step and x is the number of synthetic steps.

The use of natural and unnatural amino acids as building blocks will result in peptide libraries which may contain peptide sequences (receptor molecules) with affinity towards the analyte of interest. This is one of the approaches that are used to produce a random set of peptide molecules using chemical synthesis with predetermined sequence variations on solid-phase supports (Geysen et al. 1986). Conventional combinatorial chemistry seeks to improve the rate of success in drug and catalyst discovery through the generation of a large number of molecules which can be screened for their activity and affinity (Lavastre and Morkrn 1999; Cousins et al. 2000). Therefore, a large number of compounds can be generated in a short time using this methodology. Various types of resins such as Wang Trityl and PEGA and Merrifield are used to prepare the libraries. Screening these libraries using high-throughput methods will result in a few compounds, which may require further investigation as possible lead compounds (Crabtree 1999). Positive results can also be visualised using microscopy when tagged analytes (fluorescence dyes etc.) are used. However, the screening procedure is time and cost-inefficient and may be in some cases based on an existing library that might not contain the optimal ligand, since the produced molecular library is only containing a finite number of possibilities and the best fit can only be chosen from that confined set of molecules, depending on the make of the library. Hence, the size of a library is not decisive but more the need for appropriate diversity within the combinatorial library for a given analyte screening. Therefore, a method that can screen for an indefinite number of variants of peptide receptors, task specific for a target analyte, is required.

A synthetic peptide library for identifying ligand-binding activity has been discussed by Lam et al. (1991). These molecular libraries could be subjected to high-throughput screening (HTS) to find a perfect hit. However, HTS has generated large numbers of hits in many different biological screening assays. The major problem is to prioritise hits for further study. The use of molecular modelling as a combined approach with combinatorial chemistry to facilitate a more efficient receptor discovery process is now widely applied (Antel 1999). This helps in rationalising the design and guide the selection of the building blocks to be used in the combinatorial synthesis and produce smaller libraries with higher HTS. Molecular modelling will also visualise and evaluate the interactions between the target analyte and the potential receptor. This can provide structural informations that can direct the design of the combinatorial library and the artificial peptide.

The use of combinatorial chemistry for the production of receptors for biosensors is still in its early stages. The technology can be expensive and needs investment in equipments and personnel. However, future developments may make the technology more accessible for this application.

6.4 Dynamic Combinatorial Library

Dynamic combinatorial library (DCL) is an evolved field of combinatorial library that is viewed as a collection of compounds which are assembled dynamically by a range of building blocks in the presence of a molecular target (Huc and Lehn 1997). The concept has been reviewed in the literature (Lehn 1999; Klekota and Miller 1999; Ganesan 1998; Huc and Nguyen 2001; Corbett et al. 2006).

Huc and Lehn (1997) have proposed the DCL templating as moulding (macrocyclic or macromolecular structures assemble of high affinity building blocks around a template) and casting (template with cavity or molecular traps where the high affinity building blocks assemble them selves inside these cavities). Figure 6.3, show the dynamic combinatorial library and how the favoured compounds could be selected by the molecular target during the process of synthesis. Cousins et al. (2000) viewed it as a collection of transient compounds that are reversible assemblies of a collection of building blocks. The library members interconvert by an equilibrium processes and stabilisation of a given member results in the thermodynamic redistribution of the equilibrium, thus amplifying its concentration and favouring the formation of that structure above the other possibilities that co-exist in solution. This process has three different features when compared to conventional combinatorial library:

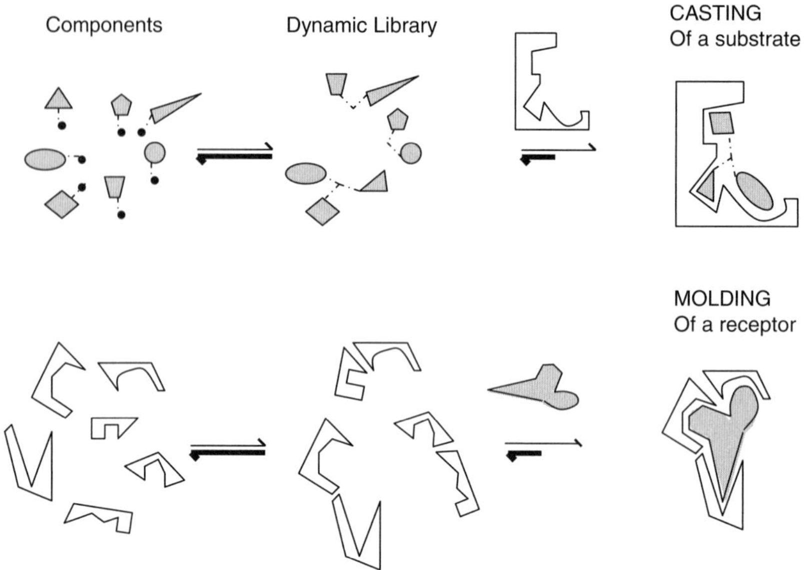

Fig. 6.3 Schematic of dynamic combinatorial synthesis (Cousins et al. 2000)

1. In situ target driven selection. The molecular target screens the library building blocks while they are synthesised.
2. Synthesis is carried out in aqueous media using reversible reactions under mild reaction conditions, especially when a protein or a bio-molecule is used as a molecular target.
3. Require the set up of efficient and accurate methods for identification and structural characterisation of active compounds in a DCL.

Dynamic combinatorial library relies on the in situ target-driven selection. During the synthesis process the high affinity building blocks of the compound library used with the molecular target will bind to the target and the equilibrium of the reaction will be shifted and products formation can be subsequently amplified. In a one assay all of the library members can be screened. By the nature of the selection, the molecules identified in a template mixture should be good receptors/hosts for the template with the desired recognition properties (Cousins et al. 2000). This is a significant development when compared to the traditional approach, in which library screening is normally carried out after the synthesis, therefore the number of screenings is large.

The molecular target type and concentration plays an important role in the synthesis process. Molecular targets in DCL play an important role in screening library members and control the equilibrium of reactions. Many types of molecular targets have already been investigated in the literature and these include: proteins

(Hochgurtel et al. 2002, 2003; Wang et al. 2003a; Zameo et al. 2005; Krasinski et al. 2005), nucleic acids (Karan and Miller 2001), peptide (Hioki and Still 1998) organic ions, such as biphenyl, halocarbons, crown ethers, transition state analogues (Yamanoi et al. 2001; Kubota et al. 2002; Yoshizawa et al. 2003; Furlan et al. 2000; Brisig et al. 2003). Inorganic ions have also been used as molecular targets such as anions, alkali metal ions and alkyl ammonium ions (Furlan et al. 2001; Saur and Severin 2005; Roberts et al. 2003; Brady and Sanders 1997a, b).

As an example of a casting process, Swann et al. (1997) reported the generation of a peptides library by incubating the non-specific protease (thermolysin) with peptide fragments (YGGFL, Tyr-Gly-Gly and Phe-Leu) in the presence of a monoclonal antibody able to recognise the peptide (Tyr-Gly-Gly-Phe-Leu). Experiments conducted on the resulting products indicated that the antibody acted as a macromolecular sink for the peptide. This is a powerful new approach for peptides with important biological identity generation. However, progress in such a system requires the availability of suitable peptidase/ligase enzyme cocktail. Immobilising the template on solid support or resins has also become a method of creating dynamic libraries and isolating high affinity receptors.

A dynamic combinatorial library of pseudo-peptide macrocycles have been generated using amino acids functionalised with both an aldehyde and a hydrazide functionality (Cousins et al. 1999; Poulsen et al. 2000). Libraries of cyclic hydrazones were then formed in the presence of dilute acid. Diversification of these libraries by the introduction of further building blocks and manipulation of equilibrium through templating are also being explored by these groups. An alkylammonium ions N-methyl quinuclidinium iodide (NMQ) and acetylcholine iodide (Ach) were used to screen a pseudo-peptide dynamic combinatorial library (Cousins et al. 2001).

Dynamic combinatorial library synthesis is normally carried out in aqueous media using reversible reactions when bio-molecules are used either as the target or the building blocks. In most interactions, the favoured products were amplified by the molecular target due to recognition and also reversible reactions where the equilibrium of the reaction shifts to generate more products. Reversible reactions used in DCLs include trans-esterification (Brady and Sanders 1997a, b), C=N double bond exchange (Saur and Severin 2005; Roberts et al. 2003), disulphide exchange (Hioki and Still 1998; Otto and Kubik 2003), olefin metathesis (Hamilton et al. 1998) and non-covalent hydrogen bonding (Calama et al. 1998) and click chemistry (Krasinski et al. 2005; Manetsch et al. 2004).

Aqueous-based Multi-Component Reaction (ab MCR) is a new method which is used as a one pot combinatorial library synthesis for the discovery of novel lead structures targeting an analyte. Dynamic combinatorial library approach is developing rapidly, playing a vital role in synthetic chemistry and producing peptides as drugs and for other health based applications, but its use for affinity receptors development is still limited and not well known. This is a new field of receptor discovery and its potential use for the development of peptides for biosensor applications is still to be realised.

6.5 Phage Display Technology

Peptide phage display is an in vitro selection technique where peptides with desired binding properties towards a target molecule can be extracted from a large pool of variants. Smith (1985) was the first to describe the expression of exogenous peptides on the surface of filamentous bacteriophage. Since then he and co-workers published many articles regarding the technology (Smith 1991; Smith and Scott 1993), also researches in the literature have seen the display of peptides and antibodies on coat proteins of phage greatly expanding the applications of the technology (Arap 2005). To date, there are laboratory manuals describing the technology (Barbas et al. 2000; Webster 2001). A peptide phage library can contain around 1010–1011 different clones [New England BioLabs Inc.]. Large libraries have also been produced using a bacteriophage vector. Many reviews have also been published regarding the technology (O'Neil and Hoess 1995; Azzazy and Highsmith 2002; Szardenings 2003). Many literatures also report on the identification of peptides with specific affinity for a range of analytes (Kay et al. 1993; Koivunen et al. 1999; Noda et al. 2001; Wang et al. 2003a, b). In this book, Chapter 18 has been dedicated to Phage display technology and therefore, the author referee the reader to the chapter for further information on the technology.

6.6 Rational Design Approach Using Computational Methods

Molecular modelling (computational modelling) can be defined as the application of computers to generate, manipulate, calculate and predict realistic molecular structures and associated properties based on quantum equations and classical physics (Cramer 1983). The use of molecular modelling approach to rationally design affinity ligands and guide the selection of the building blocks to be used in combinatorial chemistry or peptide synthesis is one of the most recent advances used in receptor discovery. The use of computer aided design for rapid developments and optimisation of the final receptor structure helps in cutting cost, time and effort in producing better receptors. It is a tool used to implement conformational study of molecules such as drugs, proteins, macromolecules etc. (Hagiwara et al. 1993; Rustici et al. 1993). Molecular modelling, serves as a bridge between theory and experiment to compare experimental results of a system, with theoretical predictions of a model, to help understanding and interpreting experimental observations and make correlations between them based on atomic, molecular and macroscopic properties (Tsai 2002). Molecular modelling is usually applied to provide structural information for the target analyte that can direct the design of combinatorial libraries and rational design of the artificial receptor. It will also visualise and evaluate the interactions between the target analyte and the potential receptor.

Recently, significant progress has been made in the use of the computer-assisted molecular design (CAMD), which represents the application of computational modelling techniques as tools for simulations of biomolecular interaction in the area of drug discovery, ligand–protein binding and facilitated structure-based ligand design (Patel et al. 1998). Many programs have also been evolved with user interfaces which are simple allowing chemist's access to such tools rather than computer specialists. Since the 1980s many de novo design computer programs have been employed within the pharmaceutical industry to aid drug design. Drugs for HIV-1 protease inhibitors, for example, have been successfully designed and now marketed (Wlodawer and Vondrasek 1998). This field is influenced by the revolution taking place in peptide science which is also influenced by the 3-D structure guided-drug discovery. Emerging databases of X-ray and NMR structures are providing high resolution molecular maps of molecular targets (Sawyer 1999; Bursavich and Rich 2002; Rich 2002). A variety of molecular modelling programs, are now available commercially (Kollman 1993; Lybrand 1995; Jones and Willett 1995). Most of the available software are drug design orientated and based on macromolecules. Software for the de novo design include QUANTA/CHARMN/MCSS/HOOK, INSIGHTII/Ludi, CERIUS 2 (MSI, San Diego, USA), SYBYL (Tripos, San Diego, USA) etc. The Protein Data Bank, worldwide archive of structural data of biological macromolecules (PDB; http://www.rcsb.org/pdb/) (Brookhaven National Laboratories, USA) (Berman et al. 2000) is used to download the structural data for biomolecules. By studying the type of forces involved in a specific target reaction (specific analyte interacting with its target biological molecule), the results can be used to guide the design of the artificial receptor. Predication of ligands that are expected to bind strongly to key regions of biologically important molecules is an important process used for selecting the best design. Current research aims at a better understanding of the physical nature of molecular recognition in protein-ligand complexes and also at the development and application of new computational tools that exploit the current knowledge on structural and energetic aspects of protein-ligand interactions in the design of novel ligands (Gunsteren et al. 2008). These approaches are based on the exponentially growing amount of information available regarding the geometry of protein structures, properties of small organic molecules exposed to a structured molecular environment and advances in computer simulation technology (Böhm 1992, 1994, 1995, 1996; Böhm and Klebe 1996; Schmuck 2001). Most commercially available software are drug design orientated and based on macromolecules. The design of biomolecules that have specific binding characteristics towards a selected target molecule is challenging and a schematic is shown in Fig. 6.4. Peptide screening based on a computational approach is innovative especially for receptor design against small compounds.

6.6.1 Molecular Mechanics

Molecular mechanics (Force fields) is one of the theoretical methods available to predict the geometry of molecules and uses classical mechanics to represent these molecules (Höltje et al. 2008). Computational modelling uses molecular mechanics

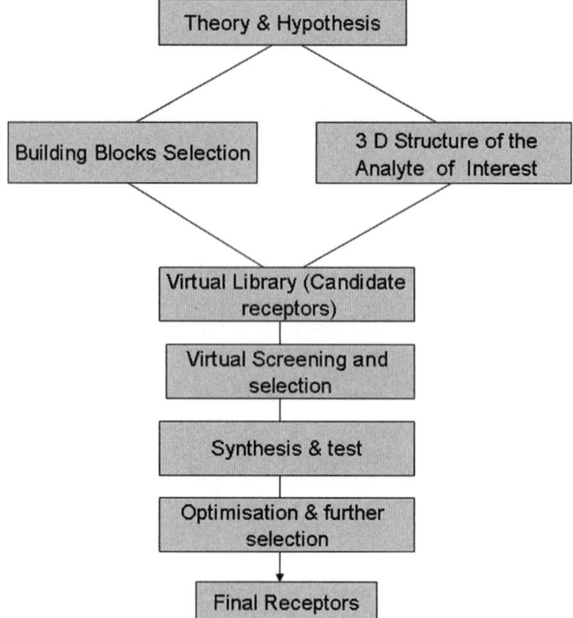

Fig. 6.4 Schematic for the use of molecular modelling in receptor discovery

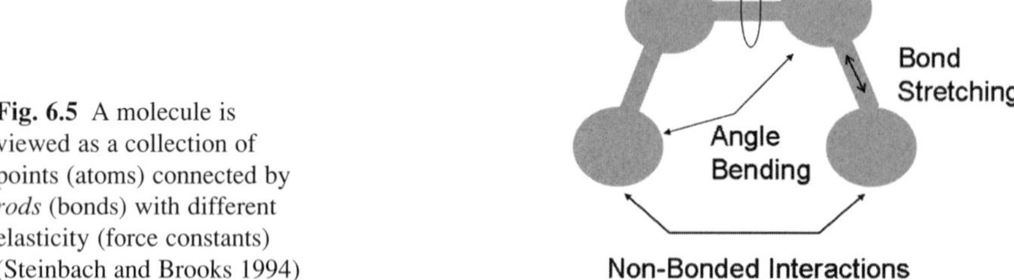

Fig. 6.5 A molecule is viewed as a collection of points (atoms) connected by *rods* (bonds) with different elasticity (force constants) (Steinbach and Brooks 1994)

to generate and present molecular data including geometries (bond lengths, bond angles, torsion angles), energies (heat of formation, activation energy, etc.), electronic properties (moments, charges, ionisation potential, electron affinity), spectroscopic properties (vibrational modes, chemical shifts) and bulk properties (volumes, surface areas, diffusion, viscosity) (Richon 1994).

In this concept, the molecules are viewed as collections of spheres, which are connected by independent flexible springs representing Bonds (Fig. 6.5) (Steinbach 2005). Hookes' law is used to model the elastic forces between the spheres. The calculation of the structure and energy of molecules is based on nuclear motions. Electrons are not considered explicitly, but assumed that they will find their optimum distribution once the positions of the nuclei are known (Richon 1994).

Force fields are used in molecular modelling to reproduce structural properties such as the energy of the structure in a particular conformation (Chianella et al. 2002). However, these energy values can not be considered absolute physical quantities. The molecular energy is calculated term by term, comparing bond parameter values that are taken from either experimental data or from ab initio (quantum mechanics) calculations. A simplified molecular mechanics representation is shown in the equation below:

$$E_{\text{Potential Energy}} = E_{\text{stretching}} + E_{\text{bending}} + E_{\text{torsional}} + E_{\text{electrostatic}} + E_{\text{van der Waals}}.$$

Apostolakis and Caflisch (1999) discussed the estimation of (relative) binding affinities, which is the most challenging of ligand design and reviewed three methods for the estimation of binding energies: (1) quantitative structure-activity relationships (QSAR); (2) empirical energy functions and (3) free energy calculations based on molecular dynamics simulations.

Examples of molecular modelling using force field include TRIPOS (Clark et al. 1989), MM2 (Allinger (1977), MM3 (Allinger et al. 1989, 1990), and MM4 (Allinger et al. 1996), AMBER (Cornell et al. 1995), MAXIMIN (Labanowski et al. 1986). Christensen and Jorgensen (1997), presented a general strategy for the energy minimisation of proteins using MAXIMIN and the software SYBYL 5.5. The influence of salvation, dielectric function and constant were investigated in the energy minimised structures by comparing them with the available X-ray structures data. The availability of high quality X-ray structures for proteins and other molecules are important for the calculations used in these models.

6.6.2 Building Blocks

For peptide receptor construction the building blocks used in the molecular modelling program include a library of amino acids which may constitute the 20 natural amino acids or also unnatural amino acids. This will depend on the availability and cost of these monomers. The structures will be imported from the PDB and built as the library of monomers to be used during the molecular interaction study.

6.6.3 Leapfrog

In computer-aided molecular discovery, models of the receptor site is generated as an experimental or homology of receptor cavity, or as a pharmacophore. This will then be turned into a series of molecules recommended for synthesis. De novo tools automatically generate such series of ligands. Leapfrog is a “de novo” ligand design program which creates ligands in the active site of the molecules of interest (template) (Bertelli et al. 2001). This uses a function called genetic search algorithm and can screen through the library in an "intelligent" approach, Leapfrog (Tripos) utilises this principle. First a program called GRID (Goodford 1985; Moon and Howe 1991) searches the receptor site on the template for interaction sites by using a small probe of; methyl, hydroxyl or carbonyl-oxygen. This data is then used to describe the desirable orientations of the peptide ligand (Goodman 1998). Then the ligand is built by placing fragments (Building blocks, amino acids in this case) and linking them together to produce the peptide sequence. The genetic algorithm calculates a binding score of the total contributions from steric, electrostatic and hydrogen bonding interactions, and records the score at each step of the virtual reaction. A slight variation of the ligand is produced and the ligand is scored again, if the new peptide ligand has a more favourable interaction then this will be stored and the first ligand is dismissed (Payne and Glen 1993). Genetic algorithms are frequently employed in docking programs, such as Flexidock (Tripos). Further discussion on recent progressions in molecular docking has been reviewed by Krovat et al. (2005) and Alberts et al. (2005). Scoring of docking processes is still regarded as one of the major challenges in the field of molecular docking. The purpose of the scoring procedure is to identify the correct binding pose by its lowest energy value, and the ranking of receptor-ligand complexes according to their binding affinities (Muegge and Rarey 2001).

New ligand structures are evaluated mainly on their binding energy and these will be saved in a database of molecular spreadsheet with their binding energy. This will be used as the list of best peptides to be synthesised for testing. The synthesised peptides are then examined for their affinity using a range of molecular interaction tools. There are disadvantages of genetic algorithms as the scoring function frequently stores a significant number of false positive structures and some of the structures selected by de novo programs are synthetically difficult to produce (Schneider and Böhm 2002; Bohacek and McMartin 1997).

The use of QSAR, which is a toxicity-modelling algorithm based on quantitative structure-activity relationships (QSAR), neuronal networks, or artificial intelligence concepts (Selassie 2003) is another alternative. QSAR is an area of computational research which builds atomistic or virtual models to predict quantities such as the binding affinity or pharmacokinetic parameters where structural features can be correlated with biological activity. A new approach named Quasar (Quasi-atomistic receptor-modelling) combines receptor modelling and QSAR technique based on a genetic algorithm by mapping an unknown or hypothetical receptor (such as the peptide receptor) in 3D and quantitatively calculating the affinity of small molecules binding to it (Vedani and Dobler 2002, 2003).

The use of rational design to produce peptide receptors for sensors applications is of great interest especially for small compounds detection and analysis. The technology has also been applied to design molecular imprinted polymers (MIPs) for toxins and environmental pollution detection (Subrahmanyam et al. 2001; Chianella et al. 2002).

6.7 Molecular Imprinting

Molecular – imprinting polymerisation has been the focus of intense research interest in the past 10 years and to date is being considered in a wide range of application areas, e.g. in the preparation of selective separation materials and as sensing layers in sensor devices (Tothill et al. 2001; Chianella et al. 2006; Piletsky et al. 2006a, b). MIPS (Molecularly Imprinted Polymers) are usually produced by forming highly cross-linked organic polymer around the molecule of interest (analyte target/template). After polymerisation, the analyte/template is removed using organic solvent leaving the void which was occupied by the analyte free. MIPS relies upon the presence of complementary interactions (non-covalent or reversible covalent) between sites on the analyte structure and the functional monomers used as the building blocks in the polymerisation process. Recognition of proteins using molecular imprinting polymers is relatively new area. Attempts at printing of amino acids derivatives (Kempe 1996) small peptides (Rachkov and Minoura 2000) and proteins (Vaidya et al. 2001) have been undertaken with limited success. As MIPS synthesis is covered as a chapter in this book, the author refers the readers to that specific chapter and also highlight that this section will only deals with amino acids as building blocks in MIPs construction.

Molecular imprinting approach has been recently applied for peptide receptors construction. In the case of peptide receptors, this approach involves formation of a complex between functional monomers (amino acids in this case) and the target molecule (template/analyte) in an appropriate solution, and cross-links the complex by polymerisation process. Following the removal of the template by washing, binding sites (imprints) are left within the polymer with a structure, which is complementary to the template molecule. The theory of imprinting polymerisation and practical aspects of MIP application are covered in several reviews (Mosbach 1994; Mosbach and Ramstrom 1996; Piletsky et al. 2001). In theory, MIPs can be prepared for any kind of substances including drugs, nucleic acids, proteins and cells. MIPs can also be prepared using rational design procedures as described above (Takeuchi and Matsui 1998). The construction of virtual library of functional monomers against a target molecule followed by the selection of the highest binding score monomers as the building blocks has been undertaken by many studies (Chianella et al. 2002; Lotierzo et al. 2004). The software facilitates calculations using different dielectric constants to reflect the polarity of the environment (solvents) in which the polymers have to be prepared or used.

Attempts have been made to use the MIP procedure but using amino acids as the monomers (Giraudi et al. 2002, 2003). As with MIP protocols the target is surrounded by amino acid monomers and then polymerised. The results show that there is an advantage that these amino acid polymers can be used in an aqueous media, however there are two distinct disadvantages, there is doubt over their stability and they cannot be fixed to a solid surface. Also the major disadvantage is the same as that of MIPs, it requires a template (pure form of the analyte) to produce the synthetic polypeptides, which can in cases be a significant cost.

6.8 Application in Sensors

Peptides are one of the most abundant molecular receptors used by nature to carry out a range of processes based on affinity interaction (Zeger et al. 1995). Therefore, the use of peptides in man made systems based on affinity reaction should be possible. However, designing new affinity receptors is usually challenging. This is due to our evolving understanding of the underlying molecular recognitions and the interactions involved in these systems. As the use of peptides as receptor molecules is new, very limited informations are therefore available in the literature. However, research in to the use of peptides in sensors and diagnostics is taking two directions; use of the peptide receptors in solid phase column to replace antibodies for sample concentration and purification before diagnosis; the use of peptides as sensing receptors for analyte detection. Peptides can be acquired either from commercial sources since there are peptide libraries available commercially from the drug discovery industry such as the commercially available Food and Drug Administration (FDA) approved drugs and natural products; commercially available phage display libraries, or libraries can be designed specifically against a target analyte using computational methods. If a commercial peptide library is going to be used, then a high throughput screening method is required to select the best affinity compounds. This is a time consuming and expensive procedure but may result in good affinity receptors. Using the rational design approach is a more direct method and it is a time and cost effective procedure. As the library will be directed against the analyte, a high affinity and specificity receptors may result from this procedure.

Analytes or designed peptides which posses specific properties such as optical (e.g. fluorescence), or electrochemical can be used to develop label free detection system. Hence, detection can be achieved using conductometric, amperometric, optical or piezoelectric. Direct detection of inert peptides and analytes can also be realised by using conformational changes to the peptide upon biding with the analyte leading to a change in conductivity or surface potential. Design of signalling peptides and their use in sensors is also a novel area of discovery. Using amino acids in the building of the peptide receptor that are able to aid in the immobilisation of the receptor on the sensor surface or detection of the affinity interaction by tagging the peptide sequence, will facilitate the use of these peptides. The use of X-ray crystallographic screening (Nienaber et al. 2001; Blundell et al. 2002) and NMR-based screening (Hajduk et al. 1999; Diercks et al. 2001) have also facilitated the discovery of high affinity receptors such as peptides.

Although essential progress has been made in the field of peptides for drugs and other applications, such as cosmetics, very few practical examples of sensors based on peptides have been reported in the literature (Parker 2008). It is anticipated that with further success in designing of effective peptides this situation will change. The sensitivity and selectivity levels achieved must compete however, with antibodies to justify their practical application in the diagnostic field. Recently there has been many documented reports where peptides or proteins have been specifically designed with folding (Jones 1994; Jin et al. 1997), binding (Fontana et al. 1993; Forst et al. 1995) or cleavage (Schneider et al. 1995) properties. Generation of peptide ligands using combinatorial chemistry and applying them as receptors in diagnostic devices such as biosensors has been reported previously (Chen et al. 1998; Day 1999; Tothill 2003). Other peptides have been synthesised to react with molecules of interest such as; nucleic acids (Gutte et al. 1979), DDT (Moser et al. 1983), estradiol (Giraudi et al. 2000, 2003), estrogen (Tozzi et al. 2002) and Mycotoxins (Tozzi et al. 2003b; Parker 2008; Heurich 2008). Examples of peptides is the 13-residue core Fc-binding peptide sequence (Asp-Cys-Ala-Trp-His-Leu-Gly-Glu-Leu-Val-Trp-Cys-Thr) which has been found to inhibit the binding of Protein A to Fc with a K_i of 25 nM (DeLano et al. 2000). Other applications where peptides have been used are for material separation and purification. Poly (amino acids) such as poly (L-arginine), poly (L-lysine), poly (L-tyrosine) and poly (L-glutamic acid), have been used for functionalisation of microporous materials in order to separate the charged compounds and also for DNA purification (Hollman et al. 2004; Liu et al. 2001).

Valero et al. (2000) developed a peptide capable of binding to a monoclonal antibody of the foot-and-mouth disease virus. The peptide was based on the amino acids of the antigenic site of the foot-and-mouth disease virus. To adapt this approach to the work considered for small analytes would require a macromolecular (e.g. antibody) structure file in which the binding pocket had been identified. The use of peptides as ion channel mimics has been reported by Katayama et al. (2000), where a gold electrode was coated with a peptide and placed into an anionic solution. Upon the addition of the analyte of interest (cyclic AMP) the analyte binds to the peptide which in turns blocks the surface of the electrode to the anionic solution. This blocking action causes a change in the response on the electrode. Figure 6.6, shows the principle of this device. Tozzi et al. (2003b) used eight amino acids to create a combinatorial library with 64 different combinations. Each dipeptide was screened for the best binding properties towards the aflatoxin. To evolve the peptide further, the best binding dipeptide (LeuLeu) was added to each well on the micro titre plate and then the plate was filled again with each of the eight amino acids in row and columns to form another 64 different combinations this time

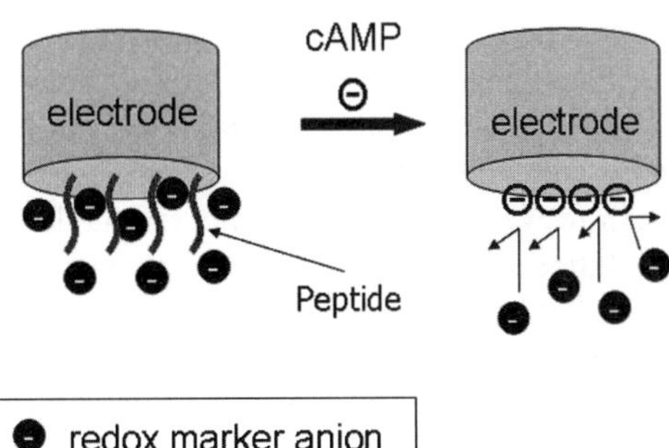

Fig. 6.6 Schematic diagram of using peptide as ion channel mimic (Katayama et al. 2000)

consisting of tetrapeptides. It was reported that the best tetrapeptide was LeuLeuAlaArg, which had a reported selectivity to be equal to the commercial antibody. Mascini et al. (2004), report on a five residue peptide optimised using the de novo program leapfrog to act as a receptor for the detection of dioxins. However, the authors used four residues of the peptide from a previous study (Kobayashi et al. 1999) to construct their peptide sequence. The four residue peptide was reported by Kobayashi et al. (1999) to have affinity for dioxins. Therefore, the sequence was not prepared from first principles by the leapfrog program. Additionally, the peptide developed was covalently attached at both ends, therefore not projecting into solution but activating as a reactive surface. Nakamura et al. (2005) developed a pentapeptide capable of binding to dioxin using combinatorial synthesis approach. The drawback of this technique is that it can generate millions of peptides, which have to go through the screening process. A high throughput method by which the peptide library produced can be screened, must be available. If the target compound is fluorescent, then this is less of a problem otherwise some other method has to be used to identify the peptide with the required affinity such as the use of SPR or QCM technology which will be highly time consuming.

At Cranfield University (UK) we have been pioneering the work on developing peptide receptors for small molecules using a combined approach of computational methods and combinatorial chemistry or peptide synthesis. The use of molecular modelling have been implemented in the design of receptors for anabolically active illegal androgens such as boldenone and stanozolol (El-Hajji et al. 2003a, b) and also for mycotoxins analysis (Parker 2008; Heurich 2008). This approach is a well-tested methodology for the design of receptor molecules modelled into binding pockets of macromolecules but has not been used for the design of relatively "small" peptide receptors for small molecular weight ligands previously. Heurich (2008) used computational approach to design a peptide receptor for Ochratoxin A (Fig. 6.7). A virtual library of functional monomers was first constructed using amino acids and then was screened against the target toxin. The highest scoring affinity peptides were then synthesised using peptide synthesis and tested using SPR (BIacore 3000). Two peptides showed good affinity towards the toxin. Full specificity studies are being conducted to evaluate the potential use of the peptides in sensor applications.

To date there is a wide spread interest in the scientific community in Europe to develop peptide receptors that can recognise small compounds. Many projects funded by the European Community are looking at this research area. Generating peptides against small compounds are being realised. The specificity and sensitivity of peptides as receptors are still being investigated to ensure that adequate informations are collected to make appropriate comparisons with antibodies. The question is, however, if today's computer modelling technology can simulate nature's evolutionary development of systems capable of distinguishing one molecule from a range of similar compounds. Therefore, further advances in computational technology is required to advance this technology and aid in producing peptides that are more specific and sensitive to the target compounds.

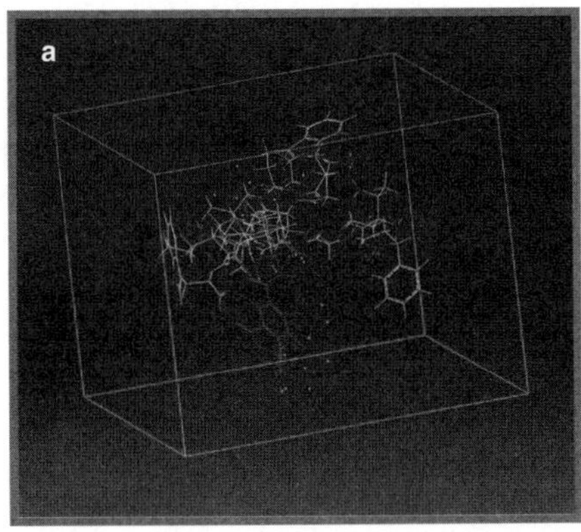
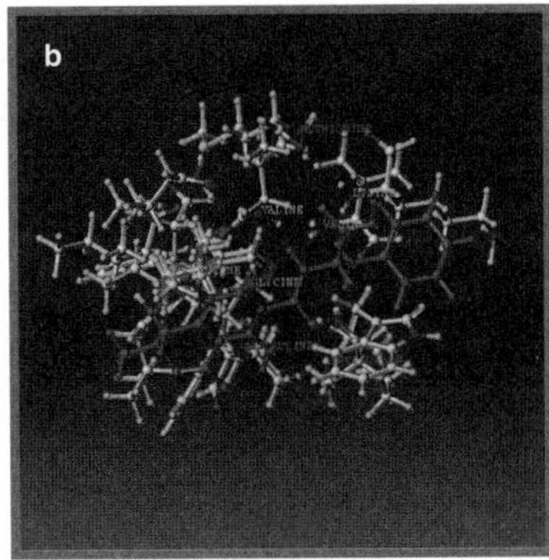

Fig. 6.7 (**a**) Electrostatic screening of the ochratoxin A template (seen in *purple*). *Coloured dots* are depicting the sites of interaction tried by the LeapFrog tool. (**b**) Ochratoxin A template (seen in *purple*) interacting with distinct amino acids (from left to right: isoleucine, valine, glycine, methionine, lysine and tryptophan) (Heurich 2008)

6.9 Future Perspective

Peptide based sensors will have their own niche for potential applications.

1. Extreme environment, (pH, pressure, temperature, organic solvents). Peptides have better stability than antibodies for extreme environment. However, this needs to be studied for each application as these factors can affect not only the stability of the peptide receptors but also their binding capabilities
2. On-line continuous monitoring
3. Sterilisable sensors for food and medical application
4. Space and military applications
5. Multisensory (microarrays and biochips)

These applications can cover the whole range of the biotechnology arena from medical, food to environmental application. The formation of arrays of sensors to diagnose one sample for several analytes is a very attractive approach. The demands for detection of a range of analytes are increasing. Also, their application in miniaturised sensors and microarrays are increasing. Using advanced fabrication technology such as photolithography, self-assembly and micro-control print offer routes to high density arrays. The drive force behind this comes from the high throughput screening program within the drug discovery businesses. The development of artificial receptors remains an important challenge which is to produce receptors that compete with the natural molecules with respect to sensitivity and stability.

The formulation of sensor arrays to diagnose one sample for several analytes is very attractive and therefore, microarray and nanoarray sensor systems are increasingly being favoured. The main issue here is the on-chip multi sensing and electronic signal amplification and processing. Achieving this also requires the use of small molecular weight receptors to be used as sensing layers to improve sensitivity and reduce steering hindrance and therefore peptides will be attractive for this application. The use of computational intelligence techniques in conjunction with these sensing devices such as Neural Networks (ANNs), mathematical modelling, chemometrics and data mining will help in moving the technology to the market.

6.10 Concluding Remarks

The need for new materials for use as stable sensor surfaces has resulted in a divers research activity in the developments of new receptor technology. By exploring material sciences and medical diagnostics, new and novel approaches have emerged to synthesis new artificial receptors. This is evident by the number of patents and publications in this area. The advantages of designing new receptors for a range of toxins in silico, are great, as many of these toxins have immunosuppressive, mutagenic and possibly carcinogenic effects and therefore being able to neither use animal resources nor handling them will be of benefit. However, there are still many challenges which need to be resolved. The range of molecular modelling programs available for drug discovery is increasing, but better understanding of small molecules interaction is needed. As this is still a new area of research, higher investments is needed to realise the potential of this technology.

References

Alberts IL, Todorov NP, Dean PM (2005) Receptor flexibility in de novo ligand design and docking. J Med Chem 48(21):6585–6596

Allinger NL (1977) Conformational analysis 130. MM2. A hydrocarbon force field utilizing V1 and V2 torsional terms. J Am Chem Soc 99:8127–8134

Allinger NL, Yuh YH, Lii JH (1989) Molecular mechanics. The MM3 force field for hydrocarbons I. J Am Chem Soc 111:8551–8566

Allinger NL, Li F, Yan L (1990) Molecular mechanics. The MM3 force field for alkenes. J Comput Chem 11:848–867

Allinger NL, Chen JA, Lii JH (1996) An improved force field (MM4) for saturated hydrocarbons. J Comput Chem 17:642–668

Antel J (1999) Integration of combinatorial chemistry and structure-based drug design. Curr Opin Drug Discov Devel 2:224–233

Apostolakis J, Caflisch A (1999) Computational ligand design. Comb Chem High Throughput Screen 2(2):91–104

Arap MA (2005) Phage display technology: applications and innovations. Genet Mol Biol 28:1–9 ISSN 1415-4757

Azzazy HM, Highsmith WE Jr (2002) Phage display technology: clinical applications and recent innovations. Clin Biochem 35:425–445

Barbas CF, Burton DR, Scott JK, Silverman GJ (2000) Phage display: a laboratory manual. Cold Spring Harbor Laboratory Press, Cold Spring Harbor, NY, pp 1–24

Berg JM, Tymoczko JL, Stryer L (2006) Biochemistry, 6th edn. W.H. Freeman and Co Ltd, New York, NY. ISBN ISBN 978-0716767664

Berman HM, Westbrook J, Feng Z, Gilliland G, Bhat TN, Weissig H, Shindyalov IN, Bourne PE (2000) The protein data bank. Nucleic Acids Res 28:235–242

Bertelli M, El-Bastawissy E, Knaggs MH, Barratt MP, Hanau S, Gilbert IH (2001) Selective inhibition of 6-phosphogluconate dehydrogenase from *Trypanosoma brucei*. J Comput Aided Mol Des 15:465–475

Blundell TL, Jhotti H, Abell C (2002) High-throughput crystallography for lead discovery in drug design. Nat Rev Drug Discov 1:45–54

Bohacek RS, McMartin C (1997) Modern computational chemistry and drug discovery: structure generating programs. Curr Opin Chem Biol 1:157–161

Böhm HJ (1992) The computer program Ludi: a new method for the de novo design of enzyme inhibitors. J Comput Aided Mol Des 6:61–78

Böhm HJ (1994) The development of a simple empirical scoring function to estimate the binding constant for a protein-ligand complex of known three-dimensional structure. J Comput Aided Mol Des 8:243

Böhm HJ (1995) Site-directed structure generation by fragment-joining. Perspect Drug Discov Des 3:21–33

Böhm HJ (1996) Towards the automatic design of synthetically accessible protein ligands: peptides, amides and peptidomimetics. J Comput Aided Mol Des 10:265–272

Böhm HJ, Klebe G (1996) What can we learn from molecular recognition in protein-ligand complexes for the design of new drugs? Angew Chem Int Ed Engl 35:2588–2614

Brady PA, Sanders JKM (1997a) Selection approaches to catalytic systems. Chem Soc Rev 26:326–337

Brady PA, Sanders JKM (1997b) J Chem Soc Perkin Trans 1:3237–3253

Brisig B, Sanders JKM, Otto S (2003) Angew Chem Int Ed 42(11):1270–1273

Bursavich MG, Rich DH (2002) Designing non-peptides peptidomimetics in the 21st century: inhibitors targeting conformational ensembles. J Med Chem 31:541–558

Calama MC, Hulst R, Fokkens R, Nibbering, NMM, Timmerman P, Reinhoudt DN (1998) Libraries of non-covalent hydrogen-bonded assemblies; combinatorial synthesis of supramolecular systems. Chem Commun 1021–1022

Chen B, Bestetti G, Turner APF (1998) The synthesis and screening of a combinatorial library for affinity ligands for glycosylated haemoglobin. Biosens Bioelectron 13:779–785

Chianella I, Lotierzo M, Piletsky S, Tothill IE, Chen B, Turner APF (2002) Rational design of a polymer specific for microcystin-LR using a computational approach. Anal Chem 74: 1288–1293

Chianella I, Karim K, Piletska E, Preston C, Piletsky SA (2006) Computational design and synthesis of molecularly imprinted polymers with high binding capacity for pharmaceutical applications-model case: adsorbent for abacavir. Anal Chim Acta 559:73–78

Christensen IT, Jorgensen FS (1997) Molecular mechanics calculations of proteins. Comparison of different energy minimization strategies. J Biomol Struct Dyn 15:473–488

Clark M, Cramer RD, Vanopdenbosh N (1989) Validation of the general-purpose Tripos 5.2 force-field. J Comput Chem 10:982–1012

Corbett PT, Leclaire J, Vial L, West KR, Wietor J-L, Sanders JKM, Otto S (2006) Dynamic combinatorial chemistry. Chem Rev 106:3652–3711

Cornell WD, Cieplak P, Bayly CI, Gould IR, Merz KM Jr, Ferguson DM, Spellmeyer DC, Fox T, Caldwell JW, Kollman PA (1995) A second generation force field for the simulation of proteins, nucleic acids and organic molecules. J Am Chem Soc 117:5179–5197

Cousins GRL, Poulsen S-A, Sanders JKM (1999) Dynamic combinatorial libraries of pseudo-peptide hydrazone macrocycles. Chem Commun 1575–1576

Cousins GRL, Poulsen S-A, Sanders JKM (2000) Molecular evolution: dynamic combinatorial libraries, autocatalytic networks and the quest for molecular function. Curr Opin Chem Biol 4:270–279

Cousins GRL, Furlan RLE, Ng YF, Redman JE, Sanders JKM (2001) Angew Chem Int Ed Engl 40:423–428
Crabtree RH (1999) Combinatorial and raid screening approaches to homogeneous catalyst discover and optimization. Chem Commun 1611–1616
Cramer RD (1983) Computer graphics in drug design. Pharm Int 106–107
Day R (1999) The development of a synthetic receptor specific to glycosylated haemoglobine for biosensing application. PhD thesis, Cranfield University
DeLano WL, Ultsch MH, de Vos AM, Wells JA (2000) Convergent solutions to binding at a protein-protein interface. Science 287:1279–1283
Diercks T, Coles M, Kessler H (2001) Applications of NMR in drug discover. Curr Opin Chem Biol 5:285–291
Eguchi M, Kahn M (2002) Design, synthesis and application of peptide secondary structure mimetics. Mini Rev Med Chem 2:447–462
El-Hajji S, Gannon G, Chen B, Tothill IE (2003a) Development of artificial receptor for androgen detection, affinity interactions 2003. 15th international conference, 27 July–1 August, Cambridge, UK
El-Hajji S, Gannon G, Chen B, Tothill IE (2003b) Artificial receptor development for androgen residues. Synthetic receptor conference, 15–17 October, Lisbon, Portugal
Fontana W, Stadler PF, Tarazona P, Weinberger ED, Schuster P (1993) Phys Rev E 47:2083–2099
Forst CV, Reidys C, Weber J (1995) In: Moran F (ed) Advances in artificial life, vol 929. Springer, Berlin, pp 3628–4147
Furlan RLE, Cousins GRL, Sanders JKM (2000) Chem Commun 18:1761–1762
Furlan RLE, Ng YF, Otto S, Sanders JKM (2001) J Am Chem Soc 123(36):8876–8877
Ganesan A (1998) Strategies for the dynamic integration of combinatorial synthesis and screening. Angew Chem Int Ed Engl 37:2828–2831
Geysen HM, Rodda SJ, Mason TJ (1986) A priori delineation of a peptide which mimics a discontinuous antigenic determinant. Mol Immunol 23:709–715
Giraudi G, Giovannoli C, Tozzi C, Baggiani C, Anfossi L (2000) Estradiol binding synthetic polypeptides. Chem Commun 13:1135–1136
Giraudi G, Giovannoli C, Tozzi C, Baggiani C, Anfossi L (2003) Molecular recognition properties of peptide mixtures obtained by polymerisation of amino acids in the presence of estradiol. Anal Chim Acta 481:41–53
Goodford PJ (1985) A computational procedure for determining energetically favourable binding sites on biologically important macromolecules. J Med Chem 28:849–857
Goodman JM (1998) Chemical applications of molecular modelling. Royal Society of Chemistry Press, London. ISBN ISBN 0-85404-579-1
Gunsteren WF, Dolenc J, Mark AE (2008) Molecular simulation as an aid to experimentalists. Curr Opin Struct Biol 18(2):149–153
Gutte B, Däumigen M, Wittschieber E (1979) Design, synthesis and characterisation of a 34-residue polypeptide that interacts with nucleic acids. Nature 281:650–655
Hagiwara D, Miyake H, Murano K, Morimoto H, Murai M, Fujii T, Nakanishi I, Matsuo M (1993) Studies on neurokinin antagonist. 3, Design and structure-activity relationships of new branched tripeptide N^a-(substituted L-apartyl, L-ornithyl, or L-lysyl)-N-methyl-N-(phenylmethyl)-L-phenylalaninamides as substance P antagonist. J Med Chem 36:2266–2278
Hajduk PJ, Meadows RP, Fesik SW (1999) NMR-based screening in drug discovery. Q Rev Biophys 32:211–240
Hamilton DG, Feeder N, Teat SJ, Sanders JKM (1998) Reversible synthesis of π-associated [2] catenanes by ring-closing metathesis: towards dynamic combinatorial libraries of catenanes. New J Chem 22:1019–1021
Heurich M (2008) Development of an affinity sensor for ochratoxin A. PhD thesis, Cranfield University, England, UK
Hioki H, Still WC (1998) Chemical evolution: a model system that selects and amplifies a receptor for the tripeptide (d)Pro(L)Val(D)Val. J Org Chem 63(4):904–905

Hochgurtel M, Kroth H, Piecha D, Hofmann MW, Nicolau C, Krause S, Schaaf O, Sonnenmoser G, Eliseev AV (2002) Proc Natl Acad Sci USA 99(6):3382–3387

Hochgurtel M, Biesinger R, Kroth H, Piecha D, Hofmann MW, Krause S, Schaaf O, Nicolau C, Eliseev AV (2003) J Med Chem 46(3):356–358

Hollman AM, Scherrer NT, Cammers-Goodwin A, Bhattacharyya D (2004) Separation of dilute electrolytes in poly(amino acid) functionalized microporous membranes: model evaluation and experimental results. J Memb Sci 239:65–79

Höltje HD, Sippl W, Rognan D, Folkers G (2008) Molecular modeling: basic principle and application, 3rd edn. Wiley-VCH, Weinheim, p 310. ISBN: 978-3-527-31568-0

Huc I, Lehn JM (1997) Virtual combinatorial libraries: dynamic generation of molecular and supramolecular diversity by self-assembly. Proc Natl Acad Sci USA 94:2106–2110

Huc I, Nguyen R (2001) Dynamic combinatorial chemistry. Comb Chem High Throughput Screen 4:53–74

Jin AY, Leung FY, Weaver DFJ (1997) J Comput Chem 18:1971–1984

Jones DT (1994) Protein Sci 3:567–574

Jones G, Willett P (1995) Docking small-molecule ligands into active sites. Curr Opin Biotechnol 6:652–656

Karan C, Miller BL (2001) J Am Chem Soc 123:7455–7456

Katayama Y, Ohuchi Y, Higashi H, Kudo Y, Maeda M (2000) The design of cyclic AMP-recognizing oligopeptides and evaluation of its capability for cyclic AMP recognition using and electrochemical system. Anal Chem 72:4671–4674

Kay BK, Adey NB, He YS, Manfredi JP, Mataragnon AH, Fowlkes DM (1993) An M13 phage library displaying random 38-amino-acid peptides as a source of novel sequences with affinity to selected targets. Gene 128:59–65

Kempe M (1996) Antibody-mimicking polymers as chiral stationary phases in HPLC. Anal Chem 68:1948–1953

Klekota B, Miller BJ (1999) Dynamic diversity and small molecule evolution: a new paradigm for ligand identification. Trends Biotechnol 17:205–209

Kobayashi S, Kitadai M, Sameshima K, Ishii Y, Tanaka A (1999) A theoretical investigation of the conformational changing of dioxins in the binging site of dioxin receptor model; role of absolute hardness – electronegativity activity diagrams for biological activity. J Mol Struct 475:203–217

Koivunen E, Arap W, Rajotte D, Lahdenranta J, Pasqualini R (1999) Identification of receptor ligands with phage display peptide libraries. J Nucl Med 40:883–888

Kollman P (1993) Free energy calculations-applications to chemical and biological phenomena. Chem Rev 93:2395–2417

Krasinski A, Radic Z, Manetsch R, Raushel J, Taylor P, Sharpless KB, Kolb HC (2005) J Am Chem Soc 127:6686–6692

Krovat EM, Steindl T, Langer T (2005) Recent advances in docking and scoring. Curr Comput Aided Drug Des 1:93–102

Kubota Y, Sakamoto S, Yamaguchi K, Fujita M (2002) Proc Natl Acad Sci USA 99:4854–4856

Labanowski J, Motoc I, Naylor CB, Mayer D, Dammkoehler RA (1986) Three-dimensional quantitative structure-activity relationships. 2. Conformational mimicry and topographical similarity of flexible molecules. Quant Struct-Act Rel 5:138–152

Lam KS, Renil M (2002) From combinatorial chemistry to chemical microarray. Curr Opin Chem Biol 6:353–358

Lam KS, Salmon SE, Hersh EM, Hruby VJ, Kazmierski WM, Knapp RJ (1991) A new type of synthetic peptide library for identifying ligand binding activity. Nature 354:82–84

Lavastre O, Morkrn JP (1999) Discovery of novel catalysts for allylic alkylation with a visual colorimetric assay. Angew Chem Int Ed Engl 38:3163–3165

Lehn J-M (1999) Dynamic combinatorial chemistry and virtual combinatorial libraries. Chem Eur J 5:2455–2463

Liu G, Molas M, Grossmann GA, Pasumarthy M, Perales JC, Cooper MJ, Hanson RW (2001) Biological properties of poly-L-lysine-DNA complexes generated by cooperative binding of polycation. J Biol Chem 276:34379–34387

Lotierzo M, Henry OYF, Piletsky S, Tothill I, Cullen D, Kania M, Hock B, Turner APF (2004) Surface plasmon resonance sensor for domoic acid based on grafted imprinted polymer. Biosens Bioelectron 20:145–152

Lybrand TP (1995) Ligand-protein docking and rational drug design. Curr Opin Struct Biol 5:224–228

Ma JS (2003) Unnatural amino acids in drug discovery. Chim Oggi 21:65–68

Manetsch R, Krasinski A, Radic Z, Raushel J, Taylor P, Sharpless KB, Kolb HC (2004) J Am Chem Soc 126:12809–12818

Mascini M, Macagnano A, Monti D, Del Carlo M, Paolesse R, Chen B, Warner P, D'Amico A, Di Natale C, Compagnone D (2004) Piezoelectric sensors for dioxins: a biomimetic approach. Biosens Bioelectron 20:1203–1210

Moon JB, Howe WJ (1991) Computer design of bioactive molecules: a method for receptor-based de novo ligand design. Proteins 11:314–328

Mosbach K (1994) Molecular imprinting. Trends Biochem Sci 19:9–14

Mosbach K, Ramstrom O (1996) The emerging technique of molecular imprinting and its future impact on biotechnology. Biotechnology 14:165–170

Moser R, Thomas RM, Gutte B (1983) An artificial crystalline DDT-binding polypeptide. FEBS Lett 157:247–251

Muegge I, Rarey M (2001) Small molecule docking and scoring. Rev Comput Chem 17:1–60

Nakamura C, Inuyama Y, Shirai K, Sugimoto N, Miyake J (2001) Detection of porphyrin using a short peptide immobilised on a surface plasmon resonance sensor chip. Biosens Bioelectron 16:1095–1100

Nakamura C, Inuyama Y, Goto H, Obataya I, Kaneko N, Nakamura N, Santo N, Miyake J (2005) Dioxin-binding pentapeptide for use in a high-sensitivity on-bead detection assay. Anal Chem 77:7750–7757

Nienaber VL, Richardson PL, Klighofer V, Bouska JJ, Giranda VL, Greer J (2001) Discovering novel ligands for macro-molecules using X-ray crystallographic screening. Nat Biotechnol 18:1105–1108

Noda K, Yamasaki R, Hironaka Y, Kitagawa A (2001) Selection of peptides that bind to the core oligosaccharide of R-form LPS from a phage-displayed heptapeptide library. FEMS Microbiol Lett 205:349–354

O'Neil KT, Hoess RH (1995) Phage display: protein engineering by directed evolution. Curr Opin Struct Biol 4:443–449

Otto S, Kubik S (2003) J Am Chem Soc 125:7804–7805

Parker C (2008) Development of an affinity sensor for the detection of aflatoxin m1 in milk. PhD thesis, Cranfield University, England, UK

Patel S, Stott IP, Bhakoo M, Elliot P (1998) Patenting computer-designed peptides. J Comput Aided Mol Des 12:543–556

Payne AWR, Glen RC (1993) Molecular recognition using a binary genetic search algorithm. J Mol Graph 11:74–91

Piletsky S, Karim K, Piletska EV, Day CJ, Freebairn KW, Legge C, Turner APF (2001) Recognition of ephedrine enentiomers by molecular imprinted polymers designed using a computational approach. Analyst 126:1826–1830

Piletsky S, Piletska EV, Sergeyeva TA, Nicholls I, Weston D, Turner A (2006a) Synthesis of biologically active molecules by imprinting polymerization. Biopolymers Cell 22:63–68

Piletsky S, Turner NW, Laitenberger P (2006b) Molecularly imprinted polymers in clinical diagnostics – future potential and existing problems. Med Eng Phys 28:971–977

Poulsen S-A, Gates P, Cousins GRL, Sanders JKM (2000) Electrospray ionization Fourier transform ion cyclotron resonance mass spectrometry of dynamic combinatorial libraries. Rapid Commun Mass Spectrom 14:44–48

Rachkov A, Minoura NJ (2000) Recognition of oxytocin and oxytocin-related peptides in aqueous media using a molecularly imprinted polymer synthesized by the epitope approach. J Chromatogr A 889:111–118

Rich DH (2002) Discovery of nonpeptide, peptidomimetic peptidase inhibitors that target alternate enzyme active site conformations. Biopolymers 66:115–125

Richon AB (1994) An introduction to molecular modelling. Mathematech 1:83

Roberts SL, Furlan RLE, Otto S, Sanders JKM (2003) Org Biomol Chem 1(9):1625–1633

Rustici M, Bracci L, Lozzi L, Neri P, Santucci A, Soldani P, Spreafico A, Niccolai N (1993) A model of the rabies virus glycoprotein active site. Biopolymers 33:961–969

Sadler K, Tam JP (2002) Peptide dendrimers: applications and synthesis. J Biotechnol 90:195–229

Saur I, Severin K (2005) Chem Commun 11:1471–1473

Sawyer TK (1999) Peptidomimetic and nonpeptide drug discovery: chemical nature and biological targets. In: Reid R (ed) Drugs and the pharmaceutical sciences, vol 101. Marcel Dekker, New York, NY, pp 81–114

Sawyer TK, Chorev M (2003) Peptide revolution: genomics, proteomics and therapeutics. Biotechiques 34(3):594–596, 598–599

Schmuck C (2001) Von der molekularen erkennung zum design neuer wirkstoffe. Chem Unserer Zeit 6:356–366

Schneider G, Böhm HJ (2002) Virtual screening and fast automated docking methods. Drug Discov Today 7:64–70

Schneider G, Schuchhardt J, Wrede P (1995) Biol Cybern 73:3

Selassie CD (2003) History of quantitative structure-activity relationships. In: Abraham DJ (ed) Burgers medicinal chemistry and drug discovery, vol 1. Wiley, New York, NY, pp 1–48

Smith GP (1985) Filamentous fusion phage: novel expression vectors that display cloned antigens on the virion surface. Science 228:1315–1317

Smith GP (1991) Surface presentation of protein epitopes using bacteriophage expression systems. Curr Opin Biotechnol 2:668–673

Smith GP, Scott JK (1993) Libraries of peptides and proteins displayed on filamentous phage. Methods Enzymol 217:228–257

Steinbach PJ (2005) Introduction to macromolecular simulation. Center for molecular modeling, center for information technology, National Institutes of Health. http://cmm.info.nih.gov/modeling/. Accessed 12, Nov. 2009

Steinbach PJ, Brooks BR (1994) Protein simulation below the glass-transition temperature: dependence on cooling protocol. Chem Phys Lett 226:447

Subrahmanyam S, Piletsky SA, Piletska EV, Karim K, Chen B, Day R, Turner APF (2001) Bite-and-switch' approach using computationally designed molecularly imprinted polymers for sensing of creatinine. Biosens Bioelectron 16:631–637

Sundaram R, Dakappagari NK, Kaumaya PT (2002) Synthetic peptides as cancer vaccines. Biopolymers 66:200–216

Swann PG, Casanova RA, Desai A, Frauenhoff MM, Urbancic M, Slomczynska U, Hopfinger A, Breton GC, Venton D (1997) Nonspecific protease-catalysed hydrolysis/synthesis of a mixture of peptides: product diversity and ligand amplification by a molecular trap. Biopolymers 40:617–625

Szardenings M (2003) Phage display of random peptide libraries: applications, limits, and potential. J Recept Signal Transduct Res 23:307–349

Takeuchi T, Matsui J (1998) Recognition of drugs and herbicides: strategy in selection of functional monomers for noncovalent molecular imprinting. ACS Symp Ser 703:119–134

Toko K (2001) Biomimetic sensor technology. Meas Sci Technol 12:221

Tothill IE (2001) Biosensors developments and potential applications in the agricultural diagnosis sector. Comput Electron Agr 30:205–218

Tothill IE (2003) On-line immunochemical assays for contaminants analysis. In: Tothill IE (ed) Rapid and on-line instrumentation for food quality assurance. Woodhead, Cambridge, ISBN: 1-85573-674-8

Tothill IE, Turner APF (1998) Biosensors: new developments and opportunities in the diagnosis of livestock diseases. Towards livestock disease diagnosis and control in the 21st century. International Atomic Energy Agency, pp 79–94

Tothill IE, Piletsky S, Magan N, Turner APF (2001) New biosensors. In: Kress-Rogers E, Brimelow CJB (eds) Instrumentation and sensors for the food industry, 2nd edn. Woodhead and CRC Press LLC, Cambridge, p 836

Tozzi C, Anfossi L, Giraudi G, Giovannoli C, Baggiani C, Vanni A (2002) Chromatographic characterisation of an estrogen-binding affinity column containing tetrapeptides selected by a combinatorial-binding approach. J Chromatogr A 966:71–79

Tozzi C, Anfossi L, Giovannoli C (2003a) Affinity chromatography techniques based on the immobilisation of peptides exhibiting specific binding activity. J Chromatogr B 797:289–304

Tozzi C, Anfossi L, Baggiani C, Giovannoli C, Giraudi G (2003b) A combinatorial approach to obtain affinity media with biding properties towards the aflatoxins. Anal Bioanal Chem 375:994–999

Tsai SC (2002) An introduction to computational biochemistry. Wiley, New York, NY

Vaidya AA, Lele BS, Kulkami MG, Mashelkar RA (2001) Creating a macromolecular receptor by affinity imprinting. J Appl Polym Sci 81:1075–1083

Valero M, Camarero JA, Haack T, Mateu MG, Domingo E, Giralt E, Andreu D (2000) Native-like cyclic peptide models of a viral antigenic site: finding a balance between rigidity and flexibility. J Mol Recognit 13:5–13

Vedani A, Dobler M (2002) 5d qsar: the key for simulating induced fit? J Med Chem 45:2139–2149

Vedani A, Dobler MQ (2003) Quantification of wide-range ligand binding to the estrogen receptor – a combination of receptor-mediated alignment and 5d QSAR. Helv Chim Acta 45:2139–2149

Wang Q, Chan TR, Hilgraf R, Fokin VV, Sharpless KB, Finn MG (2003a) J Am Chem Soc 125:3192–3193

Wang S, Humphreys ES, Chung SY, Delduco DF, Lustig SR, Wang H, Parker KN, Rizzo NW, Subramoney S, Chiang YM, Jagota A (2003b) Peptides with selective affinity for carbon nanotubes. Nat Mater 2:196–2000

Webster R (2001) Filamentous phage biology. In: Phage Barbas CF, Burton DR, Scott JK, Silverman GJ (eds) Phage display: a laboratory manual. Cold Spring Harbor Laboratory Press, Cold Spring Harbor, NY, pp 1.1–1.37

Welling GW, Geurts T, van Gorkum J, Damhof RA, Drijfhout J (1990) Synthetic antibody fragment as ligand in immunoaffinity chromatography. J Chromatogr A 512:337–343

Wlodawer A, Vondrasek J (1998) Inhibitors of HIV-1 protease: a major success of structure-assisted drug design. Annu Rev Biophys Biomol Struct 27:249–284

Yamanoi Y, Sakamoto Y, Kusukawa T, Fujita M, Sakamoto S, Tamaguchi K (2001) J Am Chem Soc 123(5):980–981

Yoshizawa M, Nagao M, Umemoto K, Biradha K, Fujita M, Sakamoto S, Yamaguchi K (2003) Chem Commun 15:1808–1809

Zameo S, Vauzeilles B, Beau JM (2005) Angew Chem Int Ed 44(6):965–969

Zeger ND, Boersma WJA, Claassen E (eds) (1995) Immunological recognition of peptides in medicine and biology. CRC Press, Boca Raton, FL, p 297. ISBN ISBN 0849389674

Chapter 7
Carbohydrates as Recognition Receptors in Biosensing Applications

Yann Chevolot, Sébastien Vidal, Emmanuelle Laurenceau, François Morvan, Jean-Jacques Vasseur, and Eliane Souteyrand

Abstract Carbohydrates are involved in crucial physiological and pathological events. One can take advantage of carbohydrate-based interaction for drug discovery, diagnosis, antibiotics, vaccine, etc. This chapter deals with biosensors and microarrays that take advantage of carbohydrates-based interactions with a special interest in devices that are designed for medical applications. A large overview of glycochemistry, followed by the biological role of carbohydrates, is given. Carbohydrate-based biosensors are then described with special emphasis on surface chemistry and signal transduction. Finally, medically relevant applications illustrate the use of carbohydrates as recognition receptors in biosensing.

Keywords Glycoarray · Carbohydrate · Glycomics · Microarray · Biosensors

Abbreviations

Ab	Antibody
Ac	Acetyl
AIDS	Acquired immune deficiency syndrome or acquired immunodeficiency syndrome
AZT	Azidothymidine or Zidovudine
BAW	Bulk Acoustic Wave
bFGF	Basic fibroblast growth factor
BSA	Bovine Serum Albumin

Y. Chevolot (✉)
Institut des Nanotechnologie de Lyon UMR CNRS 5270, Université de Lyon, Ecole Centrale de Lyon, 36 Avenue Guy de Collongue, 69134 Ecully Cedex, France
e-mail: Yann.Chevolot@ec-lyon.fr

M. Zourob (ed.), *Recognition Receptors in Biosensors*,
DOI 10.1007/978-1-4419-0919-0_7,

Bz	Benzoyl
CD4	Cluster of differentiation 4
CD44	Cluster of differentiation 44
CFG	Consortium for functional glycomics
CMP-Neu5Ac	Cytidine monophosphate-N-acetylneuraminic-acid
CNS	Central nervous system
ConA	Concanavalin A
CSPG	Chondroitin-sulfate proteoglycan
Cy3	Indodicarbocyanine 3
Cy5	Indodicarbocyanine 3
DC-SIGN	Dendritic cell-specific intercellular adhesion molecule-3-Grabbing Non-integrin
DDI	DNA directed immobilization
DNA	Deoxyribonucleic acid
DPI	Dual polarisation interferometry
DV	Dengue virus
EC	*Erythrina crista-galli* lectin
ENFET	Enzymatic field effect transistor
EQCM	Electrochemical quartz crystal microbalance
FDA	Food and drug administration
FET	Field effect transistor
FITC	Fluorescein isothiocyanate
GAG	Glycosaminoglycans
GIPL	Glycoinositol phospholipid
GlcNAc	*N*-Acetylglucosamine
GPI	Glycosylphosphatidylinositol
GSLs	Glycosphingolipids
HA	Hemagglutinin
HIV	Human immunodeficiency virus
HIV-RT	Human immunodeficiency virus reverse transcriptase
HSA	Human serum albumin
HSPG	Heparan-sulfate proteoglycan
ICAM	Inter-cellular adhesion molecule
ISFET	Ion sensitive field effect transistor
MALDI-ToF	Matrix-assisted laser desorption/ionization time of flight
MOSFET	Metal oxide semiconductor field effect transistor
MR	Mannose receptor
MRI	Magnetic resonance imaging
MS	Mass spectroscopy
MTT	3-(4,5-dimethylthiazol-2-yl)-2,5-diphenyltetrazolium bromide
NA	Neuramidase
NG2	Nerve/Glial antigen 2
NHS	*N*-Hydroxysuccinimide
NIS	*N*-iodosuccinimide

NMR	Nuclear magnetic resonance
NPC	Neuronal precursor cells
NSC	Neuronal stem cells
PCR	Polymerase chain reaction
PECAM	Platelet Endothelial cell adhesion molecule
PEG	Poly(ethylene glycol)
QCM	quartz crystal microbalance
QCM-D	quartz crystal microbalance with dissipation monitoring
QD	Quantum dot
RNA	Ribonucleic acid
SAM	Self assembled monolayer
SARS	Severe acute respiratory syndrome
SDS	Sodium dodecyl sulfate
SLe	Sialyl lewis
SPR	Surface plasmon resonance
Tf	Trifluoromethanesulphonate
TMS	Trimethylsilyl
TSM	Thickness shear mode
VCAM	Vascular cell adhesion molecule
WGA	Wheat-germ agglutinin

7.1 Introduction

Glycosylation of proteins is a common post-translational modification, and eukaryotic cells are coated with complex carbohydrates associated with lipids or proteins forming the so-called glycocalix. Carbohydrates are the most abundant type of biomolecules on earth in term of biomass. Their biological role spans from the stabilisation of protein structure to their involvement in crucial biological events such as the development, growth or survival of a cell/organism (Varki 1993; Sharon and Lis 2004; Varki et al. 1999; Dwek 1996; Bertozzi and Kiessling 2001; Bishop et al. 2007). For instance, protein–carbohydrate interactions are involved in intracellular trafficking of proteins, cell adhesion, cell recognition, cell differentiation and the development of the neuronal network. They are also involved in pathological events such as inflammation, metastasis (Galonic and Gin, 2007) and host–pathogen interactions (Imberty et al. 2004; Seeberger and Werz 2007). For example, influenza virus takes advantage of the carbohydrate motifs present on the surface of the respiratory track endothelial cells for adhering to these cells (Skehel and Wiley 2000). The switch from avian influenza to human influenza sub-type is related to the subtle change of its ability to recognise the glycosylated motif of human respiratory track endothelial cell instead of the avian ones. The elucidation of these interactions is a critical issue not only in terms of fundamental research but also in terms of applied research for the development of diagnosis or new therapeutics.

Biophysical approaches such as calorimetry (Dam and Brewer 2002), nuclear magnetic resonance (Sandström et al. 2004) and crystallography (Berhe et al. 1999) have been successfully employed and have brought many valuable pieces of information. However, these techniques are not straightforward as well as are time- and material-consuming.

When dealing with carbohydrates, several limitations arise because of their chemical nature. Firstly, carbohydrate structures are extremely diverse. It has been calculated that a hexamer can lead to more than 10^{12} structures (Laine 1994). Secondly, carbohydrate synthesis is one of the most difficult and laborious synthetic work, and carbohydrates cannot be amplified enzymatically like DNA (by polymerase chain reaction) or cloned such as proteins. The consequence is that carbohydrate material is difficult to obtain in large amounts at affordable prices. Finally, carbohydrate interactions, especially with proteins, tend generally to be weak unless a multivalent interaction is involved, taking advantage of the so-called cluster effect (Lee and Lee 1995). Consequently, analysis of carbohydrate-mediated interactions requires analytical techniques with very low detection limit.

In this context, biosensors and microarrays have several advantages. They require very little amounts of biological material. In the case of microarrays, high-throughput analysis can be performed and they usually have very low detection limits (Feizi et al. 2003; Wang 2003; Jelinek and Kolusheva 2004). Almost all biosensors are based on a biological binding/recognition element (ligand) that performs a specific binding or biochemical reaction with a target. This interaction is then converted into a signal through a transducer.

The requirements for modern biosensors are sensitivity, selectivity and reproducibility. These parameters depend on a combination of many factors among which surface chemistry, ligand immobilisation, target physico-chemical properties and signal transduction are key factors.

This chapter concerns carbohydrate based biosensor in a broad sense, that is devices comprising bound bioactive carbohydrate ligands responsible for a specific interaction at the surface of a material. Here, the ligand considered will be a saccharide or an array of saccharides and more precisely mono or oligosaccharides.

At this stage, it is worth giving a few definitions. In the following, a probe is a molecule (a biomolecule) that is immobilised on the surface of a material in order to interact with a target. The target is the biological molecule to be detected in a biological fluid.

In order to avoid confusion, a biosensor is a device that can perform directly the detection of a biochemical interaction. This means that the biological interaction is measured in real time (Rasooly 2005). Surface plasmon resonance (SPR) or electrochemical detection belong to this class of biosensors. Another class of biosensors relies on indirect detection such as fluorescence detection (Rasooly 2005).

A biochip or microarray is a special class of biosensor as it allows for the simultaneous detection of multiple interactions. Common to both cases, the ligands are immobilised on the surface of a material (plane or particles) that can be active as a transducer (electrochemically active or optically active for SPR) or inert (glass in the case of fluorescence).

In this chapter, we are concerned with carbohydrate-based biosensing. Therefore, we will first give an overview on the physico-chemical properties of carbohydrates, as these are critical for obtaining and designing the probes. We will then focus on their biological role. We will then give an overview of carbohydrate-based biosensing from their immobilisation, as this is critical for obtaining reliable devices and to the signal transducer. Finally, we will give some applications described in the literature.

7.2 General Aspects of Glycochemistry

7.2.1 Introduction

Carbohydrates were named after their composition as *hydrates of carbon,* but they can be also called saccharides from greek σάκχαρον meaning sugar. They are polyfunctional biomolecules with the general formula $C_n(H_2O)_n$ (e.g. $n = 6$ for glucose). The biological aspects of carbohydrates have long been underestimated because of their complexity in comparison to the three other classes of biomolecules. Polypeptides (proteins) and polynucleotides (e.g. DNA or RNA) are linear polymers, whereas carbohydrates are either linear or branched polymers, thereby increasing drastically the number of possible combinations (Laine 1994; Werz et al. 2007). Recent advances in glycobiology have demonstrated the numerous roles of carbohydrates in a series of biological processes such as storage and transport of energy, signal transduction, fertilisation, immune responses, pathogenesis, blood clotting, virus infection and cell-to-cell communication (Bertozzi and Kiessling 2001; Sharon and Lis 2004; Dwek 1996; Varki 1993; Varki et al. 1999). These fundamental discoveries found several applications in the design of carbohydrate-based drugs and vaccines against various diseases (Werz and Seeberger 2005b) such as malaria (Hölemann and Seeberger, 2006; Seeberger and Werz 2007), cancer (Galonic and Gin, 2007; Livingston and Ragupathi 2006) and AIDS (McReynolds and Gervay-Hague, 2007; Scanlan et al. 2007). The ever-increasing number of chemists, biochemists and biologists who are involved in research with carbohydrates is testimony to the fact that there is still a lot more work to be achieved in glycochemistry and glycobiology.

7.2.2 Structural Aspects and Chemistry of Carbohydrates

Monosaccharides are classified according to the number of carbon atoms, which ranges from 4 (tetroses), 5 (pentoses), 6 (hexoses) and up to 10 atoms. Aldoses (Fig. 7.1) and ketoses (Fig. 7.2) are polyhydroxylated aldehydes or ketones, respectively, which can cyclise intramolecularly into hemiacetals. The nucleophilic attack of a hydroxyl group on the carbonyl will provide two diastereoisomeric hemiacetals

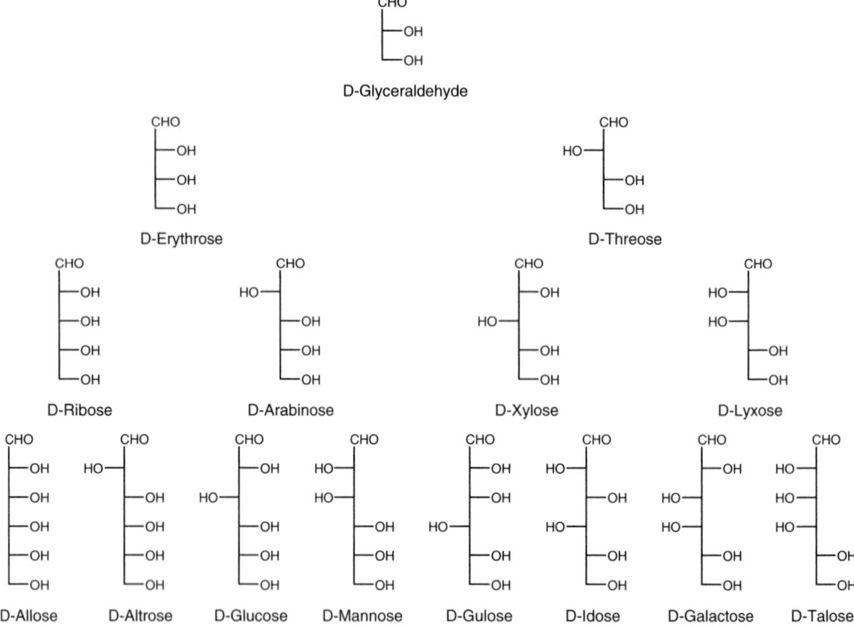

Fig. 7.1 Structures of acyclic forms of D-aldoses

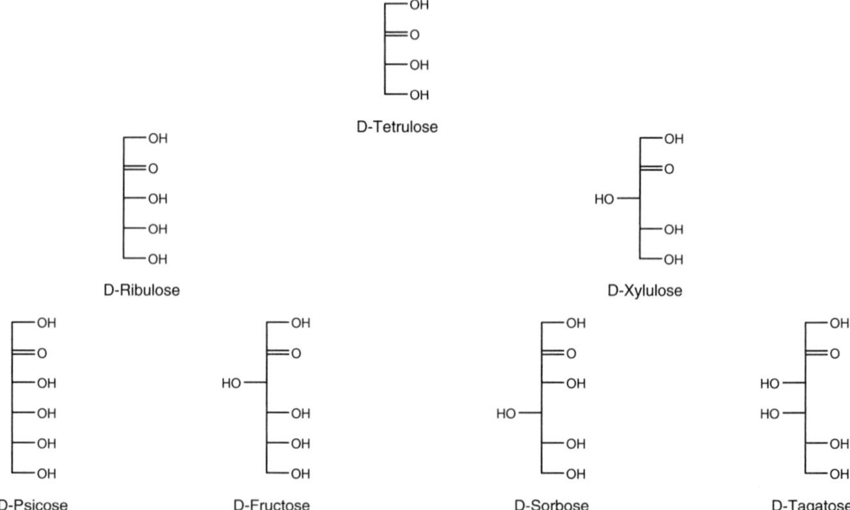

Fig. 7.2 Structures of acyclic forms of D-ketoses

called α- and β-anomers. Cyclisation can occur between O-4 and O-5 to form a five-membered ring designated as furanose or a six-membered pyranose ring, respectively (Fig. 7.3). D-Aldoses are derived from D-glyceraldehyde and their

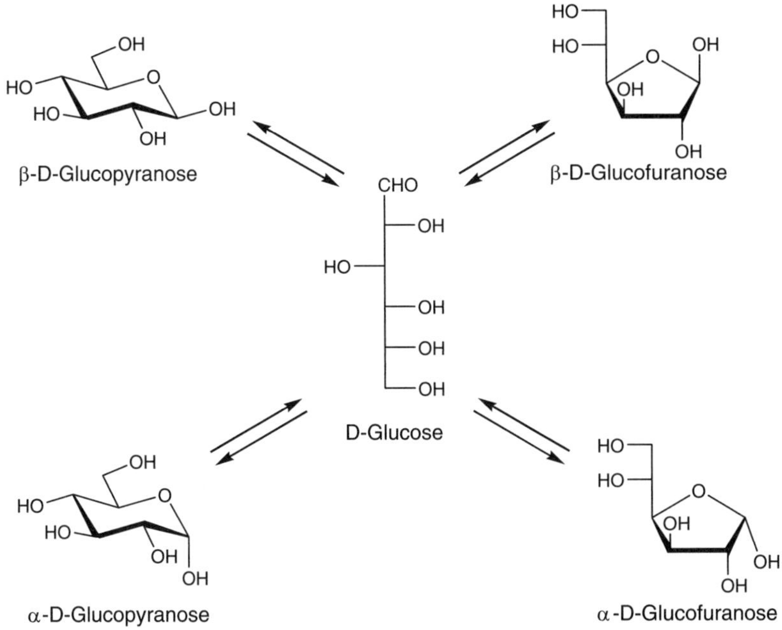

Fig. 7.3 Cyclic and acyclic forms of D-glucose

enantiomers (L-aldoses) from L-glyceraldehyde. Atom numbering starts at the most oxidised end of the carbon chain (e.g. carbonyl group for aldoses) and increases along the carbon chain.

The cyclic structures of monosaccharides can be represented according to several projections (Fig. 7.4). The Fischer projection was the first one to be applied to acyclic forms of sugars, where the carbon chain is drawn vertically with the most oxidised carbon atom (carbonyl) at the top. Later, Haworth designed a projection for a more realistic description of cyclic forms of carbohydrates. The chair conformation provides the most accurate description of the geometry of the carbohydrate ring and should be preferred to other projections. Mills and zig-zag projections are useful when considering the stereochemistry of each carbon atom along the chain.

7.2.2.1 Mutarotation

Carbohydrates are single stereoisomers in their crystalline form. Nevertheless, optical rotation of a single anomer in aqueous solution slowly changes with time to reach a constant value. Mutarotation (Pigman and Isbell 1968; Isbell and Pigman 1969) refers to this equilibration process where an α-anomer would equilibrate to a mixture of α- and β-anomers through the ring-opened aldehyde form and re-cyclisation into its opposite anomer (Fig. 7.5). This equilibrium is catalysed by acids or bases. Carbohydrates are therefore present as a mixture of acyclic and

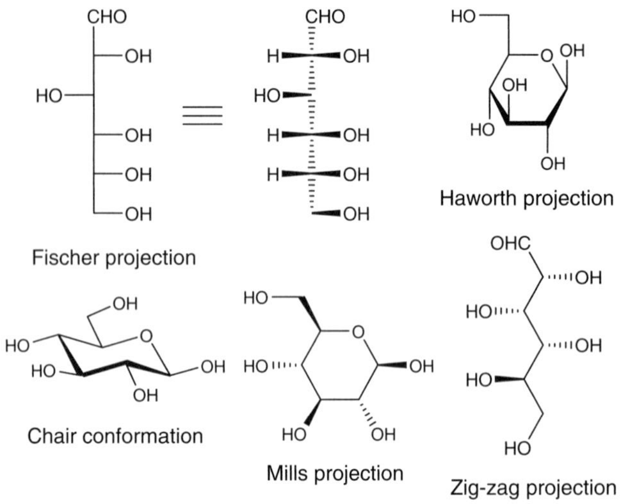

Fig. 7.4 Projections used for the description of carbohydrates

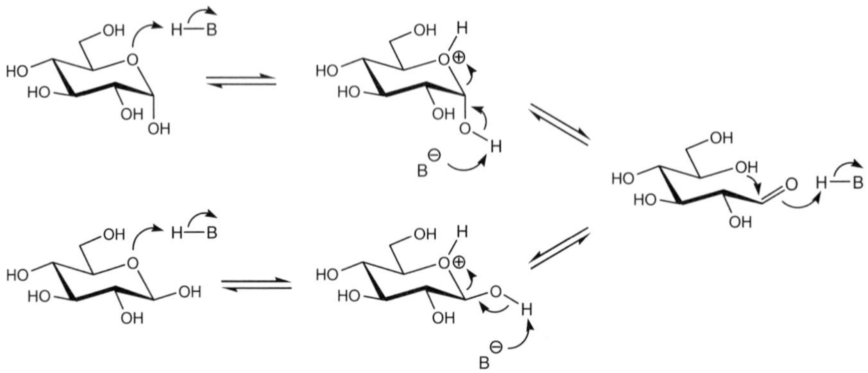

Fig. 7.5 Mechanism of mutarotation

five- or six-membered cyclic forms with two different anomers in equilibrium in aqueous solutions with ratios that can be determined by NMR spectroscopy (Table 7.1).

7.2.2.2 The Anomeric Effect and the *Exo*-anomeric Effect

The anomeric effect (Edward and Puskas 1962) was studied by Lemieux (Praly and Lemieux 1987) as the tendency for an electronegative substituent at the anomeric centre to prefer the axial orientation instead of the less hindered equatorial position (Edward and Puskas 1962; Praly and Lemieux 1987). Two explanations can be invoked based on dipolar moments or hyperconjugation (Fig. 7.6). Equatorial

Table 7.1 Composition of D-aldoses in aqueous solution at equilibrium (Angyal 1984)

Carbohydrate	Temperature (°C)	α-Pyranose (%)	β-Pyranose (%)	α-Furanose (%)	β-Furanose (%)	Acyclic form (%)
D-Ribose	31	21.5	58.5	6.5	13.5	0.05
D-Arabinose	31	60	35.5	2.5	2	0.03
D-Xylose	31	36.5	63.0	< 1	< 1	0.02
D-Lyxose	31	70	28	1.5	0.5	0.03
D-Allose	31	14	77.5	3.5	5	0.01
D-Altrose	22	27	43	17	13	0.04
D-Glucose	31	38	62	–/–	0.14	0.02
D-Mannose	44	64.9	34.2	0.6	0.3	0.005
D-Gulose	22	16	81	–/–	3	–/–
D-Idose	31	38.5	36	11.5	14	0.2
D-Galactose	31	30	64	2.5	3.5	0.02
D-Talose	22	42	29	16	13	0.03
D-Fructose	31	2.5	65.0	6.5	25.0	0.8

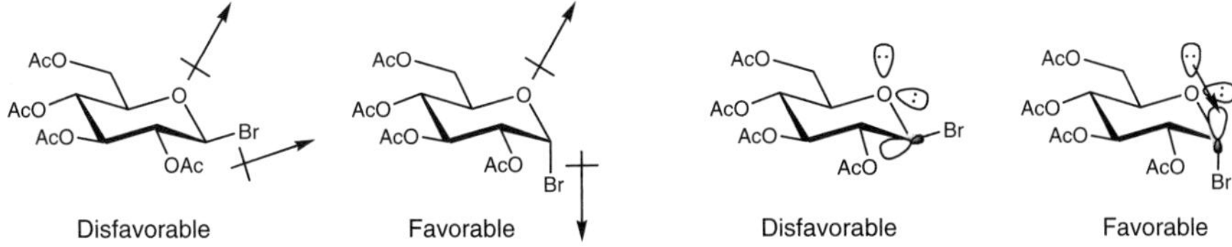

Fig. 7.6 The anomeric effect through two explanations

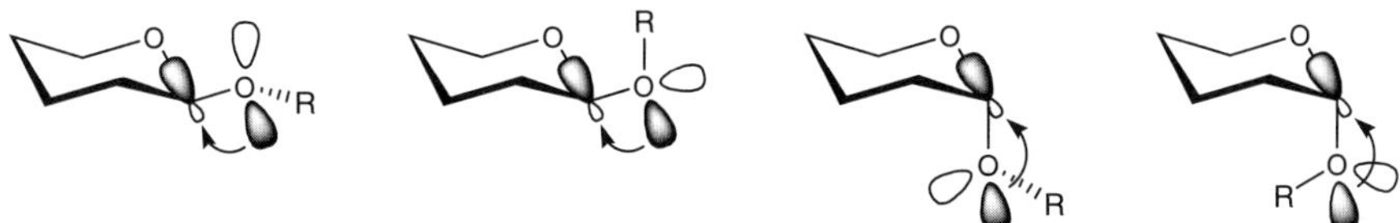

Fig. 7.7 Conformations stabilized by the *exo*-anomeric effect

configuration of an electronegative anomeric substituent results in the formation of two similar dipoles repelling each other and destabilizing the molecule. On the contrary, axially orientated substituents create roughly opposed dipoles accounting for a lower energy state. The second explanation to the anomeric effect is the presence of a stabilizing interaction through hyperconjugation between an unshared electron pair of the endocyclic oxygen atom and the σ* orbital of the anomeric bond (e.g. C–Br).

The orientation of the substituent at the anomeric centre (aglycon) can be described by the *exo*-anomeric effect (Fig. 7.7) (Praly and Lemieux 1987; Perez and Marchessault 1978; Thoegersen and Lemieux 1982) In this case, the exocyclic anomeric oxygen atom's lone pair is interacting with the anti-bonding orbital of the endocyclic C–O bond. The *exo*-anomeric effect can be observed for both α- and β-anomers.

7.2.3 Glycosylation Methods

The formation of a glycosidic bond occurs through the nucleophilic displacement of a leaving group at the anomeric centre of a glycosyl donor by an alcohol of a glycosyl acceptor (e.g. ROH) to provide the corresponding *O*-glycoside (Fig. 7.8) (Fügedi 2006; Schmidt 1986; Veeneman 1998; Whitfield and Douglas 1996). This reaction is usually promoted by a third partner (promoter) increasing the reactivity of the glycosyl donor.

The substitution at the 2-position plays an important role in the stereochemical outcome of the glycosylation process (Fig. 7.9). When a non-participating group is present at the 2-position, the major glycoside obtained is the α-anomer because of the stabilization through the anomeric effect. However, a participating group at the 2-position creates a cationic *bis*-acetal intermediate hindering the α-face of the pyranose ring and therefore favours the nucleophilic attack of the glycosyl acceptor from the β-face to give rise to the 1,2-*trans* glycoside. During this process, the alcohol could also form an orthoester which can be rearranged into the desired 1,2-*trans* glycoside under Lewis acid catalysis (Kochetkov et al., 1967, 1971).

Each glycosylation method can be distinguished through the glycosyl donor/promoter system involved (Table 7.2). Acetalation of a reducing sugar with an

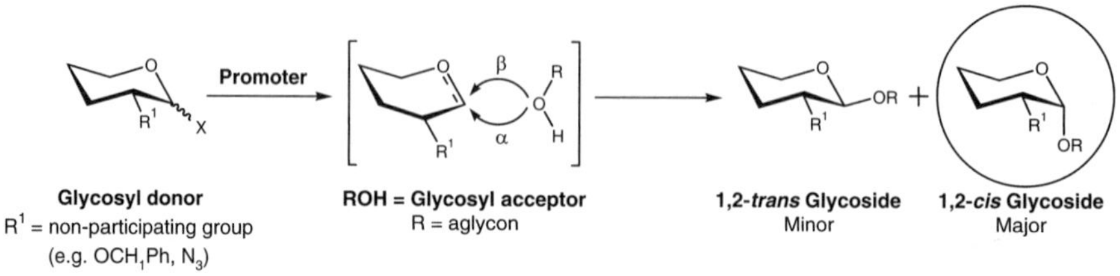

Fig. 7.8 General mechanisms for glycosylation

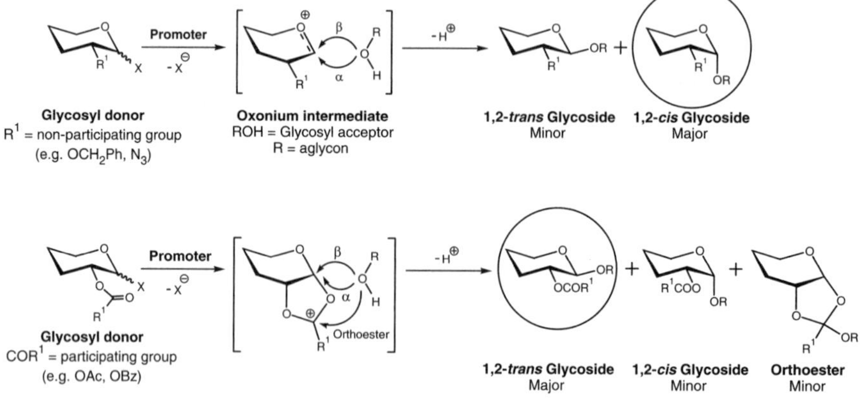

Fig. 7.9 General strategies used for the formation of 1,2-*cis* or 1,2-*trans* glycosides

Table 7.2 General glycosylation methods for the synthesis of *O*-glycosides

Promoter
ROH
- HX
Glycosyl donor
OR^1 X
OR^1 OR
R^1O OR

Glycosyl donor (name)	Leaving group(X=)	Promoters	Comments
Hemiacetals (Fischer)	OH	H^+, $BF_3{\cdot}Et_2O$, $FeCl_3$	Provides the α-anomer from the native carbohydrate
Glycosyl halides (Koenigs–Knorr)	Cl, Br	Ag_2O, AgOTf, HgO, $HgBr_2$, $Cu(OTf)_2$	Requires expensive silver or toxic mercury salts
Glycosyl fluorides	F	$SnCl_2$/AgOTf, $LiClO_4$, Cu $(OTf)_2$	Glycosyl fluorides are glycosidase substrates
1-*O*-acyl glycoside (Helferich)	OAc	$BF_3{\cdot}Et_2O$, $SnCl_4$,TMSOTf	Mixture of anomers can be used as donor
Trichloroacetimidate (Schmidt)	$OC(NH)CCl_3$	$BF_3{\cdot}Et_2O$, I_2, TMSOTf, AgOTf, TfOH	Usually requires the addition of molecular sieves
Arylthioglycosides	SAr	MeOTf,NIS/TfOH, I_2	High stability to chemical reactions and time
Glycosyl sulfoxides and sulfones (Kahne)	$S(O)_nAr$ ($n = 1$ or 2)	TMSOTf, Tf_2O, I_2	Requires the corresponding arylthioglycoside
n-Pentenyl glycosides (Fraser-Reid)	$O(CH_2)_3$ $CH = CH_2$	NBS, NIS/TfOH, iodonium dicollidine perchlorate (IDCP)	Fairly stable to functional group manipulations
Glycosyl phosphites	$OP(OR)_2$	TMSOTf, $BF_3{\cdot}Et_2O$, I_2, NIS/TfOH	Diethyl and dibenzyl phosphites commonly used
Glycosyl phosphates	$OP(O)(OR)_2$	TMSOTf, $BF_3{\cdot}Et_2O$, TfOH	Applied to both solution and solid phase synthesis
Selenoglycosides	SeAr	AgOTf/K_2CO_3, IDCP, NIS, I_2	Can be selectively activated in the presence of arylthioglycosides

alcohol in the presence of a protic acid promoter was used as one of the first glycosylation method by Fischer (Fischer 1893; Fischer 1895). It is probably the most appropriate method for the synthesis of simple glycosides (e.g. methyl, benzyl, allyl). Glycosylation was then performed from glycosyl halides in the presence of heavy metal salts by Koenigs and Knorr (Igarashi 1977; Koenigs and Knorr 1901). Glycosyl fluorides can also be used as glycosyl donors and are much more stable than their chlorinated or brominated analogues (Mukaiyama et al. 1981). They can be activated under specific conditions compatible with most protecting groups. Peracetylated carbohydrates are particularly convenient for large-scale synthesis and can be also involved in a glycosylation as glycosyl donors but their activation sometimes requires drastic conditions (Helferich and Schmitz-Hillebrecht 1933). Glycosyl trichloroacetimidates were then introduced by Schmidt and are activated under acidic conditions with either Brønsted or

Lewis acids usually in the presence of molecular sieves and at low temperature (Schmidt and Michel 1980; Pougny and Sinay 1976). Arylthioglycosides are highly stable even when stored for long periods on shelves and can also accommodate a large number of chemical reactions on the pyranose ring. They can be activated by thiophilic reagents under mild conditions, and formation of the oxonium intermediate proceeds through the addition of the electrophilic promoter to the sulphur atom (Garegg 1997; Ferrier et al. 1973; Fügedi et al. 1987). Activation can also take place under single electron transfer, electrochemical or even high pressure conditions. *n*-Pentenyl glycosides were introduced by Fraser-Reid as glycosyl donors and are activated through halogenation of the terminal double bond followed by intramolecular cyclisation to provide the oxonium intermediate and a halomethyl-2-furane (Fraser-Reid et al. 1988; Fraser-Reid et al. 1992). Glycosyl phosphites (Kondo et al. 1992; Martin and Schmidt 1992; Zhang and Wong 2000) react with alcohols at low temperature in the presence of Lewis acids affording the corresponding glycosides, and glycosyl phosphates (Hashimoto et al. 1989; Vankayalapati et al. 2002) can be activated by protic acids. Activation of selenoglycosides (Mehta and Pinto 1991; Mehta and Pinto 1993) can be achieved under similar conditions as for thioglycosides, by photo-induced or single electron transfer or iodine. They can be selectively activated in the presence of thioglycosides.

All these glycosylation methods have found numerous applications for the synthesis of natural products or for the selective formation of the desired anomer. One should also consider the influence of protecting groups on the pyranose ring, which sometimes affect the glycosylation outcome.

7.2.4 Chemo-enzymatic Glycosylation Methods

Among the powerful methods of glycosylation listed above, only Fischer glycosylation can accommodate a non-protected carbohydrate where the hydroxyl functions are not masked by protecting groups. Glycosyltransferases and glycosidases are natural enzymes capable of creating or hydrolyzing glycosidic bonds in an aqueous medium, using native sugars and with high stereo- and regioselectivities. Their use as "glycosylating agents" or promoters has attracted a lot of interest in the scientific community and a series of applications are now available for the synthesis of complex oligosaccharidic structures (Qian et al. 2001). This technology requires both biochemical and chemical skills since enzymes sometimes need to be mutated for obtaining the expected result. Sialylation is one of the most difficult glycosylation because of the poor stability of activated sialic acid derivatives. The biochemical engineering of sialyltransferases can readily provide a large series of sialylated derivatives ranging from the glycosylation of simple alcohols to mono- or disaccharides (Drouillard et al. 2006; Blixt et al. 2005; Huang et al. 2006b; Dumon et al. 2006; Muthana et al. 2007; Yu et al. 2006).

7.2.5 Glycoconjugates

Carbohydrates play an important role in a large series of biological processes and are usually present in nature either in their native form or as glycoconjugates where a sugar residue is associated to a lipid (glycolipids), a protein (glycoprotein) or a peptide (glycopeptides), along with several other natural products containing sugars. The chemical conjugation of carbohydrate moieties to proteins or polyvalent species will create neoglycoproteins or neoglycoconjugates, respectively, which can find numerous applications in biology and biochemistry. After introducing a functionalised linker at the anomeric position of a carbohydrate through standard glycosylation methods, the other end of this arm will be used for its conjugation to the desired multivalent scaffold (Fig. 7.10).

The conjugation of carbohydrate derivatives to multivalent species has been achieved through a wide range of chemical ligation techniques. When considering

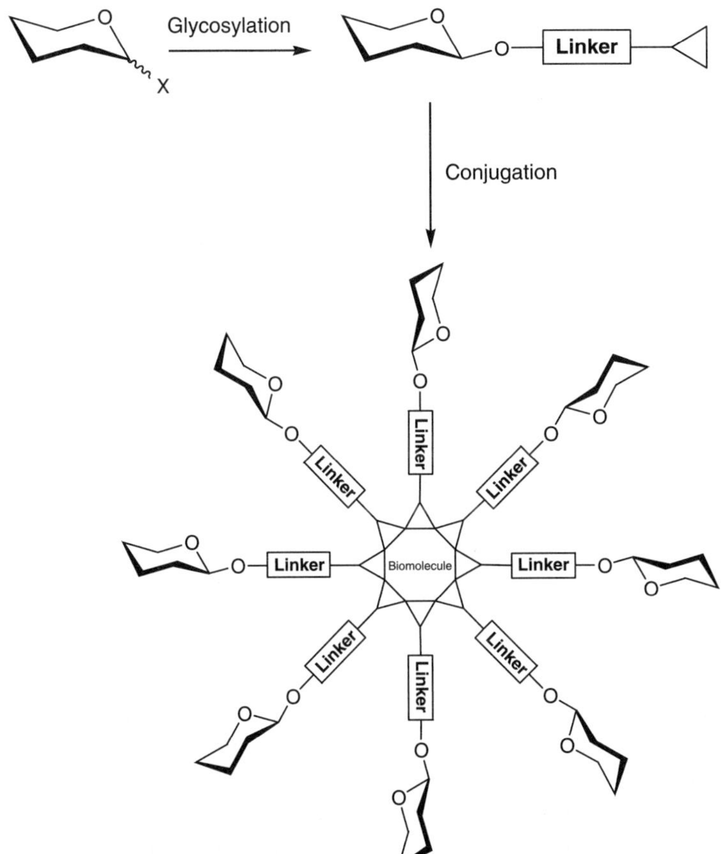

Fig. 7.10 Schematic representation of the synthesis of neoglycoproteins

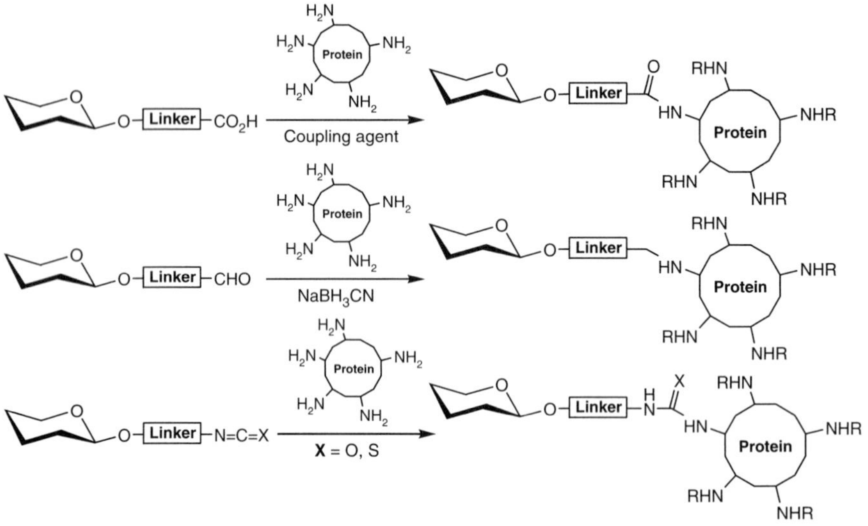

Fig. 7.11 Examples of conjugation of carbohydrates to proteins through amidation, reductive amination and urea/thiourea ligation

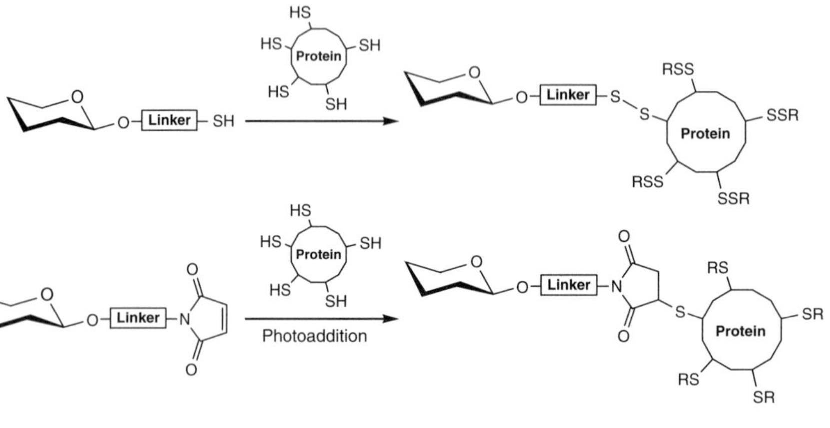

Fig. 7.12 Examples of conjugation of carbohydrates to proteins through thiol functions

the formation of neoglycoproteins, the chemist will take advantage of the reactive amino group from the lysine residues present at the surface of proteins and therefore use a series of reactions involving an amine function (Fig. 7.11). Similarly, neoglycoproteins can be prepared by reacting the thiol groups of cystein residues of proteins with thiol or maleimide functions (Fig. 7.12).

The above-mentioned techniques can be also applied to multivalent non-natural scaffolds for the preparation of neoglycoconjugates. In these cases, the nature of functionalities accessible is much broader. 1,3-Dipolar cycloaddition between an

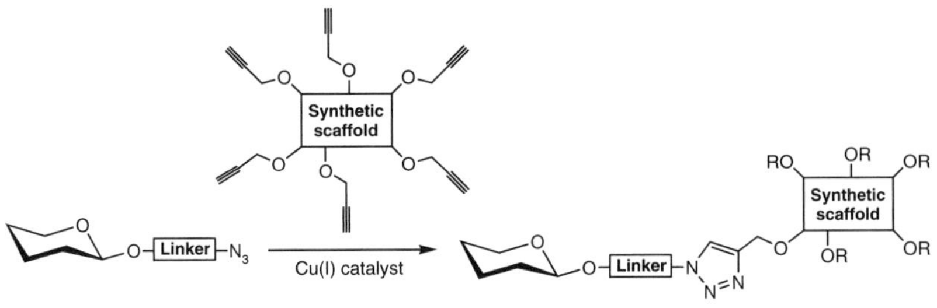

Fig. 7.13 Conjugation of carbohydrates to proteins through 1,3-dipolar cycloaddition

alkyne and an azido derivative to form 1,2,3-triazoles, developed by Huisgen (Huisgen et al., 1965, 1969) in the late 1960s, was recently introduced into the realm of techniques for conjugation and rapidly found a large number of applications in glycobiology (Fig. 7.13). This methodology is now accessible due to the improved Cu(I)-catalysed processes developed more recently (Kolb et al. 2001; Rostovtsev et al. 2002; Bock et al. 2006; Tornoe et al. 2002; Gil et al. 2007; Bouillon et al. 2006; Morvan et al. 2007). This chemical ligation technique is highly attractive since the reactive species involved are pretty much unreactive to the chemical functions present in biological media, the reaction does not create any by-product and they can be performed in water, at low temperature and on reasonable time scales.

7.2.6 *Examples of Naturally Occurring Carbohydrates*

7.2.6.1 Structures of Some Mono- and Disaccharides

A large number of monosaccharides incorporating a single carbohydrate residue are present along with several disaccharides composed of two sugar moieties (Fig. 7.14). Lactose is a disaccharide present in milk and fructose in fruits, while saccharose is the most common food sweetener. Ribose is a major component of RNA-forming nucleotide building blocks. Xylose is also called the *wood sugar* as it is one of the main components of trees and plants along with the disaccharide cellulose. Fucose is a 6-deoxy sugar found in mammalian, insect and plant cell surface.

7.2.6.2 Nucleosides

RNA and DNA can be differentiated through the nature of the carbohydrate involved in their structural subunits (nucleosides). Ribose is the sugar involved in RNA, while DNA is based on 2-deoxy-D-ribose lacking one hydroxyl group at the 2-position. The four constituents of RNA are based on a ribose residue linked to a

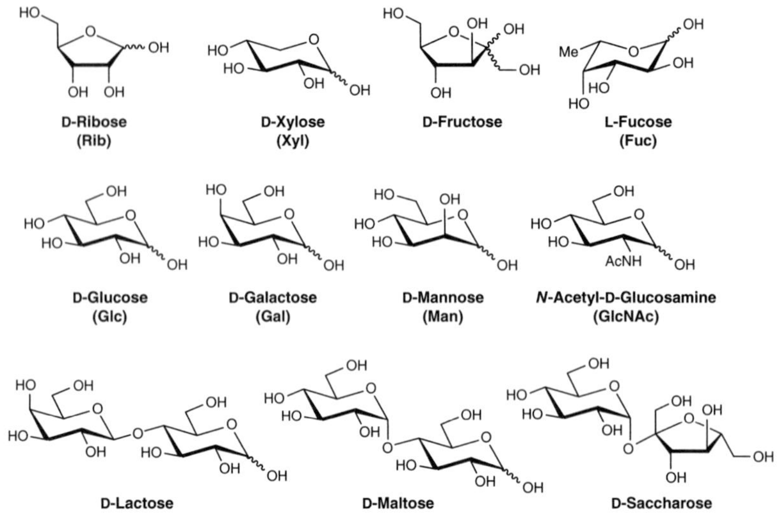

Fig. 7.14 Selection of naturally occurring mono- and disaccharides. Conventional abbreviations are mentioned in parentheses when appropriate

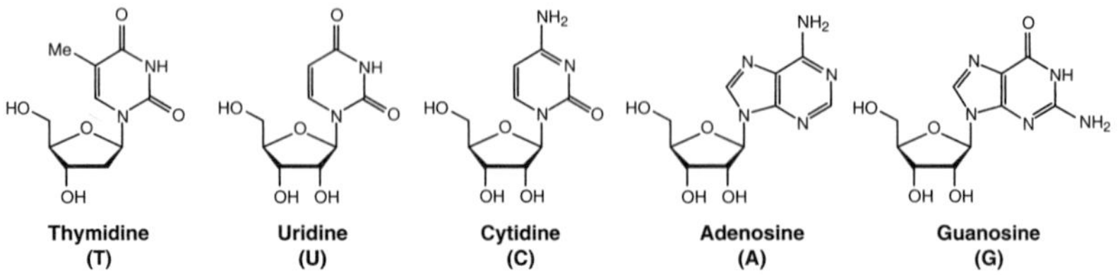

Fig. 7.15 Thymidine and the four natural nucleosides as RNA polynucleotide repeating units

pyrimidine: uracil (U) or cytosine (C), or to a purine: adenine (A) or guanine (G). In the DNA series, thymine (T) replaces uracil while the three other nucleobases are similar (Fig. 7.15).

7.2.6.3 Oligosaccharides

Carbohydrates can be associated through glycosidic linkages to form oligosaccharides possessing a wide range of biological aspects. The Lewis blood group antigens are oligosaccharides present on the surface of red blood cells and distinguish the human blood groups. These oligomers differ by the substitution (R) at the 3-position of the non-reducing galactosyl residue (Fig. 7.16).

Sialylated oligosaccharides are involved in a series of viral infection events. The substitution of lactose by sialic acid either at the $3'$ or $6'$ position differentiates between avian and human flu, respectively (Fig. 7.17) (Shinya et al. 2006).

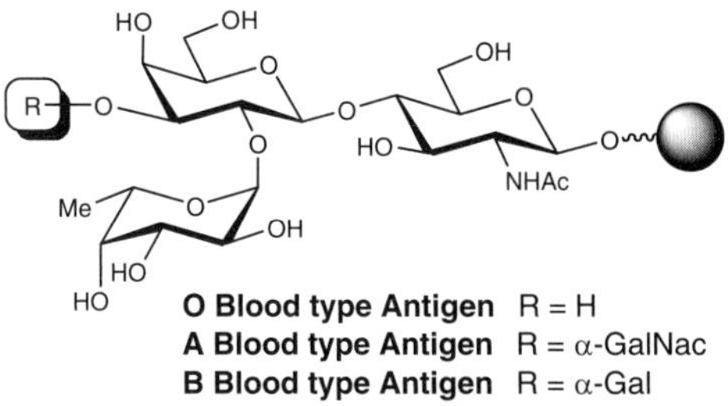

Fig. 7.16 Structure of the ABO blood antigens

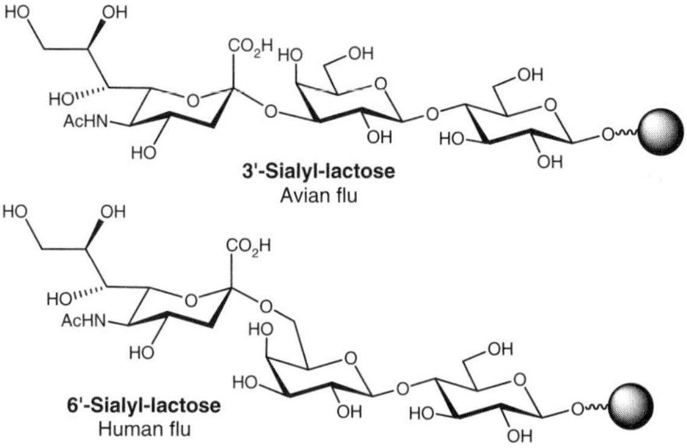

O Blood type Antigen R = H
A Blood type Antigen R = α-GalNac
B Blood type Antigen R = α-Gal

3'-Sialyl-lactose
Avian flu

6'-Sialyl-lactose
Human flu

Fig. 7.17 Structure of the 3′- and 6′-sialyl-lactose oligosaccharides involved in *influenza virus* infection

Oligosaccharides can also adopt cyclic forms conferring a very different set of chemical and physical properties. Cyclodextrins are natural cyclic oligomers composed of six or more glucose units linked through an α-1,4-glycosidic bond (Fig. 7.18) (Nepogodiev and Stoddart 1998). Their inside cavity is highly hydrophobic and can form host–guest complexes with a series of molecules of various interests in the pharmaceutical industry (drug release) or for environmental applications (interactions with toxic compounds).

7.2.6.4 Polysaccharides

Polymeric carbohydrates are also widely present in nature. Cellulose is a β(1→4)-glucose polymer with a linear shape present in plants, while amylose is the corresponding α(1→4)-polyglucoside and is organized in a helical structure (Fig. 7.19). Polysaccharides are also employed for energy storage, like glycogen which is a branched polysaccharide based on the same α(1→4)-glucose polymeric chain as in amylose. Chitin is a polymeric β(1→4)-*N*-acetyl glucosamine and is the main component of the exoskeletons of crustaceans and insects.

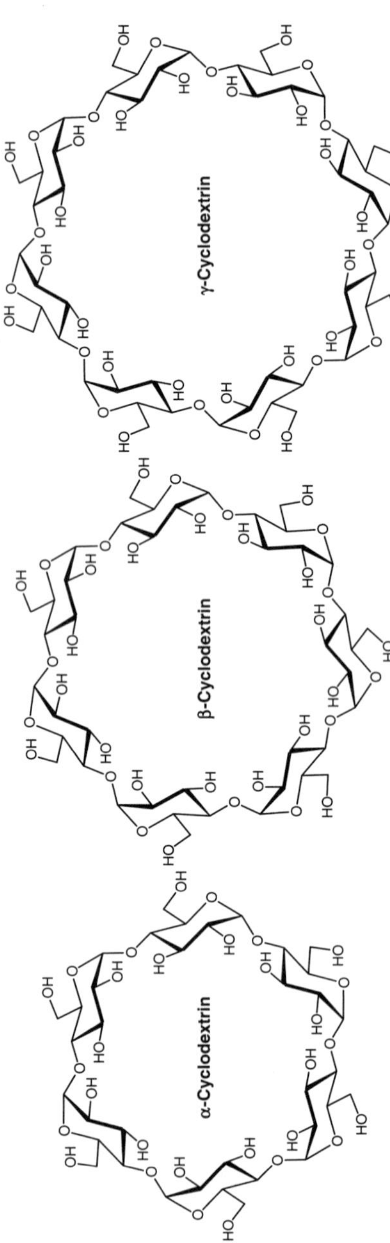

Fig. 7.18 Structure of the α-, β- and γ-cyclodextrins

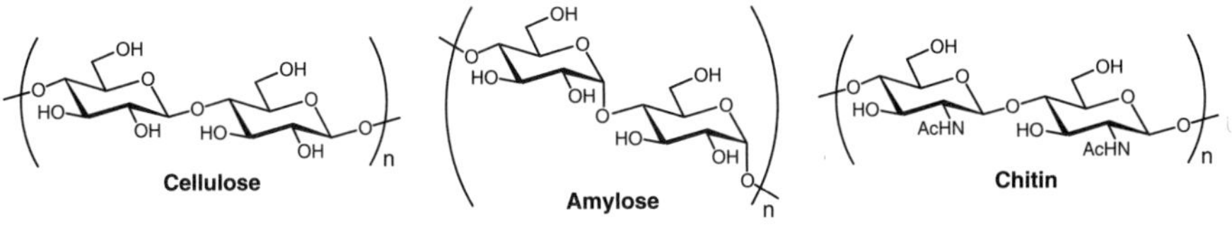

Fig. 7.19 Structure of a selection of polysaccharides

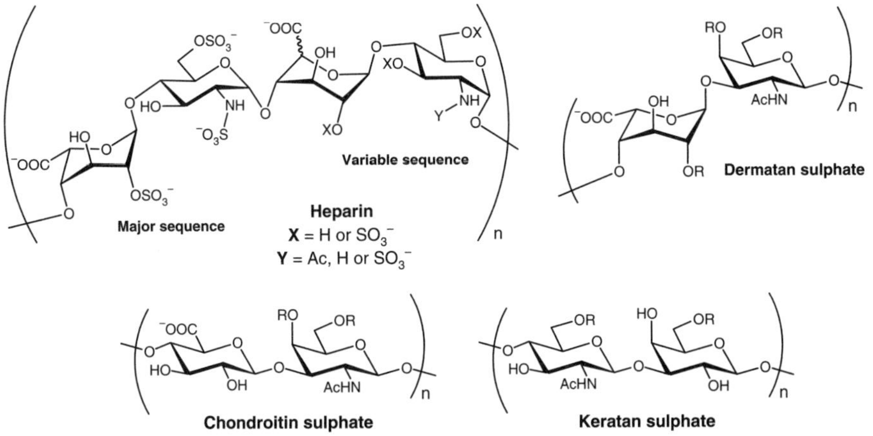

Fig. 7.20 Structure of heparin and other glycosaminoglycans (R = potentially sulphated sites)

7.2.6.5 Glycosaminoglycans

Heparin and heparan sulphate are the most complex members of the glycosaminoglycans (GAGs) family, which also includes chondroitin sulphate, keratin sulphate and dermatan sulphate (Fig. 7.20) (Bishop et al. 2007; Capila and Linhardt 2002). These highly negatively charged polysaccharides are displayed on the extracellular matrix or on the membrane of the cell. Heparin is an anticoagulant and is used for the treatment of heart diseases. It is composed of a tetrasaccharidic repeating unit with a major sequence highly conserved throughout the polymer chain and a variable sequence in which the sulphation/acetylation sites are not systematically functionalised. Heparan sulphate is a less sulphated version of heparin where the major sequence of heparin is not sulphated and the amino group acetylated.

7.2.6.6 Carbohydrate-Based Drugs from Natural and/or Synthetic Sources

A series of natural products containing carbohydrates have been isolated and their structure elucidated. Some of them are now used as drugs for the treatment of various diseases. Nojirimycin is produced by *Streptomyces* strains (Inouye et al. 1966). It is used as an antibiotic and inhibits α-glucosidases, thereby preventing the

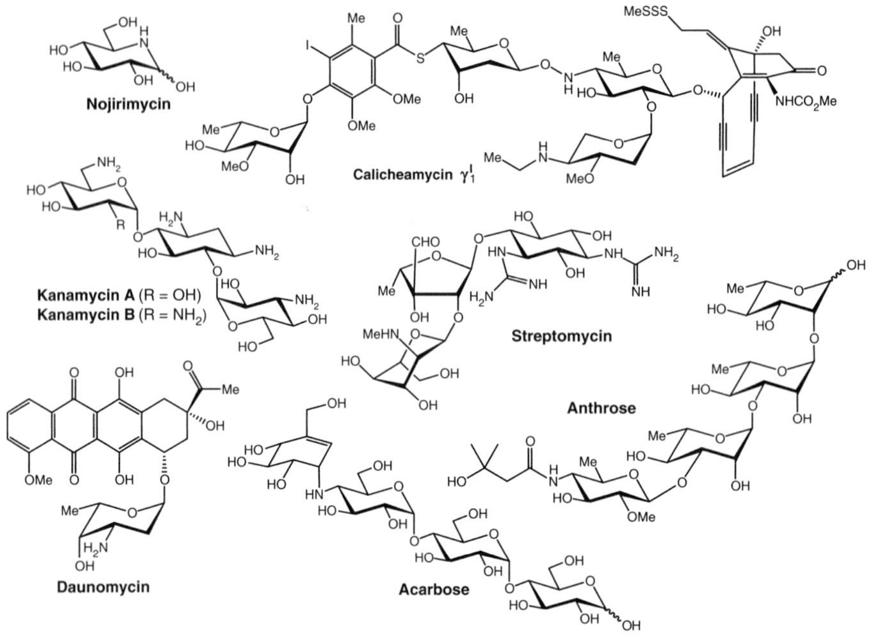

Fig. 7.21 Structures of natural products incorporating carbohydrate, one or more residue(s)

normal glycosylation of proteins (Fig. 7.21). Calicheamycin γ_1^I is isolated from *Micromonospora echinospora* and is used as an antibiotic (Nicolaou and Smith 1992). It binds to the minor groove of DNA showing a high degree of base pair sequence specificity which can be attributed to the oligosaccharide residues. Kanamycins are antibiotics affecting the 30S ribosomal subunit and preventing the correct translation of RNA (Maeda et al. 1957). Indeed, the proteins are biosynthesized but fold incorrectly, thus destroying the bacterium. Daunomycin is an anti-cancer chemotherapy agent of the anthracycline family and isolated from *Streptomyces peucetius* (Arcamone et al. 1964). Acarbose is an inhibitor of α-glucosidase used for the treatment of type 2 diabetes mellitus (Junge et al. 1984). Streptomycin is an antibiotic drug obtained from the actinobacterium *Streptomyces griseus* and inhibits bacterial growth by damaging cell membranes and blocking protein synthesis (Majumdar and Kutzner 1962). Anthrose is found on the surface of *Bacillus anthracis*. This antigen is a very promising target for the development of an anti-anthrax vaccine (Tamborrini et al. 2006).

A series of nucleoside analogues have been approved by the Food and Drug Administration (FDA) for the treatment of HIV infection (AIDS) (Fig. 7.22). Zidovudine (AZT) was one of the first nucleoside approved by FDA and acts as a chain terminator for HIV reverse transcriptase (HIV-RT). Although resistance to AZT frequently develops, the association of Lamivudine (3TC) and AZT can cause AZT-resistant HIV strains to revert to AZT-sensitive strains. Abacavir is a carbocyclic nucleoside analogue significantly less toxic than AZT and appears to have a

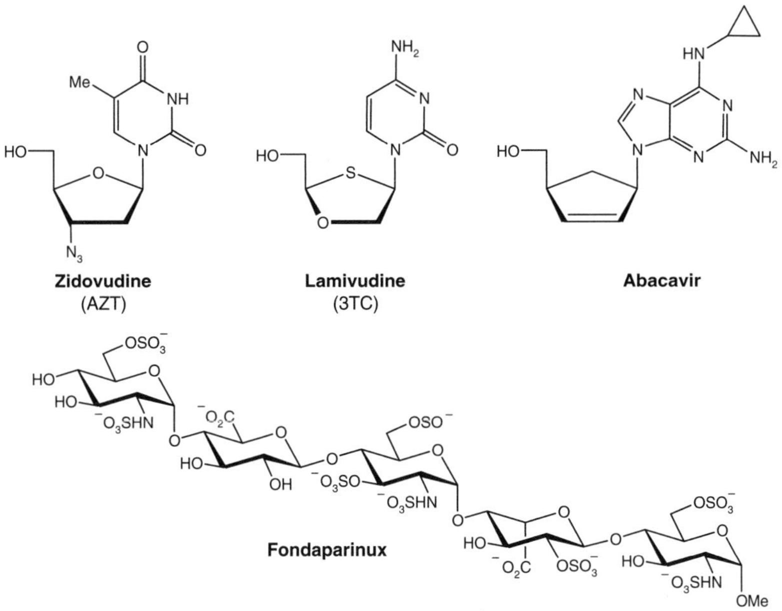

Fig. 7.22 Structures of synthetic carbohydrate-based drugs

stronger anti-HIV effect than any of the approved nucleoside analogues. The association of AZT, 3TC and Abacavir in a tri-therapy strategy improves considerably the life expectancy of AIDS (or HIV-infected) patients.

Fondaparinux has been recently marketed by Sanofi-Aventis as an heparan sulphate analog for the treatment of heart diseases (Fig. 7.22) (Petitou and van Boeckel 2004).

7.3 Biological Role

The carbohydrates are probably best known for their role in energy metabolism. But like proteins, some carbohydrates perform structural functions, providing scaffolding for bacterial and plant cell walls (cellulose, pectin), connective tissues such as cartilage in animals, and exoskeleton shells (chitin) in arthropods. Carbohydrates are also covalently bound with proteins (glycoproteins) or lipids (glycolipids) on cell surfaces. The glycoproteins and glycolipids play important roles in cell–cell recognition processes and in signal transduction

7.3.1 The Glycocalix and Extracellular Matrix Polysaccharides

The surface of most types of cells is covered by a carbohydrate-rich coat, the glycocalyx, which plays an important functional role in cell–cell and cell–matrix interactions. These specific contacts are realized by using attractive forces between cell-surface oligosaccharides and proteins (lectins) (Hakomori 1991; Hakomori 2002; Rojo et al. 2002). It is now accepted that these interactions strongly depend on divalent cations and show high specificity and low affinity. However, its mechanism is not yet clarified because of the difficulty in analysing weak affinity interactions. The main challenge to understand and control interactions between carbohydrates and lectins is the low affinity of the interactions. Nature overcomes this problem by a polyvalent presentation of ligands and receptors at the cell surface (Mammen et al. 1998).

Studies of the terminal carbohydrate residues of the glycocalyx of immune cells have shown that cell–cell and cell–matrix interactions, which play an important functional role in various processes of the immune response, are mediated at least in part by these carbohydrate residues (Hounsell 2000). The use of lectin histochemistry in lymphatic organs of chickens demonstrated that carbohydrate residues are differentially expressed in the cells of the immune system of chickens (Jorns et al. 2003). For example, in the thymus, mannose as well as *N*-acetyl glucosamine (GlcNAc)-specific lectins labelled macrophages, epithelial reticulum cells and lymphocytes within the cortex. In the bursa of Fabricius, the brush border of the lining epithelium, the macrophages and the endothelium were labelled by mannose-specific lectins.

Because glycoconjugates, including proteoglycans, glycoproteins and glycolipids, are mainly localized in the plasma membrane and in the extracellular matrix of cells, they are considered as useful cell-surface biomarkers and play crucial roles in signal transduction and cell adhesion. To illustrate this, the vascular endothelium is presented in the following.

The endothelial glycocalyx is considered to be connected to the endothelium through several "backbone" molecules, mainly proteoglycans and also glycoproteins. Proteoglycans consist of a core protein to which one or more glycosaminoglycan chains are linked. There are five types of glycosaminoglycan chains: heparan sulfate, chondroitin sulfate, dermatan sulfate, keratan sulfate and hyaluronan (or hyaluronic acid). They are linear polymers of disaccharides with variable lengths that are modified by sulfation and/or (de)acetylation to variable extents. The disaccharides are each composed of a uronic acid and a hexosamine. In vasculature, heparan sulfate proteoglycans represent 50–90% of the total amount of proteoglycans present in the glycocalyx (Pries et al. 2000). Hyaluronan, another important glycosaminoglycan, differs from others because it is not linked to a core protein. Its exact link to the cell membrane is not known, but it can be bound to the receptor CD44 (Nandi et al. 2000).

Besides the proteoglycans, well-defined glycoproteins bearing small (2–15 sugar residues) and branched carbohydrate side chains connect the glycocalyx to the endothelial cell membrane. These endothelial cell adhesion molecules play a major role in cell recruitment from the bloodstream and in cell signalling. The three families of cell adhesion molecules present in the endothelial glycocalyx are the selectin family, the integrin family and the immunoglobulin superfamily. Selectins are carbohydrate-binding glycoproteins. These glycoproteins were designated as E-, P- and L-selectin, according to the cell type on which the selectin was found (i.e. endothelium, platelets, and lymphocytes, respectively). E-selectin and P-selectin are both involved in leukocyte–endothelial cell interactions (Sperandio 2006). The carbohydrate ligands that are recognised by these selectins have been identified. E-selectin recognizes sialyl Lewis X (sLe^X) on the surface of neutrophils. P-selectin also binds sLe^X on neutrophils or leukocytes with a lower affinity. L-selectin weakly recognises sLe^X on endothelial cells, but the affinity is higher with a sulphate group on the 6-position of Gal or, perhaps more likely, on the 6-position of the GlcNAc residue. Integrins are heterodimeric molecules composed of non-covalently bound α (18 different α-subunits) and β (8 β-subunits) subunits. Integrins are found on many cell types, including endothelial cells, leukocytes, and platelets. These integrins, such as $\alpha2\beta1$, $\alpha5\beta1$ and $\alpha6\beta1$, bind to multiple extracellular matrix ligands and are responsible for interactions with laminin, fibronectin, and collagen. Many studies have focused and still focus on the interactions between these integrins and the sub-endothelial matrix during angiogenesis (Ruegg and Mariotti 2003). The best known immunoglobulin superfamily of glycoproteins includes intercellular adhesion molecule 1 and 2 (ICAM-1 and -2), vascular cell adhesion molecule 1 (VCAM-1) and platelet/endothelial cell adhesion molecule 1 (PECAM-1). They act as ligand for integrins on leukocytes and platelets and are crucial mediators of leukocyte homing to the endothelium and subsequent diapedesis. In addition, glycoproteins of the endothelial glycocalyx display functionality in coagulation, fibrinolysis and homeostasis.

Leukocyte adhesion to endothelial cells is a complex process involving capture of free-flowing leukocytes from the bloodstream, rolling on the endothelial surface, deceleration and, eventually, leukocyte immobilization (firm adhesion). Studies have shown that disruption of heparan sulphate proteoglycans of the endothelial glycocalyx stimulates firm adhesion of leukocytes (Constantinescu et al. 2003). Indeed, heparan sulfate proteoglycans are abundant on the endothelial cell surface, are highly negatively charged due to the presence of numerous sulfate groups, and bind numerous plasma proteins, contributing to the glycocalyx charge and thickness (Esko 1999; Yanagishita and Hascall 1992). The integrity of vascular endothelium depends on the number of negatively charged molecules at the endothelial cell surface. In myocardial tissue oedema, the decrease of microvascular fluid is associated with a reduction of negatively charged molecules (Gotloib et al. 1992).

Glycoconjugate antigens are also utilized as unique markers in the developing brain and central nervous system (CNS) (Yu and Yanagisawa 2006; Yanagisawa and Yu 2007; Zhang 2001). Indeed, basic fibroblast growth factor (bFGF) is important for sustaining multipotency and enhancing proliferation of neural stem cells (NSCs)/neural precursor cells (NPCs). The signal of bFGF is mediated by receptors localized in the cell surface, but additional co-factors are also required. One of these co-factors is heparan sulfate proteoglycan (HSPG; Fig. 7.1). In addition to HSPGs, chondroitin sulfate proteoglycans (CSPGs) are also expressed in NSCs/NPCs. Nerve/glial antigen 2 (NG2) is a CSPG expressed in heterogeneous cell populations, including glial precursor cells and/or NSCs (Belachew et al. 2003). Furthermore, NG2 displays high affinity for bFGF and platelet-derived growth factor (PDGF)-AA, which are mitogens of NSCs/NPCs and oligodendrocyte progenitor cells (Goretzki et al. 1999).

Another class of carbohydrate-bearing molecules are glycosphingolipids (GSLs) which are composed of a core lipid moiety (ceramide) linked to a carbohydrate chain (Yu et al. 2004). GSLs having one or more sialic acid residues are named gangliosides. GSLs are most abundant in nervous tissues and are localized on the plasma membrane of cells. Studies have shown that certain GSLs are expressed in NSCs/NPCs like GD2 in human and mouse NSCs (Klassen et al. 2001) and GD3, GT1b and GQ1b in mouse NPCs(Yanagisawa et al. 2005).

Because of the biological significance of NSCs in neurogenesis, glycoconjugates are expected to play a significant role in developing clinical uses of NSCs for the treatment of a variety of neurodegenerative disorders. First, glycoconjugates can be used as biomarkers to define and isolate NSCs and their progenies. They also may be used for enriching NSCs. In addition, glycoconjugates may be useful for inducing differentiation of NSCs into specific lineages.

7.3.2 Carbohydrates in Host–Pathogen Interactions and Metastasis

Carbohydrates can serve as key elements in various molecular recognition processes, including bacterial and viral infections, cell adhesion in inflammation and metastasis.

The glycosylated phosphatidyl inositols (GPIs) are ubiquitous eukaryotic glycolipids covalently linked to the C-terminus of membrane-associated proteins. They provide an alternative mechanism for the attachment of proteins to the plasma membrane. However, studies have shown that glycoinositol phospholipids (GIPLs) not linked to either protein or polysaccharide are major cell-surface glycolipids in trypanosomatids, and are closely related to the structure of the GPI–protein anchor (Ferguson 1999). The GIPL of the parasite *Trypanosoma cruzi*, the causative agent of Chagas' disease, contains a unique glycan chain expressing slight variations among distinct *T. cruzi* isolates, and linked through a non-*N*-acetylated glucosamine residue to an inositol phosphorylceramide, mainly composed of an *N*-lignoceroyl-dihydrosphingosine (Carreira et al. 1996). The β-D-Gal*f*-(1→3)-α-D-Man*p* substructure is involved in its antigenicity (Mendonça-Previato et al. 1983). Biological effects of *T. cruzi* GIPL and its isolated glycan and lipid moieties were investigated on cells of the immune system. Dos Reis et al. (2002) found that GIPLs are bioactive against every cell type that has been tested.

Leishmania spp. are protozoan parasites that cause a spectrum of diseases in humans, ranging from self-limiting cutaneous infections to visceral leishmaniasis. Control of this disease, which affects more than 20 million people worldwide, has been hampered by the absence of a vaccine, limitation with frontline drugs and the increased transmission as a result of co-infections with HIV (Croft et al. 2006). The promastigote stages of *Leishmania* are coated by a thick glycocalyx composed of glycosyl phosphatidyl inositol (GPI)-anchored proteins, the GPI-anchored phosphoglycan, lipophosphoglycan (LPG) and GIPLs (Naderer et al. 2004). Analysis of *Leishmania* mutants lacking individual or multiple surface components indicates that only LPG plays a critical role in macrophage infection, while loss of other surface components has relatively little affect on promastigote virulence in the mammalian host (McConville et al., 2007). The plasma membrane of *Leishmania* amastigotes is unusual in several other respects. It contains relatively high levels of externally exposed phosphatidyl serine, which may provide a mechanism for entering host cells via apoptotic cell receptors without activating microbicidal processes or proinflammatory responses (Wanderley et al. 2006).

The study of N-linked glycosylation has become an area of intense interest because of its importance in viral virulence and immune evasion for several prominent human pathogens such as Hendra (Carter et al. 2005), severe acute respiratory syndrome coronavirus (SARS-CoV) (Oostra et al. 2006), influenza, hepatitis C (Goffard et al. 2005) and West Nile (Hanna et al. 2005). HIV and influenza, two clear threats to human health, have been shown to rely on expression of specific oligosaccharides to evade detection by the host immune system.

Influenza viruses express two envelope proteins that are involved in virulence, neuraminidase (NA) and hemagglutinin (HA). Influenza hemagglutinin binds to sialic acids attached to membrane glycosphingolipids, proteoglycans and glycoproteins. It was shown that the linkage between sialic acid and the second sugar (usually galactose) could determine whether an influenza virus binds to avian or human cells (Carroll et al. 1981; Rogers and Paulson 1983), and that a single mutation in the hemagglutinin (HA) was sufficient to change binding from α-2–3′ to α-2–6′ sialyl-lactose and host cells from avian to human, or vice versa (Rogers et al. 1983; Rogers et al. 1985). Now it is generally accepted that human influenza viruses bind to α-2–6′ sialic acids and avian viruses bind to α-2–3′ sialic acids (Baum and Paulson 1990; Matrosovich et al. 1997), while swine influenza viruses are reported to bind mainly to α-2–6′ sialic acids and also to α-2–3′ sialic acid receptors (Ito et al. 1998; Rogers and D'Souza 1989).

Infection by dengue virus (DV), a mosquito-borne flavivirus leading to haemorrhagic fever in humans, has undergone a global resurgence in the past 40 years such that almost half the world's population are currently living at risk in dengue-endemic areas. The four serotypes of DV (DV1–DV4) bind to a number of opsonic and non-opsonic receptors on cells of the mononuclear phagocyte lineage including DC-SIGN (Navarro-Sanchez et al. 2003; Tassaneetrithep et al. 2003) and glycosaminoglycans (Chen et al. 1997). Recently, Miller et al. (Miller et al. 2008) have shown that the mannose receptor (MR) is a functional receptor for DV infection of human macrophage. MR and DC-SIGN both contain lectin domains, but differ distinctly in terms of ligand specificity. MR binds to terminal mannose, fucose and *N*-acetyl glucosamine, whereas DC-SIGN binds to mannose within high-mannose oligosaccharides and fucosylated glycans (Taylor et al. 2005; Feinberg et al. 2001).

Carbohydrate antigen sialyl Lewis a (CA19-9) is the most frequently applied serum tumour marker for diagnosis of cancers in the digestive organs. Moreover, malignant transformation is associated with changes in the glycosylation of cell-surface proteins and lipids. In tumour cells, alterations in cellular glycosylation may play a key role in their metastatic behaviour. Indeed, GD2 ganglioside is a major constituent of neuroectodermal tumors (Livingston 1995). Studies suggest that peptides mimicking GD2 ganglioside inhibit tumour growth through antibody and/or CD4+ T-cell-mediated mechanisms.

Oligosaccharide expression is highly relevant to many cancers. Many studies show that alterations in N-linked oligosaccharides of tumour cells are associated with carcinogenesis, invasion and metastasis (Varki 1993; Dwek 1996). Numerous studies have demonstrated a correlation between hyaluronan expression and the malignant properties of various kinds of cancer. Furthermore, hyaluronan oligosaccharides have been shown to inhibit several tumour cell types via disruption of receptor–hyaluronan interaction. So hyaluronan oligosaccharides could have potent antitumour effects due in part to the abrogation of hyaluronan-rich cell-associated matrices (Hosono et al. 2007).

Recent work has highlighted the importance of altered cell–matrix interactions in the acquisition of malignant characteristics of glioblastoma (Yamada and Araki 2001). Several studies have provided evidence that hyaluronan–cell interactions may play a role in glioma invasiveness. For example, addition of exogenous hyaluronan enhances glioma cell migration and invasion in vitro, and this effect is blocked by antibodies or anti-sense oligonucleotides to CD44, a cell-surface hyaluronan receptor (Merzak et al. 1994; Koochekpour et al. 1995; Okada et al. 1996; Tsatas et al. 2002).

7.4 Carbohydrate-Based Biosensors

Saccharide probes have been immobilised on a variety of substrates ranking from plane substrates such as polymer (Fukui et al. 2002), glass (Blixt et al. 2004), diamond (Chevolot et al. 1999) or gold (Seo et al. 2007; Miura et al. 2002; Smith et al. 2003) to nanoparticles or quantum dots (Larsen et al. 2006; Robinson et al. 2005; Huang et al. 2005). Both plane and spherical substrates are important to consider, as both can be used for biosensing.

7.4.1 Obtaining Saccharide Probes

Custom-synthesised peptides or nucleic acids are commercially available, and powerful synthetic tools such as automatic synthesisers are on the market. Furthermore, they can be amplified by PCR or cloning. On the other hand, oligosaccharides are tedious to synthesise and cannot be amplified. They are of synthetic or natural origin. Consequently, they are often available in limited quantities and expensive in the market.

Also, as mentioned in the introduction, since they have a very high structural diversity because of the fact that they can be linear or branched and the anomeric carbon can be α or β, they can be in the chair or boat conformation and the carbon involved in the glycoside linkage can vary.

Furthermore, all the functional groups are hydroxyls with special reactivity for the one carried by the anomeric carbon atom and for the primary alcohol. This means that the synthesis of oligosaccharides requires selective protections and deprotections of the desired hydroxyl function for coupling of monosaccharides through a glycosyl bond. Some authors have developed automated solid-phase synthesis, while some others have used a "one-pot" approach (Feizi et al. 2003; Werz and Seeberger 2005a; Seeberger 2008; Plante et al. 2001). Still others take advantage of enzymes (glycosidases or glycosyltranferases) to perform the synthesis of oligosaccharides. Glycosidases are enzymes that under usual conditions hydrolyse the glycosidic bond. However, as this reaction is reversible, under defined conditions, the enzyme can catalyse the formation of the linkage. Glycosyltransferases are enzymes that catalyse the glycosidic linkage formation between two saccharides. They are more efficient than glycosidases, giving better yields. However, the number of different enzymes is limited and expensive. The natural substrate of the enzyme is usually expensive as well. These reasons have motivated some authors to engineer whole bacteria with genes coding for the different enzymes required for oligosaccharides synthesis (Fort et al. 2005).

Nevertheless, synthesis of a considerable number of complex carbohydrates has been achieved but often in limited amounts and at high cost.

In the following section, the different methods employed for surface immobilisation of carbohydrates are described, followed by a paragraph on signal transduction and finally some applications.

7.4.2 Surface Physicochemistry: Non-specific Adsorption and Immobilisation

One can picture the saccharide–protein interaction as a key–lock interaction. In consequence, like protein arrays (Gordus and Mac Beath 2006), carbohydrate arrays require the preservation of the 3-D conformations of the saccharide ligand in order to perform specific biomolecular recognition with their receptor. In addition, their accessibility and orientation are major issues especially in the case of immobilised probes where the degrees of freedom are fewer than in solution. Therefore, the surface chemistry employed for immobilisation (the chemical reaction for immobilisation, but also the structure of the spacer arm) of the saccharide and the physico-chemical properties of the surface (and ligand) are major issues in maintaining the biological activity, as they will play key roles not only in the preservation of the 3-D conformation but also in the probe availability. For example, weak interactions between the surface and the probe can influence the orientation of the probe. Surface organisation of the probe (dense packing which depends among other things on the spacer arm structure) can also influence the probe availability. Organisation of the water molecules at the surface of the material may also play a role in the thermodynamics of biological recognition at the surface of the material.

The immobilisation should

- proceed with good and reproducible yield for a given probe;
- proceed with the same yield independently of the probe (especially in the case of glycoarray);
- be oriented in order to preserve the "key–lock" interaction between the probe and the target and to obtain a reproducible device;
- allow the accessibility of the probe by the target although the degree of freedom is reduced; and
- allow quality control of immobilised probes.

The performances (lower detection limit, reliability, robustness, etc.) of the micro-devices will greatly depend upon this step, as the target capture by the probe will depend on the probe availability (not taking into account diffusion and microfluidic-related phenomena (Raynal et al. 2004)).

The detection technique will determine the choice of the substrate to be used and, therefore, the subsequent coupling reactions. For example, gold is the substrate of choice for SPR detection, whereas glass is often preferred with fluorescent detection. In one case, mercapto-based coupling agents have been used (Schreiber 2000), whereas silane-based coupling agents are preferred with glass.

Furthermore, the background signal depends on non-specific adsorption of the target. It should be reduced to avoid high background.

In the following, we will discuss the various strategies employed for reducing non-specific adsorption, as well as the various strategies for immobilising carbohydrates, on surfaces of materials, more specifically mono and oligosaccharides.

Finally, the collective behaviour of the molecules such as surface organisation of probe molecules can influence the affinity of a target for a probe as discussed in section "Surface Cluster Effect or Not."

7.4.2.1 Non-specific Adsorption

A protein in solution adsorbs on a surface if the total Gibbs free energy is negative. In a simple medium where only one protein is present, the entropic factor (Norde 1985; Norde et al. 1986) seems to be the driving force for protein adsorption. Entropy is affected by the dehydration of both the surface and the protein, rearrangement of the protein structure and rearrangement of charged groups. The dehydration effect on hydrophobic surfaces can be viewed as the disruption of an ordered ice-like layer of water at the surface of the materials by the incoming protein. This disruption is due to surface–protein hydrophobic interactions that are stronger than the surface–water and protein–water interactions. This leads to an increase in entropy. On hydrophilic surfaces, the driving force is the change in protein structure (Norde et al. 1986).

In case of complex media, such as biological fluids, where different proteins coexist, the adsorption phenomenon can be regarded as a competitive adsorption (Vroman effect). The composition of the adsorbed layer may be very different from that of the bulk solution (Vroman 1962; Bamford et al., 1992).

From the above considerations, the following conclusions can be made:

- Protein adsorption on a surface depends on a combination of factors such as the solution physico-chemical characteristics (pH, ionic strength, temperature, composition, etc.), on the physico-chemical properties of the protein(s) and on the physico-chemical properties of the material surface (surface energy, roughness, surface chemistry, etc.).
- Protein adsorption can be reduced by two major means:
- By pre-adding a protein that will readily adsorb on the surface and displace the protein that one wants to analyse: BSA (bovine serum albumin), HSA (human serum albumin), casein, etc. The choice of one blocking protein over another will depend on the protein to be analysed.
- By developing a surface on which protein adsorption is thermodynamically not favoured. Neutral hydrophilic polymers such as polyethylene glycol (PEG) (Desai and Hubbell 1991; Jeon et al. 1991; Lee et al. 1997), polyhydroxyethyl methacrylate, polyvinyl alcohol, poly-*N*-vinyl 2-pyrrolidone, etc. seem to be ideal in this case. Some authors have described the use of polymers mimicking the lipid double layer (Ishihara et al. 1998; Ruiz et al. 1999).

In the case of carbohydrate-based biosensors, Table 7.3 summarises the blocking strategies commonly described in the literature. It has to be ensured that

Table 7.3 Reducing non-specific adsorption requiring passivation of the surface

Blocking	Substrate	Target	Refs
Denhardt	Glass	ConA, tetraglonolobus; EC	Zhou and Zhou(2006)
PVP K30	Polymer	Glycosyltransferase R	Seibel et al.(2006)
BSA 0.5%	Polymer	Lectin	Seibel et al.(2006)
BSA 1%	Code link-activated glass slides	Fibroblast growth factor	de Paz et al.(2006)
BSA 1%	Maleimide-modified glass slides	ConA tetraglonolobus; EC	Brun et al.(2006)
BSA 3%	NHS ester glass slides	Haemaglutinin	Stevens et al.(2006)
BSA 1%	PhotoChips	Plant lectin	Angeloni et al.(2005)
BSA 3 %	Epoxide-modified glass slides	Lectin	Manimala et al.(2006)
PVA	Hydrogel-modified glass slides	Ab	Dyukova et al.(2005)
BSA 3%	Glass	Lectin	Pei et al.(2007)
BSA0.005%	Alcyne-modified Gold	Lectin	Zhang et al.(2006)
BSA 0.1%	Polystyrene		Chevolot et al.(2001)
No blocking	NHS ester glass slides	Globo H/Ab	Huang et al.(2006a)
	Glass	Lectin and Ab	Bryan et al.(2004)
Casein/BSA	Nitrocellulose	Micronemal protein	Blumenschein et al. (2007)

The table summarises some strategies described in the literature the blocking protein does not have glycosylation motifs that can compete with the saccharide ligands and may interfere with the target molecule. Some authors have demonstrated that BSA can hide the probe immobilised at the surface from the target (Lesaicherre et al. 2002; Mac Beath and Schreiber, 2000).

7.4.2.2 Immobilisation of Saccharide Ligands

The immobilisation of the saccharides at the surfaces of the biosensor has been achieved through three main strategies:

- Physisorption
- Covalent immobilisation
- Biochemical-based interaction

7.4.2.2.1 Physisorption

Physisorption (Fukui et al. 2002; Wang et al. 2002; Satoh et al. 1999; Stoll et al. 2000; Palma et al. 2006) of carbohydrate ligands relies on the weak interactions between the molecules to be immobilised and the surface. This is a convenient technique, as it does not require surface functionalisation. Nitrocellulose or polystyrene are mainly used as the substrates.

In the case of low molecular weight species such as oligosaccharides, their interactions with the surface are too weak and require the modification of the saccharide with an anchoring tail for increasing the interaction, e.g. a lipid, a fluorescent tag (Jaipuri et al. 2008), a protein or a polymer such as polyacrylamide. Such a tail can be introduced by the condensation of the azido functional group with sp2- and sp3-bearing molecules (Fazio et al. 2002), by aminoreduction (Fukui et al. 2002) or by the condensation of amine with isocyanate (Fazio et al. 2004). The group of Wang at Stanford University and the group of Feizi at Imperial College, London, have developed neoglycoconjugates bearing such tails for their immobilisation on nitrocellulose membranes or nitrocellulose-coated glass slides. Oligosaccharides (2–20-mers) are linked to amino phospholipid 1,2-dihexadecyl-*sn*-glycero-3-phosphoethanolamine or its anthracene derivative by reductive amination. The anthracene derivative allows quality control of the relative surface density of the probes (Palma et al. 2006). The group of Wong has derivatised oligosaccharides with a C14 alkyl chain through $Cu^{(I)}$-catalysed Huisgen 1,3-dipolar cycloaddition. A similar strategy was used by Huang et al. (2005). These two groups have demonstrated in various publications that this strategy is very efficient, allowing the immobilisation of femtomoles of compounds and with a very good signal-to-noise ratio.

7.4.2.2.2 Covalent Immobilisation

The covalent immobilisation of the saccharide has been achieved by reacting with functionalised surfaces with unmodified saccharides or functionalised glycosides (saccharide bearing a functionalised aglycon).

7.4.2.2.2.a Immobilisation of Underivatised Saccharides (Fig. 7.23)

Activation of the hydroxyl functions can be achieved with a compound leading to good leaving groups under nucleophilic attack. Activating agents (CNBr, CNCl, divinylsulphone, isocyanate, etc.) have been reported for immobilisation of oligo- and polysaccharides. However, divinylsulphone may lead to intra-reactions between two adjacent hydroxyl groups.

Jouan et al. (1996) have described the immobilisation of saccharides by divinylsulfone cross-linking of amino groups and oligosaccharide hydroxyl groups. The main drawback is that the hydroxyl groups react randomly with the amino groups.

Reductive amination takes advantage of the reaction of the reducing end of the carbohydrate (or aldehydes obtained after oxidation under mild conditions agent such as nitrous acid) with an amine or an hydrazine followed by the reduction of the Schiff base by hydrides. Many examples have been reported. However, it requires long reaction times and the yields are low. To circumvent these drawbacks, other authors have immobilised saccharides through hydrazide (Lee and Shin 2005; Satoh and Matsumoto 1999) or oxime (Zhou and Zhou 2006) linkages. Thiazolidine and glycosyl amide bonds have also been described (Larsen et al. 2006). The hydrazide linkage is poorly stable at low pH, while oximes are stable over a wider range of pH (Larsen et al. 2006). Nevertheless, the immobilisation of unmodified saccharides is advantageous with complex carbohydrates from natural sources. However, the main drawbacks are that the immobilisation reactions are limited to reactions involving the reducing end of the saccharides (and consequently cannot be used with non-reducing sugars) and they often lead to the opening of the reducing end ring while it may be crucial to maintain the structure of the molecule (Larsen et al. 2006).

7.4.2.2.2.b Immobilisation of Saccharide Derivatised with a Functional Aglycon

It can be advantageous to derivatise the saccharide with a functionalised aglycon allowing for the subsequent immobilisation. Two main strategies can be envisioned. One is the use of an aglycon bearing a function that can react directly with the surface, such as thiolated or disulfide aglycons, which permits the direct immobilisation of the glycoside onto the metal surfaces (gold in particular, Fig. 7.24). The second is chemically modifying the surface with a hetero cross-linker bearing a functional group for reaction with the surface and a second functional group for reaction with the aglycon (Fig. 7.24).

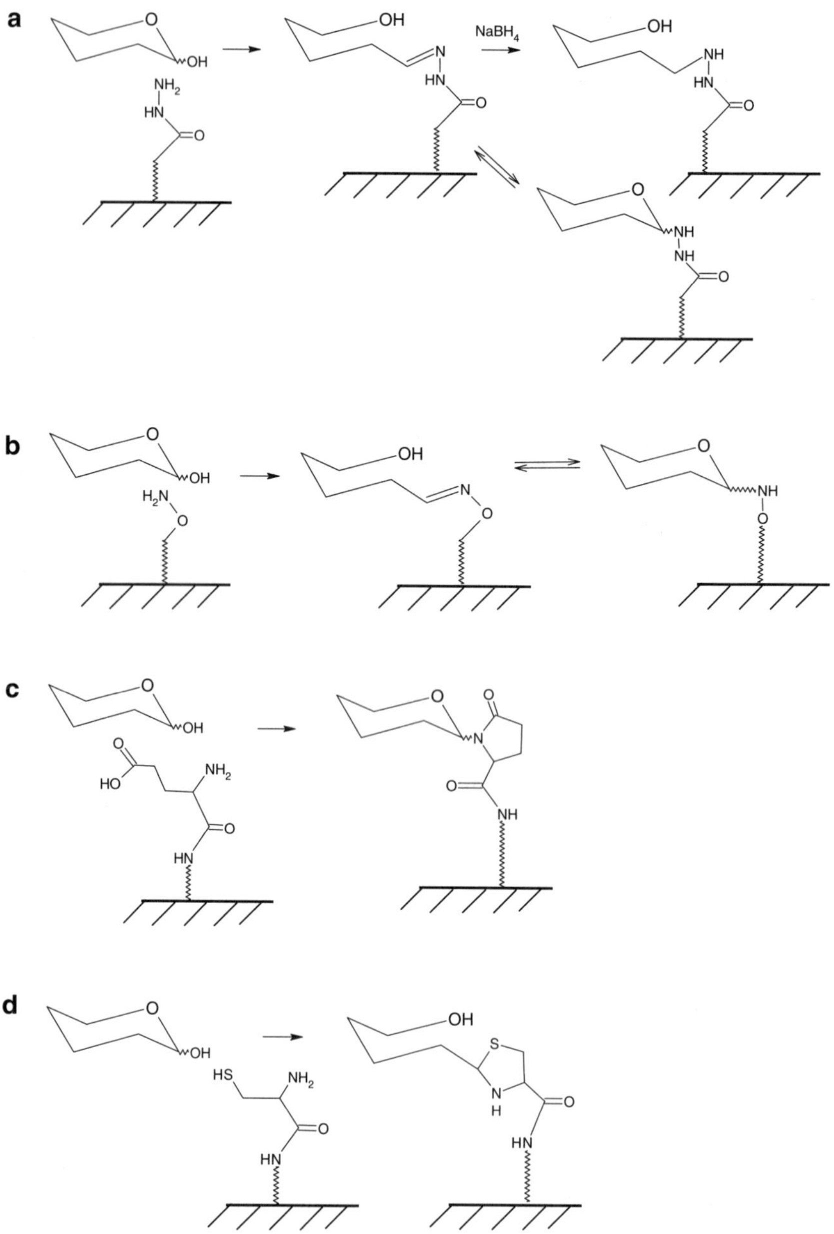

Fig. 7.23 Immobilisation of carbohydrate through their reductive end: **a** base Schiff or reductive amination, **b** oxime, **c** glycosyl amide bound formation and **d** thiazolidine. Immobilisation of underivatised saccharides usually take advantage of the reducing end of the saccharides as anchoring point, though immobilisation by the O-3, O-6 and N-2 have been described (Larsen et al. 2006; Seo et al. 2007)

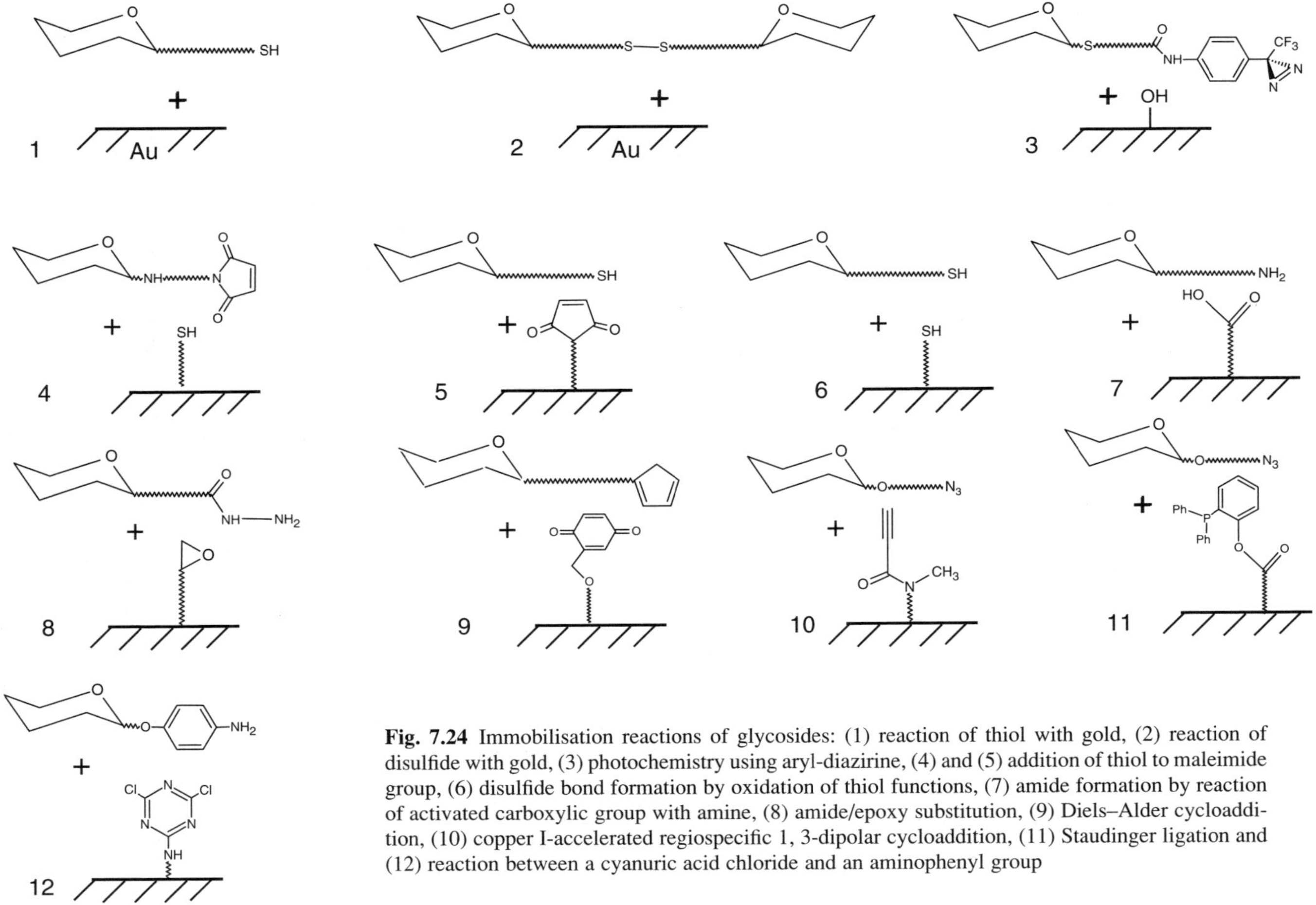

Fig. 7.24 Immobilisation reactions of glycosides: (1) reaction of thiol with gold, (2) reaction of disulfide with gold, (3) photochemistry using aryl-diazirine, (4) and (5) addition of thiol to maleimide group, (6) disulfide bond formation by oxidation of thiol functions, (7) amide formation by reaction of activated carboxylic group with amine, (8) amide/epoxy substitution, (9) Diels–Alder cycloaddition, (10) copper I-accelerated regiospecific 1, 3-dipolar cycloaddition, (11) Staudinger ligation and (12) reaction between a cyanuric acid chloride and an aminophenyl group

Direct reaction. Thiol functions have strong affinity for noble metals such as gold, silver, indium, etc. (see the corresponding chapter 2). Voltametric studies (Wink et al. 1997) have shown that upon adsorption thiols are deprotonated, leading to a covalent bond between the thiolate and the gold metal. Disulfide can also strongly bind to gold; however, the resulting films are less compact and less ordered. Many authors have taken advantage of these interactions for the immobilisation of thiol- (or disulfide-) containing biomolecules. In the case of carbohydrates, a mercapto aglycon should be introduced into the molecule.

Miura et al. (2002) have reported the immobilisation of Gb3 mimics (a saccharide ligand of the shigatoxins 1 and 2) bearing a disulfide aglycon on a gold substrate. The surface density of the immobilised Gb3 mimics was evaluated with quartz micro-balance measurements. The area per molecules was estimated to be 54 $Å^2$, leading to an estimated surface density of approximately 2×10^{14} molecules per cm^2. Ratner et al. have studied the affinity of the protein cyanovirin-N with mannosyl oligosaccharides. Immobilisation on gold substrates for SPR-binding affinity measurements was achieved with thiol derivatives of mannosyl oligosaccharides (Ratner et al. 2004).

Gold nanoparticles are often functionalised with saccharide ligands by mean of thiol reaction with gold (Chen et al. 2005). Robinson et al. (2005) have reported the synthesis of *N*-acetyl glucosamine encapsulated quantum dots (QDs). CdSe/ZnS core shell QDs were synthesized and subsequently encapsulated with pyridine. They were then allowed to react with disulfide-functionalised GlcNAc in the presence of $NaBH_4$, leading to GlcNAc-modified QDs. They have demonstrated by NMR an IR spectroscopy the reaction between the sulphur atom and the QD and the presence of the GlcNAc.

The synthesis of mannose-containing CdSe/ZnS QD has also been reported using mannose-containing phosphine oxide (Larsen et al. 2006). CdS QD modified with maltose or Lewis X was achieved using thiol or disulfide glycosides in the presence of sodium sulfide and cadmium nitrate (Larsen et al. 2006).

Other authors (Chevolot et al. 2001; Chevolot et al. 1999) have derivatised mono- and di-saccharides with photoactivatable aryl-diazirine. Immobilisation was performed by irradiation at 350 nm, leading to a reactive carbene, which reacted with the substrate. The advantage of this chemistry is the versatility of the reaction with respect to the substrate and the possibility of using photolithography for obtaining specific patterns of carbohydrate domains on the surface. In such a way, the immobilisation of galactose was performed on silicon, silicon nitride, diamond and polystyrene. Time-of-flight secondary ion mass spectroscopy (ToF-SIMS) analysis (Leonard et al. 1998; Léonard et al. 1998; Léonard et al. 2001) demonstrated that the hydroxyl groups present on diamond surfaces were involved in the immobilisation reaction. The resulting surface density was estimated to be in the range of 10^{12}–10^{13} molecules per cm^2. Square features of 25 μm were obtained using photolithography.

Surface immobilisation of saccharides can be achieved by the surface co-polymerisation of saccharide-containing monomers with monomers (without saccharides). Dubois et al. (2005) have developed an electrochemical approach based on a polypyrrole-coated electrode displaying pendant carbohydrates.

Reaction with functionalised surfaces. Several covalent immobilisations of oligosaccharides through reaction between an aglycon and a functionalised surface are described in the literature. The introduction of the aglycon can be performed by various means (refer to the corresponding paragraph on glycoside formation 7.2.3).

Surface functionalisation with the desired function such as carboxylic acid, amine, thiol and maleimide is achieved with a hetero-bifunctional cross-linker or coupling reagent. These molecules possess two functions: one that reacts with the surface and the other with the function of the aglycon. If necessary, the latter function can be protected during the reaction with the substrate. The coupling reaction may require the activation of one function, such as NHS activation of a carboxylic function, for reaction with an amine (see the corresponding chapter 2).

The usual yields are between a few percent and 10–20%, leading to a surface density between 10^{12} and 10^{13} molecules per cm^2. Therefore, high biomolecule concentrations are used for the immobilisation (mM range). To increase the yield, the saccharide solution is allowed to dry. This can lead to spot non-homogeneity, such as donut-like spots. This can be circumvented by adding moistening agents and surfactants (Dugas et al. 2005) and/or with good control of the evaporating conditions (relative humidity, temperature).

After performing the coupling reaction, the unreacted surface function should be disabled to perform further reaction. This is done by reacting the surface with a molecule bearing the same function as the aglycon but at a higher concentration. Usually this molecule is commercially available and cheap. This operation is sometime called "capping" of the surface. For example, NHS-activated surfaces after reaction with an amine-bearing carbohydrate are allowed to react with ethanolamine to deactivate the unreacted NHS esters (Fig. 7.25).

Immobilisation of carbohydrate have been achieved by taking advantage of

- thiol reactivity towards double bonds such as maleimido groups (Park et al. 2004; Brun et al. 2006; Ratner et al. 2004; Houseman et al. 2003) or towards thiol groups under oxidative condition, leading to the formation of a disulphide bond (Shin 2007);
- amine reactivity (or hydrazine) towards carboxylic (Blixt et al. 2004) (Consortium for functional glycomic (CFG)), towards aldehydes (Biskup et al. 2005) or towards epoxy groups (Lee and Shin 2005);
- cycloaddition reaction such as Diels–Alder (Houseman and Mrksich 2002) by copper I-accelerated regiospecific 1,3-dipolar cycloaddition (Huang et al. 2006a; Bryan et al. 2004; Seibel et al. 2006);
- Staudinger ligation (Kohn et al. 2003);
- the reaction between a cyanuric acid chloride and an aminophenyl group (Schwarz et al. 2003); or
- functional macromolecules such as glycoproteins, neoglycoproteins or polysaccharides.

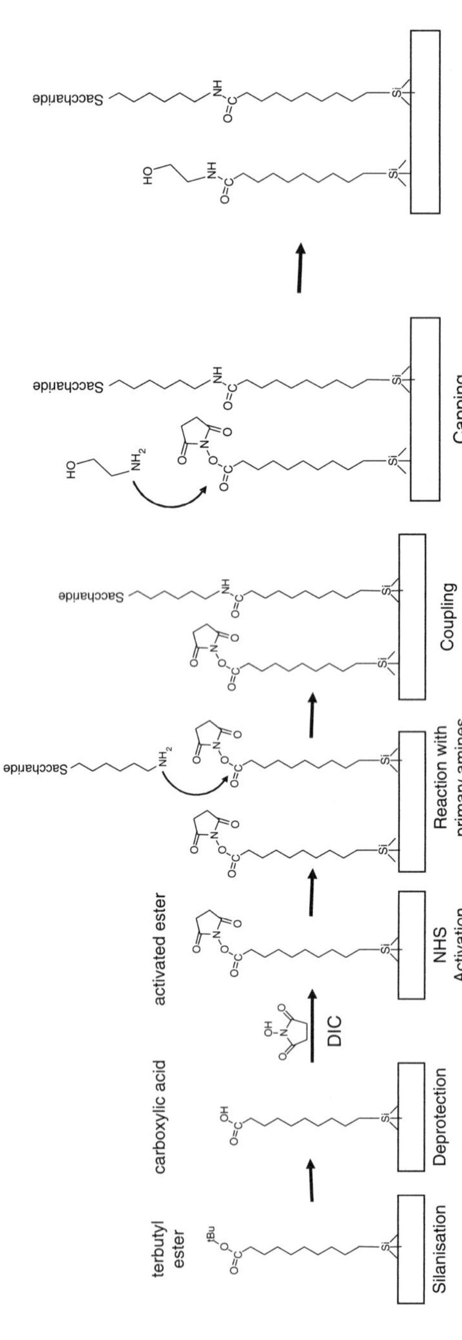

Fig. 7.25 Coupling of amine-derivatised saccharide with ester-activated surface. After the coupling reaction, the unreacted activated esters are capped with ethanolamine

Biskup et al. (2005) have reported the synthesis of amino, aldehyde and carboxylic acid glycoside for their immobilisation on amino- or formyl-functionalised glass slides.

A tremendous effort from the CFG led to a library of 200 (now 400) amino glycosides for subsequent immobilisation on NHS ester-activated glass slides. These slides have been used for a variety of applications such as antibody to viral protein affinity profiling.

The group of Seeberger in Zurich has reported the immobilisation of amine-derivatised heparin oligosaccharides onto NHS ester surfaces (de Paz et al. 2006), thiol oligomannosyl onto maleimide-functionalised surface (Brun et al. 2006) or aminoglycosides with disuccinimydyl carbonate or disuccinimidyl tetrapolyethyleneglycol cross-linker (Disney and Seeberger 2004a).

For 1,3-cycloaddition as well as Staudinger ligation, one takes advantage of azido-functionalised carbohydrates. In the latter case, Schwartz et al. have generated amine-functionalised surfaces using fourth-generation PMAM, a fourth-generation dendrimer, for increasing the number of reaction sites (amines). The surfaces were further treated to form the phosphane group, allowing subsequent Staudinger ligation (Kohn et al. 2003).

Macromolecules have been used as spacers for the immobilisation of carbohydrates taking advantage of functional groups such as amines, carboxylic acids, thiols or diazirines.

For example, BSA glycoconjugates and glycoproteins were immobilised on epoxy-derivatised glass slides (Manimala et al. 2006; Ratner et al. 2004; Adams et al. 2003). The saccharides were covalently link to BSA by reaction of thiol-modified carbohydrates with maleimide-derivatised BSA (Adams et al. 2003; Ratner et al. 2004) or by reaction of NHS ester-modified carbohydrate with BSA (Manimala et al. 2006; Manimala et al. 2005). The neoglycoprotein was then immobilised on NHS ester-activated surfaces (Ratner et al. 2004; Adams et al. 2003) or epoxy-derivatised glass slides (Manimala et al. 2005). Seeberger's group has used these strategies for immobilising carbohydrates on plane substrates as well as on microspheres (Adams et al. 2003).

Angeloni et al. (2005) described the immobilisation of oligosaccharides after surface functionalisation with Optodex. The latter is a dextran molecule bearing aryl-diazirine, allowing surface immobilisation by converting the diazirine into a reactive carbene under light activation.

As one can see, many different reactions have been employed for achieving the immobilisation of saccharides. However, it is difficult to draw a general conclusion, as the surface densities of the probes are rarely given and the washing protocols are often not the same. Dugas et al. (2004) have demonstrated that the final surface density of aminolinker-derivatised oligonucleotides immobilised on the NHS-activated ester surface is greatly dependent on the washing step. A quick rinsing with water led to 10^{14} molecules per cm^2, whereas rinsing with hot water led to 10^{13} molecules per cm^2. A further rinsing with SDS led to 3×10^{11} molecules per cm^2. This means that a significant amount of the probes can be adsorbed rather than covalently linked depending on the washing steps. In consequence, surface-oriented immobilisation can also be effected.

Linker effect. Affinities of targets for surface-immobilised probes depend on the saccharide structures and also on the aglycon structure. Miura et al. (2002) have demonstrated that shigatoxins 1 and 2 affinities for immobilised Gb3 mimics (a saccharide ligand of the shigatoxins 1 and 2) were affected by the aglycon alkyl chain length. Gb3 mimics were derivatised with a disulfide bearing 2 or 10 carbon alkyl chains as a spacer. The resulting molecules were self-assembled onto gold surfaces and the affinity of Shigatoxins 1 and 2 were studied. They found that Shigatoxin 1 binds to 2-carbon-derivatised Gb3 five times more strongly than 10-carbon-derivatised Gb3, while Shigatoxin 2 binds more strongly to 10-carbon-derivatised Gb3. Some authors (Sato et al. 1998) have demonstrated that the availability of the carbohydrate probe can be spoiled by steric hindrance due to compact packing. It may that the observed difference is related to the surface organisation of the probe. Indeed, it was demonstrated that the C10 long spacer forms compact layers compared to the C3 long spacer (Grubor et al. 2004).

Huang et al. (2006a) compared the affinity of antibodies for Globo H analogues. Globo H is a glycosphingolipid which is highly expressed on a wide range of tumoral cell lines. They synthesised four analogues of increasing complexity of the saccharide moiety of Globo H with two different aglycons bearing either an amine function or an azido function. The amino derivatives were immobilised on NHS ester-functionalised surfaces, while the azido derivatives were immobilised on alkyne-bearing surfaces through 1,3 dipolar cycloaddition. They found that the affinity of the mouse antibodies (MBr1 or VK-9) for the derivatives linked through an amide function was higher than those linked through a 1,2,3-triazol ring. They interpreted the observation as a difference of solubility between the two types of derivatives, a difference in the coupling yield or the better suitability of the shorter linker for binding.

The overall affinity of the target for the immobilised probe is a complex process, which should be viewed as the combination of the contribution of probe structure, probe surface density, probe availability (which is related to the probe orientation and to the physico-chemical properties of the spacer), the surface physico-chemical properties of the substrate and the target physico-chemical properties (Yeung and Leckband 1997).

7.4.2.2.3 Biochemical Interaction Based Immobilisation

Bochner et al. (2005) reported the use of streptavidin–biotin interaction for the immobilisation of sialic acids. One hundred and eighty biotinylated glycosides were immobilised on streptavidin-coated polystyrene titer plates and on streptavidin-coated gold surfaces for SPR measurements (CFG). Linhardt's group immobilised biotinylated heparin for studying the affinity and the kinetic of factor P/heparin interaction (Muñoz et al. 2005). Heparin was conjugated with biotin by the reaction of sulfo-NHS-LC-biotin with the free amino groups of unsubstituted glucosamine residues in heparin.

Chevolot et al. reported the use of DNA hybridisation for surface immobilisation of glycomimetics (Bouillon et al. 2006; Chevolot et al. 2007). Glycomimetics were synthesised as follows: azido glycosides were attached to pendant propargyl residues on a phosphorylated scaffold by means of solid-supported 1,3-dipolar cycloaddition. The architecture (distance between two saccharides residues, hydrophilic–lipophilic balance) was designed by adjusting the linker between two pendant propargyl residues. Each glycomimetic was tagged with a DNA sequence bearing a cyanine 3 (Cy3) dye. The resulting molecules were immobilised on a DNA array (Dugas et al. 2004) bearing complementary sequence by means of DNA/DNA hybridisation. The fluorescent dye borne by the oligonucleotide permits a quality control of the immobilised molecules. Hybridisation was performed with a 1 μM solution of Cy3 –DNA glycomimetic, leading to a probe surface density of 1–4 $\times 10^{10}$ molecules per cm^2 as estimated by Dugas et al. with radiolabelled oligonucleotides immobilised under the same conditions. The resulting analytical device had a lower detection limit in the nanomolar range (albeit the use of a 1 μM solution of glycomimetic), which is similar to what is described in the literature where typical immobilisation solution concentrations are in the millimolar range.

In the field of protein array, Niemeyer et al. (Wacker et al. 2004) have used a similar strategy called DNA directed immobilisation (DDI). They also found that immobilisation by means of hybridisation give lower detection limit of the device, which is improved, as opposed to direct covalent immobilisation. They hypothesised that the improvement was due to the rigid structure of the DNA duplex and to the reversibility of the hybridisation allowing the denser packing of the probes.

Another advantage of the DDI strategy is that biological recognition between the glycomimetics and the lectin can be first performed in solution before immobilisation of the complex on the solid support by hybridisation, circumventing the limitations due to the surface proximity.

7.4.2.2.4 Surface Cluster Effect or Not?

Individual protein–carbohydrate interactions have binding constants in the millimolar to low micromolar range, with apparent low specificity. However, under multivalent interactions in a Velcro-type mode, these interactions can be enhanced. This is the so-called cluster effect as defined by Lee in 1994(Lee and Lee, 1994, 1995): “binding affinity enhancement exhibited by a multivalent carbohydrate ligand over and beyond that expected from the concentration increase resulting from its multivalency.”

Two different mechanisms have been proposed to explain the cluster effect (Pohl and Kiessling 1999):

- Due to multivalency, the local concentration of the ligand is high at the receptor-binding site allowing an increase of 5- to 10-fold in binding affinities.
- Due to intra molecular binding, which is the chelate effect, a multivalent ligand binds with greater affinity for a multivalent receptor than the monovalent ligand, similar to the mechanism described in inorganic chemistry. It leads to an exponential increase of the affinity.

Several reviews have been published on this topic (Lundquist and Toone 2002; Mammen et al. 1998; Turnbull and Stoddart 2002).

In the case of biosensors, packing of carbohydrate at the surface of a material should permit taking advantage of the cluster as opposed to solution assays. For example, in the case of neoglycolipid-based glycol array (Jelinek and Kolusheva 2004; Stoll et al. 2000), lipid clustering and surface oligosaccharide organisation allowed higher affinity of proteins for the saccharides. On the contrary, Sato et al. observed a decreased WGA affinity by the high probe surface. They studied the affinity of WGA (Wheat Germ Agglutinin) for different glycolipids and found that the maximum affinity of WGA for GM 3 and GM4 was for a monolayer containing 20% GM (Sato et al. 1998). Similar results were found by Ebara (Ebara and Okahata 1994) with concanavalin A and glycolipids. This phenomenon was attributed to steric hindrance due to the compact packing of the carbohydrate heads in gangliosides monolayers (not mixed) limiting the access to WGA.

These apparent contradictory results suggest that the surface-based cluster effect is dependent on surface density of the probes and their surface organisation. This latter parameter is dependent on interactions between the probes through van der Waals interactions, polar, ionic forces and hydrogen bonds. Usually, when immobilising carbohydrates via a functional aglycon with a functionalised surface, the surface densities of the probes are below that of a densely packed monolayer. On the other hand, direct immobilisation of glycosides leads to denser layers.

7.4.3 Transduction

The specific biological interaction between the probe and the target can be studied by optical, gravimetric or electrochemical detection and more recently by mass spectroscopy (MS) or by the use of nanoparticles-based assays.

The detection of biomolecular interactions may require the pre-labelling of the target prior to detection or may be label free and exploit the change of physico-chemical properties induced by the biological recognition.

In the first case, strategy consists in labelling the molecule to be detected (directly or via sandwiches) or grafting different markers on probes (Cy3) and targets (Cy5) for discriminating the specific molecular interaction in a better way and requires appropriated techniques for detecting the signal associated with the label. The markers can be fluorescent, radioactive or electroactive species. Detection by fluorescence is a semi-quantitative and sensitive technique with high spatial resolution but the quenching and blenching phenomena must be monitored. Nowadays, read-out and analyses are made with standard commercial fluorescence scanners. Detection using radioactive labels such as ^{125}I, ^{32}P or ^{33}P is a more quantitative method but the weak spatial resolution and both the inherent safety and waste disposal problems have directed studies towards alternative solutions. Spe-cificity of the recognition signal must be ensured by careful washing protocols to remove non-specifically labelled probes adsorbed out of the surface without breaking the specific binding. A drawback is that the detection is performed a posteriori, out of the liquid environment and in dry conditions, which hides the analysis of the kinetics of the interactions. Advantage of these methods is that they are particularly well adapted for the parallel and massive detection analysis of numerous spots, as for instance with the DNA biochips performed by Affymetrix containing more 400,000 spots per cm^2.

In the second case, the detection strategy is based on the measure of the change of properties such as temperature, mass, electrical, optical and chemical (pH) parameter. The necessary development of new and reliable tools designed for detecting low parameter change has considerably increased on the two last decades. The release of commercial systems such as microcalorimeter, quartz crystal microbalances (QCM), field effect transistors (FETs), surface plasmon resonance (SPR) and, more recently, platforms based on piezoelectric acoustic sensors (principally bulk acoustic wave (BAW)) and thickness shear mode (TSM) sensors are leading a large number of studies determining binding specificities, affinities, kinetics and conformational changes associated with a molecular recognition event. The investigations concerned a large variety of interactions involving small molecular weight ligands, carbohydrates, proteins, nucleic acids, viruses, bacteria, cells as well as lipidic and polymeric interfaces. Beyond the direct and in situ interaction detection, the main advantage of this strategy is its ability to obtain kinetics informations on molecular interactions.

7.4.3.1 Optical Detection

7.4.3.1.1 Fluorescence

Fluorescence is one of the most commonly used detection techniques in the field of microarrays. It relies on fluorescent dyes (see Table 7.4). These molecules are capable, under illumination with a monochromatic light corresponding to their absorption maximum, of reaching an excited state. They return to the ground state by emitting a photon. The energy of the outcoming photon is less than that of the absorbed photon due to loss of energy by a non-radiative process. The difference between the maxima (in wavelength) in the excitation spectrum and the emission spectrum is called the Stokes shift. The ratio between absorbed photon and emitted photon is the fluorescence quantum yield. Table 7.4 gives some of the commonly used fluorescent dyes. Fluorescein and related molecules (Alexa, Oregon green, rhodamine, etc.) are four-ring structures with conjugated double

Table 7.4 Some of the physical properties of commonly used fluorophores (Schena 2003; http://www.invitrogen.com/)

Fluorescent probe	Exitation (nm)	Emission (nm)	Fluorescence quantum yield	ε
Fluorescein	490	519	0.71	67,000
Rhodamine 6G	525	555	0.9	85,000
Tetramethylrhodamine	555	580		176,000
Indodicarbocyanine Cy3	548	562	0.14	150,000
Indodicarbocyanine Cy5	646	664	0.15	250,000
Alexa 488	495	519	0.92	
Alexa 568	578	603	0.69	
Alexa 647	647	668	0.33	
Texas red	596	620	0.51	85,000
Bodipy650/665	646	660		

This is not an exhaustive list. Data were collected according to Invitrogen catalogue (for Alexa and microarray analysis)

bounds. FITC (fluorescein isothiocyanate) is affordable but has several drawbacks (Schena 2003). It is sensitive to self-quenching and photobleaching, and the fluorescent signal is influenced by pH (Schena 2003). The parent molecules of FITC (Alexa, Texas red, Oregon green, etc.) have additional polar, aliphatic and/or hetero cyclic groups that influence their excitation and emission spectra. These molecules are in general photostable. Furthermore, Alexa dyes are insensitive to pH change over a broad range, have high fluorescence quantum yield and a long half-life of the excited state (Schena 2003).

A second family of fluorescent dyes is of the indocarbocyanine series. The most used ones are Cy3 and Cy5. The cyanine dyes are bright, are stable to photobleaching, and are commercially available with chemical function allowing their conjugation with biomolecules. Their fluorescent signal is affected by ozone exposure as low as 5–10 ppb for Cy5 and above 100 ppb for Cy3 (Fare et al. 2003).

A third family of organic dyes used in microarrays applications is the BODIPY (boron-dipyrromethene) series. Next to these organic dyes, QDs are emerging as an alternative. Huang et al. (2005) have compared the lower detection limits for a FITC-labelled lectin plant and a QD-labelled lectin plant (Con A) interacting with immobilised D-glucopyrannoside. They found that the QD-label-based method was at least a 100 times more sensitive than the FITC approach in detecting ConA interaction.

A biosensing experiment relies on (a) the labelling of the target or (b) the detection of the target through a second interaction using a sandwich technique with antibodies, for example (Fig. 7.26).

Dyes have been modified for allowing their direct conjugation with molecules bearing amine, thiol, etc. In a fluorescent experiment, after incubation of the device with the protein of interest, the substrate is washed and dried and the remaining fluorescent signal is detected with a microarray scanner.

One of the main advantages of fluorescence is that the technique is very sensitive, allowing the detection of plant lectin with concentration as low as a few tens of picomolars (Manimala et al. 2006). However, generally no real-time experiment can be performed. Furthermore, the washing step is compulsory and

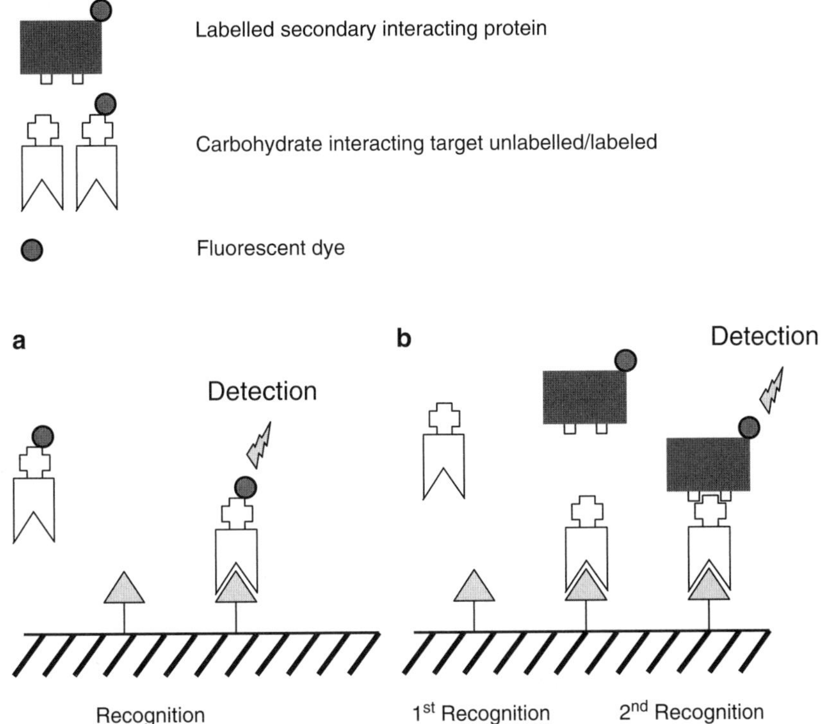

Fig. 7.26 Principle of fluorescent detection: (**a**) When target is previously labelled, then the recognition interaction is directly detected. (**b**) Sandwich technique consists in a first recognition step between probe and unlabelled target following by a secondary recognition step between the formed complex and a labelled interacting protein

critical in terms of background levels. It is tricky to find an optimum between removing non-specifically adsorbed protein and maintaining a protein that interacts weakly with the carbohydrates.

Instruments on the market (Zeptosens (http://www.zeptosens.com/en/; Disney and Seeberger 2004b)) take advantage of evanescent field for fluorescent dye excitation. By this means, real-time experiments can be performed without washing the surface. Indeed, evanescent field decreases exponentially with the distance to the surface. Consequently, only the fluorescent dyes next to the surface are excited.

Fluorescent-labelled plant lectins have been widely used as model protein for the validation of carbohydrate microarrays. The best lower detection limits, depending on the considered model, are in the nanomolar range.

Stevens et al. (2006) have used a double sandwich for probing the interaction of the influenza virus hemagglutinins of different sub-types for more than 200 saccharides displayed on a microarray. Due to the low affinity of the hemagglutinin–carbohydrate interaction, they developed an approach based on two antibodies. The hemagglutinins (HA) bear a His6 tag. First, hemagglutinins are allowed to

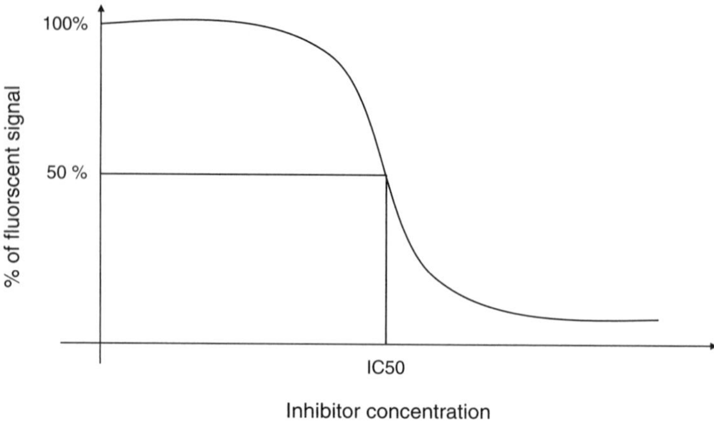

Fig. 7.27 Basic principle of an IC_{50} experiment

interact with the microarray. Next, the slides are incubated with a labelled anti-His6 mouse antibody followed by the incubation with a second labelled antibody directed against mouse antibody. With this approach, the signal is amplified, allowing the probing of weak interactions.

Usually, fluorescence give semi-quantitative results. The group of Shin (Shin 2007) has developed a procedure for quantitative analysis by determining the IC_{50} based on inhibition experiments (Fig. 7.27). In brief, the microarray was incubated with increasing amounts of a soluble inhibitor. The concentration that produces 50% of inhibition is called the IC_{50}.

An original glycoarray was developed using internally encoded microspheres (Adams et al. 2003). Each population of microspheres has a unique carbohydrate molecule covalently attached and a unique spectral signature due to an internal fluorescent dye. The microspheres are randomly spread over an imaging optical fibre. Due to the spectral signature, each microsphere is identified at the surface of the fibre with regard to the immobilised carbohydrate and position. Interactions with labelled Concanavalin A, a plant lectin, and Cyanovirin A, an HIV-inactivating protein issued from a cyanobacterium, are performed and analysed by fluorescent imaging.

7.4.3.1.2 Surface Plasmon Resonance and Surface Plasmon Resonance imaging

Surface plasmons are surface electromagnetic waves propagating at the metal/dielectric interface. Surface plasmons can be excited by an incident polarised light especially by means of an evanescent field. The total internal reflectance of the incident light will be affected as a function of the incident or reflection angle leading to a sharp minimum.

The position of this minimum is dependent on the refractive index and thickness of the surface layer. Change in the surface refractive index on the surface of the metal will thus be probed by either measuring the change in total internal reflectance or by the change in the minimum position.

Two configurations can be used: namely, the Otto set-up and the Kretschmann set-up. The Kretschmann set-up is the most widely used. In this set-up, a thin layer of gold, typically 50 nm, is evaporated on a glass substrate (with an intermediate adhesive layer such as chromium). An incident polarised light travels across the glass, and an evanescent field penetrates through the metal film and excites the surface plasmons of the outer side of the metal film.

Its use for biosensing began in the late 1980s (Lundquist and Toone 2002) and was further implemented in the early 2000s with the first commercially available surface plasmon resonance imaging (SPRi) apparatus.

SPR for detecting carbohydrate-based interactions has been reported with antibodies (Zou et al. 1999; Young et al. 1999), membrane proteins, viruses proteins (Ratner and Seeberger 2007), chemokine (de Paz et al. 2007), fibroblast growth factor (de Paz et al. 2007), plant lectin (Smith et al. 2003; Satoh and Matsumoto 1999), hSiglec7 (Karamanska et al. 2008) and enzyme (Park and Shin 2007).

SPR offers many advantages: firstly, it is a label-free detection technique; secondly, it provides rates (K_{on} and K_{off}) allowing the determination of affinity constants; and thirdly, it is a real-time measurement. The main drawback relates to its lower detection limit, which is not as good as that of fluorescence. Also, the on and off rates seem to be affected by the flow vector (mass transport) and by the orientation of the molecules at the surface of the sensor (Lundquist and Toone 2002). Recent developments have focused on coupling SPR plates with matrix assisted laser desorption/ionization-time of flight (MALDI-TOF) mass spectrometry (Lundquist and Toone 2002).

7.4.3.1.3 Dual Polarization Interferometry

Yates (Yates et al. 2006) has pointed out another optical technique based on dual polarization interferometry (DPI), which measures real-time changes in the mass and the geometry of molecules bound to a surface. DPI measures in phase both TE (transverse electric) and TM (transverse magnetic) polarizations of light. Using Maxwell's equations, one unique value for thickness and for refractive index of the layer of molecules is true for both phases of light. DPI gives absolute measurements and allows full characterisation in terms of mass, thickness and refractive index of the layer to be formed. It allows measurements in real time of probe coverage and probe orientation and then to follow in real time the biological interaction. The measurement by DPI immobilization of streptavidin on a biotinylated surface followed by the capture of the reducing end biotinylated heparin oligosaccharides is given as an example.

7.4.3.2 Gravimetric Detection: Quartz Crystal Microbalance

Among the analytical tools available to investigate the biological interactions, QCM holds a special place because this direct measurement does not require probe-labelling and allows monitoring the binding affinity. In their recent review, Cooper and Singleton (2007) register a total of 1,404 publications referencing QCM in the Web of Sciences database, of which 569 specifically involved molecular recognition studies in the period 2001–2005.

In 1959, QCM was developed by G. Sauerbrey (Sauerbrey 1959) for correlating changes in the oscillation frequency of a piezoelectric crystal with the mass deposited on it. He simultaneously developed a method for measuring the characteristic frequency (resonant frequency) and its changes by using the crystal as the frequency-determining component of an oscillator circuit. His method continues to be used as the primary tool in QCM experiments for the conversion of frequency to mass and is valid in nearly all applications under vacuum or gas.

The Sauerbrey's equation is given below:

$$\Delta f = \frac{-2\Delta m f_0^2}{A\sqrt{\rho_q \mu_q}} = -\frac{2f_0^2}{A\,\rho_q v_q}\Delta m$$

where f_0 is resonant frequency (Hz), Δf is the frequency change (Hz), Δm is the mass change (g), A is the piezoelectrically active crystal area (m^2), ρ_q is the density of quartz (ρ_q = 2.648 g/cm^3), ρ_q the shear modulus of quartz for AT-cut crystal ($\mu_q = 2.947 \times 10^{11}$ g/cm s^2) and v_q is the transverse wave velocity in quartz (3,340 m/s).

QCM measures the mass per unit area by measuring the change in frequency of a quartz crystal resonator. The resonance is disturbed by the addition or removal of a small mass on the surface of the acoustic resonator. When a thin film is attached to the sensor crystal, the frequency decreases. If the film is thin and rigid, the decrease in frequency is proportional to the mass of the film. In this way, the QCM operates as a very sensitive balance. The mass of the adhering layer is calculated by using the Sauerbrey's relation. For instance, with a crystal of resonant frequency 6 MHz, and considering an active crystal surface of 0.2 cm^2 and a frequency resolution of 1 Hz, we can detect a change of 2.46 ng. For a long time, the main application of QCM consisted in monitoring thin-film deposition in vacuum or in gas phase (King, 1964). When QCM was used in liquid environments (Bruckenstein and Shay 1985), the number of applications increased dramatically, in particular in the biology field. For instance, an adsorbed protein monolayer on a surface electrode of 1 cm^2 adds a mass of a few hundred nanograms (Muratsugu et al. 1993). But, the Sauerbrey equation developed for oscillation in air must be applied only to rigid masses attached to the crystal. Besides the mass, other physico-chemical parameters (such as viscosity, roughness, temperature) can induce a change in the resonant frequency. Thus, in liquid environment, Kanazawa and Gordon (1985) showed that quartz crystal microbalance measurements can be performed, in which case a viscosity-related decrease in the resonant frequency will be observed:

$$\Delta f = -f_0^{3/2}(\eta_l\rho_l/\pi\rho_q\mu_q)^{1/2}$$

where ρ_l is the density of the liquid and η_l is the viscosity of the liquid.

The advantage of QCM in biological measurements is connected with its ability to monitor the changes of small masses in real time, and such measurements can be performed by use of the native or synthetic molecules without any supplementary labelling.

Specifically, a QCM is constituted of a thin quartz crystal resonator disc attached with metal electrodes plated onto the upper and lower faces of the disc. Gold is usually used as the electrode material. When an alternating electric field is applied across the quartz crystal between the upper and lower electrodes, a mechanical oscillation of the characteristic frequency (f) is generated in the crystal. This resonant frequency shifts as a function of the additional bound material occurring at the upper surface. For adapting this system to molecular recognition, the metal surface is specifically biofunctionalised with a selective, passivating layer on which the analyte-specific receptor is immobilized. Appropriate flow channels with monitored microfluidics allow controlled introduction of the analyte. Nowadays, commercial QCM integrate user-friendly software for characterising the biomolecular interactions at the surface. A large variety of surface coatings have been proposed, allowing several optimized chemical treatments and biomolecule as the receptor grafting depending on the aimed interactions. Apart from the standard gold material, which is the most commonly used sensor surface because Au is chemically inert and easy to modify in different ways, we can find in product catalogues (Q Sense) electrode materials such as silicon dioxide whose main advantage is the ability to form lipid bilayers which act as a platform for further immobilisation of molecules; stainless steel surfaces used for studies of biofilm formation; titanium-coated surfaces mainly used for implant research; metals and oxides (aluminium trioxide, platinum, tantalum, tantalum nitride, tungsten, copper, chromium, iridium, iron, silicon carbide, iron carbide, silver and cobalt); and more recently commercially available polymers (polysterene, PMMA) deposited by spin coating.

QCM-D. In biochemical applications, as macromolecules (proteins, polymers, etc.) deposited on the surface do not constitute a thin rigid layer, their viscoelastic behaviour leads to a vibrational energy loss by dissipation. That is why, in addition to measuring the frequency, the dissipation is often measured to help with the analysis.

The dissipation (D) is defined as:

$$D = \frac{E_{\text{lost}}}{2\pi E_{\text{stored}}}$$

where E_{lost} is the energy lost (dissipated) during one oscillation cycle and E_{stored} is the total energy stored in the oscillator.

The QMC-D consists in measuring the dissipation in the crystal by recording the response of a freely oscillating crystal at its resonance frequency. This also gives the opportunity to jump between the fundamental frequency and overtones (e.g. 15,

25 and 35 MHz). By working at multiple frequencies and applying a viscoelastic model (the so-called Voight model), the adhering film can be characterized in detail; viscosity, elasticity and correct thickness may be extracted even for soft films when certain assumptions are made.

The QCM can also be combined with other surface-analytical instruments. The electrochemical QCM (EQCM) is particularly advanced (Schumacher 1990; Buttry and Ward 1992). Using the EQCM, one can determine the ratio of mass deposited at the electrode surface during an electrochemical reaction to the total charge passed through the electrode. This ratio is called the "current efficiency." This methods allows one to compare the deposited mass value estimated from current flowing between the working electrode (which is also the resonant electrode of QCM) and the counter electrode with the value deduced from the frequency shift of the QCM. The shrewdness of the method concerns the double use of the electrode (as the resonant QCM electrode and as the electrochemical working electrode) and the possible electrical cutting of the two functions (electrochemical and piezoelectric transducer) thanks the two different ranges of frequency.

The binding between two molecules can be quantitatively described by two parameters: the association equilibrium constant and the association rate. QCM measurements give access to these parameters. But we must keep in mind that theses values are strong depending on the experimental conditions (temperature, solution ionic strength and pH value). They also reflect structural properties of the binding sites. For instance, conformation of proteins on the surface affect drastically their recognition activities, the spacer between the surface and receptor appears as crucial factor and, in the case of lectin–carbohydrate interaction, the number of sugar monomers in the oligosaccharide chain strongly influence the interaction. Thus, uncontrolled immobilisation of receptor on biosensor surface could lead to meaningless results. The large number of parameters influencing the experimental results make a direct comparison of values (sensitivity, assay time, detection limit, association rate)reported in the literature very difficult . In their review, Cooper and Singleton (2007) have collected several tables comparing QCM assays and results for DNA sequences, for virus and phage, for selected pathogenic bacteria and for small-molecule detection. This drawback is also noticed by Lebed et al. *(*2006*)***.** Therefore, they discuss carefully the lectin–carbohydrate affinity values obtained by QCM measurement and compare them with others' previously published results.

Nevertheless, this technique used in the similar and controlled conditions allows comparison of the binding behaviour of a set of monosialogangliosides (GM1, GM2, GM3, GM4) towards wheat germ agglutinin as described by Sato et al. *(*1998*)* as far back as 1998. More recently, Pei et al. (2007) demonstrated the great interest in the use of QCM sensor for rapid identification of glycosyldisulfide lectin inhibitors from a dynamic combinatorial library. Gold-plated quartz crystals were drop-coated with polystyrene before being covered with mannan in phosphate-buffered saline (PBS), and complete blocking was achieved with BSA to obtain a stable baseline. The plant lectin concanavalin A was injected as target species alone or in presence of potential inhibitor library components. Between each screening of libraries, the Con A binding to the surface must be carefully released.

Yates et al. (2006) affirm that QCM-D provides a unique means to explore the change in mass and the molecular flexibility of oligosaccharides, GAG (glyco-aminoglycans) chains and entire proteoglycans and their complexes with proteins thanks to the ability to form membranes mimetics on the sensor surface in an environment that approximates that of the cell surface.

Nagase et al. (2003) used a lectin coated on the silver face of a piezoelectric crystal biosensor for oligosaccharide analysis. They used it for the detection of dissolved sugars and identification of erythrocytes, and obtained a detection limit approaching 100 cells.

Miura et al. (2004) used the QCM for investigating the biological functions of their synthetic glycoconjugate polymers with respect to lectins and Shiga toxin-1 (Stx-1) produced by the pathogen *Escherichia coli*. QCM detector was covered by a Gb3C10 SAM. First they demonstrated that the addition of Stx-1 to aqueous solutions induced a frequency reduction depending on the concentration. The binding affinity of Stx-1 to Gb3C10 SAM was estimated to be 1.2×10^8 (M^{-1}). Secondly, upon addition of their glycoconjugates, they exhibited a notable inhibitory activity for the adhesion of Stx-1 to the Gb3C10 SAM (degree of inhibition going from 15% up to 70% versus glycoconjugate).

7.4.3.3 Electrochemical Detection

The use of electrochemical methods involves the molecular recognition-induced redox reaction. This is particularly true for enzymatic systems. The most basic amperometric enzyme electrode concerns glucose, which was first oxidized within an immobilized layer of glucose oxidase and then determined at a conducting (platinum or carbon) electrode by measuring the current resulting either from the oxidation of hydrogen peroxide or reduction of diatomic oxygen consumed by the enzymatic reaction (Updike and Hicks 1967).

An electrochemical biosensor for determination of the tumour marker-bound sialic acid (b-SIA) was reported by Aubeck et al. (1993). This sensor consisted of a copolymer-immobilized bilayer containing the enzyme sialidase, placed in contact with a H^+ selective indicator membrane. The release of sialic acid following enzymatic cleavage resulted in a local pH change monitored by the proton-sensitive indicator electrode.

Potentiometric detection can be used if the binding of biomolecules induces large non-equilibrium potentiometric responses. One type of device named ISFET (ion sensitive field effect transistor) sensors uses this event and is derived from MOSFET (metal oxide semiconductor field effect transistor), the most important component in integrated circuits. In 1970, Bergveld et al. (Van Kerhof et al. 1993) showed the sensitivity towards ions H^+ of a MOSFET without metallic grid when the blocking silicon layer is directly in contact with the solution.

ISFET is made of a p-type silicon substrate with two areas of n-doping (called source and drain) on which metallic electrodes are applied. The channel between the source and the drain is covered of a thin blocking SiO_2 layer. Metallisation of this layer forms the gate, which is the electrode monitoring the conductivity in the channel. In the case of ISFET, an ion-sensitive membrane is put in place of the metallic layer. This structure is intrinsically blocked, so no current flows between source and drain because whatever the potential applied at this two areas, one of the p-n junction is in reverse polarisation. When a positive voltage V_G is applied between the gate and the source, the majority of holes tend to be repelled while the minority electrons of the p-substrate move to the channel. For $V_G > V_T$ (V_T is the threshold voltage), the electron density becomes larger than the hole density and the formation of n-type channel ensures electrical continuity between the source and the drain; the current flows from source to the drain. I_D is a function of V_G and V_D voltages. If $V_D < V_G - V_T$, the FET is in the linear range; for $V_D > V_G - V_T$, FET is in the saturation range.

$V_T = V_{T0} + y_0$, where V_{T0} is relative to the ISFET characteristics and y_0 is the potential between the sensitive membrane and the electrolytic solution.

Taking into account the equilibrium between the ion analyte in solution and its chemical conformation inside the sensitive membrane, y_0 is dependent of the activity of the ions to be detected. The following figure gives the shape of I_D variation for ISFET pH with an alumina sensitive membrane.

The high concentration of negative charges on heparin makes the detection of the variations in charge density in porous membranes mounted on ISFET easy. Thus, heparin biosensors based on the ISFET concept exhibit very high sensitivity thresholds of between 0.1 and 1 units per millilitres (Van Kerhof et al. 1993).

A similar transducer can be used for detecting enzymatic reactions. In fact, catalysed enzymatic reactions of hydrolase or oxidase type, respectively, consume or produce protons. The presence of the resulting products from the enzymatic reactions can be detected through the pH variation induced at the ISFET surface on which specific enzymes have been grafted. In this case, ISFET is called ENFET (enzymatic field effect transistor). Numerous molecular species can be detected by ENFET: for example, urease, glucose, organophosphorous pesticides. This type of transducer was also adapted for the detection of DNA hybridization. In this case, the SiO_2 thin film was coated with an APTS (3-aminopropyltriethoxysilane) layer on which single-strand DNA was immobilised. This FET put in contact with a solution containing complementary DNA strands shows a strong drain current resulting from the field effect induced by the arrival of the negative charge borne by DNA strands under hybridisation (Souteyrand et al. 1997). This device was called GENFET. To exploit this field effect on a microarray, Souteyrand et al. (2000) described a technique based on the optoelectrochemical impedance measurement in which electrical excitation is replaced by modulated light for scanning the surface of a electrolyte/DNA modified dielectric layer/semiconductor and for detecting directly the hybridisation on spots.

7.4.3.4 Mass Spectrometry

Mass Spectroscopy used as a detector has the advantages of being label free and of allowing the determination of the target structure after interaction in a complex medium. Bryan et al. (2004) reported the use of cleavable linker (disulfide) for subsequent electron spray ionisation MS. The use of porous silicon as a matrix for MALDI-TOF analysis was also proposed. Chen et al. (2005) used carbohydrate encapsulated gold nanoparticles for capturing proteins. After purification of reacted material from the unreacted one by centrifugation, the proteins were subjected to protease digestion. The remaining peptides were analysed by MALDI-TOF for the determination of the amino acid involved in the binding. Enzymatic reaction by enzymes can also be monitored by MALDI-TOF MS (Park and Shin 2007; Su and Mrksich 2002; Laurent et al. 2008).

7.4.3.5 Nanoparticles

The use of saccharide-modified nanoparticles (polymeric, metallic or semiconductor) for biosensing purpose is attractive, as it allows purification from a complex mixture of reacted products (by centrifugation, or magnetic field) and the multivalent display of carbohydrate. They can be used advantageously for in vivo bioimaging in transmission electron microscopy or confocal microscopy, or for glycoprofiling by mass spectroscopy (after purification and release of reacted molecules) (Larsen et al. 2006).

Carbohydrate-modified core magnetic nanoparticles could be advantageously employed for in vivo imaging by magnetic resonance imaging (MRI). Indeed, in the literature (Harris et al. 2006; Perez et al. 2002; Perez et al. 2003; Perez et al. 2004), the detection of biomolecules using super paramagnetic iron nanoparticles (and MRI) is reported. The assay is based on modified nanoparticles that self-assemble in the presence of the targeted biomolecule. The spin–spin relaxation time of surrounding water molecules is then enhanced and can be imaged by MRI, allowing the detection of the target even in complex media.

The multivalent nature of carbohydrate interactions should be suitable for this type of biosensors. However, to our knowledge, this has not been reported for probing carbohydrate–protein interaction. This would be useful for in vivo imaging in diagnosis where abnormal carbohydrate–protein interactions take place.

7.5 Biosensors: Applications

7.5.1 Antibody/Antigen

Carbohydrate biosensing can be applied to the design of sugar-based vaccine or for antibody profiling for diagnosis (IgMs profiling for tumor detection, natural autoantibodies profiling, etc.).

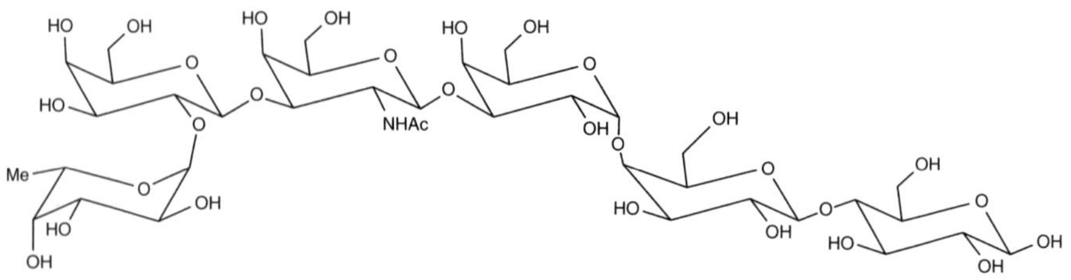

Fig. 7.28 Structure of Globo H

The cell sphingolipid Globo H possesses a hexasaccharide moiety. It is an antigenic carbohydrate such as T antigen, which is highly expressed on a variety of tumoral cell lines. Huang et al. (2006a) reported the design of a glycoarray based on Globo H and truncated Globo H ligands for understanding Globo H interaction with antibodies MBr1 and VK-9. MBr1 and VK-9 are mouse monoclonal antibodies targeting the Globo H epitope. The epitopes (including Globo H) were synthesised using "one-pot" programmable synthesis and modified with an aglycon bearing an amino or an azido function. The glycosides were covalently immobilised on a glass support using ester-activated or alkyne-modified glass slides. Affinities for MBr1 and VK-9 were probed with FITC-labelled secondary antibodies. It was shown that the monoclonal antibodies bind the terminal tetrasaccharides as well as the natural hexasaccharides and that the fucose residue is compulsory for binding. However, serum from patients with breast cancer binds both the fucosylated and defucosylated pentasaccharide, probably due to the polyclonal nature of the serum. According to the authors, their carbohydrate platform was more effective and more sensitive than ELISA for probing the affinity of antibodies for Globo H (Fig. 7.28) with the use of very minute amounts of the material.

7.5.2 *Enzyme/Carbohydrates*

Another application of carbohydrate microarray is to understand the structural features that govern substrate specificity of enzymes processing carbohydrates. The group of Mathieu (Chevolot et al. 2001) demonstrated that α-2,6-sialyltransferase was active on lactose-modified polystyrene. Detection was performed using radiolabelled CMP-Neu5Ac. The group of Shin (Shin 2007) showed, using *N*-GlcNAc and *O*-Fuc-α-modified glass slide, that LacNac could be successfully synthesised on chip using β-1,4-galactosyltransferase. They further illustrated that complex oligosaccharides could be synthesised using sequential β-1,4-galactosyltransferase followed by α-2,3′-sialyltransferase. Each step of the synthesis was characterised using FITC-labelled specific lectin.

Aminoglycosides are a broad-spectrum class of antibiotics due to their interaction with bacterial 30S ribosome. Their efficiency has decreased as a result of the emergence of resistant strains. The resistance mechanism is usually due to enzymatic modification of the molecules such as aminoglycoside acetyl transferases. Disney et al. have used aminoglycoside microarrays for probing the affinity of three acetyl transferases (from *Salmonella enterica* and from Mycobacterium tuberculosis) towards aminoglycoside antibiotics (Disney and Seeberger 2004a; Disney et al. 2004). They also studied new aminoglycosides as inhibitors of these enzymes. Based on their microarray experiment, they identified new aminoglycosides with low affinity towards the acetyl transferases, meaning they are less susceptible to enzymatic transformation by the bacteria, as well as being a potent inhibitor.

7.5.3 Carbohydrates/Lectins

Plant lectins are widely used usually as a model to assess biosensor/microarray in terms of specificity. Here, we will focus on three examples where carbohydrate biosensing was used advantageously with the perspectives of medical applications for

- analysing ligand affinity of Dectin-1, a C-type leukocytes lectin involved in phagocytosis and inflammation (Palma et al. 2006);
- analysing the specificity of hemagglutinins from modern and pandemic influenza virus (Stevens et al. 2006); and
- determining the affinity profile of GP120-binding protein. GP120 is a hyper-mannosylated protein of the HIV envelope. These mannose residues are involved in the binding of the virus to the host cells (Adams et al. 2004).

Plama et al. studied the binding of Dectin-1, a C-type leukocytes lectin involved in phagocytosis and inflammation that binds both β-1,3 and β-1,3/β-1,6 rich β-glucans (Palma et al. 2006). They developed a bioassay based on neoglycolipids oligosaccharides immobilised through physisorption onto nitrocellulose membranes. Oligosaccharides comprise mammalian oligosaccharides and oligosaccharides issued from the depolymerisation of natural polysaccharides. Fluorescence was used for detection. They showed that Dectin-1 binds exclusively β-1,3-linked glucose oligosaccharides. Ten to eleven glucose residues per oligosaccharides were compulsory for good binding.

GP120 is a hyper-mannosylated protein of the HIV envelope. The mannose residues are involved in the entry of the virus into the host cells by interacting with CD4, a chemokine family receptor present at the surface of T-lymphocytes, macrophages, dendritic cells, etc. Furthermore, the mannose residues are involved in protecting the virus from antibodies directed against GP120 by hiding the targeted epitope. This mannose oligosaccharide has been considered a potential target for the design of vaccine (it interacts with the 2G12 antibody (Calarese et al. 2005)) or as potential target for prophylactic measures. In the latter case, proteins interacting with the hyper-mannosylated moiety of GP120 would prevent the entry of the virus into the host cell. Adams et al. designed a glyco array bearing different mannose and studied the affinities of 2G12, CD4, Scytovirin (a potential anti-HIV protein), Cyanovirin N (CVN, a cyano bacteria protein with potential anti-HIV properties) and the dendritic cell lectin DC SIGN involved the internalisation of the virus. Detection was performed with FITC labelling (Adams et al. 2004). It was shown that each protein has specific requirements concerning the structure of the mannosyl oligosaccharide. For example, an α-1–2 linkage alone is necessary for the interaction of 2G12. The binding profile of the Scytovirin showed that the terminal α-1–2 mannose linkage at the terminal non-reducing end is required.

Stevens et al. took advantage of the CFG platform for studying the affinities of hemagglutinin of different strands of the influenza virus (Stevens et al. 2006). Influenza virus is an enveloped virus with a segmented negative RNA genome. Its envelope bears two proteins: hemagglutinin (HA, around 5,000 copies) and neuraminidase (3–500 copies). HAs interact with the sialic acids present at the surface of the host cell, permitting adhesion of the virus. It is the specificity of the HA towards the sialic acid linkage that determines the host specificity of the influenza virus. Human influenza viruses preferentially bind the α-2–6 linkage present at the surface of the epithelial cells of the upper respiratory track and lungs, while the avian ones recognise the α-2–3 linkage present in epithelial cells of the intestine of birds. Therefore, the subtle switch from avian to human subtype is related to the HA affinity to α-2–6.

Stevens et al. used a glycoarray bearing 200 oligosaccharides for studying the specificity of HAs from modern and pandemic influenza virus for α-2,6′/α-2′,3 linkage, and also the influence of sulfation, fucosylation and sialylation at position 2. They showed that the species barrier of the New York 1918 variant (human pandemic variant) could be reverted to an avian virus by a single mutation of the Asp190Glu. The species barrier from avian to human pandemic, as defined by the specificity of 1918 viruses (New York and South Carolina), can be jumped by the likely avian progenitor by only changing two amino acids of their HA.

7.5.4 Whole Cells

Using whole cells, the biosensor takes advantage of the natural multivalency of carbohydrate binding on the cell surface. Indeed, in nature, multivalency is often achieved by the self-association of lectine at the cellular surface upon interaction with carbohydrates (Nimrichter et al. 2004). Several authors have reported the use of whole-cell detection by means of carbohydrate/whole cell interactions.

Disney and Seeberger (2004b) have reported the use of microarrays as a means for probing bacteria (*E. coli*) affinities for immobilised oligomannose. *E. coli* strains were stained with a fluorescent cell-permeable nucleic acid dye and incubated on the microarray. They demonstrated that the platform allows discriminating strains according to their affinities for mannose residues. This issue is critical since the ability of screening cells' affinity for different carbohydrate affinities is related to their pathogenicity. The ability of the microarray to sense multiple carbohydrate-based interactions in one single experiment allows profiling pathogen–carbohydrate affinities. This should permit establishing a "finger print" like identification of the pathogen present in a complex medium. They also showed that they could identify *E. coli* from complex media such as a serum containing erythrocytes. Anti-adhesive molecule activities were evaluated using such arrays. They compared the IC_{90}s of mannose, *p*-nitro mannose and mannose-containing polymers.

Chevolot et al. (2001)tested the effect of galactose and a lactose-derivatised surface on primary rat hepatocytes. They found that their viability and xenobiotic metabolism were not affected, while neutral red uptake and MTT (3-(4,5-Dimethylthiazol-2-yl)-2,5-diphenyltetrazolium bromide) activity increased with galactose surface density. Neutral red uptake is related to the lysosomal activity. They hypothesised that this may related to the natural lysosomal activity of hepatocytes in correlation with the asialoglycoprotein receptor.

Adhesion of primary chicken hepatocytes and CD4+ human T cell onto glycoarrays (bearing 200 and 45 structures, respectively) were studied by Nimrichter et al. (2004) using fluorescent staining of the cells. Primary chicken hepatocytes express a C-type lectin that binds to *N*-acetyl glucosamine (GlcNac) present at the non-reducing terminal position of oligosaccharides. They found that the cells did bind as expected onto spots bearing GlcNAc at the terminal non-reducing end independently of the linkage and orientation. The hepatocytes did not adhere onto spots bearing galactose residues at the terminal non-reducing end. CD4+ human T cell from healthy individual express affinity towards Lewis X structure mediated by an L-selectin. On a 45-structure-bearing chip, they showed that the presence of the fucosyl residues was compulsory for obtaining good adhesion. They also showed that the adhesion was influenced by the staining protocol. According to the authors, the main limitations are the control of detachment forces, the low compatibility of cells with the robotics and high-throughput liquid and long-term storage. Adhesion of pathogens such as *E. coli* (Nimrichter et al. 2004) or viruses (Amonsen et al. 2007) was also reported.

7.6 Concluding Remarks

Carbohydrate biosensors have demonstrated to be powerful tools for studying glycobiolgy. They should find applications in fundamental research, diagnosis or drug discovery. However, availability of carbohydrate ligands in terms of synthesis, biosynthesis or extraction from natural sources remains a key issue. Furthermore, fundamental understanding should be addressed. For example, the influence of surface chemistry on the subsequent protein–carbohydrate interaction has not been widely studied, even though it may influence the overall results.

Most reported biosensors have so far focused on carbohydrate–protein interactions. However, other interactions may be of crucial interest such as carbohydrate–carbohydrate interactions, DNA–carbohydrate interactions, etc. Finally, to our knowledge, very few integrated (i.e., with integrated microfludic) systems have been reported. Such systems could in theory improve applications by reducing the amount of needed material, decreasing the experimental time, etc.

References

Adams EW, Ratner DM, Bokesch HR, Mc Mahon JB, O' Keefe BR, Seeberger PH (2004) Oligosaccharide and glycoprotein microarrays as tools in HIV glycobiology; glycan-dependent gp120/protein interactions. Chem Biol 11:875–881

Adams EW, Ueberfeld JM, Ratner DM, O' Keefe BR, Walt DR, Seeberger PH (2003) Encoded fiber-optic microsphere arrays for probing protein–carbohydrate interactions. Ang Chem Inter Ed Engl 42:5317–5320

Amonsen M, Smith DF, Cummings RD, Air GM (2007) Human parainfluenza viruses hPIV1 and hPIV3 bind oligosaccharides with α2-3-linked sialic acids that are distinct from those bound by H5 avian influenza virus hemagglutinin. J Virol 81:8341–8345

Angeloni S, Ridet JL, Kusy N, Gao H, Crevoisier F, Guinchard S, Kochhar S, Sigrist H, Sprenger N (2005) Glycoprofiling with micro-arrays of glycoconjugates and lectins. Glycobiology 15:31–41

Angyal SJ (1984) The composition of reducing sugars in solution. Adv Carbohydr Chem Biochem 42:15–68

Arcamone F, Franceschi G, Orezzi P, Cassinelli G, Barbieri W, Mondelli R (1964) Daunomycin I. The structure of daunomycinone. J Am Chem Soc 86:5334–5335

Aubeck R, Eppelsheim C, Bräuchle C, Hampp N (1993) Potentiometric thick-film sensor for the determination of the tumour marker bound sialic acid. Analyst 118:1389–1392

Bamford CH, Cooper SL, Tsurutta T (eds) (1992) The Vroman effect: festschrift in honour of the 75th birthday of Dr. Leo Vroman. VSP, Utrecht

Baum LG, Paulson JC (1990) Sialyloligosaccharides of the respiratory epithelium in the selection of human influenza virus receptor specificity. Acta Histol Supp 40:35–38

Belachew S, Chittajallu R, Aguirre AA, Yuan X, Kirby M, Anderson S, Gallo V (2003) Postnatal NG2 proteoglycan-expressing progenitor cells are intrinsically multipotent and generate functional neurons. J Cell Biol 161:169–186

Berhe S, Schofield L, Schwarz RT, Gerold P (1999) Conservation of structure among glycosylphosphatidylinositol toxins from different geographic isolates of *Plasmodium falciparum*. Mol Biochem Parasitol 103:273–278

Bertozzi CR, Kiessling LL (2001) Chemical glycobiology. Science 291:2357–2364

Bishop JR, Schuksz M, Esko JD (2007) Heparan sulphate proteoglycans fine-tune mammalian physiology. Nature 446:1030–1037

Biskup MB, Müller JU, Weingart R, Schmidt RR (2005) New methods for the generation of carbohydrate arrays on glass slides and their evaluation. ChemBioChem 6:1007–1015

Blixt O, Head S, Mondala T, Scanlan C, Huflejt ME, Alvarez R, Bryan MC, Fazio F, Calarese D, Stevens J, Razi N, Stevens DJ, Skehel JJ, Van Die I, Burton DR, Wilson IA, Cummings R, Bovin NV, Wong C-H, Paulson JC (2004) Printed covalent glycan array for ligand profiling of diverse glycan binding proteins. Proc Natl Acad Sci USA 10:17033–17038

Blixt O, Vasiliu D, Allin K, Jacobsen N, Warnock D, Razi N, Paulson JC, Bernatchez S, Gilbert M, Wakarchuk W (2005) Chemoenzymatic synthesis of 2-azidoethyl-ganglio-oligosaccharides GD3, GT3, GM2, GD2, GT2, GM1, and GD1a. Carbohydr Res 340:1963–1972

Blumenschein TM, Friedrich N, Childs RA, Saouros S, Carpenter EP, Campanero-Rhodes MA, Simpson P, Chai W, Koutroukides T, Blackman MJ, Feizi T, Soldati-Favre D, Matthews S (2007) Atomic resolution insight into host cell recognition by *Toxoplasma gondii*. EMBO J 26:2808–2820

Bochner BS, Alvarez RA, Mehta P, Bovin NV, Blixt O, White JR, Schnaar RL (2005) Glycan array screening reveals a candidate ligand for Siglec-8. J Biol Chem 280:4307–4312

Bock VD, Hiemstra H, Van Maarseveen JH (2006) Cu^{I}-catalyzed alkyne-azide "Click" cycloadditions from a mechanistic and synthetic perspective. Eur J Org Chem 2006:51–68

Bouillon C, Meyer A, Vidal S, Jochum A, Chevolot Y, Cloarec JP, Praly JP, Vasseur JJ, Morvan F (2006) Microwave assisted "click" chemistry for the synthesis of multiple labeled-carbohydrate oligonucleotides on solid support. J Org Chem 71:4700–4702

Bruckenstein S, Shay M (1985) An in situ weighing study of the mechanism for the formation of the adsorbed oxygen monolayer at a gold electrode. J Electroanal Chem Interface Electrochem 188:131–136

Brun MA, Disney MD, Seeberger PH (2006) Miniaturization of microwave-assisted carbohydrate functionalization to create oligosaccharide microarrays. ChemBioChem 7:421–424

Bryan MC, Fazio F, Lee H-K, Huang C-Y, Chang A, Best MD, Calarese DA, Blixt O, Paulson JC, Burton D, Wilson IA, Wong C-H (2004) Covalent display of oligosaccharide arrays in microtiter plates. J Am Chem Soc 126:8640–8641

Buttry DA, Ward MD (1992) Measurement of interfacial processes at electrode surfaces with the electrochemical quartz crystal microbalance. Chem Rev 92:1355–1379

Calarese DA, Lee H-K, Huang C-Y, Best MD, Astronomo RD, Stanfield RL, Katinger H, Burton DR, Wong C-H, Wilson IA (2005) Dissection of the carbohydrate specificity of the broadly neutralizing anti-HIV-1 antibody 2G12. Proc Natl Acad Sci USA 102: 13372–13377

Capila I, Linhardt RJ (2002) Heparin-protein interactions. Ang Chem Inter Ed Engl 41:390–412

Carreira JC, Jones C, Wait R, Previato JO, Mendonça-Previato L (1996) Structural variation in the glycoinositolphospholipids of different strains of *Trypanosoma cruzi*. Glycoconjugate J 13:955–966

Carroll SM, Higa HH, Paulson JC (1981) Different cell-surface receptor determinants of antigenically similar influenza virus hemagglutinins. J Biol Chem 256:8357–8363

Carter JR, Pager CT, Fowler SD, Dutch RE (2005) Role of N-linked glycosylation of the Hendra virus fusion protein. J Virol 79:7922–7925

Chen Y-J, Chen S-H, Chien Y-Y, Chang Y-W, Liao H-K, Chang C-Y, Jan M-D, Wang K-T, Chun-Cheng Lin C-C (2005) Carbohydrate-encapsulated gold nanoparticles for rapid target-protein identification and binding-epitope mapping. ChemBioChem 6:1169–1173

Chen Y, Maguire T, Hileman RE, Jonathan R, Fromm JR, Esko JD, Linhardt RJ, Marks RM (1997) Dengue virus infectivity depends on envelope protein binding to target cell heparan sulfate. Nat Med 3:866–871

Chevolot Y, Bouillon C, Vidal S, Morvan F, Meyer A, Cloarec J-P, Jochum A, Praly J-P, Vasseur J-J, Souteyrand E (2007) DNA-based carbohydrate biochips: a platform for surface glyco-engineering. Ang Chem Inter Ed Engl 46:2398–2402

Chevolot Y, Bucher O, Léonard D, Mathieu H-J, Sigrist H (1999) Synthesis and characterization of a photoactivatable glycoaryldiazirine for surface glycoengineering. Bioconjugate Chem 10:169–175

Chevolot Y, Martins J, Milosevic N, Léonard D, Zeng S, Malissard M, Berger EG, Maier P, Mathieu H-J, Crout DHG, Sigrist H (2001) Immobilisation on polystyrene of diazirine derivatives of mono and disaccharides: biological activities of modified surfaces. Bioorg Med Chem 9:2943–2953

Constantinescu AA, Vink H, Spaan JAE (2003) Endothelial cell glycocalyx modulates immobilization of leukocytes at the endothelial surface. Arterioscl Throm Vas 23:1541–1547

Cooper MA, Singleton VT (2007) A survey of the 2001 to 2005 quatrz microbalance biosensor literature: applications of acoustic physics to the analysis of biomolecular interactions. J Mol Recognit 20:154–184

Croft SL, Sundar S, Fairlamb AH (2006) Drug resistance in leishmaniasis. Clin Microbiol Rev 19:111–126

Dam TK, Brewer CF (2002) Thermodynamic studies of lectin-carbohydrate interactions by isothermal titration calorimetry. Chem Rev 102:387–429

De Paz JL, Moseman EA, Noti C, Polito L, Von Andrian UH, Seeberger PH (2007) Profiling heparin-chemokine interactions using synthetic tools. ACS Chem Biol 2:735–744

De Paz JL, Noti C, Seeberger PH (2006) Microarrays of synthetic heparin oligosaccharides. J Am Chem Soc 128:2766–2767

Desai NP, Hubbell JA (1991) Solution technique to incorporate polyethylene oxide and other water-soluble polymers into surfaces of polymeric biomaterials. Biomaterials 12:144–153

Disney MD, Seeberger PH (2004a) Aminoglycoside microarrays to explore interactions of antibiotics with RNAs and proteins. Chem Eur J 100:3308–3314

Disney MD, Seeberger PH (2004b) The use of carbohydrate microarrays to study carbohydrate-cell interactions and to detect pathogens. Chem Biol 11:1701–1707

Disney MD, Magnet S, Blanchard JS, Seeberger PH (2004) Aminoglycoside microarrays to study antibiotic resistance. Ang Chem Inter Ed Engl 43:1591–1594

Dosreis GA, Peçanha LMT, Bellio M, Previato JO, Mendonça-Previato L (2002) Glycoinositol phospholipids from *Trypanosoma cruzi* transmit signals to the cells of the host immune system through both ceramide and glycan chains. Microbes Infect 4:1007–1013

Drouillard S, Driguez H, Samain E (2006) Large-scale synthesis of H-antigen oligosaccharides by expressing *Helicobacter pylori* α-1, 2-fucosyltransferase in metabolically engineered *Escherichia coli* Cells. Ang Chem Inter Ed Engl 45:1778–1780

Dubois M-P, Gondran C, Renaudet O, Dumy P, Driguez H, Fort S, Cosnier S (2005) Electrochemical detection of *Arachis hypogaea* (peanut) agglutinin binding to monovalent and clustered lactosyl motifs immobilized on a polypyrrole film. Chem Commun:4318–4320

Dugas V, Broutin J, Souteyrand E (2005) Droplet evaporation study applied to DNA chip manufacturing. Langmuir 21:9130–9136

Dugas V, Depret G, Chevalier B, Nesme X, Souteyrand E (2004) Immobilization of single-stranded DNA fragments to solid surfaces and their repeatable specific hybridization: covalent binding or adsorption? Sens Actuators B Chem 101:112–121

Dumon C, Bosso C, Utille J-P, Heyraud A, Samain E (2006) Production of Lewis x tetrasaccharides by metabolically engineered *Escherichia coli*. ChemBioChem 7:359–365

Dwek RA (1996) Glycobiology: toward understanding the function of sugars. Chem Rev 96:683–720

Dyukova VI, Dementieva EI, Zubtsov DA, Galanina OE, Bovin NV, Rubina AY (2005) Hydrogel glycan microarrays. Anal Biochem 347:94–105

Ebara Y, Okahata Y (1994) A kinetic study of Concanavalin A binding to glycolipid monolayers by using a quartz-crystal microbalance. J Am Chem Soc 116:11209–11212

Edward JT, Puskas I (1962) Stereochemical studies: IV. Acid-catalyzed equilibration of α- and β-forms of 3-alkoxy-4-oxa-5α-cholestanes. The anomeric effect. Can J Chem 40:711–717

Esko J (1999) Proteoglycans and glycosaminoglycans. In: Varki A, Cummings R, Esko J, Freeze H, Hart GW, Marth J (eds) Essentials of glycobiology. Cold Spring Harbor Laboratory Press, New York, pp 145–161

Fare TL, Coffey EM, Dai H, He YD, Kessler DA, Kilian KA, Koch JE, Leproust E, Marton MJ, Meyer MR, Stoughton RB, Tokiwa GY, Wang Y (2003) Effects of atmospheric ozone on microarray data quality. Anal Chem 75:4672–4675

Fazio F, Bryan MC, Lee HK, Chang A, Wong CH (2004) Assembly of sugars on polystyrene plates: a new facile microarray fabrication technique. Tetrahedron Lett 45:2689–2692

Fazio F, Bryan MC, Blixt O, Paulson JC, Wong C-H (2002) Synthesis of sugar arrays in microtiter plate. J Am Chem Soc 124:14397–14402

Feinberg H, Mitchell DA, Drickamer K, Weis WI (2001) Structural basis for selective recognition of oligosaccharides by DC-SIGN and DC-SIGNR. Science 294:2163–2166

Feizi T, Fazio F, Chai W, Wong C-H (2003) Carbohydrate microarrays—a new set of technologies at the frontiers of glycomics. Curr Opin Struct Biol 13:637–645

Ferguson MA (1999) The structure, biosynthesis and functions of glycosylphosphatidylinositol anchors, and the contributions of trypanosome research. J Cell Sci 112:2799–2809

Ferrier RJ, Hay RW, Vethaviyasar N (1973) A potentially versatile synthesis of glycosides. Carbohydr Res 27:55–61

Fischer E (1893) Ueber die glucoside der alkohole. Ber Deut Chem Gesell 26:2400–2412

Fischer E (1895) Ueber die verbindungen der zucker mit den alkoholen und ketonen. Ber Deut Chem Gesell 28:1145–1167

Fort S, Birikaki L, Dubois MP, Antoine T, Samain E, Driguez H (2005) Biosynthesis of conjugatable saccharidic moieties of GM(2) and GM(3) gangliosides by engineered *E-coli*. Chem Commun 2558–2560

Fraser-Reid B, Konradsson P, Mootoo DR, Udodong U (1988) Direct elaboration of pent-4-enyl glycosides into disaccharides. J Chem Soc Chem Commun 823–825

Fraser-Reid B, Udodong U, Wu Z, Ottosson H, Merritt JR, Rao CS, Roberts C, Madsen R (1992) *n*-Pentenyl glycosides in organic chemistry: a contemporary example of serendipity. Synlett 927–942

Fügedi P (2006) Glycosylation methods. In: Levy DE, Fügedi P (eds) The organic chemistry of sugars. Taylor & Francis, New York, pp 89–180

Fügedi P, Garegg PJ, Lönn H, Norberg T (1987) Thioglycosides as glycosylating agents in oligosaccharide synthesis. Glycoconj J 4:97–108

Fukui S, Feizi T, Galustian C, Lawson AM, Chai W (2002) Oligosaccharide microarrays for high-throughput detection and specificity assignments of carbohydrate-protein interactions. Nat Biotechnol 20:1011–1017

Galonic DP, Gin DY (2007) Chemical glycosylation in the synthesis of glycoconjugate antitumour vaccines. Nature 446:1000–1007

Garegg PJ (1997) Thioglycosides as glycosyl donors in oligosaccharide synthesis. Adv Carbohydr Chem Biochem 52:179–205

Gil MV, Arévalo MJ, López O (2007) Click chemistry - What's in a name? Triazole synthesis and beyond. Synthesis 1589–1620

Goffard A, Callens N, Bartosch B, Wychowski C, Cosset F-L, Montpellier C, Dubuisson J (2005) Role of N-linked glycans in the functions of hepatitis C virus envelope glycoproteins. J Virol 79:8400–8409

Gordus A, Mac Beath G (2006) Circumventing the problems caused by protein diversity in microarrays: implications for protein interaction networks. J Am Chem Soc 128:13668–13669

Goretzki L, Brug MA, Grako KA, Stallcup WB (1999) High-affinity binding of basic fibroblast growth factor and platelet-derived growth factor-AA to the core protein of the NG2 proteoglycan. J Biol Chem 274:16831–16837

Gotloib L, Shostak A, Galdi P, Jaichenko J, Fudin R (1992) Loss of microvascular negative charges accompanied by interstitial edema in septic rats heart. Circ Shock 36:45–56

Grubor NM, Shinar R, Jankowiak R, Porter MD, Small GJ (2004) Novel biosensor chip for simultaneous detection of DNA-carcinogen adducts with low temperature fluorescence. Biosens Bioelectron 19:547–556

Hakomori S-I (1991) Carbohydrate-carbohydrate interaction as an initial step in cell recognition. Pure Appl Chem 63:473–482

Hakomori S-I (2002) The glycosynapse. Proc Natl Acad Sci USA 99:225–232

Hanna SL, Pierson TC, Sanchez MD, Ahmed AA, Murtadha MM, Doms RW (2005) N-Linked glycosylation of west nile virus envelope proteins influences particle assembly and infectivity. J Virol 79:13262–13274

Harris TJ, Von Maltzahn G, Derfus AM, Ruoslahti E, Bhatia SN (2006) Proteolytic actuation of nanoparticle self-assembly. Ang Chem Inter Ed Engl 45:3161–3165

Hashimoto S-I, Honda T, Ikegami S (1989) A rapid and efficient synthesis of 1,2-*trans*-β-linked glycosides *via* benzyl- or benzoyl-protected glycopyranosyl phosphates. J Chem Soc Chem Commun 685–687

Helferich B, Schmitz-Hillebrecht E (1933) Eine neue methode zur synthese von glykosiden der phenole. Ber Deut Chem Gesell (A and B Series) 66:378–383

Hölemann A, Seeberger PH (2006) Synthetic carbohydrate-based antimalarial vaccines and glycobiology. In: Bewley CA (ed) Protein-carbohydrate interactions in infectious diseases. Royal society of Chemistry, Cambridge, UK, pp 151–174

Hosono K, Nishida Y, Knudson W, Knudson CB, Naruse T, Suzuki Y, Ishiguro N (2007) Hyaluronan oligosaccharides inhibit tumorigenicity of osteosarcoma cell lines MG-63 and LM-8 in vitro and in vivo via perturbation of hyaluronan-rich pericellular matrix of the cells. Am J Pathol 171:274–286

Hounsell EF (2000) Glycobiology of the immune system. In: Ernst B, Hart GW, Sinay P (eds) Carbohydrate in chemistry and biology. Wiley-VCH, New-York

Houseman BT, Mrksich M (2002) Carbohydrate arrays for the evaluation of protein binding and enzymatic modification. Chem Biol 9:443–454

Houseman BT, Gawalt ES, Mrksich M (2003) Maleimide-functionalized self-assembled monolayers for the preparation of peptide and carbohydrate biochips. Langmuir 19:1522–1531

Huang CY, Thayer DA, Chang AY, Best MD, Hoffmann J, Head S, Wong CH (2006a) Carbohydrate microarray for profiling the antibodies interacting with Globo H tumor antigen. Proc Natl Acad Sci USA 103:15–20

Huang G-H, Liu T-C, Liu M-X, Mei X-Y (2005) The application of quantum dots as fluorescent label to glycoarray. Anal Biochem 340:52–56

Huang K-T, Wu B-C, Lin C-C, Luo S-C, Chen C, Wong C-H, Lin C-C (2006b) Multi-enzyme one-pot strategy for the synthesis of sialyl Lewis X-containing PSGL-1 glycopeptide. Carbohydr Res 341:2151–2155

Huisgen R, Fraunberg K, Sturm HJ (1969) Cycloadditionen von pyridyl- und pyrimidyl-aziden. Tetrahedron Lett 10:2589–2594

Huisgen R, Knorr R, Möbius L, Szeimies G (1965) 1.3-Dipolare cycloadditionen, XXIII. einige beobachtungen zur addition organischer azide an CC-dreifachbindungen. Chem Ber 98:4014–4021

Igarashi K (1977) The Koenigs-Knorr reaction. Adv Carbohydr Chem Biochem 34:243–283

Imberty A, Wimmerová M, Mitchell EP, Gilboa-Garber N (2004) Structures of the lectins from *Pseudomonas aeruginosa*: insights into the molecular basis for host glycan recognition. Microbes Infect 6:221–228

Inouye S, Tsuruoka T, Niida T (1966) The structure of nojirimycin, a piperidinose sugar antibiotic. J Antibiotics Ser A 19:288–292

Isbell HS, Pigman W (1969) Mutarotation of sugars in solution: part II. Adv Carbohydr Chem Biochem 24:14–65

Ishihara K, Nomura H, Mihara T, Kurita K, Iwasaki Y, Nakabayashi N (1998) Why do phospholipid polymers reduce protein adsorption? J Biomed Mater Res 39:323–330

Ito T, Nelson J, Couceiro SS, Kelm S, Baum LG, Krauss S, Castrucci MR, Donatelli I, Kida H, Paulson JC, Websterr G, Kawaoka Y (1998) Molecular basis for the generation in pigs of influenza A viruses with pandemic potential. J Virol 72:7367–7373

Jaipuri FA, Collet BYM, Pohl NL (2008) Synthesis and quantitative evaluation of Glycero-d-manno-heptose binding to concanavalin A by fluorous-tag assistance. Ang Chem Inter Ed Engl 47:1707–1710

Jelinek R, Kolusheva S (2004) Carbohydrate biosensors. Chem Rev 104:5987–6015

Jeon SI, Lee JH, Andrade JD, De Gennes PG (1991) Protein–surface interactions in the presence of poly-ethylene oxideI. Simplified theory. J Colloid Interface Sci 142:149–158

Jorns J, Mangold U, Neumann U, Van Damme EJM, Peumans WJ, Pfuller U, Schumacher U (2003) Lectin histochemistry of the lymphoid organs of the chicken. Anat Embryol 207:85–94

Jouan A, Hatakeyama T, Murakami K, Miyamoto Y, Yamasaki N (1996) An assay for lectin activity using microtiter plate with chemically immobilized carbohydrates. Anal Biochem 237:188–192

Junge B, Heiker F-R, Kurz J, Muller L, Schmidt DD, Wunsche C (1984) Untersuchungen zur struktur des α-glucosidaseinhibitors acarbose. Carbohydr Res 128:235–268

Kanazawa KK, Gordon IJG (1985) The oscillation frequency of a quartz resonator in contact with liquid. Anal Chim Acta 175:99–105

Karamanska R, Clarke J, Blixt O, Macrae JI, Zhang JQ, Crocker PR, Laurent N, Wright A, Flitsch SL, Russell DA, Field RA (2008) Surface plasmon resonance imaging for real-time, label-free analysis of protein interactions with carbohydrate microarrays. Glycoconj J 25:69–74

King WH Jr (1964) Piezoelectric sorption detector. Anal Chem 36:1735–1739

Klassen H, Schwartz MR, Bailey AH, Young MJ (2001) Surface markers expressed by multipotent human and mouse neural progenitor cells include tetraspanins and non-protein epitopes. Neurosci Lett 312:180–182

Kochetkov NK, Bochkov AF, Sokolavskaya TA, Snyatkova VJ (1971) Modifications of the orthoester method of glycosylation. Carbohydr Res 16:17–27

Kochetkov NK, Khorlin AJ, Bochkov AF (1967) A new method of glycosylation. Tetrahedron 23:693–707

Koenigs W, Knorr E (1901) Ueber einige derivate des traubenzuckers und der galactose. Ber Deut Chem Gesell 34:957–981

Kohn M, Wacker R, Peters C, Schröder H, Soulère L, Breinbauer R, Niemeyer CM, Waldmann H (2003) Staudinger ligation: a new immobilization strategy for the preparation of small-molecule arrays. Ang Chem Inter Ed Engl 42:5830–5834

Kolb HC, Finn MG, Sharpless KB (2001) Click chemistry: diverse chemical function from a few good reactions. Ang Chem Inter Ed Engl 40:2004–2021

Kondo H, Ichikawa Y, Wong C-H (1992) β-Sialyl phosphite and phosphoramidite: synthesis and application to the chemoenzymic synthesis of CMP-sialic acid and sialyl oligosaccharides. J Am Chem Soc 114:8748–8750

Koochekpour S, Pilkington GJ, Merzak A (1995) Hyaluronic acid/CD44H interaction induces cell detachment and stimulates migration and invasion of human glioma cells in vitro. Int J Cancer 63:450–454

Laine RA (1994) A calculation of all possible oligosaccharide isomers both branched and linear yields $1.05 \times 10(12)$ structures for a reducing hexasaccharide: the Isomer Barrier to development of single-method saccharide sequencing or synthesis systems. Glycobiology 4:759–767

Larsen K, Thygesen MB, Guillaumie F, Willats WGT, Jensen KJ (2006) Solid-phase chemical tools for glycobiology. Carbohydr Res 341:1209–1234

Laurent N, Voglmeir J, Wright A, Blackburn J, Pham NT, Wong SCC, Gaskell SJ, Flitsch SL (2008) Enzymatic glycosylation of peptide arrays on gold surfaces. ChemBioChem 9:1–5

Lebed K, Kulik AJ, Forró L, Lekka M (2006) Lectin–carbohydrate affinity measured using a quartz crystal microbalance. J Colloid Interface Sci 299:41–48

Lee JH, Jeong BJ, Lee HB (1997) Plasma protein adsorption and platelet adhesion onto comb-like PEO gradient surfaces. J Biomed Mater Res 34:105–114

Lee MR, Shin I (2005) Fabrication of chemical microarrays by efficient immobilization of hydrazide-linked substances on epoxide-coated glass surfaces. Angew Chem Int Ed Engl 44:2881–2884

Lee RT, Lee YC (1994) Neoglycoconjugates: Preparation and Applications. In: Lee RT, Lee YC (eds) Enhanced biochemical affinities of multivalent neoglycoconjugates. Academic Press, San Diego, pp 23–50

Lee YC, Lee RT (1995) Carbohydrate-protein interactions: basis of glycobiology. Acc Chem Res 28:321–327

Leonard D, Chevolot Y, Bucher O, Haenni W, Sigrist H, Mathieu HJ (1998) ToF-SIMS and XPS study of photoactivatable reagents designed for surface glycoengineering - part 2. N-[m-(3-(trifluoromethyl)diazirine-3-yl)phenyl]-4-(-3-thio(-1-D-galactopyrannosyl)-maleimidyl) butyramide (MAD-Gal) on diamond. Surf Interface Anal 26:793–799

Léonard D, Chevolot Y, Bucher O, Sigrist H, Mathieu H-J (1998) ToF-SIMS and XPS study of photoactivatable reagents designed for surface glycoengineering. Part I. N-(m-(3-(trifluoromethyl)diazirine-3-yl)phenyl)-4-maleimido-butyramide (mad) on silicon, silicon nitride and diamond. Surf Interface Anal 26:783–792

Léonard D, Chevolot Y, Heger F, Martins J, Crout DHG, Sigrist H, Mathieu HJ (2001) ToF-SIMS and XPS study of photoactivatable reagents designed for surface glycoengineering: part III:

5-carboxamidopentyl N-[m-[3-(trifluoromethyl) diazirin-3-yl] phenyl b-D-galactopyranosyl]-(1->4)-1-thio-b-D-glucopyranoside (lactose aryl diazirine) on diamond. Surf Interface Anal 31:457–464

Lesaicherre M-L, Uttamchandani M, Chena GYJ, Yao SQ (2002) Developing site-specific immobilization strategies of peptides in a microarray. Bioorg Med Chem Lett 12:2079–2083

Livingston PO (1995) Approaches to augmenting the immunogenicity of melanoma gangliosides: from whole melanoma cells to ganglioside-KLH conjugate vaccines. Immunol Rev 145:147–166

Livingston PO, Ragupathi G (2006) Cancer vaccines targeting carbohydrate antigens. Hum Vaccin 2:137–143

Lundquist JJ, Toone EJ (2002) The cluster glycoside effect. Chem Rev 102:555–576

Mac Beath G, Schreiber SL (2000) Printing proteins as microarrays for high-throughput function determination. Science 289:1760–1763

Maeda K, Ueda M, Yagishita K, Kawaji S, Kondo S, Murase M, Takeuchi T, Okami Y, Umezawa H (1957) Studies of kanamycin. J Antibiotics Ser A 10:228–231

Majumdar SK, Kutzner HJ (1962) Studies on the biosynthesis of streptomycin. Appl Environ Microbiol 10:157–168

Mammen M, Choi S-K, Whitesides GM (1998) Polyvalent interactions in biological systems: implications for design and use of multivalent ligands and inhibitors. Ang Chem Inter Ed Engl 37:2754–2794

Manimala JC, Roach TA, Li Z, Gildersleeve JC (2006) High-throughput carbohydrate microarray analysis of 24 lectins. Ang Chem Inter Ed Engl 45:3607–3610

Manimala JC, Li Z, Jain A, Vedbrat S, Gildersleeve JC (2005) Carbohydrate array analysis of anti-Tn antibodies and lectins reveals unexpected specificities: implications for diagnostic and vaccine development. ChemBioChem 6:2229–2241

Martin TJ, Schmidt RR (1992) Efficient sialylation with phosphite as leaving group. Tetrahedron Lett 33:6123–6126

Matrosovich MN, Gambaryan AS, Teneberg S, Piskarev VE, Yamnikova SS, Lvov DK, Robertson JS, Karlsson K-A (1997) Avian influenza A viruses differ from human viruses by recognition of sialyloligosaccharides and gangliosides and by a higher conservation of the HA receptor-binding site. Virology 233:224–234

McConville MJ, De Souza D, Saunders E, Likic VA, Naderer A (2007) Living in a phagolysosome; metabolism of leishmania amastigotes. Trends Parasitol 23:368–375

McReynolds KD, Gervay-Hague J (2007) Chemotherapeutic interventions targeting HIV interactions with host-associated carbohydrates. Chem Rev 107:1533–1552

Mehta S, Pinto MB (1991) Phenylselenoglycosides as novel, versatile glycosyl donors. Selective activation over thioglycosides. Tetrahedron Lett 32:4435–4438

Mehta S, Pinto MB (1993) Novel glycosidation methodology. The use of phenyl selenoglycosides as glycosyl donors and acceptors in oligosaccharide synthesis. J Org Chem 58:3269–3276

Mendonça-Previato L, Gorin PAJ, Braga AF, Scharfstein J, Previato JO (1983) Chemical structure and antigenic aspects of complexes obtained from epimastigotes of *Trypanosoma cruzi*. Biochemistry 22:4980–4987

Merzak A, Koocheckpour S, Pilkington GJ (1994) CD44 mediates human glioma cell adhesion and invasion in vitro. Cancer Res 54:3988–3992

Miller JL, Dewet BJM, Martinez-Pomares L, Radcliffe CM, Dwek RA, Rudd PM, Gordon S (2008) The mannose receptor mediates dengue virus infection of macrophages. PLOS Pathog 4:e17

Miura Y, Sasao Y, Dohi H, Nishida Y, Kobayashi K (2002) Self-assembled monolayers of globotriaosylceramide (Gb3) mimics: surface-specific affinity with shiga toxins. Anal Biochem 310:27–35

Miura Y, Wada N, Nishida Y, Mori H, Kobayashi K (2004) Chemoenzymatic synthesis of glycoconjugate polymers starting from nonreducing disaccharides. J Polym Sci Part A Polym Chem 42:4598–4606

Morvan F, Meyer A, Jochum A, Sabin C, Chevolot Y, Imberty A, Praly J-P, Vasseur J-J, Souteyrand E, Vidal S (2007) Fucosylated pentaerythrityl phosphodiester oligomers

(PePOs): automated synthesis of DNA-based glycoclusters and binding to *Pseudomonas aeruginosa* lectin (PA-IIL). Bioconjugate Chem 18:1637

Mukaiyama T, Murai Y, Shoda S-I (1981) An efficient method for glucosylation of hydroxy compounds using glycopyranosyl fluoride. Chem Lett 10:431–432

Muñoz EM, Yu H, Hallock J, Edens RE, Linhardt RJ (2005) Poly(ethylene glycol)-based biosensor chip to study heparin–protein interactions. Anal Biochem 343:176–178

Muratsugu M, Ohta F, Miya Y, Hosokawa T, Kurosawa S, Kamo N, Ikeda H (1993) Quartz crystal microbalance for the detection of microgram quantities of human serum albumin: relationship between the frequency change and the mass of protein adsorbed. Anal Chem 65:2933–2937

Muthana S, Yu H, Huang S, Chen X (2007) Chemoenzymatic synthesis of size-defined polysaccharides by sialyltransferase-catalyzed block transfer of oligosaccharides. J Am Chem Soc 129:11918–11919

Naderer T, Vince JE, Mcconville MJ (2004) Surface determinants of leishmania parasites and their role in infectivity in the mammalian host. Curr Mol Med 4:649–665

Nagase T, Nakata E, Shinkai S, Hamachi I (2003) Construction of artificial signal transducers on a lectin surface by post-photoaffinity-labeling modification for fluorescent saccharide biosensors. Chem Eur J 9:3660–3669

Nandi A, Estess P, Siegelman MH (2000) Hyaluronan anchoring and regulation on the surface of vascular endothelial cells is mediated through the functionally active form of CD44. J Biol Chem 275:14939–14948

Navarro-Sanchez E, Altmeyer R, Amara A, Schwartz O, Fieschi F, Virelizier J-L, Arenzana-Seisdedos F, Desprès P (2003) Dendritic-cell-specific ICAM3-grabbing non-integrin is essential for the productive infection of human dendritic cells by mosquito-cell-derived dengue viruses. EMBO Rep 4:723–728

Nepogodiev S, Stoddart JF (1998) Chemistry of cyclic oligosaccharides. In: Boons G-J (ed) Carbohydrate chemistry. Blackie Academic & Professional, New York, pp 322–383

Nicolaou KC, Smith AL (1992) Molecular design, chemical synthesis, and biological action of enediynes. Acc Chem Res 25:497–503

Nimrichter L, Gargir A, Gortler M, Alstock RT, Shtevi A, Weisshauss O, Fire E, Dotan N, Schnaar RL (2004) Intact cell adhesion to glycan microarrays. Glycobiology 14:197–203

Norde W (1985) Probing protein adsorption via microcalorimetry. In: Andrade JD (ed) Surface and interfacial aspects of biomedical polymers, vol. 2. Proteins adsorption. Plenum publishing, New York, pp 263–294

Norde W, Macritchie F, Nowicka G, Lyklema J (1986) Protein adsorption at solid-liquid interfaces: reversibility and conformation aspects. J Colloid Interface Sci 112:447–456

Okada H, Yoshida J, Sokabe M, Wakabayashi T, Hagiwara M (1996) Suppression of CD44 expression decreases migration and invasion of human glioma cells. Int J Cancer 66:255–260

Oostra M, De Haan CAM, De Groot RJ, Rottier PJM (2006) Glycosylation of the severe acute respiratory syndrome coronavirus triple-spanning membrane proteins 3a and M. J Virol 80:2326–2336

Palma AS, Feizi T, Zhang Y, Stoll MS, Lawson AM, Diaz-Rodrguez E, Campanero-Rhodes MA, Costa J, Gordon S, Brown GD, Chai W (2006) Ligands for the β-Glucan Receptor, Dectin-1, Assigned Using "Designer" Microarrays of Oligosaccharide Probe (Neoglycolipids) Generated from Glucan Polysaccharides. J Biol Chem 281:5771–5779

Park S, Lee MR, Pyo SJ, Shin I (2004) Carbohydrate chips for studying high-throughput carbohydrate-protein interactions. J Am Chem Soc 126:4812–4819

Park S, Shin I (2007) Carbohydrate microarrays for assaying galactosyltransferase activity. Org Lett 9:1675–1678

Pei Z, Yu H, Theurer M, Walden A, Nilsson P, Yan M, Ramstrom O (2007) Photogenerated carbohydrate microarrays. ChemBioChem 8:166–168

Perez JM, Josephson L, O' Loughlin T, Hogemann D, Weissleder R (2002) Magnetic relaxation switches capable of sensing molecular interactions. Nat Biotechnol 20:816–820

Perez JM, Simeone FJ, Saeki Y, Josephson L, Weissleder R (2003) Viral-induced self-assembly of magnetic nanoparticles allows the detection of viral particles in biological media. J Am Chem Soc 125:10192–10193

Perez JM, Simeone FJ, Tsourkas A, Josephson L, Weissleder R (2004) Peroxidase substrate nanosensors for MR imaging. Nano Lett 4:119–122

Perez S, Marchessault RH (1978) The exo-anomeric effect: experimental evidence from crystal structures. Carbohydr Res 65:114–120

Petitou M, Van Boeckel CAA (2004) A synthetic antithrombin III binding pentasaccharide is now a drug! What comes next? Ang Chem Inter Ed Engl 43:3118–3133

Pigman W, Isbell HS (1968) Mutarotation of sugars in solution: part I. Adv Carbohydr Chem Biochem 23:11–57

Plante OJ, Palmacci ER, Seeberger PH (2001) Automated solid-phase synthesis of oligosaccharides. Science 291:1523–1527

Pohl NL, Kiessling LL (1999) Scope of multivalent ligand function: lactose-bearing neoglycopolymers by ring-opening metathesis polymerization. Synthesis 1515–1519

Pougny J-R, Sinay P (1976) Reaction d'imidates de glucopyranosyle avec l'acetonitrile. Applications synthetiques. Tetrahedron Lett 17:4073–4076

Praly J-P, Lemieux RU (1987) Influence of solvent on the magnitude of the anomeric effect. Can J Chem 65:213–223

Pries AR, Secomb TW, Gaehtgens P (2000) The endothelial surface layer. Pflug Archiv Eur J Physiol 440:653–666

Qian X, Sujino K, Ratcliffe RM, Palcic MM (2001) Glycosyltransferases in oligosaccharide synthesis. In: Wang PG, Bertozzi CR (eds) Glycochemistry: principles, synthesis and applications. Marcel Dekker, New York, pp 535–566

Rasooly A (2005) Biosensor technologies. Methods 37:1–3

Ratner DM, Adams EW, Su J, O'keefe BR, Mrksich M, Seeberger PH (2004) Probing protein-carbohydrate interactions with microarrays of synthetic oligosaccharides. ChemBioChem 5:379–383

Ratner DM, Seeberger PH (2007) Carbohydrate microarrays as tools in HIV glycobiology. Curr Pharm Des 11:173–183

Raynal F, Plaza F, Beuf A, Carrière P, Souteyrand E, Martin J-R, Cloarec J-P, Cabrera M (2004) Study of a chaotic mixing system for DNA chip hybridization chambers. Phys Fluids 16:63–66

Robinson A, Fang J-M, Chou P-T, Liao K-W, Chu R-M, Lee S-J (2005) Probing lectin and sperm with carbohydrate-modified quantum dots. ChemBioChem 6:1899–1905

Rogers GN, Paulson JC, Daniels RS, Skehel JJ, Wilson IA, Wiley DC (1983) Single amino acid substitutions in Influenza haemagglutinin change receptor binding specificity. Nature 304:76–78

Rogers GN, D'souza BL (1989) Receptor binding properties of human and animal H1 influenza virus isolates. Virology 173:317–322

Rogers GN, Daniels RS, Skehel JJ, Wiley DC, Wang XF, Higa HH, Paulson JC (1985) Host-mediated selection of influenza virus receptor variants. Sialic acid-alpha 2, 6Gal-specific clones of A/duck/Ukraine/1/63 revert to sialic acid-alpha 2, 3Gal-specific wild type in ovo. J Biol Chem 260:7362–7367

Rogers GN, Paulson JC (1983) Receptor determinants of human and animal Influenza virus isolates: differences in receptor specificity of H3 hemagglutinin based on species of origin. Virology 127:361–373

Rojo J, Morales JC, Penades S (2002) Carbohydrate-carbohydrate interactions in biological and model systems. Host-guest chemistry. Springer-Verlag, Berlin, pp 45–92

Rostovtsev VV, Green LG, Fokin VV, Sharpless KB (2002) A stepwise Huisgen cycloaddition process: copper(I)-catalyzed regioselective ligation of azides and terminal alkynes. Ang Chem Inter Ed Engl 41:2596–2599

Ruegg C, Mariotti A (2003) Vascular integrins: pleiotropic adhesion and signaling molecules in vascular homeostasis and angiogenesis. Cell Mol Life Sci 60:1135–1157

Ruiz L, Fine E, Vörös J, Makohliso SA, Léonard D, Johnston DS, Textor M, Mathieu HJ (1999) Phosphorylcholine-containing polyurethanes for the control of protein adsorption and cell attachment via photoimmobilized laminin oligopeptides. J Biomater Sci Pol Ed 10: 931–955

Sandström C, Berteau O, Gemma E, Oscarson S, Kenne L, Gronenborn AM (2004) Atomic mapping of the interactions between the antiviral agent cyanovirin-N and oligomannosides by saturation-transfer difference NMR. Biochemistry 43:13926–13931

Sato T, Serizawa T, Ohtake F, Nakamura M, Terabayashi T, Kawanishi Y, Okahata Y (1998) Quantitative measurements of the interaction between monosialoganglioside monolayers and wheat germ agglutinin (WGA) by a quartz-crystal microbalance. Biochim Biophys Acta 1380:82–92

Satoh A, Fukui E, Yoshino S, Shinoda M, Kojima K, Matsumoto I (1999) Comparison of methods of immobilization to enzyme-linked immunosorbent assay plates for the detection of sugar chains. Anal Biochem 275:231–235

Satoh A, Matsumoto I (1999) Analysis of interaction between lectin and carbohydrate by surface plasmon resonance. Anal Biochem 275:268–270

Sauerbrey GZ (1959) Verwendung von schwingquarzen zur wägung dünner schichten und zur mikrowägung. Z Phys 155:206–222

Scanlan CN, Offer J, Zitzmann N, Dwek RA (2007) Exploiting the defensive sugars of HIV-1 for drug and vaccine design. Nature 446:1038–1045

Schena M (2003) Microarray analysis. Wiley, Hoboken

Schmidt RR (1986) New methods for the synthesis of glycosides and oligosaccharides - Are there alternatives to the Koenigs-Knorr method? [New Synthetic Methods (56)]. Ang Chem Inter Ed Engl 25:212–235

Schmidt RR, Michel J (1980) Facile synthesis of α- and β-O-glycosyl imidates; preparation of glycosides and disaccharides. Ang Chem Inter Ed Engl 19:731–732

Schreiber F (2000) Structure and growth of self-assembling monolayers. Prog Surf Sci 65: 151–256

Schumacher R (1990) The quartz microbalance: a novel approach to the in-situ investigation of interfacial phenomena at the solid/liquid junction. Ang Chem Inter Ed Engl 29:329–343

Schwarz M, Spector L, Gargir A, Shtevi A, Gortler M, Altstock RT, Dukler AA, Dotan N (2003) A new kind of carbohydrate array, its use for profiling antiglycan antibodies, and the discovery of a novel human cellulose-binding antibody. Glycobiology 13:749–754

Seeberger PH, Werz DB (2007) Synthesis and medical applications of oligosaccharides. Nature 446:1046–1051

Seeberger PH (2008) Automated oligosaccharide synthesis. Chem Soc Rev 37:19–28

Seibel J, Hellmuth H, Hofer B, Kicinska A-M, Schmalbruch B (2006) Identification of new acceptor specificities of glycosyltransferase with the aid of substrate microarrays. ChemBioChem 7:310–320

Seo JH, Adachi K, Lee BK, Kang DG, Kim YK, Kim KR, Lee HY, Kawai T, Cha HJ (2007) Facile and rapid direct gold surface immobilization with controlled orientation for carbohydrates. Bioconjugate Chem 18:2197–2201

Sharon N, Lis H (2004) History of lectins: from hemagglutinins to biological recognition molecules. Glycobiology 14:53R–62R

Shin I (2007) Carbohydarte microarray for high-throughput analysis of carbohydrate protein interaction. In: Bewley CA (ed) Protein-carbohydrate interactions in infectious diseases. RCS, Cambridge, pp 221–247

Shinya K, Ebina M, Yamada S, Ono M, Kasai N, Kawaoka Y (2006) Avian flu: influenza virus receptors in the human airway. Nature 440:435–436

Skehel JJ, Wiley DC (2000) Receptor binding and membrane fusion in virus entry: the influenza hemagglutinin. Annu Rev Biochem 69:531–569

Smith EA, Thomas WD, Kiessling LL, Corn RM (2003) Surface plasmon resonance imaging studies of protein-carbohydrate interactions. J Am Chem Soc 125:6140–6148

Souteyrand E, Chen C, Cloarec J-P, Nesme X, Simonet P, Navarro I, Martin J-R (2000) Comparaison between the electrochemical and optoelectrochemical impedance measurement for the detection of DNA hybridization. Appl Biochem Biotechnol 89:246–251

Souteyrand E, Cloarec J-P, Martin J-R, Wilson C, Lawrence I, Mikkelsen S, Lawrence MF (1997) Direct detection of the hybridization of specific DNA sequences by field effect. J Phys Chem 101:2980–2985

Sperandio M (2006) Selectins and glycosyltransferases in leukocyte rolling in vivo. FEBS J 273:4377–4389

Stevens J, Blixt O, Glaser L, Taubenberger JK, Palese P, Paulson JC, Wilson IA (2006) Glycan microarray analysis of the hemagglutinins from modern and pandemic influenza viruses reveals different receptor specificities. J Mol Biol 355:1143–1155

Stoll MS, Feizi T, Loveless RW, Chai W, Lawson AM, Yuen C-T (2000) Fluorescent neoglycolipids improved probes for oligosaccharide ligand discovery. Eur J Biochem 267: 1795–1804

Su J, Mrksich M (2002) Using mass spectrometry to characterize self-assembled monolayers presenting peptides, proteins, and carbohydrates. Ang Chem Inter Ed Engl 41:4715–4718

Tamborrini M, Werz DB, Frey J, Pluschke G, Seeberger PH (2006) Anti-carbohydrate antibodies for the detection of anthrax spores. Ang Chem Inter Ed Engl 45:6581–6582

Tassaneetrithep B, Burgess TH, Granelli-Piperno A, Trumpfheller C, Finke J, Sun W, Eller MA, Pattanapanyasat K, Sarasombath S, Birx DL, Steinman RM, Schlesinger S, Marovich MA (2003) DC-SIGN (CD209) mediates dengue virus infection of human dendritic cells. J Exp Med 197:823–829

Taylor PR, Gordon S, Martinez-Pomares L (2005) The mannose receptor: linking homeostasis and immunity through sugar recognition. Trends Immunol 26:104–110

Thoegersen H, Lemieux RU (1982) Further justification of the exo-anomeric effect. Conformational analysis based on nuclear magnetic resonance. Can J Chem 60:44–57

Tornoe CW, Christensen C, Meldal M (2002) Peptidotriazoles on solid phase: [1, 2, 3]-triazoles by regiospecific copper(I)-catalyzed 1, 3-dipolar cycloadditions of terminal alkynes to azides. J Org Chem 67:3057–3064

Tsatas D, Kanagasundaram V, Kaye A, Novak U (2002) EGF receptor modifies cellular responses to hyaluronan in glioblastoma cell lines. J Clin Neurosci 9:282–288

Turnbull WB, Stoddart JF (2002) Design and synthesis of glycodendrimers. Rev Mol Biotechnol 90:231–255

Updike SJ, Hicks GP (1967) The enzyme electrode. Nature 214:986–988

Van Kerhof JC, Bergveld P, Schasfoort RBM (1993) Development of an ISFET based heparin sensor using the ion-step measuring method. Biosens Bioelectron 8:463–472

Vankayalapati H, Jiang S, Singh G (2002). Glycosylation based on glycosyl phosphates as glycosyl donors. Synlett 16–25

Varki A (1993) Biological roles of oligosaccharides: all of the theories are correct. Glycobiology 3:93–130

Varki A, Cummings R, Esko J, Freeze H, Hart GW, Marth J (1999) Essentials of glycobiology. Cold Spring Harbor Laboratory Press, New York

Veeneman GH (1998) Chemical synthesis of *O*-glycosides. In: Boons G-J (ed) Carbohydrate chemistry. Blackie Academic & Professional, New York, pp 98–174

Vroman L (1962) Effect of adsorbed proteins on the wettability of hydrophilic and hydrophobic solids. Nature 196:476–477

Wacker R, Schröder H, Niemeyer CM (2004) Performance of antibody microarrays fabricated by either DNA-directed immobilization, direct spotting, or streptavidin–biotin attachment: a comparative study. Anal Biochem 330:281–287

Wanderley JL, Moreira MEC, Benjamin A, Bonomo AC, Barcinski MA (2006) Mimicry of apoptotic cells by exposing phosphatidylserine participates in the establishment of amastigotes of *Leishmania (L) amazonensis* in mammalian hosts. J Immunol 176:1834–1839

Wang D (2003) Carbohydrate microarrays. Proteomics 3:2167–2175

Wang D, Liu S, Trummer BJ, Deng C, Wang A (2002) Carbohydrate microarrays for the recognition of cross-reactive molecular markers of microbes and host cells. Nat Biotechnol 20:275–281

Werz DB, Seeberger PH (2005a) Carbohydrates as the next frontier in pharmaceutical research. Chemistry 11:3194–3206

Werz DB, Ranzinger R, Herget S, Adibekian A, Von Der Lieth C-W, Seeberger PH (2007) Exploring the structural diversity of mammalian carbohydrates (glycospace) by statistical databank analysis. Chem Biol 2:685–691

Werz DB, Seeberger PH (2005b) Total synthesis of antigen *Bacillus anthracis* tetrasaccharide—creation of an anthrax vaccine candidate. Ang Chem Inter Ed Engl 44:6315–6318

Whitfield DM, Douglas SP (1996) Glycosylation reactions: present status future directions. Glycoconj J 13:5–17

Wink T, Van Zuilen SJ, Bult A, Van Bennekom WP (1997) Self-assembled monolayers for biosensors. Analyst 122:43R–50R

Yamada KM, Araki M (2001) Tumor suppressor PTEN: modulator of cell signaling, growth, migration and apoptosis. J Cell Sci 114:2375–2382

Yanagisawa M, Taga TKN, Ariga T, Yu RK (2005) Characterization of glycoconjugate antigens in mouse embryonic neural precursor cells. J Neurochem 95:1311–1320

Yanagisawa M, Yu RK (2007) The expression and functions of glycoconjugates in neural stem cells. Glycobiology 17:57R–74R

Yanagishita M, Hascall VC (1992) Cell-surface heparan-sulfate proteoglycans. J Biol Chem 267:9451–9454

Yates EA, Terry CJ, Rees C, Rudd TR, Duchesne L, Skidmore MA, Lévy R, Thanh NTK, Nichols RJ, Clarke DT, Fernig DG (2006) Protein–GAG interactions: new surface-based techniques, spectroscopies and nanotechnology probes. Biochem Soc Trans 34:427–430

Yeung C, Leckband D (1997) Molecular level characterization of microenvironmental influences on the properties of immobilized proteins. Langmuir 13:6746–6754

Young NM, Gidney MAJ, Gudmundsson BME, Mackenzie CR, To R, Watson DC, Bundle DR (1999) Molecular basis for the lack of mimicry of Brucella polysaccharide antigens by Ab2g antibodies. Mol Immunol 36:339–347

Yu H, Chokhawala HA, Huang S, Chen X (2006) One-pot three-enzyme chemoenzymatic approach to the synthesis of sialosides containing natural and non-natural functionalities. Nat Protocols 1:2485–2492

Yu RK, Bieberich E, Xia T, Zeng GC (2004) Regulation of ganglioside biosynthesis in the nervous system. J Lipid Res 45:783–793

Yu RK, Yanagisawa M (2006) Glycobiology of neural stem cells. CNS Neurol Disord Drug Targets 5:415–423

Zhang SC (2001) Defining glial cells during CNS development. Nat Rev Neurosci 2:840–843

Zhang Y, Luo S, Tang Y, Yu L, Hou K-Y, Cheng J-P, Zeng X, Wang PG (2006) Carbohydrate-protein interactions by "clicked" carbohydrate self-assembled monolayers. Anal Chem 78:2001–2008

Zhang Z, Wong C-H (2000) Glycosylation methods: use of phosphites. In: Ernst B, Hart GW, Sinay P (eds) Carbohydrates in chemistry and biology. Wiley-VCH, Weinheim, pp 117–134

Zhou X, Zhou J (2006) Oligosaccharide microarrays fabricated on aminooxyacetyl functionalized glass surface for characterization of carbohydrate–protein interaction. Biosens, Bioelectron

Zou W, Mackenzie R, Therien L, Hirama T, Yang Q, Gidney MA, Jennings HJ (1999) Conformational epitope of the type III group B Streptococcus capsular polysaccharide. J Immunol 163:820–825

Chapter 8
Nucleic Acid Diagnostic Biosensors

Barry Glynn and Louise O'Connor

Abstract On the basis of the numbers of scientific articles published using different approaches, use of biosensors is the fastest growing technology for pathogen detection. Biosensor use still lags behind the established technologies such as polymerase chain reaction (PCR) and conventional microbiology but has already surpassed the techniques including gel electrophoresis and ELISA based immunoassays. This chapter provides an overview of nucleic acid diagnostics technologies for pathogen detection with a special emphasis on biosensor technologies including optical, piezoelectric, and electrochemical detection formats.

Keywords Biosensors · Nucleic acid diagnostics · Microbiology · PCR

Abbreviations

DNA	Deoxyribonucleic Acid
RNA	Ribonucleic Acid
cDNA	Complementary DNA
rRNA	Ribosomal RNA
tmRNA	Transfer messenger RNA
mRNA	Messenger RNA
NAD	Nucleic Acid Diagnostics

B. Glynn (✉)
Molecular Diagnostics Research Group, National Centre for Biomedical Engineering Science, National University of Ireland, Galway, Ireland
e-mail: barry.glynn@nuigalway.ie

M. Zourob (ed.), *Recognition Receptors in Biosensors*,
DOI 10.1007/978-1-4419-0919-0_8,

PCR	Polymerase Chain Reaction
ELISA	Enzyme-linked immunosorbent assay
LiPA	Line Probe Assay
FRET	Fluorescence Resonance Energy Transfer
MRSA	Methicillin-resistant *Staphylococcus aureus*
SPR	Surface Plasmon Resonance
SAM	Self Assembling Monolayer
QCM	Quartz Crystal Microbalance
GM	Genetically Modified
CCD	Charge-Coupled Device
NADH	Nicotinamide Adenine Dinucleotide
HCV	Hepatitis C Virus
CFU	Colony Forming Unit
BWA	Bio-Warfare Agents

8.1 Introduction

The use of nucleic acid sequences for specific diagnostics applications has followed a somewhat linear pattern of development since the early 1970s. Optimised methods for restriction enzyme digestion as well as reverse transcription protocols were followed by Southern, Northern, and dot blotting as well as DNA sequencing in the late 1970s. The exponential growth of molecular biology was precipitated by the 1985 description of the polymerase chain reaction (PCR) and routine laboratory manipulation of sufficient quantities of DNA for diagnostics. Alternative DNA and RNA amplification protocols were followed. The second explosion in molecular biology occurred with the invention of real-time quantitative PCR and the development of oligonucleotide microarrays and DNA chips. This trend continues today with the development of enhanced transduction elements for improved sensitivity of nucleic acid detection and the possibility of "direct" nucleic acid target detection from samples without *in vitro* amplification.

Diagnostic biosensors are devices and technologies that use a biologically derived material immobilised on a detection platform to measure the presence of one or more analytes in a sample of interest (Mascini et al. 2005). Even though biosensors can take many shapes and forms, in this chapter we solely focus on the use of nucleic acid (RNA and DNA) as the biosensor recognition element in combination with a variety of physiochemical transducers including optical, piezoelectric, and electrochemical. Examples of the use of such biosensors for microbial detection will also be described. In order for the reader to gain a better understanding of the evolution of diagnostic nucleic acid biosensors in the area of microorganism detection, a description on the technologies which contributed to the development of nucleic acid diagnostics (NAD) will also be described.

8.1.1 Development of Nucleic Acid Diagnostics

The detection of specific segments of nucleic acid by complementary probe sequences provides the basis for all nucleic acid probe based biosensors including those for infectious disease monitoring. The specificity of the interaction is dependant on the ability of different nucleotides only to form hydrogen bonds with the appropriate (complementary) partner.

In order to be detected, most nucleic acid hybridisations require in the order of 1×10^6 target molecules to be present. Currently, most detection technologies cannot improve upon this requirement for sensitivity; therefore, some level of pre-amplification of the target nucleic acid sequences is necessary. Commonly used amplification technologies include PCR which will be described in this chapter.

A new area of research worldwide has been the improvement of sensitivity levels of biosensors to potentially be competitive in a "post-PCR" diagnostics market. Key to the success of these programmes is improvements in sensor performance that makes them comparable in sensitivity to amplified methods. Alternatively, these technologies must demonstrate advantages in either reduced assay time or instrument cost in order to be commercially viable.

8.1.2 Properties of a Diagnostic Target

Nucleic acid-based diagnostics (NAD) have several advantages over conventional technologies including faster turnaround time and improved sensitivity and specificity for the detection of organisms of interest. NAD systems are available in a variety of formats ranging from simple nucleic acid probe hybridisation systems, to tests incorporating amplification of a specific genomic target utilising one of many available *in vitro* amplification technologies.

As mentioned previously, the basis of any NAD test is a specific nucleic acid target sequence, unique to the microorganism of interest. Within each organism, there are unique sequences, and these can be exploited to determine the presence of that organism in a sample of interest. DNA is a very stable molecule, which makes it particularly suitable as a molecular target for NAD tests as it can be isolated relatively simply from a variety of complex biological samples (Grennan et al. 2001; Maher et al. 2003; O'Connor et al. 2000).

Ideally, a candidate NAD target should be present in the cell at relatively high copy number while being sufficiently heterologous at the sequence level to allow for differentiation of the pathogen at both genus and species levels. To date, NAD tests have utilised a wide variety of genomic targets including multi-copy ribosomal RNA (rRNA) genes, genes encoding toxins or virulence factors, and genes involved in cellular metabolism. For example, we have successfully employed the 16S/23S rRNA intergenic spacer region in bacteria as a molecular target for the development of a variety of NAD tests for a variety of micro organisms (Grennan et al. 2001; Maher et al. 2003; O'Connor 2003).

8.1.3 DNA and RNA targets

DNA is not the only nucleic acid suitable for NADs. RNA is an alternative nucleic acid target which has the potential to be used in NAD tests. It is a labile molecule that is quickly and easily degraded, particularly once the organism is killed. This property makes handling RNA much more difficult than DNA but means RNA has the advantage of enabling viable organisms to be distinguished from non-viable organisms. rRNA remains the "gold standard" target for pathogen identification and has been exploited in a range of NADs for identification of food pathogens (Pfaller 2001). Gen-Probe Incorporated (CA, USA) have patented the use of rRNA as a target technology and have used it in a range of tests (www.gen-probe.com).

At NUI, Galway, we have also investigated the possibility of using functional, high copy number RNAs other than rRNA for the rapid, sensitive, and specific detection of bacterial pathogens. The bacterial *ssrA* gene and its encoded transcript, tmRNA, present in all bacterial phyla has many of the desired properties of a NAD target (Schonhuber et al. 2001). tmRNA is present in all bacterial species at relatively high transcript numbers (Williams 2002), for example *E. coli* has previously been shown to have approximately 5×10^2 tmRNA molecules per cell. This high-copy number of tmRNA is advantageous in developing NADs (Glynn et al. 2007; Lee et al. 1978).

8.1.4 Nucleic Acid Diagnostics Methods

Currently used formats for the detection of signature sequences within micro-organisms of interest can be divided into two categories. These are direct detection methods, which utilise a probe to hybridise to the target sequence of interest, and those employing an amplification technology to firstly amplify the target sequence of interest followed by the use of specific probe for detection of the amplified target.

Direct detection systems are available in multiple formats. They generally do not require sophisticated equipment, are simple to perform and have detection limits of 10^4–10^5 bacterial cells (Todd et al. 1999). Direct nucleic acid probe hybridisation tests are widely applied in food testing laboratories for confirmation of the identity of organisms following culture-based isolation of the food borne pathogen of concern (Fung 2002). Commercially available NADs based on direct nucleic acid probe technology include Accuprobe® (Gen-Probe, CA, USA) available for detection of *Campylobacter* spp. and *L. monocytogenes*, and Gene-Trak® (Neogen, MI, USA) for *E. coli*, *Salmonella*, and *Listeria* spp. detection. Both test systems require that the food samples are culture-enriched before sample analysis. This is then followed by a simple lysis procedure to release target nucleic acid material before the probe analysis is performed. The Gene-Trak system comprises a pathogen-specific capture probe and a detection probe with an attached chromogenic substrate. Accuprobe uses a single DNA probe to hybridise to rRNA, unbound probe is degraded chemically and the chromogenic label on the bound probe is detected using a luminometer.

The second category of NADs employs an amplification technology. While a number of amplification technologies are available the most commonly used *in vitro* technology employed in NADs is PCR. This category of NADs can use either conventional or real-time PCR which has become increasingly popular in recent years.

8.1.5 Polymerase Chain Reaction

The first report of specific DNA amplification using the PCR was in 1985. The basic method was developed by Kary Mullis (Mullis and Faloona 1987). Since this, the field of application has expanded enormously to include clinical, veterinary, food, and environmental areas.

PCR is an *in vitro* method for the synthesis of DNA sequences enzymatically. Two oligonucleotide primers hybridise to opposite strands of the target DNA and flank the region of interest. Through a series of cycles involving template denaturation, primer annealing, and extension of annealed primers by the enzyme DNA polymerase, there is an exponential accumulation of specific fragment. Products generated in one cycle can serve as a template in the next cycle. This means that the number of copies of the target DNA approximately doubles every cycle. Therefore, 20 cycles of PCR yields in the region of million-fold amplification.

Traditionally, detection of amplified target was carried out by analysing the PCR product using agarose gel electrophoresis with fragment length being used as an indicator for identification. While this method has served its purpose well, it is labour intensive and not particularly sensitive. In addition, accuracy of amplification depends on the specificity of the primers. Impure nucleic acid target or the presence of contaminating background DNA can influence the specific binding of the primers resulting in non-specific amplification and misinterpretation of the results. Because of this, most tests which are amplification based use hybridisation probes for the specific and sensitive detection of amplified product. Such probes can be immobilised onto a solid support or be free in solution and can be radioactively, colorimetric, or fluorescently labelled. When this hybridisation is carried out under stringent conditions it can be highly specific. Several variations of this technology exist including conventional Southern blot (Southern 1975), microtitre plate based hybridisation assays (Van Der Pol et al. 2000) and line probe assays (LiPA Innogenetics Belgium) (Tortoli et al. 2003).

In recent years, the basic PCR technique described above has been combined with fluorescently labelled probe hybridisation in the same reaction, allowing real time monitoring of the amplification. This development known as real time PCR has been responsible for the increase in popularity of molecular assays in diagnostic laboratories. Most real time methods are based on the principle of fluorescence resonance energy transfer (FRET). FRET occurs when two fluorescently labelled molecules are in close proximity to each other and the energy from an excited donor molecule is transferred to an acceptor molecule.

Real time PCR platforms utilise different detection formats. The most commonly used chemistry is the hydrolysis or TaqMan probe (Fitzmaurice et al. 2004). Other detection formats include molecular beacons and HybProbes. Molecular beacons are probes which form a stem-loop structure when in solution and emit a signal only when bound to their specific target (Templeton et al. 2003). An example of a commercial assay using molecular beacons is the methicillin-resistant *Staphylococcus aureus* (MRSA) from Cepheid which is performed on the Smartcycler instrument. The hybridisation probe or HybProbe format was developed specifically for use on the LightCycler® instrument (Roche). In this detection format, two separate probes are designed to hybridise to next to each other on the target DNA. HybProbes are most commonly used for melting curve analysis which is based on the probes dissociating from the target DNA at a certain temperature. Numerous protocols are described in the literature for the use of hybridisation probes (O'Connor et al. 2005).

8.2 Biosensors

The first published demonstration of an enzyme/electrode combination for biological sensing was reported in 1967 (Updike and Hicks 1967). Since this time there has been much advancement in the biosensor field. Biosensor technologies developed for the specific detection of microorganisms have included those which utilise metabolism, antibody, and DNA based systems. While immunodiagnostic biosensors are currently faster and more robust (Iqbal et al. 2000) than nucleic acid based methods, the latter are more specific and sensitive, especially when combined with an *in vitro* amplification step. Indeed, novel biosensors have been described which utilise advanced visualisation and signal amplification technologies allowing the possibility of monitoring single molecular interactions in real-time (Yao et al. 2003). The next generation of biosensors will have application in all sectors of the molecular diagnostics market. It is envisaged that for applications in microbial diagnostics, the ideal device would be a self-contained, automated system capable of organism detection directly from different sample types without pre-enrichment and also be capable of differentiating live from dead cells (Ivnitski et al. 2000). Figure 8.1 shows a general layout of a nucleic acid based biosensor.

8.2.1 Biosensor Assay formats

Two basic formats of diagnostic biosensors are generally available; direct and indirect assays (Fig. 8.2a, b) (Baeumner 2003). In the case of direct assays, the recognition element (oligonucleotide probe) is attached to the transducer and attachment or capture of the target element is detected directly. This format of detection is the basis of many devices that detect changes in mass, refractive index, or impedance. Indirect biosensor formats require addition of at least one additional

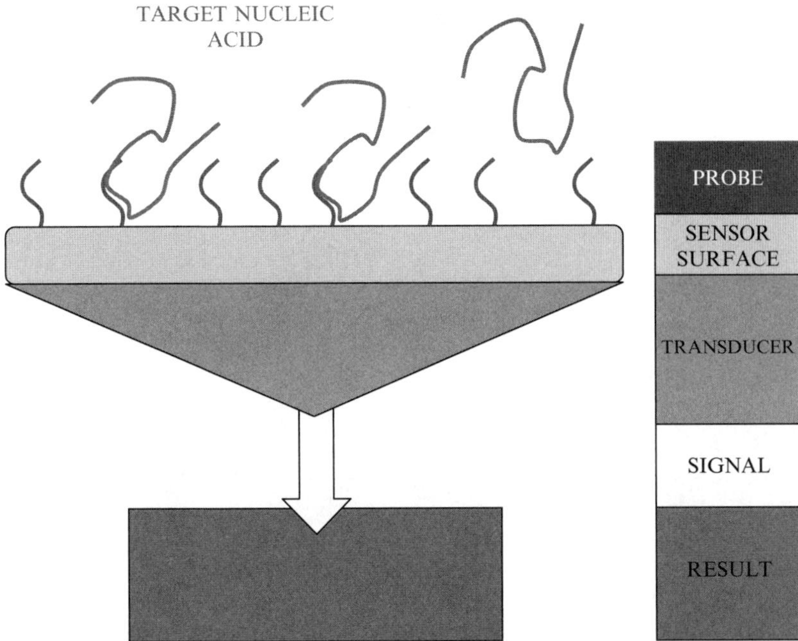

Fig. 8.1 General organisation of a nucleic acid based biosensor. Specific capture oligonucleotides are immobilised on a sensor surface. Hybridisation of complementary target nucleic acid material is converted to a result by means of a signal transducer. Hybridisation of unlabelled target nucleic acid causes a change in the physical properties of the transducer. These include changes to crystal vibration frequency, surface plasmon resonance, or electrical conductivity. In some cases signal is transmitted directly to a result. In other cases (electrochemical for example) some degree of signal amplification may be incorporated between the transducer and the end result

recognition element to the assay before the target molecule can be detected. These forms of assays are well characterised for immunoassays and form the basis of the well established sandwich assay format. Sandwich assays have also been described in nucleic acid diagnostics (Baeumner et al. 2004) (Fig. 8.2a).

Indirect assays may ultimately prove to be the most useful for dealing with complex sample types such as foodstuffs and some clinical sample types where non-specific absorption to biosensors surfaces is seen as a significant problem. The extra layer of specificity introduced by an additional recognition element (labelled oligonucleotide probes) goes some way towards reducing these spurious detections.

8.2.2 Biological Recognition Elements and Immobilisation Methods

The three main classes of recognition element used for biosensor detection of pathogens are (1) enzymes, (2) antibodies, and (3) nucleic acid. Note that enzymes

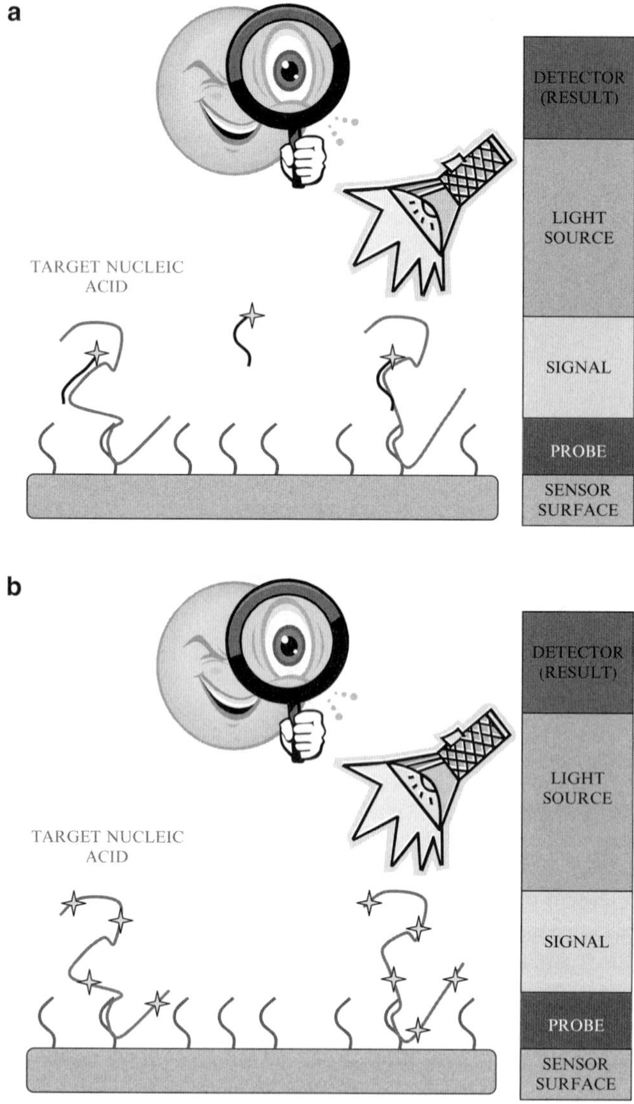

Fig. 8.2 (**a**) Biosensor format for indirect detection of unlabelled nucleic acid targets by means of a secondary "sandwich" hybridisation. In this example the target nucleic acid material is unlabelled. Detection is by means of a fluorescent label attached to a second oligonucleotide probe (reporter probe) which is designed to hybridise to the target nucleic acid at some position other than that used for the capture oligonucleotide. Illumination and detection are the same as for the detection of labelled target. (**b**) Biosensor format for detection of labelled nucleic acid targets. Fluorescent reporter molecules are introduced into the target nucleic acid target either by use of labelled PCR primers or by a post-amplification labelling process. Following appropriate wash steps to remove unbound target material, fluorescence is detected typically by excitation of the label by a laser or LED light source and detection of emissions

are generally included in biosensor platforms as labels rather than as the detection element. Some biosensor formats can detect their target molecules without labels. These methods include surface plasmon resonance (SPR), piezoelectric, and impedance based biosensors (Guan et al. 2004).

The immobilisation step is critical in the development of any class of biosensor. The key issue is that the biological molecule retains its functionality and is not inhibited sterically, or chemically rendered inert by the immobilisation process. Three common approaches for immobilisation include absorption on gold, avidin-biotin immobilisation, and self-assembling monolayers (SAMs). Absorption on gold is typically quick and simple. Unfortunately, it is also the least reliable method as it is non-controlled. Non-specific absorption can also be a significant problem especially when both the target and the capture probe have similar chemical properties (as in the case of nucleic acids). Avidin-biotin immobilisation is almost as straightforward as absorption on gold with the advantage of being more easily controlled. The high affinity between the avidin and biotin also allows the sensor surface to be washed free of any bound analyte without removing the immobilised capture oligonucleotide. For preparation of SAMs, the sensor surface is first pre-coated with reactive groups such as thiol or disulphide (Su and Li 2004). The nucleic acid probe is then chemically attached to the reactive layer. The advantage of this approach is that it is highly tunable to the specific requirements of the end-user.

8.3 Biosensor Formats for Nucleic Acid Diagnostics

8.3.1 Optical Based Systems

Optical based detection systems are the most commonly used form of biosensor. These can be divided into systems that use labels, typically fluorescence and those that are unlabelled. Labelled systems will be discussed in detail in the microarray section of this chapter. The most common form of unlabelled optical detection is based upon SPR.

8.3.2 Surface Plasmon Resonance

SPR-based biosensors have been developed by a number of companies (including Biacore International and Texas Instruments) to monitor biomolecular interactions on a surface in real-time (Rand et al. 2002). An advantage of SPR-based systems is that no labelling of target molecule is required. Figure 8.3 shows representative examples of label free detection using SPR and piezoelectric biosensors.

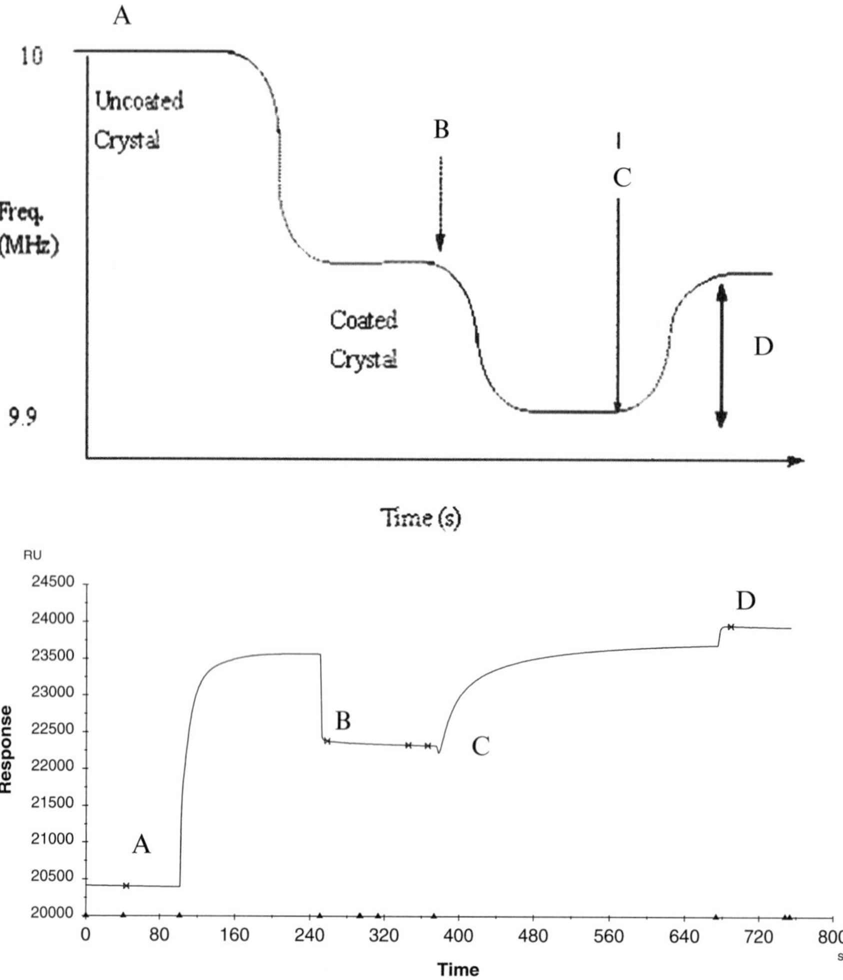

Fig. 8.3 Examples of direct detection biosensor results. Top panel shows plot of frequency response of a QCM device, bottom panel is resonance unit plot from Biacore 2000 SPR instrument. In both cases the instrument shows a baseline reading before any oligonucleotide is introduced (A) Immobilisation of capture oligonucleotide is measured by the instrument (B) Target nucleic acid was then introduced into the sample chamber and hybridisation of target material to the capture oligonucleotide is measured (C) The level of hybridisation is measured at point D as a drop in frequency for the piezoelectric device, or as an increase in resonance units for the SPR instrument. In both cases it is possible to remove the bound material and return to sensor to pre-hybridisation state (B) ready for re-use

The Biacore system uses SPR which occurs in thin metal (gold) films under total internal reflection. The gold transducer surface is modified with dextran onto which the oligonucleotide probe is attached, typically by avidin to biotin linkages. Hybridisation takes place within a continuous flow system and oligonucleotide hybridisations are monitored in real-time. Typical analysis takes about 5 min. Probe surfaces are very easily regenerated allowing many repeat hybridisations to be performed on the same surface.

Monitoring of SPR to measure DNA–DNA and DNA–RNA hybridisations has been demonstrated. Procedures are well established for the immobilisation of oligonucleotide probes to device surfaces and hybridisation of target nucleic acid to the bound probes. However, the applications of this technology for pathogen identification have yet to be fully realised (Mannelli et al. 2005; Wang et al. 2004). To date, the potential of SPR-based biosensors has been demonstrated by the detection of genetically modified (GM) soybeans and maize following PCR amplification of transgenic and wild-type sequences (Feriotto et al. 2002).

Combination of SAM with thiol deposition of specific oligonucleotide probes in an array has been described for manufacture of SPR biosensors with applications in the food industry. These SPR microarrays could combine the multiplex possibilities of microarrays with the label free detection of SPR. Sensitivity remains poor as 1 μM target concentration would correspond to 1.5×10^{13} target molecules in typical reaction volumes. However, combination of such a system with *in vitro* amplification should not be ruled out (Nelson et al. 2001, Spadavecchia et al. 2005).

Much progress will be made in the coming years in the integration of high throughput parallel SPR detectors with integrated lab-on-a-chip systems (Hoa et al. 2007).

8.3.3 Piezoelectric Biosensors

The piezoelectric principle describes the property of some materials, typically crystals, to generate an electrical potential in response to a mechanical force. The addition of mass to the sensor surface will cause a measurable change in the resonance frequency of the crystal proportional to the added mass. Piezoelectrical biosensors may be used for direct, label-free detection of specific nucleic-acid targets. These systems combine real-time readout of hybridisation measurements with relative simplicity of use.

Piezoelectric biosensors consist of specific oligonucleotides immobilised on the surface of an appropriate crystal (typically quartz). This sensor is placed in a solution containing potential target nucleic acids which bind to their complementary oligonucleotides. The mass at the surface of the crystal therefore increases and the resonance frequency of the quartz oscillation will decrease proportionally.

Piezoelectric biosensors are difficult to regenerate the following hybridisation. However, economy of scale may serve to make these detectors essentially disposable. Other difficulties remain to be overcome including lack of specificity and sensitivity as well as considerable interference at the sensor surface with buffer components and media.

The most common type of piezoelectric biosensor is the quartz crystal microbalance (QCM). QCMs have been described for the sequence specific capture of oligonucleotides corresponding to pathogenic micro algae species *Alexandrium minutum* (Dinophycaea) with a detection limit of 20 μg/ml (Lazerges et al. 2006). This is still outside the useful sensitivity range required, but, if such a system could be integrated with an *in vitro* amplification or a high copy number RNA target then it may have genuine application in the area of environmental monitoring. The current research efforts are thus laying the ground work for the future integration of QCM sensors into integrated devices for PCR based DNA detection (Wu et al. 2007).

Signal amplification of QCM response using nano-particles as "mass-enhancers" has also been reported (Mao et al. 2006). The authors report the detection of as little as 267 *E. coli* o157:H7 with the sensor when integrated with PCR but without any sample pre-enrichment. Oligonucleotide detection was in the region of 1 pM.

8.3.4 Electrochemical Biosensors

These devices work by detecting current or potential changes caused by binding reactions occurring on or near the sensor surface. Different devices are classed depending on their observed parameter. Devices that measure current are amperometric, those that measure electrical potential are potentiometric, and impedance based devices are impedometric. In principle, nucleic acid can be incorporated into each of these types of device.

Electrochemical devices are an attractive technology in that they are label free, are not affected by the turbidity or other optical properties of the sample matrix and also offer the potential for signal amplification. Some progress has been made in demonstrating the selectivity of these devices for nucleic acid based detection of micro organisms (Lazcka et al. 2007). However, sensitivity remains poor and these devices remain susceptible to non-specific interference. Solutions to these problems may involve integration of these devices with upstream sample preparation and clean up methods, possibly combined with micro-fluidic based sample concentration. Electrochemical nucleic acid detection technologies are amenable to direct electrical readout (Peng et al. 2006). Methods of electrochemical detection have included use of metal complexes, organic redox indicators, enzymes, nanoparticles, and nano-tubes.

Several experimental approaches utilising electrochemical properties for the biosensor detection of nucleic acid targets have been described. For example, the waterborne pathogen *Cryptosporidium* was detected by an oligonucleotide immobilised onto a carbon-paste transducer (Wang et al. 1998; Wang et al. 2008) that employed a sensitive chronopotentiometric transduction mode for hybridisation monitoring. Initially sensitivity was poor, requiring μg/ml levels of target DNA. However reactions could be performed in as little as 3 min. Sensitivity was improved to ng/ml levels by increased hybridisation time (typically 30 min). Such a system offers scope to monitor for a large number of microorganism on a small array type device.

8.3.5 Other Detection Formats

In order for a specific hybridisation event to be detected, oligonucleotide probes need to be labelled. Different types of reporter molecules have been described including radioactive isotopes, flurophores, enzymes, or antibody recognition sites (haptens). Radioisotopes were the original label of choice but have decreased in popularity in recent years. Fluorescent detection will become more important as detection systems (CCD etc) improve especially in regards to sensitivity. Enzyme labels are useful for applications that require signal amplification. A useful alternative to hapten binding is biotin-strepavidin complex formation. Typically biotin is attached to the oligonucleotide in a sandwich type hybridisation. A second labelling reagent containing strapavidin-enzyme complex is added post hybridisation which binds to the biotinylated probe with subsequent detection.

8.3.6 Signal Amplification

As mentioned previously, a common failing with hybridisation based nucleic acid biosensor assays is their poor sensitivity. Even the most sensitive systems require in the order of $1 \times 10^{9-10}$ molecules for detection. Sensitivity is further complicated in real samples by target loss in the sample preparation process or due to interference from other nucleic acid molecules present in the sample.

Electrochemical methods are attractive as they are directly adaptable to direct electronic readout perhaps coupled with signal amplification. The amplification of electrochemical detection signal by use of nano-particles has shown particular promise (Peng et al. 2006). Other interesting signal amplification methods try to utilise the catalytic properties of the captured DNA itself (de-los-Santos-Alvarez et al. 2006). In this approach, electro-oxidized adenines within the DNA itself catalyse the oxidation of NADH. This reaction is further improved in the presence of Ca^{2+} ions leading to a detection limit of 33 fmol DNA without any labelling. Liposome signal amplification has been incorporated into an electrochemical microfluidic biosensor for nucleic acid target detection by using an integrated potentiostat (Kwakye et al. 2006). Sensitivity was in the range of 0.01 μM while functionality was demonstrated using Dengue virus RNA.

Novel biosensors with advanced visualisation and signal amplification technologies have created the possibility of monitoring single molecular interactions in real-time (Yao et al. 2003). The application of gold nano-particles for detection of nucleic acid targets on a biosensor platform has been demonstrated to have a sensitivity of 3×10^6 target molecules when evaluated with dilutions of PCR products (Storhoff et al. 2004). A study combining biosensor-based nucleic acid detection, with nucleic acid concentration using DNA probes bound to supramagnetic nanoparticles prior to detection on the biosensor achieved detection of 2 molecules of HCV cDNA target in a 2.5 million-fold excess of non-target DNA (Fuentes et al. 2006).

8.4 Microarrays

Microarrays are a class of planar biosensor, typically consisting of numerous (10^2–10^3) oligonucleotide probes deposited in a grid on a glass surface (Fig. 8.4). Hybridisidations are performed in small volumes and molecular interactions are detected using fluorescent labels and an appropriate scanner (Affymetrix, Genescan etc). For use as in diagnostics biosensors the principle advantage of these detection formats is the very high probe density that can be incorporated into a single assay. Microarrays for microbial identification have been used in environmental and clinical settings amongst others.

Environmental microarrays (also employed in industrial settings) are used to build a picture of the microbial community present in a given setting. These assays should detect many potential microbes at relatively high taxonomic levels, i.e. species, genus, or even taxon. If possible, these assays should also offer a degree of quantification of target organisms. In contrast clinical microarrays (applied in medical, veterinary, food, and bio-defence scenarios) are required to rapidly and reliably detect the presence of specific species and strain types of only a few potentially interesting microbes.

The ability to immobilise large numbers of oligonucleotide probes onto an array allows for simultaneous multi-organism detection in a single experiment. For example Wilson et al. (2002) were able to simultaneously detect 18 pathogenic microbes using a sequencing microarray. When combined with a PCR step, the limit of detection of this assay for *B. anthracis* DNA was 10 fg (Equivalent to less than 10 cells).

Clinically relevant species identified on microarrays have included *Chlamydia* and *Chlamydophila* spp. (Sachse et al. 2005). The authors reported the possibility of direct detection of pathogens from clinical tissue although their then reported detection limit was in the order of 6×10^6 target molecules. However, this is still

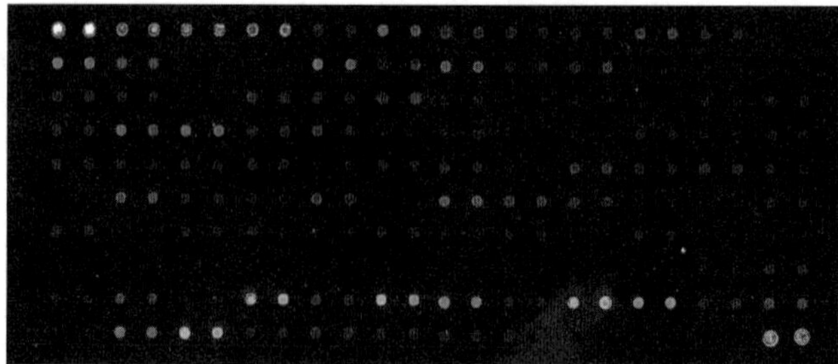

Fig. 8.4 Detection of labelled nucleic acid targets on a low-density diagnostics microarray. Diagnostics microarrays typically contain less immobilised probes than whole genome or gene-expression arrays. Probes can be included on the array that are capable of differentiating between different species, or in the example shown here, multiple probes for the same target organism can be immobilised in order to rapidly identify the optimum probe sets for that particular microorganism

far superior to sensitivities reported for gel electrophoresis based detection (1×10^{11}). As mentioned earlier, the bacterial rRNA genes have many of the properties associated with an ideal diagnostics target. It is no surprise therefore to find in the literature many instances of the microarray based detection of this nucleic acid diagnostic target. The advantage of this target on a microarray is the possibility of differentiating between very large numbers of potentially present species in a single assay. The high intracellular levels of rRNA in bacteria also make it an interesting target for microarray detection without *in vitro* amplification (Small et al. 2001). Identification of bacterial species was possible from of 0.5 μg of total RNA prepared from soil without any PCR step in a 2 h hybridisation. As 16s rRNA has a large amount of secondary structured, which could potentially prevent oligonucleotide hybridisation, the authors designed their labelled reporter probes to have a "chaperone" effect on the microarray thus improving assay sensitivity. 23s rRNA has been used for the direct detection of *S. aureus* in a flow through microarray (Anthony et al. 2005) using 2 labelled 20 mer oligonucleotides as reporters and 60 mer immobilised oligonucleotides for target capture. Approximately 16,000 cfu were required for detection. Detection levels were 100 times lower when the authors attempted direct detection of mRNA targets. However, these results are still very impressive and point the way towards extremely sensitive bacterial detection without *in vitro* amplification.

A difficulty associated with the successful penetration of microarray detection systems into the market place is the extremely high cost of the associated infrastructure (hybridisation stations, detector, and spotters). This is potentially offset by the possibility of highly parallel detection and the massive number of potential oligonucleotides deployed. However, only one sample is measured at a time.

As microarray technology matures, these planar arrays are being supplemented by further evolutions including microbead and suspension microarray formats. Suspension microarrays are a modification of the original planar microarrays with the differentiating probes being immobilised onto the surface of polystyrene microbeads containing internal fluorescent dyes (Ahn and Walt 2005; Borucki et al. 2005).

A microarray-based assay incorporating signal amplification and suspension microarray technologies has been reported for the identification and sub-typing of *L. monocytogenes* from genomic DNA (Borucki et al. 2005). Micro bead type arrays have been developed for identification of Salmonella spp. Detection sensitivities of 1×10^3 CFU ml^{-1} were reported when the *Salmonella invA2* and *spvB* genes were targeted (Ahn and Walt 2005). A multiplex PCR for *E. coli* O157:H7 (*eaeA*, *hlyA*, *stx1*, and *stx2* genes) and *Salmonella* (*invA* gene) combined with a suspension microarray detection system had a similar sensitivity for each species (Straub et al. 2005).

The developments in biosensor and microarray technologies have generated promising results to date. However, for application in diagnostics several important issues remain to be resolved. While biosensors are designed for the optimum detection of minute quantities of target nucleic acid, it is important that the sample material be free of contaminants before application to the biosensor surface (Fung 2002). This is of particular relevance to the food industry for example where regulatory standards require the detection of 1 viable target cell in 25 g of foodstuff such as ground beef. Current biosensor technologies are unable to detect such a low bacterial load from a food matrix without either sample amplification (using culture-based enrichment or target amplification by PCR) or extensive sample purification techniques. These major hurdles have to be overcome before biosensors and microarrays will provide "real-time" detection of pathogens in food samples.

8.5 Use of Nucleic Acid Biosensors for Rapid Pathogen Detection

Conventional methods for the detection of pathogenic microorganisms are typically time consuming, often requiring at least 72 h for a confirmed result. The results of microbiological screening can be difficult to interpret and are generally not available in a time-scale compatible with clinical management of an infection. Alternative, molecular biology methods are more sensitive, but analysis time can remain lengthy. Future methods such as nucleic acid biosensor technology are required that can rapidly detect low concentrations of pathogenic organisms in food, clinical, or environmental samples without pre-enrichment or other complex and time-consuming steps. Some of the fields where biosensor technology could potentially be of use are described below. The field of bio-threat monitoring is one which has received particular attention in recent times. Technologies currently in use such as the Razor (Idaho Technologies), the Bio-Seeq Plus (Smiths Detection), and the GeneXpert system (Cepheid) represent significant advances in that they allow detection within a 30-minute timeframe. Future devices will be required to operate within the "Detect to Protect" window which requires results within 5 min. There has, therefore, been a recent surge in developing technologies that bridge the gap between the 30 min "Detect to Treat" and the 5 min "Detect to Protect" time criteria. When considering the detection of biological threat agents, it is necessary to consider that these may exist as bacteria (both vegetative and as spores) viruses and toxins. In a real-sense, nucleic acid based biosensor detection of bio-threat agents is an extension of the technologies employed for environmental monitoring with the same basic considerations. Further considerations specific to "field-deployable" assays include low energy, user friendliness and sufficient robustness to work in a variety of challenging environments and not necessarily a laboratory. Military applications of nucleic acid biosensors are associated with the rapid ultra sensitive detection of selected bio-warfare agents (BWAs). Detection devices are required to be portable, robust, rapid and user friendly. To date direct detection biosensors have been described that take advantage of changes at a solid/liquid interface that measure pH, oxygen consumption, ion concentration, potential difference, current, or optical properties. Of these, only optical biosensors have so far shown genuine utility as nucleic acid based biosensors. These work by measuring very small changes in the refractive index which occur when target material is captured by an oligonucleotide probe bound to a solid phase surface. Undoubtedly, this is a field where advances in nucleic acid biosensors could have significant impact.

8.6 Future Developments

As of 2007 use of biosensor technology for pathogen detection ranked fourth behind PCR, conventional microbiology and ELISA based methods. However, biosensors are the fastest growing technology in this sector. Optical and electrochemical sensors each account for over 30% of this market with piezoelectric coming third at 16% with the remaining sensors making up the other 16% (Lazcka et al. 2007). The reason that conventional methods have remained on-stream for so long is their relatively high sensitivity and specificity. Emerging technologies such as biosensors promise to shorten the time to result but must match the performance of pre-existing technologies. Only then will secondary considerations such as overall unit cost, user friendliness, and ease of operation become relevant.

The driving force behind the need for pathogen detection biosensors is that the turnaround time for conventional approaches such as microbiology are too slow. For example, even a negative *campylobacter* result can take 4–9 days while confirmation of a positive takes from 14–16 days. This is a real issue in time critical applications such as clinical microbiology and processes such as found in the food industry.

Biosensor systems for microbial diagnostics currently exist in a marketplace dominated by the PCR. In order to gain significant market share in the future, these systems need to demonstrate comparable sensitivity and specificity as the current state of the art *in vitro* amplification assays. When this quantum leap in sensitivity is achieved, the secondary advantages of biosensor systems (speed, sensitivity, unit cost) will come to the fore. Several systems are in development that do not require *in vitro* amplification. These instead rely on advances in detection systems either by deployment of advanced and novel dyes or by means of signal amplification. Nanosphere Inc (Northbrook IL, USA) has succeeded in developing two automated systems that combine advanced optical detection with micro fluidic sample handling. The *Verigene Autolab* and the *Verigene Mobile* use DNA probes conjugated to gold nano-particles for detection in combination with the company's proprietary Bio-barcode technology. Nanosphere has demonstrated detection of DNA and RNA targets at sensitivities comparable with those associated with PCR. The *Autolab* system may also be used for identification of amplified DNA samples while multiplexing is achieved by using a mixture of different oligonucleotide-nanoparticle conjugates in the same assay. The *Verigene Mobile* is capable of detection of non-amplified product in a total assay time of less than 1 h and is expected to reach the market in 2008.

8.6.1 *Micro and Nano Scale Biosensors*

Technology development in all areas has witnessed an inexorable trend towards miniaturization. Microprocessor manufacture has placed computing power comparable to the Apollo moon shots within a pocket calculator sized device. This trend also occurs in biosensor development. Micro and nano scale fabrication methods are being applied to biosensors and in some cases their application leads to a change in the way these devices are designed and manufactured. Development of rapid detection systems for pathogens progresses towards the integrated devices utilising solid state technology that do not require introduction of reagents or the separation of unbound probe molecules. Electrochemical and optical detection methods are attractive in this context as they couple signal generated by hybridisation with instrument readout. These assay formats also do not require the use of labelled reporter molecules such as enzymes or flurophores. Further advantages of micro-scale devices include cheaper unit cost due to mass production, much reduced assay cost due to several orders of magnitude less reagent used and a concurrent lowering of potential environmental impact due to lower reagent levels. Micro-fluidic devices have improved mixing rates resulting in shorter analysis times. Multi-analyte detection on the same device is also possible.

8.7 Concluding Remarks

The incorporation of nucleic acid diagnostic biosensors has benefited from close integration with the physical sciences. The area of micro and nano biosensors involves the application of nanotechnology, micro-fluidics, bioinformatics and molecular biology. Only when these elements are in place, it is possible to consider the required detection of picomolar concentrations of analyte in nanolitre volumes in a matter of seconds, without target amplification. The satisfaction of these challenging preconditions will be necessary for rollout of a true point of care molecular biosensor.

System engineers will play a key role in future biosensor development as advances in filtration, flow-injection, micro-fabrication, and detection are combined. In order to be suitable for routine applications, biosensor devices must have the capability to distinguish target sequences in a multi-organism background often in the presence of complex sample matrix. Devices should be rapid, sensitive, adaptable (allow interchangeable probe cartridges), in a user friendly and inexpensive configuration.

Several systems have shown promise in laboratory settings while failing when exposed to real-world materials. A problem facing biosensors for the direct detection of microorganisms is the lack of sensitivity of assays in real samples. This is a particular issue in the case of pathogens such as *E. coli* O157:H7 or *Salmonella* where the infectious dose is 10 cells and water samples should have no more than 4 coliforms per 100 ml and drinking water should have no coliforms at all.

Therefore, a biosensor for water analysis would require a minimum sensitivity of 1 cell from 100 ml water.

The advancements described above together with the increasing awareness of the key criteria for consideration in developing NAD assays provide potential for new bio analytical test methods that will enable multi-parameter testing and at-line monitoring for microbial contaminants.

References

Ahn S, Walt DR (2005) Detection of *Salmonella* spp. Using microsphere-based, fiber-optic DNA microarrays. Anal Chem 77:5041–5047

Anthony RM, Schuitema AR, Oskam L, Klatser PR (2005) Direct detection of *Staphylococcus aureus* mRNA using a flow through microarray. J Microbiol Methods 60:47–54

Baeumner AJ (2003) Biosensors for environmental pollutants and food contaminants. Anal Bioanal Chem 377:434–445

Baeumner AJ, Jones C, Wong CY, Price A (2004) A generic sandwich-type biosensor with nanomolar detection limits. Anal Bioanal Chem 378:1587–1593

Borucki MK, Reynolds J, Call DR, Ward TJ, Page B, Kadushin J (2005) Suspension microarray with dendrimer signal amplification allows direct and high-throughput subtyping of *Listeria monocytogenes* from genomic DNA. J Clin Microbiol 43:3255–3259

De-los-Santos-Alvarez P, De-los-Santos-Alvarez N, Lobo-Castanon MJ, Miranda-Ordieres AJ, Tunon-Blanco P (2006) Amplified label-free electrocatalytic detection of DNA in the presence of calcium ions. Biosens Bioelectron 21:1507–1512

Feriotto G, Borgatti M, Mischiati C, Bianchi N, Gambari R (2002) Biosensor technology and surface plasmon resonance for real-time detection of genetically modified roundup ready soybean gene sequences. J Agric Food Chem 50:955–962

Fitzmaurice J, Glennon M, Duffy G, Sheridan JJ, Carroll C, Maher M (2004) Application of real-time PCR and RT-PCR assays for the detection and quantitation of vt1 and vt2 toxin genes in *E. coli* O157:H7. Mol Cell Probes 18:123–132

Fuentes M, Mateo C, Rodriguez A, Casqueiro M, Tercero JC, Riese HH, Fernandez-Lafuente R, Guisan JM (2006) Detecting minimal traces of DNA using DNA covalently attached to superparamagnetic nanoparticles and direct PCR-ELISA. Biosens Bioelectron 21:1574–1580

Fung DY (2002) Predictions for rapid methods and automation in food microbiology. J AOAC Int 85:1000–1002

Glynn B, Lacey K, Reilly J, Barry T, Smith T, Maher M (2007) Quantification of bacterial tmRNA using in vitro transcribed RNA standards and two-step qRT-PCR. Res J Biol Sci 2:564–570

Grennan B, O'Sullivan NA, Fallon R, Carroll C, Smith T, Glennon M, Maher M (2001) PCR-ELISAs for the detection of *Campylobacter jejuni* and *Campylobacter coli* in poultry samples. Biotechniques 30:602–606 608–610

Guan JG, Miao YQ, Zhang QJ (2004) Impedimetric biosensors. J Biosci Bioeng 97:219–226

Hoa XD, Kirk AG, Tabrizian M (2007) Towards integrated and sensitive surface plasmon resonance biosensors: A review of recent progress. Biosens Bioelectron 23:151–160

Iqbal SS, Mayo MW, Bruno JG, Bronk BV, Batt CA, Chambers JP (2000) A review of molecular recognition technologies for detection of biological threat agents. Biosens Bioelectron 15:549–578

Ivnitski D, Abdel-Hamid I, Atanasov P, Wilkins E, Stricker S (2000) Application of electrochemical biosensors for detection of food pathogenic bacteria. Electroanalysis 12:317–325

Kwakye S, Goral VN, Baeumner AJ (2006) Electrochemical microfluidic biosensor for nucleic acid detection with integrated minipotentiostat. Biosens Bioelectron 21:2217–2223

Lazcka O, Del Campo FJ, Munoz FX (2007) Pathogen detection: A perspective of traditional methods and biosensors. Biosens Bioelectron 22:1205–1217

Lazerges M, Perrot H, Antoine E, Defontaine A, Compere C (2006) Oligonucleotide quartz crystal microbalance sensor for the microalgae *Alexandrium minutum* (Dinophyceae). Biosens Bioelectron 21:1355–1358

Lee SY, Bailey SC, Apirion D (1978) Small stable RNAs from *Escherichia coli*: Evidence for the existence of new molecules and for a new ribonucleoprotein particle containing 6s RNA. J Bacteriol 133:1015–1023

Maher M, Finnegan C, Collins E, Ward B, Carroll C, Cormican M (2003) Evaluation of culture methods and a DNA probe-based PCR assay for detection of *Campylobacter* species in clinical specimens of feces. J Clin Microbiol 41:2980–2986

Mannelli I, Minunni M, Tombelli S, Wang R, Michela Spiriti M, Mascini M (2005) Direct immobilisation of DNA probes for the development of affinity biosensors. Bioelectrochemistry 66:129–138

Mao X, Yang L, Su XL, Li Y (2006) A nanoparticle amplification based quartz crystal microbalance DNA sensor for detection of *Escherichia coli* O157:H7. Biosens Bioelectron 21:1178–1185

Mascini M, Tombelli S, Palchetti I (2005) New trends in nucleic acid based biosensors: University of Florence (Italy), 25–28 October 2003. Bioelectrochemistry 67:131–133

Mullis KB, Faloona FA (1987) Specific synthesis of DNA in vitro via a polymerase-catalyzed chain reaction. Methods Enzymol 155:335–350

Nelson BP, Grimsrud TE, Liles MR, Goodman RM, Corn RM (2001) Surface plasmon resonance imaging measurements of DNA and RNA hybridization adsorption onto DNA microarrays. Anal Chem 73:1–7

O'Connor L (2003) Detection of *Listeria monocytogenes* using a PCR/DNA probe assay. Methods Mol Biol 216:185–192

O'Connor L, Joy J, Kane M, Smith T, Maher M (2000) Rapid polymerase chain reaction/DNA probe membrane-based assay for the detection of *Listeria* and *Listeria monocytogenes* in food. J Food Prot 63:337–342

O'Connor L, Lahiff S, Casey F, Glennon M, Cormican M, Maher M (2005) Quantification of als1 gene expression in *Candida albicans* biofilms by RT-PCR using hybridisation probes on the Lightcycler. Mol Cell Probes 19:153–162

Peng H, Soeller C, Cannell MB, Bowmaker GA, Cooney RP, Travas-Sejdic J (2006) Electrochemical detection of DNA hybridization amplified by nanoparticles. Biosens Bioelectron 21:1727–1736

Pfaller MA (2001) Molecular approaches to diagnosing and managing infectious diseases: Practicality and costs. Emerg Infect Dis 7:312–318

Rand AG, Ye JM, Brown CW, Letcher SV (2002) Optical biosensors for food pathogen detection. Food Technology 56:32–39

Sachse K, Hotzel H, Slickers P, Ellinger T, Ehricht R (2005) DNA microarray-based detection and identification of *Chlamydia* and *Chlamydophila* spp. Mol Cell Probes 19:41–50

Schonhuber W, Le Bourhis G, Tremblay J, Amann R, Kulakauskas S (2001) Utilization of tmRNA sequences for bacterial identification. BMC Microbiol 1:20

Small J, Call DR, Brockman FJ, Straub TM, Chandler DP (2001) Direct detection of 16s rRNA in soil extracts by using oligonucleotide microarrays. Appl Environ Microbiol 67:4708–4716

Southern EM (1975) Detection of specific sequences among DNA fragments separated by gel electrophoresis. J Mol Biol 98:503–517

Spadavecchia J, Manera MG, Quaranta F, Siciliano P, Rella R (2005) Surface plamon resonance imaging of DNA based biosensors for potential applications in food analysis. Biosens Bioelectron 21:894–900

Storhoff JJ, Marla SS, Bao P, Hagenow S, Mehta H, Lucas A, Garimella V, Patno T, Buckingham W, Cork W, Muller UR (2004) Gold nanoparticle-based detection of genomic DNA targets on microarrays using a novel optical detection system. Biosens Bioelectron 19:875–883

Straub TM, Dockendorff BP, Quinonez-Diaz MD, Valdez CO, Shutthanandan JI, Tarasevich BJ, Grate JW, Bruckner-Lea CJ (2005) Automated methods for multiplexed pathogen detection. J Microbiol Methods 62:303–316

Su XL, Li Y (2004) A self-assembled monolayer-based piezoelectric immunosensor for rapid detection of *Escherichia coli* O157:H7. Biosens Bioelectron 19:563–574

Templeton KE, Scheltinga SA, van der Zee A, Diederen BM, van Kruijssen AM, Goossens H, Kuijper E, Claas EC (2003) Evaluation of real-time PCR for detection of and discrimination between *Bordetella pertussis, Bordetella parapertussis*, and *Bordetella holmesii* for clinical diagnosis. J Clin Microbiol 41:4121–4126

Todd EC, Szabo RA, MacKenzie JM, Martin A, Rahn K, Gyles C, Gao A, Alves D, Yee AJ (1999) Application of a DNA hybridization-hydrophobic-grid membrane filter method for detection and isolation of verotoxigenic *Escherichia coli*. Appl Environ Microbiol 65:4775–4780

Tortoli E, Mariottini A, Mazzarelli G (2003) Evaluation of Inno-LIPA *Mycobacteria* V2: Improved reverse hybridization multiple DNA probe assay for Mycobacterial identification. J Clin Microbiol 41:4418–4420

Updike SJ, Hicks GP (1967) The enzyme electrode. Nature 214:986–988

Van Der Pol B, Quinn TC, Gaydos CA, Crotchfelt K, Schachter J, Moncada J, Jungkind D, Martin DH, Turner B, Peyton C, Jones RB (2000) Multicenter evaluation of the Amplicor and automated Cobas Amplicor CT/NG tests for detection of *Chlamydia trachomatis*. J Clin Microbiol 38:1105–1112

Wang J, Fernandes JR, Kubota LT (1998) Polishable and renewable DNA hybridization biosensors. Anal Chem 70:3699–3702

Wang R, Tombelli S, Minunni M, Spiriti MM, Mascini M (2004) Immobilisation of DNA probes for the development of SPR-based sensing. Biosens Bioelectron 20:967–974

Wang R, Zhang L, Feng Y, Ning C, Jian F, Xiao L, Zhao J, Wang Y (2008) Molecular characterization of a new genotype of *Cryptosporidium* from American minks (*Mustela vison*) in China. Vet Parasitol 154:162–166

Williams KP (2002) The tmRNA website: Invasion by an intron. Nucleic Acids Res 30:179–182

Wilson WJ, Strout CL, DeSantis TZ, Stilwell JL, Carrano AV, Andersen GL (2002) Sequence-specific identification of 18 pathogenic microorganisms using microarray technology. Mol Cell Probes 16:119–127

Wu VC, Chen SH, Lin CS (2007) Real-time detection of *Escherichia coli* O157:H7 sequences using a circulating-flow system of quartz crystal microbalance. Biosens Bioelectron 22: 2967–2975

Yao G, Fang X, Yokota H, Yanagida T, Tan W (2003) Monitoring molecular beacon DNA probe hybridization at the single-molecule level. Chemistry 9:5686–5692

Chapter 9
Tissue-Based Biosensors

Victor Acha, Thomas Andrews, Qin Huang, Dhiraj K. Sardar, and Peter J. Hornsby

Abstract Biosensors based on tissue structures in living animals can be used to detect and measure hormones, drugs, and toxins. The potential use of tissue-based biosensors extends to such diverse fields of biomedical science as physiology, pharmacology, and biodefense. In general, tissue-based biosensors can be formed from genetically modified cells or by direct genetic modification in order to introduce biosensor proteins into a tissue in the animal. Biosensor cells transduce the concentration of the molecule being detected into a physical signal, which can be precisely measured. Biophotonics provides the most versatile basis for tissue-based biosensors. Light output from biosensor cells can be in the form of fluorescence or bioluminescence, and, of these two, bioluminescence offers advantages of not requiring an input source of light and having a more favorable signal to noise ratio in living animals than fluorescence. Protein-protein interactions can be used to detect almost any molecule, by means of fusion proteins that can be used to generate resonance energy transfer. Bioluminescence resonance energy transfer (BRET) has the potential to be used for the measurement of a wide variety of molecules in living animals. Two examples are discussed here: a tissue-based biosensor for the hormone vasopressin, and a biosensor for rapamycin, both based on BRET. The possible extension of tissue-based biosensors to human subjects will require solutions to several problems, particularly the mode of detection of the physical output, and demonstration of the safety of the genetic modifications needed to introduce biosensor proteins into cells in vivo.

Keywords Biosensors · Tissues · Animals · Biomedicine · Hormones · Drugs · Toxins · Bioluminescence · Fluorescence · Resonance energy transfer · Luciferase · Photon counting · Imaging · BioMEMS · Vasopressin · Rapamycin · Human subjects

P. Hornsby (✉)
Sam and Ann Barshop Institute for Longevity and Aging Studies, University of Texas Health Science Center, San Antonio, TX78245, USA
e-mail: hornsby@uthscsa.edu

M. Zourob (ed.), *Recognition Receptors in Biosensors*,
DOI 10.1007/978-1-4419-0919-0_9, © Springer Science+Business Media, LLC 2010

Abbreviations

BRET	Bioluminescence resonance energy transfer
PLGA	Poly(D,L-lactic-co-glycolic acid)
PAR-1	Protease activated receptor-1
IκBα	Nuclear factor of kappa light polypeptide gene enhancer in B-cells inhibitor, alpha
TNFα	Tumor necrosis factor, alpha
FLuc	Firefly luciferase
RLuc	*Renilla* luciferase
CCD	Charge coupled device
FRET	Fluorescence resonance energy transfer
GFP	Green fluorescent protein
CrFK	Crandell feline kidney
BioMEMS	Bio-microelectromechanical systems

9.1 Introduction

Cells, tissues, or even whole organisms can be used as biosensors to detect or measure substances of importance. The substances detected/measured might be of environmental/biodefense importance, such as hazardous bacteria, or of medical importance, such as insulin and other hormones in the body. The molecular basis for most biosensors are proteins that interact with the substance that is being detected. Biosensor proteins can be incorporated into physical (nonliving) sensors, or into living cells, tissues and organisms by genetic modification. In some cases, the endogenous nongenetically modified properties of cells and tissues can be exploited to produce biosensors (Pancrazio 2001; Reininger-Mack et al. 2002).

Tissue-based biosensors, as the name implies, are biosensors based on tissue structures. The biosensor proteins can be incorporated into tissues either via the transplantation of cells that have been genetically modified to express them, or by the direct introduction of appropriate genes into the tissues. Some tissue-based biosensors that have been developed are designed to operate in vitro (Pancrazio 2001; Reininger-Mack et al. 2002), but the work under consideration in this chapter concerns the use of tissue-based biosensors in living animals or human subjects (Fig. 9.1). Cells within the tissue transduce the concentration of the substance being monitored to a physical signal (e.g., electrical, photonic, or magnetic).

At the current state of development of the field, we face a variety of questions concerning the optimal design and the use of tissue-based biosensors. Published work has established the proof of principle for tissue-based biosensors, but their potential has been largely unexplored to date.

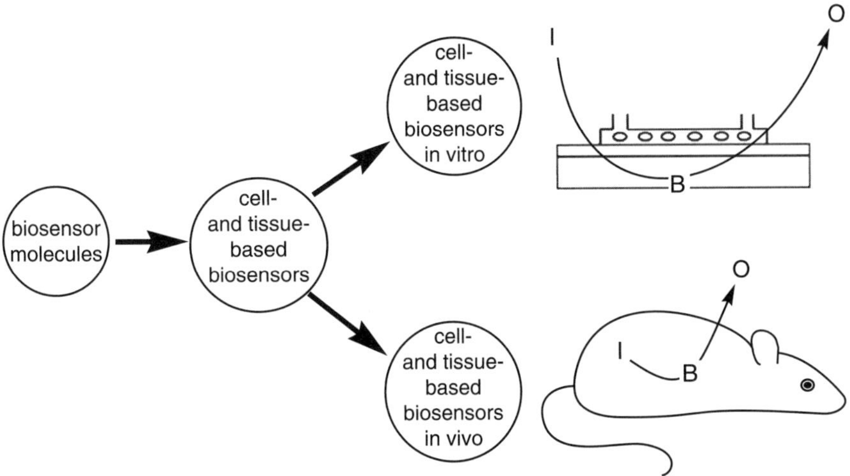

Fig. 9.1 Tissue-based biosensors are an extension of the concept of biosensor molecules (proteins) that can be incorporated into living cells. Biosensor cells (B) can be used in devices in vitro that transduce an input, I, the concentration of molecule to which the biosensor responds, to an output, O, comprising a physical signal that can be measured within the device or by some external method. Similarly, biosensor cells can be transplanted into a living animal, or alternatively biosensor molecules can be incorporated into the endogenous tissues of the animal. Biosensor tissues in the living animal also transduce an input, the concentration of the sensed molecule, into a physical output. The output can measured externally to the animal or partly internally and partly externally

9.2 Tissue-Based Biosensors in Experimental Animals

Research in this field has been concerned with the development of tissue-based biosensors for living animals based on genetically modified cells. The aim has been to be able to use the defined properties of biosensor proteins, tested and validated in systems in vitro, in living animals, principally mice. These tissue-based biosensors can be used in a variety of applications. Some of these are: (i) *Physiology*. Substances of physiological relevance, such as hormones, are typically measured in blood samples. Repeated blood sampling cannot be performed in small animals such as mice because the blood volume would become depleted; and unless performed with indwelling catheters blood sampling involves stressing the animal, which can severely affect the experiment. Levels of hormones change very quickly, making a way to continuously and noninvasively monitor hormones very desirable. The implementation of tissue-based biosensors would be very valuable in research in which hormonal changes are critical, such as the effects of caloric restriction on aging in rodents (Nelson et al. 1995; Heydari et al. 1996). (ii) *Drug discovery and testing*. Tissue-based biosensors can be used to monitor how novel drugs act in an in vivo setting; they can be used both to ascertain how potently the drug affects a specific molecular pathway and to monitor its pharmacokinetics. (iii) *Biodefense*.

A mouse samples the environment in the same way as an at-risk human does, by breathing the air. The lungs serve as a highly efficient delivery system to bring substances from the environment into contact with biosensor tissues that can detect a pathogen and transduce the concentration of the pathogen into a physical signal. Moreover, processing or activation of agents that may be necessary for pathogenic effects will occur in the mouse as they would in a human who is exposed to the same agent.

9.3 Incorporating Biosensor Molecules Into Tissues

Biosensor proteins may be incorporated into tissue-based biosensors either by transplantation of genetically modified cells into an experimental animal, thereby creating a customized tissue, or by genetic modification of the animal's own tissues. The latter may be accomplished by viral or nonviral vectors that are introduced into a tissue. For example, an implant formed from an appropriate material could be coated with a vector encoding the biosensor proteins. When inserted into tissues, the vector-coated implant would accomplish the genetic modification of the surrounding cells. Transfection may be achieved by using commonly used implant materials, such as poly(D,L-lactic-co-glycolic acid) (PLGA), that has been processed to form a modified surface (Zheng et al. 2000). Such an implant could both achieve the necessary genetic modification of adjacent cells and could also house a physical sensor for the output of the genetically modified cells (electrical, photonic, or magnetic). Biosensor tissues formed by genetic modification of the animal's own tissues would have the advantage that the body would not react to the biosensor tissue as foreign, and the tissue would not be subjected to rejection and would not encounter biocompatibility issues (although it should be noted that the implant itself could still elicit a foreign body reaction or be subject to other biocompatibility issues). Such implants may be envisaged as a future development in this field, and could represent the way in which the technology may be introduced into human use (see below).

The above discussion assumes that differentiated tissue cells will be genetically modified. A more powerful technique could comprise the genetic modification of stem cells, thereby creating a continuously renewing source of biosensor cells. For example, hematopoietic stem cells could be modified, resulting in genetically modified white blood cells. The output from such cells could be detected at many sites in the body, depending on the physical nature of the output. Light could be detected from areas of the skin where the blood comes into close proximity to the surface, such as the external ear. In a dark environment, light emission from the skin of the animal can be detected using appropriate photonics apparatus (photon counting) in a completely noninvasive way. Such methods in essence would make the entire animal into a biosensor.

Implants that can genetically modify cells that they contact, and biosensors based on genetic modification of stem cells, are examples of exciting future developments of tissue-based biosensor technology. Most of the current examples of this technology involve biosensor cells transplanted into experimental animals to form artificial tissues. The following discussion in the chapter surveys the present state of the art in this field.

9.4 Biophotonics-Based Biosensors and Biosensors Based on Other Physical Outputs

Probably the simplest physical output that a biosensor tissue could produce is electrical, which can be measured very simply. The proof of principle for this concept has been provided in a pioneering series of experiments by Edelberg and colleagues. They showed that cardiomyocytes can be implanted into the external ear of the mouse, and there they can survive and form beating heart-like structures (Edelberg et al. 1998; Christini et al. 2001; Edelberg et al. 2002; Tang et al. 2003). By measuring the electrical activity of the transplanted cardiomyocytes these investigators were able to monitor circulating levels of adrenergic hormones and monitor the bioactivity of circulating drugs. For example, they engineered cardiomyocytes to monitor the local coagulation potential by overexpression of the human thrombin receptor, protease activated receptor-1 (PAR-1). PAR-1 engineered cells implanted in vivo detected local increases in thrombin by changes in chronotropic activity (Tang et al. 2003).

Whereas biosensors based on electrical output have the advantage of simplicity of measurement of the output, biosensors based on principles of biophotonics are attractive because in theory they can be designed to detect almost any biological or chemical substance, and because light may be measured with a high degree of precision. However, current techniques for light measurement require a dark environment and relatively complex equipment.

The output from biosensors based on biophotonics principles can be either fluorescence or bioluminescence. Although both these outputs from a biosensor tissue could achieve sensitive measurement of detected molecules, fluorescence is more complex because it requires both an input and an output light path. Nevertheless, biosensors that operate in living animals (mice) have been very valuable. Some examples follow: (i) Transgenic mice have been generated which express functional fluorescent calcium indicator proteins under control of a tetracycline-responsive promoter (Hasan et al. 2004). (ii) Synaptically targeted pHluorins were expressed in the olfactory sensory neurons of transgenic mice. Using these mice, it was possible to investigate spatial patterns of glomerular activity after odorant stimulation within the olfactory bulb in vivo (Bozza et al. 2004). (iii) A calcium-sensing protein was expressed and analyzed in smooth muscle of transgenic mice (Ji et al. 2004). However, current methods of imaging of tissues in mice using fluorescence methods involve the imaging of exposed tissues under a microscope, and the animal must be anesthetized or immobilized. Here, we focus mainly on bioluminescence, which escapes many of these restrictions. Bioluminescence requires only an output path, because light production is endogeneous.

A key factor favoring bioluminescence in comparison to fluorescence is a superior signal to noise ratio. In a detailed study of autofluorescence and autobioluminescence in living mouse tissues, detection of bioluminescence required far fewer cells than detection of fluorescence. When cells that express both fluorescent and bioluminescent proteins were injected subcutaneously, 3.8×10^5 cells were required to obtain a fluorescence signal above background, but only 400 cells were required to obtain a bioluminescence signal above background (Troy et al. 2004).

Various molecular mechanisms can be employed to link the molecule being detected to the output of light in the form of bioluminescence. For example, the gene encoding the bioluminescent protein may be placed under the control of a promoter that is regulated by the molecule being detected. This would be appropriate for detection of a molecule that acts via a transcriptional response (e.g., a hormone like thyroid hormone or estrogen). Alternatively, the level may be under translational regulation, or under post-translational regulation, e.g., by proteolysis. The latter can be used to link the level of a biosensor protein to an intracellular pathway that is regulated by proteolysis. For example, a fusion protein of IκBα and FLuc is degraded by the proteasome under the control of an intracellular pathway stimulated by TNFα (Gross and Piwnica-Worms 2005). Thus, a cell expressing this fusion protein can serve as a biosensor for TNFα and other factors that activate the same intracellular pathway. In yet another variant, the gene encoding a bioluminescent protein is split into two sections, each encoding a half-protein. Neither half-protein is bioluminescent, but when the two halves are brought together in the presence of a detected molecule bioluminescence is restored (Paulmurugan et al. 2002).

In all these cases, the usefulness of the biosensor is greatly improved by coexpression of a second bioluminescent protein that acts as an internal control. For example, the biosensor protein can be based on one type of luciferase (e.g., firefly luciferase, FLuc) and a second luciferase (e.g., *Renilla* luciferase, RLuc) can serve as an internal control. FLuc and RLuc emit light at distinct wavelengths, and have separate substrates for light production, thereby enabling light output from the two luciferases to be measured simultaneously. The internal control protein should be expressed from a constitutive promoter so that its levels are invariant. The ratio [biosensor protein light output/control protein light output] provides a much more accurate measure because it becomes insensitive to the distance between the source of the bioluminescence and the light detector or other variables. Therefore, the use of two luciferases, because it enables the distance from the light source to the detector to vary, facilitates the use of bioluminescence-based biosensors in conscious freely moving animals. This topic is developed in more detail.

In addition to molecular processes that depend on changing the intracellular level of the bioluminescent protein as the result of action of the detected molecule, a particularly powerful form of biophotonics-based biosensor is based on protein-protein interactions, which can be linked to light output via bioluminescence resonance energy transfer (BRET), as described in detail below. Because BRET produces a ratio, it does not require an independent internal control bioluminescent protein.

9.5 Measuring Light Output from Bioluminescence-Based Biosensor Tissues in Living Animals

There are several challenges involved in optimizing the performance of tissue-based biosensors that use bioluminescence. A severe challenge arises from poor transmission of light through tissues, including the skin, which absorb light strongly at wavelengths <600 nm (Weissleder and Ntziachristos 2003). Several approaches are possible to overcome this limitation, including the development of bioluminescent proteins that emit light at longer wavelengths; the use of more sensitive detection devices and the use of skin/body wall windows that permit light to reach the detector without impedance by intervening tissues.

Another set of challenges arises from the need for a substrate to be supplied to the bioluminescence-emitting protein. In order to produce light, luciferases need oxygen and a specific substrate, and in some cases also ATP. For FLuc the substrate is D-luciferin; for RLuc the substrate is coelenterazine. These molecules are cell permeable and are usually delivered by intraperitoneal or intravenous injection for observation of bioluminescence in experimental animals (Bhaumik and Gambhir 2002).

Ideally, substrates would be directly and continuously supplied to biosensor cells in the living animal under the investigator's control. A pioneering example of in vivo bioluminescence monitoring via continuous supply of luciferase substrate in conscious freely moving animals is a study in which FLuc was expressed in the brain under the control of a circadian rhythm-responsive promoter. Bioluminescence was measured continuously using an indwelling fiber optic probe (Yamaguchi et al. 2001). Continuous delivery of D-luciferin by micro-osmotic pumps was also shown to enable real-time bioluminescence imaging of FLuc activity in vivo (Gross et al. 2007). Thus, long-term administration of luciferase substrate in vivo can be performed without encountering apparent toxicity.

Another potential mode of substrate delivery is topical, i.e., directly via the skin to biosensor cells located under the surface. This has not yet been used extensively in published experiments; however, coelenterazine can serve as a substrate when applied topically to the surface of the eye in mice infected with herpes simplex virus genetically engineered to express RLuc (Luker et al. 2002).

Desirably, light output from the biosensor tissues would be monitored continuously, without direct contact with the animal. A frequently employed method is to use CCD (charge coupled device) camera-based imaging to detect bioluminescence (De and Gambhir 2005). Photon counting using a photomultiplier provides a more sensitive and quantitative method, although it does not provide spatial information on the distribution of the bioluminescence (Huang et al. 2007b). In currently published experiments, both methods require that animals (mice) be anesthetized for light measurement. However, in this lab we have used the skin/body wall window model, described below, to measure bioluminescence in unanesthetized animals placed in a custom restrainer. Ideally, light measurements should be made in freely moving conscious animals in a minimally invasive way, and ultimately it should be possible to use telemetry for monitoring. For example, light output from biosensor tissues could be converted to a radio signal by a self-contained implanted miniature device that would enable truly remote and noninvasive monitoring.

The use of skin/body wall windows provides an intermediate step toward such completely noninvasive monitoring. Windows can be positioned over the site of transplantation of biosensor cells, in conjunction with a fiber optic that transmits light from the cells to a photon counter (Fig. 9.2). Windows enable measurement of light emission from bioluminescence-emitting cells transplanted into the kidney, as illustrated in the figure, or at other sites in the body. The plastic window, which has excellent light transmission properties, obviates the problems resulting from light absorption by the skin and other tissues (Huang et al. 2007b).

Using skin/body wall windows we assessed whether it was possible to simultaneously measure RLuc and FLuc activities in a biosensor tissue. Cells expressing both proteins were transplanted into the kidney of immunodeficient mice and were

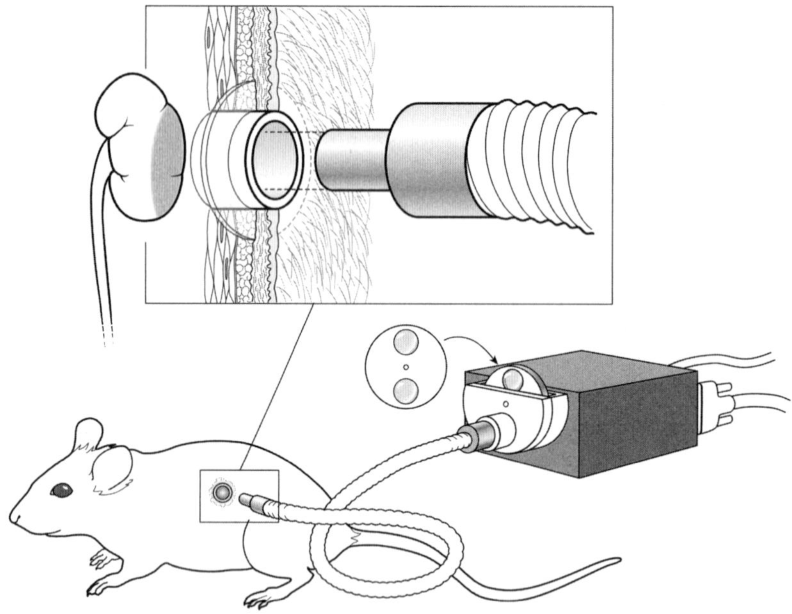

Fig. 9.2 Diagrammatic representation of a mouse with a skin/body wall window and biosensor cells transplanted under the capsule of the kidney. The gray shading represents the transplanted cells. A plastic window is fitted in the skin and body wall adjacent to the kidney. A woven polyester collar attached to the window enables the window to be tightly integrated into the tissues of the mouse without loss of integrity of the skin around the edge of the window. When the mouse is anesthetized, the end of a fiber optic light guide may be inserted into the opening of the window. Following injection of luciferase substrates into the peritoneal cavity of the animal, light emitted from the bioluminescence-emitting cells is detected by a photon counter. A filter wheel interposed in the light path enables the measurement of light at defined wavelengths. Data from the photon counter is recorded on a computer, which also controls the filter wheel. Reproduced with permission from Huang et al. (2007b), copyright 2007 SPIE

permitted to establish a vascular supply, as previously described for other cell types (Thomas et al. 1997; Thomas and Hornsby 1999; Sun et al. 2004). Insertion of the window into the skin/body wall and transplantation of bioluminescent cells into the kidney was performed in a single surgical procedure. The size and shape of the window allowed one end of a fiber optic light guide to be inserted within it when the mouse was immobilized under anesthesia. The light guide was used to transmit light to a photon counter (Fig. 9.2). Optical filters were used to discriminate the shorter wavelength light emitted by RLuc and the longer wavelength light emitted by FLuc. Animals were used for experiments at 20–30 days following surgery, when the transplanted cells had become vascularized beneath the kidney capsule. When luciferase substrates (D-Luciferin or coelenterazine) were injected separately, light emission was detected at the appropriate wavelength and not at the other (inappropriate) wavelength. When both substrates were injected together, light was detected at both wavelengths in proportion to the levels of the two luciferases in the biosensor tissue. These experiments provide the proof of principle that one luciferase can serve as a biosensor protein (e.g., by being under the regulation of a specific promoter, or by fusion to a protein that is under proteolytic regulation, etc.) while the other luciferase serves as an internal control.

9.6 Tissue-Based Biosensors Based on Bioluminescence Resonance Energy Transfer (BRET)

Biosensors based on protein-protein interactions are very versatile and useful. The detection of most molecules can be adapted to take advantage of a protein-protein interaction. The key feature of such interactions is that resonance energy transfer can occur between the two proteins when they are in close proximity (~10 nm). In one form of this phenomenon, each of the two interacting proteins is a fusion protein with a fluorescent protein. In fluorescence resonance energy transfer (FRET), light energy is transferred between two different fluorescent protein molecules without emission of a photon (Li et al. 2006). Bioluminescence resonance energy transfer (BRET), by contrast, involves one bioluminescent protein and one fluorescent protein (Massoud et al. 2007). The bioluminescent protein is capable of converting some form of chemical energy into light energy. The wavelength of the light emitted from the bioluminescent protein is suitable for excitation of the fluorescent protein, although (as in FRET) the energy is transferred between the two molecules without emission of a photon. Because of the versatility of BRET, like FRET, it can be adapted via fusion proteins to almost any protein-protein interaction and thereby to the detection of a huge variety of molecules.

Although BRET has yet to be used extensively in living animals, it holds great promise for the future. BRET can be used in ways very similar to FRET. The essential and important difference is that BRET requires no source of light. Thus for a tissue-based biosensor, methods of measuring the emitted light are needed, but not a method for illuminating the biosensor, thus greatly simplifying the apparatus needed. BRET is not affected by variations in the size of the transplant or the efficiency of light collection from the transplant because it measures a *ratio* rather than absolute levels. The following is a quote from a recent review: "Unlike FRET, BRET does not require excitation of the donor with an external light source and offers several advantages in this respect. BRET assays show no photo-bleaching or photoisomerization of the donor protein, no photodamage to cells, and no light scattering or auto-fluorescence from cells or microplates (when used in vitro), which can be caused by incident excitation light. In addition, one main advantage of BRET over FRET is the lack of emission arising from direct excitation of the acceptor. This reduction in background should permit the detection of interacting proteins at much lower concentrations than is possible for FRET. The use of BRET technology for imaging protein-protein interactions in living subjects is currently being validated, with a view to effective applications in the evaluation of new pharmaceuticals" (quoted from Massoud et al. 2007).

The fluorescent protein emits light at a longer wavelength than the bioluminescent protein. Thus the amount of light emitted at the longer wavelength, as a fraction of the total light emitted, gives a measure of the fraction of the molecules with close apposition of the two proteins. This measures the extent of the protein-protein interaction thereby measures the concentration of the agent that caused the conformational change, e.g., a ligand such as a hormone or toxin. BRET has been used in the study of a large number of protein-protein reactions, particularly with plasma membrane receptors (Pfleger and Eidne 2005).

9.7 Examples of Tissue-Based Biosensors That Use BRET

9.7.1 Example 1: Vasopressin

Many hormones and other molecules exert their biological actions via G-protein coupled plasma membrane receptors. By linking the receptor as a fusion protein to RLuc and by linking the signal tansduction protein β-arrestin to modified green fluorescent protein (GFP) the concentration of the hormonal or other ligand can be linked to BRET. The principle of the interaction of the two fusion proteins and the mechanism by which they respond to a ligand such as vasopressin is shown in Fig. 9.3. RLuc and modified GFP (GFP2) require an analog of coelenterazine for energy transfer to take place (coelenterazine 400a, also termed DeepBlueC, a trademark of Packard Instrument Co.). Moreover, whereas the interaction of RLuc with native coelenterazine results in light emission at 475 nm, interaction of RLuc with coelenterazine 400a results in light emission at 395 nm, which is then shifted to 510 nm when BRET occurs between RLuc and GFP2 (Hamdan et al. 2005). Optical filters may be used to distinguish light from free RLuc and light emitted by BRET resulting from GFP2/RLuc interaction.

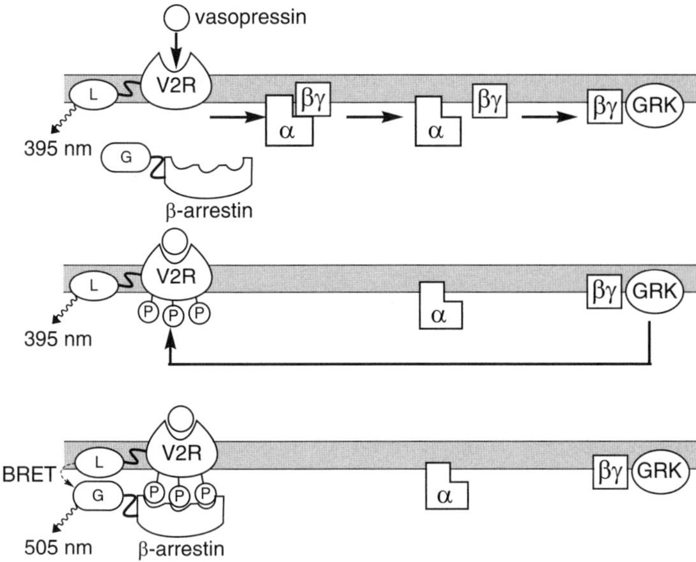

Fig. 9.3 Principle of a cell-based biosensor for vasopressin that can be used in cultured cells and in cells transplanted in mice. The ligand, the hormone vasopressin, binds to a fusion protein V2R-L (V2R = V2 vasopressin receptor; L = RLuc). If luciferase substrate (coelenterazine 400a) is supplied, light is emitted by RLuc at 395 nm. When vasopressin binds to the receptor, the associated G protein (α, β, and γ subunits) is activated, leading to dissociation of the $\beta\gamma$ subunit. This in turn activates G protein receptor kinase (GRK), which phosphorylates the receptor. The phosphorylated receptor is recognized by and bound by a fusion protein G-β-arrestin (G = GFP^2). The binding brings the RLuc and GFP^2 fusion proteins into close proximity (<10 nm) allowing bioluminescence energy transfer (BRET) to occur. Light is emitted at a longer wavelength, 510 nm, in proportion to the number of vasopressin receptors that are bound by β-arrestin, which is proportional to the fraction of receptors that have been activated by vasopressin, which in turn is proportional to the concentration of vasopressin in the external environment of the biosensor cell

Based on our prior experience with transplantation of various cells to form tissues in the subrenal capsule space (Huang et al. 2007a), we set out to examine whether biosensor cells could survive in this site over an extended period of time following transplantation into immunodeficient mice (Huang et al. 2007b). Crandell feline kidney (CrFK) cells were transfected with pSV2A/L-AΔ5′, a plasmid encoding FLuc and *neo*, and stable transfectants were selected with G418. These cells were then transfected with plasmids pGFP2-β-arrestin2 and pV2-vasopressin-receptor-hRLuc (Packard Bioscience, Meriden, CT) or a modified version of this plasmid encoding hRLuc mutated at 8 codons (Loening et al. 2006). This pair of fusion proteins, GFP2-β-arrestin2 and V2-vasopressin-receptor-hRLuc, constitute a biosensor for the hormone vasopressin (Bertrand et al. 2002). hRLuc is a humanized form of RLuc; GFP^2 is a modified form of GFP (Breit et al. 2006).

In order to visualize the cells in the kidney, a plastic window was fitted in the skin of the mice over this organ. The window allows the measurement of light emission from the transplanted cells without distortion by intervening layers of tissues. When RLuc substrate (coelenterazine 400a) is administered to the animal, light emission from the transplanted cells is detected at two wavelengths giving a measure of the extent of BRET. We demonstrated that, both in vitro and in the living animal, these cells respond to the administration of vasopressin by an increase in BRET, thus providing proof of principle that BRET can be measured accurately in biosensor tissues in vivo (Huang et al. 2007b) (Fig. 9.4).

9.7.2 Example 2: Rapamycin

In a second example of the use of biosensors based on BRET, we engineered cells to respond to a drug by adapting fusion proteins to produce BRET in the presence of rapamycin (Fig. 9.5). The principle of this biosensor is similar to that in published experiments (De and Gambhir 2005) although the fusion proteins were independently derived in this lab. As shown in the figure, the addition of rapamycin caused an immediate ~fivefold increase in BRET. Presumably, the difference in timing between this result and the increase in BRET following addition of vasopressin to vasopressin-responsive biosensor cells (Fig. 9.4) results from the fact that the latter takes place following several different intracellular steps, whereas this is not the case for the rapamycin, which directly brings together the two interacting proteins.

9.8 Potential Uses of Tissue-Based Biosensors in Human Medicine

Tissue-based biosensors could provide a powerful technology in human medical diagnosis, if several rather severe problems could be solved. Presumably, any adaptation of this technology to human subjects would use the patient's own cells. Whereas it is easy to speculate that an implant of the type described earlier in this chapter could be adapted to provide huge variety of biomedical readouts, this has to be accomplished with minimal inconvenience and with no risk to the patient. Genetic modification of the patient's cells could be accomplished using a vector-coated implant, as described earlier. Potentially, such an implant could be small enough to be surgically implanted with little inconvenience, yet would both produce the necessary genetic modification and also house the detector mechanism for the output of the biosensor cells. Such a biosensor would allow noninvasive monitoring of hormones, drugs or toxins.

Despite the versatility of biosensors based on biophotonics principles, such biosensors are unlikely to be very adaptable to human patients because of the need for supply of a substrate for light production, the need for a dark environment in the vicinity of the biosensor cells, and the requirement for sensitive measurement systems. Probably, problems associated with measurement can be solved, based on

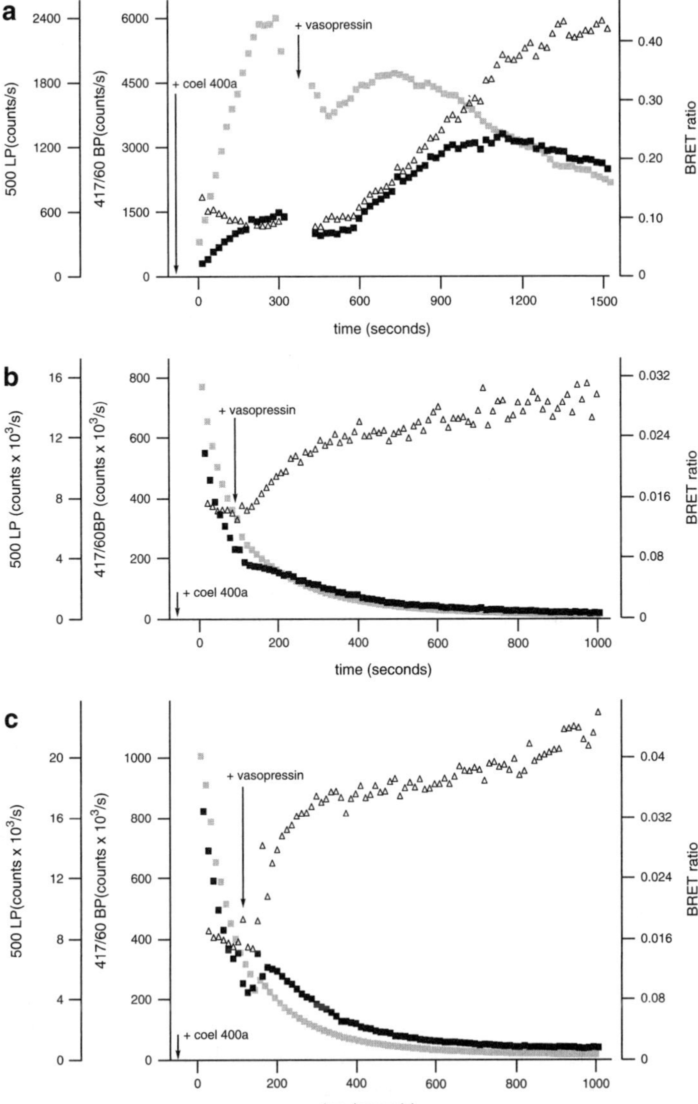

Fig. 9.4 Observations on BRET in biosensor cell transplants and in cell suspensions in vitro. (**a**) Vasopressin-sensitive biosensor cells were transplanted beneath the kidney capsule of immunodeficient mice. Following an intraperitoneal injection of 2 mg/kg coelenterazine 400a, light emission was measured using 417/60 bandpass (BP) and 500 longpass (LP) filters (gray and black symbols respectively). The triangles show the ratio of light detected using the two filters [BRET ratio = (500 LP)/(417/60 BP)]. After ~350 s light emission acquisition was stopped temporarily and vasopressin (0.5 mg/kg) was injected intraperitoneally into the animal. Light emission using the two filters was then measured for another ~1,100 s. (**b**) Vasopressin-sensitive biosensor cells were studied in vitro. A suspension of 100,000 cells was used in an incubation with coelenterazine 400a at 25°C. Light emission using 417/60 BP and 500 LP filters and the BRET ratio are plotted. After 100 s vasopressin was added to the cells without interrupting light acquisition. (**c**) The experiment was performed as described in (b) except that the cell suspension was maintained at 37°C throughout the experiment. Reproduced with permission from Huang et al. 2007b

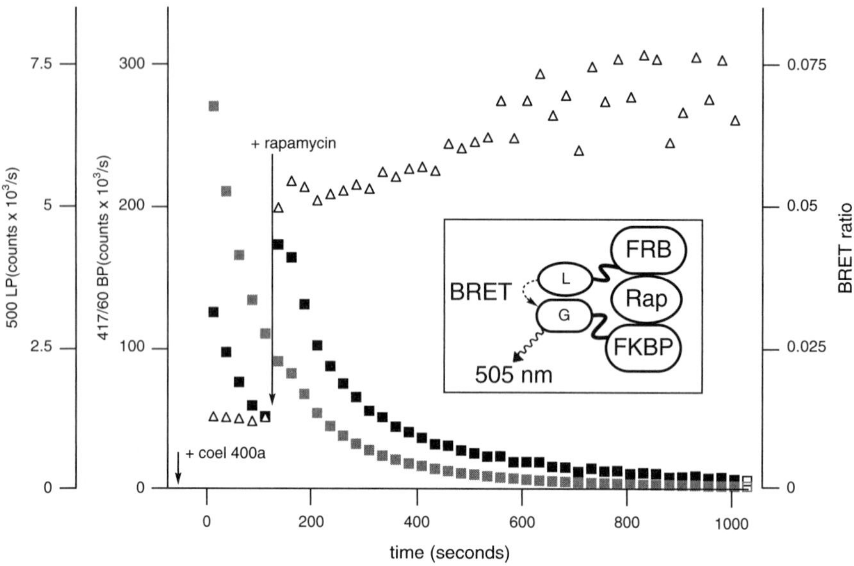

Fig. 9.5 A cell-based biosensor for a drug, rapamycin. *Inset*: The principle of the biosensor is illustrated. The ligand, rapamycin (Rap) binds and brings together two fusion proteins: G-FKBP (G = GFP^2, FKBP = FK506 binding protein) and FRB-L (FRB = FKBP and rapamycin binding domain of mammalian target of rapamycin protein; L = RLuc). When luciferase substrate, coelenterazine 400a, is supplied, light is emitted by RLuc at 395 nm. When the two fusion proteins are brought together by the presence of rapamycin, BRET occurs and light is emitted at the longer wavelength, 510 nm, in proportion to the number of rapamycin-bound protein pairs, which is turn proportional to the concentration of rapamycin (intracellular and extracellular, as rapamycin rapidly crosses the plasma membrane into the cytoplasm). *Graph*: Example of the response of cells expressing these biosensor proteins. Cells were suspended in a suitable buffer and coelenterazine 400a was added to begin the reaction. Light emission was measured using 417/60 bandpass (BP) and 500 longpass (LP) filters (*gray* and *black* symbols respectively). The triangles show the ratio of light detected using the two filters [BRET ratio = (500 LP)/(417/60 BP)]. After ~150 s 1 μM rapamycin was added to the cells without interrupting light acquisition

the probability that electronic devices will continue to shrink, while increasing in power, at a rate that exceeds the rate of advances in biomedical science. This depends on the likely rapid progress in development in BioMEMS (biomedical micro-electro-mechanical systems) (Gourley 2005). A system in which the output of the biosensor tissue is linked telemetrically to a wearable monitor device is certainly within reach in a few years. However, greater problems stem from the use of light as the physical output of the biosensor. Despite the versatility of biophotonics, biosensors based on electrical output are more likely to be implemented in human subjects.

The safety of such biosensors is a major issue and needs to be well established before they could be in general use. Indeed, any diagnostic device that involves genetic modification of patient's cells will have difficulty being justified in view of the possible adverse consequences of such genetic modifications. Gene therapy achieved by retroviral modification of hematopoietic stem cells was associated by a lymphoproliferative disorder caused by insertional mutagenesis (Hacein-Bey-Abina et al. 2003).

9.9 Concluding Remarks

The use of tissues as biosensors, derived by transplantation of biosensor cells or by genetic modification of tissues in situ, has great promise for biomedical science. Tissue-based biosensors could provide in vivo readouts for hormones, toxins, drugs and other molecules. While the current state of development of the field limits use of these biosensors to experimental animals, the future application of this technology to human subjects must await the resolution of several practical and safety issues.

References

Bertrand L, Parent S, Caron M, Legault M, Joly E, Angers S, Bouvier M, Brown M, Houle B, Menard L (2002) The BRET2/arrestin assay in stable recombinant cells: a platform to screen for compounds that interact with G protein-coupled receptors (GPCRS). J Recept Signal Transduct Res 22:533–541

Bhaumik S, Gambhir SS (2002) Optical imaging of Renilla luciferase reporter gene expression in living mice. Proc Natl Acad Sci USA 99:377–382

Bozza T, McGann JP, Mombaerts P, Wachowiak M (2004) In vivo imaging of neuronal activity by targeted expression of a genetically encoded probe in the mouse. Neuron 42:9–21

Breit A, Gagnidze K, Devi LA, Lagace M, Bouvier M (2006) Simultaneous activation of the δ opioid receptor (δOR)/sensory neuron-specific receptor-4 (SNSR-4) hetero-oligomer by the mixed bivalent agonist bovine adrenal medulla peptide 22 activates SNSR-4 but inhibits δOR signaling. Mol Pharmacol 70:686–696

Christini DJ, Walden J, Edelberg JM (2001) Direct biologically based biosensing of dynamic physiological function. Am J Physiol 280:H2006–H2010

De A, Gambhir SS (2005) Noninvasive imaging of protein-protein interactions from live cells and living subjects using bioluminescence resonance energy transfer. FASEB J 19:2017–2019

Edelberg JM, Aird WC, Rosenberg RD (1998) Enhancement of murine cardiac chronotropy by the molecular transfer of the human β2 adrenergic receptor cDNA. J Clin Invest 101:337–343

Edelberg JM, Jacobson JT, Gidseg DS, Tang L, Christini DJ (2002) Enhanced myocyte-based biosensing of the blood-borne signals regulating chronotropy. J Appl Physiol 92:581–585

Gourley PL (2005) Brief overview of BioMicroNano technologies. Biotechnol Prog 21:2–10

Gross S, Piwnica-Worms D (2005) Real-time imaging of ligand-induced IKK activation in intact cells and in living mice. Nat Methods 2:607–614

Gross S, Abraham U, Prior JL, Herzog ED, Piwnica-Worms D (2007) Continuous delivery of D-luciferin by implanted micro-osmotic pumps enables true real-time bioluminescence imaging of luciferase activity in vivo. Mol Imaging 6:121–130

Hacein-Bey-Abina S, Von Kalle C, Schmidt M, McCormack MP, Wulffraat N, Leboulch P, Lim A, Osborne CS, Pawliuk R, Morillon E, Sorensen R, Forster A, Fraser P, Cohen JI, de Saint Basile G, Alexander I, Wintergerst U, Frebourg T, Aurias A, Stoppa-Lyonnet D, Romana S,

Radford-Weiss I, Gross F, Valensi F, Delabesse E, Macintyre E, Sigaux F, Soulier J, Leiva LE, Wissler M, Prinz C, Rabbitts TH, Le Deist F, Fischer A, Cavazzana-Calvo M (2003) LMO2-associated clonal T cell proliferation in two patients after gene therapy for SCID-X1. Science 302:415–419

Hamdan FF, Audet M, Garneau P, Pelletier J, Bouvier M (2005) High-throughput screening of G protein-coupled receptor antagonists using a bioluminescence resonance energy transfer 1-based beta-arrestin2 recruitment assay. J Biomol Screen 10:463–475

Hasan MT, Friedrich RW, Euler T, Larkum ME, Giese G, Both M, Duebel J, Waters J, Bujard H, Griesbeck O, Tsien RY, Nagai T, Miyawaki A, Denk W (2004) Functional fluorescent Ca2+ indicator proteins in transgenic mice under TET control. PLoS Biol 2:e163

Heydari AR, You S, Takahashi R, Gutsmann A, Sarge KD, Richardson A (1996) Effect of caloric restriction on the expression of heat shock protein 70 and the activation of heat shock transcription factor 1. Dev Genet 18:114–124

Huang Q, Chen M, Liang S, Acha V, Liu D, Yuan F, Hawks CL, Hornsby PJ (2007a) Improving cell therapy – experiments using a cell transplantation model in immunodeficient mice. Mech Age Dev 128:25–30

Huang Q, Yow RM, Schneider EC, Acha V, Sardar DK, Hornsby PJ (2007b) Bioluminescence measurements in mice using a skin window. J Biomed Optics 12:054012

Ji G, Feldman ME, Deng KY, Greene KS, Wilson J, Lee JC, Johnston RC, Rishniw M, Tallini Y, Zhang J, Wier WG, Blaustein MP, Xin HB, Nakai J, Kotlikoff MI (2004) Ca^{2+}-sensing transgenic mice: postsynaptic signaling in smooth muscle. J Biol Chem 279:21461–21468

Li IT, Pham E, Truong K (2006) Protein biosensors based on the principle of fluorescence resonance energy transfer for monitoring cellular dynamics. Biotechnol Lett 28:1971–1982

Loening AM, Fenn TD, Wu AM, Gambhir SS (2006) Consensus guided mutagenesis of Renilla luciferase yields enhanced stability and light output. Protein Eng Des Sel 19:391–400

Luker GD, Bardill JP, Prior JL, Pica CM, Piwnica-Worms D, Leib DA (2002) Noninvasive bioluminescence imaging of herpes simplex virus type 1 infection and therapy in living mice. J Virol 76:12149–12161

Massoud TF, Paulmurugan R, De A, Ray P, Gambhir SS (2007) Reporter gene imaging of protein-protein interactions in living subjects. Curr Opin Biotechnol 18:31–37

Nelson JF, Karelus K, Bergman MD, Felicio LS (1995) Neuroendocrine involvement in aging: evidence from studies of reproductive aging and caloric restriction. Neurobiol Aging 16: 837–843

Pancrazio J (2001) Preface. Special issue of Biosensors and Biolectronics on cell and tissue based biosensors. Biosens Bioelectron 16:427–428

Paulmurugan R, Umezawa Y, Gambhir SS (2002) Noninvasive imaging of protein-protein interactions in living subjects by using reporter protein complementation and reconstitution strategies. Proc Natl Acad Sci USA 99:15608–15613

Pfleger KD, Eidne KA (2005) Monitoring the formation of dynamic G-protein-coupled receptor-protein complexes in living cells. Biochem J 385:625–637

Reininger-Mack A, Thielecke H, Robitzki AA (2002) 3D-biohybrid systems: applications in drug screening. Trends Biotechnol 20:56–61

Sun B, Huang Q, Liu S, Chen M, Hawks CL, Wang L, Zhang C, Hornsby PJ (2004) Progressive loss of malignant behavior in telomerase-negative tumorigenic adrenocortical cells and restoration of tumorigenicity by human telomerase reverse transcriptase. Cancer Res 64: 6144–6151

Tang L, Christini DJ, Edelberg JM (2003) Genetically engineered biologically based hemostatic bioassay. Ann Biomed Eng 31:159–162

Thomas M, Hornsby PJ (1999) Transplantation of primary bovine adrenocortical cells into *scid* mice. Mol Cell Endocrinol 153:125–136

Thomas M, Northrup SR, Hornsby PJ (1997) Adrenocortical tissue formed by transplantation of normal clones of bovine adrenocortical cells in *scid* mice replaces the essential functions of the animals' adrenal glands. Nature Med 3:978–983

Troy T, Jekic-McMullen D, Sambucetti L, Rice B (2004) Quantitative comparison of the sensitivity of detection of fluorescent and bioluminescent reporters in animal models. Mol Imaging 3:9–23

Weissleder R, Ntziachristos V (2003) Shedding light onto live molecular targets. Nat Med 9:123–128

Yamaguchi S, Kobayashi M, Mitsui S, Ishida Y, van der Horst GT, Suzuki M, Shibata S, Okamura H (2001) View of a mouse clock gene ticking. Nature 409:684

Zheng J, Manuel WS, Hornsby PJ (2000) Transfection of cells mediated by biodegradable polymer materials with surface bound polyethyleneimine. Biotechnol Progress 16:254–257

Chapter 10
Biosensing with Plants: Plant Receptors for Sensing Environmental Pollution

S.K. Basu and I. Kovalchuk

Abstract Chemical pollution, associated with anthropogenic activities, is one of the worst forms of environmental pollution impacting unique global ecosystems and changing the fragile and vulnerable environment all over the planet. It is essential and important to monitor and detect harmful and toxic pollutants in our natural environment. Plants provide us with an unique opportunity to biologically monitor and detect pollutants in the environment at a molecular level. This is a new approach and is also known as plant biomonitoring, phytobiomonitoring, or phytomonitoring. In this chapter, we investigate the importance and implications of plants for an efficient biomonitoring and biosensing of specific environmental pollutants and the emergence of technologies associated with phytomonitoring. We also investigate mechanisms of ligand–receptor interactions in plants and evaluate the possibility of using these systems for biomonitoring.

Keywords Biological monitoring · Biomonitoring · Biosensors · Environment · Ligand–receptor interaction · Ligands · Phytomonitoring · Phytosensing · Receptor kinases · Plant

Abbreviations

ABAR	ABA-binding protein
ABAs	Abscisic acids
ARC1	Arm Repeat Containing1

I. Kovalchuk (✉)
Department of Biological Sciences, Hepler Hall, University of Lethbridge, 4401 University Drive, Lethbridge, AB, Canada T1K 3M4
e-mail: igor.kovalchuk@uleth.ca

M. Zourob (ed.), *Recognition Receptors in Biosensors*,
DOI 10.1007/978-1-4419-0919-0_10,

BAK1	Bri1-associated receptor kinase 1
BRI1	Brassinosteroid-insensitive 1
CALUX	Chemical-activated luciferase gene expression
CaM	Calmodulin
CCH	Copper Chaperone
CLV3	CLAVATA 3
GAs	Gibberellic acids
GID1	GA-insensitive dwarf mutant
GLR	Glutamate receptor-like genes
LRRs	Leucine-rich repeats
MRP4	Multi-drug resistance-associated protein
Nr	Never ripe
PBB	Plant-based biomonitoring
PSK	Peptide phytosulfokine
RAN	Responsive to Antagonist1
RLK	Receptor-like kinase
RLKs	Receptor-like kinases
RPK	Receptor protein kinase
RPK1	Receptor-like protein kinase
SAM	Shoot apical meristem
SCR	S-locus cysteine-rich protein
SI	Self-Incompatibility
SLG	S-locus glycoprotein
SP11	S-locus Protein 11
SRK	S-locus receptor kinase
TRP	Transient receptor potential
TRPA1	Transient receptor potential A1
WAK	Cell wall-associated kinase

10.1 Introduction

Environment pollution is one of the worst impacts that human civilization imposed on itself (Schwartz and Jones 1997; Guieysse et al. 2001; Datta Banik et al. 2007; Lee and Yang 2008). One of the most detrimental forms of environmental pollution is chemical pollution or pollution induced by the widespread and indiscriminate use of different complex organic compounds and inorganic elements and compounds (Holst and Nogel 1997; Guieysse et al. 2001; Lee and Yang 2008). Several factors are attributed towards rapid dispersal of global chemical pollution. Among those are: non-scientific management of industrial areas and sites across the planet; lack of proper processing and monitoring of industrial effluents; indiscriminate use of such agro-chemicals as fertilizers, non-sanitized manures, pesticides and related chemicals for enhancing agricultural productivity; rapid rate of urbanization and industrialization, promotion of mega-scale industrial agriculture, destruction of unique ecosystems and habitats due to agro-industrial expansion and growth of human population (Schwartz and Jones 1997; Holst and Nogel 1997; Yoon et al. 2002; Kovalchuk and Kovalchuk 2003; Lee and Yang 2008).

Nowadays, a wide range of environmentally sensitive chemicals are known to seriously impact living organisms, both prokaryotes and eukaryotes. They are toxicogens, allergens, mutagens, teratogens, clastogens, xenobiotics, different agro-industrial chemicals, industrial effluents, pharmaceuticals, plastics, radioactive waste products, fossil fuel and their associated end products (Powell 1997; Schwartz and Jones 1997; Guieysse et al. 2001; Gadzala-Kopciuch et al. 2004). Most of these chemical pollutants have both short- and long-term detrimental effects on the environment (Kovalchuk and Kovalchuk 2003; Gadzala-Kopciuch et al. 2004).

Analysis of toxic pollution levels is an important task for environmental health management (Choi and Gu 2003). This type of monitoring can be done using various physical, chemical and biological agents (Choi and Gu 2003; Green and Knutzen 2003; Kovalchuk and Kovalchuk 2003, 2008). Physical- and chemical-based methods allow a fairly precise measurement of the level of contaminants, but they lack the flexibility of biological agents. Moreover, using biological monitors is more relevant for environment and human health, as they are less harmful for the environment than more intrusive chemical and/or physical approaches, and are more reliable, cheaper and faster (Gadzala-Kopciuch et al. 2004). There are a large number of chemical approaches that are currently used, such as Gas Chromatography, High Performance Liquid Chromatography, Mass Spectrometry, Atomic Absorption, Emission Spectra and other intermixing approaches. However, the cost associated with elaborate instrumentation is quite high compared to bioanalytical approaches, such as the use of enzymes, aptamers, antibodies, peptides, microbial, animal or plant cells (Asari et al. 2004).

Kandimalla and Ju (2005) suggested that environmental bioreceptors could be simply classified into three separate classes (1) biocatalysts that depend on specific enzymes to catalyze key biochemical reactions and generate a specific, easily identifiable/detectable product; (2) bioaffinity receptors that depend on aptamers, such as DNA/RNA oligonucleotides, peptides and polypeptides, and antibodies to promptly recognize and/or identify a target receptor and then associatively bind to it; and (3) microbial or cell-based systems that depend on using microbes as biological identifying agents for a specific pollutant. However, it is important to note that microbial bioassays are comparatively cheaper than enzyme bioassays, and they operate over a wider range of temperature and pH (Gadzala-Kopciuch et al. 2004).

Biological monitoring, also known as biomonitoring, can be done with various bacterial and eukaryotic organisms (Powell 1997; Kovalchuk and Kovalchuk 2003; Gadzala-Kopciuch et al. 2004). Of particular importance and relevance is biomonitoring with higher eukaryotes (Kovalchuk and Kovalchuk 2008). In this respect, plants are the best and most suitable monitors, since they enter the food chain, they are sedentary organisms and hence are ethically more convenient to use. Moreover, plants can be modified to sense a specific type of pollutant (Schwartz and Jones

1997; Kovalchuk and Kovalchuk 2003, 2008). Fairly recently, the use and development of transgenic plant lines for rapid detection of environmental pollution and contamination using highly efficient molecular screening techniques have been increasingly investigated (Hohn et al. 1999; Kovalchuk et al. 1998a; Kovalchuk et al. 2000a, b; Kovalchuk and Kovalchuk 2003, 2008).

These novel biological monitoring approaches using specially designed plants have been often referred to as plant biomonitoring or phytobiomonitoring or phytomonitoring (Powell 1997; Basu et al. in press). The term "bioanalytics," as used by Gadzala-Kopciuch et al. (2004), referred collectively to all bioassays, biological tests, biosensors and bioreceptor- based studies and investigations. Recent investigations of ligand–receptor interactions in plants also hold promise for developing future strategies and technologies for efficient environmental monitoring (biosensing) using signals generated by plants under specific stress environments. These genetically modified plants can be designed to pick up and transmit specific signals against specific environmental pollutants in their immediate surroundings through specific ligand–receptor interactions. A simple classification of the proposed approaches for environmental biomonitoring is presented in Fig. 10.1.

In this chapter, we specifically investigate the role of plants in efficient biomonitoring and biosensing of their natural environment, in response to specific environmental contaminants and pollutants as well as technologies associated with efficient phytobiomonitoring. We also investigate mechanisms of ligand–receptor interactions in plants with emphasis on their possible future potential as phytobiosensors.

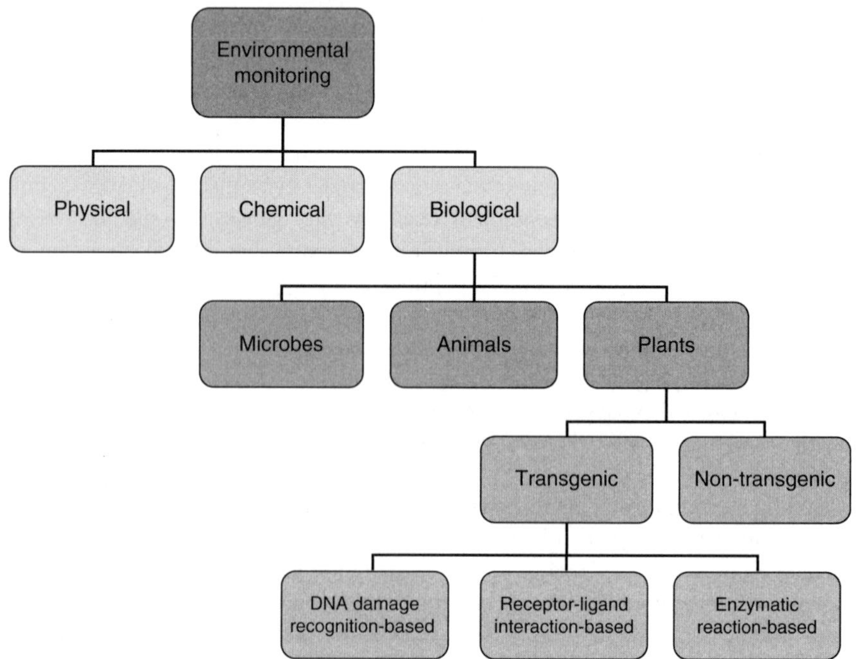

Fig. 10.1 A simple classification of the proposed environmental biomonitoring approach

10.2 Biomonitoring

Environmental pollution has been a constant companion of human progress and endeavor all across the globe. Harmful pollutants released by indiscriminate anthropogenic sources entering food chains and food webs are toxic to the ecology of a particular habitat and detrimental to the overall health of the environment (Powell 1997; Datta Banik et al. 2007). Hence, it is necessary to be able to detect toxic pollutants in our immediate environment and also assess their level and concentration in the ecosystem, so that the appropriate abatement technology and strategy could be adopted well in time to reduce the damage intensity (Lebel et al. 1993; McCormick and Garins 1997; Powell 1997; Gadzala-Kopciuch et al. 2004; van der Auwera et al. 2008).

Thus, biomonitoring is an efficient and effective management tool for detecting a particular toxicant, a xenobiotic or a group of pollutants and assessing their possible concentration in the environment. Several studies have been conducted using different biological organisms for effective monitoring, such as microbes (Wikström et al. 1999; Guieysse et al. 2001; Wikström et al. 2001; Choi and Gu 2003), micro- and macro-algae (McCormick and Garins 1997; Chen 2001; Stauber et al. 2002), lichens (Bielczyk 1993), animals (Vetter et al. 1996; Gimeno et al. 1997; Berecka et al. 2003; Green and Knutzen 2003) and most notably plants (Seidman et al. 1965, Powell 1997; Schwartz and Jones 1997; Hohn et al. 1999; Kovalchuk et al. 1999a, b; Kovalchuk et al. 2000a, b, c, d; Filkowski et al. 2003; Gadzala-Kopciuch et al. 2004; Kovalchuk and Kovalchuk 2003, 2008).

The CALUX-bioassay (chemical-activated luciferase gene expression) has become quite popular as a dependable biomonitoring tool for detecting a number of toxic pollutants, such as polychlorinated dioxins and furans (see Sakai and Takigami 2003). This highly efficient bioassay is capable of detecting the dioxin-like activity, including unknown substances in highly complex toxic industrial waste samples and a wide variety of environmental samples (Sakai and Takigami 2003). The CALUX-bioassay has also been successfully utilized as a biomonitoring tool in waste wood recycling for developing animal bedding in Japan and as a safety pre-screening measure for detecting traces of toxic dioxin-like compounds in wood residues (Asari et al. 2004).

Another example is the use of Cytochrome P450s as biomarkers in the case of biomonitoring of polycyclic aromatic hydrocarbons, aromatic amines, benzene, toluene, and nitrosamines associated with tobacco smoking (see Lee and Yang 2008).

Plants have been used for biomonitoring for a long time. The effective use of onion (*Allium cepa*) root-tips for detecting the mutagenic influence of pollutants has been well documented. The assay is based on the detection of chromosomal abnormalities and the corresponding mitotic index in cells of onion root-tips. In the past, this model was used for studying soil samples from different parts of Ukraine, contaminated as a result of Chernobyl nuclear incident in 1986 (Kovalchuk et al. 1998a, b). The most practical advantages of using this assay are the low cost and simplicity of establishment. In brief, onion seeds are germinated on Whatman paper soaked in contaminated water or water extracts from contaminated soil. Root-tips of germinated seeds are squished and stained in 2% orcein dissolved in 45% acetic acid, mounted on a slide and observed under a microscope. The measured parameters include seed germination percentage, length of formed roots, percentage of cells undergoing mitosis and percentage of various chromosomal aberrations.

Despite its simplicity, the *Allium cepa* test does not have high sensitivity. Another disadvantage is that one needs to bring samples to the laboratory and prepare water extracts. Under ideal conditions, one would rely on the use of a "portable" biosensor. In this respect, planting seeds of a sensitive biomonitor in the polluted area seems to be a valid idea. Such a biomonitor should have high sensitivity, rapid growth, and it should be tolerant to high levels of contamination.

In this respect, thale cress *Arabidopsis thaliana* (the *Brassicacae* family) is an ideal candidate; it has a short reproductive season (2–3 months) and is quite tolerant to pollutants. For better visualization of the damaging influence of pollutants, this plant has been transgenically modified. Several reporter genes introduced into the Arabidopsis genome were able to monitor genotoxicity. These reporters are based on the so-called visual markers – enzymes (gene products) that are able to cleave particular substrates, and this enzymatic reaction can be easily observed. Two major visual markers were used in the past, GUS (based on the uidA gene) and LUC (based on the luciferase gene). The former codes for the enzyme β-glucuronidase, and the latterfor the luciferase. The cleavage of the substrate X-gluc and luciferince by β-glucuronidase and luciferase, respectively, results in the production of an easily detectable product. In the case of β-glucuronidase, it is a blue precipitate, and in the case of luciferase, it is a light-emitting product.

Transgenic assays employing the aforementioned markers are based on two reporter constructs, mutation-based and recombination-based. The former is based on the modification of a single nucleotide in a coding region of the gene in such a way that it changes a coding triplet into the one that either encodes a stop codon or amino-acid that inactivates enzymatic ability. The latter is much more complex, it consists of two truncated non-functional copies of the transgene.

Both reporters were integrated into the plant genome. Activity of the point mutation reporter gene can only be restored via a point mutation reverting the stop-codon or non-sense mutation to the endogenous nucleotide. The reported recombination activity can be restored via a homologous recombination event that utilizes regions of homology as a template. Cells in which restoration of the transgene structure occurred can be easily monitored. Whereas the point mutation assay monitors all types of DNA damage that result in point mutation, the recombination assay monitors all sorts of strand breaks that result in homologous recombination events. Thus, the assays are more effective if used in parallel.

Indeed over the last decade, a number of studies reported substantial success in using these plants for environmental biomonitoring (Kovalchuk et al. 1998a, b, 1999a, b, 2000a, b; Ries et al. 2000; Kovalchuk et al. 2001a, b; Besplug et al. 2004; Li et al. 2006; van der Auwera et al. 2008).

10.2.1 What Is Ligand–Receptor Interaction?

A ligand can be defined as a specific substance that is capable of binding and forming a thermodynamically stable complex with other biomolecules (mostly macromolecules, such as receptor proteins) through either ionic and hydrogen bonding or Van der Waals forces to perform a specific biologically important function (Nelson and Cox 2000; Gadzala-Kopciuch et al. 2004). The word ligand is derived from the Latin word *ligare* which means *to bind* (Keller et al. 1998). A ligand is a specific signal-inducing molecule, such as an inhibitor, activator, substrate or neurotransmitter that binds to macromolecules such as proteins irreversibly via intermolecular forces mentioned above. The strength of binding is referred to as affinity. It depends on the strength of intermolecular forces governing the binding processes (Fig. 10.2). Ligand binding to receptor biomolecule, such as a receptor protein, usually results in changes in the three-dimensional structure conformation of a receptor protein, which in turn governs and directs a specific biological function of target biomolecules (Keller et al. 1998; Friefelder 1999). The affinity strength is an important aspect of ligand–receptor interaction, as stronger affinity indicates longer association of a ligand with its corresponding receptor protein and subsequent utilization of higher binding energy for a necessary conformational change in the receptor protein that directs and governs its specific biological action (Keller et al. 1998; Friefelder 1999; Nelson and Cox 2000). Several types of ligands are reported in biochemistry, only a few representative types are mentioned below:

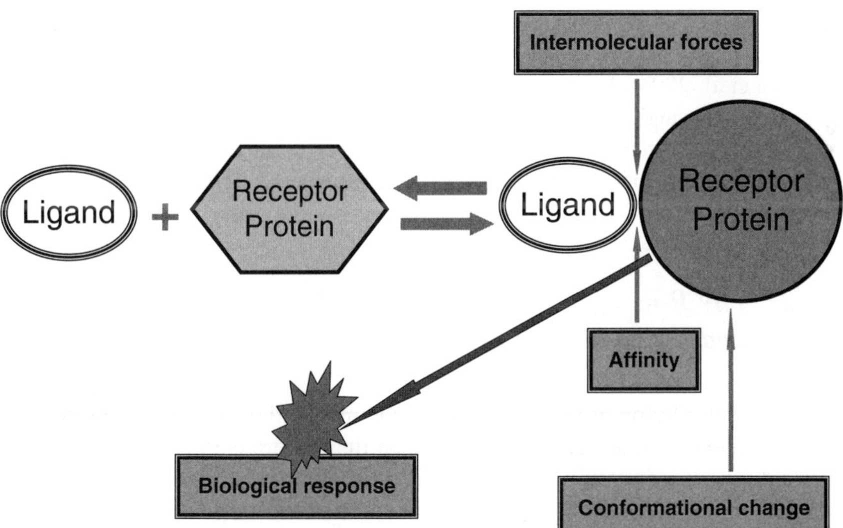

Fig. 10.2 A schematic representation of ligand-receptor protein complex formation and subsequent conformational changes of the receptor protein

1. High-affinity ligand – A low concentration of ligand, sufficient to bind to the maximum number of ligand binding sites, available in a specific receptor molecule and capable of initiating a biological response.
2. Low-affinity ligand – A high concentration of ligand necessary to bind to the maximum number of ligand binding sites, available in a specific receptor molecule to initiate the maximum possible biological response.
3. Selective ligands – Bind specifically to a few selected receptors only.
4. Non-selective ligands – Bind non-specifically to any receptors.
5. Agonist – Ligand binding to a specific receptor molecule induces a conformational change of this particular receptor and initiates a biological response in this receptor.
6. Antagonist – Ligand binding to a specific receptor molecule fails to initiate any biological response.

10.2.1.1 What Are Biosensors (Bioreceptors)?

It is interesting to mention that biosensors (also known as bioreceptors) are a specific group of chemical sensors that are activated and driven strictly by biological mechanisms (Keller et al. 1998; Gadzala-Kopciuch et al. 2004). They could be enzymes, hormones, antibodies (plantibodies in case of plants), peptides or polypeptides, proteins, animal or plant tissue, plant or animal or even microbial cells (Kandimalla and Ju 2005). While receptor-based biosensors are associated mostly with ligand–receptor interaction, another type of biosensors called catalytic bioreceptors are dependent on biocatalysts (enzymes) for subsequent recognition and binding with specific chemical compounds (Keller et al. 1998; Gadzala-Kopciuch et al. 2004). Biosensors are characterized by their extremely high specificity, sensitivity and selectivity that help them to identify and detect concentration levels of a wide variety of environmental toxicants, irritants, pollutants and xenobiotics (Gadzala-Kopciuch et al. 2004; Kandimalla and Ju 2005).

10.2.2 The Role of Hormones in Plant Development and Stress Signaling

In multi-cellular living organisms, cellular communications or cross talks between cells or within different cell organelles inside the cell are extremely important for the maintenance of several important biochemical and physiological functions, such as immune response, signal transduction, defense response against specific pathogens, uptake and transport of important ions and biomolecules necessary for cellular respiration, nutrition and metabolism (Bleecker and Kende 2000; Matsubayashi et al. 2001; Shiu and Bleecker 2001a, b; Chow and McCourt 2006).

Extensive studies conducted on animals have revealed wealth of information regarding different molecules and receptors involved in intricate intercellular and intracellular communications (Shiu and Bleecker 2001a, b; Matsubayashi 2003). Several signaling peptides have been identified, characterized and investigated till date; most of them are characterized by the presence of specific receptors attached to the plasma membrane of receiving cells (Matsubayashi et al. 2001; Shiu and Bleecker 2001a, b; Yin et al. 2002; Chow and McCourt 2006).

In plants too, cellular communications have been intermittently associated with different bioactive compounds referred to as plant hormones and phytohormones (Matsubayashi 2003). Till date, all major plant hormones, such as auxins, cytokinins, gibberellic acids, abscisic acids, ethylene, specific steroid molecules (brassinosteroids/brassinolides) and a multitude of plant peptide hormones, have been found to be associated with cell signaling and cell-cell communications (Bleecker and Kende 2000; Ryan 2000; Matsubayashi et al. 2001; Yin et al. 2002; Chow and McCourt 2006).

Numerous studies, from classical to detailed molecular investigations, have been conducted on several aspects of different plant hormone receptors (see Bleecker et al. 1998; Ryan and Pearce 1998; Bleecker and Kende 2000; Ryan 2000; Matsubayashi et al. 2001; Yin et al. 2002; Heyl and Schmulling 2003; Tichtinsky et al. 2003; Chow and McCourt 2006). Only major breakthroughs have been highlighted in the following subsections dealing with specific phytohormones.

10.2.2.1 Cytokinin

Cytokinin is a major plant hormone playing an important role in the process of cell division, regulation of organellar growth, stress response, disease resistance and abscission, to some extent (Sexton and Roberts 1982; Chow and McCourt 2006). Major progress has been made in uncovering the mechanism of cytokinin perception and signaling (see Heyl and Schmulling 2003). Three different receptors have been identified in the case of cytokinin. The first is *CKI1*, the Arabidopsis receptor histidine kinase gene (Kakimoto 1996). The second is the Arabidopsis histidine kinase *CRE1/AHK4* gene complex (Inoue et al. 2001; Suzuki et al. 2001). Yamada et al. (2001) reported that the Arabidopsis AHK4 histidine kinase is a cytokinin-binding receptor that transduces cytokinin signal across the plant cell membrane. Ueguchi et al. (2001) also confirmed that AHK4 is actively involved in the cytokinin-signaling pathway as a "direct receptor molecule." The next generation of receptors, mainly Arabidopsis histidine phosphotransferase proteins and Arabidopsis response regulators (AHK2 and AHK3), were reported with the availability of the Arabidopsis genome sequence (Heyl and Schmulling 2003).

Riefler et al. (2006) conducted a detailed study on the loss of function with three Arabidopsis histidine kinases (cytokinin receptors), namely, *AHK2*, *AHK3* and *CRE1/AHK4*. They reported partly redundant and repetitive functions for all cytokinin receptors and prominent roles for the *AHK2/AHK3* combination in quantitative regulation of organ growth in plants, with opposite controlling mechanisms exhibited in case of roots and shoots.

10.2.2.2 Gibberellic Acids (GAs)

Gibberellic acids (GAs) are plant hormones performing a number of important functions in plants that are critical for their efficient growth and development (Sexton and Roberts 1982; Chow and McCourt 2006). Although there have been a number of speculations and unconfirmed reports regarding suspected membrane-bound and soluble GA receptors, no concrete proof has been provided yet. Ueguchi-Tanaka et al. (2005) for the first time reported the isolation and characterization of a new GA-insensitive dwarf mutant rice (*gid1*). The authors reported that this suspected rice gene (*OsGID1*) is a soluble receptor conducting GA signal transduction in rice. Nakajima et al. (2006) cloned three GA receptor genes (*AtGID1a*, *AtGID1b* and *ATGID1c*) from Arabidopsis and found each of them to be an ortholog of the rice GA receptor gene (*OsGID1*). Biochemical analyses were conducted to establish the fact that all the three genes functioned as GA receptors in Arabidopsis. GA binding activities of three recombinant proteins were established through binding assays, and similar ligand specificity was observed among all the three proteins, with the highest affinity being for GA_4.

10.2.2.3 Abscisic Acids (ABAs)

Abscisic acids (ABAs) are also major plant hormones that are associated with a stress response mechanism of plants (Sexton and Roberts 1982; Chow and McCourt 2006). Hong et al. (1997) isolated an Arabidopsis cDNA clone for the receptor-like protein kinase gene (*RPK1*). The corresponding peptide consisted of four basic components of a receptor kinase: an amino terminal sequence, a domain with five-extracellular LRR sequences, a membrane-spanning domain and a cytoplasmic protein kinase domain. The gene expression in floral parts, stems, leaves and roots were induced within 1 h after ABA treatment. The authors found that this particular gene is overall responsive to all major stresses, such as dehydration, high salt and low temperature. The researchers also found that this dehydration-inducible gene expression is independent of ABA, and that the gene is possibly associated with the ABA pathway for signal transduction. Razem et al. (2006) reported the RNA-binding FCA proteins to be ABA receptors. FCAs are associated with important functions in plants, such as RNA metabolism and regulation of flowering time.

Abscission is a natural process associated with plants. It is attributed to the actions of three different plant hormones ABA, GA and cytokinin, although accumulation of the gaseous hormone ethylene and several cell-wall degrading enzymes are also reported to be associated with abscission (Sexton and Roberts 1982). However, the regulatory pathways associated with abscission are not clear. Jinn et al. (2000) were the first who reported that HAESA, the Arabidopsis LRR receptor kinase, is actively involved in the process of developmental regulation of floral organ abscission. The researchers detected that HAESA is a plasma membrane serine/threonine protein kinase.

Such genes and their corresponding protein receptors have great potential in phytosensing, since it is responsive to fairly divergent environmental parameters discussed above and can easily provide a record of environmental deterioration of a particular site or locality.

In later studies, Zhang et al. (2002) reported purification and identification of a 42-kilodalton ABA-binding protein (ABAR) from the epidermis of broad bean leaves, presumably involved in stomatal signaling. This receptor-protein (CHLH) gene encodes for the H subunit of magnesium-chelatase and plays an important role in chlorophyll synthesis as well as in plastid-to-nucleus signaling. The same research group also demonstrated that ABAR/CHLH is an ubiquitous protein that specifically binds to ABA in Arabidopsis and regulates a number of important physiological functions associated with seed germination, growth following germination and stomatal dynamics (Shen et al. 2006). Since ABAR/CHLH is expressed in both green and non-green tissues of Arabidopsis, the researchers suggested that this complex is probably capable of perceiving an ABA signal at all plant levels (Shen et al. 2006).

10.2.2.4 Ethylene

Ethylene, a gaseous hormone, is associated with a number of physiological and biochemical activities related to natural plant growth and development (Bleecker and Kende 2000). This gaseous hormone is detected by a number of plant receptors belonging to the ETR1 protein family from Arabidopsis. The *ETR1* gene was the first member of the receptor family genes that was cloned by Chang et al. (1993). In 1998, Sakai et al. reported the third response gene from the same gene family. The newly cloned *ETR2* gene was found to be similarly associated with ethylene signaling like the previously reported *ETR1* and *ETR1*-like *ERS* genes in Arabidopsis (Sakai et al. 1998). Around five different *ETR1* genes (*ETR1*, *ETR2*, *ERS1*, *ERS2* and *EIN4*) have been reported till date in Arabidopsis (Bleecker and Kende 2000; Chow and McCourt 2006). A specific mutation induced in any of these genes results in ethylene insensitivity in plants. Several models have been proposed for signal transduction involving genetic dominance of different mutations in different *ETR1* genes (Bleecker et al. 1998; Bleecker and Kende 2000). A simple flowchart mechanism of ethylene ligand–receptor interaction is represented in Fig. 10.3.

Wilkinson et al. (1997), in their study on mutated forms of *ETR1*, reported that the *etr1-1* gene of Arabidopsis encoding a mutated receptor is also responsible for a significant delay in fruit ripening, abscission and floral senescence. The researchers concluded that the capability of this mutated gene to perform in heterologous plants indicates that the ethylene response pathway is highly conserved and amenable to genetic modifications. Such genes could certainly play an important role in developing a ligand-based model for phytosensing.

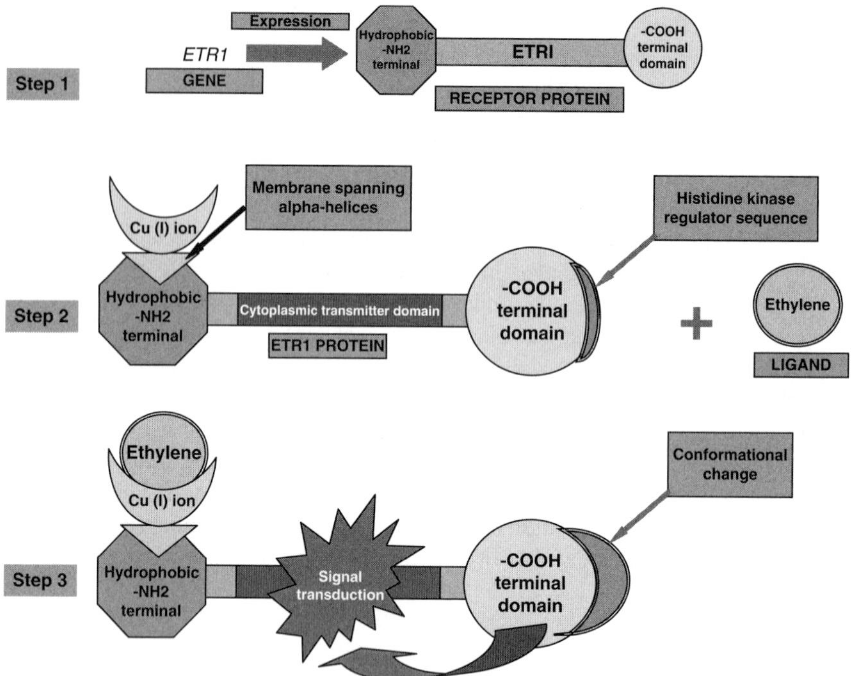

Fig. 10.3 A simple flowchart diagram representing the mechanism of ethylene genes involved in ligand–receptor interaction (after Bleecker et al. 1998). Step 1. Expressions of *ETR1* gene. Step 2. Receptor activation via binding of a transition metal ion to N-terminal alpha-helical domain of the ETR1 protein. Step 3. Signal transduction based on binding of ethylene molecule to a transition metal ion and subsequent conformational changes in the sensor domain

Tomato plants have the Arabidopsis homolog *ETR1* ethylene receptors as a big family of five members represented as *LeETR 1-5* (Tieman et al. 2000). The US researchers carried out an investigation to study ethylene responses of the *LeETR* gene family along with a tomato *Never ripe* (*Nr*) mutant type. The tomato ethylene receptors, NR and LeETR4, were found to be negative regulators of ethylene response, and they also showed a profile of functional compensation within a multi-gene family. Despite major differences in their corresponding structures, divergent receptors were found to be functionally redundant (Tieman et al. 2000).

10.2.2.5 Polypeptide Hormones

Currently, four major peptide hormones have been reported (Matsubayashi 2003). They include: a wound response hormone systemin, a sulfated peptide plant hormone phytosulfokine (PSK) associated with cellular differentiation and dedifferentiation, the apical meristema tissue-growth regulating CLAVATA subfamily of polypeptides and the self-incompatibility hormones or *S*-Locus signaling hormones (Ryan et al. 2002), including the *S*-determinant *S*-locus cysteine-rich protein (SCR) also called *S*-locus Protein 11 (SP11). The first two were isolated based on their biological activities, and their receptors have been isolated and purified from the plasma membranes on the basis of their highly specific ligand–receptor interactions. The latter two were identified genetically, while analyzing the function of specific ligands for previously cloned receptors (Bleecker and Kende 2000; Matsubayashi et al. 2001; Matsubayashi 2003; Chow and McCourt 2006). All these four signaling peptides have now been reported to be associated with specific cellular communication via intricate ligand–receptor interactions (see Matsubayashi 2003).

Systemin

Systemin is an 18-amino acid plant polypeptide hormone that is secreted from a larger prohormone protein (200-amino acid precursors) known as prosystemin by the process of proteolytic cleavage (Ryan and Pearce 1998; Ryan 2000). This hormone was isolated from injured tomato leaves and is capable of inducing approximately 20 plant defense genes (Ryan and Pearce 1998). Systemin, isolated from tomato, has been found to be inactive in tobacco, suggesting that systemin receptor(s) and other recognition factors are absent in the latter (Ryan et al. 2002). Biochemical approach has been adapted for the identification of systemin receptors, and radio-labeled systemin was finally shown (Meindl et al. 1998; Scheer and Ryan 1999) to bind to a suspected receptor in the cell membrane (~160Kd protein named SR160). Subsequent purification and protein sequencing were done to identify the complete cDNA sequence of SR160, and it was found to be most closely related to BRI1 involved in brassinosteroid signaling (Yin et al. 2002).

Phytosulfokine (PSK)

Matsubayashi et al. (2002) used ligand-based affinity chromatography for purifying a 120 kD membrane protein interacting with a sulfated peptide plant hormone, phytosulfokine (PSK) from microsomal fractions of carrot. The authors reported retrieving the corresponding cDNA fragment encoding a 1021-amino acid receptor kinase that contains LRR, a 36-amino acid-island, separated from a single transmembrane domain and a cytoplasmic kinase domain. The researchers suggested that PSK, a small five-amino acid peptide phytohormone (a plant signal for cellular dedifferentiation and multiplication), and the receptor kinase described above are a perfect ligand-receptor pair.

CLAVATA3 (CLV3)

CLV3 (CLAVATA 3) is an extracellular signaling polypeptide hormone in plants. It is about 49-amino acid in length and is associated with cell fate of the shoot apical meristem (SAM) of Arabidopsis (see Barinaga 1999; Ryan et al. 2002). The proteins associated with *CLV1* (encoding a putative receptor kinase) (Clark et al. 1997) and *CLV2* (encoding a small polypeptide) (Ryan et al. 2002) constitute a unique receptor-ligand association that subsequently regulates a delicate balance between cell proliferation and dedifferentiation in the Arabidopsis SAM (Franssen 1998). Clark et al. (1997) for the first time demonstrated that *CLV1* expression in the Arabidopsis inflorescence is associated with the activity of meristematic tissues. *CLV3* is produced in a different meristematic region compared to *CLV1* and is an extracellular signal that travels inside the cell to exert its effects on the *CLV1* intracellular domain (Clark et al. 1997; Barinaga 1999). Researchers have also found two CLV1-containing protein complexes.

Another important finding was the detection of an enzyme KAPP that binds to CLV1 and has the ability to inactivate the receptor by detaching a phosphate group from a receptor molecule (Barinaga 1999). According to Clark (2001), this phosphate removal may either provide a necessary threshold to fire the pathway or switch *CLV1* off. Later, Brand et al. (2000) reported overexpressing *CLV3* in transgenic Arabidopsis plants so that meristem cell accumulation and fate is directly proportional to the level of *CLV3* activity. The researchers also showed that the *CLV3* signaling mechanism is executed through a *CLV1/CLV2* receptor kinase complex. The authors also demonstrated that the *CLV* pathway works by repression of a specific transcription factor (*WUSCHEL*). Matsubayashi et al. (2001) suggested that CLV1 and CLV3 are peptide ligands and constitute a receptor kinase pair.

Trotochaud et al. (2000) demonstrated a series of interactions among members of this protein family. The demonstrations included: co-immunoprecipitation of CLV3 (detected as a ~25 kD soluble multimer complex) and CLV1; binding of yeast CLV1 and CLV2 with plant-derived CLV3 involving CLV1 kinase activity; in vivo CLV3 association with active CLV1; and bonding of more than 75% CLV3 to CLV1.

S-Locus Signaling Hormones

This subfamily of peptide hormones is represented by 47–60 amino acid long intercellular signaling peptides, collectively referred to as *S*-locus cysteine-rich protein (SCR)/*S*-locus Protein 11 (SP11) (Ryan et al. 2002). These peptides have been reported to play a pivotal role in the phenomenon of pollen Self-Incompatibility (SI) among Brassicaceae (syn. Cruciferae) family members. The SI system or the phenomenon of self-pollen rejection in plants is an important regulatory mechanism naturally evolved to restrict inbreeding in different plant species. This rejection of self-pollen is specifically determined by the *S*-locus (Franklin-Tong 2002; Kachroo et al. 2002). During the SI process, pollen of a specific crucifer lands on a stigma of the same plant, and self-recognition switch is triggered suspending the pollination mechanism. SI causes complete dehydration of the stigma papillae, and stigmatic aquaporin molecules have been recognized as important partners in the dehydration step (Franklin-Tong 2002; Kachroo et al. 2002; Ryan et al. 2002).

Detailed molecular investigation of the SI system in *Brassica* has led to the discovery of two distinct stigmatic components, namely the *S*-locus receptor kinase (SRK) and the *S*-locus glycoprotein (SLG) and a pollen component, SCR/SP11, at the *S*-locus (Franklin-Tong 2002). Stein et al. (1996) reported identifying the gene product of SRK_6 in the stigma of the crucifer *B. oleracea* and suggested that it shares similar characteristics with an integral membrane protein. By expression of SRK_6 in the transgenic tobacco line, the researchers established the existence of the membrane-localized transmembrane protein kinase, with the ability of intercellular recognition. They found that SRK_6 have an overlap time period and similar functional aspects to that of the previously reported *SLG* gene. Finally, the authors indicated that SRK probably functions in the papillary cells of the stigma as a distinct cellular receptor.

The SCR/SP11 interaction with SRK on cell walls of the stigmatic papillae has been reported. SRK has a characteristic receptor-like kinase structure with an extracellular cysteine-major domain at the N-terminal and intermediate transmembrane region (Ryan et al. 2002). It has been postulated that interaction of SCR/SP11 and SRK initiates a continuous signal transduction pathway in the stigma that subsequently results in the fast depletion of pollen growth at the pre- or post-pollen germination stage (see Franklin-Tong 2002). However, two independent research groups (Kachroo et al. 2001; Takayama et al. 2001) have now convincingly identified the long sought-after ligand (pollen)–receptor (stigma) interaction. They reported that inhibition of pollen was compatible and successfully achieved by coating the stigma with *S*-locus signaling peptides (SCR/SP11), however a success was recorded in the case of a compatible pollen–stigma interaction for eliciting the SI response.

Takayama et al. (2001) artificially synthesized the SP11 peptide and reported detecting eight cysteine residues that form four disulfide bridges in the molecular structure of SP11. The fact that they have convincingly demonstrated the biologically active nature of this artificially synthesized SP11 peptide is the most promising aspect of their work. Future SI-related research and plant ligand–receptor interaction studies would be undoubtedly enriched and promoted by these latest works.

Plant cell proteins, like SCR/SP11, are also similarly located on the pollen grain. Their role in SI, however, is not essential. It is also important to note that although SLG has a strong similarity with SRC in the extracellular domain, the former is non-essential for the SI reaction (Franklin-Tong 2002). The SCR/SP11 (ligand in the pollen) interaction with SRK (receptor in the stigma) triggers phosphorylation of the kinase domain in SRK, if the corresponding S-alleles match with each other. This in its turn phosphorylates another ARC1receptor (another protein receptor kinase) which possibly regulates aquaporins responsible for dehydration of the pollen. A simple representation of a SI model described above is represented in Fig. 10.4. ARC1 is also associated with SI. However, since it is not a peptide plant hormone, it will be discussed in the section on receptor-like kinases (RLKs).

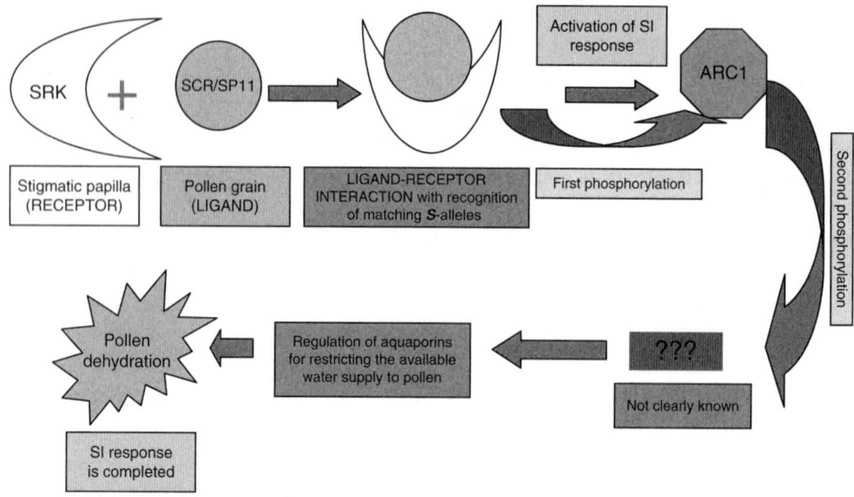

Fig. 10.4 A simple diagram representing the SI reaction in the Brassicaceae member (after Franklin-Tong 2002; Ryan et al. 2002). Binding of SCR/SP11 (ligand in the pollen) to SRK (receptor in the stigma) triggers phosphorylation of the kinase domain in SRK, if the corresponding S-alleles match with each other. This in its turn phosphorylates another ARC1 receptor which possibly regulates aquaporins responsible for dehydration of the pollen

10.2.3 Receptor-Like Kinases

Plant genomes represent a wide variety of RLKs (Tichtinsky et al. 2003) that are structurally related to tyrosine and serine/threonine RLK families reported for animals (Cock et al. 2002). Plant RLKs have been shown to represent receptors for different plant peptides (Torii 2000; Tichtinsky et al. 2003; Boller 2005). RLKs are characterized as transmembrane proteins with putative amino-terminal extracellular domains and intricate carboxy-terminal intracellular kinase domains. They play an important role in receiving and transmitting extracellular signals into plant cells (Vinagre et al. 2006).

Plant RLKs belong to a big gene family of ~610 members representing ~2.5% Arabidopsis protein-coding sequences. They have a strong similarity with animal receptor tyrosine kinases (Shiu and Bleecker 2001a, b). Detailed phylogenetic analyses indicate that plant RLKs have evolved independent of their animal counterparts (Cock et al. 2002; Tichtinsky et al. 2003). However, Shiu and Bleecker (2001a, b), suggested that RLKs from Arabidopsis represent a monophyletic gene family related to animal receptor kinases. A comparison between animal Raf kinases, RLKs from plants and animal receptor tyrosine kinases exhibited a common ancestry with serine/threonine/tyrosine kinases. According to Boller (2005), the need to identify several peptide signals in self/non-self recognition and perception in plants may have initiated a wide array of diversity and branching in reported RLKs from the plant kingdom. Cock et al. (2002), however, reported some common characteristic aspects of both plant and animal RLKs, such as a single membrane-pass structure; receptor activation associated with trans-autophosphorylation and dimerization processes and the presence of phosphatases and inhibitors down-regulating the activity of target receptors.

Distribution patterns of four major RLK subfamilies in plants indicate a probable duplication, followed by reshuffling of the Arabidopsis genome as well as production of tandem repeats. The researchers also suggested that the RLK gene family may have been close to the present-day gene family prior to the divergence of terrestrial plants into distinctly different lineages (Lease et al. 1998; Torii 2000; Tichtinsky et al. 2003). A wide diversity has been observed in RLK extracellular domains: some of them are common to both the plant and animal kingdoms, while others are specific to plants only. Several of these motifs are involved in a wide range of functions such as protein–protein interactions or protein–cell wall associations (components like glycoproteins) (Lease et al. 1998; Torii 2000).

Another notable characteristic of RLKs is the occurrence of leucin-rich repeats (LRR) in the extracellular domain. Almost one-third of all RLKs reported till date have 1–32 LRR in the extracellular domain. These are often involved in different protein–protein interactions. RLK-LRR has been subdivided into 12 subfamilies on the basis of the relationship between amino acids in their kinase domains (Devart and Clark 2004). The unique number and locations of LRRs in their extracellular domain structure characterize an individual member of each subfamily. LRRs constitute less than half the components of the extracellular domain in several reported RLKs (Shiu and Bleecker 2001a, b; Devart and Clark 2004).

RLK activities in a cell membrane, similar to that in animals, have been reported in Arabidopsis by Schaller and Bleecker (1994). The authors found that plant protein kinases were: ~115–135 kDa in molecular weight; associated with a membrane and glycosylated; showed a preference for divalent manganese ions instead of more commonly reported divalent magnesium ions (similar to another RLK cloned from Arabidopsis *AtTMK1*); and phosphorylated only at serine and threonine residues, suggesting their putative role as cell receptors.

Recent investigations have detected and identified a number of receptors and sensors in a cell wall that performs important functions, such as cell wall elongation and extension, rigidity of mechanical tissues and a number of co-ordinated sensing and feedback mechanisms to regulate growth and development of a cell wall (see Humphrey et al. 2007). In a fairly recent publication, Hématy and Höfte (2008) reported discovering three novel plasma membrane-bound receptor kinases related to the animal *CrRLK1* in plants. The authors suggested that they had detected highly conserved extracellular kinase domains in each of the three receptors, each of them being able to bind specific ligands that could possibly be involved in synthesis of reactive oxygen species.

These three new receptor-kinases are: (1) THESEUS1 (reported to be activated in cellulose-deficient mutants and is suspected to be involved as a cell wall integrity sensor, preventing cellular elongation in Arabidopsis); (2) FERONIA (detected in the female gametophyte of Arabidopsis only within synergid cells, essential for growth inhibition of compatible pollen tubes and the associated sperm delivery process) and (3) *AmRLK* (participating in the regulation of the polar conical outgrowth of epidermal cells in *Antirrhinum* petals).

Recently, Hématy et al. (2007) reported the suspected functioning of THESUS1 in cellular elongation in further detail. The authors reported that a mutation in the *THE1* gene and a subsequent overexpression of the functional fusion protein THE1-GFP in Arabidopsis had no impact on non-mutated control plants. However, attenuation, prolonged growth inhibition and ectopic lignification of seedlings were observed in the case of lines mutated in cellulose synthase *CESA6*. In addition, the T-DNA insertion mutant of *THE1* also showed similar symptoms. These results clearly demonstrate the involvement of *THE1* in the process of cellular growth and elongation in plants and possibly its function as a putative receptor for cell wall integrity.

Vinagre et al. (2006) reported the identification of a specific RLK, called the *SHR5* gene, with the suspected involvement in the signal transduction process associated with sugarcane plants and endophytic bacterial interactions. The authors reported down-regulation of the gene in plants specifically associated with beneficial endophytic bacteria. Subsequent reduction in the level of gene expression in case of more compatible plant–microbe interactions has also been noted. The researchers concluded that the gene-specific mRNA levels must be inversely correlated with the efficiency of the symbiotic association.

Such studies could give us an indirect idea about environmental quality. The distortion of symbiotic association between a host plant and microbes would result in a different level of gene expression. This can be linked to a specific, easily detected visual marker. A simple example of phytosensing of the environment could thus be achieved.

Another research group (Voets et al. 2007) devoted to molecular and cellular research in animals reported for the first time charge-neutralizing mutations in the transmembrane segment 4 (S4) and the S4–S5 connector of human TRPM8, a transient receptor potential (TRP) channel superfamily member. Mutations were capable of reducing the channel gating charge, suggesting TRPM8 to be a part of the voltage sensor of the neuron. The authors detected significant alterations in thermal, voltage-sensing and methanol affinity by inducing specific mutations in this region. They proposed a model with synergistic effects of three above-mentioned factors on TRPM8 gating. Similar thermo-voltage sensitive receptors in plants have not been currently reported. If detected, these receptors would be ideal candidates to be used as phytosensors.

Charpenteau et al. (2004) first reported the detection of a calmodulin (CaM) binding protein RLK in Arabidopsis named *AtCaMRLK*. This particular RLK had all four essential components of a receptor kinase, namely, an amino terminal, a domain consisting of seven leucine-rich repeats (LRR), a single putative membrane-spanning segment and a protein kinase domain (Fig. 10.5). Researchers showed that a region of 23 amino acids localized at the kinase domain of the target receptor actually binds to CaM in a calcium-dependent mechanism. The authors reported that the specific CaM binding domain was highly conserved across many

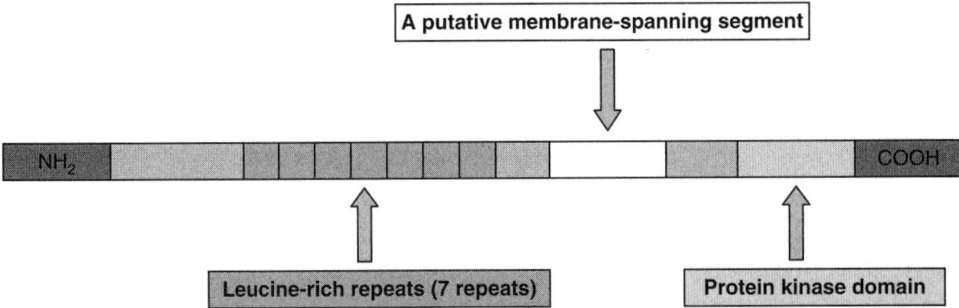

Fig. 10.5 A diagram representing the structure of a typical receptor kinase *AtCaMRLK* (after Charpenteau et al. 2004)

members of the plant RLK family. This also indicates a possible role of calcium/CaM in the regulation of RLK pathways in plants.

Koizumi et al. (2001) first reported the cloning and amino acid sequences of *AtIre1-1* and *AtIre1-2*, two Arabidopsis homologs of yeast and mammalian *Ire1*. In yeast and mammals, the *Ire1* is a transmembrane receptor protein kinase (RPK) with a sensor domain located in the lumen of endoplasmic reticulum. This gene is a first member in the signal transduction process from the endoplasmic reticulum to the nucleus for regulating an unfolded protein response reaction. The researchers used these two homolog genes to study their direct involvement in unfolded protein response signaling in plants. They found the Ire1-2 protein to be associated with auto-transphosphorylation. These receptors could also be used for biosensoring. Any pollutant that disrupts the signal transduction process or auto-transphosphorylation could be essentially picked up in the environment by specifically designed phytosensing plants.

10.2.4 Cell Wall Associated Kinase and WAK-Like Kinase

The cell wall associated kinase (WAK) and the WAK-like kinase (WAKL) represent a distinct group or subfamily under RLK, characterized by its 26 members serving as signal molecules with a direct communication link between the cell wall and cell cytoplasm (Verica et al. 2003). Earlier, researchers have associated this gene subfamily with bacterial pathogenic responses and heavy metal tolerances (He et al. 1998; Sivaguru et al. 2003). He et al. (1998) provided strong evidence to the fact that there is a strong direct linkage or association between signals transmitted from the extracellular matrix by the target protein kinase (WAK1) and the cascade that follows pathogen recognition. The researchers suggested that this response is triggered only if Arabidopsis plants are infected with bacterial pathogens. On the other hand, Sivaguru et al. (2003) reported the aluminum-induced organ-specific expression of the *WAK1* gene and cell type-specific localization of WAK1 proteins in Arabidopsis. The most interesting outcome of this investigation was that the authors reported that transgenic plants overexpressing the *WAK1* gene had an increased heavy metal tolerance to aluminum, making it an interesting candidate for future phytosensing research. The most characteristic aspect of this subfamily is the presence of epidermal growth factor-like repeats within the extracellular domain and their compact cell wall association (He et al. 1996; Wagner and Kohorn 2001; Verica et al. 2003). Such studies provide us with an important tool, as they are amenable to necessary genetic modifications in future to investigate their possibilities and effectiveness in environmental phytosensing.

Hou et al. (2005) reported a functional role of WALK4, a member of the WAK/WAKL subfamily from Arabidopsis, with respect to mineral responses. The authors detected that WAKL4 has strong cell wall association like other family members, and they also reported WAKL4 expression to be induced by a host of ions such as sodium, potassium, copper, nickel and zinc. Impairment of the WAKL4 promoter resulted in hypersensitivity to all ions mentioned above. The authors observed that a T-DNA insertion at 40 bp upstream of the *WALK4* start codon had no impact under normal growth conditions, but it could significantly alter expression levels under the influence of diverse heavy metal salts. They established that WALK4 was necessary for up-regulation of zinc transporter genes at the time of this metal deficiency. They also reported that the WALK4 T-DNA resulted in zinc accumulation being reduced in the plant shoot system.

10.2.5 Leucine-Rich Repeats: LRR-RLK Subfamily

According to Shiu and Bleecker (2001b), it is quite reasonable to find out that the LRR-RLK subfamily is associated intermittently with plant signaling, as this subfamily represents over 200 genes in the Arabidopsis family. It should also be noted that extracellular LRR motifs bind to the brassinosteroid. Chow and McCourt (2006) reported that the LRR motifs represent the specific and highly organized protein–protein interaction domain rather than a motif interacting randomly with other biomolecules. Several specific functions have been attributed to the LRR domain (see Chow and Mccourt 2006). Spaink (2002) reported that LRR motifs are capable of identifying carbohydrate microbial factors associated with nodule formation and nitrogen fixation. It has been reported by Schuster and Nelson (2000) that in animals the same group of proteins is associated with the immune system and is referred to as Toll-like receptors. Clark (2001) described a similar protein subfamily with the common intracellular serine/threonine kinase domain that was able to initiate a signaling cascade within the plant cell. Spaink (2002) suggested that the LRR receptor subfamily is specifically involved in plant–microbe interactions, and the LRR receptor subfamily represents the largest group of receptors in the plant family, since around 174 members have been reported from Arabidopsis alone.

Any mutation affecting the Arabidopsis *BRI1* (*Brassinosteroid-Insensitive 1*) gene encoding for a specific brassinosteroid cell surface receptor results in stunted plant growth and delay in development. Li et al. (2001) reported the identification of a new locus, *brs1-1D* via activation tagging of a weak *bri1* allele. This *BRS1* gene is supposedly involved in the secretion of a serine carboxypeptidase that plays an important role in the early events of *BRI1* signaling. Brasionosteroids control plant growth and development through the formation of a protein complex among LRR receptor-like protein kinases and *BRI1*. Li et al. (2002) used the activation-tagging approach to screen a dominant genetic *bri1* suppressor, *bak1-1D* (*bri1-associated receptor kinase 1-1Dominant*), encoding a distinctly different LRR receptor kinase. The authors reported development of bigger organs in Arabidopsis lines because of overexpression of *BAK,* the semi-dwarf phenotypic character, and reduced brassinosteroid sensitivity expressed in a null allele of the same gene. They detected that BAK1 is a serine/threonine protein kinase that is capable of in vivo and in vitro interactions with BRI1. This study strongly suggests BAK1 to be an integrated component of brassinosteroid signaling in higher plants like Arabidopsis.

Nam and Li (2002) also identified BAK1 using yeast two-hybrid screening methods. The authors found that in yeast, BAK1/BRI1 interaction is activated through the process of transphosphorylation, and both of them exhibit similar patterns of gene expression and subcellular localizations. Based on their mutation analyses study, the authors suggested that both BAK1 and BRI1 act synergistically to regulate brassinosteroid signaling.

Another receptor kinase-like protein encoded by the rice disease resistance gene *Xa21*, which confers resistance against the pathogenic bacteria *Xanthomonas oryzae* pv. *oryzae* race 6, was discovered by Song et al. (1995). This specific protein was detected to have both LRR motif as well as a serine/threonine kinase-like domain. The authors suggested that this protein might have a role in the identification of the cellular surface of a pathogen ligand and also in the simultaneous activation of an intracellular defense response. Sun et al. (2004) reported another LRR-receptor kinase-like protein that was constitutively expressed and encoded by a pathogen resistance gene, *Xa26*, conferring resistance to *X. oryzae* pv. *oryzae* in rice.

While screening for proteins interacting with the kinase domain of the SRK, Gu et al. (1998) isolated a plant protein, arm repeat containing1 (ARC1). The authors detected the specific phosphorylation-dependent binding of ARC1 with kinase domains of SRK-910 and A14 but not with Arabidopsis RLKs. The researchers observed that ARC1 expression is restricted to the floral stigma only, which is a target site for the floral incompatibility response. In the following year, Stone et al. (1999) reported that suppression of ARC1 mRNA levels in the *B. napus* W1 line is associated with the partial breakdown of SI resulting in seed production, which proves that ARC1 acts as a positive effector of the *B. napus* SI response. Later, He et al. (2000) devised a novel assay for studying mechanisms of receptor kinase activation and signaling in plants. The researchers designed a chimeric construct by fusing BRI1 to the serine/threonine kinase domain of *Xa21* (a disease resistance receptor of rice) that can initiate cellular defense in rice cells after treatment with brassinosteroids. The authors suggested that the BRI1 extracellular domain perceives brassiosteroids, indicating a standard mechanism of signaling for LRR receptor kinases. This system could facilitate the detection of several new LRR kinases ligands reported to be the biggest in the plant kingdom.

10.3 Possible Implications of Ligand–Receptor Interaction Studies on Future Phytosensing Research

Recently, a simple biosensor that allows the subtype-selective nuclear hormone receptor binding to be detected by growth phenotypic alterations in Gram-negative bacteria *Escherichia coli* has been reported by Skretas and Wood (2005). This has great potential for use in plants. Himelblau and Amasino (2000) reported that two Arabidopsis genes (homologs to yeast and mammalian copper transporting genes) are supposedly involved in copper sequestration. These are the *copper chaperone* (*CCH*) and the *responsive to antagonist1* (*RAN1*) genes. The authors mentioned that these two genes are involved in an intricate pathway that is also essential for developing functional ethylene receptors and could thus have an important role in biosensing under stress environments.

In another fairly recent investigation, Keinänen et al. (2007) reported isolation of genes that are up-regulated by copper from a copper-tolerant birch tree (*Betula pendula*). Around ~31% of cDNA fragments isolated were up-regulated by copper, of which ~51% exhibited an over twofold induction. The authors were able to identify 21 copper-induced genes, including plasma intrinsic protein 2 and the glutamine synthetase and multi-drug resistance-associated protein (MRP4). These genes showed greater copper up-regulation in copper-tolerant lines compared to non-tolerant lines with significantly distinct clonal differences.

Several studies utilized animal-based receptors in plants. Expression of human nuclear receptors in plants for the discovery of plant-derived ligands has been achieved by Doukhanina et al. (2007). The animal glucorticoid receptor-mediated transcriptional induction system in transgenic Arabidopsis and tobacco has been developed by Aoyama and Chua (1997). An estrogen receptor-based transactivator mediated highly inducible gene expression in Arabidopsis has been designed by Zuo et al. (2000).

The possibilities of utilizing plants for carbon and nitrogen sensing in the environment have been investigated by Coruzzi and Zhou (2001) using "matrix effects." The role of elicitors and elicitor receptors in the case of plant-microbial pathogen interaction has been reviewed in depth by Dixon and Lamb (1990). In addition, plant–microbe associations in ligand–receptor interactions for symbiosis and nodule formation have been extensively studied by Spaink (2002). In a later study, Spaink (2004) also highlighted the significance of specific recognition of bacteria by plant LysM domain receptor kinases.

Kim et al. (2001) reported the identification of a mammalian ionophoric glutamate receptor gene, *AtGluR2*, in Arabidopsis that alters calcium utilization, when it is overexpressed, but overexpression of the *AtGluR2* gene does not impact the process of calcium uptake. Transgenic plants exhibited hypersensitivity to sodium and potassium ionic stresses which was recovered by subsequent supplementation with calcium. The authors suggested that the protein expressed by the gene AtGluR2 possibly plays a significant role in calcium nutrition of plants, and is presumably involved in regulating the ion allocation among different calcium sinks during normal growth and development as well as under stress-induced conditions.

Very recently, Andersson et al. (2008), using a mouse model, reported the transient receptor potential A1 (TRPA1) to be a sensory receptor for multiple products of oxidative stress. The authors detected that TRPA1 is actually expressed by a subset of neurons, where it serves as a very useful biosensor for environmental irritants. They observed that a number of exogenous compounds (irritants) such as acrolein, allyl isothiocyanate, cinnamaldehyde etc. are involved in activating TRPA1 by the covalent modification of cysteine residues. The authors showed that TRPA1 is activated by a number of thiol-reactive molecular species. Overall, this work showed that several chemical species synthesized during oxidative stress can successfully induce activation of TRPA1 in sensory neurons of animals.

Similar studies have not yet been reported in plant species. However, it is quite likely that similar systems are also available in plants, and, if detected, could be effectively applied in phytosensing.

10.4 Concluding Remarks

Overall, it is important to note that biomonitoring or phytomonitoring is an important and essential technology for detecting and analyzing the presence and concentrations of specific environmental pollutants. The greatest advantage of this technology is that it is cheaper to use on a wider scale and at the industrial level, compared to other physical and chemical methods available. Further advantages include: flexibility, in the sense that it can be used in situ compared to other competitive methods, and simplicity of using sensors for detection, scoring and recording the influence of a pollutant. In addition, phytosensors provide the possibility of long-term monitoring.

The advantages of the phytomonitoring technology are summarized in Fig. 10.6. In future, this technology can be used as an economic, convenient and quick kit for growing seeds of specifically designed transgenic plants that are sensitive to specific pollutants at a particular site for quick and easy detection of toxic pollutant(s) and their corresponding concentration levels.

However, it is also important to remember that to monitor new toxic pollutants introduced into the environment, researchers need to continuously enrich and modify the available technologies to effectively and efficiently cater for the needs of the future. It could also possibly generate a micro-industry within or under the rapidly emerging plant biotechnology and genetic engineering mega-industry (Fig. 10.7). It has a future possibility of generating a number of jobs in the biotech

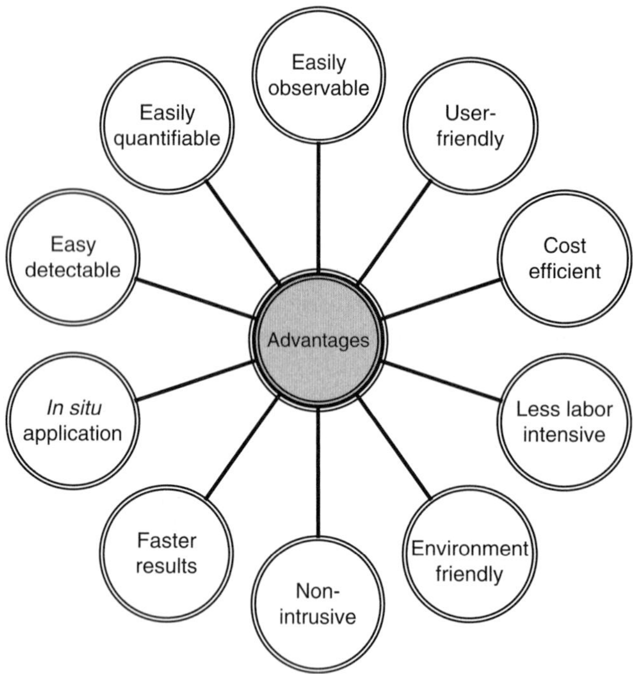

Fig. 10.6 Advantages of phytomonitoring

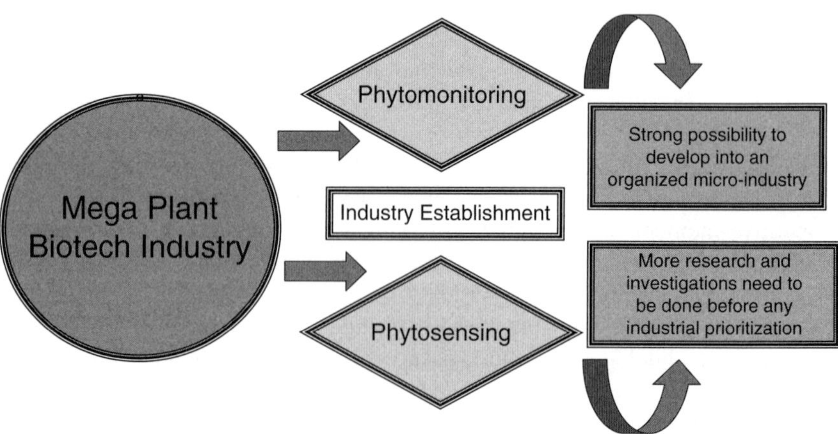

Fig. 10.7 A flowchart representing the future status of environmental biomonitoring approaches

market and developing a niche for the perennial and evergreen market of environmental monitoring and management on local, regional and global scales.

We foresee a boom for the phytomonitoring technology: this industry will be developing in underdeveloped and developing parts of the globe. Plant-based biomonitoring (PBB) has much greater potential compared to other expensive alternatives used at present in developing and underdeveloped countries because of its less sophisticated requirements and a lower need for expensive laboratory-based tests and laboratory infrastructure, and also due to lesser requirements for highly trained technical personnel for processing and recording the information. Simple kit-based seeds grown on toxic sites could easily provide the urgent and important information necessary for adopting appropriate environmental management initiatives.

An essential peak to conquer will be the development of plant-based systems responding to a multitude of toxic pollutants, systems with the ability to provide information on a quantifiable scale using simple observations and recordings. In short, phytomonitoring does carry the possibility of becoming an important tool and technology for sustainable environmental health management because of its eco-friendly nature and ability to be easily transformed into a regular household item for detection of even negligible amounts of indoor pollutants in the long run.

On the contrary, biosensing or phytosensing technology based on plant signal transduction and ligand–receptor interaction is still in infancy. Although extensive research demonstrated physico-biochemical, genetical and molecular biological aspects of ligand–receptor interactions, their possible role in biosensing or phytosensing needs to be critically investigated in much more detail. To the best of our knowledge, only a handful number of laboratories around the world are currently looking into the possibility of utilizing ligand–receptor interactions in phytosensing. There is no primary literature available, making it even more challenging and attractive for future investigators to venture into this area of research.

References

Andersson DA, Gentry C, Moss S, Bevan S (2008) Transient receptor potential A1 is a sensory receptor for multiple products of oxidative stress. J Neurosci 28(10):2485–2494

Aoyama T, Chua N-H (1997) A glucocorticoid-mediated transcriptional induction system in transgenic plants. Plant J 11(3):605–612

Asari M, Takatuski H, Yamazaki M, Azuma T, Takigami H, Sakai S (2004) Waste wood recycling as animal bedding and development of biomonitoring tool using CALUX assay. Environ Int 30(5):639–649

Barinaga M (1999) Key molecular signals in plants. Science 283:1825–1826

Basu SK, Kovalchuk I, Eudes F (in press) Biotech crops for ecology and environment. In: Kole C, Michler CH, Abbott AG, Hall TC (eds) Biotech plants: utilization and deployment, vol 2. Springer, Berlin

Berecka B, Gdzala-Kopcihch R, Barosze-Wicz J, Buszewski B (2003) The determination of nonphenol polyethoxylates in the environmental samples using coupled chromatography techniques. Chem Anal 48:413–428

Besplug J, Filkowski J, Burke P, Kovalchuk I, Kovalchuk O (2004) Atrazine induces homologous recombination but not point mutation in the transgenic plant-based biomonotoring assay. Arch Environ Cont Toxic 46(3):296–300

Bielczyk U (1993) Lichens scale. Aura 3:22

Bleecker AB, Kende H (2000) Ethylene: a gaseous signal molecule in plants. Annu Rev Cell Dev Biol 16:1–18

Bleecker AB, Esch JJ, Hall AE, Rodríguez FI, Binder BM (1998) The ethylene-receptor family from Arabidopsis: structure and function. Philos Trans R Soc Lond B Biol Sci 353 (1374):1405–1412

Boller T (2005) Peptide signalling in plant development and self/non-self perception. Curr Opin Cell Biol 17(2):116–122

Brand U, Fletcher JC, Hobe M, Meyerowitz EM, Simon R (2000) Dependent of stem cell fate I Arabidopsis on a feedback loop regulated by CLV3 activity. Science 289:617–619

Chang C, Kwok SF, Bleecker AB, Meyerowitz EM (1993) Arabidopsis ethylene-response gene *ETR1*: similarity of product to two-component regulators. Science 262:539–544

Charpenteau M, Jaworski K, Ramirez BC, Tretyn A, Ranjeva R, Ranty B (2004) A receptor-like kinase from *Arabidopsis thaliana* is a calmodulin-binding protein. Biochem J 379(3): 841–848

Chen Y-C (2001) Immobilized microalga Secnedesmus quadicauda (Chlorophyta, Chlorococcales) for long-tem storage and for application of water quality control in fish culture. Aquaculture 195:71–80

Choi SH, Gu MB (2003) Toxicity biomonitoring of degradation byproducts using freeze-dried recombinant bioluminescent bacteria. Anal Chim Acta 481(2):229–238

Chow B, McCourt P (2006) Plant hormone receptors: perception is everything. Genes Dev 20:1998–2008

Clark SE (2001) Cell signalling at the shoot meristem. Nat Rev Mol Cell Biol 2:276–284

Clark SE, Williams RE, Meyerowitz EM (1997) The CLAVATA1 gene encodes a putative receptor kinase that controls shoot and floral meristem assize in Arabidopsis. Cell 89:575–585

Cock JM, Vanoosthuyse V, Gaude T (2002) Receptor kinase signalling in plants and animals: distinct molecular systems with mechanistic similarities. Curr Opin Cell Biol 14(2):230–236

Coruzzi GM, Zhou L (2001) Carbon and nitrogen sensing and signaliong in plants: emerging 'matrix effects'. Curr Opin Plant Biol 4:247–253

Datta Banik S, Basu SK, De AK (2007) Environment concerns and perspectives. A P H Publishing Corporation, New Delhi

Devart A, Clark SE (2004) LRR-containing receptors regulating plant development and defense. Development 131:251–261

Dixon RA, Lamb CJ (1990) Molecular communication in interactions between plants and microbial pathogens. Annu Rev Plant Physiol Plant Mol Biol 41:339–367

Doukhanina EV, Apuya NR, Yoo H-D, Wu C-Y, Davidow P, Krueger S, Flavell RB, Hamilton R, Bobzin SC (2007) Expression of human nuclear receptors in plants for the discovery of plant-derived ligands. J Biomol Screen 12(3):385–395

Filkowski J, Besplug J, Burke P, Kovalchuk I, Kovalchuk O (2003) Recombination- and point mutation-monitoring transgenic plants reveal the genotoxicity of commonly used herbicides 2, 4-D and dicamba. Mutat Res 542(1–2):23–32

Franklin-Tong VE (2002) Receptor-ligand interaction demonstrated in Brassica self-incompatibility. Trends Genet 18(3):113–115

Franssen HJ (1998) Plants embrace a stepchild: the discovery of peptide growth regulators. Curr Opin Biol 1(5):384–387

Friefelder DM (1999) Molecular biology. Science Books International, USA

Gadzala-Kopciuch R, Berecka B, Bartoszewicz J, Buszewski B (2004) Some considerations about bioindicators in environmental monitoring. Pol J Env Stud 13(5):453–462

Gimeno S, Komen H, Vanderbosch PWM, Browmer T (1997) Disruption of sexual differences in genetic male common carp (Cyprinus carpio) exposed to an alkylphenol, during different life stages. Environ Sci Technol 31(10):2884–2890

Green NW, Knutzen J (2003) Organohalogens and metals in marine fish and mussels and some relationships to biological variables at reference localities in Norway. Mar Pollut Bull 46(3): 362–374

Gu T, Mazzurco M, Sulaman W, Matias DD, Goring DR (1998) Binding of an arm repeat protein to the kinase domain of the S-locus receptor kinase. Proc Natl Acad Sci USA 95: 382–387

Guieysse B, Wikström P, Forsman M, Mattiasson B (2001) Biomonitoring of continuous microbial community adaptation towards more efficient phenol-degradation in a fed-batch bioreactor. Appl Microbiol Biotechnol 56:780–787

He ZH, Fujiki M, Kohorn BD (1996) A cell-wall associated, receptor like protein kinase. J Biol Chem 271:19789–19793

He ZH, He D, Kohorn BD (1998) Requirement for the induced expression of a cell wall associated receptor kinase for survival during the pathogen response. Plant J 14:55–63

He Z, Wang Z-Y, Li J, Zhu Q, Lamb C, Ronald P, Chory J (2000) Perception of brassinosteroids by the extracellular domain of the receptor kinase BRI1. Science 288:2360–2363

Hématy K, Höfte H (2008) Novel receptor kinases involved in growth regulation. Curr Opin Plant Biol 11(3):321–328

Hématy K, Sado PE, Van Tuinen A, Rochange S, Desnos T, Balzergue S, Pelletier S, Renou JP, Höfte H (2007) A receptor-like kinase mediates the response of Arabidopsis cells to the inhibition of cellulose synthesis. Curr Biol 17(14):R541–R542

Heyl A, Schmulling T (2003) Cytokinin signal perception and transduction. Curr Opin Plant Biol 6:480–488

Himelblau E, Amasino RM (2000) Delivering copper within plant cells. Curr Opin Plant Biol 3:205–210

Hohn B, Kovalchuk I, Kovalchuk O (1999) Transgenic plants sense radioactive contamination. BioWorld 6:13–15

Holst RW, Nogel DJ (1997) Radiation effects on plants. In: Wang W, Gorusch JW, Hughes JS (eds) Plants for environmental studies. Lewis Publishers, New York, pp 38–79

Hong SW, Jon JH, Kwak JM, Nam HG (1997) Identification of a receptor-like protein kinase gene rapidly induced by abscisic acid, dehydration, high salt, and cold treatments in Arabidopsis thaliana. Plant Physiol 113:1203–1212

Hou X, Tong H, Selby J, Dewitt J, Peng X, He ZH (2005) Involvement of a cell wall-associated kinase, WAKL4, in Arabidopsis mineral responses. Plant Physiol 139(4):1704–1716

Humphrey TV, Bonetta DT, Goring DR (2007) Sentinels at the wall: cell wall receptors and sensors. New Phytol 176(1):7–21

Inoue T, Higuchi M, Hshimoto Y, Seki M, Kobayashi M, Kato T, Tabata S, Shinozaki K, Kakimoto T (2001) Identification of *CRE1* as cytokinin receptor from Arabidopsis. Nature 409:1060–1063

Jinn TL, Stone JM, Walker JC (2000) HAESA, an Arabidopsis leucine-rich repeat receptor kinase, controls floral orgam abscission. Genes Dev 14:108–117

Kachroo A, Schopfer CR, Nasrallah ME, Nasrallah JB (2001) Allele-specific receptor complex interaction in *Brassica* self-incompatibility. Science 293:1824–1826

Kachroo A, Nasrallah ME, Nasrallah JB (2002) Self-incompatibility in the Brassicaceae: receptor-ligand signalling and cell-to-cell communication. Plant Cell 8:429–445

Kakimoto T (1996) CKI1, a histidine kinase homolog implicated in cytokinin signal transduction. Science 274:982–985

Kandimalla VB, Ju HX (2005) Binding of acetylcholinesterase to multiwall carbon nanotube-cross-linked chitosan composite for flow-injection amperometric detection of an organophosphorous insecticide. Chem Eur J 12:1074–1080

Keinänen SI, Hassinen VH, Kärenlampi SO, Tervahauta AI (2007) Isolation of genes up-regulated by copper in a copper-tolerant birch (*Betula pendula*) clone. Tree Physiol 27(9):1243–1252

Keller R, Mermet JM, Otto M, Widemer HW (1998) Analytical chemistry. Willey, Wienheim

Kim SA, Kwak JM, Jae S-K, Wang M-H, Nam HG (2001) Overexpression of the AtGluR2 gene encoding an Arabidopsios homolog of mammalian glutamate receptors impairs calcium utilization and sensitivity to ionic stress in transgenic plants. Plant Cell Physiol 42(1):74–84

Koizumi N, Martinez IM, Kimata Y, Kohno K, Sano H, Chrispeels MJ (2001) Molecular characterization of two Arabidopsis Ire1 homologs, endoplasmic reticulum-located transmembrane protein kinases. Plant Physiol 127(3):949–962

Kovalchuk I, Kovalchuk O (2003) A new use for transgenic plants-environmental biomonitors. Biotechnol Genet Eng Rev 20:23–45

Kovalchuk I, Kovalchuk O (2008) Transgenic plants as sensors of environmental pollution genotoxicity. Sensors 8:1539–1558

Kovalchuk I, Kovalchuk O, Arkhipov A, Hohn B (1998a) Transgenic plants are sensitive bioindicators of nuclear pollution caused by the Chernobyl accident. Nat Biotechnol 16:1054–1059

Kovalchuk O, Kovalchuk I, Arkhipov A, Telyuk P, Hohn B, Kovalchuk L (1998b) The Allium cepa chromosome aberration test reliably measures genotoxicity of soils of inhabited areas in the Ukraine contaminated by the Chernobyl accident. Mutat Res 415(1–2):47–57

Kovalchuk I, Kovalchuk O, Hohn B (1999a) Transgenic plants as bioindicators of environmental pollution. Review. AgBiotechNet 1(October):ABN 030

Kovalchuk O, Kovalchuk I, Titov V, Arkhipov A, Hohn B (1999b) Radiation hazard caused by the Chernobyl accident in inhabited areas of Ukraine can be monitored by transgenic plants. Mutat Res 446:49–55

Kovalchuk I, Kovalchuk O, Hohn B (2000a) Genome-wide variation of the somatic mutation frequency in transgenic plants. EMBO J 19:4431–4438

Kovalchuk I, Kovalchuk O, Fritsch O, Hohn B (2000b) Des bio-indicators de pollution radioactive. Biofutur 205:48–51

Kovalchuk O, Arkhipov A, Barylyak I, Karachov I, Titov V, Hohn B, Kovalchuk I (2000c) Plants experiencing chronic internal exposure to ionizing radiation exhibit higher frequency of homologous recombination than acutely irradiated plants. Mutat Res 449:47–56

Kovalchuk O, Arkhipov A, Dubrova Yu, Hohn B, Kovalchuk I (2000d) Wheat mutation rate after Chernobyl. Nature 407:583–584

Kovalchuk I, Kovalchuk O, Hohn B (2001a) Biomonitoring of genotoxicity of environmental factors with transgenic plants. Trends Plant Sci 6:306–310

Kovalchuk O, Titov V, Hohn B, Kovalchuk I (2001b) A sensitive transgenic plant system to detect toxic inorganic compounds in the environment. Nat Biotechnol 19:568–572

Kovalchuk O, Kovalchuk I, Telyuk P, Hohn B, Titov V, Kovalchuk L (2003) Monitoring of the genotoxicity of drinking water from the inhabited areas of Ukraine affected by the Chernobyl accident using plant bioassays. Bull Env Cont Toxicol 70:847–853

Lease L, Ingham E, Walker JC (1998) Challenges in understanding RLK function. Curr Opin Plant Biol 1:388–392

Lebel EG, Masson J, Bogucki A, Paszkowski J (1993) Stress-induced intrachromosomal recombination in plant somatic cells. Proc Natl Acad Sci USA 90:422–426

Lee H-S, Yang M (2008) Applications of CYP-450 expression for biomonitoring in environmental health. Environ Health Prev Med 13(2):84–93

Li J, Lease KA, Tax FE, Walker JC (2001) BRS1, a serine carboxypeptdase, regulates BRI1 signaling in Arabidopsis thaliana. Proc Natl Acad Sci USA 98(10):5916–5921

Li J, Wen J, Lease KA, Doke JT, Tax FE, Walker JC (2002) BAK1, an Arabidopsis LRR receptor-like protein kinase, interacts BRI1 and modulates brassinosteroid signalling. Cell 110:213–222

Li L, Jean M, Belzile F (2006) The impact of sequence divergence and DNA mismatch repair on homeologous recombination in Arabidopsis. Plant J 45:908–916

Matsubayashi Y (2003) Ligand-receptor pairs in plant peptide signalling. J Cell Sci 116(19): 3863–3870

Matsubayashi Y, Yang H, Sakagami Y (2001) Peptide signals and their receptors in higher plants. Trends Plant Sci 6(12):573–577

Matsubayashi Y, Ogawa M, Morita A, Sakagami Y (2002) An LPR receptor kinase involved in perception of a peptide plant hormone, phytosulfokine. Science 296:1470–1472

McCormick PV, Garins J (1997) Algal indicators of aquatic ecosystem condition and change. In: Wang W, Gorusch JW, Hughes JS (eds) Plants for environmental studies. Lewis Publishers, New York, pp 177–207

Meindl T, Boller T, Felix G (1998) The plant wound hormone systemin binds with the N-terminal part to its receptor but needs the C-terminal part to activate it. Plant Cell 10:1561–1570

Nakajima M, Shimada A, Takashi Y, Kim Y-C, Park Y-C, Ueguchi-Tanaka M, Suzuki H, Katoh E, Iuchi S, Kobayashi M, Maeda T, Matsuoka M, Yamaguchi I (2006) Identification and characterization of Arabidopsis gibberellin receptors. Plant J 46:880–889

Nam KH, Li J (2002) BRIK/BAK1, a receptor kinase pair mediating brassinosteroid signalling. Cell 110:203–212

Nelson DL, Cox MM (2000) Lehninger principles of biochemistry, 3rd edn. Worth Publishres, New York

Powell RL (1997) The use of vascular plants as "field" biomonitors. In: Wang W, Gorusch JW, Hughes JS (eds) Plants for environmental studies. Lewis Publishers, New York, pp 335–365

Razem FA, El-Kereamy A, Abrams SR, Hill RD (2006) The RNA-binding protein FCA is an abscisic acid receptor. Nature 439:290–294

Riefler M, Novak O, Strnad M, Schmülling T (2006) Arabidopsis cytokinin receptor mutants reveal functions in shoot growth, leaf senescence, seed size, germination, root development, and cytokinin metabolism. Plant Cell 18(1):40–54

Ries G, Heller W, Puchta H, Sandermann H, Seidlitz HK, Hohn B (2000) Elevated UV-B radiation reduces genome stability in plants. Nature 406:98–101

Ryan CA (2000) The systemin signalling pathway: differential activation of plant defensive genes. Biochim Biophys Acta 1477(1–2):112–121

Ryan CA, Pearce G (1998) SYSTEMIN: a polypeptide signal for plant defensive genes. Annu Rev Cell Dev Biol 14:1–17

Ryan CA, Pearce G, Scheer J, Moura DS (2002) Polypeptide hormones. Plant Cell 14:S251–S264

Sakai S, Takigami H (2003) Integrated biomonitoring of dioxin-like compounds for waste management and environment. Ind Health 41:205–214

Sakai H, Hua J, Chen QG, Chang C, Medrano LJ, Bleecker AB, Meyerowitz EM (1998) ETR2 is an ETR1-like gene involved in ethylene signalling in Arabidopsis. Proc Natl Acad Sci USA 95:5812–15817

Schaller GE, Bleecker AB (1994) Receptor-like kinase activity in membranes of *Arabidopsis thaliana.* FEBS Lett 339(3):313–314

Scheer JM, Ryan CA (1999) A 160-kD systemin receptor on the surface of *Lycopersicon peruvianum* suspension-cultured cells. Plant Cell 11:1525–1536

Schuster JM, Nelson PS (2000) Toll receptors: an expanding role in our understanding of human disease. J Leucoc Biol 67:767–773

Schwartz OJ, Jones LW (1997) Bioaccumulation of xenobiotic organic chemicals by terrestrial plants. In: Wang W, Gorusch JW, Hughes JS (eds) Plants for environmental studies. Lewis Publishers, New York, pp 417–449

Seidman G, Hindawai IJ, Heck WW (1965) Environmental conditions affecting the use of plants as indicators of air pollution. J Air Pollut Cont Assoc 5:168–170

Sexton R, Roberts JA (1982) Cell biology of abscission. Annu Rev Plant Physiol 33:133–162

Shen Y-Y, Wang X-F, Wu F-Q, Du S-Y, Cao Z, Shang Y, Wang X-L, Peng C-C, Yu X-C, Zhu S-Y, Fan R-C, Xu Y-H, Zhang D-P (2006) The Mg-chelatase H subunit is an abscisic acid receptor. Nature 443:823–826

Shiu S-H, Bleecker AB (2001a) Receptor-like kinases from Arabidopsis from a monophyletic gene family related to animal receptor kinases. Proc Natl Acad Sci USA 98(2):10763–10768

Shiu S-H, Bleecker AB (2001b) Plant-receptor like kinase gene family: diversity, function, and signalling. Sci STKE 113:RE22

Sivaguru M, Ezaki B, He Z-H, Tong H, Osawa H, Baluska F, Volkman D, Matsumoto H (2003) Aluminium-induced gene expression and protein localization of a cell wall-associated receptor kinase in Arabidopsis. Plant Physiol 132:2256–2266

Skretas G, Wood DW (2005) Rapid detection of subtype-selective nuclear hormone receptor binding with bacterial genetic selection. Appl Envir Microbiol 71(12):8995–8997

Song W-Y, Wang G-L, Chen L-L, Kim H-S, Pi L-Y, Holsten T, Gardner J, Wang B, Zhai W-X, Zhu L-H, Fauquet C, Ronald P (1995) A receptor kinase-like protein encoded by the rice disease resistance gene, Xa21. Science 270:1804–1806

Spaink P (2002) A receptor in symbiotic dialogue. Nature 417:910–911

Spaink HP (2004) Specific recognition of bacteria by plant LysM domain receptor kinases. Trends Microbiol 12(5):201–204

Stauber JL, Franklin NM, Adams MS (2002) Application of flow cytometry to ecotoxicity testing using microalgae. Trends Biotechnol 20:141–143

Stein JC, Dixit R, Nasrallah ME, Nasrallah JB (1996) SRK, the stigma-specific S locus receptor kinase of Brassica, is targeted to the plasma membrane in transgenic tobacco. Plant Cell 8:429–445

Sun X, Cao Y, Yang Z, Xu C, Li X, Wang S, Zhang Q (2004) Xa26, a gene conferring resistance to Xanthomonas oryzae pv. oryzae in rice, encodes an LRR receptor kinase-like protein. Plant J 37:517–527

Suzuki T, Miwa K, Ishikawa K, Yamada H, Aiba H, Mizumo T (2001) The Arabidopsis sensor His-kinase, AHK4 can respond to cytokinins. Plant Cell Phys 42:107–113

Takayama S, Shimosato H, Shiba H, Funato M, Che F-S, Watanabe M, Iwano M, Isogai A (2001) Direct ligand-receptor complex interaction controls *Brassica* self-incompatibility. Nature 413:534–538

Tichtinsky G, Vannsthuyse V, Cock JM, Gaude T (2003) Making inroads into plant receptor kinase signalling pathways. Trends Plant Sci 8(5):231–237

Tieman DM, Taylor MG, Ciardi JA, Klee HJ (2000) The tomato ethylene receptors NR and LeETR4 are negative regulators of ethylene response and exhibit functional compensation within multigene family. Proc Natl Acad Sci USA 97(10):5663–5668

Torii KU (2000) Receptor kinase activation and signal transduction in plants: an emerging picture. Curr Opin Plant Biol 3:361–367

Trotochaud AE, Jeong S, Clark SJ (2000) CLAVATA3, a multimeric ligand for the CLAVATA1 receptor-kinase. Science 289:613–617

Ueguchi C, Sato S, Kato T, Tabata S (2001) The AHK4 gene involved in the cytokinin-signaling pathway as a direct receptor molecule in Arabidopsis thaliana. Plant Cell Physiol 42(7):751–755

Ueguchi-Tanaka M, Ashikari M, Nakajima M, Itoh H, Katoh E, Kobayashi M, Chow T, Hsing Y, Kitano H, Yamaguchi I (2005) GIBBERELLIN INSENSITIVE DWARF1 encodes a soluble receptor for gibberellin. Nature 437:693–698

Van der Auwera G, Baute J, Bauwens M, Peck I, Piette D, Pycke M, Asselman P, Depicker A (2008) Development and application of novel constructs to score C:G-t0T:A transitions and homologous recombination in Arabidopsis. Plant J 45:908–916

Verica JA, Chae L, Tong H, Ingmire P, He ZH (2003) Tissue-specific and developmentally regulated expression of a cluster of tandemly arrayed cell wall-associated kinase-like kinase genes in Arabidopsis. Plant Physiol 133:1732–1746

Vetter W, Luckas B, Heidemann G, Skirnisson K (1996) Organochlorine residues in marine mammals from Northern Hemisphere-a consideration of the composition of organochlorine residues in the blubber of marine mammals. Sci Total Environ 186(1–2):29–40

Vinagre F, Vargas C, Schwarcz K, Cavalcante J, Noguiera EM, Baldani JI, Ferreira PCG, Hemerly AS (2006) SHR5: a novel plant receptor kinase involved in plant-N_2-fixing endophytic bacteria association. J Exp Bot 57(3):559–569

Voets T, Owsianik G, Janssens A, Talavera K, Nilius B (2007) TRPM8 voltage sensor mutants reveal a mechanism for integrating thermal and chemical stimuli. Nat Chem Biol 3(3):174–182

Wagner T, Kohorn B (2001) Wall-associated kinases are expressed throughout plant development and are required for cell expansion. Plant Cell 13:303–318

Wikström P, Andersson A-C, Forsman M (1999) Biomonitoring complex microbial communities using random amplified polymorphic DNA and principal component analysis. FEMS Microbiol Ecol 28(2):131–139

Wikström P, Johansson T, Lundstedt S, Hägglund L, Forsman M (2001) Phenotypic biomonitoring using multivariate flow cytometric analysis of multi-stained microorganisms. FEMS Microbiol Ecol 34(3):187–196

Wilkinson JQ, Lanahan MB, Clark DG, Bleecker AB, Chang C, Meyerowitz EM, Klee HJ (1997) A dominant mutant receptor from Arabidopsis confers ethylene insensitivity in heterologous plants. Nat Biotechnol 15:444–447

Yamada H, Suzuki T, Terada K, Takei K, Ishikawa K, Miwa K, Yamashino T, Mizuno T (2001) The Arabidopsis AHK4 histidine kinase is a cytokinin-binding receptor that transduces cytokinin signals across the membrane. Plant Cell Physiol 42(9):1017–1023

Yin Y, Wu D, Chory J (2002) Plant receptor kinases: systemin receptor identified. Proc Natl Acad Sci USA 99(14):9090–9092

Yoon JM, Oh B-T, Just CL, Schnoor JL (2002) Uptake and leaching of octahydro-1, 3, 5, 7-tetranitro-1, 3, 5, 7-tetrazocine by hybrid poplar trees. Environ Sci Technol 36:4649–4655

Zhang DP, Wu ZY, Li XY, Zhao ZX (2002) Purification and identification of a 42-kilodalton abscisic acid-specific binding protein from the epidermis of broad bean leaves. Plant Physiol 128:714–725

Zuo J, Niu Q-W, Chua N-H (2000) An estrogen based transactivator XVE mediates highly inducible gene expression in transgenic plants. Plant J 24(2):265–273

Chapter 11
Bacteriophage-Based Biosensors

Mohammed Zourob and Steven Ripp

Abstract Bacteriophages, or phages for short, are viruses that infect bacteria and are considered the main regulators of microbial balance on Earth. These viruses are extremely specific, and their long-term survivability and ability to reproduce quickly in suitable hosts play a major role in the preservation of a dynamic equilibrium amid the diverse variety of bacterial species in the Earth's ecosystem.

The many advantages of bacteriophages make them valuable tools for the detection and identification of bacterial pathogens as well as potentially promising recognition elements in biosensors as they have distinct advantages over other biorecognition receptors. Among these advantages are the specificity of the interaction of this type of virus with its target host cell, its ability to lyse and kill its host, and its capacity to multiply during the infection process. In addition, they are ubiquitous in nature, harmless to humans, economically and easily produced, have a far longer shelf life as they withstand harsh environments, reducing the environmental limitations and enabling regeneration of the biosensor surface. They can also be immobilized onto transducing devices in much the same manner as antibodies or DNA probes. For these reasons, the present chapter will provide an overview of the techniques, methods, and assays that employ bacteriophages in biosensing applications.

Keywords Bacteriophage · Biosensor · Pathogen · Phage · Reporter

M. Zourob (✉)
GROUPE GDG ENVIRONNEMENT LTÉE, 430, rue Saint-Laurent, Trois-Rivières, QC, Canada G8T 6H3,
e-mail: mohammed.zourob@gdg.ca

M. Zourob (ed.), *Recognition Receptors in Biosensors*,
DOI 10.1007/978-1-4419-0919-0_11, © Springer Science+Business Media, LLC 2010

Abbreviations

DNA	Deoxyribonucleic acid
RNA	Ribonucleic acid
PCR	Polymerase chain reaction
mRNA	Messenger RNA
RT-PCR	Real-time polymerase chain reaction
ELISA	Enzyme-linked immunosorbant assay
ATP	Adenosine triphosphate
AMP	Adenosine monophosphate
PP1	Protein phosphatase 1
AK	Adenylate kinase
ADP	Adenosine diphosphate
b-PAPG	p-aminophenyl-b-d-galactopyranoside
HPLC	High-performance liquid chromatography
SACs	Signal-amplifying cells
MRSA	Methicillin-resistant *Staphylococcus aureus*
cELISA	Competitive enzyme-linked immu nosorbent assay
LPS	Lipopolysaccharide
TMB	3,3′,5,5′-tetramethylbenzidine peroxidase
MALDI-MS	Matrix-assisted laser desorption/ionization mass spectrometry
RLU	Relative light output
PolyDADMAC	Polydiallyldimethylammonium chloride
Ni (II)	Nickel (II)
NTA	Nitrilotriacetic acid
SPR	Surface plasmon resonance
SPE	Screen printed electrode
EDC	1-ethyl-3-[3-dimethylaminopropyl] carbodiimide hydrochloride
TOF-SIMS	Time-of-flight secondary ion mass spectroscopy
SEPTIC	SEnsing of phage-triggered ion cascade
IS	Impedance spectroscopy
SEM	Scanning electron microscopy
LB	Langmuir–Blodgett
OFRR	Opto-fluidic ring resonator
pM	Pico Molar
Sulfo-NHS	N-hydroxysulfosuccinimide
BSA	Bovine serum albumin
DAPI	4′,6-diamidino-2-phenylindole
Gfp	Green fluorescent protein
PMTs	Photomultiplier tubes (PMTs)
CCD	Charge-coupled device
OHHL	N-3-(oxohexanoyl)-l-homoserine lactone
MGIT	Mycobacterial growth indicator tube
VBNC	Viable but nonculturable

11.1 Introduction

Bacteriophages, or phages for short, are viruses that infect bacteria and are considered the main regulators of microbial balance on Earth as they total an estimated 10^{32} entities (Fig. 11.1) (Kutter and Sulakvelidze 2005). Bacteriophages are found in abundance in the soil, manure and feces, thermal vents, and water. These viruses are extremely specific, and their long-term survivability and ability to reproduce quickly in suitable hosts play a major role in the preservation of a dynamic equilibrium amid the diverse variety of bacterial species in the Earth's ecosystem (Kutter and Sulakvelidze 2005).

The many advantages of bacteriophages make them valuable tools for the detection and identification of bacterial pathogens as well as potentially excellent vehicles for diagnostics and therapeutics of bacterial disease. Among these advantages are the specificity of the interaction of this type of virus with its target host cell, its ability to lyse and kill its host, and its capacity to multiply during the infection process.

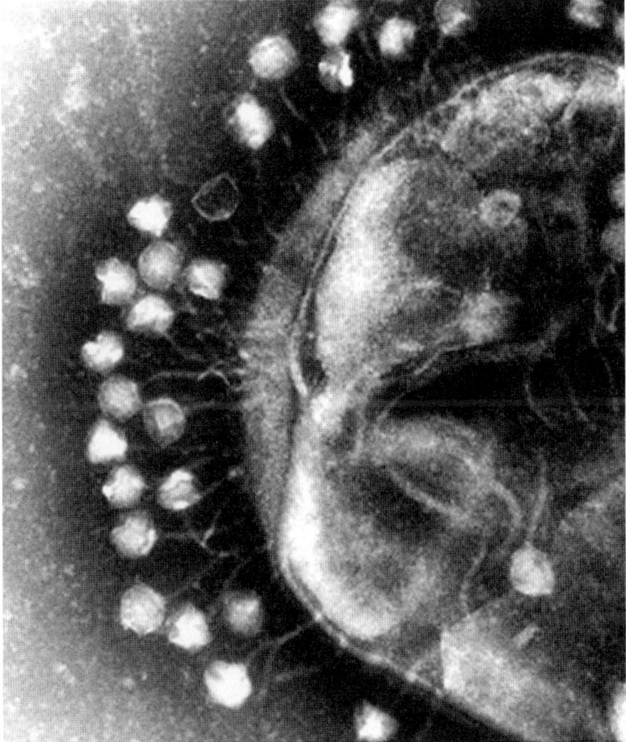

Fig. 11.1 Multiple bacteriophage attached to the outer surface of a bacterial cell (from http://www.Wikipedia.org)

Modern phage research is attributed to Frederick W. Twort who began serious bacteriological work in 1915. Twort was trying to grow Caccinia virus on agar media when he noticed the growth of many mucoid and glassy colonies of micrococci, which he identified as being bacterial contamination. However, when examined under the microscope, the colonies seemed to have disintegrated into small granules that could be dyed red with Giemsa stain (Twort 1915).

In 1917, Félix d'Herelle discovered microbes able to lyse bacteria and provoke their death, and thus called them "bacteriophages" (from Greek "to eat" bacteria) (D'Herelle 1917). D'Herelle later noted that phages were natural resistant agents to bacteria as phage titers increased in patients with infectious diseases when they started to convalesce (D'Herelle 1919a). In 1919 he went on to test phages as prophylaxis against the infection of chickens by *Bacillus gallinarum*. The results suggested that chickens treated with phages died less and recovered faster, and that subsequent infections by *B. gallinarum* were prevented (D'Herelle 1919b).

In 1954 Cherry et al. used a broad host range phage to detect the presence of *Salmonella enterica* and noted that the replication of phages in a plated bacterial culture resulted in the development of clear zones of cell lysis referred to as plaques (Fig. 11.2) (Cherry et al. 1954).

Phages are parasites that use the bacterial machinery to direct their replication in the target host. Each phage particle, called a virion, encloses its genome in a protein

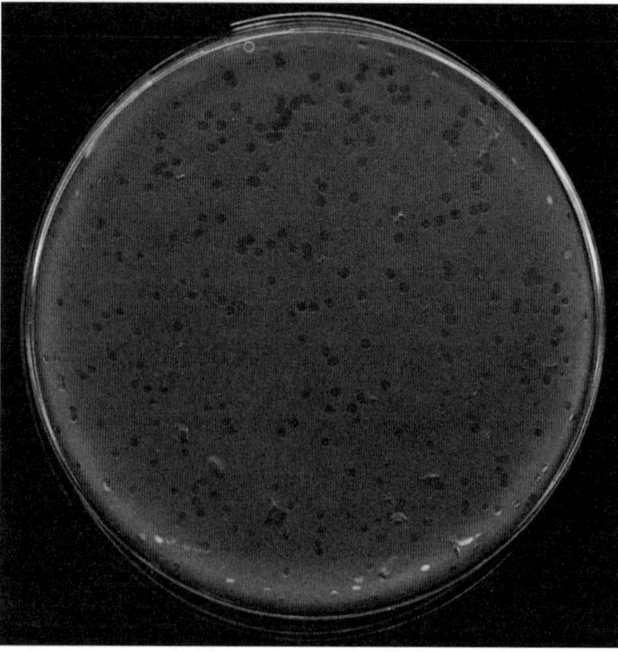

Fig. 11.2 Bacteriophage plaques formed on a lawn of bacterial host cells (from http://www.Wikipedia.org)

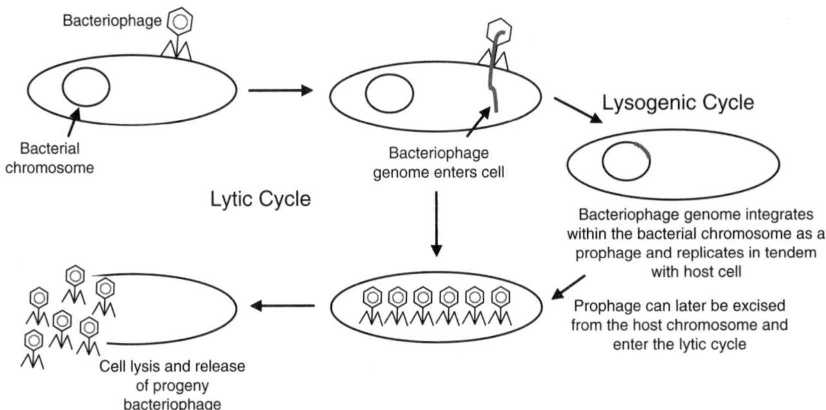

Fig. 11.3 The lytic and lysogenic cycles of bacteriophage replication

or lipoprotein coat called a capsid. The capsid and the nucleic acid together form the nucleocapsid (Kutter and Sulakvelidze 2005). Phages are grouped in 12 families and multiple genera based on their morphology, nucleic acid homology and type, and serology (Mandeville et al. 2003). Moreover, phages can be further divided into two categories depending on their life cycles and means of propagation (Fig. 11.3) (Mandeville et al. 2003). The first group comprises the virulent (or lytic, or productive) phages that are only capable of lytic propagation consisting of infection of the host bacterium by the phage, replication of the phage genome, production of the phage structural components, and release of the newly assembled phages. More often than not, this process results in bacterial lysis and death. The second category comprises the temperate (or reductive) phages that are capable of lytic as well as lysogenic propagation where the genome will assume a quiescent state called a prophage, rendering the coexistence of the bacteriophage and host cell possible. This is achievable by incorporating the phage genome into the chromosome of the bacterium by a site-specific integration procedure (Mandeville et al. 2003). Advantages of temperate phages consist in shielding their hosts from infection by other phages, and inducing considerable changes in the properties of their host-bacteria, such as restriction systems and resistance to antibiotics (Kutter and Sulakvelidze 2005).

Tailed phages of the Caudovirales represent over 95% of the bacteriophages described in the literature. Half of the mass of the virion is made of double-stranded DNA while the other half is made of protein. The head is icosahedral and is constructed from multiple copies of one or two specific proteins (Kutter and Sulakvelidze 2005).

Tail morphology defines the three main families of tailed phages. The *Siphoviridae* family comprises approximately 60% of phages, the latter possessing long flexible tails. Twenty-five percent are *Myoviridae* with contractile tails, and the remaining 15% are *Podoviridea* with tiny, thick tails (Kutter and Sulakvelidze 2005). *Archaephages*, as their name suggests, infect *Archae* bacteria and have polymorphic shapes.

Tailless phages are divided into ten families differentiated by phage shape, the presence or absence of a lipid coat, the possession of single- or double-stranded DNA or RNA, whether or not they are released from their host cells by lysis, and whether or not they are segmented (Kutter and Sulakvelidze 2005).

Bacteriophages are promising recognition elements in biosensors as they have distinct advantages over antibodies as recognition receptors (Shabani et al. 2008, 2009; Balasubramanian et al. 2007). First, they are ubiquitous in nature and highly specific to bacteria while remaining harmless to humans. Second, phage production can be less complicated and less expensive than antibody production. Third, phages have a far longer shelf life as they withstand harsh environments, reducing the environmental limitations and enabling regeneration of the biosensor surface. Fourth, they can also be immobilized onto transducing devices in much the same manner as antibodies or DNA probes. For these reasons, the present chapter will review the developments of bacteriophages as recognition receptors in biosensors.

11.2 Detection by Phage Amplification

Phages have stirred much attention in recent years as a means to detect bacteria in food and water and for disease surveillance. The reason remains that requirements for pathogenic bacterial detection are rapidity, sensitivity, specificity, and cost. Historically, conventional culture methods were and are still used for bacterial detection. These methods are, however, time consuming and require enrichment steps in order to visualize colonies on agar plates. Over the years, mass spectrometry and biochemical detection systems have been developed to ameliorate and increase the speed and sensitivity of culture methods but these methods retain negative aspects such as high cost and difficulty in handling.

Polymerase chain reaction (PCR) and immunoassays are other approaches designed to ameliorate the sensitivity of detection. However, PCR techniques detect the presence of nucleic acid and thus cannot differentiate living cells from dead cells. It is, however, possible to couple PCR and reverse transcription to detect mRNA, but high cost render RT-PCR difficult for regular use. Immunoassays such as ELISA are simple and rapid but they can have low sensitivity for detection of pathogens.

As mentioned above, bacteriophages bind specifically to their host and are optimized in a fashion that makes them ideal tools for the detection of bacterial pathogens. They are able to distinguish between live and dead cells, are robust, and easy to produce. Many detection techniques have been developed to exploit the specific binding ability of bacteriophages. These include the detection of intracellular components released from the bacterium after its lysis, the production of fluorescently labeled phages or fluorescently labeled antibodies directed against specifically adsorbing phage, inhibition of metabolism and growth, cell-wall recognition, and use of reporter phages. All of these methods will be discussed in this chapter.

11.3 Detection of Released Intracellular Components During Phage Lysis

The lysis of bacterial cells by bacteriophages culminates in the release of intracellular materials of the host bacterium and phage progeny. Two proteins are involved in the destruction of the bacterial cell wall. The first protein is a holin that perforates the cytoplasmic membrane by forming small pores in it. The second protein is an endolysin that will degrade the peptidoglycan provoking the release of a variety of detectable intracellular markers that can be easily measured (Schmelcher and Loessner 2008). The specificity of this detection method relies on the use of host-specific lytic phage.

11.3.1 Measurement of Adenosine Triphosphate Release

The most popular phage lysis method measures bacterial intracellular adenosine triphosphate (ATP). The amount of ATP in an average, actively metabolizing bacterial cell is quite consistent (approximately 10^{-15} g/cell) (Rees and Dodd 2006). The amount of ATP metabolized by a bacterial cell decreases rapidly after cell death. It is possible to determine the total number of viable cells by detecting the released intracellular ATP by the firefly luciferase/luciferin enzyme system after cell disrubtion using detergent-based agents (Stanley 1989). The amount of light produced can be determined by small luminometers and the quantity of ATP in the assay is directly proportional to the amount of light produced by the luciferase/luciferin reaction:

$$\mathrm{ATP} + \mathrm{luciferin} + \mathrm{O_2} \xrightarrow{\text{Firefly luciferase} + \mathrm{Mg^{2+}}} \mathrm{AMP} + \mathrm{PP_1} + \mathrm{oxyluciferin} + \mathrm{CO_2} + h\nu\ (\text{light})$$

However, detection of specific pathogens is not possible using this method.

Specificity can be achieved by specific cell lysis, which can be accomplished by intact phage particles or purified recombinant phage endolysins (Schmelcher and Loessner 2008). Blasco et al. (1998), for example, developed the phage adenylate kinase (AK) assay. AK is an enzyme that directs the conversion of ADP to ATP and AMP:

$$\mathrm{ADP} + \mathrm{ADP} \xrightarrow{\text{Firefly luciferase} + \mathrm{Mg^{2+}}} \mathrm{ATP} + \mathrm{AMP}$$

When a cell bursts open after a phage infection event, AK is released as part of the cellular milieu. The *Escherichia coli*-specific phage NCIMB 10359 or the *Salmonella*-specific phage Newport was added to bacterial cultures, and, if suitable host cells were present, infection, lysis, and release of AK occurred. ADP was then added to drive the AK-mediated reaction toward the generation of ATP, and resulting ATP pools were detected using a commercially available firefly luciferase assay. Detection limits approached 10^4 cfu/mL of *E. coli* or *Salmonella* within an assay time of less than 2 h. These less than optimum detection limits were attributed to the high unspecific ATP background in such samples. Wu et al. (2001) later optimized assay incubation times and phage concentrations to increase detection limits to 10^3 cfu/mL. Squirrel et al. (2002) improved the sensitivity of the assay further by capturing and concentrating the pathogens of interest using immunomagnetic beads coated with antibodies before performing the phage-mediated bioluminescence AK assay.

Stewart et al. (1996) employed Ply 118 purified recombinant phage endolysins for specific lysing of *Listeria monocytogenes* and coupled cellular lysis to a bioluminescence-based ATP assay. Schuch et al. (2002) similarly detected *Bacillus anthracis* using the PlyG recombinant endolysins. They reported detection limits as low as 100 spores in 60 min (after incubation with germination solution) and 2.5×10^3 spores after 10 min.

11.3.2 Measurement of Enzymes and Other Cytoplasmic Markers

Other intracellular constituents can be used to monitor phage–cell lysis such as β-galactosidase and the endolysins that are produced as late gene products during phage replication to release progeny phage from the bacterial cells. Neufeld et al. (2003) used electrochemistry to measure amperometric changes in solution due to phage-mediated cell lysis. Infection of *E. coli* by a lytic version of phage lambda ultimately leads to the release of cellular components, such as the enzyme β-d-galactosidase. β-d-galactosidase can be measured amperometrically with a potentiostat via the addition of the substrate *p*-aminophenyl-β-d-galactopyranoside (β-PAPG) to yield the product *p*-aminophenol, which is oxidized at the carbon anode. *E. coli* could be detected within 6–8 h at a detection limit of 1 cfu/100 mL. Yemini et al. (2007) used the same principle to detect *Bacillus cereus*, where lysis by phage B1-7064 instigated cellular release of the enzyme α-glucosidase, as well as *Mycobacterium smegmatis* using phage D29 and the cellular release of β-glucosidase. Theoretically and advantageously, any phage/host combination can be detected using this method as long as the phage is lytic and an appropriate electrochemically detectable enzymatic marker is released by the target cell. However, a single phage would likely be insufficient to infect across the spectrum of target bacterial cells desired and false negative signals arising from cross-infections or naturally lysing cells would have to be accounted for. Neufeld et al. (2005) addressed some of these concerns in their assays using a phage-encoded alkaline phosphatase enzyme that had to be delivered to the host cell in order to be expressed. Thus, only after an active infection event would the enzyme be synthesized and then later released by the cell during lysis. This assay could detect a single *E. coli* cfu/mL in less than 3 h in both pure and mixed cultures.

11.3.3 Measurement of Phage Progeny

Since phage-infected cells ultimately synthesize and then release often times large numbers of progeny phage, a phage infection event can be assayed for simply by monitoring for a release and accumulation of progeny phage. The most basic format of these assays involves combining wild-type lytic phage with a bacterial culture. If bacterial hosts infectable by the phage are present, then progeny phage are produced and the resulting amplification in phage numbers serves as the signal for that infection occurrence. The technique was first demonstrated by Hirsh and Martin (1983) who used phage Felix-01 to detect *Salmonella*. Felix-01 was added to a sample, and if *Salmonella* were present, infection occurred and progeny phage was produced. The resulting amplification in phage numbers was detected by high-performance liquid chromatography (HPLC). Detection limits were fairly poor at 10^6 *Salmonella* per gram or mL sample. Stewart et al. (1998) demonstrated a more sensitive phage amplification assay again using phage Felix O-1 for detecting *Salmonella* as well as phages NCIMB 10116 and 10884 for detecting *Pseudomonas aeruginosa*. Phages were combined with bacterial cultures and incubated for up to 25 min to allow for infection to occur. Residual phages, i.e., those that had not participated in an active infection, were then destroyed via the addition of a virucide, in this case, pomegranate rind. After 3 min, virucidal activity was neutralized and "helper cells" were added to provide fresh hosts for the new progeny phage to infect. A standard top agar plaque plate was then prepared, where each plaque ideally represented a single phage/host infection event. By enumerating plaques, an estimate of the number of phages present could be calculated. Within 4 h, this assay could detect as few as 600 *S. typhimurium* or 40 *P. aeruginosa* cells/mL.

Favrin et al. (2001) used phage SJ2 for the detection of *Salmonella* Enteritidis. *Salmonella* cells were initially concentrated using immunomagnetic separation and then incubated with phage SJ2 for 10 min. The paramagnetic beads allowed the *Salmonella* cells to be easily washed and recovered while simultaneously removing free phage from the sample. Upon continued incubation, *Salmonella* cells infected with phage now lysed and released their phage progenies. Healthy *Salmonella* "signal-amplifying cells" (SACs) were then added to serve as new hosts for the progeny phage, and their subsequent infection could be monitored simply by measuring optical density, where a decrease in optical density indicated that signal-amplifying cell concentrations were declining due to infection and lysing by phage while an increase in optical density indicated an unaffected and growing population of signal-amplifying cells (Fig. 11.4). The assay was tested in artificially contaminated skim milk powder, chicken rinses, and ground beef, with an average detection limit of 3 cfu/g or mL within a total assay time of 20 h, inclusive of requisite preenrichment incubations. It has also been applied toward the detection of *E. coli* O157:H7 using phage LG1 and anti-*E. coli* paramagnetic beads, with a detection limit of 2 cfu/g ground beef within an assay time of 23 h (Favrin et al. 2003).

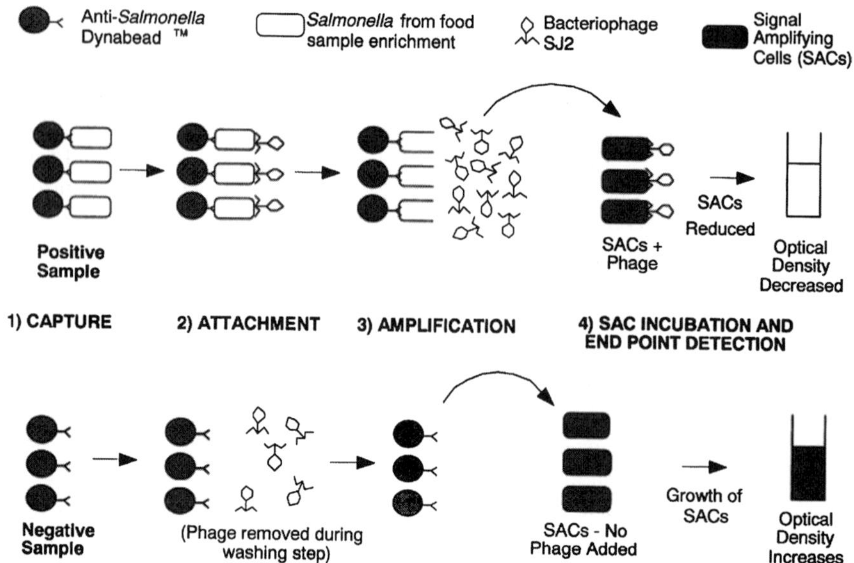

Fig. 11.4 A phage amplification assay incorporating immunomagnetic separation. *Salmonella* cells present in the sample are captured with polyclonal antibody-coated paramagnetic beads. The addition of *Salmonella*-specific phage results in infection, cell lysis, and the amplified release of progeny phage. The addition of healthy signal-amplifying cells (SACs) provides new hosts for these phage, and subsequent infection (or noninfection if *Salmonella* was not initially present) is monitored by optical density. (With permission from Favrin et al. (2003), copyright 2008, American Society for Microbiology)

Jassim and Griffiths (2007) developed a phage amplification assay that used fluorochromic stains to monitor target host cell viability associated with phage infection and lysis. SYTO9 dye stains viable cells fluorescent green while propidium iodide stains dead cells fluorescent red. The propidium iodide penetrates only those cells with damaged cell membranes, as occurs when a host bacterium is infected with a phage. Therefore, after performing the phage amplification assay, each dye is added to the sample and the ratio of green-to-red fluorescent cells is used to indicate the occurrence of phage infection and, therefore, the presence of target bacterial cells. Using phage NCIMB 10116, target *P. aeruginosa* cells could be detected down to 10 cfu/mL in pure culture within 4 h with no preenrichment.

Ulitzur and Ulitzur (2006) used novel phage mutant repair mechanisms to identify endpoint plaque formation due to infected target bacteria. Phage carrying amber mutations (phage Felix-O1 for *Salmonella*), ultraviolet light-irradiated mutations (phage OE for *E. coli*), or temperature-sensitive mutations (phage AR1 for *E. coli* O157:H7) were constructed. These phages could not form plaques on their host cells unless their mutations were repaired by recombination or complementation, thereby negating the need to wash and/or centrifuge the assay samples to remove free phage. For example, two temperature-sensitive phage mutants were allowed to coinfect their *E. coli* O157:H7 targets at the permissive temperature (37°C), then incubated at their restrictive temperature (42°C) to prevent further infection cycles. Subsequent plaque formation was, therefore, only possible if the mutation had been repaired since any remaining mutant phage, due to their temperature sensitivity, could not plaque at 42°C, and the number of plaques thus reflected the number of *E. coli* host cells in the sample. Detection was achieved down to 1 cell/mL in a 3.5 h assay. Similar coinfection strategies with the other phage mutants yielded detection limits of 10 or less target cells/mL in 3–5 h assay formats.

One of the most applied phage amplification assays is for the clinical detection of *Mycobacterium tuberculosis* using the commercially available FASTPlaqueTB™ kit (Mole and Maskell 2001). The phage, referred to as Actiphage™, is added to the sample for 1 h followed by the addition of a virucide (Virusol™). Since these tests are highly standardized for this specific phage/host combination, the virucidal approach works exceedingly well. After a 5-min incubation, the virucide is neutralized and a rapidly growing mycobacterial cell suspension is added (Sensor™ cells) to promote additional phage infection and amplification. Resulting phage are enumerated on plaque plates. Two large-scale studies verified the detection of 65–83% of confirmed *M. tuberculosis* infections in sputum samples within 2 days using this assay (Pai et al. 2005). Although direct microscopic methods can achieve results in 2 h with corresponding sensitivity, the simplicity of phage amplification, especially when packaged in kit form, and the ability to perform these assays with no additional costly equipment and minimal user training, offers significant benefits, particularly in economically disadvantaged countries and low resource environments. A FASTPlaque-Response™ kit is also available for establishing rifampicin resistant *M. tuberculosis*. The sample is preincubated in the presence or absence of rifampicin antibiotic and then subjected to the phage amplification assay. If the cells are resistant to rifampicin, the number of plaques enumerated will be similar in both samples. If the cells are sensitive to rifampicin, the number of plaques in the rifampicin-treated sample will be less than the rifampicin free sample. Susceptibility testing for various other antituberculosis drugs (isoniazid, ethambutol, streptomycin, pyrazinamide, ciprofloxacin) can additionally be achieved (Mole and Maskell 2001).

Microphage Inc., based in Colorado, USA, (http://www.microphage.com) developed a phage amplification assay for methicillin-resistant *Staphylococcus aureus* (MRSA). The clinical sample is incubated with MRSA-specific phages. The subsequent increase in yield of phages due to the presence of MRSA in the sample is detected via immunological identification of a phage-specific protein. Blood samples can be processed in as little as 5 h. Their detection assays can also be used for antibiotic susceptibility testing.

Several research groups have developed unique technologies to monitor the amplification of progeny phage after infection events. Guan et al. (2006), for example, combined phage amplification with a competitive enzyme-linked immunosorbent assay (cELISA) to detect *Salmonella* Typhimurium using phage BP1 and a biotinylated version of BP1. *Salmonella* cultures were incubated with wild-type BP1 phage and resulting phage supernatants added to ELISA microtiter plates coated with *Salmonella* Typhimurium smooth lipopolysaccharide (LPS) to which the phage attached. The biotinylated version of the phage was additionally added, which could be detected by the colorimetric substrate 3,3′,5,5′-tetramethylbenzidine peroxidase (TMB). If excess wild-type BP1 phage were present, due to the availability of suitable *Salmonella* host cells, then few biotinylated phage would attach and a weak colorimetric signal would be detected. If no target *Salmonella* were present, then BP1 replication would not occur and excess biotinylated phage would bind to the smooth LPS to yield an intense yellow color. In another technique, Madonna et al. (2003) used matrix-assisted laser desorption/ionization mass spectrometry (MALDI-MS) to identify the molecular weight signature of the phage capsid protein. *E. coli* in pure culture was concentrated by immunomagnetic separation and then infected with phage MS2. Analysis of 1 μL of this sample by MALDI-MS was sufficient to detect the MS2 capsid protein, providing a detection limit of approximately 10^4 *E. coli* cells/mL in an assay time of 2 h.

11.4 Direct Detection Through Cell Wall Recognition

11.4.1 Phage Immobilization

Bacteriophage specificity can be used to capture bacterial targets, and by adhering bacteriophages directly to a transducer interface, fully standalone biosensors can be created. Bacteriophages have been immobilized onto transducer interfaces using various immobilization techniques including physical adsorption (Bennett et al. 1997; Sun et al. 2001; Singh et al. 2009; Gervais et al. 2007; Petrenko 2008; Balasubramanian et al. 2006; Nanduri et al. 2007a, b; Lakshmanan et al. 2007a, b; Petrenko 2008; Johnson et al. 2008; Wan et al. 2007; Li et al. 2009), Langmuir-Blodgett deposition (Olsen et al. 2006; Guntupalli et al. 2008), and covalent immobilization (Shabani et al. 2009; Zhu et al. 2008; Handa et al. 2008). Bennett et al. (1997) reported a method to separate and concentrate *Salmonella* from food materials by using a biosorbent coated with phages physically attached to a solid phase. In their approach, they soaked two different polystyrene surfaces, microplates and dipsticks, into a 5×10^{10} pfu/mL Sapphire-specific lytic phage solution, followed by washing in order to remove unbound phage, and blocking the remaining adsorption sites. These biosorbents were then incubated with *Salmonella* and mixed bacterial cultures, and after washing, the specific recovery of *Salmonella* cells was assessed either by PCR or by epifluorescence microscopy using acridine orange as a dye for labeling the bacterial nucleic acids. In both cases, *Salmonella* could be separated from mixed bacterial cultures, and 9 of 11 *Salmonella* strains gave positive signals. One of the strains not detected by the test, *Salmonella enterica* serovar Arizonae CRA 1568, is known to synthesize an incomplete lipopolysaccharide molecule that results in an inability of the phage to adsorb to this strain. However, the detection limit for this method was rather high. With a PCR detection step, a concentration of 10^5 cfu/mL was required to generate a positive signal, and a concentration of 10^7 cfu/mL was necessary in the initial culture to ensure that this amount of cells was captured by the biosorbent. This represents a capture efficiency of 1%, which is clearly not sufficient.

Sun et al. (2001) described specific immobilization of phages, exploiting the high affinity of biotin to streptavidin, as opposed to physical adsorption. In this work, phage SJ2 for *Salmonella* was biotinylated using sulfosuccinimido-biotin, which reacts with primary amino groups of the phage coat proteins. Afterward, the biotinylated phages were incubated with streptavidin-coated magnetic beads. This phage-based biosorbent was applied to capture target cells of *S. enteritidis* by magnetic separation, using magnetic beads coated with nonbiotinylated phage as a negative control. The capture efficiency was determined by using a recombinant bioluminescent *Salmonella* strain as a target organism and measuring relative light output (RLU) after capture. With this technique, approximately 20% of the target cells could be recovered when a culture of 2×10^6 cfu/mL was used, which was a significant improvement when compared to the physical adsorption method.

Gervais et al. (2007) demonstrated streptavidin-mediated attachment of bacteriophage genetically modified to display biotin on their capsid. The biotinylated phages were immobilized on streptavidin-coated gold surface electrodes fabricated in microtitre chips. This approach yielded a phage attachment density of 4.4 phages/μm^2, representing a 15-fold increase over simple physisorption (Fig. 11.5). Impedance was used to monitor cell growth. The immobilized phage lyse their target cells and inhibit or slow cell growth in comparison to unmodified wells. The main limitation of this method is that the phage requires genetic modification which can be a time-consuming process.

Singh et al. (2009) reported the immobilization of wild-type T4 phage on gold surfaces through hydrogen bonding using different approaches. The gold surfaces were modified by hydrophilic interaction with sugars (dextrose and sucrose) and

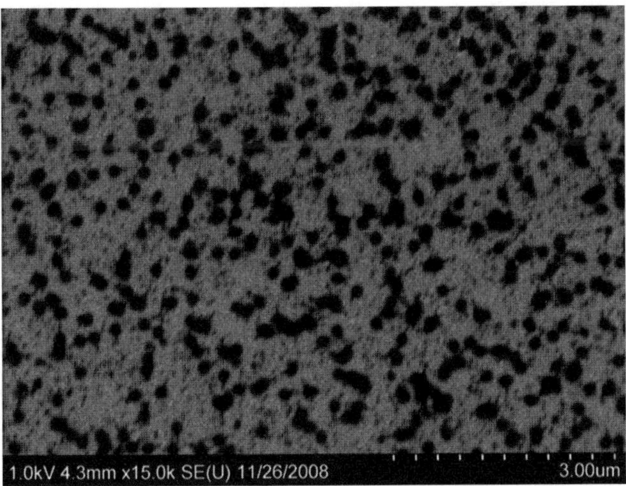

Fig. 11.5 T4 phage immobilized on a gold-coated surface. (Provided courtesy of Dr. S. Evoy-Alberta University)

through thiol linkage with the amino acids histidine and cysteine. In addition, gold surfaces modified with amino acids where further activated with glutaraldehyde. The attachment density of phages with glutaraldehyde activation was 18 $\pm$ 0.15 phages/μm^2. This represented a 4-fold improvement over the values obtained with biotinylated phages, and a 37-fold improvement compared with simple physisorbtion. Subsequently, the *E. coli* host bacterial capture density improved by 9-fold compared with physically adsorbed phages.

Jabrane et al. (2008, 2009) studied the effect of charged coated papers on the immobilization of phages. They precoated a low charge density cationic layer composed of PolyDADMAC onto the paper surface and showed maintenance of T4 phage bioactivity and increased phage spatial orientation perpendicular to the surface of the paper. They explained this by the fact that the capsid of the T4 phage carries a negative net charge while the tail fibers are slightly cationic.

Udit et al. (2008) expressed hexahistidine sequences from bacteriophage Qβ via coexpression of the wild-type and hexahistidine-modified coat proteins in *E. coli*. The resulting polyvalent hexahistidine moieties impart strong affinity for immobilized metal ions. SPR gold surfaces were modified to display Ni (II)-NTA moieties for immobilizing the expressed Qβ bacteriophage. A dissociation constant for binding to Ni–NTA of approximately 10 nM was measured by surface plasmon resonance under noncompetitive, physiological conditions. Affinity chromatography over immobilized metal columns was also demonstrated to purify the phages from both crude cell lysates and after chemical derivatization.

T4 bacteriophages were immobilized onto screen printed electrode (SPE) microarrays via 1-ethyl-3-[3-dimethylaminopropyl] carbodiimide hydrochloride (EDC) by formation of amide bonds between the phage and the electrochemically generated carboxylic groups at the carbon surface (Shabani et al. 2008, 2009). The attachment of the phage was investigated using time-of-flight secondary ion mass spectroscopy (TOF-SIMS). Surface chemical mapping (secondary ion spectra) was acquired from an area of 40 $\times$ 40 μm, using at least three different positions per electrode, to verify the immobilization process and confirm the attachment of phage at the electrode surface. Fig. 11.6 shows 40 $\times$ 40 μm^2 intensity maps of negative and positive fragments. It is clear from the intensity map that CN^- and CNO^- fragments are present on the EDC- and T4-modified surface, showing gradually higher intensities. Also, the presence of K^+ is a good indication of the presence of biological entities such as cells and viruses (Nygren et al. 2007), which is only observed after T4 immobilization. The relative intensity map for total ion reflects a homogenous distribution for each surface following the modification processes. This same research group immobilized gamma bacteriophage as probes to electrochemically functionalized SPE microarrays for the detection of *Bacillus anthracis* Sterne vegetative cells (Shabani et al. 2009). The carbon electrodes were initially functionalized through cyclic-voltammetric reduction of a nitroaryl diazonium moiety, followed by further reduction of nitro groups to amino groups, and finally activated with glutaraldehyde. Functionalization of the carbon electrodes and the binding of Gamma phage were verified by X-ray photoelectron spectroscopy and TOF-SIMS, respectively.

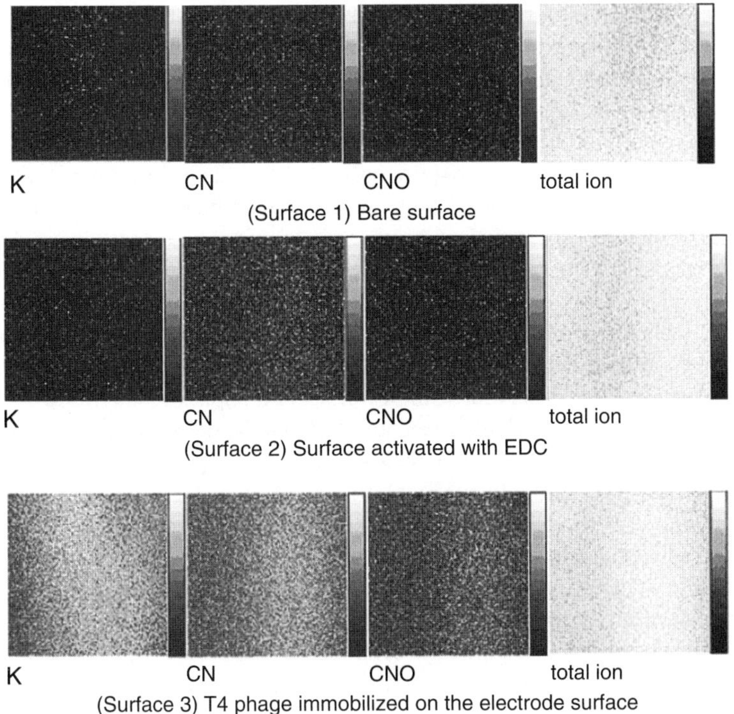

Fig. 11.6 Approximately 40 × 40 μm^2 intensity maps of various positive and negative ions from each surface during the modification process, bare (Surface 1), EDC (Surface 2), phage T4 (Surface 3). Ion intensity is scaled individually to show maximum counts as white and zero counts in black

11.4.2 Affinity Detection

11.4.2.1 Electrical

An electrooptical analyzer (ELUS EO) technique was used to characterize the biospecific binding between the bacterium *E. coli* XL-1 and the phage M13K07, a filamentous phage of the family *Inoviridae* (Bunin et al. 2004; Guliy et al. 2007). The operating principle of the analyzer is based on the polarizability of microorganisms, which depends strongly on their composition, morphology, and phenotype. The principle of analysis of the interaction of *E. coli* with phage M13K07 is based on recording changes of optical parameters of bacterial suspensions upon phage adsorption to cells, nucleic acid injection, phage amplification, and cell lysis. The interaction of *E. coli* with phage M13K07 induced a strong and specific electrooptical signal as a result of substantial changes in the electrooptical properties of the mixed bacterial suspension (*E. coli* K-12 and *Azospirillum brasilense* Sp7).

Kish et al. (2005) (Dobozi-King et al. 2005; Biard and Kish 2005) proposed a method called SEnsing of Phage-Triggered Ion Cascade (SEPTIC) for the rapid detection and identification of bacteria via the electrical field caused by the stochastic emission of ions during phage infection. The method is based on the massive transitory ion leakage that occurs at the moment of phage DNA injection into the host cell. The ion fluxes require only that the cells be physiologically viable (i.e., have energized membranes) and can occur within seconds after mixing the cells with sufficient concentrations of phage particles. These fluxes were detected using a nano-well, a lateral, micrometer-sized capacitor of titanium electrodes with a gap size of 150 nm, and used it to measure the electrical field fluctuations in microliter samples containing phage and bacteria. In mixtures where the analyte bacteria were sensitive to the phage, large stochastic waves with various time and amplitude scales were observed, with power spectra approximately following a 1/*f* 2 law from 1–10 Hz. The calculations indicated that the detection and identification of a single bacterium could be achieved with wild-type phages within a time window of 10 min. The advantage of this method is fast, immobilization-free detection but the cells must be physiologically viable.

The bacteriophage/host cell relationship can be used to not only detect target bacteria but to detect and identify bacteriophage as well. In the dairy fermentation industry, for example, the infection of bacterial starter cultures by contaminating bacteriophages results in significant economic losses. To detect these bacteriophages, the research group of Muñoz-Berbel et al. (2008) developed an impedimetric approach that monitored capacitive changes in a bacterial biofilm upon exposure to a bacteriophage. *E. coli* WG5 were immobilized on an electrode surface and PhiX174 bacteriophages were introduced to the sensor surface. Impedimetric changes occurring at the microelectrode surface due to bacteriophage infection and degradative lysis of the *E. coli* biofilm were measured over a 24-h period by nonfaradic impedance spectroscopy (IS) in urban sewage. The method was sensitive enough to detect phage concentrations of approximately 10^4 pfu/mL. Impedimetric detection was later adapted to an on-chip biosensor format and tested in milk samples (García-Aljaro et al. 2009). Detection of bacteriophages occurred down to 1 pfu/mL but responses were nonlinear, making this assay more of a presence/absence indicator than a quantitative test.

An impedance technique was employed for specific, direct and label-free detection of bacteria using bacteriophages as recognition receptors immobilized covalently onto functionalized SPE microarrays (Shabani et al. 2007; Shabani et al. 2008, 2009). T4 bacteriophages were immobilized onto SPE microarrays via 1-ethyl-3-[3-dimethylaminopropyl] carbodiimide hydrochloride (EDC) by formation of amide bonds between the phage and the electrochemically generated carboxylic groups at the carbon surface. The immobilized T4 phages were then used to specifically detect *E. coli* bacteria. Scanning electron microscopy (SEM) was used to verify the presence of surface-bound bacteria (Fig. 11.7). Impedance measurements (Nyquist plots) show shifts of the order of 10^4 Ω due to the binding of *E. coli* bacteria to the T4 phages (Fig. 11.8). No significant change in impedance was observed for control experiments using immobilized T4 phage in the presence of

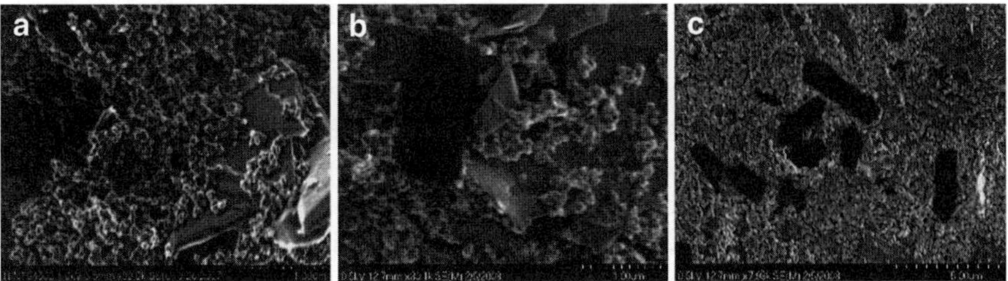

Fig. 11.7 SEM images of bacteria bound to phage-modified surface. (**a**) T4 phage immobilized to surface, (**b**) bacteria bound to immobilized T4 phage (*high resolution*), (**c**) bacteria bound to immobilized T4 phage (*low resolution*).

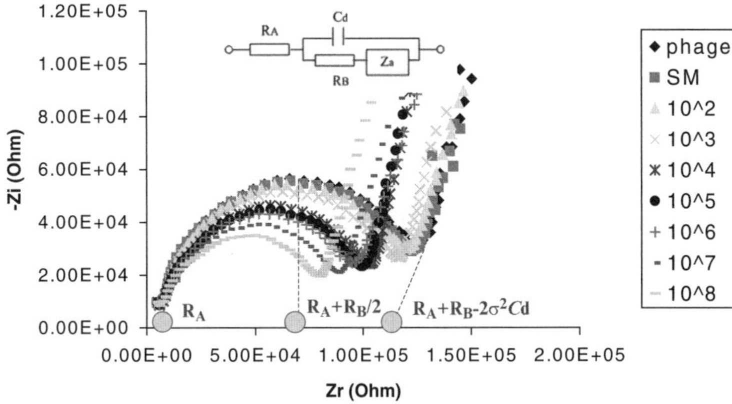

Fig. 11.8 Nyquist plots for a T4-modified surface in the presence of *E. coli* at different concentrations

Salmonella. Concentration–response curves yielded a detection limit of 10^4 cfu/mL in 50 μL samples. Impedimetric techniques were also employed for the detection of *Bacillus anthracis Sterne* vegetative cells using *Gamma* phage as probes attached to electrochemically functionalized SPE microarrays (Shabani et al. 2009). Concentration–response curves yielded a detection limit of 10^3 cfu/mL in 40-μL samples.

11.4.2.2 Optical Sensors

Bacteriophages as recognition receptors were employed in a number of optical biosensors including surface plasmon resonance sensors (García-Aljaro et al. 2008; Nanduri et al. 2007a, b; Balasubramanian et al. 2007), optical light microscope systems (Guntupalli et al. 2008), optical microresonators (Zhu et al. 2008), colorimetric microtiter plates (Knezevic and Petrovic 2008), and ELISA and atomic force microscopy (Handa et al. 2008).

Surface plasmon resonance (SPR) is a well-established technique for the measurement of molecules binding to surfaces and the quantification of binding constants between surface-immobilized ligand and analyte in solution. A two-channel microfluidic SPR sensor device was used to detect the presence of somatic coliphages, a group of bacteriophages that have been proposed as fecal pollution indicators in water, using their host, *E. coli* WG5, as a target for their selective detection (García-Aljaro et al. 2008). *E. coli* WG5 was immobilized on gold sensor chips using avidin–biotin and exposed to bacteriophages extracted from wastewater. SPR was able to monitor the binding of bacteriophages to their bacterial hosts, and demonstrated detection of as few as 1 PFU/mL of bacteriophage after an incubation period of 120 min, which equates to an approximate limit of detection of around 10^2 PFU/mL. The bacteriophage–bacterium interaction appeared to cause a structural change in the surface-bound bacteria, possibly due to collapse of the cell, which was observed as an increase in mass density on the sensor chip. Balasubramanian et al. (2006) employed an SPR-based SPREETA™ sensor for label-free detection of *Staphylococcus aureus* using lytic phage as highly specific and selective biorecognition elements. The lytic phage was physically adsorbed on the gold surface by flowing the phage solution overnight on the gold surface followed by washing. *S. aureus* bacteria were then exposed to the immobilized phages. The detection limit was determined to be 10^4 cfu/mL. Detection specificity was investigated by an inhibition assay while selectivity was examined with *Salmonella typhimurium*. The inhibition assay was performed by preincubating different concentrations of phages with a known concentration of *S. aureus* for 1 h to block the receptor sites from binding and the solutions were introduced through the sensor. It was found that the response of the sensor decreased as the concentration of the phage increased. This technique can be employed for rapid and label-free detection of different bacterial pathogens. SPR was employed for the detection of β-galactosidase (β-gal) using specific landscape phage 1G40 physically adsorbed on the gold surface of SPR SPREETA™ sensor chips (Nanduri et al. 2007a, b). Wild-type phage F8-5 was used in the reference channel as nonspecific to the β-gal phage. It was shown that concentrations of β-gal between 10^{-12} M and 10^{-7} M could be readily detected, with the linear part of the calibration curve between 10^{-9} M and 10^{-6} M (Fig. 11.9). When β-gal was preincubated with different concentrations of free 1G40 phage prior to exposure to the biosensor, concentration-dependent inhibition was observed, indicating the biosensor had high specificity toward β-gal. Batch mode and flow through modes were employed to study the binding of β-gal to immobilized phage on the SPR sensor surface. Experiments using a flow through mode provided more consistent results in the full dose range and showed higher sensitivity as opposed to the batch mode studies. The mean K_d and binding valences for the flow through mode studies was 1.3 ± 0.001 nM and 1.5 ± 0.03, in comparison to 26 ± 0.003 nM and 2.4 ± 0.01 for the batch mode studies. The average thickness of phage 1G40 adlayer deposited through flow through and batch mode was 3 ± 0.002 nm and 0.66 ± 0.001 nm, respectively.

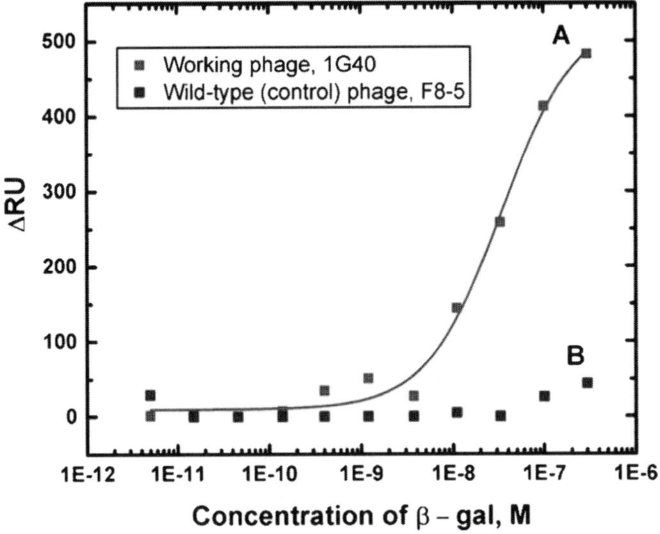

Fig. 11.9 Dose–responses of a phage-immobilized SPR sensor in a flow through mode. Curve A represents responses obtained from the working phage, 1G40 immobilized on the SPR in one of the two channels of the sensor against graded concentrations of β-gal (0.0032–210 nM). B represents the responses obtained from the wild-type phage, F8-5 immobilized on the second channel. (Reproduced from Naduri et al. (2007) with permission of Elsevier)

S. aureus-specific bacteriophage were used as a probe for the detection of methicillin-resistant *S. aureus* (MRSA) in aqueous solution using a CytoViva^{TM} optical light microscope system (Guntupalli et al. 2008). Biorecognition phage monolayers were transferred to glass substrates using Langmuir–Blodgett (LB) technique and were exposed individually to MRSA in solution at logarithmic concentrations ranging from 10^6 to 10^9 cfu/mL, and observed for real-time binding using a CytoViva^{TM} optical light microscope system. Results indicate that LB monolayers possessed high levels of elasticity (K), measuring 22 and 29 mN/m for 10^9 and 10^{11} pfu/mL phage concentrations, respectively. Near-instantaneous MRSA–phage binding produced $33 \pm 5\%$, $10 \pm 1\%$, $1.1 \pm 0.1\%$, and $0.09 \pm 0.01\%$ coverage of the substrate that directly correlated to a decrease in MRSA concentrations of 10^9, 10^8, 10^7, and 10^6 cfu/mL. The exclusive selectivity of phage monolayers was verified with *Salmonella enterica* subsp. *enterica* serovar *typhimurium* (*S. typhimurium*) and *Bacillus subtilis*.

An opto-fluidic ring resonator (OFRR) was used to monitor the binding of filamentous phage R5C2, displaying peptides to streptavidin specifically, and employed as a model receptor (Zhu et al. 2008). The experimental detection limit was approximately 100 pM streptavidin and the $K_{d(apparent)}$ was 25 pM. Specificity was verified using the RAP5 phage, which is not specific to streptavidin, as the negative control. Sensing surface regeneration results show that the phage maintained functionality after surface regeneration, which greatly improves the sensors' reusability.

Colorimetric microtiter plates based on crystal violet staining and measurement of optical density were used for the selection of the most effective *P. aeruginosa* phages for inhibition of biofilm formation and its eradication (Knezevic and Petrovic 2008). Four different bacteriophages isolated from municipal water were tested in this study with *P. aeruginosa* PA-4u and ATCC 9027 as the bacterial hosts. Biofilm formation could be inhibited by over 50% under active growth conditions but relatively little effect was observed when mature biofilms were exposed to phage.

A monolayer of bacteriophage P22 whose tailspike proteins specifically recognize *Salmonella* serotypes was covalently bound to glass substrates through a bifunctional cross-linker 3-aminopropyltrimethoxysilane deposited using chemical vapor deposition. The amino-functionalized glass surface was activated with *N*-hydroxysulfosuccinimide (Sulfo-NHS) and 1-ethyl-3-[3-dimethylaminopropyl] carbodiimide hydrochloride (EDC) prior to phage immobilization (Handa et al. 2008). The specificity of the binding of *S. typhimurium* to the immobilized bacteriophage monolayer was studied by enzyme-linked immunosorbent assay (ELISA) and atomic force microscopy. *E. coli* and a Gram-positive *Listeria monocytogenes* bacterium were also studied as control bacteria. The durability of the P22 phage coating in ambient air was tested to show that 50% of immobilized phage particles were still capable of recognizing *S. typhimurium* after remaining in a dry state for 7 days.

11.4.2.3 Mass Sensitive Sensors

Bacteriophages as bioprobes were employed in a number of mass sensitive sensors based on quartz crystal microbalance (Nanduri et al. 2007a, b), surface acoustic wave (Olsen et al. 2006), magnetoelastic (Lakshmanan et al. 2007a, b; Petrenko 2008; Johnson et al. 2008; Wan et al. 2007), and magnetostrictive millimeter cantilever (Li et al. 2009) sensors.

Landscape phage 1G40 was immobilized by physical adsorption on the surface of a quartz crystal microbalance and used for detection of β-galactosidase from *E. coli* (Nanduri et al. 2007a, b). The sensor had a detection limit of a few nanomoles and a response time of ~100 s over the range of 0.003–210 nM. The binding dose–response curve had a typical sigmoid shape and the signal was saturated at a β-galactosidase concentration of about 200 nM. No response was observed in mixed solutions even when the concentration of BSA exceeded the concentration of β-galactosidase by a factor of approximately 2,000.

Surface acoustic wave biosensors were demonstrated for the detection of *Salmonella typhimurium* using filamentous phage physically adsorbed onto a piezoelectric transducer (Olsen et al. 2006). Quantitative deposition studies of the phages onto the transducer surface indicated that approximately 3×10^{10} phage particles/cm^2 could be irreversibly adsorbed for 1 h at room temperature to prepare working biosensors. The biosensors possessed a rapid response time of <180 s, had a low-detection limit of 10^2 cells/mL, and were linear over a range of 10^1–10^7

cells/mL with a sensitivity of 10.9 Hz per order of magnitude of *S. typhimurium* concentration.

Millimeter magnetoelastic ribbon sensors were demonstrated in buffer for the detection of *S. typhimurium* (Lakshmanan et al. 2007a, b; Petrenko 2008) and *Bacillus anthracis* spores using specific filamentous bacteriophage. Phage were immobilized onto the surface of the sensors by physical adsorption. The phage-immobilized magnetoelastic sensors were exposed to *S. typhimurium* cultures at concentrations ranging from 5×10^1 to 5×10^8 cfu/mL, and the corresponding changes in resonance frequency were monitored. It was experimentally shown that the sensitivity of the magnetoelastic sensors was higher for sensors with smaller physical dimensions. An increase in sensitivity from 159 Hz/decade for a 2 mm sensor to 770 Hz/decade for a 1 mm sensor was observed. SEM was used to verify the frequency changes caused by *S. typhimurium* binding to phage immobilized on the sensor surface. A detection limit on the order of 10^3 cfu/mL was obtained for a sensor with dimensions of $1 \times 0.2 \times 0.015$ mm (Lakshmanan et al. 2007a, b). A magnetoelastic biosensor immobilized with filamentous bacteriophage for the detection of *Salmonella typhimurium* in water and fat-free milk was also demonstrated using sensors with dimensions of $2 \times 0.4 \times 0.015$ mm (Lakshmanan et al. 2007a, b). A detection limit of 5×10^3 cfu/mL, with a sensitivity of 159 Hz/decade, was obtained for the sensors tested in water samples, as compared to 118 Hz/decade in fat-free milk. The dissociation constant K_d and the binding valencies for *S. typhimurium* spiked water samples was 136 and 2.4, respectively, while the K_d and binding valencies in spiked fat-free milk samples was 149 and 2.5 cfu/mL, respectively.

Fabricated microscale magnetoelastic particles ($500 \times 100 \times 4$ μm) composed of an amorphous iron–boron binary alloy was demonstrated for *B. anthracis* spore detection using landscape affinity-selected filamentous phages physically adsorbed (Johnson et al. 2008; Wan et al. 2007; Petrenko 2008). The biosensors were tested in *B. anthracis* spore solutions with concentrations from 10^3 to 10^8 cfu/mL. A detection limit of 10^3 cfu/mL, with a sensitivity of 6.5 kHz/decade, was observed. SEM images were used for visual correlation of the sensors' responses (Fig. 11.10). Specificity of the biosensor was measured after exposure to mixed spore solutions containing *B. anthracis* Sterne, *Bacillus cereus*, and *Bacillus megaterium* strains. Binding kinetics demonstrated that the K_d and the binding valency for the sensor in *B. anthracis* spore solutions was 193 cfu/mL and 2.32, respectively (Wan et al. 2007).

Magnetostrictive millimeter (length of 2.7 mm and width of 1 mm) cantilevers were recently employed for *B. anthracis* spore detection using physically adsorbed f8/8 landscape phage (Li et al. 2009). It was found that the resonance frequency changes with time continuously and eventually reaches its saturated value. Captured spore cells were observed using SEM. It was demonstrated that the shift in the resonance frequency was related to the total mass of spore cells on the sensor surface. The results showed good dose–response relationships between 10^6 and 10^8 cfu/mL. The reported detection limit was 10^5 cfu/mL.

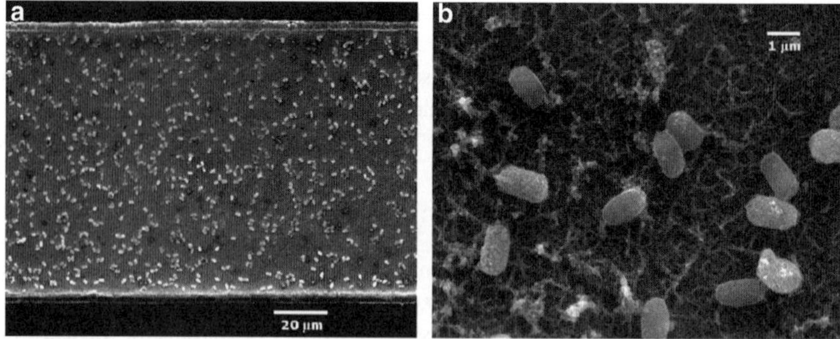

Fig. 11.10 SEM image of a gold-coated magnetoelastic sensor coated (**a**) with filamentous phage (**b**) attachment of the spores to the immobilized phage. (Reproduced from Johnson et al. (2008) with permission of Elsevier)

Fig. 11.11 *Escherichia coli* O157:H7 cells labeled with fluorescent bacteriophages (With permission from Goodridge et al. (1999b), copyright 2008 Elsevier)

11.4.3 Fluorescently Labeled Phages

Labeled phage relies on the initial attachment of the phage to its host cell to signal host cell presence. Goodridge et al. (1999a, b) demonstrated this technique by staining the DNA of the *E. coli* O157:H7 specific phage LG1 with the fluorescent dye YOYO-1. Upon attachment of this fluorescently labeled phage to a suitable *E. coli* O157:H7 host, a fluorescent halo could be visualized around the cell when viewed with an epifluorescent microscope (Fig. 11.11). When confronted with a mixed bacterial culture, immunomagnetic separation was first used to separate target *E. coli* cells from the mix, and then fluorescently labeled LG1 phage was added. Flow cytometry permitted rapid enumeration of phage attached *E. coli* cells. Detection limits in artificially contaminated ground beef and raw milk approached approximately 2 cfu/g after a 6 h enrichment and 10 cfu/mL after a 10-h enrichment, respectively. Kenzaka et al. (2006) demonstrated a similar procedure using DAPI

(4′,6-diamidino-2-phenylindole)-labeled phage T4 to detect *E. coli*. Using epifluorescence microscopy, they successfully enumerated *E. coli* within 30 min in water samples obtained from a fecally contaminated canal in Thailand. Mosier-Boss et al. (2003) labeled P22 phage with a fluorescent SYBR gold dye. In this case, the labeled DNA was visualized after phage-mediated injection into target *Salmonella* Typhimurium host cells.

11.5 Indirect Detection

11.5.1 Detection Based on Inhibition of Metabolism and Growth

Generally, lytic phages infect host cells leading to inhibition of growth of cells and killing cells when phages are present at high enough concentrations. A number of studies was demonstrated for monitoring the metabolic activity and microbial growth of microbial populations including impedimetric (Okigbo et al. 1985; Andrews et al. 1997; Mclntyre and Griffiths 1997; Chang et al. 2002), conductivity (Pugh and Arnott 1987; Carminat and Neviani 1991; Sevensson 1994; Chang et al. 2002), and optically (Preer et al. 1971).

Chang et al. (2002), based on the knowledge that growth in a microbial culture could be monitored electrochemically by measuring changes in electrical parameters occurring as complex growth media substrates were broken down into smaller highly charged molecules such as acids, hypothesized that the presence of phage within a bacterial culture, provided that suitable host cells were present, would impede culture growth and, therefore, directly affect growth media composition. Thus, by comparing conductance measurements between phage-supplemented and phage-free samples or between phage-specific and nonphage-specific bacterial cultures, one could easily screen samples for the presence of phage-specific pathogens. Phage AR1 and its *E. coli* O157:H7 host were used to demonstrate the technique. Pure cultures of *E. coli* O157:H7 or non-O157:H7 cells at 10^6 cfu/mL with or without phage AR1 addition were placed in test tubes fitted with platinum electrodes and conductance measurements were taken every 6 min. Resulting conductance curves could discriminate between *E. coli* O157:H7 and non-O157:H7 cultures within a 24-h period.

11.6 Detection by Reporter Phages

Reporter phages are genetically endowed with reporter genes that transmit an easily visualized signal that correlates with productive phage attachment or infection events. Due to ease of measurement, signals are most often optical, taking the form of bioluminescence, fluorescence, or colorimetric via the luciferase (*lux* and *luc*), green fluorescent protein (*gfp*), and β-galactosidase (*lacZ*) reporter genes, respectively. More uncommon signals such as ice formation (*inaW*) have been applied as well. The reporter gene is placed within the phage's genome, and upon infection of an appropriate bacterial host, is transferred to that host whereupon only then is it expressed. Interfacing the assay with an appropriate transducer capable of measuring the signal realizes the complete detection scheme, which can be carried out in formats that range from single sample test tubes to high-throughput microtiter plates to lab-on-a-chip devices.

11.6.1 Bioluminescent Reporter Phages (lux **and** *luc)*

Bioluminescence is the production and emission of light by a living organism and luciferase is the general term for the enzyme that mediates the light-emitting reaction. Luciferases derived from bacteria are denoted as *lux* while those derived from insects, primarily the firefly (*Photinus pyralis)*, are denoted as *luc* (the firefly luciferase is sometimes also referred to as *fflux*). Both *lux* and *luc* have been very well-characterized and routinely applied in reporter gene assays (Daunert et al. 2000; Viviani 2002). Signal detection is typically achieved using photomultiplier tubes (PMTs), charge-coupled device (CCD) cameras, luminometers and micro-luminometers, or even something as simple as X-ray film or the naked eye.

11.6.1.1 *lux* Reporter Phages

Bioluminescent bacteria are the most abundant and widely distributed of the light-emitting organisms on Earth and can be found in both aquatic and terrestrial environments. The majority of these microorganisms are classified into three genera: *Vibrio*, *Photobacterium*, and *Photorhabdus* (*Xenorhabdus*). Their luciferase consists of two proteins, LuxA and LuxB, that cooperatively generate 490 nm light from the oxidation of a long chain fatty aldehyde in the presence of a reduced riboflavin phosphate ($FMNH_2$) and oxygen. Three other proteins in the operon, LuxC, LuxD, and LuxE, regenerate the aldehyde substrate required for this reaction. Reporter systems can contain only the *luxAB* genes or the entire *luxCDABE* gene cassette. *luxAB* alone generates a bioluminescent signal only after addition of an aldehyde substrate, normally *n*-decanal, due to the absence of the *luxCDE* genes that would normally perform this function. Including the full *luxCDABE* operon permits autonomous generation of light independent of any substrate additions. Phage reporters so far developed utilize only the *luxAB* genes since incorporation of the entire *lux* operon is difficult due to its size and potential interference with headful packaging constraints of the phage. Ulitzur and Kuhn (Ulitzur and Kuhn 1987) first demonstrated *luxAB* reporter phage detection using phage λ Charon 30 to detect *E. coli* down to 10–100 cells/mL in artificially contaminated milk or urine within 1 h. Kodikara et al. (1991) later used this assay to detect *E. coli* in swab samples from slaughterhouse surfaces and swine carcasses at a detection limit of 10 cells/cm^2 or g after a 4-h preenrichment. Waddell and Poppe (2000) inserted the *luxAB* genes into phage ΦV10 for the specific detection of *E. coli* O157:H7 within 1 h at concentrations of 10^6 cfu/mL.

Besides *E. coli* hosts, phage reporters have been designed for additional targets as well. Chen and Griffiths (1996) constructed several P22-based *luxAB* reporter phage specific for the A, B, and D_1 serotypes of *Salmonella enterica* and directly detected 10^8 cfu/mL within 1–3 h. By adding a 6 h preincubation step, detection of as few as 10 cfu/mL were possible. Reporter phages were also injected directly into poultry eggs artificially contaminated with *Salmonella* at a lower inoculum of 63 cfu/mL. By imaging the entire egg with a CCD camera, points of bioluminescent light corresponding to phage-infected *Salmonella* cells could be visualized after *n*-decanal addition, permitting localized identification of infected areas within the unadulterated egg. Thouand et al. (2008) recently adapted these reporter phage to a more commercially viable kit-like format consisting of four steps; (1) addition of the *luxAB* reporter phage to a 12–14-h enriched culture, (2) a 1 h incubation sufficient to allow contact between phage and host *Salmonella*, (3) addition of fresh growth medium followed by a 2–4-h incubation, and (4) addition of the *n*-decanal reagent and measurement of light. They showed that bacterial concentrations enriched above 10^6 cfu/mL could be detected within 16 h in various artificially contaminated poultry feed, feces, and litter samples. Loessner et al. (1997) created an A511 *luxAB* reporter phage for *Listeria monocytogenes*. Phage A511 infects approximately 95% of the *L. monocytogenes* serovars responsible for human listeriosis. Cheeses, chocolate pudding, cabbage, lettuce, ground beef, liverwurst, milk, and shrimp were each inoculated with *L. monocytogenes* from 0.1 to 1,000 cfu/g and allowed to preincubate for 20 h. The reporter phages and *n*-decanal substrate were then added and detection limits of 1 cell/g were demonstrated within less than 24 h. Foods such as ground beef that contained more complex microflora yielded detection limits of approximately 10 cfu/g. A total of 348 naturally contaminated meats, poultry, dairy products, and other environmental samples were additionally assayed in parallel with standard plating techniques, and *Listeria* positive samples were found to correspond between the two methods. With standard plating requiring 72–96 h, the less than 24 h A511 *luxAB* reporter phage assay was shown to be highly efficient while generating statistically comparable results.

Using *lux* but in a different approach, Brigati et al. (2007) designed a phage reporter assay for the detection of *E. coli* O157:H7 that did not rely on the addition of *n*-decanal for signal generation. This allowed the assay to be fully autonomous and capable of continuous monitoring of target pathogens. Besides the *luxCDABE* genes discussed above, the *lux* operon also contains a *luxI* and *luxR* gene that are involved in a cell-to-cell communication network referred to as quorum sensing. The *luxI* gene product synthesizes the chemical *N*-3-(oxohexanoyl)-l-homoserine lactone (OHHL) that interacts with the *luxR* gene which, in turn, activates the *luxCDABE* gene cassette to generate bioluminescence. The *luxI* gene was inserted into phage PP01 and was expressed upon infection of an *E. coli* O157:H7 host,

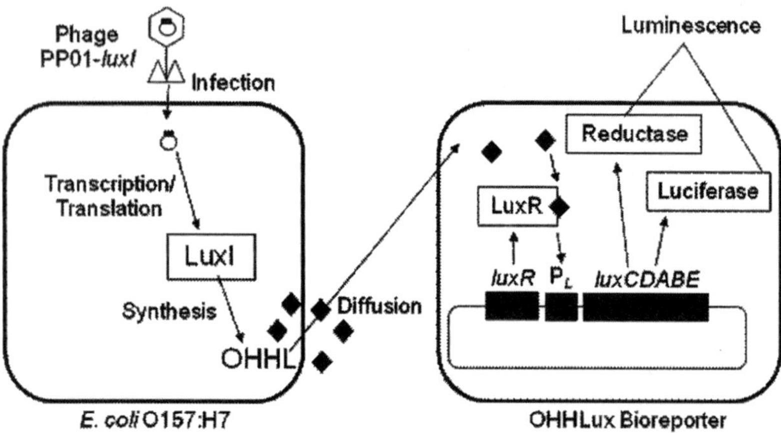

Fig. 11.12 Bacteriophage-based bioluminescent assay format for the detection of *Escherichia coli* O157:H7. The PP01 reporter phage carries a *luxI* gene that, upon infection of its *E. coli* host, forces synthesis of the chemical OHHL. A bioluminescent bioreporter cell sensitive to OHHL subsequently yields a bioluminescent light response to denote the presence of phage infective *E. coli* O157:H7 cells

thereby obligating the *E. coli* cell to generate OHHL (Fig. 11.12). Within the assay there was also included a bioluminescent bioreporter cell containing the *luxRCDABE* genes and its exposure to the OHHL chemical instigated bioluminescent light emission. Thus, *E. coli* O157:H7 could be detected indirectly via its phage-derived synthesis of a signature chemical analyte. The assay was demonstrated in artificially contaminated apple juice and tap water with detection limits of 1 cfu/mL within 22 and 12.5 h, respectively, using a 96-well microtiter plate format. Monitoring of an artificially contaminated spinach leaf rinsate was performed with a CCD camera that allowed real-time, continuous visualization of the bioluminescent signal, permitting detection of 1 cfu/mL within 6 h after a 2-h preincubation (Ripp et al. 2008).

11.6.1.2 *luc* Reporter Phages

The *luc* firefly luciferase catalyzes the two-step conversion of D-luciferin to oxyluciferin to generate a 560-nm bioluminescent light signal. Analogous to the addition of *n*-decanal in *luxAB* reporter assays, the D-luciferin substrate must also be added exogenously during *luc*-based reporter assays. The use of *luc* in reporter phage has primarily been associated with *Mycobacterium* specific phages and the detection and assessment of drug susceptibility in *M. tuberculosis* (Carriere et al. 1997; Sarkis et al. 1995). Susceptibility to antimycobacterial drugs can be determined by comparing bioluminescent output kinetics from reporter phage added to antibiotic-free or antibiotic-amended *Mycobacterium* cultures. If the drug is effective, fewer host *Mycobacterium* cells are available for phage infection and less light is therefore emitted as compared to the antibiotic-free control. These assays can require several days to complete, but far surpass the several week time periods required using conventional antibiotic susceptibility assays. Bardarov et al. (2003) used *luc* reporter phage phAE142 to assay sputum samples for *M. tuberculosis* presence and establish antibiotic susceptibility profiles. *M. tuberculosis* could be detected within 7 days at concentrations greater than 10^4 cfu/mL, while antibiotic susceptibility profiles could be obtained within 3 days. This compares to conventional growth assays such as the Mycobacterial Growth Indicator Tube (MGIT) that requires 9 days until results and antibiotic resistant tests that require up to 12 days. The phage assay is inexpensive and simple to perform, and has been marketed as a diagnostic tool valuable to resource-poor countries. To demonstrate, Banaiee et al. (2008) provided the assay to laboratory technicians in a South African hospital setting and showed an approximate 98% agreement between it and conventional BACTEC methods. A phage, referred to as Che12, showing greater specificity toward *M. tuberculosis* has recently been described and genetically incorporated with the *luc* gene (Kumar et al. 2008)

11.6.2 Fluorescent Reporter Phages (*gfp*)

The green fluorescent protein (*gfp*) reporter gene emits a 508-nm fluorescent signal when activated by ultraviolet or blue light. Funatsu et al. (2002) first demonstrated *gfp* signaling in phage lambda for broad-based detection of *E. coli*. Reporter phage genomically incorporated with the *gfp* gene were combined with a mixed culture of *E. coli* and *Mycobacterium smegmatis* for 4 h and then observed under an epifluorescent microscope. *E. coli* cells that had been infected with the phage fluoresced while noninfectable *M. smegmatis* cells did not. Tanji et al. (2004) constructed another *E. coli*-specific *gfp* reporter phage using a phage T4 modified to inactivate its lytic activity. Thus, these phage were no longer able to kill their hosts, which is an advantage in an enumeration assay where destroying the target one wishes to detect is not altogether favorable. This reporter phage, referred to as T4e⁻/GFP, was added to a mixed culture of *E. coli* and *P. aeruginosa* where it was shown to differentiate between the two microbes when observed under the epifluorescent microscope (Fig. 11.13). Miyanaga et al. (2006) later used T4e⁻/GFP reporter

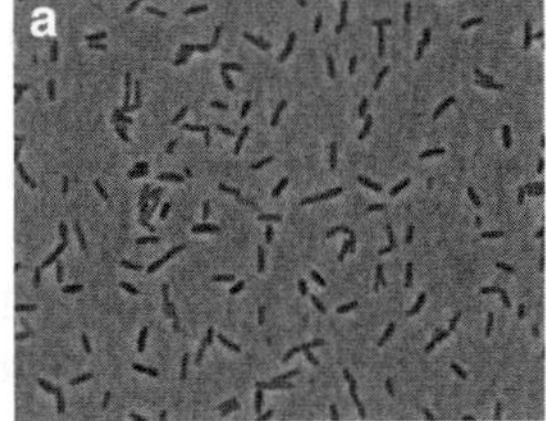

Fig. 11.13 (**a**) A microscopic image of an *Escherichia coli*/*Pseudomonas aeruginosa* bacterial mix, and (**b**) a same field-of-view fluorescent microscopic image showing fluorescent *E. coli* cells after addition of the *E. coli*-specific T4e⁻/GFP reporter phage. (With permission from Tanji et al. (2004), copyright 2008 Elsevier)

phage to detect *E. coli* in sewage influent. Since the host range of T4e^{-}/GFP was not inclusive of all *E. coli* present in the sewage samples, only approximately 8% of the total *E. coli* population was detected, which was not unexpected since the use of a single reporter phage severely confines target acquisitions. Namura et al. (2008) addressed this by isolating two additional *E. coli*-specific phage from sewage and genetically modifying them with *gfp* inserts. Together, these reporter phages demonstrated a host range covering nearly 50% of the *E. coli* sewage isolates.

Oda et al. (2004) created a much narrower host range *gfp* reporter phage using the *E. coli* O157:H7-specific phage PP01 which was capable of discriminating between *E. coli* O157:H7 and *E. coli* K12 cells within 10 min based on fluorescence emission. Sensitivity was later improved by once again inactivating the lytic activity of the phage (Awais et al. 2006), providing clearer and sharper epifluorescent images that translated into easier and more direct identification of *E. coli* O157:H7 in mixed cultures. These assays additionally discriminated between healthy and stressed cells, where healthy cells emitted bright green fluorescence and metabolically stressed cells emitted faded fluorescent signals. This permitted simultaneous differentiation of healthy cells from viable but nonculturable (VBNC) cells, which cannot be accomplished with conventional plating methods without supplemental and time-consuming steps.

11.6.3 Colorimetric Reporter Phages (lacZ)

The *lacZ* gene encodes a β-galactosidase enzyme that catalyzes the hydrolysis of β-galactosides. With the addition of a fluorescent, luminescent, or chemiluminescent substrate, the reaction endpoint can be optically observed in a typical reporter gene fashion. Goodridge and Griffiths (2002) inserted the *lacZ* reporter gene into phage T4 for the detection of *E. coli*. After addition of a chemiluminescent substrate to phage-infected *E. coli* cells, they were able to achieve detection down to 100 cfu/mL in pure culture within 12 h. The *lacZ* reporter gene has, however, seen very little additional use in reporter phage assays since *lux*, *luc*, and *gfp* assays typically maintain greater sensitivities.

11.6.4 Ice Nucleation Reporter Phages (inaW)

A more unusual application of reporter phages involves the *inaW* ice nucleation gene. Its protein product, InaW, integrates within the bacterial outer cell membrane where it acts as a catalyst for ice crystal formation within a temperature range of −2 to −10°C. The signaling endpoint for *inaW* is, therefore, ice crystal formation at supercooled temperatures. Wolber and Green (1990) inserted the *inaW* gene into *Salmonella* phage P22 to create the BIND^{TM} (Bacterial Ice Nucleation Diagnostic) assay. Upon infection of *Salmonella* by the reporter phage, the *inaW* gene is expressed and samples freeze faster when supercooled as compared to *Salmonella* or other cells not expressing *inaW*. To easily visualize ice formation, the BIND^{TM} assay used an indicator dye that turned orange if freezing occurred and fluorescent green if it did not, thereby providing a simple colorimetric or fluorescent signaling endpoint. When tested in raw eggs and milk artificially contaminated with *Salmonella*, detection limits of less than 10 cells/mL were demonstrated. The BIND^{TM} assay, although once sold as a commercial kit, is no longer marketed.

11.7 Other Detection Methods Using Phages

11.7.1 Phage-Conjugated Quantum Dots

Edgar et al. (2006) developed a unique phage-conjugated quantum dot method for the detection of *E. coli*. Quantum dots are fluorescent probes consisting of colloidal semiconductor nanocrystals that exhibit high quantum yield and excellent photostability. Phage T7 was engineered to display a biotinylation peptide on its major capsid protein. Upon infection, progeny phage synthesized within the *E. coli* cell became biotinylated. After lytic release from the cell, these biotinylated phage could be detected by a streptavidin functionalized quantum dot designed to attach to and label only biotinylated phage. Thus, if no suitable hosts were present, then no biotinylated phages were produced and the functionalized quantum dot, having nothing to attach to, would simply be washed away. Fluorescence microscopy permitted visualization of a single quantum dot conjugated phage. Typical detection limits in laboratory mixed culture approached 10 cells/mL within an assay time of 1 h. Testing of river water samples demonstrated detection of 20 *E. coli* cells/mL within 1 h.

11.8 Conclusions and Future Remarks

Biosensing relies on specific analyte recognition by an appropriate receptor. A good recognition receptor should be highly specific, stable, and versatile, and few receptors meet these prerequisites as well as bacteriophages do. Considering the global abundance of phage with which we share this planet, appropriate phage/host specificity ranges should theoretically exist for any bacterial host desired. Their ease of production, long shelf life, and conduciveness to immobilization further warrants their application in biosensing regimens. With applications ranging from the detection of food and waterborne pathogens to military and homeland defence-related biological threat identification to in vivo diagnostic surveillance of disease states, the innovative simplicity, specificity, and sensitivity of bacteriophage-mediated biosensors cannot go unnoticed.

References

Andrews DMA, Gharbia SE, Shah HN (1997) Characterization of a novel bacteriophage in *Fusobacterium varium, Clinical Infectious Diseases* 25, Supplement 2. Proceedings of the 1996 Meeting of the Anaerobe Society of the Americas, Sep 1997, pp S287–S288

Awais R, Fukudomi H, Miyanaga K, Unno H, Tanji Y (2006) A recombinant bacteriophage-based assay for the discriminative detection of culturable and viable but nonculturable *Escherichia coli* O157:H7. Biotechnol Prog 22:853–859

Balasubramanian S, Sorokulova IB, Vodyanoy VJ, Simonian AL (2007) Lytic phage as a specific and selective probe for detection of *Staphylococcus aureus* – A surface plasmon resonance spectroscopic study. Biosens Bioelectron 22:948–955

Banaiee N, January V, Barthus C, Lambrick M, RoDiti D, Behr MA, Jacobs WR, Steyn LM (2008) Evaluation of a semi-automated reporter phage assay for susceptibility testing *Myobacterium tuberculosis* isolates in South Africa. Tuberculosis 88:64–68. doi:10.1016/j.tube.2007.08.006

Bardarov S, Dou H, Eisenach K, Banaiee N, Ya S, Chan J, Jacobs WR, Riska PF (2003) Detection and drug-susceptibility testing of *M. tuberculosis* from sputum samples using luciferase reporter phage: comparison with the Mycobacteria Growth Indicator Tube (MGIT) system. Diagn Microbiol Infect Dis 45:53–61

Bennet AR, Davids FG, Valhodimou S, Banks JG, Betts RP (1997) The use of bacteriophage-based systems for the separation and concentration of Salmonella. J App Microbiol 83:259–265

Blasco R, Murphy MJ, Sanders MF, Squirrell DJ (1998) Specific assays for bacteria using phage mediated release of adenylate kinase. J Appl Microbiol 84:661–666

Biard JR, Kish LB (2005) Enhancing the sensitivity of the septic bacterium detection method by concentrating the phage-infected bacteria via DC electrical current. Fluctuation Noise Lett 5: L153–L158

Brigati JR, Ripp S, Johnson CM, Jegier P, Sayler GS (2007) Bacteriophage-based bioluminescent bioreporter for the detection of *Escherichia coli* O157:H7. J Food Prot 70:1386–1392

Bunin VD, Ignatov OV, Guliy OI, Zaitseva IS, O'Neil D, Ivnitski D (2004) Electrooptical analysis of the *Escherichia coli*–phage interaction. Anal Biochem 328:181–186

Carminat D, Neviani E (1991) Application of the Conductance Measurement Technique for Detection of *Streptococcus salivarius* ssp. *thermophilus* Phages. J Dairy Sci 74:1472–1476

Carriere C, Riska PF, Zimhony O, Kriakov J, Bardarov S, Burns J, Chan J, Jacobs WR (1997) Conditionally replicating luciferase reporter phages: improved sensitivity for rapid detection and assessment of drug susceptibility of *Mycobacterium tuberculosis*. J Clin Microbiol 35:3232–3239

Chang TC, Ding HC, Chen S (2002) A conductance method for the identification of *Escherichia coli* O157:H7 using bacteriophage AR1. J Food Prot 65:12–17

Chen J, Griffiths MW (1996) *Salmonella* detection in eggs using *lux*$^+$ bacteriophages. J Food Prot 59:908–914

Cherry WB, Davis BR, Edwards PR, Hogan RB (1954) A simple procedure for the identification of the genus *Salmonella* by means of a specific bacteriophage. J Lab Clin Med 44:51–55

Corbitt AJ, Bennion N, Forsythe SJ (2000) Adenylate kinase amplification of ATP bioluminescence for hygiene monitoring in the food and beverage industry. Lett Appl Microbiol 30:443–447

Daunert S, Barrett G, Feliciano JS, Shetty RS, Shrestha S, Smith-Spencer W (2000) Genetically engineered whole-cell sensing systems: coupling biological recognition with reporter genes. Chem Rev 100:2705–2738

D'Herelle F (1917) Sur un microbe invisible antagoniste des bacilles dysentériques. Comptes Rend Acad Sci Paris 165:373–375

D'Herelle F (1919a) Du rôle du microbe filtrant bacteriophage dans la fièvre thyphoide. Compte Rend Acad Sci Paris 168:631–634

D'Herelle F (1919b) Sur une épizooric de thyphose aviaire. Comptes Rend Acad Sci Paris 169:817–819

Dobozi-King M, Seo S, Kim JU, Young R, Cheng M, Kish LB (2005) Rapid detection and identification of bacteria: SEnsing of Phage-Triggered Ion Cascade (SEPTIC). J Biol Phys Chem 5:3–7

Easter MC, Gibson DM (1985) Rapid and automated detection of *Salmonella* by electrical measurements. J Hyg (Lond) 94:245–262

Edgar R, McKinstry M, Hwang J, Oppenheim AB, Fekete RA, Giulian G, Merril C, Nagashima K, Adhya S (2006) High-sensitivity bacterial detection using biotin-tagged phage and quantum-dot nanocomplexes. Proc Natl Acad Sci USA 103:4841–4845

Favrin SJ, Jassim SA, Griffiths MW (2003) Application of a novel immunomagnetic separation-bacteriophage assay for the detection of *Salmonella enteritidis* and *Escherichia coli* O157: H7 in food. Int J Food Microbiol 85:63–71

Favrin SJ, Jassim SA, Griffiths MW (2001) Development and optimization of a novel immunomagnetic separation-bacteriophage assay for detection of *Salmonella enterica* serovar enteritidis in broth. Appl Environ Microbiol 67:217–224

Funatsu T, Taniyama T, Tajima T, Tadakuma H, Namiki H (2002) Rapid and sensitive detection method of a bacterium by using a GFP reporter phage. Microbiol Immunol 46:365–369

García-Aljaro C, Muñoz-Berbel X, Jenkins A, Blanch AR, Muñoz FX (2008) Surface plasmon resonance assay for real-time monitoring of somatic coliphages in wastewaters. Appl Environ Microbiol 74:4054–4058

García-Aljaro C, Muñoz-Berbel X, Muñoz FJ (2009) On-chip impedimetric detection of bacteriophages in dairy samples. Biosens Bioelectron 24:1712–1716

Gervais L, Gel M, Allain B, Tolba M, Brovko L, Zourob M, Mandeville R, Griffiths M, Evoy S (2007) Immobilization of biotinylated bacteriophages on biosensor surfaces. Sens Actuat B 12:615–621

Goodridge L, Chen J, Griffiths MW (1999a) Development and characterization of a fluorescent-bacteriophage assay for detection of *Escherichia coli* O157:H7. Appl Environ Microbiol 65:1397–1404

Goodridge L, Chen J, Griffiths MW (1999b) The use of fluorescent bacteriophage assay for detection of *Escherichia coli* O157:H7 in inoculated ground beef and raw milk. Int J Food Microbiol 47:43–50

Goodridge L, Griffiths MW (2002) Reporter bacteriophage assays as a means to detect foodborne pathogenic bacteria. Food Res Int 35:863–870

Griffiths M, Goodridge L, McIntyre L, Ilenchuk TT (2003) Diagnostic and therapeutic applications of lytic phages. Anal Lett 36:3241–3259

Guan JW, Chan M, Allain B, Mandeville R, Brooks BW (2006) Detection of multiple antibiotic-resistant *Salmonella enterica* serovar Typhimurium DT104 by phage replication-competitive enzyme-linked immunosorbent assay. J Food Protect 69:739–742

Guliy OI, Bunin VD, O'Neil D, Ivnitski D, Ignatov OV (2007) A new electro-optical approach to rapid assay of cell viability. Biosens Bioelectron 23:583–587

Guntupalli R, Sorokulova I, Krumnow A, Pustovyy O, Olsen E, Vodyanoy V (2008) Real-time optical detection of methicillin-resistant Staphylococcus aureus using lytic phage probes. Biosens Bioelectron 24:151–154

Handa H, Gurczynski S, Jackson MP, Auner G, Walker J, Mao G (2008) Recognition of *Salmonella typhimurium* by immobilized phage P22 monolayers. Surf Sci 602:1392–1400

Hirsh DC, Martin LD (1983) Rapid detection of *Salmonella* spp by using Felix-01 bacteriophage and high-performance liquid-chromatography. Appl Environ Microbiol 45:260–264

Jabrane T, Jeaidi J, Dubé M, Mangin PJ (2008) Gravure printing of enzymes and phages. Adv Print Media Technol 35:279–288

Jabrane T, Dubé M, Mangin PJ (2009) Bacteriophage immobilization on paper surface: effect of cationic pre-coat layer. In: Proceedings of Canadian PAPTAC 95th Annual Meeting 2009, Montreal, Canada, pp 31–314

Jassim SAA, Griffiths MW (2007) Evaluation of a rapid microbial detection method via phage lytic amplification assay coupled with live/dead fluorochromic stains. Lett Appl Microbiol 44:673–678

Johnson ML, Wan J, Huang S, Cheng Z, Petrenko VA, Kim D-J, Chen I-H, Barbaree JM, Hong JW, Chin BA (2008) A wireless biosensor using microfabricated phage-interfaced magnetoelastic particles. Sens Actuat A 144:38–47

Kenzaka T, Utrarachkij F, Suthienkul O, Nasu M (2006) Rapid monitoring of *Escherichia coli* in southeast Asian urban canals by fluorescent-bacteriophage assay. J Health Sci 52:666–671

Kish LB, Cheng M, Kim JU, Seo S, King MD, Young R, Der A, Sshmera G (2005) Estimation of detection limits of the phage-invasion based identification of bacteria. Fluctuat Noise Lett 5: L105–L108

Kodikara CP, Crew HH, Stewart GSAB (1991) Near on-line detection of enteric bacteria using *lux* recombinant bacteriophage. FEMS Microbiol Lett 83:261–266

Knezevic P, Petrovic O (2008) A colorimetric microtiter plate method for assessment of phage effect on *Pseudomonas aeruginosa* biofilm. J Microbiol Meth 74:114–118

Kumar V, Loganathan P, Sivaramakrishnan G, Kriakov J, Dusthakeer A, Subramanyam B, Chan J, Jacobs WR, Rama NP (2008) Characterization of temperate phage Che12 and construction of a new tool for diagnosis of tuberculosis. Tuberculosis 88:616–623. doi:10.1016/j.tube.2008.02.007

Kutter E, Sulakvelidze A (2005) Bacteriophages: biology and applications. CRC Press, Boca Raton, p 510

Lakshmanan RS, Guntupalli R, Hu J, Kim D-J, Petrenko VA, Barbaree JM, Chin BA (2007a) Phage immobilized magnetoelastic sensor for the detection of *Salmonella typhimurium.* J Microbiol Meth 71:55–60

Lakshmanan RS, Guntupalli R, Hu J, Petrenko VA, Barbaree JM, Chin BA (2007b) Detection of *Salmonella typhimurium* in fat free milk using a phage immobilized magnetoelastic sensor. Sens Actuat B 126:544–550

Li S, Fu L, Barbaree JM, Cheng Z-Y (2009) Resonance behavior of magnetostrictive micro/milli-cantilever and its application as a biosensor. Sens Actuat B 137:692–699

Loessner MJ, Rudolf M, Scherer S (1997) Evaluation of luciferase reporter bacteriophage A511::*luxAB* for detection of *Listeria monocytogenes* in contaminated foods. Appl Environ Microbiol 63:2961–2965

Mandeville R, Griffiths M, Goodridge L, McIntyre L, Ilenchuk TT (2003) Diagnostic and Therapeutic Applications of Lytic Phages. Anal Lett 3241–3259

McIntyre L, Griffiths MW (1997) A bacteriophage-based impedemtric method for the detection of pathogens in diary products. J Diary Sci 80:107

Miyanaga K, Hijikata TF, Furukawa C, Unno H, Tanji Y (2006) Detection of *Escherichia coli* in the sewage influent by fluorescent labeled T4 phage. Biochem Eng J 29:119–124

Madonna AJ, Van Cuyk S, Voorhees KJ (2003) Detection of *Escherichia coli* using immunomagnetic separation and bacteriophage amplification coupled with matrix-assisted laser desorption/ionization time-of-flight mass spectrometry. Rap Commun Mass Sp 17:257–263

Mole RJ, Maskell WOC (2001) Phage as a diagnostic - the use of phage in TB diagnosis. J Chem Technol Biotechnol 76:683–688

Mosier-Boss PA, Lieberman SH, Andrews JM, Rohwer FL, Wegley LE, Breitbart M (2003) Use of fluorescently labeled phage in the detection and identification of bacterial species. Appl Spectrosc 57:1138–1144

Muñoz-Berbel X, García-Aljaro C, Muñoz FJ (2008) Impedimetric approach for monitoring the formation of biofilms on metallic surfaces and the subsequent application to the detection of bacteriophage. Electrochimica Acta 53(2008):5739–5744

Namura M, Hijikata T, Miyanaga K, Tanji Y (2008) Detection of *Escherichia coli* with fluorescent labeled phages that have a broad host range to *E. coli* in sewage water. Biotechnol Prog 24:481–486

Nanduri V, Balasubramanian S, Sista S, Vodyanoy VJ, Simonian AL (2007a) Highly sensitive phage-based biosensor for the detection of β-galactosidase. Anal Chim Acta 589:166–172

Nanduri V, Sorokulova IB, Samoylov AM, Simonian AL, Petrenko VA, Vodyanoy V (2007b) Phage as a molecular recognition element in biosensors immobilized by physical adsorption. Biosens Bioelectron 22:986–992

Neufeld T, Mittelman AS, Buchner V, Rishpon J (2005) Electrochemical phagemid assay for the specific detection of bacteria using *Escherichia coli* TG-1 and the M13KO7 phagemid in a model system. Anal Chem 77:652–657

Neufeld T, Schwartz-Mittelmann D, Biran EZ, Rishpon RJ (2003) Combined phage typing and amperometric detection of released enzymatic activity for the specific identification and quantification of bacteria. Anal Chem 75:580–585

Nygren H, Hagenhoff B, Malmberg P, Nilsson M, Richter K (2007) Bioimaging TOF-SIMS: high resolution 3D imaging of single cells. Microsc Res Tech 70:969–974

Oda M, Morita M, Unno H, Tanji Y (2004) Rapid detection of *Escherichia coli* O157:H7 by using green fluorescent protein-labeled PP01 bacteriophage. Appl Environ Microbiol 70:527–534

Olsen EV, Sorokulova IB, Petrenko VA, Chen I-H, Barbaree JM, Vodyanoy VJ (2006) Affinity-selected filamentous bacteriophage as a probe for acoustic wave biodetectors of *Salmonella typhimurium*. Biosens Bioelectron 21:1434–1442

Okigbo LM, Ooberg CJ, Richardson GH (1985) Lactic culture activity tests using ph and impedance instrumentation. J Dairy Sci 68:2521–2526

Pai M, Kalantri S, Pascopella L, Riley LW, Reingold AL (2005) Bacteriophage-based assays for the rapid detection of rifampicin resistance in *Mycobacterium tuberculosis*: a meta-analysis. J Infect 51:175–187

Petrenko VA (2008) Landscape phage as a molecular recognition interface for detection devices. Microelectronics J 39:202–207

Preer JR Jr, Preer LB, Rudman B, Jurand A (1971) Isolation and composition of bacteriophage-like particles from kappa of killer paramecia. Mol Gen Genet 111:202–208

Pugh SJ, Arnott ML (1987) Automated conductemetric detection of Salmonella in confectionary products. In rapid microbiological methods for foods, beverages & pharmaceuticals. Soc Appl Bacteriol Tech Series 185

Rees CE, Dodd CE (2006) Phage for rapid detection and control of bacterial pathogens in food. Adv Appl Microbiol 59:159–186

Ripp S, Jegier P, Johnson CM, Brigati J, Sayler GS (2008) Bacteriophage-amplified bioluminescent sensing of *Escherichia coli* O157:H7. Anal Bioanal Chem 391:507–514

Sarkis GJ, Jacobs WR, Hatfull GF (1995) L5 luciferase reporter mycobacteriophages: a sensitive tool for the detection and assay of live mycobacteria. Mol Microbiol 15:1055–1067

Sanders MF (1995) A rapid bioluminescent technique for the detection and identification of *Listeria monocytogenes* in the presence of *Listeria innocua*. In: Campbell AK, Kricka LJ, Stanley PE (eds) Bioluminescence and chemiluminescence: fundamental and applied aspects. Wiley, Chichester, U.K., pp 454–457

Schmelcher M, Loessner MJ (2008) Chapter 27 page 727. In Zourob M, Elwary S, Turner A (eds) Principles of bacterial detection. Springer, New York, p 970

Schuch R, Nelson D, Fischetti VA (2002) A bacteriolytic agent that detects and kills *Bacillus anthracis*. Nature 418:884–889

Sevensson UK (1994) Conductimetric analyses of bacteriophage infection of two groups of bacteria in DL-Lactococcal starter cultures. J Dairy Sci 77:3524–3531

Silley P, Forsythe S (1996) Impedance microbiology–a rapid change for microbiologists. J Appl Bacteriol 80:233–243

Singh A, Glass N, Tolba M, Brovko L, Griffiths M, Evoy S (2009) Immobilization of bacteriophages on gold surfaces for the specific capture of pathogens. Biosens Bioelectron 24:3645–3651

Shabani A, Zourob M, Allain B, Lawrence M, Mandeville R (2007) Electrochemical detection of bacteria using bacteriophage. Signals Syst Electron 2:45–47

Shabani A, Zourob M, Allain B, Marquette CA, Lawrence MF, Mandeville R (2008) Bacteriophage-modified microarrays for the direct impedimetric detection of bacteria. Anal Chem 80:9475–9482

Shabani A, Zourob M, Marquette CA, Lawrence MF, Mandeville R (2009) Carbon microarrays for the direct impedimetric detection of bacillus anthracis using gamma phage as probe. International Journal of Environmental Analytical Chemistry (Submitted)

Squirrel DJ, Price RL, Murphy MJ (2002) Rapid and specific detection of bacteria using bioluminescence. Anal Chim Acta 457:109–114

Sun W, Brovko L and Griffiths M (2001) Use of bioluminescent *Salmonella* for assessing the efficiency of constructed phage-based biosorbent. J Ind Microbiol Biotechol 27:126–128

Stanley PE (1989) A review of bioluminescent ATP techniques in rapid microbiology. J Biolumin Chemilumin 4:375–380

Stewart GSAB, Jassim SAA, Denyer SP, Newby P, Linley K, Dhir VK (1998) The specific and sensitive detection of bacterial pathogens within 4 h using bacteriophage amplification. J Appl Microbiol 84:777–783

Stewart GSAB, Loessner MJ, Scherer S (1996) The bacterial *lux* gene bioluminescent biosensor revisited. ASM News 62:297–301

Tanji Y, Furukawa C, Na SH, Hijikata T, Miyanaga K, Unno H (2004) *Escherichia coli* detection by GFP-labeled lysozyme-inactivated T4 bacteriophage. J Biotechnol 114:11–20

Thouand G, Vachon P, Liu S, Dayre M, Griffiths MW (2008) Optimization and validation of a simple method using P22:: luxAB bacteriophage for rapid detection of *Salmonella enterica* serotypes A, B, and D in poultry samples. J Food Prot 71:380–385

Twort F (1915) An investigation on the nature of ultramicroscopic viruses. Lancet 11:1241

Udit AK, Brown S, Baksh MM, Finn MG (2008) Immobilization of bacteriophage Qβ on metal-derivatized surfaces via polyvalent display of hexahistidine tags. J Inorg Biochem 102: 2142–2146

Ulitzur S, Kuhn J (1987) Introduction of *lux* genes into bacteria, a new approach for specific determination of bacteria and their antibiotic susceptibility. In: Scholmerich J et al (eds) Bioluminescence and chemiluminescence: new perspectives. Wiley, New York, pp 463–472

Ulitzur N, Ulitzur S (2006) New rapid and simple methods for detection of bacteria and determination of their antibiotic susceptibility by using phage mutants. Appl Environ Microbiol 72:7455–7459

Viviani VR (2002) The origin, diversity, and structure function relationships of insect luciferases. Cell Mol Life Sci 59:1833–1850

Wan J, Johnson ML, Guntupalli R, Petrenko VA, Chin BA (2007) Detection of *Bacillus anthracis* spores in liquid using phage-based magnetoelastic micro-resonators. Sens Actuat B 127:559–566

Waddell TE, Poppe C (2000) Construction of mini-Tn10*luxABcam*/*Ptac*-ATS and its use for developing a bacteriophage that transduces bioluminescence to *Escherichia coli* O157:H7. FEMS Microbiol Lett 182:285–289

Wolber PK, Green RL (1990) Detection of bacteria by transduction of ice nucleation genes. Trends Biotechnol 8:276–279

Wu Y, Brovko L, Griffiths MW (2001) Influence of phage population on the phage-mediated bioluminescent adenylate kinase (AK) assay for detection of bacteria. Lett Appl Microbiol 33:311–315

Yemini M, Levi Y, Yagil E, Rishpon J (2007) Specific electrochemical phage sensing for *Bacillus cereus* and *Mycobacterium smegmatis*. Bioelectrochemistry 70:180–184

Zhu H, White IM, Suter JD, Fan X (2008) Phage-based label-free biomolecule detection in an opto-fluidic ring resonator. Biosens Bioelectron 24:461–466

Part III
Synthetic and Engineered Receptors

Chapter 12
Antibody Engineering for Biosensor Applications

Neal A.E. Hopkins

Abstract Antibodies are a well-established class of affinity reagents used extensively in detection and diagnostic applications. However, while these proteins have been developed to provide considerable functionality in conventional applications (e.g. sandwich ELISA), their exploitation in emergent biosensor technologies requires careful consideration.

Historically, diagnostic technologies have been the primary consumer of antibody products. As a result, antibody development has been skewed toward satisfying the requirements of conventional diagnostic assay formats (ostensibly sandwich assays). The resulting catalogue of antibody products provides a convenient resource for potential biosensor exploitation. However, when applied outside of their traditional context (on novel surface/transducer interfaces) they can behave unpredictably and undesirably (e.g. loss of activity or specificity).

The surface immobilisation of a recognition element within a biosensor is an invariant feature of biosensor assay design. The resulting interface represents a complex system composed of interdependent technical challenges affecting assay stability, specificity, and sensitivity. It is likely that a holistic approach to interface development is required as the development of individual interface components is unlikely to deliver the technical advances required for the practical exploitation/commercialization of biosensor technologies.

The use of antibodies in biosensor applications requires a detailed understanding of their inherent properties and the interface to which they are to be tethered. This chapter illustrates the core properties of antibody structure and function and their significance in biosensing applications. The different classes of antibody reagents available to biosensor developers are discussed with a focus on recombinant antibody technologies. The opportunities available in biosensor development, regarding assay and interface design, are briefly considered. Finally, strategies for

---

N.A.E. Hopkins
Detection Department, Defence Science and Technology Laboratory, Porton Down, Salisbury, SP4 0JQ, UK
e-mail: NAHOPKINS@mail.dstl.gov.uk

M. Zourob (ed.), *Recognition Receptors in Biosensors*,
DOI 10.1007/978-1-4419-0919-0_12, © Springer Science+Business Media, LLC 2010

the bespoke engineering and application of antibodies in biosensor technologies are discussed in detail.

Keywords Antibody engineering · Biosensor interface · Stability · Affinity · Specificity · Sensitivity · Immobilisation · Monoclonal · Polyclonal · Recombinant antibody · Domain antibody

Abbreviations

ELISA	Enzyme-linked immunosorbent assay
Ig	Immunoglobulin
IgG	Immunoglobulin gamma
CDR	Complementarity determining region
hCG	Human chorionic gonadotropin
PCR	Polymerase chain reaction
PDB	Protein databank
DARPin	Designed ankyrin repeat proteins
PSA	Prostate specific antigen
PSMA	Prostate specific membrane antigen
SPR	Surface Plasmon resonance
AFM	Atomic force microscopy
QCM-D	Quartz crystal microbalance with dissipation
PEG	Polyethylene glycol (identical to PEO)
PEO	Polyethylene oxide (identical to PEG)
NSB	Non-specific binding
MA	Methacrylate
NHS	N-hydroxysuccinimide
2-MEA	2-mercaptoethylamine
tRNA	Transfer ribonucleic acid
DNA	Deoxyribonucleic acid
mRNA	Messenger ribonucleic acid
HRP	Horseradish peroxidase

12.1 Antibodies – Nature's Own Biosensor

Antibodies (or immunoglobulins) are a key part of the higher eukaryotic immune system. These proteins enable the immune system to specifically identify and eliminate harmful materials. Since their discovery in the late nineteenth century the mechanisms of both antibody generation and action have been the subject of continuous scientific enquiry. The exploitation of this research has lead to the development of powerful technologies that have changed society.

Emil von Behring and Shibasaburō Kitasato are credited with the first crude purifications of antibodies (Behring and Kitasato 1890). They discovered that serum preparations taken from an animal exposed to tetanus had a protective effect when injected into a different animal, one which had not been exposed to the disease. While their antitoxin preparation obviously had some interesting properties it was nearly another 50 years before the primary agent of immunity was identified. In 1939 Arne Tiselius and Elvin Kabat identified the "immunoglobulin gamma" protein known today as IgG (Tiselius and Kabat 1939).

Considering the role antibodies play in the immune response they may well be regarded as nature's own biosensor. Their role in nature is to specifically identify foreign materials by binding tightly to them. In binding their "antigen", antibodies signal to other components of the immune system that they have identified something that needs to be destroyed (see (Nezlin and Ghetie 2004) for a review of non antigen related antibody interactions). The complex mechanisms used by animals to generate antibodies and the mechanisms by which antigenic materials are removed, have been well studied. Interested readers may find the reviews of Peled et al. (2008) and Delves and Roitt (2000) useful.

Since their identification in the mid twentieth century, IgG antibodies have been exploited by many different technologies. Their application as therapeutic agents, as demonstrated by Behring and Kitasato, continues today and will likely become their most lucrative commercial exploitation (Visiongain 2008). However, antibodies have found application in many other technology areas with considerable effect. These versatile proteins provide the same functionality to biosensor technologies as they provide to the immune system, the specific detection and identification of materials.

12.2 Antibody Structure and Function

Biosensors require recognition elements with both high specificity and affinity. Considering their role in nature, antibodies are an obvious candidate and have been extensively used as affinity reagents in diagnostic applications. However, alternative recognition elements are increasingly available (Ngundi et al. 2006; Kolmar and Skerra 2008) and in order that they may be constructively compared their structures and the mechanisms by which they function should be understood.

12.2.1 Immunoglobulin Gamma and the Ig Fold

The immunoglobulin gamma (IgG) protein introduced in Sect. 12.1 is the most well studied member of the immunoglobulin protein family. It is also the most widely

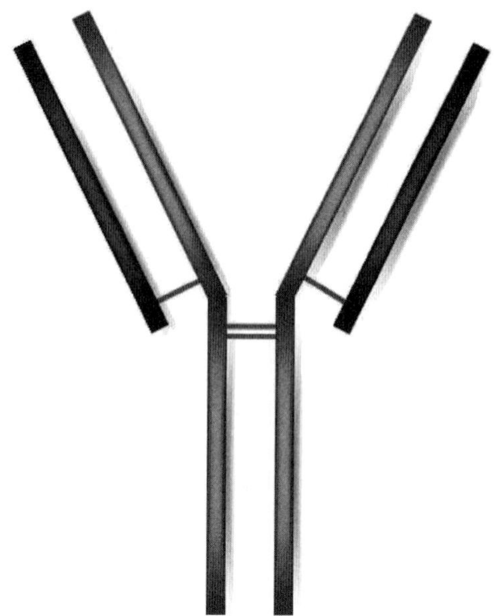

Fig. 12.1 IgG schematic showing its four composite polypeptide chains. H chains (long), L chains (short), and inter chain disulphide bonds

exploited structure in technology applications such as biosensors. Its basic structure is outlined in Fig. 12.1 and is composed of four polypeptide chains of two identical pairs; two "H" or heavy chains (50 kDa) and two "L" or light chains (25 kDa). These rudimentary names relate to the size of the composite peptide chains, which assemble into a homodimer (150 kDa).

Each of the polypeptide chains forms a series of discrete domains of comparable size (Fig. 12.2); four for the H chain (known as V_H, C_H1, C_H2 and C_H3) and two for the L (known as V_L and C_L). These domains form a "Y" shaped molecule held together through covalent disulphide bonds and noncovalent hydrophobic interactions. The arms of the "Y" are known as "Fab" regions and the stem as the "Fc" region. The Fab regions contain the antigen binding sites, while the Fc is primarily involved in host effector functions.

The two domains at the tips of each "Y" arm are "V" or variable domains. This V_H–V_L dimer displays the antigen binding site (also known as the paratope), which is unique to each antibody. As such the properties of these variable domains vary considerably between different antibodies. The remaining eight domains of the IgG are known as "C" or constant domains and provide a highly conserved stabilizing structure supporting the V domains. Many immunoglobulins are glycoproteins and so, in addition to the H and L polypeptide chains, they utilise sugars (oligosaccharides) to augment structural and functional behaviour. These oligosaccharides are covalently bound to a conserved location within the C_H2 domains (shown schematically in Fig. 12.2) and separate the two C_H2 domains of the Fc (Raju 2008).

All 12 domains of the IgG molecule are immunoglobulin (Ig) folds and share sequence and structural homology. This domain unit of around 100 amino acid residues occurs extensively in nature, serving many different functions across major

Fig. 12.2 IgG schematic showing individual Ig folds of the H chain (N- to C-terminus V_H, C_H1, C_H2 and C_H3) and the L chain (N- to C-terminus V_L and C_L in red). The oligossacharides orient between C_H2 domains of the Fc region

phylogenic branches. These include complex intracellular functions, e.g. striated muscle contraction and DNA transcriptional regulation (Tskhovrebova and Trinick 2004; Nagata et al. 1999), and extracellular functions, e.g. cell adhesion proteins (Gong and Chatterjee 2003). The abundant utility of the Ig fold illustrates the versatility of this structural motif. It is likely that both divergent and convergent evolutionary processes have generated this diversity owing to the considerable disparity between sequence homology and structural commonality (Halaby and Mornon 1998). Within this immunoglobulin superfamily the composite folds of immunoglobulins, such as those of the IgG, have both sequence and structural homology.

In the immunoglobulin family (Sect. 12.2.2), each Ig fold domain has a characteristically compact structure consisting of two β sheets, which may be likened to two slices of bread in a sandwich (Fig. 12.3). Each slice is composed of antiparallel β strands, which are connected by flexible loops that may vary in length (Fig. 12.4). The strands alternate between hydrophobic and hydrophilic residues with side chains held perpendicular to the plane of the sheet. Hydrophobic residues are orientated toward the core of the β sandwich, while hydrophilic residues project into its environment. The sandwich is stabilized by hydrophobic interactions between the sheets, hydrogen bonding between antiparallel β strands, and a characteristic disulphide bridge, which braces the two sheets together. This pleated β core is highly conserved in immunoglobulins and is the modular structural unit common to all classes of immunoglobulin (e.g. IgG, IgNARv, V_HH – see Sect. 12.2.2).

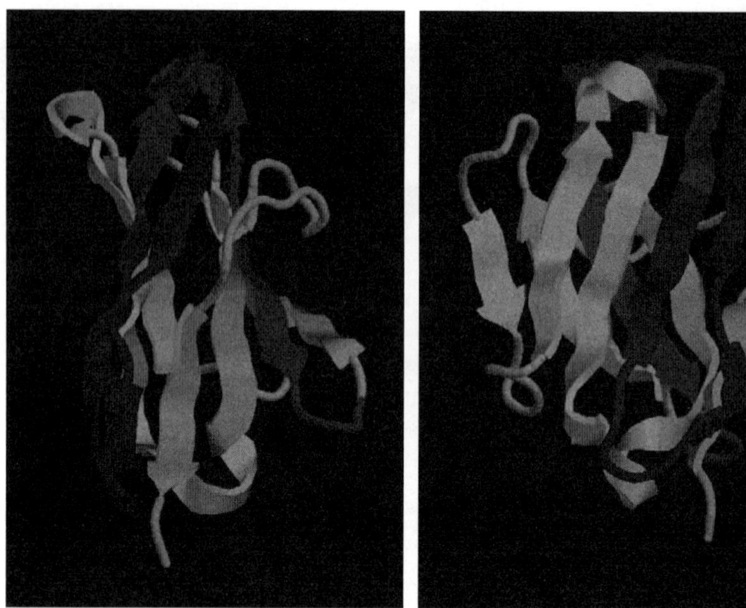

Fig. 12.3 Schematic depiction of a human V_H Ig fold domain derived using PDB file 1T2J and RasMol. Left: side view of the anti-parallel β sheet sandwich. Right: 90° rotation showing CDR peptide loops linking anti-parallel β strands at the top of the domain unit

A space filling model of the IgG structure showing the peptide chains and their composite Ig domains is shown in Fig. 12.5. This structure is shown in the same orientation in Fig. 12.6 as a ribbon diagram showing the primary peptide chain as it forms each Ig fold of the IgG structure.

The physical dimensions of the core Ig fold are approximately $2.5 \times 3.5 \times 5$ nm. Whole IgG antibodies are not constrained materials (discussed further in Sect. 12.2.3) and defined dimensions may be misleading. However, bearing this in mind, they may be generally of dimensions $15 \times 7 \times 3.5$ nm (Saerens et al. 2008).

The conserved hydrophobic core of the Ig fold typically has a high intrinsic thermodynamic stability making it particularly suited to its role as a modular protein unit. This stability results from the β sandwich core structure and the disulphides which brace them. These disulphides are not required for domain folding (Lappalainen et al. 2008) but serve to condense the β sandwich (Halaby et al. 1999) and thereby improve domain stability (Thies et al. 2002). In some animals, such as camelids and cartilaginous fish, disulphide bracing is used more extensively in the V_H domains resulting in further improvements in domain thermodynamic stability (Harmsen and De Haard, 2007; Dooley et al. 2003; Lipovsek et al. 2007). The stability of the Ig fold monomer generally persists in the multimeric structures (discussed further in Sect. 12.2.2). Indeed, of all serum proteins the IgG is thought to be the most stable (Haab et al. 2005).

For biosensing applications the most important parts of the IgG protein are its antigen binding sites. Antigen binding at the V domains is mediated by their

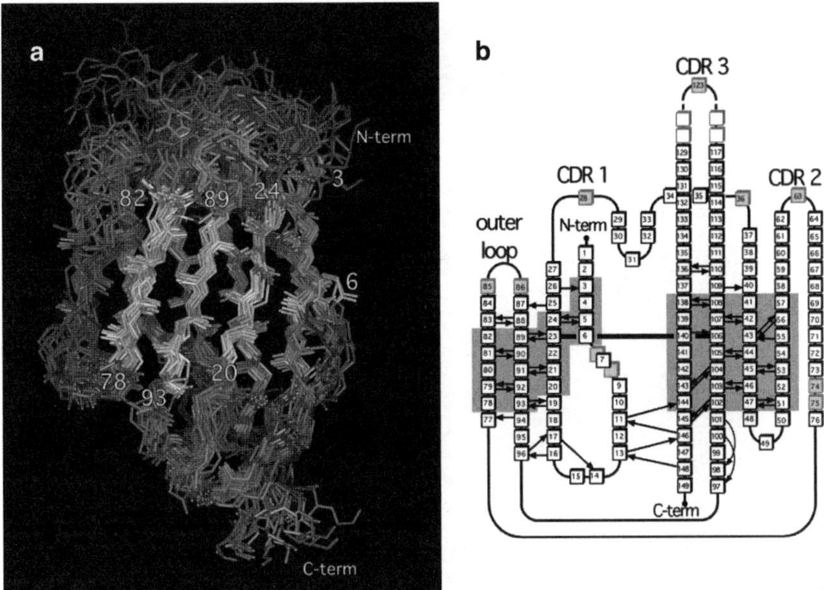

Fig. 12.4 (**a**) Representative structures of disparate Ig fold domains from $V_L\iota$ (PDB entries 1MFA, 2FB4, and 8FAB *pink*), $V_L\kappa$ (PDB entries 1A2Y, 1F58, 1FLR, 1HIL, 25C8, and 2FBJ *magenta*), V_H (PDB entries 1A2Y, 1A6V, 1F58, 1FLR, 1MFA, 1MRC, and 2HMI *cyan*), T Cell Receptor Va (PDB entries 1A07, 1B88, 1BD2, 1KB5, 1NFD, and 1TCR *orange*), T Cell Receptor Vb (PDB entries 1A07, 1BD2, 1KB5, 1NFD, 1TCR green) and T Cell Receptor Vd (PDB entry 1TVD *red*) were aligned by a least-squares fit of the C^{α} atoms of residues 3–7, 20–24, 41–47, 51–57, 78–82, 89–93, 102–108 and 138–144 (indicated in *white*). (**b**) Structure and main-chain hydrogen bonding pattern of immunoglobulin variable domains. Arrows indicate hydrogen bonds that are present in the majority of the structures in all types of immunoglobulin variable domains. The loop and turn regions that accommodate gaps are indicated in *yellow*. *Grey* areas underlie the residues whose C^{α} positions were used for least-squares superposition of the structures (With permission of Honegger and Pluckthun (2001)), copyright 2008 Elsevier

complementarity determining regions or CDRs. These are flexible loops that connect the β strands of the hydrophobic core (Fig. 12.4b). These loops form a continuous surface on one side of the V_H and V_L dimer. This binding surface faces out into the solvent environment available to bind its complementary antigen. In biosensor applications the other eight constant domains of the IgG are not explicitly required (further detailed below) though they can have some beneficial properties, such as V_H–V_L dimer stabilisation (Rothlisberger et al. 2005).

12.2.2 Conventional and Recombinant Antibodies

The IgG (introduced in detail above) is not the only structure/scaffold available for use in biosensors. While all antibody scaffolds are multimers of the modular Ig

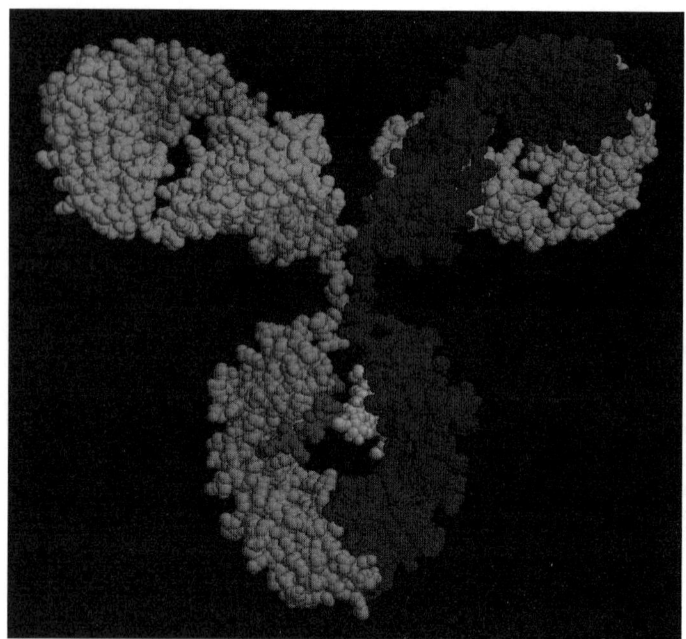

Fig. 12.5 Space filling model of a human IgG1 antibody derived using the PDB file from (Clark 1997) and RasMol. The structure is coloured by chain showing H chains, L chains, and oligosaccharides

fold, the composition of these domain units and their quaternary arrangement can yield antibodies with quite different properties (Mallender et al. 1996; Solar and Gershoni 1995).

In biosensing applications the term conventional generally refers to well-established affinity reagents developed for use in diagnostic assays. These materials tend to be polyclonal and monoclonal antibodies and are primarily of the IgG class. Historically these materials have been derived using *in vivo* animal sources which will be described in more detail in Sect. 12.3.1. Theoretically any naturally occurring antibody structure (some of which are illustrated in Fig. 12.7) may be used as a conventional antibody, but these are practically limited to those of the IgG class. This is mainly as a result of the methods used to isolate and produce monoclonal antibodies and the generally higher affinity that IgG reagents have over their structural cousins. One important aspect of conventional antibodies is their defined and static properties. These antibodies represent a finished product and are not generally amenable to protein engineering.

While the conventional Ig structures of laboratory animals were defined in the 1980s, other immune structures are still being identified across a variety of animal sources. In 1993, the heavy chain antibodies of camelids were reported (Hamers-casterman et al. 1993) (Fig. 12.8). This was followed in 1995 by another heavy chain structure found in cartilaginous fish (Greenberg et al. 1995). These are of particular interest for exploitation in recombinant approaches though they may

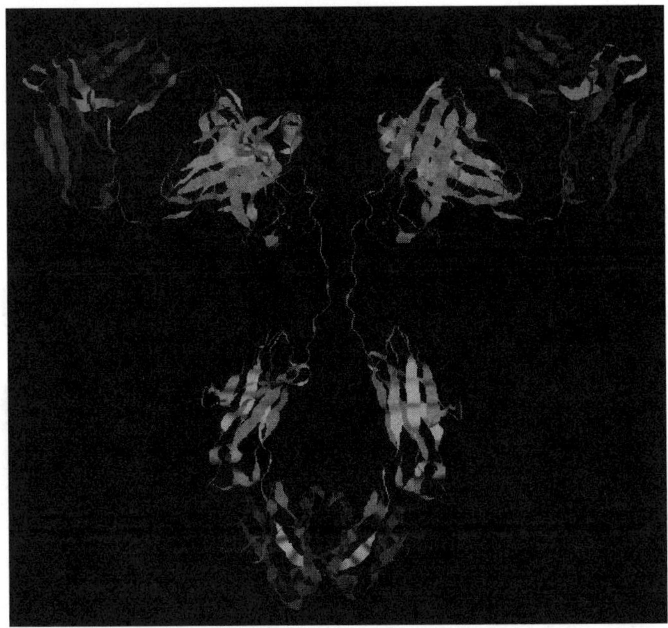

Fig. 12.6 Ribbon schematic of the human IgG1 antibody in Fig. 12.5, derived using the PDB file from (Clark 1997) and RasMol. Colour graduating from N-terminal V domains (top) to C-terminal C_H3 domains (bottom)

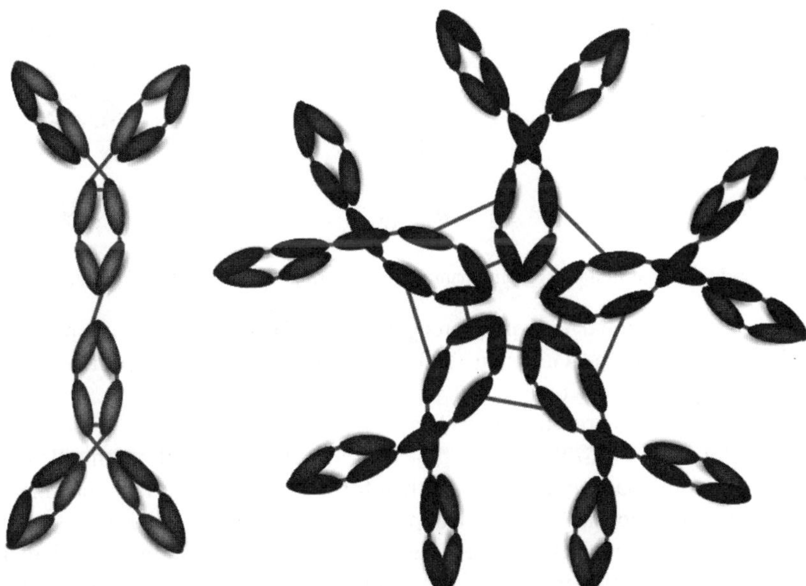

Fig. 12.7 Left: Schematic of tetravalent IgA. Right: Pentavalent IgM. Disulphide linkages connect the H chains of these multimers. Other immunoglobulin structures (e.g. IgD and IgE) also occur but are not included here as they have similar molecular architecture to IgG

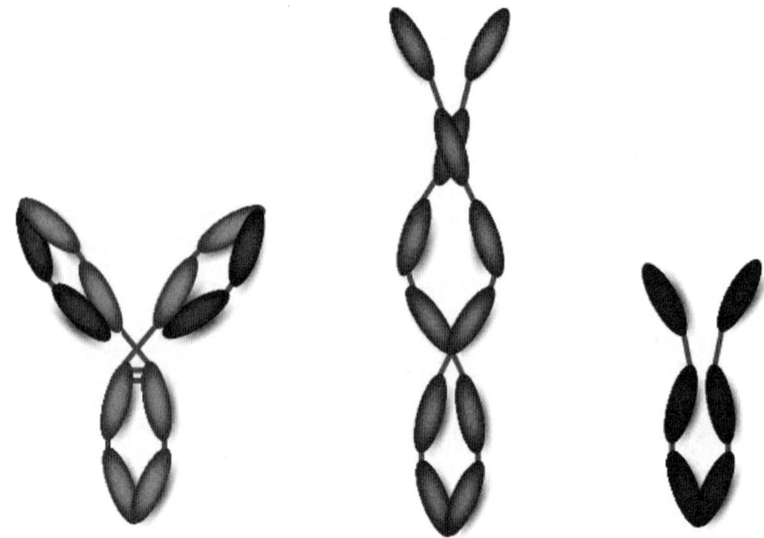

Fig. 12.8 Schematic of recently discovered antibody structures and the conventional IgG. Left: IgG for reference (H chain and L chain). Middle: IgNARv antibody from cartilaginous fish (H chain). Right: Heavy chain antibody from camelids (H chain)

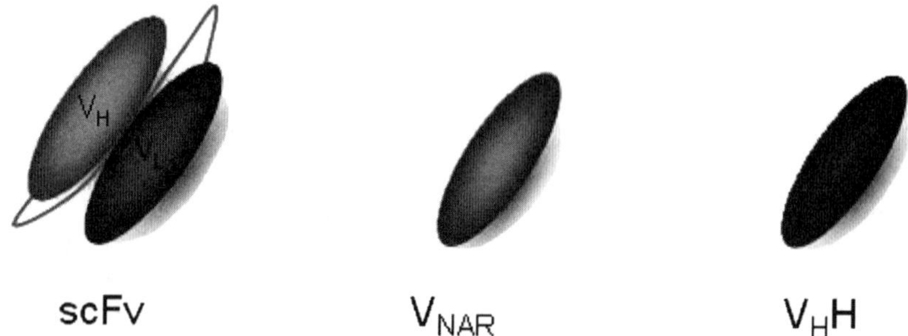

Fig. 12.9 Schematic of minimal recombinant antibody formats from structures in Fig. 12.8

be used as conventional polyclonal reagents, e.g. the heavy chain antibodies of camelids (Anderson and Goldman, 2008). It is an intriguing prospect that other structures may yet be discovered.

The term recombinant does not relate to any defined group of antibody structures; it refers instead to the use of biotechnology to genetically manipulate the antibody. Recombinant antibodies are therefore not fixed materials and may be genetically engineered for particular applications. This enables the use of practically any antibody scaffold and even the minimization of these binding elements through the removal of extraneous constant domains. Recombinant antibody structures may be any of those above (e.g. Figs. 12.7 and 12.8) but of particular interest to biosensors are those of minimal structure (Fig. 12.9).

The scFv recombinant antibody format was first reported over 20 years ago (Bird et al. 1988; Huston et al. 1988) and is the most accessible format owing to the ready conversion of existing monoclonal hybridomas into this format (Orlandi et al. 1989). The scFv is a structurally minimized monovalent version of the conventional IgG format in which the variable (Fv) domains are genetically removed and fused into a single peptide chain. As a result of their wide accessibility much of the literature concerning recombinant technology is biased toward this structure.

Recombinant structures of the recently discovered heavy chain antibodies are composed of only the V_H domain (Fig. 12.9). These structures are primarily derived from cartilaginous fish (IgNARv) and camelids (V_HH) though additional domains have been engineered from mammalian V_H domains. The V_HH and IgNARv recombinant domains are Ig folds but they have some identifying structural features/motifs. The camelid V_HH is analogous to the V_H of the conventional IgG as it uses three CDR loops to mediate binding. These domains are characterized by their extended CDR3 loops (relative to the mammalian domains). This characteristic loop serves two functions; the first obviously relates to antigen binding but the second is to help stabilize the domain. This is accomplished through the shielding of the putative hydrophobic V_L interface. This interface also carries some characteristic mutations (known as the "V_HH tetrad") which increase the hydrophilicity of the redundant V_L interface. In contrast the IgNARv (which is thought to be an evolutionary relative of the IgG) has only two CDR loops which are analogous to the CDR1 and CDR3 of the camelid domain. The VL interface is entirely absent in the IgNARv domain and the CDR3 loop (though again of longer length relative to the mammalian CDR3) is purely engaged in antigen binding.

The antibody engineering knowledge base continues to grow and with it our ability to specifically engineer these materials to incorporate novel properties. This is well illustrated by the synthetic human V_H domains which are being developed as versatile binding partners with considerable inherent stability. While the V_H domains of heavy chain antibodies provide a naturally robust domain scaffold this human V_H has been purpose built. The human V_H domain 4D5 (similar to that in Fig. 12.3) has particularly high inherent stability resulting from its Ig fold core. This domain is so stable that it can readily maintain its structure without any disulphide bonds. However, while the thermodynamic stability of these engineered V_H domains is high (Barthelemy et al. 2008), its propensity to aggregate in its denatured state is also high. Through recombinant engineering this propensity has been reduced and the resulting scaffold demonstrates many of the properties of its V_HH/IgNARv cousins (Famm et al. 2008;Christ et al. 2007).

12.2.3 The Nature of Antibody Binding

When exploiting an affinity reagent in biosensor technologies, the mechanism by which it functions (i.e. captures the analyte of interest) may be of significance. Different recognition elements have differing binding mechanisms (Yu et al. 2007) and, while all interactions are obviously governed by the laws of thermodynamics, there may be significant variation in their suitability for use in a particular technology (Brod et al. 2008). Important aspects of antibody functionality include their inherent dynamic flexibility, complementarity of paratope shape/contour, binding thermodynamics, avidity, attractive range, and resistance to mechanical disruption. Some or all of these factors may be of significance depending on the intended biosensor application.

12.2.3.1 Dynamic Flexibility and Antibody Stability

Polypeptides are complex polymers and on every structural level (i.e. primary to quaternary) they should be considered as dynamic materials. It is generally the case that proteins with greater flexibility suffer greater instability (i.e. propensity to denature and aggregate). In this regard the Ig fold (introduced in Sect. 12.2.1) represents one of the most structurally constrained and stable protein motifs (Mirny and Shakhnovich 1999;Halaby et al. 1999). This does not mean the unit is static but that it is more compact and energetically constrained, reducing its intrinsic dynamic movement. However, while Ig fold monomers are structurally constrained, multimers (such as the IgG) may have considerable inter-domain flexibility (Fig. 12.10).

Antibodies have evolved to operate primarily in solution and exhibit considerable dynamic flexibility. The importance of this dynamic flexibility and its relation to *in vivo* antibody function is poorly understood and remains a subject

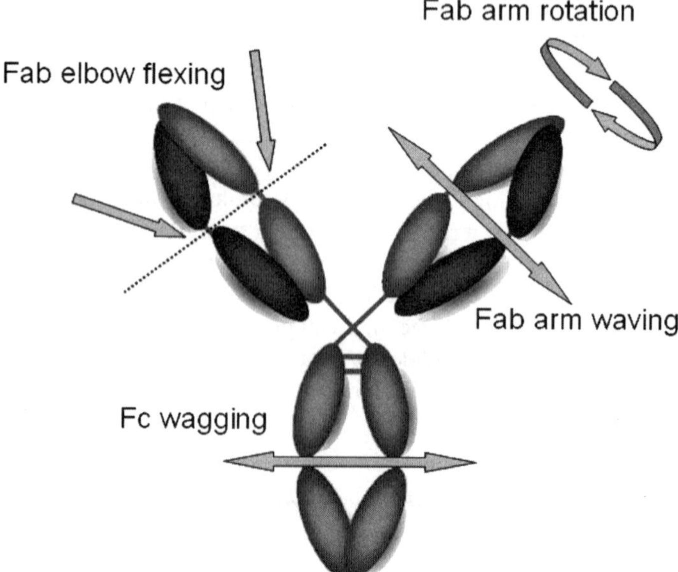

Fig. 12.10 Schematic IgG showing the structures inherent dynamic inter-domain flexibility (based on a diagram from the home pages of Mike Clark (www.path.cam.ac.uk)

of continuing academic interest (Kamerzell and Middaugh 2007; Thorpe and Brooks 2007; Thielges et al. 2008). It may be expected that the stability of such inherently dynamic structures may be compromised (Kamerzell et al. 2008). Indeed the most mobile and flexible part of the IgG molecule is the most frequently identified location of peptide cleavage (typically within the hinge region or C_H2 domain) (Gaza-Bulseco and Liu 2008; Xiang et al. 2007).

The first crystal structure of a native whole antibody was reported by Harris et al. (1992). Such studies have provided considerable insights into the fundamentals of Ig fold structure and even antibody–antigen complexes. Few of the reported structures can, however, provide any substantive comment on how antibodies appear in their native solvent environment. Other whole antibody structures have provided some insight into hinge flexibility. Saphire et al. (2002) reported the crystallization of a native antibody with Fab regions in two significantly different orientations relative to the Fc region. This work provides an intriguing snap shot of the potential dynamic gyration of Fab regions. However, crystal structures do not necessarily relate directly to native solvated conformation as they present a time averaged view of an artificially immobilized state. Crystal structures should therefore be regarded as high confidence models of protein structure and not accepted as *de facto* solvent structures.

Other practical approaches are better able to quantify natural Fab–Fc flexibility. Bongini et al. exploited cryo-electron tomography to illustrate gross parameters of Fab–Fc gyration in a murine IgG2a monoclonal antibody. In this particular study the average Fab angle relative to the Fc was ~110° with a standard deviation of ~30° (Bongini et al. 2004). However, the Fab–Fc dynamic is merely the tip of the iceberg with regard to antibody motion dynamics. Other comparatively modest dynamics such as intra- and inter-chain motions may affect variable domains and their binding interactions (Tam et al. 2007; Adachi et al. 2003). The role of flexibility *in vivo* and the biological significance of the Fc saccharide are not well understood but are likely to be of importance in IgG binding, stability, and the recruitment of *in vivo* effector functions (Mimura et al. 2000; Prabakaran et al. 2008; Liu et al. 2008b). The perturbation of antibody motion dynamics and even modest interference in inter-domain flexibilities may be translated directly into reduced binding activity (Pellequer et al. 1999). Motion dynamics may, therefore, be of considerable importance when accommodating these proteins in biosensor technologies.

12.2.3.2 Binding Surface Complementarity

In order for binding to occur it is important that interacting surfaces are contoured to complement each other (Janin et al. 2008). The closer the interaction between two surfaces the greater the potential affinity of the binding interaction (solvent entropy may be maximized and enthalpic contacts closely coordinated – discussed in greater detail in Sect. 12.2.3.3). For antibodies, the potential surface area available for antigen interaction may be in excess of 2000Å^2 (Lo Conte et al. 1999).

The natural diversity of paratope structures (antibody binding sites) may be defined and grouped into structurally related families. Chothia et al. were able to interpret early crystallization data to identify the repetitive structural conformations of CDR loops and group them into canonical classes (Chothia and Lesk 1987). This provides an interesting insight into one of the structural constraints of antibody stability and diversity. While antibody variable domain scaffolds can achieve considerable diversity in binding function, this appears to follow some structural rules to maintain the integrity of the Ig hydrophobic core.

The whole of the antibody interface is not necessarily utilized in every binding interaction. Binding energy can be distributed unevenly across the interface (Persson and Ohlin 2007). For IgG structures binding haptens (low molecular weight antigens, e.g. biotin or fluorescein), for example, the binding interface is usually between the variable domains (Persson and Ohlin 2007; Pozharski et al. 2005). The single V_H domains of camelids and cartilaginous fish are also capable of binding low molecular weight species though the binding site is obviously different as the V_L domain is absent (Anderson and Goldman 2008).

The complementary binding interface may not be a static surface, i.e. the CDRs which compose the paratope may not be tightly constrained/ordered structures. Antibodies with comparable antigen affinity may have rigid or flexible binding sites depending on the local interactions of the CDR loops. Flexible binding sites bind antigens through a slightly different mechanism compared to more constrained binding sites (Kourentzi et al. 2008). Instead of a "lock and key" interaction where one element simply coordinates with the other, flexible binding sites exhibit a degree of induced fit in response to antigen. This results in the remoulding of the binding site. In practical use this plasticity may increase the probability of promiscuous antibody binding as the binding site may also remould itself to accommodate materials other than its cognate antigen. Interestingly, *in vivo* antibody selection processes (i.e. affinity maturation) have been found to progressively reduce plasticity of antibody binding sites as affinity and stability are improved (Sinha et al. 2002).

Considering the importance of surface complementarity in binding interactions, modest distortions of binding domains may skew binding thermodynamics and rather than promoting binding may well repel it (van Oss 2003). The plasticity of antibody structures is often poorly appreciated when they are applied in sensing technologies.

12.2.3.3 Antigen Binding Thermodynamics

In order to understand protein binding interactions it is important to briefly consider the forces which fold and stabilize proteins generally. All amino acids within a protein contribute to its thermodynamic behaviour with their enthalpic and entropic contributions. These properties do not manifest themselves independently; rather they are activated by the solvent environment (which in nature is invariably aqueous). To fold a polypeptide from primary chain to organized quaternary structure may seem both entropically and enthalpically expensive considering that the primary peptide is chaotically distributed and polar groups are well solvated. However, the overall entropy change, driven by the removal of trapped hydration shells surrounding hydrophobic residues, is favourable (Lappalainen et al. 2008). For a comprehensive discussion of structural protein thermodynamics, readers may refer to the book by Cooper (1999).

When two proteins interact to specifically bind each other, the process occurs in the same hydrodynamic environment which drives protein folding (Cooper et al. 2007). However, the majoritatively entropic force driving protein folding cannot impart any degree of selectivity to binding. Interactions which occur via this mechanism generally result in poorly controlled and promiscuous interactions (Sinha et al. 2002; Calabrese et al. 2008; Chang et al. 2008). In antibody interfaces it is enthalpic contributions which endow the binding site with the required specificity (Moreira et al. 2007a), though the stability of the bound complex may well be ascribed to entropic contributions (Sundberg et al. 2000).

While it is convenient to consider enthalpy and entropy separately, this considerably oversimplifies protein binding interfaces. Amino acids are diverse and while they may be conveniently divided into different functional categories (aliphatic, hydrophilic etc.) there are residues which are capable of multitasking. Tyrosine has been identified as an indicative residue of putative protein binding sites. Clackson and Wells, (1995) demonstrated that these residues can make significant individual contributions to the free energy of protein binding and termed such features "hot-spot" residues (for a recent review see Moreira et al. (2007b)). These versatile residues can contribute to binding in several ways including hydrogen bonding, as well as van der Waals, hydrophobic, and amino-aromatic (cation-π) interactions (Shiroishi et al. 2007). In antibody–antigen complexes tyrosine has special significance (Rubinstein et al. 2008). In point mutation studies the loss of such residues from binding sites has been shown to significantly reduce antibody affinity (Shiroishi et al. 2007). Tyrosine also occurs frequently in the binding sites of synthetic protein engineering scaffolds developed through random mutagenic protein engineering (e.g. ankyrin repeat proteins (Sennhauser and Grutter 2008) and fibronectin FnIII scaffolds (Gilbreth et al. 2008)). This reinforces the general importance of this versatile residue in high affinity protein interactions.

Obviously not all significant antibody interactions can be ascribed to tyrosine. Binding is a complex process and it is important to remember that both entropy and enthalpy are components of all interactions. In protein interactions, the components of free energy may counteract one another. Brummell et al. (1993) demonstrated that mutations in antibody CDR loops may improve enthalpic contacts while reducing the contribution of entropic forces and *vice versa*. In this study, the relationship between entropy and enthalpy was found to be directly proportional. It is interesting to note that high affinity binding interfaces exploiting primarily entropic or strongly charged enthalpic forces (i.e. significantly biased to one or the other), generally result in more promiscuous binding interactions (Birtalan et al., 2008). This suggests that protein surfaces mediating high affinity yet specific interactions accomplish this through the titration of entropic and enthalpic extremes.

All antibody binding contacts are not necessarily direct; solvent may actively participate in binding interactions (Rodier et al. 2005). Indeed, while some antibody interactions maximize entropy through complete surface dehydration, others trap water molecules at considerable entropic cost. The reduction in entropy is offset by the coordination of trapped water in favourable and specific enthalpic bridging bonds (Yokota et al. 2003).

12.2.3.4 Affinity, Avidity, Attraction, and Mechanical Resistance

Binding site affinity (K_D) is a mathematical expression combining two properties of binding site behaviour; association rate K_a (i.e. propensity of an antibody to bind to its antigen) and dissociation rate K_d (the propensity of the bound complex to break down) where $K_D = K_d/K_a$ (a review is available concerning protein–protein interaction kinetics (Schreiber 2002)). Antibody binding affinity and its explicit relation to biosensor functionality is discussed in more detail in Sect. 12.6.3.1.

Most naturally occurring antibodies (Fig. 12.8) present two binding sites but some present more (e.g. IgA and IgM – see Fig. 12.7). The effect of increasing valency (binding site density/number) is to increase the functional affinity of the resulting multimer. This is a molecular property known as avidity. Multimerisation does not change the affinity of the composite binding sites, but it does change the overall affinity of the molecular multimer. It does this by increasing the probability that a binding collision will occur and that it will reoccur should the antigen dissociate (Yoshimoto et al. 2008). In addition, by increasing the number of bonds made between the multimer and antigen the probability of complete dissociation of the molecular species is greatly reduced. IgM for example is well known for its lower affinity binding sites relative to IgG (Strandh et al. 1998), but as a multimer it can perform adequately as an affinity reagent (Campbell and Mutharasan 2008).

The thermodynamic forces driving antibody binding are not entirely passive, i.e. binding events are not governed solely by diffusion as may be expected. The enthalpic and entropic properties of antibody binding sites have an attractive force associated with them which are distance dependant. While these forces are significantly affected by the ionic composition of the solvent, in physiological conditions (or comparable ionic environments) these forces may act over tens of nanometres. Electrostatic charge effects act over the longest distance and are of primary importance in the initial association phase of antibody binding (Sinha et al. 2002). These forces coordinate the binding surfaces rotating them into optimal orientations for association. The docking phase occurs when the binding surfaces of the associated pair are within one nanometre of each other (Leckband et al. 2000). At this range entropic forces of surface dehydration are effective and stabilize the bound complex (van Oss 2003).

The mechanical strength of binding interactions may also be of significance, especially when considering flow cell design and microfluidic engineering (e.g. shear forces from solvent flow). Single molecule atomic force studies of antibody interactions have generally found that forces in the region 50–100 pN are enough to dissociate the bound complex (though the rate at which the force is applied may be significant) (Verbelen et al. 2007). The prediction of antibody behaviour in this respect is not straightforward as binding affinity is not indicative of force resistance. For example, the biotin streptavidin bond has an affinity of ~10^{-15} M^{-1} (Green, 1975) and a mechanical resistance of ~250 pN (2.5×10^{-10}N) (Zhou et al. 2006). Compared to antibody–antigen interactions this value seems disproportionately low. Berquand et al. tested an antibody lysozyme complex with an affinity of ~10^{-9} M^{-1} with an associated binding resistance of ~50 pN (Berquand et al. 2005b). Given the considerable disparity between affinity and mechanical resistance, affinities are clearly no indication of mechanical strength. A correlation between force resistance and thermally induced binding dissociation has been reported (Schwesinger et al. 2000). This relationship suggests that entropic contributions may be of greater significance in force resistance. High avidity interactions increase the mechanical stability of captured analytes with linear proportionality (Berquand et al. 2005a). Such surfaces may greatly improve the rate of binding in high flow applications (Catimel et al. 1997).

12.2.4 General Thermodynamic Stabilities of Antibody Scaffolds

The most widely used conventional and recombinant antibody reagents are the IgG and scFv respectively. In academic studies these formats have been engineered to carry identical variable domains and their relative stabilities have been determined (Pantoliano et al. 1991). Stability variation in V domain identical antibodies is entirely attributable to the manner in which they are presented. The IgG and scFv will be used here to illustrate the potential variability of antibody scaffold stability.

The V_H–V_L domain association of both IgG and scFv is reversible and has an intrinsic affinity of between 10^{-9} and 10^{-6} M^{-1} (Pluckthun 1992). In the IgG structure the variable domain dimer is constrained by the closely associated C_L and C_H1 domains (to which they are covalently bound). In contrast, the V_H–V_L dimer of the scFv has no supporting structure and the domains associate by means of the comparatively weak hydrophobic interaction acting over the inter-domain surface area of between 583 and 874Å^2 (Padlan 1994). As the V_H–V_L interaction is not covalent it may be influenced by solvent ionic strength and pH (Arndt et al. 1998). The inter-domain tether of the scFv (between the V_H–V_L dimer) effectively restricts the freedom of the dissociated domains, increasing the probability that they may reassociate. This linker has been found to considerably improve scFv stability. However, while this pushes the equilibrium of heterogeneous structural states (Fig. 12.11) toward the associated form it does not entirely prevent dissociation. It is this inevitability that makes scFv vulnerable to promiscuous interactions as the hydrophobic regions are transiently exposed. At its most benign this promiscuity results in domain swapping, a

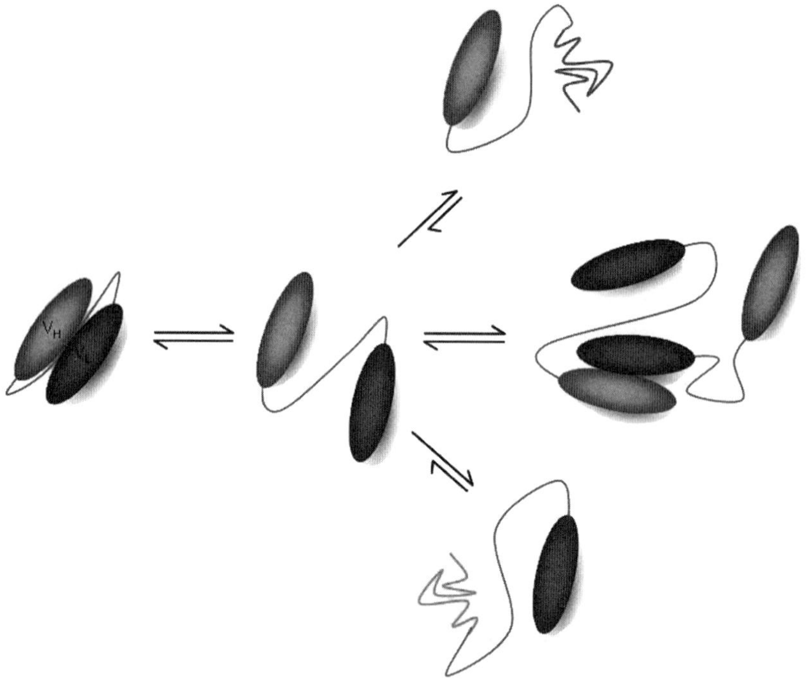

Fig. 12.11 Schematic showing single chain antibody inter-domain dynamics. The association of the twin variable domains is noncovalent, which leads to their inevitable dissociation. While transient this dissociation exposes the hydrophobic domain interface which may subsequently participate in domain-swapping or lead to denaturation of the V_H or V_L domain and irreversible aggregation

phenomenon in which the dissociated domains may interact with other dissociated scFv molecules. However, at its most severe, this may result in the aggregation and permanent denaturation of scFv or their promiscuous uncontrolled binding. It is this fundamental instability which has limited the exploitation of the scFv scaffold.

While the weak inter-domain affinity of the variable domains is the most fundamental limitation of the scFv scaffold (Reiter et al. 1994) this may be exacerbated by the poor inherent stability of one of the domains (Worn and Pluckthun 1999). This may well be offset by a cooperative stabilizing effect of its partner domain (as the entropic cost of maintaining the domains' integrity is reduced (Worn and Pluckthun 1998)) though this is again dependant on the affinity of the dimer interaction. For this reason many of the early investigators of the scFv format quickly moved onto the Fab format to improve the general stability of the recombinant product (Hoogenboom et al. 1991; Demarest et al. 2006). This, however, doubles the size of the product and complicates the production methods required (e.g. titrating the potentially differential expression of H and L chains).

Recently the scFab format has been reported where the composite Fab chains are fused in a similar fashion to the scFv, the only difference being that it is twice the size with a longer linker (Hust et al. 2007). Whether this format provides any tangible benefit remains to be seen though this appears to be a convenient method for the production of the Fab scaffold.

Considering the unpredictability of multi-domain interactions contributing to scaffold instability (e.g. IgG, scFv, and Fab), and the relative robustness of the minimal Ig fold, smaller structures of single Ig domains have been investigated. Naturally occurring heavy chain antibodies (Sect. 12.2.2) use only one variable domain in their binding interactions. The V_H domains of these antibodies have evolved to operate in isolation which necessitates a high level of inherent stability. Despite the structural differences of the V_HH and IgNARv heavy chain variable domains (Sect. 12.2.2), their antigen binding properties and thermodynamic stabilities are comparable. A stabilizing feature of both IgNARv and V_HH is the use of additional intra-domain disulphide bonds. These serve to constrain the dynamics of the domain unit and therefore have a considerable stabilizing effect. As a result, these single domain antibodies exhibit unusually high thermodynamic stability. In addition, once denaturing conditions are removed (e.g. cooling following exposure to elevated temperature) the domains typically recover well and refold spontaneously into their native structural conformation. It should be stated here that the stability of IgNARv and V_HH domains are themselves variable and depend on peptide primary sequence; however, the inherent properties afforded by their structural features do provide a more inherently robust domain (Omidfar et al. 2007).

12.3 Antibody Technologies

12.3.1 Conventional Antibodies

12.3.1.1 Polyclonal Antibodies

Polyclonal antibodies are the most basic antibody affinity reagent available. They are produced by an animal in response to an administered immunogen (the antigen) and represent a snap-shot of its immune response at the time of exsanguination (blood extraction). The extracted blood is separated into a crude serum fraction containing all soluble proteins and insoluble debris (as Emil von Behring had done in 1890). This serum can then be purified to enrich the antibody fraction (e.g. by protein G affinity chromatography). The resulting purified fraction is a complex mixture of many different antibodies including antibodies of interest (i.e. those binding the antigenic epitopes of the immunogen) and, inevitably, any other antibodies present in the animal at the time of exsanguination. It is possible to further improve this fraction through antigen affinity purification where the antigen is used to selectively purify only those antibodies which bind to it. This drastically increases the cost of the antibody as a reagent as yields can be low and the process time consuming (Petrenko and Vodyanoy 2003). It does, however, improve the antigen binding activity of the fraction though the material will still be polyclonal as it contains antibodies which bind all antigenic epitopes of the immunogen.

The use of polyclonal antibodies in analytical capabilities peaked in the late 1970s with demand outstripping supply. In some laboratories, the death of a polyclonal donor (e.g. a rabbit) was a significant blow to the capability of the laboratory. In such times it might have been necessary to produce many batches of a replacement polyclonal antibody before a preparation of sufficient activity was isolated.

The hCG case study: Human chorionic gonadotropin (hCG) is a hormone produced at elevated levels in pregnancy and has been exploited for decades as a biomarker for pregnancy testing. The 36.7 kDa protein is a heterodimer consisting α and β subunits. The β subunit is unique to hCG; however, the α-subunit is identical to other hormones such as luteinizing hormone, follicle-stimulating hormone, and thyroid-stimulating hormone. *In vivo* these hormones occur together and an exclusively polyclonal-based hCG assay will bind all these hormones through the shared alpha subunit. The sensitivity of the assay is therefore reduced as the assay specific signal must compete with the background hormonal noise.

The issue of heterogeneity in epitope binding is the principle limitation of polyclonal antibodies as illustrated by the hCG case study. This is of increasing severity when more complex systems are interrogated (e.g. detection of bacteria). The likelihood of common elements occurring between analyte and background interferents increases with the complexity of the analyte (Zhao and Liu, 2005). Bacterial cells, for example, are considerably heterogeneous on account of the huge variety of proteins, saccharides, and lipids they comprise. The immunogenic differences between closely related bacteria (e.g. *Salmonella enterica* serotype Typhimurium and *Salmonella enterica* serotype Enteritidis) are negligible and polyclonal antibodies generated using whole cell preparations can bind both strains (Barlen *et al.*, 2007). However, the practical implications in strain identification may be significant if action is required, for example, to locate and mitigate food poisoning hazards (Barlen et al. 2007). While polyclonal cross reactivity is, to some extent, tolerable in the hCG assay, cross reactivity with more complex analytes (e.g. whole cells) can preclude their use.

It is primarily the complexity of the immunogen which generates variability and low specificity in polyclonal antibody products. When simple molecules (i.e. haptens) are used as immunogens, antibodies with virtually identical kinetic properties can be produced, even when different animals of the same species are used (Ostler et al. 2002). When polyclonal antibodies are applied in detection systems, they can out perform their more technologically advanced and expensive counterparts, monoclonal and recombinant antibodies (Le Berre and Kane 2006). However, with complex antigens their variability is too great which can easily defeat the assay. This complexity may be considerably reduced through the substitution of the native immunogen with synthetic materials which represent unique portions of the natural material (Reeves and Lummis 2006). Such an approach may simultaneously reduce the cross reactivity of the antibody and improve batch to batch reproducibility, especially when combined with subsequent antigen affinity purification (Bayry et al. 1999).

Despite the issues surrounding polyclonal antibodies they are, in the short term, the most economical affinity reagents. They are often employed in proof of principle studies where the intended goal is not an analytical assay but merely the demonstration that a certain technology is capable of detecting a material. Their heterogeneous nature provides an advantage in this application as their varying stability can afford some degree of tolerance to often unrefined or crude application methods. This can ensure that proof of principle experiments are achievable without extensive assay optimization and interface development. Such results obscure the fact that further work is essential to deliver the anticipated capability. This is one of the principal barriers preventing new technologies entering commercial markets.

12.3.1.2 Monoclonal Antibodies

As outlined above, polyclonal antibodies are considerably heterogeneous, which can provide modest advantages in certain applications, but they have considerable limitations in the longer term, the most obvious being their finite supply and binding promiscuity. Polyclonal materials are the net product of all B cell lymphocytes within the donor animal. Each lymphocyte produces a single antibody of defined and static properties (i.e. every antibody secreted by the cell is identical). These antibody secreting cells can be isolated from donors and grown outside the animal into monoclonal cell cultures. However, the continuous culture of these cells results in the rapid loss of antibody secreting ability and so the process is not viable as a production method.

In 1975, researchers in the UK delivered a step change in antibody technology. Georges Kohler and César Milstein published a method for the creation of B cells which retained antibody secreting capacity in continuous culture (Kohler and Milstein 1975). This process involves the fusion of an antibody secreting cell with a myeloma cell (plasma cancer cell) to combine the properties of each. The resulting cell fusion is capable of secreting monospecific antibody and is continuously culturable. The discovery of this reproducible and scalable method for the production of monoclonal antibodies was a considerable advancement for which they were awarded the Nobel prize (Garfield 1985).

Kohler and Milstein's technology advance was not protected by patent but gifted (perhaps inadvertently) to the scientific community. The commercial revenue of such a patent could have been considerable but its free access accelerated the development and exploitation of antibody technologies. Unfortunately the resulting boom in exploitation resulted in considerable disappointment in certain applications. This might have been inevitable owing to the considerable commercial possibilities of a technology which was not fully understood. It was anticipated that monoclonal technology would deliver "silver bullets" with which to specifically target disease but it quickly became apparent that antibodies raised in one species (e.g., murine) were not suited to treating disease in another (e.g., humans). Such treatments were defeated by phenomena such as the HAMA (human anti mouse antibody) response and these early therapeutic antibodies returned to the research lab for further development.

The hCG case study: Monoclonal antibody technology enabled the production of hCG specific antibodies which bind only the specific β subunit of the hormone. When deployed in hCG assays these new reagents provided a level of specificity to screen out other hormones reducing assay noise. The resulting assay can diagnose pregnancy at an earlier stage compared to a wholly polyclonal assay (Chard, 1992). While both assays can deliver the correct result, the sensitivity and reliability of detection in the monoclonal assay provides a considerable commercial advantage in the consumer diagnostics market.

Following the monoclonal technology boom, expectations levelled out and by the late 1980s murine monoclonal antibodies had established a firm place in diagnostic laboratories (replacing polyclonal antibodies). These new materials enabled the development of new technologies which have changed our approach to health care diagnostics and disease control. Monoclonal antibodies enabled strain differentiation in proteomically similar disease causing organisms. This has aided the effective monitoring and treatment of these conditions (e.g. strain differentiation in hepatitis infection).

Monoclonal antibodies are not a diagnostic panacea as the technology has some potentially significant limitations. For example, it is not possible to control the specificity of the antibodies produced. An immunogen is still delivered to the animal in order to generate the antibody and as a result some key limitations may result. For example, the animal is likely to produce antibodies to the most abundant and antigenic epitopes available, a phenomenon known as immunodominance. This makes good sense for the animal as it will be able to combat the antigen most effectively, but these epitopes may not be unique to the antigen and so the antibodies will not be specific (Morozova and Morozov 2008). While immunization with complex antigens may produce monoclonal reagents of desirable specificity (Hearty et al. 2006), this is a hit and miss affair. As with polyclonal antibody production, the response of the animal may be steered to some degree through the manipulation of the immunogen, e.g. use of specific recombinant proteins or representative synthetic peptides. This approach is obviously limited by the quality of information known regarding the natural immunogen and the ability of synthetic approaches to adequately mimic it.

Another potential limitation of monoclonal technology comes from the B cell extraction and fusion process. An estimated 10^{15} B cell lymphocytes may be available in the immunized animal (Tonegawa 1983), many of which will bind the immunogen, but there will inevitably be a background of those which do not. The extraction and fusion process may result in the isolation of hundreds of successful hybridoma cell lines but this represents a considerable loss of potential binders from the animal donor (those 10^{15} potential lymphocytes). This drastically reduces the chance of generating a useful reagent as only a fraction of a percent of the potential antibody producing cells make it to the fusion and selection process. The hundreds of hybridomas that do make it require intensive efforts to screen them for antibody production (as not all fusions result in antibody secreting cell lines) and binding specificity (as not all secretors produce useful antibody). Having identified those hybridomas which do secrete antibody and bind the immunogen, these must then be screened for specificity with any known interferents. The epitope identified by a specific monoclonal antibody may occur in other materials or have partial similarity with other epitopes. While these side reactions may not be of significant affinity, they can result in elevated assay noise leading to a reduction in sensitivity or assay false positives or negatives (Michaud et al. 2003). Even when a suitable hybridoma has been identified it is still possible for cell lines to lose their viability or cease antibody secretion with the potential loss of the diagnostic assay or associated capability. This may often be attributable to a certain technology practitioners' poor cell culture techniques as hybridoma cell lines are an artificial cell type composed of a genetic mosaic of myeloma and B cell lymphocytes. As a result they can become increasingly heterogeneous over time and without careful subcloning and seed stock maintenance these cultures can lose their clonal integrity (Bibila and Robinson 1995).

The process of producing useful monoclonal antibody hybridomas requires great skill and considerable time. It therefore requires considerable investment with no guarantee that the product will be of use. However, once a useful hybridoma has been identified it may be used to produce grams of specific antibody on an industrial scale. The maturity and widespread commercialization of mammalian cell culture makes the production process economical. However, this technology is ultimately limited by the selection process of useful hybridoma clones and the skill of those producing and maintaining them.

An additional issue of monoclonal antibodies is that their inherent structural stabilities are notoriously variable (Liu et al. 2008a; Ionescu et al. 2008). While the stability of the modular Ig fold is high, this variability occurs as a result of the inherent variability required of V domains in antigen binding. This results in heterogeneous affinity, stability, and durability in antibodies against different analytes which may have significant implications in technology development (explored further in Sect. 12.4).

12.3.2 Recombinant Antibodies

Recombinant antibodies began to develop into a technology domain in the late 1980s (Skerra and Pluckthun 1988). Somewhat surprisingly, this technology has yet to realize its potential and is notable primarily through its absence as an applied commercial technology (Haab 2006). While this is likely the result of a complicated patent environment (not seen with monoclonal antibody exploitation, Sect. 12.3.1.2), it also indicates there are technical barriers yet to be overcome.

12.3.2.1 Basic Principles

Conventional antibodies (discussed above) have static properties and should an antibody fail to function appropriately in a given technology (e.g. ELISA), it is usually discarded and another sought. Recombinant antibody technology fundamentally changes this long standing tradition. In principle, every facet of recombinant antibody behaviour may be tailored, to fit the antibody around the application. This is the basic tenet of recombinant technology.

The process of recombinant antibody development is outlined in Fig. 12.12 which may equally describe *in vivo* antibody development (though the terminology is different). The *in vitro* mimicry of natural antibody selection/development has been a goal of protein engineers since the conception of recombinant antibody technology. This has been driven by the ethical need to remove/reduce the use of animals in antibody production, but there are also considerable technological advantages which will be discussed in detail in Sects. 12.3.2.2–12.4.

In vivo antibody development can be divided into two phases. In the first phase a precursor antibody is selected from a diverse library of different antibody structures. This is analogous to the upper and right boxes of Fig. 12.12

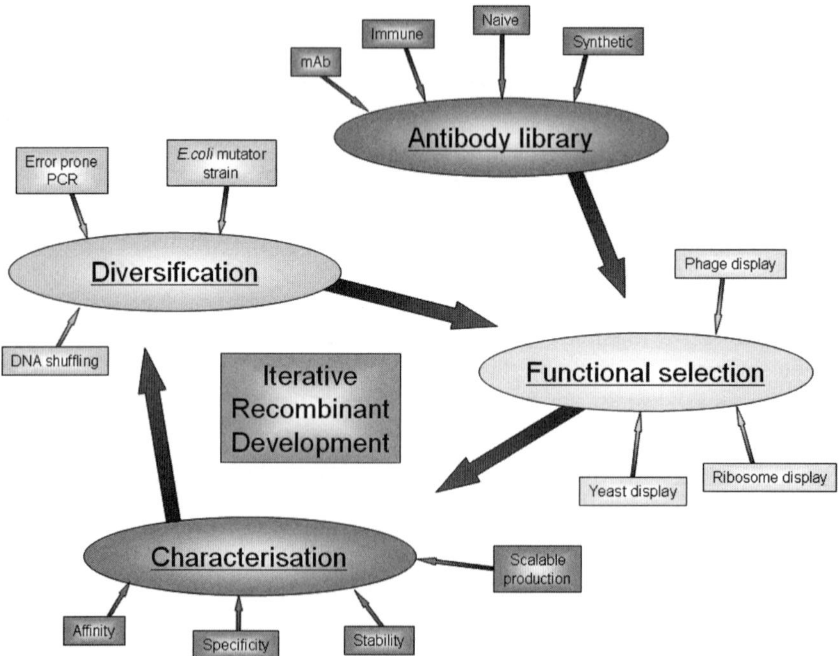

Fig. 12.12 The cycle of iterative recombinant antibody development. Once a lead has been identified through functional selection (right boxes) from a variety of potential sources (upper boxes), it may be improved through an iterative process of characterization (lower boxes), diversification (left boxes) and reselection to develop desirable properties

where antibodies are displayed on circulating B cell lymphocytes, which are effectively waiting to bind antigenic materials. In the second phase, this precursor is matured through controlled mutagenesis into a product with more useful *in vivo* functionality (e.g. stability, specificity and affinity), analogous to the lower and left boxes in Fig. 12.12.

The success and versatility of *in vivo* antibody generation relies on three core processes. Firstly, the considerable diversity within the structural library of 10^{15} lymphocytes ensures there is a reasonable probability that several candidate antibody structures will be selected for mutational improvement. Secondly, these candidate structures are positively selected through their affinity to antigenic materials, i.e. it is antigen binding that stimulates their continued development. Thirdly, somatic hypermutation which follows this initial binding interaction creates sub-libraries of the parental antibody structure. This involves the controlled introduction of mutations into the antibody structure generating a sub-library of related antibody structures with numerous subtle variations. These sub-libraries are then subjected to a repeat selection process on the basis of their *in vivo* functionality, primarily antigen affinity (Rajewsky 1996).

The considerable complexity of *in vivo* processes continues to confound research scientists in their pursuit of a truly parallel *in vitro* method. However, considerable practical advances have been made which, when taken together, enable the development of products with equal or greater utility relative to those derived *in vivo*. Some of the *in vitro* methods developed to mimic this complex process are included in Fig. 12.12 as satellites around the main technical processes. Some of these will be discussed in more detail in the following sections.

12.3.2.2 The Power of Recombinant Selection and Reengineering

The natural immune system can utilize a vast diversity of unique antibody structures estimated to be in the region of 10^{15} unique structures (Tonegawa 1983). Assuming such diversity is available outside of these *in vivo* processes (discussed further in Sect. 12.3.2.3), how can such diversity be practically interrogated *in vitro* to select antibodies of interest?

Recombinant affinity selection methods are generally referred to as bio-panning. These have been specifically developed to accommodate and interrogate libraries of structural diversity. The natural process of B cell pre-somatic clonal selection (introduced Sect. 12.3.2.1) is a form of bio-panning. Each lymphocyte cell represents a discrete package displaying a unique antibody which is tethered to the cell's membrane. The cell contains the genetic script to construct the antibody and so the genotype and phenotype of each unique antibody are contained within this discrete cellular unit. *In vitro* bio-panning display vectors seek to achieve the same objective but substitute the mammalian cell for something more robust. The most widely used display vector system is phage display first reported in 1985 (Smith 1985). Phage display has been well reviewed (Petrenko and Vodyanoy 2003; Conrad and Scheller

2005; Sidhu and Koide 2007) but a basic introduction will be given here to illustrate the bio-panning concept.

Antibody phage display involves the genetic fusion of the minimal antibody combining site (e.g. the V_H–V_L domains) to a coat protein of a virus (e.g. pIII protein of the M13 *Escherichia coli* bacteriophage) in such a fashion as it is displayed on the virion surface (McCafferty et al. 1990). The phage then provides a discrete package combining the genotype and phenotype of the antibody fragment. This format enables the affinity separation of virions of interest from the structural library of different immunoglobulin fragments (see Fig. 12.13 below).

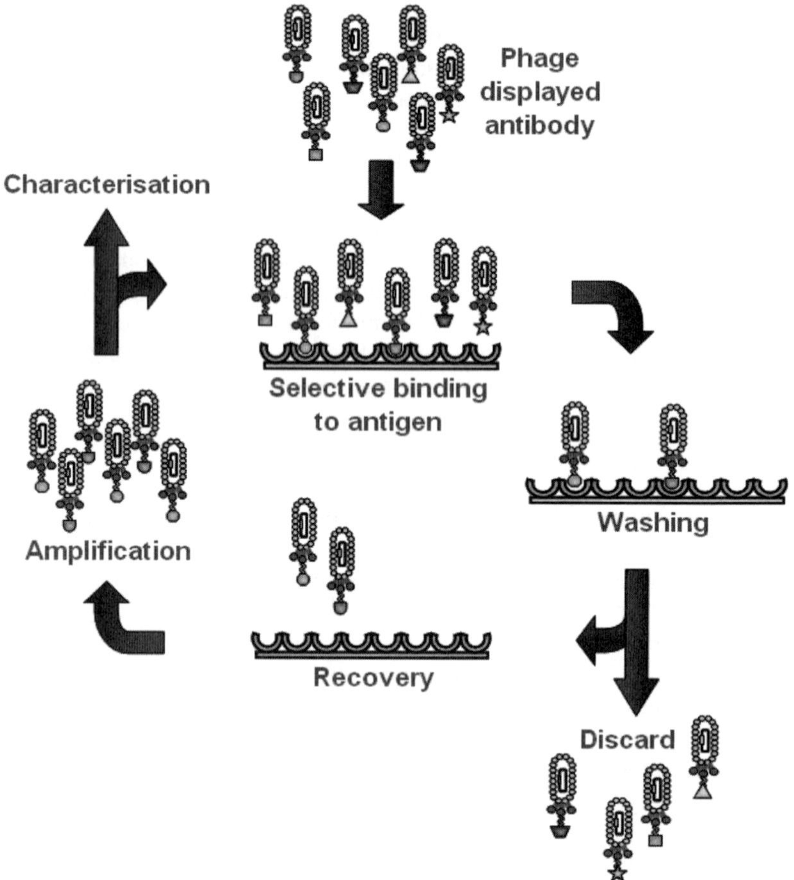

Fig. 12.13 Schematic of phage display bio-panning. Antibodies are fused to the coat protein of the phage and incubated with a surface immobilized analyte of interest (*top*). Unbound phage particles are washed from the surface (*right*) and those which bound to the analyte are eluted (bottom) and amplified in the viral host, e.g. *E. coli* (*left*). The antibody may be cloned out of the phage and characterized or the cycle may be repeated to enrich the content of the library for high affinity/specific binders

While the phage vector is a convenient and relatively robust system (Jung et al. 1999), the diversity which may be practically accommodated is generally between 10^6 and 10^{11} unique antibody structures (Leong and Chen 2008). Maximal diversity is a function of the display system used and other bio-panning systems have been developed in the pursuit of more accommodating display vectors. In the case of phage display, diversity is constrained by the transformation efficiency practically achievable in *E. coli* (which is essential to the process). Phage display is, at best, ten thousand times smaller than the potential diversity of antibody structures (i.e. 10^{15} B cell somatic possibilities). This adversely affects the probability that a useful antibody may be isolated from such a library. New display vector systems such as ribosome display and mRNA display (Lipovsek and Pluckthun 2004) accommodate greater diversity and can be used to practically interrogate libraries of between 10^{12} and 10^{13} (Hoogenboom 2005). Relative to the phage display system, this is a considerable improvement, but it is still considerably lower than theoretical *in vivo* antibody diversity.

While library size fundamentally limits the possibility of producing immediately useful antibodies it is possible to improve candidate materials akin to *in vivo* processes. Once an antibody is isolated it may be mutated into a sub-library through a variety of means (e.g. error prone PCR and DNA shuffling (Luginbuhl et al. 2006)). This sub-library may then be subjected to reselection. Ribosome and mRNA display systems use PCR as an integral component in the recovery of candidate materials at the end of the bio-panning process. This enables the use of error prone PCR to diversify the library incrementally through several rounds of bio-panning (He and Taussig 2005). This mimics the natural process of somatic hypermutation.

Several diversification strategies may be employed simultaneously. For example, error prone PCR may be used to introduce point mutations in the coding sequence. The resulting mutations may then be recombined through the use of DNA-shuffling. If such an approach is used in repeated selection rounds, beneficial mutations may be added to and recombined in many different combinations. Such an approach has been termed "sexual PCR" as it accelerates mutation propagation and the selection of cooperative or additive mutations.

The greatest advantage of using *in vitro* bio-panning methods is that selective pressures, which dominate *in vivo* antibody development, may be removed and new pressures exerted, which are designed for specific applications (Choi et al. 2008; Demarest and Glaser 2008). This process enables a library of structures to be subjected to engineered selective pressures in order that antibodies with desirable properties may be specifically isolated (e.g. stability, specificity, and affinity). There are practical limits on the stringency of these designed selective pressures as some display vector systems are more robust than others. For example, ribosome display is well suited to affinity engineering as the selective pressure is relatively passive. Indeed, the greater diversity accommodated in this method generally returns greater improvements in affinity (Dufner et al. 2006). However, if stability is the selective pressure, ribosome display may not be suitable as the required arrangement of mRNA, ribosome and protein is potentially the weakest link and not the antibody structure. As a result any environmental stress exerted as a selective pressure is likely to destroy the display vector before it affects the antibody. For stability improvement phage display has been found to be generally useful (Jung et al. 1999).

With conventional antibodies it is the *in vivo* selection process used to create them which fundamentally limits their technological capability. The selective pressures acting on *in vivo* development have been honed by evolution to generate materials for use in the controlled environment of circulating animal serum. Despite the relative robustness of conventional antibodies (compared to other serum proteins) when taken out of their natural context, e.g. applied to a biosensor surface, they may fail to operate appropriately. This is not surprising considering the thermodynamics of the new environment (e.g. a biosensor surface) are likely to be far removed from that which they have been selected to operate. Recombinant antibodies change this fundamental limitation of conventional antibody reagents.

Recombinant antibodies are the only antibody technology capable of combining selective pressures in the selection process. As discussed in Sect. 12.3.1.2 monoclonal antibodies are only screened for specificity once they have been through many time consuming (and therefore expensive) processes without any guarantee that the products will be suitable for a given application.

12.3.2.3 Sources of Recombinant Antibodies

Having considered how we might interrogate diverse structural libraries, where does library diversity originate from?

The most accessible recombinant antibodies are those derived by converting existing monoclonal hybridoma cell lines into the recombinant scFv format. This method takes advantage of the *in vivo* maturation process and so the *in vivo* screening against extensive natural diversity has already occurred. The conversion of monoclonal antibodies is a particularly attractive option considering the substantial investments made in monoclonal generation over the last 30 years. However, the process is not always successful and care should be taken in such a process as the context in which the variable domains are presented may change (i.e. the loss of the constant domains changes the inter-domain dynamic which in turn affects binding characteristics) (Torres and Casadevall 2008). This may affect stability, affinity, and even specificity which can negate the point of using the material in the first place (Solar and Gershoni 1995; Schenk et al. 2007; Tam et al. 2007).

An approach which is generally more reliable is to harvest the genetic component of an animal's immune response to an immunogen (i.e. extract the genetic script from the animals B cells). Orlandi et al. (1989) first reported a method for the harvest of genetic information from antibody producing cells and its subsequent manipulation through molecular biology. In theory this method could harvest the whole immune response of the animal by extracting the mRNA (the antibody's genetic script) from antibody secreting cells. Unlike the few hundred B cells evaluated in conventional monoclonal technology, this method enables the screening of a more representative proportion of the animal immune response. Once the genetic information is extracted, it can be assembled into a display vector system. Libraries generated in this way are biased toward the antigen used to immunize the animal. These are called immune libraries as the B cells are a product of natural immune selection and affinity maturation. This has practical advantages as the process of *in vivo* affinity maturation has effectively enriched the content of the library and, therefore, improves the probability that an antibody of good specificity/affinity may be isolated. Immune libraries are a useful way to ensure that a display system of potentially restricted library diversity (e.g. phage display, Sect. 12.3.2.2) contains useful products. However, they are still a product of natural processes.

Other libraries of animal origin have been produced using B cell progenitor cells (Nissim et al. 1994). These so called naïve libraries provide a start point for the development of antibody products but they tend to require substantial improvement to develop their affinity and other facets of their behaviour. This results from the generally suppressed diversity of such libraries and the associated reduced probability of finding something immediately useful. The principle advantage of naïve libraries is that once the library is constructed no more animals are required to produce antibodies. The resulting naïve library (sometimes referred to as a "single pot" library) can be used repeatedly to isolate antibodies against unrelated materials. However, the practicality of having to invest considerable effort in substantially improving every antibody isolated is not cost effective and the materials isolated from these low diversity libraries may not represent an ideal starting point. They are merely the best a limited library could offer.

Considering antibody binding diversity is primarily limited to the CDRs, what has prevented the use of random mutagenesis in the generation of vast synthetic mutational libraries? This approach is limited by our current understanding of protein structure and function, and specifically how to synthetically introduce the variation required into the antibody scaffold without compromising its ability to fold into a stable and robust entity. Previous attempts at synthetic library production through the random mutation of CDR regions have successfully generated libraries of considerable diversity. However, the functional activity of such libraries (i.e. the proportion of binders in the library which retain structural integrity while providing diversity in binding capability) is not acceptable. For example, Tanha et al. completely randomized the H-CDR3 loop of a recombinant llama domain and affinity screened the synthetic library against peptide binding antibodies. The results demonstrated that there was a biased distribution of feature location within the CDR toward its N-terminus (Tanha et al. 2001). This illustrates that the locations of residues tolerant to mutation (in this scaffold and for this "epitope") are clustered together and the other residues within H-CDR3 may perform additional functions perhaps relating to structural integrity.

Semi-synthetic libraries have been produced as a "half-way house" where a naïve library has been generated and then synthetically diversified to increase its functionality (Goldman et al. 2006). Other semi-synthetic approaches have harvested genomic germline sequences (i.e. pre-somatic rearrangement) and synthetically shuffled the natural repertoire of CDR gene sequences (Soderlind et al. 2000). This approach produced a more functionally active library compared to the approach of Goldman et al., which suggests that we have much to learn regarding the diversification of binding identity while retaining the structural identity of the parent scaffold. It is, therefore, reasonable to believe that the composition of CDR loops, which evolution has tailored expressly for this application, is not entirely random (Lo Conte et al. 1999; AlLazikani et al. 1997).

Currently the most reliable (and therefore most commercially favoured) route for recombinant antibody development relies on the use of immune libraries. It would appear that considerable work is required to define the locations of mutational tolerance in CDR regions. Though some structural parameters have been understood for some time (AlLazikani et al. 1997) it was only recently that such parameters have been used in the statistical design of synthetic libraries (Rothe et al. 2008). Whether these new hi-tech libraries deliver the much anticipated bounty remains to be seen.

12.3.2.4 Recombinant Antibody Production

For any of the advantages of recombinant technologies to be realized in detection and diagnostic capabilities, the product must be amenable to scaled industrial production. While conventional antibody technologies have reached a high level of maturity (e.g. extensive commercial support industries) (Birch and Racher 2006), this is not currently the case for recombinant antibody technology. While this is a likely result of patent complications limiting technology exploitation, it is exacerbated by the complexities of the immunoglobulin fold as a protein domain unit. Discussed below are two of the principal challenges facing scalable recombinant antibody production. For a more comprehensive coverage of the challenges in this area recent reviews are available (Schirrmann et al. 2008; Leong and Chen 2008; Schirrmann et al. 2008; Arbabi-Ghahroudi et al. 2005; Harmsen and De Haard 2007).

The process of making recombinant proteins *per se* should be an economical process as prokaryotic expression systems (most notably *E. coli*) are a considerably cheaper technology option than those required in monoclonal antibody production (e. g. mammalian cell culture). However, some of the properties of the immunoglobulin fold motif present technical challenges which generally prevent prokaryotic production on an industrial scale. The most convenient production method is to produce the protein as a soluble native structure. However, the disulphides within Ig folds (required for domain stability) are not compatible with conventional cytoplasmic *E. coli* production (Bessette et al. 1999). While *E. coli* is capable of processing disulphide bonded proteins in soluble form (through periplasmic secretion), the yields produced may not be suitable for the production of industrially meaningful quantities (e.g. gram to kilogram). Low yields are also attributable to the absence of native folding chaperones and disulphide isomerases in these recombinant hosts, which affects both the quality and quantity of material produced (Hu et al. 2007; Miller et al.

2005; Hu et al. 2007). In twin domain arrangements (such as the scFv) this is further exacerbated by a core property of this class of recombinant scaffold – domain swapping (as discussed in Sect. 12.2.4). This dynamic situation can lead to the protein exploring conformations which lead to irreversible denaturation and aggregation. Conventional antibody formats like the IgG use twin variable domains but the stabilizing influence of the constant domains is substantial (Rothlisberger et al. 2005).

It is possible to produce the antibodies as insoluble inclusion bodies in *E. Coli,* which result from nascent peptide aggregation. While this is an acceptable way to produce substantial quantities of protein by mass, the harvested material requires extensive processing to refold it into the desired product (Sinacola and Robinson 2002). This process is complicated by the necessity of appropriate disulphide formation and must be optimized for every antibody. Though some scFv reagents may be successfully produced these reports tend to be the exceptions to the rule and as a result this method is not amenable to commercial production (Donini et al. 2003). More sophisticated eukaryotic expression systems (e.g. *Pichia pastoris*) are capable of producing more reliable protein yields and it is anticipated that these systems will have the capability to deliver a more universal and, therefore, industrially scalable production process (Miller et al., 2005; Damasceno et al. 2007; Cai et al. 2009; Wan et al. 2008; Li et al. 2008).

12.3.3 Alternative Protein Scaffolds

Owing to the technical challenges facing Ig fold exploitation protein engineers have sought more industrially amenable scaffolds to perform the same function (examples of which are shown in Fig. 12.14). Immunoglobulins are not the only protein structure exploited in animal adaptive immunity. Parallel evolutionary processes have generated alternative scaffolds serving the same function (Cooper and Alder 2006). The hagfish for example has the oldest adaptive immune system (currently known), which exploits "leucine rich repeat" motifs completely unrelated to the Ig fold. These accommodate binding interface diversity and have potential for recombinant exploitation (Pancer and Mariuzza 2008).

Natural immune systems are not the only source of affinity materials. Protein scaffolds may be engineered synthetically and it is interesting to note that researchers leading recombinant antibody technology in the 1990s have been developing this technology area. A popular feature of these alternative scaffolds is the absence of disulphide bonds (e.g. the α-helical "affibody" and ankyrin repeat protein (Li et al. 2006)) though some scaffolds are homologous to the Ig fold (e.g. neocarzinostatin or "monobodies" based on the FnIII domain of fibronectin).

While for commercial reasons such investigations are typically geared to therapeutic markets (Nuttall and Walsh 2008), their utility in biosensor technologies may be of interest as these technologies aim to address many of the same practical/commercial challenges facing immunoglobulin scaffolds. However, considering their simplicity, the structural integrity of these materials may be of concern as

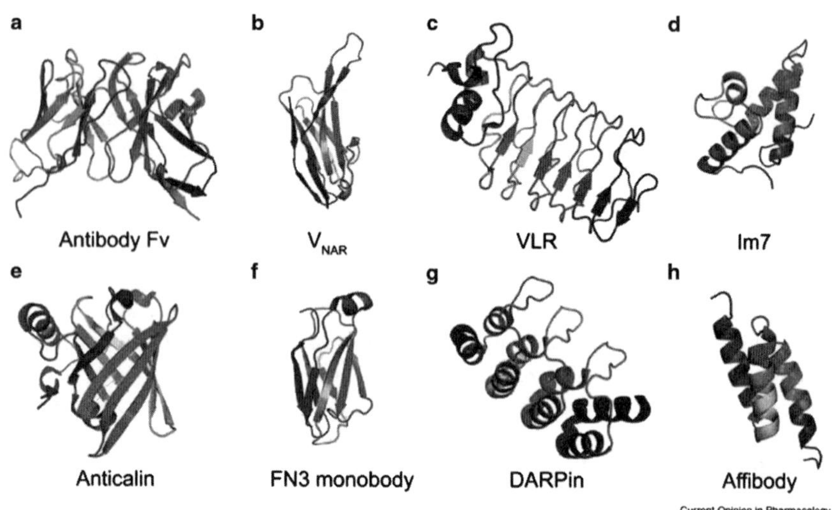

Fig. 12.14 Cartoon illustrating diversity of protein folds utilized as binding-scaffolds. (**a**) Antibody Fv fragment (PDB 1NMA); (**b**) VNAR single domain antibody (PDB 1VES); (**c**) lamprey variable lymphocyte receptor (PDB 2O6R); (**d**) Im7 immunity protein (PDB 1AYI); (**e**) Anticalin (PDB 1LNM); (**f**) FN3 monobody (PDB 2OCF); (**g**) DARPin (PDB 2JAB) and (**h**) affibody (PDB 2B88). Underlying folds are based on: b-sheet and b-barrel configurations (**a** and **e**); single immunoglobulin domain frameworks (**b** and **f**); modular repeat units (**c** and **g**); and helical bundles (**d** and **h**) (With permission of Nuttall and Walsh (2008)), copyright 2008 Elsevier

modest temperature fluctuation or matrix variation (e.g. ionic concentration) may disrupt the proteins binding interface and compromise their utility.

12.3.4 Limitations of Protein Recognition Elements

Structure/stability of proteins relates directly to their thermodynamic environment. As protein recognition elements are frequently evolved/designed in aqueous environments they are predominantly reliant of these environments to maintain their structural identity. So, while it is technically possible to design assays for the detection of practically any material using protein based recognition elements, it is analyte solubility in these aqueous environments which may potentially limit their utilization (Suzuki et al. 2003).

Solvent additives can be employed to improve analyte solubility (Whelan et al. 1993) and other solvent matrices may be used to perform the immunoassay (e.g. ionic liquids (Baker et al. 2006)), but water is the ideal solvent for proteins. Water is also frequently involved in antibody binding interactions and its removal may compromise binding energetics in addition to protein stability (Siegel et al. 2008).

Protein thermodynamics are also strongly affected by temperature. *In vivo* selected antibodies, such as monoclonals, are generated by the animal to function in physiological conditions inside the donor. This limits the temperature to 37°C and deviation from this may reduce stability or specificity. In practice monoclonal IgG antibodies are fairly resistant to thermal denaturation, but if antigen binding is expected to occur at elevated temperature there will likely be some suppression of sensitivity resulting from affinity reduction (Lipschultz et al. 2002).

12.4 Opportunities in Biosensor Development

Following the conception of biosensor technology in the mid 1960s, patent offices around the world have processed technology claims for a myriad of devices. However, to date few of these contenders have satisfied a commercial market (with some notable exceptions such as the glucose oxidase sensor). Some technologies have been developed and "pushed" into the commercial market with little consideration for their concept of use or the commercial competition; such enterprises have been short lived (Luong et al. 2008). If any technology is to be "pulled through" to a commercial market, considerable advantages over conventional methods need to be consistently demonstrated (Luong et al. 2008). The appetite of society to identify, monitor, and control materials may well continue to grow, but it is ultimately the affordability of these technologies which determines whether society chooses to exploit them.

Conventional diagnostics (e.g. high throughput hospital diagnostics) are a mature technology and it is a challenging objective to identify any incremental change which may be readily implemented to improve the economy and sensitivity of these processes. However, some considerable compromises have been made in these processes and these will be used to illustrate potential opportunities for future biosensor technologies. Particular attention is paid to the associated costs of analysis, the speed of data output, and the multiplexity of sample analysis.

12.4.1 Assay Economics

The financial burden of conventional analytical capabilities is attributable mainly to the skilled labour and complex reagents required (e.g. enzyme/fluorophore conjugates). In order that these costs are not prohibitive, compromises have been made in the design of conventional analytical capabilities.

Conventional sandwich assays (i.e. dual antibody assays) are well established and provide considerable utility through the careful partnering of antibody reagents (e.g. in enzyme linked immunosorbent assay (ELISA) or hand-held immunochromatographic assay). By using a substrate immobilized antibody to capture a target analyte and another to report the binding event a high level of specificity can be achieved (Fig. 12.15). The success of such approaches relies on the overlaying of antibody specificity where each antibody may cross-react with alternate materials to its partner. Such assays provide a useful capability in batch based diagnostic assays.

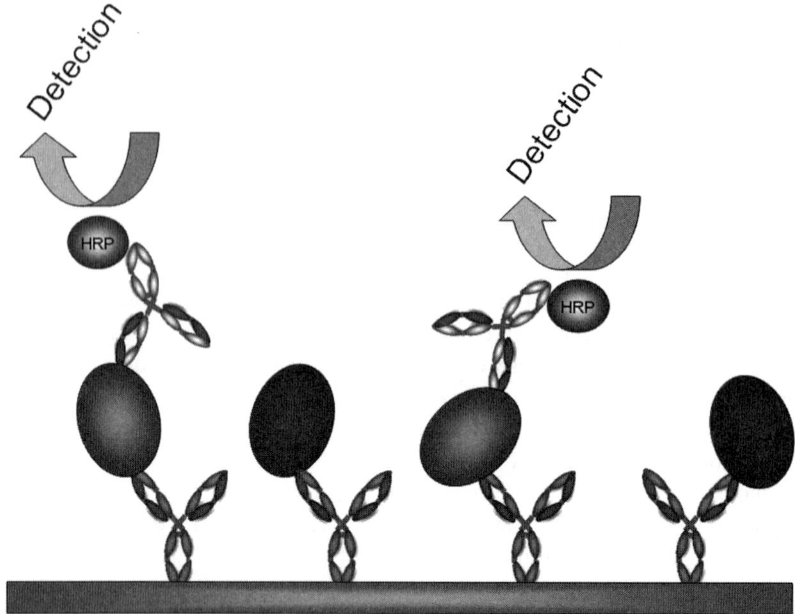

Fig. 12.15 Schematic of conventional enzyme linked immunosorbent assay format. The surface immobilized capture antibody binds both analyte and cross-reactant. The secondary antibody horseradish peroxidase (HRP) conjugate only binds the analyte. Assay specificity results from the combined stringency of both antibody binding specificities. HRP enzyme activity provides signal amplification (through the production of a coloured product) disclosing the presence of the analyte

For example, the centralized screening of diagnostic samples in hospitals enables the utilization of high throughput sample processing and analysis (e.g. ELISA). Sandwich assays require several reagents per analyte and the process takes a comparatively long period of time as it requires sample collection, processing, and analysis. Sandwich assays are not readily applied to continuous monitoring sensors (e.g. "in line" water quality monitoring for live data output) as they require the sequential addition of reagents and buffers, separated by washing steps.

Through the use of capture assays (i.e. single antibody assays) biosensors can reduce the need for reagents and provide continuous "real-time" data feeds for use in monitoring applications (e.g. air and water quality). A capture assay uses a single immobilized antibody for each analyte and so reduces reagent consumption (Fig. 12.16). As no additional reagents are required the assay can run continuously to provide real time data, for example, in environmental monitoring.

The sensitivity of capture assays may be reduced when compared to sandwich assays as the use of enzyme conjugates (e.g. horse radish peroxidise) to report the primary binding event enables signal amplification and, thereby, increases sandwich assay sensitivity. Capture assays rely solely on the sensitivity of the transduction mechanism to report binding events. It is also important to note that specificity requirements of antibodies deployed in capture assays are more

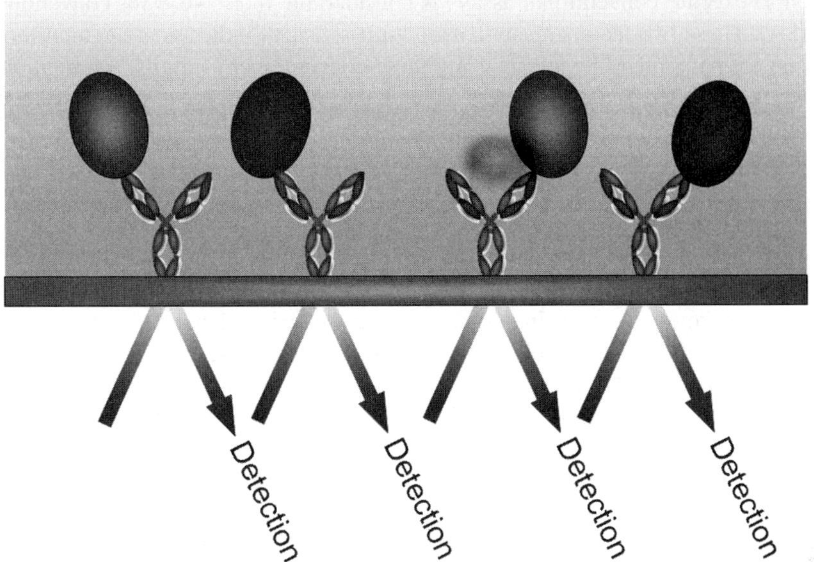

Fig. 12.16 Schematic of surface plasmon resonance (SPR) capture assay format (evanescent field decaying with distance from the surface). Surface immobilized antibodies may bind materials other than the analyte of interest and due to the detection method (in this case it is SPR) these are registered as positive binding events. Assay specificity depends entirely on the antibody used as the capture element

stringent. Any antibody cross reactivity (i.e. nonspecific interactions) will be more evident in the resulting capture assay. At best this may result in elevated background noise in the assay reducing sensitivity, and at worst the assay will report false positives. The production and selection of reagents for capture assays must therefore be carefully considered (Michaud et al. 2003).

Capture assays do, however, have considerable potential advantages. The reduction in reagent consumption reduces the cost per assay. In addition, the sensing interface or chip may be reused for multiple analyses. This requires binding site regeneration or cleaning between assays. These chips would ideally be the only consumable in the system which would reduce the skill and, therefore, the training required to operate the system or conduct routine maintenance. Capture assays also have great advantages in mulitiplexed assay design (discussed in greater detail in Sect. 12.4.3).

12.4.2 Speed of Analysis

The speed of analysis is currently limited by two challenging factors. Firstly, the sample matrices used in conventional diagnostics are limited. Complex samples (e.g. whole blood) require some degree of sample processing. Secondly, the time taken to execute conventional assays is considerable (e.g. 3–6 h for conventional ELISA). These two factors mean that any time sensitive indicators (e.g. monitoring sepsis) are not suited to analysis using these conventional methods. As mentioned above, centralized sample processing and the development of high throughput screening have helped to reduce the cost of conventional analysis. As these limitations are inherent to the established technologies, there is a gap in the market for biosensor technologies capable of sensitively interrogating complex matrices with rapid/automated data interpretation.

Considering the hCG pregnancy test, the hCG hormone is present in the urine and blood of pregnant women (Chard 1992). These two matrices provide hCG in different metabolic forms but, more importantly, the complexity of matrix composition is significantly different (e.g. cells, proteins, saccharides, salts, urea, etc.). Home test kits (immunochromatographic assays) probe the comparatively simple matrix of urine while the complex matrix of whole blood requires processing before running the hCG assay or it would fail to operate appropriately (Decramer et al. 2008).

It is the biosensor interface which limits the exploitation of current technologies in this kind of application. While it has been widely demonstrated that many biosensor technologies are, in principle, capable of rapid analysis it is often the process of sample handling and preparation which delay analysis (Latterich et al. 2008). In order for a biosensor to operate in complex matrices, the interface through which it operates must be engineered to retain specificity, sensitivity, and stability within the sample. This is, arguably, a more challenging undertaking than producing a rapid assay format *per se*. For example, healthcare diagnostics often require sample fractionation prior to analysis. This may serve several practical functions; partial purification to reduce sample complexity, removal of assay interferants, and elevation of analyte concentration assisting in its detection. However, in future healthcare applications (e.g. "point of care" diagnostics) time consuming sample processing should not be required which necessitates the interrogation of complex sample matrices (e.g. whole blood).

The requirements of continuous environmental monitoring are potentially as challenging. Continuous monitoring of sewage or industrial effluents for environmental pollutants will experience considerable variation in sample complexity across a wide range of known and potentially unknown materials. A biosensor must be capable of responding appropriately to these materials or the sensor output is of no value to decision makers.

12.4.3 Multiplexity of Analysis

Single analyte assays, such as hCG chromatographic pregnancy test strips, provide the user with a simple yes/no answer to a specific question. This kind of immunoassay delivers the right level of information in a convenient and simple format. There is little conceivable improvement possible in this type of application as the needs of the user are well addressed (Chard 1992). Future biosensor technologies must aim for simultaneous quantification of many targets for multi-factorial analysis.

Genomic sequencing studies (e.g. the Human Genome Project) have enabled the correlation of genotype to phenotype, which has obvious application in pre-emptive healthcare (Kopf and Zharhary 2007). Such approaches are useful to determine congenital predisposition to well studied disease states, but whether or not an individual has such a condition can only be determined phenotypically (Ingvarsson et al. 2008). For example, an individual's predisposition to prostate cancer does not indicate that the individual will develop the life threatening condition. They are merely more likely to do so in their lifetime which suggests close monitoring of the condition may be pertinent. In prostate cancer, higher than average serum concentrations of PSA (prostate specific antigen) correlate strongly with the disease. However, this biomarker alone does not enable diagnosis of the condition directly and additional invasive procedures are currently required. A multiplexed assay could assess a number of biomarkers characteristic of a disease condition. For example, a combined assay for PSA and PSMA (prostate specific membrane antigen) improves the stringency of the prostate cancer assay (Ramachandran et al. 2008). The more the biomarkers probed in an assay the greater the confidence in the result, which in turn leads to more successful and cost effective health care.

Multiplexed assays could be further developed for the screening of many potential conditions simultaneously in pre-emptive proteome profiling (Latterich et al. 2008). In such approaches it is foreseeable that diseases yet to appear as physical manifestations may be detected and treated in a preventative manner (Ellmark et al. 2008). Data interpretation is of great importance in such an application. As biomarkers like PSA are not explicit indicators of disease, a quantitative or at least ratiometric (i.e. abundance of one analyte relative to another) approach must be employed. Sandwich assays are capable of incorporating array features for internal calibration enabling a degree of quantitative concentration determination (Kingsmore 2006). Capture assays, in comparison, are less amenable to deriving quantitative data because of their design, but this is primarily a factor relating to chip to chip variability, which means (with current production techniques) that one chip from one production batch may behave quite differently from another. Capture data are, however, readily suited to ratiometric analysis where assay signals may be compared to one another to identify trends deviating from "normal" values. This approach is particularly suitable for diagnostic testing of human sample matrices where many potentially suitable reference materials may be available (e.g. serum albumin).

Multiplexed arrays could also have potential in autonomous home care much in the same way as the glucose oxidase biosensor has enabled diabetics to self medicate. This technology has improved the quality of life for diabetic individuals and reduced the financial burden to health care providers. Currently the most common method for monitoring organ transplant rejection is through biopsy, which is both invasive and costly and does not enable active intervention as it is reactionary rather than pre-emptive (McManus et al. 2006). Post operative transplant care may be managed more effectively and more comfortably in the home using biomarkers. For example, immune suppressant medication regimes could be modified and responsively tailored by the patients to their individual requirements (Gwinner 2007).

While conventional sandwich assays can already probe several analytes simultaneously, there are practical issues of multiplexing this assay format. The potential of assay interference (i.e. nonspecific assay noise) resulting from the cross reactions of reporter antibodies is significantly increased when sandwich assays are multiplexed. It has been suggested that such assays have a practical ceiling of 25 analytes (Kingsmore 2006). In contrast, capture assays have no upper limit other than the practical limitation of the number of assays which can be addressed by a particular transduction method.

While a multiplexed assay is obviously of great potential benefit to many applications, it has the potential to raise the costs of biosensor chips considerably. Multiplexed chips generally increase in cost in proportion to the number of antibodies applied (Haab 2006). In Sect. 12.4.1, it was suggested that the reuse of biosensor chips would reduce the cost of analysis. In multiplexed assays this becomes more important to make the process economically and functionally favourable. However, antibodies are by nature variable proteins and their stability in biosensor assays can be affected by many factors. The most significant of these factors, in multiplexed applications, is the need to clean the chip, or regenerate the binding sites, at the end of one assay in preparation for the next. It is possible to engineer an assay to accommodate the specific needs of one particular antibody (e.g. ionic buffer composition (Einhauer and Jungbauer 2001)); however, when a multiplexed array is considered the antibodies must have universal stability/functional properties. Conventional antibodies provide considerable challenges in this regard as demonstrated by the literature studies (Steinhauer et al. 2002; Quinn et al. 1999). For example, regenerating solutions used to remove bound analyte to enable biochip reuse (e.g. 10 mM potassium hydroxide) disrupt the bonding between antibody and analyte. However, if the solution also inactivates any of the antibodies on the biochip it is no longer of any use (Nath et al. 2008). Antibody standardization is a key technical challenge in pursuit of multiplexed immunoassays. Arguably this may never be realized using conventional monoclonal antibodies as their selection is an opportunistic process. The selection processes of recombinant antibody technologies are more flexible and enable the selection of desirable properties such as universal properties (e.g. regeneration).

12.5 The Biosensor Interface

The theoretical sensitivity of a biosensor is ultimately determined by the physical properties of the transduction method (Luppa et al. 2001). However, it is the interface between sensor and sample matrix that determines whether the sensor output is of any practical value (Lahiri et al. 1999). Regardless of the biosensor transduction method used, the immobilization of recognition elements to an interface within these systems is invariable (Steinhauer et al. 2002). Successful development of this interface is likely the key challenge facing the commercial exploitation of any new technology. This is well illustrated by commercial surface plasmon resonance (SPR) biosensor systems which have been supplying the research market since the late 1980s (examples of which are given in a review (Baird and Myszka 2001)). The development of these sensors alone would not have led to the general adoption of this technique in research labs. The successful commercialization of biochips for use in such sensors has ensured the rapid and widespread adoption of the technology.

Considering the market opportunities for biosensor development (Sect. 12.4) the potential demands of the interface are considerable. To illustrate the importance of interface design it is necessary to consider what technical challenges may be encountered. As each biosensor may require bespoke interface development, it is only possible to illustrate some of the main principles of interface design rather than any specific solutions.

12.5.1 Surface Adsorption of Antibodies – An Interface Case Study

Many different methods have been used to immobilize antibodies but the most straightforward, and therefore extensively used in proof of principle studies, is physical adsorption. While adsorption is simple to carry out, it is not capable of consistently delivering high quality, reproducible surfaces/biochips. This case study focuses on biochip sensitivity and stability resulting from this widely used method.

The passive physical adsorption of protein may, in principle, be used to coat any surface. While the precise mechanisms may not be well understood (Rabe et al. 2008) it is the most convenient and straight forward method of applying antibodies to surfaces. The method is simple in that it only requires the incubation of an aqueous protein solution over a surface and in time the protein adsorbs to it. The forces acting on a protein in adsorption processes are the same hydrodynamic enthalpic and entropic forces driving protein folding and antigen–antibody binding (Sect. 12.2.3). In this process the biosensor interface must be considered an extension of the proteins energetic system. This system should include all components of the interface including the recognition element, the surface to which it is tethered, and the three dimensional (3D) environment surrounding it (i.e. sample matrix, solvent, and any material used to tether the recognition element).

A protein adopts a folded state as a result of favourable entropy (Sect. 12.2.3). The tethering of a protein to a surface should, therefore, stabilize the protein by reducing the entropic cost of maintaining the folded structure (Knotts et al. 2008). This may result in a modest improvement in the stability of the bound protein, but this effect is protein dependant and even varies with the location of the tethering point within the protein (Knotts et al. 2008). This unpredictable variation is attributable to enthalpic changes within the immobilized protein which may offset the entropic gains in stability (Knotts et al. 2008). These forces are relatively modest and are only evident in proteins adsorbed to weakly interacting surfaces. The strength of protein surface interactions can have a more profound impact on protein structure and function.

The strength of surface interaction and mechanism by which proteins adhere to surfaces depend on the chemical nature of the substrate (Hemmersam et al. 2008). If the surface is hydrophobic (e.g. aliphatic) the process is primarily entropic, driven by the solvent environment. In aqueous environments hydrophobic surfaces maintain a hydration shell of energetically constrained water, at considerable energetic cost. This energetically trapped water may be readily displaced by materials such as proteins. Surfaces coated in this manner can yield densely packed, strongly bound protein, but this is achieved at the expense of protein structural identity/integrity. The surface forces binding the protein modify its energetic environment and can promote substantial conformational changes (Baujard-Lamotte et al. 2008). Surface derived entropic forces may continue to act on the adsorbed protein and over time the protein may rearrange itself to sate surface entropy by spreading itself over as much of the surface as possible (Figueira and Jones 2008). The structural rearrangement is dependent on the inherent stability of the adsorbed protein, but even the most stable scaffolds are prone to some deformation (Gwenin et al. 2007a; Figueira and Jones 2008). Generally, the deformed protein remodels itself to expose its hydrophilic residues to solvent and to hide its hydrophobic residues at the surface interface. The net effect is to convert the surface into a hydrophilic interface by coating itself with protein.

In comparison, enthalpically mediated adsorption is mild and potentially reversible. When surfaces are polar/charged (e.g. hydroxylated) they more actively participate in solvent dynamics (Bolduc and Masson 2008). Relative to a hydrophobic surface they are energetically engaged with the bulk solvent, reducing the net entropic cost of surface hydration. This does not entirely remove the entropic drive to exclude water from the interface, but it does reduce it. On these less entropically compromised surfaces enthalpic forces are more dominant and, by modifying the charge state of the protein, surface interactions may be controlled (Chen et al. 2003). Any residual hydrophobic forces may be attenuated to some degree through the use of surfactants such as the widely used "Tween p20" (Wang et al. 2006b).

A protein's charge state, and subsequent electrostatic behaviour, is determined by the pH and ionic composition of its solvent environment. Proteins have a net charge resulting from the sum of the electrostatic contributions of their solvent exposed amino acids. The pH at which this net charge is equal to zero (i.e. all charge is cancelled out by equal and opposite charge) is known as the isoelectric point or pI. At this point a protein has no net charge but will likely have an uneven charge distribution across its surface giving it some degree of polarity resulting from charge clusters. A protein's charge state may, therefore, be controlled to provide some degree of control over its adsorption to hydrophilic surfaces

(Wang et al. 2008). This can facilitate protein self assembly into ordered, two dimensional (2D) monolayers (Andolfi et al. 2004). Unlike hydrophobic immobilization (which is frequently irreversible), hydrophilic surfaces adsorb protein through electrostatic interactions which may be reversed by modifying pH or ionic concentration (Su et al. 1998).

Antibodies adsorbed onto two-dimensional solid substrates can be probed using a variety of physical methods including AFM (atomic force microscopy), neutron reflection spectroscopy, SPR (surface Plasmon resonance), and QCM-D (quartz crystal microbalance with dissipation) (Lu et al. 2007). Using neutron reflection spectroscopy and AFM, Xu et al. (2006) demonstrated that an IgG antibody adsorbs to hydrophilic surfaces in the same plane as the surface. The same observation has been made with other antibodies and surfaces (San Paulo and Garcia 2000; Thomson 2005; Martinez et al. 2008; Lee et al. 2008). This may well not be surprising as the most rational thermodynamic place for the antibody to be is in plane with the surface (as entropy may be maximized through solvent exclusion from both the surface and the globular protein). While such a surface may not be expected to retain binding activity (as protein flexibility is considerably compromised) this approach can, rather surprisingly, generate surfaces with good activity (Xu et al. 2007). This is a testament to the potential durability and robustness of immunoglobulin scaffolds.

Using the above adsorption method it could be reasonably assumed that assay sensitivity and dynamic range may be elevated by maximizing the amount of antibody adsorbed to the surface. However, in such two-dimensional interfaces this is somewhat simplistic as steric hindrance (from both antigen and antibody binding site crowding) can restrict binding interactions (Xu et al. 2007). However, optimizing binding site density (i.e. reducing surface coverage) to improve binding site accessibility reveals a greater proportion of the underlying surface. This may erroneously interact with unintended sample matrix components and thereby reduce the sensitivity or specificity of the assay. Blocking of such areas with traditional blocking agents (e.g. bovine serum albumin – BSA) only serves to sterically block the binding sites once more (Reimhult et al. 2008; van Oss 2003).

In addition to sterically compromised binding sites, the stability of adsorbed sites can be affected as they may become distorted as a result of forces exerted by the underlying substrate. Biochip consumables should ideally have a long shelf life and be stable up to and including the point of use. Considering the importance of surface contour complementarity in binding interactions (Sect. 12.2.3.2) any remoulding of the variable domains will be detrimental to their activity (Chou et al. 2004). In addition, the mechanism of physical adsorption is non-covalent and material may be lost or dynamically substituted by other proteins (Noh and Vogler 2007). In protracted use this may lead to the gradual loss of antibody from the surface which will reduce sensitivity and erode confidence in biosensor assay data. The conditions required in chip cleaning and binding site regeneration may also reduce chip activity (Chou et al. 2004) and promote the removal of antibody from the surface or increase the promiscuity of its binding (Dimitrov et al. 2007).

It is clear from this case study that antibody adsorption may have considerable limitations regarding the technological aspirations of biosensors, but it is important to note that these problems are not insurmountable and depend on the end application. Some commercially viable diagnostic tests exploit the simple process of physical adsorption to great effect. Pregnancy test strips are constructed through the adsorption of antibodies to both gold colloid and nitrocellulose membranes and function effectively. However, these devices are stored in desiccated form and are reconstituted for single use. In addition, the sample matrix is either relatively simple (e.g. urine) or sample processing is undertaken to prepare the sample for analysis (e.g. blood). This is quite a different concept compared to the complex and interrelated technical obstacles introduced in Sect. 12.4.

12.5.2 The Necessity of Holistic Interface Development

Bespoke engineering of biosensor interfaces can potentially provide many benefits including assay sensitivity, stability, dynamic range, speed of execution, biochip shelf life, reduced false positives and negatives, recycling/reuse of bioconsumables, environmental durability, and reduced batch variability. Considering the potential benefit to biosensor technology it is interesting to note that interface engineering has been relatively neglected in comparison to the investments made in pursuit of patentable sensor technologies. The significant financial and temporal investments required to overcome these interrelated technical challenges are likely the reasons for the apparent lack of advancement in this regard. If any technology is to exploit current market opportunities these challenges must be addressed simultaneously, which requires a clear understanding of all facets of the interface and sample matrix behaviour.

Obviously, the precise requirements of surface bound antibodies remain to be determined, but considering their general properties (discussed in Sect. 12.2) some educated guesswork is possible. Antibody flexibility has been implicated in both binding function and stability. Compromising this dynamic flexibility through surface adsorption can influence their ability to bind antigen (Briand et al. 2007; Boujday et al. 2008). Polymers provide a useful tool to separate the antibody from the interface to effectively restore its solvated state. In SPR biosensors the evanescent field (the sensing area) decays exponentially with increasing distance from the 2D gold substrate. This sensitive zone is estimated to extend over 300 nm from the surface. Hydrogels, such as those based on the saccharide dextran, have been used to great effect to maximally populate this 3D evanescent field. Not only is the activity of bound antibody improved when applied in this solvated matrix, but the resulting stability of the antibodies can also be improved. Polymer chemistries are incredibly diverse and it is conceivable that any properties a biosensor interface requires may be achieved through synthetic approaches (Goddard and Hotchkiss 2007; Zhang et al. 2008). However, in some sensor technologies the interface must be 2D and these approaches are not suitable (Perkins and Squirrell 2000).

Nonspecific binding (NSB) can be a considerable challenge with 2D surfaces as the energetic forces discussed in Sect. 12.5.1 may act on any material presented to them (e.g. whole blood).

To address this 2D challenge the Whitesides group developed a self assembling monolayer interface based on oligoethylene glycol (Lahiri et al. 1999). The fundamental success of their approach related to their choice of polymer, oligoethylene glycol (OEG also known as polyethylene glycol or PEG depending on its molecular weight). PEG is considered a special or unique polymer and may be used in both aqueous and organic solvents. In aqueous environments PEG has some particularly advantageous properties owing to its elemental composition and bond arrangement. The polymer backbone consists of a simple repeating unit of saturated [carbon–carbon–oxygen]. The oxygen bond arrangement of this unit adopts a gauche conformation with bond angles similar to those in the surrounding water. This property, combined with the polarity of the oxygen, results in extensive interactions with aqueous solvent closely mimicking the enthalpic and entropic equilibria of the bulk solvent. The lack of any explicit charge in PEG is a further boon considering that one of the principle forces promoting nonspecific surface fouling is electrostatic (e.g. hydroxyl mediated binding). The use of PEG as a replacement for conventional blocking agents has also been suggested to reduce the effects of steric crowding in antibody binding occlusion (Reimhult et al. 2008). However, care must be taken when employing more elaborate polymer arrangements as the thermodynamic environment created may well have unintended properties such as reduced binding activity as a direct result of increased polymer entropy (Mehne et al. 2008).

While the Whitesides interface provides continuous close coverage of the gold substrate, if combinations of different surface topologies are required for different analytes (e.g. proteins and cells (Perkins and Squirrell 2000)) then this method is not readily suited to the production of zoned interface areas, i.e. a combination of 2D and 3D interfaces on the same bio-chip. Rubina et al. (2003) reported the zoned polymerization of methacrylate (MA) promoted by ultra-violet irradiation. In this approach the methacrylate monomer may carry any desired functionality (e.g. oligoethylene glycol (Zhang et al. 2008)) which may then be incorporated into the polymer. This functionality may include the antibody itself. In such a process antibody immobilization and polymer synthesis may be combined in light mediated arraying (a recent review is available (Rubina et al. 2008)). Such fabrication methods also enable both antibody functionalized hydrogel and planar interfaces to be deposited in discrete locations on the same arrayed biochip interface.

As a point of interest, just as evolution has developed antibodies for specific and sensitive detection, evolution has overcome the challenge of specific and controlled surface interactions. Mammalian cell communication is complex and has evolved to interact through specific mechanical and proteomic processes to modulate cell mobility, gene expression, hormone release, and any other functions of multicellular life. Considering these natural systems function with much greater complexity, biosensor design may benefit from understanding the natural processes behind both specific and nonspecific binding interactions (Robert et al. 2008). This may well be a useful source of inspiration in the pursuit of more universally successful biosensor interfaces (Wu et al. 2007).

12.6 Bespoke Antibody Engineering for Biosensors

Conventional diagnostic antibodies are a mature technology and have been developed around specific technologies (primarily sandwich assays such as ELISA or immunochromatographic assays). In principle, the breadth of antibodies developed for these conventional assays may be readily applied to biosensors. Any such application is, however, a fundamental change in their use and when real samples are interrogated in developmental systems, this generally results in poor assay performance (stability, specificity, and sensitivity).

It is frequently the case that new biosensor technologies use conventional antibodies to demonstrate their potential capabilities (e.g. potential sensitivity and specificity). This is, however, merely the beginning of assay development which will likely require the production of bespoke reagents designed for the technology (Iqbal et al. 2000). While recombinant protein technologies enable the engineering of antibodies in this regard it is of little value without a clear understanding of what properties are required of the antibodies. For biosensor technologies to deliver useful capabilities in any application the development of new recognition elements must be a synergistic process.

12.6.1 Stability

A desirable level of stability may be difficult to define. However, considering that the antibody is likely to be a limiting feature of a technology (e.g. biochip shelf life), there may well be no upper limit. As such, the greater the stability of an antibody biochip the more financially viable the product may be (e.g. reduced requirement of climate control in storage). The application of antibodies in any technology (e.g. therapeutics or diagnostics) places considerable demands on their stability and successful development in one technology area may, in principle, be exploited in another. The IgG scaffold has a good track record in this regard (Sect. 12.2.4).

Antibody stability relates directly to the composite Ig folds, specifically their arrangement and domain features (Sect. 12.2.4). The relatively homologous structures within the IgG class can exhibit considerable variation in stability (Liu et al. 2008a). This variation is attributable to their most variable portion, the V domains (Ionescu et al. 2008). By minimizing the antibody binding element to its variable domains, the varying stability observed across monoclonals can be more extreme across recombinant formats such as the scFv. In the case of the scFv, this relates to the fact that much of the stabilizing force provided by the constant domains is removed (see Sect. 12.2.4). This section primarily focuses on the engineering of antibody structures of particular interest to biosensing (i.e. minimal binding domains). For further reading with regard to IgG stability engineering there are articles available concerning therapeutic development which may be relevant (Wang et al. 2007; Liu et al. 2006; Gaza-Bulseco and Liu 2008; Garber and Demarest 2007).

As proteins are dynamic flexible structures, they actively explore all energetic conformations afforded to them by the solvating environment (Sect. 12.2.3). So, an inherently flexible domain may explore more potential conformations, which generally increase the probability that it may adopt aggregation prone states (Espargaro et al. 2008). The same process, which drives protein folding into a discrete domain unit, also drives disordered domains to aggregate. This frequently leads to the loss of domain structural identity and associated binding properties. A method for general stability comparison between different antibodies has recently been suggested, which quantifies free thiol content under mildly denaturing conditions (Lacy et al. 2008). The rationale behind this method is that should Ig folds rely on disulphides to stabilize them, then these domains are likely to be of low inherent stability. This is a fair assumption and, if a panel of antibodies are available, this may assist in their down selection. However, when a particularly desirable antibody suffers from instability this may be improved through protein engineering using two approaches; direct modification of its inherent thermodynamic behaviour or indirect modification by influencing the thermodynamics of its immediate environment. Below, these approaches have been divided into sections covering mutational approaches to directly improve stability and indirect fusion proteins commonly used to stabilize antibodies and other synthetic polymers, which have been employed to the same effect.

12.6.1.1 Improving Stability Through Mutagenesis and Protein Design

Different degrees of stability are, to some degree, inherent to the different scaffolds available (Sect. 12.2.4) (Demarest and Glaser 2008). However, regardless of the scaffold choice, once a useful antibody has been identified it may be improved through either rational or evolutionary mutagenic approaches. These approaches focus on directly improving the thermodynamic stability inherent to the antibody. While this may be attractive, it does require considerable investment of effort and is, therefore, an expensive process.

High thermodynamic stability generally necessitates a tightly condensed hydrophobic core braced with disulphides, extensive hydrogen bonding, and Van der Waals interactions. In addition, the solvent interface of the protein must be dynamically engaged with the solvent to reduce nonspecific interactions (Sect. 12.2.3.3). With multi-domain scaffolds it may also be beneficial to improve the inter-domain interactions to reduce dissociation and subsequent promiscuity.

In reviewing the considerable literature covering stabilizing mutations in scFv antibodies, it quickly becomes apparent that there is little consensus regarding successful point mutations. Considering the sequence homology of these materials, this is both surprising and frustrating. However, while this precludes the use of these data as a tool for the identification of desirable residue substitutions, they may be generally indicative of destabilizing residues. Through primary sequence comparison with consensus sequences a probability that a given residue is destabilizing may be calculated (Monsellier and Bedouelle 2006). This approach exploits evolutionary selection which, it may be reasonably assumed, has consistently selected domains of greater stability. The success of this approach relies heavily on class dependant homology alignment and the success rate has been suggested to be in the region of 60% (Monsellier and Bedouelle 2006). Substituted residues between classes or even sub-classes (i.e. different Vκ domain sub groups) may have different effects and so careful consideration of comparative sequences is required (Willuda et al. 1999). These selected point mutations dynamically interact with their surrounding environment (whether solvent or peptide) and the outcome of such interactions may not be readily predicted. As such, the rational design of antibody domains is limited to the quality of known information regarding sequence diversity and their structural/environmental context.

The principal strength of consensus mutagenesis is in rectifying gross sequence abnormalities. For example, errors in PCR amplifications may be carried through to selected antibodies and otherwise be undetected. Residue #6 of the V_H domain is either glutamine or glutamic acid and this residue participates in the complex hydrogen bonding network in the core of V_H (Langedijk et al. 1998). This location is often randomized as part of the PCR primer set designed for B cell gene extraction from animal sources. As such, it can be artificially mutated and stability may be improved by simply restoring the amino acid to its consensus residue through point mutation.

Another homology based approach is loop grafting where the binding identity of the unstable donor antibody (contained within the CDR peptide loops) is transplanted into a structurally homologous host. While this approach has generated some robust materials (Jung and Pluckthun 1997), its success requires much to be known about both the donor and the host on a structural level. Such information is difficult to infer through primary sequence comparison alone. Without this detailed understanding the subtleties of protein behaviour may be disrupted and the binding properties of the antibody may well be modified. A related approach which may be more generically applied is the insertion of natural CDR diversity into a stable domain host to construct a library from which to select stable antibodies (Soderlind et al. 2000). This approach developed by the Borrebaeck group has been used to great effect (Ingvarsson et al. 2008).

The ideal approach to improve antibody stability is through a random evolutionary approach as this may identify unique, antibody specific mutations. The strength of this approach is that it requires no prior knowledge and will likely generate materials with novel mutations which could not have been identified in any other way. As outlined in Sect. 12.3.2.2, recombinant selection processes enable the screening of diverse populations of antibodies. Once an antibody has been identified, a sub-library may be constructed where it is diversified through various genetic methods (Sect. 12.3.2.2 and a recent review (Benhar 2007)). This sub-library may then be reselected using selective pressures appropriate to the desired application. In such an approach it is generally advisable to remove the disulphides from the Ig folds, either chemically or genetically (Jermutus et al. 2001; Brockmann et al. 2005). This automatically exerts a selective pressure on the core thermodynamic stability of the Ig folds. Selection may be further skewed by incubating the disulphide free mutant library at elevated temperatures or with organic solvents and surfactants, each of which reduces the stabilizing power of solvent entropy. By repeating the affinity selection process only those mutants of improved function will be recovered. This method has been used to improve the stabilities of many scFv antibodies and provides a broadly applicable method for antibody improvement. However, this does require that the display vector used (e.g. phage display) is robust enough to survive the process, which may practically limit the variety or severity of selective pressures available. *In vitro* vector systems (such as mRNA or ribosome display) are not very tolerant of aggressive selection strategies (Worn and Pluckthun 2001). Comparatively, phage display is chemically and thermally robust, which enables a greater breadth of conditions to be considered (Famm and Winter 2006; Jung et al. 1999).

An alternative approach to stabilize variable domains is through disulphide bond cross bracing. Disulphide bonding may be exploited to reduce motion dynamics either within individual domains (Hagihara et al. 2007) or between scFv domains (Young et al. 1995). However, as with consensus mutagenic approaches above, there is little agreement on which to base a generic strategy for the required cysteine substitutions. These substitutions are again likely to be class and sub-class specific and those which have been successfully reported have generally been undertaken with considerable understanding of antibody structure (e.g. their crystal structure). Inter-domain V_H–V_L disulphide bonds have been shown to improve resistance to thermal aggregation, denaturants, nonpolar and polar solvents, surfactants, and proteases (Dooley et al. 1998). However, while the product may be useful, these stable structures are difficult to synthesize owing to their additional structural complexity (Young et al. 1995; Worn and Pluckthun 1999).

It is important to note that any change to the antibody variable domains in pursuit of greater stability may well change its fundamental binding characteristics. This serves to emphasize the importance of using affinity selection (e.g. by bio-panning) to verify the specificity of the resulting products.

12.6.1.2 Stabilising Fusion Partners

Considering the investment of effort required in mutagenic stabilization of individual antibodies more generic strategies have been sought. To this end many proteins have been genetically fused to recombinant antibody fragments. In the case of the scFv format, it may be argued that the constant domains of the IgG provide a very

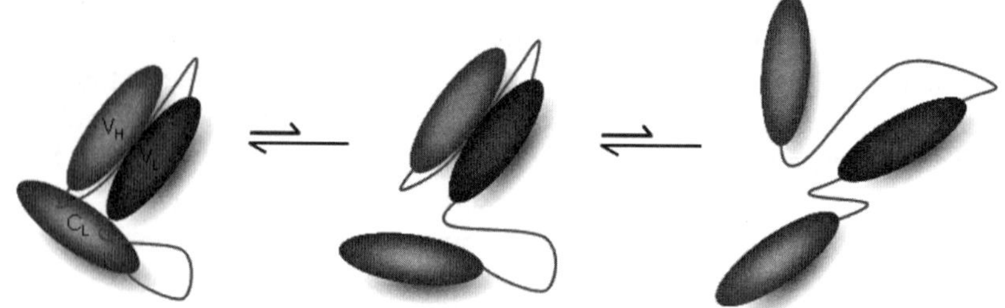

Fig. 12.17 Schematic showing a suggested arrangement of the kappa constant domain fused to the V_H–V_L dimer. Fully dissociated domains are prone to denaturation as shown in Fig. 12.11 though the stabilizing mechanism of the constant kappa domain has yet to be defined

useful fusion and (as a generic strategy) their native IgG structure is hard to improve upon (Demarest et al. 2006; Rothlisberger et al. 2005). However, minimized recombinant antibody scaffolds are of specific interest to biosensor applications (Sect. 12.2.2) and the expression systems available for their production have potential to reduce the cost of supply (Sect. 12.3.2).

In the case of the recombinant scFv format, the fusion of a single constant domain to the V_H–V_L scaffold (e.g. constant kappa light chain) has proven an effective way to stabilize the dimer interface or at least skew the structural equilibrium toward an associated state (Fig. 12.17) (Hayhurst 2000). Surprisingly the kappa fusion has been found to have a generalized stabilizing effect with proteins unrelated to Ig folds (Palmer et al. 2006). Other fusions, such as thioredoxin, have been used to facilitate cytoplasmic expression. The precise mechanisms by which these divergent fusion proteins improve the stability of these structurally unrelated substrates are not well understood and appear not to be specific. For example, thioredoxin was thought to provide some disulphide interchange activity, but mutations which remove its active site residues did not reduce the so called "chaperone" effect (Jurado et al. 2006). It is more likely that the stabilizing phenomena relate to steric effects of molecular crowding and other modest influences imparted through localized entropic and nonspecific enthalpic modification (Knotts et al. 2008).

Considering the mild and unpredictable interactions of fusion proteins, it is difficult to predict the effects of an N- or C-terminal fusion. For example, the maltose binding protein fused to the N-terminus of a scFv appears to improve its stability, but fusion to the C-terminus is poorly tolerated (Hayhurst 2000). Such fusions are always context dependant and, in the case of antibody variable domains, N-terminal fusions can actually occlude the binding site reducing its activity (Martin et al. 2006).

Once a successful fusion has been identified, it may be generally applicable within an antibody class (e.g. kappa fusion to scFv C-terminus) (Hayhurst 2000). Nonspecific stabilization (e.g. thioredoxin) is generally not as effective as specific evolutionarily developed (e.g. C_H1/C_L dimer) stabilization (Randles et al. 2008) and so as a generic approach protein fusion may only ever provide modest stability improvements.

12.6.1.3 Polymer Conjugation

The immediate thermodynamic environment of a protein may be affected greatly by simply conjugating polymers to the finished product. This strategy has primarily been developed for therapeutic applications for which a recent review is available (Pasut and Veronese 2007). While the solution stability of the protein may be improved (which may well equate to better thermal stability (Treethammathurot et al. 2008)), it is not possible to apply these same approaches to biosensor application as binding interactions are frequently compromised (Kubetzko et al. 2005). However, the general principle of polymer stabilization may well be useful in the context of interface development (van Oss 2003). The use of polymers in biosensor interface engineering may enable the stabilization of immobilized antibody (discussed further in Sects. 12.5.2 and 12.6.4).

12.6.2 Specificity

This attribute of antibody binding is obviously the primary reason such materials are applied in biosensors. However, while the specificity of one antibody in a given technology may be adequate, this may not be readily transferable to another. For example, the transition from sandwich assay to capture assay format is required for many reasons (outlined in Sect. 12.4) and is particularly punishing of any cross reactivity between antibody and matrix components. As such the demands future technologies will make on the specificity of their reagents may be considerable.

The generation of antibodies with the required level of specificity for capture assays is challenging. Defining the level of specificity required must be the first part of this challenge. For example, whole blood may be considerably complex but its components can be experimentally defined. Problematic elements may be identified and natural variations within a population may be quantified. In other applications it may be impossible to identify or predict what materials may be encountered. For example, in real time air quality monitoring it is desirable to quantify pollen allergens (Takahashi et al. 2001). To define the variability of this bio-aerosol within a variable background of vehicle exhaust particulates and fungal spore detritus is practically impossible considering that even year to year trends are unique (Kaarakainen et al. 2008). Where problematic matrix components may be identified, this knowledge may be used to adapt antibody selection strategies accordingly. However, this does rely on there being some degree of flexibility in the antibody production process. The sections below reveal the key difference between conventional and recombinant technologies in this regard.

12.6.2.1 Conventional Antibody Screening

Polyclonal antibodies are poorly controlled materials owing to the use of unpredictable *in vivo* animal immune responses to produce them. The only opportunity to influence these products is through the design of the immunogen. Through the use of designed features (e.g. synthetic peptide epitopes identified as unique to the native antigen) the animal immune response may be steered to produce a more specific material. The resulting polyclonal material may generally contain a more specific antibody as a result, but there will always be complications with polyclonal antibodies no matter how ingenious be the immunogen (Michaud et al. 2003). It is primarily antigen heterogeneity which results in polyclonal cross-reactivity and unpredictability. Polyclonal antibodies may, however, be more reliably depended upon to produce useful reagents when considering simple immunogens such as haptens (Le Berre and Kane 2006).

While monoclonal antibodies are a defined material, they may also suffer cross-reactivity issues in a similar fashion to polyclonals (though with reduced severity) owing to their production from similar immunization methods as used in polyclonal production. The monoclonal production process separates the practical processes of specificity screening from initial hybridoma-fusion selection. This results in the generation of many monoclonal cell lines before determining whether they are of sufficient specificity. This makes the process unpredictable and, as with polyclonals, the only way to steer the process is to deliver an engineered immunogen to the animal. This is not to say that the hybridoma production process is entirely inflexible but just that the mature commercialized aspects of this technology are relatively inflexible. Indeed, a recent study reported a method for the effective enrichment of B cell lymphocytes from the animal host prior to myeloma fusion (Adekar et al. 2008). Though this innovative approach improves the probability of generating a useful product it is a complex process requiring great skill and will ultimately still require production through mammalian cell culture.

12.6.2.2 Recombinant Antibody Selection

The only antibody technology capable of readily coupling antibody selection with specificity screening is recombinant (Love et al. 2008). Affinity bio-panning using display systems (e.g. phage display) enables the exploitation of *a priori* information in affinity panning rounds. This may enable the active removal of cross reacting antibodies from a library for the selective isolation of truly specific antibodies. This generally involves coating the affinity panning substrates with the specific analyte of interest and including the nonspecific (i.e. cross reactive) materials free in solution. As a result, the cross reacting antibody species in the library are bound by the nonspecific materials in solution and are easily removed from the library on washing. Love et al. (2008) exploited this process to great effect generating antibodies to *Bacillus anthracis* protein EA1. Compared to their conventional monoclonal counterparts the resultant scFvs improved both specificity and sensitivity of the biosensor assay. The antibodies exhibited exquisite specificity identifying *B. anthracis* but not responding to other very closely related *Bacilli* sp.

As discussed in Sect. 12.3.2.2, the diversity of the library is of considerable importance for many aspects of antibody design (e.g. affinity and specificity). Considering that antibody affinity may well correlate with its level of binding specificity (Morozova and Morozov 2008) it is reasonable to assume that greater library diversity will likely increase the specificity (along with the affinity) of isolated antibodies. However, this may place unrealistic demands on library design and diversity may need to be rationalized in order that it may be accommodated in display vectors. For example, synthetic libraries have been developed which are designed for activity against specific target structures (e.g. peptides (Cobaugh et al. 2008)). This relies on the detailed understanding of how canonical CDR frameworks and Ig fold conformations relate to antigen binding. This requires that many assumptions are made and it is generally better to harness as much diversity as practically possible in the selected library.

In silico methods have been used to genetically predict residues at the binding site of well defined antibodies/antigens that can be mutated/designed to improve their specificity. This approach requires a considerable amount of information to be known about both the structure of the antibody binding site and its interaction with antigen/cross-reactants. This approach has been successfully demonstrated not only in the modification of pre-existing antibodies (Winkler et al. 2000; McCarthy and Hill 2001) but also in the design of new structures (Wang et al. 2006a; Chang et al. 2007). However, the costs associated with accruing the required information, and general recalcitrance of protein crystallography, make it unsuitable for general use.

The most efficient (i.e. cost effective) route to generate specific antibodies is the use of recombinant antibody technology. Recombinant technology is the only way of coupling specificity screening to the selection process. Compared with both poly and monoclonal technologies the probability of generating a product of the required specificity for a given application is greatly increased. It is interesting to note that all serious pharmaceutical discovery laboratories currently use recombinant selection processes for lead identification in therapeutic antibody programmes (Benhar 2007).

12.6.3 Sensitivity

The sensitivity of a biosensor is obviously dependant upon many factors; however, those relating directly to the binding element are affinity and avidity.

12.6.3.1 Affinity Maturation

As suggested in Sect. 12.3, *in vivo* antibody generation occurs as a two step process where a lead is first identified and then mutated into a more functional product. A similar process may be undertaken using designed *in vitro* methods. The affinities of some antibodies, such as those isolated from naïve or semi-synthetic libraries, may be lower than practically useful (as discussed in Sect. 12.3.2). The Nissim library, for example, was one of the first naïve libraries and has been used extensively to identify many scFv materials (Nissim et al. 1994). The affinity of isolated recombinant antibodies generally increases in proportion to the initial diversity of the recombinant library (Benhar 2007). The Nissim single pot library is of modest diversity (10^8 unique clones) and provides antibodies of modest affinity. However, these recombinant materials can be *in vitro* engineered to improve their affinity to a more desirable/functional level.

This maturation may be practically undertaken using the mutational diversification methods introduced in Sect. 12.3.2.2 where a candidate antibody is mutated to produce a sub-library of subtly divergent sequences and, therefore, a structural diversity with related specificity but modified binding affinity. This sub-library may then be subjected to affinity based bio-panning for the isolation of higher affinity antibodies which retain specificity for the analyte. However, what level binding affinity is required or desirable?

It is a common expectation that the higher the affinity of a binding interaction the greater the sensitivity of the assay. This maxim holds true for conventional diagnostic assays such as the ELISA, as the captured analyte must be retained on the antibody coated surface through many washing steps. However, as a biosensor may detect binding events in "real time" (e.g. SPR) the demands on affinity may be reduced. In this context low affinity antibody binding may be useful, though this may be accompanied by a reduction in binding specificity (Sect. 12.2.3). When investigating antibody affinity, it is generally observed that greater affinity translates into greater specificity in the resulting assay (Michaud et al. 2003). So, while high affinity is not a technological necessity, it may have certain functional benefits.

It is worth considering that there may be an upper limit beyond which affinity is not useful. Extreme affinities can render chip regeneration impossible as the conditions required to disrupt the bound complex are so aggressive that proteins stand little chance of surviving the cleaning process (binding site regeneration). This makes extreme affinities less practically useful in multiplexed arrays (Sect. 12.4.3).

It has been suggested that a useful antibody binding affinity (K_D) for biosensor applications may be in the region of K_D 10^{-8} M^{-1} (in the order of 10 nM) derived from a high association rate (K_a) of ~10^5 M^{-1} s^{-1} and moderate dissociation rate (K_d) of ~10^{-3} M^{-1} s^{-1} (Saerens et al. 2008). This enables a rapid response to analyte (e.g. within a 10 min analysis period) but will also enable simple surface regeneration for repeat analysis (Saerens et al. 2008). In practice a higher affinity will not present any problem as long as the association rate is high and the surface may be regenerated with no loss of activity.

The key part of the *in vitro* affinity maturation process, as with all bio-panning selection processes, is the appropriate exertion of an effective selective pressure. Within the phage display system selective pressures may be readily applied through the modification of the affinity panning process. For example, with "off rate selection" the analyte adsorbed substrate is coated with the phage displayed antibody and subsequently incubated in a solution with free antigen (Forsman et al.,

2008). This actively selects those antibody structures with reduced dissociation rates. The stringency of selection relates directly to the duration of this incubation step as antibodies with lower affinities are eluted from the surface more rapidly and prevented from rebinding by free antigen in solution. As affinity is progressively improved through such processes, it becomes more difficult to exert an effective selective pressure and this practical limitation is a function of the vector system employed (Foote and Eisen 2000).

While off-rate selection generates antibodies of greater net affinity, the exerted selective pressure actively steers selection toward antigen–antibody complexes with low dissociation rates (K_d). This attribute may be of considerable interest in therapeutic or conventional diagnostic applications (Luginbuhl et al. 2006), but in biosensing it is primarily the association rate (K_a) which is of most interest as it directly affects the speed with which an assay responds to analyte (Sinha et al. 2002). This is another illustration of the divergent requirements of conventional detection technologies and emergent biosensor technologies. In nature the affinity and, therefore, half life of antibody interactions is actively selected to be no greater than 12 min (Batista and Neuberger 1998), i.e. affinities below this level are actively selected against, but there is no natural capacity to select for antibodies of any greater affinity. Such a time constraint balances the selective pressure between association and dissociation rates and so conventional *in vivo* antibody isolation methods (i.e. monoclonal and polyclonal antibody technologies) generate antibodies which are optimized neither for ELISA sandwich assays (where dissociation rates are most significant) nor for biosensor capture assays (where association rates are most significant).

As observed in stability maturation (Sect. 12.6.1) random mutagenesis quite often generates the most significant improvement in antibody function and this is usually the case with affinity maturation. However, as explained above, improvement in net K_D is not explicitly desirable and it is the K_a constituent which is of greater interest. For K_a improvement it is primarily electrostatic mutations which are required (Sect. 12.2.3). Exerting an appropriate selective pressure to discriminate between mutations relating to association or dissociation may be complex and unworkable. It is likely that the detailed thermodynamic screening of mutations will be required and while not impossible, this may be an onerous task.

It is frequently observed in experimental studies that the effects of enthalpically beneficial mutations have an equal and opposite effect on the entropic component of binding energy (Brummell et al. 1993). Therefore, while the net affinity of interaction (K_D) does not change, binding is driven by different energetic processes. For example, mutational studies have been undertaken to determine the effects of framework residues within the vernier zone (CDR flanking residues) (Makabe et al. 2008). It is possible that this balance between entropy and enthalpy may be exploited to specifically improve the electrostatically driven association rate specifically for biosensing applications.

As with specificity engineering, it is possible to exploit crystal structure data through *in silico* modeling to design affinity enhancing mutations (Barderas et al. 2008). Indeed, this has been used to successfully tune the electrostatics of several antibody interactions; however, somewhat surprisingly, this has only a 60% success rate (Lippow et al. 2007). While this enables true affinity engineering of the binding interaction it requires the generation of expensive crystallographic data, which are not generally available for the essentially limitless diversity of diagnostic analytes.

A more general strategy to hone the electrostatics of antibody binding interactions may be to target specific locations. Indeed, with semi-biased approaches the mutations may be steered toward the refinement of binding electrostatics. For example, once putative hot spot residues are identified (through consensus sequence comparison or mutational studies) they may be specifically targeted for mutagenesis (Ho et al. 2005). Another simplified mutagenic strategy is to use a restricted repertoire of amino acids in mutated locations (e.g. tyrosine, serine, glycine, or arginine (Birtalan et al. 2008)).

12.6.3.2 Multimerization Strategies

Binding site density (surface avidity) can translate directly into assay sensitivity and dynamic range. While this is closely related to affinity, it contributes in its own right to the functionality of the biosensor surface. This is especially the case when considering multivalent analytes (e.g. cells and viruses) as the force resistance of the bound material to the shear forces of fluid dynamics increases in proportion to the number of surface interactions (Sect. 12.2.3.4). However, the valency of the antibody as a molecular entity (e.g. bivalent IgG) is not of significance *per se* when it is immobilized at a sensing interface. The implications of molecular valency are considerable when antibodies are in solution, as evidenced by the differences of affinity and avidity of IgM relative to IgG. However, when preparing a biosensor surface this property is likely only to be advantageous if the immobilization process is crude and uncontrolled as the deformation of one domain/region does not necessarily result in total inactivation of all binding sites.

Various multimerization strategies have been employed in designed recombinant reagents, many of which have been developed for therapeutic applications (e.g. cell targeting for drug delivery). Examples include "streptabodies" based on streptavidin and biotinylated antibodies (Cloutier et al. 2000), p53 tetramerization domain fusion to scFv (Kubetzko et al. 2006), scFv of various inter-domain linker lengths (Desplancq et al. 1994), and decavalent sdAb (Stone et al. 2007). These approaches may be applied equally in biosensors, but few of them are covalent strategies and as such may reorganize themselves in immobilization (Sect. 12.5.1).

In preference, the biosensor interface should be regarded as a scaffold onto which the antibody may multimerize. In this way, valency is maximized and extraneous artefacts associated with immobilization (e.g. partial deformation, instability, and nonspecific binding) are minimized. This places considerable demands on the immobilization methods employed, but this approach is generally more amenable to reproducibility and surface stability.

12.6.4 Immobilization

Having produced a stable antibody of sufficient specificity and affinity it must be immobilized sympathetically on the biosensor transducer. The process of immobilizing the antibody at the interface is critical to every facet of biosensor function/performance (Sect. 12.5). It is essential that both the application (e.g. point of care) and interface (e.g. hydrogel) required of an assay are understood before any attempt is made to develop an immobilization strategy. In addition, it is ultimately the cost of the process which will dictate which process is utilized in commercial development; therefore, any solution developed must be both technically sound and economically scalable.

The process of antibody immobilization will be divided into three sections: physical/chemical engineering (primarily conventional antibodies), affinity immobilization (a combination of both recombinant and conventional antibodies) and genetic engineering (recombinant antibodies). While polymers have already been introduced as potentially stabilizing protein fusions (Sect. 12.6.1.3) or generally useful nonspecific binding shields at the biosensor interface (Sect. 12.5.2), these will not be considered here in any detail as their required properties will be application dependant.

12.6.4.1 Coupling Through Primary Amines

The success of the Biacore system is due in large part to the success and relative simplicity of the amine coupling process by which antibodies (and proteins generally) may be immobilized to their commercially available surfaces (e.g. CM5 hydrogel or C1 planar surfaces). This chemistry requires simple surface processing to derivatize tethered carboxyl groups (e.g. the CM5 carboxymethyl dextran hydrogel) with the amine reactive N-hydroxy succinimide (NHS) ester. This ester is readily displaced by amine groups which occur across the surface of hydrophilic proteins. In commercial SPR systems this process has become the standard method for the immobilization of protein and is the one by which all other immobilization methods are measured.

Despite the widespread use of amine coupling, it is not a panacea for interface production as it results in considerable heterogeneity in the orientation of tethered proteins. As lysine residues occur with reasonable frequency across the surface of proteins, the point of immobilization may vary accordingly. The structural implications of changing the function/behaviour of these lysines can be significant. When considering antibody structures with little or no structural redundancy (e.g. scFv antibodies composed of only variable domains), their structure is important and lysines within these domains may participate in intra and inter-domain cross bracing salt bridges (Arndt et al. 1998). The removal of such stabilizing interactions has been shown to reduce the solvent phase stability of these reagents (Hugo et al. 2002; Bosshard et al. 2004; Gvritishvili et al. 2008). In coupling a protein by these groups the stability of the bound structures may be reduced as stabilizing interactions are permanently disrupted. In addition, the random distribution of binding site orientations will reduce binding site availability especially if reactive lysines are located within the binding site (Kortt et al. 1997). To compound this issue the reaction is not explicitly lysine specific as is frequently assumed. Peptide-mapping mass spectrometry studies have identified other reactive nucleophilic groups (primarily the hydroxyls of serine and tyrosine) which can have reactivity to the NHS chemistry (Swaim et al. 2004; Kalkhof and Sinz 2008).

The heterogeneity of surface immobilized structures which are inherently large and multivalent (e.g. IgG) may be tolerable owing to the inherent redundancy of the constant domains and multiple binding sites. Amine coupling of these structures has a proven track record of producing surfaces of reasonable utility in biosensors. However, while these surfaces are of practical use their actual activity (i.e. active binding capacity of immobilized antibody binding sites) is usually below 10% (Goodchild et al. 2006). This figure provides much room for improvement with implications for assay sensitivity and dynamic range.

Any amine based approach for antibody derivatization will suffer the same problems experienced above (e.g. biotinylation). Considering that antibodies may be fluorescently labeled and conjugated to other proteins very successfully through a multitude of commercial amine crosslinkers, what is it about biosensor surface immobilization which appears to inactivate the antibody? It is fair to say that any antibody derivatized through its primary amines will result in a variety of products some of which may be inactivated to a similar extent as found in biosensor immobilization. The difference is in how these materials are then employed in binding assays. When incubated as a fluorescent reporter conjugate in conventional sandwich assays, the active functional conjugate is effectively affinity purified out of the bulk solution. Any reduced activity of the bulk conjugate can be easily offset by modifying assay parameters (e.g. conjugate concentration or volume). In such an application poor activity is never such a limiting issue as it is in biosensor interface immobilization.

12.6.4.2 Enzymatic and Chemical Engineering

The chemical diversity of proteins makes controlling their covalent reactivity a demanding challenge. There are, however, some features of the IgG which may be exploited toward this end. The saccharide bound to the C_H2 domain of the Fc region (Fig. 12.2) and the inter-chain dithiols provide useful reactivities for site specific immobilization.

Oxidation of transient saccharide cis-diols may be undertaken using sodium periodate to yield aldehydes. These aldehydes have reactivity toward nucleophiles and so may react with amines, but they have greater reactivity toward hydrazide functionalities (e.g. adipic acid dihydrazide). The resulting schiff-base requires stabilization through reduction (e.g. with sodium cyanoborohydride). Interestingly, Shmanai et al. (2001) found when exploiting this method that assay sensitivity was

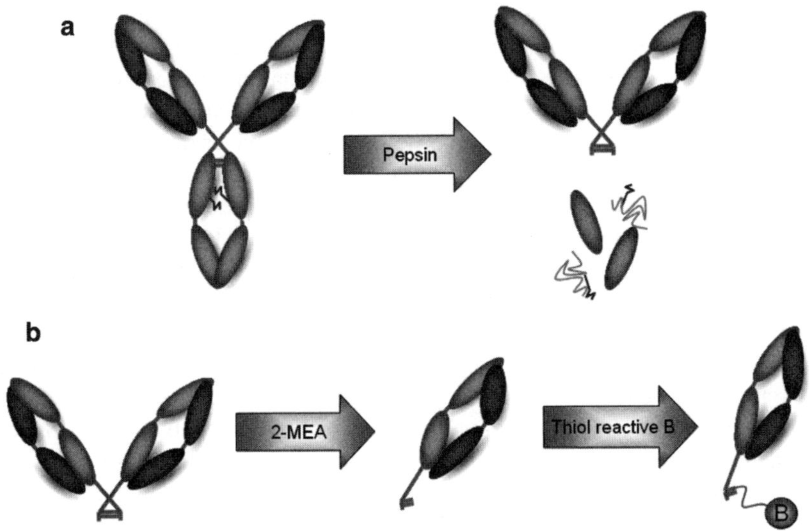

Fig. 12.18 (**a**) Schematic of the semi-specific proteolytic cleavage of the IgG using Pepsin. The resulting $F(ab)_2$ structure consists of two Fab arms are bound together with disulphide linkages. (**b**) $F(ab)_2$ may be reduced and chemically targeted for site specific functionalization/immobilization. For example, 2-mercaptoethylamine (2-MEA) may be used to reduce the disulphides and thiol reactive "B" reagent (e.g. biotin maleimide) used to specifically react to it

only improved two fold. Considering the investment of time in this multistep process, and the consumption of reagents, this method is not economically viable.

IgG may be selectively cleaved using the enzyme pepsin to yield a $F(ab)_2$ structure bound together through disulphide bonds (Fig. 12.18a). These may be semi-selectively activated through reduction with mild reducing agents such as 2-MEA (2-mercaptoethylamine) to yield Fab' as shown in Fig. 12.18b. The reactivity of the thiol functionality is more controllable than that of the sugar derived aldehydes and can be used in a variety of ways. Catimel et al. (1997) used a dithiol exchange reaction (using a surface bound pyridyl leaving group) to immobilized Fab' via its thiol to a dextran hydrogel. In comparison to amine coupling this method improved binding activity twofold, as was found by Shmanai et al. when immobilizing via the saccharide.

To minimize the proceeding steps in specific thiol generation (for example, for Fab' the IgG must be digested, purified, reduced, repurified, functionalized, and repurified) thiols have also been introduced through the chemical reduction (Cho et al. 2007) and the light mediated reduction (Duroux et al. 2008) of whole IgG. The resulting products may be less costly to produce, but there is likely to be a considerable compromise as the products are destined to be heterogeneous with implications for stability, sensitivity, and specificity of interactions.

In addition to the somewhat underwhelming improvements in sensitivity, the physical properties of the Fab' protein can be changed considerably on the removal of the Fc region, which can affect the surface densities readily achieved in such orientation studies (Goodchild et al. 2006). Considering the complexities of varying antibody properties (which necessitates the optimization of the process for each antibody) the surfaces produced do not provide the anticipated reward. On a commercial basis none of these approaches are acceptable as they are complicated and costly, and do not provide a proportional improvement in assay capability.

Other methods have been investigated such as site specific enzymatic biotinylation of C-terminal lysine residues of the Fc region by carbamoyl transferase. However, a similar outcome is experienced as above; the activity of the immobilized binding sites is only doubled. It is curious that all of these approaches generally return a two fold improvement in binding site activity. However, all of these studies have used a hydrogel to immobilize the antibody. This serves to illustrate an important point; the architecture of the interface has considerable implications for the degree of control required when immobilizing antibodies. With hydrogels (such as the carboxymethyl dextran discussed) stringent control over immobilization is not necessarily required, even for small binding elements like single domain antibodies (Saerens et al. 2005). When antibodies are applied to planar surfaces, however, considerable control is required in order that they may interact with their analyte and maintain structural integrity (Saerens et al. 2005). For example Fab' monolayers can be produced on planar surfaces mediated by the site specific biotinylation of the reduced hinge thiol. The resulting surface achieved a binding site density of 85% of the theoretical maximum (calculated as closely packed Fab' cylinders in a monolayer bound to streptavidin) with an activity of over 90% (Peluso et al. 2003). Although this required the investment of considerable effort, it did return a tenfold improvement in sensitivity. Overall, sophisticated chemistries may be academically interesting, but the properties of the resulting systems may not be economically exploitable even if they do improve key assay parameters (Derwinska et al. 2008).

12.6.4.3 Natural and Synthetic Affinity Interactions

Evolution has developed natural proteins which bind antibodies with high affinity (primarily derived from bacterial pathogens). These may be exploited as intermediary proteins with which to immobilize antibodies to biosensor surfaces. While this appears to swap the problem of immobilizing one protein for another (i.e. that of the antibody for an intermediary) there are some potential advantages: binding valency and their potential for generic use. These natural antibody binding proteins (e.g. Protein A) tend to display several binding sites, which in nature results in high avidity interactions. When used as an intermediary in biosensor immobilization the partial inactivation is less of a problem as antibody binding sites are often still available and the amine coupling of such proteins is an adequate method for general use. In the study above, where Catimel et al. investigated the orientated immobilization of Fab', they compared this with Protein A mediated immobilisation. While the Fab' method yielded a twofold improvement in binding activity, the Protein A method returned a threefold improvement (Catimel et al. 1997). This was again

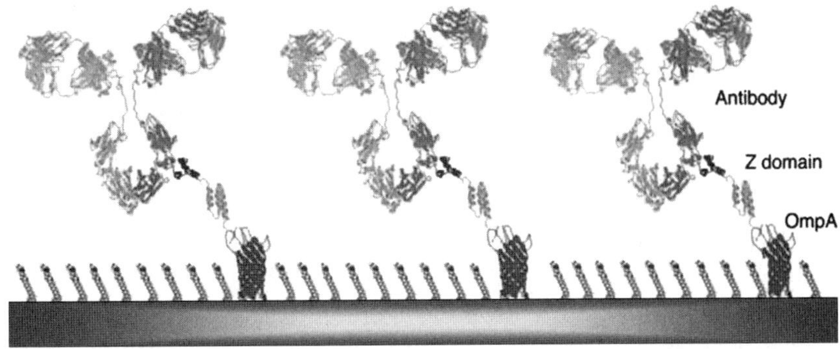

Fig. 12.19 Schematic (not to scale). The Orla-18 OmpA Protein A derived fusion scaffold. The protein OmpAZ has a single cysteine residue in periplasmic turn four allowing the protein to bind in a specific orientation to the gold. The space between protein molecules is filled with thioPEG. The OmpAZ protein has two SpA Z domains that can bind IgG antibodies in their constant regions (With permission of Le Brun et al. (2008)), copyright 2008 Springer

undertaken using a hydrogel, but when using planar surfaces the effect is more rewarding.

Recombinant engineering of IgG binding proteins has enabled the development of some interesting tools for planar interface engineering. One of the principle advantages of producing proteins recombinantly is that they may be engineered according to the needs of a given application. Protein G has been recombinantly engineered to include free cysteine residues (i.e. unpaired thiols) which may be used as chemical handles for its immobilization. The preferential reactivity of this thiol toward gold enables the formation of an orientated monolayer of Protein G across the gold surface (Bae et al. 2005; Fowler et al. 2007).

In a related approach, the Orla™ protein scaffold is fused to antibody binding proteins (e.g. Protein A/G/L) (Le Brun et al. 2008). This self assembling scaffold is a beta barrel structure based on the *E. coli* OmpA membrane protein (Fig. 12.19). Site specific cysteines enable the covalent tethering of this carrier scaffold to a gold surface. The hydrophobic regions of the beta barrel promote its self assembly into an ordered structure across the surface, which projects the fusion protein into solution. The activities of antibodies coupled to these kinds of surface are considerably greater than their amine immobilized counterparts though their utility is limited to various isotypes of IgG (i.e. murine IgG1 has relatively weak affinity for Protein A and so is of little use with this isotype). The improved activity may well be attributable to the flexibility Protein A affords to the bound IgG in addition to the preferential orientation of the binding sites (Caruso et al. 1996).

The methods above require the use of additional proteins which must be produced and purified, and which can only add to the complexity and cost of their use. The advantage is that once a method is successfully identified it may be applied generally to antibodies of the IgG class (bearing in mind isotype specificities). However, there are considerable advantages in chemically synthetic approaches as the associated costs of chip fabrication may be greatly reduced.

The engineering of a synthetic peptide as a high affinity binder of the Fc region of IgG has recently been demonstrated in biosensing applications (Jung et al. 2008). This peptide exhibits isotype specificity (much as Protein A does), but it may be handled robustly and may be incorporated in conventional alkane thiol self assembled monolayers. An alternative synthetic approach has been to use a primary amine coordinating functionality known as the "calixcrown" (Lee et al. 2003). This may be expected to generate surfaces with the same heterogeneity as covalent amine immobilization outlined in Sect. 12.6.4.1. However, as this is an affinity based interaction the immobilized material is not held in a fixed state but in a dynamic equilibrium. This has one intriguing advantage; the associated antibodies are preferentially coordinated via amines with the strongest interactions (Ly et al. 2008). Immobilized antibody appears to be preferentially coordinated via the Fc region providing a more generic approach to Fc biased immobilization with all the advantages of being entirely synthetic.

No matter how high the affinity of an interaction, it must always have a half life and (as discussed in Sect. 12.2.3.4) the mechanical resistance is of considerable importance in assays where shear forces may be experienced. In applications where the gradual loss of material from the sensing interface is not likely to defeat the assay (e.g. single shot assays for point of care use) affinity immobilization may be a viable option for scalable production. However, in biosensor applications such a process is unlikely to meet the technical aspirations outlined in Sect. 12.4.

12.6.4.4 Genetic Engineering

Protein engineering at the genetic level provides great opportunities for exquisite control over antibody immobilization. A genetically engineered immobilisation process should ideally be capable of promoting simple covalent immobilization without the need for laborious antibody purification or processing. For this reason affinity interactions are useful to effectively purify the antibody from crude fractions onto the surface and once the interaction associates it should ideally be switched in some way to form a covalent bond. Practically combining these two properties can be complicated. At its simplest, genetic engineering may be possible by the insertion of specific reactive residues (e.g. lysines or cysteines) or affinity motifs (e.g. His tag or biotin mimic), but many sophisticated approaches have been demonstrated such as the fusion of antibodies to proteins of engineered function (e.g. enzyme derivatives and protein splicing) and selective enzyme substrates (e.g. Biotin ligase BirA and transglutaminase).

The simplest approach involves the insertion of specific residues for covalent immobilization, but this is restricted to a great extent by the tolerance of such features in the expression hosts used to produce the antibody (Sect. 12.3.2.4). For example, a free cysteine inserted at the C-terminus enables the orientated immobilization of scFv antibodies (Torrance et al. 2006) though this is poorly tolerated in scalable expression methods (Schmiedl et al. 2000). An alternative approach is to attempt to bias conventional amine coupling through the insertion of serial

C-terminal lysines, but this also is poorly tolerated by expression hosts (Saerens et al. 2005). The precise location of these inserted residues may be investigated to improve their tolerance (Albrecht et al. 2006; Shen et al. 2005). However, considering the expense involved in identifying such locations, and with no guarantee that the identified solution will be transferable to other recombinant antibodies, this approach is likely to be restricted for use in enzyme technologies (Ryan and Fagain 2007; Gwenin et al. 2007b).

Several novel chemistries have been developed/adapted to site specifically immobilize proteins in general (reviewed by (Jonkheijm et al. 2008)). However, for these to be brought to bear in antibody applications a precursor must be covalently bound to the antibody, and herein lies their limitation. While chemistries such as the "click" chemistry (Wang et al. 2003) and the related Staudinger ligation (Watzke et al. 2006), provide useful targeted functionalities they still require the insertion of chemical groups into the antibody structure. This is a complex challenge and may easily result in heterogeneous immobilization as observed in amine based approaches. Some research groups have accomplished such site specific insertions through the novel use of non-natural amino acids in the biosynthetic production of the materials (Goerke and Swartz 2009). However, while such approaches are academically interesting, they are likely to be costly (low *in vitro* cell free yields), time consuming (purification after synthesis is still required), and reagent intensive (e.g. artificial tRNAs) in scaled production processes. Unless site-specific functionalities can be readily introduced to antibodies (which is not likely to be a scalable process as illustrated above), such chemistries have limited utility in site-specific antibody immobilization.

The second most accessible route to genetically tailor immobilization involves the use of affinity tags which are often inserted as C-terminal fusions for the purpose of purification (Wingren et al. 2005). The "hexa-his" tag for example is well tolerated in recombinant expression systems and because of its abundant use as a purification tag, nitrilotriacetic acid (NTA) support matrices have been developed for its use in biosensors. The affinity of the interaction between hexa-his tag and NTA coordinated nickel (Ni^{2+}) ions is in the region of $10^{-6}\,M^{-1}$ (Porath et al. 1975; Maly et al. 2002). The association may be strengthened by the use of twin His tags on the same protein fused in series at the C-terminus of the scFv antibody (Steinhauer et al. 2006). However, the affinity of His tags is fundamentally too low to be generally useful.

Many other affinity tags have been reported but the highest affinity tag must be the biotin acceptor peptide (BAP) (Schatz 1993). The addition of this 15 residue peptide to the antibody along with the coexpression of the *E. coli* biotin ligase, BirA, in the expression host (Cloutier et al. 2000; Scholler et al. 2006) leads to the site specific biotinylation of a specific lysine residue within the tag. The exploitation of this tag has been abundant owing to its high affinity interaction with streptavidin. The suitability of this method for scaled antibody production is questionable, however, owing to the coexpression of BirA and the fact that the immobilization method is affinity based.

Other tags have been identified with more straightforward production/expression pathways but these are inevitably of lower affinity (e.g. the streptavidin binding peptides StrepTagII (Schmidt and Skerra 2007) and Nano-tag (Lamla and Erdmann 2004)). Of the other affinity based peptide tags, the designed E-coil/K-coil leucine zipper provides a useful screening tool for the facile assessment of surface immobilized antibodies in the earliest stages of recombinant antibody development (De Crescenzo et al. 2003). In this method one leucine rich helix is immobilized on the biosensor surface while the other is fused to the antibody C-terminus. The interaction between these two affinity partners enables combined purification and immobilization from crude extracts negating the need for expensive pilot scale production and purification. Such methods may assist in the cost effective screening of selected recombinant antibody leads for surface immobilized applications.

The combined properties of affinity purification followed by covalent immobilization may seem contradictory or overly complex, but some designed enzymes are progressing toward this technically challenging objective. One rather elegant solution to this challenge involved the fusion of cutinase to a scFv, which enabled its affinity association with its surface tethered phosphate ligand (Kwon et al. 2004). Having associated specifically with its substrate, the enzyme attempts to process the phosphate through ester-hydrolysis and in the process is covalently trapped (Hodneland et al. 2002). Other solutions have been reported with a similar objective such as human O6-alkylguanine-DNA alkyltransferase, which has been rapidly commercialized as the SNAP-tag^{TM} (Iversen et al. 2008). Another potential contender is the recently reported phosphopantetheinyl transferase fusion method (Wong et al. 2008). While potentially attractive, enzyme fusions increase the size of the recombinant structure. The enzymes themselves may have compromising stabilities or poor resistance to nonspecific interactions.

An interesting alternative to enzyme fusion is the use of protein trans-splicing as developed by Camarero et al. (reviewed in (Camarero, 2008)). In this approach natural protein trans-splicing mechanisms are utilized and a natural self-splicing domain (e.g. the DnaE intein from *Synechocystis*) is split into two fragments; one fragment is linked to the surface and the other to the recombinant protein. These two fragments have an affinity for each other and associate readily. Once associated, a spontaneous splicing reaction occurs which releases both intein fragments, and covalently and site specifically couples the protein of interest to the surface. One attractive feature of this process is the complete removal of the extraneous intein components from the surface, leaving only the protein of interest.

It is interesting to note that the only protein fusion method introduced above (i.e. enzyme or intein based) to have been applied to antibodies is the cutinase fusion. The reasons for the apparent lack of utilization of the others in antibody immobilization may be numerous. Whether any of these new methods can meet this technical challenge, and whether they are amenable to scaled production in the required expression hosts in economic quantity, remains to be seen. However, it is surly only a matter of time before biotechnology finds a way to solve this challenging objective.

12.6.5 Other Engineering Strategies for Biosensors

Antibody engineering should be considered as part of the much larger discipline of nano-engineering. As our understanding of protein sequence, structure, and function grows, so too does our ability to engineer new materials using biopolymers. Antibodies are one of the most well studied classes of proteins and this has enabled the engineering of these structures to incorporate functionalities in addition to specific binding interactions. Some of these nano-engineering strategies of specific interest to biosensing are briefly introduced below.

The specificity of antibody CDR loops may be transferred to other proteins (e.g. green fluorescent protein (Kiss et al. 2006) and even synthetic molecular carriers (Timmerman et al. 2007). "Binding Bodies" have been reported which display CDR peptides on benzene derived carrier groups (Timmerman et al. 2005). The affinities of these structures are likely to be compromised in relation to their parental antibody structures (primarily due to their increased molecular entropy (Riechmann and Winter 2006)); therefore, the benefits in production (e.g. cost effective synthesis) or functionality (e.g. fluorescence) must be significant to offset the potential loss in sensitivity.

The process of array production has been combined with protein synthesis by the Taussig group (He et al. 2008). This novel nano-engineering approach assembles all the components required to *in vitro* synthesize the antibody and tether it to the surface in one single synthetic step. In this process coding DNA/mRNA, ribosome and all necessary substrates/catalysts are delivered to the surface within the small volume of an arrayed spot. This method exploits cell free expression technology (recently reviewed (He 2008)). This entirely *in vitro* approach could break the linear relationship between chip cost and array size as array fabrication does not require the complex and time consuming processes of protein expression and purification.

Antibodies may be nano-engineered into self-disclosing binding elements capable of independently reporting the binding event through fluorescence. Several groups have successfully demonstrated this principle covering both haptens (Medintz et al. 2005) and proteins (Renard et al. 2002). The engineering and verification in such approaches must be extensive to ensure that specific and sensitive binding properties are retained, but this demonstrates an attractive principle; the binding element becomes the biosensor.

The multi-domain structure of the antibody binding site can be exploited as an engineering feature. Blenner et al. substituted the serine-glycine linker in the 4D5flu scFv antibody with an elastin like peptide (Blenner and Banta, 2008). These peptides are responsive to environmental cues (e.g. temperature and ionic concentration) and through the modification of the antibody's environment the inter-domain V_H–V_L association could be modified. The precise mechanism of affinity switching in this case was a result of a scFv dimer and not due to the intra-molecular properties of the monomer. However, it does demonstrate that through modifying the intra/inter-molecular dynamics the binding interactions may be engineered to mechanically switch binding on or off.

As biopolymer materials (such as antibodies) are developed with ever increasing stability, specificity, and sensitivity, their exploitation in novel engineering strategies and the development of truly nanoscale devices become an achievable reality.

12.7 Concluding Remarks

The discovery of polyclonal antibodies in the 1930s ultimately led to the generation of ground breaking technologies in diagnostic healthcare. The migration from polyclonal to monoclonal technology in the late 1970s provided many advantages, without much requirement to change the underlying technologies. In the detection and diagnostic domains the situation remains largely unchanged today.

Considering the advantages recombinant antibody technologies offer, it is somewhat surprising that this technology remains poorly exploited. While this may have resulted from a variety of factors (e.g. patent restrictions, early scaffold instabilities, and the convention and simplicity of monoclonal technology), in some technology applications recombinant antibody exploitation is likely to be fundamental to their success, and biosensors are one of these applications.

Conventional capabilities based on monoclonal and polyclonal antibodies have some inherent factors which compromise their utility (e.g. speed of response). These limitations provide good opportunities for the development of biosensor technologies and it is surely a matter of time before successful systems are delivered into a commercial market. However, the delivery of these new capabilities requires the development of a new wave of recognition receptors. What these new recognition receptors are, however, remains to be seen.

Synthetic materials (e.g. aptamers and peptides) do not need to be produced in biological systems and so the cost of their production is assumed to be lower than that of antibodies. Considering the relative maturity of these synthetic technologies this is perhaps not surprising. While recombinant protein production is a mature technology, the field of protein engineering (of for example single domain antibodies) is an evolving technology. Historically, the scFv recombinant antibody was difficult to produce with some inherent compromising properties; however, as knowledge of protein structure and function increases, these early technical challenges are being overcome.

The cost of these various affinity reagents is only one aspect to consider in the commercialization of sensor technologies. The most useful recognition element will provide a compromise between cost and functionality. It is a gambling certainty that recombinant antibodies can provide all the functionality that is demanded of them in biosensing applications (e.g. stability, specificity, and sensitivity). As antibody engineering technology matures, the costs associated with their production may be expected to reduce. It remains to be seen which class of recognition receptor has the greatest functionality.

References

Adachi M, Kurihara Y, Nojima H, Takeda-Shitaka M, Kamiya K, Umeyama H (2003) Interaction between the antigen and antibody is controlled by the constant domains: normal mode dynamics of the HEL-HyHEL-10 complex. Protein Sci 12:2125–2131

Adekar SP, Jones RM, Elias MD, Al-Saleem FH, Root MJ, Simpson LL, Dessain SK (2008) Hybridoma populations enriched for affinity-matured human IgGs yield high-affinity antibodies specific for botulinum neurotoxins. J Immunol Methods 333:156–166

Albrecht H, DeNardo GL, DeNardo SJ (2006) Monospecific bivalent scFv-SH: effects of linker length and location of an engineered cysteine on production, antigen binding activity and free SH accessibility. J Immunol Methods 310:100–116

AlLazikani B, Lesk AM, Chothia C (1997) Standard conformations for the canonical structures of immunoglobulins. J Mol Biol 273:927–948

Anderson GP, Goldman ER (2008) TNT detection using llama antibodies and a two-step competitive fluid array immunoassay. J Immunol Methods 339:47–54

Andolfi L, Bruce D, Cannistraro S, Canters GW, Davis JJ, Hill HAO, Crozier J, Verbeet MP, Wrathmell CL, Astier Y (2004) The electrochemical characteristics of blue copper protein monolayers on gold. J Electroanal Chem 565:21–28

Arbabi-Ghahroudi M, Tanha J, MacKenzie R (2005) Prokaryotic expression of antibodies. Cancer Metastasis Rev 24:501–519

Arndt KM, Muller KM, Pluckthun A (1998) Factors influencing the dimer to monomer transition of an antibody single-chain Fv fragment. Biochemistry 37:12918–12926

Bae YM, Oh BK, Lee W, Lee WH, Choi JW (2005) Study on orientation of immunogrlobulin G on protein G layer. Biosens Bioelectron 21:103–110

Baird CL, Myszka DG (2001) Current and emerging commercial optical biosensors. J Mol Recognit 14:261–268

Baker SN, Brauns EB, McCleskey TM, Burrell AK, Baker GA, 2006. Fluorescence quenching immunoassay performed in an ionic liquid. Chem Commun (Camb): 2851–2853.

Barderas R, Desmet J, Timmerman P, Meloen R, Casal JI (2008) Affinity maturation of antibodies assisted by in silico modeling. Proc Natl Acad Sci USA 105:9029–9034

Barlen B, Mazumdar SD, Lezrich O, Kampfer P, Keusgen M (2007) Detection of salmonella by surface plasmon resonance. Sensors 7:1427–1446

Barthelemy PA, Raab H, Appleton BA, Bond CJ, Wu P, Wiesmann C, Sidhu SS (2008) Comprehensive analysis of the factors contributing to the stability and solubility of autonomous human V-H domains. J Biol Chem 283:3639–3654

Batista FD, Neuberger MS (1998) Affinity dependence of the B cell response to antigen: a threshold, a ceiling, and the importance of off-rate. Immunity 8:751–759

Baujard-Lamotte L, Noinville S, Goubard F, Marque P, Pauthe E (2008) Kinetics of conformational changes of fibronectin adsorbed onto model surfaces. Colloids Surf B Biointerfaces 63:129–137

Bayry J, Prabhudas K, Bist P, Reddy GR, Suryanarayana VVS (1999) Immune affinity purification of foot and mouth disease virus type specific antibodies using recombinant protein adsorbed to polystyrene wells. J Virol Methods 81:21–30

Behring E, Kitasato S (1890) Ueber das Zustandekommen der Diphtheria-Immunitat und der Tetanus-Immunitat bci thicrcn. Dtsch Med Wochenschr 16:1113–1114

Benhar I (2007) Design of synthetic antibody libraries. Expert Opin Biol Ther 7:763–779

Berquand A, Xia N, Castner DG, Clare BH, Abbott NL, Dupres V, Adriaensen Y, Dufrene YF (2005) Antigen binding forces of single antilysozyme Fv fragments explored by atomic force microscopy. Langmuir 21:5517–5523

Bessette PH, Aslund F, Beckwith J, Georgiou G (1999) Efficient folding of proteins with multiple disulfide bonds in the *Escherichia coli* cytoplasm. Proc Natl Acad Sci USA 96:13703–13708

Bibila TA, Robinson DK (1995) In pursuit of the optimal fed-batch process for monoclonal-antibody production. Biotechnol Prog 11:1–13

Birch JR, Racher AJ (2006) Antibody production. Adv Drug Deliv Rev 58:671–685

Bird RE, Hardman KD, Jacobson JW, Johnson S, Kaufman BM, Lee SM, Lee T, Pope SH, Riordan GS, Whitlow M (1988) Single-chain antigen-binding proteins. Science 242:423–426

Birtalan S, Zhang YN, Fellouse FA, Shao LH, Schaefer G, Sidhu SS (2008) The intrinsic contributions of tyrosine, serine, glycine and arginine to the affinity and specificity of antibodies. J Mol Biol 377:1518–1528

Blenner MA, Banta S (2008) Characterization of the 4D5Flu single-chain antibody with a stimulus-responsive elastin-like peptide linker: a potential reporter of peptide linker conformation. Protein Sci 17:527–536

Bolduc OR, Masson JF (2008) Monolayers of 3-mercaptopropyl-amino acid to reduce the nonspecific adsorption of serum proteins on the surface of biosensors. Langmuir 24:12085–12091

Bongini L, Fanelli D, Piazza F, De los Rios P, Sandin S, Skoglund U (2004) Freezing immunoglobulins to see them move. Proc Natl Acad Sci USA 101:6466–6471

Bosshard HR, Marti DN, Jelesarov I (2004) Protein stabilization by salt bridges: concepts, experimental approaches and clarification of some misunderstandings. J Mol Recognit 17: 1–16

Boujday S, Bantegnie A, Briand E, Marnet PG, Salmain M, Pradier CM (2008) In-depth investigation of protein adsorption on gold surfaces: correlating the structure and density to the efficiency of the sensing layer. J Phys Chem B 112:6708–6715

Briand E, Gu C, Boujday S, Salmain M, Herry JM, Pradier CM (2007) Functionalisation of gold surfaces with thiolate SAMs: topography/bioactivity relationship – a combined FT-RAIRS, AFM and QCM investigation. Surf Sci 601:3850–3855

Brockmann EC, Cooper M, Stromsten N, Vehniainen M, Saviranta P (2005) Selecting for antibody scFv fragments with improved stability using phage display with denaturation under reducing conditions. J Immunol Methods 296:159–170

Brod E, Nimri S, Turner B, Sivan U (2008) Electrical control over antibody–antigen binding. Sens Actuators B Chem 128:560–565

Brummell DA, Sharma VP, Anand NN, Bilous D, Dubuc G, Michniewicz J, MacKenzie CR, Sadowska J, Sigurskjold BW, Sinnott B, Young NM, Bundle DR, Narang SA (1993) Probing the combining site of an anticarbohydrate antibody by saturation mutagenesis – role of the heavy-chain Cdr3 residues. Biochemistry 32:1180–1187

Cai HW, Chen LH, Wan L, Zeng LY, Yang H, Li SF, Li YP, Cheng JQ, Lu XF (2009) High-level expression of a functional humanized anti-CTLA4 single-chain variable fragment antibody in *Pichia pastoris*. Appl Microbiol Biotechnol 82:41–48

Calabrese MF, Eakin CM, Wang JM, Miranker AD (2008) A regulatable switch mediates self-association in an immunoglobulin fold. Nat Struct Mol Biol 15:965–971

Camarero JA (2008) Recent developments in the site-specific immobilization of proteins onto solid supports. Biopolymers 90:450–458

Campbell GA, Mutharasan R (2008) Near real-time detection of cryptosporidium parvum oocyst by IgM-functionalized piezoelectric-excited millimeter-sized cantilever biosensor. Biosens Bioelectron 23:1039–1045

Caruso F, Rodda E, Furlong DN (1996) Orientational aspects of antibody immobilization and immunological activity on quartz crystal microbalance electrodes. J Colloid Interf Sci 178:104–115

Catimel B, Nerrie M, Lee FT, Scott AM, Ritter G, Welt S, Old LJ, Burgess AW, Nice EC (1997) Kinetic analysis of the interaction between the monoclonal antibody A33 and its colonic epithelial antigen by the use of an optical biosensor – a comparison of immobilisation strategies. J Chromatogr A 776:15–30

Chang CEA, McLaughlin WA, Baron R, Wang W, McCammon JA (2008) Entropic contributions and the influence of the hydrophobic environment in promiscuous protein–protein association. Proc Natl Acad Sci USA 105:7456–7461

Chang H, Qin WS, Li Y, Zhang JY, Lin Z, Lv M, Sun YX, Feng JN, Shen BF (2007) A novel human scFv fragment against TNF-alpha from de novo design method. Mol Immunol 44:3789–3796

Chard T (1992) Pregnancy tests – a review. Hum Reprod 7:701–710
Chen SF, Liu LY, Zhou J, Jiang SY (2003) Controlling antibody orientation on charged self-assembled monolayers. Langmuir 19:2859–2864
Cho IH, Paek EH, Lee H, Kang JY, Kim TS, Paek SH (2007) Site-directed biotinylation of antibodies for controlled immobilization of solid surfaces. Anal Biochem 365:14–23
Choi DH, Katakura Y, Ninomiya K, Shioya S (2008) Rational screening of antibodies and design of sandwich enzyme linked immunosorbant assay on the basis of a kinetic model. J Biosci Bioeng 105:261–272
Chothia C, Lesk AM (1987) Canonical structures for the hypervariable regions of immunoglobulins. J Mol Biol 196:901–917
Chou SF, Hsu WL, Hwang JM, Chen CY (2004) Development of an immunosensor for human ferritin, a nonspecific tumor marker, based on surface plasmon resonance. Biosens Bioelectron 19:999–1005
Christ D, Famm K, Winter G (2007) Repertoires of aggregation-resistant human antibody domains. Protein Eng Des Sel 20:413–416
Clackson T, Wells JA (1995) A hot-spot of binding-energy in a hormone-receptor interface. Science 267:383–386
Clark M (1997) Antibody engineering: IgG effector mechanisms. Chem Immunol 65:88–110
Cloutier SM, Couty S, Terskikh A, Marguerat L, Crivelli V, PugniΦres M, Mani J-C, Leisinger H-J, Mach JP, Deperthes D (2000) Streptabody, a high avidity molecule made by tetramerization of in vivo biotinylated, phage display-selected scFv fragments on streptavidin. Mol Immunol 37:1067–1077
Cobaugh CW, Almagro JC, Pogson M, Iverson B, Georgiou G (2008) Synthetic antibody libraries focused towards peptide ligands. J Mol Biol 378:622–633
Conrad U, Scheller J (2005) Considerations on antibody-phage display methodology. Comb Chem High Throughput Screen 8:117–126
Cooper A (1999) Thermodynamics of protein folding and stability. Protein: a comprehensive treatise 2:217–270
Cooper A, Cameron D, Jakus J, Pettigrew GW (2007) Pressure perturbation calorimetry, heat capacity and the role of water in protein stability and interactions. Biochem Soc Trans 35:1547–1550
Cooper MD, Alder MN (2006) The evolution of adaptive immune systems. Cell 124:815–822
Damasceno LM, Anderson KA, Ritter G, Cregg JM, Old LJ, Batt CA (2007) Cooverexpression of chaperones for enhanced secretion of a single-chain antibody fragment in Pichia pastoris. Appl Microbiol Biotechnol 74:381–389
De Crescenzo G, Litowski JR, Hodges RS, O'Connor-McCourt MD (2003) Real-time monitoring of the interactions of two-stranded de novo designed coiled-coils: effect of chain length on the kinetic and thermodynamic constants of binding. Biochemistry 42:1754–1763
Decramer S, de Peredo AG, Breuil B, Mischak H, Monsarrat B, Bascands JL, Schanstra JP (2008) Urine in clinical proteomics. Mol Cell Proteomics 7:1850–1862
Delves PJ, Roitt IM (2000) Advances in immunology: the immune system – second of two parts. N Eng J Med 343:108–117
Demarest SJ, Chen G, Kimmel BE, Gustafson D, Wu J, Salbato J, Poland J, Elia M, Tan XQ, Wong K, Short J, Hansen G (2006) Engineering stability into *Escherichia coli* secreted Fabs leads to increased functional expression. Protein Eng Des Sel 19:325–336
Demarest SJ, Glaser SM (2008) Antibody therapeutics, antibody engineering, and the merits of protein stability. Curr Opin Drug Discov Devel 11:675–687
Derwinska K, Sauer U, Preininger C (2008) Adsorption versus covalent, statistically oriented and covalent, site-specific IgG immobilization on poly(vinyl alcohol)-based surfaces. Talanta 77:652–658
Desplancq D, King DJ, Lawson ADG, Mountain A (1994) Multimerization behavior of single-chain Fv variants for the tumor-binding antibody B72.3. Protein Eng 7:1027–1033
Dimitrov JD, Lacroix-Desmazes S, Kaveri SV, Vassilev TL (2007) Transition towards antigen-binding promiscuity of a monospecific antibody. Mol Immunol 44:1854–1863

Donini M, Morea V, Desiderio A, Pashkoulov D, Villani ME, Tramontano A, Benvenuto E (2003) Engineering stable cytoplasmic intrabodies with designed specificity. J Mol Biol 330:323–332

Dooley H, Flajnik MF, Porter AJ (2003) Selection and characterization of naturally occurring single-domain (IgNAR) antibody fragments from immunized sharks by phage display. Mol Immunol 40:25–33

Dooley H, Grant SD, Harris WJ, Porter AJ (1998) Stabilization of antibody fragments in adverse environments. Biotechnol Appl Biochem 28:77–83

Dufner P, Jermutus L, Minter RR (2006) Harnessing phage and ribosome display for antibody optimisation. Trends Biotechnol 24:523–529

Duroux M, Skovsen E, Neves-Petersen MT, Duroux L, Gurevich L, Borrebaeck CAK, Wingren C, Petersen SB (2008) Light-induced immobilisation of biomolecules as an attractive alternative to micro-droplet dispensing-based arraying technologies (vol 7, pg 3491, 2007). Proteomics 8:1113

Einhauer A, Jungbauer A (2001) Affinity of the monoclonal antibody M1 directed against the FLAG peptide. J Chromatogr A 921:25–30

Ellmark P, Hogerkorp CM, Ek S, Belov L, Berglund M, Rosenquist R, Christopherson RI, Borrebaeck CAK (2008) Phenotypic protein profiling of different B cell sub-populations using antibody CD-microarrays. Cancer Lett 265:98–106

Espargaro A, Castillo V, de Groot NS, Ventura S (2008) The *in vivo* and *in vitro* aggregation properties of globular proteins correlate with their conformational stability: the SH3 case. J Mol Biol 378:1116–1131

Famm K, Hansen L, Christ D, Winter G (2008) Thermodynamically stable aggregation-resistant antibody domains through directed evolution. J Mol Biol 376:926–931

Famm K, Winter G (2006) Engineering aggregation-resistant proteins by directed evolution. Protein Eng Des Sel 19:479–481

Figueira VBC, Jones JP (2008) Viscoelastic study of the adsorption of bovine serum albumin on gold and its dependence on pH. J Colloid Interf Sci 325:107–113

Foote J, Eisen HN (2000) Breaking the affinity ceiling for antibodies and T cell receptors. Proc Natl Acad Sci USA 97:10679–10681

Forsman A, Beirnaert E, asa-Chapman MMI, Hoorelbeke B, Hijazi K, Koh W, Tack V, Szynol A, Kelly C, McKnight A, Verrips T, de Haard H, Weiss RA (2008) Llama antibody fragments with cross-subtype human immunodeficiency virus type 1 (HIV-1)-neutralizing properties and high affinity for HIV-1 gp120. J Virol 82:12069–12081

Fowler JM, Stuart MC, Wong DKY (2007) Self-assembled layer of thiolated protein G as an immunosensor scaffold. Anal Chem 79:350–354

Garber E, Demarest SJ (2007) A broad range of Fab stabilities within a host of therapeutic IgGs. Biochem Biophys Res Commun 355:751–757

Garfield E (1985) The 1984 Nobel-prize in medicine is awarded to Jerne, Niels, K., Milstein, Cesar, and Kohler, Georges, J.F., for their contributions to immunology. Curr Contents 8:3–18

Gaza-Bulseco G, Liu HC (2008) Fragmentation of a recombinant monoclonal antibody at various pH. Pharm Res 25:1881–1890

Gilbreth RN, Esaki K, Koide A, Sidhu SS, Koide S (2008) A dominant conformational role for amino acid diversity in minimalist protein–protein interfaces. J Mol Biol 381:407–418

Goddard JM, Hotchkiss JH (2007) Polymer surface modification for the attachment of bioactive compounds. Prog Polym Sci 32:698–725

Goerke AR, Swartz JR (2009) High-level cell-free synthesis yields of proteins containing site-specific non-natural amino acids. Biotechnol Bioeng 102:400–416

Goldman ER, Anderson GP, Liu JL, Delehanty JB, Sherwood LJ, Osborn LE, Cummins LB, Hayhurst A (2006) Facile generation of heat-stable antiviral and antitoxin single domain antibodies from a semisynthetic llama library. Anal Chem 78:8245–8255

Gong NL, Chatterjee S (2003) Platelet endothelial cell adhesion molecule in cell signaling and thrombosis. Mol Cell Biochem 253:151–158

Goodchild S, Love T, Hopkins N, Mayers C (2006) Engineering antibodies for biosensor technologies. Adv Appl Microbiol 58:185–226

Green NM (1975) Avidin. Adv Protein Chem 29:85–133

Greenberg AS, Avila D, Hughes M, Hughes A, Mckinney EC, Flajnik MF (1995) A new antigen receptor gene family that undergoes rearrangement and extensive somatic diversification in sharks. Nature 374:168–173

Gvritishvili AG, Gribenko AV, Makhatadze GI (2008) Cooperativity of complex salt bridges. Protein Sci 17:1285–1290

Gwenin CD, Jones JP, Kalaji M, Lewis TJ, Llewellyn JP, Williams PA (2007a) Viscoelastic change following adsorption and subsequent molecular reorganisation of a nitroreductase enzyme on a gold surface: a QCM study. Sens Actuators B Chem 126:499–507

Gwenin CD, Kalaji M, Williams PA, Jones RM (2007b) The orientationally controlled assembly of genetically modified enzymes in an amperometric biosensor. Biosens Bioelectron 22:2869–2875

Gwinner W (2007) Renal transplant rejection markers. World J Urol 25:445–455

Haab BB (2006) Applications of antibody array platforms. Curr Opin Biotechnol 17:415–421

Haab BB, Geierstanger BH, Michailidis G, Vitzthum F, Forrester S, Okon R, Saviranta P, Brinker A, Sorette M, Perlee L, Suresh S, Drwal G, Adkins JN, Omenn GS (2005) Immunoassay and antibody microarray analysis of the HUPO Plasma Proteome Project reference specimens: systematic variation between sample types and calibration of mass spectrometry data. Proteomics 5:3278–3291

Hagihara Y, Mine S, Uegaki K (2007) Stabilization of an immunoglobulin fold domain by an engineered disulfide bond at the buried hydrophobic region. J Biol Chem 282:36489–36495

Halaby DM, Mornon JPE (1998) The immunoglobulin superfamily: an insight on its tissular, species, and functional diversity. J Mol Evol 46:389–400

Halaby DM, Poupon A, Mornon JP (1999) The immunoglobulin fold family: sequence analysis and 3D structure comparisons. Protein Eng 12:563–571

Hamers-casterman C, Atarhouch T, Muyldermans S, Robinson G, Hamers C, Songa EB, Bendahman N, Hamers R (1993) Naturally-occurring antibodies devoid of light-chains. Nature 363:446–448

Harmsen MM, De Haard HJ (2007) Properties, production, and applications of camelid single-domain antibody fragments. Appl Microbiol Biotechnol 77:13–22

Harris LJ, Larson SB, Hasel KW, Day J, Greenwood A, Mcpherson A (1992) The 3-dimensional structure of an intact monoclonal-antibody for canine lymphoma. Nature 360:369–372

Hayhurst A (2000) Improved expression characteristics of single-chain Fv fragments when fused downstream of the *Escherichia coli* maltose-binding protein or upstream of a single immunoglobulin-constant domain. Protein Expr Purif 18:1–10

He MY (2008) Cell-free protein synthesis: applications in proteomics and biotechnology. N Biotechnol 25:126–132

He MY, Stoevesandt O, Taussig MJ (2008) In situ synthesis of protein arrays. Curr Opin Biotechnol 19:4–9

He MY, Taussig MJ (2005) Ribosome display of antibodies: expression, specificity and recovery in a eukaryotic system. J Immunol Methods 297:73–82

Hearty S, Leonard P, Quinn J, O'Kennedy R (2006) Production, characterisation and potential application of a novel monoclonal antibody for rapid identification of virulent Listeria monocytogenes. J Microbiol Meth 66:294–312

Hemmersam AG, Rechendorff K, Foss M, Sutherland DS, Besenbacher F (2008) Fibronectin adsorption on gold, Ti-, and Ta-oxide investigated by QCM-D and RSA modelling. J Colloid Interf Sci 320:110–116

Ho M, Kreitman RJ, Onda M, Pastan I (2005) *In vitro* antibody evolution targeting germline hot spots to increase activity of an anti-CD22 immunotoxin. J Biol Chem 280:607–617

Hodneland CD, Lee YS, Min DH, Mrksich M (2002) Selective immobilization of proteins to self-assembled monolayers presenting active site-directed capture ligands. Proc Natl Acad Sci USA 99:5048–5052

Honegger A, Pluckthun A (2001) Yet another numbering scheme for immunoglobulin variable domains: an automatic modeling and analysis tool. J Mol Biol 309:657–670

Hoogenboom HR (2005) Selecting and screening recombinant antibody libraries. Nat Biotechnol 23:1105–1116

Hoogenboom HR, Griffiths AD, Johnson KS, Chiswell DJ, Hudson P, Winter G (1991) Multi-subunit proteins on the surface of filamentous phage – methodologies for displaying antibody (Fab) heavy and light-chains. Nucleic Acids Res 19:4133–4137

Hu XJ, O'Hara L, White S, Magner E, Kane M, Wall JG (2007) Optimisation of production of a domoic acid-binding scFv antibody fragment in *Escherichia coli* using molecular chaperones and functional immobilisation on a mesoporous silicate support. Protein Expr Purif 52: 194–201

Hugo N, Lafont V, Beukes M, Altschuh D (2002) Functional aspects of co-variant surface charges in an antibody fragment. Protein Sci 11:2697–2705

Hust M, Jostock T, Menzel C, Voedisch B, Mohr A, Brenneis M, Kirsch MI, Meier D, Dubel S (2007) Single chain Fab (scFab) fragment. BMC Biotechnol 7:14

Huston JS, Levinson D, Mudgetthunter M, Tai MS, Novotny J, Margolies MN, Ridge RJ, Bruccoleri RE, Haber E, Crea R, Oppermann H (1988) Protein engineering of antibody-binding sites – recovery of specific activity in an anti-digoxin single-chain Fv analog produced in *Escherichia coli*. Proc Natl Acad Sci USA 85:5879–5883

Ingvarsson J, Wingren C, Carlsson A, Ellmark P, Wahren B, Engstrom G, Harmenberg U, Krogh M, Peterson C, Borrebaeck CAK (2008) Detection of pancreatic cancer using antibody microarray-based serum protein profiling. Proteomics 8:2211–2219

Ionescu RM, Vlasak J, Price C, Kirchmeier M (2008) Contribution of variable domains to the stability of humanized IgG1 monoclonal antibodies. J Pharm Sci 97:1414–1426

Iqbal SS, Mayo MW, Bruno JG, Bronk BV, Batt CA, Chambers JP (2000) A review of molecular recognition technologies for detection of biological threat agents. Biosens Bioelectron 15: 549–578

Iversen L, Cherouati N, Berthing T, Stamou D, Martinez KL (2008) Templated protein assembly on micro-contact-printed surface patterns. Use of the SNAP-tag protein functionality. Langmuir 24:6375–6381

Janin J, Bahadur RP, Chakrabarti P (2008) Protein–protein interaction and quaternary structure. Q Rev Biophys 41:133–180

Jermutus L, Honegger A, Schwesinger F, Hanes J, Pluckthun A (2001) Tailoring *in vitro* evolution for protein affinity or stability. Proc Natl Acad Sci USA 98:75–80

Jonkheijm P, Weinrich D, Schroder H, Niemeyer CM, Waldmann H (2008) Chemical strategies for generating protein biochips. Angew Chem Int Ed 47:9618–9647

Jung S, Honegger A, Pluckthun A (1999) Selection for improved protein stability by phage display. J Mol Biol 294:163–180

Jung S, Pluckthun A (1997) Improving in vivo folding and stability of a single-chain Fv antibody fragment by loop grafting. Protein Eng 10:959–966

Jung YW, Kang HJ, Lee JM, Jung SO, Yun WS, Chung SJ, Chung BH (2008) Controlled antibody immobilization onto immunoanalytical platforms by synthetic peptide. Anal Biochem 374:99–105

Jurado P, de Lorenzo V, Fernandez LA (2006) Thioredoxin fusions increase folding of single chain Fv antibodies in the cytoplasm of *Escherichia coli*: evidence that chaperone activity is the prime effect of thioredoxin. J Mol Biol 357:49–61

Kaarakainen P, Meklin T, Rintala H, Hyvaerinen A, Karkkainen P, Vepsalainen A, Hirvonen MR, Nevalainen A (2008) Seasonal variation in airborne microbial concentrations and diversity at landfill, urban and rural sites. Clean – Soil Air Water 36:556–563

Kalkhof S, Sinz A (2008) Chances and pitfalls of chemical cross-linking with amine-reactive N-hydroxysuccinimide esters. Anal Bioanal Chem 392:305–312

Kamerzell TJ, Middaugh CR (2007) Two-dimensional correlation spectroscopy reveals coupled immunoglobulin regions of differential flexibility that influence stability. Biochemistry 46:9762–9773

Kamerzell TJ, Ramsey JD, Middaugh CR (2008) Immunoglobulin dynamics, conformational fluctuations, and nonlinear elasticity and their effects on stability. J Phys Chem B 112:3240–3250

Kingsmore SF (2006) Multiplexed protein measurement: technologies and applications of protein and antibody arrays. Nat Rev Drug Discov 5:310–320

Kiss C, Fisher H, Pesavento E, Dai MH, Valero R, Ovecka M, Nolan R, Phipps ML, Velappan N, Chasteen L, Martinez JS, Waldo GS, Pavlik P, Bradbury ARM (2006) Antibody binding loop insertions as diversity elements. Nucleic Acids Res 34:e132

Knotts TA, Rathore N, de Pabloz JJ (2008) An entropic perspective of protein stability on surfaces. Biophys J 94:4473–4483

Kohler G, Milstein C (1975) Continuous cultures of fused cells secreting antibody of predefined specificity. Nature 256:495–497

Kolmar H, Skerra A (2008) Alternative binding proteins get mature: rivalling antibodies. FEBS J 275:2667

Kopf E, Zharhary D (2007) Antibody arrays – an emerging tool in cancer proteomics. Int J Biochem Cell Biol 39:1305–1317

Kortt AA, Oddie GW, Iliades P, Gruen LC, Hudson PJ (1997) Nonspecific amine immobilization of ligand can be a potential source of error in BIAcore binding experiments and may reduce binding affinities. Anal Biochem 253:103–111

Kourentzi K, Srinivasan M, Smith-Gill SJ, Willson RC (2008) Conformational flexibility and kinetic complexity in antibody–antigen interaction. J Mol Recognit 21:114–121

Kubetzko S, Balic E, Waibel R, Zangemeister-Wittke U, Pluckthun A (2006) PEGylation and multimerization of the anti-p185(HER-2) single chain Fv fragment 4D5 – effects on tumor targeting. J Biol Chem 281:35186–35201

Kubetzko S, Sarkar CA, Pluckthun A (2005) Protein PEGylation decreases observed target association rates via a dual blocking mechanism. Mol Pharmacol 68:1439–1454

Kwon Y, Han ZZ, Karatan E, Mrksich M, Kay BK (2004) Antibody arrays prepared by cutinase-mediated immobilization on self-assembled monolayers. Anal Chem 76:5713–5720

Lacy ER, Baker M, Brigham-Burke M (2008) Free sulfhydryl measurement as an indicator of antibody stability. Anal Biochem 382:66–68

Lahiri J, Isaacs L, Tien J, Whitesides GM (1999) A strategy for the generation of surfaces presenting ligands for studies of binding based on an active ester as a common reactive intermediate: a surface plasmon resonance study. Anal Chem 71:777–790

Lamla T, Erdmann VA (2004) The nano-tag, a streptavidin-binding peptide for the purification and detection of recombinant proteins. Protein Expr Purif 33:39–47

Langedijk AC, Honegger A, Maat J, Planta RJ, van Schaik RC, Pluckthun A (1998) The nature of antibody heavy chain residue H6 strongly influences the stability of a V-H domain lacking the disulfide bridge. J Mol Biol 283:95–110

Lappalainen I, Hurley MG, Clarke J (2008) Plasticity within the obligatory folding nucleus of an immunoglobulin-like domain. J Mol Biol 375:547–559

Latterich M, Abramovitz M, Leyland-Jones B (2008) Proteomics: new technologies and clinical applications. Eur J Cancer 44:2737–2741

Le Berre M, Kane M (2006) Biosensor-based assay for domoic acid: comparison of performance using polyclonal, monoclonal, and recombinant antibodies. Anal Lett 39:1587–1598

Le Brun AP, Holt SA, Shah DS, Majkrzak CF, Lakey JH (2008) Monitoring the assembly of antibody-binding membrane protein arrays using polarised neutron reflection. Eur Biophys J 37:639–645

Leckband DE, Kuhl TL, Wang HK, Muller W, Herron J, Ringsdorf H (2000) Force probe measurements of antibody–antigen interactions. Methods 20:329–340

Lee SK, Kim HC, Cho SJ, Jeong SW, Jeon WB (2008) Binding behavior of CRP and anti-CRP antibody analyzed with SPR and AFM measurement. Ultramicroscopy 108:1374–1378

Lee Y, Lee EK, Cho YW, Matsui T, Kang IC, Kim TS, Han MH (2003) ProteoChip: a highly sensitive protein microarray prepared by a novel method of protein, immobilization for application of protein–protein interaction studies. Proteomics 3:2289–2304

Leong SSJ, Chen WN (2008) Preparing recombinant single chain antibodies. Chem Eng Sci 63:1401–1414
Li JN, Mahajan A, Tsai MD (2006) Ankyrin repeat: a unique motif mediating protein–protein interactions. Biochemistry 45:15168–15178
Li TJ, Cheng JR, Hu BS, Liu Y, Qian GL, Liu FQ (2008) Construction, production, and characterization of recombinant scFv antibodies against methamidophos expressed in Pichia pastoris. World J Microbiol Biotechnol 24:867–874
Lipovsek D, Lippow SM, Hackel BJ, Gregson MW, Cheng P, Kapila A, Wittrup KD (2007) Evolution of an interloop disulfide bond in high-affinity antibody mimics based on fibronectin type III domain and selected by yeast surface display: molecular convergence with single-domain camelid and shark antibodies. J Mol Biol 368:1024–1041
Lipovsek D, Pluckthun A (2004) *In-vitro* protein evolution by ribosome display and mRNA display. J Immunol Methods 290:51–67
Lippow SM, Wittrup KD, Tidor B (2007) Computational design of antibody-affinity improvement beyond in vivo maturation. Nat Biotechnol 25:1171–1176
Lipschultz CA, Yee A, Mohan S, Li YL, Smith-Gill SJ (2002) Temperature differentially affects encounter and docking thermodynamics of antibody–antigen association. J Mol Recognit 15:44–52
Liu HC, Gaza-Bulseco G, Faldu D, Chumsae C, Sun J (2008a) Heterogeneity of monoclonal antibodies. J Pharm Sci 97:2426–2447
Liu HC, Gaza-Bulseco G, Sun J (2006) Characterization of the stability of a fully human monoclonal IgG after prolonged incubation at elevated temperature. J Chromatogr B Analyt Technol Biomed Life Sci 837:35–43
Liu HC, Gaza-Bulseco G, Xiang T, Chumsae C (2008b) Structural effect of deglycosylation and methionine oxidation on a recombinant monoclonal antibody. Mol Immunol 45:701–708
Lo Conte L, Chothia C, Janin J (1999) The atomic structure of protein–protein recognition sites. J Mol Biol 285:2177–2198
Love TE, Redmond C, Mayers CN (2008) Real time detection of anthrax spores using highly specific anti-EA1 recombinant antibodies produced by competitive panning. J Immunol Methods 334:1–10
Lu JR, Zhao XB, Yaseen M (2007) Protein adsorption studied by neutron reflection. Curr Opin Colloid Interface Sci 12:9–16
Luginbuhl B, Kanyo Z, Jones RM, Fletterick RJ, Prusiner SB, Cohen FE, Williamson RA, Burton DR, Pluckthun A (2006) Directed evolution of an anti-prion protein scFv fragment to an affinity of 1 pM and its structural interpretation. J Mol Biol 363:75–97
Luong JHT, Male KB, Glennon JD (2008) Biosensor technology: technology push versus market pull. Biotechnol Adv 26:492–500
Luppa PB, Sokoll LJ, Chan DW (2001) Immunosensors – principles and applications to clinical chemistry. Clin Chim Acta 314:1–26
Ly T, Liu ZJ, Pujanauski BG, Sarpong R, Julian RR (2008) Surveying ubiquitin structure by noncovalent attachment of distance constrained bis(crown) ethers. Anal Chem 80:5059–5064
Makabe K, Nakanishi T, Tsumoto K, Tanaka Y, Kondo H, Umetsu M, Sone Y, Asano R, Kumagai I (2008) Thermodynamic consequences of mutations in Vernier zone residues of a humanized anti-human epidermal growth factor receptor murine antibody-528. J Biol Chem 283:1156–1166
Mallender WD, Carrero J, Voss EW (1996) Comparative properties of the single chain antibody and Fv derivatives of mAb 4-4-20 – relationship between interdomain interactions and the high affinity for fluorescein ligand. J Biol Chem 271:5338–5346
Maly J, Illiano E, Sabato M, De Francesco M, Pinto V, Masci A, Masci D, Masojidek J, Sugiura M, Franconi R, Pilloton R (2002) Immobilisation of engineered molecules on electrodes and optical surfaces. Mater Sci Eng C-Biomim Supramol Syst 22:257–261
Martin CD, Rojas G, Mitchell JN, Vincent KJ, Wu JH, McCafferty J, Schofield DJ (2006) A simple vector system to improve performance and utilisation of recombinant antibodies. BMC Biotechnol 6:46

Martinez NF, Lozano JR, Herruzo ET, Garcia F, Richter C, Sulzbach T, Garcia R (2008) Bimodal atomic force microscopy imaging of isolated antibodies in air and liquids. Nanotechnology 19:384011

McCafferty J, Griffiths AD, Winter G, Chiswell DJ (1990) Phage antibodies – filamentous phage displaying antibody variable domains. Nature 348:552–554

McCarthy BJ, Hill AS (2001) Altering the fine specificity of an anti-Legionella single chain antibody by a single amino acid insertion. J Immunol Methods 251:137–149

McManus CA, Rose ML, Dunn MJ (2006) Proteomics of transplant rejection. Transplant Rev 20:195–207

Medintz IL, Goldman ER, Lassman ME, Hayhurst A, Kusterbeck AW, Deschamps JR (2005) Self-assembled TNT biosensor based on modular multifunctional surface-tethered components. Anal Chem 77:365–372

Mehne J, Markovic G, Proll F, Schweizer N, Zorn S, Schreiber F, Gauglitz G (2008) Characterisation of morphology of self-assembled PEG monolayers: a comparison of mixed and pure coatings optimised for biosensor applications. Anal Bioanal Chem 391:1783–1791

Michaud GA, Salcius M, Zhou F, Bangham R, Bonin J, Guo H, Snyder M, Predki PF, Schweitzer BI (2003) Analyzing antibody specificity with whole proteome microarrays. Nat Biotechnol 21:1509–1512

Miller KD, Weaver-Feldhaus J, Gray SA, Siegel RW, Feldhaus MJ (2005) Production, purification, and characterization of human scFv antibodies expressed in *Saccharomyces cerevisiae*, *Pichia pastoris*, and *Escherichia coli*. Protein Expr Purif 42:255–267

Mimura Y, Church S, Ghirlando R, Ashton PR, Dong S, Goodall M, Lund J, Jefferis R (2000) The influence of glycosylation on the thermal stability and effector function expression of human IgG1-Fc: properties of a series of truncated glycoforms. Mol Immunol 37:697–706

Mirny LA, Shakhnovich EI (1999) Universally conserved positions in protein folds: reading evolutionary signals about stability, folding kinetics and function. J Mol Biol 291: 177–196

Monsellier E, Bedouelle H (2006) Improving the stability of an antibody variable fragment by a combination of knowledge-based approaches: validation and mechanisms. J Mol Biol 362:580–593

Moreira IS, Fernandes PA, Ramos MJ (2007a) Hot spot occlusion from bulk water: a comprehensive study of the complex between the lysozyme HEL and the antibody FVD1.3. J Phys Chem B 111:2697–2706

Moreira IS, Fernandes PA, Ramos MJ (2007b) Hot spots-A review of the protein–protein interface determinant amino-acid residues. Proteins 68:803–812

Morozova TY, Morozov VN (2008) Force differentiation in recognition of cross-reactive antigens by magnetic beads. Anal Biochem 374:263–271

Nagata T, Gupta V, Sorce D, Kim WY, Sali A, Chait BT, Shigesada K, Ito Y, Werner MH (1999) Immunoglobulin motif DNA recognition and heterodimerization of the PEBP2/CBF Runt domain. Nat Struct Biol 6:615–619

Nath N, Hurst R, Hook B, Meisenheimer P, Zhao KQ, Nassif N, Bulleit RF, Storts DR (2008) Improving protein array performance: focus on washing and storage conditions. J Proteome Res 7:4475–4482

Nezlin R, Ghetie V (2004) Interactions of immunoglobulins outside the antigen-combining site. Academic, San Diego

Ngundi MM, Kulagina NV, Anderson GP, Taitt CR (2006) Nonantibody-based recognition: alternative molecules for detection of pathogens. Expert Rev Proteomics 3:511–524

Nissim A, Hoogenboom HR, Tomlinson IM, Flynn G, Midgley C, Lane D, Winter G (1994) Antibody fragments from a single pot phage display library as immunochemical reagents. EMBO J 13:692–698

Noh H, Vogler EA (2007) Volumetric interpretation of protein adsorption: competition from mixtures and the Vroman effect. Biomaterials 28:405–422

Nuttall SD, Walsh RB (2008) Display scaffolds: protein engineering for novel therapeutics. Curr Opin Pharmacol 8:609–615

Omidfar K, Rasaee MJ, Kashanian S, Paknejad M, Bathaie Z (2007) Studies of thermostability in *Camelus bactrianus* (Bactrian camel) single-domain antibody specific for the mutant epidermal-growth-factor receptor expressed by Pichia. Biotechnol Appl Biochem 46:41–49

Orlandi R, Gussow DH, Jones PT, Winter G (1989) Cloning immunoglobulin variable domains for expression by the polymerase chain-reaction. Proc Natl Acad Sci USA 86:3833–3837

Ostler EL, Resmini M, Brocklehurst K, Gallacher G (2002) Polyclonal catalytic antibodies. JImmunol Methods 269:111–124

Padlan EA (1994) Anatomy of the antibody molecule. Mol Immunol 31:169–217

Palmer E, Liu H, Khan F, Taussig MJ, He MY (2006) Enhanced cell-free protein expression by fusion with immunoglobulin C kappa domain. Protein Sci 15:2842–2846

Pancer Z, Mariuzza RA (2008) The oldest antibodies newly discovered. Nat Biotechnol 26: 402–403

Pantoliano MW, Bird RE, Johnson S, Asel ED, Dodd SW, Wood JF, Hardman KD (1991) Conformational stability, folding, and ligand-binding affinity of single-chain-Fv immunoglobulin fragments expressed in *Escherichia coli*. Biochemistry 30:10117–10125

Pasut G, Veronese FM (2007) Polymer-drug conjugation, recent achievements and general strategies. Prog Polym Sci 32:933–961

Peled JU, Kuang FL, Iglesias-Ussel MD, Roa S, Kalis SL, Goodman ME, Scharff MD (2008) The biochemistry of somatic hypermutation. Annu Rev Immunol 26:481–511

Pellequer JL, Chen SWW, Roberts VA, Tainer JA, Getzoff ED (1999) Unraveling the effect of changes in conformation and compactness at the antibody V-L–V-H interface upon antigen binding. J Mol Recognit 12:267–275

Peluso P, Wilson DS, Do D, Tran H, Venkatasubbaiah M, Quincy D, Heidecker B, Poindexter K, Tolani N, Phelan M, Witte K, Jung LS, Wagner P, Nock S (2003) Optimizing antibody immobilization strategies for the construction of protein microarrays. Anal Biochem 312:113–124

Perkins EA, Squirrell DJ (2000) Development of instrumentation to allow the detection of microorganisms using light scattering in combination with surface plasmon resonance. Biosens Bioelectron 14:853–859

Persson H, Ohlin M (2007) Exploring central and peripheral diversity in antibody evolution. Mol Immunol 44:2729–2736

Petrenko VA, Vodyanoy VJ (2003) Phage display for detection of biological threat agents. J Microbiol Meth 53:253–262

Pluckthun A (1992) Monovalent and bivalent antibody fragments produced in *Escherichia coli* – engineering, folding and antigen-binding. Immunol Rev 130:151–188

Porath J, Carlsson J, Olsson I, Belfrage G (1975) Metal chelate affinity chromatography, a new approach to protein fractionation. Nature 258:598–599

Pozharski E, Moulin A, Hewagama A, Shanafelt AB, Petsko GA, Ringe D (2005) Diversity in hapten recognition: structural study of an anti-cocaine antibody M82G2. J Mol Biol 349:570–582

Prabakaran P, Vu BK, Gan JH, Feng Y, Dimitrov DS, Ji XH (2008) Structure of an isolated unglycosylated antibody C(H)2 domain. Acta Crystallogr D Biol Crystallogr 64:1062–1067

Quinn J, Patel P, Fitzpatrick B, Manning B, Dillon P, Daly S, O'Kennedy R, Alcocer M, Lee H, Morgan M, Lang K (1999) The use of regenerable, affinity ligand-based surfaces for immunosensor applications. Biosens Bioelectron 14:587–595

Rabe M, Verdes D, Zimmermann J, Seeger S (2008) Surface organization and cooperativity during nonspecific protein adsorption events. J Phys Chem B 112:13971–13980

Rajewsky K (1996) Clonal selection and learning in the antibody system. Nature 381:751–758

Raju TS (2008) Terminal sugars of Fc glycans influence antibody effector functions of IgGs. Curr Opin Immunol 20:471–478

Ramachandran N, Srivastava S, LaBaer J (2008) Applications of protein microarrays for biomarker discovery. Proteomics Clin Appl 2:1444–1459

Randles LG, Batey S, Steward A, Clarke J (2008) Distinguishing specific and nonspecific interdomain interactions in multidomain proteins. Biophys J 94:622–628

Reeves DC, Lummis SCR (2006) Detection of human and rodent 5-HT3B receptor subunits by anti-peptide polyclonal antibodies. BMC Neurosci 7:27

Reimhult K, Petersson K, Krozer A (2008) QCM-D analysis of the performance of blocking agents on gold and polystyrene surfaces. Langmuir 24:8695–8700

Reiter Y, Brinkmann U, Kreitman RJ, Jung SH, Lee B, Pastan I (1994) Stabilization of the Fv fragments in recombinant immunotoxins by disulfide bonds engineered into conserved framework regions. Biochemistry 33:5451–5459

Renard M, Belkadi L, Hugo N, England P, Altschuh D, Bedouelle H (2002) Knowledge-based design of reagentless fluorescent biosensors from recombinant antibodies. J Mol Biol 318: 429–442

Riechmann L, Winter G (2006) Early protein evolution: building domains from ligand-binding polypeptide segments. J Mol Biol 363:460–468

Robert P, Sengupta K, Puech PH, Bongrand P, Limozin L (2008) Tuning the formation and rupture of single ligand-receptor bonds by hyaluronan-induced repulsion. Biophys J 95:3999–4012

Rodier F, Bahadur RP, Chakrabarti P, Janin J (2005) Hydration of protein–protein interfaces. Proteins 60:36–45

Rothe C, Urlinger S, Lohning C, Prassler J, Stark Y, Jager U, Hubner B, Bardroff M, Pradel I, Boss M, Bittlingmaier R, Bataa T, Frisch C, Brocks B, Honegger A, Urban M (2008) The human combinatorial antibody library HuCAL GOLD combines diversification of all six CDRs according to the natural immune system with a novel display method for efficient selection of high-affinity antibodies. J Mol Biol 376:1182–1200

Rothlisberger D, Honegger A, Pluckthun A (2005) Domain interactions in the Fab fragment: a comparative evaluation of the single-chain Fv and Fab format engineered with variable domains of different stability. J Mol Biol 347:773–789

Rubina AY, Dementieva EI, Stomakhin AA, Darii EL, Pan'kov SV, Barsky VE, Ivanov SM, Konovalova EV, Mirzabekov AD (2003) Hydrogel-based protein microchips: manufacturing, properties, and applications. Biotechniques 34:1008–1014

Rubina AY, Kolchinsky A, Makarov AA, Zasedatelev AS (2008) Why 3-D? Gel-based microarrays in proteomics. Proteomics 8:817–831

Rubinstein ND, Mayrose I, Halperin D, Yekutieli D, Gershoni JM, Pupko T (2008) Computational characterization of B-cell epitopes. Mol Immunol 45:3477–3489

Ryan BJ, Fagain CO (2007) Arginine-to-lysine substitutions influence recombinant horseradish peroxidase stability and immobilisation effectiveness. BMC Biotechnol 7:86

Saerens D, Frederix F, Reekmans G, Conrath K, Jans K, Brys L, Huang L, Bosmans E, Maes G, Borghs G, Muyldermans S (2005) Engineering camel single-domain antibodies and immobilization chemistry for human prostate-specific antigen sensing. Anal Chem 77: 7547–7555

Saerens D, Huang L, Bonroy K, Muyldermans S (2008) Antibody fragments as probe in biosensor development. Sensors 8:4669–4686

San Paulo A, Garcia R (2000) Protein biochips: on the threshold of success. Biophys J 78: 1599–1605

Saphire EO, Stanfield RL, Crispin MDM, Parren PWHI, Rudd PM, Dwek RA, Burton DR, Wilson IA (2002) Contrasting IgG structures reveal extreme asymmetry and flexibility. J Mol Biol 319:9–18

Schatz PJ (1993) Use of peptide libraries to map the substrate-specificity of a peptide-modifying enzyme – a 13 residue consensus peptide specifies biotinylation in *Escherichia coli*. Biotechnology 11:1138–1143

Schenk JA, Sellrie F, Bottger V, Menning A, Stocklein WFM, Micheel B (2007) Generation and application of a fluorescein-specific single chain antibody. Biochimie 89:1304–1311

Schirrmann T, Al-Halabi L, Dubel S, Hust M (2008) Production systems for recombinant antibodies. Front Biosci 13:4576–4594

Schmidt TGM, Skerra A (2007) The Strep-tag system for one-step purification and high-affinity detection or capturing of proteins. Nat Protoc 2:1528–1535

Schmiedl A, Breitling F, Winter CH, Queitsch I, Dubel S (2000) Effects of unpaired cysteines on yield, solubility and activity of different recombinant antibody constructs expressed in *E. coli*. J Immunol Methods 242:101–114

Scholler N, Garvik B, Quarles T, Jiang S, Urban N (2006) Method for generation of in vivo biotinylated recombinant antibodies by yeast mating. J Immunol Methods 317:132–143

Schreiber G (2002) Kinetic studies of protein-protein interactions. Curr Opin Struct Biol 12:41–47

Schwesinger F, Ros R, Strunz T, Anselmetti D, Guntherodt HJ, Honegger A, Jermutus L, Tiefenauer L, Pluckthun A (2000) Unbinding forces of single antibody–antigen complexes correlate with their thermal dissociation rates. Proc Natl Acad Sci USA 97:9972–9977

Sennhauser G, Grutter MG (2008) Chaperone-assisted crystallography with DARPins. Structure 16:1443–1453

Shen Z, Stryker GA, Mernaugh RL, Yu L, Yan HP, Zeng XQ (2005) Single-chain fragment variable antibody piezoimmunosensors. Anal Chem 77:797–805

Shiroishi M, Tsumoto K, Tanaka Y, Yokota A, Nakanishi T, Kondo H, Kumagai I (2007) Structural consequences of mutations in interfacial Tyr residues of a protein antigen–anti body complex – the case of HyHEL-10-HEL. J Biol Chem 282:6783–6791

Shmanai V, Nikolayeva T, Vinokurova L, Litoshka A (2001) Oriented antibody immobilization to polystyrene macrocarriers for immunoassay modified with hydrazide derivatives of poly(meth) acrylic acid. BMC Biotechnol 1:4

Sidhu SS, Koide S (2007) Phage display for engineering and analyzing protein interaction interfaces. Curr Opin Struct Biol 17:481–487

Siegel RW, Baugher W, Rahn T, Drengler S, Tyner J (2008) Affinity maturation of tacrolimus antibody for improved immunoassay performance. Clin Chem 54:1008–1017

Sinacola JR, Robinson AS (2002) Rapid refolding and polishing of single-chain antibodies from *Escherichia coli* inclusion bodies. Protein Expr Purif 26:301–308

Sinha N, Mohan S, Lipschultz CA, Smith-Gill SJ (2002) Differences in electrostatic properties at antibody–antigen binding sites: implications for specificity and cross-reactivity. Biophys J 83:2946–2968

Skerra A, Pluckthun A (1988) Assembly of a functional immunoglobulin-Fv fragment in *Escherichia coli*. Science 240:1038–1041

Smith GP (1985) Filamentous fusion phage – novel expression vectors that display cloned antigens on the virion surface. Science 228:1315–1317

Soderlind E, Strandberg L, Jirholt P, Kobayashi N, Alexeiva V, Aberg AM, Nilsson A, Jansson B, Ohlin M, Wingren C, Danielsson L, Carlsson R, Borrebaeck CAK (2000) Recombining germline-derived CDR sequences for creating diverse single-framework antibody libraries. Nat Biotechnol 18:852–856

Solar I, Gershoni JM (1995) Linker modification introduces useful molecular-instability in a single-chain antibody. Protein Eng 8:717–723

Steinhauer C, Wingren C, Hager ACM, Borrebaeck CAK (2002) Single framework recombinant antibody fragments designed for protein chip applications. Biotechniques: 38–45

Steinhauer C, Wingren C, Khan F, He MY, Taussig MJ, Borrebaeck CAK (2006) Improved affinity coupling for antibody microarrays: Engineering of double-(His)(6)-tagged single framework recombinant antibody fragments. Proteomics 6:4227–4234

Stone E, Hirama T, Tanha J, Tong-Sevinc H, Li SH, MacKenzie CR, Zhang JB (2007) The assembly of single domain antibodies into bispecific decavalent molecules. J Immunol Methods 318:88–94

Strandh M, Ohlin M, Borrebaeck CAK, Ohlson S (1998) New approach to steroid separation based on a low affinity IgM antibody. J Immunol Methods 214:73–79

Su TJ, Lu JR, Thomas RK, Cui ZF, Penfold J (1998) The adsorption of lysozyme at the silica-water interface: a neutron reflection study. J Colloid Interf Sci 203:419–429

Sundberg EJ, Urrutia M, Braden BC, Isern J, Tsuchiya D, Fields BA, Malchiodi EL, Tormo J, Schwarz FP, Mariuzza RA (2000) Estimation of the hydrophobic effect in an antigen–antibody protein–protein interface. Biochemistry 39:15375–15387

Suzuki S, Otani T, Iwasaki S, Ito K, Omura H, Tanaka Y (2003) Monitoring of 15 pesticides in rainwater in Utsunomiya, eastern Japan, 1999–2000. J Pestic Sci 28:1–7

Swaim CL, Smith JB, Smith DL (2004) Unexpected products from the reaction of the synthetic cross-linker 3, 3′-dithiobis(sulfosuccinimidyl propionate), DTSSP with peptides. J Am Soc Mass Spectr 15:736–749

Takahashi Y, Ohashi T, Nagoya T, Sakaguchi M, Yasueda H, Nitta H (2001) Possibility of real-time measurement of an airborne *Cryptomeria japonica* pollen allergen based on the principle of surface plasmon resonance. Aerobiologia 17:313–318

Tam FCH, Ma CH, Leung DTM, Sutton B, Lim PL (2007) Carrier-specificity of a phosphorylcholine-binding antibody requires the presence of the constant domains and is not dependent on the unique VH49 glycine or VH30 threonine residues. J Immunol Methods 321: 152–163

Tanha J, Xu P, Chen ZG, Ni F, Kaplan H, Narang SA, MacKenzie CR (2001) Optimal design features of camelized human single-domain antibody libraries. J Biol Chem 276: 24774–24780

Thielges MC, Zimmermann J, Yu W, Oda M, Romesberg FE (2008) Exploring the energy landscape of antibody–antigen complexes: protein dynamics, flexibility, and molecular recognition. Biochemistry 47:7237–7247

Thies MJW, Talamo F, Mayer M, Bell S, Ruoppolo M, Marino G, Buchner J (2002) Folding and oxidation of the antibody domain C(H)3. J Mol Biol 319:1267–1277

Thomson NH (2005) The substructure of immunoglobulin G resolved to 25 kDa using amplitude modulation AFM in air. Ultramicroscopy 105:103–110

Thorpe IF, Brooks CL (2007) Molecular evolution of affinity and flexibility in the immune system. Proc Natl Acad Sci USA 104:8821–8826

Timmerman P, Beld J, Puijk WC, Meloen RH (2005) Rapid and quantitative cyclization of multiple peptide loops onto synthetic scaffolds for structural mimicry of protein surfaces. Chembiochem 6:821–824

Timmerman P, Puijk WC, Meloen RH (2007) Functional reconstruction and synthetic mimicry of a conformational epitope using CLIPS (TM) technology. J Mol Recognit 20:283–299

Tiselius A, Kabat EA (1939) An electrophoretic study of immune sera and purified antibody preparations. J Exp Med 69:119–131

Tonegawa S (1983) Somatic generation of antibody diversity. Nature 302:575–581

Torrance L, Ziegler A, Pittman H, Paterson M, Toth R, Eggleston I (2006) Oriented immobilisation of engineered single-chain antibodies to develop biosensors for virus detection. J Virol Methods 134:164–170

Torres M, Casadevall A (2008) The immunoglobulin constant region contributes to affinity and specificity. Trends Immunol 29:91–97

Treethammathurot B, Ovartlarnporn C, Wungsintaweekul J, Duncan R, Wiwattanapatapee R (2008) Effect of PEG molecular weight and linking chemistry on the biological activity and thermal stability of PEGylated trypsin. Int J Pharm 357:252–259

Tskhovrebova L, Trinick J (2004) Properties of titin immunoglobulin and fibronectin-3 domains. J Biol Chem 279:46351–46354

van Oss CJ (2003) Long-range and short-range mechanisms of hydrophobic attraction and hydrophilic repulsion in specific and aspecific interactions. J Mol Recognit 16:177–190

Verbelen C, Gruber HJ, Dufrene YF (2007) The NTA-HiS(6) bond is strong enough for AFM single-molecular recognition studies. J Mol Recognit 20:490–494

Visiongain (2008) Therapeutic monoclonal antibodies report 2008–2023. Visiongain

Wan L, Cai HW, Yang H, Lu YR, Li YY, Li XW, Li SF, Zhang J, Li YP, Cheng JQ, Lu XF (2008) High-level expression of a functional humanized single-chain variable fragment antibody against CD25 in Pichia pastoris. Appl Microbiol Biotechnol 81:33–41

Wang Q, Chan TR, Hilgraf R, Fokin VV, Sharpless KB, Finn MG (2003) Bioconjugation by copper(I)-catalyzed azide-alkyne [3+2] cycloaddition. J Am Chem Soc 125:3192–3193

Wang ST, Feng JN, Guo JW, Guo LM, Li Y, Sun YX, Qin WS, Hu MR, Han GC, Shen BF (2006a) A novel designed single domain antibody on 3-D structure of ricin A chain remarkably blocked ricin-induced cytotoxicity. Mol Immunol 43:1912–1919

Wang W, Singh S, Zeng DL, King K, Nema S (2007) Antibody structure, instability, and formulation. J Pharm Sci 96:1–26

Wang XQ, Wang YN, Xu H, Shan HH, Lub JR (2008) Dynamic adsorption of monoclonal antibody layers on hydrophilic silica surface: a combined study by spectroscopic ellipsometry and AFM. J Colloid Interf Sci 323:18–25

Wang ZH, Viana AS, Jin G, Abrantes LM (2006b) Immunosensor interface based on physical and chemical immunogylobulin G adsorption onto mixed self-assembled monolayers. Bioelectrochemistry 69:180–186

Watzke A, Kohn M, Gutierrez-Rodriguez M, Wacker R, Schroder H, Breinbauer R, Kuhlmann J, Alexandrov K, Niemeyer CM, Goody RS, Waldmann H (2006) Site-selective protein immobilization by Staudinger ligation. Angew Chem Int ed 45:1408–1412

Whelan JP, Kusterbeck AW, Wemhoff GA, Bredehorst R, Ligler FS (1993) Continuous-flow immunosensor for detection of explosives. Anal Chem 65:3561–3565

Willuda J, Honegger A, Waibel R, Schubiger PA, Stahel R, Zangemeister-Wittke U, Pluckthun A (1999) High thermal stability is essential for tumor targeting of antibody fragments: engineering of a humanized anti-epithelial glycoprotein-2 (epithelial cell adhesion molecule) single-chain Fv fragment. Cancer Res 59:5758–5767

Wingren C, Steinhauer C, Ingvarsson J, Persson E, Larsson K, Borrebaeck CAK (2005) Microarrays based on affinity-tagged single-chain Fv antibodies: sensitive detection of analyte in complex proteomes. Proteomics 5:1281–1291

Winkler K, Kramer A, Kuttner G, Seifert M, Scholz C, Wessner H, Schneider-Mergener J, Hohne W (2000) Changing the antigen binding specificity by single point mutations of an anti-p24 (HIV-1) antibody. J Immunol 165:4505–4514

Wong LS, Thirlway J, Micklefield J (2008) Direct site-selective covalent protein immobilization catalyzed by a phosphopantetheinyl transferase. J Am Chem Soc 130:12456–12464

Worn A, Pluckthun A (1998) Mutual stabilization of V-L and V-H in single-chain antibody fragments, investigated with mutants engineered for stability. Biochemistry 37:13120–13127

Worn A, Pluckthun A (1999) Different equilibrium stability behavior of ScFv fragments: identification, classification, and improvement by protein engineering. Biochemistry 38:8739–8750

Worn A, Pluckthun A (2001) Stability engineering of antibody single-chain Fv fragments. J Mol Biol 305:989–1010

Wu L, Xiao BT, Jia XL, Zhang Y, Lu SQ, Chen J, Long M (2007) Impact of carrier stiffness and microtopology on two-dimensional kinetics of P-selectin and P-selectin glycoprotein ligand-1 (PSGL-1) interactions. J Biol Chem 282:9846–9854

Xiang T, Lundell E, Sun Z, Liu H (2007) Structural effect of a recombinant monoclonal antibody on hinge region peptide bond hydrolysis. J Chromatogr B Analyt Technol Biomed Life Sci 858:254–262

Xu H, Zhao XB, Grant C, Lu JR, Williams DE, Penfold J (2006) Orientation of a monoclonal antibody adsorbed at the solid/solution interface: a combined study using atomic force microscopy and neutron reflectivity. Langmuir 22:6313–6320

Xu H, Zhao XB, Lu JR, Williams DE (2007) Relationship between the structural conformation of monoclonal antibody layers and antigen binding capacity. Biomacromolecules 8:2422–2428

Yokota A, Tsumoto K, Shiroishi M, Kondo H, Kumagai I (2003) The role of hydrogen bonding via interfacial water molecules in antigen–antibody complexation – the HyHEL-10-HEL interaction. J Biol Chem 278:5410–5418

Yoshimoto K, Hoshino Y, Ishii T, Nagasaki Y (2008). Binding enhancement of antigen-functionalized PEGylated gold nanoparticles onto antibody-immobilized surface by increasing the functionalized antigen using alpha-sulfanyl-omega-amino-PEG. Chemical Communications: 5369–5371.

Young NM, MacKenzie CR, Narang SA, Oomen RP, Baenziger JE (1995) Thermal stabilization of a single-chain Fv antibody fragment by introduction of a disulphide bond. FEBS Lett 377:135–139

Yu JP, Jiang YX, Ma XY, Lin Y, Fang XH (2007) Energy landscape of aptamer/protein complexes studied by single-molecule force spectroscopy. Chem Asian J 2:284–289

Zhang Z, Zhang M, Chen SF, Horbetta TA, Ratner BD, Jiang SY (2008) Blood compatibility of surfaces with superlow protein adsorption. Biomaterials 29:4285–4291

Zhao ZJ, Liu XM (2005) Preparation of monoclonal antibody and development of enzyme-linked immunosorbent assay specific for *Escherichia coli* O157 in foods. Biomed Environ Sci 18:254–259

Zhou J, Zhang LZ, Leng YS, Tsao HK, Sheng YJ, Jiang SY (2006) Unbinding of the streptavidin-biotin complex by atomic force microscopy: a hybrid simulation study. J Chem Phys 125:104905

Chapter 13
Genetically Engineered Proteins as Recognition Receptors

Jonathan D. Dattelbaum

Abstract The design of fluorescent protein biosensors for the detection of analytes has applications in diverse fields, from molecular life sciences to military readiness. The engineering of biosensors has progressed, from using naturally occurring proteins to the redesign of polypeptide sequences using extensive computational and experimental screening to create entirely new binding molecules. Bacterial periplasmic binding proteins have high affinity and specificity for a variety of biochemically important sugars, amino acids, and anions. The mutation of these proteins and the labeling with fluorescent probes has been utilized to create many reagentless biosensors. Further advances combined genetic fusions of these proteins with derivatives of green fluorescent proteins to create FRET-based biosensors which are expressed in living cells and report on metabolic processes in real-time. A number of proteins and polypeptides have also been used as scaffolds and reengineered to bind new ligands. Computational tools, such as ROSETTA and DEZYMER, have successfully predicted specific mutations to reconstruct protein scaffolds producing novel binders. A complementary approach is to select a highly stable scaffold, randomize predetermined sites, and create phage or ribosome display libraries which may be screened for specificity toward a target analyte. While both of these methods have individually led to the production of novel polypeptides with nanomolar binding constants, the combination of screening libraries and utilizing computational tools is likely to provide the greatest progress in the future of biosensor design.

Keywords Fluorescence · Biosensor · Periplasmic binding proteins · Affinity receptor · Green fluorescent protein · FRET · Rational design · Environmentally-sensitive probes · Protein engineering · Ligand binding · Rosetta · DEZYMER · Protein scaffold

---

J.D. Dattelbaum
Department of Chemistry, University of Richmond, Richmond, VA 23173, USA
e-mail: jdattelb@richmond.edu

M. Zourob (ed.), *Recognition Receptors in Biosensors*,
DOI 10.1007/978-1-4419-0919-0_13,

Abbreviations

ABD-F	7-Fluorobenz-2-oxa-1,3-diazole-4-sulfonamide
Acrylodan	6-Acryloyl-2-dimetholaminonaphthalene
cAMP	Cyclic-adenosine-5′-monophosphate
Cys	Cysteine
Dapoxyl	(2-Bromoacetamidoethyl)sulfonamide
FAD	Flavin adenine dinucleotide
FRET	Fluorescence resonance energy transfer
IANBD	Ester N-(2-(iodoacetoxy)ethyl)-N-methyl)amino-7-nitrobenz-2-oxa-1,3-diazole
IAANS	2-(4′- Iodoacetamido)anilino)naphthalene-6-sulfonatic acid
IAEDANS	5-((((2-Iodacetyl)amino)ethyl)amino)naphthalene-1-sulfonic acid
Lys	Lysine
MDCC	7-Diethylamino-3-((((2-aleimidyl)ethyl)amino)carbonyl)coumarin
Met	Methionine
NADH	Nicotinamide adenine dinucleotide
SPR	Surface plasmon resonance
Trp	Tryptophan

13.1 Introduction

The analytical detection of target molecules is a common theme in molecular life sciences. Traditional chemical techniques like liquid–liquid extractions and HPLC may not be appropriate for the analysis of many biochemically interesting metals, small molecules, and proteins, particularly when in vivo imaging is desirable. To provide tools compatible with living systems, many groups utilize biological molecules because of evolutionarily refined levels of specificity and selectivity for a target. Fluorescent protein biosensors are one class of sensing modules which rely on polypeptides as the recognition biomolecule and fluorescence as the signal transduction mechanism. Often the term reagentless is also used to describe this group, particularly when receptor proteins are used as the basis for detection. This term is meant to reflect the absence of additional substrates or other chemicals beyond the protein biosensor. Because of the enormous advances in molecular biology over the past 20 years, the genetic manipulation to alter overexpressed proteins is now a matter of experimental routine. The goal in protein biosensor design is to engineer a polypeptide to possess desirable affinity for an analyte (either a small molecule or another polypeptide) and to monitor binding by a change in some fluorescence characteristic. This chapter reviews the use of wild-type proteins as sensing receptors for small molecules as well as current

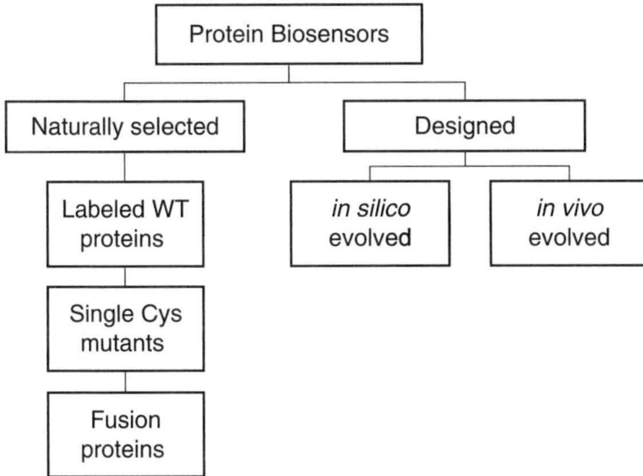

Fig. 13.1 Outline of strategies used to design fluorescent protein biosensors

computational and experimental approaches to develop an expanded toolkit of polypeptides to detect physiologically interesting target analytes (Fig. 13.1).

13.1.1 Fluorescence in Biological Sensing

Fluorescence is a versatile and highly sensitive technique used for over a hundred years to characterize ecological and biological systems. As early as 1877, Baeyer reportedly used fluorescein to search for underground water routes linking the Danube and Rhine Rivers through the Alps (Berlman 1971). In the middle of the last century, Gregorio Weber (1952a, b) prominently expanded the use of fluorescence methods to study biomolecules, especially binding interactions. Traditionally, Trp or Trp analogs have been shown to be powerful fluorescent probes for studying protein dynamics and ligand binding (Lakowicz 1999). However, the UV excitation and emission wavelengths for Trp limit its usefulness in biological systems due to interference from ubiquitous biomolecules such as NADH, FAD, and pyridoxal phosphate. Therefore, a number of commercially available extrinsic probes have been synthesized with amine and sulfhydryl reactive moieties used for covalent attachment to Lys and Cys residues. Consequently, to measure ligand binding, the fluorescence properties of these extrinsic dyes establish the mechanisms commonly used in biosensor design (e.g. environmental sensitivity, static quenching, FRET or anisotropy) (Lakowicz 1999). Upon receptor-analyte binding, there are at least three ways to correlate a change in protein structure with a change in fluorescence emission properties including: (1) direct contact between the analyte and fluorescent probe, (2) a change in protein conformation which alters the microenvironment of a fluorescent probe located away from the analyte binding site, and (3) changes in protein oligomerization state that alter the probe environment or interactions.

13.1.2 Determination of Apparent Dissociation Constant (K_d) Using Fluorescence

The dissociation constant (K_d) is an essential factor which provides a quantitative evaluation of the specificity between receptor and analyte. Determination of individual ligand K_d values may be accomplished by monitoring the ligand-dependent emission during a titration of the labeled protein-probe conjugate. For receptor proteins with a single and independent binding site, the equilibrium dissociation constant may be written following the law of mass action,

$$[PL] \rightleftarrows [P] + [L] \tag{13.1}$$

$$K_d = \frac{[P][L]}{[PL]} \tag{13.2}$$

where $[P]$ is the concentration of free protein, $[L]$ is the concentration of free ligand and $[PL]$ is the concentration of bound complex. The fraction of bound (f_b) complex may be determined by titration of the labeled protein with ligand and used to determine $[PL]$,

$$f_b = \frac{[PL]}{[P]_{tot}} = \frac{F - F_f}{F_b - F_f} \tag{13.3}$$

where F_f and F_b are the fluorescence intensities measured in the absence of ligand and in the presence of saturating ligand, respectively, and $[P]_{tot}$ is the total labeled protein concentration. Using this value and the total starting ligand concentration, the concentration of free ligand $[L]$ is now calculated from

$$[L] = [L]_{tot} - [PL] \tag{13.4}$$

Rearranging (13.2) for $[PL]$ and combining with (13.3) provides the Langmuir isotherm (13.5) which is plotted (f_b against $[L]$) and fit to solve for K_d.

$$f_b = \frac{n[L]}{(K_d + [L])} \tag{13.5}$$

For this derivation, the number of independent binding sites (n) was 1, whereas n may equal any integer as long as no cooperativity exists between sites (Klotz 1997). Finally, while it is preferable to make separate samples for each ligand concentration, dilution of the labeled protein may be necessary due to limited material. In this case, the observed fluorescence should be adjusted according to

$$F = F_{obs} \times (\mathrm{Vol}_{final}/\mathrm{Vol}_{initial}) \tag{13.6}$$

The fitted data can then be used as a standard curve to determine the unknown ligand concentrations, one order of magnitude above and below the K_d value. Any program with the ability to enter user-defined equations is capable of providing reasonable solutions to this equation. Applications developed by OriginLab (Northampton, MA) or SigmaPlot (San Jose, CA) are two of the most commonly used software packages in the literature. If a shift in emission maxima is observed upon the addition of ligand, a more complex analysis is needed as outlined by de Lorimier et al. (2002).

13.2 Naturally Selected Biosensors

13.2.1 Fluorescent-Labeled Proteins as Biosensors

In forming a general timeline for progressive strategies utilized in biosensor design, three stages along the path to creating genetically engineered fluorescent protein biosensors will be discussed: chemical labeling of wild-type proteins, site-directed mutagenesis for single amino acid alterations and fusion proteins for entirely encodable proteins (Fig. 13.1). This classification is meant to provide an organizing framework to describe important work reported in the recent literature. The first stage in biosensor design occurred by covalently modifying fluorescent probes to wild-type proteins without genetic manipulations. Because Nature has produced an impressive variety of polypeptides with the capability to bind a plethora of ligands, the genetic manipulations in this stage generally consist of cloning, expressing, and purifying naturally occurring proteins without significant alterations to the amino acid sequence or to the natural ligand binding ability. One of the earliest fluorescent protein biosensors used fluorescein-iodoacetamide attached to Met residues on the *Escherichia coli* galactose binding protein to detect and investigate sugar binding (Zukin et al. 1977). An elegant protein biosensor was engineered to image cAMP levels in single fibroblast cells using cAMP-dependent protein kinase (Adams et al. 1991). The sensor was based on a FRET signal from the cAMP-dependent oligomerization of fluorescein-labeled catalytic domains and rhodamine-labeled regulatory domains which make up the quaternary structure of this holoenzyme (Fig. 13.2). The authors determined conditions to label only one or two amine-reactive probes per subunit, albeit on unspecified Lys residues. The binding of cAMP to the regulatory subunit causes a dissociation of the catalytic and regulatory domains. Because FRET is a distance-dependent process, energy transfer is diminished upon cAMP binding and is correlated with intracellular cAMP levels in the 0.01–10 μM range. Due to the abundance of Lys residues, the labeling of amino groups in proteins is widely applicable although not readily controllable in terms of the number of probes added or position located. A similar argument may be made against exclusive labeling of naturally positioned Cys residues as well.

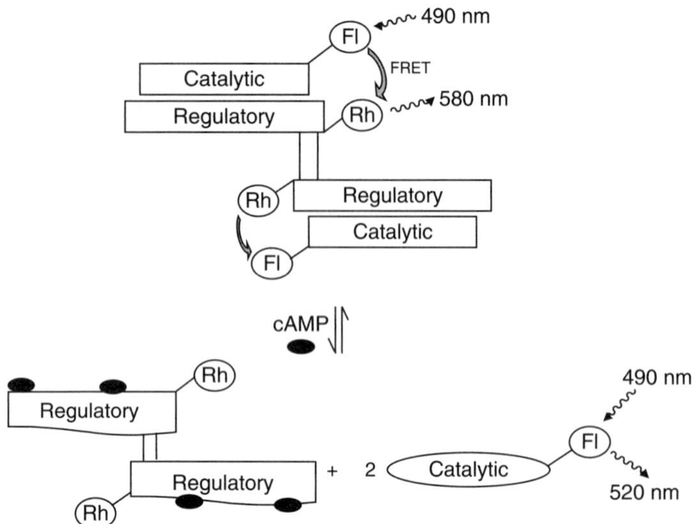

Fig. 13.2 FRET-based protein biosenor used to detect intracellular cAMP. Fluorescein (Fl) and rhodamine (Rh) were covalently attached to the subunits of cAMP-dependent protein kinase. (Reprinted with permission from Adams et al. (1991) copyright, 2008 Nature publishing group)

13.2.2 Genetically Modified Single Cys Biosensors

The second stage of designing protein biosensors used site-specific mutagenesis of overexpressed proteins in order to attach extrinsic probes at defined positions. Routine laboratory methods to alter DNA sequences may be performed using overlapping PCR mutagenesis (Ho et al. 1989) as well as using comprehensive kits from several vendors, e.g., QuickChange, Stratagene Inc. This stage was ushered in by the work of Cass and colleagues on a family of wild-type *E. coli* periplasmic binding proteins (PBPs) (Zhou and Cass 1991; Sohanpal et al. 1993; Gilardi et al. 1994). These proteins are highly soluble and are extensively investigated as the core of many biosensor designs (Dwyer and Hellinga 2004).

E. coli possesses a number of structurally similar PBPs that bind to a diverse set of ligands including amino acids, sugars, lipids, and anions for bacterial chemotaxis and nutrient uptake (Higgins 1992). Data using CD and fluorescence methods demonstrated that two such proteins, ribose and glucose/galactose binding proteins, undergo significant conformation alterations and increased stability following ligand binding (Zukin et al. 1979). Detailed X-ray crystallographic structural determinations of many PBPs in both the ligand bound and unbound states confirmed these observations as a general rule for this family of proteins (Table 13.1). Typically, these proteins consist of a single polypeptide chain folded into two distinct domains, connected by a hinge region forming a cleft for ligand binding (Fig. 13.3). Because PBPs contain few or no Cys residues, many groups use site-directed mutagenesis to construct proteins with unique Cys residues which

Table 13.1 Representative periplasmic binding proteins genetically altered for use as biosensors

PBP	Ligand	PDB ID		References
		Open form	Closed form	
AraF	Arabinose	8abp	1abe	Quiocho and Vyas (1984), Vermersch et al. (1991)
LivJ	Leucine/isoleucine/valine	1z15	1z16	Trakhanov et al. (2005)
MglB	Glucose/galactose	2fw0	2fvy	Borrok et al. (2007)
RbsB	Ribose	1urp	2dri	Bjorkman et al. (1994), Bjorkman and Mowbray (1998)
ArgT	Lysine/arginine/ornithine	2lao	1lst	Oh et al. (1993)
HisJ	Histidine		1hsl	Yao et al. (1994)
GlnH	Glutamine	1ggg	1wdn	Hsiao et al. (1996), Sun et al. (1998)
MalE	Maltose	1omp	1anf	Sharff et al. (1992), Quiocho et al. (1997)
Sbp	Sulfate		1sbp	He and Quiocho (1993)
PstS	Phosphate	1oib	1ixh	Yao et al. (1996), Wang et al. (1997)

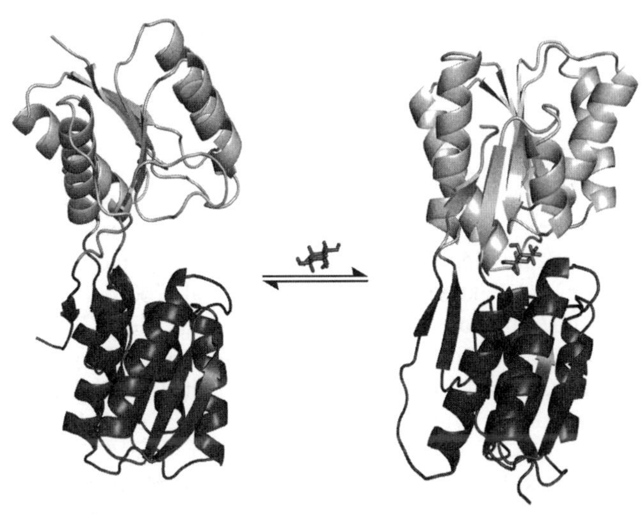

Fig. 13.3 Extensive conformational change observed for the *E. coli* ribose binding protein in the absence (PDB ID 1URP) and presence (PDB ID 2DRI) of ribose. From Vercillo et al. (2007)

serve as an attachment point for thiol-reactive fluorophores. As a result of the large ligand-dependent conformational changes observed for most PBPs, environmentally-sensitive fluorescent probes are successfully employed to correlate ligand binding and protein conformation. Examples of commercially available fluorophores which display altered emission depending on microenvironment polarity include dansyl aziridine, ABD-F, IANBD ester, Acrylodan, IAEDANS, IAANS, MDCC, and Dapoxyl (Fig. 13.4). Each of these probes may be purchased with

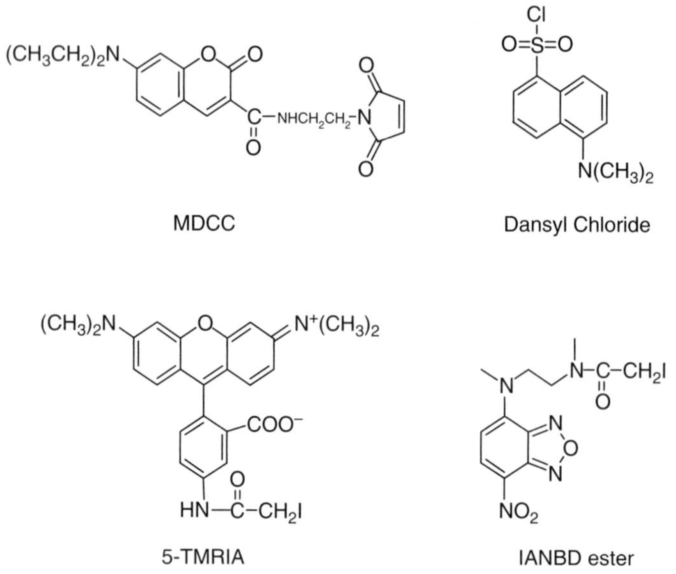

MDCC

Dansyl Chloride

5-TMRIA

IANBD ester

Fig. 13.4 Structures of environmentally sensitive fluorescent probes commonly utilized for detecting ligand binding

a maleimide, alkyl halide, or haloacetamide moiety capable of reaction with the sulfhydryl group in Cys residues. Many examples exist where genetically engineered PBPs are used as the recognition receptor and environmentally-sensitive probes as the signaling mechanism, including biosensors for amino acids (Dattelbaum and Lakowicz 2001; Staiano et al. 2006; Chino et al. 2007; Sakaguchi et al. 2007); maltose (Gilardi et al. 1994; Marvin et al. 1997; Dattelbaum et al. 2005); ribose (deLorimier et al. 2002; Vercillo et al. 2007); glucose/galactose (Marvin and Hellinga 1998; Tolosa et al. 1999; Salins et al. 2001; Cuneo et al. 2006; Thomas et al. 2006; Khan et al. 2008); phosphate (Brune et al. 1994; Salins et al. 2004); sulfate (Shrestha et al. 2002), and nickel (Salins et al. 2002).

The success of this design strategy relies on choosing the most advantageous location for Cys insertion. Amino acids selected for probe positioning are categorized as endosteric, peristeric or allosteric to describe the location of the fluorophore relative to the ligand binding pocket. Endosteric and peristeric refer to amino acids within and surrounding the ligand binding site, respectively, and may be identified through visual inspection of protein structures (Brune et al. 1994; Gilardi et al. 1994). For these sites, fluorescence changes are generally attributed to direct contact between probe and ligand. Allosteric sites may be located anywhere in the protein, and as the name implies, fluorophores attached to these locations report on ligand binding as a result of significant conformational alterations in the presence and absence of ligand. Both visual inspection (Dattelbaum and Lakowicz 2001) and computational modeling (Marvin et al. 1997; deLorimier et al. 2002)

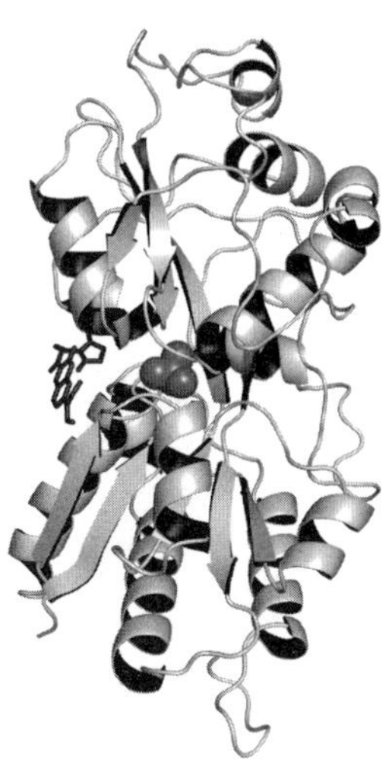

Fig. 13.5 *E. coli* phosphate binding protein (PDB ID 1A54) mutant A197C with bound phosphate (*spheres*) and covalently modified with the environmentally sensitive probe, MDCC (Reprinted with permission from Brune et al. (1994), copyright 2008 American Chemical Society)

of crystal structures in the open and closed state have been used successfully to create allosteric-type fluorescent sensing proteins.

An early example of using extrinsic probes and PBPs was performed on the phosphate binding protein (PhBP) by Webb and colleagues (Brune et al. 1994). Using site-directed mutagenesis, they constructed six individual mutants such that each protein product would encode a single cysteine residue. Peristeric and endosteric sites were chosen by visual inspection and deductive reasoning using the recently established closed crystal X-ray structure of PhBP. Each mutant was purified directly from the periplasmic space using osmostic shock and labeled with seven different environmentally sensitive probes. The fluorescence response of each labeled mutant in the presence and absence of phosphate was performed to establish an empirical matrix of amino acid location and fluorescent probe. Forty-two different combinations were attempted and yielded one exceptional phosphate sensing protein conjugate. An X-ray structure of PhBP:A197C labeled with MDCC (Fig. 13.5) shows the endosteric placement of the environmentally sensitive probe. Upon the addition of inorganic phosphate, a large increase in fluorescence signal (405% or fivefold) and 10 nm blue shift in emission maximum are observed for this labeled mutant (Fig. 13.6). A complete titration with increasing amounts of ligand was used to calculate an apparent K_d value of 0.03 μM, consistent with data established by equilibrium dialysis experiments. When genetic alterations are used to modify a protein, the affect of such changes to ligand binding and overall stability should be investigated. The authors assessed the impact of potential inhibitors containing phosphate or phosphate-like groups and found that only arsenate produced a significant signaling response from this mutant. However, arsenate binds with 100-fold less efficiency than phosphate which is a regrettable but acceptable source of error for cellular sensing. Following the construction and characterization phases, this phosphate sensing protein is used to investigate the

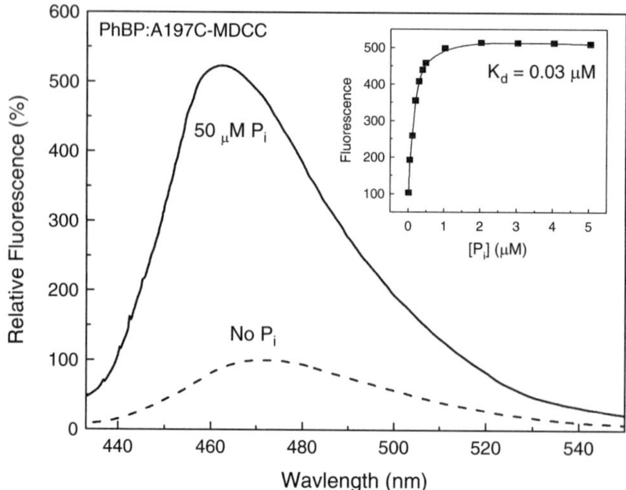

Fig. 13.6 Fluorescence emission spectra of MDCC covalently bound to PhBP:A197C in the presence and absence of 50 μM P_i. The inset shows the complete titration and fitted data used to determine K_d (Brune et al. 1994) (Reprinted with permission from Brune et al. (1994), copyright 2008 American Chemical Society)

kinetics of inorganic phosphate release during a single turn of the actomyosin subfragment ATPase reaction.

One of the principal observations from the PhBP data presented is that both the position and the specific environmentally sensitive probe combine to produce the observed changes in fluorescence emission. While not necessarily surprising, the difficult task of matching probe and position is a challenge still unresolved in the current literature where no complete theoretical framework exists. While global protein conformational changes are essential to this sensing strategy, the emission response of environmentally sensitive fluorophores is affected by alterations to factors which control the probe microenvironment including solvent accessibility, direct interactions with ligand, direct interactions with surrounding amino acids, or probe mobility. While a few groups have endeavored to identify the underlying mechanism leading to the spectral alterations in the absence and presence of ligand (e.g. (Gilardi et al. 1997; Dattelbaum and Lakowicz 2001; Dattelbaum et al. 2005)), the process of identifying the best environmentally sensitive probe for signal transduction at a specific site relies primarily on serendipitous connections.

13.2.3 Glucose Binding Protein Biosensors

The need for fluorescent biosensors, useful to healthcare professionals as well as individual patients, has led to much interest in the design of a glucose biosensor (Moschou et al. 2004; Newman and Turner 2005). Initial interest in using fluorescence methods to sense glucose focused on using the lectin, Concanavailin A (Mansouri and Schultz 1984; Meadows and Schultz 1988). More recently, the *E. coli* glucose/galactose binding protein (GGBP), which also belongs to the PBP superfamily, has been utilized as the basis for designing a fluorescent protein biosensor. Marvin and Hellinga (1998) constructed single-Cys GGBP mutant by analogy with previous work performed on the maltose binding protein (MBP). To do this, they selected two different classes of mutational sites. Like the PhBP, two sites within the glucose binding pocket (endosteric sites) were chosen as sites of fluorophore attachment where they hoped for a fluorescence change by direct contact with the ligand. The second grouping of sites was located in the hinge region of GGBP where no direct contact is made with the glucose molecule (allosteric sites). It was hypothesized that glucose-dependent hinge-bending conformational movements within the overall protein would result in signal transduction via an allosteric mechanism which relies on a network of noncovalent interactions connecting the probe microenvironment with the glucose binding site. Two different environmentally sensitive probes were used, IANBD amide and acrylodan. As with the PhBP, variations were observed in fluorescence response depending on the site of attachment and probe identity (Fig. 13.7). The probe at position H152C, an endosteric mutation, gave the largest fluorescence response with a K_d value of 20 μM for glucose (Fig. 13.8). Because the wild-type dissociation constant is much smaller (~1 μM), the fluorophore seems to inhibit glucose access to the binding pocket.

A computational method for selecting the site of fluorophore attachment has been formulated by Hellinga and colleagues (deLorimier et al. 2002). In general, the C_α–C_α displacement is calculated for every amino acid between overlapping open- and closed-states of PBPs to rapidly identify areas of the protein that undergo significant conformational changes. These represent a rationally selected set of

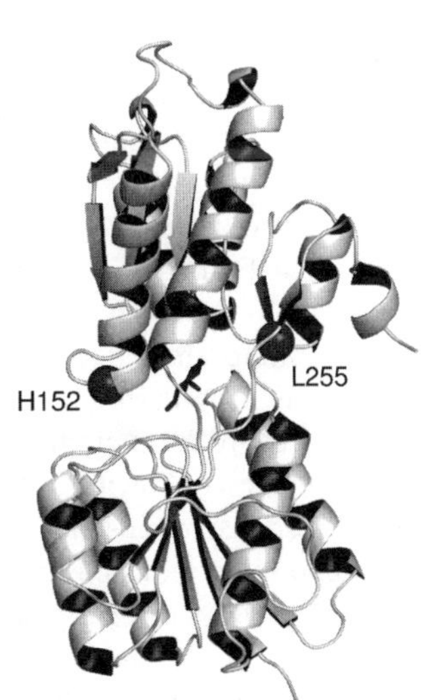

Fig. 13.7 *E. coli* glucose binding protein (PDB ID 1GLG) showing bound glucose and attachment sites for conjugating fluorophores (*spheres*). Typical endosteric and allosteric mutation sites are represented at positions H152 and L255, respectively. Adapted from Marvin and Hellinga (1998)

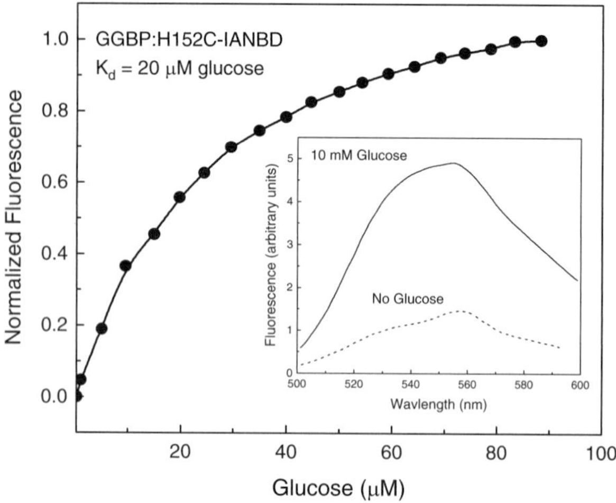

Fig. 13.8 Fluorescence emission data for the binding of glucose to the GGBP H152C-IANBD conjugate. The titration data were normalized using (13.3). Inset shows changes in the emission spectra in the absence (*dash*) and presence (*solid*) of 10 mM glucose. Redrawn from Marvin and Hellinga (1998) (Reprinted with permission from Marvin and Hellinga (1998), copyright 2008 American Chemical Society)

amino acid locations likely to experience environmental changes which may be detected using appropriate fluorescent probes. This method is particularly useful for identifying the allosteric-type of sites for probe attachment which are not easily exposed from visual inspection of protein structures.

13.2.3.1 Ratiometric Glucose Sensing Using GGBP

GGBP was also investigated as the basis for a biosensor for in vivo glucose measurements using environmentally sensitive probes to detect glucose binding (Tolosa et al. 1999; Salins et al. 2001; Ge et al. 2003; Ge 2004; Thomas et al. 2006). A genetically engineered GGBP:Q26C mutant was labeled with the environmentally sensitive probe, IA-ANS, and was used to measure glucose concentrations in a feed batch fermentation system to measure glucose consumption by yeast cells (Ge et al. 2003). Even though this design was a significant first step to developing a practical biosensor, the quantitative response relies on fluorescence emission at a single wavelength which may vary greatly, particularly with power fluctuations experienced by the excitation source, and may increase the error associated with the measurement (Lakowicz 1994). Ge et al. introduced an internal standard into the ligand-responsive biosensor by establishing conditions to covalently attach a second fluorescent molecule to the N-terminus of GGBP (Marvin and Hellinga 1998; Ge et al. 2004). Care must be taken in choosing this second probe molecule which needs to be nonresponsive to the ligand being measured and possess an emission maximum away from the actual reporting fluorophore. This strategy effectively allowed a ratiometric approach to quantitating glucose. In this work, acrylodan was attached to position L255C for the glucose-dependent response and a ruthenium metal-ligand charge transfer complex was attached to the amino-terminus as the nonreporting internal standard. Using this sensor, glucose concentrations in the 0.4–1.4 μM range were detectable.

13.2.3.2 In Vivo Glucose Detection

The ideal glucose measurement would be conducted in blood without the need for drawing blood or blood serum. There are at least three issues which add to the complexity of designing a glucose sensor for in vivo applications using GGBP. First, it is well established that most biological systems possess some level of interfering fluorescence termed "the ubiquitous blue fluorescence." Because water and the primary mammalian blood proteins and pigments have low absorption coefficients in the near-infrared region, fluorescent reporter molecules with excitation and emission in the 600–1,000 nm spectral region should be strongly considered for transdermal sensing (Doornbos et al. 1999). Secondly, blood glucose levels are considered normal in the 1–5 mM range which would completely saturate GGBP ($K_d = 1$ μM) unless significant dilutions are performed. The GGBP binding constant must be altered by site-directed mutagenesis of the amino acids involved in glucose binding. Thirdly, the specificity of the mutated GGBP needs to be explored in detail because of the significant quantities of potential interferents, such as other sugars, found in blood plasma (West 1990).

13.2.3.3 Excitonic Detection of Glucose

Der and Dattelbaum (2008) try to address all three issues by constructing a glucose biosensor using molecular exciton theory whereby two rhodamine dyes are attached to opposite domains of GGBP (Fig. 13.9). As a result of ground state dimerization between the two fluorescent probes, a splitting of the excited state energy level produces measurable changes to both the absorption and emission spectra (Kasha 1963). To select sites for Cys insertion using the known GGBP closed conformation, pairs of amino acids were selected on opposite sides of the ligand binding pocket which are separated by an approximate length of two rhodamine molecules. The authors hypothesized that dimer formation (and hence, quenching) would be altered by the known glucose-dependent conformational change. Because two identical thiol-reactive dyes are used, conditions for protecting one site followed by laborious purification are not needed. The largest fluorescence response was obtained by engineering mutations at residues R96C and D168C, where a 40% increase in rhodamine emission was observed upon the addition of glucose. This mutation takes advantage of the large rotational motion of GGBP to pull the two probes apart following glucose binding. Finally, alanine-scanning mutagenesis of amino acids within the glucose binding pocket was performed to successfully

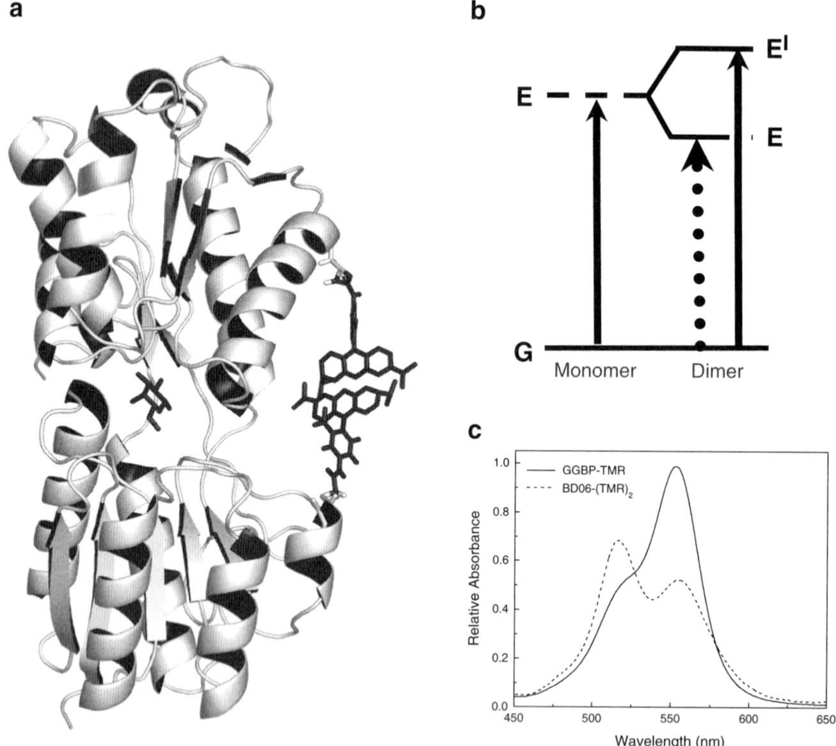

Fig. 13.9 Model of glucose biosensor designed using exciton theory. (**a**) GGBP closed structure (PDB 1GLG) with bound glucose and TMR probes attached via Cys residues to the upper and lower domains. (**b**) Electronic energy level splitting predicted by molecular exciton theory following dimer formation. (**c**) Absorption spectra of single- and double-labeled TMR probes attached to GGBP

increase the GGBP K_d to ~1 mM which is a more physiologically relevant value. The benefit of this method is the dependence on proximity between the two probes and not on an ambiguous environmental factor which may not be understood when designing the biosensor. While some investigators have identified the exact mechanism of ligand-dependent emission changes (Dattelbaum et al. 2005), these insights are most often discovered after a useful probe–position pair is identified. The purpose of using excitonic quenching is to provide a straightforward theoretical balance for building biosensors based on proteins with large conformational changes.

13.2.3.4 Near IR Detection of Glucose

Thomas et al. (2006) peristerically mutated GGBP to create several new single and double Cys mutants to covalently attach Nile Red, which is an environmentally sensitive reporter and emits in the near-IR region. GGBP E149C/A213C/L238S displayed a 50% increase in fluorescence upon the addition of glucose. The authors determined that the probe bound to position E149C was most likely responsible for the observed response. Additionally, because these mutations are located along the rim of the glucose binding pocket, the dissociation constant was significantly increased to 7 mM, well within the normal blood glucose range.

13.2.4 Design of GFP Fusion Biosensors

The third stage in the design of genetically encoded receptor molecules encompasses the fusion of naturally selected proteins with fluorescent proteins for ratiometric measurements based on FRET. The original green fluorescent protein (GFP) was isolated from the jellyfish *Aequorea victoria* more than 40 years ago and has been used in an impressive array of applications in life sciences over the last two decades (Tsien 1998). The discovery of similar fluorescent proteins with expanded emission properties throughout the visible region and into the near-IR made possible true genetically encoded fluorescent biosensors (Wang et al. 2008). Tsien and colleagues led the effort to design different FRET-based biosensors and they, as well as others, have constructed sensors to measure many analytes including cAMP (DiPilato et al. 2004), Ca^{2+} (Romoser et al. 1997), proteases (Mitra et al. 1996; Luo et al. 2001), serine/threonine kinases (Zhang et al. 2001; Sato et al. 2002; Wang et al. 2005) and Cdc42 (Itoh et al. 2002; Nalbant et al. 2004), to name but a few. Of particular note is the sensor for calcium which is designed around the calcium-binding domain of calmodulin with the cyan fluorescent protein (CFP) and yellow fluorescent protein (YFP) translationally fused to the N- and C-termini, respectively (Miyawaki et al. 1999). This sensor has been used to study the release of calcium in a variety of cells in vivo and a more highly engineered version has become commercially available from Invitrogen/Molecular Probes (Carlsbad, CA). While the next section focuses on the use of fluorescent proteins in ligand binding using FRET, an in-depth discussion of the numerous additional applications employing fluorescent proteins may be found in Wang et al. (2008).

13.2.4.1 FRET Biosensors Using PBPs

The design of FRET-based genetically encoded biosenors utilizing the PBP family began with maltose (Fehr et al. 2002) and glucose (Ye and Schultz 2003) (Fig. 13.10). Frommer and colleagues have continued to design a family of genetically encoded biosensors based on PBPs and have made many in vivo measurements (Fehr et al. 2005; Lalonde et al. 2005; Looger et al. 2005; Gu et al. 2006; Dulla et al. 2008). The typical arrangement is to clone the enhanced CFP (ECFP) upstream of the binding protein gene and to add the enhanced YFP (EYFP) gene downstream, as well as a 6xHIS tag to facilitate purification of the expressed fusion construct. The experiment is performed by exciting the donor molecule, ECFP

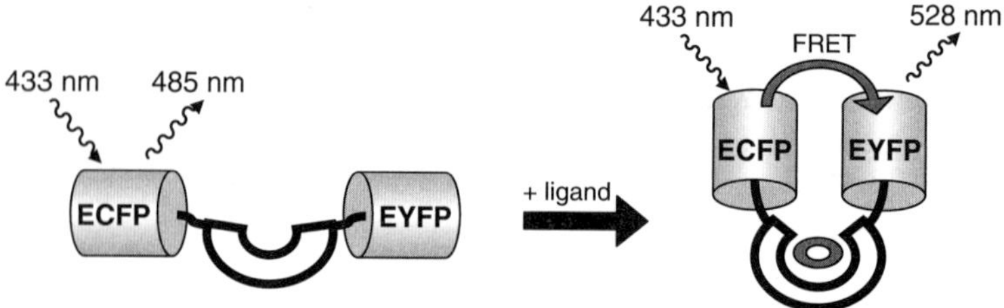

Fig. 13.10 Design strategy of genetically encoded FRET biosensors following ligand-dependent conformational changes

(~433 nm), which undergoes radiationless Förster transfer to the acceptor, EYFP, and emits light at ~530 nm. However, FRET is not a complete process and emission is also observed from ECFP at ~485 nm. A ratio of emission intensities 530/485 is measured with increasing concentration of ligand and used to design a standard curve for quantitative ligand determination. Using this signaling strategy, Fehr et al. (2002) constructed a translational fusion of ECFP–MBP–EYFP, although no maltose-dependent signaling was observed. To provide proper spatial rearrangement of the two GFP variants, five amino acids were truncated from the amino-terminus of the mature MBP to allow better separation of donor and acceptor in the open and closed states. Wild-type MBP binds tightly to maltose (~2 mM in this study) and a W62A mutation, which alters a direct contact between protein and ligand that produced a good signaling response and an apparent K_d of 226 μM. Using confocal microscopy, this engineered FRET-based maltose biosensor provided accurate images of cytosolic maltose concentrations in single yeast cells. Further redesign was performed to limit FRET losses as a result of conformation entropy by altering the linkers used to connect the fluorescent proteins to the N- and C-termini or by genetic insertion of fluorescent protein sequences within the binding protein gene (Deuschle et al. 2005). The addition of internally fused fluorophores produced the highest degree of restricted motion which correlated well with the predicted enhancement in dynamic FRET signaling of glucose and glutamate concentrations. Building on these results, Ha et al. (2007) rationalized that a rigid linker mimicking the helical structure found at the C-terminus of ECFP would also provide tighter coupling between conformational movements and signaling. They demonstrate 10-fold gains in the FRET response and utilize this nanosensor to image maltose levels in actively respiring yeast cells.

13.3 Designed Evolution

13.3.1 In Silico Evolution: Biosensors by Design

While nature has evolved a plethora of binding proteins which biochemists are able to engineer for the development of biosensors, expansion of the current set of protein folds and structures is ongoing using computational (so-called "rational") and combinatorial (so-called "irrational") methodologies. Because the primary objective of protein engineering is to identify new binders for other proteins or small ligands, brute force is utilized by both methods albeit in very different ways. Computationally driven processes require vast amounts of computing power to screen in silico for the lowest energy interactions which then must be constructed, expressed and screened at the laboratory bench. Combinatorial approaches usually take advantage of the rapid screening capabilities of phage display and ribosome display but a reliable and facile assay must be available. Because neither method is assured of a successful outcome, they work best when used complementary with each other to undertake a process of rescreening and redesign to achieve a new binding protein scaffold or structure.

13.3.2 Computational Design Using ROSETTA

Because the design of new binding receptors using computational methods is a large, rapidly advancing field, the following section is meant to provide a brief introduction to some of the major points in this area and the reader is directed to more dedicated reviews on the subject (Butterfoss and Kuhlman 2006). In general, there are two types of algorithms utilized to efficiently sample the conformational flexibility of amino acid sequence space: stochastic (e.g. Monte Carlo and Genetic algorithms) and deterministic (e.g. dead end elimination) (Voigt et al. 2000). In either case, the program searches for the global minimized energy conformation utilizing a specific force field equation, which is an energy function that describes both covalent and non-covalent thermodynamic parameters (Boas and Harbury 2007). Stochastic methods tend to be faster but may become trapped in local energy minima whereas deterministic approaches tend to require more computing power, and by definition always find the global energy minima. However, the global minimum is defined by the force field and may not provide the experimentally best protein–protein or protein–ligand interaction. Depending on the parameters used, several solutions may be obtained from a calculation. An optimized and efficient method to rank order the output sequences may be important in choosing the experimentally constructed structures. The Rosetta software suite (http://www.rosettacommons.org/) is an example of a stochastic program which provides a variety of computational tools available to academics for free and to companies for fee-based services (Das and Baker 2008). While some differences exist among the many calculations performed by the program, the Rosetta modules relevant to this discussion typically use a Monte Carlo sampling approach to optimize the side-chain conformation and investigate the free energy landscape of binding interactions to small molecules and proteins. A diverse set of published results utilizing this tool may be found in the literature as well as at the software website.

13.3.3 Computational Design Using DEZYMER

A second example of powerful protein design software is DEZYMER (Hellinga and Richards 1991), which currently employs a dead end elimination deterministic algorithm (Looger and Hellinga 2001). To date, most of the published results using DEZYMER involve the redesign of binding interactions between protein-small molecules or protein-metal ions (Benson et al. 1998, 2002; Looger et al. 2003; Allert et al. 2004; Yang et al. 2005); however, the principles should be applicable to protein–protein interactions as well. In this algorithm, a ligand of interest is docked within a predefined cage coordinate system. While most of the protein is treated as a rigid scaffold, a predetermined number of amino acids directly interacting with the ligand are systematically altered in sequence space. Repeated rounds of DEE at each amino acid rejects rotamers that do not meet the energetic criteria and ultimately yields a list of mutations in the binding pocket needed to reengineer the protein to bind the new ligand. While the global minimum energy conformation is obtained for any particular ligand pose, it is frequently necessary to rotate the ligand in 3D space to compile a list of globally minimized structures and then apply a rank ordering based on energetic criteria chosen by the user. Using DEZYMER, Hellinga and colleagues first created a number of new binding proteins for lactose, TNT and serotonin (Looger et al. 2003). Allert et al. (2004) then used DEZYMER to predict 12 amino acid mutations within the glucose binding protein (GGBP) to redesign the ligand binding pocket for pinacolyl methyl phosphonic acid (PMPA), the primary hydrolytic product of the nerve gas agent, soman. The authors constructed 14 of the DEZYMER output sequences and successfully obtained eight new PMPA binders to the GGBP scaffold with a dissociation constant ranging from 45 nM to 10 μM. Of the six remaining designs, four showed no binding to PMPA as determined by the fluorescence assay used and the last two either did not express or precipitated out of solution prior to analysis. They demonstrate that this method may be expanded to other periplasmic binding protein scaffolds by redesigning the ribose binding protein with PMPA specificity below 100 nM. While no structural data is given, alanine-scanning mutagenesis confirmed binding interactions predicted by the program. Computational tools should prove valuable in drug development as the algorithms continue to undergo refinement and computing power continues to increase.

13.4 In Vivo Evolution: Receptor Proteins by Combinatorial Screening

A complementary method to the computational design of novel protein and ligand binding functionality is to screen libraries of genetically randomized polypeptide domains. Because phage and ribosome displays are primarily used to screen large libraries raised for a particular target, this area may be thought of as in vivo-directed evolution of binding partners. Immunoglobulins or their fragments continue to provide the primary source of novel binding motifs for research laboratories. However, these proteins typically require complex methods for expression and isolation, as well as possessing limited stability with prolonged storage. There is a clear need for non-antibody scaffolds which address these issues and here we will describe only polypeptide-based systems. In theory, any spontaneously folding motif may be utilized for design of new ligand binding molecules, but it is wise to fully consider the thermodynamic and kinetic aspects as well. In particular, domain stability is a key issue because the protein sequence must be able to handle the proportionately large number of mutations required for the successful redesign or insertion of binding function. While these methods provide high throughput screening of binding to any target, detailed biochemical investigations to determine dissociation constants and specificity towards other target molecules are rarely performed, yet offer crucial information required to fully characterize the binding sequences. The following sections describe some of the polypeptide scaffolds which have been used for the design of new binding proteins (Fig. 13.11).

13.4.1 Trp Cage Motif

The Trp Cage is a 20 amino acid polypeptide which rapidly folds into a complex and highly organized tertiary structure (Neidigh et al. 2002). The isolation of this miniprotein is the result of detailed investigations into exendin-4, a small (39 amino

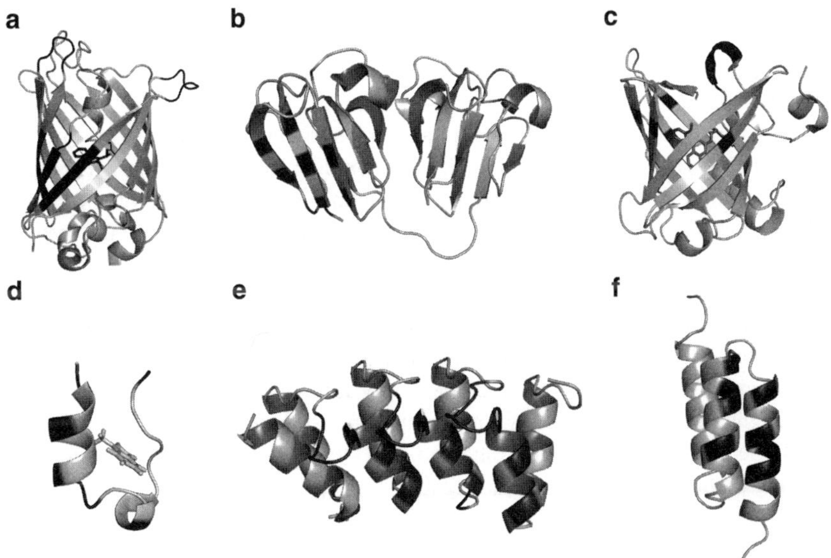

Fig. 13.11 Representative scaffolds screened for designing novel binding capabilities. Amino acids mutated during library development are shown in black. (**a**) Green fluorescent protein (PDB ID 2HGD), (**b**) g-crystallin (PDB ID 2JDG), (**c**) lipocalin (PDB ID 1N0S), (**d**) Trp cage (PDB ID 1L2Y), (**e**) ankyrin repeat (PDB ID 2R5Q), (**f**) affibody domain (1H0T)

acid) glucagon-like peptide hormone originally isolated from saliva obtained from the Gila monster, *Heloderma suspectum*. The primary feature of this motif is a hydrophobic cluster of Pro and Phe residues surrounding both faces of a single Trp residue. While fairly stable ($T_m = 43.5°C$), several positions have been identified which do not harm the folding capabilities. Herman et al. (2007) constructed a T7 phage library via fusion to gp10B to express the Trp Cage randomized at these seven different amino acid locations. Following several rounds of panning against streptavidin, the library yielded 10 clones with binding capabilities to this protein. Sequence analysis of the clones identified a specific His-Pro-Gln motif commonly associated with specific binding to strepavidin (Giebel et al. 1995). The Trp Cage library was also screened for peptides capable of binding to human bronchial epithelial cells. While panning is performed for only one cell line in this report, the authors note that the method may be generalized for the identification of peptides which target any cell line.

13.4.2 γ-B-Crystallin

The all β-sheet protein, γ-B-crystallin, is a single polypeptide chain with two domains of similar Greek key motifs consisting of about 40 amino acids each (Bloemendal et al. 2004). This protein persists at extremely high concentrations (860 mg/mL) in the eyes of invertebrates and provides much of the refracting power for vision. Interestingly, this protein exists as a highly stable monomer, contains no disulfide bridges, has no known affinity for other proteins, and possesses no known enzymatic activity. Using sequence alignment of 84 crystallin homologs and surface accessibility, Ebersbach et al. (2007) identified eight amino acids (2, 4, 6, 15, 17, 19, 36, and 38) as solvent accessible locations for randomization and for the design of a new ligand binding site. The authors call the new binding proteins developed from the γ-B-crystallin scaffold, Affilins. After constructing an M13 phage library of γ-B-crystallin fused to coat protein III, an in-depth spectroscopic analysis of the resulting binders to estradiol/testosterone, IgG (F_c fragment only), and the proform of human nerve growth factor (PrNGF) was performed. Using surface plasmon resonance measurements, Affilins were found with sub micromolar dissociation constants for each of the diverse molecules. The authors also demonstrated a broad range of temperature stability (56–80°C) and selectivity ratios of 5:1 for Affilin binding to estradiol over testosterone. X-ray data comparing the wild type γ-B-crystallin and one of the F_c binding domains showed that scaffold structure remained intact following mutagenesis of the Affilin rigid β-sheet domains.

13.4.3 Min-23 (a Knottin)

Min-23 is a short, highly stable ($T_m = 100°C$) peptide containing a cystine-stabilized β-sheet motif (Heitz et al. 1999). This 23 amino acid peptide is a minimal segment derived from the squash trypsin inhibitor EETI-II (*Ecballium elaterium* trypsin inhibitor) belonging to the Knottin family of peptides (Gelly et al. 2004). Souriau et al. (2005) created chimeras of Min-23 to demonstrate that an exposed β-turn loop may be altered by peptide insertion and still allow proper folding. A phage library was constructed by replacing residues 17–20 with ten randomly selected amino acids and screened for binding to seven different proteins ranging in size up to 70 kDa. SPR was used to determine binding constants for some of the mutants and were reported to be of the order of 10^{-8} M. It is important to note that in order to express the cystine-containing polypeptides in bacterial cells, translational fusions with maltose binding protein are typically constructed to increase solubility and expression yields.

13.4.4 Scorpion Toxins and Defensins

These two classes of peptides share identical structural motifs and are typically stabilized by 3–5 disulfide bonds folded into multistranded antiparallel β-strands with an α-helix in parallel (White et al. 1995; Selsted and Ouellette 2005). The first reported redesign encompassed the addition of a metal binding site into charybdotoxin, a 37 amino acid toxin from the scorpion *Leiurus quinquestriatus hebraeus* (Vita et al. 1995). Three histidine residues were site-specifically engineered into two of the β-strands (a.a. 25–29 and 32–36) to mimic the Zn^{2+} binding site of carbonic anhydrase. Additional mutations were added in this region to stabilize or ease van der Waals contacts. Metal binding was monitored using UV and fluorescence spectroscopy upon Cu^{2+} binding with a reported K_d of 4×10^{-8} M. Binding affinities for other metals including Zn^{2+}, Cd^{2+}, Ni^{2+} and Mn^{2+} were determined to have lower affinities and found to mimic that of carbonic anhydrase. In another attempt to utilize scorpion toxin scaffolds, nine amino acids were transplanted into the scyllatoxin to mimic HIV-1 envelope protein (gp120) binding exhibited by CD4. A small 27 amino acid miniprotein was produced and was found to inhibit infection of $CD4^{+}$ cells by different virus isolates (Vita et al. 1999).

Defensins are a widely distributed family of 30–54 amino acid peptides investigated for their inhibitory activity of microbes, fungi, parasites and tumor cells (White et al. 1995). Functional studies have concluded that lytic activity is due to membrane insertion and pore formation to disrupt the normal cellular activity. Zhoa et al. (2004) designed a conformationally constrained phage display library of insect defensin A through reconstruction of 29 amino acid peptide and randomized the seven positions found on the β-strands (a.a. 23–26 and 32–34). Four to five rounds of biopanning were performed and selected for defensins with binding affinity against four different proteins: TNFa, TNF receptor 1, TNF receptor 2 and monoclonal antibody against BMP-2. While binding was demonstrated, further biophysical analysis and determination of the association constants will be helpful for comparing the efficacy of defensins as a design scaffold. Structural studies of defensin interactions with enzymes like α-amylase contribute to our understanding of defensin inhibitory activity as well as improving our ability to design additional binding libraries (Lin et al. 2007). A third class of antimicrobial peptides with this same β-strand structure is the γ-thionins (Stec 2006), however, to date, no redesign of this peptide as a scaffold can be found in the literature.

13.4.5 Lipocalins

The rigid and highly conserved lipocalin fold consists primarily of an eight stranded β-barrel containing a ligand binding site and a C-terminal α-helical tail (Flower 1996). This family of proteins has less than 20% sequence homology and has been found in plants, animals and bacteria to bind a diverse array of physiologically important molecules. As a result of this broad specificity, lipocalins are known with binding affinity for nonnatural hazardous chemicals, such as diphenylamine, dimethyl-phthalate, resorcinol and dinitrotoluene. D'Auria and colleagues developed a competitive ligand-binding assay for these explosive components using truncated odorant binding protein, a monomeric lipcalin found in animals. (Ramoni et al. 2007). In this assay, analyte binding displaced a fluorescent aminoanthracene probe from the binding pocket, and the binding constant for each hazardous material was determined. A detailed study of the ligand binding pocket from tear lipocalin was also performed using the fluorescent probe, ANS, which found evidence of energy transfer to nearby Trp residues (Gasymov et al. 2008). The last two examples demonstrate potential methods for developing reliable assays for ligands once new binding molecules have been found. Skerra and colleagues have developed phage display libraries utilizing the billin-binding protein, a lipocalin from the butterfly, *Pieris brassicae* (Beste et al. 1999). Randomization of 16–17 amino acids found in four loop regions, connecting the β-strands, led to the panning of lipocalins (~174 amino acid single polypeptide) with binding specificity for fluorescein (35 nM) and digoxigenin (30 nM) (Schlehuber et al. 2000; Vopel et al. 2005). The authors renamed the engineered lipocalins, "anticalins" for its new ligand binding ability. Further X-ray crystallographic studies of mutant Dig16A revealed that the β-barrel remains intact and only small structural deviations are observed in the highly mutated ligand binding loop region (Korndorfer et al. 2003).

13.4.6 Staphylococcal Protein A Domain

A 58 amino acid (~6.5 kDa) segment of domain B from staphylococcal protein A may be isolated as a stable three helix bundle (Nilsson et al. 1987). Following substitutions of Gly29Ala and Ala1Val, this polypeptide is renamed the Z-domain which possesses less binding activity for antibody (F_{ab}) fragments than the wild-type and increases the stability of helix 2 to allow for faster folding of the polypeptide (Nord et al. 1997). The randomization of 13 amino acids on helices 1 and 2 of the Z-domain followed by construction of phagemid libraries produces a redesigned scaffold, termed "affibodies," with specificities for human, bacterial, and viral target proteins (Gronwall et al. 2007). The structural and thermodynamic basis for affibody binding has been investigated and confirmed the importance of helix 1 and helix 2 in ligand binding (Dogan et al. 2006; Lendel et al. 2006). An interesting application of affibodies has been in imaging tumor cells which over-express specific epidermal growth factor receptors (EGFR1). Friedman et al. performed directed evolution of five carefully selected positions to yield second generation EGFR1-binding affibodies with 5–10 nM affinities (Friedman et al. 2007, 2008). These affibodies were labeled with Indium-111 and showed specific binding to EGFR-expressing tumor cells within 4 h post-injection in murine models (Orlova et al. 2007; Tran et al. 2007).

13.4.7 Ankyrin Repeat Proteins

Ankyrin repeat proteins are built from tightly joined repeating units (ca. 33 amino acids) which consist of a β-turn followed by two antiparallel α-helices and a loop reaching the turn of the next repeat (Sedgwick and Smerdon 1999; Kohl et al. 2003). Found in all phyla, typical ankyrin repeat proteins contain four to six structural units capable of domain stabilization as well as presenting an expanded surface for target binding. Mimicking this useful natural strategy, Plückthun and colleagues use ribosome display to construct designed ankyrin repeat proteins (DARPins) containing two to three repeating units and an N-terminal cap to prevent aggregation (Stumpp et al. 2003; Binz et al. 2004). Six surface exposed sites in each ankyrin repeating unit are randomized to create highly stable and soluble DARPins resulting in high binding affinity to target proteins, including human epidermal growth factor 2 (91 pM) (Zahnd et al. 2006, 2007), as well as serving as protease (Kawe et al. 2006; Schweizer et al. 2007) and kinase (Binz et al. 2004) inhibitors. Because of the successful application as protein inhibitors, the authors note some interesting potential for modulation of protein functionality in vivo in addition to the original goal of creating simple binders.

13.4.8 Green Fluorescent Protein

The fluorescent protein family has grown rapidly over the last decade to include more than 30 different members most described structurally by an 11 strand β-barrel enclosing a central α-helix (Phillips 2006). In the original GFP, three consecutive amino acids (Ser65-Tyr66-Gly67) on the helix undergo spontaneous cyclization and oxidation to form a new chromophore with good quantum yield and green luminescence (Ward 2006). While GFP and its variants have been widely used to design genetically encoded FRET-based sensors for many targets, the insertion and randomized library approach, which works well with other scaffolds described above, has not been as effective for the fluorescent protein family members. An early report by Abedi et al. (1998) found several exposed loop regions connecting β-strands into which 20 amino acid linkers may be added and still retain GFP emission. Despite this early success, only a few groups have been able to successfully graft new binding domains into the GFP structure and couple binding events with the intrinsic fluorescence. Ding and colleagues insert pentapeptides directly into specific locations within β-strands to design intrinsically fluorescent biosensors for endotoxin and for gram-negative bacteria expressing these lipopolysaccharides (Goh et al. 2002a, b). Biosensors have been engineered by utilizing circular permutation of GFP and domain insertion of polypeptides able to chelate specific metals for environmental and biomedical applications. The transposition of Ca^{2+} binding domain from calmodulin detected physiologically relevant levels of this important metal (Baird et al. 1999). After Imoto and colleagues added the M13 fragment of myosin light chain kinase to this system, several groups have added to the redesign for better signal-to-noise characteristics allowing Ca^{2+} imaging in a variety of cell lines (Nakai et al. 2001; Ohkura et al. 2005; Souslova et al. 2007). Also, a designed coiled-coil domain has been successfully grafted into circularly permutated GFPs between amino acids 190–191 with binding affinities for Cu^{2+} and Zn^{2+} in a concentration-dependent manner (Mizuno et al. 2007). Further, some fluorescent protein family members possess intrinsic transition metal binding proteins and many laboratories continue to optimize the signaling response (Richmond et al. 2005; Sumner et al. 2006). The design of new fluorescent proteins which couple intrinsic fluorescence with affinity for metals, small molecules or larger polypeptides should continue to be a highly active area of research in the next decade.

Several other single chain polypeptide scaffolds have been investigated and mutants identified with high levels of binding affinity to a variety of targets; these scaffolds include human fibronectin tenth type III (FN3) domain (Koide et al. 1998; Richards et al. 2003; Karatan et al. 2004; Huang et al. 2006), basic pancreatic trypsin inhibitor (BPTI) (Kiczak et al. 2001), Kuntiz domain (Markland et al. 1996a, b), tendamistat (McConnell and Hoess 1995), Src homology domains 2 and 3 (Malabarba et al. 2001), PDZ domains (Schneider et al. 1999; Ferrer et al. 2005), carbohydrate binding domain (CBD, knottin family) (Smith et al. 1998; Lehtio et al. 2002), cytotoxic t lymphocyte-associated antigen (CTLA-4) (Nuttall et al. 1999; Hufton et al. 2000) and thermostable carbohydrate-binding module (CBM4-2) (Gunnarsson et al. 2004, 2006). Finally, the design of repeating unit that scaffolds with the ability to bind two (or more) target molecules has been accomplished using the human A-domain, termed avimers, found in various cell surface receptors (Silverman et al. 2005).

13.5 Concluding Remarks

The future of biosensor design will continue along the transdisciplinary application of techniques in biochemistry, biophysics, molecular mechanics, and spectroscopic engineering. The production of novel polypeptides with binding selectivity for natural and non-natural ligands will only be possible through utilizing a variety of scaffolds for sensor design. Physiologically important small molecules and polypeptides present a diverse problem for generating interaction surfaces and one scaffold is unlikely to satisfy this complexity. More work is also needed towards the creation of new strategies to detect target binding. While fluorescence is a highly sensitive technique, improved methods to genetically incorporate fluorescent probes will alleviate the need to rely solely on bulky fluorescent proteins or on purification and chemical modification in vitro. Protocols have been developed to expand the genetic code and incorporate non-natural amino acids into proteins (Hendrickson et al. 2004). The design of engineered tRNA–tRNA synthatase orthogonal pairs capable of inserting chromophores, particularly with near-IR spectral properties, into a protein scaffold would be an exciting advance to the biosensing community. Alternate methods to fluorescence detection may also be appropriate. While surface plasmon resonance is widely used in laboratories for screening scaffold binding to targets, the current SPR experiment requires a flow cell which may be difficult to miniaturize and operate in clinical settings, for example. One technique that may be miniaturized into a portable, field-ready sensor is a quartz crystal microbalance which is also capable of detecting target binding with high signal-to-noise ratios (Marx 2003). For these reasons, the experimental design of viable protein biosensors will continue to increase our knowledge and bridge areas of biological investigations.

References

Abedi MR, Caponigro G, Kamb A (1998) Green fluorescent protein as a scaffold for intracellular presentation of peptides. Nucleic Acids Res 26:623–630

Adams SR, Harootunian AT, Buechler YJ, Taylor SS, Tsien RY (1991) Fluorescence ratio imaging of cyclic amp in single cells. Nature 349:694–697

Allert M, Rizk SS, Looger LL, Hellinga HW (2004) Computational design of receptors for an organophosphate surrogate of the nerve agent soman. Proc Natl Acad Sci USA 101:7907–7912

Baird GS, Zacharias DA, Tsien RY (1999) Circular permutation and receptor insertion within green fluorescent proteins. Proc Natl Acad Sci USA 96:11241–11246

Benson DE, Haddy AE, Hellinga HW (2002) Converting a maltose receptor into a nascent binuclear copper oxygenase by computational design. Biochemistry 41:3262–3269

Benson DE, Wisz MS, Liu W, Hellinga HW (1998) Construction of a novel redox protein by rational design: conversion of a disulfide bridge into a mononuclear iron-sulfur center. Biochemistry 37:7070–7076

Berlman IB (1971) Fluorescence specta of aromatic molecules. Academic, New York

Beste G, Schmidt FS, Stibora T, Skerra A (1999) Small antibody-like proteins with prescribed ligand specificities derived from the lipocalin fold. Proc Natl Acad Sci USA 96:1898–1903

Binz HK, Amstutz P, Kohl A, Stumpp MT, Briand C, Forrer P, Grutter MG, Pluckthun A (2004) High-affinity binders selected from designed ankyrin repeat protein libraries. Nat Biotechnol 22:575–582

Bjorkman AJ, Binnie RA, Zhang H, Cole LB, Hermodson MA, Mowbray SL (1994) Probing protein–protein interactions. The ribose-binding protein in bacterial transport and chemotaxis. J Biol Chem 269:30206–30211

Bjorkman AJ, Mowbray SL (1998) Multiple open forms of ribose-binding protein trace the path of its conformational change. J Mol Biol 279:651–664

Bloemendal H, de Jong W, Jaenicke R, Lubsen NH, Slingsby C, Tardieu A (2004) Ageing and vision: structure, stability and function of lens crystallins. Prog Biophys Mol Biol 86: 407–485

Boas FE, Harbury PB (2007) Potential energy functions for protein design. Curr Opin Struct Biol 17:199–204

Borrok MJ, Kiessling LL, Forest KT (2007) Conformational changes of glucose/galactose-binding protein illuminated by open, unliganded, and ultra-high-resolution ligand-bound structures. Protein Sci 16:1032–1041

Brune M, Hunter JL, Corrie JET, Webb MR (1994) Direct, real-time measrement of rapid inorganic phosphate release using a novel fluorescent probe and its application to actomyosin subfragment atpase. Biochemistry 33:8262–8271

Butterfoss GL, Kuhlman B (2006) Computer-based design of novel protein structures. Annu Rev Biophys Biomol Struct 35:49–65

Chino S, Sakaguchi A, Yamoto R, Ferri S, Sode K (2007) Branched-chain amino acid biosensing using fluorescent modified engineered leucin/isoleucine/valine binding protein. Int J Mol Sci 8:513–525

Cuneo MJ, Changela A, Warren JJ, Beese LS, Hellinga HW (2006) The crystal structure of a thermophilic glucose binding protein reveals adaptations that interconvert mono and di-saccharide binding sites. J Mol Biol 362:259–270

Das R, Baker D (2008) Macromolecular modeling with rosetta. Ann Rev Biochem 77:363–382

Dattelbaum JD, Lakowicz JR (2001) Optical determination of glutamine using a genetically engineered protein. Anal Biochem 291:89–95

Dattelbaum JD, Looger LL, Benson DE, Sali KM, Thompson RB, Hellinga HW (2005) Analysis of allosteric signal transduction mechanisms in an engineered fluorescent maltose biosensor. Protein Sci 14:284–291

deLorimier RM, Smith JJ, Dwyer MA, Looger LL, Sali KM, Paavola CD, Rizk SS, Sadigov S, Conrad DW, Loew L et al (2002) Construction of a fluorescent biosensor family. Protein Sci 11:2655–2675

Der BS, Dattelbaum JD (2008) Construction of a reagentless glucose biosensor using molecular exciton luminescence. Anal Biochem 375:132–140

Deuschle K, Okumoto S, Fehr M, Looger LL, Koszhukh L, Frommer WB (2005) Construction and optimization of a family of genetically encoded metabolite sensors by semirational protein engineering. Protein Sci 14:2304–2314

DiPilato LM, Cheng X, Zhang J (2004) Fluorescent indicators of camp and epac activation reveal differential dynamics of camp signaling within discrete subcellular compartments. Proc Natl Acad Sci USA 101:15081–15086

Dogan J, Lendel C, Hard T (2006) Thermodynamics of folding and binding in an affibody: affibody complex. J Mol Biol 359:1305–1315

Doornbos RMP, Lang R, Aalders MC, Cross FW, Sterenborg HJCM (1999) The determination of *in vivo* human tissue optical properties and absolute chromophore concentrations using spatially resolved steady-state diffuse reflectance spectroscopy. Phys Med Biol 44:967–981

Dulla C, Tani H, Okumoto S, Frommer WB, Reimer RJ, Fluguenard JR (2008) Imaging of glutamate in brain slices using fret sensors. J Neurosci Methods 168:306–319

Dwyer MA, Hellinga HW (2004) Periplasmic binding proteins: a versatile superfamily for protein engineering. Curr Opin Struct Biol 14:495–504

Ebersbach H, Fiedler E, Scheuermann T, Fiedler M, Stubbs MT, Reimann C, Proetzel G, Rudolph R, Fiedler U (2007) Affilin-novel binding molecules based on human γ-b-crystallin, an all β-sheet protein. J Mol Biol 372:172–185

Fehr M, Frommer WB, Lalonde S (2002) Visualization of maltose uptake in living yeast cells by fluorescent nanosensors. Proc Natl Acad Sci USA 99:9846–9851

Fehr M, Okumoto S, Deuschle K, Lager I, Looger LL, Persson J, Kozhukh L, Lalonde S, Frommer WB (2005) Development and use of fluorescent nanosensors for metabolite imaging in living cells. Biochem Soc Trans 33:287–290

Ferrer M, Maiolo J, Kratz P, Jackowski JL, Murphy DJ, Delagrave S, Inglese J (2005) Directed evolution of pdz variants to generate high-affinity detection reagents. Protein Eng Des Sel 18:165–173

Flower DR (1996) The lipocalin protein family: structure and function. Biochem J 318:1–14

Friedman M, Nordberg E, Hoiden-Guthenberg I, Brisimar H, Adams GP, Nilsson FY, Carlss J, Stahl S (2007) Phage display selection of affibody molecules with specific binding to the extracellular domain of the epidermal growth factor receptor. Protein Eng Des Sel 20: 189–199

Friedman M, Orlova A, Johansson E, Eriksson TLJ, Hoiden-Guthenberg I, Tolmachev V, Nilsson FY, Stahl S (2008) Directed evolution to low nanomolar affinity of a tumor-targeting epidermal growth factor receptor-binding affibody molecule. J Mol Biol 376:1388–1402

Gasymov OK, Abduragimov AR, Glasgow BJ (2008) Ligand binding site of tear lipocalin: contribution of a trigonal cluster of, charged residues probed by 8-anilino-1-naphthalenesulfonic acid. Biochemistry 47:1414–1424

Ge X, Tolosa L, Rao G (2004) Dual-labeled glucose binding protein for ratiometric measurements of glucose. Anal Chem 76:1403–1410

Ge X, Tolosa L, Simpson J, Rao G (2003) Genetically engineered binding proteins as biosensors for fermentation and cell culture. Biotechnol Bioeng 84:723–731

Gelly JC, Gracy J, Kaas Q, Le-Nguyen D, Heitz A, Chiche L (2004) The knottin website and database: a new information system dedicated to the knottin scaffold. Nucleic Acids Res 32: D156–D159

Giebel LB, Cass RT, Milligan DL, Young DC, Arze R, Johnson CR (1995) Screening of cyclic peptide phage libraries identifies ligands that bind streptavidin with high affinities. Biochemistry 34:15430–15435

Gilardi G, Mei G, Rosato N, Agro AF, Cass AE (1997) Spectroscopic properties of an engineered maltose binding protein. Protein Eng 10:479–486

Gilardi G, Zhou LQ, Hibbert L, Cass AE (1994) Engineering the maltose binding protein for reagentless fluorescence sensing. Anal Chem 66:3840–3847

Goh YT, Frecer V, Ho B, Ding JL (2002a) Rational desing of green fluorescent protein mutants as biosensor for bacterial endotoxin. Protein Eng Des Sel 15:493–502

Goh YT, Ho B, Ding JL (2002b) A novel fluorescent protein-based biosensor for gram-negative bacteria. Appl Environ Microbiol 68:6343–6352

Gronwall C, Jonsson A, Lindstrom S, Gunneriusson E, Stahl S, Herne N (2007) Selection and characterization of affibody ligands binding to alzheimer amyloid beta peptides. J Biotech 128:162–183

Gu H, Lalonde S, Okumoto S, Looger LL, Scharff-Poulsen AM, Grossman AR, Kossmann J, Jakobsen I, Frommer WB (2006) A novel analytical method for in vivo phosphate tracking. FEBS Lett 580:5885–5893

Gunnarsson LC, Karlsson EN, Albrekt AS, Andersson M, Holst O, Ohlin M (2004) A carbohydrated binding module as a diversity-carrying scaffold. Protein Eng Des Sel 17:213–221

Gunnarsson LC, Karlsson EN, Andersson M, Holst O, Ohlin M (2006) Molecular engineering of a thermostable carbohydrate-binding module. Biocatal Biotransformation 24:31–37

Ha JS, Song JJ, Lee YM, Kim SJ, Sohn JH, Shin CS, Lee SG (2007) Design and application of highly responsive fluorescence resonance energy transfer biosensors for detection of sugar in living saccharomyces cerevisiae cells. Appl Environ Microbiol 73:7408–7414

He JJ, Quiocho FA (1993) Dominant role of local dipoles in stabilizing uncompensated charges on a sulfate sequestered in a periplasmic active transport protein. Protein Sci 2:1643–1647

Heitz A, Le-Nguyen D, Chiche L (1999) Min-21 and min-23, the smallest peptides that fold like a cystine-stabilized β-sheet motif: design, solution structure, and thermal stability. Biochemistry 38:10615–10625

Hellinga HW, Richards FM (1991) Construction of new ligand binding sites in proteins of known structure. I. Computer-aided modeling of sites with pre-defined geometry. J Mol Biol 222:763–785

Hendrickson TL, de Crecy-Lagard V, Schimmel P (2004) Incorporation of nonnatural amino acids into proteins. Annu Rev Biochem 73:147–176

Herman RE, Badders D, Fuller M, Makienko EG, Houston ME, Quay SC, Johnson PH (2007) The trp cage motif as a scaffold for the display of a randomized peptide library on bacteriophage t7. J Biol Chem 282:9813–9824

Higgins CF (1992) Abc transporters: from microorganisms to man. Annu Rev Cell Biol 8:67–113

Ho SN, Hunt HD, Horton RM, Pullen JK, Pease LR (1989) Site-directed mutagenesis by overlap extension using the polymerase chain reaction. Gene 77:51–59

Hsiao CD, Sun UJ, Rose J, Wang B-C (1996) The crystal structure of glutamine-binding protein from *Escherichia coli*. J Mol Biol 262:225–242

Huang J, Koide A, Nettle KW, Greene GL, Koide S (2006) Conformation-specific affinity purification of proteins using engineered binding proteins: application to the estrogen receptor. Protein Expr Purif 47:348–354

Hufton SE, van Neer N, van den Beuken T, Desmet J, Sablon E, Hoogenboom HR (2000) Development and application of cytotoxic t lymphocyte-associated antigen 4 as a protein scaffold for the generation of novel binding ligands. FEBS Lett 475:225–231

Itoh RE, Kurokawa K, Ohba Y, Yoshizaki H, Mochizuki N, Masuda M (2002) Activation of rac and cdc42 video imaged by fluorescent resonance energy transfer-based single-molecule probes in the membrane of living cells. Mol Cell Biol 22:6582–6591

Karatan E, Merguerian M, Han Z, Scholle MD, Koide S, Kay BK (2004) Molecular recognition properties of fn3 monobodies that bind the src sh3 domain. Chem Biol 11:835–844

Kasha M (1963) Energy transfer mechanisms and the molecular exciton model for molecular aggregates. Radiat Res 20:55–71

Kawe M, Forrer P, Amstutz P, Pluckthun A (2006) Isolation of intracellular proteinase inhibitors derived from designed ankyrin repeat proteins by genetic screening. J Biol Chem 281:40252–40263

Khan F, Gunudi L, Pickup JC (2008) Fluorescence-based sensing of glucose using engineered glucose/galactose-binding protein: a comparison of fluorescence resonance energy transfer and environmentally sensitive dye labeling strategies. Biochem Biophys Res Commun 365: 102–106

Kiczak L, Kasztura M, Koscielska-Kasprzak K, Kadlez M, Otlewsdi J (2001) Selection of potent chymotrypsin and elastase inhibitors from m13 phage library of basic pancreatic trypsin inhibitor (bpti). Biochim Biophys Acta 17:153–163

Klotz IM (1997) Ligand-receptor energetics: a guide for the perplexed. Wiley, New York

Kohl A, Binz HK, Forrer P, Stumpp MT, Pluckthun A, Grutter MG (2003) Designed to be stable: crystal structure of a consensus ankyrin repeat protein. Proc Natl Acad Sci USA 100:1700–1705

Koide A, Bailey CW, Huang X, Koide S (1998) The fibronectin type iii domain as a scaffold for novel binding proteins. J Mol Biol 284:1141–1151

Korndorfer IP, Schlehuber S, Skerra A (2003) Structural mechanism of specific ligand recognition by a lipocalin tailored for the complexation of digoxigenin. J Mol Biol 330:385–396

Lakowicz JR (1994) Emerging biomedical applications of time-resolved fluorescence spectroscopy. In: Lakowicz JR (ed) Topics in fluorescence spectroscopy, vol 4, Probe design and chemical sensing. Plenum, New York, p 501

Lakowicz JR (1999) Principles of fluorescence spectroscopy. Plenum, New York

Lalonde S, Ehrhardt DW, Frommer WB (2005) Shining light on signaling and metabolic networks by genetically encoded biosensors. Curr Opin Plant Biol 8:574–581

Lehtio J, Teeri TT, Nygren PA (2002) Alpha-amylase inhibitors selected from a combinatorial library of a cellulose binding domain scaffold. Proteins 41:316–322

Lendel C, Dogan J, Hard T (2006) Structural basis for molecular recognition in an affibody: affibody complex. J Mol Biol 359:1293–1304

Lin KF, Lee PH, Hsu MP, Chen CS, Lyu PC (2007) Structure-based protein engineering for α-amylase inhibitory activity of plant defensin. Proteins 68:530–540

Looger LL, Dwyer MA, Smith JJ, Hellinga HW (2003) Computational design of receptor and sensor proteins with novel functions. Nature 423:185–190

Looger LL, Hellinga HW (2001) Generalized dead-end elimination algorithms make large-scale protein side-chain structure prediction tractable: implications for protein design and structural genomics. J Mol Biol 307:429–445

Looger LL, Lalonde S, Frommer WB (2005) Genetically encoded fret sensors for visualizing metabolites with subcellular resolution in living cells. Plant Physiol 138:555–557

Luo KQ, Yu VC, Pu Y, Chang DC (2001) Application of the fluorescence resonance energy transfer method for studying the dynamics of caspase-3 activation during uv-induced apoptosis in living hela cells. Biochem Biophys Res Commun 283:1054–1060

Malabarba MG, Milia E, Faretta M, Zamponi R, Pelicci PG, Di Fiore PP (2001) A repertoire library that allows the selection of synthetic sh2s with altered binding specificities. Oncogene 20:5186–5194

Mansouri S, Schultz JS (1984) A minature optical glucose sensor based on affinity binding. Biotechnolgy 2:885–890

Markland W, Ley AC, Lee SW, Ladner RC (1996a) Iterative optimization of high-affinity protease inhibitors using phage display.1. Plasmin. Biochemistry 35:8045–8057

Markland W, Ley AC, Ladner RC (1996b) Iterative optimization of high-affinity protease inhibitors using phage display. 2. Plasma kallikrein and thrombin. Biochemistry 35:8058–8067

Marvin JS, Corcoran EE, Hattangadi NA, Zhang JV, Gere SA, Hellinga HW (1997) The rational design of allosteric interactions in a monomeric protein and its applications to the construction of biosensors. Proc Natl Acad Sci USA 94:4366–4371

Marvin JS, Hellinga HW (1998) Engineering biosensors by introducing fluorescent allosteric signal transducers: construction of a novel glucose sensor. J Am Chem Soc 120:7–11

Marx KA (2003) Quartz crystal microbalance: a useful tool for studying thin polymer films and complex biomolecular systems at the solution-surface interface. Biomacromolecules 4: 1099–1120

McConnell SJ, Hoess RH (1995) Tendamistat as a scaffold for confomationally constrained phage peptide libraries. J Mol Biol 250:460–470

Meadows D, Schultz JS (1988) Fiber-optic biosensors based on fluresence energy transfer. Talanta 35:145–150

Mitra RD, Silva CM, Youvan DC (1996) Fluorescence resonance energy transfer between blue-emitting and red-shifted excitation derivatives of the green fluorescent protein. Gene 173:13–17

Miyawaki A, Griesbeck O, Heim R, Tsien RY (1999) Dynamic and quantitative ca2+ measurements using improved cameleons. Proc Natl Acad Sci USA 96:2135–2140

Mizuno T, Murao K, Tanabe Y, Oka M, Tanaka T (2007) Metal-ion-dependent gfp emission in vivo by combining a circularly permutated green fluorescent protein with an engineered metal-ion-binding coiled-coil. J Am Chem Soc 129:11378–11383

Moschou EA, Sharma BV, Deo SK, Daunert S (2004) Fluorescence glucose detection: advances towards the ideal *in vivo* biosensor. J Fluoresc 14:535–547

Nakai J, Ohkura M, Imoto K (2001) A high signal-to-noise ca^{2+} probe composed of a single green fluorescent protein. Nat Biotechnol 19:137–141

Nalbant P, Hodgson L, Kraynov V, Toutchkine A, Hahn KM (2004) Activation of endogenous cdc42 visualized in living cells. Science 305:1615–1619

Neidigh JW, Fesinmeyer RM, Andersen NH (2002) Designing a 20-residue protein. Nat Struct Biol 9:425–430

Newman JD, Turner AP (2005) Home blood glucose biosensors: a commerical perspective. Biosens Bioelectron 20:2435–2453

Nilsson B, Moks T, Jansson B, Abrahmsen L, Elmblad A, Holmgren E, Henrichson C, Jones TA, Uhlen M (1987) A synthetic igg-binding domain based on staphylococcal protein-a. Protein Eng Des Sel 1:107–113

Nord K, Gunneriusson E, Ringdahl J, Stahl S, Uhlen M, Nygren PA (1997) Binding proteins selected from combinatorial libraries of an alpha-helical bacterial receptor domain. Nat Biotechnol 15:772–777

Nuttall SD, Rousch MJ, Irving SE, Hufton SE, Hoogenboom HR, Hudson PJ (1999) Design and expression of soluble ctla-4 variable domain as a scaffold for the display of functional polypeptides. Proteins 36:217–227

Oh BH, Pandit J, Kang CH, Nikaido K, Gokcen S, Ames GF, Kim SH (1993) Three-dimensional structures of the periplasmic lysine/arginine/ornithine-binding protein with and without a ligand. J Biol Chem 268:11348–11355

Ohkura M, Matsuzaki M, Kasai H, Imoto K, Nakai J (2005) Genectically encoded bright ca^{2+} probe applicable for dynamic ca^{2+} imaging of dendritic spines. Anal Chem 77:5861–5869

Orlova A, Feldwisch J, Abrahmsen L, Tolmachev V (2007) Affibody molecules for molecular imaging and therapy for cancer. Cancer Biother Radiopharm 22:573–584

Phillips GN (2006) The three-dimensional structure of green fluorescent protein and its implications for function and design. Methods Biochem Anal 47:67–82

Quiocho FA, Spurlino JC, Rodseth LE (1997) Extensive features of tight oligosaccharide binding revealed in high-resolution structures of the maltodextrin transport/chemosensory receptor. Structure 5:997–1015

Quiocho FA, Vyas NK (1984) Novel stereospecificity of the l-arabinose-binding protein. Nature 310:381–386

Ramoni R, Bellucci S, Grycznyski I, Grycznyski Z, Grolli S, Staiano M, De Bellis G, Micciulla F, Pastore R, Tiberia A et al (2007) The protein scaffold of the lipocalin odorant-binding protein is suitable for the design of new biosensors for the detection of explosive components. J Phys Condens Matter 19:395012 (7pp) doi: 10.1088/0953-8984/19/39/395012

Richards J, Miller M, Abend J, Koide A, Koide S, Dewhurst S (2003) Engineered fibronectin type iii domain with a rgdwxe sequence binds with enhanced affinity and specificity to human $v\beta3$ integrin. J Mol Biol 326:1475–1488

Richmond TA, Takahashi TT, Shimkhada R, Bernsdorf J (2005) Engineered metal binding sites on green fluorescence protein. Biochem Biophys Res Commun 268:462–465

Romoser VA, Hinkle PM, Persechini A (1997) Detection in living cells of ca^{2+}-dependent changes in the fluorescence emission of an indicator composed of two green fluorescent protein variants linked by a calmodulin-binding sequence. J Biol Chem 272:13270–13274

Sakaguchi A, Ferri S, Tsugawa W, Sode K (2007) Novel fluorescent sensing systme for alpha-fructosyl amino acids based on engineered fructosyl amino acid binding protein. Biosens Bioelectron 22:1933–1938

Salins LL, Deo SK, Daunert S (2004) Phosphate binding protein as the biorecognition element in a biosensor for phosphate. Sens Actuators B Chem 97:81–89

Salins LL, Goldsmith ES, Ensor CM, Daunert S (2002) A fluorescence-based sensing system for the environmental monitoring of nickel using the nickel binding protein from *Escherichia coli*. Anal Bioanal Chem 372:174–180

Salins LL, Ware RA, Ensor CM, Daunert S (2001) A novel reagentless sensing system for measuring glucose based on the galactose/glucose-binding protein. Anal Biochem 294:19–26

Sato M, Ozawa T, Inukai K, Asano T, Umezawa Y (2002) Fluorescent indicators for imaging protein phosphorylation in single living cells. Nat Biotechnol 20:287–294

Schlehuber S, Beste G, Skerra A (2000) A novel type of receptor protein, based on the lipocalin scaffold, with specificity for digoxigenin. J Mol Biol 297:1105–1120

Schneider S, Buchert M, Georgiev O, Catimel B, Halford M, Stacker SA, Baechi T, Moelling K, Hovens CM (1999) Mutagenesis and selection of pdz domains that bind new protein targets. Nat Biotechnol 17:170–175

Schweizer A, Roschitzki-Voser H, Amstutz P, Briand C, Gulotti-Georgieva M, Prenosil E, Binz HK, Capitani G, Baici A, Pluckthun A et al (2007) Inhibition of caspase-2 by a designed ankyrin repeat protein: specificity, structure, and inhibition mechanism. Structure 15:625–636

Sedgwick SG, Smerdon SJ (1999) The ankyrin repeat: a diversity of interactions on a common structural framework. Trends Biochem Sci 24:311–316

Selsted ME, Ouellette AJ (2005) Mammalian defensins in the antimicrobial immune response. Nat Immunol 6:551–557

Sharff AJ, Rodseth LE, Spurlino JC, Quiocho FA (1992) Crystallographic evidence of a large ligand-induced hinge-twist motion between the two domains of the maltodextrin binding protein involved in active transport and chemotaxis. Biochemistry 31:10657–10663

Shrestha S, Salins LL, Ensor CM, Daunert S (2002) Rationally designed fluorescently labeled sulfate-binding protein mutants: evaluation in the development of a sensing system for sulfate. Biotechnol Bioeng 78:517–526

Silverman J, Lu Q, Bakker A, To W, Duguay A, Alba BM, Smith R, Rivas A, Li P, Le H et al (2005) Multivalent avimer proteins evolved by exon shuffling of a family of human receptor domains. Nat Biotechnol 23:1556–1561

Smith GP, Patel SU, Windass JD, Thornton JM, Winter G, Griffiths AD (1998) Small binding proteins selected from a combinatorial repertoire of knottins displayed on phage. J Mol Biol 277:317–332

Sohanpal K, Watsuji T, Zhou LQ, Cass AEG (1993) Reagentless fluorescence sensors based upon specific binding proteins. Sens Actuators B Chem 11:547–552

Souriau C, Chiche L, Irving R, Hudson P (2005) New binding specificities derived from min-23, a small cystine-stabilized peptidic scaffold. Biochemistry 44:7143–7155

Souslova EA, Belousov VV, Lock JG, Stromblad S, Kasparov S, Bolshakov A, Pinelis VG, Labas YA, Lukyanov S, Mayr LM et al (2007) Single fluorescent protien-based ca^{2+} sensors with increased dynamic range. BMC Biotechnol 7:37

Staiano M, Scognamiglio V, Mamone G, Rossi M, Parracino A, Rossi M, D'Auria S (2006) Glutamine-binding protein from escherichia coli specifically binds a wheat gliadin peptides. 2. Resonance energy transfer studies suggest a new sensing approach for an easy detection of wheat gliadin. J Proteome Res 5:2083–2086

Stec B (2006) Plant thionins – the structural perspective. Cell Mol Life Sci 63:1370–1385

Stumpp MT, Forrer P, Binz HK, Pluckthun A (2003) Designing repeat proteins: modular leucine-rich repeat protein libraries based on the mammalian ribonuclease inhibitor family. J Mol Biol 332:471–487

Sumner JP, Westerber NM, Stoddard AK, Hurst TK, Cramer M, Thompson RB, Fierke CA, Kopelman R (2006) Dsred as a highly sensitive, selective, and reversible fluorescence-based biosensor for both cu^{+} and cu^{2+} ions. Biosens Bioelectron 21:1302–1308

Sun Y-J, Rose J, Wang B-C, Hsiao C-D (1998) The structure of glutamine-binding protein complexed with glutamine at 1.94 a resolution: compaisons with other amino acid binding proteins. J Mol Biol 278:219–229

Thomas KJ, Sherman DB, Amiss TJ, Andalus SA, Pitner B (2006) A long-wavelength fluorescent glucose biosensor based on bioconjugates of galactose/glucose binding protein and nile red derivatives. Diabetes Technol Ther 8:261–268

Tolosa L, Gryczynski I, Eichhorn LR, Dattelbaum JD, Castellano FN, Rao G, Lakowicz JR (1999) Glucose sensor for low-cost lifetime-based sensing using a genetically engineered protein. Anal Biochem 267:114–120

Trakhanov S, Vyas NK, Luecke H, Kristensen DM, Ma J, Quiocho FA (2005) Ligand-free and -bound structures of the binding protein (livj) of the escherichia coli abc leucine/isoleucine/valine transport system: trajectory and dynamics of the interdomain rotation and ligand specificity. Biochemistry 44:6597–6608

Tran T, Engfeldt T, Orlova A, Sandstrom M, Feldwisch J, Abrahmsen L, Wennborg A, Tolmachev V, Karlstrom AE (2007) Tc-99 m-maeee-z(her2: 342), an affibody molecule-based tracer for the detection of her2 expression in malignant tumors. Bioconjug Chem 18:1956–1964
Tsien RY (1998) The green fluorescent protein. Annu Rev Biochem 67:509–544
Vercillo NC, Herald KJ, Fox JF, Der BS, Dattelbaum JD (2007) Analysis of ligand binding to a ribose biosensor using site-directed mutagenesis and fluorescence spectroscopy. Protein Sci 16:362–368
Vermersch PS, Lemon DD, Tesmer JJ, Quiocho FA (1991) Sugar-binding and crystallographic studies of an arabinose-binding protein mutant (met108leu) that exhibits enhanced affinity and altered specificity. Biochemistry 30:6861–6866
Vita C, Drakopoulou I, Vizzavona J, Rochette S, Martin L, Menez A, Roumestand C, Yang YS, Ylisastigui L, Benjouad A et al (1999) Rational engineering of a miniprotein that reproduces the core of the cd4 site interacting with hiv-1 envelope glycoprotein. Proc Natl Acad Sci USA 96:13091–13096
Vita C, Roumestand C, Toma F, Menez A (1995) Scorpion toxins as natureal scaffolds for protein engineering. Proc Natl Acad Sci USA 92:6404–6408
Voigt CA, Gordon B, Mayo SL (2000) Trading accuracy for speed: a quantitative comparison of search algorithms in protein sequence design. J Mol Biol 299:789–803
Vopel S, Muhlbach H, Skerra A (2005) Rational engineering of a fluorescein-binding anticalin for improved ligand affinity. Biol Chem 386:1097–1104
Wang Y, Botvinick EL, Zhao Y, Berns MW, Usami S, Tsien RY, Chien S (2005) Visualizing the mechanical activation of src. Nature 434:1040–1045
Wang Y, Shyy JY, Chien S (2008) Fluorescence proteins, live-cell imaging, and mechanobiology: seeing is believing. Ann Rev Biomed Eng 10:1–38
Wang Z, Luecke H, Yao N, Quiocho FA (1997) A low energy short hydrogen bond in very high resolution structures of protein receptor–phosphate complexes. Nat Struct Biol 4:519–522
Ward WW (2006) Biochemical and physical properties of green fluorescent protein. Methods Biochem Anal 47:39–65
Weber G (1952a) Polarization of the fluorescence of macromolecules 1. Theory and experimental methods. Biochem J 51:145–155
Weber G (1952b) Polarization of the fluorescence of macromolecules 2. Fluorescent conjugates of ovalbumin and bovine serum albumin. Biochem J 51:155–167
West JB (ed) (1990) Best and taylor's physiological basis of medical practice. Williams & Wilkins, Baltimore, MD
White SH, Wimley WC, Selsted ME (1995) Structure, function, and membrane integration of defensins. Curr Opin Struct Biol 5:521–527
Yang W, Wilkins AL, Ye Y, Liu Z, Li S, Urbauer JL, Hellinga HW, Kearnery A, van der Merwe PA, Yang JJ (2005) Design of a calcium-binding protein with desired structure in a cell adhesion molecules. J Am Chem Soc 127:2085–2093
Yao N, Ledvina PS, Choudhary A, Quiocho FA (1996) Modulation of a salt link does not affect binding of phosphate to its specific active transport receptor. Biochemistry 35: 2079–2085
Yao N, Trakhanov S, Quiocho FA (1994) Refined 1.89-a structure of the histidine-binding protein complexed with histidine and its relationship with many other active transport/chemosensory proteins. Biochemistry 33:4769–4779
Ye K, Schultz JS (2003) Genetic engineering of an allosterically based glucose indicator protein for continuous glucose monitoring by fluorescence resonance energy transfer. Anal Chem 75:3119–3127
Zahnd C, Pecorari F, Straumann N, Wyler E, Pluckthun A (2006) Selection and characterization of her2 binding-designed ankyrin repeat proteins. J Biol Chem 281:35167–35175
Zahnd C, Wyler E, Schwenk JM, Steiner D, Lawrence MC, McKern NM, Pecorari F, Ward CW, Joos TO, Pluckthun A (2007) A designed ankyrin repeat protein evolved to picomolar affinity to her2. J Mol Biol 369:1015–1028

Zhang J, Ma Y, Taylor SS, Tsien RY (2001) Genetically encoded reporters of protein kinase a activity reveal impact of substrate tethering. Proc Natl Acad Sci USA 98:14997–145002

Zhao A, Xue Y, Zhang J, Gao B, Feng J, Mao C, Zheng L, Liu M, Wang F, Wang H (2004) A conformation-constrained peptide library based on insect defensin a. Peptides 25:629–635

Zhou LQ, Cass AE (1991) Periplasmic binding protein based biosesnors. 1. Preliminary study of maltose binding protein as sensing element for maltose biosensor. Biosens Bioelectron 6: 445–450

Zukin RS, Hartig PR, Koshland DE (1979) Effect of an induced conformational change on the physical properties of two chemotactic receptor molecules. Biochemistry 18:5599–5605

Zukin RS, Hartig PR, Koshland DE Jr (1977) Use of a distant reporter group as evidence for a conformational change in a sensory receptor. Proc Natl Acad Sci USA 74:1932–1936

Chapter 14
Biosensing Systems Based on Genetically Engineered Whole Cells

Anjali Kumari Struss, Patrizia Pasini, and Sylvia Daunert

Abstract Genetically engineered whole cells as biosensing systems in biosensors have been employed, in the past two decades, for the detection of a variety of analytes. In addition to being rapid, specific/selective, and sensitive, these whole-cell-based sensing systems provide information pertaining to the analyte bioavailability. This information is particularly important to study the effect of harmful/toxic chemicals on living systems. The whole cells used for designing and developing cell-based sensing systems can be either prokaryotic or eukaryotic in nature. These intact prokaryotic or eukaryotic cells can be genetically engineered to recognize the analytes of interest and respond with the production of a measurable signal in a dose-dependent manner. Generally, prokaryotic bacterial whole-cell sensing systems are developed by introducing a plasmid construct with a reporter gene fused to a promoter, which is induced by a target analyte through a regulatory protein. Similarly, a receptor, which is activated by a target analyte, is coupled with a reporter gene for the development of genetically modified eukaryotic cell-based biosensing systems. The most commonly used reporter proteins in whole-cell biosensing include luminescent proteins, such as bacterial and firefly luciferases; green fluorescent protein along with its variants; and β-galactosidase. The analytes that can be detected using genetically manipulated whole-cell sensing systems range from general toxicants and cell stress factors to specific analytes, such as metals, metalloids, organic pollutants, sugars, drugs, and bacterial signaling molecules. In order to develop self-contained sensing devices based on recombinant whole-cell sensing systems, preservation, miniaturization, and portability are important issues that need to be addressed.

Keywords Biosensors · Genetic engineering · Whole-cells · Reporter genes · Luminescence · Regulatory protein · Receptor · Promoter

S. Daunert (✉)
Department of Chemistry, University of Kentucky, Lexington, KY 40506-0055, USA
e-mail: daunert@uky.edu

M. Zourob (ed.), *Recognition Receptors in Biosensors*,
DOI 10.1007/978-1-4419-0919-0_14,

Abbreviations

Lux	Bacterial luciferase
$FMNH_2$	Reduced flavin mononucleotide
FMN	Flavin mononucleotide
Luc	Firefly luciferase
ATP	Adenosine triphosphate
Ruc	Renilla luciferase
GFP	Green fluorescent protein
V. fischeri	*Vibrio fischeri*
EDCs	Endocrine disrupting compounds
RE	Response element
S. cerevisiae	*Saccharomyces cerevisiae*
hAR	Human androgen receptor
ARE	Androgen response element
CALUX	Chemical-activated luciferase expression
AhR	Aryl hydrocarbon receptor
DRE	Dioxin-responsive element
CCD	Charge-coupled device
PDMS	Poly-(dimethylsiloxane)
BBIC	Bioluminescent-bioreporter integrated circuit
IC	Integrated circuit

14.1 Introduction

A biosensor is a sensing device that integrates a biological sensing element with a signal transducer. The biological sensing element is capable of recognizing a chemical change in its environment and the transducer produces a measurable signal in response to the change, in a dose-dependant manner. The biological sensing elements can be designed employing tissues, whole cells, organelles, or specific biomolecules. These biomolecules may range from enzymes, antibodies, receptors, nucleic acids, and ion channels to lipid bilayers. The biological sensing element confers selectivity to a biosensor, whereas the sensitivity is determined by the signal produced by the transducer. The transduction signal can be based on either mass, calorimetric, piezoelectric, magnetic, electrochemical, or optical principles. Although isolated biomolecules as sensing elements in biosensors confer high specificity/selectivity and fast detection times, these fail to provide information regarding the bioavailability of target analytes in a sample (Daunert et al. 2000). Additionally, the extraction and purification of biomolecules are tedious and expensive, and their stability is a major concern. Thus, whole cells and tissues serve as better models of sensing elements when there is a need for understanding the interaction of analytes present in a sample with cells of higher organisms. Although tissues as biological recognition elements provide valuable physiological and functional information, they are relatively less stable, difficult to grow and manage, and have limited applications compared with whole cells.

14.2 Whole-Cell-Based Sensing Systems

These are biosensing systems that employ whole cells as the biological recognition element. These whole-cell sensing systems usually contain a plasmid that is genetically engineered by coupling a sensing element with a reporter gene. Generally, in inducible-operon-based whole-cell sensing systems, the sensing element in the plasmid is comprised of an operon promoter region and a regulatory gene, which encodes for a regulatory protein. Additionally, the expression of the reporter system is under the transcriptional control of the operon promoter. In the presence of the target analyte, the regulatory protein recognizes the analyte and forms a regulatory protein–analyte complex. The resulting complex then binds to the promoter, thus triggering the expression of all the genes that are under the transcriptional control of the promoter, which includes the reporter gene. The expressed reporter protein produces a measureable signal as a function of the concentration of analyte present in the test sample. Consequently, the amount of analyte present in a test sample can be assayed directly by measuring the intensity of the signal produced by the reporter and employing appropriate calibration plots. In most eukaryotic whole-cell sensing systems, receptors are coupled with a reporter system. Specifically, the sensing element is comprised of a receptor, which recognizes and binds the analyte, and specific gene sequences, which are activated by the receptor–analyte complex and control expression of the reporter gene. In the next section, we discuss the reporters that are commonly used in the design of whole-cell sensing systems.

14.3 Luminescent Reporter Genes

Reporter genes employed in whole-cell sensing encode for proteins that produce a measurable signal in the presence of target analytes. The concentration of target analytes present in a test sample can be correlated with the measurable signal produced by reporter proteins. As mentioned in an earlier section, the sensitivity of any biosensing system is determined by the signal produced by the transducer, specifically, the reporter protein in the case of whole-cell sensing systems. Thus, a reporter gene should have certain characteristics to be employed effectively in a biosensing system. First of all, it should reflect the physical or chemical change that occurs in the presence of a target analyte (Daunert et al. 2000). The signal intensity should be such that it is easily quantifiable even in the presence of endogenous proteins or enzyme activities (Daunert et al. 2000; Feliciano et al. 2006a).

Specifically, the signal-to-noise ratio should be high for the analyte to be successfully detected and quantified. Moreover, the procedure for the measurement of the signal produced by the reporter protein should to be simple and fast (Daunert et al. 2000). The expression of the reporter protein should be controlled only by the regulatory gene of interest, and the amount of reporter protein expressed should correspond to the amount of reporter gene transcribed (Bronstein et al. 1996; Feliciano et al. 2006a). There should not be any other protein that is similar to the reporter protein in the whole cells used (Bronstein et al. 1996). The reporter protein should also be environmentally safe and nontoxic to the sensing cells (Feliciano et al. 2006a). Furthermore, a suitable reporter should be selected depending on the cell type used for biosensing, gene expression, nature of target analytes, desired sensitivity, dynamic range and response time, availability of a measurement instrument, and the desired application (Daunert et al. 2000; Feliciano et al. 2006a). In that regard, luminescent reporter genes proved to be suitable to meet these requirements. Particularly, the sensitivity and rapidity of luminescence measurements made the genes encoding for luminescent proteins the reporter genes of choice in the design and development of whole-cell sensing systems. The next section presents a brief summary of the reporter genes that are most commonly used in the design of whole-cell sensing systems, including bioluminescent, chemiluminescent, and fluorescent reporter genes.

Enzymes that catalyze a light-emitting reaction are generically known as luciferases and their substrates are referred to as luciferins (Meighen 1993; Baldwin et al. 1995). The most commonly used luciferases are from the bacteria *Vibrio harveyi*, *Vibrio fischeri*, *Photorhabdus luminescens*, *Photobacterium leiognathi*, and *Xenorhabdus luminescens*, the firefly *Photinus pyralis*, and the sea pansy *Renilla reniforms* (Meighen 1993; Feliciano et al. 2006a; Galluzzi and Karp 2006). Bacterial luciferase (Lux) is encoded by *luxA* and *luxB* genes of the *luxCDABE* cassette. The *luxC*, *luxD*, and *luxE* genes of the *lux* operon of the bioluminescent bacteria code for the enzymes involved in the synthesis and recycling of the long-chain aldehyde substrate for the luciferase. Although *luxA* and *luxB* genes are sufficient for the production of bioluminescence in bacteria, the presence of *luxCDE* voids the need for external addition of substrate for the luciferase (Meighen 1993; Galluzzi and Karp 2006). Thus, the use of the whole operon, *luxCDABE*, in a plasmid construct is advantageous, especially for real-time monitoring applications. Bacterial luciferase catalyzes the oxidation of reduced flavin mononucleotide ($FMNH_2$) and a long-chain aldehyde to flavin mononucleotide (FMN) and the corresponding carboxylic acid in the presence of molecular oxygen (Table 14.1). The wavelength and quantum yield of the light emitted are 490 nm and ~0.1, respectively (Köhler et al. 2000). Bacterial luciferase allows one of the lowest limits of detection (down to attomole levels) for cell-based biosensors (Van Dyk and Rosson 1998). However, the application of bacterial luciferases as reporter proteins is limited because of their heat-labile nature (Naylor 1999). On the other hand, the eukaryotic firefly luciferase (Luc) is commonly employed as a reporter protein in mammalian cell-based sensing systems (Billard and DuBow 1998). The firefly luciferase converts a heterocyclic carboxylic acid (D-luciferin) to

Table 14.1 Reporter proteins, their catalyzed reactions and methods of detection

Reporter protein	Detection method
Bacterial luciferase $FMNH_2$ + R-CHO + O_2 → FMN + R-COOH + H_2O + $h\nu$ (490 nm)	Bioluminescence
Firefly luciferase Firefly luciferin + ATP + O_2 + Mg^{2+} → oxyluciferin + AMP + P_i + $h\nu$ (560 nm)	Bioluminescence
Sea pansy luciferase Sea pansy coelenterazine + O_2 → coelenteramide + CO_2 + $h\nu$ (480 nm)	Bioluminescence
Aequorin Jellyfish coelenterazine + O_2 + Ca^{2+} → coelenteramide + CO_2 + $h\nu$ (~465 nm)	Bioluminescence
Green fluorescent protein Post-translational formation of an internal chromophore Excitation at ~395 nm and emission at 509 nm	Fluorescence
β-Galactosidase Hydrolysis of β-galactosides	Chemiluminescence Fluorescence Colorimetric Electrochemical

oxyluciferin in the presence of ATP, Mg^{2+}, and molecular oxygen with light emission at 560 nm. The quantum yield of the bioluminescent reaction is reported to be approximately tenfold higher than that of the bacterial luciferase reactions. Additionally, it also offers a broad dynamic range (up to 7–8 orders of magnitude) of measurement (Naylor 1999). An interesting feature of eukaryotic luciferases is their different emission maximum wavelengths ranging from 547 to 593 nm, which allows for their usage in multiple analyte detection in a single assay (Tauriainen et al. 1999). As for bacterial luciferases, there is no endogenous activity of the firefly luciferase in mammalian cells (Naylor 1999; Daunert et al. 2000). However, the use of firefly luciferase as a reporter protein has a drawback because of the need for external addition of substrate and ATP for the production of bioluminescence (Billard and DuBow 1998; Yagi 2007). Another luciferase, the sea pansy luciferase (Ruc), catalyzes a bioluminescent reaction involving oxidative decarboxylation of the substrate coelenterazine (Lorenz et al. 1991). In vivo, this reaction is followed by energy transfer to an accessory protein, specifically the green fluorescent protein (GFP), resulting in the emission of green light (509 nm). However, in vitro, in the absence of GFP, it results in the emission of blue light (480 nm) and can be employed as such for analytical purposes. Similar to the firefly luciferase, the sea pansy luciferase allows limits of detection down to subattomole levels and a broad dynamic range (five orders of magnitude) of measurement (Gu et al. 2004; Feliciano et al. 2006a). Use of the sea pansy luciferase along with the firefly luciferase has gained popularity in assays where dual reporter proteins are required (Gu et al. 2004; Feliciano et al. 2006a; Fan and Wood 2007). However, both of these reporter proteins require an external addition of substrate, which makes their usage in field applications impractical (Gu et al. 2004; Feliciano et al. 2006a).

Aequorin is a bioluminescent calcium-binding protein from the jellyfish *Aequorea victoria* (Ohmiya and Hirano 1996). Upon calcium binding, in the presence of molecular oxygen, it undergoes a conformational change that determines the oxidation of its chromophore, coelenterazine, and the consequent light output. Aequorin emits light at ~465 nm and the quantum yield of the bioluminescent reaction is 0.16–0.18. As a labeling protein, it offers high sensitivity (down to subattomole levels) with low background interferences (Feliciano et al. 2006a). Although aequorin is used as a label in immunoassays and gene probe assays, and as a probe to detect intracellular Ca^{2+} levels, it has scarcely been employed as a reporter in cell-based sensing applications (Shimomura 2005; Feliciano et al. 2006a).

The green fluorescent protein is another protein from the jellyfish *Aequorea victoria* (Müller-Taubenberger and Anderson 2007). In vivo, GFP acts as an accessory protein and fluoresces because of an energy transfer from the Ca^{2+} activated aequorin in the jellyfish, emitting green light at 509 nm (Miyawaki 2002). The main advantage of GFP as a reporter protein is that it does not require any external additions of substrate or cofactors (Naylor 1999). In the wild-type GFP, the fluorescent light can be generated by using long UV wavelengths, since the excitation wavelength of GFP is ~395 nm (Billard and DuBow 1998). The quantum yield of the reaction is ~0.88 (Köhler et al. 2000). GFP has been mutated to obtain various GFP variants with altered spectral properties and structural stabilities, and improved signal intensity (Billard and DuBow 1998; Müller-Taubenberger and Anderson 2007; Yagi 2007). These GFP variants include mutants with blue, cyan, and yellow emission wavelengths, thus facilitating their use for imaging and multianalyte detection studies. Disadvantages associated with GFP include its relatively low sensitivity, interference due to background signal from sample components, requirement of molecular oxygen, and potential toxicity toward cells (Daunert et al. 2000; Feliciano et al. 2006a; Yagi 2007). Also, the formation of the active fluorophore of GFP takes longer than the core protein, thus limiting its online applications (Gu et al. 2004).

β-Galactosidase is a protein encoded by the *lacZ* gene of *Escherichia coli* (Gu et al. 2004). This enzyme has extensively been employed as a reporter and a variety of substrates are available for measuring its activity. Depending on the type of substrate chosen, the signal produced can be detected using colorimetric, electrochemical, or chemiluminescent methods (Yagi 2007). The most commonly used substrates are: *O*-nitrophenyl β-D-galactopyranoside for colorimetric detection, *p*-aminophenyl β-D-galactopyranoside for the electrochemical assay, and 1,2-dioxetane derivatives for chemiluminescence measurements. The chemiluminescence-based detection of β-galactosidase activity allows the highest sensitivity, enabling a limit of detection in the order of femtograms of enzyme, and a dynamic range up to 4–5 orders of magnitude (Bronstein et al. 1996; Gu et al. 2004; Yagi 2007). The major limitations of using β-galactosidase as a reporter protein lie in the requirements of adding a substrate and lysing the cells to make the enzyme available to the substrate (Gu et al. 2004). Moreover, β-galactosidase endogenous activity has been reported in mammalian cell sensing systems (Naylor 1999).

14.4 Advantages and Disadvantages of Whole-Cell-Based Sensing Systems

There are several advantages in using whole-cell-based sensing systems, for instance, their ability to withstand environmental conditions such as wide ranges of temperatures, pH, and ionic strengths, especially with respect to isolated proteins (Daunert et al. 2000; Feliciano et al. 2006a). Moreover, whole cells can be prepared easily and rapidly in laboratory conditions and are relatively inexpensive and easy to handle compared with conventional physicochemical analytical methods. These whole-cells can be freeze-dried or frozen in protective media and stored for a long time (Galluzzi and Karp 2006). The most significant advantage of using whole-cell sensing systems lies in their ability to provide information on the bioavailability, general toxicity, and genotoxicity of the analyte (Belkin 2003). Conventional analytical methods are incapable of detecting the amount of analyte that is actually available to living cells, and therefore tend to overestimate the amount that is harmful to the cells (Hansen and Sørensen 2001). Additionally, advances in molecular biology and recombinant DNA technology can be exploited to construct various custom-engineered plasmids that are specific to analytes of interest. Thus, whole-cell sensing systems can be utilized to detect analytes in fields as diverse as medicine, molecular biology, biochemistry, biotechnology, environmental science, defense, and food processing, among others. Furthermore, whole-cell sensing systems facilitate easy, fast, cost-effective, sensitive, selective, and accurate detection of target analytes and are amenable to high-throughput screening, miniaturization and automation (Daunert et al. 2000; Feliciano et al. 2006a; Harms et al. 2006). Thus, they serve as promising tools for field applications.

Although whole-cell sensing systems serve as a great tool for detecting analytes in various types of samples, there are also a few disadvantages. For instance, there are complexities associated with using living whole cells, such as potential interferences due to cell molecules, metabolic pathways, and possible redundancies of transcriptional control in the sensing cells (Daunert et al. 2000; Galluzzi and Karp 2006; Harms et al. 2006). Furthermore, high background signals may arise when fluorescence is used as the detection principle, because of the presence of fluorescent biomolecules in the cells. Additionally, environmental factors and components of the sample matrix may have adverse effects on the sensing potential of the sensing system. To circumvent this problem, a bacterial whole-cell-based sensing system, with an internal correction mechanism of the analytical response, was developed in our laboratory (Mirasoli et al. 2002). It was achieved by introducing an additional reporter gene under control of a constitutive promoter, which provided a reference signal of the analytical performance of the sensing cells. Furthermore, whole-cell sensing systems have longer response times, compared with those based on isolated biomolecules, since it takes time for the analytes to pass through the barrier of the cell wall and reach the recognition elements as well as for the cell machinery to produce the reporter protein in response to the analytes (D'Souza 2001). There is also an added variability in response among batches of the prepared sensing cells (Galluzzi and Karp 2006). Moreover, the detection range offered by whole-cell sensing systems is relatively narrow compared with cell-free assays.

14.5 Bacterial Whole-Cell Sensing Systems

Microbial cells, particularly prokaryotic microorganisms such as bacteria, have extensively been used as whole-cell sensing systems. There are several benefits in using genetically engineered bacterial whole cells as tools for sensing analytes because of their inherent fast growth rate, low maintenance, availability of a variety of growth substrates, and cost-effectiveness. Most importantly, bacteria are amenable to genetic manipulation, and hence, they can easily be custom-engineered for sensing a specific analyte. Moreover, bacterial whole-cell sensing systems generally do not require either expensive instrumentation or highly trained personnel. Examples of bacteria that are commonly employed in the design and development of whole-cell sensing systems include *Escherichia coli*, *Bacillus subtilis*, *Rhizobium leguminosarum*, and *Pseudomonas* species.

It is significant to note that the detection of analytes using whole-cell sensing systems is generally based on assays where the expression of the reporter protein is either constitutive or inducible (Gu et al. 2004). In constitutive systems, such as that originally observed in *Vibrio fischeri* and subsequently employed for toxicity evaluation purposes, the bioluminescent luciferase reporter protein is constantly expressed (Hansen and Sørensen 2001) (Fig. 14.1). However, in the presence of toxic chemicals/substances the bioluminescence signal is decreased/inhibited. The inhibition in the bioluminescent signal produced by the luciferase protein is proportional to the overall toxic/inhibitory effect of the sample present in the local environment of the whole sensing cells. The main limitation of such constitutive systems is their lack of specificity for an analyte. Although these sensing systems provide information about the overall toxicity of a sample, they fail to provide insight into the molecular nature of the sample components or into their mechanisms of action on the system.

In inducible operon-based whole-cell sensing systems, the sensing cells generally contain an engineered plasmid that expresses regulatory and reporter proteins. The expression of a reporter protein is under the control of a specific promoter. The promoter, incorporated in the genetically modified plasmid, is specific to an analyte or a group of analytes that are of interest. The promoter may be under negative or positive control of the regulatory protein. In the first case, in the absence of analyte, the regulatory protein binds to the promoter and inhibits gene transcription; in the presence of analyte, the regulatory protein binds the analyte and releases itself from the promoter, thus activating transcription of the gene sequences that are under the control of the promoter. In the second case, in the absence of analyte, the regulatory protein is not bound to the promoter and gene transcription is inhibited; in the presence of analyte, the regulatory protein binds the analyte, then binds the

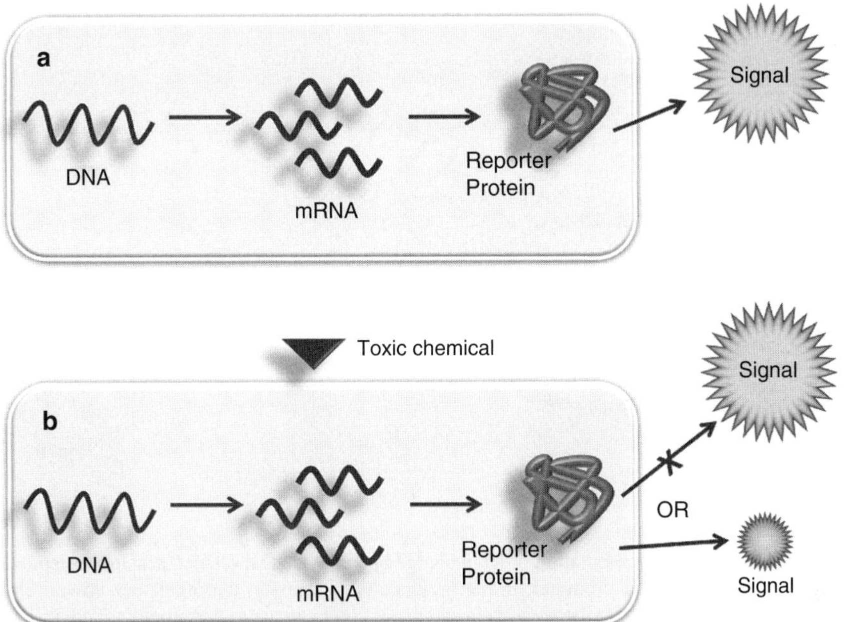

Fig. 14.1 Schematic representation of a constitutive operon-based whole-cell sensing system. (**a**) In the absence of toxic chemicals the bioluminescent reporter protein is constantly expressed. (**b**) In the presence of toxic chemicals/substances the bioluminescence signal is decreased/ inhibited

promoter, and activates transcription of the gene sequences that are under the control of the promoter. In both the cases, these gene sequences include the reporter gene. Thus, the presence of an analyte(s) elicits an increase in the signal produced by the reporter protein that is expressed under the control of the promoter specific to the analyte(s). A schematic of an inducible-operon-based sensing system with positive control of the promoter is shown in Fig. 14.2. Commonly used reporter genes for the development of whole-cell sensing systems are discussed in an earlier section. Inducible operon-based whole-cell sensing systems are specific to an analyte or selective for a class of analytes. Additionally, they may provide insight into the mechanisms of action of the sample components on the sensing system.

In both constitutive and inducible operon-based whole-cell sensing systems, the amount of reporter protein expressed is dependant on the concentration of analyte (s) present in the test sample. Thus, a change in the measureable signal of the reporter protein can be correlated with the presence and amount of analyte(s) present in the local environment of the whole-cell sensing system. A variety of analytes can be detected using these genetically engineered whole bacterial cells. These analytes may range from general toxicants or stress factors to specific analytes or classes of analytes.

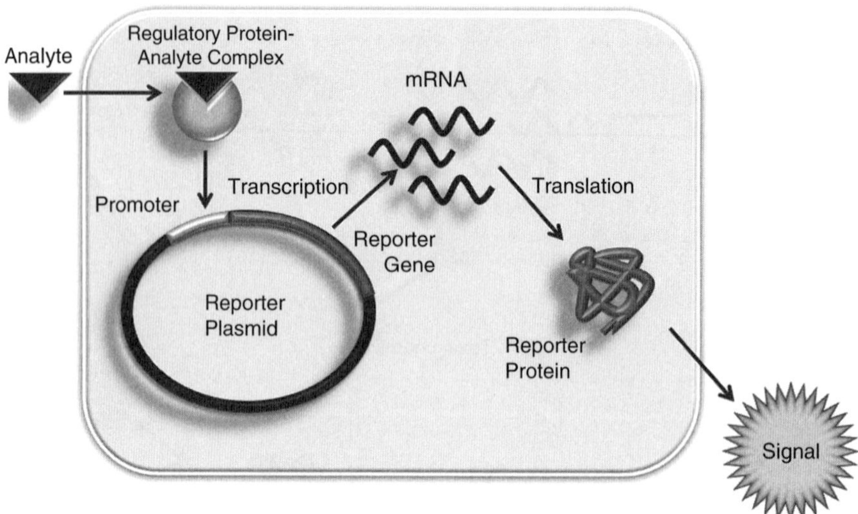

Fig. 14.2 Schematic representation of an inducible operon-based whole-cell sensing system with positive regulation of the promoter. In the presence of an analyte, the regulatory protein binds the analyte, and then the regulatory protein–analyte complex binds to the promoter and activates gene transcription. The expression of the reporter system is under the transcriptional control of the operon promoter

14.5.1 General Toxicants

Whole-cell sensing systems may be employed to evaluate the overall toxic effect of compounds and/or samples. In this case, the detection of toxic chemicals using whole-cell sensing systems is based on the general toxicity of a sample to the sensing cells rather than the chemical identity of a compound (Belkin 2003). In most studies, the reporter gene cassette *luxCDABE*, from the marine bacterium *V. fischeri*, is employed for studying the general toxicity of a sample. One of such commercial systems, the Microtox® toxicity test, is based on the measurement of the decreased light emission by the marine bacterium *V. fischeri* in the presence of toxic samples (Bulich and Isenberg 1981; Qureshi et al. 1998). As discussed in an earlier section, this bacterium is naturally bioluminescent, but in the presence of a sample that contains toxic chemicals/substances, the naturally produced light is decreased or inhibited. The major drawback of using such a system is that the inhibition in bioluminescence can be caused by any condition that can cause a decrease in the metabolic activity of the sensing cells. For instance, these conditions include temperature, pH, and salt and metal ion concentrations among others (Qureshi et al. 1998; Sørensen et al. 2006). Other microorganisms have been used for general toxicity evaluation, the main reason being that *V. fischeri* is not an ideal representative of environments other than the marine one. These microorganisms are either naturally constitutively luminescent or can be genetically manipulated to luminesce constitutively. Few examples of such microorganisms are *Escherichia coli* (Horsburgh et al. 2002; Tiensing et al. 2002), *Pseudomonas fluorescens* (Tiensing et al. 2002), *Photobacterim leiognathi* (Ulitzur et al. 2002), *Synechocystis* PCC6803 (Shao et al. 2002), and *Janthinobacterium lividum* (Cho et al. 2004). Along with these prokaryotic microorganisms, naturally bioluminescent eukaryotic fungi, specifically, *Armillaria mellea* and *Mycena citricolor*, have also been employed for toxicity studies (Weitz et al. 2002). Additionally, the yeast *Saccharomyces cerevisiae* has also been cited for such analyses (Hollis et al. 2000). These nonspecific whole-cell sensing systems have been used to evaluate the toxicity of a variety of samples, ranging from industrial effluents, wastewater, drinking water, sediments, and soil to industrial organic and inorganic samples (Qureshi et al. 1998; Horsburgh et al. 2002; Shao et al. 2002; Tiensing et al. 2002; Weitz et al. 2002; Cho et al. 2004; Gu et al. 2004).

14.5.2 Stress Factors

These are factors that may cause damage to cells. They may range from heat shock, cold shock, oxidative damage, DNA damage, osmotic stress, and growth inhibition to membrane synthesis, among others (Gu et al. 2004; Vollmer et al. 2004). The cells respond by activating processes that protect against the stress factors, and repair the damages caused by them (Sørensen et al. 2006). These responses are genetically encoded in stress-response regulons. The whole-cell sensing systems developed and used to detect these stress factors are semispecific in nature, since their promoters are inducible by different stimuli/stress factors (Köhler et al. 2000; Vollmer et al. 2004). Additionally, since these promoters belong to global regulatory circuits, the stress response is generally linked to complex cascades.

The SOS regulon/DNA damage repair is one of the well-studied stress responses in bacteria (Fig. 14.3). Single-stranded DNA fragments may result when the bacterial double-stranded chromosomal DNA is damaged. The resulting single-stranded DNA fragments in the bacteria then activate the protein RecA (Gu et al. 2004). The activated RecA protein then binds to the single-stranded DNA fragment and the resulting protein–DNA complex causes the autocatalytic cleavage of the repressor protein LexA (Yagi 2007). The cleavage of the LexA repressor protein results in the expression of SOS regulon genes that are downstream of the promoters that are repressed by the protein. The expression of these genes results in proteins that are involved in DNA repair processes. Whole-cell sensing systems bearing plasmids that incorporate a reporter gene with the *recA* gene and its promoter have been developed by several researchers for detecting genotoxicity (Nunoshiba and Nishioka 1991; Vollmer et al. 1997; Gu et al. 2000; Polyak et al. 2001; Kostrzynska et al. 2002; Bartolome et al. 2003; Mitchell and Gu 2004a, b; Norman et al. 2005). Following similar approaches, a number of whole-cell sensing systems have also been developed where plasmids are

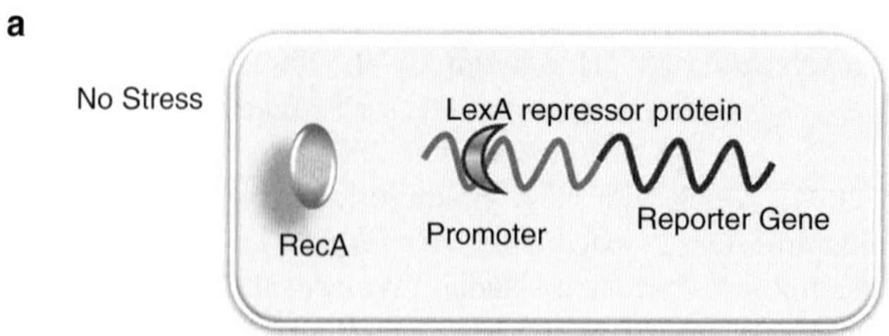

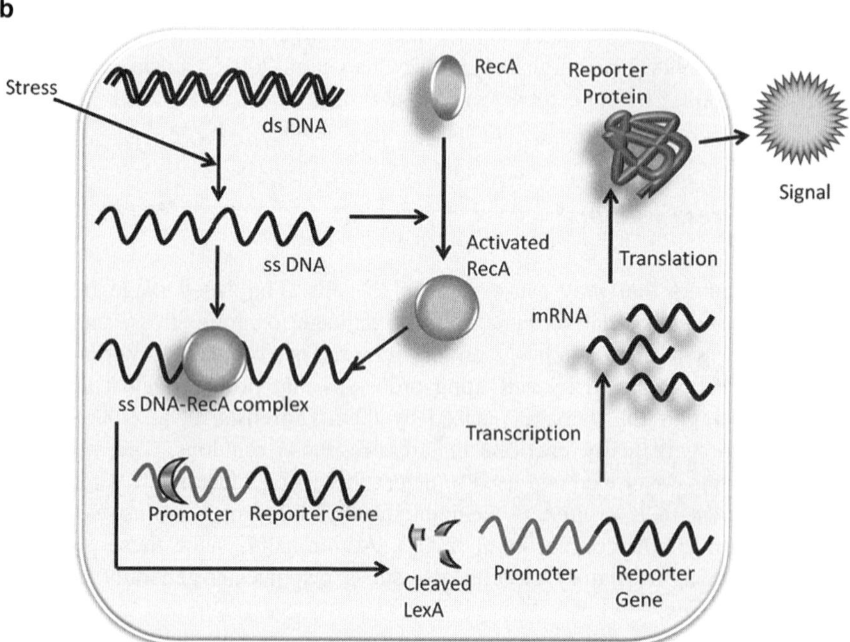

Fig. 14.3 Schematic representation of DNA damage stress-inducible biosensing system. (**a**) In the absence of DNA damage, the repressor protein LexA is bound to the promoter region, thus preventing the expression of reporter protein. The DNA is present in the cells in its natural double stranded (ds) form. (**b**) In the presence of DNA damage stress, single stranded DNA (ss DNA) fragments may result, which then activate the protein RecA. The activated RecA protein–single stranded DNA complex then causes the autocatalytic cleavage of the repressor protein LexA. This cleavage results in the expression of SOS regulon genes that are downstream of the promoters that are repressed by the LexA protein. In genetically modified whole-cell-based sensing systems the reporter gene is under control of this promoter. Thus, in the presence of DNA damage stress, the reporter protein produces a measurable signal in a dose-dependant manner

constructed by coupling reporter genes to promoters that are inducible by various stress factors (Gu et al. 2004; Vollmer et al. 2004; Gu 2005; Sørensen et al. 2006). Examples of stress-inducible whole-cell sensing systems are reported in Table 14.2.

Table 14.2 Semi-specific/stress-inducible biosensing systems

Stress	Regulatory gene	Reporter gene	Host organism	References
Heat shock	*grpE*	*rfp*	*E. coli*	Hever and Belkin (2006)
		gfp-luciferase	EPC	Molina et al. (2002)
		luxCDABE	*E. coli*	Van Dyk et al. (1995), Gu and Gil (2001), Premkumar et al. (2001, 2002a, b), Choi and Gu (2002, 2003), Gu et al. (2002a), Molina et al. (2002), Kim and Gu (2003), Pedahzur et al. (2004), Lee et al. (2005)
	rpoH	*gfp*	*E. coli*	Cha et al. (1999)
Cold shock	*cspA*	*lacZ*	*E. coli*	Vasina et al. (1998), Shapiro and Baneyx (2002, 2007)
		luc	*E. coli*	Shapiro and Baneyx (2007)
Oxidative damage	*katG*	*luxCDABE*	*E. coli*	Premkumar et al. (2001, 2002b), Choi and Gu (2002, 2003), Kim and Gu (2003), Kim et al. (2003), Lee and Gu (2003), Mitchell and Gu (2004a, b), Pedahzur et al. (2004), Lee et al. (2005)
	sodA	*luxCDABE*	*E. coli*	Lee and Gu (2003), Lee et al. (2005)
	micF	*luxCDABE*	*E. coli*	Premkumar et al. (2001), Pedahzur et al. (2004)
DNA damage	*recA*	*gfp*	*E. coli*	Kostrzynska et al. (2002), Premkumar et al. (2002a), Bartolome et al. (2003), Kuang et al. (2004), Mitchell and Gu (2004a, b), Norman et al. (2005), Hever and Belkin (2006)
		luxCDABE	*S. typhimurium*	Davidov et al. (2000), Baumstark-Khan et al. (2005)
		luxCDABE	*E. coli*	Vollmer et al. (1997), Min et al. (1999), Davidov et al. (2000), Gu et al. (2000, 2002a), Polyak et al. (2001), Premkumar et al. (2001, 2002b), Choi and Gu (2002, 2003), Kim and Gu (2003), Kim et al. (2003), Pedahzur et al. (2004), Lee et al. (2005)
		lacZ	*E. coli*	Vollmer et al. (1997)
		luxCDABE	*P. aeruginosa*	Elasri and Miller (1999)
	umuD	*luc*	*E. coli*	Tani et al. (2004), Maehana et al. (2006)
	cda	*gfp*	*E. coli*	Norman et al. (2005, 2006)
	cda	*gfp*	*S. typhimurium*	Østergaard et al. (2007)
	cda	*luxCDABE*	*E. coli*	Ptitsyn et al. (1997)

(*continued*)

Table 14.2 (continued)

Stress	Regulatory gene	Reporter gene	Host organism	References
	nrdA	*luxCDABE*	*E. coli*	Hwang et al. (2008)
	RAD54	*gfp*	*S. cerevisiae*	Walmsley et al. (1997), Afanassiev et al. (2000), Knight et al. (2004)
	RNR2	*gfp*	*S. cerevisiae*	Afanassiev et al. (2000)
	GADD45α	*egfp*	TK6	Hastwell et al. (2006)
Osmotic stress	*osmY*	*lacZ*	*E. coli*	Biran et al. (1999)
		luxCDABE	*E. coli*	Van Dyk et al. (1998)
Growth inhibition	*uspA*	*luxCDABE*	*E. coli*	Van Dyk et al. (1995)
Membrane synthesis/ damage	*fabA*	*luxCDABE*	*E. coli*	Premkumar et al. (2001, 2002b), Choi and Gu (2002, 2003), Gu et al. (2002a), Kim and Gu (2003), Kim et al. (2003), Pedahzur et al. (2004), Lee et al. (2005)

E. coli(Escherichia coli); *S. typhimurium* (*Salmonella typhimurium*); *P. aeruginosa* (*Pseudomonas aeruginosa*); *S. cerevisiae* (*Saccharomyces cerevisiae*); *EPC* Epithelioma papulosum of cyprini, a fish cell line; *TK6* Thymidine kinase heterozygote, a human lymphoblastoid cell line

14.5.3 Specific Analytes or Groups of Analytes

A good amount of work has been performed by several research groups, including ours, toward designing and developing whole-cell sensing systems for the detection of specific analytes or classes of analytes. These whole-cell sensing systems are specific to their target analytes or selective for classes of analytes, in contrast to the systems that are designed for the detection of general toxicants or stress factors. The plasmids of such whole-cell sensing systems are based on genetic constructs that contain an inducible promoter, and regulatory and reporter genes, in which the expression of the reporter protein is under the control of the promoter. The presence of the target analyte in a sample induces the promoter thus causing the expression of the genes downstream, which also include the reporter gene(s). There is a plethora of whole-cell sensing systems that have been designed to specifically or selectively detect various analytes, including metals, metalloids, organic pollutants, sugars, bacterial signaling molecules, and drugs, among others. A representative list of whole-cell sensing systems for inorganic and organic compounds is shown in Table 14.3.

The possibility of developing specific/selective bacterial cell sensing systems is related to the presence of certain operons, induced by selected compounds, in bacteria found in nature. As an example, bacteria in metal-containing environments have evolved with protective operons that confer resistance toward the metals present (Ramanathan et al. 1997a). The genes in these operons encode proteins involved in a variety of metal detoxification mechanisms, thus allowing the bacteria to withstand levels of metals that would be toxic otherwise. Importantly, the

Table 14.3 Specific/selective biosensing systems

Analyte	Regulatory unit	Reporter gene	References
Inorganics			
Nitrate	*nar*	*gfp*	Taylor et al. (2004)
	nblA	*luxAB*	Mbeunkui et al. (2002)
	nar	*luxCDABE*	Prest et al. (1997)
Phosphate	*pho*	*luxCDABE*	Dollard and Billard (2003)
Metals/metalloids			
Arsenite, arsenate and antimonite	*ars*	*luc*	Tauriainen et al. (1997, 1998)
		luxAB	Ramanathan et al. (1997b), Stocker et al. (2003)
		gfp	Roberto et al. (2002), Stocker et al. (2003), Rothert et al. (2005)
		lacZ	Stocker et al. (2003), Date et al. (2007)
Cadmium, lead and antimonite	*cad*	*luc*	Tauriainen et al. (1998)
Cadmium, lead and zinc	*znt*	*rs-gfp*	Shetty et al. (2003)
Cadmium, lead, mercury and zinc	*znt*	*luxCDABE*	Riether et al. (2001)
Chromate	*chr*	*luxCDABE*	Peitzsch et al. (1998)
Cobalt and nickel	*cnr*	*luxCDABE*	Peitzsch et al. (1998), Tibazarwa et al. (2001)
Copper	*copA*	*luxCDABE*	Holmes et al. (1994), Vulkan et al. (2000), Riether et al. (2001)
	cup1	*gfp*	Shetty et al. (2004)
	cup1	*lacZ*	Lehmann et al. (2000)
Copper and silver	*copA*	*luxCDABE*	Riether et al. (2001), Stoyanov et al. (2003)
Iron	*pvd*	*gfp*	Joyner and Lindow (2000)
Mercury	*mer*	*luc*	Virta et al. (1995), Ivask et al. (2001), Hakkila et al. (2002)
		luxCDABE	Selifonova et al. (1993), Holmes et al. (1994), Lyngberg et al. (1999), Hansen and Sørensen (2000), Rasmussen et al. (2000), Hakkila et al. (2002, 2004)
		luxAB	Omura et al. (2004)
		gfp	Hansen and Sørensen (2000), Hakkila et al. (2002)
		lacZ	Hansen and Sørensen (2000)
	znt	*lacZ*	Biran et al. (2003)
Mercury and organomercurials	*mer*	*luc*	Ivask et al. (2001), Hakkila et al. (2004)
Zinc	*smtB*	*egfp*	Date et al. (2007)
Zinc and trace levels of copper and cadmium	*smtA*	*luxCDABE*	Erbe et al. (1996)

(*continued*)

Table 14.3 (continued)

Analyte	Regulatory unit	Reporter gene	References
Organics			
Alkanes	*alk*	*luxAB*	Sticher et al. (1997)
Benzene, toulene, ethylbenzene, xylene (BTEX)	*sep*	*luxCDABE*	Phoenix et al. (2003)
	xyl	*luc*	Kobatake et al. (1995), Ikariyama et al. (1997), Willardson et al. (1998)
	tbu	*gfp*	Stiner and Halverson (2002)
	tod	*luxCDABE*	Applegate et al. (1998)
	lac	*luxCDABE*	Gil et al. (2002)
Aromatic compounds	*sep*	*luxCDABE*	Phoenix et al. (2003)
	xyl	*luc*	Kobatake et al. (1995)
Chlorobenzoates	*fcbA*	*luxCDABE*	Rozen et al. (1999)
Naphthalene/ salicylate	*nah*	*luxCDABE*	King et al. (1990), Heitzer et al. (1994), Ripp et al. (2000)
		lacZ	Cebolla et al. (1997)
Phenolic compounds	*mop*	*luxCDABE*	Abd-El-Haleem et al. (2002)
	dmp	*lacZ*	Wise and Kuske (2000)
		luxAB	Shingler and Moore (1994)
		luxCDABE	Leedjärv et al. (2006)
Polychlorinated biphenyls	*bph*	*luxCDABE*	Layton et al. (1998)
		lacZ	Xu et al. (2005)
Hydroxylated polychlorinated biphenyls	*hbp*	*luxAB*	Turner et al. (2007)
Dihydroxylated (chloro-) biphenyls	*clc*	*lacZ*	Feliciano et al. (2006b)
Chlorocatechols	*clc*	*lacZ*	Guan et al. (2000, 2002)
Amino acids, malate	*his*	*luxAB*	Corthier et al. (1998)
	ald	*luxAB*	Corthier et al. (1998)
	mlc	*luxAB*	Corthier et al. (1998)
	Δ*lysA*	*gfp*	Chalova et al. (2008)
Lactose	*lacI*	*bfp2*	Shrestha et al. (2001)
L-Arabinose	*araC*	*gfp*	Shetty et al. (1999)
		gfpuv	Shrestha et al. (2001)
Tetracyclines	*tetA*	*luxCDABE*	Korpela et al. (1998), Kurittu et al. (2000a, 2000b)
		luc	Kurittu et al. (2000a)
N-Acyl homoserine lactones	*rhl*	*luxCDABE*	Winson et al. (1998), Kumari et al. (2006)
	las	*luxCDABE*	Winson et al. (1998), Kumari et al. (2006)
	lux	*luxCDABE*	Winson et al. (1998), Yan et al. (2007)
		gfp	Andersen et al. (2001)
	tra	*lacZ*	Fuqua and Winans (1996), Shaw et al. (1997)

expression of the resistance genes is tightly regulated and induced by the presence of specific metals in the environment of the cell. The promoter and regulatory genes of these protective operons, from such bacteria in nature, can be cloned and used for the development of inducible promoter based whole-cell sensing systems by coupling to a reporter gene. Bacterial operons involved in metabolic and signaling pathways have also been exploited for the construction of bacterial cell biosensing systems.

14.6 Eukaryotic Whole-Cell Sensing Systems

Although bacterial whole cells have several advantages as sensing systems, their being prokaryotes makes them less relevant to eukaryotes. For instance, it was demonstrated by Glover et al. that chemicals, which are known to be toxic to eukaryotes and were also detected as such by genetically modified eukaryotic yeast sensing cells, were assessed as nontoxic by prokaryotic biosensing systems (Hollis et al. 2000). In addition, physiologically relevant molecules, such as certain organic environmental pollutants and drugs, exert their effects on humans and other mammals by binding to cell receptors and activating specific signaling pathways. Therefore, genetically modified eukaryotic cells, which express such receptors and a reporter gene under control of specific receptor-activated gene sequences, are particularly suited for the design and development of whole-cell sensing systems for the detection of these molecules. Additionally, these sensing systems can provide information pertaining to the biological activity and bioavailability of target compounds in eukaryotic cells. Endocrine disrupting compounds (EDCs) are a group of such chemicals for which eukaryotic cell sensing systems have been developed. EDCs are often found as environmental contaminants and include harmful substances with diverse chemical structures, which mimic or interfere with endocrine/hormonal compounds, such as androgens and estrogens, by binding to their respective receptors, thus disrupting their natural physiological function and, consequently, having adverse effects on humans and animals (Birnbaum 1995; Gu et al. 2002b). Although the toxic effects of EDCs have been demonstrated using bacterial whole-cell sensing systems, these failed to provide physiologically relevant information pertaining to their estrogenic effects (Gu et al. 2002b). Thus, eukaryotic microorganisms, such as yeast, and mammalian cells may serve as better tools to detect these compounds and understand their effects on higher organisms at the gene level.

The eukaryotic cells employed for sensing may naturally express the receptors of interest. Alternatively, yeast or mammalian cells may be genetically modified in order to express such receptors (Walmsley and Keenan 2000). These receptors are then coupled with a reporter system for the development of genetically engineered eukaryotic whole-cell sensing systems. This is accomplished by inserting in the cells a plasmid containing a reporter gene and the gene sequence of a specific response element (RE) that controls the expression of the reporter protein. In the

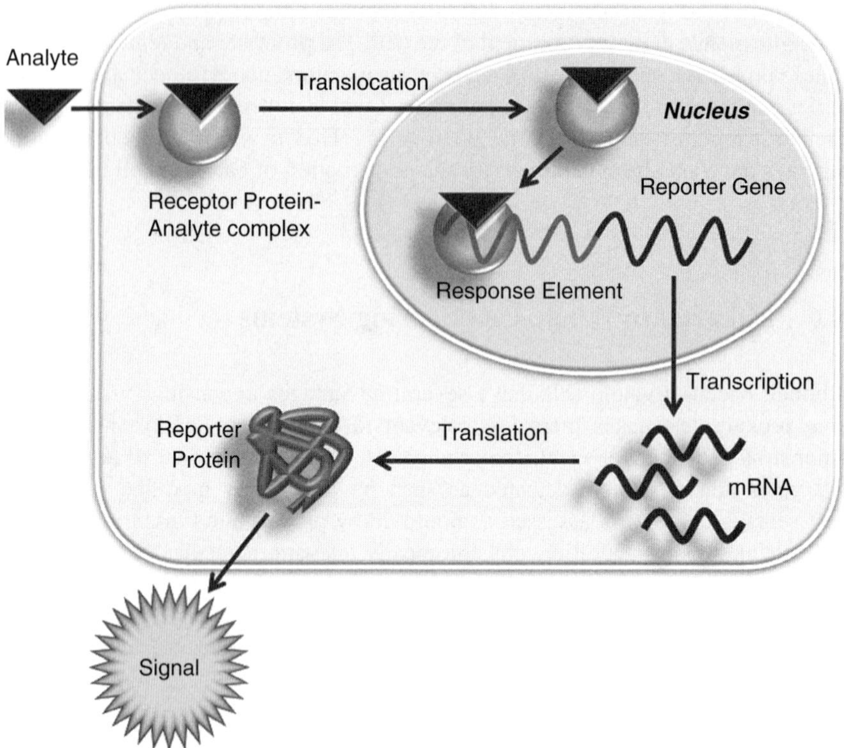

Fig. 14.4 Schematic representation of a receptor-based eukaryotic cell sensing system. In the presence of a ligand, the receptor binds the ligand, resulting in a ligand–receptor complex. The complex then moves into the cell nucleus and binds to the RE DNA sequences, resulting in the expression of the reporter protein in a dose-dependent manner

presence of a receptor ligand, the receptor binds the ligand and moves into the cell nucleus. In the nucleus, the ligand–receptor complex then binds to the RE DNA sequences, resulting in the expression of the reporter protein in a dose-dependent manner (Fig. 14.4). Importantly, receptors from a given mammalian organism can be expressed in yeast or mammalian cells from different species. This allowed for the development of genetically modified sensing systems based on yeast or nonhuman mammalian cells that carry human receptors.

Although both genetically modified yeast and mammalian cell sensing systems may provide functionally and physiologically relevant information, there are certain practical advantages in using yeast cells. For instance, yeast cells grow faster and are easier to culture and manage than mammalian cells. Specifically, yeast is less prone to contamination, thus requiring less strict sterile culture conditions; unlike mammalian cells, it does not need several replication cycles before reaching optimal metabolic activity; furthermore, recombinant yeast has longer shelf-life. Additionally, yeast cells do not have any human-relevant endogenous receptors and thus, lack complex interactions of the receptor of interest with other receptors, which may occur in mammalian cells (Routledge and Sumpter 1996; Legler et al. 2002). The absence of endogenous hormones also reduces the background signal, consequently increasing the signal-to-noise ratio (Legler et al. 2002). On the other hand, assays using recombinant mammalian cell lines can be more sensitive than yeast-based systems. In fact, lower permeability of analytes through the thick cell walls of yeast cells is normally observed, although this problem could be overcome by employing permeabilized yeast cells (Gonchar et al. 1998; Smutok et al. 2007). Most importantly, mammalian cells may provide better insight into the mechanisms of action of the test compounds.

14.6.1 Yeast Cells

Genetically manipulated yeast cells as sensing systems share most of the advantages with bacterial whole cells, for instance, high growth rate and ease of genetic manipulation (Walmsley and Keenan 2000). Moreover, the yeast cells are robust and can be grown on a variety of substrates and easily immobilized. The genome of the yeast commonly used in the development of eukaryotic biosensing systems, specifically, *Saccharomyces cerevisiae*, has also been completely sequenced (Goffeau et al. 1996). Due to their thick cell wall and complex eukaryotic protective mechanisms, yeast cells can withstand extremes of pH, osmotic shock, temperature, and solvents exposure (Parry 1999).

Various yeast-cell-based sensing systems expressing receptors, such as human estrogen, androgen, and olfactory receptors, have been designed and utilized for the detection of a variety of chemicals (Routledge and Sumpter 1996; Minic et al. 2005; Roda et al. 2006; Nakama et al. 2007; Radhika et al. 2007).

As an example, Michelini and coworkers employed recombinant *S. cerevisiae* to develop a whole-cell sensing system for androgen-like compounds. Yeast cells were genetically modified to express the human androgen receptor (hAR) and contain the gene sequence androgen response element (ARE) that controlled the expression of the reporter protein (Michelini et al. 2005). The reporter gene used in their study was firefly luciferase. In the presence of androgenic compounds, the hAR bound to the compound moves into the nucleus and binds the ARE DNA sequences, resulting in the expression of the reporter protein. The activity of the luciferase protein was simply measured by the light emitted after addition of its substrate D-luciferin. The yeast-based sensing system responded to test compounds in a dose-dependent manner and exhibited a limit of detection down to subnanomolar levels.

14.6.2 Mammalian Cells

Genetically modified mammalian cells as biosensing systems may be used to provide responses relevant to humans, for instance, information regarding the bioavailability of target analytes in a sample and their effects (Pancrazio et al. 1999).

Whole-cell sensing systems based on genetically modified mammalian cells have been developed to detect and study the biochemical effects of compounds, including environmental chemicals and drugs, that interact with cell receptors, such as the estrogen, androgen, aryl hydrocarbon, and G protein-coupled receptor, among others Table 14.4.

In one of the first studies, Murk and coworkers developed a chemical-activated luciferase expression (CALUX) assay to detect dioxin-like compounds in sediments and pore water (Murk et al. 1996). They used a recombinant rat hepatoma cell line, which naturally expressed the aryl hydrocarbon receptor (AhR) and was genetically modified by transfection with a plasmid construct bearing the dioxin-responsive element (DRE) sequences and the firefly luciferase gene under transcriptional control of the DREs. Dioxin-like compounds bind to the AhR and then the resulting dioxin-like compound/AhR complex moves to the cell nucleus where it binds to DREs and activates the reporter expression in a dose-dependent fashion (Brown et al. 1992; Murk et al. 1996). Thus, the dioxin-like activity can be calculated by measuring the light emitted by the firefly luciferase reporter protein after addition of its substrate D-luciferin. A detection limit in the order of femtomoles of the reference dioxin compound was reported. The CALUX assay was demonstrated to be a rapid, sensitive tool for the assessment of relative toxicity of AhR-active compounds in sediments and pore water. The CALUX system is now commercially available as a kit, which is commonly used for analyzing the dioxin-like toxic activity of various environmental samples (http://www.biodetectionsystems.com). On a similar line, CALUX systems have also been developed for the detection of estrogen- and androgen-like compounds (Sonneveld et al. 2005, 2006).

14.7 Integration of Genetically Engineered Whole-Cell Sensing Elements in Biosensors

The capabilities of whole-cell sensing systems would be exploited to their full extent if they could be routinely employed for on-site monitoring. For that, miniaturization and portability are critical issues. Work has been done in the direction of integrating genetically engineered whole cells in a biosensor, in order to achieve self-contained sensing devices. Toward this goal, a variety of approaches have been utilized where genetically modified whole cells are supported on solid surfaces. These approaches include the whole sensing cells immobilization on portable devices, such as optical fibers and high-throughput platforms, including multiwell microtiter plates, microfluidics platforms, microarrays, and integrated circuits. For instance, recombinant whole cells have been immobilized using matrices such as agar medium (Gil et al. 2000, 2002; Mbeunkui et al. 2002; Mitchell and Gu 2006), "washed agar" (Mbeunkui et al. 2002), agarose (Mbeunkui et al. 2002; Tani et al.

Table 14.4 Receptor-based eukaryotic cell sensing systems

Cell type	Receptor	Reporter gene	Analyte	References
Yeast cell				
S. cerevisiae	hER	*lacZ*	Estrogen	Klein et al. (1998)
	hER	*lacZ*	Estrogen-like compounds	Routledge and Sumpter (1996), Coldham et al. (1997), Fang et al. (2000), Roda et al. (2006)
	hAR	*luc*	Androgen-like compounds	Michelini et al. (2005)
	hER-α	*lacZ*	Organic biocides	Nakama et al. (2007)
	hAR, hER-α, hER-β	*luc*	Androgen- and estrogen-like compounds	Leskinen et al. (2005)
	AhR	*luc*	Dioxin-like compounds	Leskinen et al. (2008)
	hER, hAR, hPR	*lacZ*	Endocrine disruptive chemicals	Gaido et al. (1997)
	17OR	*luc*	Helional (an odorant)	Minic et al. (2005)
	Olfr226	*gfp*	2,4-Dinitrotoluene	Radhika et al. (2007)
Mammalian cell line				
HGELN, MVLN	hER-α	*luc*	Natural and synthetic estrogens	Gutendorf and Westendorf (2001), Villeneuve et al. (2002)
H4IIE	AhR	*luc*	Dioxin-like compounds	Murk et al. (1996), Villeneuve et al. (2002), Leskinen et al. (2008)
MELN	ER-α	*luc*	Estrogen-like compounds	Pillon et al. (2005)
CHO-K1	AR	*luc*	Androgen- and anti-androgen-like compounds	Satoh et al. (2005)
MDA-kb2	AR, GR	*luc*	Androgen- and anti-androgen-like compounds	Satoh et al. (2005)
CHO-K1	AR	*luc*	Endocrine disruptive chemicals	Satoh et al. (2008)
MVLN	hER-α	*luc*	Endocrine disruptive chemicals	Satoh et al. (2008)
CV-1	AR	*luc*	Androgen-like compounds	Takeyoshi et al. (2003)
Hepa 1c1c7	AhR	*luc*	Dioxin-like compounds	Joung et al. (2007)
U2-OS	AR, ER-α	*luc*	Androgen- and estrogen-like compounds	Sonneveld et al. (2005, 2006)

hER human estrogen receptor; *hAR* human androgen receptor; *hER-α* human estrogen receptor alpha; *hER-β* human estrogen receptor beta; *AhR* aryl hydrocarbon receptor; *GR* glucocorticoid receptor; *hPR* human progestrone receptor; *17OR* rat 17 olfactory receptor; *Olfr226* rat olfactory receptor; *HGELN* transgenic human cell line; *MVLN* recombinant human breast carcinoma cells; *H4IIE* rat hepatocarcinoma cell line; *MELN* transgenic human cell line; *CHO-K1* chinese hamster ovary cell line; *MDA-kb2* human breast cancer cell line; *CV-1* cercopithecus aethiops kidney cell line; *Hepa 1c1c7* mouse hepatic cells; *U2-OS* human osteoblastic osteosarcoma cell line

2004; Maehana et al. 2006), alginates (Heitzer et al. 1994; Corbisier et al. 1999; Elasri and Miller 1999; Davidov et al. 2000; Ripp et al. 2000; Polyak et al. 2001; Leth et al. 2002; Hakkila et al. 2004), sol–gel derived silicates (Premkumar et al. 2002a, b; Mitchell and Gu 2006) and polycarbonate or dialysis membranes (Ikariyama et al. 1997). Gu and coworkers demonstrated that immobilizing recombinant bacterial whole cells in a thin agar layer with the addition of glass beads enhanced their sensitivity to toxic gaseous chemicals (Gil et al. 2002). It was shown that the addition of glass beads improved gas diffusion in the matrix resulting in an increased sensitivity of the whole-cell sensing system. Premkumar et al. used an antibody-based immobilization of recombinant *E. coli* cells on gold or silica glass surfaces (Premkumar et al. 2001). The immobilization was achieved by glutaraldehyde-based anchoring of nonspecific anti-*E. coli* antibodies on the gold/silica glass surface. The binding of the recombinant *E. coli* cells on these antibodies attached to the surface resulted in immobilized whole sensing cells. Stocker et al. used genetically engineered *E. coli*, with β-galactosidase as the reporter protein, responsive to arsenite and arsenate (Stocker et al. 2003). The sensing bacterial cells, in a drying protectant solution, were dried on a paper strip. The paper strip was then kept in the test solution for 30 min for the development of color. This protocol provided a fast and convenient method for the visual monitoring of the target analyte in a dose-dependant manner, in a test sample, as shown in Fig. 14.5. Leth et al. immobilized recombinant *Alcaligenes eutrophus* cells in an alginate matrix compatible with fiber optics for the detection of copper ions (Leth et al. 2002). In another approach, the recombinant whole cells were directly immobilized at the tip of fiber optic cables. Such an approach allows the integration of whole cell sensing elements in the measurement system itself. Ikariyama et al. immobilized recombinant *E. coli* cells on the fiber-optic end with either dialysis membrane or polycarbonate membrane (Ikariyama et al. 1997). Using such fiber optics cell-based biosensor, they successfully monitored benzene-related aromatic compounds. Polyak et al. alginate-immobilized recombinant *E. coli* cells, responsive to genotoxicants, on the exposed core of a fiber optic end (Polyak et al. 2001). Hakkila et al. alginate-immobilized recombinant bacterial cells onto the tip of an optical fiber for the detection of mercury (Hakkila et al. 2004). Walt and coworkers immobilized recombinant *E. coli* cells on the face of an optical fiber that contained a high-density array of microwells (Biran and Walt 2002). The distal end of an optical fiber bundle was etched to yield an array of microwells in such a manner that each microwell accommodates a single cell. It was previously demonstrated by the same group that

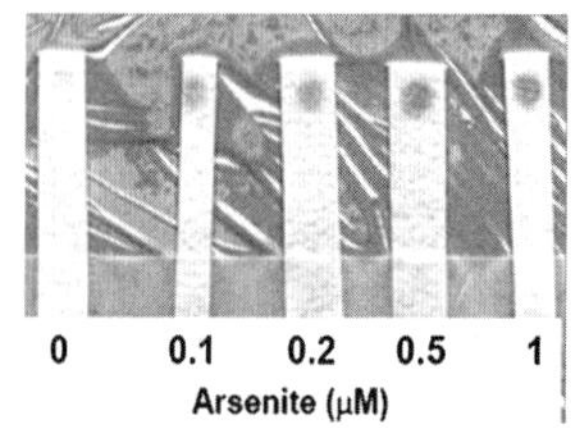

Fig. 14.5 Paper strip biosensor with whole sensing cells exposed to various concentrations of arsenite standard for the visualization of color

such an approach allows for the identification of a specific cell location within the array and correlation of the responses of these cells to the changes in their microenvironment (Taylor and Walt 2000). They fabricated microwell arrays capable of simultaneously monitoring genetically modified whole cells with different reporter genes. Such a microwell array system allowed for the monitoring of expression of reporter gene by a single cell. The light signal produced by individual cells at the end of the optical fiber was captured by a CCD camera. Using the same approach, they developed biosensors with recombinant whole cells in an array on an optical fiber for the detection of mercury (Biran et al. 2003) and genotoxins (Kuang et al. 2004). Van Dyk et al. developed a novel solid-phase system consisting of 689 bacterial strains with nonredundant functional genes, each one of these strains containing the *luxCDABE* cassette as reporter (Van Dyk et al. 2001). These 689 nonredundant functional genes, referred to as LuxArray 1.0, represented 27% of the predicted transcriptional units in *E. coli*. The LuxArray 1.0 reporter strains were printed in high density to porous nylon membranes on agar plates. After 6 h of incubation at 37°C, the membranes were moved to new prewarmed plates containing either only agar medium or agar medium with nalidixic acid, an inhibitor of DNA gyrase. After further incubation, images were collected using a CCD camera for each array, which allowed for a simultaneous monitoring of the expression of the reporter protein in each strain in the presence or absence of nalidixic acid. The effect of nalidixic acid on these 689 reporter strains was monitored by correlating with the pixel density of the images captured by the CCD camera. In another study, Mbeunkui et al. immobilized recombinant cyanobacteria, strain *Synechocystis* PCC 6803, in the wells of a 24-well microtiter plate using "washed agar." This system was used to monitor nitrogen bioavailability in aquatic ecosystems (Mbeunkui et al. 2002). In our lab, we developed a miniaturized whole-cell sensing system for the detection of arsenite/antimonite by incorporating the sensing cells into a microfluidics, microcentrifuge platform in the shape and size of a compact disk, which can be operated in the field (Rothert et al. 2005). This system, which employed recombinant *E. coli*, with gfp_{uv} as the reporter gene, allowed for a rapid and sensitive detection of the target analytes, as well as for the reduction of reagent and sample consumption. Potential for portability and high-throughput screening of samples is envisioned. Tani and coworkers immobilized recombinant *E. coli* cells, using agarose in a grid-like three-dimensional microfluidic network on a chip format (Tani et al. 2004; Maehana et al. 2006). The genetically engineered bacteria, with firefly luciferase as the reporter gene, that were employed were responsive to stress factors. The three-dimensional assembly was fabricated using two poly-(dimethylsiloxane) (PDMS) layers with microchannels separated by a silicon substrate with perforated microwells. The microchannels in the two PDMS layers were connected by the perforated wells on the silicon chip. First, sensing cells in an agarose matrix were introduced into the microchannels of the first PDMS layer and were immobilized in the microchannels as well as the perforated microwells of the Si chip. Second, test samples with substrate for the firefly luciferase protein were introduced in the microchannels of the second PDMS layer, which was sealed on the other side of the silicon chip. The channels of both the PDMS layers connected with each other and with the immobilized bacterial strain at the perforated well. The interaction between the samples and the sensing strains then was detected by measuring the bioluminescence at each of the microwells. In this way, several assays could be performed using various sensing strains at each of the microwells in a single three-dimensional chip. Gil et al. developed a portable sensor kit using recombinant bioluminescent *E. coli* cells for the detection of toxic gases (Gil et al. 2000). They immobilized the sensing bacteria in a polypropylene test tube, using solid agar medium. The end side of the tube was attached to the distal end of a fiber optic probe, which was then connected to a luminometer. In the presence of the chosen toxic chemical, vapors of benzene, the biosensor response was demonstrated to be dose-dependent. Sayler and coworkers developed a portable biosensor, using recombinant *Pseudomonas fluorescens* strain HK44 (Ripp et al. 2000). These bacteria represent the first recombinant microorganisms approved for field testing in the United States and are used to monitor polyaromatic hydrocarbons. The biosensing system was mainly employed for monitoring these organic chemicals in soil. The biosensing unit consisted of an assembly that suspended the recombinant cells immobilized in alginate matrix below a fiber optic cable, which was then connected to a photomultiplier-based system. The fiber optic probe was lowered into acrylonitrile-butadiene styrene pipes that extended to various depths into the soil. In another study, Gu and coworkers developed a portable biosensor, using various strains of freeze-dried recombinant *E. coli* cells (Choi and Gu 2002). They utilized one constitutive and four inducible strains for the detection of general toxicity and stress factors, respectively. The portable biosensor consisted of a vial containing a recombinant bacteria strain, a small light-proof test chamber, and a luminometer. The vial containing the sensing bacteria was connected to the end of a fiber optic cable and was inserted through a hole into the test chamber. The other side of the fiber optic probe was connected to a luminometer. As a further approach, Simpson et al. constructed bioluminescent-bioreporter integrated circuits (BBIC) by interfacing recombinant bioluminescent *Pseudomonas putida* with a complementary metal oxide semiconductor photodiode-integrated circuit (IC) (Simpson et al. 1998; Ripp et al. 2004). The bioluminescent bacteria were immobilized on the IC. In the presence of target chemicals, the bacteria produced luminescence, which was detected, processed, and communicated by the IC part of the BBIC. Such an instrument, which combines a sensing element interfaced on a small measurement device, itself provides for a self-sufficient portable biosensor. This voids the need of bulky instrumentation, thus having great potential for various field-applications (Nivens et al. 2004).

14.8 Concluding Remarks

Genetically engineered whole cells as biosensing systems allow for rapid, specific/selective, and sensitive detection of target analytes, along with providing information on their bioavailability. Additionally, the use of bioluminescent, chemiluminescent, and fluorescent reporter genes in the design of these systems enhances their sensitivity and rapidity. There are numerous whole-cell sensing systems that have been designed and developed to detect various analytes and stress conditions, or to evaluate the overall toxicity of samples. The use of whole-cell sensing systems in laboratory settings has advanced significantly since their first application almost two decades ago (King et al. 1990). These biosensing systems hold great potential as an ideal platform for bio-monitoring applications in the fields of biotechnology, pharmacology, and medical and environmental sciences. In addition to the integration of the sensing cells into miniaturized and portable devices, cell preservation is an important issue that still needs to be addressed. Toward this goal, several preservation methods have been proposed and employed to maintain the sensing cells viable and functional during storage and transport. These methods include freeze-drying, vacuum drying, continuous culture, and encapsulation in either organic or inorganic polymers, such as hydrogels and sol–gels, respectively (Bjerketorp et al. 2006). Recently, in our laboratory, we developed an alternative method of preservation, storage, and transport of sensing bacteria, which is based on the use of spore-forming bacteria as host microorganisms and the generation of spores as preservation elements (Date et al. 2007). Spores are naturally hardy, dormant forms of life, which can be germinated and brought back to metabolically active cells in appropriate conditions. This method was demonstrated to be simple, inexpensive, robust, and appropriate for the long-term storage of sensing bacteria. Moreover, it is expected to facilitate the integration of living cells into portable analytical devices as well.

References

Abd-El-Haleem D, Ripp S, Scott C, Sayler GS (2002) A luxCDABE-based bioluminescent bioreporter for the detection of phenol. J Ind Microbiol Biotechnol 29:233

Afanassiev V, Sefton M, Anantachaiyong T, Barker G, Walmsley R, Wölfl S (2000) Application of yeast cells transformed with GFP expression constructs containing the RAD54 or RNR2 promoter as a test for the genotoxic potential of chemical substances. Mutat Res/Genet Toxicol Environ Mutagen 464:297–308

Andersen JB, Heydorn A, Hentzer M, Eberl L, Geisenberger O, Christensen BB, Molin S, Givskov M (2001) gfp-Based *N*-acyl homoserine-lactone sensor systems for detection of bacterial communication. Appl Environ Microbiol 67:575–585

Applegate BM, Kehrmeyer SR, Sayler GS (1998) A chromosomally based tod-luxCDABE whole-cell reporter for benzene, toluene, ethybenzene, and xylene (BTEX) sensing. Appl Environ Microbiol 64:2730–2735

Baldwin TO, Christopher JA, Raushel FM, Sinclair JF, Ziegler MM, Fisher AJ, Rayment I (1995) Structure of bacterial luciferase. Curr Opin Struct Biol 5:798–809

Bartolome AJ, Ulber R, Scheper T, Sagi E, Belkin S (2003) Genotoxicity monitoring using a 2D-spectroscopic GFP whole cell biosensing system. Sens Actuators B Chem 89:27–32

Baumstark-Khan C, Cioara K, Rettberg P, Horneck G (2005) Determination of geno- and cytotoxicity of groundwater and sediments using the recombinant SWITCH test. J Environ Sci Health A Tox Hazard Subst Environ Eng 40:245–263

Belkin S (2003) Microbial whole-cell sensing systems of environmental pollutants. Curr Opin Microbiol 6:206–212

Billard P, DuBow MS (1998) Bioluminescence-based assays for detection and characterization of bacteria and chemicals in clinical laboratories. Clin Biochem 31:1–14

Biran I, Walt DR (2002) Optical imaging fiber-based single live cell arrays: a high-density cell assay platform. Anal Chem 74:3046–3054

Biran I, Klimentiy L, Hengge-Aronis R, Ron EZ, Rishpon J (1999) On-line monitoring of gene expression. Microbiology 145:2129–2133

Biran I, Rissin DM, Ron EZ, Walt DR (2003) Optical imaging fiber-based live bacterial cell array biosensor. Anal Biochem 315:106–113

Birnbaum LS (1995) Developmental effects of dioxins and related endocrine disrupting chemicals. Toxicol Lett 82–83:743–750

Bjerketorp J, Hakansson S, Belkin S, Jansson JK (2006) Advances in preservation methods: keeping biosensor microorganisms alive and active. Curr Opin Biotechnol 17:43–49

Bronstein I, Martin CS, Fortin JJ, Olesen CE, Voyta JC (1996) Chemiluminescence: sensitive detection technology for reporter gene assays. Clin Chem 42:1542–1546

Brown MM, McCready TL, Bunce NJ (1992) Factors affecting the toxicity of dioxin-like toxicants: a molecular approach to risk assessment of dioxins. Toxicol Lett 61:141–147

Bulich AA, Isenberg DL (1981) Use of the luminescent bacterial system for the rapid assessment of aquatic toxicity. ISA Trans 20:29–33

Cebolla A, Sousa C, de Lorenzo V (1997) Effector specificity mutants of the transcriptional activator NahR of naphthalene degrading pseudomonas define protein sites involved in binding of aromatic inducers. J Biol Chem 272:3986–3992

Cha HJ, Srivastava R, Vakharia VN, Rao G, Bentley WE (1999) Green fluorescent protein as a noninvasive stress probe in resting *Escherichia coli* cells. Appl Environ Microbiol 65:409–414

Chalova V, Zabala-Díaz I, Woodward C, Ricke S (2008) Development of a whole cell green fluorescent sensor for lysine quantification. World J Microbiol Biotechnol 24:353–359

Cho J-C, Park K-J, Ihm H-S, Park J-E, Kim S-Y, Kang I, Lee K-H, Jahng D, Lee D-H, Kim S-J (2004) A novel continuous toxicity test system using a luminously modified freshwater bacterium. Biosens Bioelectron 20:338–344

Choi SH, Gu MB (2002) A portable toxicity biosensor using freeze-dried recombinant bioluminescent bacteria. Biosens Bioelectron 17:433–440

Choi SH, Gu MB (2003) Toxicity biomonitoring of degradation byproducts using freeze-dried recombinant bioluminescent bacteria. Anal Chim Acta 481:229–238

Coldham NG, Dave M, Sivapathasundaram S, McDonnell DP, Connor C, Sauer MJ (1997) Evaluation of a recombinant yeast cell estrogen screening assay. Environ Health Perspect 105:734–742

Corbisier P, van der Lelie D, Borremans B, Provoost A, de Lorenzo V, Brown NL, Lloyd JR, Hobman JL, Csöregi E, Johansson G, Mattiasson B (1999) Whole cell- and protein-based biosensors for the detection of bioavailable heavy metals in environmental samples. Anal Chim Acta 387:235–244

Corthier G, Delorme C, Ehrlich SD, Renault P (1998) Use of luciferase genes as biosensors to study bacterial physiology in the digestive tract. Appl Environ Microbiol 64:2721–2722

D'Souza SF (2001) Microbial biosensors. Biosens Bioelectron 16:337–353

Date A, Pasini P, Daunert S (2007) Construction of spores for portable bacterial whole-cell biosensing systems. Anal Chem 79:9391–9397

Daunert S, Barrett G, Feliciano JS, Shetty RS, Shrestha S, Smith-Spencer W (2000) Genetically engineered whole-cell sensing systems: coupling biological recognition with reporter genes. Chem Rev 100:2705–2738

Davidov Y, Rozen R, Smulski DR, Van Dyk TK, Vollmer AC, Elsemore DA, LaRossa RA, Belkin S (2000) Improved bacterial SOS promoter: lux fusions for genotoxicity detection. Mutat Res/ Genet Toxicol Environ Mutagen 466:97–107

Dollard M-A, Billard P (2003) Whole-cell bacterial sensors for the monitoring of phosphate bioavailability. J Microbiol Methods 55:221–229

Elasri MO, Miller RV (1999) Study of the response of a biofilm bacterial community to UV radiation. Appl Environ Microbiol 65:2025–2031

Erbe JL, Adams AC, Taylor KB, Hall LM (1996) Cyanobacteria carrying ansmt-lux transcriptional fusion as biosensors for the detection of heavy metal cations. J Ind Microbiol Biotechnol 17:80–83

Fan F, Wood KV (2007) Bioluminescent assays for high-throughput screening. Assay Drug Dev Technol 5:127–136

Fang H, Tong W, Perkins R, Soto AM, Prechtl NV, Sheehan DM (2000) Quantitative comparisons of in vitro assays for estrogenic activities. Environ Health Perspect 108:723–729

Feliciano J, Pasini P, Deo SK, Daunert S (2006a) Photoproteins as reporters in whole-cell sensing. Wiley-VCH Verlag GmbH & Co. KGaA, Weinheim, Germany

Feliciano J, Shifen X, Xiyuan G, Lehmler H-J, Bachas LG, Daunert S (2006b) ClcR-based biosensing system in the detection of cis-dihydroxylated (chloro-)biphenyls. Anal Bioanal Chem 385:807–813

Fuqua C, Winans SC (1996) Conserved cis-acting promoter elements are required for density-dependent transcription of *Agrobacterium tumefaciens* conjugal transfer genes. J Bacteriol 178:435–440

Gaido KW, Leonard LS, Lovell S, Gould JC, Babaï D, Portier CJ, McDonnell DP (1997) Evaluation of chemicals with endocrine modulating activity in a yeast-based steroid hormone receptor gene transcription assay. Toxicol Appl Pharmacol 143:205–212

Galluzzi L, Karp M (2006) Whole cell strategies based on lux genes for high throughput applications toward new antimicrobials. Comb Chem High Throughput Screen 9:501–514

Gil GC, Mitchell RJ, Tai Chang S, Bock Gu M (2000) A biosensor for the detection of gas toxicity using a recombinant bioluminescent bacterium. Biosens Bioelectron 15:23–30

Gil GC, Kim YJ, Gu MB (2002) Enhancement in the sensitivity of a gas biosensor by using an advanced immobilization of a recombinant bioluminescent bacterium. Biosens Bioelectron 17:427–432

Goffeau A, Barrell BG, Bussey H, Davis RW, Dujon B, Feldmann H, Galibert F, Hoheisel JD, Jacq C, Johnston M, Louis EJ, Mewes HW, Murakami Y, Philippsen P, Tettelin H, Oliver SG (1996) Life with 6000 genes. Science 274:563–567 546

Gonchar MV, Maidan MM, Moroz OM, Woodward JR, Sibirny AA (1998) Microbial O_2- and H_2O_2-electrode sensors for alcohol assays based on the use of permeabilized mutant yeast cells as the sensitive bioelements. Biosens Bioelectron 13:945–952

Gu M (2005) Environmental biosensors using bioluminescent bacteria. In: Environmental chemistry. pp 691–698. http://dx.doi.org/10.1007/3-540-26531-7_63

Gu BM, Gil GC (2001) A multi-channel continuous toxicity monitoring system using recombinant bioluminescent bacteria for classification of toxicity. Biosens Bioelectron 16:661–666

Gu MB, Min J, LaRossa RA (2000) Bacterial bioluminescent emission from recombinant *Escherichia coli* harboring a recA::luxCDABE fusion. J Biochem Biophys Methods 45:45–56

Gu MB, Gil GC, Kim JH (2002a) Enhancing the sensitivity of a two-stage continuous toxicity monitoring system through the manipulation of the dilution rate. J Biotechnol 93: 283–288

Gu MB, Min J, Kim EJ (2002b) Toxicity monitoring and classification of endocrine disrupting chemicals (EDCs) using recombinant bioluminescent bacteria. Chemosphere 46:289–294

Gu M, Mitchell R, Kim B (2004) Whole-cell-based biosensors for environmental biomonitoring and application. In: Biomanufacturing. pp 269–305. http://dx.doi.org/10.1007/b13533

Guan X, Ramanathan S, Garris JP, Shetty RS, Ensor M, Bachas LG, Daunert S (2000) Chlorocatechol detection based on a clc operon/reporter gene system. Anal Chem 72:2423–2427

Guan X, D'Angelo E, Luo W, Daunert S (2002) Whole-cell biosensing of 3-chlorocatechol in liquids and soils. Anal Bioanal Chem 374:841–847

Gutendorf B, Westendorf J (2001) Comparison of an array of in vitro assays for the assessment of the estrogenic potential of natural and synthetic estrogens, phytoestrogens and xenoestrogens. Toxicology 166:79–89

Hakkila K, Maksimow M, Karp M, Virta M (2002) Reporter Genes lucFF, luxCDABE, gfp, and dsred have different characteristics in whole-cell bacterial sensors. Anal Biochem 301:235–242

Hakkila K, Green T, Leskinen P, Ivask A, Marks R, Virta M (2004) Detection of bioavailable heavy metals in EILATox-Oregon samples using whole-cell luminescent bacterial sensors in suspension or immobilized onto fibre-optic tips. J Appl Toxicol 24:333–342

Hansen LH, Sørensen SJ (2000) Versatile biosensor vectors for detection and quantification of mercury. FEMS Microbiol Lett 193:123–127

Hansen LH, Sørensen SJ (2001) The use of whole-cell biosensors to detect and quantify compounds or conditions affecting biological systems. Microb Ecol 42:483–494

Harms H, Wells M, van der Meer J (2006) Whole-cell living biosensors – are they ready for environmental application? Appl Microbiol Biotechnol 70:273–280

Hastwell PW, Chai L-L, Roberts KJ, Webster TW, Harvey JS, Rees RW, Walmsley RM (2006) High-specificity and high-sensitivity genotoxicity assessment in a human cell line: validation of the GreenScreen HC GADD45a-GFP genotoxicity assay. Mutat Res/Genet Toxicol Environ Mutagen 607:160–175

Heitzer A, Malachowsky K, Thonnard JE, Bienkowski PR, White DC, Sayler GS (1994) Optical biosensor for environmental on-line monitoring of naphthalene and salicylate bioavailability with an immobilized bioluminescent catabolic reporter bacterium. Appl Environ Microbiol 60:1487–1494

Hever N, Belkin S (2006) A dual-color bacterial reporter strain for the detection of toxic and genotoxic effects. Eng Life Sci 6:319–323

Hollis RP, Killham K, Glover LA (2000) Design and application of a biosensor for monitoring toxicity of compounds to eukaryotes. Appl Environ Microbiol 66:1676–1679

Holmes DS, Dubey SK, Gangolli S (1994) Development of biosensors for the detection of mercury and copper ions. Environ Geochem Health 16:229–233

Horsburgh AM, Mardlin DP, Turner NL, Henkler R, Strachan N, Glover LA, Paton GI, Killham K (2002) On-line microbial biosensing and fingerprinting of water pollutants. Biosens Bioelectron 17:495–501

Hwang ET, Ahn J-M, Kim BC, Gu MB (2008) Construction of a nrdA::luxCDABE fusion and its use in *Escherichia coli* as a DNA damage biosensor. Sensors 8:1297–1307

Ikariyama Y, Nishiguchi S, Koyama T, Kobatake E, Aizawa M, Tsuda M, Nakazawa T (1997) Fiber-optic-based biomonitoring of benzene derivatives by recombinant *E. coli* bearing luciferase gene-fused TOL-plasmid immobilized on the fiber-optic end. Anal Chem 69: 2600–2605

Ivask A, Hakkila K, Virta M (2001) Detection of organomercurials with sensor bacteria. Anal Chem 73:5168–5171

Joung KE, Chung YH, Sheen YY (2007) DRE-CALUX bioassay in comparison with HRGC/MS for measurement of toxic equivalence in environmental samples. Sci Total Environ 372: 657–667

Joyner DC, Lindow SE (2000) Heterogeneity of iron bioavailability on plants assessed with a whole-cell GFP-based bacterial biosensor. Microbiology 146:2435–2445

Kim BC, Gu MB (2003) A bioluminescent sensor for high throughput toxicity classification. Biosens Bioelectron 18:1015–1021

Kim BC, Park KS, Kim SD, Gu MB (2003) Evaluation of a high throughput toxicity biosensor and comparison with a *Daphnia magna* bioassay. Biosens Bioelectron 18:821–826

King JMH, DiGrazia PM, Applegate B, Burlage R, Sanseverino J, Dunbar P, Larimer F, Sayler GS (1990) Rapid, sensitive bioluminescent reporter technology for naphthalene exposure and biodegradation. Science 249:778–781

Klein KO, Baron J, Barnes KM, Pescovitz OH, Cutler GB Jr (1998) Use of an ultrasensitive recombinant cell bioassay to determine estrogen levels in girls with precocious puberty treated with a luteinizing hormone-releasing hormone agonist. J Clin Endocrinol Metab 83:2387–2389

Knight AW, Keenan PO, Goddard NJ, Fielden PR, Walmsley RM (2004) A yeast-based cytotoxicity and genotoxicity assay for environmental monitoring using novel portable instrumentation. J Environ Monit 6:71–79

Kobatake E, Niimi T, Haruyama T, Ikariyama Y, Aizawa M (1995) Biosensing of benzene derivatives in the environment by luminescent *Escherichia coli*. Biosens Bioelectron 10:601–605

Köhler S, Belkin S, Schmid RD (2000) Reporter gene bioassays in environmental analysis. Fresenius J Anal Chem 366:769–779

Korpela MT, Kurittu JS, Karvinen JT, Karp MT (1998) A recombinant escherichia coli sensor strain for the detection of tetracyclines. Anal Chem 70:4457–4462

Kostrzynska M, Leung KT, Lee H, Trevors JT (2002) Green fluorescent protein-based biosensor for detecting SOS-inducing activity of genotoxic compounds. J Microbiol Methods 48:43–51

Kuang Y, Biran I, Walt DR (2004) Living bacterial cell array for genotoxin monitoring. Anal Chem 76:2902–2909

Kumari A, Pasini P, Deo SK, Flomenhoft D, Shashidhar H, Daunert S (2006) Biosensing systems for the detection of bacterial quorum signaling molecules. Anal Chem 78:7603–7609

Kurittu J, Karp M, Korpela M (2000a) Detection of tetracyclines with luminescent bacterial strains. Luminescence 15:291–297

Kurittu J, Lonnberg S, Virta M, Karp M (2000b) A group-specific microbiological test for the detection of tetracycline residues in raw milk. J Agric Food Chem 48:3372–3377

Layton AC, Muccini M, Ghosh MM, Sayler GS (1998) Construction of a bioluminescent reporter strain to detect polychlorinated biphenyls. Appl Environ Microbiol 64:5023–5026

Lee HJ, Gu MB (2003) Construction of a sodA::luxCDABE fusion *Escherichia coli*: comparison with a katG fusion strain through their responses to oxidative stresses. Appl Microbiol Biotechnol 60:577–580

Lee JH, Mitchell RJ, Kim BC, Cullen DC, Gu MB (2005) A cell array biosensor for environmental toxicity analysis. Biosens Bioelectron 21:500–507

Leedjärv A, Ivask A, Virta M, Kahru A (2006) Analysis of bioavailable phenols from natural samples by recombinant luminescent bacterial sensors. Chemosphere 64:1910–1919

Legler J, Dennekamp M, Vethaak AD, Brouwer A, Koeman JH, van der Burg B, Murk AJ (2002) Detection of estrogenic activity in sediment-associated compounds using in vitro reporter gene assays. Sci Total Environ 293:69–83

Lehmann M, Riedel K, Adler K, Kunze G (2000) Amperometric measurement of copper ions with a deputy substrate using a novel *Saccharomyces cerevisiae* sensor. Biosens Bioelectron 15:211–219

Leskinen P, Michelini E, Picard D, Karp M, Virta M (2005) Bioluminescent yeast assays for detecting estrogenic and androgenic activity in different matrices. Chemosphere 61:259–266

Leskinen P, Hilscherova K, Sidlova T, Kiviranta H, Pessala P, Salo S, Verta M, Virta M (2008) Detecting AhR ligands in sediments using bioluminescent reporter yeast. Biosens Bioelectron 23:1850–1855

Leth S, Maltoni S, Simkus R, Mattiasson B, Corbisier P, Klimant I, Wolfbeis OS, Csöregi E (2002) Engineered bacteria based biosensors for monitoring bioavailable heavy metals. Electroanalysis 14:35–42

Lorenz WW, McCann RO, Longiaru M, Cormier MJ (1991) Isolation and expression of a cDNA encoding *Renilla reniformis* luciferase. Proc Natl Acad Sci USA 88:4438–4442

Lyngberg OK, Stemke DJ, Schottel JL, Flickinger MC (1999) A single-use luciferase-based mercury biosensor using *Escherichia coli* HB101 immobilized in a latex copolymer film. J Ind Microbiol Biotechnol 23:668–676

Maehana K, Tani H, Kamidate T (2006) On-chip genotoxic bioassay based on bioluminescence reporter system using three-dimensional microfluidic network. Anal Chim Acta 560:24–29

Mbeunkui F, Richaud C, Etienne AL, Schmid R, Bachmann T (2002) Bioavailable nitrate detection in water by an immobilized luminescent cyanobacterial reporter strain. Appl Microbiol Biotechnol 60:306–312

Meighen EA (1993) Bacterial bioluminescence: organization, regulation, and application of the lux genes. FASEB J 7:1016–1022

Michelini E, Leskinen P, Virta M, Karp M, Roda A (2005) A new recombinant cell-based bioluminescent assay for sensitive androgen-like compound detection. Biosens Bioelectron 20:2261–2267

Min J, Kim EJ, LaRossa RA, Gu MB (1999) Distinct responses of a recA::luxCDABE Escherichia coli strain to direct and indirect DNA damaging agents. Mutat Res/Genet Toxicol Environ Mutagen 442:61–68

Minic J, Persuy M-A, Godel E, Aioun J, Connerton I, Salesse R, Pajot-Augy E (2005) Functional expression of olfactory receptors in yeast and development of a bioassay for odorant screening. FEBS J 272:524–537

Mirasoli M, Feliciano J, Michelini E, Daunert S, Roda A (2002) Internal response correction for fluorescent whole-cell biosensors. Anal Chem 74:5948–5953

Mitchell RJ, Gu MB (2004a) Construction and characterization of novel dual stress-responsive bacterial biosensors. Biosens Bioelectron 19:977–985

Mitchell RJ, Gu MB (2004b) An *Escherichia coli* biosensor capable of detecting both genotoxic and oxidative damage. Appl Microbiol Biotechnol 64:46–52

Mitchell RJ, Gu MB (2006) Characterization and optimization of two methods in the immobilization of 12 bioluminescent strains. Biosens Bioelectron 22:192–199

Miyawaki A (2002) Green fluorescent protein-like proteins in reef Anthozoa animals. Cell Struct Funct 27:343–347

Molina A, Carpeaux R, Martial JA, Muller M (2002) A transformed fish cell line expressing a green fluorescent protein–luciferase fusion gene responding to cellular stress. Toxicol In Vitro 16:201–207

Müller-Taubenberger A, Anderson K (2007) Recent advances using green and red fluorescent protein variants. Appl Microbiol Biotechnol 77:1–12

Murk AJ, Legler J, Denison MS, Giesy JP, van de Guchte C, Brouwer A (1996) Chemical-activated luciferase gene expression (CALUX): a novelin vitrobioassay for ah receptor active compounds in sediments and pore water. Fundam Appl Toxicol 33:149–160

Nakama A, Funasaka K, Shimizu M (2007) Evaluation of estrogenic activity of organic biocides using ER-binding and YES assay. Food Chem Toxicol 45:1558–1564

Naylor LH (1999) Reporter gene technology: the future looks bright. Biochem Pharmacol 58: 749–757

Nivens DE, McKnight TE, Moser SA, Osbourn SJ, Simpson ML, Sayler GS (2004) Bioluminescent bioreporter integrated circuits: potentially small, rugged and inexpensive whole-cell biosensors for remote environmental monitoring. J Appl Microbiol 96:33–46

Norman A, Hansen LH, Sørensen SJ (2005) Construction of a ColD cda promoter-based SOS-green fluorescent protein whole-cell biosensor with higher sensitivity toward genotoxic compounds than constructs based on recA, umuDC, or sulA promoters. Appl Environ Microbiol 71:2338–2346

Norman A, Hansen LH, Sørensen SJ (2006) A flow cytometry-optimized assay using an SOS-green fluorescent protein (SOS-GFP) whole-cell biosensor for the detection of genotoxins in complex environments. Mutat Res/Genet Toxicol Environ Mutagen 603:164–172

Nunoshiba T, Nishioka H (1991) 'Rec-lac test' for detecting SOS-inducing activity of environmental genotoxic substances. Mutat Res/DNA Repair 254:71–77

Ohmiya Y, Hirano T (1996) Shining the light: the mechanism of the bioluminescence reaction of calcium-binding photoproteins. Chem Biol 3:337–347

Omura T, Kiyono M, Pan-Hou H (2004) Development of a specific and sensitive bacteria sensor for detection of mercury at picomolar levels in environment. J Health Sci 50:379–383

Østergaard TG, Hansen LH, Binderup M-L, Norman A, Sørensen SJ (2007) The cda Geno-Tox assay: a new and sensitive method for detection of environmental genotoxins, including nitroarenes and aromatic amines. Mutat Res/Genet Toxicol Environ Mutagen 631:77–84

Pancrazio JJ, Whelan JP, Borkholder DA, Ma W, Stenger DA (1999) Development and application of cell-based biosensors. Ann Biomed Eng 27:697–711

Parry JM (1999) Use of tests in yeasts and fungi in the detection and evaluation of carcinogens. IARC Sci Publ (146):471–485

Pedahzur R, Polyak B, Marks RS, Belkin S (2004) Water toxicity detection by a panel of stress-responsive luminescent bacteria. J Appl Toxicol 24:343–348

Peitzsch N, Eberz G, Nies DH (1998) Alcaligenes eutrophus as a bacterial chromate sensor. Appl Environ Microbiol 64:453–458

Phoenix P, Keane A, Patel A, Bergeron H, Ghoshal S, Lau PCK (2003) Characterization of a new solvent-responsive gene locus in *Pseudomonas putida* F1 and its functionalization as a versatile biosensor. Environ Microbiol 5:1309–1327

Pillon A, Servant N, Vignon F, Balaguer P, Nicolas J-C (2005) In vivo bioluminescence imaging to evaluate estrogenic activities of endocrine disrupters. Anal Biochem 340:295–302

Polyak B, Bassis E, Novodvorets A, Belkin S, Marks RS (2001) Bioluminescent whole cell optical fiber sensor to genotoxicants: system optimization. Sens Actuators B Chem 74:18–26

Premkumar JR, Lev O, Marks RS, Polyak B, Rosen R, Belkin S (2001) Antibody-based immobilization of bioluminescent bacterial sensor cells. Talanta 55:1029–1038

Premkumar JR, Sagi E, Rozen R, Belkin S, Modestov AD, Lev O (2002a) Fluorescent bacteria encapsulated in sol–gel derived silicate films. Chem Mater 14:2676–2686

Premkumar RJ, Rosen R, Belkin S, Lev O (2002b) Sol–gel luminescence biosensors: encapsulation of recombinant *E. coli* reporters in thick silicate films. Anal Chim Acta 462:11–23

Prest AG, Winson MK, Hammond JRM, Stewart GSAB (1997) The construction and application of a lux-based nitrate biosensor. Lett Appl Microbiol 24:355–360

Ptitsyn LR, Horneck G, Komova O, Kozubek S, Krasavin EA, Bonev M, Rettberg P (1997) A biosensor for environmental genotoxin screening based on an SOS lux assay in recombinant *Escherichia coli* cells. Appl Environ Microbiol 63:4377–4384

Qureshi AA, Bulich AA, Isenberg DL (1998) Microtox toxicity test systems – where they stand today. CRC Press, Washington, DC

Radhika V, Proikas-Cezanne T, Jayaraman M, Onesime D, Ha JH, Dhanasekaran DN (2007) Chemical sensing of DNT by engineered olfactory yeast strain. Nat Chem Biol 3:325–330

Ramanathan S, Ensor M, Daunert S (1997a) Bacterial biosensors for monitoring toxic metals. Trends Biotechnol 15:500–506

Ramanathan S, Shi W, Rosen BP, Daunert S (1997b) Sensing antimonite and arsenite at the subattomole level with genetically engineered bioluminescent bacteria. Anal Chem 69:3380–3384

Rasmussen LD, Sørensen SJ, Turner RR, Barkay T (2000) Application of a mer-lux biosensor for estimating bioavailable mercury in soil. Soil Biol Biochem 32:639–646

Riether K, Dollard MA, Billard P (2001) Assessment of heavy metal bioavailability using *Escherichia coli* zntAp::lux and copAp::lux-based biosensors. Appl Microbiol Biotechnol 57:712–716

Ripp S, Nivens DE, Ahn Y, Werner C, Jarrell IJ, Easter JP, Cox CD, Burlage RS, Sayler GS (2000) Controlled field release of a bioluminescent genetically engineered microorganism for bioremediation process monitoring and control. Environ Sci Technol 34:846–853

Ripp SA, Daumer KA, Garland JL, Simpson ML, Sayler GS (2004) Remote sensing of microbial volatile organic compounds with a bioluminescent bioreporter integrated circuit. Environ Monit Remediation III 5270:208–213

Roberto FF, Barnes JM, Bruhn DF (2002) Evaluation of a GFP reporter gene construct for environmental arsenic detection. Talanta 58:181–188

Roda A, Mirasoli M, Michelini E, Magliulo M, Simoni P, Guardigli M, Curini R, Sergi M, Marino A (2006) Analytical approach for monitoring endocrine-disrupting compounds in urban waste water treatment plants. Anal Bioanal Chem 385:742–752

Rothert A, Deo SK, Millner L, Puckett LG, Madou MJ, Daunert S (2005) Whole-cell-reporter-gene-based biosensing systems on a compact disk microfluidics platform. Anal Biochem 342:11–19

Routledge EJ, Sumpter JP (1996) Estrogenic activity of surfactants and some of their degradation products assessed using a recombinant yeast screen. Environ Toxicol Chem 15:241–248

Rozen Y, Nejidatl A, Gartemann K-H, Belkin S (1999) Specific detection of *p*-chlorobenzoic acid by *Escherichia coli* bearing a plasmid-borne fcbA'::lux fusion. Chemosphere 38: 633–641

Satoh K, Nonaka R, Ohyama K-, Nagai F (2005) Androgenic and antiandrogenic effects of alkylphenols and parabens assessed using the reporter gene assay with stably transfected CHO-K1 cells (AR-EcoScreen System). J Health Sci 51:557–568

Satoh K, Nonaka R, Ohyama KI, Nagai F, Ogata A, Iida M (2008) Endocrine disruptive effects of chemicals eluted from nitrile-butadiene rubber gloves using reporter gene assay systems. Biol Pharm Bull 31:375–379

Selifonova O, Burlage R, Barkay T (1993) Bioluminescent sensors for detection of bioavailable Hg(II) in the environment. Appl Environ Microbiol 59:3083–3090

Shao CY, Howe CJ, Porter AJR, Glover LA (2002) Novel cyanobacterial biosensor for detection of herbicides. Appl Environ Microbiol 68:5026–5033

Shapiro E, Baneyx F (2002) Stress-based identification and classification of antibacterial agents: second-generation *Escherichia coli* reporter strains and optimization of detection. Antimicrob Agents Chemother 46:2490–2497

Shapiro E, Baneyx F (2007) Stress-activated bioluminescent *Escherichia coli* sensors for antimicrobial agents detection. J Biotechnol 132:487–493

Shaw PD, Ping G, Daly SL, Cha C, Cronan JE Jr, Rinehart KL, Farrand SK (1997) Detecting and characterizing *N*-acyl-homoserine lactone signal molecules by thin-layer chromatography. Proc Natl Acad Sci USA 94:6036–6041

Shetty RS, Ramanathan S, Badr IHA, Wolford JL, Daunert S (1999) Green fluorescent protein in the design of a living biosensing system for L-arabinose. Anal Chem 71:763–768

Shetty RS, Deo SK, Shah P, Sun Y, Rosen BP, Daunert S (2003) Luminescence-based whole-cell-sensing systems for cadmium and lead using genetically engineered bacteria. Anal Bioanal Chem 376:11–17

Shetty RS, Deo SK, Yue L, Daunert S (2004) Fluorescence-based sensing system for copper using genetically engineered living yeast cells. Biotechnol Bioeng 88:664–670

Shimomura O (2005) The discovery of aequorin and green fluorescent protein. J Microsc 217:3–15

Shingler V, Moore T (1994) Sensing of aromatic compounds by the DmpR transcriptional activator of phenol-catabolizing *Pseudomonas* sp. strain CF600. J Bacteriol 176:1555–1560

Shrestha S, Shetty RS, Ramanathan S, Daunert S (2001) Simultaneous detection of analytes based on genetically engineered whole cell sensing systems. Anal Chim Acta 444:251–260

Simpson ML, Sayler GS, Applegate BM, Ripp S, Nivens DE, Paulus MJ, Jellison GE (1998) Bioluminescent-bioreporter integrated circuits form novel whole-cell biosensors. Trends Biotechnol 16:332–338

Smutok O, Dmytruk K, Gonchar M, Sibirny A, Schuhmann W (2007) Permeabilized cells of flavocytochrome b2 over-producing recombinant yeast *Hansenula polymorpha* as biological recognition element in amperometric lactate biosensors. Biosens Bioelectron 23:599–605

Sonneveld E, Jansen HJ, Riteco JAC, Brouwer A, van der Burg B (2005) Development of androgen- and estrogen-responsive bioassays, members of a panel of human cell line-based highly selective steroid-responsive bioassays. Toxicol Sci 83:136–148

Sonneveld E, Riteco JAC, Jansen HJ, Pieterse B, Brouwer A, Schoonen WG, van der Burg B (2006) Comparison of in vitro and in vivo screening models for androgenic and estrogenic activities. Toxicol Sci 89:173–187

Sørensen SJ, Burmølle M, Hansen LH (2006) Making bio-sense of toxicity: new developments in whole-cell biosensors. Curr Opin Biotechnol 17:11–16

Sticher P, Jaspers MC, Stemmler K, Harms H, Zehnder AJ, van der Meer JR (1997) Development and characterization of a whole-cell bioluminescent sensor for bioavailable middle-chain alkanes in contaminated groundwater samples. Appl Environ Microbiol 63:4053–4060

Stiner L, Halverson LJ (2002) Development and characterization of a green fluorescent protein-based bacterial biosensor for bioavailable toluene and related compounds. Appl Environ Microbiol 68:1962–1971

Stocker J, Balluch D, Gsell M, Harms H, Feliciano J, Daunert S, Malik KA, Van der Meer JR (2003) Development of a set of simple bacterial biosensors for quantitative and rapid measurements of arsenite and arsenate in potable water. Environ Sci Technol 37:4743–4750

Stoyanov JV, Magnani D, Solioz M (2003) Measurement of cytoplasmic copper, silver, and gold with a lux biosensor shows copper and silver, but not gold, efflux by the CopA ATPase of *Escherichia coli*. FEBS Lett 546:391–394

Takeyoshi M, Kuga N, Yamasaki K (2003) Development of a high-performance reporter plasmid for detection of chemicals with androgenic activity. Arch Toxicol 77:274–279

Tani H, Maehana K, Kamidate T (2004) Chip-based bioassay using bacterial sensor strains immobilized in three-dimensional microfluidic network. Anal Chem 76:6693–6697

Tauriainen S, Karp M, Chang W, Virta M (1997) Recombinant luminescent bacteria for measuring bioavailable arsenite and antimonite. Appl Environ Microbiol 63:4456–4461

Tauriainen S, Karp M, Chang W, Virta M (1998) Luminescent bacterial sensor for cadmium and lead. Biosens Bioelectron 13:931–938

Tauriainen S, Virta M, Chang W, Karp M (1999) Measurement of firefly luciferase reporter gene activity from cells and lysates using *Escherichia coli* arsenite and mercury sensors. Anal Biochem 272:191–198

Taylor LC, Walt DR (2000) Application of high-density optical microwell arrays in a live-cell biosensing system. Anal Biochem 278:132–142

Taylor CJ, Bain LA, Richardson DJ, Spiro S, Russell DA (2004) Construction of a whole-cell gene reporter for the fluorescent bioassay of nitrate. Anal Biochem 328:60–66

Tibazarwa C, Corbisier P, Mench M, Bossus A, Solda P, Mergeay M, Wyns L, van der Lelie D (2001) A microbial biosensor to predict bioavailable nickel in soil and its transfer to plants. Environ Pollut 113:19–26

Tiensing T, Strachan N, Paton GI (2002) Evaluation of interactive toxicity of chlorophenols in water and soil using lux-marked biosensors. J Environ Monit 4:482–489

Turner K, Xu S, Pasini P, Deo S, Bachas L, Daunert S (2007) Hydroxylated polychlorinated biphenyl detection based on a genetically engineered bioluminescent whole-cell sensing system. Anal Chem 79:5740–5745

Ulitzur S, Lahav T, Ulitzur N (2002) A novel and sensitive test for rapid determination of water toxicity. Environ Toxicol 17:291–296

Van Dyk TK, Rosson R (1998) Photorhabdus luminescens luxCDABE promoter probe vectors. Methods Mol Biol 102:985–996

Van Dyk TK, Smulski DR, Reed TR, Belkin S, Vollmer AC, LaRossa RA (1995) Responses to toxicants of an *Escherichia coli* strain carrying a uspA'::lux genetic fusion and an *E. coli* strain carrying a grpE'::lux fusion are similar. Appl Environ Microbiol 61:4124–4127

Van Dyk TK, Ayers BL, Morgan RW, Larossa RA (1998) Constricted flux through the branched-chain amino acid biosynthetic enzyme acetolactate synthase triggers elevated expression of genes regulated by rpos and internal acidification. J Bacteriol 180:785–792

Van Dyk TK, DeRose EJ, Gonye GE (2001) LuxArray, a high-density, genomewide transcription analysis of *Escherichia coli* using bioluminescent reporter strains. J Bacteriol 183:5496–5505

Vasina JA, Peterson MS, Baneyx F (1998) Scale-up and optimization of the low-temperature inducible cspA promoter system. Biotechnol Prog 14:714–721

Villeneuve DL, Khim JS, Kannan K, Giesy JP (2002) Relative potencies of individual polycyclic aromatic hydrocarbons to induce dioxinlike and estrogenic responses in three cell lines. Environ Toxicol 17:128–137

Virta M, Lampinen J, Karp M (1995) A luminescence-based mercury biosensor. Anal Chem 67:667–669

Vollmer AC, Van Dyk TK (2004) Stress responsive bacteria: biosensors as environmental monitors. In: Poole RK (ed) Advances in microbial physiology. Academic, London, pp 131–174

Vollmer AC, Belkin S, Smulski DR, Van Dyk TK, LaRossa RA (1997) Detection of DNA damage by use of *Escherichia coli* carrying recA'::lux, uvrA'::lux, or alkA'::lux reporter plasmids. Appl Environ Microbiol 63:2566–2571

Vulkan R, Zhao FJ, Barbosa-Jefferson V, Preston S, Paton GI, Tipping E, McGrath SP (2000) Copper speciation and impacts on bacterial biosensors in the pore water of copper-contaminated soils. Environ Sci Technol 34:5115–5121

Walmsley RM, Keenan P (2000) The eukaryote alternative: advantages of using yeasts in place of bacteria in microbial biosensor development. Biotechnol Bioprocess Eng 5:387–394

Walmsley RM, Billinton N, Heyer WD (1997) Green fluorescent protein as a reporter for the DNA damage-induced gene *RAD54* in *Saccharomyces cerevisiae*. Yeast 13:1535–1545

Weitz HJ, Campbell CD, Killham K (2002) Development of a novel, bioluminescence-based, fungal bioassay for toxicity testing. Environ Microbiol 4:422–429

Willardson BM, Wilkins JF, Rand TA, Schupp JM, Hill KK, Keim P, Jackson PJ (1998) Development and testing of a bacterial biosensor for toluene-based environmental contaminants. Appl Environ Microbiol 64:1006–1012

Winson MK, Swift S, Fish L, Throup JP, Jorgensen F, Chhabra SR, Bycroft BW, Williams P, Stewart GSAB (1998) Construction and analysis of luxCDABE-based plasmid sensors for investigating *N*-acyl homoserine lactone-mediated quorum sensing. FEMS Microbiol Lett 163:185–192

Wise AA, Kuske CR (2000) Generation of novel bacterial regulatory proteins that detect priority pollutant phenols. Appl Environ Microbiol 66:163–169

Xu S, D'Angelo E, Ghosh D, Feliciano J, Deo SK, Daunert S (2005) Detection of polychlorinated biphenyls employing chemical dechlorination followed by biphenyl whole cell sensing system. Toxicol Environ Chem 87:287–298

Yagi K (2007) Applications of whole-cell bacterial sensors in biotechnology and environmental science. Appl Microbiol Biotechnol 73:1251–1258

Yan L, Allen MS, Simpson ML, Sayler GS, Cox CD (2007) Direct quantification of *N*-(3-oxo-hexanoyl)-L-homoserine lactone in culture supernatant using a whole-cell bioreporter. J Microbiol Methods 68:40–45

Chapter 15
Photosynthetic Proteins Created by Computational and Biotechnological Approaches in Biosensing Applications

Maria Teresa Giardi

Abstract Chromophore molecules, such as chlorophylls and bacteriochlorophylls, bacteriorhodopsins, and quinones, are accommodated with photosynthetic proteins in photochemically active Reaction Centers (RCs) with interchromophore distances and gaps in the energy levels that ensure light-induced unidirectional electron transfer through lipid membranes. The structure and function of photosynthetic proteins differ across the photosynthetic evolutionary scale, allowing for their application in a range of technologies, particularly for the construction of photo-optical electrical biodevices. Practical monitoring programs require rapid, simple, and low-cost screening procedures for the detection of harmful chemicals in the environment. Traditional chemical methods of pollution monitoring such as gas and high-performance liquid chromatography, atomic absorption, and mass spectrometry are sensitive and effective. However, photosynthetic biosensors better reflect the real physiological impact of active compounds present in the sample because even low concentrations of pollutants affect the living organisms by altering physiological processes. Here, we provide a general description of the fundamental and technical research in this sector and an overview of biochips and biosensors based on photochemical activity that have been developed for the bioassay of pollutants, and applied in photodevices for light capturing and life support.

Keywords Photosystem II · Reaction centers · Technological applications · Biosensing · Evolution scale · Biosensors · Biodevices · Pesticides · Heavy metals · Herbicides · Water monitoring · Agrofood monitoring · Space applications

---

M.T. Giardi
Department of Molecular Design and of Agrofood, Area of Research of Rome, Institute of Crystallography, National Council of Research, Via Salaria Km 29.300, Monterotondo scalo, Rome 00016, Italy
e-mail: mariateresa.giardi@mlib.ic.cnr.it

M. Zourob (ed.), *Recognition Receptors in Biosensors*,
DOI 10.1007/978-1-4419-0919-0_15,

Abbreviations

RC	Reaction Center
PSI and PSII	Photosystem 1 and 2
NAD/NADH	Nicotinamide Adenin Dinucleotide/Nicotinamide Adenin Dinucleotide reduced
NADP, NADPH	Nicotinamide Adenin Dinucleotide Phosphate/Nicotinamide Adenin Dinucleotide Phosphate reduced
Q_A, Q_B	Primary and secondary quinone electron acceptor
FeS	Iron–sulfur
Chl_2	Chlorophyll dimer
$BChl_2$	Bacteriochlorophyll dimer
Φ	Pheophytin
BΦ	Bacterio-Pheophytin
$\Phi \cdot Q$	Pheophytin–Quinone
PCR	Polymerase Chain Reaction
LHC	Light Harvesting Complex
OEC	Oxygen Evolving Complex
Chl *a*	Chlorophyll *a*
MIP	Molecularly Imprinted Polymer
PCy	Plastocyanin
Fd	Ferredoxin
PVA-SbQ	Poly(vinylalcohol) bearing styrylpyridinium polymers
BSA	Bovine Serum Albumin
GA	Glutaraldehyde
DCPIP	2,6-Dichlorophenolindophenol
DCBQ	2,5-Dichlorobenzoquinone
C_{10}, C_{50}	Analyte concentration inhibiting 10 or 50% signal, respectively
HRP	Horseradish Peroxidase
NTA	Nitrilotriacetic acid
ITO/Au	Indium-tin oxide/gold
CBRNs	Chemical, Biologic, Radiological, and Nuclear compounds

15.1 The Emergence of Technology Based on Photosynthetic Proteins

The potential for producing a new generation of technological devices that integrate the knowledge coming from various fields–chemistry, biology, computer science, electronics, engineering–is currently attracting increasing attention. This trend has led to the emergence of a new technological science (Giardi and Pace 2005; Giardi and Piletska 2006) based on the use of protein components that function as

Fig. 15.1 A sensor (**a**) and a biochip (**b**) for multielectrode systems for bacterial RCs

electronic parts (Fig. 15.1). In this context, bioelectronics is a field of molecular electronics that investigates the use of native, as well as modified biological molecules in electronic or photonic devices. This approach is considered promising since evolution has often solved problems such as creating and optimizing thermal stability and cyclicity, as the organism must be able to function in the extreme and resource-limited environments. Several proteins have been explored for device applications, and, among others, photosynthetic protein complexes are presently the most largely utilized (Table 15.1). Chromophore molecules, such as chlorophylls and bacteriochlorophylls, bacteriorhodopsins and quinones, are arranged with photosynthetic proteins in photochemically active Reaction Centers (RCs) with inter-chromophore distances and gaps in the energy levels that ensure unidirectional electron transfer. In green plants, algae, and cyanobacteria, photosystem I (PSI) produces NADPH, which provides reducing equivalents for carbon dioxide fixation in the reactions of the Calvin cycle; photosystem II (PSII) catalyzes one of the most thermodynamically demanding reactions in biology – the conversion of light energy into activated oxidants capable of converting water into oxygen (Fig. 15.2). Photosynthetic bacteria are more primitive, dating from the time when the Earth had a reducing atmosphere, and, as a consequence, they have simpler photoconversion pathways with only one photosystem. The primary energy and electron-transfer processes in bacteria have been studied extensively and have served as the principal inspiration for biomimetic studies over the past 15 years. RC-biotechnology exploits the characteristics of pigment-protein complexes, the RCs being located within the membranes of plants, algae, cyanobacteria, and bacteria. All of these complexes share a common feature, i.e., their solar energy-triggered photochemistry. However, they differ in their structures and energy conversion modes, thus providing a choice of photochemical microreactors that can be tailored to the needs of diverse applications. From the point of view of analytical applications, the advantage of using RC-technology stems from the ability of the complexes to specifically recognize certain chemicals or particular physico-chemical conditions, and to generate physiological responses that can be easily monitored as generated electric current through amperometric, potentiometric, and optical systems. Moreover, the photosynthetic complexes are naturally abundant and readily available. Engineered RCs have the potential to provide the

Table 15.1 Example of different performances of biosensors developed from various photosynthetic organisms and transduction systems for herbicide detection

Biosensor characteristics	Photosystem	Detected herbicides
Electrochemical		
Clark's oxygen electrode and Screen-printed electrode.	PSII particles from *Synechoccus elongatus*	Diuron, atrazine, simazine, ioxynil, bromoxynil, dinoseb
Multi-biosensor with three flow cells, illumination at 800, 650, 430 nm. Screen-printed electrode	Isolated chloroplasts from *Spinacia oleracea L.*	Diuron, atrazine, simazine, deethylterbuthylazine
Multi-biosensor with four flow cells, illumination at 650 nm. Screen-printed electrode	Thylakoid from *Spinacia oleracea L., Senecio vulgaris,* and its mutant resistant to atrazine	Various uree, diamine, triazine, and phenolic compounds
Optical		
Fluorescence analyzer	Thylakoid from *Spinacia oleracea L.*	Diuron, Linuron, metribuzin, prometrine, atrazine, cyanazine
Chemiluminescent micro-fluidic sensor	Thylakoid from *Spinacea oleracea L.* and higher plants	Atrazine, diuron
Multiarray biosensor. Four light sources, photodiode detector	PSII wild type, mutants from atrazine-resistant plants *Chlamydomonas reinhardtii* wild-type and mutants selected by bioinformatics	Diamine, triazine, urea, and phenolic compounds
Spectroscopic		
Photoelectronic biosensor	RC from *Rhodobacter sphaeroides*	Phenanthroline

core for numerous innovative devices. In fact, single change in the amino acid sequences of photosynthetic proteins can induce new properties, such as an increase in stability, in the specificity of the binding of analytes, and in the efficiency of photochemical processes.

This review provides an overview of currently existing biosensing applications (Fig. 15.3) of photosynthetic proteins from various organisms across the evolutionary scale, including RCsII of photosystem II from plants, algae, and cyanobacteria, and RCs from bacteria.

15.2 Key Issues in Designing Photosystems-Based Biosensors

Biosensors are analytical devices which consist of a sensing element called a biomediator (enzymes, tissues, living cells) and a transducer that translates the chemical signal into an electrical signal for further processing (Fig. 15.4). When designing a biosensing device, several steps need to be taken in order to develop its optimal potential (Fig. 15.5).

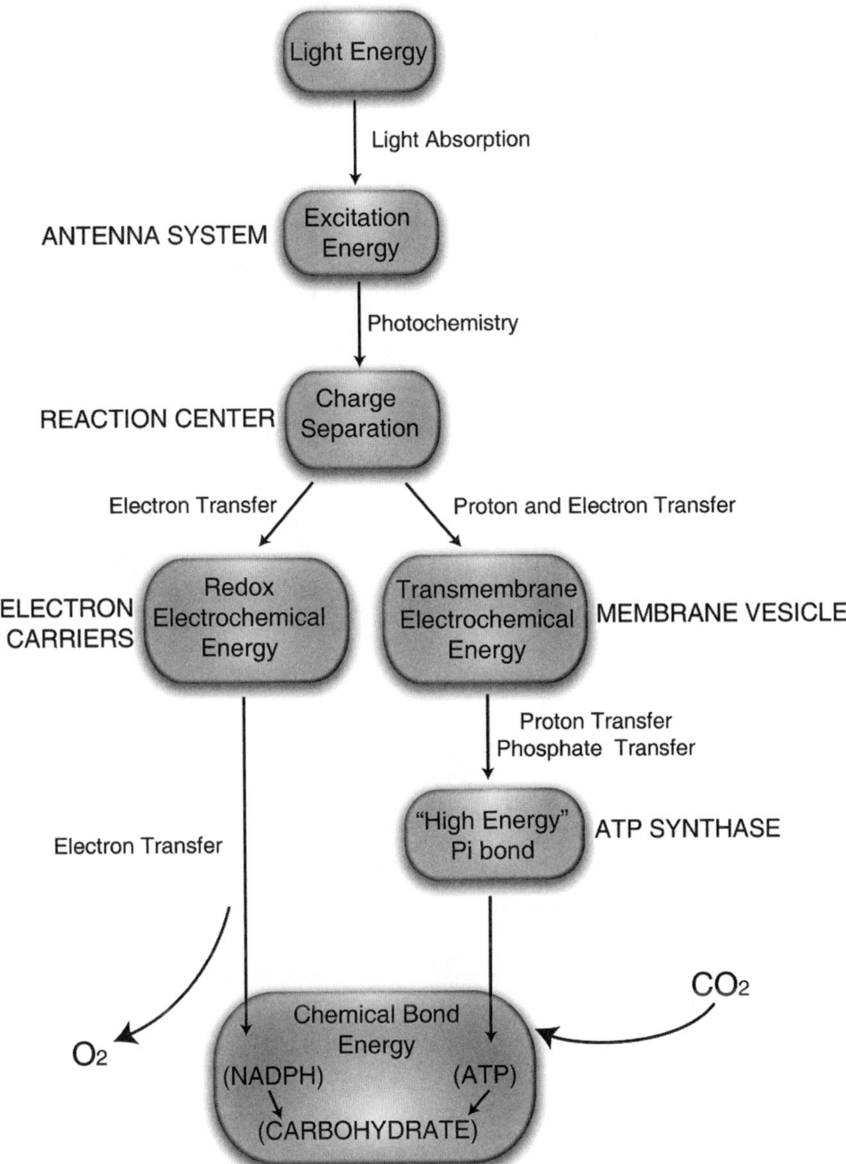

Fig. 15.2 Schematic representation of light-triggered chemistry in photosystem II (PSII). Light energy is absorbed by light harvesting complexes; excitation energy is transferred from this antenna to the "core" of the PSII complex, the Reaction Center (RC), where charge separation takes place. Electron, proton, and phosphate transfer reactions and oxygen evolution (oxidation of water), resulting in an accumulation of reducing equivalents (NADPH), and ADP phosphorylation to ATP. The accumulated electrochemical energy results in carbohydrate synthesis by CO_2 reduction

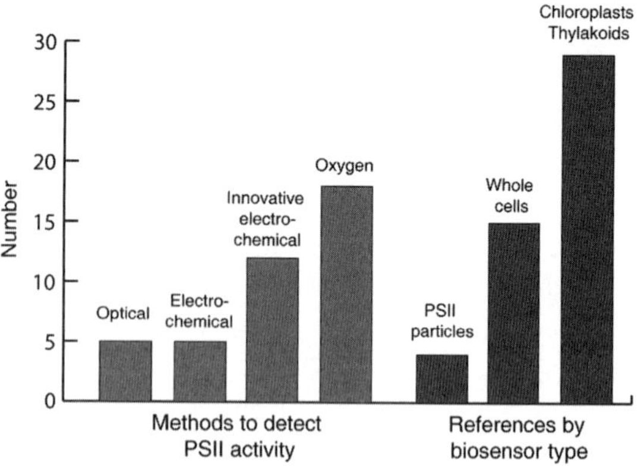

Fig. 15.3 Relevance of different detection methods and biological photosynthetic material for photosystems-based applications

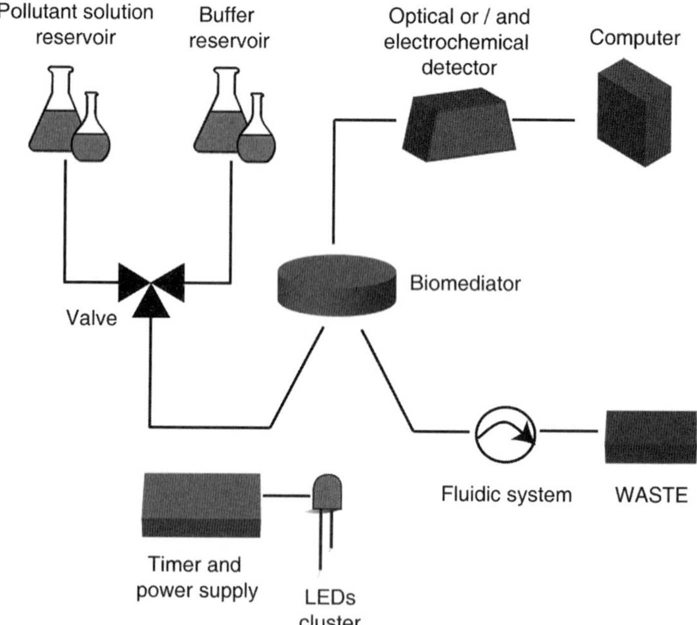

Fig. 15.4 Schematic representation of a biosensor based on photosystem II or reaction center. The immobilized biomediator interacts alternatively with the pollutant-containing aqueous solution, or with the regenerating buffer. This alternative flow is ensured by a three-way valve and a fluidic system. The LED activation of the biomediator is controlled by means of a timer. The biomediator-generated signal is detected either optically or electrochemically by a transducer connected to a computer

First of all, the most suitable photosynthetic organism, either a wild-type or a mutant, needs to be carefully selected (see Sect. 15.3). At this stage, it is important to take into account the physiology of the photosynthetic organism. For instance,

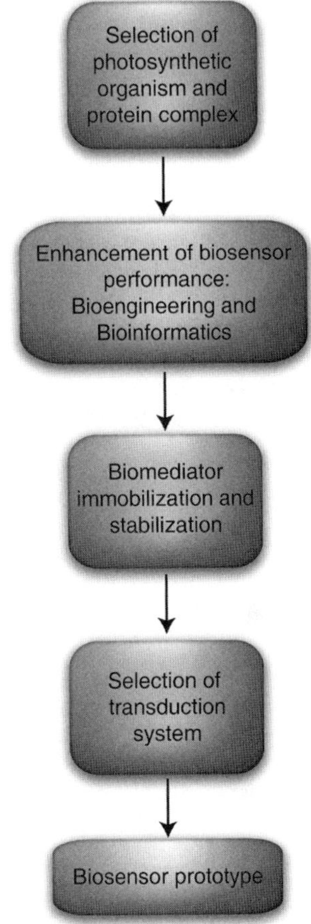

Fig. 15.5 Sequence of steps for the design and construction of a photosynthetic-based biosensor

choosing a thermophilic cyanobacterium as a biomediator affords better biosensor stability for monitoring purposes. Furthermore, some key features of biomediators can be enhanced by means of bioengineering and bioinformatics approaches, which have been exploited for the improvement of sensitivity and/or selectivity toward analytes, and tolerance to damaging factors such as exposure to pesticides, temperature, or radiation. Mutations of the PSII complex that produce specific properties which can be carried out; e.g., a modification of a single amino acid on D1 protein (i. e., the core protein of Photosystem II that binds the quinone Q_B) generates resistance toward pollutant compounds. From this perspective, bioinformatics can be applied to screen for optimal molecular modifications of the biomediator (Sect. 15.8).

Further aspects to be addressed in biosensor design and construction are biomediator isolation, immobilization and stabilization by biochemical and chemical techniques (Sect. 15.4). Photosynthetic activities can be measured either by electrochemical, or by optical systems (Fig. 15.6) (Sects. 15.6, 15.7). These activities are sensitive to the presence of contaminants, such as pesticides and heavy metals,

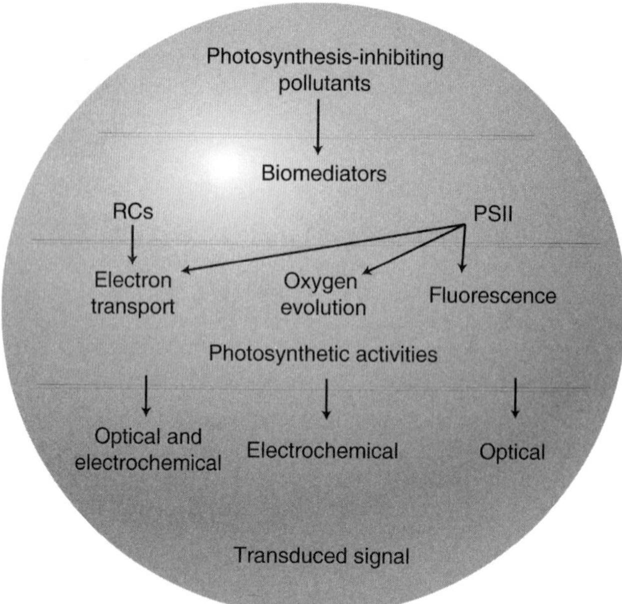

Fig. 15.6 Different signals generated by the activity of the photosynthetic biomediators. From top to bottom: The photosynthesis-inhibiting pollutant binds to the biomediator and impairs the physiological activities; as a consequence, the signals produced by the photosynthetic activities of biomediators are inhibited, and, finally, this inhibition is detected through the transduction system for the specific signal

which inhibit the specific signal in a concentration-dependent manner, thus allowing for analyte determination. The sensitivity of the photosystem toward a given analyte can be quantified through the concentration of the compound of interest resulting in a predetermined percentage of signal inhibition. For instance, C_{10} or C_{50} are the concentrations of a chemical compound that causes 10% or 50% inhibition, respectively.

A suitable transduction system for the detection of the chosen signal and subsequent data analysis is, then developed, and finally all the items mentioned above converge into biosensor prototypes for specific applications (e.g., field portable, benchtop, miniature-size prototype etc.). Results obtained with biosensing devices based on immobilized biomediators are presented in Sects. 15.6 and 15.7, where different types of biosensors are described in more detail.

15.3 General Remarks on the Reaction Centers of Photosystems

The RCs of photosynthetic organisms are remarkably conserved and appear to function through similar mechanisms of energy transfer (Fig. 15.7). However, the pathway used for these processes varies depending on the specific molecular

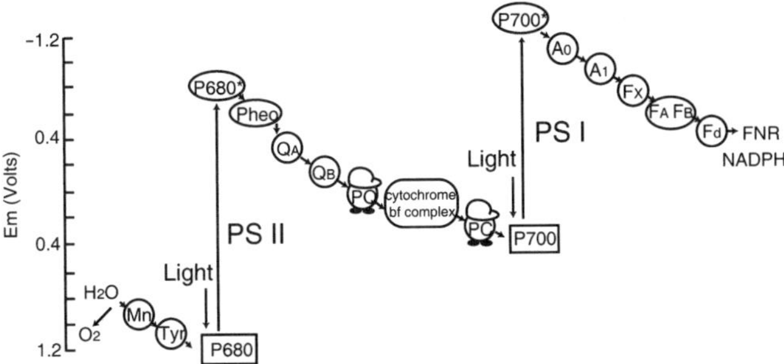

Fig. 15.7 Schematic representation of light-induced electron chain in eukarya and oxygen-evolving bacteria. The process starts at PSII, where light-induced charge separation, with formation of the pair P680$^+$-Pheo$^-$, triggers the electron flow to P700, through a series of quinone (Q_A and Q_B) and plastoquinone (PQ) acceptors. A second light-induced electron flow from reduced P700, through a series of acceptors (iron-sulfur proteins and ferredoxin), occurs in PSI, where NADP$^+$ is finally reduced to NADPH

constituents carrying out electron transfer present in different organisms. The results of phylogenetic studies indicate that all RCs fall into two basic categories: those with pheophytin as the primary electron acceptor and quinone as the secondary acceptors, the 'pheophytin–quinone ($\Phi{\cdot}Q$) type' (e.g., photosystem II), and those with iron sulfur clusters as early acceptors, the 'iron–sulfur (FeS) type' (Fig. 15.8). The two RC types are distributed among the photosynthetic organisms of both the eukarya and the bacteria. The eukarya, plants, and algae, and the bacteria that perform oxygen-evolving photosynthesis, cyanobacteria, contain two photosystems. The anoxygenic photosynthetic bacteria possess only one photosystem and are susceptible in the presence of oxygen to the inhibition of photosynthetic pigment biosynthesis.

In the ($\Phi{\cdot}Q$) type RC of the anoxygenic photosynthetic bacteria, as well as in the photosystem II (PSII) of the eukarya and in the oxygenic photosynthetic bacteria, the electron is transferred from an external donor to the primary electron donor, a bacteriochlorophyll dimer ($BChl_2$) or a chlorophyll dimer (Chl_2). After light absorption, the electron flows through the chain from the bacterio-pheophytin (BΦ) or the pheophytin (Φ) to the quinone transfer components. The quinone Q_B in the Reaction Center II is bound to a protein called D1. In the (FeS) type RC of the anoxygenic photosynthetic bacteria, energy is transferred from the primary electron donor ($BChl_2$) through a series of acceptors and three iron-sulfur clusters FeS-X, FeS-A and FeS-B to reduce the final acceptors NAD or NADP.

PSII is the multisubunit pigment-protein complex embedded in the thylakoid membranes of higher plants, algae and cyanobacteria (Fig. 15.9) that comprises a Reaction Center II, of which two polypeptides -D1 and D2- are extremely relevant in view of biosensor development for their role in the binding of pollutants such as herbicides.

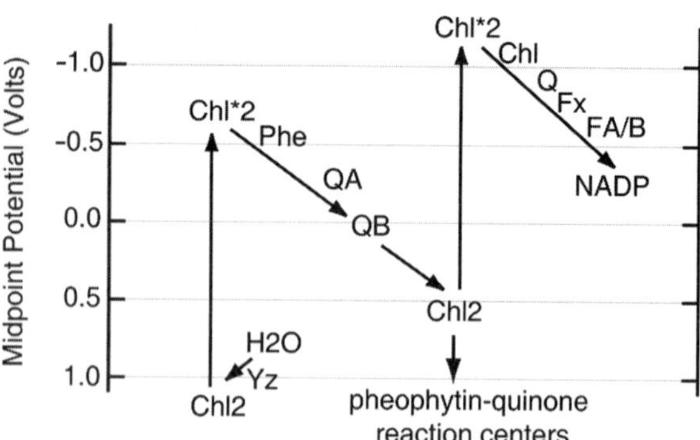

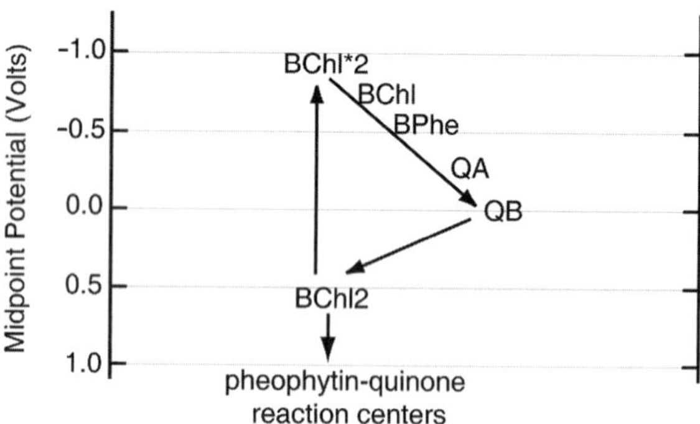

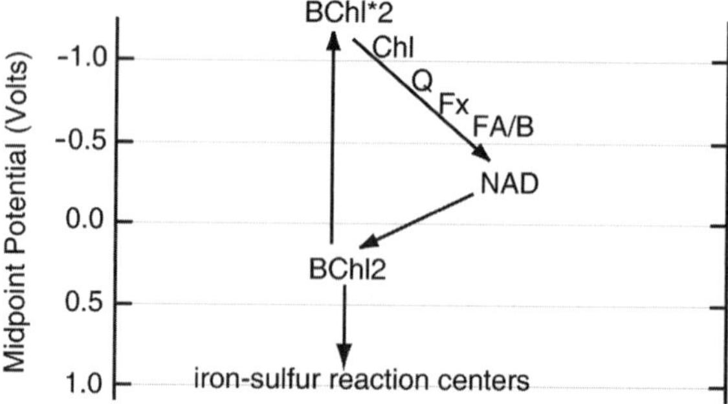

Fig. 15.8 Electron transfer schemes in pheophytin-quinone (Φ-*Q*) (cyanobacteria and higher plants; purple bacteria and chloroflexaceae), and iron-sulfur (Fe-S) (green sulfur bacteria and halobacteria) Reaction Centers (RCs)

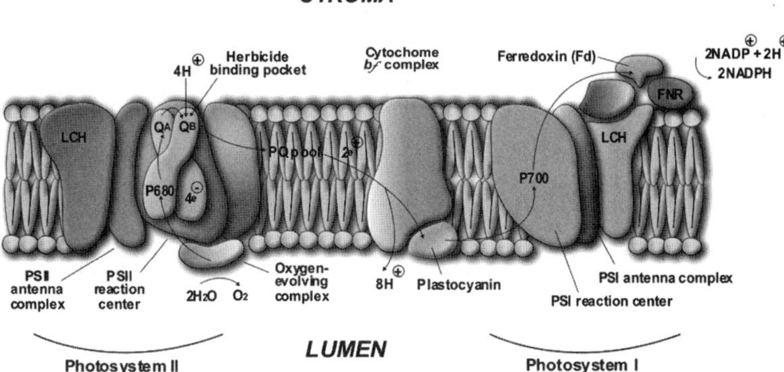

Fig. 15.9 Representation of photosystems I and II (PSI and PSII) complexes, embedded in the thylakoid membranes of higher plants, algae, and cyanobacteria. Water oxidation to oxygen and NADP reduction to NADPH occur at separate sites, the PSII and PSI respectively. The electron flow is light-triggered at both sites, by excitation of P680 and P700. Electrons are carried from PSII to PSI by a pool of plastoquinones and plastocyanin. The herbicide-binding pocket is located at PSII

15.4 Immobilization Procedures for Biomediator Stabilization

The first biosensing systems developed for monitoring polluted compounds and testing phytotoxicity used intact cells of algae, cyanobacteria, and diatoms. These systems are still frequently used to measure either changes in photocurrent, inhibition of electron transport with artificial mediators, or changes in chlorophyll fluorescence. Notwithstanding their sensitive detection at parts per billion (ppb) levels of chemical compounds, a drawback is that the cell envelope can be impermeable to electrolytes. In recent years, progress has been made in the isolation of RC and PSII particles. It is now possible, by detergent solubilization, to isolate preparations from plant thylakoid and cyanobacterial membranes that are quite stable and pure, thus avoiding the cell wall obstacle and permitting the detection of low levels of herbicides in the water and soil. These preparations are capable of light-induced oxygen evolution and electron transfer both in the absence and in the presence of artificial electron acceptors. However, the practical applications of biosensors based on PSII preparations are prone to be limited by their instability, particularly when illuminated. In vivo, the cellular environment and mechanisms of repair contribute to the preservation of the structures and photosynthetic activities against external factors such as temperature or light; however, the activity of the photosynthetic material decreases quickly after isolation at a rate that depends mainly on storage and operating conditions. During operation of the biomediator-based device, its functional stability may be affected by adverse operating conditions such as temperature, ionic strength, pH, illumination and the presence of electron donors/acceptors. Problems related to functional instability of biological preparations can be overcome by means of immobilization on a physical support.

Table 15.2 Main immobilization procedures for photosynthetic materials

Physical immobilization		Chemical immobilization	
Adsorption	Gel inclusion	Reticulation	Coreticulation
Paper filter	*Polysaccharide gel:* agar, agarose, carragenan, alginate	Glutaraldehyde	Glutaraldehyde-gelatin
Alumina filter	*Protein gel:* gelatin		Glutaraldehyde-collagen
Glass microfibre filter	*Synthetic gel:*		Glutaraldehyde-bovine serum albumine
Diethylaminoethyl-cellulose	Polyacrylamide, Polyurethane, Photocrosslinkable resin Vinyl Poly(vinylalcohol), Poly (vinyalcohol)-styrylpyridinum groups		

Immobilization methods are identified either as chemical or physical procedures depending on whether covalent bonds are established or not (Table 15.2). Physical methods involve adsorption of the photosynthetic material on a support or its inclusion in a natural or synthetic gel. Adsorption methods use various supports, such as filter paper discs, alumina filter discs, glass microfiber filters, and columns containing diethylaminoethyl-cellulose. This immobilization technique is sufficiently mild as to preserve the native activity of the material. Moreover, the adsorption on an inert or ion exchanger support is a simple, economic methodology. However, the interaction forces involved are weak, hence release of the biocatalyst may be observed during operating conditions of the biosensor.

Embedding the photosynthetic material in a natural or synthetic polymer gel results in a three-dimensional distribution of the biocatalyst. Among the natural polysaccharides, gels, one of the most widely used is alginate, while agar, agarose, carragenan are seldom used. The advantage in employing alginate gel lies in the possibility to operate at low temperature while mixing the biological material, whereas other gels need heating. Gelatin is a protein gel widely used for the entrapment of biological material. Synthetic gels include polyacrylamide gel, polyurethane prepolymer, photo-crosslinkable resin prepolymer, vinyl monomers, poly(vinylalcohol) polymers, poly(vinylalcohol) bearing styrylpyridinium polymers (PVA-SbQ). Polyurethane gel can be used for both physical and chemical methods of immobilization, because of its ability to either include the biocatalyst, or to react with some of its functional groups.

In general, chemical immobilization methods are scarcely used with photosynthetic biomolecules because of the loss of activity due to the denaturing effect of the bonding agents. Glutaraldehyde, however, has been fully shown to preserve photosynthetic activity in chloroplasts. The denaturing effect of the reticulating agent can be decreased by the addition of proteins such as gelatin, collagen, or BSA during the polymerization. These shield the biocatalyst from being excessively modified through covalent bonding.

A comparative study of sensitivity to herbicides was performed on thylakoid membranes after two types of immobilization procedures: the physical PVA-SbQ entrapment, and the chemical coreticulation with BSA-Glutaraldehyde (BSA-GA). The two approaches were similar in terms of detection of some classes of herbicides. The physicochemical properties of the materials potentially enabled the insoluble BSA-GA matrix to withstand longer incubation periods in aqueous solution. This feature may prove advantageous in the detection of chemicals with a slower rate of inhibition. On the other hand, preservation of the PVA-SbQ immobilized material in the dry state prevents microbial contamination of the samples.

The photosynthetic material can be coupled to an electrode by suitably modifying its metal surface through the self-assembled monolayer approach. In a recent investigation, an isolated PSII from the cyanobacterium *Thermosynechoccus elongatus* with a genetically introduced histidine tag was attached onto gold electrodes modified with thiolates bearing Ni(II)-nitrilotriacetic (NTA) acid groups (Fig. 15.10). An organic surface of predefined composition was obtained by self-assembling of mixed thyolates (i.e., 1-octanethyol and 16-mercaptohexanedecanoic acid) onto gold, where the thiol end is anchored onto the metal surface. The carboxy end is, then, coupled through an amide bond to a Ni(II)- activated Nα',Nα''-bis (carboxymethyl)-l-lysine that interacts with the histidine tag.

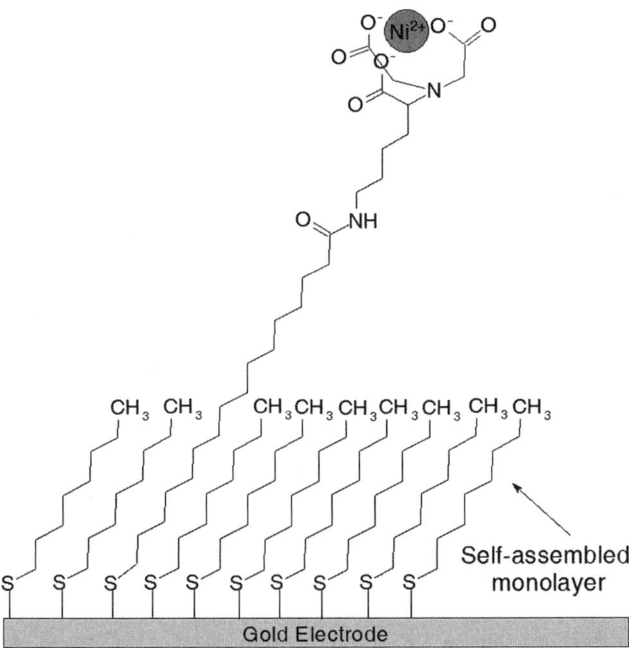

Fig. 15.10 Gold electrodes modified with thiolates bearing Ni(II)-nitrilotriacetic acid groups. Isolated photosystems from the cyanobacterium *T. elongatus* with a genetically introduced histidine tag are immobilized through interaction with the Ni-activated acidic groups

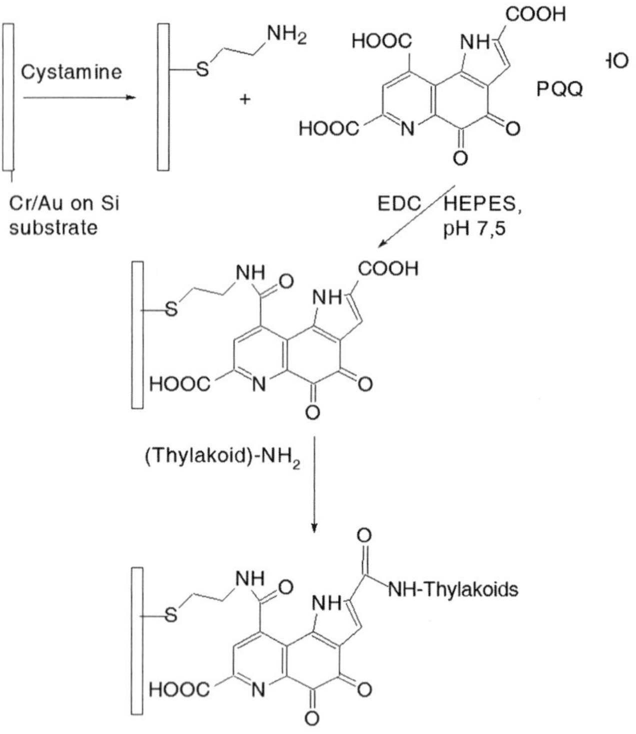

Fig. 15.11 Immobilization of thylakoids by means of self-assembled monolayers of cystamine and pyrroloquinoline quinone (PQQ). Cystamine is anchored to the electrode surface through thiolate bonds. PQQ is then coupled to the amino group of anchored cystamine, and provides the carboxy group for the immobilization of thylakoids

A further example Lam et al. (2006) is provided by the immobilization of photosynthetic thylakoids directly onto a gold electrode surface by means of self-assembled monolayers of cystamine and pyrroloquinoline quinone (Fig. 15.11).

Thylakoid membranes have been immobilized onto magnetic beads (Varsamis et al. 2008) to be utilized in a micro-fluidic sensor for herbicide monitoring (see Sect. 15.7 for a description of the device). The amino groups of thylakoids were coupled through a GA bridge (Fig. 15.12) to magnetic particles, obtained by coating a layer of magnetite and amino-polystyrene onto monodispersed particles.

15.5 Transduction Systems

Transduction systems that generate signals from the photo-induced activity of the biomediators make use, either directly or indirectly, of their redox and optical properties. The photo-induced electron flow results in oxidation of water to oxygen, and the concomitant reduction of an electron acceptor. In the Hill reaction, the

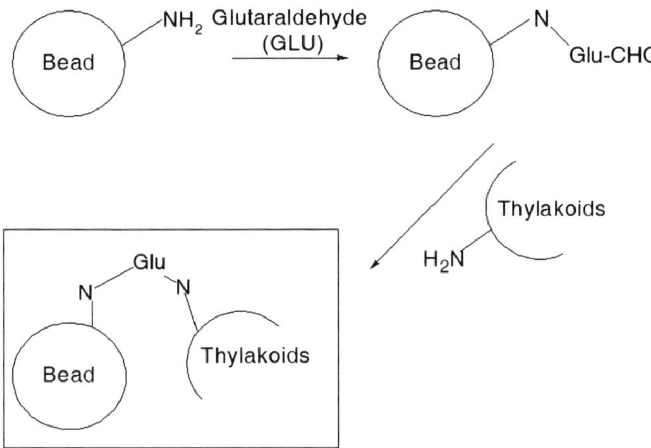

Fig. 15.12 Immobilization of thylakoids onto amino-polystyrene magnetic beads. Glutaraldehyde is the bifunctional linker that reacts with amino groups of both the beads and the thylakoids

electron acceptor can be an artificial one, such as the dye 2,6-dichlorophenolindophenol (DCPIP), and an electrochemical signal can be generated by reoxidizing it at a working electrode. This is the basis for amperometric and potenziometric methods. These processes are sensitive in a concentration-dependent manner to binding of photosynthetic herbicides, or to other contaminants, such as heavy metals, thus allowing for analyte determination. In fact, as early as the 1950s, selection of the most effective herbicides took advantage of the fact that they can inhibit the Hill reaction in isolated chloroplasts. Furthermore, since the binding of the polluted compounds to the photosynthetic protein is not covalent, it has been shown that biomediator regeneration with 10 mM phosphate buffer can be achieved, with approximately 10% decrease in the inhibition after one regeneration.

Optical transduction systems are based on the modifications of fluorescence properties of Photosystems, induced by binding of phytotoxic chemicals, and their disruption of the photo-induced electron flow.

Amperometric and fluorescence-based transduction systems have been integrated into modular devices, where both the types of measurements can be carried out either alternatively or simultaneously. The possibility of integrating different signals, obtained from a single sample, can afford versatile biosensors for the simultaneous detection of different classes of analytes.

More detail and results about different transduction systems are to be found in the description of specific devices in the following sections.

15.6 Electrochemical Detection of Pollutants

Stable and sensitive biosensors for pesticide detection use PSII particles from thermophilic species, for example the thermophilic cyanobacterium *Synechococcus elongatus*. Two types of biosensors for the detection of photosynthetic herbicides

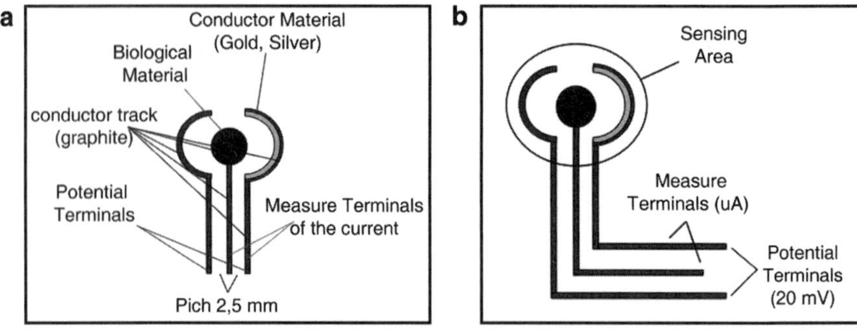

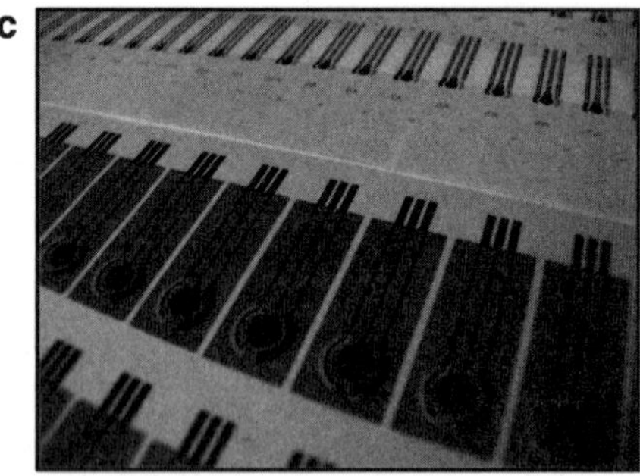

Fig. 15.13 Screen-printed electrodes (SPEs) for photosystem-based biosensors. (**a**), (**b**) Constructive scheme of SPEs; (**c**) Actual SPEs

have been constructed using these particles: a Clark's oxygen electrode (Laberge et al. 1999; Koblizek et al. 1998) and a screen-printed carbon–silver electrode (Koblizek et al. 2002) (Fig. 15.13). The Clark's electrode traces oxygen produced by PSII particles, whereas the positively polarized carbon-silver electrode registers the reduction of an artificial electron acceptor as electrical current. The signal-to-noise ratio was maximized by mounting the biosensor onto a flow system, which resulted in a high concentration of PSII oxygen evolution in the microenvironment of the measuring probe. Moreover, because the herbicides can be washed out, the biomediator can be reused for several samples. As for the device (Fig. 15.14) based on the screen-printed electrode, its performance was characterized by a relatively good stability (half-life of 24 h), and the limits of detection of several herbicides are reported in Table 15.3, together with those found with the biosensor based on Clark's electrode.

In the other developed electrochemical detection systems, the pool of plastoquinones that acts as electron acceptor in nature is replaced with artificial mediators for electrochemical monitoring of the electron flow in the light-induced process. Typically, DCPIP is used as the electron acceptor, but other mediators have also been tested, i.e., potassium ferricynide, and duroquinone. Since the electron flow through PSII is affected by the presence of herbicides, the reduction of the electrochemical mediators is, in turn, also proportionally affected, and can be

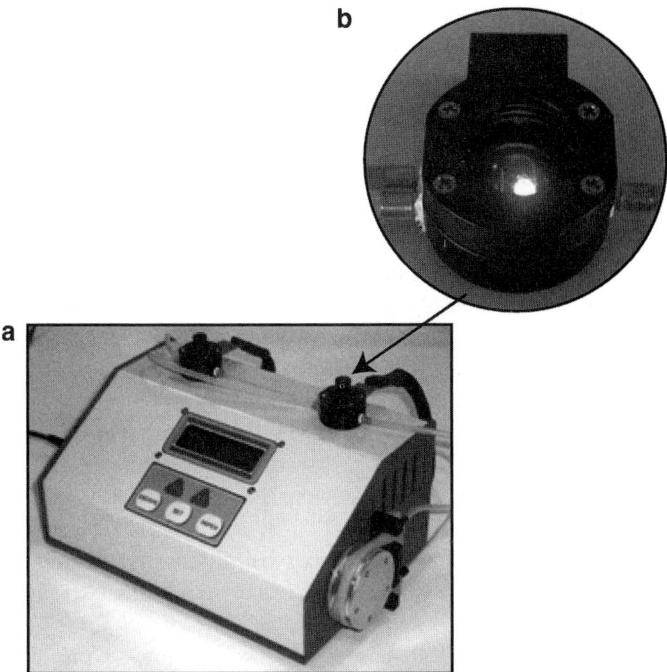

Fig. 15.14 (**a**) A Bioamperometer based on screen-printed electrodes. (**b**) Detail of the bioamperometer: flow cell with the optical system that activates the photosynthetic biomediator

Table 15.3 Limits of detection of two classes of herbicides with two types of biosensors, based on isolated PSII particles from thermophilic cyanobacterium *S. elongatus*

Herbicide	Clark electrode Limit of Detection (M)	Screen-printed sensor Limit of Detection (M)
Classical		
Diuron	5×10^{-10}	1×0^{-9}
Atrazine	2×10^{-9}	2×10^{-9}
Simazine	1×10^{-8}	4×10^{-9}
Phenolic		
Ioxynil	9×10^{-9}	$\sim 10^{-7}$
Bromoxynil	2×10^{-7}	$\sim 10^{-6}$
Dinoseb	6×10^{-8}	$\sim 10^{-6}$

detected either optically or amperometrically. In a different experimental setup (Touloupakis et al. 2005), a multibiosensor was designed and constructed for the amperometric detection of herbicides by the photosynthetic thylakoid from *Spinacia oleracea L.*, *Senecio vulgaris* and its mutant resistant to atrazine, immobilized either with BSA-GA, or gelatin. The detection units consisted of screen-printed sensors composed of a graphite working electrode, onto which the immobilized thylakoids were deposited, and an Ag/AgCl reference electrode on a polymeric substrate. The device was composed of four flow cells, each containing a different immobilized thylakoid, with independent illumination at 650 nm to activate electron transfer in PSII. The goal was to obtain a biosensor in which each of the four simultaneously-working flow cells would specifically recognize a different class of herbicides, by taking advantage of the peculiar properties of various PSII preparations. It will be shown in Sect. 15.8 that the mutation of a single amino acid on the D1 protein of PSII is able to induce resistance toward a herbicide. For example, the replacement of a serine moiety with a glycine one induces resistance to atrazine but not to the other herbicides; the biomediator is, then, resistant (i.e., "nonbinding") to the class of triazinic compounds and, therefore, selective for those classes of herbicides other than triazine (phenolic, ureidic, and diazinic herbicides). Since "nonbinding" of triazinic compounds translates into a "nonresponse" to their presence, a PSII featuring such selectivity allows for selective analysis of nontriazinic herbicides. By combining two or more selective biomediators in a series of multiarrayed flow cells, and by suitable data processing, different pesticide subclasses can be detected individually in a single analysis (Giardi et al. 2009, Scognamiglio et al. 2009, Rea et al. 2009, unpublished data). Table 15.4 summarizes the recognition activity of differently immobilized biomediators, isolated from the wild type organism, and from mutants from soil treated with atrazine for five years.

The multiarray design of PSII-based biosensors being so promising, it has undergone further refinements in terms of cell construction. Recent measurements were carried out on a portable bio-amperometer equipped with two noninterfering sensing cells, each illuminated in a programmable way by two LEDs at 470 and 660 nm, for optimal biomediator illumination. The biological reaction cell of 4–6 μl volume, to which the biomediator and the disposable screen-printed electrode are introduced, is separated from the optical compartment by a glass window. The flow system, based on rotary peristaltic pumps, allows for both static (i.e., no electrolyte flow), and dynamic measurements (i.e., with electrolyte circulation at a preadjusted

Table 15.4 Herbicide recognition activity of differently immobilized PSIIs, either wild-type or mutant

Plant species	Immobilization	Half-life (h)		Detected herbicide subclasses	Range of recognition (M)
		pH 7.5	pH 6.5		
S. oleracea	BSA-Glu	15.5	54.2	Urea, diamine, triazine	10^{-8}–10^{-6}
S. oleracea	Gelatin	7.3	24.3	Urea, diamine, triazine, phenolic	10^{-8}–10^{-6}
S. oleracea	$CdCl_2$	13.6	39.2	Urea, diamine, triazine, phenolic	10^{-9}–10^{-6}
Senecio vulgaris	$CdCl_2$	7.0	24.4	Urea, diamine, triazine	10^{-7}–10^{-5}
Senecio mutant	$CdCl_2$	5.2	12.3	Urea	10^{-7}–10^{-5}
A. retroflexus	$CdCl_2$	9.6	13.6	Urea, diamine, triazine, phenolic	10^{-8}–10^{-5}
Amaranthus mutant	$CdCl_2$	7.3	21.0	Urea, diamine	10^{-7}–10^{-5}

flow rate). Such a design afforded a flexible, small, and modular tool for the analysis of herbicides and, potentially, of agro-food, pharmaceutical, and biomedical samples. In fact, the equipment was also tested as a tyrosinase-based amperometric biosensor for the detection of phenolic compounds.

PSII is particularly affected by cations of toxic metals, and, therefore, should provide an excellent tool for their rapid, simple, and low-cost screening. The principle of measurement is based on the photocurrent from the reoxidation at the working electrode of a photoreduced electron acceptor. This can be either hydrogen peroxide produced by the photosynthetic system by reduction of ambient oxygen, or an artificial electron acceptor, such as 2,5-dichlorobenzoquinone (DCBQ). As was extensively discussed in Sect. 15.4, in order to increase the stability of isolated photosynthetic membranes, or submembrane fractions, several immobilization techniques have been developed, and the activities of free and immobilized membranes have been compared. Thylakoid membranes immobilized in a crosslinked BSA-GA matrix were used to detect heavy metal chlorides as well as other contaminants such as sulfite, nitrite, and herbicides. Immobilized thylakoids were stable under storage conditions. The photosynthetic activity was measured as the initial rate of oxygen evolution with DCBQ as the artificial electron acceptor. When comparing the data for $CuCl_2$ and $HgCl_2$, it was shown that, after an incubation time of 10 min, the concentrations of the $CuCl_2$, which inhibited 50% of oxygen production (C_{50}) were 100 μM with free thylakoids and 40 μM with immobilized photosynthetic membranes. In the case of $HgCl_2$, C_{50} was 60 μM for the immobilized membrane, and 40 μM for free thylakoids. The latter result was explained in terms of the decreased accessibility of the main inhibitory site of the mercury ions, which is located at the oxygen evolving complex on the lumenal side of the thylakoids, due to the higher microviscosity of the crosslinked preparations. Besides BSA-GA, various photosynthetic materials were also entrapped in poly PVA-S_bQ (Avramescu et al. 1999; Rouillon et al. 2000), and were used to detect various heavy metals (Table 15.5). Concentrations of heavy metals (Cr, Pb, Cd, Zn, Ni, Cu) which produce a 10% inhibition of the activity (C_{10}) of submembrane fractions immobilized in PVA-S_bQ are in the range 20 ÷ 445 mg/L (Hg and Cu, respectively). Immobilization of photosynthetic materials in PVA-S_bQ results in a roughly tenfold decreased sensitivity toward $HgCl_2$, both with cyanobacteria and PSII submembrane fractions, since immobilization matrices provide a barrier between the sensing element and the toxicants.

The micromolar (corresponding to ~ 0.1 ppm) range detection limits of the PSII biosensors toward heavy metals is not acceptable for the analysis of environmental water, since they are extremely toxic to humans at ppb levels. However, these biosensors are suitable for monitoring certain metals in sewage sludge, where concentrations range from about 3 to 800 mg/Kg, since C_{10} for a number of metal salts ranges from 20 to 445 mg/kg. One approach that is being investigated in order to increase the sensitivity of the biosensors for metals detection consists of manipulating the photosynthetic proteins to increase the number of thiol groups, so as to improve the metal-binding ability of the protein.

Table 15.5 Photosynthetic biomediators to detect heavy metals

Biological material	Immobil. method	Detection Method	Heavy metals	C_{10} of $HgCl_2$ (μM)
Spinach thylakoid Membranes	PVA-S_bQ entrapment	Electrochemical	$HgCl_2$, $CuCl_2$, $PbCl_2$, $NiCl_2$, $ZnCl_2$, $CuCl_2$	200
Synechococcus sp. PCC 7942 whole cells	PVA-S_bQ entrapment	Electrochemical	$HgCl_2$	60
Synechocystis sp. PCC 6803 whole cells	PVA-S_bQ entrapment	Electrochemical	$HgCl_2$	50
Spinach photosystem II submembranes	PVA-S_bQ entrapment	Electrochemical	$HgCl_2$, $Hg(NO_3)_2$, $CuCl_2$, $PbCl_2$, $NiCl_2$, $ZnCl_2$, $CrCl_3$	50
Spinach thylakoid Membranes	BSA-GA Cross-linking	Electrochemical	$PbCl_2$, $CdCl_2$	
Spinach thylakoid Membranes	BSA-GA Cross-linking	Oxygen evolution	$HgCl_2$, $CuCl_2$	
Chlorella vulgaris whole cells	Alumina filter disc adsorption	Oxygen evolution	$Hg(NO_3)_2$, $CuSO_4$	

15.7 Optical Transducers for the Detection of Herbicides

Chlorophyll *a* (chl *a*) fluorescence induction is the light intensity dependent on polyphasic changes in chl *a*, giving a fluorescence emission when a dark-adapted leaf or a suspension of isolated chloroplasts or thylakoids is illuminated with continuous light. The fluorescence intensity rises from an initial low value F_0, which is observed when all PSII RCs are oxidized, to the highest one F_p, when all PSII RCs are reduced. Therefore, the fluorescence intensity varies as a function of time, and, by plotting $F(t)/F_0$ vs. time, one can calculate the area (FA) contained between the curve and the horizontal line at F_p (Fig. 15.15). When the inhibitor herbicide displaces the secondary quinone Q_B from its binding pocket, an accumulation of the primary quinone Q_A occurs, and the fluorescence emission reaches its maximum in a shorter time. As a consequence, the area between the $F(t)/F_0$ curve and the F_p horizontal line is dramatically reduced. Fluorescence monitoring of the photosynthesis inhibition of herbicides has been exploited for the development of optical biosensors. Thylakoids from *Spinacea Oleracea L.* were isolated and immobilized in a BSA-GA matrix (Euzet et al. 2005), and prepared in the form of pastilles to be contacted with the test solution (i.e., buffer, with or without herbicides, or natural water). After 10 min contact with the test solution, the pastilles were dried and introduced in the measuring chamber of a fluorescence analyzer. The measurement of fluorescence showed the greatest sensitivity toward

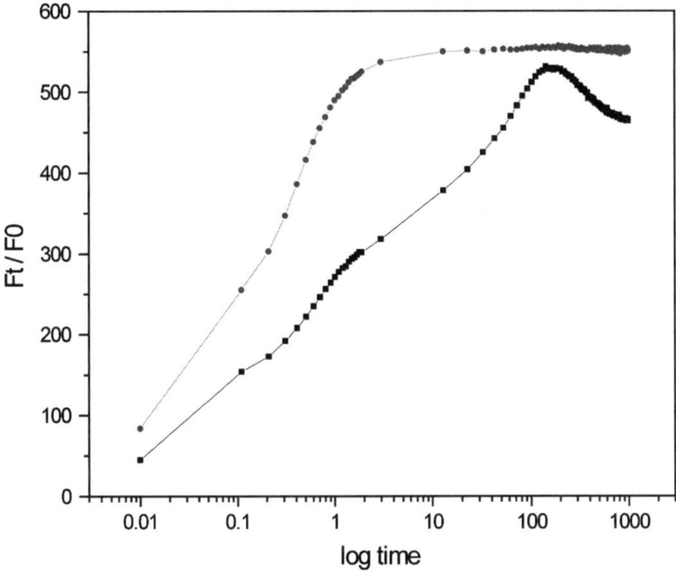

Fig. 15.15 Fast chlorophyll *a* fluorescence induction curves as a function of time measured on dark-adapted spinach illuminated at 650 nm (*bottom curve*), and in the presence of a herbicide (*upper curve*). The measured parameter for quantification is the area comprised between a curve and the straight line corresponding to its maximum value, or correlated parameters

diuron, followed by metribuzin > atrazine > cyanazine. Concentrations of herbicides inducing 10% (C_{10}) or 50% (C_{50}) inhibition were in the order of 0.35–10 μg/L, and 200–400 μg/L, respectively (corresponding to 0.35–400 ppb overall). These values are not acceptable as detection limits in environmental waters as such, since current EU and EPA regulation is more stringent (max 0.1 ppb for a single herbicide in the EU; in the ppb range in the US). However, suitable detection limits were reached through preconcentration of small volumes of sample by elution through a molecularly imprinted polymer (MIP) cartridge (Breton et al. 2006). MIPs have been developed to mimic, to some degree, biological receptors: a template (the analyte, such as cyanazine), the polymer precursors (monomers, such as methacrylic acid), and a crosslinker are dissolved in an appropriate solvent, and UV-promoted polymerization is carried out. Removal of the template after polymerization leads to a polymer containing cavities that act as specific recognition sites for the analyte. Results showed that not only does the MIP cartridge affords better sensitivity, but it allows for selectivity as well. In fact, triazines show a higher retention onto the cartridge than diuron and metribuzin.

The spectroscopic properties of the bacterial ($\Phi \cdot Q$) RC type can be exploited for the development of an optical transducer for use in photosynthetic herbicide detection. The absorption band of the bacteriochlorophyll dimer is more intense than its corresponding excited form. The time required for the bacteriochlorophyll dimer to return to the stationary state strictly depends on the electron transfer to the secondary ubiquinone acceptor Q_B. In the presence of herbicides, this electron transfer is impaired as the herbicide prevents Q_B from binding to its pocket. Consequently, the only possible recombination path involves directly the primary ubiquinone Q_A, with a reduced lifetime. A photoelectronic system was realized for monitoring temporal changes in absorption of the RC complex from purple bacterium *Rhodobacter sphaeroides*, following a flash lamp excitation. The introduction of a mathematical model made it possible to determine the concentration of photosynthetic herbicides.

A new type of biosensor utilizes a chemiluminescence detection system, based on the horseradish peroxidase (HRP)-catalyzed chemiluminescence of luminol/hydrogen peroxide (Varsamis et al. 2008). The latter is produced by photoinduction, through the PSII. The luminescent signal was investigated under illumination of thylakoids extracted from higher plants and it was observed that H_2O_2 production increased in a time – and light intensity – dependent manner. A chemiluminescence microfluidic sensor for herbicide monitoring has been proposed, in which PSII and HRP are immobilized onto magnetic beads, using GA as crosslinker (see Sect. 15.4). The PSII-loaded beads are magnetically anchored to the illuminated region of a flow channel, while the HRP-loaded beads are anchored in the vicinity of the detector optical fibre. The flow channel is made from a silicon elastomer, while the whole construct is sandwiched between the two blocks of machined Perspex. Luminol is added to the sample prior to its passage into the flow chamber, and the beads are recovered by washing them off with water/buffer after switching off of the magnetic field.

The presence of atrazine and diuron herbicides in the thylakoid samples was shown to reduce the measured hydrogen peroxide in a concentration-dependent manner, thus allowing for herbicide determination.

Similarly to amperometric biosensors, multiarray design has been applied to fluorescence-based instruments (Fig. 15.16). Multiarray biosensors (Giardi et al. 2005; Tibuzzi et al. 2007a; Tibuzzi et al. 2008) were used in order to investigate the response of wild-type, PSII-D1 protein mutants obtained by site-directed mutagenesis, and atrazine-resistant plants to a series of herbicides. Each cell in a multiarray biosensor (Fig. 15.17) consists of two independent sub-modules, i.e., a fixed compartment lodging the optical devices, and a removable upper section containing the biosample, that can be easily disposed of after testing, and reloaded with different biosensors for new analyses. The optical compartment may consist of four excitation light sources and the photodiode detector for fluorescence emission capture. The multiarray setup allows advantage to be taken of the selectivity of different mutant biomediators toward different classes of herbicides. The simultaneous response of the differently selective biomediators during the analysis of a given sample allows for identification and quantitation of different herbicide subclasses. In this respect, the use of a series of *Chlamydomonas reinhardtii* mutants (described in Sect. 15.8) in multiarray detection systems proved particularly useful.

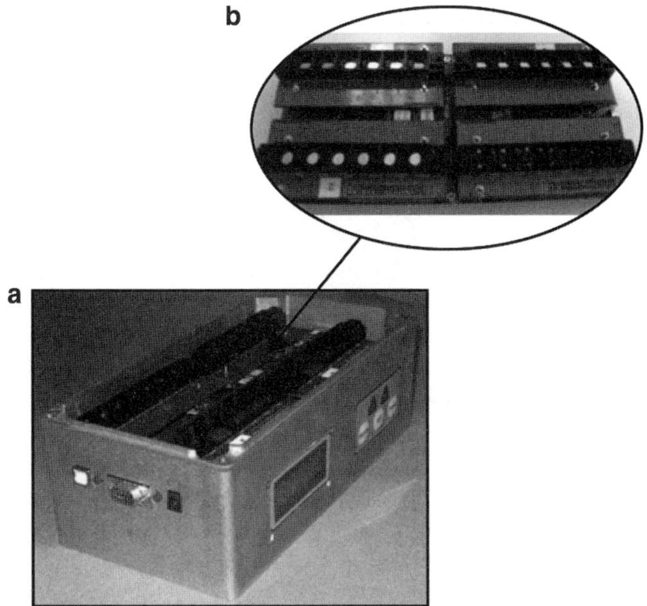

Fig. 15.16 (a) Multifluorimetric biosensor. (b) A detail of the arrays of cells in the biodevice

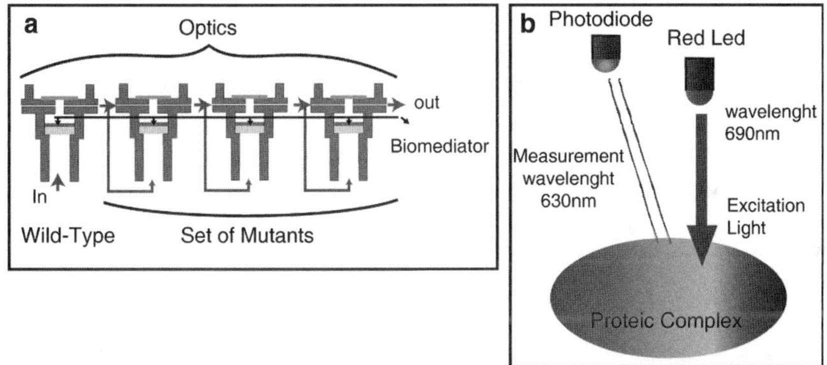

Fig. 15.17 (a) Multiarray of cells for the simultaneous fluorescence measurement of response to herbicide of a set of wild-type and mutant photosystems. (b) Schematic representation of the measuring cell

15.8 Enhancing Biomediator Properties by Bioengineering and Bioinformatics

Recent progress both in chloroplast engineering and in crystal structure analysis has considerably increased our manipulative possibilities as well as our knowledge of structure-function relationships in photosynthetic proteins (Barber 2006; Mattoo

2006). Therefore, the development of more stable, more specific and more sensitive photosystems-based biosensors now appears feasible, by making use of various mutagenesis techniques and in vivo expression of modified photosystem complexes. Even redesigning photosystems in such a way as to confer novel properties to the sensor element appears to be within reach. For example, by applying a molecular "LEGO" approach, aminoacids or small protein modules with the desired properties can be fused with PSII without compromising its function.

A clear example of the importance of such an approach is given by the modification of D1 protein of PSII. The D1 subunit has been identified as the "herbicide binding protein" by competition of such compounds in the plastoquinone Q_B binding site. Abundant biochemical and genetic evidence has accumulated which places the binding niche between amino acids 211 and 275. The study of specific amino acid substitutions (Wlski et al. 2006) in the D1 reaction center subunit provides a powerful tool for biosensing applications. For example, mutation at position 264 from Ser to Gly in the D1 protein, which is the only one observed in nature, turned out to be the molecular basis for triazine resistance in *Amaranthus hybridus*. While engineering the D1 protein in prokaryotes has been a comparatively simple task, manipulation of D1 in chloroplasts of green algae or higher plants has taken quite some time. The eucharyotic green alga *C. reinhardtii* has been particularly instrumental in engineering the D1 subunit of PSII. Its manipulation in terms of selection for photosynthetic growth represents a very robust method that can be easily extended to screen for tolerance to potentially damaging factors. A PCR-based mutagenesis protocol (Fig. 15.18) has been developed to successfully introduce various point mutations into the D1 protein. Single point mutations can have dramatic effects on the inhibitor binding, resulting in either resistance or supersensitivity. In case of resistance, the C_{50}-value of the inhibitor is higher in the mutant, when compared with the "susceptible" wild type.

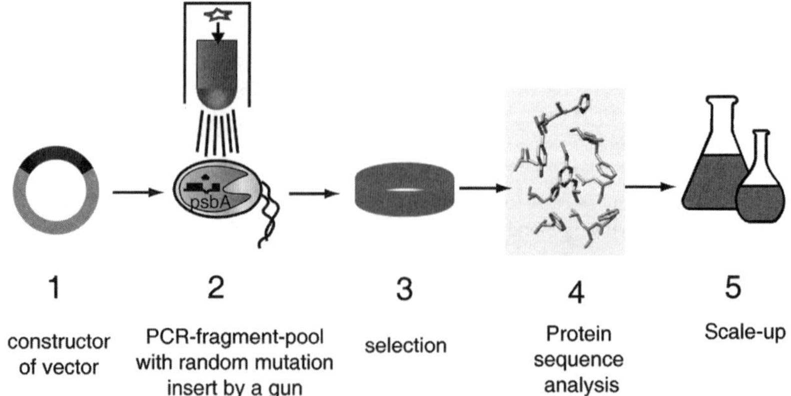

Fig. 15.18 Production of foreign proteins in *C. reinhardtii* chloroplasts. (1) Ligation of transgene into a vector containing chloroplast DNA sequences for homologous recombination; (2) Particle gun transformation of vector DNA into cells immobilized on filters; (3) Growth of transformants on selective media; (4) Screening for protein expression; (5) Cultivation of transformants in larger volumes of liquid media

In most of the mutants obtained so far, resistance is high against triazines and triazinones, and much less pronounced against ureas. The Phe255→Tyr mutant is an exception to this rule, because both triazinones, metamitron, and metribuzin exhibit supersensitivity.

The mutations Ser264→Lys and Ile have been generated by site-directed mutagenesis. Since Lys bears a positive charge, different affinities toward herbicides were expected. In fact, the resulting mutants exhibit different degrees of resistance to some herbicides (triazines atrazine, ametryn, and prometryn), they both exhibit high resistance against the carbamate herbicide phenisopham, while supersensitivity for phenolic herbicides is observed in the Lys mutant, and the Ile mutant only shows supersensitivity against bromoxynil.

Changes in binding affinities for herbicides in the Leu275→Phe mutant are marginal. This mutant is resistant to triazinones, and supersensitivity is observed against the phenolic herbicide ioxynil.

A set of D1 mutants of *C. reinhardtii* has been obtained, in which Phe206 is exchanged by site-directed mutagenesis against Ile, Glu, Ser, Ala, His, Asn, Val, and Lys. The Ser and Lys mutants all exhibited supersensitivity to atrazine and diuron. These data show that Phe206 is either part of the herbicide binding niche, or its substitution has a long range effect on it. These new molecular findings could prove relevant for the development of more sensitive herbicide biosensors.

As an alternative method, the reconstitution of a system consisting of an overexpressed D1 protein derived from selected mutated organisms and chromophore quinones could enhance the sensitivity and specificity of herbicide detection.

Bioinformatics has been recently utilized to address D1 protein modifications. This relatively recent discipline has been developed on the knowledge of biology, statistics and informatics to manage and elaborate the enormous amount of data generated by biochemical, genetic and biomolecular technologies. Based on predictive computational approaches, it is used to model the structure and intermolecular interactions of complex biological molecules such as proteins. Recent investigations aim at predicting suitable mutations within both the D1 and D2 proteins for the improvement of specificity, sensitivity, and binding affinity toward herbicides for biosensing applications. Due to the importance of the Q_B pocket in photosynthesis and in agrofood applications, models have been built using the methods of molecular mechanics and notions of contact-surface between atoms which account for the experimentally functional state of Dl protein mutants in the region of Q_B. The D1 protein-herbicide interactions can be investigated by the combination of homology-based protein modelling and virtual mutagenesis. *T. elongatus* exhibits a high similarity with *C. reinhardtii* in terms of aminoacid sequences. Thus, the three-dimensional structure of D1 and D2 proteins from *T. elongatus* were used as templates to model the three-dimensional structure of the corresponding proteins from *C. reinhardtii*. The affinities of different mutants toward atrazine have been predicted by molecular docking. This is the computational study of intermolecular forces between two or more structures, such as protein-protein, or a protein and a small ligand. In the latter case, the protein molecule, due to its size, is usually treated as a rigid body. Results obtained by docking atrazine onto Phe255→Tyr and

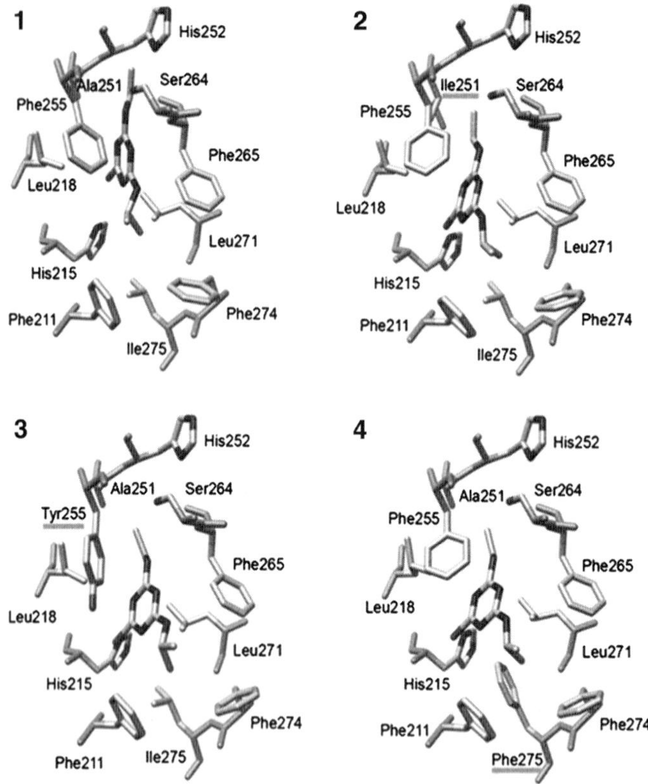

Fig. 15.19 Docking of atrazine to the binding site of D1 protein (**1**) and with single aminoacid substitutions (**2**) Ile251, (**3**) Tyr255, (**4**) Phe275

Leu275→Phe mutant structures (Fig. 15.19) indicate that the first mutation leads to a strong increase in the binding energy value for atrazine (lower affinity), while the latter leads to a decreased binding energy value (higher affinity). This is in line with the experimental evidence that mutation of Phe255 to Tyr confers resistance to atrazine, but not to metribuzin, while mutation of Leu275 to Phe confers resistance to metribuzin, but not to atrazine. Analysis of the interactions of atrazine with wild type as well as mutant proteins allows the residues which play an important role in herbicide binding, to be identified (Rea et al. 2009).

15.9 Innovative Applications of Photosystem-Based Technologies

The principles outlined in the previous sections for detection of pollutants by photosystem-based biosensors (i.e., electron flow in the RC as the source of an electrochemical or optical signal) can be exploited for other biosensing needs, thus expanding the application of such biosensors to agro-food, pharmaceutical, and biomedical fields.

For instance, in a preliminary study (Tibuzzi et al. 2007b), the possibility of determining the antioxidant power of solutions with amperometric transduction was explored. Different immobilized RCs from either cyanobacteria, algae, and higher plants were used as biomediators on positive-biased screen-printed electrodes in order to detect the current variations caused by the concomitant presence in solution of H_2O_2 and antioxidants at different concentrations.

Analytical applications are not the only ones where monitoring the activity of photosystems is of interest. In particular, photosynthetic organisms are being investigated for their oxygen-evolving properties in view of innovative applications, such as oxygen and biomass production in space (Javanmardian and Palsson 1992), and gas exchange in artificial organs (Zwischenberger et al. 2006). With a view to human space exploration, in order to use algae and cyanobacteria as bioreactors for life-support in space, the effect of space radiation on PSII needed to be assessed. For this reason, various algae and isolated PSII particles were exposed to gamma rays in the laboratory, and to actual space radiation during space missions (Angelini et al. 2001; Giardi et al. 2002; Esposito et al. 2002, Rea et al. 2008). The functionality of PSIIs under such stress was monitored through their fluorescence parameters, which meant building a computerized automatic optical biodevice for real-time measurements in space (Fig. 15.20). The exposure of the biological sample to ionizing radiation in space leads to a loss of functionality due to inactivation of enzymes, based on Radiation Target theory. These studies indicate that this damage is correlated to the radiation dose in a reproducible way, so that the device could be used to detect ionizing radiation.

The feasibility of using immobilized microalgal cells as an oxygen supply system for encapsulated pancreatic islets has been recently assessed (Bloch et al. 2006). Immunoisolation of pancreatic islets is necessary in order to prevent rejection after transplantation to patients with insulin-dependent diabetes mellitus, type I. Pancreatic islets are placed in a compartment protected from the recipient's immune system by hydrogels or by semipermeable membranes. However, immunoisolation leads to loss of vasculature, resulting in oxygen and nutrients

Fig. 15.20 (**a**) A biodevice for monitoring of photosynthetic activity of different types of cells and of PSII in space. (**b**) Upper open view of the automated apparatus

deficiency, loss of cells functionality, and, eventually, cell death. To overcome this problem, several approaches have been investigated, including induction of neovascularization, introduction of oxygen carriers into the immunoisolated matrix, oxygen generation by electrolysis of water. As an alternative, the green thermophylic alga *Chlorella sorokiniana*, immobilized in alginate beads, was found to be able to compensate for oxygen consumption of islets and provide optimal insulin secretion from encapsulated islets perfused with oxygen-free medium.

A further example of an innovative application of photosynthetic organisms for life support is the prototype of an "artificial lung", i.e., a para-corporeal device where a high temperature strain of the algae *Chlorella pyrenoidosa* is coupled to the blood stream through a gas transfer membrane to serve as the O_2 producer and CO_2 remover.

The RC of photosynthetic bacteria can be isolated easily from ready available cultures and is particularly stable against denaturation, which makes bacterial RC advantageous for technological applications, when compared with the photosynthetic material extracted from plants. Besides the development of devices for pollutant monitoring, bacterial RCs have been proven suitable for use in photochemical cells that promote the conversion of visible light energy into electrical or chemical energy. These are processes where the extracted functional biomolecules perform simplified and alternative versions of photosynthesis, converting light energy into new exploitable forms of energy.

In view of the realization of electrical systems, the orientation and the conformation of the photosynthetic complex at the surface of electrodes are the important factors because they influence the interfacial electron speed. The predominant orientation of the RC can be controlled by the surface properties of the physical support. The protein component of the purple bacteria RC consists of three subunits: L, M (membrane-embedded) and H (hydrophilic, located at the membrane surface) polypeptides. The hydrophilic subunit drives the orientation of the RC with respect to the electrode, unless immobilization is oriented, otherwise by chemical modifications of the support and of the protein. Oriented immobilization was achieved in a protein film linked to bifunctional crosslinking agents. Two different bifunctional reagents, 4-aminothiophenol and 2-mercaptoethylamine, generated opposite orientations showing dramatic differences in their electrochemical characteristics. Histidine tagging, either on the H, or on the M subunits of the RC from *R. sphaeroides* was also used to determine protein orientation on a nickel-nitrilotriacetic acid support. The immobilized protein showed the orientation-dependent photochemical properties that have been proved useful for herbicide detection. Furthermore, when the M subunit was histidine-tagged, the protein was estimated to conserve a quantum yield and a power conversion efficiency of 12%. This property was exploited for the realization of a photovoltaic cell, by immobilizing the oriented monolayer on an indium-tin oxide/gold (ITO/Au) electrode (Badura et al. 2006).

Using the same procedure, a histidine-tagged Photosystem I extracted from plants was immobilized onto a Ni^{2+}-NTA functionalized ITO/Au electrode. The PSI is a much larger complex than the bacterial RC, and its activity could be preserved only by stabilizing it with surfactant peptides. The isolated components of the PSI were chemically modified and reassembled to create a simple system that still converted light energy into chemical energy. The induced photochemical process resulted in the catalytic evolution of molecular hydrogen by the reduction of hexachloroplatinate ions.

Further technological improvements of these innovative technologies will be needed prior to their practical application.

15.10 Why make Photosystems-Based Biosensors for Environmental Monitoring?

Photosynthetic-based biosensors provide an alternative to traditional chemical methods for environmental monitoring programs, in that they are suitable to carry out rapid, simple and low-cost in-field screening for the detection of harmful chemicals. These include pesticides, heavy metals, and in particular the so-called "threat agents", or CBRNs (i.e., Chemical, Biological, Radiological, and Nuclear compounds), which are chemicals originating from several sources.

Pesticides (e.g., diazines, ureas, triazines, phenols) originating from sources such as agricultural run-off can contaminate soils, and surface and ground waters. Worldwide, millions of tons of pesticides are applied annually, of which an estimated amount of 76.4 million pounds of atrazine are used in the US alone (www.epa.gov/pesticides/factsheets/atrazine_background.htm). Such a massive use of pesticides can affect the health of ecosystems and, hence, of man. Organophosphates and carbamates can also have severe neurotoxic effects on humans and infants in particular. For these reasons, the intake of commercial pesticide preparations in the environment as well as their trace amounts in surface, ground and drinking waters, and in food and soil are strongly regulated. Regulations are now being strictly enforced in Europe with a maximum pesticide concentration of 1 μg/l per pesticide in drinking water. The most recent EU directive concerning pesticides (Commission Regulation (EC) No 149/2008 of 29th January 2008) establishes the maximum residue levels of pesticides on food and feed of plant and animal origin at 0.01–0.05 mg/kg depending on the pesticide and food species. The increasing concern for pesticides intake arising in the world necessitates the development of reliable control procedures and sensor systems.

Heavy metals (e.g., mercury, arsenic, chromium, lead) are toxic or poisonous at low concentrations, which cannot be degraded, and tend to bioaccumulate. Their toxic effects arise from their action on enzymes of the metabolic chain. Heavy metal poisoning could result, for instance, from drinking water contamination, high ambient air concentrations near emission sources, or intake via the food chain. Heavy metals can enter water supply systems from streams, lakes, rivers, and groundwater contaminated by the industrial and consumer waste, or even from acidic rain breaking down soils.

The so-called CBRN threat agents are compounds which have gained increasing importance since the public concern for terrorist attacks has spread around the world. Detecting CBRN agents has recently become one of the most important and crucial objectives for all National Security Agencies in the world, in particular in USA and Europe. Reports of various agencies indicate that the presence of mutagenic compounds in different habitats is a common phenomenon rather than an exception.

The Environmental Mutagen Information Centre database contains over 20,000 citations to literature on agents tested for mutagenic activity (www.nlm.nih.gov/pubs/factsheets/emicfs.html). The problem of the presence of harmful chemicals in natural habitats is important because they may be capable, even at extremely low concentrations, of inducing serious diseases, including cancer, and may lead to fertility problems and negative genetic changes in future generations. Given the large number of potentially harmful chemicals, the challenge of determining mutagenic contamination in environmental samples with quick and preliminary assays, should be met, before a detailed, time-consuming analysis can be carried out.

The list of potential harmful agents comprises a wide range of chemical and biological characteristics thus requiring highly flexible analytical systems and diverse detection capabilities. Also, a diffused environmental monitoring requires that analytical systems be field capable and relatively inexpensive.

15.11 Concluding Remarks

In perspective, biosensors may play a key role in early warning and protection of public health. In fact, photosynthetic-based biosensors can alleviate the need for costly and often environmentally unsound analyses in two ways: (a) The developed biosensor allows rapid field detection of pollutants and assessment of their phytotoxicity. Therefore, it is suitable for prescreening environmental samples, providing an indication of which samples require further analysis by standard, more costly and time consuming techniques, such as chromatography and immunological methods. This should avoid the use of large quantities of organic solvents and the need for expensive equipment to extract polluted compounds; (b) the use of an array of PSII-based biosensors that are each highly specific to a given class of pollutant, in conjunction with appropriate data elaboration, will enable the user to generate an "identikit" of the pollutants present in samples.

References

Angelini G, Ragni P, Esposito D, Giardi P, Pompili ML, Moscardelli R, Giardi MT (2001) A device to study the effect of space radiation on photosynthetic organisms. 1st international workshop on space radiation research and 11th Annual NASA space radiation health investigators' workshop Arona (Italy), May 27–31, 2000. Phys Med 17:267–268

Avramescu A, Rouillon R, Carpentier R (1999) Potential use of a cyanobacterium *Synechocystis sp*. Immobilized in poly(vinylalcohol): Application to the detection of pollutants. Biotechnol Tech 13:559–562

Badura A, Esper B, Ataka K, Grunwald C, Wöll C, Kuhlmann J, Heberle J, Rögner M (2006) Light-driven water splitting for (bio-)Hydrogen production: Photosystem 2 as the central part of a bioelectrochemical device. Photochem photobiol 82:1385–1390

Barber J (2006) Photosystem II: An enzyme of global significance. Biochem Soc Trans 34:619–631

Bloch K, Papismedov E, Yavriyants K, Vorobeychik M, Beer S, Vardi P (2006) Immobilized microalgal cells as an oxygen supply system for encapsulated pancreatic islets: A feasibility study. Artif Organs 30:715–718

Breton F, Euzet P, Piletsky SA, Giardi MT, Rouillon R (2006) Integration of photosynthetic biosensor with molecularly imprinted polymer-based solid phase extraction cartridge. Anal Chim Acta 569:50–57

Esposito D, Pace E, Margonelli A, Rizzuto M, Giardi P, Giardi MT (2002) A biodosimeter for the detection of ionizing radiation utilizing isolated enzymes. Radiat Prot Dosimetry 99:303–305

Euzet P, Giardi MT, Rouillon R (2005) A crosslinked matrix of thylakoids coupled to the fluorescence transducer in order to detect herbicides. Anal Chim Acta 539(1–2):263–269

Giardi MT, Pace E (2005) Photosynthetic proteins for technological applications. Trends Biotechnol 23:257–263

Giardi MT, Piletska EV, (2006) Biotechnological applications of photosynthetic proteins: Biosensors, Biochips and Biodevices. Biotechnology intelligence unit. Landes Bioscience/ Eurekah.com Sprinter Science + Business Media, LLC, New York

Giardi MT, Guzzella L, Euzet P (2005) Detection of herbicide subclasses by an optical multibiosensor based on an array of photosystem II mutants. Environ Sci Technol 39: 5378–5384

Giardi MT, Scognamiglio V, Rea G, Rodio G, Antonacci A, Lambreva M, Pezzotti G, Johanningmeier U (2009) Optical biosensors for environmental monitoring based on computational and biotechnological tools for engineering the photosynthetic D1 protein of *Chlamydomonasreinhardtii*. Biosens Bioel 25:294–300

Giardi MT, Zanini A, Esposito D, Fasolo F, Torzillo G, Faraloni C (2002) Biodevices for the detection of space radiation effect on photosynthetic organisms. Proceedings of the Second European Workshop on Exo/Astrobiology, Graz, Austria, 16–19 September 2002 (ESA November 2002), p 455–456

Javanmardian M, Palsson BO (1992) Design and operation of an algal photobiorector system. Adv Space Res 12:231–235

Koblizek M, Masojidek J, Komenda J, Kucera T, Pilloton R, Mattoo A, Giardi MT (1998) A sensitive photosystem II-based biosensor for detection of a class of herbicides. Biotechnol Bioeng 60:664–669

Koblizek M, Maly J, Masojidek J, Komenda J, Kucera T, Giardi MT, Mattoo AK, Pilloton R (2002) A biosensor for the detection of triazine and phenylurea herbicides designed using photosystem II coupled to a screen-printed electrode. Biotechnol Bioeng 78:110–116

Laberge D, Chartrand J, Rouillon R, Carpentier R (1999) In vitro phytotoxicity screening test using immobilized spinach thylakoids. Environ Toxicol and Chem 18:2851–2858

Lam K, Irwin EF, Healy KE, Lin L (2006) Bioelectrocatalytic self-assembled thylakoids for micro-power and sensing applications. Sensors Actuators B Chem 117:480–487

Mattoo A (2006) The D1 Protein: Past and future perspectives. Book series: advances in photosynthesis and respiration, Vol 21. Book Photoprotection, photoinhibition, gene regulation, and environment. Springer, Netherlands

Rea G, Esposito D, Damasso M, Serafini A, Margonelli A, Faraloni C, Torzillo G, Zanini A, Bertalan I, Johanningmeier U, Giardi MT (2008) Ionizing radiation impacts photochemical quantum yield and oxygen evolution activity of photosystem II in photosynthetic microorganisms. J Radiat Biol 84:867–877

Rea G, Polticelli F, Antonacci A, Scognamiglio V, Katiyar P, Kulkarni SA, Johanningmeier U, Giardi MT (2009) Structure-based design of novel *Chlamydomonas reinhardtii*. D1-D2 photosynthetic proteins for herbicide monitoring. Protein Sci 18:2139–51

Rouillon R, Boucher N, Gingras Y, Carpentier R (2000) Potential for the use of photosystem II submembrane fractions immobilized in poly(vinylalcohol) to detect heavy metals in solution or in sewage sludge. J Chem Technol Biotechnol 75:1003–1007

Scognamiglio V, Raffi D, Lambreva M, Rea G, Tibuzzi A, Pezzotti G, Johanningmeier U, Giardi MT (2009) *Chlamydomonas reinhardtii* genetic variants as probes for fluorescence sensing system in detection of pollutants. Anal Bioanal Chem 394:1081–7

Tibuzzi A, Pezzotti G, Lavecchia T, Rea G, Giardi MT (2008) A portable light-excitation equipped bio-amperometer for electrogenic biomaterials to support the technical development of most biosensors. Sens Transducers J 88:9–20

Tibuzzi A, Rea G, Pezzotti G, Esposito D, Giardi MT (2007a) A new miniaturized multiarrays biosensor system for fluorescence detection. J Phys Condens Matter 19:395006 12pp

Tibuzzi A, Rea G, Bianconi D, Giardi MT, Pezzotti G (2007) A flexible modular bio-instrument for amperometric detection of bioenergetic material: a RC-based measurement of antioxidant power. Techical Conference: Nanotech in Life Science & Medicine, Symposium: Bio Sensors & Diagnostics, Area: Biosensor-General, Santa Clara, CA, USA

Touloupakis E, Giannoudi L, Piletsky SA, Guzzella L, Pozzoni F, Giardi MT (2005) A multi-biosensor based on immobilized Photosystem II on screen-printed electrodes for the detection of herbicides in river water. Biosens Bioelectron 20:1984–1992

Varsamis DG, Touloupakis E, Morlacchi P, Ghanotakis DF, Giardi MT, Cullen C (2008) Development and testing of an optical (chemiluminescence) micro-fluidic sensor for herbicides monitoring. Talanta 77:42–47

Wlski S, Johanningmeier U, Hertel S, Oettmeier W (2006) Herbicide binding in various mutants of the photosystem II D1 protein of Chlamydomonas reinhardii. Pestic Biochem Physiol 84: 157–164

Zwischenberger BA, Clemson LA, Zwischenberger JB (2006) Artificial lung: Progress and prototypes. Expert Rev Med Devices 3:485–497

Chapter 16
Oligonucleotides as Recognition and Catalytic Elements

Keith E. Herold and A. Rasooly

Abstract Oligonucleotides function as recognition elements for both nucleic acids and proteins in many ways, starting from their basic function as genetic material recognizing the complementary sequences of RNA or DNA, through their role as genetic regulatory elements in the form of antisense DNA/RNA, interfering RNA (RNAi) as well as enzymatic activities such as trans-cleaving ribozymes. Oligonucleotides can also serve as recognition elements for proteins in the form of aptamers. All of these functions are derived from the basic primary, secondary, and tertiary structures along with their combinatorial nature, which allows vast variability within even short sequences.

For biotechnology, the utility of oligonucleotides goes far beyond their basic function as a carrier of genetic information. Oligonucleotides are widely used as recognition elements for a large number of DNA and protein manipulation technologies, including numerous methods for DNA amplification, manipulation (e.g., mutation insertion or repair), and DNA sequence recognition and analysis. They are used for manipulation of gene expression including the use of ribozymes to catalyze RNA digestion, and siRNA and anti-sense RNA/DNA used for gene silencing in both in vitro and in vivo. In addition, they serve as an alternative to antibodies as ligands for protein detection.

The term oligonucleotide (or oligo) refers to a short segment of DNA or RNA commonly synthesized today by polymerizing nucleotide precursors using automated synthesizers. Although most oligos used in molecular biology are short in the range of 20–30 bases (or mer), the term is used here also to refer to longer sequences (e.g., ~200 bases) and to ribozymes, which are traditionally not viewed as oligos.

In this manuscript, we describe the basic structure of oligonucleotides relevant to general utility in molecular biology and biotechnology. This utility includes function as recognition elements, basic utility for DNA amplification, recognition and

K.E. Herold (✉)
Fischell Department of Bioengineering, University of Maryland, College Park, MD 20742, USA
e-mail: herold@umd.edu

M. Zourob (ed.), *Recognition Receptors in Biosensors*,
DOI 10.1007/978-1-4419-0919-0_16, © Springer Science+Business Media, LLC 2010

manipulation, application for the regulation of gene expression and function and utility as protein recognition elements.

Keywords Oligonucleotides · Recognition · Catalytic · RNA · DNA · Aptamers

Abbreviations

DNA	Deoxyribonucleic acid
RNA	Ribonucleic acid
RNAi	interfering RNA
mRNA	messenger RNA
rRNA	ribosomal RNA
tRNA	transfer RNA
ssDNA	single-stranded DNA
dsDNA	double-stranded DNA
PCR	Polymerase chain reaction
NN	Nearest-neighbor
DMSO	Dimethyl sulfoxide
Bp	Base pair
SNP	Single nucleotide polymorphisms
LNA	Locked nucleic acids
LATE-PCR	Linear-after-the-exponential-PCR
MSP	Methylation-specific PCR
CpG	—C—phosphate—G—
MLPA	Multiplex ligation-dependent probe amplification
Q-PCR	Quantitative PCR
RT-PCR	Reverse transcription PCR
TAIL-PCR	Thermal asymmetric interlaced PCR
WGA	Whole genome amplification
DOP-PCR	Segenerate oligonucleotide primed PCR
MDA	Multiple-displacement amplification
ChIP	Chromatin immunoprecipitation
cDNA	complementary DNA
BLAST	Basic Local Alignment Search Tool
aRNA	Antisense RNA
miRNAs	microRNAs
siRNAs	small interfering RNAs
RISC	induced silencing complex
siRNA	short interfering RNA
shRNA	short hairpin RNA
esiRNAs	endoribonuclease-prepared short interfering RNAs
SELEX	systematic evolution of ligands by exponential enrichment
QCM	Quartz crystal microbalance
SPR	Surface plasmon resonance
CNT-FETs	Carbon nanotube field-effect transistors

16.1 Introduction

Natural nucleic acids, including RNA (ribonucleic acids) and DNA (deoxyribo nucleic acids), serve various biological functions in the cell and can also be applied as biological recognition elements in biodetection against various targets. Although both RNA and DNA can form double-stranded helices, RNA is usually single stranded because it is separated from its DNA template during synthesis. DNA is usually found in a double-stranded helical form, bound to the template from which it is synthesized. These characteristics are associated with differing functions in the cell and they impact our ability to use these molecules. DNA is primarily a carrier of genetic information, exhibiting long-term stability and interactions with a small set of enzymes that regulate transcription and duplication during mitosis. RNA has multiple functions in the cell, including mRNA (messenger RNA) that acts as a short-term carrier of genetic information, rRNA (ribosomal RNA) that helps form ribosomes, tRNA (transfer RNA) that direct amino acids during protein synthesis on the ribosome, and a number of small RNA molecules involved in gene regulation (anti-sense RNA), RNA transcript splicing (ribozymes) and degradation of RNA molecules through RNA interference (RNAi).

Nucleic acids bind to other molecules in several distinct ways. The most obvious is Watson–Crick binding between complementary sequences, which can occur between two like molecules (both RNA or DNA), between one DNA and one RNA molecule, or between two segments of a single molecule (a hairpin). This type of binding is a fundamental aspect of life, involved in carrying genetic information, transcription to mRNA, and translation to protein. A second type involves proteins that bind to nucleic acids by adopting an electrostatic surface profile that mates with the target molecule. A third type of nucleic acid binding, similar to protein binding, involves single-stranded RNA molecules, which fold into a compact form (an aptamer or a ribozyme) that presents an electrostatic binding surface that can interact with other molecules. All of these types of nucleic acid binding can be harnessed for recognition of nucleic acids and proteins used for experimental and clinical applications.

16.2 DNA Hybridization

Our understanding of the structure of DNA was advanced considerably in 1953 when Watson and Crick published a paper (Watson and Crick 1953) about the double helical structure. In the range of temperatures where biological cells are viable, DNA is normally found in cells as a double-stranded helix (dsDNA) held together by hydrogen bonds between the complimentary bases. Elevated temperatures (i.e., above the melting temperature), can open these hydrogen bonds, resulting in "denatured" single-stranded DNA (ssDNA), which upon cooling can anneal to re-form the dsDNA. If DNA from two genetically distinct sources is first denatured, then mixed and cooled, double-stranded DNA hybrids can form between some sections of the genetically distinct molecules. The concept of nucleic acid hybridization (Rich 1960), which originally referred to DNA-RNA hybridization, is used today to refer to the process of alignment of complimentary bases, winding of the two strands into a helical configuration and hydrogen binding between the strands (whether RNA or DNA). The concept of the annealing temperature (or melting temperature), at which DNA molecules transition between double-stranded and single-stranded forms was introduced around 1960 (Doty et al. 1960; Marmur and Lane 1960). The annealing process is a dynamic hybrid formation process, which involves many intermediate, metastable states. The transition is dominated by random thermal fluctuations that break all but the most energetically favorable hybrids. The foundation for many current nucleic acid technologies is the Southern blot method (Southern 1975), which employs gel electrophoresis, for size separation, followed by blotting to a membrane and hybridization of a labeled probe sequence. Microarray technology expands the basic approach by reducing the size of the array elements used for probes and increasing the number of probes used for hybridization (Ekins et al. 1989, 1990; Ekins and Chu 1999) enabling "multi-analyte" analysis of nucleic acids using "microspots" distributed on an inert solid support and detected by fluorescent labels.

Several other major technical milestones in engineering DNA as a recognition element include: (1) advances in DNA synthesis chemistry that enable routine synthesis of oligos of specified sequence (Behlke and Devor 2007), (2) the discovery of PCR (polymerase chain reaction) (Mullis 1990), and (3) the discovery of DNA sequencing techniques (Sanger and Coulson 1975). Although the basic chemistry of DNA sequencing is very similar to that developed in 1975, the technology and bioinformatics aspects of sequencing have undergone rapid change that is currently accelerating (Hall 2007). The current state of the art includes the availability of the genome sequence of many organisms, the low cost availability of custom DNA sequences and the ability to amplify DNA sequences, using PCR, to increase the sensitivity of many detection methods.

The use of short sections of DNA (oligonucleotides, or oligos) as recognition elements is a key aspect in several microarray technologies due to the exquisite sequence specificity of hybridization. Hybridization is destabilized significantly when only a few mismatched bases exist. Thus, a system based on DNA hybridization as a recognition element can be highly specific with low rates of false signals. Although there are many variations on the theme, the microarray configuration involving surface tethered oligos highlights the potential. When target DNA molecules are labeled (e.g., with a fluorescent group) and then brought in contact with the microarray surface at annealing conditions, a fluorescent signal at a particular location on the surface indicates that the sample contains the complementary sequence.

DNA hybridization also plays a central role in PCR where oligo primers are employed to define, via hybridization to the target DNA, a region of the target that is to be amplified (i.e., create multiple millions of copies). The primers are designed to bind to the target to define and initiate extension of the target section via DNA polymerase. Amplification of the target via PCR provides a method to increase sensitivity of an assay because of the high fidelity that PCR provides. Primer design is critical to the specificity and amplification yield of a PCR reaction. Primer design includes many considerations, some of which overlap with probe design. A starting point for design of oligos for such applications is an understanding of the factors that impact melting temperature.

16.2.1 Melting Temperature

The melting temperature, T_m, of a DNA hybrid is defined as the temperature at which one-half of the molecules are double-stranded and the remaining half are single-stranded. The melting temperature depends on various factors including the length of the DNA, the sequence (strong correlation to GC content), salt concentration, and DNA concentration. Duplex stability is a local property that varies considerably with local sequence composition. DNA melting occurs over a range of temperatures with local melting of AT rich regions occurring before GC rich regions. The state of a DNA solution can be measured through the effect of duplex state on many different solution variables including viscosity, light absorption, and others. The kinetics of denaturation and renaturation can confound melting curve measurements unless a sufficiently slow ramp rate is used to allow the system to remain close to equilibrium. Sequences can exhibit widely differing kinetics depending on the presence of significant metastable intermediate states.

A melting curve can be measured using a spectrometer with a temperature ramping capability to measure the change in absorbance as the DNA denatures. An example melting curve plotting optical density at 260 nm is shown in Fig. 16.1. At low temperature, the absorbance is low consistent with double-stranded DNA. The melting temperature, ΔH and ΔS values were obtained from a thermodynamic analysis of the data. The transition from double-stranded to single-stranded is termed "DNA melting." The melting characteristics of a particular sample depend strongly on the length of the complementary region, the cation concentration and the base sequence. Many melting temperature models have been proposed and the best models today are the nearest-neighbor (NN) models. The NN models assume that the sequence dependence can be correlated using a free energy function involving the identity and orientation of neighboring base pairs.

16.2.2 Melting Temperature Prediction

The nearest neighbor (NN) melting temperature model (Breslauer et al. 1986; Sugimoto et al. 1996; SantaLucia 1998) can be summarized as

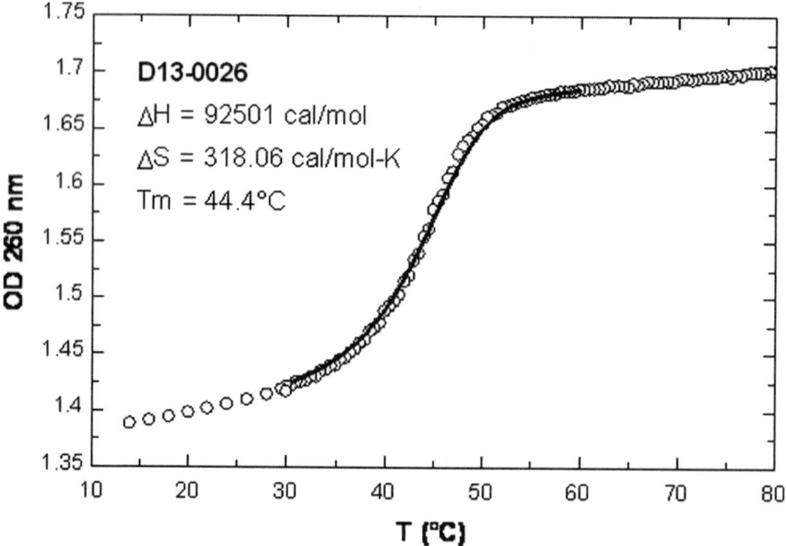

Fig. 16.1 Melting curve for DNA measured using an AVIV spectrometer. The sequence is 16 bp long (CAACTTGATATTAATA) with a single mismatch at (C) at position 8 in the otherwise complementary sequence. [DNA] = 5.5 μM each, [NaCl] = 1 M. Solid line is a melting model, open circles are data taken at 0.5 K increments except below 30°C where 2 K increments were used. Calculated melting temperature for this sequence without mismatch is 51.6°C and with mismatch is 41.9°C (Herold and Rasooly 2003). Ramp rate is 1.3 K/min

$$T_{\mathrm{m}} = \frac{\Delta H}{\Delta S + R \ln C_{\mathrm{t}}} + 16.6 \log_{10}[x_{\mathrm{salt}}]$$

where:

$$\Delta H = \sum_i n_i \Delta H_i + \Delta H_{\mathrm{init\ GC}} + \Delta H_{\mathrm{init\ AT}} + \Delta H_{\mathrm{sym}}$$

$$\Delta S = \sum_i n_i \Delta S_i + \Delta S_{\mathrm{init\ GC}} + \Delta S_{\mathrm{init\ AT}} + \Delta S_{\mathrm{sym}}$$

and where ΔH_i and ΔS_i are the enthalpy and entropy of the $n-1$ pairs of neighboring nucleotides contained in the sequence of interest. R is the universal gas constant (1.987 cal/mol K). C_{t} is the total molar concentration of single strands when oligos are self complementary or it is equal to ¼ of this concentration in the case of non self-complementary sequences. The model assumes an equal concentration of the complementary strands. The NN model was implemented using experimentally derived parameters (SantaLucia 1998) and is available, along with a number of other nucleotide analysis tools (Herold and Rasooly 2003).

Other solution variables that can affect melting temperature include pH and ionic strength. An increase in pH tends to destabilize a DNA duplex due to deprotonation of bases. This effect is minimized in most oligo applications by use of an appropriate pH buffer that limits changes in pH. The ionic strength measures the electrostatic properties of the aqueous-based solvent. DNA duplexes bind cations, and thus, higher ionic strength (i.e., higher salt concentration) tends to stabilize duplexes, leading to an increase in T_m with higher salt concentration as indicated in the NN melting temperature model.

Various organic solvent additives are known to decrease stability of duplexes including formamide (Casey and Davidson 1977), dimethyl sulfoxide (DMSO) (Chester and Marshak 1993), ethanol and methanol. These reagents form complexes with single-stranded DNA and thus disrupt duplex formation. They are sometimes added to an aqueous DNA solution to change the DNA melting characteristics. Formamide, at concentrations up to 50%, is known to lower the melting temperature of DNA by approximately 0.7 K for each 1% of formamide. Formamide is sometimes used in microarray hybridization to create stringent hybridization conditions, which minimize mismatch and non-specific annealing of target molecules to the tethered oligos while allowing the recognition reaction to proceed at lower temperatures (55–65°C).

The effect of sequence length can be understood by looking at the melting temperature for a large number (10^5) of random sequences as shown in Fig. 16.2. Figure 16.2a is a plot of the distribution of melting temperatures for random sequences of a particular length, with each trace for a different length sequence. The average melting temperature is indicated by the peak value showing that the melting temperature rises with the sequence length. The sensitivity of the mean melting temperature to length is larger for shorter oligos. Another aspect that can be seen in Fig. 16.2a is that the width of the distribution decreases as the length increases. This means that the influence of sequence on melting temperature is more significant for short oligos. These observations highlight a general rule that the melting temperature of short oligos is more sensitive, as compared to long oligos, to changes in any reaction variable.

Another view of the variables that affect the melting temperature is seen in Fig. 16.2b, where both curves are for 16 bp length oligos. The lower curve is for an ensemble of 10^5 random sequences as in Fig. 16.2a. The other curve is calculated for an ensemble of 10^5 random 16 bp sequences that all have the same GC content of 50% (obtained by random shuffling of an input sequence). The peak values for both distributions in Fig. 16.2b occur at the same temperature. Constraining the GC content to a fixed value narrows the distribution and increases the peak height considerably. This result shows that although GC content is an important variable, it is not sufficient to accurately predict DNA melting temperature. Applications that require closely matched melting temperatures, including primer pairs and groups of microarray probes, require the accuracy of the NN model.

Another feature that motivates the use of longer oligo sequences is the possibility of a random sequence match. Assuming equal probabilities for each of the four

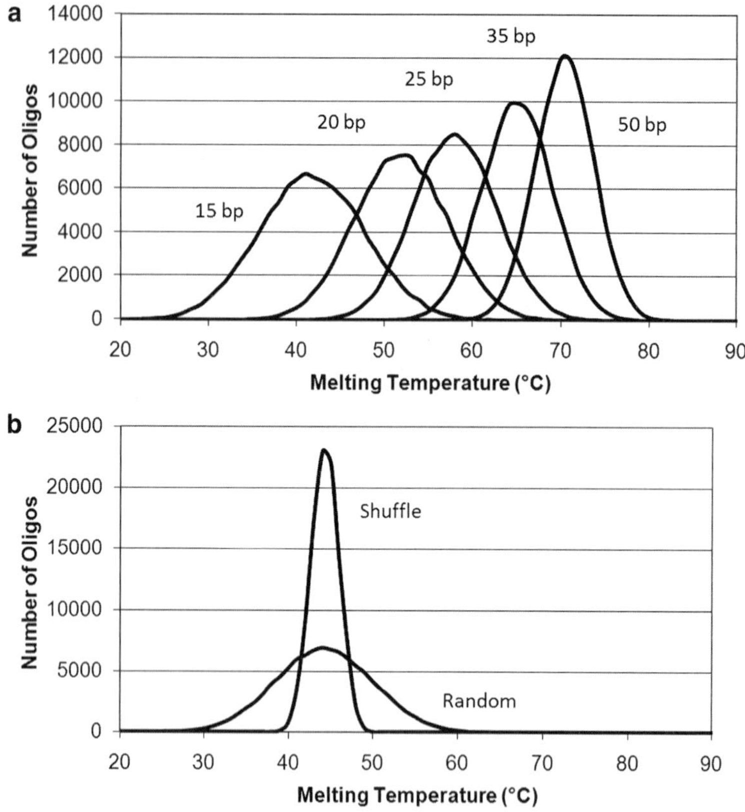

Fig. 16.2 DNA melting temperature calculated using a nearest neighbor model and an ensemble of 10^5 random sequences. (**a**) Random distributions for five different oligo lengths. (**b**) Random and GC-fraction constrained distributions of 16 bp length

bases (i.e., $P = 0.25$) at any given position in a nucleotide sequence, the probability of a random sequence matching at a particular position can be written as

$$P = \left(\frac{1}{4}\right)^{(N-m)} \quad \frac{N!}{m!\,(N-m)!}$$

where N is the length of the oligo and m is the number of allowed mismatch locations in the hybrid. To obtain the probability of a random match to a long target sequence, the value must be multiplied by the sequence length as $P_{\text{tot}} = PL$ where L is the target length. Although the probability of a random exact match to an oligo of 20 bp is only 4.5×10^{-6} when the target is genomic DNA from a typical bacterium ($L = 5 \times 10^6$ bp) it rises to 2.7×10^{-3} for a genome similar to the human genome ($L = 3 \times 10^9$ bp). Mismatches increase the probability still further. The probability of a random match can be minimized by: (1) using long oligo sequences (an increase of 5 bp decreases the probability by 10^3 in the range of the example), (2) using short target sequences, or (3) checking the oligo sequence against the known sequence of the target to avoid oligos with multiple potential hybridization sites.

16.2.3 Effect of Mismatched Bases on Melting Temperature

Hybridization between two sequences that are complementary with the exception of a mismatch at one or a few locations will occur if the binding strength of the complementary regions is sufficient to offset the steric hindrance caused by the mismatches. However, any mismatch will cause a decrease in melting temperature that can be significant. This sensitivity to mismatch is the basis for the high specificity of DNA hybridization. In general, mismatches have a larger destabilizing effect if they are away from the ends of a duplex since an interior mismatch must be anchored on two sides by compensating binding forces. The magnitude of the effect of a mismatch on the melting temperature depends strongly on the length of the duplex with larger effects seen in shorter sequences (He et al. 1991; Allawi and SantaLucia 1997, 1998a, b, c, d; Peyret et al. 1999). Also, some mismatches are more stable than the others; for example, G binds more tightly to C but it also binds with A and T. An example showing both the length and base specific effects on mismatch stability is provided in Table 16.1. Mismatch stability influences specificity of binding, impacting all hybridization applications. Some applications, such as microarray SNP identification, must be able to discriminate single base differences. In such cases, probe length must be optimized to trade off between mismatch temperature sensitivity (enhanced with short probes) and overall specificity (enhanced with long probes). However, for other hybridization applications, complementarity to within a few base pairs indicates a match and thus such applications can use longer probes effectively.

Discrimination of DNA sequences with a single mismatched base sometimes lacks the required specificity. One approach to address this for microarray analysis is tiling, the use of short overlapping oligonucleotide probes (Hacia et al. 1999; Cutler et al. 2001; Kallioniemi 2001) that provide high specificity through overlapping sequences. Several computer programs are available for probe design including a program enabling tiling design (Herold and Rasooly 2003).

Table 16.1 Example melting temperature calculations (Herold and Rasooly 2003) with mismatched base at position 7 (bold). The corresponding complementary base is A which exhibits the highest stability

Sequence	Opposing base (position 7)			
	G	A	T	C
GATCGATCGATC	21.4	**31.5**	19.4	17.0
GATCGATCGATCGATCGATCGATC	53.3	**56.0**	51.1	51.0

Non-natural nucleosides, such as inosine (Watkins and SantaLucia 2005) and LNA (locked nucleic acids) (McTigue et al. 2004) can be used in oligos to exploit several unique features. Inosine binds equally well with A, T or C bases and thus provides a method for designing degenerate probes (or primers) that can hybridize with a group of sequences instead of with a single sequence. One application where this is of interest is PCR amplification of bacterial sequences that exhibit some genetic divergence. If degenerate primers can be designed that anneal equally well to a range of targets, then a single primer set can be used as the amplification step in a detection system for a broad range of microorganisms. LNA nucleosides are locked in the sense that they are constrained geometrically in comparison to standard DNA nucleosides. Single-stranded LNA exhibits a curved backbone that is better prepared for helix formation as compared to DNA. Thus, the entropy of single-stranded LNA is lower than DNA, leading to higher melting temperatures for the same oligo length. LNA is used in critical applications, such as SNP detection, where high sensitivity is required.

16.2.4 Hybridization Interference Through Self-Annealing

Watson–Crick binding can create unwanted self-annealed structures if care is not taken in the oligo design process. Such self-annealing interactions include hairpins and self-annealing between two oligos. Hairpins are caused by self-annealing between the folded-back ends of an oligo with a loop. Self-annealing between two oligos can occur when an oligo sequence includes two segments that are the reverse complement of each other. In general, sequences that can form hairpins can also self-anneal between oligo molecules. As with the stability of any duplex, the stability of such interfering structures depends on the number of complementary bases and the particular sequence. Hairpin structures are further stabilized by the tethered proximity of the complementary sequence. The dynamic equilibrium in solution includes all of the possible molecular conformations with the concentration of each determined by the relative stability. Since oligos that exist in one of the self-annealed forms are not available for hybridization, these structures interfere with recognition reactions. The best option for dealing with self-annealing is to avoid it by using oligos that do not possess self-annealing sequences. For typical oligos, it is a simple matter to write a computer program to examine the sequence and determine the potential for self-annealing (Herold and Rasooly 2003). This type of check can be integrated into an oligo search algorithm for selection of probes and primers. Another option for dealing with self-annealing is to use stringent solution conditions such as high temperature and addition of denaturing agents. This approach is often used for microarray hybridization reactions.

Another type of interference can occur when oligos of different sequence are present in solution (e.g., PCR primers). Annealing between such oligos can reduce the activity of those oligos for their main function. In the case of PCR primers, inter-primer annealing can cause the formation of interfering PCR products called primer-dimers which are formed when the 3′ ends of the primers anneal followed by extension of the double-stranded molecule by polymerase. Primer-dimers reduce the yield of the primary PCR reaction by reducing the activity of the primers, consuming PCR reactants and occupying the polymerase. Primer-dimers are a concern in any PCR reaction, and are more likely when multiple oligos are present in the reaction such as in multiplex PCR where they often limit the number of primer sets that can be used in a single reaction. Primer-dimers can often be avoided through careful primer design where the sequences are analyzed by computer to flag the primer-dimer potential (Herold and Rasooly 2003).

16.2.5 Hybridization to Probes Immobilized on a Solid Matrix

The kinetics of hybridization to probes immobilized on a solid matrix depends on several additional factors such as the DNA probe surface density and the length of the immobilized DNA. The effect of the immobilization substrate was studied and it was shown that hybridization is largely independent of immobilization substrate (Stillman and Tonkinson 2000, 2001). Hybridization kinetics on a solid matrix can be described as a function of the immobilized DNA density following (Fixe et al. 2004):

$$V_0 = \frac{V_{\max} I}{K + I}$$

In this equation, V_0 is the initial hybridization rate, $V_{\max}$ is the maximum calculated hybridization rate, I is the surface density of immobilized DNA probes and K is the probe surface density where the hybridization rate is half of the maximum (Stillman and Tonkinson 2001). Experimentally, plotting the initial rate of hybridization for different immobilized capture probe densities demonstrates that the hybridization rate is dependent on the DNA surface concentration.

Hybridization specificity determines the non-specific binding to the probe measured as “background level.” Stringent hybridization conditions are conditions that are largely unfavorable to hybridization, minimizing less-specific or random binding to the probes while capturing only the most highly complementary sequences to the probes. Several parameters affect hybridization stringency including: temperature, salt concentration, formamide concentration, probe size, target concentration, hybridization chamber configuration, and time. High stringency (e.g., high hybridization temperature and low salt in hybridization buffers), permits very specific hybridization only between very similar sequences while low stringency conditions (e.g., lower temperature and high salt), allows hybridization when the sequences that are less similar

Another approach to reduce non-specific annealing of the target DNA to the probe is to add non-complementary carrier DNA to the hybridization buffer. This dilutes the target DNA for the same basic solution conditions and, thus, increases stringency. Another factor for hybridization is time. A typical hybridization may take 12 to 24 h, with shorter times (a few hours) used for shorter probes, shorter targets and higher target concentration.

To increase the strength of hybridization, especially when the target is long double-stranded DNA and the probes are short oligoprobes, fragmentation of the target DNA is done prior to hybridization. Common methods for DNA fragmentation include physical shearing (e.g., sonic fragmentation), chemical shearing, and randomly cleavage by a nuclease enzyme such as DNase I, partial digestion with a restriction enzyme, which cuts relatively frequently within the genome (e.g., four base cutter), or double digestion with two restriction enzymes.

16.3 Application of Oligonucleotides as Recognition Elements for Nucleotide Analysis

In molecular biology, oligonucleotides can be used as recognition elements in a wide range of hybridization based assays for DNA and RNA. The original suite of hybridization assays, introduced starting in 1975, include Southern blot hybridization, northern blot hybridization, dot blot, random priming for DNA labeling. A second generation of hybridization based technologies, which has since revolutionized molecular biology, includes polymerase chain reaction (PCR), whole genome amplification and microarray analysis.

16.3.1 Original Hybridization-Based Assays

16.3.1.1 Southern Blot Technology

Southern blotting technology (Southern 1975) was the first method which enabled the detection and characterization of specific DNA sequences in the presence of other DNA sequences. In this method, the sample is digested by restriction enzyme (s), size separated using agarose gel electrophoresis and transferred (blotting) onto a filter membrane. Labeled oligonucleotide probes (or other DNA) are then hybridized to the membrane and imaged to indicate the presence of the target. The advantages of oligonucleotide probes include that they are easy to synthesize (including control of length) in a single-stranded, labeled form. Southern blotting laid the foundation for many other similar nucleic acid analysis methodologies including northern blot hybridization, dot blot and DNA microarray.

16.3.1.2 Northern Blot Hybridization

Northern blot hybridization (Alwine et al. 1977) is similar to Southern blot except that the target is RNA. Northern blot techniques were used originally for studying gene expression.

16.3.1.3 Dot Blot

Dot blot (or Slot blot) (Thomas 1980) techniques are hybridization methods where the RNA or DNA samples to be detected are not separated electrophoretically prior to hybridization. Instead, the labeled target mixture is spotted through circular (or slot) templates directly onto the membrane followed by hybridization and target detection. Avoiding the nucleic acid separation and blotting (used by Southern and northern hybridization) simplifies DNA and RNA analysis (although the size of the DNA fragment or RNA transcript is not determined by this method) and enables more accurate quantitation.

16.3.1.4 Random Priming

Random priming (Feinberg and Vogelstein 1983) is a common method for DNA labeling using "random" oligonucleotide sequences to prime DNA synthesis. The method is based on the hybridization of combinatorial oligonucleotides of all possible sequences to the denatured template DNA to be labeled. The complementary DNA strand is then synthesized by a "Klenow" fragment of DNA Polymerase I, using the random oligonucleotides as primers and labeled nucleotides for the synthesis, thus labeling the newly synthesized complementary DNA.

16.3.2 Oligonucleotides as Recognition Elements in Polymerase Chain Reaction (PCR)

Polymerase chain reaction (PCR) (Saiki et al. 1985; Mullis 1990) is a powerful technique for amplifying DNA that has many variations. One constant in all of these variations is the need to design primers that will anneal to the target. Once the primers are annealed to the target, DNA polymerase starts from the 3′ end of the primer and extends the second strand of the molecule, base by base, using the target strand as a template. Because the polymerase starts from the 3′ end of the primer, stable binding at the 3′ end is much more important than binding at the 5′ end. For most PCR applications, the objective is to amplify a specific sequence. Thus, binding specificity is a primer design goal. Because G (guanine) binds well to C, A and T, the presence of G in a primer, particularly near the ends, can promote non-specific binding. Various approaches can be used to avoid this problem including (1) setting a maximum binding strength for the primer ends, or (2) looking for primers with end triplets CAT or CTA (Sergeev et al. 2005).

Non-specific annealing of the primers to the target is usually an unwanted phenomenon. Such non-specific annealing occurs readily when the annealing temperature is set too low. Typical practice is to use an annealing temperature 5–10 K below the nominal primer melting temperature as a starting point. Non-specific annealing shows up on a gel as a smear instead of a sharp band.

In the thermocycling process of PCR, the denaturation step uses a temperature that is sufficiently high to disrupt the double-stranded structure of most target DNA within a few seconds. In the absence of primers, slow cooling to the annealing temperature would allow renaturation. However, the primers outcompete the native strands due to their small size and excess concentration. The result is that renaturation of the native strands of DNA is blocked by the primers. Annealing of the primers is rapid, with the reaction dynamics controlled more by the heat capacity of the aqueous solution. Once the primers anneal, the extension reaction starts. Raising the reaction temperature to around 72°C simply provides the optimum temperature for the extension. The speed of extension depends on the particular polymerase used and the reaction conditions but a useful design estimate is 100 bp/s.

In addition to degenerate primers based on unusual bases such as inosine, another type of degeneracy is to use a mix of primers that include a few or many sequences. Oligos with random sequences can be obtained commercially and are used in a number of DNA amplification methods.

The many variations of PCR technology use oligonucleotides as recognition elements and hybridization properties in various ways:

- *Allele-specific PCR*: A diagnostic or cloning technique to identify or utilize single nucleotide polymorphisms (SNP) uses oligonucleotide primers whose 3′ end terminates at the SNP position.
- *Assembly PCR*: A DNA synthesis method for long DNA sequences which uses overlapping segments generated from different primer sequences. Assembly PCR is used for the assembly of sequences such as genes or an entire plasmid.
- *Asymmetric PCR*: Used to amplify one strand of DNA more than the other by using an excess of one primer leading to linear amplification in late PCR cycles and generating single-stranded DNA for hybridization (or other applications). A modification of this technology Linear-After-the-Exponential-PCR (LATE-PCR), uses a limiting primer with a higher melting temperature.
- *Methylation-specific PCR (MSP)*: A method used to detect methylation of CpG islands in genomic DNA. The method is based on sodium bisulfite treatment which converts unmethylated cytosine bases to uracil, which is recognized by DNA PCR primers as thymine. Thus, one primer set can recognize DNA with cytosines for amplification of methylated DNA, and another recognizes uracil to amplify unmethylated DNA.
- *Multiplex ligation-dependent probe amplification (MLPA)*: A method to detect copy number variations in genomic sequences (deletions or duplications) or SNPs, which permits multiple targets to be amplified with only a single primer pair. MLPA is based on the ligation of two adjacent annealing oligonucleotides (only when there is perfect homology with the target sequence) followed by quantitative PCR amplification of the ligated products. The resulting amplification products are separated normally by gel electrophoresis. MLPA tests are designed in a way that the length of each amplification product is unique so several sites can be analyzed simultaneously.

- *Multiplex PCR*: The use of multiple, unique primer sets within a single PCR reaction to amplify several targets simultaneously. Annealing temperatures for each of the primer sets must be optimized to work correctly within a single reaction, and amplicon sizes (i.e., their base pair length) should be sufficiently different to allow identification by gel electrophoresis.
- *Nested PCR*: Two sets of nested primers are used in two successive PCR reactions. The first reaction generates a long PCR amplicon, which becomes the target for the second PCR reaction. Nested PCR is used to increase the specificity of DNA amplification since the final product depends on a match to four primers instead of two as in non-nested PCR.
- *Quantitative PCR (Q-PCR)*: A method used to measure the quantity of a PCR product (often in real-time) using fluorescent dyes, such as Sybr Green which preferably binds to dsDNA generated during amplification, or by using a fluorescent reporter and a quencher in close proximity on a DNA probe (TaqMan). During PCR, the exonuclease activity of Taq polymerase degrades that proportion of the probe that has annealed to the template separating the fluorescent reporter from the quencher so more PCR amplification will result in less quenching and more light emission.
- *Reverse Transcription PCR (RT-PCR)*: A PCR-based method used to amplify RNA. This method uses the reverse transcriptase enzyme, which synthesizes cDNA from an RNA template. The reverse transcription step is followed by PCR amplification of the cDNA. The method is used mainly for transcription analysis but may become more important as RNA techniques increase in importance.
- *Thermal asymmetric interlaced PCR (TAIL-PCR)*: A method for amplification of unknown sequences flanking a known sequence using a nested pair of primers (with differing annealing temperatures) for the known sequence and a degenerate primer for the unknown sequence.

In addition to thermocycling-based amplification of DNA, isothermal methods for DNA amplification were developed including a Helicase-dependent amplification. This isothermal amplification technique uses the enzyme DNA Helicase for DNA unwinding, instead of thermal denaturation as required by PCR.

16.3.3 Whole Genome Amplification (WGA) Technologies

In addition to PCR and other methods which amplifies specific sequences, whole genome amplification (WGA) can be used to amplify for the whole genome in a single reaction.

When the amount of DNA is limited (e.g., less than 1 μg) WGA protocols can be used to obtain enough target DNA for analysis. There are three WGA amplification approaches, all based on extension of random primers (6–8 mers) annealed at random locations along the genome, resulting in amplification (10–10,000 fold) of the entire genome with varying levels of bias.

- Random sequences (such as hexamers or octamers) can be used to create random priming locations as in degenerate oligonucleotide primed PCR (DOP-PCR) (Telenius et al. 1992; Barbaux et al. 2001) which was one of the first methods used for WGA amplification. DOP-PCR typically uses Taq DNA polymerase. DOP-PCR is designed to provide piece-wise amplification of a long target sequence (such as a whole genome). DOP-PCR uses a mix of primers containing (typically) a random hexamer flanked by 5′ and 3′ anchor sequences. The DOP primers anneal at many locations along the target resulting in random-length overlapping amplicons, which have been shown to provide an approach to whole genome amplification, although the amplification is somewhat non-uniform along the length (bias). In DOP-PCR, the anchor sequences (on the order of 6–8 bases long) combined with the attached random hexamer provides statistically predictable frequency of binding along a long target sequence. Liberal use of G in the anchor sequences further encourages random amplification.
- Multiple-displacement amplification (MDA) (Luthra and Medeiros 2004; Paez et al. 2004; Gadkar and Rillig 2005) uses the highly processive bacteriophage Phi29 DNA polymerase and random exonuclease-resistant primers (to protect them from degradation during the amplification) in an isothermal amplification reaction. In contrast to the Klenow fragment, the Phi29 polymerase generates long DNA products (>10 kb in length) with higher amplification. Another advantage of Phi29 polymerase over the Klenow fragment is the higher proofreading activity resulting in lower mis-incorporation rates compared to other DNA polymerases.
- OmniPlex converts randomly fragmented genomic DNA into a library of DNA fragments, which can be amplified by PCR (Barker et al. 2004a, b). This approach consists of initial DNA fragmentation followed by adapter ligation to form a genomic library which can be amplified by PCR using primers complementary to the adapter sequence. This technique results in a range of fragment lengths (between 150 and 2,000 bp) depending weakly on the amount of sample.

16.3.4 Microarray Applications

Microarray technologies include some of the most important applications for oligonucleotides as recognition elements (Sergeev et al. 2006a, b). Microarray technologies are a high-throughput large-scale nucleic acid identification system based on hybridization which enables simultaneous identification of a large number (up to several hundred thousand) of labeled target molecules (e.g., DNA or cDNA). that the target molecules anneal specifically to the immobilized probes on the array which are spatially ordered. The probes are immobilized (chemically bonded) in discrete spots on a solid matrix, such as a microscope slide. The bound targets can be detected via their label and identified by their binding position on the array (Fig. 16.3). The technology is based on early work on nucleic acid hybridization on a nitrocellulose filter which was first developed in 1965 for DNA-RNA

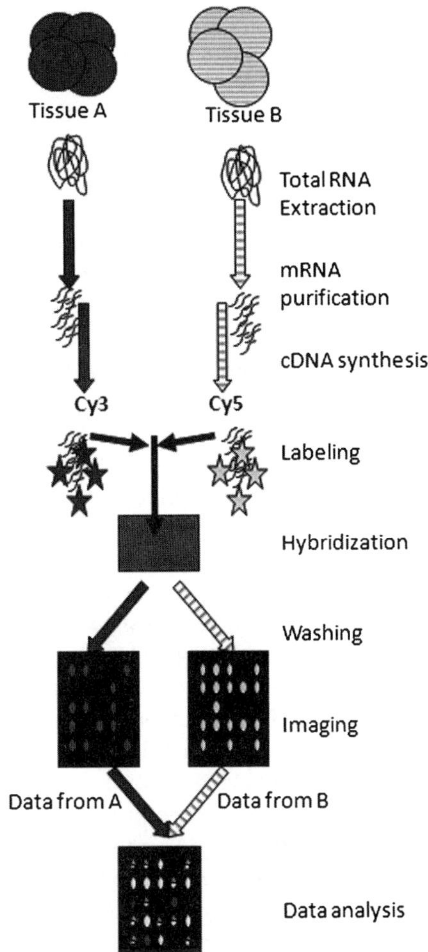

Fig. 16.3 Microarray differential gene expression experiment steps Processing of cellular extracts from two tissue samples (A and B) to obtain a mixture of total mRNA, which is then purified and converted to cDNA by reverse transcriptase. The cDNA in each sample is labeled with a fluorescent dye unique to that sample (Cy3 and Cy5) and the samples are then mixed and hybridized to the microarray. After washing, the array is scanned using excitation for both dyes and data analysis is performed. The data for the two dyes is combined and for each of the microarray spots, the relative intensity of each of the dyes is measured and the relative abundance of each mRNA is calculated

hybridization (Gillespie and Spiegelman 1965), followed by in situ hybridization (Gall and Pardue 1969) and Southern blot (Southern 1975) (used to detect specific DNA sequences) and northern blot analysis for RNA analysis (Alwine et al. 1977). Current microarray designs typically utilize a solid matrix for probe immobilization such as silicon or glass. However, several alternative approaches appear in the literature such as using porous aluminum oxide (Wu et al. 2004), polyacrylamide pads (Guschin et al. 1997), microsphere beads (Yang et al. 2001), solid state electronics (Westin et al. 2001), nitrocellulose and PVDF (polyvinylidene fluoride, a piezoelectric material).

One of the most common applications of microarray technology is for differential gene expression analysis. In such experiments, cellular extracts are taken from two tissue samples (A and B in Fig. 16.3) and the purified mRNA from each sample is converted to cDNA by reverse transcriptase. The cDNA in each sample is then labeled with a unique fluorescent dye (e.g., Cy3 and Cy5). The two labeled samples are then mixed and hybridized to the microarray. After washing, the array is scanned using excitation for both dyes and data analysis of the fluorescent signals is performed. The data for the two dyes combine and for each of the microarray spots, the relative intensity of each of the dyes is measured and the relative abundance of each mRNA is calculated to determine differential gene expression.

In addition to gene expression, microarrays are used for genetic analysis. Such analyses include microbial identification and identification of protein-binding locations ChIP-chip (chromatin immunoprecipitation).

Most DNA microarrays employ single-stranded probes in the length range 18–60 nt. Short probes tend to avoid non-specific annealing while longer probes tend to provide a stronger signal because they bind with higher affinity. Several technologies are used for manufacturing microarrays including robotic spotting of pre-synthesized probes, and various in-situ syntheses of probes on the chip surface such as lithographic techniques, which involve synthesis of oligonucleotides on the surface of the array in multiple cycles where each cycle adds one base (nucleotide).

Spotting technologies have become more widely used with the development of contact printing arrayers, designed at Pat Brown's laboratory (Schena et al. 1995), which made the printing technology freely available. The probes including oligonucleotides (of any length), cDNA, whole genomes, antibodies, proteins or any other biomolecules are pre-synthesized and then deposited onto the array surface in aqueous solution as a small droplet. The probes adsorb to the surface and can be subsequently covalently linked. In contrast, in-situ methods (mainly lithographic techniques) synthesize the probes directly on the array surface by computer controlled photochemistry. Due to synthesis errors which accumulate with length, on-array synthesis is practical mainly for short probes. Spotting methods are practical for low density arrays consisting of 10's or 100's of probes, requiring only a computer controlled robotic spotting system. A laboratory with a robotic spotter can make custom arrays at minimum cost. Lithographic methods (Pease et al. 1994) are more practical for high-density arrays (which are costly to manufacture) such as the genome arrays that are currently available for some organisms. The high density microarray technology was originally used for quantitative monitoring of gene expression of *Arabidopsis* using a cDNA microarray (Schena et al. 1995) and human gene expression analysis monitored by hybridization to a high-density (16,000 spots) oligonucleotide array (Lockhart et al. 1996). Today, high density arrays are used for a range of applications in basic research and in the drug discovery industry.

In microarray analysis, the hybridization step is at the heart of the whole process. The DNA probes on the microarray substrate and the complementary labeled DNA (or RNA) target must anneal to form a double-stranded tethered molecule. Following annealing, unbound target is then washed off the array leaving labeled target tethered to the surface at positions indexed to known probe sequences.

The conditions used for the hybridization reaction and wash conditions can strongly influence the microarray signal strength. In general, consideration of these variables leads to tradeoffs between signal strength and signal specificity. Important variables include temperature, time, buffer ionic strength, target concentration and the presence of denaturing agents or detergents. The hybridization temperature is chosen to allow the target to bind but to minimize non-specific binding.

Common hybridization protocols utilize static incubation of the target in the hybridization solution with the array in a hybridization chamber. Hybridization kinetics are governed by diffusion from the liquid to the surface. The hybridization chamber is typically a small cassette. After loading the sample, the hybridization chamber is sealed and placed in an incubator or a temperature-controlled water bath. The speed of the hybridization reaction can be increased through mixing. Another factor affecting hybridization is the array feature size, which influences nucleic acid surface capture in DNA microarrays (Dandy et al. 2007). The wash steps are designed to disrupt and dislodge any target molecules that are weakly bound (such as non-specific binding) to the array and to leave only target molecules that are tightly bound (i.e., that are complementary to the probes).

For high throughput systems, robotic hybridization stations have been developed to handle multiple chips automatically, including incubation and washing steps. Such automation is expected to reduce the variability of microarray experiments.

16.3.4.1 Design of Oligonucleotides as Recognition Elements

In general, DNA probes on an array should be designed with similar melting temperatures to enable uniform hybridization affinity across the array. Probes should be designed without significant hairpin and self-annealing potential since these properties can interfere with target hybridization. Alternatively, use of stringent hybridization conditions can minimize such interference. Hybridization interference due to the tethered end can be minimized by including a short linker section that allows the probe to stand off from the array surface.

The design of oligonucleotides as recognition elements relies on using bioinfomatics tools including genomics databases, DNA homology search tools, secondary structure analysis, and calculation of melting temperature. The software tools commonly used for cross-homology testing of probes against a reference database include BLAST (Basic Local Alignment Search Tool) (Altschul et al. 1990), which enables rapid sequence comparison, and finds alignments that optimize a measure of local similarity. Prediction of secondary structures can be based on a thermodynamic approach, related to the melting temperature determination methods discussed above.

Several computer programs such as Oligo Design, a computer program for development of probes for oligonucleotide microarrays (Herold and Rasooly 2003), Array Designer (Premier Biosoft Intl) (http://www.premierbiosoft.com/dnamicroarray/index.html), ArrayOligoSel (Leiske et al. 2006) and ArrayOligoSelector (Bozdech et al. 2003) are available for the design of oligonucleotides as probes.

The criteria for selecting probes in such programs include: (i) probe length, (ii) target Tm, (iii) maximum stability of hairpins, (iv) maximum stability of self-dimers, and (v) the size of repeats. Some of these programs incorporate interrogation the designed sequences in a BLAST search against a genomic database (e.g., NCBI). Such interrogation may result in the identification of significant homologies and repeat regions within the target genome which may be automatically flagged and avoided. Such computer programs are capable of designing thousands of highly specific PCR primers and microarray probes for these applications with similar melting temperatures to avoid temperature-induced variations in hybridization.

These algorithms can generate multiple BLAST results that complicate the selection of the probes. Some of the algorithms do not take into account the impact of mismatch position within the probe which reduces the accuracy of the melting temperature calculations. Other factors which may complicate probe design including low GC content (Kuroda et al. 2001) for microbial genomes and frequent conserved repeats which may lead to cross-hybridization (Talla et al. 2003) resulting in erroneous target identification.

Some of these limitations may be overcome by an algorithm (Charbonnier et al. 2005) which incorporates filtering of oligonucleotide probe against libraries of annotated probes recognizing highly conserved targets shared by different genomes. This approach for the design of whole-genome microarrays can be performed within a single genome or between several different genomes (e.g., strains/organisms).

16.4 Oligonucleotides as Genetic Regulatory Elements

Three types of nucleic acids are used to silence gene expression including antisense DNA/RNA, interfering RNA (RNAi) and trans-cleaving ribozymes. All three utilize antisense sequences for annealing to specific RNA targets. However, mechanistically, each type is different. RNA/DNA antisense sequences interfere with translation by binding to mRNA interfering with cellular processing of the mRNA. RNAi disrupts translation by binding to mRNA and employing cellular machinery for degradation of the target mRNA. Trans-cleaving ribozymes act as enzymes catalyzing mRNA cleavage.

16.4.1 Antisense Oligonucleotides

Antisense RNA (aRNA) and DNA are single-stranded oligonucleotides complementary to mRNA. The small (13–25 nt) single-stranded oligomers bind via Watson–Crick base pairing to their target and thus can potentially interfere with several steps of RNA processing related to export, stability, and message translation including splicing, polyadenylation, blocking of the translational apparatus and induction of cleavage of the target RNA.

Most commonly, antisense DNA oligonucleotides base-pair with an mRNA target and mediate its destruction by RNase H, an enzyme that digests the RNA in a double-stranded DNA/RNA hybrid. This gene-silencing approach avoids translation of one particular protein and has potential uses in research and therapy. Antisense oligonucleotides are also being developed for manipulation of alternative splicing ratios from a single gene (Kole and Sazani 2001; Sazani et al. 2003; Vacek et al. 2003; Sazani and Kole 2003a, b). This can be used to silence mutations that cause aberrant splicing, thus restoring normal function of a defective gene. This is important since close to 50% of genetic disorders are caused by mutations that cause defects in pre-mRNA splicing (Sazani and Kole 2003a). Thus, targeting splicing with antisense oligonucleotides significantly extends the clinical potential of these compounds.

Antisense oligonucleotides have significant potential to provide effective therapies for a wide variety of diseases, including viral infections, cancer, inflammatory and cardiovascular diseases. In general, there are two major requirements for oligonucleotides that target splicing: (i) the drugs must not activate RNaseH, which would destroy the mRNA before splicing, and (ii) they have to be able to effectively compete with natural splicing factors.

16.4.1.1 Antisense RNA (aRNA)

The majority of RNA-based antisense oligonucleotides work as steric blockers, physically preventing or inhibiting progression of the splicing or the translational machinery (e.g., translation initiation or elongation). A major disadvantage of the application of "naked" antisense RNA oligonucleotides is their insufficient stability against nucleases (Drude et al. 2007).

One strategy to overcome the stability problem is the development of a suitable application system, which protects the oligonucleotide, enhances the penetration and enables targeting the compartment of action using liposomes (Zimmer et al. 1999) or nanoparticles as colloidal drug carrier systems (Zobel et al. 1999). Another strategy is the structural modification of natural oligonucleotides leading to more stable artificial structures.

"First generation" antisense: To increase nuclease resistance, several modifications were used in "first generation" antisense oligonucleotides including replacing one of the non-bridging oxygen atoms of the phosphodiester backbone with sulfur (Campbell et al. 1990). This thio modification was found to increase the half life of an antisense oligonucleotide from minutes to about 8–9 h. The disadvantages of phosphothioate oligonucleotides include relatively low binding affinity to complementary sequences, and non-specific binding to proteins (Guvakova et al. 1995) leading to low specificity in target recognition as well as high toxicity (Brown et al. 1994).

"Second generation" antisense: The toxicity is reduced with "second generation" antisense oligonucleotides that contain phosphorothioate segments at the 3′ and 5′ ends and a central segment with modified nucleotides, 2′-MOE (2′-O-methoxy ethyl), that do not support RNaseH activity. These modified antisense

molecules show higher target affinity and lower protein affinity (Agrawal et al. 1997).

"Third generation" antisense: The "third generation" antisense oligonucleotides use a modification at the sugar moiety by introduction of a 6-aminohexyl group (Urban and Noe 2003), which enhances the stability of the modified oligonucleotides against the catalytic degradation effect of ribozymes. This modification provides high target affinity while showing (Urban and Noe 2003) little toxicity and greatly enhanced stability against degradation by nucleases.

16.4.1.2 DNA Antisense

The majority of DNA antisense technologies consist of deoxyribonucleotide analogs functioning via an RNaseH dependent mechanism. Binding of the antisense nucleotide to its target results in the recruitment of RNaseH leading to degradation of the target RNA by RNaseH (cleaving the RNA part of RNA/DNA hybrids) (Zamaratski et al. 2001).

Similar to RNA antisense, replacement of natural structures by artificial nucleotides is a general strategy to increase stability of the nucleic acids. Peptide type nucleic acids (PNA) have had a significant impact on DNA antisense drug development (Nielsen et al. 1991, 1994; Demidov et al. 1994; Nielsen 1999, 2000, 2001a, b, 2004), because they are neutral molecules (Nielsen et al. 1994; Wittung et al. 1994; Urban and Noe 2003) capable of hybridization with DNA and RNA including the formation of triple helices. PNA sequences show increased melting temperature (i.e., higher affinity). The oligomers are resistant to endo- and exo-nuclease mediated degradation as well as protease digestion (Demidov et al. 1994). Nevertheless, all drug candidates presently in clinical development consist of only minor modifications of natural oligonucleotides.

16.4.2 RNA Interference (RNAi)

RNA interference (RNAi) is a mechanism for gene expression inhibition (Fire et al. 1998; Kennerdell and Carthew 1998) by degradation of specific RNA transcripts interfering with mRNA translation (Lopez et al. 1998), or epigenetic silencing (Lippman et al. 2003; Ekwall 2004) through RNA-directed DNA methylation. RNAi is one of the most rapidly growing applications today for oligonucleotides as recognition elements. Originally discovered in plants, where RNAi plays a role in the immune response against viruses, it was later discovered in yeast, fungi, nematodes and more recently mammalian cells (Napoli et al. 1990; Sharp 1999; Elbashir et al. 2001a). To avoid long-double-stranded-RNA-dependent interferon response, short-interfering siRNAs are widely used for RNA interference in mammalian cells. The selective interference of RNAi makes it a very important functional genomics research tool enabling controlled gene silencing, which can help identify gene function. RNAi technology has broad potential medical applications including antiviral therapies and inhibition of gene expression in cancerous cells among a multitude of potential uses.

16.4.2.1 The Various Types of RNAi and RNAi Processing

In RNAi, short sequences of RNA serve as recognition elements (by homologous base pairing (Fire et al. 1998; Kennerdell and Carthew 1998)), which recognize target mRNAs and directs their degradation. In mammalian cells, two main types of short RNAs, microRNAs (miRNAs) and small interfering RNAs (siRNAs) are thought to be loaded into the same RNA-induced silencing complex (RISC) (Hammond et al. 2000) which includes a catalytic RNase component called Argonaute (Okamura et al. 2004; Rand et al. 2005). Sources of siRNA include long strands of exogenous dsRNA (typically of virial origin) that are cleaved into siRNA by Dicer and custom-made dsRNA designed in the laboratory to silence particular genes. siRNA sequences have complementarity to the target where they guide mRNA degradation or translation silencing (Okamura et al. 2004). For miRNA, 70-nt stem-loop pre-miRNA in the nucleus are processed by the RNase III Drosha enzyme (Lee et al. 2003; Han et al. 2006) which exists as part of the "microprocessor complex," which also contains the double-stranded RNA binding protein Pasha (Yeom et al. 2006). The pre-miRNA is further processed to mature miRNAs in the cytoplasm by the ribonuclease II Dicer (Macrae et al. 2006) and the guide strand selected by the Argonaute protein and is then incorporated into RISC. There are several pathways and types of RNAi, and the most useful types for studying gene expression and for potential therapeutic applications are:

Short Interfering RNA (siRNA): siRNA precursors are long, double-stranded RNA (dsRNA) molecules (Vermeulen et al. 2005) cleaved into short fragments of 20–25 (Zamore et al. 2000) by the ribonuclease II Dicer (Macrae et al. 2006) in the cytoplasm. One strand called the *guide strand* is selected by the Argonaute proteins (e.g., AGO1, AGO2) (Okamura et al. 2004; Rand et al. 2005) (the catalytic RNase component of RISC) and is then incorporated into RISC causing degradation of similar sequences. This mechanism is used by plants as a defense against RNA viruses and can also be harnessed in the laboratory for gene silencing by designing the guide strand to target particular mRNA sequences.

Short hairpin RNA (shRNA): shRNA is a sequence of RNA folded in a hairpin including a double-stranded stem. shRNA is cleaved by the cellular machinery into siRNA and can be used to silence gene expression via RNA interference. Several shRNA expression vectors have been engineered using plasmid systems (Brummelkamp et al. 2002; Paddison et al. 2002; Yu et al. 2002) retroviral (Devroe and Silver 2002), adenoviral (Huang et al. 2004) and lentiviral (Abbas-Terki et al. 2002) vectors. shRNA expression in mammalian cells can, sometimes, cause the cell to mount an interferon response. Interestingly, this type of interferon response is not observed with miRNA.

Micro RNA (miRNA): miRNA precursors are processed to short, 70-nt stem-loop pre-miRNA by the nuclease Drosha and the double-stranded RNA binding protein Pasha. The pre-miRNA is further processed by Dicer and the guide strand selected by the Argonaute protein is then incorporated into RISC. There is a tight connection between the miRNA and RNAi molecular machineries. It was shown in both mammalian cells and plants that siRNAs can function as miRNAs (Doench et al. 2003) and both are capable of targeting mRNA for degradation or translation silencing depending on the degree of complementarity between the small RNA and its mRNA target (Okamura et al. 2004).

Endoribonuclease-prepared short interfering RNAs (esiRNAs): An alternative approach for generating siRNAs, which can overcome the variable silencing ability inherent in other methods is to prepare multiple siRNAs from a cDNA template targeting the gene of interest. After PCR amplification of the cDNA, a corresponding long dsRNA is prepared using T7 RNA polymerase. Digestion of this dsRNA molecule into siRNAs with the bacterial enzyme RNase III results in endoribonuclease-prepared siRNAs or esiRNAs (Yang et al. 2002). esiRNAs represent a population of siRNAs which target different regions of the gene of interest. Due to redundancy, esiRNAs provide a higher probability of gene silencing, saving time and money for the researcher. Comparative analyses with chemically synthesized siRNAs (Kittler et al. 2007) demonstrated a high-silencing efficacy of esiRNAs and a 12-fold reduction of down-regulated off-target transcripts as detected by microarray analysis. esiRNA libraries offer an efficient, cost-effective and specific alternative to previously available mammalian RNAi resources.

16.4.2.2 Short Interfering RNA (siRNA) Design

siRNA technologies have significant potential as tools in functional genomics and as therapeutic agents. Thus, design of effective siRNA has many applications. Some of the requirements for developing effective siRNA applications are described (Reynolds et al. 2004; Santoyo et al. 2005; Matveeva et al. 2007) including:

- Using sequences shorter than 30 nt to avoid interferon response and the activation of the host protein kinase PKR immune response used against virus infection.
- siRNAs with 3′ overhanging UU dinucleotides are the most effective (Elbashir et al. 2001b).
- siRNAs with 30–50% GC content are more active than those with a higher G/C content; avoid sequences with >50% G + C content (Reynolds et al. 2004; Matveeva et al. 2007)
- Off-target effects (the interactions between siRNA and random mRNA transcripts causing RNAi knockdown of non-targeted genes) can be avoided by using sequence analysis tools (Du et al. 2005; Qiu et al. 2005; Yamada and Morishita 2005; Qiu and Lane 2006; Rual et al. 2007).

- Avoid stretches of >4 T's or A's in the target sequence when designing sequences to be expressed from an RNA Pol III promoter because 4–6-nt poly (T) acts as a termination signal for RNA Pol III (Du et al. 2005; Qiu et al. 2005; Yamada and Morishita 2005; Qiu and Lane 2006; Rual et al. 2007).
- Select multiple siRNA target sites at different positions along the length of the gene because some regions of the mRNA may be bound by regulatory proteins or highly structured.
- Avoid sequences that share a high homology (>16 contiguous base pairs) with other coding sequences.
- Targeted regions for siRNA should be located 50–100 nt downstream of the start codon (ATG) to avoid regions where regulatory proteins may bind (Elbashir et al. 2002).
- For improved results, use sequence motifs AA(N19)TT or NA(N21), or NAR (N17)YNN, where R is purine (A, G), Y is pyrimidine (C, U) and N is any nucleotide (Santoyo et al. 2005).
- Avoid targeting intron sequences because RNAi works only in the cytoplasm and not in the nucleus.
- siRNA experiments should include appropriate controls including a negative control of a non-complementary sequence with thermodynamically similar properties as the siRNA sequence.

siRNA design tools: Numerous algorithms have been developed to predict the most effective target sequences for a given target gene. Several examples of such tools include:

- *The Rational siRNA Design Template*: Software which can be used to design multiple siRNA sequences for a given target sequence and which also calculates the molecular weights and melting temperature. The software is based on the Rational siRNA design algorithm (Reynolds et al. 2004; Jiang et al. 2007).
- *TROD: T7 RNAi Oligo Designer*: The program provides tools for selecting and narrowing the number of candidate siRNA sequences. RNA T7 RNA polymerase (Dudek and Picard 2004) has a strong preference for a G as initiation nucleotide. The software generates a prioritized list of ready-to-order DNA oligonucleotide sequences, with links to perform a BLAST analysis.
- *RFRCDB-siRNA*: Instead of directly predicting the gene-silencing activity of siRNAs, it takes siRNAs as queries to search against the siRNA-centric database (Jiang et al. 2007).
- *E-RNAi*: This is a program with capability to access published predesigned dsRNAs (Arziman et al. 2005).
- *DEQOR*: This program uses a scoring system to predict regions in a gene that show high silencing capacity (e.g., part of the exon sequence and GC content) based on the base pair composition and the corresponding siRNAs with high silencing potential for chemical synthesis applying BLAST searches (Henschel et al. 2004).
- *RNA Workbench (RNAWB)*: This program takes into consideration free energies of the duplex (Varekova et al. 2008). A computational tool to minimize interference with the functions of unrelated genes was developed (Ui-Tei et al. 2007).

16.4.2.3 Oligonucleotide Synthesis and Delivery for RNAi

Oligonucleotide synthesis for RNAi: Several methods used for siRNA and shRNA synthesis include exogenous chemical synthesis and endogenous synthesis via genetically engineered vectors.

Chemical synthesis of siRNA using solid-phase support chemistry similar to that used to generate DNA primers for PCR and probes for microarrays is possible. Chemically synthesized double-stranded RNA is then transfected into cells where it interacts with the endogenous RISC complex to cause RNAi. The problem with this approach is that the RNAi effect lasts only until the transfected siRNA is degraded by normal cellular nuclease mechanisms.

Expression vectors and viral vectors include retroviral (Devroe and Silver 2002), adenoviral (Huang et al. 2004) and lentiviral (Abbas-Terki et al. 2002) vectors for siRNA or shRNA. Plasmid systems (Brummelkamp et al. 2002; Paddison et al. 2002; Yu et al. 2002) utilize RNA polymerase III (Pol III) or Pol II promoters (Paule and White 2000; Schramm and Hernandez 2002). These systems use vectors to introduce DNA (for siRNA or shRNA) into the cell where it is transcribed continuously into siRNA or shRNA, which then interacts with RISC causing RNAi. Thus, these systems synthesize the oligonucleotides, which cause RNAi in vivo.

PCR can be used to generate siRNA or shRNA expression cassettes followed by in vitro transcription of siRNA or shRNA. The short length Pol III promoters (less than 300 bp) can be used to generate expression cassettes using PCR to amplify a linear fragment of double-stranded DNA containing the necessary promoters and the siRNA or shRNA sequence (Huang et al. 2004). The cassette itself or the purified in vitro transcript of the cassette can serve as the source of nucleic acid for RNAi. This system works similar to the vector-based approach except the DNA is transfected into the cells.

Another method typically starts from cDNA and involves in vitro transcription of long strands of dsRNA. The dsRNA is then processed with purified mammalian Dicer or the *E. coli* enzyme RNase III to digest the nucleic acid into a population of siRNA duplexes. The result is multiple siRNA oligos targeted to the gene of interest (esiRNA).

siRNA delivery into cells: There are several methods for siRNA and/or shRNA delivery into cells (Aigner 2007; de Fougerolles 2008; Judge and Maclachlan 2008; Pirollo and Chang 2008; Pushparaj et al. 2008; Shim and Kwon 2008; Wolff and Rozema 2008) including liposome-based transfection, electroporation, calcium phosphate co-precipitation, microinjection and vector delivery techniques of plasmid vectors designed to express siRNA molecules. Although lipid-based transfection is commonly used for adherent cells, cells in suspension are often more difficult to transfect. Electroporation techniques, generally, have higher rates of delivery. There are a variety of plasmid and viral vectors used to produce siRNA or shRNA molecules. As a drug-delivery system for siRNA, one question that arises is whether the disease target lends itself to systemic or local administration (e.g., tumors, skin lesions, and mucosal tissue). Of all the various delivery methods, the most established methods are plasmids and viral vectors. DNA-based vectors for shRNA expression offer several advantages over chemical and in vitro synthesized siRNA. Vector-based construction is less expensive than chemical synthesis of siRNA, the selection of transfected cells is possible via antibiotic selection and the option of inducible shRNA transcription is also available.

siRNA stability in the cells: in vivo unmodified, naked siRNAs are rapidly degraded by endo- and exonucleases in cells, blood or serum. Chemical modifications can be introduced into the RNA duplex structure to protect it and to enhance its in vivo biological stability without adversely affecting siRNA activity. Chemical modifications to the backbone, base, or sugar of the RNA have been developed to enhance siRNA stability and activity (Bumcrot et al. 2006) *delaying* degradation from minutes to hours. Using in vitro synthesized modified siRNA combined with an appropriate delivery system, in vivo siRNA gene-silencing activity can be sustained for several days (Bumcrot et al. 2006). In addition, chemical modifications in the guide strand can reduce most off-target effects (Jackson et al. 2006) and increase thermal stability within the siRNA duplex for improved potency of siRNA (Khvorova et al. 2003; Schwarz et al. 2003; Reynolds et al. 2004; Jiang et al. 2007).

Plasmid vectors: Delivery of siRNA into mammalian cells can be achieved via the transfection of plasmid vectors that participate in transcription in the cell. Plasmid systems (Brummelkamp et al. 2002; Paddison et al. 2002; Yu et al. 2002) utilize RNA polymerase III (Pol III) or Pol II promoters to initiate and terminate RNA transcripts at well-defined positions to drive the expression of either siRNA or short hairpin RNA transcripts. Plasmid vectors often use two RNA Polymerase III promoters (U6 and H1) to drive transcription (Paule and White 2000; Schramm and Hernandez 2002; Lambeth et al. 2005) enabling the synthesis of small, highly abundant RNA transcripts. The target sequence is cloned in both sense and antisense orientations with a small spacer group in between (short hairpin RNA or shRNA). The resulting transcribed RNA hairpin can be recognized and cleaved by Dicer into siRNA. RNA duplexes may also be transcribed without hairpin structures and directly processed by the RISC. A bi-functional siRNA construct that enables its detection by quantitative PCR and which retains RNAi activity for selective knock-down of targeted mRNA in mammalian cells is described (Jiang et al. 2005). The main limitation of plasmid systems is relatively poor transfection efficiency.

Viral vectors: Several viral vector systems have been used including; retroviral (Devroe and Silver 2002), adenoviral (Huang et al. 2004) and lentiviral (Abbas-Terki et al. 2002) vectors. Retroviral vectors have been used for siRNA delivery overcoming the problem of poor transfection of plasmid-based systems. Viral vectors are highly efficient delivery systems for nucleic acids; however, they require an active cell division for gene transduction and their clinical application is hindered by the induction of toxic immune responses and inadvertent gene expression changes following random integration into the host genome. Retroviral vectors and adeno-associated viruses are used as vectors for siRNA delivery into mammalian cells.

16.4.2.4 siRNA Libraries

In addition to targeted siRNA, high-throughput technologies can be used to facilitate multi-target siRNA analysis for large scale genomic studies. The format for such assays is often microtiter-based using libraries of siRNA or shRNA molecules. The libraries may be based on validated siRNA sequences or based on randomly produced sequences. This strategy may be used to identify drug targets or new signaling pathways. Various approaches for generating and selecting such libraries were reported.

Enzymatic strategies: Digestion of long dsRNA with endoribonucleases in vitro by using bacterial RNase III or recombinant Dicer generates a heterogeneous population of siRNAs capable of interacting with multiple sites on the target mRNA (Buchholz et al. 2006). In such methods, long dsRNA can be converted in vitro into siRNAs with a length of 18–25 base pairs (bp).

Genome-wide data sets for the production of endoribonuclease-prepared short interfering RNAs (esiRNAs) for human, mouse and rat were developed and used to predict the optimal region for esiRNA synthesis for every protein-coding gene of these three species generating 16,242 esiRNAs (Kittler et al. 2007). A database called RiDDLE is available for retrieval of target sequences and primer information (Kittler et al. 2007)

A short hairpin RNA (shRNA) library from cDNA was developed as a tool for high-throughput loss-of-function genetic screens in mammalian cells (Wu et al. 2007). Enzymatically digested cDNA fragments are equalized by suppression PCR based on subtractive hybridization.

siRNA target sequence selection system (Kasim et al. 2006) based on siRNA expression vector libraries carrying siRNA expression fragments (100 random siRNA expression plasmids) originating from fragmentized target genes, followed by a group selection screening system, was used for screening of siRNA target sequences.

Libraries of synthetic siRNA: An alternative approach to the enzymatic strategies is the use of synthetic libraries. Libraries of synthetic siRNA and microRNAs (miRNA) were used to probe the tumor necrosis factor-related apoptosis-inducing ligand (TRAIL) induced apoptosis pathway (Ovcharenko et al. 2007).

16.4.3 Ribozymes

Ribozymes are RNA molecules which catalyze chemical reactions. The name derives from “ribonucleic acid enzyme” and they are also referred to as catalytic RNA. The substrate for ribozymes is often either RNA or DNA. Ribozymes are involved in intron splicing where they cleave the phosphodiester backbone of their target at a specific cutting site. Ribozymes have also been shown to catalyze aminotransferase activity of the ribosome (Supattapone 2004).

Ribozymes, which traditionally are not viewed as oligos, were included in this chapter because short (e.g., 13 or 19-mer) oligoribonucleotides *can* act as a ribozyme for the specific cleavage of oligoribonucleotide substrate (Uhlenbeck 1987; Jeffries and Symons 1989). Also, ribozyme building blocks can be prepared from short oligoribonucleotides including more complex chemical synthesis and assembly of an artificially branched hairpin ribozyme with structural modifications (Ivanov et al. 2004) with new catalytic structures that cannot be prepared enzymatically (Slim et al. 1991) including incorporation of polyamines to stabilize ribozymes against chemical and enzymatic degradation (Marsh et al. 2004).

RNA catalytic activities were discovered in the 1980s by groups studying RNase P and intron splicing. While studying RNase P, which consists of an RNA molecule with associated proteins, it was concluded that the RNA molecule is the catalytic subunit of the enzyme (Guerrier-Takada et al. 1983). While studying RNA splicing in *Tetrahymena,* it was shown that the "control" treatment (not containing the nuclear extract) also resulted in RNA splicing, suggesting that splicing is independent of any protein (i.e., self-splicing RNA) (Kruger et al. 1982). Five major classes of ribozymes have been described based on their sequences and three-dimensional structures (Bunnell and Morgan 1998): the *Tetrahymena* group I, RNase P, hammerhead, hairpin ribozyme, and the hepatitis delta virus ribozyme.

In terms of activity, ribozymes fold into specific three-dimensional structures to form catalytic centers. Many ribozymes function by binding to a target RNA sequence through Watson–Crick base pairing and then cleaving the phosphodiester backbone of the target at a specific cutting site. Ligation of the new RNA target fragments follow the trans-splicing of the specific RNA target.

The hammerhead ribozymes are the best understood ribozymes. Hammerhead ribozymes are small catalytic RNAs that undergo self-cleavage of their own backbone to produce two RNA products. Thus, hammerhead ribozymes do not catalyze multiple turnovers like an enzyme. As other ribozymes, the hammerhead ribozymes are antisense RNA forming a stem-loop structure in which some of the nucleotides within the sequence selectively form Watson–Crick base pairs with others (stem), while the rest remain single stranded (loop). Hammerhead ribozymes contain three base-paired stems and a highly conserved core of residues required for cleavage.

Ribozymes can be used as molecular scissors enabling the study of gene function by manipulating mRNA. The discovery of ribozymes led to the development of new synthetic ribozymes with self-cleaving RNAs. Methods for in vitro selection of self-cleaving RNAs from random-sequence RNAs were developed (Tang and Breaker 2000) including ribozymes with novel structures. Ribozyme RNA catalysis has potential applications in the regulation of gene expression and in protein synthesis. However, using ribozymes as molecular scissors is *not* simple. The cell is producing a large number of RNAs from an even larger number of genes so the engineering of ribozyme specificity is complicated. In addition, the experimental systems for ribozymes are more complex than other gene-silencing approaches such as siRNA.

16.5 Oligonucleotides as Ligands-Aptamer Binding

For biodetection, the recognition element (ligand) is the element which binds to the analyte of interest (target molecule), and the two main factors important for every ligand are affinity and specificity. The measure of ligand affinity is the equilibrium dissociation constant (K_D) for binding between the ligand and the target. Antibodies are the best natural ligands with equilibrium dissociation constants (K_D) on the order of 10^{-9} M. The specificity of antibodies can be quite high for monoclonal antibodies with lesser specificity for polyclonal antibodies. In general, antibodies are *not* easy to synthesize in large quantities because they require either a live animal host (polyclonal) or a cell culture (monoclonal). Neither of these methods is well suited for high throughput or multianalyte applications, where a very large number of diverse ligands have to be developed.

In recent years, there have been several important advances in the development of new types of "synthetic" ligands that mimic antibodies in that they are derived from a large molecular library and can bind to almost any target by rearranging the molecular sequence. Selection of high-affinity ligands is based on relatively simple in vitro screening. Included are aptamers, peptides, scaffolded-peptides, combined binding agents derived from low-affinity ligands and combinatorial chemistry ligands. With the exception of aptamers, the major drawback of many of these ligands is the low K_D's (~10^{-6}–10^{-7} M), which are 2–3 orders of magnitude less than antibodies.

Aptamers (Ellington and Szostak 1990; Tuerk and Gold 1990) are single-stranded nucleic acid molecules, either DNA or RNA, that fold into a three-dimensional secondary structure with binding affinity for a target molecule (including dyes, proteins, peptides, drugs, organic and inorganic molecules or even whole cells). Aptamers are selected in vitro for their specific binding affinity to target molecules by a method called SELEX (systematic evolution of ligands by exponential enrichment) starting from a large library of random sequences. SELEX starts from a large library of molecules (complexity ~10^{10}–10^{16}), either RNA or DNA, that contain a central region of random sequence (N nucleotides), which acts as the binding element, flanked by constant sequences used for PCR amplification. In vitro selection of high-affinity aptamer sequences involves cyclic steps of binding, separation and amplification leading to binding affinities that rival antibodies. After multiple cycles have been performed, the candidate aptamers are then typically cloned and sequenced. Once the sequence of a high-affinity aptamer is known, it can be easily synthesized. See chapter 17 for more details about aptamers and their use in biosensor.

16.5.1 Aptamer Structure and Design

The first aptamers were RNA selected for organic dyes (Ellington and Szostak 1990) and T4 Polymerase (Tuerk and Gold 1990). In vitro selection schemes for single-stranded DNA were developed later for organic dyes (Ellington and Szostak 1992) and human thrombin (Bock et al. 1992). Thus, both DNA and RNA aptamers can be used and several detailed protocols for DNA and RNA libraries appear in the literature (Ulrich et al. 2004).

RNA aptamers: For the development of a RNA aptamer, a small quantity of a large library of DNA sequences is first synthesized including a randomized central region flanked by PCR primers. PCR is then used to generate the required amount of DNA. A T7-promoter is normally incorporated into the 5′ constant sequence of the DNA library to allow generation of the RNA library by in vitro transcription (Ulrich et al. 2004). After selection, the purified RNA is reverse transcribed to cDNA, followed by PCR amplification to give double-stranded DNA, which serves as template for in vitro transcription to RNA for the next SELEX cycle. A restriction site (e.g., *Eco*RI and *Hin*dIII) can be included in the flanking sequences to allow the insertion of the group of aptamers resulting from SELEX into a bacterial vector for cloning followed by sequencing to arrive at individual high-affinity aptamers.

The complexity and the randomness of the library is an important aspect in arriving at high-affinity aptamers (Ellington and Szostak 1990). To address the stability of RNA, which is an issue due to RNAse activity in biological samples, libraries are made nuclease-resistant by incorporating modified bases during the in vitro transcription step (e.g., 2′-F-dCTP and 2′-F-dUTP) (Heidenreich et al. 1995; Schurer et al. 2001; Ulrich et al. 2004) or by protecting the RNA during the selection using an RNAse inhibitor (Kusser 2000; Boiziau and Toulme 2001).

DNA aptamers: DNA libraries can be constructed by solid phase synthesis, followed by purification and randomness verification. To obtain larger quantities, the library can be PCR amplified following single-strand purification (using biotinylated forward primer and streptavidin capturing) (Kusser 2000; Boiziau and Toulme 2001).

RNA aptamers versus DNA aptamers: Both ssRNA and ssDNA libraries have been used in SELEX experiments and each approach offers advantages and disadvantages. In terms of selection, some ligands lead to the isolation of more DNA sequences while others show more affinity to RNA sequences. Moreover, RNA and DNA aptamers with corresponding sequences do not show similar binding characteristics (Ellington and Szostak 1992). However, although the sequences of the highest affinity aptamers for RNA and DNA differ, the final binding affinity obtained from SELEX is similar for both.

In terms of applications, the main advantage of DNA (over RNA) aptamers is the greater stability shown by DNA. DNA aptamers do not need modifications for in vivo use, whereas RNA aptamers must be stabilized against nuclease activity. On the other hand, as DNA production results in a double-stranded molecule, DNA aptamer development requires additional steps such as denaturation and purification of the sense strand (e.g., using a triple-biotinylated antisense primer) which adds cost and complexity to the aptamer development process. Unlike RNA, the quantity of the DNA pool used for selection is limited by the yield of the antisense strand obtained following purification, whereas the in vitro transcription reaction for generation of RNA aptamers represents an amplification step.

RNA aptamers have been reported to form more flexible structures than DNA aptamers (Ulrich et al. 2004) and RNA aptamers can be expressed inside cells (Good et al. 1997), which may be of great importance in developing therapeutic aptamers. So, each type of aptamer offers some advantages.

Aptamer library complexity: An important aspect of aptamer selection is that the length of the randomized tract of nucleotides has a primary role in determining the initial abundance of high-affinity structures. A typical library today consists of ~10^{14}–10^{15} randomized sequences of length *N* between two fixed sequences that are used for PCR amplification. In such libraries, the number of combinations is 4^N so with 30 random bases, the number of combination is $4^{30} = 10^{18}$. With such a large number of molecules, it is expected that some will exhibit high affinity binding to any particular target. The synthesis of the oligoucleotides is not perfect and a fraction will exhibit errors which accumulate with each cycle of oligonucleotide synthesis. Chemical deprotection synthesis has a coupling efficiency of approximately 98% for each cycle. For 25 base oligonucleotides acid-mediated deprotection yield is ~60%. Errors that accumulate in the randomized sequence contribute to library complexity but errors in the flanking primer sections reduce the efficiency of the entire SELEX process.

Long sequences cause problems in synthesis because errors are dependent on the number of bases. Thus, longer sequences imply a lower synthesis yield. Errors in the randomized section are not a concern but errors in the PCR primer flanking sequences are critical. As long as the synthesis yield is sufficient to support PCR, the amplification step can be used to compensate for low yield. However, a concern arises then about the complexity of the resulting library. If the synthesis process *skews* the randomization of the central section of the sequence, then the resulting library may randomly miss the high affinity sequences that are sought. Minimization of these effects leads to short randomized sections in the order of 25 nt. Another factor that affects the design length is the practical problem of the number of molecules in the randomized library. Since that number (4^N) increases rapidly as *N* increases, the use of longer randomized lengths requires larger reaction volumes (to maintain an acceptable library concentration) and larger synthesis surface area to achieve a complete starting library. In many cases, relatively short aptamers (with a randomized length of ~25 nt) have been found to work well and to result in high binding affinity. In an experiment for selection of aptamers to isoleucine (Legiewicz et al. 2005), the outcomes of 11 independent selections of 16, 22, 26, 50, 70, and 90 nt targeting isoleucine were compared. The simplest isoleucine binding aptamer was present as a domain in the majority of the resulting aptamers. In this study, it was found that the number of SELEX cycles needed is lower for shorter sequences.

Oligonucleotide folds are typically between 25 and 50 nt long, so libraries with *N* greater than approximately 50 nt may contain sequences comprising two independent folds, complicating the process of in vitro selection. Moreover, random pools containing more than 90 random positions are difficult to synthesize and are likely to form self-aggregates that precipitate (Bartel and Szostak 1993; Lorsch and Szostak 1994).

The probability of finding a high-affinity target-binding sequence depends on many factors such as the stringency of the screen, the desired K_D, the target and the library complexity. Using a library with $N \sim 100$ (Ellington and Szostak 1990) for selection targeting a relatively simple dye, it was estimated that approximately 10% of the library exhibited measurable affinity for the dye. In another study using a library with $N = 36$, it was estimated that perhaps one in 10^{11} sequences contained adenosine-binding motifs (Burke and Gold 1997), which is a more complex target than the dye.

Aptamer affinity: The ability of a given aptamer sequence to survive the in vitro selection process depends on two requirements, thermal stability and target affinity. Aptamers achieve their high-binding affinity and specificity by adopting defined three-dimensional structures upon binding their target and essentially becoming encapsulated by the ligand as shown by aptamer-target crystal structure (Huang et al. 2003). DNA aptamers are often based on the guanine quartet (quadruplex structures are highly stable DNA or RNA structures formed on G-rich sequences) as the fundamental secondary structure unit (Smirnov and Shafer 2000). In general, guanine quadruplex structures, stabilized by guanine quartets, can be constructed from one, two, or four oligonucleotide strands (Williamson 1994).

Thermodynamic stability of a series of aptamer sequences was studied as a function of modifications in the various loop domains of the aptamer, as well as the effect of adding an additional quartet (Smirnov and Shafer 2000). Changes in aptamer loop sequence can have a significant impact on aptamer stability. With protein as the target, aptamer binding can be extraordinarily tight and the resulting aptamers are very specific (Brody et al. 1999). Aptamer dissociation constants in the pM to low nM range have been reported by various groups (Hermann and Patel 2000; Misono and Kumar 2005). Aptamers in the length range 30–70 are typical (Carothers et al. 2006a, b). The binding site of the aptamer can be ascertained by affinity measurements over a set of sequence variations at each position. Aptamers with the same binding site sequence can exhibit higher target affinity when the binding site support structure is optimized (i.e., when the aptamer is longer).

For a particular application, binding specificity may be more important than binding affinity. Binding-specificity is a function of the size and shape of the binding pocket and the range of related targets that must be distinguished. Current aptamer selection schemes optimize for affinity. It was shown recently that affinity does not correlate well with specificity (Carothers et al. 2006a, b). Specificity can be significantly enhanced by increasing the size of the binding pocket or by binding to two (or more) adjacent epitopes (using proximity-dependent DNA ligation assays) (Fredriksson et al. 2002).

16.5.2 Aptamer Selection

As discussed above, aptamers are selected with repeated cycles of in vitro selection often referred to as SELEX "Systematic Evolution of Ligands by

Exponential Enrichment" or "in vitro evolution" (Ellington and Szostak 1990; Tuerk and Gold 1990), which is an in vitro selection process consisting of multiple cycles of two sequential steps, selection and amplification. SELEX can efficiently reduce a complex randomized library of nucleic acid sequences (complexity ~(10^{10}–10^{16}) to a small subset of one or a few sequences that bind tightly to the target.

The selection of high-affinity aptamers requires 8–20 cycles of separation and amplification. The first cycle typically starts from a large pool of random sequences. Each sequence in the pool is flanked on each end by PCR primer binding sequences that enable the subsequent amplification step. Enrichment of aptamers with high affinity for the target is accomplished via one of several possible separation schemes such as chromatography, capillary electrophoresis or surface plasmon resonance. After removing low-affinity species, the high-affinity species are eluted from the target and amplified via PCR resulting in a concentrated pool of aptamer candidates with which to start another cycle of enrichment. Recently, this process has been automated (Cox et al. 2002a, b) such that high-affinity aptamers to a new target can be generated within a few days. Various alternative SELEX approaches have been described (Tombelli et al. 2005; Mairal et al. 2008) including conventional selection by partitioning based on affinity, photoSELEX selection, yielding increased specificity and affinity, using photoinduced covalent bonds with the target molecule via substitution of one of the nucleic acid bases in the library with a photoactivatable nucleotide (Koch et al. 2004), and capillary electrophoresis SELEX, resulting in high-affinity aptamers in a few cycles (Mosing et al. 2005).

16.5.3 Non-SELEX Selection of Aptamers

The SELEX approach involves multiple rounds of alternating steps of partitioning and PCR amplification. Non-SELEX selection of aptamers is a process that involves repetitive steps of partitioning with no amplification steps (Berezovski et al. 2006). Such a method, called non-equilibrium capillary electrophoresis of equilibrium mixtures (NECEEM), was used for selection of aptamers. In the first step of NECEEM, a randomized DNA library is mixed with the target protein and incubated to form the equilibrium mixture. DNA molecules with high affinity (potential aptamers) bind to the target, while those with low affinity (non-aptamers) do not bind. The resulting DNA-protein complexes are purified by non-equilibrium capillary electrophoresis. It was shown that three steps of NECEEM-based partitioning are sufficient to improve the affinity of a DNA library to a target protein by more than four orders of magnitude. Moreover, NECEEM-based selection took only 1 h in contrast to several days required for a typical SELEX procedure.

16.5.4 Aptamer Applications

Aptamers have many potential applications in biotechnology, pharmacology and biosensing. Each of these applications is based on the high affinity and the high specific recognition capabilities of aptamers (Hermann and Patel 2000). While antibodies can exhibit similar binding characteristics, generation of antibodies is complicated by the need for live animals (polyclonal antibodies) or cell culture (monoclonal antibodies) leading to high cost. Other factors favoring aptamers include longer shelf-life and simpler chemical modification. Furthermore, aptamers may be the only practical choice for targets that are weakly immunogenic and/or toxic. Discrimination ability of aptamers has been demonstrated in the ability to separate enantiomers (stereoisomers) (Michaud et al. 2003, 2004). Clinical and pharmacological research has included aptamer-based therapeutic agents against cancer, prion, HIV, and hepatitis. Recently, FDA approval of the first aptamer-based (Ng and Adamis 2006; Ng et al. 2006) therapeutic is a milestone in aptamer development. This drug (Pegaptanib) is designed to bind to, and inactivate, VEGF which is implicated as an angiogenic factor in age-related macular degeneration. Aptamers are attractive as therapeutics because they exhibit minimal immunogenic reaction. However, nucleases in the blood stream are quite effective such that the half-life of a simple aptamer is on the order of 1 min. Various chemical modifications of aptamers have increased the half-life to as much as 8 h.

In the last few years, a number of studies were published using aptamers as the recognition element in a biosensor. For many of these studies, it was necessary to immobilize the aptamer to a sensor surface (Balamurugan et al. 2006, 2008). Aptamer biosensor studies include QCM (Liss et al. 2002) SPR (Wang et al. 2007, 2008)] fluorescent detection (Rankin et al. 2006), cantilever (Savran et al. 2004), carbon nanotube field-effect transistors (CNT-FETs) (Maehashi et al. 2007), electrochemical detection (Kim et al. 2007; Yoon et al. 2008) and Quantum-dot electrochemical detection (Hansen et al. 2006). Related work includes aptamer beacons (Hamaguchi et al. 2001; McCauley et al. 2003).

16.6 Concluding Remarks

Oligonucleotide recognition elements form the basis of many current biology, biotechnology, biomedical and biosensing protocols. The power of these techniques stems from a number of key characteristics of polynucleotides, which have been harnessed into technologies for amplification and sequencing. Oligonucleotide recognition has a long history and a bright future based on the explosion of new techniques in the literature.

Oligonucleotides play a major role in biotechnology as recognition elements for nucleic acids and proteins. In addition to their basic biological function as genetic material, they have a role as genetic regulatory elements and as recognition elements for proteins. In this manuscript, we have described various aspects of oligonucleotides as recognition elements including their basic structure, various aspects of their biotechnology utility including DNA amplification, recognition and manipulation, their application for the regulation of gene expression, and their function and utility as proteins recognition elements.

The classical view of the role of nucleic acid as solely genetic material is *changing* rapidly. In the last few years, their role as genetic regulatory elements has become better understood and appreciated. Their capability to alter gene expression and to affect translation is multidimensional, including several antisense DNA/RNA modalities, various interfering RNA (RNAi) mechanisms and pathways, and their enzymatic activities as trans-cleaving ribozymes. All these mechanisms utilize their basic capabilities for annealing to specific RNA targets derived from their basic primary, secondary and tertiary structures.

Although each of these mechanisms relies on recognition capabilities, each genetic regulatory mechanism is completely different. While RNA/DNA antisense sequences interfere with several steps of RNA processing related to export, stability, and message translation including splicing, polyadenylation, blocking of the translational apparatus and induction of cleavage of the target RNA by RNaseH, RNAi disrupts translation by binding to mRNA and employing cellular machinery for degradation of the target RNA and ribozymes act as enzymes catalyzing RNA cleavage.

In addition to the base pairing interactions with other DNA and RNA molecules, specific short sequences within the DNA (or present as free oligonucleotides in experimental systems), are key elements in genetic regulation of processes such as DNA replication and gene expression, where such sequences serve as highly specific protein and other regulatory factor binding or recognition sites. Nucleic acid-protein interactions play a crucial role in the regulation of replication, transcription and translation in the cell.

Modern biotechnology harnesses oligonucleotide natural capabilities as recognition elements where oligonucleotides are widely used for a large number of DNA and protein manipulation technologies from DNA amplification and manipulation through their use to alter gene expression and their use as an alternative to antibodies as ligands for protein detection.

Most of the work on the role of oligonucleotides in regulation of processes was done in just the last few years. As we gain a better understanding of these functions and their interaction with other cell modalities, it is expected their utility in the rapidly developing biotechnology field will continue to grow.

References

Abbas-Terki T, Blanco-Bose W et al (2002) Lentiviral-mediated RNA interference. Hum Gene Ther 13(18):2197–2201

Agrawal S, Jiang Z et al (1997) Mixed-backbone oligonucleotides as second generation antisense oligonucleotides: in vitro and in vivo studies. Proc Natl Acad Sci USA 94(6):2620–2625

Aigner A (2007) Nonviral in vivo delivery of therapeutic small interfering RNAs. Curr Opin Mol Ther 9(4):345–352

Allawi HT, SantaLucia J Jr (1997) Thermodynamics and NMR of internal G.T mismatches in DNA. Biochemistry 36(34):10581–10594

Allawi HT, SantaLucia J Jr (1998a) NMR solution structure of a DNA dodecamer containing single G*T mismatches. Nucleic Acids Res 26(21):4925–4934

Allawi HT, SantaLucia J Jr (1998b) Nearest-neighbor thermodynamics of internal A.C mismatches in DNA: sequence dependence and pH effects. Biochemistry 37(26):9435–9444

Allawi HT, SantaLucia J Jr (1998c) Thermodynamics of internal C.T mismatches in DNA. Nucleic Acids Res 26(11):2694–2701

Allawi HT, SantaLucia J Jr (1998d) Nearest neighbor thermodynamic parameters for internal G.A mismatches in DNA. Biochemistry 37(8):2170–2179

Altschul SF, Gish W et al (1990) Basic local alignment search tool. J Mol Biol 215(3):403–410

Alwine JC, Kemp DJ et al (1977) Method for detection of specific RNAs in agarose gels by transfer to diazobenzyloxymethyl-paper and hybridization with DNA probes. Proc Natl Acad Sci USA 74(12):5350–5354

Arziman Z, Horn T, et al (2005) E-RNAi: a web application to design optimized RNAi constructs. Nucleic Acids Res 33(Web Server issue):W582–W588

Balamurugan S, Obubuafo A et al (2006) Designing highly specific biosensing surfaces using aptamer monolayers on gold. Langmuir 22(14):6446–6453

Balamurugan S, Obubuafo A et al (2008) Surface immobilization methods for aptamer diagnostic applications. Anal Bioanal Chem 390(4):1009–1021

Barbaux S, Poirier O et al (2001) Use of degenerate oligonucleotide primed PCR (DOP-PCR) for the genotyping of low-concentration DNA samples. J Mol Med 79(5–6):329–332

Barker DL, Hansen MS et al (2004a) Two methods of whole-genome amplification enable accurate genotyping across a 2320-SNP linkage panel. Genome Res 14(5):901–907

Barker KS, Crisp S et al (2004b) Genome-wide expression profiling reveals genes associated with amphotericin B and fluconazole resistance in experimentally induced antifungal resistant isolates of Candida albicans. J Antimicrob Chemother 54(2):376–385

Bartel DP, Szostak JW (1993) Isolation of new ribozymes from a large pool of random sequences [see comment]. Science 261(5127):1411–1418

Behlke MA, Devor EJ (2007) Chemical synthesis of oligonucleotides. http://www.idtdna.com/support/technical/TechnicalBulletinPDF/Chemical_Synthesis_of_Oligonucleotides.pdf

Berezovski MV, Musheev MU, Drabovich AP, Jitkova JV, and Krylov SN (2006) *Non-SELEX: selection of aptamers without intermediate amplification of candidate oligonucleotides.* Nat Protoc 1(3):p.1359–1369

Bock LC, Griffin LC et al (1992) Selection of single-stranded DNA molecules that bind and inhibit human thrombin. Nature 355(6360):564–566

Boiziau C, Toulme JJ (2001) A method to select chemically modified aptamers directly. Antisense Nucleic Acid Drug Dev 11(6):379–385

Bozdech Z, Zhu J et al (2003) Expression profiling of the schizont and trophozoite stages of Plasmodium falciparum with a long-oligonucleotide microarray. Genome Biol 4(2):R9

Breslauer KJ, Frank R et al (1986) Predicting DNA duplex stability from the base sequence. Proc Natl Acad Sci USA 83(11):3746–3750

Brody EN, Willis MC et al (1999) The use of aptamers in large arrays for molecular diagnostics. Mol Diagn 4(4):381–388

Brown DA, Kang SH et al (1994) Effect of phosphorothioate modification of oligodeoxynucleotides on specific protein binding. J Biol Chem 269(43):26801–26805

Brummelkamp TR, Bernards R et al (2002) A system for stable expression of short interfering RNAs in mammalian cells. Science 296(5567):550–553

Buchholz F, Kittler R et al (2006) Enzymatically prepared RNAi libraries. Nat Methods 3(9):696–700

Bumcrot D, Manoharan M et al (2006) RNAi therapeutics: a potential new class of pharmaceutical drugs. Nat Chem Biol 2(12):711–719

Bunnell BA, Morgan RA (1998) Gene therapy for infectious diseases. Clin Microbiol Rev 11 (1):42–56
Burke DH, Gold L (1997) RNA aptamers to the adenosine moiety of S-adenosyl methionine: structural inferences from variations on a theme and the reproducibility of SELEX. Nucleic Acids Res 25(10):2020–2024
Campbell JM, Bacon TA et al (1990) Oligodeoxynucleoside phosphorothioate stability in subcellular extracts, culture media, sera and cerebrospinal fluid. J Biochem Biophys Methods 20 (3):259–267
Carothers JM, Davis JH et al (2006a) Solution structure of an informationally complex high-affinity RNA aptamer to GTP. RNA 12(4):567–579
Carothers JM, Oestreich SC et al (2006b) Aptamers selected for higher-affinity binding are not more specific for the target ligand. J Am Chem Soc 128(24):7929–7937
Casey J, Davidson N (1977) Rates of formation and thermal stabilities of RNA:DNA and DNA: DNA duplexes at high concentrations of formamide. Nucleic Acids Res 4(5):1539–1552
Charbonnier Y, Gettler B et al (2005) A generic approach for the design of whole-genome oligoarrays, validated for genomotyping, deletion mapping and gene expression analysis on Staphylococcus aureus. BMC Genomics 6:95
Chester N, Marshak DR (1993) Dimethyl sulfoxide-mediated primer Tm reduction: a method for analyzing the role of renaturation temperature in the polymerase chain reaction. Anal Biochem 209(2):284–290
Cox JC, Hayhurst A et al (2002a) Automated selection of aptamers against protein targets translated in vitro: from gene to aptamer. Nucleic Acids Res 30(20):e108
Cox JC, Rajendran M et al (2002b) Automated acquisition of aptamer sequences. Comb Chem High Throughput Screen 5(4):289–299
Cutler DJ, Zwick ME et al (2001) High-throughput variation detection and genotyping using microarrays. Genome Res 11(11):1913–1925
Dandy DS, Wu P et al (2007) Array feature size influences nucleic acid surface capture in DNA microarrays. Proc Natl Acad Sci USA 104(20):8223–8228
de Fougerolles AR (2008) Delivery vehicles for small interfering RNA in vivo. Hum Gene Ther 19 (2):125–132
Demidov VV, Potaman VN et al (1994) Stability of peptide nucleic acids in human serum and cellular extracts. Biochem Pharmacol 48(6):1310–1313
Devroe E, Silver PA (2002) Retrovirus-delivered siRNA. BMC Biotechnol 2:15
Doench JG, Petersen CP et al (2003) siRNAs can function as miRNAs. Genes Dev 17(4):438–442
Doty P, Marmur J et al (1960) Strand separation and specific recombination in deoxyribonucleic acids: physical chemical studies. Proc Natl Acad Sci USA 46(4):461–476
Drude I, Dombos V et al (2007) Drugs made of RNA: development and application of engineered RNAs for gene therapy. Mini Rev Med Chem 7(9):912–931
Du Q, Thonberg H et al (2005) A systematic analysis of the silencing effects of an active siRNA at all single-nucleotide mismatched target sites. Nucleic Acids Res 33(5):1671–1677
Dudek P, Picard D (2004) TROD: T7 RNAi Oligo Designer. Nucleic Acids Res 32(Web Server issue):W121–123
Ekins R, Chu F et al (1990) Multispot, multianalyte, immunoassay. Ann Biol Clin (Paris) 48 (9):655–666
Ekins R, Chu F et al (1989) High specific activity chemiluminescent and fluorescent markers: their potential application to high sensitivity and 'multi-analyte' immunoassays. J Biolumin Chemilumin 4(1):59–78
Ekins R, Chu FW (1999) Microarrays: their origins and applications. Trends Biotechnol 17 (6):217–218
Ekwall K (2004) The roles of histone modifications and small RNA in centromere function. Chromosome Res 12(6):535–542
Elbashir SM, Harborth J et al (2001a) Duplexes of 21-nucleotide RNAs mediate RNA interference in cultured mammalian cells. Nature 411(6836):494–498

Elbashir SM, Harborth J et al (2002) Analysis of gene function in somatic mammalian cells using small interfering RNAs. Methods 26(2):199–213

Elbashir SM, Martinez J et al (2001b) Functional anatomy of siRNAs for mediating efficient RNAi in Drosophila melanogaster embryo lysate. Embo J 20(23):6877–6888

Ellington AD, Szostak JW (1990) In vitro selection of RNA molecules that bind specific ligands. Nature 346(6287):818–822

Ellington AD, Szostak JW (1992) Selection in vitro of single-stranded DNA molecules that fold into specific ligand-binding structures. Nature 355(6363):850–852

Feinberg AP, Vogelstein B (1983) A technique for radiolabeling DNA restriction endonuclease fragments to high specific activity. Anal Biochem 132(1):6–13

Fire A, Xu S et al (1998) Potent and specific genetic interference by double-stranded RNA in Caenorhabditis elegans. Nature 391(6669):806–811

Fixe F, Dufva M et al (2004) One-step immobilization of aminated and thiolated DNA onto poly (methylmethacrylate) (PMMA) substrates. Lab Chip 4(3):191–195

Fredriksson S, Gullberg M et al (2002) Protein detection using proximity-dependent DNA ligation assays. Nat Biotechnol 20(5):473–477

Gadkar V, Rillig MC (2005) Application of Phi29 DNA polymerase mediated whole genome amplification on single spores of arbuscular mycorrhizal (AM) fungi. FEMS Microbiol Lett 242(1):65–71

Gall JG, Pardue ML (1969) Formation and detection of RNA-DNA hybrid molecules in cytological preparations. Proc Natl Acad Sci USA 63(2):378–383

Gillespie D, Spiegelman S (1965) A quantitative assay for DNA-RNA hybrids with DNA immobilized on a membrane. J Mol Biol 12(3):829–842

Good PD, Krikos AJ et al (1997) Expression of small, therapeutic RNAs in human cell nuclei. Gene Ther 4(1):45–54

Guerrier-Takada C, Gardiner K et al (1983) The RNA moiety of ribonuclease P is the catalytic subunit of the enzyme. Cell 35(3 Pt 2):849–857

Guschin D, Yershov G et al (1997) Manual manufacturing of oligonucleotide, DNA, and protein microchips. Anal Biochem 250(2):203–211

Guvakova MA, Yakubov LA et al (1995) Phosphorothioate oligodeoxynucleotides bind to basic fibroblast growth factor, inhibit its binding to cell surface receptors, and remove it from low affinity binding sites on extracellular matrix. J Biol Chem 270(6):2620–2627

Hacia JG, Fan JB et al (1999) Determination of ancestral alleles for human single-nucleotide polymorphisms using high-density oligonucleotide arrays. Nat Genet 22(2):164–167

Hall N (2007) Advanced sequencing technologies and their wider impact in microbiology. J Exp Biol 210(Pt 9):1518–1525

Hamaguchi N, Ellington A et al (2001) Aptamer beacons for the direct detection of proteins. Anal Biochem 294(2):126–131

Hammond SM, Bernstein E et al (2000) An RNA-directed nuclease mediates post-transcriptional gene silencing in Drosophila cells. Nature 404(6775):293–296

Han J, Lee Y et al (2006) Molecular basis for the recognition of primary microRNAs by the Drosha-DGCR8 complex. Cell 125(5):887–901

Hansen JA, Wang J et al (2006) Quantum-dot/aptamer-based ultrasensitive multi-analyte electrochemical biosensor. J Am Chem Soc 128(7):2228–2229

He L, Kierzek R et al (1991) Nearest-neighbor parameters for G.U mismatches: [formula; see text] is destabilizing in the contexts [formula; see text] and [formula; see text] but stabilizing in [formula; see text]. Biochemistry 30(46):11124–11132

Heidenreich O, Kang SH et al (1995) Ribozyme-mediated RNA degradation in nuclei suspension. Nucleic Acids Res 23(12):2223–2228

Henschel A, Buchholz F, et al (2004) DEQOR: a web-based tool for the design and quality control of siRNAs. Nucleic Acids Res 32(Web Server issue):W113–W120

Hermann T, Patel DJ (2000) Adaptive recognition by nucleic acid aptamers. Science 287 (5454):820–825

Herold KE, Rasooly A (2003) Oligo Design: a computer program for development of probes for oligonucleotide microarrays. Biotechniques 35(6):1216–1221

Huang A, Chen Y et al (2004) Functional silencing of hepatic microsomal glucose-6-phosphatase gene expression in vivo by adenovirus-mediated delivery of short hairpin RNA. FEBS Lett 558 (1–3):69–73

Huang DB, Vu D et al (2003) Crystal structure of NF-kappaB (p50)2 complexed to a high-affinity RNA aptamer. Proc Natl Acad Sci USA 100(16):9268–9273

Ivanov SA, Volkov EM et al (2004) Chemical synthesis of an artificially branched hairpin ribozyme variant with RNA cleavage activity. Tetrahedron 60(41):9273–9281

Jackson AL, Burchard J et al (2006) Position-specific chemical modification of siRNAs reduces "off-target" transcript silencing. RNA 12(7):1197–1205

Jeffries AC, Symons RH (1989) A catalytic 13-mer ribozyme. Nucleic Acids Res 17(4): 1371–1377

Jiang M, Arzumanov AA et al (2005) A bi-functional siRNA construct induces RNA interference and also primes PCR amplification for its own quantification. Nucleic Acids Res 33(18):e151

Jiang P, Wu H et al (2007) RFRCDB-siRNA: improved design of siRNAs by random forest regression model coupled with database searching. Comput Methods Programs Biomed 87 (3):230–238

Judge A, Maclachlan I (2008) Overcoming the innate immune response to small interfering RNA. Hum Gene Ther 19(2):111–124

Kallioniemi OP (2001) Biochip technologies in cancer research. Ann Med 33(2):142–147

Kasim V, Taira K et al (2006) Screening of siRNA target sequences by using fragmentized DNA. J Gene Med 8(6):782–791

Kennerdell JR, Carthew RW (1998) Use of dsRNA-mediated genetic interference to demonstrate that frizzled and frizzled 2 act in the wingless pathway. Cell 95(7):1017–1026

Khvorova A, Reynolds A et al (2003) Functional siRNAs and miRNAs exhibit strand bias. Cell 115(2):209–216

Kim YS, Jung HS et al (2007) Electrochemical detection of 17beta-estradiol using DNA aptamer immobilized gold electrode chip. Biosens Bioelectron 22(11):2525–2531

Kittler R, Surendranath V et al (2007) Genome-wide resources of endoribonuclease-prepared short interfering RNAs for specific loss-of-function studies. Nat Methods 4(4):337–344

Koch TH, Smith D et al (2004) Kinetic analysis of site-specific photoaptamer-protein cross-linking. J Mol Biol 336(5):1159–1173

Kole R, Sazani P (2001) Antisense effects in the cell nucleus: modification of splicing. Curr Opin Mol Ther 3(3):229–234

Kruger K, Grabowski PJ et al (1982) Self-splicing RNA: autoexcision and autocyclization of the ribosomal RNA intervening sequence of Tetrahymena. Cell 31(1):147–157

Kuroda M, Ohta T et al (2001) Whole genome sequencing of meticillin-resistant Staphylococcus aureus. Lancet 357(9264):1225–1240

Kusser W (2000) Chemically modified nucleic acid aptamers for in vitro selections: evolving evolution. J Biotechnol 74(1):27–38

Lambeth LS, Moore RJ et al (2005) Characterisation and application of a bovine U6 promoter for expression of short hairpin RNAs. BMC Biotechnol 5:13

Lee Y, Ahn C et al (2003) The nuclear RNase III Drosha initiates microRNA processing. Nature 425(6956):415–419

Legiewicz M, Lozupone C et al (2005) Size, constant sequences, and optimal selection. RNA 11(11):1701–1709

Leiske DL, Karimpour-Fard A et al (2006) A comparison of alternative 60-mer probe designs in an in-situ synthesized oligonucleotide microarray. BMC Genomics 7:72

Lippman Z, May B et al (2003) Distinct mechanisms determine transposon inheritance and methylation via small interfering RNA and histone modification. PLoS Biol 1(3):E67

Liss M, Petersen B et al (2002) An aptamer-based quartz crystal protein biosensor. Anal Chem 74(17):4488–4495

Lockhart DJ, Dong H et al (1996) Expression monitoring by hybridization to high-density oligonucleotide arrays. Nat Biotechnol 14(13):1675–1680

Lopez PJ, Marchand I et al (1998) Translation inhibitors stabilize Escherichia coli mRNAs independently of ribosome protection. Proc Natl Acad Sci USA 95(11):6067–6072

Lorsch JR, Szostak JW (1994) In vitro selection of RNA aptamers specific for cyanocobalamin. Biochemistry 33(4):973–982

Luthra R, Medeiros LJ (2004) Isothermal multiple displacement amplification: a highly reliable approach for generating unlimited high molecular weight genomic DNA from clinical specimens. J Mol Diagn 6(3):236–242

Macrae IJ, Zhou K et al (2006) Structural basis for double-stranded RNA processing by Dicer. Science 311(5758):195–198

Maehashi K, Katsura T et al (2007) Label-free protein biosensor based on aptamer-modified carbon nanotube field-effect transistors. Anal Chem 79(2):782–787

Mairal T, Ozalp VC et al (2008) Aptamers: molecular tools for analytical applications. Anal Bioanal Chem 390(4):989–1007

Marmur J, Lane D (1960) Strand separation and specific recombination in deoxyribonucleic acids: biological studies. Proc Natl Acad Sci USA 46(4):453–461

Marsh AJ, Williams DM et al (2004) The synthesis and properties of oligoribonucleotide-spermine conjugates. Org Biomol Chem 2(14):2103–2112

Matveeva O, Nechipurenko Y et al (2007) Comparison of approaches for rational siRNA design leading to a new efficient and transparent method. Nucleic Acids Res 35(8):e63

McCauley TG, Hamaguchi N et al (2003) Aptamer-based biosensor arrays for detection and quantification of biological macromolecules. Anal Biochem 319(2):244–250

McTigue PM, Peterson RJ et al (2004) Sequence-dependent thermodynamic parameters for locked nucleic acid (LNA)-DNA duplex formation. Biochemistry 43(18):5388–5405

Michaud M, Jourdan E et al (2004) Immobilized DNA aptamers as target-specific chiral stationary phases for resolution of nucleoside and amino acid derivative enantiomers. Anal Chem 76 (4):1015–1020

Michaud M, Jourdan E et al (2003) A DNA aptamer as a new target-specific chiral selector for HPLC. J Am Chem Soc 125(28):8672–8679

Misono TS, Kumar PK (2005) Selection of RNA aptamers against human influenza virus hemagglutinin using surface plasmon resonance. Anal Biochem 342(2):312–317

Mosing RK, Mendonsa SD et al (2005) Capillary electrophoresis-SELEX selection of aptamers with affinity for HIV-1 reverse transcriptase. Anal Chem 77(19):6107–6112

Mullis KB (1990) The unusual origin of the polymerase chain reaction. Sci Am 262(4):56–61, 64–65

Napoli C, Lemieux C et al (1990) Introduction of a chimeric chalcone synthase gene into petunia results in reversible co-suppression of homologous genes in trans. Plant Cell 2(4):279–289

Ng EW, Adamis AP (2006) Anti-VEGF aptamer (pegaptanib) therapy for ocular vascular diseases. Ann NY Acad Sci 1082:151–171

Ng EW, Shima DT et al (2006) Pegaptanib, a targeted anti-VEGF aptamer for ocular vascular disease. Nat Rev Drug Discov 5(2):123–132

Nielsen PE (1999) Applications of peptide nucleic acids. Curr Opin Biotechnol 10(1):71–75

Nielsen PE (2000) Peptide nucleic acids: on the road to new gene therapeutic drugs. Pharmacol Toxicol 86(1):3–7

Nielsen PE (2001a) Peptide nucleic acids as antibacterial agents via the antisense principle. Expert Opin Investig Drugs 10(2):331–341

Nielsen PE (2001b) Targeting double stranded DNA with peptide nucleic acid (PNA). Curr Med Chem 8(5):545–550

Nielsen PE (2004) PNA Technology. Mol Biotechnol 26(3):233–248

Nielsen PE, Egholm M et al (1991) Sequence-selective recognition of DNA by strand displacement with a thymine-substituted polyamide. Science 254(5037):1497–1500

Nielsen PE, Egholm M et al (1994) Peptide nucleic acid (PNA). A DNA mimic with a peptide backbone. Bioconjug Chem 5(1):3–7

Okamura K, Ishizuka A et al (2004) Distinct roles for Argonaute proteins in small RNA-directed RNA cleavage pathways. Genes Dev 18(14):1655–1666

Ovcharenko D, Kelnar K et al (2007) Genome-scale microRNA and small interfering RNA screens identify small RNA modulators of TRAIL-induced apoptosis pathway. Cancer Res 67 (22):10782–10788

Paddison PJ, Caudy AA et al (2002) Short hairpin RNAs (shRNAs) induce sequence-specific silencing in mammalian cells. Genes Dev 16(8):948–958

Paez JG, Lin M et al (2004) Genome coverage and sequence fidelity of phi29 polymerase-based multiple strand displacement whole genome amplification. Nucleic Acids Res 32(9):e71

Paule MR, White RJ (2000) Survey and summary: transcription by RNA polymerases I and III. Nucleic Acids Res 28(6):1283–1298

Pease AC, Solas D et al (1994) Light-generated oligonucleotide arrays for rapid DNA sequence analysis. Proc Natl Acad Sci USA 91(11):5022–5026

Peyret N, Seneviratne PA et al (1999) Nearest-neighbor thermodynamics and NMR of DNA sequences with internal A.A, C.C, G.G, and T.T mismatches. Biochemistry 38(12): 3468–3477

Pirollo KF, Chang EH (2008) Targeted delivery of small interfering RNA: approaching effective cancer therapies. Cancer Res 68(5):1247–1250

Pushparaj PN, H'Ng SC et al (2008) Refining siRNA in vivo transfection: silencing SPHK1reveals its key role in C5a-induced inflammation in vivo. Int J Biochem Cell Biol 40 (9):1817–1825

Qiu S, Adema CM et al (2005) A computational study of off-target effects of RNA interference. Nucleic Acids Res 33(6):1834–1847

Qiu S, Lane T (2006) RNA string kernels for RNAi off-target evaluation. Int J Bioinform Res Appl 2(2):132–146

Rand TA, Petersen S et al (2005) Argonaute2 cleaves the anti-guide strand of siRNA during RISC activation. Cell 123(4):621–629

Rankin CJ, Fuller EN et al (2006) A simple fluorescent biosensor for theophylline based on its RNA aptamer. Nucleosides Nucleotides Nucleic Acids 25(12):1407–1424

Reynolds A, Leake D et al (2004) Rational siRNA design for RNA interference. Nat Biotechnol 22 (3):326–330

Rich A (1960) A Hybrid Helix Containing Both Deoxyribose and Ribose Polynucleotides and Its Relation to the Transfer of Information between the Nucleic Acids. Proc Natl Acad Sci USA 46 (8):1044–1053

Rual JF, Klitgord N et al (2007) Novel insights into RNAi off-target effects using C. elegans paralogs. BMC Genomics 8:106

Saiki RK, Scharf S et al (1985) Enzymatic amplification of beta-globin genomic sequences and restriction site analysis for diagnosis of sickle cell anemia. Science 230(4732):1350–1354

Sanger F, Coulson AR (1975) A rapid method for determining sequences in DNA by primed synthesis with DNA polymerase. J Mol Biol 94(3):441–448

SantaLucia J Jr (1998) A unified view of polymer, dumbbell, and oligonucleotide DNA nearest-neighbor thermodynamics. Proc Natl Acad Sci USA 95(4):1460–1465

Santoyo J, Vaquerizas JM et al (2005) Highly specific and accurate selection of siRNAs for high-throughput functional assays. Bioinformatics 21(8):1376–1382

Savran CA, Knudsen SM et al (2004) Micromechanical detection of proteins using aptamer-based receptor molecules. Anal Chem 76(11):3194–3198

Sazani P, Astriab-Fischer A et al (2003) Effects of base modifications on antisense properties of 2′-O-methoxyethyl and PNA oligonucleotides. Antisense Nucleic Acid Drug Dev 13(3): 119–128

Sazani P, Kole R (2003a) Therapeutic potential of antisense oligonucleotides as modulators of alternative splicing. J Clin Invest 112(4):481–486

Sazani P, Kole R (2003b) Modulation of alternative splicing by antisense oligonucleotides. Prog Mol Subcell Biol 31:217–239

Schena M, Shalon D et al (1995) Quantitative monitoring of gene expression patterns with a complementary DNA microarray. Science 270(5235):467–470

Schramm L, Hernandez N (2002) Recruitment of RNA polymerase III to its target promoters. Genes Dev 16(20):2593–2620

Schurer H, Stembera K et al (2001) Aptamers that bind to the antibiotic moenomycin A. Bioorg Med Chem 9(10):2557–2563

Schwarz DS, Hutvagner G et al (2003) Asymmetry in the assembly of the RNAi enzyme complex. Cell 115(2):199–208

Sergeev N, Distler M et al (2006a) Microarray analysis of Bacillus cereus group virulence factors. J Microbiol Methods 65(3):488–502

Sergeev N, Herold KE et al (2006b) Microarray analysis of microbial pathogens. Encyclopedia Rapid Microbiol Methods 3:363–405

Sergeev N, Rasooly A et al (2005) Oligonucleotide design for PCR primers and microarray probes. Encyclopedia Med Genomics Proteomics 2:938–942

Sharp PA (1999) RNAi and double-strand RNA. Genes Dev 13(2):139–141

Shim MS, Kwon YJ (2008) Controlled delivery of plasmid DNA and siRNA to intracellular targets using ketalized polyethylenimine. Biomacromolecules 9(2):444–455

Slim G, Pritchard C et al (1991) Synthesis of site-specifically modified oligoribonucleotides for studies of the recognition of TAR RNA by HIV-1 tat protein and studies of hammerhead ribozymes. Nucleic Acids Symp Ser 24:55–58

Smirnov I, Shafer RH (2000) Effect of loop sequence and size on DNA aptamer stability. Biochemistry 39(6):1462–1468

Southern EM (1975) Detection of specific sequences among DNA fragments separated by gel electrophoresis. J Mol Biol 98(3):503–517

Stillman BA, Tonkinson JL (2000) FAST slides: a novel surface for microarrays. Biotechniques 29 (3):630–635

Stillman BA, Tonkinson JL (2001) Expression microarray hybridization kinetics depend on length of the immobilized DNA but are independent of immobilization substrate. Anal Biochem 295 (2):149–157

Sugimoto N, Nakano S et al (1996) Improved thermodynamic parameters and helix initiation factor to predict stability of DNA duplexes. Nucleic Acids Res 24(22):4501–4505

Supattapone S (2004) Prion protein conversion in vitro. J Mol Med 82(6):348–356

Talla E, Tekaia F et al (2003) A novel design of whole-genome microarray probes for Saccharomyces cerevisiae which minimizes cross-hybridization. BMC Genomics 4(1):38

Tang J, Breaker RR (2000) Structural diversity of self-cleaving ribozymes. Proc Natl Acad Sci USA 97(11):5784–5789

Telenius H, Carter NP et al (1992) Degenerate oligonucleotide-primed PCR: general amplification of target DNA by a single degenerate primer. Genomics 13(3):718–725

Thomas PS (1980) Hybridization of denatured RNA and small DNA fragments transferred to nitrocellulose. Proc Natl Acad Sci USA 77(9):5201–5205

Tombelli S, Minunni M et al (2005) Analytical applications of aptamers. Biosens Bioelectron 20 (12):2424–2434

Tuerk C, Gold L (1990) Systematic evolution of ligands by exponential enrichment: RNA ligands to bacteriophage T4 DNA polymerase. Science 249(4968):505–510

Uhlenbeck OC (1987) A small catalytic oligoribonucleotide. Nature 328(6131):596–600

Ui-Tei K, Naito Y et al (2007) Guidelines for the selection of effective short-interfering RNA sequences for functional genomics. Methods Mol Biol 361:201–216

Ulrich H, Martins AH et al (2004) RNA and DNA aptamers in cytomics analysis. Cytometry A 59 (2):220–231

Urban E, Noe CR (2003) Structural modifications of antisense oligonucleotides. Farmaco 58 (3):243–258

Vacek M, Sazani P et al (2003) Antisense-mediated redirection of mRNA splicing. Cell Mol Life Sci 60(5):825–833

Varekova RS, Bradac I, et al (2008) http://www.rnaworkbench.com: A new program for analyzing RNA interference. Comput Methods Programs Biomed 90(1):89–94
Vermeulen A, Behlen L et al (2005) The contributions of dsRNA structure to Dicer specificity and efficiency. RNA 11(5):674–682
Wang J, Lv R et al (2008) Characterizing the interaction between aptamers and human IgE by use of surface plasmon resonance. Anal Bioanal Chem 390(4):1059–1065
Wang Z, Wilkop T et al (2007) Surface plasmon resonance imaging for affinity analysis of aptamer-protein interactions with PDMS microfluidic chips. Anal Bioanal Chem 389 (3):819–825
Watkins NE Jr, SantaLucia J Jr (2005) Nearest-neighbor thermodynamics of deoxyinosine pairs in DNA duplexes. Nucleic Acids Res 33(19):6258–6267
Watson JD, Crick FH (1953) Molecular structure of nucleic acids; a structure for deoxyribose nucleic acid. Nature 171(4356):737–738
Westin L, Miller C et al (2001) Antimicrobial resistance and bacterial identification utilizing a microelectronic chip array. J Clin Microbiol 39(3):1097–1104
Williamson JR (1994) G-quartet structures in telomeric DNA. Annu Rev Biophys Biomol Struct 23:703–730
Wittung P, Nielsen PE et al (1994) DNA-like double helix formed by peptide nucleic acid. Nature 368(6471):561–563
Wolff JA, Rozema DB (2008) Breaking the bonds: non-viral vectors become chemically dynamic. Mol Ther 16(1):8–15
Wu H, Dinh A et al (2007) Generation of shRNAs from randomized oligonucleotides. Biol Proced Online 9:9–17
Wu Y, de Kievit P et al (2004) Quantitative assessment of a novel flow-through porous microarray for the rapid analysis of gene expression profiles. Nucleic Acids Res 32(15):e123
Yamada T, Morishita S (2005) Accelerated off-target search algorithm for siRNA. Bioinformatics 21(8):1316–1324
Yang D, Buchholz F et al (2002) Short RNA duplexes produced by hydrolysis with Escherichia coli RNase III mediate effective RNA interference in mammalian cells. Proc Natl Acad Sci USA 99(15):9942–9947
Yang L, Tran DK et al (2001) BADGE, Beads Array for the Detection of Gene Expression, a high-throughput diagnostic bioassay. Genome Res 11(11):1888–1898
Yeom KH, Lee Y et al (2006) Characterization of DGCR8/Pasha, the essential cofactor for Drosha in primary miRNA processing. Nucleic Acids Res 34(16):4622–4629
Yoon H, Kim JH et al (2008) A novel sensor platform based on aptamer-conjugated polypyrrole nanotubes for label-free electrochemical protein detection. Chembiochem 9(4):634–641
Yu JY, DeRuiter SL et al (2002) RNA interference by expression of short-interfering RNAs and hairpin RNAs in mammalian cells. Proc Natl Acad Sci USA 99(9):6047–6052
Zamaratski E, Pradeepkumar PI et al (2001) A critical survey of the structure-function of the antisense oligo/RNA heteroduplex as substrate for RNase H. J Biochem Biophys Methods 48 (3):189–208
Zamore PD, Tuschl T et al (2000) RNAi: double-stranded RNA directs the ATP-dependent cleavage of mRNA at 21 to 23 nucleotide intervals. Cell 101(1):25–33
Zimmer A, Aziz SA et al (1999) Synthesis of cholesterol modified cationic lipids for liposomal drug delivery of antisense oligonucleotides. Eur J Pharm Biopharm 47(2):175–178
Zobel HP, Zimmer A et al (1999) Evaluation of aminoalkylmethacrylate nanoparticles as colloidal drug carrier systems. Part I: Synthesis of monomers, dependence of the physical properties on the polymerization methods. Eur J Pharm Biopharm 47(3):203–213

Chapter 17
Aptamers: Versatile Tools for Reagentless Aptasensing

Eva Baldrich

Abstract From time to time, the reporting of a "new" bio-component, molecule or reagent contributes to revolutionize a certain field. This happened, for example, when antibodies (Ab) were applied for the first time to detection in the 1950s, opening the doors to the possibility of immuno-diagnosis (Yallow and Berson 1959), or when the development of monoclonal antibodies (MAb) in 1975 imparted a significant improvement in the specificity of the existing assays (Kohler and Milstein 1975). With aptamers' arrival onto the scene, we may be witnessing another step in bio-sensing evolution: molecules of a striking chemical simplicity and production that behave as efficiently as antibodies and favor the development of innovative approaches, such as aptabeacons and aptamer arrays.

In this chapter, we will summarize the recent advances in aptamer and aptasensor development. The first part of the text is devoted to provide a wide background on aptamer biochemistry, including the description of the aptamer selection technology, advantages and drawbacks of aptamers versus traditional biorecognition components, aptamer folding and stabilization, or the molecular basis of the aptamer–target interaction. The second part of the chapter, on the other hand, focuses on the different aptasensors and aptasensing strategies reported to date.

Keywords Aptamer · Aptasensor · Reagentless detection · Sensor development

Abbreviations

3-D	Three-dimensional
Ab	Antibody

E. Baldrich
Institut de Microelectrònica de Barcelona - Centro Nacional de Microelectrónica (IMB-CNM), CSIC, Campus Universidad Autónoma de Barcelona, 08193 - Bellaterra, Barcelona, Spain
e-mail: Eva.baldrich@imb-cnm.csic.es

M. Zourob (ed.), *Recognition Receptors in Biosensors*,
DOI 10.1007/978-1-4419-0919-0_17,

AMP	Adenosine monophosphate, also known as 5′-adenylic acid
ATP	Adenosine 5′-triphosphate
bFGF	Basic fibroblast growth factor
BSA	Bovine serum albumin
CE	Capillary electrophoresis
CE-SELEX	Capillary electrophoresis SELEX
CV	Cyclic voltammetry
Cy3	Fluorophore of the cyanine dye family
DNA	Deoxyribonucleic acid
DPV	Differential pulse voltammetry
ELAA	Enzyme-linked aptamer assay
ELISA	Enzyme-linked immunoassay
ELONA	Enzyme-linked oligonucleotide assay
FET	Field effect transistor
FITC	Fluorescein, one of the most used fluorophores
FRET	Förster/fluorescence resonance energy transfer
HCV	Hepatitis C virus
HIV	Human immunodeficiency virus
IES	Faradaic impedance spectroscopy
IgE	Human immunoglobulin E
Ip	Iso-electric point
IMPDH	Inosine monophosphate dehydrogenase
ISFET	Ion-selective field-effect transistor
LOD	Limit of detection
MAb	Monoclonal antibody
MB	Molecular beacon
NIS	Non-faradic impedance spectroscopy
NMR	Nuclear magnetic resonance
PCR	Polymerase chain reaction
PDGF	Platelet-derived growth factor
QCM	Quartz crystal microbalance
RNA	Ribonucleic acid
SAM	Self-assembled monolayer
SAW	Surface acoustic wave sensor
SELEX	Systematic evolution of ligands by exponential enrichment
SPR	Surface plasmon resonance
TBA	Thrombin-binding aptamer
UV	Ultraviolet
VEGF	Vascular endothelial growth factor
XRC	X-ray crystallography

17.1 Introduction

Aptamers were reported for the first time in the early 1990s and nearly simultaneously by three different laboratories in La Jolla, Boston and Bolder (USA) (Ellington and Szostak 1990; Robertson and Joyce 1990; Tuerk and Gold 1990).

They were described as artificial nucleic acid ligands, selected in vitro from DNA/RNA random pools thanks to their high affinity for non-nucleic acid targets of the researcher's choice. Aptamers were thus chemically related to nucleic acid probes, but behaved more like Ab. The newly described oligonucleotides were soon referred to as "aptamers," coined from the Latin *aptus*, meaning "fitting" (Jayasena 1999). Since their inception, an important number of aptamers has been selected against a great variety of targets, showing surprising versatility compared to other bio-recognition components. For example, aptamers have been selected not only against small molecules, drugs, peptides and hormones (Brambs et al. 2005; Gebhardt et al. 2000; Haller and Sarnow 1997; Levesque et al. 2007; Mann et al. 2005; Shoji et al. 2007; Vater et al. 2003; Wilson and Szostak 1999; Win et al. 2006), but also against bigger and more complex targets such as proteins, whole spores and even whole cells (Ikanovic et al. 2007; Lee et al. 2005; Mallikaratchy et al. 2006; Mendonsa and Bowser 2004; Ohuchi et al. 2006; Shangguan et al. 2006; Spiridonova et al. 2003). In the later case, the authors demonstrated that aptamers could be used to target a disease state, such as cancer, without prior knowledge of the molecular changes associated with its developing. Interestingly, during the last years, an increasing number of aptamers have been selected against pathogens, including prions, viruses, bacteria and other parasites (Bellecave et al. 2003; Ellingham et al. 2006; Gopinath et al. 2006; Homann et al. 2006; Hwang et al. 2004; Jones et al. 2006; Kikuchi et al. 2003; King et al. 2007; Kulbachinskiy et al. 2004; Misono and Kumar 2005; Mosing et al. 2005; Pan et al. 2005; Ramos et al. 2007; Rhie et al. 2003; Sando et al. 2007; Sekiya et al. 2006; Weiss et al. 1997). This makes us raise the question on whether aptamers will be useful and used in the future as a substitute of more established components for therapeutics and/or diagnostics.

In this respect, most of the reported aptamers have shown high affinity and specificity for their targets, equaling or even surpassing the efficiency of their equivalent Ab. If we also consider the relatively simple techniques and apparatus required for their isolation, the number of alternative molecules that can be screened per experiment and their chemical simplicity, we will understand that aptamers have been proposed by several authors as ideal *alternative* candidates to the traditional antibodies for use in analytical devices and techniques (Brody and Gold 2000; Clark and Remcho 2002; Hesselberth et al. 2000; Jayasena 1999; Osborne et al. 1997; Tombelli et al. 2005b).

17.2 Some General Notes on Aptamer Technology

17.2.1 Aptamer Selection: SELEX and Automated SELEX

17.2.1.1 The Principle of Traditional SELEX

Aptamers are selected from single-stranded DNA or RNA combinatorial libraries via an in vitro iterative process of adsorption, recovery and re-amplification known as SELEX, an acronym for *S*ystematic *E*volution of *L*igands by *Ex*ponential

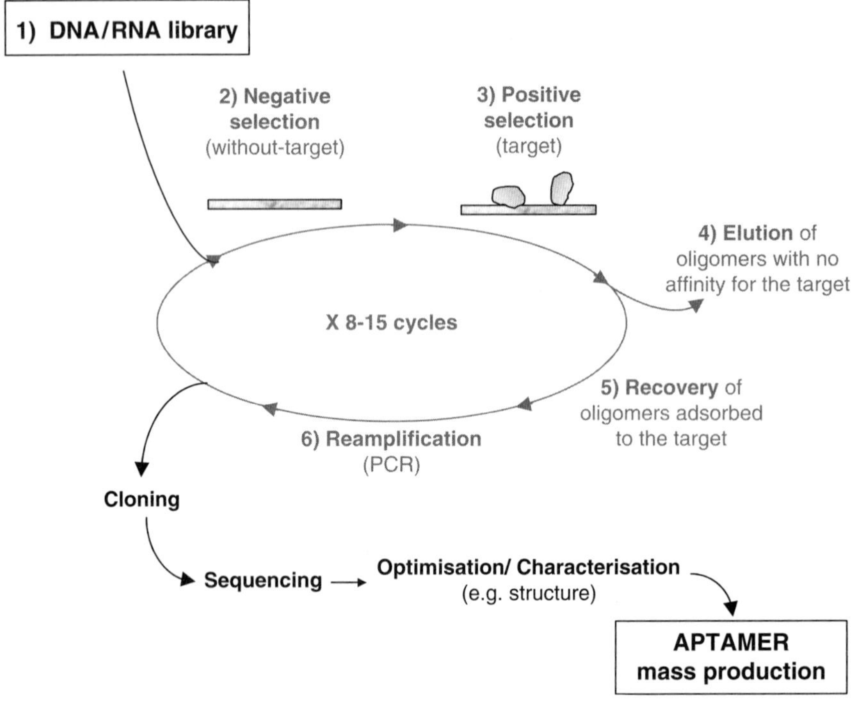

Fig. 17.1 Scheme of the SELEX procedure. The DNA/RNA library is submitted to 8–15 rounds of SELEX. The aptamer candidates are then cloned, sequenced, and characterized. Mass production of the best-performing aptamers allows assay development

enrichment (Ellington and Szostak 1990; Robertson and Joyce 1990; Tuerk and Gold 1990) (Fig. 17.1). The procedure begins with the production of a DNA-random library using a standard DNA synthesizer. The library size, in terms of number of different molecules, depends on their length and typically consists in 10^{13}–10^{15} for nucleotides of 70–150 bases. Each molecule contains a central region of random sequence, flanqued by two short fixed primers for later PCR amplification. When RNA aptamers want to be selected, the DNA library is first submitted to *in vitro* transcription in order to produce the corresponding RNA library. DNA libraries/aptamers are cheaper to produce and are particularly useful for applications requiring high stability, such as biosensing, environmental monitoring and therapeutics. RNA libraries, on the other hand, often yield aptamers with higher binding affinities and can be actively transcribed in vivo from suitable templates, anticipating their potential application in therapeutics. As a third alternative, aptamers can be selected from modified libraries bearing chemical functionalities not present in natural nucleic acids. Chemically modified oligonucleotides are resistant to degradation by nucleases and display novel physico–chemical and folding properties, and to a certain extent expand the existing genetic code (Eaton 1997; Kusser 2000; Piccirilli et al. 1990).

The SELEX procedure continues with the incubation of the DNA/RNA library with the target under study, which may be free in solution or immobilized on a physical support, such as an affinity column, microtiter plate, magnetic particles or nanoparticles (Bruno and Kiel 2002; Burke and Gold 1997; Cox and Ellington 2001; Moreno et al. 2003; Nieuwlandt et al. 1995; Rhodes et al. 2000; Stoltenburg et al. 2005; Tuerk and Gold 1990; Zhang and Anderson 1998). If there exist sequences in the library able to bind to the target, they can be separated from the library pool by simple elution, wash, magnetic separation, centrifugation, or filtration, among others (Bruno and Kiel 2002; Burke and Gold 1997; Cox and Ellington 2001; Moreno et al. 2003; Nieuwlandt et al. 1995; Rhodes et al. 2000; Stoltenburg et al. 2005; Zhang and Anderson 1998). The adsorbed species, and thus aptamer candidates, can now be recovered, re-amplified by PCR taking advantage of the fixed primers, and submitted to additional selection cycles, usually 8–15, with incubation and washing conditions of increasing stringency. In order to improve aptamer specificity, SELEX often includes rounds of negative selection, to eliminate non-specific binders (e.g. nucleotides binding to the physical support, surface-blocking agent, or ions present in the SELEX solution) and/or counter selection to remove nucleotides binding to molecules similar to the target (enzyme isotypes; proteins from different species or states) (Conrad et al. 1994; Kimoto et al. 2002). At this point, the aptamer candidates can be cloned, sequenced and further studied, in order to identify the consensus motif and 3-D structure responsible for target binding. The final aim is to produce the aptamer minimal consensus motif, whose behavior is predictable and production is as cheap as possible. The production of aptamers offers an additional advantage versus other capture elements: aptamers can be submitted to post-SELEX modification in order to, for example, improve their kinetics, nuclease resistance and/or tissue distribution. This can be accomplished by introducing modifications in and around the binding interface, followed by the study of the aptamer performance (Eaton et al. 1997). For additional information on SELEX, the reader is directed to the numerous reviews published (Gopinath 2007; Hamula et al. 2006; Kopylov and Spiridonova 2000; Nieuwlandt 2000; Uphoff et al. 1996; Wilson and Szostak 1999).

17.2.1.2 Improved SELEX

A typical SELEX experiment carried out manually may take weeks or even months. For this reason, several groups have worked in the development of faster and easier aptamer production strategies. Capillary electrophoresis (CE) is an excellent tool for the resolution of bound and unbound nucleic acids in solution, based on the mobility shift induced by the complex formation. CE is characterized by its speed, resolution capacity, and minimal sample dilution and manipulation required. Consistently, CE-SELEX has generated high affinity aptamers against several proteins and enzymes after only 2–4 cycles of selection carried out in just three working days, not requiring target immobilization or manipulation (Berezovski et al. 2005; Drabovich et al. 2006; Mendonsa and Bowser 2004; Mosing et al. 2005; Tang et al. 2006). The efficiency of this methodology, however, seems limited to target molecules of a certain size, with the only exception reported for the aptamer against the small ricin toxin (Tang et al. 2006). Surface plasmon resonance (SPR), regularly used to monitor bio-molecule interactions in real time, has also been successfully applied to SELEX (Misono and Kumar 2005). SPR-SELEX can be carried out on line, recording the kinetics behavior of the candidate aptamers even before their isolation. In a completely different approach, selection of aptamers against resofurin, a fluorescent molecule, was accomplished on microarray chips, which contained hundreds of random oligonucleotides (Asai et al. 2004). Sequences able to bind the target were next submitted to point mutation, followed by additional binding-mutation cycles. Three complete cycles were sufficient to produce affinity binders as, according to the authors, the best performing sequences do not have to compete with less efficient binders during the PCR steps carried in classical SELEX.

Among the SELEX variants described for specific purposes, "Toggle SELEX" was reported as a tool to generate aptamers which simultaneously recognizes related targets (Bianchini et al. 2001; White et al. 2001). With this purpose, SELEX is carried out by "toggling" between targets during alternating rounds of selection. The procedure allowed the authors to produce an aptamer that recognized and inhibited both human and porcine thrombin with dissociation constants (K_d) of around 1 nM. The system is useful for the production of therapeutic aptamers, which may not progress from bench top to clinical trials without having demonstrated first their efficacy in animal models. A different strategy is based on the production of aptamer pools, which could be treated as the equivalent to polyclonal Ab versus the use of isolated species (Bruno and Kiel 1999; Moreno et al. 2003). At least one company, AptaRes (http://www.aptares.net/), offers fast production of "polyclonal aptamers," which include aptamers of different affinities against various epitopes of the chosen target, synthesized using a one-round procedure. Alternatively, authors have searched for procedures able to generate aptamers with higher affinities. "Covalent SELEX" exploits oligonucleotides modified with reactive groups able to crosslink with the target, thereby preventing dissociation of the complex. For example, PhotoSELEX uses photoactivatable nucleotides which, following exposition to UV light, form a permanent covalent bond with the target. This allows intensive and stringent washes, efficiently removing non-specific binders (Golden et al. 2000). PhotoSELEX has been reported to produce photoaptamers with low pM dissociation constants against various serum and viral proteins, in just 6–8 rounds of selection (Bock et al. 2004; Golden et al. 2000; Jensen et al. 1995). Finally, Berezovski has described a different approach named "Non-SELEX Selection" (Berezovski et al. 2006). In this case, the DNA library is submitted to serial partitioning steps, with no amplification in-between, using non-equilibrium capillary electrophoresis of equilibrium mixtures (NECEEM). In only one hour and three rounds of partitioning, aptamers were isolated with K_d of down to 0.2 μM for h-Ras protein as a target.

17.2.1.3 Automated SELEX

The previously described SELEX systems are, in practice, repetitive, time-consuming and *not* applicable for high-throughput selection. Automated systems should provide flexibility and versatility, allowing the handling of multiple targets in parallel, as well as the sequential selection of buffers, reagents and stringent conditions without any direct intervention of the researcher. The first SELEX automated robotic platform described was based on the use of a Beckman Biomek 2000 pipetting robot, which integrated a PCR thermal cycler, a magnetic bead separator, reagent trays and a pipette tip station, and was successfully used to select poly-T binding RNA molecules (Cox et al. 1998). The equipment was later optimized, substituting, among others, magnetic bead capture by filtering, allowing selection of aptamers against proteins and synthetic peptides in just a couple of days (Cox and Ellington 2001; Cox et al. 2002a, b). Eulberg reported on the incorporation of cooling, heating, storage and vacuum units to a RoboAmp 4200 instrument (Biotech). The robotic platform was used to produce an RNA spiegelmer against an 11 amino acid peptide neurotransmitter (Eulberg et al. 2005). More recently, Hybarger described a microchannel-based automated microfluidic SELEX (Hybarger et al. 2006), claimed by the authors as smaller, cheaper and faster than the previously reported platforms. At present, the company NascaCell (http://www.nascacell.de) commercially provides fully characterized customized-aptamers, generated with their automated SELEX platform within weeks.

17.2.2 Aptamer Stability: Spiegelmers and Chemically Modified Aptamers

One of the main concerns in relation to the real applicability of aptamers is the known *sensitivity* of nucleic acids to nuclease attack. Nucleic acids free in solution, and specially RNA, are fast digested by the ubiquitous nucleases and often show lifetimes of less than 10 min (James 2001; Kusser 2000). This may limit aptamer lifetime and use on real samples and in vivo applications. Although some DNA aptamers have shown unusual stability in solution, presumably due to their tight folding (Bock et al. 1992; Griffin et al. 1993; Hicke et al. 1996; Tasset et al. 1997; Weiss et al. 1997), a great deal of research has been devoted to *increasing* aptamer stability under such conditions (Kusser 2000).

17.2.2.1 Spiegelmers

Spiegelmers are "mirror-image aptamers," synthesized from L-nucleosides instead of the D-enantiomeric forms present in nature (Klussmann et al. 1996; Leva et al. 2002a; Nolte et al. 1996). Their stability relies on the fact that L-nucleosides are

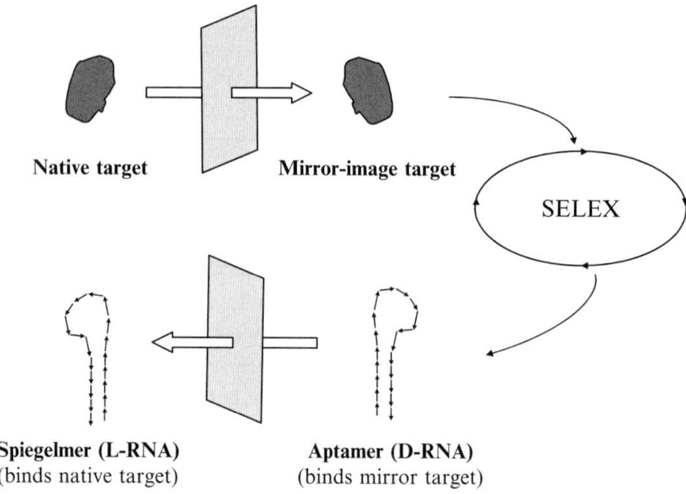

Fig. 17.2 Spiegelmers and mirror SELEX

neither found in nature nor naturally produced, and thus not efficiently digested by nucleases. Since polymerases used in aptamer production are also enantio-specific, spiegelmer selection has to be based on a "mirror-image" SELEX, which starts with the synthesis of the enantiomer of the chosen target, D-target (Fig. 17.2). This synthetic D-target is then submitted to SELEX versus a "normal" nucleic acid library in order to obtain D-aptamers. By virtue of molecular symmetry, a D-aptamer binds the mirror image of a target (D-target) as efficiently as its equivalent L-aptamer (spiegelmer) binds the native L-target. Spiegelmers are thus chemically synthesized at this point by using L-nucleosides to copy the sequences of the selected D-oligomers. The spiegelmers reported to date have shown excellent performances, with stabilities of hours to up to 10 days in serum or even in the presence of purified nucleases (Williams et al. 1997). The requirement for the synthesis of a target enantiomer, however, has limited spiegelmer selection towards only small molecules and peptides (Faulhammer et al. 2004; Helmling et al. 2004; Klussmann et al. 1996; Leva et al. 2002a; Nolte et al. 1996; Purschke et al. 2003; Vater et al. 2003; Williams et al. 1997). Spiegelmers are commercialized as pharmaceutical candidates by the German firm Noxxon Pharma AG (http://www.noxxon.net/noxxon). Chemically and structurally similar to D nucleic acids, spiegelmers induce minimal toxicity and immunogenic response in animals and appear as powerful candidates for in vivo diagnostics and therapeutics (Sooter and Ellington 2002; Vater and Klussmann 2003; Wlotzka et al. 2002).

17.2.2.2 Chemically Modified Aptamers

The alternative production of chemically modified aptamers, less or not efficiently recognized by nucleases, appears as a more universally applicable strategy in the

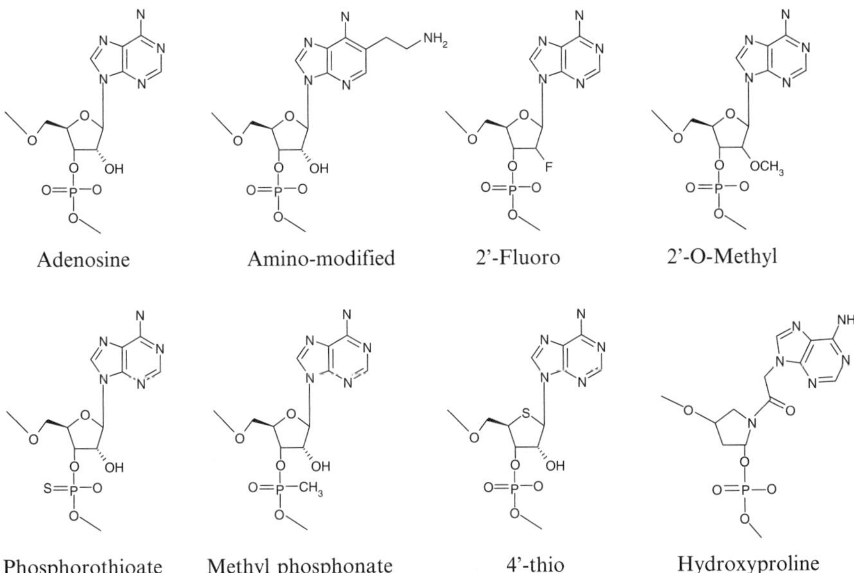

Fig. 17.3 Examples of modified nucleotides. The schemes show the molecular structure of unmodified adenosine (*top left*) and some modified molecules commonly used in production of chemically modified aptamers. None of these modified molecules is efficiently recognized by nucleases

search for stable ligands. This is generally accomplished by incorporating chemically modified nucleosides during the library synthesis, provided that they are efficiently recognized by the enzymes used along the synthesis and SELEX procedures (some examples are illustrated in Fig. 17.3). Alternatively, post-selectional modifications can be carried out by incorporating different chemical groups to some of the nucleotides taking place in the binding pocket. This strategy, however, requires accurate studies as to ensure that the modified aptamers retain binding efficiency, and often aptamer re-selection. Working with chemically modified libraries provides an additional benefit: the "new bases" contribute to generate a wider variety of 3-D structures (Shoji et al. 2007). Concerning the existing modification strategies, only a few works report successful amendment of the oligonucleotide backbone, such as the use of α-thio-substituted nucleosides (Jhaveri et al. 1998; King et al. 1998; Ruckman et al. 1998). Conversely, the modification of the 2′-OH position of the pyrimidine ribonucleotides, one of the major targets for ribonucleases, has been repeatedly exploited, even when its success depends on the purine/pyrimidine ratio in the final aptamer. For example, 2′-amino modified RNA aptamers have been produced against human neutrophil elastase, thyroid stimulating hormone, IgE, acetylcholine and tumor necrosis factor (Lee and Sullenger 1997; Lin et al. 1994, 1996; Schoetzau et al. 2003; Wiegand et al. 1996; Yan et al. 2004) and 2′-fluoro modification served in the selection of RNA aptamers towards the keratinocyte and the vascular endothelial growth factors

(VEGF) (Pagratis et al. 1997; Ruckman et al. 1998). Alternative modification strategies have also been described (Dewey et al. 1995; Ito 1997, 2001; Vaish et al. 2000; Zinnen et al. 2002) and an assortment of about 100 modified purine and pyrimidine phosphamidites are commercially available or easily synthesized (Eaton 1997; Eaton et al. 1997; Kopylov and Spiridonova 2000).

17.2.3 The Molecular Basis of the Aptamer–Target Interaction. Implications in Assay Development

Aptamer selection relies on the fact that oligonucleotides fold in solution into a variety of three-dimensional structures, which are stabilized through canonical Watson–Crick as well as unusual base pairing established between nucleotide bases (Chou et al. 2003; Keniry 2000). Every single nucleic acid sequence will potentially adopt different structures under different conditions, each of them exposing on surface slightly different combinations of electro-chemical groups (their "fingerprints"). An aptamer is selected if one or more of the structures present in the selection buffer exhibit on surface binding pockets or clefts for the specific recognition and tight binding of motifs present on the surface of the target, and thus their respective 3-D structures are complementary. The aptamer–target interaction will then be highly specific. In fact, interactions between proteins and nucleic acids occur naturally during many gene regulatory processes, and an important percentage of biochemical reactions rely on the formation of complexes between nucleic acids and other compounds. Such molecular interactions are essentially based on surface recognition principles and stabilized by hydrogen bonding, hydrophobic interaction and ionic contacts. In the case of SELEX, the difference is that the researcher chooses the target.

The fact that aptamer–target binding depends on the aptamer folding into a specific 3-D structure, rather than just its sequence, has encouraged researchers to study aptamers' folding in depth. Aptamer structure is mainly determined by X-ray crystallography (XRC) and nuclear magnetic resonance (NMR), and additionally studied by other means, such as circular dichroism spectroscopy, differential scanning calorimetry, UV melting experiments or Raman spectroscopy. Supplementary information about the target-binding event can be obtained by electrophoresis, atomic force microscopy or by NOESI magnetic resonance, but the aptamer–target complexes are commonly modeled with indirect methods, especially in the case of complex targets. In any case, aptamer folding may be affected by a variety of external factors, as for instance, incubation temperature and buffer composition. Also, certain ions are required for some aptamers to fold into the target-binding structure as well as to decrease the levels of non-target protein non-specific adsorption. For these reasons, most authors carry out their aptameric assays under similar conditions as those employed in the selection procedure. Nevertheless, it has been demonstrated that aptamer optimal working conditions are not necessarily the SELEX ones, specially if a multiplexed-assay is to be optimized, and that series of optimization experiments should be performed for each aptamer or structure type (Baldrich et al. 2004; Cho et al. 2006; Papamichael et al. 2007). Besides, aptamer folding may be impaired by steric hindrance following immobilization or modification with, for example, reporter molecules. In these cases, using long spacers and small-size labels seems essential (Balamurugan et al. 2006; Baldrich et al. 2004; Bini et al. 2007; Li et al. 2006; Tang et al. 2007a). The requirement for a denaturation step prior to assay performance, in order to improve aptamer proper folding, is unclear. The few existing studies indicate that denaturing may be unnecessary or even deleterious for certain aptamers, as it is the case for the 15-nucleotide thrombin-binding aptamer (TBA), known to fold into a quadruplex (Baldrich et al. 2004; Bini et al. 2007). In other cases, however, aptamers submitted to denaturation showed slightly better limits of detection and reproducibility, as happens for the hairpin-forming aptamer against HIV Tat protein (Minunni et al. 2004).

Conversely to what might be expected, the reported aptamers show a striking repetition of patterns among the folded structures responsible for target binding. For instance, an important number of aptamers fold into quadruplex, stem-and-loop, hairpin and three-stem helix junction structures. In relation to this, aptamer conformation can determine the assay formats to be optimized. As an example, mainly quadruplex and helix-junction structures have been formatted into reagentless aptabeacons (see later in the text). On the other hand, the types of interactions occurring between aptamer and target will condition the success of the different regeneration strategies.

17.2.3.1 Aptamers Folding into G-Quartet Stabilized Structures

The possibility that guanine-rich sequences could arrange into quadruplexes was already suggested in the early 1960s (Gellert et al. 1962). Since then, guanine-rich segments have been found in biologically significant regions of the genome (Wang 1994) and several proteins have been described to bind or promote quadruplex formation (Keniry 2000; Shafer and Smirnov 2000). A quadruplex consists of arrays of G-quartets, each one composed of four guanines arranged into a square planar configuration. The G-quartets are piled in the space, and are connected by lateral loops, which contain the additional bases of the aptamer. Quadruplexes are classified based in the number of aptamer molecules taking place in the folding (monomer, dimmer and tetramer), the positioning of their loop regions (chair and basket for lateral and diagonal loops) and the relative strand orientation (parallel, antiparallel or mixed) (Fig. 17.4) (Keniry 2000; Shafer and Smirnov 2000). The structure leaves a central empty space that accommodates certain cations, which binding is reported to promote and stabilize the quadruplex structure versus the random coil. For this reason, it has been long accepted that target binding depends on the presence of those cations and thus they were always included in the assays binding buffer. Although sodium and potassium are usually the ions of

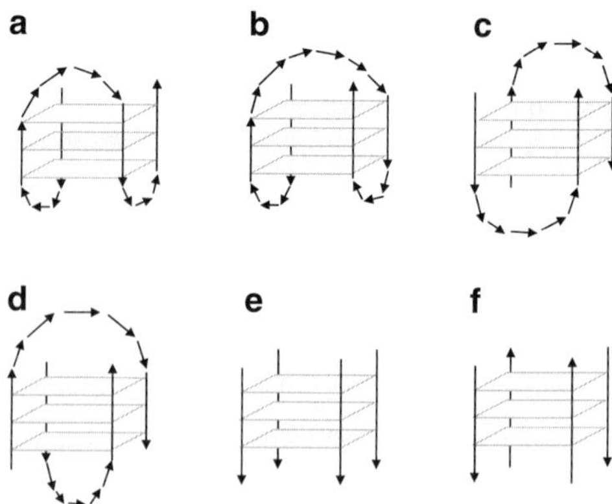

Fig. 17.4 Examples of quadruplex structures. (**a**) Monomer chair; (**b**) monomer basket; (**c**) dimmer chair; (**d**) dimmer basket; (**e**) parallel tetramer; (**f**) antiparallel tetramer

choice, NH^{4+}, T^{+}, Sr^{2+}, Ba^{2+}, and Pb^{2+} are reported to induce a similar effect. The quadruplexes formed in the presence of different cations, however, show slightly different structure and affinity for the target. This is of extreme importance when, for example, an aptamer evaluated in vitro is to be used in real samples of different ion concentration.

The best known example in this conformational category is the TBA, which was the first aptamer reported against a protein (Bock et al. 1992). Its target-binding conformation contains two G-quartets and three lateral loops arranged into a chair quadruplex structure (Fig. 17.5a) (Macaya et al. 1993; Marathias and Bolton 1999). The TBA quadruplex is stabilized by several cations, especially K^{+}, which was long believed to be essential for thrombin binding. Nevertheless, it has been demonstrated that TBA can *also* bind thrombin in the absence of stabilizing ions, especially at low temperatures (Baldrich and O'Sullivan 2005; Baldrich et al. 2004; Ho and Leclerc 2004; Nagatoishi et al. 2007). As target binding induces TBA conformational rearrangement, by promoting and/or tightening the quadruplex structure, it was soon formatted into reagentless fluorescent and electrochemical molecular beacon formats (see later in the text). Other aptamers known to fold into a quadruplex are the ones against streptavidin, human hepatocyte growth factor, hamster prion protein rPrP23-231, Staphylococcal enterotoxin B, adenosine and ATP (Huizenga and Szostak 1995; Purschke et al. 2003; Saito and Tomida 2005; Weiss et al. 1997; Wiegand et al. 1996).

17.2.3.2 Aptamers Folding into Hairpin and Stem-and-Loop Structures

A number of the aptamers known to bind small targets, as well as a few against proteins, fold into hairpin and stem-and-loop structures (Fig. 17.5b, c). In this case, a stem segment, stabilized by standard Watson–Crick base pairing, is disrupted by one or more loops. One or few base mismatches usually mark the frontier between the two structural parts. Upon complex formation, the otherwise mobile and less

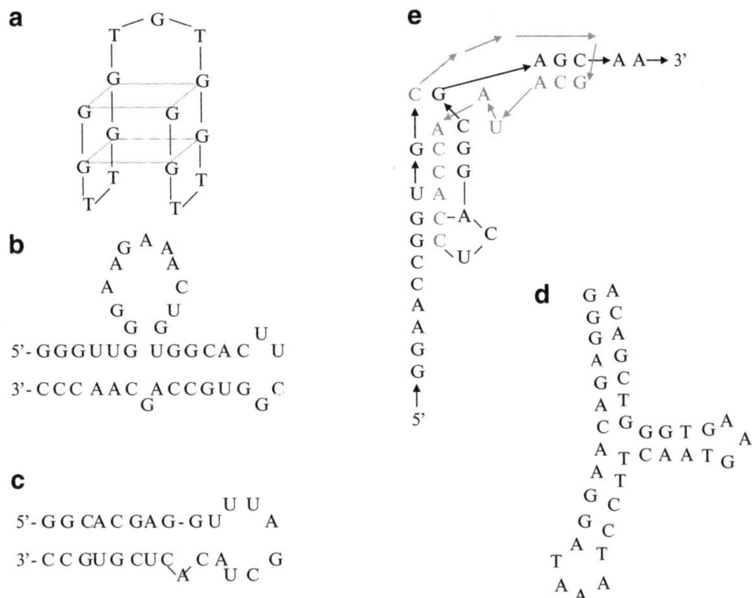

Fig. 17.5 Examples of aptamer folding. (**a**) TBA folding into a chair quadruplex. (**b**) AMP-binding RNA aptamer folding into a stem-and-loop. (**c**) Tobramycin RNA aptamer folding into a hairpin. (**d**) RNA aptamer against vitamin B12 folding into a pseudoknot. (**e**) The three-stem junction of the codeine DNA aptamer

structured loops fold into a well-organized and highly structured binding pocket, in which the target "fits" completely or partially. Target interaction and structure stabilization often require presence of magnesium ions and rely in combinations of "S-like" turns of the backbone, base zippers and/or base stacking, as well as the establishment of networks of hydrogen bonds and electrostatic interactions. These combinations of structural motifs generate highly specific binding sites.

Among the stem-and-loop forming aptamers, the AMP/ATP aptamer contains an S-shape folded asymmetrical internal loop, where the purine ring of the AMP intercalates (Dieckmann et al. 1996; Jiang et al. 1996). Flavin mononucleotide binds to its aptamer by zippering up an asymmetric loop through base mismatch and base triple formation, followed by intercalation of the target (Burgstaller and Famulok 1994; Fan et al. 1996). Binding of L-arginine and L-citrulline to their aptamers involves stacking of the target over a "platform" generated by the big aptamer loop folding (Famulok 1994). Tobramycin binds within the major groove spanning the hairpin loop-stem junction of its RNA aptamer, more of 50% of its surface remaining buried in the complex (Jiang et al. 1997; Patel et al. 1997). Other examples are the DNA aptamers against human hepatocyte growth factor, human IgE, Taq DNA polymerase and thalidomide (Saito and Tomida 2005; Shoji et al. 2007; Wiegand et al. 1996, Yakimovich et al. 2003), and the RNA aptamers against VEGF and the release factor 1 from *Thermus thermophilus* (Bozza et al. 2006;

Szkaradkiewicz et al. 2002). The molecular basis of these and more examples has been extensively summarized (Feigon et al. 1996; Patel et al. 1997; Patel and Suri 2000).

17.2.3.3 Aptamers Folding into Complex Structures

Finally, some of the known aptamers fold into more complex structures, often combining several of the previously described motifs. The binding sites and target interaction are, in these cases, determined by combinations of complementary packing of hydrophobic surfaces, hydrogen bonding and dipolar interactions between target and aptamer. For example, the RNA/DNA aptamers selected against the gonadotropin-releasing hormone, platelet-derived growth factor (PDGF), HIV-1 gp120 glycoprotein, guanine, sulforhodamine B, and tetracycline form a three-stem helix junction, shaped as a three-armed star (Fig. 17.5d) (Berens et al. 2001; Fang et al. 2001; Holeman et al. 1998; Leva et al. 2002a, b; Noeske et al. 2007; Sayer et al. 2002). In the latter case, the target is known to bind to inter-helical regions, probably via ionic interactions with the charged RNA backbone. The 3-D structure responsible for the sulforhodamine and human RNase H1 aptamers contains series of G-quartets stacked at the end of a stem segment (Pileur et al. 2003; Wilson and Szostak 1998). The aptamers towards streptavidin, hepatitis C virus RNA polymerase, and *Escherichia coli* release factor 1 arrange into series of hairpins and/or stem-and-loop segments, each domain flanked and separated from the others by single-chain fragments (Sando et al. 2007; Tahiri-Alaoui et al. 2002; Vo et al. 2003). The vitamin B12 aptamer forms a two-helix pseudoknot, stabilized by a three-stranded zipper with a perpendicularly stacking duplex (Fig. 17.5e) (Sussman et al. 2000). This is the first case in which an aptamer is demonstrated to create a pre-packed hydrophobic surface with many bumps and pockets, serving directly as target-binding site. Pseudoknots have also been described for the aptamers against HIV nucleocapsid and reverse transcriptase, biotin, and L-arginine (Chaloin et al. 2002; Geiger et al. 1996; Kim and Jeong 2003; Wilson et al. 1998a).

17.2.4 Advantages and Drawbacks Versus Other Biorecognition Molecules

When compared to antibodies, aptamers appear as ideal alternative candidates for use in existing analytical devices and techniques. Aptamers bind their cognate targets with high specificity and affinity, often with dissociation constants in the micromolar to sub-picomolar range, and discriminate between closely related molecules (Geiger et al. 1996; Haller and Sarnow 1997; Jenison et al. 1994; Leva et al. 2002a; Mannironi et al. 1997; Sassanfar and Szostak 1993; Shoji et al. 2007). For example, the L-arginine aptamer binds its target 12,000 times better than

D-arginine (Geiger et al. 1996) and the theophylline aptamer binds its target 10,000 times better than caffeine, which differs by only one methyl group, and at least tenfold more efficiently than an antibody isolated for the same purpose (Geiger et al. 1996; Zimmerman et al. 1997). The selectivity of aptamers, on the other hand, can be directed by carrying out the appropriate series of SELEX and counter-SELEX steps, as shown by reports of aptamers capable of discriminating between related protein forms and isoforms with, for example, enantioselectivity (Conrad et al. 1994; Kimoto et al. 2002; Shoji et al. 2007). Conversely, the discrimination capacity can be reduced and aptamers have been obtained that recognize ERK-2 protein both in its native and denatured forms, or that indistinctly bind human and porcine thrombin (Bianchini et al. 2001; White et al. 2001). In addition, aptamers should not generate the false positives associated to using Ab when studying rare clinical samples exhibiting heterophilic antibodies, which recognize the antibodies from other animal species used in immunoassays, or in the presence of rheumatoid factors and auto-antibodies in samples obtained from individuals with autoimmune disorders.

Aptamers possess the interesting property of being chemically produced, completely avoiding the use of animals. Once its sequence has been established, an aptamer can be synthesized at any time. DNA/RNA synthesis is easier, quicker and cheaper than Ab production, and produces highly pure molecules. Therefore, little or no batch-to-batch variation is expected in aptamer production. Apart from avoiding ethical concerns, the aptamer selection and production procedures are potentially automated in their whole length. Also for this reason, and at least in theory, aptamers can be selected under non-physiological conditions and/or towards toxins (Purschke et al. 2003; Tang et al. 2006, 2007a), something hard to accomplish when producing Ab. This is especially interesting in the case of small molecules which, due to their size, would not induce an immune response. In this context, aptamers have been successfully selected against metal ions, organic dyes, drugs, amino acids, cofactors, antibiotics, nucleotide base analogs and nucleotides (Burke and Gold 1997; Ciesiolka and Yarus 1996; Connell et al. 1993; Ellington and Szostak 1990, 1992; Famulok 1994; Geiger et al. 1996; Haller and Sarnow 1997; Harada and Frankel 1995; Hofmann et al. 1997; Huizenga and Szostak 1995; Jenison et al. 1994; Kiga et al. 1998; Lauhon and Szostak 1995; Lorsch and Szostak 1994; Mannironi et al. 1997; Sassanfar and Szostak 1993; Wallace and Schroeder 1998; Wallis et al. 1995, 1997; Wang et al. 1996; Wilson et al. 1998b). Nonetheless, aptamer selection is, in practice, restricted to several conditions as, for instance, a nucleic acid library cannot be submitted to selection at temperatures inducing the nucleotides' melting or in the absence of cations.

From an integration point of view, aptamers are smaller and less complex, with a molecular mass often one order of magnitude lower than that of Ab (5–25 kDa versus 150 kDa). They are also easier to manufacture and modify, either during or after synthesis, by incorporation of modified nucleotides or functional groups. This favors their subsequent immobilization and/or conjugation to molecular reporters. Thanks to the *unique* chemical and structural characteristics of nucleic acids, aptamers can be reversibly denatured, permitting the design of re-usable devices. In combination, these properties encourage the development of novel sensing formats such as the aptamer beacons and arrays (Bock et al. 2004; Collett et al. 2005b; Cho et al. 2006).

The main concerns in respect to the real applicability of aptamers are related to two inherent properties of nucleic acids: structural pleiomorphism and chemical simplicity. Structural pleiomorphism refers to the fact that a single DNA/RNA sequence can adopt more than one conformation. This implies that only a percentage of the aptamer molecules used in an assay might be folded into the appropriate structure. This fact can, not only reduce the assay efficiency, but also increase its production cost. The way to minimize this effect is to carefully characterize the folded structure responsible for aptamer–target binding, removing bases not involved in structural recognition that could contribute to pleiomorphism, followed by determination/optimization of the various parameters playing a role in the aptamer folding. As it happens with immunoassays, *optimization* is the only clue for optimal aptamer assay performance.

The fact that nucleic acid molecules are composed of just four different units also leads to concerns on whether such chemical simplicity will generate enough aptamer diversity. For example, all nucleic acids have a low iso-electric point (pI) and display uniform hydrophilicity, meaning that it will be difficult to isolate an aptamer against a highly hydrophobic target or towards target regions negatively charged. Also, the rigidity of the ribose ring and the phosphodiester and glycosidic bonds, as well as the characteristics of the complementary pairing, impairs conformational limitations to the oligonucleotides. In this respect, it has been proved that either the attachment of different substituents or the incorporation of base analogs to the nucleotide bases *can* increase the diversity of oligonucleotide libraries, as far as they can behave as proper substrates for the enzymes used in SELEX (Eaton 1997; Kusser 2000; Piccirilli et al. 1990). Those modifications also extend the conformational diversity and enhance the ability of aptamers to interact with their targets. Moreover, chemical groups incorporated post-SELEX can help to achieve a better aptamer–target contact in complexes of known structure.

17.3 Aptamers as Bio-recognition Elements in Biosensor Development

Aptamers were used for the first time as a bio-recognition element with analytical purposes in 1996, when Drolet and co-workers described the first enzyme-linked oligonucleotide assay (ELONA), otherwise called enzyme-linked aptamer assay (ELAA) (Drolet et al. 1996). The assay was, in fact, a mixed ELISA/ELAA sandwich, which used Ab and fluorophore-labeled aptamer as capture and reporter components, respectively. Nonetheless, the authors demonstrated for the first time that aptamers could be used for true diagnostic applications. Other mixed ELAA/ELISA sandwich assays (Baldrich et al. 2004; Rye and Nustad 2001) and double ELAA sandwich assays using aptamers as capture as well as reporting elements

(Bruno and Kiel 1999, 2002; Centi et al. 2007; Drolet et al. 1996; Ikebukuro et al. 2005) have been developed since then, and aptamers have been successfully exploited for the optimization of faster, easier and yet sensitive assays, such as the competition and displacement formats (Baldrich et al. 2004, 2005; Papamichael et al. 2007). Additionally, aptamers have been used as affinity reagents in several analytical applications, such as affinity chromatography, electrochromatography, affinity probe capillary electrophoresis or mass spectrometry, allowing the separation, purification and quantification of various analytes. The advances in these fields have already been reviewed (Mosing and Bowser 2007; Ravelet et al. 2006; Tombelli et al. 2005b).

These results demonstrated that aptamers could be used as bio-recognition elements in place of other traditional components, such as antibodies, with at least similar performances. The assay formats, discussed above, however, do *not* take advantage of some of the most exciting properties of aptamers. It is their integration to certain bio-sensing formats that demonstrates in its full depth the potential that aptamers offer in the field. Aptamer properties such as their high specificity, small size, modification and immobilization versatility, regenerability or conformational change induced by the target binding have been successfully exploited to optimize a variety of bio-sensing formats, including a number of reagentless strategies. We will summarize the recent advances in aptasensor development, starting with the more conventional gravimetric, optical, and electrochemical detection strategies, to finish with the unique molecular beacons.

17.3.1 Mass-Sensitive and Resonant Aptasensors: Specificity Online

In resonance sensing, the researcher propagates a mechanical oscillation across or along the sensor/resonator surface and monitors resonant frequency over time. If the sensor suffers an increase in mass load, for example by capture of a target by the immobilized ligand, a change in the resonance frequency is measured

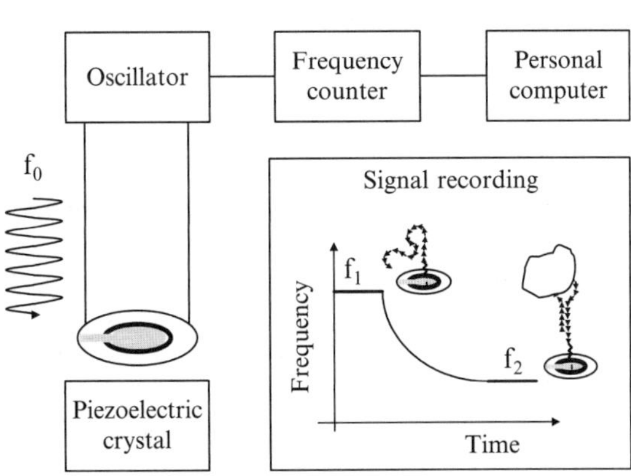

Fig. 17.6 Scheme of piezoelectric sensor performance. A wave of known frequency (f_0) is propagated along or across the sensing surface. The outlet frequency recorded in the absence of target (f_1) differs from that following target capture (f_2)

(Fig. 17.6). As resonant sensors are uniquely sensitive to the mass of the molecular species bound, they are reagentless-sensing formats. This implies that their specificity absolutely depends on the absence of cross-sensitivity towards unwanted components, and thus the method for transducer surface functionalization used to immobilize the ligand molecule. This transduction procedure, however, appears to allow incorporation of relatively thick layers, which has been translated into the optimization of various functionalization strategies, that generate extraordinary low levels of unspecific adsorption. Some examples of mass-sensitive resonant sensors are the quartz crystal resonators, the acoustic sensors and some cantilevers (Cooper and Singleton 2007; Lucklum and Hauptmann 2006).

The first example of aptameric piezoelectric transduction was provided by Liss and co-workers, who reported detection of human IgE by a DNA aptamer using a quartz-crystal microbalance (QCM) (Liss et al. 2002). The authors demonstrated the system specificity by detecting IgE in the presence of an excess of various proteins or protein mixtures, with no interference having been observed. When compared to a similarly prepared immunosensor, both formats detected IgE down to 0.5 nmol L^{-1} (3.3 ng cm^{-2}), but the aptasensor generated a tenfold higher linear detection range and survived regeneration in much better shape. Similar results were obtained by Minunni and co-workers. In this case, the authors immobilized a biotinylated RNA aptamer on streptavidin-functionalized piezoelectric quartz-crystals and studied the interaction of its target (HIV-1 Tat protein) by using a quartz crystal analyser (Minunni et al. 2004). The time needed for the whole measurement (baseline, interaction, washing, regeneration, and new baseline) was of just 25 min. When the aptasensor performance was compared to that of an immunosensor, produced using the same surface chemistry, a wider linear range characterized the antibody but higher sensitivity could be observed for the aptamer. They also managed to regenerate the aptasensor up to 15 times with no significant loss of sensitivity. The same team reproduced these results and demonstrated similar performances for QCM and SPR systems, although they observed interferences by another related HIV protein: Rev (17 and 25% of the signal generated by Tat using QCM and SPR, respectively) (Tombelli et al. 2005a). More recently, Bini and co-workers used the TBA immobilized on piezoelectric transducers to determine the different parameters influencing the analytical performances of the system (Bini et al. 2007). They observed that the length and chemical composition of the spacer used for aptamer immobilization strongly influenced detectability, with strategies separating the aptamer from the surface as much as possible generating best results in terms of signal amplitude, reproducibility, and selectivity. The sensor showed specificity for thrombin versus human serum albumin, even when assayed in serum or plasma samples, and could be regenerated up to 20 times by simple immersion in 2 M NaCl.

The application of surface acoustic wave (SAW) love-wave sensors to aptamers has also been reported. In this case, changes in the sensor surface loading alters the velocity of propagation of an acoustic wave, resulting in adjustment of either the resonance frequency of the resonator or the phase shift between the output and input signals. The device was successfully applied to detection of both thrombin and

HIV-1 Rev peptide by their aptamers, with detection limits around 75 pg cm^{-2} and a mass loading 36-fold higher for thrombin than for the non-target protein elastase (Schlensog et al. 2004). The same team presented in a later work improved measure set-up and data analysis, showing that the performance of their SAW aptamensor was comparable to that of a better characterized commercial SPR equipment (Gronewold et al. 2005). The system was additionally applied to study the interactions occurring between thrombin and different proteins involved in the blood-coagulation cascade, demonstrating the love-wave sensor as a good and reliable tool for the study of ligand/analyte interactions in real time. In a recent work, Jung and co-workers report a number of strategies to produce aptamer functionalized SAW sensors by exploiting either aminocellulose derivatives or NH2-(organo-) polysiloxanes for surface fuctionalization (Jung et al. 2007). Interestingly, the whole procedure, including bio-fuctionalization, detection and regeneration, could be integrated in a microfluidic system and followed on-line.

A different approach was developed by Savran et al. (2004), who reported on the optimization of a micromachined cantilever. The device contained two adjacent gold cantilevers: one functionalized with aptamers and the other, used as a reference, containing unrelated oligonucleotides. Target binding results here in a change in surface stress, which translates into a differential physical bending and is measured by interferometry (Fig. 17.7). The sensor was functionalized by simple self-assembly of the thiolated oligonucleotides and was able to detect the Taq polymerase target with a K_d of 15 pM and even in the presence of complex protein mixtures four times more concentrated by weight. Alternatively, oscillating cantilevers, functionalized with RNA aptamers, have been applied to the detection of HCV helicase by measuring the shift in resonance frequency induced by the target capture event (Hwang et al. 2007).

17.3.2 Optical Aptasensors

17.3.2.1 Surface Plasmon Resonance: Reagentless Study in Real Time

Surface plasmon resonance (SPR) spectroscopy allows the study of bio-molecular interactions in real-time, requiring neither aptamer nor target labeling and making

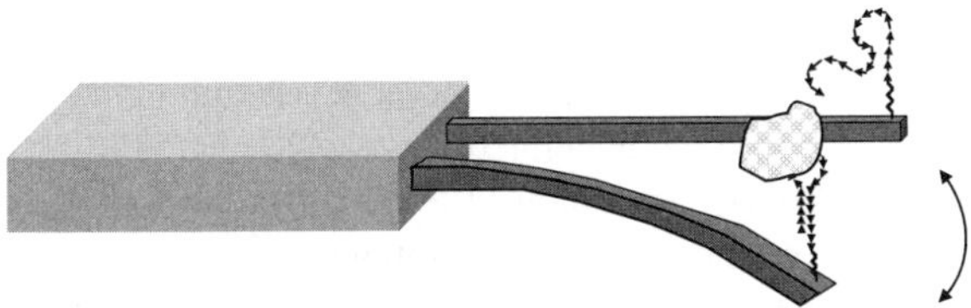

Fig. 17.7 Scheme of a surface-stress microcantilever. Target capture induces a change in surface stress, measurable in terms of physical bending

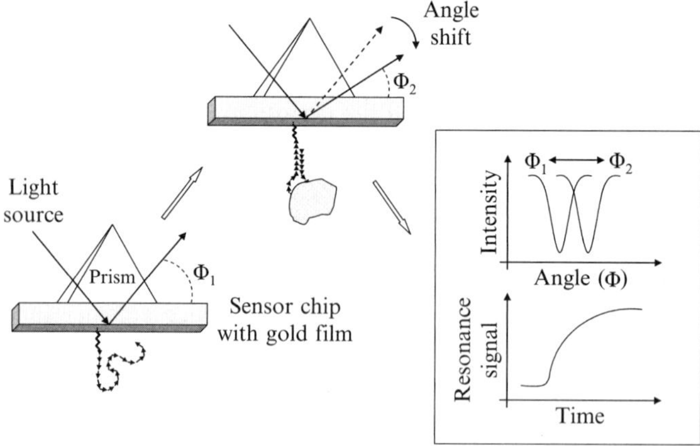

Fig. 17.8 Scheme of the performance of an SPR sensor. The beam of polarized light reaches the gold film that covers the chip and an evanescent wave is generated. Part of the energy is absorbed by the free electrons in the gold layer, generating a resonance plasmon and decreasing light reflection. Subsequent target capture onto the sensing surface correlates with additional decrease in both angle and intensity of the reflected light

use of relatively simple optical system devices (Fig. 17.8). As happens with other bio-components, SPR has been extensively used to study the binding kinetics of numerous aptamer-target couples. For example, Win reported dissociation constants in the low micromolar range for aptamers against codeine by using the target immobilized on the solid support and flowing different concentrations of the aptamers or aptamer pools studied (Win et al. 2006). The same strategy was used by Dey and co-workers to study the aptamer binding HIV-1 gp120 glycoprotein (Dey et al. 2005). Most authors, however, use the aptamer immobilized for these studies (in most cases using a biotinylated species on (strept)avidin-coated surfaces), and flow increasing concentrations of free target as to calculate the affinity and dissociation constants (Dey et al. 2005; Fukusaki et al. 2001; Hartmann et al. 1998; Leva et al. 2002a; Li et al. 2006; Sando et al. 2007). This approach also allows the study of the aptamer specificity if non-target species are flowed instead (Leva et al. 2002a; Sando et al. 2007). For instance, Leva used SPR to confirm the lack of cross-binding of the GnRH spiegelmers against various related or unrelated neuroactive peptides (Leva et al. 2002a). In the case of the spiegelmer produced against the Staphylococcal enterotoxin B, the authors studied the aptamer–target kinetics by carrying out an SPR displacement assay (Purschke et al. 2003). The assay consisted of competition for the free target between immobilized and increasing amounts of free aptamer, and generated K_d of ~420 nM. SPR, on the other hand, was also exploited by Baldrich to demonstrate the differences in TBA affinity for its native thrombin target and the HRP-labeled suboptimal, which were the basis of an aptameric enzyme-linked displacement assay (Baldrich et al. 2005).

SPR has also been used to optimize aptamer self-assembly and assay performance with bio-sensing purposes. For example, Balamurugan used SPR to show that target detectability is improved when long spacers are used to immobilize the aptamer, with better results having been obtained for oligo(ethylene oxide)-containing instead of hydrocarbon-based linkers (Balamurugan et al. 2006). They also demonstrated the requirement for co-adsorption of thiolated species, such as mercaptoethanol or mercaptohexanol, which removed aptamers non-specifically adsorbed and filled the surface unoccupied gaps. Tang and co-workers compared the efficiency of two thrombin-binding aptamers and described several SPR assay formats, including one-step direct binding, competition, and sandwiched schemes, with excellent performances for thrombin detection (Tang et al. 2007b). They also studied several factors affecting target binding such as use of a DNA spacer, the aptamer surface coverage and salt concentration. Tombelli used a biotinylated aptamer, immobilized on commercial dextran-coated chips modified with streptavidin, to detect HIV-1 Tat protein by SPR (Tombelli et al. 2005a). The assay showed a linear range between 0.12 and 2.5 ppm and an average coefficient of variation of only 7% for all the concentrations tested in triplicate, demonstrating that both the immobilization and detection procedures were reproducible. In addition, the authors claimed to have regenerated the sensor surface up to 50 times by injecting a solution containing NaOH 12 mM and 1.2% ethanol.

In a different way, Li exploited SPR imaging to determine the relative surface densities of the various aptamers immobilized in a microarray format, generated via a novel ligation approach (Li et al. 2006). SPR imaging was also used by another team to study the relative affinities of four different aptamers for IgE in a multiplexed format (Wang et al. 2007). The best-performing aptamer showed results similar to those obtained for QCM and regular SPR, with a linear range between 8.4 and 84 nM and LOD of 2 nM IgE.

17.3.2.2 Fluorescence Aptasensors: Not Having to Pre-treat Samples?

The first fluorescence aptasensor was described by Kleinjung et al. (1998). The device consisted of optical fibers derivatized with avidin, to which a biotinylated RNA aptamer against L-adenosine had been affinity-captured. Subsequent capture and fluorescence detection of increasing concentrations of FITC-labeled L-adenosine were performed in real time and allowed the evaluation of the association and dissociation rates. The performance of a competitive assay with native L-adenosine allowed selective detection in the submicromolar range, with 1,700-fold discrimination of the chiral L- and D-enantiomers. A similar strategy was reported by Lee and Walt, who covalently conjugated a thrombin-binding aptamer to the surface of microspheres, and distributed these microspheres into the microwells of optical imaging fibers to create a microarray (Lee and Walt 2000). The competitive binding assay between unlabeled and FITC-labeled thrombin allowed detection of the native species with an LOD of 1 nM and a dynamic range of nanomolar to low micromolar concentrations, in an assay performed in 15 min including the regeneration step. The authors demonstrated the possibility of using aptamer microarrays for multi-analyte detection for the first time.

The unique properties characterizing aptamers, on the other hand, allow the optimization of novel reagentless formats by exploiting optical transduction strategies, which are difficult to develop with bigger and more complex capture molecules, such as Ab. For example, using plane polarized light to excite a fluorophore and detecting linearly polarized components of the emission can give information about the size, shape, and flexibility of the fluorophore or the fluorophore-linked molecule. Fluorescence anisotropy has been used in this way to monitor DNA/protein binding, to study conformational changes in proteins, and to study the self-association of peptides and proteins. The interaction of a target with often bigger fluorophore-labeled Ab, which by definition rotate much slower in solution than small molecules, induces in most cases changes in size or conformation too small to be detected by fluorescence anisotropy. Consistently, reported Ab-based assays making use of fluorescent polarization have always used a competitive format (Wilson et al. 1998b). Aptamers, on the other hand, are relatively small molecules, so the binding of target proteins will bring a significant change in their molecular weight while frequently inducing structural reorganization. Aptamers labeled with a reporter fluorophore will, thus, suffer a detectable change in the fluorophore rotational diffusion rate induced by target binding, which is measurable as changes in the evanescent wave-induced fluorescence anisotropy (Fig. 17.9). This strategy was exploited with aptamers for the first time by Potyrailo and co-workers, who developed a low-volume aptasensor for thrombin detection (sample volume 140 pL) using fluorescein-labeled TBA covalently bound to a glass support (Potyrailo et al. 1998). The fluorescence anisotropy of the labeled aptamer was found to increase as a function of thrombin concentration, with the assay showing an LOD of 0.7 amol of thrombin and a dynamic range of three orders of magnitude, in an assay of just 10 min. The bio-sensor could be regenerated and re-used by addition of a denaturating agent to remove the previously bound thrombin. This

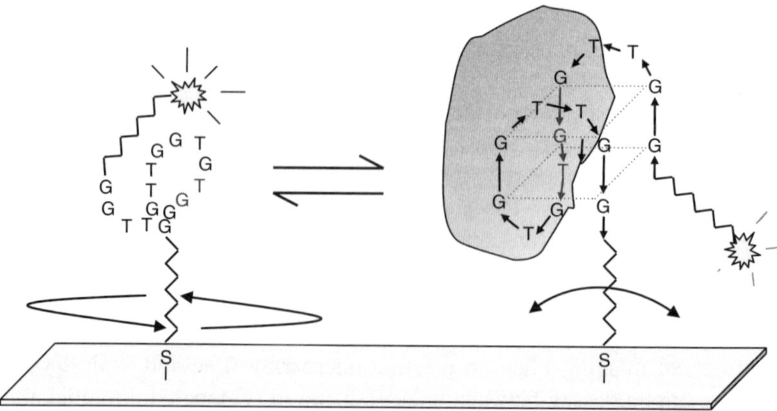

Fig. 17.9 Mechanism of action of a fluorescence aptasensor based in the changes in the evanescence wave-induced fluorescence anisotropy generated by target binding. The increase in size due to capture of a relatively big target limits the aptamer rotational diffusion and thus affects the emission of the fluorescent label

format was later formatted by McCauley into arrays composed of four different aptamers (against thrombin, bFGF, VEGF and IMPDH II), demonstrating that fluorescence anisotropy can be exploited in a multiplexed-sensing format (McCauley et al. 2003). In this work, detection was feasible in bacteria lysates or diluted human serum, with apparent low interference of their components.

Alternatively, reports of successful detection of subnanomolar concentrations of other targets, but using the labeled aptamer in solution instead of immobilized, exist at least for PDGF and IgE (Fang et al. 2001; Gokulrangan et al. 2005). In these cases, however, the ionic composition of the binding buffer influences the assay sensitivity, potentially limiting its application in the study of real samples. In a different strategy, the incorporation of fluorophores around the ATP aptamer target-binding site led to the design of "signaling aptamers." ATP binding induced in this context significant structural reorganization of the aptamer, which presumably contributed to alter the electronic environment of the attached fluorophore and translated into increase of the fluorescence intensity (Jhaveri et al. 2000; Yamana et al. 2003). This assay format allowed, not only target detection, but also the study of the aptamer–target kinetics and binding constants in a fast and easy assay format. In these last cases, it should be stressed that working in solution instead of directly on the sensing surface, apart from not being strictly considered aptasensing, does not allow regeneration and reuse of the system.

17.3.2.3 Aptamer Microarrays: Multiplexing Detection

Of special interest is the recent development of aptameric arrays as they offer the possibility of simultaneous detection of multiple targets and high-throughput screening (Fig. 17.10). Additionally, this field can take advantage to certain extent of the fabrication and spotting techniques already developed for the more consolidated DNA microarrays. The first works reporting on aptamer arrays were based in the production of optical fiber arrays (Kleinjung et al. 1998; Lee and Walt 2000). The authors functionalized the surface with a single aptamer, but these works demonstrated the possibility of using aptamers in an array format for multianalyte detection. The later incorporation into scene of photoaptamers, which remain covalently bound to their targets following photoactivated crosslinking, allowed performing highly stringent washing steps, thus efficiently removing non-specifically attached species and improving the array specificity (Bock et al. 2004; Golden et al. 2000; Petach and Gold 2002). The enhanced signal-to-noise ratios and lower limits of detection generated allowed the team from the SomaLogic company to develop a 17-plex photoaptamer array by aptamer spotting onto activated glass slides (Bock et al. 2004). The array was detected in a sandwich format using either NH-reactive fluorophores (Fig. 17.10a) or fluorophore-labeled antibodies (Fig. 17.10b), and exhibited limits of detection below 10 fM for several analytes measured in 10% serum.

Ellington and co-workers have extensively worked in the production of aptamer microarrays (Collett et al. 2005a, b; Cho et al. 2006; Kirby et al. 2004). In a first

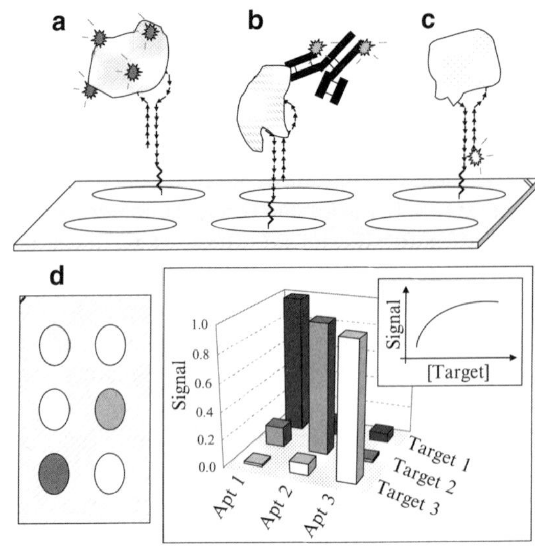

Fig. 17.10 Scheme of an aptamer array. (**a**) Detection of a fluorophore-labeled target. Detection of native target can be carried in this format by performing a competition between labeled and native targets, or by labeling the native target following its capture using a reacting fluorophore. (**b**) Sandwich detection using a fluorophore-labeled Ab. (**c**) Detection of the changes in fluorescence anisotropy of the labeled aptamer induced by target capture. (**d**) The use of different aptamers and fluorophores allows the simultaneous detection and quantization of multiple targets

approach they immobilized biotinylated aptamers on commercially available streptavidin-agarose beads, which were then deposited in the wells of micromachined flow chips (Kirby et al. 2004). They observed specific detection of three fluorophore-labeled molecules (lysozyme, ricin and HIV Rev protein) by their respective aptamers, and carried out a sandwich assay using Alexa-labeled Ab to detect native ricin with an LOD of 320 ng mL^{-1}. Interestingly, the chips could be regenerated and re-used with no loss in function. In later works, however, this team used streptavidin-coated glass slides instead, where biotinylated RNA and DNA aptamers against lysozyme, ricin, IgE and thrombin were spotted (Collett et al. 2005a, b; Cho et al. 2006). Since the four aptamers had been produced in different selection buffers, the authors defined a universal buffer in which the four of them retained affinity. In some of the cases, the assay detected efficiently the fluorophore-labeled analyte at concentrations over seven orders of magnitude (10 pg mL^{-1} to 100 μg mL^{-1}). In spite of the potential degradation of immobilized RNA aptamers by nucleases, the arrays produced showed no diminution in performance even after incubation in cell lysate for 30 min at room temperature (Collett et al. 2005a).

In a different approach, McCauley and co-workers, at the company Archemix, produced on streptavidin-coated glass slides microarrays of RNA and DNA aptamers, which had been previously labeled with fluorescein (McCauley et al. 2003). Target binding could thus be directly measured by studying changes in fluorescence polarization anisotropy (Fig. 17.10c). The authors demonstrated the specific detection and quantization of four different proteins, three of them relevant to cancer diagnosis, even in serum and complex bacterial cell lysates. Lin and co-workers reported on an alternative consisting on self-assembly of a modified TBA, which bore a fluorescent nucleotide analogue in close proximity to the target-binding site, into high-density nanoarrays (Lin et al. 2006). Target binding generates in this format a significant increase in fluorescence, measurable using confocal fluorescence microscope imaging, with thrombin being detected in the low nanomolar range. Finally, Bera Aberem and co-workers used the TBA labeled with Cy3, and hybridized it with a chromic, cationic, water-soluble polythiophene previous to spotting onto silanized glass slides (Bérn Abérem et al. 2006). The assembly resulted in quenching of the Cy3 emission by the polymer due to Förster resonance energy transfer (FRET). Subsequent target binding induced polymer displacement and significant increase of the fluorescence, while incubation with unrelated proteins generated no changes in emission.

17.3.3 Electrochemical Aptasensors

Electrochemical transduction offers the advantages of high sensitivity, fast response, low cost of production and the possibility of miniaturization. The variety of electrochemical techniques available provides, on the other hand, versatility to the detection scheme, which can be selected according to the type of sample to be studied.

A number of examples have been reported in which electrochemical detection of the aptamer–target event is carried. For example, a recent report exploits magnetic particles in combination with an ELAA sandwich assay for the electrochemical detection of thrombin (Centi et al. 2007). Chronopotentiometric-stripping detection of lysozyme, previously captured and concentrated using aptamers-bearing magnetic particles, was also reported with a detection limit of 7 nM (Kawde et al. 2005). Electrochemical detection of an ELAA displacement assay showed to be more efficient than the equivalent colorimetric detection, allowing detection of less than 1 nM of displacing thrombin in solution (Baldrich et al. 2005). In those examples, however, target capture is not directly coupled to signal transduction on sensing surfaces, and thus cannot be considered true sensors. The next sections summarize the electrochemical aptasensors reported to date, most of which exploit voltammetric or impedimetric detection. Electrochemical aptamer molecular beacons (or aptabeacons) state as a unique case in voltammetric aptasensing and will be described in detail in a later section.

17.3.3.1 Amperometry, Voltammetry and Square Wave Pulsed Voltammetric Techniques: In the Name of Label

Ikebukuro and co-workers proposed the first electrochemical aptasensor reported (Ikebukuro et al. 2005). The assay was based in a sandwich ELAA approach and used simultaneously two different aptamers shown to bind at different sites of their target thrombin. One of them was thiolated and self-assembled on gold electrodes. The aptamer-modified electrode was then incubated in the presence of either thrombin or BSA and with the second aptamer, labeled with the enzyme pyrroquinoleine quinine glucose dehydrogenase. Glucose was finally added as enzyme substrate and its hydrolysis followed amperometrically. The assay showed good correlation between current generated and thrombin concentration in the range 40–100 nM, with a detection limit of 10 nM and no cross-binding to BSA. Another sandwich format to detect thrombin was proposed by Polsky, who used platinum nanoparticles instead of enzymes in the search for catalytic labels less instable under varied thermal and environmental conditions (Polsky et al. 2006). The electrocatalytic reduction of H_2O_2 was measured by linear sweep voltamperometry and allowed detection of thrombin, but not unrelated proteins, down to 1 nM.

In spite of the good performances reported, sandwich formats require several incubation and washing steps to be carried out, turning up to be quite long assays. The requirement for addition of labeled bio-components also complicates integration in rapid and miniaturized devices. The displacement assay formats have been suggested as an interesting alternative, combining speediness and sensitivity, as well as improved functioning compared with the equivalent Ab-based displacement assays (Baldrich et al. 2005). In this respect, the colorimetric aptamer displacement assay previously described by Baldrich was formatted by Hansen and co-workers into a multiplexed electrochemical displacement (Hansen et al. 2006). The new format consisted of simultaneous self-assembly of two different thiolated aptamers onto gold electrodes, followed by capture of their respective suboptimal targets, produced by labeling thrombin and lysozyme with quantum dot nanocrystals. Subsequent incubation with the native unmodified targets induced displacement of the labeled suboptimals (Fig. 17.11a). Detection by square-wave stripping voltammetry of the labels remaining on surface allowed extremely low detection limits (0.5 pM for thrombin), consistent with multi-labeling of each suboptimal unit and the pre-concentration step associated to the electrochemical stripping transduction mode. The authors claimed that up to six different targets could be simultaneously measured per assay, taking into account the number non-overlapping metal peaks potentially exploited. A second displacement assay strategy was described by Xiao, who used a thrombin-binding aptamer immobilized on electrodes, this time hybridized with a complementary sequence labeled with methylene blue (Xiao et al. 2005b). The DNA forms, in the absence of thrombin, a rigid duplex which keeps the redox label far away from the electrode surface. Upon thrombin binding, the complementary labeled sequence is displaced and the methylene blue label can approach to the electrode surface (Figs.17.11b). The subsequent increase in current could be measured by differential pulse voltammetry (DPV), with a detection limit of 3 nM. Zuo reported a similar format to detect ATP, but this time labeling the aptamer, and not the complementary strain, with ferrocene (Zuo et al. 2007). Target binding forces the aptamer rearrangement, displacing the complementary strain while approaching the aptamer labeled end to the electrode surface. The alternative use of a DNA probe, immobilized onto an electrode modified with electropolymerized tyamine and gold nanoparticles, was recently reported by Wu as a tool to capture a ferrocene-labeled aptamer against adenosine (Wu et al. 2007). The target-

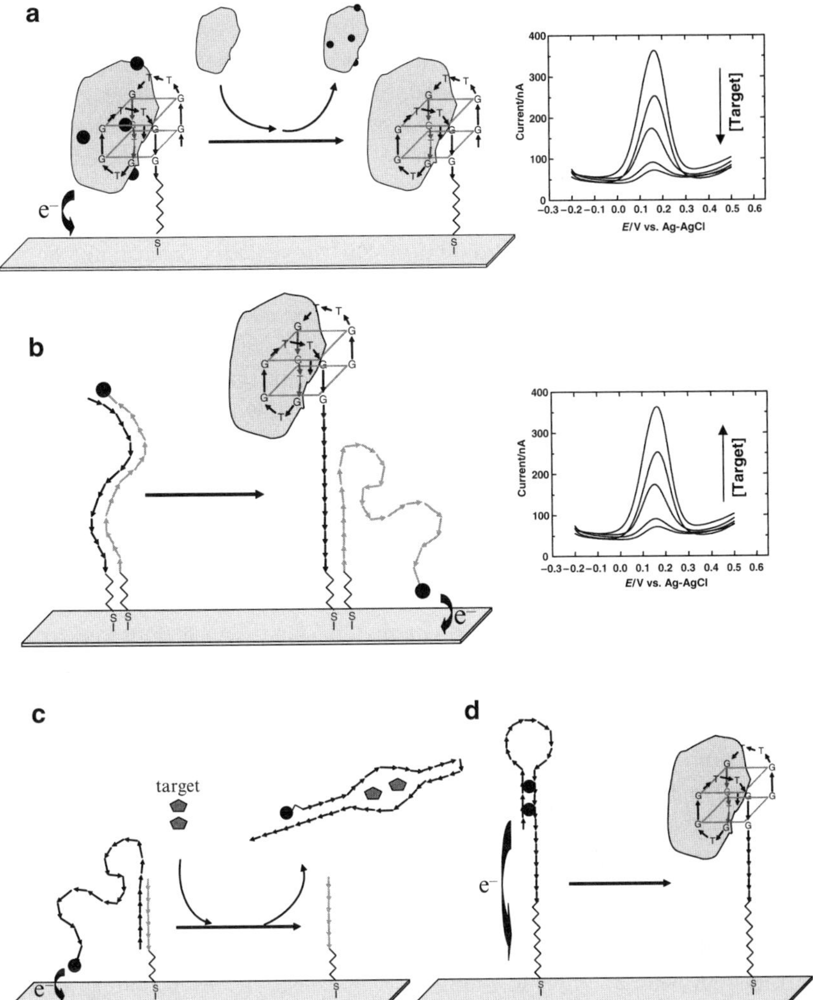

Fig. 17.11 Examples of electrochemical displacement assays. (**a**) A suboptimal target, labeled with an electroactive species, is captured and subsequently displaced by native the target present in the sample. (**b**) The aptamer is immobilized with a complementary sequence which is labeled with an electroactive species. Target capture displaces the complementary sequence, whose label can now freely reach the electrode surface. (**c**) The labeled aptamer is immobilized through base-pairing with a complementary sequence. The structure rearrangement induced by target interaction forces the two oligonucleotides apart and the labeled aptamer is washed away from the electrode surface. (**d**) The aptamer is modified as to fold into a hair-pin in the absence of target and the electroactive species is intercalated within the double-strain segment. Target capture induces hair-pin disruption and displacement of the intercalated label

induced aptamer rearrangement generated (in this case, aptamer release from the surface), and thus decrease in signal recorded, with a detection limit of 20 nM (Fig. 17.11c).

Yet, a third example of displacement strategy was reported by Bang et al. (2005). The authors immobilized via EDC/NHS chemistry an anti-thrombin aptamer, modified as to form a stem–loop beacon, on a SAM-modified gold electrode. The sensor was subsequently incubated with methylene blue, which intercalates within the double strand segment. The aptamer conformational change induced by target binding forces the two strains of the doublet apart, releasing the intercalated methylene blue and translating into a decrease of signal (Fig. 17.11d). Detection of different concentrations of thrombin by DPV revealed a linear range between 0 and 51 nM and LOD of 11 nM, with little interference by non-target proteins such as ovalbumin and casein. Cheng and co-workers used a redox-active cation instead, $[Ru(NH_3)_6]^{3+}$, which interacts with the negatively charged aptamer phosphate backbone by electrostatic interaction (Cheng et al. 2007). The system allowed lysozyme detection by cyclic voltammetry (CV) at physiological concentrations (between 0.5 and 50 μg mL^{-1}).

Finally, several authors have explored voltammetric detection of the aptamer–target binding event by detecting changes in the diffusion, and thus electron transfer, of electroactive species present in the detection solution. In the case reported by Hianik and co-workers, a thrombin-binding aptamer was immobilized on gold electrodes via avidin–biotin capture (Hianik et al. 2005). The authors measured, by DPV, the reduction of methylene blue, which appeared to be in solution too far from the electrode surface, but was efficiently detected following intercalation to the aptamer-captured thrombin. This detection strategy generated a detection limit of 10 nM thrombin with no cross-binding to unrelated proteins, and did not require chemical modification of neither the aptamer nor the target. In a related approach, Le Floch self-assembled a thrombin-binding aptamer on gold electrodes and measured by square wave voltammetry (SWV) an electroactive polymer present in solution (Le Floch et al. 2006). In the absence of target, the polymer binds to the negatively charged aptamer-coated surface and a high current peak is obtained. Thrombin capture physically blocks the sensor surface, translating into a decrease of the signal generated. The limit of detection, however, was relatively high, and attempts to improve assay sensitivity by including a step of nuclease degradation extended the assay and complicated manipulation. More recently, Kim reported on the detection of estradiol by CV and SWV, based on the interference induced by target binding in the electron flow produced by the ferrocyanide–ferricyanide redox reaction taking place in solution (Kim et al. 2007). Target capture produced decreases in current in a concentration range between 10 pM and 100 nM, with little interference of unrelated species.

17.3.3.2 Impedance Sensors: Reagentless Sensing the Interface

Impedance biosensors measure changes in either impedance or capacitance of an interface. As target capture by the aptamer-functionalized surface induces its partial blocking, resulting in an increase of the interfacial electron-transfer resistance,

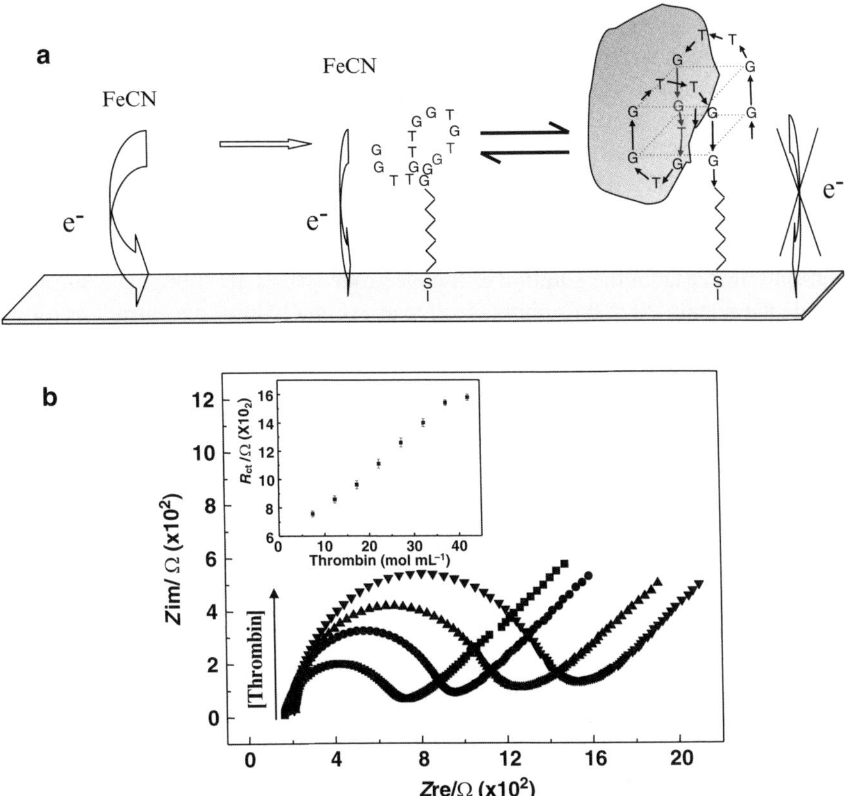

Fig. 17.12 Impedimetric aptasensor. (**a**) Electrode surface functionalization by aptamer immobilization induces its partial blocking. Subsequent capture of the aptamer target additionally contributes to decrease electron transfer across the interface. (**b**) This is noticeable in the Nyquist plot of the Faradaic impedance measurement as an enlargement of the graph semicircle portion, consistent with an electron-transfer-limited process

impedance monitoring has been successfully applied to the detection of such a binding event (Fig. 17.12). The proven success of this methodology in the development of aptasensors relates to the fact that impedimetric sensing requires no labels to be used. This allows truly reagentless detection of the target-binding event in an extremely sensitive format, which is compatible with the standard microfabrication technology. On the other hand, as aptamers are smaller in size than other capture bio-components, such as Ab, they produce less important surface blocking on their own, provide lower background signals and higher relative impedance changes due to target biding, and thus grant improved sensor sensitivity.

The first electrochemical impedance spectroscopy (IES) aptasensor was reported by Xu and co-workers (Xu et al. 2005). They detected human IgE in the presence of

$[Fe(CN)_6]^{3-/4-}$ using its hairpin aptamer, self-assembled on photolithographic arrayed gold electrodes. The aptasensor detected IgE in a concentration range between 2.5 and 100 nM, with a detection limit of 0.1 nM and averaged relative standard deviation under 8%. Compared to impedance immunosensing, aptasensing generated lower levels of non-specific, and thus background noise, while larger impedance relative signals. The same approach was applied to thrombin detection using the TBA self-assembled on gold electrodes, with detection limits in the low nanomolar range (Fig. 17.12) (Cai et al. 2006; Radi et al. 2005). The sensitivity of this format could be significantly improved by denaturing the captured thrombin using guanidine hydrochloride, causing and augmented blocking effect against electron transfer from the solution to the electrode (Xu et al. 2006). The amplified impedance signals led to thrombin detection at extremely low concentrations (down to 10 fM).

Alternatively, non-faradic impedance spectroscopy (NIS) has also been investigated in order to detect the target in saline buffer with no added electroactive species. Reports exist for at least thrombin and PDGF detection (Liao and Cui 2007, Löhndorf et al. 2005). In the first case, the authors immobilized an RNA aptamer on microfabricated devices. They then followed target binding by measuring at a fixed frequency the changes in NIS, caused by the modification of the effective dielectric constant within the 68 nm gaps that separate two gold electrodes (Löhndorf et al. 2005). The system allowed simultaneous measurement of five sensor channels, including reference channels for the study of unspecific binding events. In the second case, the aptamer was conjugated to silanized silica wafers. The system allowed specific detection in the low nanomolar range with discrimination against unrelated species, and demonstrated its feasibility for in vivo applications (Liao and Cui 2007).

17.3.3.3 Field Effect Transistors: Dressing "Nano"

Field effect transistors (FET) and related biosensors measure changes in the charge distribution within proximity to sensor surface, occurred as a consequence of the interactions between the charges of the molecules approaching and the mobile charges present at an electrolyte–insulator–semiconductor interface. The distance limit at which this interaction can be measured, the Debye length, varies as the inverse square root of the ionic strength. Accordingly, FETs exhibit their maximal sensitivity at low ionic strengths, where counter-ion shielding is reduced. Under physiological conditions, however, the size of Ab immobilized on the sensor surface (10–15 nm) forces target immuno-capture to take place beyond or close to the Debye length. Aptamers, much smaller in average, emerge as an ideal candidate to fill this gap, allowing target binding within the electrical double layer in buffer solution.

So and co-workers reported on the first attempts to optimize a single-walled carbon nanotube field-effect transistor for thrombin detection (So et al. 2005). The TBA was conjugated to carbodiimidazole-activated Tween 20, which had been coated onto the carbon nanotube walls through hydrophobic interactions. Thrombin capture was next monitored in real time by tracking the changes of the electrical double layer. The assembly showed a linear range in a thrombin concentration range of 0–100 nM, a detection limit of 10 nM, and little response against non-target elastase. The same approach was later on applied to the real-time detection of human IgE, with a detection limit of 250 pM (Maehashi et al. 2007). The aptamer-functionalized device performed better than an equivalent assembly displaying monoclonal Ab instead.

In order to produce an ion-selective field-effect transistor (ISFET) for the detection of small molecules, Zayats optimized a displacement method (Zayats et al. 2006). The sensor incorporated an adenosine-binding aptamer, conjugated to the silanized surface and hybridized to a shorter complementary oligonucleotide. Target binding induced aptamer folding into a hairpin and displacement of the complementary oligonucleotide. This affected the ISFET gate potential and its interfacial transfer resistance. The measurement could be carried out either potentiometrically or impedimetrically, with detection limits in the low micromolar range in an assay of just 4 min.

17.3.4 Aptamer Molecular Beacons: Aptabeacons

Molecular beacons (MB) were described for the first time in 1996 as nucleic acids probes, modified as to fold into a hairpin in solution (Tyagi and Kramer 1996). If a complementary nucleic acid sequence was present, its hybridization disrupted the MB hairpin. This conformational change set apart the two fluorophores, or a fluorophore-quencher pair, placed at the two ends of the MB and affected their light emission, measurable in real time. The fact that the aptamer–target binding event often correlates with a conformational rearrangement of the former, soon suggested that some aptamers could be easily formatted into aptamer MB. Reports on both fluorescent and electrochemical aptamer MB, otherwise named aptabeacons, confirmed this as one of the most promising aptasensing formats, truly setting aptamers aside from antibodies as a highly versatile capture bio-component.

17.3.4.1 Fluorescent Aptabeacons: Fast-and-Easy . . . But in Solution

A fluorescence aptabeacon contains, attached to its two ends, either two fluorophores or a fluorophore-quencher pair (Fig. 17.13). The change in conformation following target binding will translate into changes in the distance separating such two molecules, and thus changes in FRET efficiency between them. Interestingly, the subsequent adjustment in the fluorophore/s emission can be followed in real time.

When a fluorophore-quencher system is used, approximation of the two labels generates a decrease in the emission of the fluorophore, which is quenched by

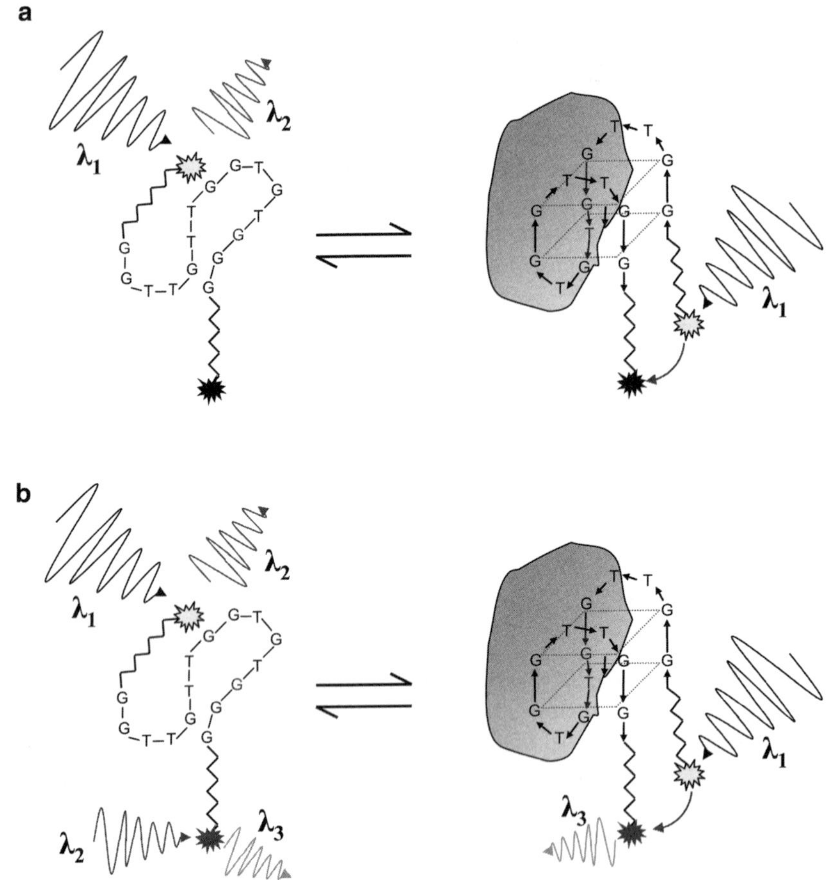

Fig. 17.13 Fluorescent aptabeacon. (**a**) In the quenching format a fluorophore and a quencher are placed close to each other following target binding; the decrease in fluorescence measured is proportional to target concentration. (**b**) In the FRET format, two fluorophores of overlapping excitation-emission spectra are used; target capture translates in decrease of emission by the donor fluorophore while increase of the acceptor emission

the quencher (Fig. 17.13a). Target presence will thus translate into increase or decrease of the fluorophore emission, for MB with a "turn-on" and "turn-off" setting respectively. The first example of a quenching aptabeacon was reported by Yamamoto and Kumar, who produced two RNA oligonucleotides derived from a previously described aptamer against HIV Tat protein (Yamamoto and Kumar 2000). One of the chains formed a hairpin in the absence of target, to which ends a fluorophore and quencher were added. Under these conditions, the hairpin was closed and the two labels were brought together, "turning off" the fluorophore fluorescence emission. In the presence of Tat, but not in the presence of other RNA binding proteins, the two oligonucleotides used underwent a conformational change and formed together a duplex that bound the target. The new conformation impaired distance between the labels and prevented FRET, leading to enhancement of the fluorophore fluorescence. Similar results were obtained by Hamaguchi, who engineered a thrombin-binding aptamer by adding a nucleotide tail to one of its ends as to form a stem-and-loop in the absence of thrombin (Hamaguchi et al. 2001). Target binding promotes aptamer folding into a quadruplex structure, detaching the fluorophore from the quencher and permitting it to fluoresce. A radically different strategy was described by Frauendorf and Jäschke, who assembled a theophylline-binding aptamer with a hammerhead ribozyme domain, and labeled the two ends of the resulting molecule with a fluorophore-quencher pair (Frauendorf and Jäschke 2001). Target binding induced again aptamer rearrangement, causing in this occasion the activation of the ribozyme, which cleaved the nucleotide strain and removed the quencher.

Alternatively, Stojanovic designed a quenching aptabeacon against cocaine, this time exploiting a switch-off format (Stojanovic et al. 2001). In this case, the aptamer folds into a three-way junction, which remains partially opened in the absence of target, thus keeping apart the fluorophore-quencher pair. Cocaine binding promoted three-way junction closing, translating into label approximation and decrease of emission. The aptabeacon detected cocaine present in serum in the micromolar range, with selectivity over its metabolites. This "turn off" format has also been exploited to develop aptabeacons against thrombin, PDGF, and L-argininamide, which fold into quadruplex, three-way junction and hairpin structures respectively (Fang et al. 2003; Li et al. 2002; Ozaki et al. 2006).

The main drawback associated with the quenching format is that the emission of some fluorophores can be *affected* by external factors, such as buffer composition, and is quenched by components including gold and guanidine nucleotides. It is thus difficult to differentiate between changes in emission due to target binding or due to changes in the labels environment, limiting this format application in the study of complex biological samples. As an alternative, the MB can be labeled with two fluorophores, instead. In this case, labels showing overlapping emission and excitation spectra will have been chosen. When the two fluorophores are distant from each other, both emit fluorescence if excited at the appropriate wave lengths. When in close proximity, the "donor" fluorophore will be excited at the corresponding wave length, but instead of emitting fluorescence, its energy emission will be absorbed by the "acceptor" fluorophore. The "acceptor" fluorophore will thus emit fluorescence without having been apparently excited. In this format, the system will measure only the fluorescence due to FRET, which takes place when the two fluorophores are placed together (Fig. 17.13b).

The first example of so-called FRET aptabeacon was reported by Li and coworkers, who took advantage of the conformational change induced by thrombin binding on the TBA (Li et al. 2002). As already discussed, thrombin binding stabilizes quite a tight quadruplex conformation for this aptamer, which places its two ends close to each other. The authors labeled the MB with a donor–acceptor fluorophore pair and thrombin biding resulted in increase of the acceptor emission simultaneous to decrease in donor intensity. Compared to the quenching format, the FRET approach showed improved sensitivity and target quantitation, detecting picomolar concentrations of thrombin. This format was later formatted into a displacement assay for the real-time study of interactions between thrombin and other proteins (Cao and Tan 2005) and a FRET aptabeacon has been reported for PDGF detection (Vicens et al. 2005). Nevertheless, and in spite of the afore mentioned promising results, some authors have reported problems to reproduce the FRET format (Baldrich and O'Sullivan 2005; Stojanovic et al. 2001). In these works, instead of FRET, a simultaneous decrease in emission by the two fluorophores was registered. This was attributed to the fact that any two fluorophores put in close proximity can act as π-overlap quenchers of each other's fluorescence.

Up to date, the fluorescent aptabeacons described perform in solution. This is partly due to the fact that fluorophore attachment onto surfaces contributes to its quenching, as well as to the difficulty to simultaneously incorporate two reporter molecules plus immobilization strategies to a single aptamer molecule. Some authors defend that working in solution allows easy and fast study of samples in real-time without the requirement for washes or additional reagents. However, this does not always take into account the fact that fluorophores and fluorescent detection are often affected by components present in, for example, biological samples. On the other hand, fluorescent MB used in solution cannot be regenerated and reused, even if they are quite expensive to produce. Nevertheless, the reporting of new fluorophores, more stable against a variety of conditions, such as changes in pH or temperature, will doubtless contribute to the development of this format.

17.3.4.2 Electrochemical Aptamer Molecular Beacons: The Jewel in the Crown

As previously described in the text, the design of aptamer molecular beacons is based in the fact that aptamers undergo spontaneous conformational change following target interaction. In the case of electrochemical aptabeacons, the aptamer is immobilized onto an electrode surface and its free extreme labeled with an electroactive label. The conformational change induced by target binding translates into adjustment in the distance between sensing surface and label, and thus increase or decrease in electron transfer rate (Fig. 17.14). In contrast with voltammetric or amperometric detection using exogenous redox labels, the use of electrochemical aptabeacons is a completely reagentless and regenerable sensing approach. Additionally, it eludes potential interference due to unspecific binding of non-target species onto the sensing surface that might affect the performance of other sensing formats such as QCM or SPR. When compared to fluorescent molecular beacons, that may produce false signals arising from biological samples endogenous fluorescence, undergo masking by optically dense contaminants, or suffer photobleaching, the electrochemical counterparts remain unaffected by the low electroactive background observed in clinical samples and benefit from the relative stability of electroactive labels. Their main limitation, however, is the requirement for large-scale

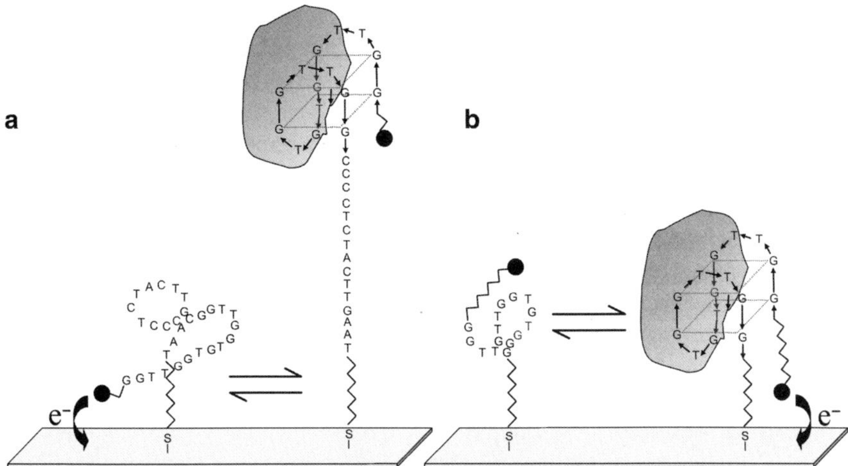

Fig. 17.14 TBA electrochemical aptabeacons. (**a**) *Turn-off* format: target binding induces decrease in electron transfer. (**b**) *Turn-on* format: target binding induces increase in electron transfer. The main difference between the two formats is the length of the immobilization linker used by the authors

aptamer rearrangement induced by target binding, which has to date limited aptabeacon optimization to a small number of aptamers, most of them folding into quadruplex or helix-junction structures.

The first electrochemical aptabeacon was reported by Xiao and co-workers, who self-assembled the TBA on gold electrodes through a long-thiolated 15-base spacer, after having modified its free end with a methylene blue moiety (Xiao et al. 2005a). The authors expect the aptamer to remain relatively unfolded in the absence of thrombin, allowing physical contact and electron exchange between the attached MB and the electrode surface. Target capture would promote aptamer folding into a G-quartet conformation, pushing the label apart (Fig. 17.14a). The subsequent decrease in signal was measured by DPV, detecting thrombin in a range from few nanomolar to low micromolar concentrations, and even in 50% serum.

Radi and co-workers followed a slightly different strategy in order to generate a “signal on” rather than “signal off” aptabeacon (Radi et al. 2006). In this case, the authors self-assembled the original 15-mer TBA with no additional bases, used ferrocene as the redox label, and carried out detection by DPV in a cation-free environment. Under these conditions, the aptamer loose conformation placed the redox label too far from the electrode surface for efficient electron transfer to take place. Thrombin binding induced G-quadruplex tightening and ferrocene physical approximation to the electrode, which translated into an increase in electron transfer and signal generation (Fig. 17.14b). The detection limit reported was of 0.5 nM with a linear range of 5–35 nM. The sensor “signal-on” functioning was confirmed in a later work using other electrochemical techniques, including potential step chronoamperometry, chronopotentiometry, square-wave voltammetry and chronopotentiometric-stripping analysis (Sánchez et al. 2006).

The same approximation was applied for the design of electrochemical aptabeacons against cocaine and PDGF, again using methylene blue labeling (Baker et al. 2006; Lai et al. 2007). In these two cases, the aptamer remains partially unfolded in the absence of target, folding into a three-stem junction following its capture. The conformational change approaches the label to the electrode surface, generating increases in current proportional to the amount of target bound. The analysis by DPV-generated detection limits of 10 μM and 500 μM cocaine in buffered saline and 50% fetal calf serum or human saliva, respectively (Baker et al. 2006; Lai et al. 2007), and detection of PDGF in both undiluted and twofold diluted serum down to 1 nM and 50 pM in that order (Lai et al. 2007).

17.4 Concluding Remarks

In less than 20 years of history, aptamers have become promising molecular tools in research areas such as diagnostics and bio-analysis, therapeutics and drug discovery, and quantitative and functional proteomics. This illustrates the most important characteristic of aptamers: their *versatility*. However, it is in the bio-sensor field where aptamers display in depth their various advantages versus other biorecognition molecules; aptasensing allows the development of miniaturized, highly sensitive and specific, reagentless and reusable devices. The above-described displacement-based sensors and aptabeacons are but a few examples. Aptamer real applicability will certainly benefit from recent improvements in the aptamer technology field, including resistance to nucleases via chemical modification and spiegelmer production, better characterization of the molecular basis of aptamer folding and target binding, and development of robotic platforms and microchips for automated SELEX.

On the other hand, the commercial use of aptamers will be limited by the number of patents protecting their different uses. For example, the USA-based *Archemix* possess the right on more than 200 patents covering all aspects of aptamer technology around the world, including aptamer composition, SELEX methodologies, and use of aptamers in therapeutics. Also, in the USA, *SomaLogic* holds via more than 150 patents the exclusive rights to aptamer-based diagnostics and ex vivo detection applications, and is responsible of the development of photoSELEX and photoaptamers. The French *Aptanomics* and the Germans *Nascacell Technologies* and *Noxxon* AG owe the patents for the use of peptide aptamers, reporter aptazymes and spiegelmers in therapeutics and drug screening. The Americans *Aptamera Inc*. and *Genta Inc*. develop aptamers targeted against cancer. The British *Isis Innovation Ltd* and USA *VITEX* hold patents involving aptamers for the diagnosis of prion diseases among others.

At present, at least three companies produce customized-aptamers against any targets selected by the customers: RNA-tec (Leuven, Belgium), AptaRes

(Luckenwalde, Germany) and Nascacell (Munchen, Germany). In spite of this and the increasing amount of aptamer assays and sensors reported, no diagnostic or analytical products based on the use of aptamers have been commercialized yet. Only one aptamer-derived product has reached the market: *Macugen*, developed by *Gilead Inc.* and commercialized by *OSI Pharmaceuticals* in the USA and *Pfizer* in Europe. *Macugen* is the first aptamer therapeutic approved to date for use in humans, which is based on the use of anti-VEGF inhibitory aptamers. A number of other drug candidates are at different stages of clinical trials at the moment.

It can be hardly expected that aptasensors displace in brief the more established immunoassays. In this respect, any aptameric device will have to perform not similarly but significantly *better* than an immuno-counterpart as to be accepted in the market. Nonetheless, aptamers possess properties that suit them to circumvent some of the limitations associated with the use of antibodies. For this reason, aptamers can be an appropriate solution in contexts where Ab are inadequate, or in combination to Ab as to offer improved performance. In this respect, aptamer integration might facilitate the detection of small and/or toxic targets not efficiently detected by Ab. The use of photoaptamers can allow the target covalent capture and thus lower detection limit. The possibility to regenerate aptamers allows the production of reusable, and thus cheaper devices. Aptamer integration in arrayed formats sets aside the possibility of multiplexed detection. Not forgetting that automatized production will approach aptamers to an increasing number of researchers and facilitate their use in different contexts. We will have to wait and see if these possibilities really materialize in the near future.

References

Asai R, Nishimura SI, Aita T, Takahashi K (2004) In vitro selection of DNA aptamers on chips using a method for generating point mutations. Anal Lett 37:645–656

Baker BR, Lai RY, Wood MS, Doctor EH, Heeger AJ, Plaxco KW (2006) An electronic, aptamer-based small-molecule sensor for the rapid, label-free detection of cocaine in adulterated samples and biological fluids. J Am Chem Soc 128(10):3138–3139

Balamurugan S, Obubuafo A, Soper SA, McCarley RL, Spivak DA (2006) Designing highly specific biosensing surfaces using aptamer monolayers on gold. Langmuir 22:6446–6453

Baldrich E, O'Sullivan CK (2005) Ability of thrombin to act as molecular chaperone, inducing formation of quadruplex structure of thrombin-binding aptamer. Anal Biochem 341:194–197

Baldrich E, Restrepo A, O'Sullivan CK (2004) Aptasensor development: elucidation of critical parameters for optimal aptamer performance. Anal Chem 76:7053–7063

Baldrich E, Acero JL, Reekmans G, Laureyn W, O'Sullivan CK (2005) Displacement enzyme linked aptamer assay. Anal Chem 77:4774–4784

Bang GS, Cho S, Kim B-G (2005) A novel electrochemical detection method for aptamer biosensors. Biosens Bioelectron 21:863–870

Bellecave P, Andreola ML, Ventura M, Tarrago-Litvak L, Litvak S, Astier-Gin T (2003) Selection of DNA aptamers that bind the RNA-dependent RNA polymerase of hepatitis C virus and inhibit viral RNA synthesis in vitro. Oligonucleotides 13:455–463

Berens C, Thain A, Schroeder R (2001) A tetracycline-binding RNA aptamer. Bioorg Med Chem 9:2549–2556

Berezovski M, Drabovich A, Krylova SM, Musheev M, Okhonin V, Petrov A, Krylov SN (2005) Nonequilibrium capillary electrophoresis of equilibrium mixtures: a universal tool for development of aptamers. J Am Chem Soc 127:3165–3171

Berezovski M, Musheev M, Drabovich A, Krylov SN (2006) Non-SELEX selection of aptamers. J Am Chem Soc 128:1410–1411

Bérn Abérem M, Najari A, Ho A-A, Gravel J-F, Nobert P, Boudreau D, Leclerc M (2006) Protein detecting arrays based on cationic polythiophene–DNA–aptamer complexes. Adv Mater 18:2703–2707

Bianchini M, Radrizzani M, Brocardo MG et al (2001) Specific oligobodies against ERK-2 that recognize both the native and the denatured state of the protein. J Immunol Methods 252:191–197

Bini A, Minunni M, Tombelli S, Centi S, Mascini M (2007) Analytical performances of aptamer-based sensing for thrombin detection. Anal Chem 79:3016–3019

Bock LC, Griffin LC, Latham JA, Vermaas EH, Toole JJ (1992) Selection of single-stranded DNA molecules that bind and inhibit human thrombin. Nature 355:564–566

Bock C, Coleman M, Collins B, Davis J, Foulds G, Gold L, Greef C, Heil J, Heilig JS, Hicke B, Hurst MN, Husar GM, Miller D, Ostroff R, Petach H, Schneider D, Vant-Hull B, Waugh S, Weiss A, Wilcox SK, Zichi D (2004) Photoaptamer arrays applied to multiplexed proteomic analysis. Proteomics 4:609–618

Bozza M, Sheardy RD, Dilone E, Scypinski S, Galazka M (2006) Characterization of the secondary structure and stability of an RNA aptamer that binds vascular endothelial growth factor. Biochemistry 45:7639–7643

Brambs HJ, Hoffmann M, Pauls S (2005) Diagnosis and interventional therapy for ductal gallstones. Radiologe 45:1004–1011

Brody EN, Gold L (2000) Aptamers as therapeutic and diagnostic agents. Rev Mol Biotechnol 74:5–13

Bruno JG, Kiel JL (1999) In vitro selection of DNA aptamers to anthrax spores with electrochemiluminescence detection. Biosens Bioelectron 14:457–464

Bruno JG, Kiel JL (2002) Use of magnetic beads in selection and detection of biotoxin aptamers by electrochemiluminescence and enzymatic methods. BioTechniques 32:178–183

Burgstaller P, Famulok M (1994) Isolation of RNA aptamers for biological cofactors by in vitro selection. Angew Chem Int Ed Engl 33:1084–1087

Burke DH, Gold L (1997) RNA aptamers to the adenosine moiety of S-adenosyl methionine: structural inferences from variations on a theme and the reproducibility of SELEX. Nucleic Acids Res 25:2020–2024

Cai H, Lee TM-H, Hsing I-M (2006) Label-free protein recognition using an aptamer-based impedance measurement assay. Sens Actuators B Chem 114:433–437

Cao Z, Tan W (2005) Molecular aptamers for real-time protein–protein interaction study. Chem Eur J 11:4502–4508

Centi S, Tombelli S, Minunni M, Mascini M (2007) Aptamer-based detection of plasma proteins by an electrochemical assay coupled to magnetic beads. Anal Chem 79:1466–1473

Chaloin L, Lehmann MJ, Sczakiel G, Restle T (2002) Endogenous expression of a high-affinity pseudoknot RNA aptamer suppresses replication of HIV-1. Nucleic Acids Res 30:4001–4008

Cheng AKH, Ge B, Yu H-Z (2007) Aptamer-based biosensors for label-free voltammetric detection of lysozyme. Anal Chem 79:5158–5164

Cho EJ, Collett JR, Szafranska AE, Ellington AD (2006) Optimization of aptamer microarray technology for multiple protein targets. Anal Chim Acta 564:82–90

Chou SH, Chin KH, Wang AHJ (2003) Unusual DNA duplex and hairpin motifs. Nucleic Acids Res 31:2461–2474

Ciesiolka J, Yarus M (1996) Small RNA-divalent domains. RNA 2:785–793

Clark SL, Remcho VT (2002) Aptamers as analytical reagents. Electrophoresis 23:1335–1340

Collett JR, Cho EJ, Lee JF, Levy M, Hood AJ, Wan C, Ellington AD (2005a) Functional RNA microarrays for high-throughput screening of antiprotein aptamers. Anal Biochem 338: 113–123

Collett JR, Eun JC, Ellington AD (2005b) Production and processing of aptamer microarrays. Methods 37:4–15

Connell GJ, Illangesekare M, Yarus M (1993) Three small ribooligonucleotides with specific arginine sites. Biochemistry 32:5497–5502

Conrad R, Keranen LM, Ellington AD et al (1994) Isozyme-specific inhibition of protein kinase C by RNA aptamers. J Biol Chem 269:32051–32054

Cooper MA, Singleton VT (2007) A survey of the 2001 to 2005 quartz crystal microbalance biosensor literature: applications of acoustic physics to the analysis of biomolecular interactions. J Mol Recognit 20:154–184

Cox JC, Ellington AD (2001) Automated selection of anti-protein aptamers. Bioorg Med Chem 9:2525–2531

Cox JC, Rudolph P, Ellington AD (1998) Automated RNA selection. Biotechnol Prog 14:845–850

Cox JC, Hayhurst A, Hesselberth J, Bayer TS, Georgiou G, Ellington AD (2002a) Automated selection of aptamers against protein targets translated in vitro: from gene to aptamer. Nucleic Acids Res 30:e108

Cox JC, Rajendran M, Riedell T, Davidson EA, Sooter LJ, Bayer TS, Schmitz-Brown M, Ellington AD (2002b) Automated acquisition of aptamer sequences. Comb Chem High Throughput Screen 5:289–299

Dewey TM, Mundt A, Crouch GJ, Zyzniewski CM, Eaton BE (1995) New uridine derivatives for systematic evolution of RNA ligands by exponential enrichment. J Am Chem Soc 117:8474–8475

Dey AK, Griffiths C, Lea SM, James W (2005) Structural characterization of an anti-gp120 RNA aptamer that neutralizes R5 strains of HIV-1. RNA 11:873–884

Dieckmann T, Suzuki E, Nakamura GK, Feigon J (1996) Solution structure of an ATP-binding RNA aptamer reveals a novel fold. RNA 2:628–640

Drabovich AP, Berezovski M, Okhonin V, Krylov SN (2006) Selection of smart aptamers by methods of kinetic capillary electrophoresis. Anal Chem 78:3171–3178

Drolet DW, Moon-McDermott L, Romig TS (1996) An enzyme-linked oligonucleotide assay. Nat Biotechnol 14:1021–1027

Eaton BE (1997) The joys of in vitro selection: chemically dressing oligonucleotides to satiate protein targets. Curr Opin Chem Biol 1:10–16

Eaton BE, Gold L, Hicke BJ, Janjicé N, Jucker FM, Sebesta DP, Tarasow TM, Zichi DA (1997) Post-SELEX combinationl optimisation of aptamers. Bioorg Med Chem 5:1087–1096

Ellingham M, Bunka DHJ, Rowlands DJ, Stonehouse NJ (2006) Selection and characterization of RNA aptamers to the RNA-dependent RNA polymerase from foot-and-mouth disease virus. RNA 12:1970–1979

Ellington AD, Szostak JW (1990) In vitro selection of RNA molecules that bind specific ligands. Nature 346:818–822

Ellington AD, Szostak JW (1992) Selection in vitro of single-stranded DNA molecules that fold into specific ligand-binding structures. Nature 355:850–852

Eulberg D, Buchner K, Maasch C, Klussmann S (2005) Development of an automated in vitro selection protocol to obtain RNA-based aptamers: identification of a biostable substance P antagonist. Nucleic Acids Res 33:e45

Famulok M (1994) Molecular recognition of amino acids by RNAaptamers: an L-citrulline binding RNA motif and its evolution into an L-arginine binder. J Am Chem Soc 116:1698–1706

Fan P, Suri AK, Fiala R, Live D, Patel DJ (1996) Molecular recognition in the FMN-RNA aptamer complex. J Mol Biol 258:480–500

Fang X, Cao Z, Beck T, Tan W (2001) Molecular aptamer for real-time oncoprotein platelet-derived growth factor monitoring by fluorescence anisotropy. Anal Chem 73:5752–5757

Fang X, Sen A, Vicens M, Tan W (2003) Synthetic DNA aptamers to detect protein molecular variants in a high-throughput fluorescence quenching assay. ChemBioChem 4:829–834
Faulhammer D, Eschgfaller B, Stark S, Burgstaller P, Englberger W, Erfurth J, Kleinjung F, Rupp J, Vulcu SD, Schroder W, Vonhoff S, Nawrath H, Gillen C, Klussmann S (2004) Biostable aptamers with antagonistic properties to the neuropeptide nociceptin/orphanin FQ. RNA 10:516–527
Feigon J, Dieckmann T, Smith FW (1996) Aptamer structures from A to Z. Chem Biol 3:61.l–61.7
Frauendorf C, Jäschke A (2001) Detection of small organic analytes by fluorescing molecular switches. Bioorg Med Chem 9:2521–2524
Fukusaki E-I, Hasunuma T, Kajiyama S-I, Kazawa A, Itoh TJ, Kobayashi A (2001) SELEX for tubulin affords specific T-rich DNA aptamers. Bioorg Med Chem Lett 11:2927–2930
Gebhardt K, Shokraei A, Babaie E, Lindquist BH (2000) RNA aptamers to *S*-adenosylhomocysteine: kinetic properties, divalent cation dependency, and comparison with anti-*S*-adenosylhomocysteine antibody. Biochemistry 39:7255–7265
Geiger A, Burgstaller P, von der Eltz H, Roeder A, Famulok M (1996) RNA aptamers that bind L-arginine with sub-micromolar dissociation constants and high enantioselectivity. Nucleic Acids Res 24:1029–1036
Gellert M, Lipsett MN, Davies DR (1962) Helix formation by guanylic acid. Proc Natl Acad Sci USA 48:2013–2018
Gokulrangan G, Unruh JR, Holub DF, Ingram B, Johnson CK, Wilson GS (2005) DNA aptamer-based bioanalysis of IgE by fluorescence anisotropy. Anal Chem 77:1963–1970
Golden MC, Collins BD, Willis MC, Koch TH (2000) Diagnostic potential of photoSELEX-evolved ssDNA aptamers. J Biotechnol 81:167–178
Gopinath SCB (2007) Methods developed for SELEX. Anal Bioanal Chem 387:171–182
Gopinath SCB, Sakamaki Y, Kawasaki K, Kumar PKR (2006) An efficient RNA aptamer against human influenza B virus hemagglutinin. J Biochem 139:837–846
Griffin LC, Toole JJ, Leung LLK (1993) The discovery and characterization of a novel nucleotide-based thrombin inhibitor. Gene 137:25–31
Gronewold TMA, Glass S, Quandt E, Famulok M (2005) Monitoring complex formation in the blood-coagulation cascade using aptamer-coated SAW sensors. Biosens Bioelectron 20:2044–2052
Haller AA, Sarnow P (1997) In vitro selection of a 7-methyl-guanosine binding RNA that inhibits translation of capped mRNA molecules. Proc Natl Acad Sci USA 94:8521–8526
Hamaguchi N, Ellington A, Stanton M (2001) Aptamer beacons for the direct detection of proteins. Anal Biochem 294:126–131
Hamula CLA, Guthrie JW, Zhang HQ, Li XF, Le XC (2006) Selection and analytical applications of aptamers. Trends Analyt Chem 25:681–691
Hansen JA, Wang J, Kawde A-N, Xiang Y, Gothelf KV, Collins G (2006) Quantum-dot/aptamer-based ultrasensitive multi-analyte electrochemical biosensor. J Am Chem Soc 128: 2228–2229
Harada K, Frankel AD (1995) Identification of two novel arginine binding DNAs. EMBO J 14:5798–5811
Hartmann R, Nørby PL, Martensen PM, Jørgensen P, James MC, Jacobsen C, Moestrup SK, Justesen J (1998) Activation of 2′-5′ oligoadenylate synthetase by single-stranded and double-stranded RNA aptamers. J Biol Chem 273:3236–3246
Helmling S, Maasch C, Eulberg D, Buchner K, Schroder W, Lange C, Vonhoff S, Wlotzka B, Tschop MH, Rosewicz S, Klussmann S (2004) Inhibition of ghrelin action in vitro and in vivo by an RNA-Spiegelmer. Proc Natl Acad Sci USA 101:13174–13179
Hesselberth J, Robertson MP, Jhaveri S, Ellington AD (2000) In vitro selection of nucleic acids for diagnostic applications. Rev Mol Biotechnol 74:15–25
Hianik T, Ostatnaé V, Zajacovaé Z, Stoikova E, Evtugyn G (2005) Detection of aptamer–protein interactions using QCM and electrochemical indicator methods. Bioorg Med Chem Lett 15:291–295

Hicke BJ, Watson SR, Koenig A, Lynott CK, Bargatze RF, Chang Y-F et al (1996) DNA aptamers block L-selectin function in vivo. J Clin Invest 98:2688–2692
Ho H-A, Leclerc M (2004) Optical sensors based on hybrid aptamer/conjugated polymer complexes. J Am Chem Soc 126:1384–1387
Hofmann H-P, Limmer S, Hornung V, Sprinzl M (1997) Ni^{2+}-Binding RNA motifs with an asymmetric purine-rich internal loop and a G-A base pair. RNA 3:1289–1300
Holeman LA, Robinson SL, Szostak JW, Wilson C (1998) Isolation and characterization of fluorophore-binding RNA aptamers. Fold Des 3:423–431
Homann M, Lorger M, Engstler M, Zacharias M, Goringer HU (2006) Serum-stable RNA aptamers to an invariant surface domain of live African trypanosomes. Comb Chem High Throughput Screen 9:491–499
Huizenga DE, Szostak JW (1995) A DNA aptamer that binds adenosine and ATP. Biochemistry 34:656–665
Hwang B, Cho JS, Yeo HJ, Kim JH, Chung KM, Han K, Jang SK, Lee SW (2004) Isolation of specific against NS3 helicase and high-affinity RNA aptamers domain of hepatitis C virus. RNA 10:1277–1290
Hwang KS, Lee S-M, Eom K, Lee JH, Lee Y-S, Park JH, Yoon DS, Kim TS (2007) Nanomechanical microcantilever operated in vibration modes with use of RNA aptamer as receptor molecules for label-free detection of HCV helicase. Biosens Bioelectron 23:459–465
Hybarger G, Bynum J, Williams RF, Valdes JJ, Chambers JP (2006) A microfluidic SELEX prototype. Anal Bioanal Chem 384:191–198
Ikanovic M, Rudzinski WE, Bruno JG, Allman A, Carrillo MP, Dwarakanath S, Bhahdigadi S, Rao P, Kiel JL, Andrews CJ (2007) Fluorescence assay based on aptamer-quantum dot binding to *Bacillus thuringiensis* spores. J Fluoresc 17:193–199
Ikebukuro K, Kiyohara C, Sode K (2005) Novel electrochemical sensor system for protein using the aptamers in sandwich manner. Biosens Bioelectron 20:2168–2172
Ito Y (1997) Modified nucleic acids for in vitro selection. Nucleic Acids Symp Ser 37:259–260
Ito Y, Suzuki A, Kawazoe N, Imanishi Y (2001) In vitro selection of RNA aptamers carrying multiple biotin groups in the side chains. Bioconjug Chem 12:850–854
James WC (2001) Aptamers. In: Meyers RA (ed) Encyclopedia of analytical chemistry. Wiley, Chichester, pp 4848–4871
Jayasena SD (1999) Aptamers, an emerging class of molecules that rival antibodies in diagnostics. Clin Chem 45:1628–1650
Jenison RD, Gill SC, Pardi A, Polisky B (1994) High-resolution molecular discrimination by RNA. Science 263:1425–1429
Jensen KB, Atkinson BL, Willis MC, Koch TH, Gold L (1995) Using in vitro selection to direct the covalent attachment of human immunodeficiency virus type 1 Rev protein to high-affinity RNA ligands. Proc Natl Acad Sci USA 92:12220–12224
Jhaveri S, Olwin B, Ellington AD (1998) In vitro selection of phosphorothiolated aptamers. Bioorg Med Chem Lett 8:2285–2290
Jhaveri SD, Kirby R, Conrad R, Maglott EJ, Bowser M, Kennedy RT, Glick G, Ellington AD (2000) Designed signaling aptamers that transduce molecular recognition to changes in fluorescence intensity. J Am Chem Soc 122:2469–2473
Jiang F, Kumar RA, Jones RA, Patel DJ (1996) Structural basis of RNA folding and recognition in an AMP-RNA aptamer complex. Nature 382:183–186
Jiang L, Suri AK, Fiala R, Patel DJ (1997) Saccharide-RNA recognition in an aminoglycoside antibiotic-RNA aptamer complex. Chem Biol 4:35–50
Jones LA, Clancy LE, Rawlinson WD, White PA (2006) High-affinity aptamers to subtype 3a hepatitis C virus polymerase display genotypic specificity. Antimicrob Agents Chemother 50:3019–3027
Jung A, Gronewold TMA, Tewes M, Quandt E, Berlin P (2007) Biofunctional structural design of SAW sensor chip surfaces in a microfluidic sensor system. Sens Actuators B Chem 124:46–52

Kawde A-N, Rodriguez MC, Lee TMH, Wang J (2005) Label-free bioelectronic detection of aptamer–protein interactions. Electrochem Commun 7:537–540

Keniry MA (2000) Quadruplex structures in nucleic acids. Biopolymers 56:123–146

Kiga D, Futamura Y, Sakamoto K, Yokoyama S (1998) An RNA aptamer to the xanthine/guanine base with a distinctive mode of purine recognition. Nucleic Acids Res 26:1755–1760

Kikuchi K, Umehara T, Fukuda K, Hwang J, Kuno A, Hasegawa T, Nishikawa S (2003) RNA aptamers targeted to domain II of hepatitis C virus IRES that bind to its apical loop region. J Biochem 133:263–270

Kim MY, Jeong S (2003) RNA aptamers that bind the nucleocapsid protein contain pseudoknots. Mol Cells 16:413–417

Kim YS, Jung HS, Matsuura T, Lee HY, Kawai T, Gu MB (2007) Electrochemical detection of 17β-estradiol using DNA aptamer immobilized gold electrode chip. Biosens Bioelectron 22:2525–2531

Kimoto M, Shirouzu M, Mizutani S et al (2002) Anti-(Raf-1) RNA aptamers that inhibit Ras-induced Raf-1 activation. Eur J Biochem 269:697–704

King DJ, Ventura DA, Brasier AR, Gorenstein DG (1998) Novel combinatorial selection of phosphothioate oligonucleotide aptamers. Biochemistry 37:16489–16493

King DJ, Safar JG, Legname G, Prusiner SB (2007) Thioaptamer interactions with prion proteins: sequence-specific and non-specific binding sites. J Mol Biol 369:1001–1014

Kirby R, Cho EJ, Gehrke B, Bayer T, Park YS, Neikirk DP, McDevitt JT, Ellington AD (2004) Aptamer-based sensor arrays for the detection and quantitation of proteins. Anal Chem 76:4066–4075

Kleinjung F, Klussmann S, Erdmann VA, Scheller FW, Fürste JP, Bier FF (1998) High-affinity RNA as a recognition element in a biosensor. Anal Chem 70:328–331

Klussmann S, Nolte A, Bald R, Erdmann VA, Fürste JP (1996) Mirror-image RNA that binds D-adenosine. Nat Biotecnol 14:1112–1116

Kohler G, Milstein C (1975) Continuous cultures of fussed cells secreting antibody of predefined specificity. Nature 256:495–497

Kopylov AM, Spiridonova VA (2000) Combinatorial chemistry of nucleic acids: SELEX. Mol Biol 34:940–954

Kulbachinskiy A, Feklistov A, Krasheninnikov I, Goldfarb A, Nikiforov V (2004) Aptamers to *Escherichia coli* core RNA polymerase that sense its interaction with rifampicin, sigma-subunit and GreB. Eur J Biochem 271:4921–4931

Kusser W (2000) Chemically modified nucleic acid aptamers for in vitro selections: evolving evolution. J Biotechnol 74:27–38

Lai RY, Plaxco KW, Heeger AJ (2007) Atamer-based electrochemical detection of picomolar platelet-derived growth factor directly in blood serum. Anal Chem 79:229–233

Lauhon CT, Szostak JW (1995) RNA aptamers that bind flavin and nicotinamide redox cofactors. J Am Chem Soc 117:1246–1257

Le Floch F, Ho HA, Leclerc M (2006) Label-free electrochemical detection of protein based on a ferrocene-bearing cationic polythiophene and aptamer. Anal Chem 78:4727–4731

Lee SW, Sullenger BA (1997) Isolation of a nuclease-resistant decoy RNA that can protect human acetylcholine receptors from myasthenic antibodies. Nat Biotechnol 15:41–45

Lee M, Walt DR (2000) A fiber-optic microarray biosensor using aptamers as receptors. Anal Biochem 282:142–146

Lee SK, Park MW, Yang EG, Yu JH, Jeong SJ (2005) An RNA aptamer that binds to the beta-catenin interaction domain of TCF-1 protein. Biochem Biophys Res Commun 327:294–299

Leva S, Burmeister J, Muhn P, Jahnke B, Fesser D, Erfurth J, Burgstaller P, Klussmann S (2002a) GnRH binding RNA and DNA spiegelmers: a novel approach toward GnRH antagonism. Chem Biol 9:351–359

Leva S, Lichte A, Burmeister J, Muhn P, Jahnke B, Fesser D, Erfurth J, Klussmann S (2002b) GnRH binding RNA and DNA Spiegelmers: a novel approach toward GnRH antagonism. Chem Biol 9:351–359

Levesque D, Beaudoin J-D, Roy S, Perreault J-P (2007) In vitro selection and characterization of RNA aptamers binding thyroxine hormone. Biochem J 403:129–138
Li JJ, Fang X, Tan W (2002) Molecular aptamer beacons for real-time protein recognition. Biochem Biophys Res Commun 292:31–40
Li Y, Lee HJ, Corn RM (2006) Fabrication and characterization of RNA aptamer microarrays for the study of protein–aptamer interactions with SPR imaging. Nucleic Acids Res 34: 6416–6424
Liao W, Cui XT (2007) Reagentless aptamer based impedance biosensor for monitoring a neuro-inflammatory cytokine PDGF. Biosens Bioelectron 23:218–224
Lin Y, Qiu Q, Gill SC, Jayasena SD (1994) Modified RNA sequence pools for in vitro selection. Nucleic Acids Res 22:5229–5234
Lin Y, Nieuwlandt D, Magallanez A, Feistner B, Jayasena SD (1996) High-affinity and specific recognition of human thyroid stimulating hormone (hTSH) by in vitro-selected 2′-amino-modified RNA. Nucleic Acids Res 24:3407–3414
Lin C, Katilius E, Liu Y, Zhang J, Yan H (2006) Self-assembled signaling aptamer DNA arrays for protein detection. Angew Chem Int Ed Engl 45:5296–5301
Liss M, Petersen B, Wolf H, Prohaska E (2002) An aptamer-based quartz crystal protein biosensor. Anal Chem 74:4488–4495
Löhndorf M, Schlecht U, Gronewold TMA, Malaveé A, Tewes M (2005) Microfabricated high-performance microwave impedance biosensors for detection of aptamer–protein interactions. Appl Phys Lett 87:1–3
Lorsch JR, Szostak JW (1994) In vitro selection of RNA aptamers specific for cyanocobalamin. Biochemistry 33:973–982
Lucklum R, Hauptmann P (2006) Acoustic microsensors-the challenge behind microgravimetry. Anal Bioanal Chem 384:667–682
Macaya RF, Schultze P, Smith FW, Roe JA, Feigon J (1993) Thrombin-binding DNA aptamer forms a unimolecular quadruplex structure in solution. Proc Natl Acad Sci USA 90:3745–3749
Maehashi K, Katsura T, Kerman K, Takamura Y, Matsumoto K, Tamiya E (2007) Label-free protein biosensor based on aptamer-modified carbon nanotube field-effect transistors. Anal Chem 79:782–787
Mallikaratchy P, Stahelin RV, Cao Z, Cho W, Tan W (2006) Selection of DNA ligands for protein kinase C-delta. Chem Commun (Camb) (30):3229–3132
Mann D, Reinemann C, Stoltenburg R, Strehlitz B (2005) In vitro selection of DNA aptamers binding ethanolamine. Biochem Biophys Res Commun 338:1928–1934
Mannironi C, Nardo AD, Fruscoloni P, Tocchini-Valentini GP (1997) In vitro selection of dopamine RNA ligands. Biochemistry 36:9726–9734
Marathias VM, Bolton PH (1999) Determinants of DNA quadruplex structural type: sequence and potassium binding. Biochemistry 38:4355–4364
McCauley TG, Hamaguchi N, Stanton M (2003) Aptamer-based biosensor arrays for detection and quantification of biological macromolecules. Anal Biochem 319:244–250
Mendonsa SD, Bowser MT (2004) In vitro selection of high-affinity DNA ligands for human IgE using capillary electrophoresis. Anal Chem 76:5387–5392
Minunni M, Tombelli S, Gullotto A, Luzi E, Mascini M (2004) Development of biosensors with aptamers as bio-recognition element: the case of HIV-1 Tat protein. Biosens Bioelectron 20:1149–1156
Misono TS, Kumar PKR (2005) Selection of RNA aptamers against human influenza virus hemagglutinin using surface plasmon resonance. Anal Biochem 342:312–317
Moreno M, Rincon E, Pineiro D, Fernandez G, Domingo A, Jimenez-Ruiz A, Salinas M, Gonzalez VM (2003) Selection of aptamers against KMP-11 using colloidal gold during the SELEX process. Biochem Biophys Res Commun 308:214–218
Mosing RK, Bowser MT (2007) Microfluidic selection and applications of aptamers. J Sep Sci 30:1420–1426

Mosing RK, Mendonsa SD, Bowser MT (2005) Capillary electrophoresis-SELEX selection of aptamers with affinity for HIV-1 reverse transcriptase. Anal Chem 77:6107–6112
Nagatoishi S, Tanaka Y, Tsumoto K (2007) Corrigendum to "Circular dichroism spectra demonstrate formation of the thrombin-binding DNA aptamer G-quadruplex under stabilizing-cation-deficient conditions" [Biochem Biophys Res Commun 352 (2007) 812–817] (doi:10.1016/j.bbrc.2006.11.088). Biochem Biophys Res Commun 354:837–838
Nieuwlandt D (2000) In vitro selection of functional nucleic acid sequences. Curr Issues Mol Biol 2:9–16
Nieuwlandt D, Wecker M, Gold L (1995) In vitro selection of RNA ligands to substance P. Biochemistry 34:5651–5659
Noeske J, Buck J, Fütig B, Nasiri HR, Schwalbe H, Wöhnert J (2007) Interplay of 'induced fit' and preorganization in the ligand induced folding of the aptamer domain of the guanine binding riboswitch. Nucleic Acids Res 35:572–583
Nolte A, Klussmann S, Bald R, Erdmann VA, Fürste JP (1996) Mirror-design of L-oligonucleotide ligands binding to L-arginine. Nat Biotechnol 14:1116–1119
Ohuchi SP, Ohtsu T, Nakamura Y (2006) Selection of RNA aptamers against recombinant transforming growth factor-beta type III receptor displayed on cell surface. Biochimie 88:897–904
Osborne SE, Matsumura I, Ellington AD (1997) Aptamers as therapeutic and diagnostic reagents: problems and prospects. Curr Opin Chem Biol 1:5–9
Ozaki H, Nishihira A, Wakabayashi M, Kuwahara M, Sawai H (2006) Biomolecular sensor based on fluorescence-labeled aptamer. Bioorg Med Chem Lett 16:4381–4384
Pagratis NC, Fitzwater T, Jellinek D, Dang C (1997) Potent 2′-amino- and 2′-fluoro-2′-deoxyribonucleotide RNA inhibitors of keratinocyte growth factor. Nat Biotechnol 15:68–73
Pan Q, Zhang XL, Wu HY, He PW, Wang FB, Zhang MS, Hu JM, Xia B, Wu JG (2005) Aptamers that preferentially bind type IVB pili and inhibit human monocytic-cell invasion by *Salmonella enterica* serovar *Typhi*. Antimicrob Agents Chemother 49:4052–4060
Papamichael KI, Kreuzer MP, Guilbault GG (2007) Viability of allergy (IgE) detection using an alternative aptamer receptor and electrochemical means. Sens Actuators B Chem 121:178–186
Patel DJ, Suri UAK (2000) Structure, recognition and discrimination in RNA aptamer complexes with cofactors, amino acids, drugs and aminoglycoside antibiotics. Rev Mol Biotechnol 74:39–60
Patel DJ, Suri AK, Jiang F, Jiang L, Fan P, Kumar RA, Nonin S (1997) Structure, recognition and adaptive binding in RNA aptamer complexes. J Mol Biol 272:645–664
Petach H, Gold L (2002) Dimensionality is the issue: use of photoaptamers in protein microarrays. Curr Opin Biotechnol 13:309–314
Piccirilli JA, Krauch T, Moroney SE, Benner SA (1990) Enzymatic incorporation of a new base pair into DNA and RNA extends the genetic alphabet. Nature 343:33–37
Pileur F, Andreola M-L, Dausse E, Michel J, Moreau S, Yamada H, Gaidamakov SA, Cazenave C (2003) Selective inhibitory DNA aptamers of the human RNase H1. Nucleic Acids Res 31:5776–5788
Polsky R, Gill R, Kaganovsky L, Willner I (2006) Nucleic acid-functionalized Pt nanoparticles: catalytic labels for the amplified electrochemical detection of biomolecules. Anal Chem 78:2268–2271
Potyrailo RA, Conrad RC, Ellington AD, Hieftje GM (1998) Adapting selected nucleic acid ligands (aptamers) to biosensors. Anal Chem 70:3419–3425
Purschke WG, Radtke F, Kleinjung F, Klussmann S (2003) A DNA Spiegelmer to staphylococcal enterotoxin B. Nucleic Acids Res 31:3027–3032
Radi AE, Sanchez JLA, Baldrich E, O'Sullivan CK (2005) Reusable impedimetric aptasensor. Anal Chem 77:6320–6323
Radi AE, Sanchez JLA, Baldrich E, O'Sullivan CK (2006) Reagentless, reusable, ultrasensitive electrochemical molecular beacon aptasensor. J Am Chem Soc 128:117–124

Ramos E, Pineiro D, Soto M, Abanades DR, Martin ME, Salinas M, Gonzalez VM (2007) A DNA aptamer population specifically detects Leishmania infantum H2A antigen. Lab Invest 87:409–416

Ravelet C, Grosset C, Peyrin E (2006) Liquid chromatography, electrochromatography and capillary electrophoresis applications of DNA and RNA aptamers. J Chromatogr A 1117:1–10

Rhie A, Kirby L, Sayer N, Wellesley R, Disterer P, Sylvester I, Gill A, Hope J, James W, Tahiri-Alaoui A (2003) Characterization of 2′-fluoro-RNA aptamers that bind preferentially to disease-associated conformations of prion protein and inhibit conversion. J Biol Chem 278:39697–39705

Rhodes A, Deakin A, Spaull J, Coomber B, Aitken A, Life P, Rees S (2000) The generation and characterization of antagonist RNA aptamers to human oncostatin M. J Biol Chem 275:28555–28561

Robertson DL, Joyce GF (1990) Selection in vitro of an RNA enzyme that specifically cleaves single-stranded DNA. Nature 344:467–468

Ruckman J, Green LS, Waugh S, Gillette WL, Henninger DD, Claesson-Welsh L, Tanjic N (1998) 2′-Fluoropyrimidine RNA-based aptamers to the 165-amino acid form of Vascular Endothelial Growth Factor (VEGF165). Inhibition of receptor binding and VEGF-induced vascular permeability through interactions requiring the exon 7-encoded domain. J Biol Chem 273: 20556–20567

Rye PD, Nustad K (2001) Immunomagnetic DNA aptamer assay. BioTechniques 30:290–295

Saito T, Tomida M (2005) Generation of inhibitory DNA aptamers against human hepatocyte growth factor. DNA Cell Biol 24:624–633

Sánchez JLA, Baldrich E, Radi AE-G, Dondapati S, Sánchez PL, Katakis I, O'Sullivan CK (2006) Electronic 'off–on' molecular switch for rapid detection of thrombin. Electroanalysis 18 (1):1957–1962

Sando S, Ogawa A, Nishi T, Hayami M, Aoyama Y (2007) In vitro selection of RNA aptamer against *Escherichia coli* release factor 1. Bioorg Med Chem Lett 17:1216–1220

Sassanfar M, Szostak JW (1993) An RNA motif that binds ATP. Nature 364:550–553

Savran CA, Knudsen SM, Ellington AD, Manalis SR (2004) Micromechanical detection of proteins using aptamer-based receptor molecules. Anal Chem 76:3194–3198

Sayer N, Ibrahim J, Turner K, Tahiri-Alaoui A, James W (2002) Structural characterization of a 2′F-RNA aptamer that binds a HIV-1 SU glycoprotein, gp120. Biochem Biophys Res Commun 293:924–931

Schlensog MD, Gronewold TMA, Tewes M, Famulok M, Quandt E (2004) A Love-wave biosensor using nucleic acids as ligands. Sens Actuators B Chem 101:308–315

Schoetzau T, Langner J, Moyroud E, Roehl I, Vonhoff S, Klussmann S (2003) Aminomodified nucleobases: functionalized nucleoside triphosphates applicable for SELEX. Bioconjug Chem 14:919–926

Sekiya S, Noda K, Nishikawa F, Yokoyama T, Kumar PKR, Nishikawa S (2006) Characterization and application of a novel RNA aptamer against the mouse prion protein. J Biochem 139:383–390

Shafer RH, Smirnov I (2000) Biological aspects of DNA/RNA quadruplexes. Biopolymers 56:209–227

Shangguan D, Li Y, Tang Z, Cao ZC, Chen HW, Mallikaratchy P, Sefah K, Yang CJ, Tan W (2006) Aptamers evolved from live cells as effective molecular probes for cancer study. Proc Natl Acad Sci USA 103:11838–11843

Shoji A, Kuwahara M, Ozaki H, Sawai H (2007) Modified DNA aptamer that binds the (R)-isomer of a thalidomide derivative with high enantioselectivity. J Am Chem Soc 129:1456–1464

So H-M, Won K, Kim YH, Kim B-K, Ryu BH, Na PS, Kim H, Lee J-O (2005) Single-walled carbon nanotube biosensors using aptamers as molecular recognition elements. J Am Chem Soc 127:11906–11907

Sooter LJ, Ellington AD (2002) Reflections on a novel therapeutic candidate. Chem Biol 9:857–858

Spiridonova VA, Rog EV, Dugina TN, Strukova SM, Kopylov AM (2003) Aptamer DNA: a new type of thrombin inhibitors. Russ J Bioorg Chem 29:450–453

Stojanovic MN, de Prada P, Landry DW (2001) Aptamer-based folding fluorescent sensor for cocaine. J Am Chem Soc 123:4928–4931

Stoltenburg R, Reinemann C, Strehlitz B (2005) FluMag-SELEX as an advantageous method for DNA aptamer selection. Anal Bioanal Chem 383:83–91

Sussman D, Nix JC, Wilson C (2000) The structural basis for molecular recognition by the vitamin B12 RNA aptamer. Nat Struct Biol 7:53–57

Szkaradkiewicz K, Nanninga M, Nesper-Brock M, Gerrits M, Erdmann VA, Sprinzl M (2002) RNA aptamers directed against release factor 1 from *Thermus thermophilus*. FEBS Lett 514:90–95

Tahiri-Alaoui A, Frigotto L, Manville N, Ibrahim J, Romby P, James W (2002) High affinity nucleic acid aptamers for streptavidin incorporated into bi-specific capture ligands. Nucleic Acids Res 30:e45

Tang JJ, Xie JW, Shao NS, Yan Y (2006) The DNA aptamers that specifically recognize ricin toxin are selected by two in vitro selection methods. Electrophoresis 27:1303–1311

Tang JJ, Yu T, Guo L, Xie JW, Shao NS, He ZK (2007a) In vitro selection of DNA aptamer against abrin toxin and aptamer-based abrin direct detection. Biosens Bioelectron 22:2456–2463

Tang Q, Su X, Loh KP (2007b) Surface plasmon resonance spectroscopy study of interfacial binding of thrombin to antithrombin DNA aptamers. J Colloid Interface Sci 315:99–106

Tasset DM, Kubik MF, Steiner W (1997) Oligonucleotide inhibitors of human thrombin that bind distinct epitopes. J Mol Biol 272:688–698

Tombelli S, Minunni M, Luzi E, Mascini M (2005a) Aptamer-based biosensors for the detection of HIV-1 Tat protein. Bioelectrochemistry 67:135–141

Tombelli S, Minunni M, Mascini M (2005b) Analytical applications of aptamers. Biosens Bioelectron 20:2424–2434

Tuerk C, Gold L (1990) Systematic evolution of ligands by exponential enrichment: RNA ligands to bacteriophage T4 DNA polymerase. Science 249:505–510

Tyagi S, Kramer FR (1996) Molecular beacons: probes that fluoresce upon hybridization. Nat Biotechnol 14:303–308

Uphoff KW, Bell SD, Ellington AD (1996) In vitro selection of aptamers: the dearth of pure reason. Curr Opin Struct Biol 6:281–288

Vaish NK, Fraley AW, Szostak JW, McLaughlin LW (2000) Expanding the structural and functional diversity of RNA: analog uridine triphosphates as candidates for in vitro selection of nucleic acids. Nucleic Acids Res 28:3316–3322

Vater A, Klussmann S (2003) Toward third-generation aptamers: Spiegelmers and their therapeutic prospects. Curr Opin Drug Discov Devel 6:253–261

Vater A, Jarosch F, Buchner K, Klussmann S (2003) Short bioactive Spiegelmers to migraine-associated calcitonin gene-related peptide rapidly identified by a novel approach: tailored-SELEX. Nucleic Acids Res 31(21):e130

Vicens MC, Sen A, Vanderlaan A, Drake TJ, Tan W (2005) Investigation of molecular beacon aptamer-based bioassay for platelet-derived growth factor detection. ChemBioChem 6:900–907

Vo NV, Oh JW, Lai MM (2003) Identification of RNA ligands that bind hepatitis C virus polymerase selectively and inhibit its RNA synthesis from the natural viral RNA templates. Virology 307:301–316

Wallace ST, Schroeder R (1998) In vitro selection and characterization of streptomycin-binding RNAs: recognition discrimination between antibiotics. RNA 4:112–123

Wallis MG, Ahsen U, Schroeder R, Famulok M (1995) A novel RNA motif for neomycin recognition. Chem Biol 2:543–552

Wallis MG, Streicher B, Wank H, Ahsen U, Clodi E, Wallace ST, Famulok M, Schroeder R (1997) In vitro selection of a viomycin-binding RNA pseudoknot. Chem Biol 4:357–366

Wang KY (1994) Tertiary structure motif of oxytricha telomere DNA. Biochemistry 33:7517–7527
Wang Y, Killian J, Hamasaki K, Rando RR (1996) RNA molecules that specifically and stoichiometrically bind aminoglycoside antibiotics with high affinities. Biochemistry 35:12338–12346
Wang Z, Wilkop T, Xu D, Dong Y, Ma G, Cheng Q (2007) Surface plasmon resonance imaging for affinity analysis of aptamer–protein interactions with PDMS microfluidic chips. Anal Bioanal Chem 389:819–825
Weiss S, Proske D, Neumann M, Groschup MH, Kretzschmar HA, Famulok M, Winnacker EL (1997) RNA aptamers specifically interact with the prion protein PrP. J Virol 71:8790–8797
White R, Rusconi C, Scardino E, Wolberg A, Lawson J, Hoffman M, Sullenger B (2001) Generation of species cross-reactive aptamers using "Toggle" SELEX. Mol Ther 4:567–573
Wiegand TW, Williams PB, Dreskin SC, Jouvin MH, Kinet JP, Tasset D (1996) High-affinity oligonucleotide ligands to human IgE inhibit binding to Fc epsilon receptor I. J Immunol 157:221–230
Williams KP, Liu X-H, Schumacher TNM, Lin HY, Ausiello DA, Kim PS, Bartel DP (1997) Bioactive and nuclease-resistant L-DNA ligand of vasopressin. Proc Natl Acad Sci USA 94:11285–11290
Wilson C, Szostak JW (1998) Isolation of a fluorophore-specific DNA aptamer with weak redox activity. Chem Biol 5:609–617
Wilson DS, Szostak JW (1999) In vitro selection of functional nucleic acids. Annu Rev Biochem 68:611–647
Wilson C, Nix J, Szostak J (1998a) Functional requirements for specific ligand recognition by a biotin-binding rna pseudoknot. Biochemistry 37:14410–14419
Wilson DH, Groskopf W, Hsu S, Caplan D, Langner T, Baumann M, Demanno D, Manderino G (1998b) Rapid, automated assay for progesterone on the Abbott AxSYM(TM) analyzer. Clin Chem 44:86–91
Win MN, Klein JS, Smolke CD (2006) Codeine-binding RNA aptamers and rapid determination of their binding constants using a direct coupling surface plasmon resonance assay. Nucleic Acids Res 34:5670–5682
Wlotzka B, Leva S, Eschgfaller B, Burmeister J, Kleinjung F, Kaduk C, Muhn P, Hess-Stumpp H, Klussmann S (2002) In vivo properties of an anti-GnRH Spiegelmer: an example of an oligonucleotide-based therapeutic substance class. Proc Natl Acad Sci USA 99:8898–8902
Wu Z-S, Guo M-M, Zhang S-B, Chen C-R, Jiang J-H, Shen G-L, Yu R-Q (2007) Reusable electrochemical sensing platform for highly sensitive detection of small molecules based on structure-switching signaling aptamers. Anal Chem 79:2933–2939
Xiao Y, Lubin AA, Heeger AJ, Plaxco KW (2005a) Label-free electronic detection of thrombin in blood serum by using an aptamer-based sensor. Angew Chem Int Ed Engl 44:5456–5459
Xiao Y, Piorek BD, Plaxco KW, Heegert AJ (2005b) A reagentless signal-on architecture for electronic, aptamer-based sensors via target-induced strand displacement. J Am Chem Soc 127:17990–17991
Xu D, Xu D, Yu X, Liu Z, He W, Ma Z (2005) Label-free electrochemical detection for aptamer-based array electrodes. Anal Chem 77:5107–5113
Xu Y, Yang L, Ye X, He P, Fang Y (2006) An aptamer-based protein biosensor by detecting the amplified impedance signal. Electroanalysis 18:1449–1456
Yakimovich OY, Alekseev YI, Maksimenko AV, Voronina OL, Lunin VG (2003) Influence of DNA aptamer structure on the specificity of binding to Taq DNA polymerase. Biokhimiya 68:274–282
Yallow RS, Berson SA (1959) Assay of plasma insulin in human subjects by immunological methods. Nature 185:1648–1649
Yamamoto R, Kumar PKR (2000) Molecular beacon aptamer fluoresces in the presence of Tat protein of HIV. Genes Cells 5:389–396
Yamana K, Ohtani Y, Nakano H, Saito I (2003) Bis-pyrene labeled DNA aptamer as an intelligent fluorescent biosensor. Bioorg Med Chem Lett 13:3429–3431

Yan X, Gao X, Zhang Z (2004) Isolation and characterization of 2′-amino-modified RNA aptamers for human TNFalpha. Genomics Proteomics Bioinformatics 2:32–42

Zayats M, Huang Y, Gill R, Ma C-A, Willner I (2006) Label-free and reagentless aptamer-based sensors for small molecules. J Am Chem Soc 128:13666–13667

Zhang F, Anderson D (1998) In vitro selection of bacteriophage φ29 prohead RNA aptamers for prohead binding. J Biol Chem 273:2947–2953

Zimmerman G, Jenison R, Wick C, Simorre J, Pardi A (1997) Nat Struct Biol 4:644–649

Zinnen SP, Domenico K, Wilson M, Dickinson BA, Beaudry A, Mokler V, Daniher AT, Burgin A, Beigelman L (2002) Selection, design, and characterization of a new potentially therapeutic ribozyme. RNA 8:214–228

Zuo X, Song S, Zhang J, Pan D, Wang L, Fan C (2007) A target-responsive electrochemical aptamer switch (TREAS) for reagentless detection of nanomolar ATP. J Am Chem Soc 129:1042–1043

Chapter 18
Phage Display Technology in Biosensor Development

Scott C. Meyer and Indraneel Ghosh

Abstract The utility of phage display technology in biosensor development lies in the ability to rapidly discover novel peptide and protein modules that serve as molecular detection domains in sensor architectures. Phage display allows for connecting a protein or peptide (phenotype) to the DNA (genotype) that encodes it, allowing users to screen billion member libraries of compounds against targets of interest, while using the genetic information of the virus for both output identification and amplification. This process known as in vitro selection mimics natural selection through the use of selective pressure followed by the amplification of the selected mutant in a bacterial host cell. Because of the robust nature of phage, the output virions from a selection can sometimes be used directly in a biosensor. More commonly, the genetic information contained within the phage can be translated to a peptide, protein, or antibody that serves as the recognition domain in a biosensor. Several modalities of phage display have been utilized for discovering receptors for biosensors, including peptide display (low and high valency), antibody display, and small protein display. These phage display approaches have greatly aided in the development of novel biosensor receptors.

Keywords Phage display · M13 filamentous bacteriophage · In vitro selection · Peptide library · Streptavidin · Avidin · Landscape phage · Antibody · Protein scaffold · Protein grafting · Zinc finger

I. Ghosh (✉)
Department of Chemistry, University of Arizona, Tucson, AZ 85721-0041, USA
e-mail: ghosh@u.arizona.edu

M. Zourob (ed.), *Recognition Receptors in Biosensors*,
DOI 10.1007/978-1-4419-0919-0_18,

Abbreviations

DNA	Deoxyribonucleic acid
GFP	Green fluorescent protein
cDNA	complementary DNA
ELISA	Enzyme-linked immunosorbent assay
FACS	Fluorescence-activated cell sorting
CMV	Cucumber mosaic virus
SEB	Staphylococcal enterotoxin B
PCR	Polymerase chain reaction
HRP	Horseradish peroxidase
MSMC	Magnetostrictive microcantilever
AFM	Atomic force microscopy
CDR	Complementarity determining regions
scFv	Single chain variable fragment
phOx	2-phenyloxazol-5-one
SPR	Surface plasmon resonance
HIV	Human immunodeficiency virus
NMR	Nuclear magnetic resonance
APP	Avian pancreatic polypeptide
IgG	Immunoglobulin G
SEER	SEquence Enabled Reassembly
MBD	Methyl binding domain

18.1 Introduction

Phage display is a versatile and robust methodology (Smith 1985) that utilizes bacteriophage particles for the determination and manipulation of protein and peptide interactions (Barbas et al. 2001). Bacteriophage are viruses that exclusively infect bacteria, such as the commonly used laboratory strains of *E. coli*. The success of phage display stems from the physical connection between phenotype (the peptide or protein attached to the phage surface) and its genotype (encoded in the phage's genetic material). The use of filamentous phage as a platform for the display of recombinant proteins and peptides was first described by Smith in 1985 and has subsequently developed into a burgeoning field. Today, phage display is commonly used in a wide variety of fields involving in vitro evolution, which includes protein-drug development, enzyme optimization, and in the development of the recognition handle in biosensors.

By necessity, here, the description of phage display will be brief, but for a detailed account, we recommend a few excellent resources, including a textbook by Barbas et al. (2001) and two thorough reviews (Kehoe and Kay 2005; Petrenko et al. 1996). Several phage platforms have been used for phage display, including T4, T7, and λ phage; however, M13 (and the related fd) filamentous phage are the most common. For the purpose of simplicity, this chapter focuses primarily on M13

Table 18.1 Genomic proteins of the M13 bacteriophage (Kehoe and Kay 2005)

Protein	Function	Size (a.a.)
pI	Phage assembly	348
pII	DNA replication	410
pIII	Structural protein (minor coat protein, 3–5 copies)	406
pIV	Phage assembly	405
pV	Single-stranded DNA binding/transport	87
pVI	Structural protein (minor coat protein, 3–5 copies)	112
pVII	Structural protein (minor coat protein, 3–5 copies)	33
pVIII	Structural protein (major coat protein, ~2,700 copies)	50
pIX	Structural protein (minor coat protein, 3–5 copies)	32
pX	DNA replication	111
pXI	Phage assembly	108

filamentous phage display, though other phage systems will be mentioned. Filamentous phage are named as such because of their dimensions: M13 phage are cylindrical in shape with dimensions of approximately 1,000 nm in length by 6 nm in diameter (Petrenko et al. 1996). Unlike some bacteriophage, M13 are secreted in a nonlytic fashion, i.e., they do not kill the host cells during excretion to the media. Each wild-type M13 phage particle consists of single-stranded genomic DNA (6407 bases) encoding 11 proteins, denoted pI – pXI (Table 18.1). Of these, the major coat protein (pVIII) is the most abundant, with 2700 copies in the wild-type M13, though that number varies with the length of the DNA contained in the phage particle (Kehoe and Kay 2005). On one end of the phage, there are three to five copies each of pVII and pIX, while at the other end, there are three to five copies of pIII and pVI (Fig. 18.1a).

In the most common forms of phage display, a recombinant protein or peptide of interest is fused to the N-terminus of either the pIII protein for monovalent or low valency phage display or pVIII for multivalent display. In both pIII and pVIII display, the wild-type function of the phage proteins can be retained, though with certain constraints. For instance, adding more than ~8 residues to the pVIII protein often disrupts phage production; however, large proteins can be displayed correctly by including copies of wild-type pVIII along with fusion proteins (Petrenko et al. 1996). Similarly, a mixture of wild-type and fusion pIII proteins can be used to achieve predominantly monovalent display as an alternative to pentavalent pIII display (Fig. 18.1b). Here, an excess of wild-type pIII is mixed with the recombinant pIII such that each phage particle displays one or zero copies of the fusion pIII protein, while the other copies are wild-type. In this way, monovalent display of the peptide or protein can be achieved.

M13 phage particles are assembled in the periplasm of *E. coli*, which is a nonreducing environment and contains several proteases (Gottesman 1996) (Fig. 18.2). This localization to the periplasmic space is mediated by N-terminal peptide signaling sequences for both pIII and pVIII proteins. The signaling sequences are subsequently proteolytically cleaved in the *E. coli* periplasm (Petrenko et al. 1996). The nonreducing environment of the periplasm allows for the formation of inter- and intramolecular disulfide bonds, which are not usually possible in the reducing environment of the cytosol. Hence, the periplasmic

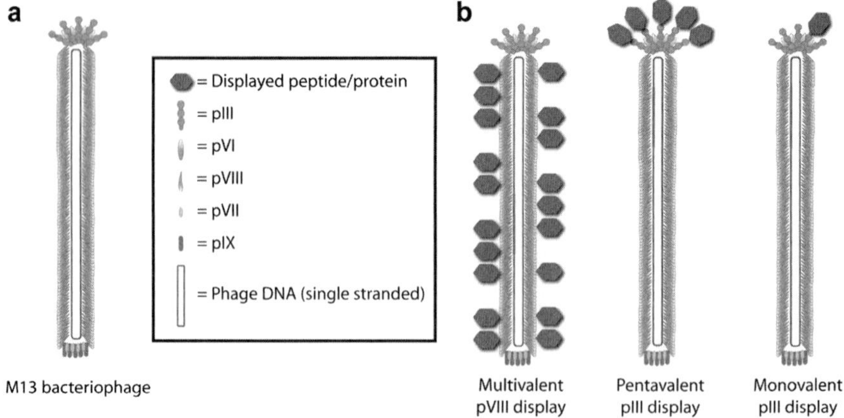

Fig. 18.1 Phage display with M13 filamentous bacteriophage. (**a**) The M13 bacteriophage consists of a protein coat made of the major coat protein, pVIII, and four minor coat proteins, pIII and pVI on one end, and pVII and pIX on the other. (**b**) In the most common forms of phage display, peptides or proteins of interest can be displayed on the surface of phage as multivalently as pVIII fusions, pentavalently as pIII fusions, or monovalently as pIII fusions. The genetic information for the displayed peptide or protein is contained within the phage particle itself

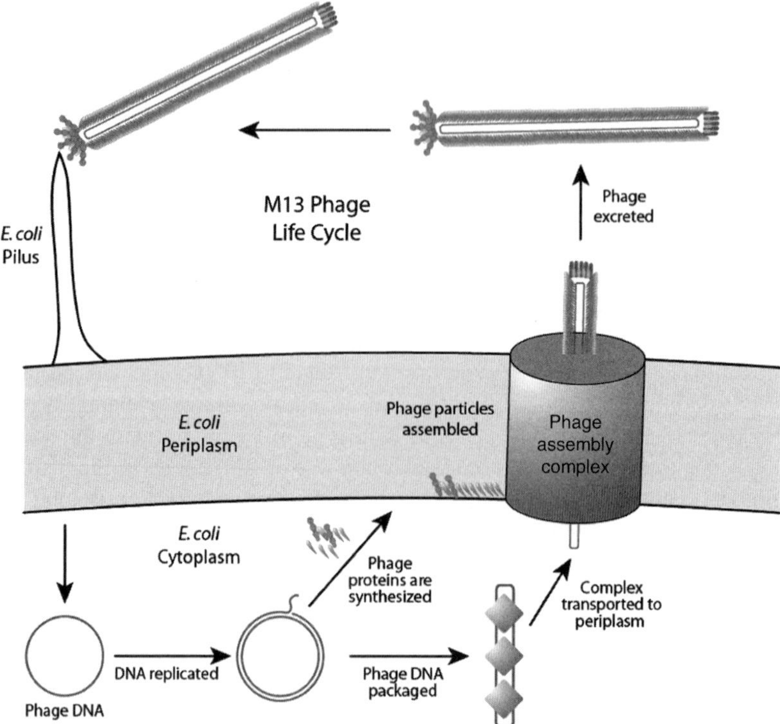

Fig. 18.2 M13 phage life cycle. The phage virions infect *E. coli* through the pilus and the single-stranded phage DNA is transported to the cytoplasm. Phage DNA is replicated and phage proteins are synthesized, all of which are then transported to the periplasm. Phage particles are assembled around the single stranded phage DNA and mature virions are non-lytically excreted, leaving the *E. coli* cell intact

assembly of the phage particles is often exploited to display cyclic peptides via the formation of a single intramolecular disulfide bond (discussed in Sect. 18.2) (O'Neil et al. 1992). Additionally, the nonreducing environment aids in the display of some eukaryotic proteins that rely on correct formation of disulfide bonds to achieve their native fold (Bass et al. 1990). As for the size limit of proteins that can be displayed on phage, an 86 kDa heterodimeric enzyme, penicillin G acylase, has been reported to be functionally displayed on both the pIII and pVIII proteins (Verhaert et al. 1999). Still, penicillin G acylase marks an extreme case, since most proteins displayed on phage are usually single chain antibodies, small proteins, or peptide libraries (Petrenko et al. 1996). Though it is a versatile and robust platform, not all proteins can be functionally displayed on phage, and it is difficult to make broad predictions about which species can be successfully displayed. For example, some variants of the green fluorescent protein (GFP) that are readily expressed in the cytosol of *E. coli* have been difficult to display functionally on phage (Paschke and Hohne 2005).

18.2 In Vitro Selections: Peptide Phage Display

18.2.1 Introduction to In Vitro Selections

The most common application of phage display in biotechnology has been in vitro selections for the development of protein or peptide ligands and receptors from biological libraries. In vitro selection, also referred to as "directed evolution", exploits the connection between phenotype and genotype, allowing one to iteratively screen and amplify protein or peptide libraries. In this way, novel protein or peptide interactions can be discovered from a pool of randomized or biologically-derived variants. In vitro selection is an important tool in the field of biosensors because it facilitates the production of novel ligand-receptor systems, thereby extending the utility of biosensors beyond known biomolecular interactions.

For in vitro selections utilizing phage display, a library of proteins or peptides is constructed through the randomization of a gene of interest, or by incorporation of a biologically obtained genomic or immunoglobulin cDNA (Petrenko et al. 1996). This DNA library is then ligated to the DNA of the phage coat protein (usually pIII for low valency phage display or pVIII for high valency display). The resulting recombinant fusion protein is then expressed on the surface of the phage particle, with the corresponding DNA encapsulated inside. For phage libraries, it is important to note that each phage particle contains the DNA for the one specific library member that is displayed on the phage coat protein. This one-to-one correspondence allows for billions of library members to be simultaneously screened, with the genetic information providing a means to amplify, isolate, and identify the selected library members of interest.

In a typical phage display selection (Fig. 18.3), the target is immobilized onto a solid support and the library-containing phage are washed over, allowing for capture

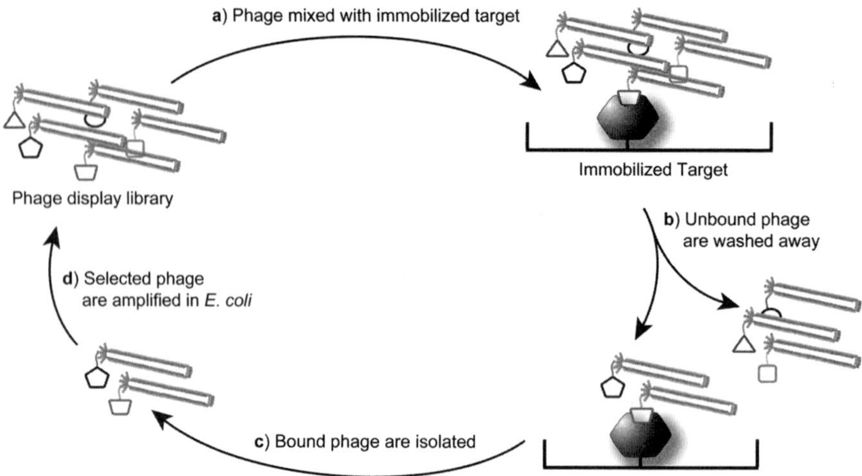

Fig. 18.3 Typical phage display selection. (**a**) The phage display library is exposed to an immobilized target of interest. (**b**) After immobilization, the weakly bound phage are washed away, leaving only the tightly bound phage. (**c**) These phage are isolated and used to infect *E. coli*. (**d**) Growing the infected *E. coli* amplifies the selected phage, which can then be resubjected to the immobilized target

of library members by the immobilized target. Weakly bound phage are washed away, leaving the more tightly bound phage to be isolated. *E. coli* are then infected with the bound phage and grown to achieve amplification of the selected library members. The newly amplified phage are then isolated and resubjected to the target in a subsequent round of selection. Through this iterative screening and amplification process, phage particles displaying variants that bind the target more tightly are favored over those variants that interact weakly or not at all. In this way, phage display selections mimic the process of natural selection, though in a controlled manner.

Typically, several rounds of selection (> 3) are necessary to identify library members that exhibit high target binding affinity. After a given round of selection, *E. coli* that are infected with the output phage can be plated with dilutions such that individual colonies can be isolated. DNA encoding individual library members is then isolated from single colonies and sequenced. When a narrow distribution of library members has been achieved, similar sequences appear in the selection output, thus providing consensus motifs. These sequences can then be cloned into an expression vector and recombinantly expressed, synthesized via solid phase peptide synthesis, or utilized directly as phage fusions in biosensor applications.

18.2.2 Peptide Phage Display

An early application of phage display was the use of randomized peptide libraries for the in vitro selection of peptide ligands. Randomized peptide libraries can be constructed at the DNA level as fusions to phage coat proteins by synthesizing oligonucleotides with a mixture of nucleotides at the desired positions of diversity. The mixture of oligonucleotides can then be manipulated via standard cloning methods to produce functional phage coat protein fusions. The most common method in peptide fusion construction is the use of cassette mutagenesis (Petrenko et al. 1996). In many cases, the phage coat proteins are manipulated either in the form of plasmids encoding the phage coat protein of interest or by direct modification of the phage genome. In either case, the randomized oligonucleotide library is extended to double-stranded DNA by Polymerase Chain Reaction (PCR) amplification with appropriate primers or with nonamplifying polymerization (for instance with the Klenow fragment of DNA polymerase I (Klenow and Hennings 1970)). The full length double stranded DNA can then be cloned into the phage DNA using appropriate restriction enzymes and subsequent ligation.

Given the degeneracy of the genetic code, complete randomization of all three positions of the codon is not necessary to achieve incorporation of all twenty naturally occurring amino acid residues (Fig. 18.4). A common strategy is to include all four nucleotides (N) in the first two positions of the codon with only G and C (S) in the third position, resulting in an NNS triplet for all randomized positions. Besides covering all twenty amino acids, this mixture excludes two of the three stop codons, leaving only the amber stop codon (TAG). Limiting the stop codons in a library is important, since a stop in the middle of a peptide will disrupt coat protein synthesis (in the case of N-terminal library fusions) or truncate the

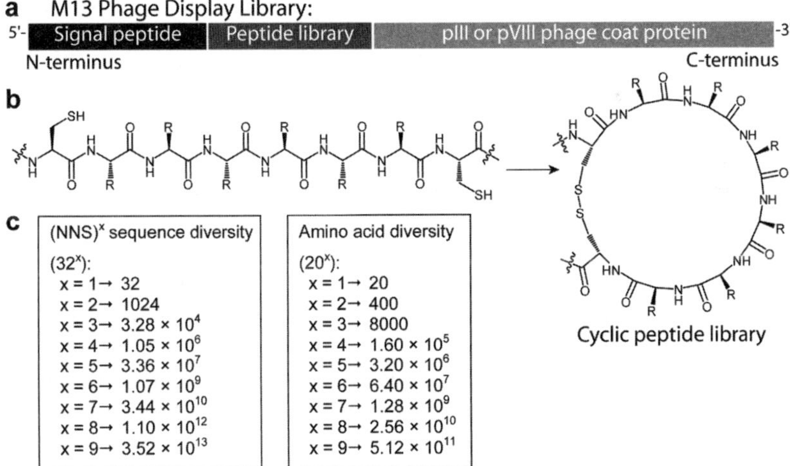

Fig. 18.4 Peptide phage display. (**a**) The M13 phage display library gene consists of an N-terminal signal peptide, followed by the randomized peptide library and the rest of the pIII gene. (**b**) A cyclic peptide library can be accessed by incorporation of two conserved cysteine residues, which form an intramolecular disulfide bond in the periplasm of *E. coli*. (**c**) The theoretical diversity of a peptide library increases rapidly as the number of amino acid residues being randomized (x) increases. However, the theoretical diversity of the amino acids increases more slowly than the genetic diversity due to the degeneracy of the genetic code

displayed peptide (for C-terminal library fusions). Randomized libraries can also be biased toward a subset of amino acid residues (e.g. hydrophobic, polar, etc.) through the use of other nucleotide mixtures. In addition to the elimination of stop codons, limiting the number of possible sequences of a library is important if complete coverage (hard library) of the theoretical library is desired. The total number of possible library members can be simply calculated by multiplying the number of degenerate nucleotides at every randomized position. For example for the NNS mixed codon strategy the total individual library members can be easily calculated by the equation below:

$$y = (N \times N \times S)^x \Rightarrow (4 \times 4 \times 2)^x = (32)^x \quad (1)$$

where y is the total theoretical library size and x is the number of amino acid residues that are being randomized. The limiting experimental factor for library size in phage display is the insertion of the recombinant library DNA into the host *E. coli* and propagation of *E. coli* cultures, which usually confines the library size to ~10^9–10^{10} members in most cases. This translates to a maximum of six completely randomized amino acid residues with the NNS mixed codon set ($x = 6$, $y = 1.07 \times 10^9$ members). If a more restricted codon set is used, a greater number of residues can be randomized while still enabling complete library coverage. However, many investigators sacrifice complete coverage in favor of randomizing a large number of residues ($x > 8$) resulting in a library that contains only a small fraction of its potential diversity (Kay et al. 1993). An incomplete library is often more than sufficient to identify useful tight-binding motifs, which can be used as is or later optimized by further phage-display selection.

In addition to the size of the peptide library, many researchers have also varied the morphology of the peptide. While the utilization of complex protein scaffolds for peptide phage display will be discussed in Sect. 18.4, an important modification of peptide morphology is the cyclization of peptides (O'Neil et al. 1992). Cyclic peptides are often more resistant to proteolysis, and have been shown to bind their targets with much higher affinity than their linear counterparts in many cases (Giebel et al. 1995; Meyer et al. 2006). This increase in binding affinity is likely due to the exclusion of unfavorable binding conformations and entropic effects. In phage display, cyclization is achieved through the incorporation of two conserved cysteine residues flanking an appropriate length of "randomized" amino acid residues. Upon phage assembly in the periplasm, the conserved cysteine residues spontaneously form a disulfide bond, thus cyclizing the displayed peptide (Fig. 18.4b).

18.2.3 Phage Display Targeting Biotin-Binding Proteins

Some of the earliest in vitro phage display selections were focused on the development of peptide ligands for streptavidin (Devlin et al. 1990), a biotin binding protein that has been utilized in the development of numerous biosensors (Hengsakul and Cass 1997; Lin et al. 1997; Samoylov et al. 2002). Streptavidin is a homotetrameric bacterial protein (species, *streptomyces*), which binds the small molecule biotin with extremely high affinity (Green 1990). There is also a eukaryotic glycoprotein analog of streptavidin, called avidin, that can be isolated from hen egg white and binds biotin with a similarly high affinity (the K_Ds for the (strept)avidin-biotin complexes are ~10^{-15} M). Researchers often exploit the high affinity of (strept) avidin-biotin interactions for the immobilization of biotin-tagged proteins, peptides, dyes, or a variety of other molecules. Furthermore, the specificity of the (strept)avidin-biotin interactions generally provides low backgrounds during immobilization (Finn and Hofmann 1990), hence the utility in biosensor applications.

The availability and utility of the biotin-binding proteins has prompted several groups to target them with phage display selections (Devlin et al. 1990; Giebel et al. 1995; Kay et al. 1993; McLafferty et al. 1993; Petrenko and Smith 2000). An early example of this, indeed one of the earliest examples of peptide phage display, was carried out by Devlin et al (1990), against streptavidin. These authors discovered a unique consensus motif containing the sequence HPQ, which has subsequently been found in various forms by several groups. Although many selections against streptavidin resulted in sequences that contained HPQ as a major consensus motif, the flanking sequences varied widely, with other motifs also being isolated (Table 18.2). Additionally, selections that were carried out with cyclic peptide libraries produced ligands that showed much higher affinity upon cyclization (Giebel et al. 1995; Zang et al. 1998).

Interestingly, several peptides that contain the HPQ consensus motif for streptavidin have been shown to have poor affinity for avidin in both its native and deglycosylated states (Kay et al. 1993). This orthogonality provided the potential for a unique system where two receptors share high affinity for a

Table 18.2 Streptavidin- and avidin-binding peptide motifs discovered by phage display

Sequence	Morphology	Reference
Streptavidin:		
HPQ	pIII, linear	Devlin et al. (1990) and Kay et al. (1993)
	pIII, cyclic	Giebel et al. (1995) and McLafferty et al. (1993)
	T7 phage (linear and cyclic)	Krumpe et al. (2006)
EPDW(F/Y)	pIII, linear	Caparon et al. (1996)
GD(F/W)XF	pIII, linear	Roberts et al. (1993)
PWXWL	pIII, linear	Roberts et al. (1993)
VP(E/D)(G/S)AF	pVIII linear, multivalent	Petrenko and Smith (2000)
Deglycosylated Avidin:		
D(R/L)A(S/T)P(Y/W)	pIII, cyclic	Meyer et al. (2006)
D(P/L)A(S/T)P(Y/W)	T7 phage, cyclic	Krumpe et al. (2006)
VPEY(T)	pVIII linear, multivalent	Petrenko and Smith (2000)

common ligand (biotin), while differentially binding a lower affinity ligand (the HPQ peptide motif). Later, peptide ligands for the deglycosylated form of avidin were discovered in both monovalent (Krumpe et al. 2006; Meyer et al. 2006) and multivalent phage display selections (Petrenko and Smith 2000). Binding studies showed that these peptides did not bind to streptavidin, thus providing a fully orthogonal system that could find utility in the biosensor field (Gaj et al. 2007).

18.2.4 Peptide Phage Display in the Detection of Toxins

The utility of phage display in biosensor applications arises from the ease of development of ligands for a given biological target, with only minimal preliminary information required. In contrast to design approaches, phage display can rapidly provide ligands for a wide variety of novel targets. This ease of development lends itself well to cases where the rapid development of diagnostic tools is essential, for instance, in the detection of environmental toxins (Petrenko and Sorokulova 2004; Turnbough 2003). A short list of toxins targeted by phage display is given in Table 18.3, though a more detailed list can be found elsewhere (Petrenko and Sorokulova 2004).

The examples of targeting toxins with phage display summarized in Table 18.3 demonstrate the ease of translating results from in vitro selections to detection platforms. For instance, peptides derived from phage display can be used for the detection of an analyte by labeling a fluorescent probe with the selected peptide (Williams et al. 2003), or by labeling the phage particles displaying the selected peptides with a fluorescent dye (Knurr et al. 2003). These tools can then be used in

Table 18.3 Peptide phage display targeting biological toxins in biosensors (Petrenko and Sorokulova 2004)

Target	Analytical Test	Sensitivity	Reference
Staphylococcal enterotoxin B (SEB)	ELISA, Fluoroimmunoassay	1.4 ng/well	Goldman et al. (2000)
Cucumber mosaic virus (CMV)	ELISA, dot blot	5 ng CMV protein	Gough et al. (1999)
Bacillus subtilis spores	FACS w/dye labeled phage	$< 10^7$ spores	Knurr et al. (2003)
Bacillus anthracis spores	FACS w/ peptide labeled fluorescent protein	$< 10^7$ spores	Williams et al. (2003)
M. paratuberculosis	Capture PCR	10–200 cells/mL	Stratmann et al. (2002)
Salmonella typhimurium	ELISA and FACS	$K_{D(apparent)} = 106$ pM	Sorokulova et al. (2005)

Fluorescence Activated Cell Sorting (FACS), microscopy, or other applications. Furthermore, phage can be utilized directly in Enzyme-Linked Immunosorbent Assay (ELISA) with an anti-phage antibody conjugated to a reporter, such as horseradish peroxidase (HRP) (Gough et al. 1999; Sorokulova et al. 2005).

All of these methods focus more on the interaction between the analyte and the phage rather than the nature of that interaction itself. In an interesting application, Belcher and colleagues used phage display to select peptides that bound various metals (Whaley et al. 2000). Subsequently, phage particles displaying these peptides were utilized to form various magnetic and semiconducting nanostructures (Mao et al. 2004). The use of the phage displayed peptides in applications ranging from materials to biosensors implies that a molecular-level analysis of the binding interactions is often unnecessary. When a detectable interaction between an analyte and a signal generating species is required, phage can frequently be used directly in the sensor itself.

18.2.5 Landscape Phage and the Detection of Salmonella typhimurium

Important examples of phage particles as components of biosensors can be found in the use of multivalent "landscape phage" developed by (Petrenko et al. 1996) (Fig. 18.5). In this application, peptides are displayed on every copy of the pVIII phage coat protein. Given the relatively rigid structure of the phage coat, the peptides are displayed in defined three dimensional orientation that can be regarded as an "organic landscape". Furthermore, by displaying the peptide

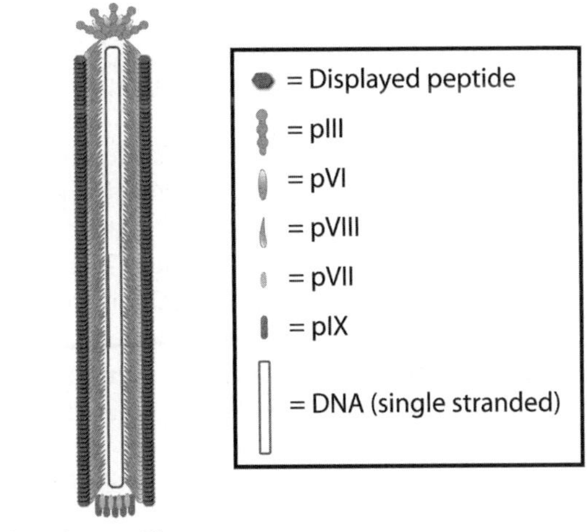

Fig. 18.5 In landscape phage display, every copy of the pVIII phage coat protein displays the peptide library member. The close-packed nature of the phage coat restricts the conformations accessible to the displayed peptide and increases avidity via multivalent interactions

on the major coat protein of the phage, one can take advantage of the avidity effect usually seen with multivalent ligand binding to a surface with multiple epitopes. Landscape phage selections provide an expedient path to the discovery of ligands for novel targets and authors have speculated that the landscape phage could serve as substitutes for antibodies (Petrenko and Smith 2000). However, recent advances in the development of antibody fragments and their ubiquity in biotechnology will provide stiff resistance. Though individual antibodies are likely to be superior to individual members of a landscape phage library, the defined orientation and extremely high multivalency of the displayed peptides have the possibility of replacing antibodies with landscape phage in some biosensor applications that require high affinity but can tolerate lower specificity (Petrenko and Smith 2000; Petrenko 2008).

One field where the multivalency of landscape phage provides a tangible advantage is in the recognition of cell surfaces. By their very nature, cell surfaces, including bacterial cells, tend to display multiple copies of various receptors. Therefore, multivalent ligand platforms, such as landscape phage, have the potential to bind to these cell surfaces tightly due to the multivalent avidity effect. With this in mind, Petrenko and coworkers have developed landscape phage targeting the disease causing bacterium *Salmonella typhimurium* (Sorokulova et al. 2005). In vitro selection with a pVIII phage library produced a number of consensus peptides. Landscape phage displaying this peptide were shown to bind *Salmonella typhimurium* better than the bound control bacteria and were active in both ELISA and FACS assays (Petrenko and Sorokulova 2004; Sorokulova et al. 2005). Subsequent work led to the development of an acoustic wave biodetector, a magnetoelastic sensor, and a magnetostrictive microcantilever (MSMC) for *salmonella* detection (Fu et al. 2007; Lakshmanan et al. 2007; Olsen et al. 2006). These examples highlight the success of the landscape phage approach since the resulting biosensors were able to detect low levels of *Salmonella typhimurium*. By contrast, another group carried out a low valency peptide phage display against the lipopolysaccharide portion of *salmonella*'s outer membrane (Kim et al. 2005) that also produced phage that selectively recognize the bacterium, though with lower signal to noise in subsequent assays.

In addition to binding motifs derived from randomized phage display libraries, many bacteria are infected by known wild-type bacteriophages. For instance, *Salmonella typhimurium* is targeted by the bacteriophage P22, which has been used to detect the bacterium in a biosensor (Handa et al. 2008). In this work, the authors immobilized P22 in a monolayer and selectively captured *salmonella* bacteria on the phage-coated surface, followed by interrogation with ELISA and AFM. Additionally, bacteriophage P22 has also been fluorescently labeled and utilized for selective imaging of *salmonella* (Mosier-Boss et al. 2003). These examples illustrate the utility and versatility of phage in the detection of a target of biological interest, namely the *Salmonella typhimurium* bacterium.

18.3 Antibody Phage Display

18.3.1 Introduction to Antibody Phage Display

Since the first demonstration by Smith in 1985, phage display was adapted to display antibodies and antibody fragments. Antibodies (immunoglobulins) are glycoprotein complexes consisting of two heavy chains and two light chains, all of which are covalently attached through specific intermolecular disulfide bonds (Fig. 18.6). As part of the eukaryotic immune response, antibodies are produced to bind to a specific target (antigen). Most of the diversity found in different antibodies is localized to three hypervariable or complementarity determining regions (CDR) on the variable domains of the heavy and light chains (V_H and V_L, respectively). These CDRs are loops that vary greatly in sequence depending on the antigen, and the variable domains (V_H and V_L) essentially provide a scaffold for antigen binding. The variable regions of antibodies can be assembled in various forms, including the Fab portion and the single chain variable fragment (scFv), which still bind their purported antigen. These truncations often facilitate

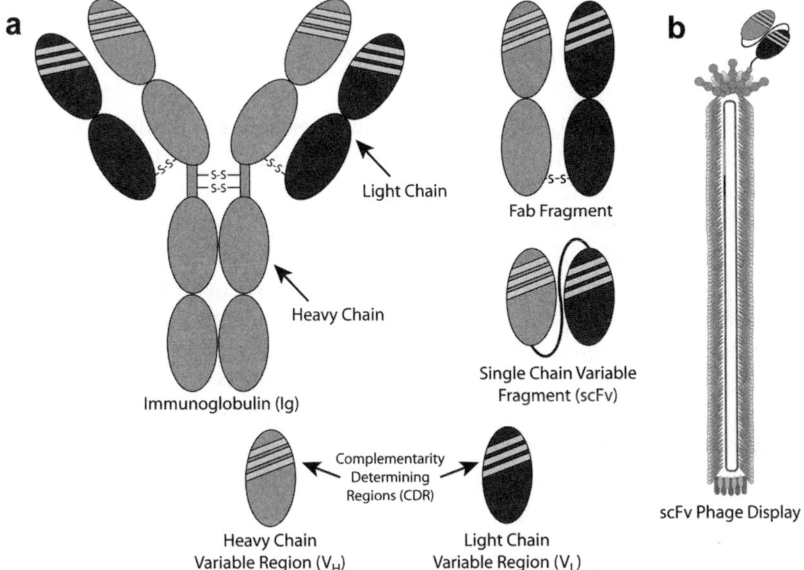

Fig. 18.6 Antibody phage display. (**a**) Immunoglobulins (Ig) consist of two heavy chains (*gray*) and two light chains (*dark gray*) covalently tethered by intermolecular disulfide bonds. The diversity of immunoglobulins is concentrated on the Complementary Determining Regions (CDR) of the heavy and light chain. The domain containing these CDRs is called the variable region. Antibodies are commonly dissected into fragments containing the requisite variable regions in either the Fab or single chain variable fragment (scFv) format. (**b**) Phage display is typically carried out with either Fab or scFv fragments of antibodies. A representative phage display format (scFv displayed monovalently on the pIII protein) is shown

the use of antibodies in certain applications, such as antibody drugs and phage display.

Typically, antibodies are isolated as mixtures of different immunoglobulin clones, which consist of many distinctive antibodies that are capable of binding the same antigen. In some cases, these polyclonal mixtures may be beneficial, however, it is often desirable to isolate one particular chemical species – a monoclonal antibody. With phage display, it is possible to generate high affinity monoclonal antibodies from polyclonal pools. Given the low valency nature of most immunoglobulins, the display modality of choice for antibodies is as pIII fusions. Display of complete antibodies on the surface of phage, while possible, is generally not desirable due to their large size and extensive disulfide bonding requirements. As an alternative, fragments of antibodies are often displayed on the surface of phage and several groups have developed methods for doing so (Barbas et al. 1991; Clackson et al. 1991; Hoogenboom et al. 1991; Krebber et al. 1997; McCafferty et al. 1990; Sheets et al. 1998). Antibody phage libraries can be constructed by PCR amplification of the variable regions of immunoglobulins from immunized animal sources, from immune human donors, or from "naïve" pools of immunoglobulins (Hoogenboom et al. 1998).

18.3.2 Early Examples of Antibody Phage Display

Several reviews of antibody phage display have been written (Bradbury and Marks 2004; Burton 1993; Griffiths 1993; Griffiths and Duncan 1998; Hoogenboom et al. 1998; Hudson 1998; Rader and Barbas 1997; Soderlind et al. 1992; Winter et al. 1994) and therefore only a few examples will be highlighted here. It should be noted that monoclonal antibody fragments derived from phage display are just as applicable to biosensors as antibodies derived using other methods (Cooper 2002). One of the first described phage display antibody enrichments was demonstrated in 1990 by Winter and coworkers using a single chain variable fragment (scFv) from the hen egg-white lysozyme antibody (McCafferty et al. 1990). A fusion between the pIII phage protein and the heavy chain and light chain variable region of the antibody (covalently tethered with a glycine and serine linker) was incorporated into the phage particle. It was then possible to isolate the antibody-containing phage from a mixture with parent phage (no antibody) after two rounds, even when outnumbered at a ratio of a million to one. This million-fold enrichment showed that phage bearing tight-binding antibodies could be isolated from a mixture at ratios reflective of a highly diverse library. The same group later isolated high affinity scFv antibody fragments for the hapten phOx (2-phenyloxazol-5-one) from a pool of antibodies derived from immunized mice (Clackson et al. 1991) showing that high affinity, monoclonal antibody fragments could be isolated from a polyclonal pool.

Subsequently, many groups have extended antibody phage display to single chain variable fragments and disulfide tethered Fab fragments (Barbas et al.

1991; Hoogenboom et al. 1991). For disulfide tethered Fab fragments, the variable portion of the light and heavy chains of the immunoglobulin proteins (one of which is fused to the pIII phage protein) are exported to the periplasm, where they dimerize and form a covalent disulfide bond, followed by incorporation into the assembling phage. Both of the chains are encoded in the phage-packaged DNA, and hence the connection between the genotype and phenotype of the displayed Fab fragments is maintained.

18.3.3 Phage Selected Antibodies in the Detection of Toxins

Though a comprehensive list of antibody phage display targets would be beyond the scope of this chapter, a short list of biological toxin targets is given in Table 18.4. These antibodies have demonstrated the utility in various biosensor technologies (Petrenko and Sorokulova 2004), most notably in ELISA and Surface Plasmon Resonance (SPR) applications. Phage particles can also be directly used in biosensors as receptors (see Sect. 18.2) and labeled phage have been used as optical markers in FACS or microscopy. The large size of the phage particles provides many potential sites for dye labeling, thus providing an advantage over the labeled antibodies alone. Additionally, it is possible to take advantage of the multiple surfaces of filamentous phage to label one surface (pVIII) with particles as large

Table 18.4 Antibody phage display targeting biological toxins in biosensors (Petrenko and Sorokulova 2004)

Target	Format	Analytical Test	Reference
Botulinum neurotoxins	Fab	ELISA, FACS, SPR	Emanuel et al. (1996) and Emanuel et al. (2000)
Human immunodeficiency virus (HIV) gp120 protein	Fab	ELISA, SPR	Barbas et al. (1994) and Burton et al. (1991)
Puumala virus	Fab	Immunoblot, ELISA	Salonen et al. (1998)
Measles virus	Fab	ELISA, Radioimmunoprecipitation	Nicacio et al. (2002)
Hepatitis C virus	scFv	ELISA	Chan et al. (1996)
Hepatitis C virus	Fab	ELISA	Plaisant et al. (1997)
Human cytomegalovirus	scFv	SPR	Pini et al. (1997)
Bacillus subtilis	scFv	ELISA	Zhou et al. (2002)
Salmonella serogroup B, O-polysaccharide	scFv	ELISA, SPR	Deng et al. (1994) and Deng et al. (1995)
Brucella melitensis	scFv	ELISA	Hayhurst et al. (2003)
Listeria monocytogenes	scFv	Bioelectrochemical sensors	Benhar et al. (2001)
Mycobacterium tuberculosis (MtKatG protein)	scFv	Bioelectrochemical sensors	Benhar et al. (2001)

as semiconductor quantum dots, while leaving another surface (pIII) free for binding to a target of choice (Chen et al. 2004).

18.3.4 Lab on a Chip Applications

Phage have also been used in chip-based technologies as receptors for various analytes (Liao et al. 2006). Chip-based biosensors take advantage of miniaturization of detection components to maximize sensitivity with small sample sizes. Phage have proven to be very useful in these applications due to the ease of generating novel receptor/ligand interactions, ease of immobilization of phage, and the potential specificity of phage displayed ligands. Phage can be immobilized in a number of different ways, including (a) physical adsorption (Nanduri et al. 2007); (b) covalent bond formation, (amide bond with phage surface lysines or disulfide bonds with free thiols on the phage (Lang et al. 2000)); and (c) noncovalent interactions between an affinity tag and a functionalized surface (e.g. a hexahistidine tag on the phage and a metal functionalized surface (Finucane and Woolfson 1999) or biotinylated phage on a streptavidin coated surface (Gervals et al. 2007)). Phage-immobilized chip assays have been used with optical sensors to detect cancerous cells both with phage displayed antibodies (Liao et al. 2006; and references cited therein) and with phage displayed peptides (Jia et al. 2007).

18.4 Protein Phage Display

18.4.1 Introduction to Protein Phage Display

Structured protein scaffolds are emerging as an alternative to antibodies, which are often difficult to be expressed and folded. Protein phage display has aided in the development of novel protein-protein and protein-nucleic acid interactions through the use of in vitro selection strategies including those coupled with rational design. These novel proteins that are discovered through phage display can serve as modules in biosensor for recognition of almost any chemical species recognized by a natural protein.

Due to limitations on library size (see Sect. 18.2), phage display proteins cannot be fully mutated, like peptide display libraries. Typically, protein libraries are constructed by synthetic randomization at specific residues (similar to the peptide library synthesis described in Sect. 18.2), by error prone PCR (Aslan et al. 2005; Pannekoek et al. 1993; Vartanian et al. 1996), or in a variety of other methods (Brakmann 2001; Kunkel 1985; Wang et al. 2006). When libraries are constructed via synthesis of a randomized oligonucleotide, the positions of diversity are usually chosen on the basis of both structural information (choosing the randomized residues to be in the correct orientation in space), and sequence considerations (residues that are close together in the amino acid sequence which are easier to simultaneously mutate). In contrast to libraries constructed by randomization of synthetic oligonucleotides, libraries constructed with error prone PCR have random mutations that can be spread over many more residues. This method takes advantage of the inherent imperfection of polymerases under certain conditions (e.g. differing concentrations of nucleotides and the presence of manganese) to control the incorporation of differential levels of mutations (1–10%) in a protein spread throughout the sequence (Vartanian et al. 1996).

18.4.2 Protein-protein Interactions in Phage Display

Many researchers have attempted to imitate antibodies by incorporating mutations between loops joining secondary structural elements of globular proteins, while others have more recently attempted to utilize the secondary structural domains, such as the α-helix or β-sheet to their benefit (Zhao and Chmielewski 2005). To date, several scaffolds have been utilized for phage display, including α-helical scaffolds (Nygren 2008), β-sheet scaffolds (Rajagopal et al. 2004; Rajagopal et al. 2006; Smith et al. 2006), loop scaffolds (Ku and Schultz 1995; McConnell and Hoess 1995) and antibody fragments (see Sect. 18.3). Protein scaffolds utilized for in vitro selections have recently been reviewed extensively (Binz et al. 2005; Hosse et al. 2006), a summary of some of these scaffolds utilized in phage-display is given in Table 18.5, and a sampling of these will be explored in more detail below.

18.4.2.1 Loop Scaffolds

In an effort to mimic antibody binding, several groups have developed various protein scaffolds for the display of peptides on loops rather than more rigid secondary structures, such as α-helices or β-sheets. With antibodies, much of the diversity is concentrated on three loops of each variable domain (hypervariable loops) (Burton 1993). The loops found in antibodies are more conformationally constrained than free peptides, but they are less constrained than α-helices or β-sheets. Therefore, loops can accommodate a large amount of sequence diversity without disruption of the overall fold of the protein scaffold. This attribute of antibodies that allows them to recognize a wide range of antigens can be mimicked with other protein scaffolds as well.

An early example of using loops as scaffolds for phage display involved tendamistat (Fig. 18.7a), a 74 amino acid β-sheet protein that is a known inhibitor of α-amylase (McConnell and Hoess 1995). Tendamistat is similar to antibodies in which there is a predominantly β-sheet structure and it contains a disulfide bond. On the other hand, a predominantly α-helical protein, cytochrome b_{562} (Fig. 18.7b),

Table 18.5 Protein scaffolds utilized in phage display (Binz et al. 2005)

Scaffold	Size (a.a.)	Fold	Randomization	Reference
Tendamistat	74	β-sandwich	Loop	McConnell and Hoess (1995)
CTLA-4	136	β-sandwich	Loop	Hufton et al. (2000)
Fibronectin	94	β-sandwich	Loop	Koide et al. (1998)
T-cell receptor	250	β-sandwich	Loop	Li et al. (2005)
Lipocalins	160–180	β-barrel	Loop	Beste et al. (1999)
Cellulose-binding domain of cellobiohydrolase I (CBD)	36	β_3	Loops/surface	Smith et al. (1998)
TEM-1 β-lactamase	265	α/β	Loop	Legendre et al. (2002)
Insect defensin A peptide	29	α/β_2	Loop	Zhao et al. (2004)
Cytochrome b_{562}	106	α-helices	Loop	Ku and Schultz (1995)
Z domain of protein A (affibody)	58	α-helices	Helix	Nygren (2008)
pVIII phage coat protein	50	Mostly helix	Helix	Petrenko et al. (2002)
Avian pancreatic polypeptide (APP)	36	α/ppII	Helix	Chin and Schepartz, (2001)
Zinc fingers	26	α/β	Helix	Bianchi et al. (1995) and Wolfe et al. (1999)
Min-23	23	β_3	β-turn	Souriau et al. (2005)
HTB1 (B1 domain of protein G)	56	α/β_4	β-sheet	Rajagopal et al. (2006) and Smith et al. (2006)

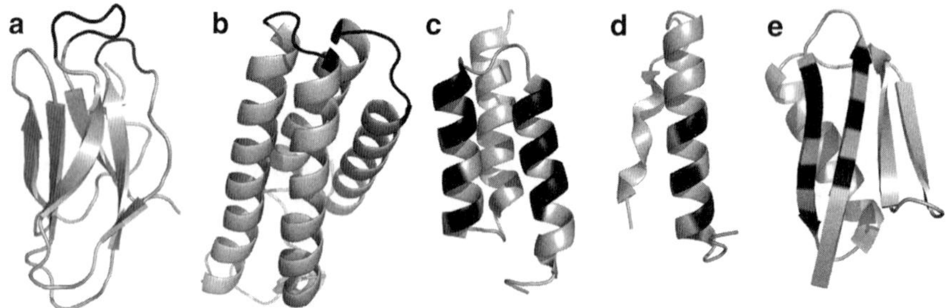

Fig. 18.7 Protein scaffolds utilized in phage display. Amino acid positions that were randomized in protein scaffold are highlighted in black. (**a**) Loop scaffold tendamistat [PDB ID: 1OK0] (**b**) Loop scaffold cytochrome b_{562} [PDB ID: 256B] (**c**) The Z-domain of *staphylococcal* protein A, used to derive affibodies [PDB ID: 2B87] (**d**) Avian Pancreatic Polypeptide (APP) [PDB ID: 1PPT] (**e**) The hyperthermophilic variant of the B1 domain of *streptococcal* protein G (HTB1) [PDB ID: 1GB4]

was also used as a loop scaffold in phage display (Ku and Schultz 1995). It is interesting to note that proteins with morphologies as different as tendamistat and cytochrome b_{562} (β-sheet vs. α-helix) can both be utilized as loop scaffolds, demonstrating the flexibility of this method.

18.4.2.2 α-Helical Scaffolds

Some of the early examples of protein scaffolds in phage display utilized a small (58 a.a.) α-helical domain derived from the Z-domain of *staphylococcal* protein A (Nord et al. 1997). These small, three-helical bundle proteins, dubbed "affibodies", contain no disulfide bonds and fold readily into native-like structures. The lack of disulfide bonds as well as the stable fold potentially provide an advantage over antibody fragments and other protein scaffolds (Nygren and Uhlen 1997). One face of the α-helical bundle is randomized (Fig. 18.7a) to display a discrete binding surface, organizing the residues into a more-or-less predetermined orientation. Affibodies have been discovered that bind various protein targets including: Taq DNA polymerase, human apolipoprotein A-1, recombinant human factor VIII, human IgA, human β-amyloid peptides, human transferrin, HIV glycoprotein 120 (gp120), and the breast cancer marker ErbB2 (Her2) (Nygren 2008). The affinities of these affibodies varied widely from the pico- to micromolar range.

Another approach to develop new protein-protein interactions is protein grafting, where the residues of a protein that are known to participate in the binding of a target (from an NMR or crystal structure) are first transplanted (grafted) onto a desired protein scaffold. Following grafting, adjacent amino acids are optimized by phage-display. This is primarily done to improve the designed interaction, both functionally and structurally, which provides a smaller peptide scaffold that is more amenable to chemical manipulation (Chin et al., 2001; Chin and Schepartz, 2001). A prototypical example of phage display in protein grafting was done by Chin and Schepartz utilizing avian pancreatic polypeptide (APP) (Fig. 18.7d) as the protein scaffold to target Bcl-2 and Bcl-X_L (Chin and Schepartz, 2001). In this work, the authors grafted residues that were known to interact with Bcl-X_L from an α-helical fragment of the protein Bak onto the α-helix of APP. Other residues on the α-helix were then randomized for a phage display selection against Bcl-2, which is known to be structurally similar to Bcl-X_L). The resulting proteins showed much higher affinity for Bcl-2 (K_D = 52 nM for one mutant than the original Bak fragment used in the design (K_D = 4.9 μM).

18.4.2.3 β-Sheet Scaffolds

β-sheet scaffolds have proven to be challenging to manipulate due to their greater potential than α-helical scaffolds to aggregate. β-sheet scaffolds can potentially form extended β-sheet structures such as those commonly found in amyloid-like fibrils (Dobson 2003). However, by utilizing a small, highly stable protein, access to β-sheet motifs can be achieved in phage display (Fig. 18.7e). For instance, a 56-residue hyperthermophilic variant of the B1 domain of *streptococcal* protein G (HTB1) (Malakauskas and Mayo 1998) has been used to target human thrombin (Rajagopal et al. 2004; Rajagopal et al. 2006) and β-amyloid peptides (Smith et al. 2006). Like the affibody scaffold, HTB1 contains no disulfide bonds, and is stable and well-folded. Furthermore, HTB1 is known to bind to human IgG with its non-β-sheet surface. This provides the opportunity to ensure that the selected mutants are properly folded, through inclusion of a preliminary "structural selection" against IgG. Only the properly folded HTB1 variants have the potential to bind IgG, thus ensuring the retention of native-like structure with IgG binding. HTB1 thereby provides a β-sheet protein scaffold upon which randomized residues can be presented in a predetermined orientation for use in phage display. Like affibodies, the small proteins discovered with this approach are stable and well-folded (Rajagopal et al. 2006), therefore lending themselves to a myriad of applications, including biosensor technology.

18.4.3 Protein-DNA Interactions in Phage Display: Zinc Finger Domains

In addition to protein-protein interactions, phage display has also been used to discover and optimize protein-DNA interactions, including one of the most important DNA binding scaffolds in biotechnology, the "zinc finger" motif (Pavletich and Pabo 1991; Wolfe et al. 2000a). Zinc fingers are modular domains that rely on zinc atoms for stability while recognizing double stranded DNA in a sequence-specific manner (Fig. 18.8). The modularity of zinc fingers can be exploited by swapping out native, designed, or selected domains to produce recombinant proteins that are theoretically capable of specifically recognizing almost any given double stranded DNA sequence (Segal et al. 2003). These designed, or "reprogrammed", zinc fingers targeted against novel DNA sequences have been made possible through a large scale phage-display effort, though the newly designed fingers can also be reoptimized via phage display to achieve the desired sequence specificity and affinity (Greisman and Pabo 1997; Segal et al. 1999; Wolfe et al. 1999; Wolfe et al. 2000b).

Zinc fingers that have been designed with the help of phage display studies have been used recently in biosensors in the form of a technology called SEquence Enabled Reassembly (SEER) (Ghosh et al. 2006; Ooi et al. 2006; Porter et al. 2007; Stains et al. 2005; Stains et al. 2006). In this method, zinc fingers are used in a split protein complementation assay to recognize double stranded DNA in a sequence-specific manner. Two zinc fingers, or other DNA binding domains, are fused to separate halves of a split signal-generating protein domain. Binding of the DNA recognition domains brings the two halves of a signal generating domain together, which refold to produce a detectable signal. SEER has been demonstrated with zinc fingers (Ooi et al. 2006; Stains et al. 2005) and a methyl binding domain (MBD) (Porter et al. 2007; Stains et al. 2006) as the DNA recognizing domains, and green fluorescent protein (GFP) (Stains et al. 2005; Stains et al. 2006), β-lactamase (Ooi et al. 2006; Porter et al. 2007), or firefly luciferase (Porter et al. 2008) as the signal generating domain (Fig. 18.8b). The broad utility of SEER is dependent upon the universality of zinc finger DNA recognition, which in turn has been greatly aided by in vitro selection in the form of phage display.

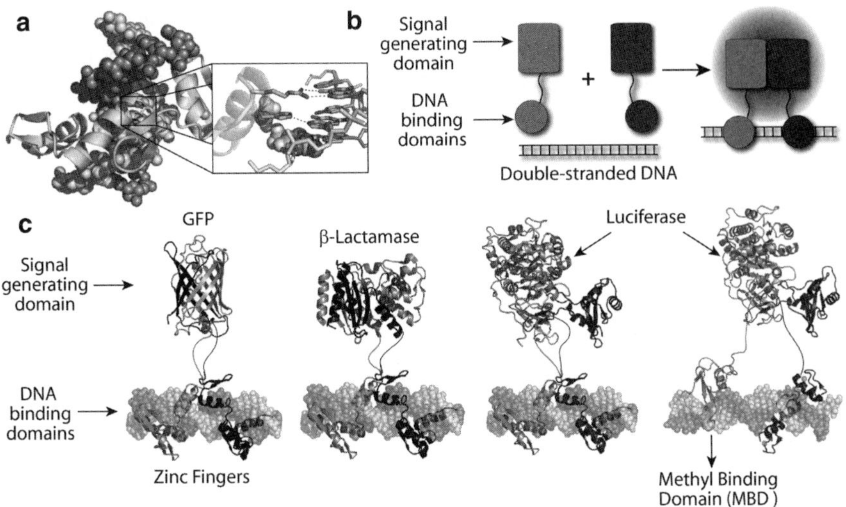

Fig. 18.8 Overview of the SEquence Enabled Reassembly (SEER) system. (**a**) Zinc fingers (natural, designed, or selected by phage-display) bind sequence-specifically to DNA. This recognition is governed by specific hydrophobic packing as well as hydrogen bonding between the DNA strand and residues on the α-helix of the zinc finger. (**b**) SEER consists of two DNA binding domains fused to separate halves of a signal generating domain. When the two DNA binding domains recognize the DNA, they bring the halves of the signal generating domain together, leading to an observable signal. (**c**) SEER has been demonstrated with zinc fingers and the Methyl Binding Domain (MBD) for DNA recognition, while split GFP, β-lactamase, and firefly luciferase have been used for signal generation

18.5 Concluding remarks

Biosensors rely on specific analyte recognition by an appropriate biological module. However, many analytes of interest do not have appropriate binding partners that can be easily integrated into detection devices. Therefore, methods such as phage display that are capable of discovering novel receptors for a user-defined analyte greatly facilitate biosensor development. Phage display has aided the discovery of novel interactions involving peptides, antibodies and proteins through the use of in vitro selection strategies. The connection between genotype and phenotype afforded by phage display allows for the iterative screening and amplification necessary to isolate binding domains from vast libraries. As we have discussed in this chapter, the peptide, protein, or antibody domains available from phage display can target virtually all biomacromolecules, whether peptide, protein, nucleic acid, carbohydrate, or even inorganic. These binding domains, with a wide range of tunable affinities and selectivities, can be rapidly incorporated into the existing biosensor architectures. Furthermore, interesting advances have allowed for the phage particles themselves to be integrated into the devices. Future advances in phage display selection methodologies may perhaps allow for the rapid discovery of new signaling domains to complement the detection domains described in this chapter. Phage display is clearly an immensely powerful technology for providing key detection modules in user-defined biosensors.

References

Aslan FM, Yu Y, Mohr SC, Cantor CR (2005) Engineered single-chain dimeric streptavidins with an unexpected strong preference for biotin-4-fluorescein. Proc Natl Acad Sci USA 102:8507–8512

Barbas CF, Kang AS, Lerner RA, Benkovic SJ (1991) Assembly of combinatorial antibody libraries on phage surfaces: The gene-III site. Proc Natl Acad Sci USA 88:7978–7982

Barbas CF, Hu D, Dunlop N, Sawyer L, Cababa D, Hendry RM, Nara PL, Burton DR (1994) In vitro evolution of a neutralizing human antibody to human immunodeficiency virus type 1 to enhance affinity and broaden strain cross-reactivity. Proc Natl Acad Sci USA 91: 3809–3813

Barbas CF, Burton DR, Scott JK, Silverman GJ (2001) Phage display: a laboratory manual. Cold Spring Harbor Laboratory Press, Cold Spring Harbor, NY

Bass S, Greene R, Wells JA (1990) Hormone phage: an enrichment method for variant proteins with altered binding properties. Proteins 8:309–314

Benhar I, Eshkenazi I, Neufeld T, Opatowsky J, Shaky S, Rishpon J (2001) Recombinant single chain antibodies in bioelectrochemical sensors. Talanta 55:899–907

Beste G, Schmidt FS, Stibora T, Skerra A (1999) Small antibody-like proteins with prescribed ligand specificities derived from the lipocalin fold. Proc Natl Acad Sci USA 96:1898–1903

Bianchi E, Folgori A, Wallace A, Nicotra M, Acali S, Phalipon A, Barbato G, Bazzo R, Cortese R, Felici F, Pessi A (1995) A conformationally homogeneous combinatorial peptide library. J Mol Biol 247:154–160

Binz HK, Amstutz P, Pluckthun A (2005) Engineering novel binding proteins from nonimmunoglobulin domains. Nat Biotechnol 23:1257–1268

Bradbury ARM, Marks JD (2004) Antibodies from phage antibody libraries. J Immunol Methods 290:29–49

Brakmann S (2001) Discovery of superior enzymes by directed molecular evolution. Chembiochem 2:865–871

Burton DR, Barbas CF, Persson MAA, Koenig S, Chanock RM, Lerner RA (1991) A large array of human monoclonal antibodies to type 1 human immunodeficiency virus from combinatorial libraries of asymptomatic seropositive individuals. Proc Natl Acad Sci USA 88:10134–10137

Burton DR (1993) Monoclonal-antibodies from combinatorial libraries. Acc Chem Res 26:405–411

Caparon MH, DeCiechi PA, Devine CS, Olins PO, Lee SC (1996) Analysis of novel streptavidin-binding peptides, identified using a phage display library, shows that amino acids external to a perfectly conserved consensus sequence and to the presented peptides contribute to binding. Mol Divers 1:241–246

Chan SW, Bye JM, Jackson P, Allain JP (1996) Human recombinant antibodies specific for hepatitis C virus core and envelope E2 peptides from an immune phage display library. J Gen Virol 77:2531–2539

Chen LM, Zurita AJ, Ardelt PU, Giordano RJ, Arap W, Pasqualini R (2004) Design and validation of a bifunctional ligand display system for receptor targeting. Chem Biol 11:1081–1091

Chin JW, Grotzfeld RM, Fabian MA, Schepartz A (2001) Methodology for optimizing functional miniature proteins based on avian pancreatic polypeptide using phage display. Bioorg Med Chem Lett 11:1501–1505

Chin JW, Schepartz A (2001) Design and evolution of a miniature Bcl-2 binding protein. Angew Chem Int Ed Engl 40:3806–3809

Clackson T, Hoogenboom HR, Griffiths AD, Winter G (1991) Making antibody fragments using phage display libraries. Nature 352:624–628

Cooper MA (2002) Optical biosensors in drug discovery. Nat Rev Drug Discov 1:515–528

Deng SJ, Mackenzie CR, Sadowska J, Michniewicz J, Young NM, Bundle DR, Narang SA (1994) Selection of antibody single-chain variable fragments with improved carbohydrate-binding by phage display. J Biol Chem 269:9533–9538

Deng SJ, Mackenzie CR, Hirama T, Brousseau R, Lowary TL, Young NM, Bundle DR, Narang SA (1995) Basis for selection of improved carbohydrate-binding single-chain antibodies from synthetic gene libraries. Proc Natl Acad Sci USA 92:4992–4996

Devlin JJ, Panganiban LC, Devlin PE (1990) Random peptide libraries: a source of specific protein binding molecules. Science 249:404–406

Dobson CM (2003) Protein folding and misfolding. Nature 426:884–890

Emanuel P, OBrien T, Burans J, DasGupta BR, Valdes JJ, Eldefrawi M (1996) Directing antigen specificity towards *botulinum* neurotoxin with combinatorial phage display libraries. J Immunol Methods 193:189–197

Emanuel PA, Dang J, Gebhardt JS, Aldrich J, Garber EAE, Kulaga H, Stopa P, Valdes JJ, Dion-Schultz A (2000) Recombinant antibodies: a new reagent for biological agent detection. Biosens Bioelectron 14:751–759

Finn FM, Hofmann K (1990) Isolation and characterization of hormone receptors. Methods Enzymol 184:244–74

Finucane MD, Woolfson DN (1999) Core-directed protein design II. Rescue of a multiply mutated and destabilized variant of ubiquitin. Biochemistry 38:11613–11623

Fu LL, Li SQ, Zhang KW, Chen IH, Petrenko VA, Cheng ZY (2007) Magnetostrictive microcantilever as an advanced transducer for biosensors. Sensors 7:2929–2941

Gaj T, Meyer SC, Ghosh I (2007) The AviD-tag, a NeutrAvidin/avidin specific peptide affinity tag for the immobilization and purification of recombinant proteins. Protein Expr Purif 56:54–61

Gervals L, Gel M, Allain B, Tolba M, Brovko L, Zourob M, Mandeville R, Griffiths M, Evoy S (2007) Immobilization of biotinylated bacteriophages on biosensor surfaces. Sens Actuators B Chem 125:615–621

Ghosh I, Stains CI, Ooi AT, Segal DJ (2006) Direct detection of double-stranded DNA: Molecular methods and applications for DNA diagnostics. Mol Biosyst 2:551–560

Giebel LB, Cass RT, Milligan DL, Young DC, Arze R, Johnson CR (1995) Screening of cyclic peptide phage libraries identifies ligands that bind streptavidin with high affinities. Biochemistry 34:15430–15435

Goldman ER, Pazirandeh MP, Mauro JM, King KD, Frey JC, Anderson GP (2000) Phage-displayed peptides as biosensor reagents. J Mol Recognit 13:382–387

Gottesman S (1996) Proteases and their targets in Escherichia coli. Annu Rev Genet 30:465–506

Gough KC, Cockburn W, Whitelam GC (1999) Selection of phage-display peptides that bind to cucumber mosaic virus coat protein. J Virol Methods 79:169–180

Green NM (1990) Avidin and streptavidin. Methods Enzymol 184:51–67

Greisman HA, Pabo CO (1997) A general strategy for selecting high-affinity zinc finger proteins for diverse DNA target sites. Science 275:657–661

Griffiths AD (1993) Production of human antibodies using bacteriophage. Curr Opin Immunol 5:263–267

Griffiths AD, Duncan AR (1998) Strategies for selection of antibodies by phage display. Curr Opin Biotechnol 9:102–108

Handa H, Gurczynski S, Jackson MP, Auner G, Walker J, Mao G (2008) Recognition of *Salmonella typhimurium* by immobilized phage P22 monolayers. Surf Sci 602:1392–1400

Hayhurst A, Happe S, Mabry R, Koch Z, Iverson BL, Georgiou G (2003) Isolation and expression of recombinant antibody fragments to the biological warfare pathogen *Brucella melitensis*. J Immunol Methods 276:185–196

Hengsakul M, Cass AEG (1997) Alkaline phosphatase-strep tag fusion protein binding to streptavidin: Resonant mirror studies. J Mol Biol 266:621–632

Hoogenboom HR, Griffiths AD, Johnson KS, Chiswell DJ, Hudson P, Winter G (1991) Multi-subunit proteins on the surface of filamentous phage: methodologies for displaying antibody (Fab) heavy and light chains. Nucleic Acids Res 19:4133–4137

Hoogenboom HR, de Bruine AP, Hufton SE, Hoet RM, Arends JW, Roovers RC (1998) Antibody phage display technology and its applications. Immunotechnology 4:1–20

Hosse RJ, Rothe A, Power BE (2006) A new generation of protein display scaffolds for molecular recognition. Protein Sci 15:14–27

Hudson PJ (1998) Recombinant antibody fragments. Curr Opin Biotechnol 9:395–402

Hufton SE, van Neer N, van den Beuken T, Desmet J, Sablon E, Hoogenboom HR (2000) Development and application of cytotoxic T lymphocyte-associated antigen 4 as a protein scaffold for the generation of novel binding ligands. FEBS Lett 475:225–231

Jia YF, Qin M, Zhang HK, Niu WC, Li X, Wang LK, Li X, Bai YP, Cao YJ, Feng XZ (2007) Label-free biosensor: a novel phage-modified light addressable potentiometric sensor system for cancer cell monitoring. Biosens Bioelectron 22:3261–3266

Kay BK, Adey NB, He YS, Manfredi JP, Mataragnon AH, Fowlkes DM (1993) An M13 phage library displaying random 38-amino-acid peptides as a source of novel sequences with affinity to selected targets. Gene 128:59–65

Kehoe JW, Kay BK (2005) Filamentous phage display in the new millennium. Chem Rev 105:4056–4072

Kim YG, Lee CS, Chung WJ, Kim E, Shin DS, Rhim JH, Lee YS, Kim BG, Chung JH (2005) Screening of LPS-specific peptides from a phage display library using epoxy beads. Biochem Biophys Res Commun 329:312–317

Klenow H, Hennings I (1970) Selective Elimination of Exonuclease Activity of Deoxyribonucleic Acid Polymerase from Escherichia-Coli-B by Limited Proteolysis. Proc Natl Acad Sci USA 65:168–175

Knurr J, Benedek O, Heslop J, Vinson RB, Boydston JA, McAndrew J, Kearney JF, Turnbough CL (2003) Peptide ligands that bind selectively to spores of *Bacillus subtilis* and closely related species. Appl Environ Microbiol 69:6841–6847

Koide A, Bailey CW, Huang XL, Koide S (1998) The fibronectin type III domain as a scaffold for novel binding proteins. J Mol Biol 284:1141–1151

Krebber A, Bornhauser S, Burmester J, Honegger A, Willuda J, Bosshard HR, Pluckthun A (1997) Reliable cloning of functional antibody variable domains from hybridomas and spleen cell repertoires employing a reengineered phage display system. J Immunol Methods 201:35–55

Krumpe LRH, Atkinson AJ, Smythers GW, Kandel A, Schumacher KM, McMahon JB, Makowski L, Mori T (2006) T7 lytic phage-displayed peptide libraries exhibit less sequence bias than M13 filamentous phage-displayed peptide libraries. Proteomics 6:4210–4222

Ku J, Schultz PG (1995) Alternate protein frameworks for molecular recognition. Proc Natl Acad Sci USA 92:6552–6556

Kunkel TA (1985) Rapid and efficient site-specific mutagenesis without phenotypic selection. Proc Natl Acad Sci USA 82:488–492

Lakshmanan RS, Guntupalli R, Hu J, Kim DJ, Petrenko VA, Barbaree JM, Chin BA (2007) Phage immobilized magnetoelastic sensor for the detection of *Salmonella typhimurium*. J Microbiol Methods 71:55–60

Lang S, Xu J, Stuart F, Thomas RM, Vrijbloed JW, Robinson JA (2000) Analysis of antibody A6 binding to the extracellular interferon gamma receptor alpha-chain by alanine-scanning mutagenesis and random mutagenesis with phage display. Biochemistry 39:15674–15685

Legendre D, Vucic B, Hougardy V, Girboux AL, Henrioul C, Van Haute J, Soumillion P, Fastrez J (2002) TEM-1 beta-lactamase as a scaffold for protein recognition and assay. Protein Sci 11:1506–1518

Li Y, Moysey R, Molloy PE, Vuidepot AL, Mahon T, Baston E, Dunn S, Liddy N, Jacob J, Jakobsen BK, Boulter JM (2005) Directed evolution of human T-cell receptors with picomolar affinities by phage display. Nat Biotechnol 23:349–354

Liao W, Guo S, Zhao XS (2006) Novel probes for protein chip applications. Front Biosci 11:186–197

Lin VSY, Motesharei K, Dancil KPS, Sailor MJ, Ghadiri MR (1997) A porous silicon-based optical interferometric biosensor. Science 278:840–843

Malakauskas SM, Mayo SL (1998) Design, structure and stability of a hyperthermophilic protein variant. Nat Struct Biol 5:470–475

Mao CB, Solis DJ, Reiss BD, Kottmann ST, Sweeney RY, Hayhurst A, Georgiou G, Iverson B, Belcher AM (2004) Virus-based toolkit for the directed synthesis of magnetic and semiconducting nanowires. Science 303:213–217

McCafferty J, Griffiths AD, Winter G, Chiswell DJ (1990) Phage antibodies: filamentous phage displaying antibody variable domains. Nature 348:552–554

McConnell SJ, Hoess RH (1995) Tendamistat as a scaffold for conformationally constrained phage peptide libraries. J Mol Biol 250:460–470

McLafferty MA, Kent RB, Ladner RC, Markland W (1993) M13 bacteriophage displaying disulfide-constrained microproteins. Gene 128:29–36

Meyer SC, Gaj T, Ghosh I (2006) Highly selective cyclic peptide ligands for NeutrAvidin and avidin identified by phage display. Chem Biol Drug Des 68:3–10

Mosier-Boss PA, Lieberman SH, Andrews JM, Rohwer FL, Wegley LE, Breitbart M (2003) Use of fluorescently labeled phage in the detection and identification of bacterial species. Appl Spectrosc 57:1138–1144

Nanduri V, Sorokulova IB, Samoylov AM, Simonian AL, Petrenko VA, Vodyanoy V (2007) Phage as a molecular recognition element in biosensors immobilized by physical adsorption. Biosens Bioelectron 22:986–992

Nicacio CD, Williamson RA, Parren PWHI, Lundkvist A, Burton DR, Bjorling E (2002) Neutralizing human Fab fragments against measles virus recovered by phage display. J Virol 76:251–258

Nord K, Gunneriusson E, Ringdahl J, Stahl S, Uhlen M, Nygren PA (1997) Binding proteins selected from combinatorial libraries of an alpha-helical bacterial receptor domain. Nat Biotechnol 15:772–777

Nygren PA, Uhlen M (1997) Scaffolds for engineering novel binding sites in proteins. Curr Opin Struct Biol 7:463–469

Nygren PA (2008) Alternative binding proteins: affibody binding proteins developed from a small three-helix bundle scaffold. Febs Journal 275:2668–2676

O'Neil KT, Hoess RH, Jackson SA, Ramachandran NS, Mousa SA, DeGrado WF (1992) Identification of novel peptide antagonists for GPIIb/IIIa from a conformationally constrained phage peptide library. Proteins 14:509–515

Olsen EV, Sorokulova IB, Petrenko VA, Chen IH, Barbaree JM, Vodyanoy VJ (2006) Affinity-selected filamentous bacteriophage as a probe for acoustic wave biodetectors of *Salmonella typhimurium*. Biosens Bioelectron 21:1434–1442

Ooi AT, Stains CI, Ghosh I, Segal DJ (2006) Sequence-enabled reassembly of beta-lactamase (SEER-LAC): a sensitive method for the detection of double-stranded DNA. Biochemistry 45:3620–3625

Pannekoek H, Vanmeijer M, Schleef RR, Loskutoff DJ, Barbas CF (1993) Functional display of human plasminogen-activator inhibitor-1 (PAI-1) on phages: novel perspectives for structure-function analysis by error-prone DNA synthesis. Gene 128:135–140

Paschke M, Hohne W (2005) A twin-arginine translocation (Tat)-mediated phage display system. Gene 350:79–88

Pavletich NP, Pabo CO (1991) Zinc Finger DNA Recognition: Crystal-Structure of a Zif268-DNA Complex at 2.1-A. Science 252:809–817

Petrenko VA, Smith GP, Gong X, Quinn T (1996) A library of organic landscapes on filamentous phage. Protein Eng 9:797–801

Petrenko VA, Smith GP (2000) Phages from landscape libraries as substitute antibodies. Protein Eng 13:589–592

Petrenko VA, Smith GP, Mazooji MM, Quinn T (2002) Alpha-Helically constrained phage display library. Protein Eng 15:943–950

Petrenko VA, Sorokulova IB (2004) Detection of biological threats. A challenge for directed molecular evolution. J Microbiol Methods 58:147–168

Petrenko VA (2008) Landscape phage as a molecular recognition interface for detection devices. Microelectronics J 39:202–207

Pini A, Spreafico A, Botti R, Neri D, Neri P (1997) Hierarchical affinity maturation of a phage library derived antibody for the selective removal of cytomegalovirus from plasma. J Immunol Methods 206:171–182

Plaisant P, Burioni R, Manzin A, Solforosi L, Candela M, Gabrielli A, Fadda G, Clementi M (1997) Human monoclonal recombinant Fabs specific for HCV antigens obtained by repertoire cloning in phage display combinatorial vectors. Res Virol 148:165–169

Porter JR, Stains CI, Segal DJ, Ghosh I (2007) Split beta-lactamase sensor for the sequence-specific detection of DNA methylation. Anal Chem 79:6702–6708

Porter JR, Stains CI, Jester BW, Ghosh I (2008) A general and rapid cell-free approach for the interrogation of protein-protein, protein-DNA, and protein-RNA interactions and their antagonists utilizing split-protein reporters. J Am Chem Soc 130:6488–6497

Rader C, Barbas CF (1997) Phage display of combinatorial antibody libraries. Curr Opin Biotechnol 8:503–508

Rajagopal S, Meza-Romero R, Ghosh I (2004) Dual surface selection methodology for the identification of thrombin binding epitopes from hotspot biased phage-display libraries. Bioorg Med Chem Lett 14:1389–1393

Rajagopal S, Meyer SC, Goldman A, Zhou M, Ghosh I (2006) A minimalist approach toward protein recognition by epitope transfer from functionally evolved beta-sheet surfaces. J Am Chem Soc 128:14356–14363

Roberts D, Guegler K, Winter J (1993) Antibody as a surrogate receptor in the screening of a phage display library. Gene 128:67–69

Salonen EM, Parren PWHI, Graus YF, Lundkvist A, Fisicaro P, Vapalahti O, Kallio-Kokko H, Vaheri A, Burton DR (1998) Human recombinant puumala virus antibodies: cross-reaction with other hantaviruses and use in diagnostics. J Gen Virol 79:659–665

Samoylov AM, Samoylova TI, Pathirana ST, Globa LP, Vodyanoy VJ (2002) Peptide biosensor for recognition of cross-species cell surface markers. J Mol Recognit 15:197–203

Segal DJ, Dreier B, Beerli RR, Barbas CF (1999) Toward controlling gene expression at will: selection and design of zinc finger domains recognizing each of the 5'-GNN-3' DNA target sequences. Proc Natl Acad Sci USA 96:2758–2763

Segal DJ, Beerli RR, Blancafort P, Dreier B, Effertz K, Huber A, Koksch B, Lund CV, Magnenat L, Valente D, Barbas CF (2003) Evaluation of a modular strategy for the construction of novel polydactyl zinc finger DNA-binding proteins. Biochemistry 42:2137–2148

Sheets MD, Amersdorfer P, Finnern R, Sargent P, Lindqvist E, Schier R, Hemingsen G, Wong C, Gerhart JC, Marks JD (1998) Efficient construction of a large nonimmune phage antibody library: the production of high-affinity human single-chain antibodies to protein antigens. Proc Natl Acad Sci USA 95:6157–6162

Smith GP (1985) Filamentous fusion phage: novel expression vectors that display cloned antigens on the virion surface. Science 228:1315–1317

Smith GP, Petrenko VA (1997) Phage Display. Chem Rev 97:391–410

Smith GP, Patel SU, Windass JD, Thornton JM, Winter G, Griffiths AD (1998) Small binding proteins selected from a combinatorial repertoire of knottins displayed on phage. J Mol Biol 277:317–332

Smith TJ, Stains CI, Meyer SC, Ghosh I (2006) Inhibition of beta-amyloid fibrillization by directed evolution of a beta-sheet presenting miniature protein. J Am Chem Soc 128:14456–14457

Soderlind E, Simonsson AC, Borrebaeck CAK (1992) Phage display technology in antibody engineering: design of phagemid vectors and in vitro maturation systems. Immunol Rev 130:109–124

Sorokulova IB, Olsen EV, Chen IH, Fiebor B, Barbaree JM, Vodyanoy VJ, Chin BA, Petrenko VA (2005) Landscape phage probes for *salmonella typhimurium*. J Microbiol Methods 63:55–72

Souriau C, Chiche L, Irving R, Hudson P (2005) New binding specificities derived from Min-23, a small cystine-stabilized peptidic scaffold. Biochemistry 44:7143–7155

Stains CI, Porter JR, Ooi AT, Segal DJ, Ghosh I (2005) DNA sequence-enabled reassembly of the green fluorescent protein. J Am Chem Soc 127:10782–10783

Stains CI, Furman JL, Segal DJ, Ghosh I (2006) Site-specific detection of DNA methylation utilizing mCpG-SEER. J Am Chem Soc 128:9761–9765

Stratmann J, Strommenger B, Stevenson K, Gerlach GF (2002) Development of a peptide-mediated capture PCR for detection of mycobacterium avium subsp paratuberculosis in milk. J Clin Microbiol 40:4244–4250

Turnbough CL (2003) Discovery of phage display peptide ligands for species-specific detection of Bacillus spores. J Microbiol Methods 53:263–271

Vartanian JP, Henry M, WainHobson S (1996) Hypermutagenic PCR involving all four transitions and a sizeable proportion of transversions. Nucleic Acids Res 24:2627–2631

Verhaert RMD, van Duin J, Quax WJ (1999) Processing and functional display of the 86 kDa heterodimeric penicillin G acylase on the surface of phage fd. Biochem J 342:415–422

Wang TW, Zhu H, Ma XY, Zhang T, Ma YS, Wei DZ (2006) Mutant library construction in directed molecular evolution. Mol Biotechnol 34:55–68

Whaley SR, English DS, Hu EL, Barbara PF, Belcher AM (2000) Selection of peptides with semiconductor binding specificity for directed nanocrystal assembly. Nature 405:665–668

Williams DD, Benedek O, Turnbough CL (2003) Species-specific peptide ligands for the detection of *bacillus anthracis* spores. Appl Environ Microbiol 69:6288–6293

Winter G, Griffiths AD, Hawkins RE, Hoogenboom HR (1994) Making antibodies by phage display technology. Annu Rev Immunol 12:433–455

Wolfe SA, Greisman HA, Ramm EI, Pabo CO (1999) Analysis of zinc fingers optimized via phage display: evaluating the utility of a recognition code. J Mol Biol 285:1917–1934

Wolfe SA, Nekludova L, Pabo CO (2000a) DNA recognition by Cys(2)His(2) zinc finger proteins. Annu Rev Biophys Biomol Struct 29:183–212

Wolfe SA, Ramm EI, Pabo CO (2000b) Combining structure-based design with phage display to create new Cys(2)His(2) zinc finger dimers. Structure 8:739–750

Zang X, Yu Z, Chu YH (1998) Tight-binding streptavidin ligands from a cyclic peptide library. Bioorg Med Chem Lett 8:2327–2332

Zhao A, Xue YN, Zhang H, Gao B, Feng JN, Mao CQ, Zheng L, Liu NG, Wang F, Wang H (2004) A conformation-constrained peptide library based on insect defensin A. Peptides 25:629–635

Zhao L, Chmielewski J (2005) Inhibiting protein-protein interactions using designed molecules. Curr Opin Struct Biol 15:31–34

Zhou B, Wirsching P, Janda KD (2002) Human antibodies against spores of the genus Bacillus: a model study for detection of and protection against anthrax and the bioterrorist threat. Proc Natl Acad Sci USA 99:5241–5246

Chapter 19
Molecularly Imprinted Polymer Receptors for Sensors and Arrays

Glen E. Southard and George M. Murray

Abstract Molecular imprinting is a process for making selective binding sites in synthetic polymers. The process may be approached by designing the recognition site or by simply choosing monomers that may have favorable interactions with the imprinting molecule. The successful application of the methodology to biochemical sensing typically requires the designed approach. The process involves building a complex of an imprint molecule and complementary polymerizable ligating monomers. At least one of the molecular complements must exhibit a discernable physical change associated with binding. This change in property can be any measurable quantity, but a change in luminescence is the most sensitive and selective analytical technique. By copolymerizing the complexes with a matrix monomer and a suitable level of crosslinking monomer, the imprint complex becomes bound in a polymeric network. The network may need to be mechanically and chemically processed to liberate the imprinting species and create the binding site. The design of the binding site requires insights into its chemistry. These insights are derived from studies of molecular recognition and self-assembly and include considerations of molecular geometry, size, and shape, as well as molecule-to-ligand thermodynamic affinity.

This chapter explores the use of molecular imprinting in the construction of sensors for biomolecules and biological constructs. It is not a review, but rather a tutorial aimed at exploring the different ways molecularly imprinted receptors can aid in the determination of biomolecules. It begins with a brief history of molecular imprinting and follows with a description of the major approaches. A description of the types of molecules used to make imprints, both inorganic and organic, is given. The different polymer morphologies are explored. Sensor transduction strategies compatible with imprinted receptors are described. A few examples of sensors based on each of the described transduction are described to illustrate the wide range of sensors that have been prepared using molecular imprinting.

G.M. Murray (✉)
Department of Materials Science and Engineering, Center for Laser Applications, University of Tennessee Space Institute, 411 B. H. Goethert Parkway MS 24, Tullahoma, TN 37388, USA
e-mail: gmurray@utsi.edu

M. Zourob (ed.), *Recognition Receptors in Biosensors*,
DOI 10.1007/978-1-4419-0919-0_19,

An attempt is made to discuss what remains to be done with imprinting and sensing in the future.

Keywords Molecularly imprinted polymers · Biosensors · Synthetic Receptors · Receptor arrays

Abbreviations

MIP	Molecularly imprinted polymer
MMA	Methyl methacrylate
EGDM	Ethylene glycol dimethacrylate
CRPs	Controlled free radical polymerizations
ITO	Indium tin oxide
QCM	Quartz crystal microbalances
SAWS	Surface acoustic wave sensors
SPR	Surface plasmon resonance
ISE	Ion selective electrode
TMV	Tobacco mosaic viruses
PEDOT	Poly(3,4-ethylenedioxythiophene)
MO	Morphine
MIA	Molecularly imprinted sorbent assay
2,4-D	2,4-dichlorophenoxyacetic acid
ZON	Zearalenone
NBDRA	2,4-dihydroxybenzoic acid 2-[methyl(7-nitro-benzo[1,2,5]oxadiazol-4-yl)amino]ethyl ester
PMRA	2,4-dihydroxy-N-pyren-1-ylmethylbenzamide
PARA	2,4- dihydroxybenzoic acid 2-[(pyrene-l-carbonyl)amino] ethyl ester
RCM	Ring-closing metathesis
MWD	Molecular weight dispersity

19.1 Introduction

This chapter seeks to explore the use of molecular imprinting in the construction of sensors for biomolecules. It is not an attempt at an exhaustive review, but rather a tutorial aimed at exploring the different ways in which molecularly imprinted receptors can aid in the determination of biomolecules. The chapter begins with a brief history of molecular imprinting and follows with a description of the major approaches. A description of the types of molecules used to make imprints, both inorganic and organic, is given. The different polymer morphologies are explored. Sensor transduction strategies compatible with imprinted receptors are described. A few examples of sensors based on each of the described transduction are described to illustrate the wide range of sensors that have been prepared using molecular imprinting.

The production of a biochemical sensor requires the realization of two goals. The first goal is the development of a sufficiently specific chemical recognition element that allows a molecule, or a class of molecules, to be selectively identified in a complex chemical matrix. The second goal, to obtain analytically sensitive signal transduction, requires a measurable change in a physical property of the material. While these two goals are not always separable, the successful design of chemical sensors in a systematic fashion requires both the goals to be achieved. Most transduction elements are based on optical, resistive, surface acoustic wave (SAW) or capacitive measurements, although a variety of other approaches is also available. These well-developed transduction methods dominate largely due to their ease of operation, high sensitivity, and low cost. However, the chemical recognition elements in these detectors lag far behind. Indeed, most reports on chemical sensors end with a statement suggesting that many other devices could be fabricated if only suitable chemical recognition units were available. The missing element is a general approach to chemical recognition that allows the rational design and assembly of materials in a stable and reusable form. One approach to solve selective recognition is to prepare molecularly imprinted polymers (MIPs).

Molecular recognition can be approached in many ways. Each method attempts to construct molecular complements that will provide a selective binding capability. Classically, a molecule with an appropriate structure is tested as a complexant with a variety of candidate compounds until an association is found. Molecular modeling applies all available chemical information to design specific site interactions. This approach allows the rational design of recognition sites and would be optimal if all the aspects of the calculations were correct. Unfortunately, it is not yet known how to do all the calculations, and the synthesis of the designed molecules can be tedious. At the other end of the spectrum is the combinatorial approach. In this approach, little is assumed to be known and an assembly or library of potential functionalities is tested for binding by a screening method. If selectivity is required, then a more complex screening process is required. This method, normally employing peptide libraries, ignores the vast amount of chemistry that is known, does not use many of the elements in the periodic table, and relies on nature to solve the riddle of recognition.

Molecular imprinting falls between the two extremes and lies more closely to the rational design approach. MIPs are made by first building a complex of a target molecule and associated ligating monomers. Some planning should be done at this stage to choose the proper complexing monomers to maximize selectivity. This is not recognized by all practitioners and has led to marginal results in many of the attempts at applying molecular imprinting to sensing. In the case of sensors, the complexing monomer may be required to have an appropriate change in properties. In conventional imprinting, the complex is formed by a self-assembly process in the expectation that the ligating atoms will find the optimal geometry. This will be true only if the ligands were chosen correctly. The

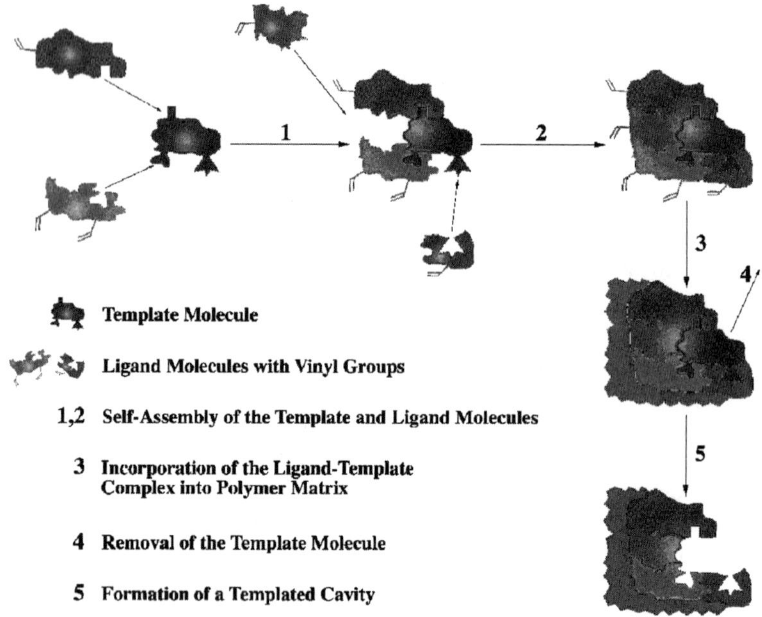

Fig. 19.1 The process involved in conventional molecular imprinting

complex is usually dissolved in a larger amount of other polymerizable molecules. The bulk of the other molecules for the polymer are typically made with crosslinking monomers. The crosslinkers are necessary to hold the complexing molecules in place after the target molecule or "template" is removed. It is also common to add a solvent to the mixture. The solvent molecules get caught up in the growing polymer and leave gaps and pores in the structure to make the target complexes more accessible after the polymer is formed. Typically, after polymerization, a chunk of plastic is obtained. This chunk is ground up into a powder to expose a maximum surface area and the target molecule is removed by washing it out with an appropriate solvent. The powder is left with specially shaped holes that have a "memory" for the target molecule and are ready to recapture that specific molecule when reexposed (Fig. 19.1).

19.2 Molecular Imprinting

19.2.1 Chronology

Molecular imprinting is considered to have originated with experiments by a Russian scientist Polyakov (1931), who was preparing silica gels for chromatography.

He prepared silica gels from sodium silicate solutions using ammonium carbonate as initiator. The gels he prepared would later selectively adsorb certain solvent molecules present during the sol gel preparation.

Conceptually, MIPs also trace their origin back to suppositions about the operation of the human immune system by Stuart Mudd in the 1930s and Linus Pauling in the 1940s. Mudd's (1932) contribution to the understanding of the immune system was to propose the idea of complementary structures. This supposition states that the reason for a specific antibody attacking a specific target or "antigen" is that the shape of the antibody provides an excellent fitting cavity for the shape of the antigen. The description is very similar to the "lock and key" analogy used to explain the action of enzymes, the molecules responsible for hastening and directing biochemical reactions. In this case, the enzyme forms the lock for a particular chemical key to fit and as this "key" is turned, the enzyme directs and hastens the production of desired products from the chemical target.

The idea of molecular imprinting was brought to the biological community by concepts developed by Linus Pauling (1940) and illustrated in his paper, "A Theory of the Structure and Process of Formation of Antibodies." Pauling described the process of antibody formation in terms of a self assembly of protein antibody around an antigen template. He followed his theoretical paper with an experimental paper where he used denatured globulins to sequester dye "antigens" in an attempt to reproduce nature in vitro (Pauling and Campbell 1942). This was followed by a paper from one of Pauling's postdoctoral fellows describing the preparation of sol gels selective for binding dyes. The paper ended with the suggestion of preparing sorbents that would resolve optical isomers, a major success of molecular imprinting. Subsequent to the early work, several papers relating to the formation of sol gel imprinting followed for a period of about 15 years (Dickey 1949, 1955).

It is recognized that the first detailed paper on molecular imprinting using synthetic organic polymers was the paper by Wulff and Sarhan (1972). This paper describes what will later be known as the covalent approach to molecular imprinting. Of particular note in this effort was that the polymer produced was selective toward an enantiomer, D-glyceric acid, an important step in bioseparations. This became one of the most impressive successes in molecular imprinting, the resolution of racemates. The investigations of Wulff and Sarhan (1972) and later those of Shea (Shea and Thompson 1978; Shea et al. 1980), Neckers (Damen and Neckers 1980a, b, c), Mosbach (Arshady and Mosbach 1981) and their coworkers indicate that polymers with specific cavities can be produced. Since this pioneering work, there has been a gradual but steady increase in the number of molecularly imprinted materials. In the last 10 years, there has been a change in emphasis from sequestering and separations media toward sensing. Ye and Haupt (2004) have recently published an instructive review on MIPs as antibody and receptor mimics for assays, sensors, and drug discovery.

19.2.2 Approaches

Classical imprinting is divided into two approaches: covalent imprinting and non-covalent imprinting. Covalent imprinting utilizes a covalently linked template molecule. This approach was developed primarily by Wulff and Shea in the 1980s. This requires functionalization of the template or template analog with a polymerizable group. The functionalization should be sufficiently stable to survive polymerization, but the bond must cleave with mild conditions, and after cleavage, must have a positive interaction with the analyte. Common linkages are esters and amides. Crosslinkers and matrix monomers with functionalities capable of favorable are also used. Polymers are formed, ground, and sieved for size and then the template is removed, typically by acid–base chemistry. The advantage of this approach is better site homogeneity in the absence of site aggregation with concomitant increased selectivity. The disadvantages include: intrinsic complexity, difficulty in complete removal of the template molecule from the polymer, and that the kinetics of exchange are generally slower.

The alternative to the covalent approach to conventional imprinting is the noncovalent or self-assembled approach. In this approach, crosslinking and matrix monomers are chosen that should have associative properties to the template. In practice, the matrix monomers typically used are methyl methacrylate (MMA) and ethylene glycol dimethacrylate (EGDM) with the addition of a selective coordinator monomer, such as acrylic acid. Again, the monomers, crosslinker, and template are mixed, polymerized, then ground, and sieved. The template is removed typically by Soxhlet extraction. The advantage to this approach is its relative simplicity. The disadvantage is the site heterogeneity and lesser selectivity. One method to mitigate the heterogeneity is to cap residual reactive groups to minimize nonspecific interactions (Umpleby et al. 2001).

A third approach, similar to the covalent approach, is the metal ion-mediated approach. In this approach, metal ion coordination or the combination of coordination and charge compensation is used to form the basis of a recognition site. As in the covalent approach, the choice of the complexing monomer is critical. When using a metal complex, the metal ion must be strongly tethered to the polymer to avoid loss of function. The advantage of this approach, especially to the application of sensing, is the ability of changes in metal ion complexation to yield measurable changes in color, luminescence, or some other physical property. The use of an appropriate coordinating tether can influence the metal's affinity for the template. In the case of ion imprinting, it is greatly desired that the binding site provides charge compensation. The imprint process needs to provide an appropriate geometry as well as a thermodynamic affinity. If the charged site utilizes a metal ion for signal transduction, that metal ion may need to have its thermodynamic affinity modified by the selection of appropriate chelating support ligands. These ligands must have a high-enough affinity to disallow the removal of the metal ion and the proper charge density to give the metal the appropriate "hardness."

19.3 Ligating Monomers

19.3.1 *Inorganic*

As discussed earlier, silica was the first extensively investigated means for preparing selective sorbents by an imprinting method. Molecular imprinting in sol gel materials has been recently reviewed (Díaz-Garcia and Laíño 2005). Sol gel imprinting benefits from the extensive inventory of functionalized metal alkoxides provided by Gelest, Inc., 11 East Steel Road, Morrisville, PA. The catalog also includes a tutorial on the production of sol gel materials. There is also a wide selection of materials that include vinyl substituents for the production of hybrid materials (Fig. 19.2).

19.3.2 *Organic Monomers*

The number of commercially available polymerizable complexants is fairly limited. However, the preparation of polymerizable complexants is a growing area in imprinting. The monomers initially used for the covalent approach were target molecules functionalized with vinyl groups via a hydrolysable group, such as an ester or amide. Therefore, the monomer depended on the target. This led to the slower kinetics exhibited by covalent MIPs.

The synthesis of polymerizable ligands for complexing metals has suffered from the lack of utilization of state-of-the-art techniques. Subsequently, investigators have either been forced to purchase readily available, but not always the most desirable, ligands, or have relied upon noncatalytic and conventional carbon–carbon bond reactions, such as Wittig or Friedel–Crafts techniques. Metal, usually palladium, catalyzed carbon–carbon bond forming reactions have demonstrated that modern cross coupling reactions have a role for readily synthesizing specialized ligands for polymerizable metal complexes that have been difficult, if not impossible, to generate by more traditional chemical reactions (Southard and Murray 2005; Southard et al. 2006) (Fig. 19.3).

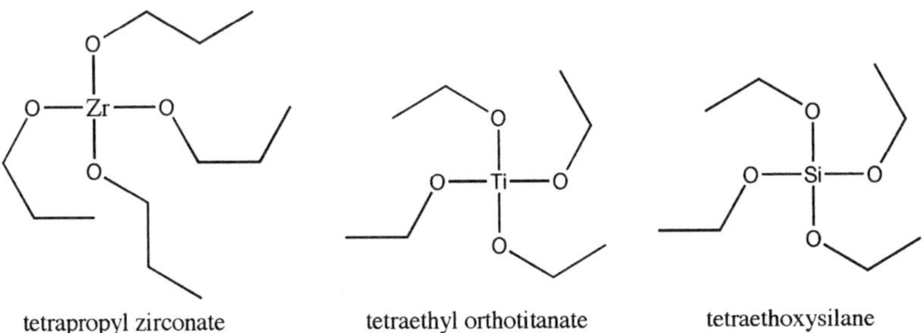

Fig. 19.2 Some metal alkoxide monomers

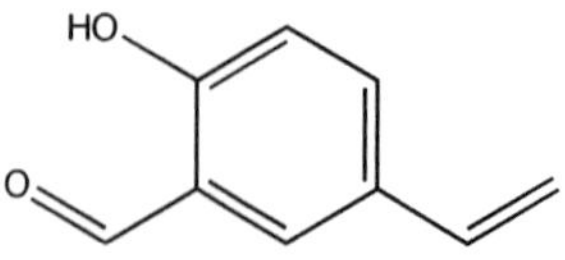

5-Vinylsalicylaldehyde

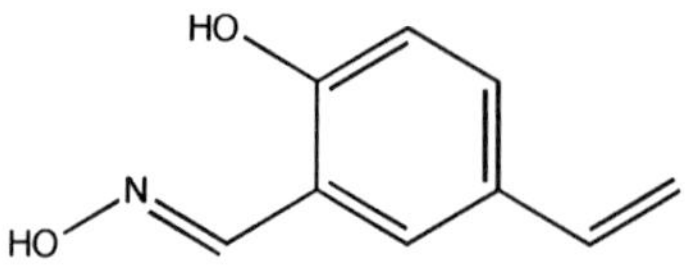

5-Vinylsalicylaldoxime

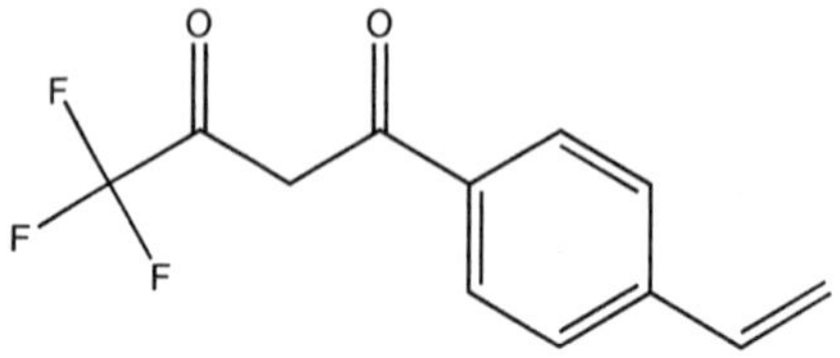

4-Vinylbenzoyltrifluoroacetone

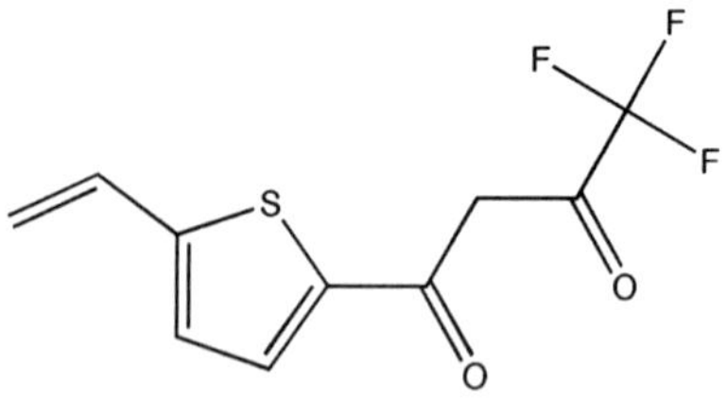

5-Vinylthenoyltrifluoroacetone

Fig. 19.3 Some recently prepared metal ion-coordinating monomers

19.4 Polymer Morphology

19.4.1 Bulk Polymers

Most molecular imprints are prepared as bulk polymers, typically using sol gel chemistry or free radical-initiated polymerization. The resultant monolith is then ground, sieved, and the template removed prior to use. While this is a reasonable approach to making selective sorbents for chemical separations, it does not lend itself well to the preparation of sensors because of inconsistent reproducibility, adhesion to substrate problems, light-scattering issues, and refractive index matching problems. However, several methods have been developed that use sieved particles in sensing applications.

One way to use ground bulk polymers as sensors is to incorporate particles of a specific size in the construction of a membrane for an ion-selective electrode (ISE). This is a simple process, since it only involves suspending the polymers in a solution of PVC in THF and casting a film on glass within the confines of a ring, also typically made of glass.

Another way to prepare sensors is to synthesize the polymer in contact with a transducer substrate. This process is a complicated one in that it requires something in the mix that will bind to the substrate. Naturally, some sol gels will bond directly to glass substrates when the surface is properly cleaned and prepared. This makes sol gel techniques of especial utility for optical transduction.

An increasing number of investigators are using emulsion or suspension polymerizations to give bulk MIPs with a near-uniform particle size distribution ranging from submicron to 500 micron spheres depending upon the conditions of polymerization (Fig. 19.4). Emulsion or suspension polymerizations are an attractive alternative to the single-phase monolith concept of creating MIPs as all of the particles,

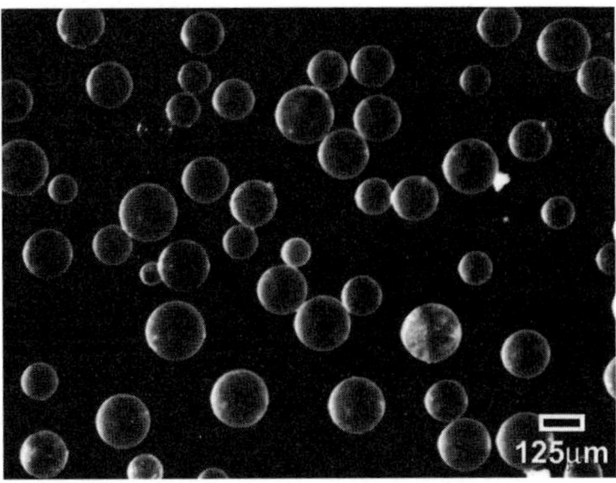

Fig. 19.4 MIP beads imprinted for dicyanoaurate prepared by suspension polymerization

instead of only a sieved fraction, may be used for sensor building. Additionally, some of the thermodynamic problems, such as auto-acceleration, with single-phase monolith polymerizations are overcome through the use of a two-phase polymerization. One of the major disadvantages of emulsion or suspension polymerizations is that the imprint molecule needs to have a high solubility in the polymer phase or it may be lost in the carrier solution. Additionally, determining an adequate formulation for the bead polymerization may require more development time than the single-phase monolith preparation.

Titirici and Sellergren (2004) prepared silica particles containing immobilized peptidic templates used for the generation of hierarchically imprinted polymers. The pores of the silica mold were filled with a mixture of monomers/initiator and polymerized, followed by dissolution of the silica template. This method leaves behind imprinted polymers with binding sites located at the surface, which are capable of recognizing larger molecules with the same immobilized epitope. The particles are of uniform size and shape, with improved chromatographic properties.

There is increasing interest in using controlled free radical polymerizations (CRPs), specifically atom transfer (Coessens et al. 2001), nitroxide-mediated (Hawker et al. 2001), and reversible addition fragmentation chain transfer (Le et al. 1998; Chen et al. 2004) polymerizations. CRPs have several advantages over traditional free-radical polymerization such as predictable rates, mitigation of auto-acceleration and narrow molecular weight distributions. They may be used to produce linear polymers, which may be isolated, characterized, and used as macro-initiators for subsequent polymerizations. They may form a number of polymer morphologies such as block, star, comb, tapered, emulsion, bead, or traditional monoliths that give great flexibility for preparing sensors with varied form factors. The resulting polymers often possess reactive end groups, which readily allow tethering to a surface or another macromolecule. Additionally, glass or other surfaces may be easily functionalized to allow CRP growth directly from the substrates surface. The disadvantages of CRPs in regard to molecular imprinting are: they require a somewhat high polymerization temperature, typically over 80°C; may require some additional synthetic steps for ligands or initiators; and each polymerization system has its own set of idiosyncrasies.

19.4.2 Polymer Films

The casting of MIPs as films may be accomplished by utilizing one of five different methods. In general, the first entails suspending in solution a MIP of uniform particle size with a soluble but inert polymer. The mixture can be cast by spin coating or evaporation to give a film. The method has several drawbacks because of the potential masking of the MIP by the carrier polymer, uneven particle distribution, and problems with reproducibility.

The second method begins as a normal bulk polymerization that is allowed to polymerize to a predetermined viscosity and before the polymerization has gelled. The viscous solution is then cast onto a substrate and the polymerization allowed to cure. Due to the high surface area resulting from the casting, the method requires a porogen with a low vapor pressure, and thus, the MIP will retain its porosity. Also, great care must be taken to give reproducible results by casting the polymerization at the correct viscosity, which will affect the thickness of the MIP coating.

The third method is to cast an uncrosslinked polymer film containing an unbound template molecule. The polymer film is either thermally or UV-initiated and the film is crosslinked by a condensation or grafting mechanism with the template bound to the polymer through the same mechanism. An interesting derivation of this technique is to press large objects, such as spores or viruses, into a soft precured polymer cast onto a hard substrate. The polymer is then cured with the objects in place under pressure.

The fourth method may be performed by electrochemical means. Electro-active monomers, such as thiophene and pyrrole or their derivatives, may be polymerized on an electrode in the presence of an electrochemically stable analyte. The method is most suitable for ISEs.

The fifth method is to grow MIP films directly from the surface of a substrate by free radical or CRP methods in the presence of vinyl-substituted template molecules. The substrate, glass, ITO, etc. is first functionalized with a group susceptible to forming free radicals or CRP conditions and then placed into a polymerization solution appropriate to the condition. The polymerization is carried out and the film washed free of the analyte. The method is useful for a large number of sensing methods, but requires some advanced polymerization techniques.

19.4.3 Surface Imprinting

The term surface imprinting has been interpreted in different ways. True imprinting requires crosslinking to retain shape recognition. This is the process used by several investigators to prepare selective recognition elements. The advantage to surface imprinting is speed accessibility afforded by the surface or near-surface recognition site. The process may employ core particles as substrates or may include the preparation of the core as part of the synthesis.

19.4.4 Soluble MIPs

Using MIP technology as a standard recognition element in the rational design of chemical sensors has shown a significant progress in the last 10 years. However, there are several residual difficulties that are yet to be overcome. One main drawback of MIPs technology is that the polymer matrix is by necessity structurally rigid. The templated resins formed are inherently intractable; they cannot be dissolved or recast without loss of the template structure and function. Furthermore, the removal of the target molecules from the template cavity usually requires relatively high temperatures and long extraction times with solvents that swell the polymer matrix. Incomplete removal of the target leads to high background interference and diminished capacity. The extraction process itself can also lead to loss of the template structure and to long-term relaxation phenomenon within the matrix which changes the materials' sensitivity over time.

Zimmerman et al. (2003) prepared a type of soluble crosslinked polymer by using a porphyrin encapsulated into a crosslinked core with multiple arms to enhance solubility (Beil and Zimmerman 2004; Beil et al. 2004). This process is possible due to CRP techniques that can form pseudo-living polymers. Modern techniques of CRP also allow the preparation of block copolymers with potentially crosslinkable substituents in specific locations. The inclusion of crosslinkable mers proximate to the binding complex in the core of a star polymer allows the formation of molecularly imprinted macromolecules that are soluble and processable. The use of discreet self-assembled macromolecular structures that incorporate the active binding environments required for MIP formation, but are not part of a bulk matrix, offers many advantages: these structures are soluble in a broad range of solvents; they can be recast into a variety of structures; and they do not lose their structural form. They can be used directly in solution, as thin films or coatings on a variety of substrates, as block polymers or powders.

19.5 Sensor Transduction

19.5.1 Mass Sensors

There are several transducers that are sensitive to changes in mass. The three most common transducers are bulk acoustic wave, SAW, and surface plasmon resonance (SPR) sensors. Mass transduction is attractive because of the sensitivity of the devices, speed of response, and good sensitivity. Mass transduction does impose more rigorous conditions for imprinting, since it is necessary to eliminate nonspecific binding. This means that the polymer matrix must be prepared from monomers that do not interact with likely interferents. This restriction is less important in separations media and is the main reason why the approaches used for separations media fare poorly when applied to sensing.

Bulk acoustic wave. Bulk acoustic wave transducers, also called quartz crystal microbalances (QCM), are mass sensitive detectors that respond through a change in the characteristic frequency of a quartz crystal oscillator. Originally for gas phase sensing, they have since been adapted to liquid sensing and used in flow cells. Quartz is one member of a family of crystals that experience the piezoelectric effect. Applying alternating current to the quartz crystal will induce oscillations. Piezoelectricity is the ability of materials, some crystals, certain ceramics, and polyvinylidene difluoride, to generate an electric potential in response to applied mechanical stress or conversely to change dimension with applied potential. With an alternating current between the electrodes of a properly cut crystal, a standing shear wave is generated. Sensing occurs through the dampening of the oscillation when the crystal is loaded by a deposit. Common equipment allows resolutions down to 1 Hz on crystals with a fundamental resonant frequency in the 4–6 MHz range.

Surface acoustic wave sensors (SAWS). SAWS are related to the QCMs in that they employ piezoelectric materials. In this case, an acoustic wave travels along the surface of a material having some elasticity, with amplitude that typically decays exponentially with the depth of the substrate. SAWs were discovered in 1887 by Lord Rayleigh. Rayleigh waves, named after their discoverer, have a longitudinal and a vertical component that can couple with a medium in contact with the surface of the device. This coupling strongly affects the frequency of the wave, allowing SAW sensors to directly sense mass. Electronic devices employing the SAW normally utilize interdigitated transducers to convert an electrical signal to an acoustic wave and back again. Like the bulk acoustic wave sensors discussed earlier, these sensors are made using piezoelectric materials. The major difference between SAWS and QCMs is that the frequency of the oscillation is higher, affording a greater sensitivity.

Surface plasmon resonance (SPR). SPR is an optical phenomenon that occurs when light interacts with a free-electron metal under specific conditions. The resonance is highly sensitive to the refractive index of the environment close to the metal surface. Binding events caused by molecular interactions near the surface cause corresponding increases in refractive index. This allows changes in refractive index to be monitored in real time. SPR is considered as one of the gold standard techniques for measuring biokinetics, affinity, and interactions (Fig. 19.5).

There are several commercial SPR vendors. Biacore has developed a variety of laboratory grade instruments that are in use in various applications from assay development to production facility quality control. ICX Technologies has produced a low-cost, simplified embodiment of an SPR system called the SPREETA. This system has developed a self-contained sensor module consisting of the essential SPR components; a light source, polarizer, prism, sensing area, and SPR angel detector. This system, shown in Fig. 19.6, reduced the assay chip costs to approximately $50.

The sensors have evolved into a more developed system called SensiQ (Fig. 19.7). The original SensiQ is a manually operated system. A more basic

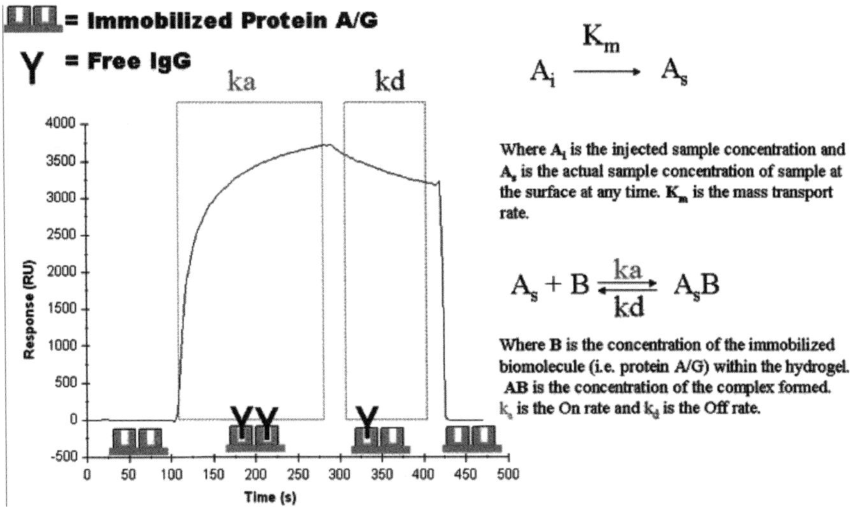

Fig. 19.5 Surface plasmon resonance interactions

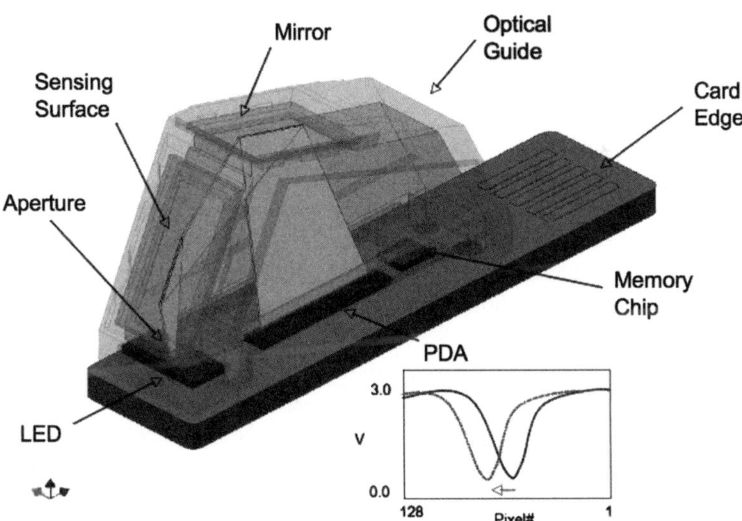

Fig. 19.6 The SPR "all in one" chip

version has been introduced, the SensiQ Discovery. An automated version is now available as the SensiQ Pioneer.

MIPS have been demonstrated for use in the capture of high mass proteins such as horseradish peroxidase (HRP MW 44 kDa) and lactoperoxidase (MW 77 kDa).

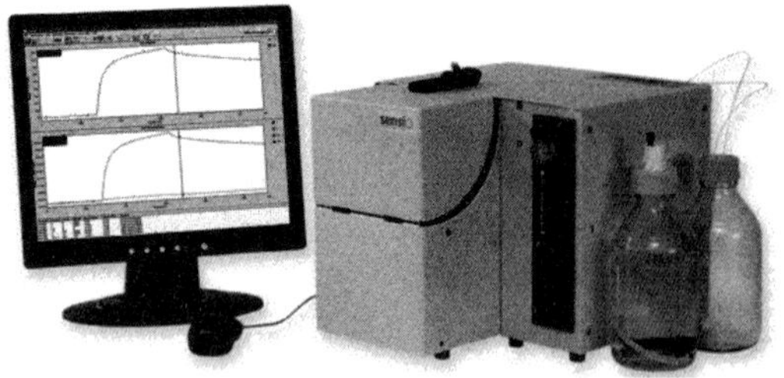

Fig. 19.7 SensiQ, surface plasmon resonance instrument

Piletsky et al. (2001) also developed a method to attach these polymers to the bottom of a microplate well.

19.5.2 Electrochemical Sensors

Conductivity. The simplest electrochemical approach to ion sensing that could employ imprinted polymers is measuring changes in polymer conductivity with ion binding. The most straightforward application would be the detection of ions. However, the binding of an analyte to a polymer could change conductivity in a variety of ways. The binding of the analyte molecule can swell the polymer and allow ions present in the polymer to migrate more easily. Binding larger and less mobile counter ion analytes to the imprinted sites can reduce the conductivity of a MIP film.

Capacitance. Capacitors are usually designed to maintain a fixed structure. Certain factors can change the structure of the capacitor; the resulting change in capacitance can be used to sense those factors. As a MIP sensor, typically, the dielectric is composed of a MIP. The binding of the analyte to the dielectric results in a change in capacitance. Both polymer swelling and changing the dielectric constant of the polymer contribute to the change in capacitance (19.1).

$$\text{Capacitance} = \frac{\text{Area} \times \text{Dielectric}}{\text{Gap}} \tag{19.1}$$

ISEs. Another approach to electrochemical transduction by chemical recognition is incorporating the imprinted polymer as the active ingredient in a membrane of an ISE. ISEs are devices that, when incorporated in an electrochemical cell with an appropriate reference electrode, produce a potential that varies predictably with the concentration of a certain ion in solution. If the response of the electrode follows theory, the response is Nernstian and is given by the Nernst Equation:

$$E = K - \frac{RT}{nF} \ln \frac{\alpha_{\text{ext}}}{\alpha_{\text{int}}}, \tag{19.2}$$

where E is the measured potential, K is, in this case, an experimentally derived constant, R is the gas constant, T is the absolute temperature, n is the ionic charge, F is Faraday's constant, and the α's are the activities of the analyte ion in the external and internal solutions. Inserting the values of the constants and changing to logarithms to the base 10, the equation simplifies to:

$$E = K - 0.0592/n \log(\alpha), \tag{19.3}$$

Hence, the useful cell potential is:

$$E_{\text{cell}} = K - 0.0592/n \log(\alpha) - E_{\text{ref}}. \tag{19.4}$$

Voltammetry. Voltammetry is an electrochemical method that involves the application of a potential at an electrode surface and measures the resulting current, using a three-electrode system. The electrodes include: a reference electrode to provide a known potential, an auxiliary electrode for the passage of current, and a working electrode involved in the electrochemical processes of the cell. The technique is performed using a potentiostat. The potentiostat drives the cell and supplies whatever current is needed between working and counter electrodes to maintain a specific voltage between working and reference electrode. Amperometry is a variant of voltammetry in which the applied potential is maintained as a constant.

19.5.3 Optical Sensors

Spectroscopic sensing requires that a chromophore must be available. It may be that the chromophore is influenced by the rebinding of the imprinted analyte or may simply be a part of the analyte. Sensors based on intrinsic analyte chromophores can benefit from molecular imprinting by both selectivity and sensitivity enhancement. In terms of selectivity, molecules similar to the analyte are likely to have similar spectroscopic parameters that could be the source of interference in a conventional sensing strategy. The inability of the interferent to bind to the imprinted polymer allows discrimination. In terms of sensitivity, the analyte molecules can be effectively concentrated from the solution by the imprinted polymer. The degree of concentration is controlled by the relevant binding constant. In fluorimetry, the target chromophore can be shielded from quenchers by the polymer matrix. The inclusion of noncomplexing monomers isolates the chromophore, eliminates cross-talk, and reduces concentration quenching. In the case of metal ion luminescence, the use of appropriate complexing ligands can enhance the intensity by several orders of magnitude through a variety of mechanisms. Examples will be given in terms of absorbance (colorimetry), fluorescence, and metal ion luminescence.

19.6 Sensor Examples

19.6.1 Mass Sensors

Bulk acoustic wave. Hayden and Dickert (2001) developed an elegant technique to form packed receptor sites on dual microbalances coated with a polymer surface, using a stamping procedure. Polymer or sol-gel solutions in a prepolymerized stage are spin cast onto the transducer surface. A stamp with a cell layer is pressed on the coated device and will embed the cells partially, depending on the layer thickness and the viscosity of the prepolymer. Receptor sites were built during the curing process because of the self-organization process of oligomeric components around the cell surface. During this polymerization step, crosslinking and densification of the prepolymerized layer occurred primarily. Best imprinting results were gathered with a constant force applied on the stamp during the polymerization of polymers. In the case of the sol-gel layers, the best results were achieved pressing the stamp on the coated microbalance for a short time. QCMs coated with yeast-imprinted polymers (yeast-IPs) from sol-gels are much less sensitive than MIPs from polyurethane. The imprinted pits are densely packed and form a honeycomb-like structure on the polymer surface. The imprints do not have a uniform diameter but a size distribution corresponding to the cell population used for the stamp. The yeast-IPs of polyurethane has pits with a usual depth of up to 1 μm. Therefore, the sensor layers have to be thicker than 1 μm to get a complete fingerprint of the cells our layers were around 2 μm.

More recently, Dickert et al. applied surface imprinting techniques on polymer-coated QCMs that have been used to detect tobacco mosaic viruses (TMVs) in aqueous media. MIPs, tailor-made by self-organization of monomers around a template (TMV), were generated directly on the gold electrodes. Imprinted trenches on the polymer surface mimicking the shape and surface functionality of the virus serve as recognition sites for readsorption, after washing out of the template. The sensors are applicable to TMV detection ranging from 100 ng mL^{-1} to 1 mg mL^{-1} within minutes. Furthermore, direct measurements without time-consuming sample preparation are possible in complex matrices such as tobacco plant sap (Dickert et al. 2004). Bolisay et al. (2006) made MIPs for TMV that were selective against Tobacco Necrosis Virus.

SAW. Most imprinted polymer-coated SAW sensors have been used for the detection of small molecules. A review of MIP coated piezo-electric transducers of both kinds has been published recently (A'vila et al. 2008). Dickert and Hayden (2002) were successful in detecting a single yeast cell-MIP interaction by using dual shear wave SAW devices, designed from a LiTaO3 substrate and gold electrodes, operating at a frequency of 428 MHz. These acoustic wave devices can be applied to liquid-phase measurements and the increased resonance frequency appreciably improves the sensitivity of the transducer.

19.6.2 Electrochemical Sensors

Conductivity. Piletsky et al. (1995) produced conductometric sensors for the pesticide atrazine. The mode of interaction with the sensor was suspected to be a function of reorganization of the polymer when the analyte was rebound, effecting the diffusion of co- and counter-ions. The construction and demonstration of a conductometric chemical sensor based on a molecularly imprinted for benzyltriphenylphosphonium ions was reported by Kriz et al. (1996). The MIP-based sensor gave a higher conductivity reading than the reference sensor when exposed to the analyte.

Capacitance. A glucose biosensor based on capacitive detection was developed using MIPs. The sensitive layer was prepared by electro-polymerization of *o*-phenylenediamine on a gold electrode in the presence of the template (glucose). Cyclic voltammetry and capacitive measurements monitored the process of electro-polymerization. Surface uncovered areas were plugged with 1-dodecanethiol to make the layer dense, and the insulating properties of the layer were studied in the presence of redox couples. The template molecules and the unbound thiol were removed from the modified electrode surface by washing with distilled water. A capacitance decrease could be obtained after injection of glucose. The electrode constructed similarly but with ascorbic acid or fructose showed only a small response compared with glucose. The stability and reproducibility of the biosensor were also investigated (Cheng et al. 2001).

ISEs. Mosbach's group prepared the first imprinted polymer ISEs for calcium and magnesium ions (Rosatzin et al. 1991). The monomer used in the fabrication of the electrode was a neutral ionophore *N,N*′-dimethyl-*N,N*′-bis(4-vinylphenyl)-3-oxapentadiamide. This was chosen as a Ca^{2+}- and Mg^{2+}-selective ionophore to produce a polymer cavity that would resemble a polyether site with imprinting the means to acquire size selection. The imprinting process gave enhancement by factors of 6 and 1.7 for Ca^{2+} and Mg^{2+} binding, respectively, over an unimprinted polymer blank as measured by batch extraction. A calibration curve (slope 27.15 mV per decade) and selectivity coefficients against alkali metals and alkaline earths for the Ca^{2+} electrode were provided. The membranes were prepared using a mixture of 32.1 wt% PVC, 65.3 wt% bis(2-ethylhexyl) sebacate, 1.85 wt% imprinted polymer, and 0.7 wt% potassium tetrakis (*p*-chlorophenyl)borate. The electrode was evaluated with metal chloride solutions in a 1:1 methanol–water mixture (v/v) with an unspecified pH.

We have prepared ISEs for Pb^{2+} and UO_2^{2+} (Murray et al. 1997). The Pb^{2+} electrode was based on a Pb^{2+}- imprinted polymer that had been characterized previously as an ion resin (Zeng and Murray 1996), and applied to the analysis of trace amounts of Pb^{2+} in seawater (Bae et al. 1998). The uranyl ion electrodes were prepared from two polymers, one employing vinylbenzoate and the other vinylsalicylaldoxime. Typical membranes were prepared by mixing 90 mg of polyvinylchloride (PVC) powder and 30 mg of imprinted polymer particles with

0.2 mL of plasticizer, usually either 2-nitrophenyl octyl ether (NPOE) or dioctyl phenylphosphonate (DOPP). The resulting electrode is filled with 1 mM $M^{n+}(NO_3^-)_n$ and 1 mM NaCl internal solution and conditioned for 24 h by soaking in a 1 mM of $M^{n+}(NO_3^-)_n$ solution. Our measurements were made with a model 350 Corning pH/ion Analyzer. Rapid response is facilitated by the use of a high-flow Ag/AgCl reference electrode, manufactured by Orion. Thus, the cell schematic is:

$$\mathrm{Hg|Hg_2Cl_2, KCl(sat.)||test\ solution|PVC\ membrane|10\ mM\ CaCl_2|AgCl|Ag.}$$

The pH electrode is simply an ISE for H^+. Glass or ISFET electrodes for pH are the best-characterized and the most successful ISEs. Many sensors employ chemical reactions to produce H^+ or OH^- ions and employ pH electrodes for signal transduction. An imprinted sensor based on this approach is the glucose sensor produced by Arnold's group (Chen et al. 1997). This polymer employs a metal ion (Cu^{2+}) in the imprinted site and makes use of the metals directed bonding to obtain enhanced chemical selectivity. The imprinted polymer was synthesized using methyl-b-glucopyranoside. The imprinting molecule is removed by replacement with OH^-. At a pH above 8, the Cu^{2+} complex has water and hydroxide bound to the two coordination sites formerly occupied by the imprint molecule. At slightly higher pH values (>9), glucose readily replaces the water and OH^-, liberating H^+. The increase in H^+ is then measured and is proportional to the amount of glucose in solution. The pH change can be measured by a pH electrode or alternatively by a colorimetric or fluorimetric dye.

Voltammetry. Morphine-sensitive devices have been constructed based on MIPs. The imprinted polymer exhibited recognition properties previously. In a first example, a method of detection based on competitive binding was used to measure morphine in the concentration range 0.1–10 pg mL^{-1}. A morphine concentration of 0.5 pg mL^{-1} gave a peak current (by oxidation) of 4 nA. The method of morphine detection involves two steps. In the first step, morphine binds selectively to the MIP in the sensor. In the second step, an electro-inactive competitor (codeine) is added in excess, whence some of the bound morphine is released. The released morphine is detected by an amperometric method. This sensor, based on an artificial recognition system, demonstrates autoclave compatibility, long-time stability, and resistance to harsh chemical environments (Kriz and Mosbach 1995). A second example, MIP particles for morphine, was synthesized, using a precipitation polymerization method. The conducting polymer, poly(3,4-ethylenedioxythiophene) (PEDOT), was utilized to immobilize the MIP particles onto the indium tin oxide (ITO) glass as a MIP/PEDOT-modified electrode. The sensitivity of the MIP/PEDOT-modified electrode with MIP particles was 41.63 A cm^{-2} mM, which is more than that with non -MIP particles or that of a single PEDOT film with no incorporated particles in detecting morphine ranging from 0.1 to 2 mM. The detection limit was 0.3 mM

(S/N = 3). The electrode can discriminate against codeine as an interfering species (Ho et al. 2005).

19.6.3 Optical Sensors

Colorimetry. One means to a chromophore is Au nanoparticles. They can be used in a variety of ways as optical tags. In one instance, the sensing mechanism was based upon the variable proximity of the Au nanoparticles immobilized in the imprinted polymer, which exhibits selective binding of a given analyte accompanied by swelling that causes a blue-shift in the plasmon absorption band of the immobilized Au nanoparticles. Adrenaline, the analyte, was detected selectively to be as low as 5 μM. The combination of molecular imprinting and the Au nanoparticle-based sensing system was shown to be a general strategy for constructing sensing materials in a tailor-made fashion because of the wide applicability of the imprinting technique and the sensing mechanism's being independent of the analyte recognition system (Matsui et al. 2004).

In one example, a MIP of morphine (MO) was prepared through thermal radical copolymerization of methacrylic acid (MAA) and EDMA in the presence of MO templates, and a molecularly imprinted sorbent assay (MIA) based on a colorimetric reporter was developed to determine the adsorption isotherm of MO–MIP binding. In practice, the MO-bound MIP was brought into contact with an aqueous mixture of Fe^{3+} and $[Fe(CN)_6]^{3-}$ so that the 3-phenolic group of MO was oxidized and $[Fe(CN)_6]^{3-}$ was reduced to $[Fe(CN)_6]^{4-}$. As a result, the MO-bound MIP was stained with Prussian blue (PB), which was attributed to the instant coprecipitation of Fe^{3+} and $[Fe(CN)_6]^{4-}$ (Ksp = 10^{-40}). Accordingly, MO–MIP binding of the blue dye could be detected by visible spectroscopy. In addition, such staining could successfully distinguish MO from codeine. Upon data analyses, a two-site binding isotherm with two dissociation constants of 6.00×10^{-5} and 1.03×10^{-3} M was found for MO–MIP binding. MIAs for non-MIP were also performed.

A more general approach to colorimetry is a dye displacement assay. After removing the target molecules from the MIP, dye-labeled target molecules can be introduced into the MIP receptor sites. After coating the dye-impregnated MIP onto a suitable surface, the displacement assay is ready to use. A solution of interest is applied onto the polymer surface; if dye is released from the MIP and a color change observed, then the target molecule is present in the test solution. The unlabeled target molecule will preferentially bind to the MIP receptor site (and displace the dye-labeled molecule), since the unlabeled target was used to form the receptor sites during the cross linking reaction (Fig. 19.8).

Another type of dye displacement array based on generic MIPs with a lesser affinity has been developed (Shimizu and Greene 2005). A colorimetric sensor array composed of seven MIPs was shown to accurately identify seven different

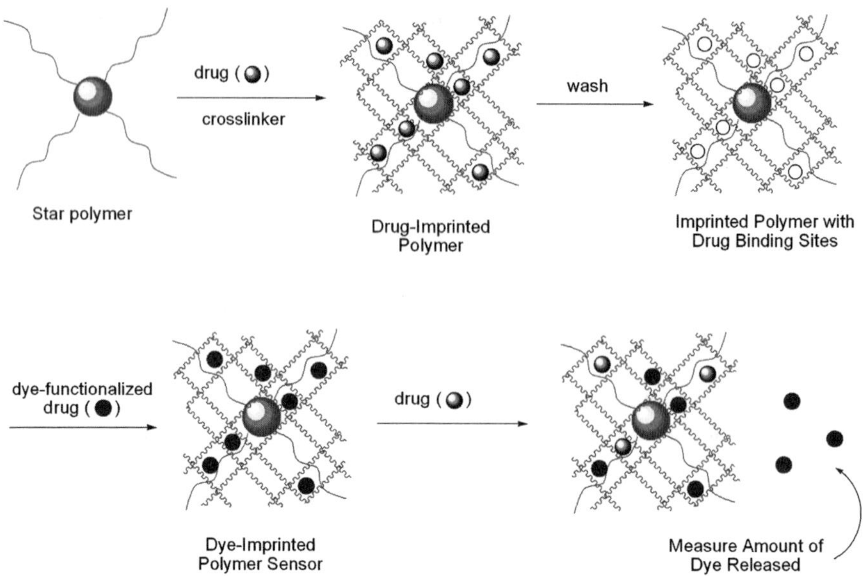

Fig. 19.8 The concept of operation of a dye labeled displacement assay based on a MIP

aromatic amines. The response patterns were classified using linear discriminant analysis with 94% classification accuracy. Analyses of the response patterns of the analytes to the imprinted polymer array show that different patterns arise from the imprinting process. The noncovalent imprinting process allowed facile preparation of polymers for the array from EGDM and methacrylic acid (80:20) in the presence of six different template molecules plus a blank nonimprinted polymer. The response of the imprinted polymer array was coupled to a colorimetric response, using a dye displacement strategy. A benzofurazan dye was selected and shown to give an accurate measure of the binding properties of the imprinted polymer array to all the seven analytes. The colorimetric response allows the inclusion of analytes without chromophores that were not among the original templates. This broadens the potential utility of the imprinted polymer sensor array strategy to a wider range of analytes and applications.

Fluorescence. A fluorescent ligand displacement assay has been developed for the herbicide 2,4-dichlorophenoxyacetic acid (2,4-D) based on a MIP. The format of the assay is analogous to competitive fluoro-immunoassay formats and uses a coumarin derivative as a nonrelated fluorescent probe. The specificity and the sensitivity of the assay are on par with a radioligand displacement assay using the same MIP and radiolabeled 2,4-dichlorophenoxyacetic acid. The assay can be used both in aqueous buffer and in organic solvents, the detection limit being about 100 nM (Haupt et al. 1998).

A further development was of an imaging assay, analogous to competitive enzyme immunoassays, developed using a MIP. Again, the antigen was 2,4-dichlorophenoxyacetic acid. It was labeled with tobacco peroxidase, and the chemiluminescence reaction with luminol was used for detection. Microtiter plates (96 or 384 wells) were coated with polymer microspheres imprinted with 2,4-D and were fixed in place by using poly(vinyl alcohol) as glue. In a competitive mode, the analyte–peroxidase conjugate was incubated with the free analyte in the microtiter plate, after which the bound fraction of the conjugate was quantified. After addition of the chemiluminescent substrates, light emission was measured in a high-throughput imaging format with a CCD camera. Calibration curves corresponding to analyte concentrations ranging from 0.01 to 100 $\mu g\ mL^{-1}$ were obtained (Surugiu et al. 2001).

A MIP-based optode has been developed for zearalenone (ZON) mycotoxin analysis. The automated flow through assay is based on the displacement of tailor-made highly fluorescent tracers by the analyte from a MIP prepared by UV irradiation of a mixture of cyclododecyl 2,4-dihydroxybenzoate (template, ZON mimic), 1-allyl piperazine (functional monomer), trimethylolpropane trimethacrylate (cross-linker), and 2,2′-azobisisobutyronitrile in acetonitrile (porogen). Three fluorescent analogs of ZON, namely 2,4-dihydroxybenzoic acid 2-[methyl(7-nitro-benzo[1,2,5] oxadiazol-4-yl)amino]ethyl ester (NBDRA), 2,4-dihydroxy-N-pyren-1-ylmethyl-benzamide (PMRA), and of 2,4-dihydroxybenzoic acid 2-[(pyrene-l-carbonyl) amino] ethyl ester (PARA), have been molecularly engineered for the assay development. The pyrene-containing tracers also inform on the characteristics of the microenvironment of the MIP binding sites. PARA was selected to optimize the ZON displacement fluorosensor and shows a limit of detection of 2.5×10^{-5} M in acetonitrile. A positive cross-reactivity has been found for β-zearalenol, but not for resorcinol, resorcylic acid, 17-estradiol, estrone, or bisphenol-A (Navarro-Villoslada et al. 2007).

Metal ion luminescence. An optical sensor for dicrotophos, an organophosphonate pesticide, was developed by using soluble MIPs (Southard et al. 2007). The soluble MIPs were prepared by reversible addition fragmentation chain transfer (RAFT) polymerization followed by ring-closing metathesis (RCM). The polymerization was done in the presence of a template to generate a processable star MIP. The core of the star polymer was a dithiobenzoate substituted tris(β-diketonate) europium(III) complex. The tris(β-diketonate)europium(III) complex served as a polymerization substrate for the three-armed RAFT-mediated star polymer and as a luminescent binding site for the dicrotophos. The star arms were AB block copolymers. Block A was either 1-but-3-enyl-4-vinylbenzene or a mixture of 1-but-3-enyl-4-vinylbenzene and styrene. Block B was styrene or MMA. The but-3-enyls of block A were reacted by RCM with a second generation Grubbs catalyst to give an intramolecularly cross-linked core (Fig. 19.9). The intramolecularly cross-linked MIP was soluble in common organic solvents. The 30% cross-linked soluble and processable star MIP was applied to the determination of dicrotophos with sub-ppb level detection limits.

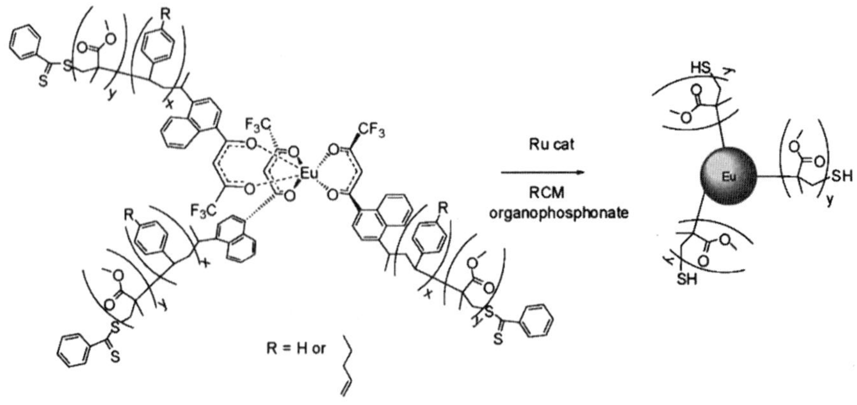

Fig. 19.9 A general scheme for cross-linking of the polymeric core using second-generation Grubbs catalyst

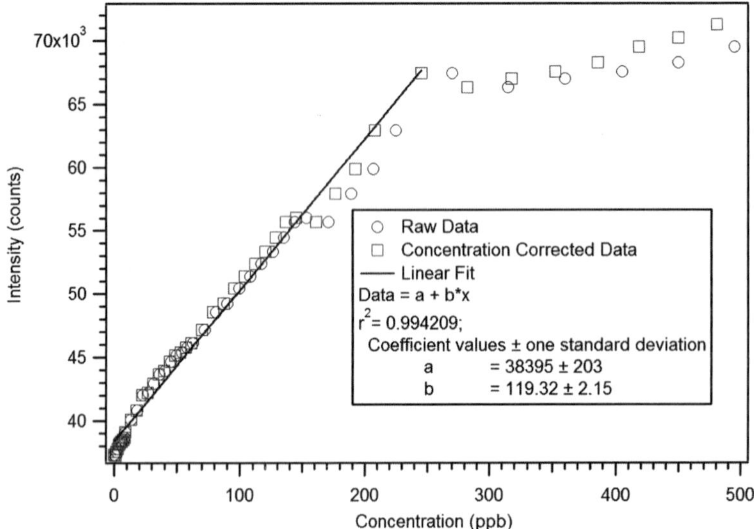

Fig. 19.10 Luminescence titration of 0.1 mg mL^{-1} soluble MIP with 0.10 mM dicrotophos

The soluble MIPs exhibited good sensitivity (LODs in the low ppb range) and a very high selectivity (no interference from near identical interferents) (Figs. 19.9 and 19.10). The polydispersities of the polymers were initially high (MWD = 3.1) because of interstar cross-linking. By cross-linking in very dilute solution, the polydispersity improved (MWD = 1.38). The process results in macromolecules with terminal thiol groups amenable to binding to gold. The soluble MIPs provided a good example of the production of synthetic antibodies, improved chemical sensors, or highly stable and efficient luminescent plastics (Fig. 19.11).

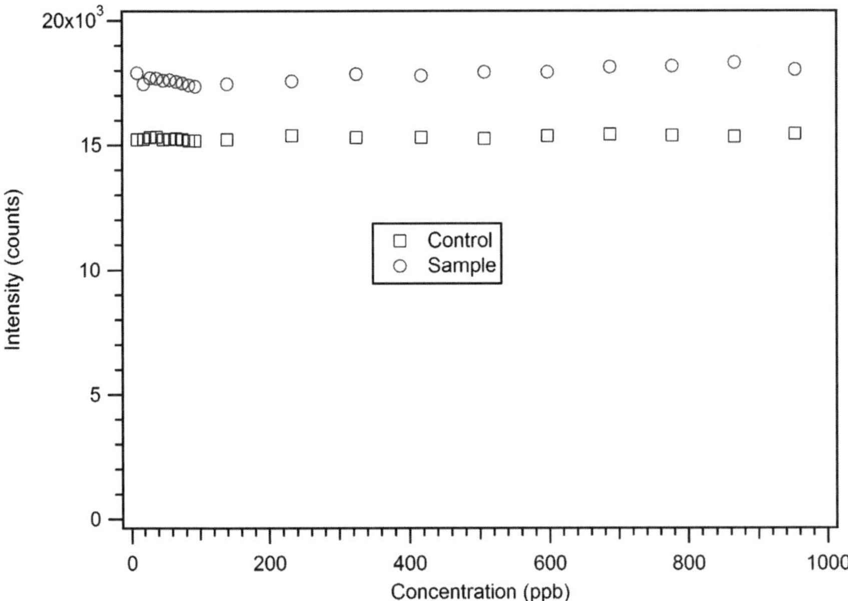

Fig. 19.11 A dichlorvos luminescence interference test with soluble MIP shows essentially no response up to 1 ppm or 1,000 times the detection limit for dicrotophos

19.7 Concluding Remarks

It is now apparent that molecular imprinting is becoming an attractive means for the preparation of molecular, supra-molecular, and biological construct recognition sites. While many sensors transduction methods offer competitive sensitivity, low cost, and ease of use, few allow for selective recognition without some chemical interface. Molecular imprinting has been used to generate numerous artificial receptors that have demonstrated selective binding of their target molecule. Research needs to be focused on developing MIPs with improved functions. The future of MIPs as receptors for bioanalytical applications is promising, both for their potential in fundamental research and for industrial applications. Given the possibility of fine-tuning the binding strength and the specificity of MIPs, these artificial targets may result in faster and more cost-effective identification of new bioactive molecules.

References

A'vila M, Zougagh M, Escarpa A, R1'os A (2008) Molecularly imprinted polymers for selective piezoelectric sensing of small molecules. Trends Anal Chem 27:55–65

Arshady R, Mosbach K (1981) Synthesis of substrate-selective polymers by host-guest polymerization. Makromol Chem 182:687–692

Bae SY, Zeng X, Murray GM (1998) A photometric method for the determination of Pb^{2+} following separation and preconcentration using a templated ion exchange resin. J Anal At Spectrom 10:1177–1181
Beil JB, Zimmerman SC (2004) Synthesis of nanosized cored star polymers. Macromolecules 37:778–787
Beil JB, Lemcoff NG, Zimmerman SC (2004) On the nature of dendrimer cross-linking by ring-closing metathesis. J Am Chem Soc 126:13576–13577
Bolisay LD, Culver JN, Kofinas P (2006) Molecularly imprinted polymers for tobacco mosaic virus recognition. Biomaterials 27:4165–4168
Chen G, Guan Z, Chen C-T, Fu L, Sundaresan V, Arnold (1997) A glucose sensing polymer. Wat Biotech 15:354–357
Chen M, Ghiggino KP, Mau AWH, Rizzardo E, Sasse WHF, Thang SH, Wilson GJ (2004) Synthesis of functionalized RAFT agents for light harvesting macromolecules. Macromolecules 37:5479–5481
Cheng Z, Wang E, Yang X (2001) Capacitive detection of glucose using molecularly imprinted polymers. Biosens Bioelectron 16:179–185
Coessens V, Pintauer T, Matyjaszewski K (2001) Functional polymers by atom transfer radical polymerization. Prog Polym Sci 26:337–377
Damen J, Neckers DC (1980a) Stereoselective syntheses via a photochemical template effect. J Am Chem Soc 102:3265–3267
Damen J, Neckers DC (1980b) Memory of synthesized vinyl polymers for their origins. J Org Chem 45:1382–1387
Damen J, Neckers DC (1980c) On the memory of synthesized vinyl polymers for their origins. Tetrahedron Lett 21:1913–1916
Díaz-Garcia ME, Laíño RB (2005) Molecular imprinting in sol gel materials: recent developments and applications. Mikrochim Acta 149:19–36
Dickert FL, Hayden O (2002) Bioimprinting of polymers and sol-gel phases: selective detection of yeasts with imprinted polymers. Anal Chem 74:1302–1306
Dickert FL, Hayden O, Bindeus R, Mann K-J, Blaas D, Waigmann E (2004) Bioimprinted QCM sensors for virus detection screening of plant sap. Anal Bioanal Chem 378:1929–1934
Dickey FH (1949) The preparation of specific adsorbents. Proc Natl Acad Sci USA 35:227–229
Dickey FH (1955) Specific adsorption. J Phys Chem 59:695–707
Haupt K, Mayes AG, Mosbach K (1998) Herbicide assay using an imprinted polymer-based system analogous to competitive fluoroimmunoassays. Anal Chem 70:3936–3939
Hawker CJ, Bosman AW, Harth E (2001) New polymer synthesis by nitroxide mediated living radical polymerizations. Chem Rev 101:3661–3688
Hayden O, Dickert FL (2001) Selective microorganism detection with cell surface imprinted polymers. Adv Mater 3:1480–1483
Ho K-C, Yeh W-M, Tung T-S, Liao J-Y (2005) Amperometric detection of morphine based on poly(3, 4-ethylenedioxythiophene) immobilized molecularly imprinted polymer particles prepared by precipitation polymerization. Anal Chim Acta 542:90–96
Kriz D, Mosbach K (1995) Competitive amperometric morphine sensor based on an agarose immobilized molecularly imprinted polymer. Anal Chim Acta 300:71–75
Kriz D, Kempe M, Mosbach K (1996) Introduction of molecularly imprinted polymers as recognition elements in conductometric chemical sensors. Sens Actuators B 33:178–181
Le TP, Moad G, Rizzardo E, Thang SH (1998) PCT Int. Appl. WO 9801478 A1 980115. Chem Abstr 128:115390
Matsui J, Akamatsu K, Nishiguchi S, Miyoshi D, Nawafune H, Tamaki K, Sugimoto N (2004) Composite of Au nanoparticles and molecularly imprinted polymer as a sensing material. Anal Chem 76:1310–1315
Mudd S (1932) A hypothetical mechanism of antibody production. J Immunol 23:423–427
Murray GM, Jenkins AL, Bzhelyansky A, Uy OM (1997) Molecularly imprinted polymers for the selective sequestering and sensing of ions. JHUAPL Tech Digest 18:432–441

Navarro-Villoslada F, Urraca JL, Moreno-Bondi MC, Orellana G (2007) Zearalenone sensing with molecularly imprinted polymers and tailored fluorescent probes. Sens Actuators B 121:67–73

Pauling L (1940) A theory of the structure and process of formation of antibodies. J Am Chem Soc 62:2643–2657

Pauling L, Campbell DH (1942) The manufacture of antibodies in vitro. J Exp Med 76:211–220

Piletsky SA, Piletskaya EV, Elgersma AV, Karube L (1995) Atrazine sensing by molecularly imprinted membranes. Biosens Bioelectron 10:959–964

Piletsky SA, Piletska EV, Bossi A, Karim K, Lowe P, Turner APF (2001) Substitution of antibodies and receptors with molecularly imprinted polymers in enzyme-linked and fluorescent assays. Biosens Bioelectron 16:701–707

Polyakov MV (1931) Adsorption properties and structure of silica gel. Zh Fiz Khim 2:799–805

Rosatzin T, Andersson L, Simon W, Mosbach K (1991) Preparation of Ca^{2+} selective sorbents by molecular imprinting using polymerizable ionophores. J Chem Soc Perkin Trans 2:1261–1265

Shea KJ, Thompson EA (1978) Template synthesis of macromolecules selective functionalization of an organic polymer. J Org Chem 43:4253–4255

Shea KJ, Thompson EA, Pandey SD, Beauchamp PS (1980) Template synthesis of macromolecules. Synthesis and chemistry of functionalized macroporous poly(divinylbenzene). J Am Chem Soc 102:3149–3155

Shimizu K, Greene MT (2005) Colorimetric molecularly imprinted polymer sensor array using dye displacement. J Am Chem Soc 127:5695–5700

Southard GE, Murray GM (2005) Synthesis of vinyl-substituted β-diketones for polymerizable metal complexes. J Org Chem 70:9036–9039

Southard GE, Van Houten KA, Murray GM (2006) Heck cross-coupling for synthesizing metal complexing monomers. Synthesis 15:2475–2477

Southard GE, Van Houten KA, Murray GM (2007) Soluble and processable phosphonate sensing star molecularly imprinted polymers. Macromolecules 40:1395–1400

Surugiu I, Danielsson B, Ye L, Mosbach K, Haupt K (2001) Chemiluminescence imaging ELISA using an imprinted polymer as the recognition element instead of an antibody. Anal Chem 73:487–491

Titirici MM, Sellergren B (2004) Peptide recognition via hierarchical imprinting. Anal Bioanal Chem 378:1913–1921

Umpleby RJ II, Rushton GT, Shah RN, Rampey AM, Bradshaw JC, Berch JK Jr, Shimizu KD (2001) Recognition directed site-selective chemical modification of molecularly imprinted polymers. Macromolecules 34:8446–8452

Wulff G, Sarhan A (1972) Use of polymers with enzyme-analogous structures for the resolution of racemates. Angew Chem Int Ed Engl 11:341–356

Ye L, Haupt K (2004) Molecularly imprinted polymers as antibody and receptor mimics for assays, sensors and drug discovery. Anal Bioanal Chem 378:1887–1897

Zeng X, Murray GM (1996) Synthesis and characterization of site selective ion exchange resins templated for lead (II) ion. Sep Sci Technol 31:2403–2418

Zimmerman SC, Zharov I, Wendland MS, Rakow NA, Suslick KS (2003) Molecular imprinting inside dendrimers. J Am Chem Soc 125:13504–13518

Chapter 20
Biomimetic Synthetic Receptors as Molecular Recognition Elements

Hans-Jörg Schneider, Soojin Lim, and Robert M. Strongin

Abstract The rapid development of supramolecular chemistry has in recent years made available synthetic host compounds for almost all possible analytes. The study of artificial receptors has led to a better understanding and thus the design of binding mechanisms, which also underly the function of biological receptors. The principles ruling sensitivity and selectivity of host–guest complexes are discussed on a quantitative basis, with emphasis on the significance of multivalent interactions. Limitations and advantages of synthetic versus biological receptors are compared. Complexation in particular of biologically relevant analytes is illustrated with representative examples, ranging from simple inorganic cations and anions over aminoacids, peptides, nucleotides, carbohydrates, to terpenes and steroids. Particular attention is given to receptors that function in natural aqueous environment and furnish optical detection signals.

Keywords Supramolecular complexes · Noncovalent interactions · Selectivity · Sensitivity · Sensors

Abbreviations

ΔG	Free energy
ΔΔG	Free energy difference
K	Equilibrium constant
PCA	Principal component analysis
CHEF	Chelation enhanced fluorescence
PET	Photoinduced electron transfer

H-J. Schneider (✉)
Universität des Saarlandes, D 66041, Saarbrücken, Germany
e-mail: h-j.schneider@mx.uni-saarland.de

M. Zourob (ed.), *Recognition Receptors in Biosensors*,
DOI 10.1007/978-1-4419-0919-0_20,

DMSO	Dimethyl sulfoxide
$CDCl_3$	Chloroform, deuterated
ATP	Adenosin triphosphate
GTP	Guanosin triphosphate
CTP	Cytidine triphosphate
UTP	Uridine triphosphate
HEPES	4-(2-hydroxyethyl)-1-piperazineethanesulfonic acid (a buffer)
NADP	Nicotinamide adenine dinucleotide phosphate
NADPH	Nicotinamide adenine dinucleotide phosphate, reduced form
CD's	Cyclodextrins

Note: Aminoacids are mentioned with the usual three character notation, too many to be explained here, see common textbooks.

20.1 Introduction

Artificial receptors can in principle be made for every analyte, in contrast to systems relying on biological elements. These are essentially restricted to receptors of their natural target molecules, although they may often be used for structurally related analytes or may be synthetically varied in order to achieve the desired goal. The drawback of many synthetic host compounds is often a smaller sensitivity and selectivity in comparison to natural receptors. Over the last decade, organic chemists, however, have learned their lessons from nature and have in the framework of supramolecular chemistry developed synthetic host compounds that in some cases perform even better than biological systems. Sensor array techniques, supported by PCA analyses, hold much promise to overcome the selectivity limitation, as do, e.g., computer-assisted multiwavelength measurements. In addition to their unlimited applicability, synthetic host compounds have the advantage of higher stability and less fatigue. They also can be equipped with any desired signaling or reporter units, such as fluorescence dyes, making the use of coupled reaction sequences unnecessary, which are typical of many enzymatic assays. Alternatively, they can be implemented in electrochemical devices, such as electrodes.

Immobilization in or on inorganic or organic polymers has become standard operation, and reduces the problem of washing out. Miniaturization down to nanoscale and thin films poses no problems, leading to higher speed and sensitivity of response. Finally, the cost of synthetic receptors can be significantly lower than those based on biological materials. In the following, an overview on the available synthetic host–guest systems for the detection of a large variety of analytes is given, restricted to examples that highlight possible applications as well as the underlying mechanisms of molecular recognition. The latter aspect is emphasized, as it contributes to our understanding of interactions also in biological systems, and will be of help for the rational design of new sensors. For a general discussion of synthetic host–guest complexes, we refer to some of the available books (Vögtle 1993; Lehn 1995;

Hamilton 1996; Lehn et al. 1996; Beer et al. 1999; Schneider and Yatsimirsky 2000; Steed and Atwood 2000; Atwood and Steed 2004; Schmuck and Wennemers 2004; Ariga and Kunitake 2005; Cragg 2005; Schrader and Hamilton 2005; Schalley 2007; Steed et al. 2007) and reviews (Hartley et al. 2000; Houk et al. 2003; Meyer et al. 2003; Paulini et al. 2005; Kruppa and König 2006; Albrecht 2007;and Oshovsky et al. 2007), which also address the noncovalent binding mechanisms (Schneider 1991; Schneider 1994; Schneider 2009; Nassimbeni 2003; Hunter 2004).

20.2 How to Achieve High Selectivity, Affinity, and Sensitivity

High selectivity, affinity, and sensitivity are the major issue for the design of receptor compounds. We will briefly discuss in general terms how these goals depend on each other, and how they can be optimized. High affinity is reached also in natural receptors by providing many interaction sites A, B, C etc. simultaneously. In the laboratory, one can synthesize host molecules with as many interaction sites as desired, with the aim to use them simultaneously by the analyte molecule. This way one can maximize the affinity, and, as will be shown, also the selectivity. In all these cases, the lock-and-key condition of optimal geometric fit between the host and the guest must be pertained. Geometric fitting is the basis of shape-dependent selectivity and will not be discussed here in detail; suffice it to note that conformation changes in the complexation process is a common feature in almost all real situations, but will always go on the expense on strain energy built up by deviation from the minimum energy in the ground state. Positive and negative cooperativity between the single binding sites may also occur, and then will lead to deviation from the simple additivity of single interactions. As a rule, the corresponding single interaction energies ΔG_A, ΔG_B, ΔG_C, etc. add up to a total free complexation free energy ΔG_t , which is the basis for the chelate effect well-known from coordination chemistry and now often referred to as multivalency (Mammen et al. 1998; Mulder et al. 2004; Badjic et al., 2005). Synthetic host–guest complexes can indeed achieve very high affinities by multivalent interactions, which may come close or even exceed the binding strength in natural complexes. This has been found early for siderophores (Buhleier et al. 1978; Shanzer et al. 1991; Raymond 1994; Heinz et al. 1999), more recently, for instance for the biotin-streptavidin complex (Scheme 20.1, (**a**), Rao et al. 1998; Rao et al. 2000). The complex with cucurbituril (Scheme 20.1 (**b**), Rekharsky et al. 2007) illustrates also that the often-assumed general decrease of mobility in terms of entropy with an increase of binding strength enthalpy (Dunitz 1995; Williams et al. 2004) does not necessarily hold. It should be noted that implementation of even more binding sites in a host than can be used at a given time by the complementary analyte molecule also leads to significantly enhanced affinities (Schneider et al. 2004; Liu et al. 2001; Abraham et al. 2002; Arduini et al. 2005), e.g., with dendrimers which can provide a large excess of binding sites for small guest compounds (Page et al. 1996).

If a given receptor R reacts with either analyte *i* or *j*, it follows

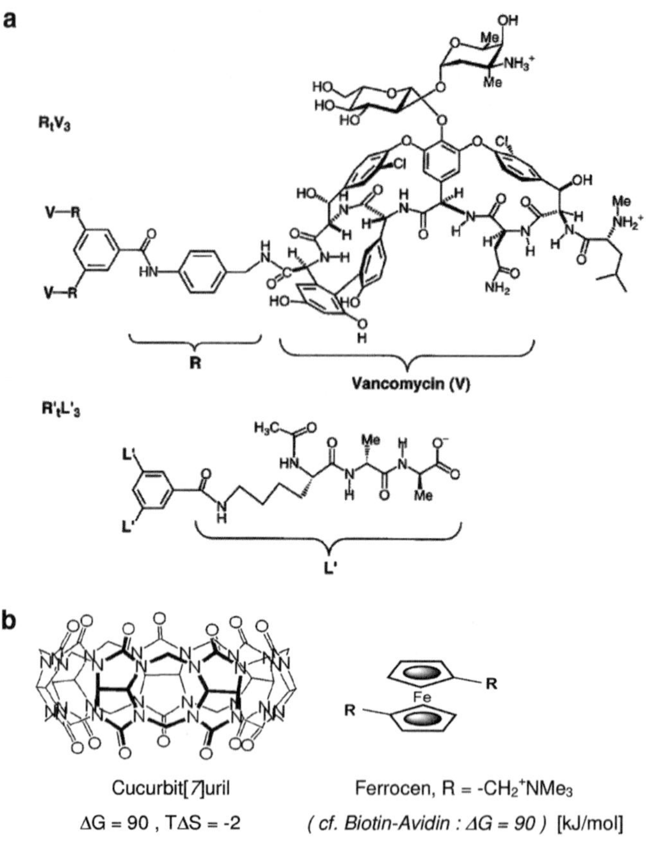

Scheme 20.1 Synthetic host-guest complexes with affinities exceeding, e.g., the biotin-streptavidin association ($K = 10^{17}$ [M^{-1}]). (**a**) a tris-vancomycinderivative (Rao et al. 1998; Rao et al. 2000)¸ (**b**) cucurbituril complex (Rekharsky et al. 2007)

$$K_{Ri}/K_{Rj} = \exp\left[n\left(\Delta\Delta G_j - \Delta\Delta G_i\right)/RT\right]$$

If a number *n* of interactions contribute, the equation will then be

$$K_{Ri}/K_{Rj} = \exp\left[n\left(\Delta\Delta G_j - \Delta\Delta G_i\right)/RT\right]$$

It is important to note that therefore the presence of multiple binding sites enhances not only affinity but also the selectivity in supramolecular complexes. In simple metal ion or anion complexes, where only one kind of forces act simultaneously on the target molecule, there should be linear correlations between total binding strength and selectivity. Figure 20.1 illustrates both the linear increase of binding strength with the number of interactions (the chelate effect) and the accompanying linear increase of selectivity (Schneider and Yatsimirsky 2008).

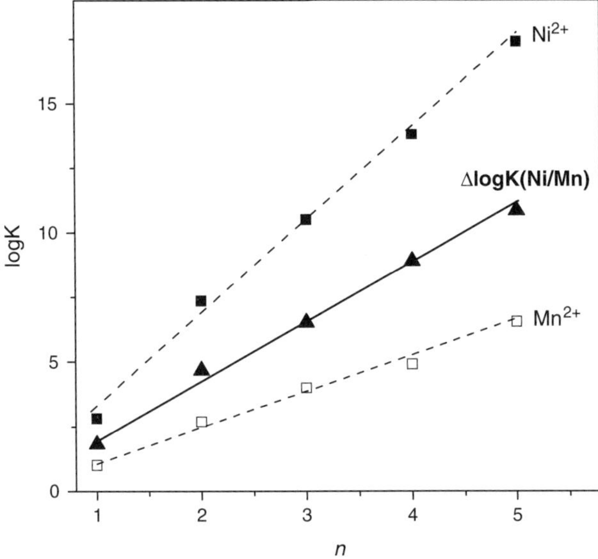

Fig. 20.1 Logarithms of stability constants of Mn^{2+} and Ni^{2+} complexes with linear polyamine ligands $H_2N(CH_2CH_2NH)_{n-1}H$ as a function of total number n of nitrogen donor atoms. As n increases from one to five the selectivity of binding of Ni^{2+} over Mn^{2+} increases in terms of K_{Ni}/K_{Mn} by nine orders of magnitude (With permission of Schneider and Yatsimirsky (2008), copyright 2008 the Royal Society of Chemistry)

Scheme 20.2 A chiral crown ether host in, in which hydrogen bonds with the terminal–NH_3 group provide the primary binding force, and the naphthylside groups provide enantioselectivity (Helgeson et al. 1973)

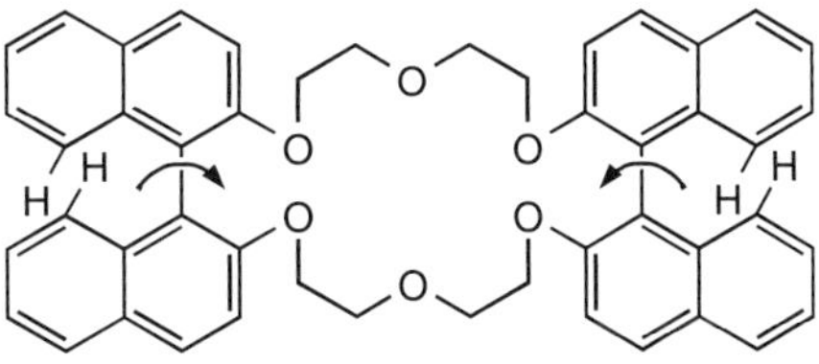

Most natural as well as synthetic host molecules use interactions of different nature and strength, where often one binding site A secures a high affinity, and other binding sites B, C, D, etc. control the selectivity. A classical example is the crown ether complex with aminoacids, in which hydrogen bonds with the terminal –NH_3 group provide the primary binding force, and the naphthylside groups provide the enantioselectivity (Scheme 20.2) (Helgeson et al. 1973).

Scheme 20.3 illustrates that two complexes may differ by a factor of, e.g., 10exp5 in strength, but will retain the same selectivity between guest molecules i and j, as long as the secondary discriminating interactions at the sites B,C, D, etc remain constant. Therefore, complexes with anisotropic analyte molecules will often not exhibit a simultaneous increase or decrease of binding selectivity and affinity. The discrimination factors ΔG_B, ΔG_C, ΔG_D, etc., including geometric

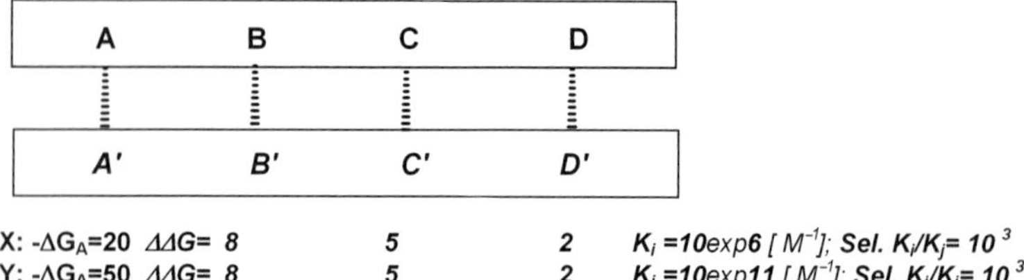

Scheme 20.3 Schematic illustration of complexes with same selectivity but different affinity: the discriminating binding free energy difference ΔΔG between to guest molecules ***i*** and ***j*** with two host molecules **X** and **Y** is the same; only the primary interaction ΔG differ (all ΔG values in kJ/mol) (Schneider and Yatsimirsky 2008)

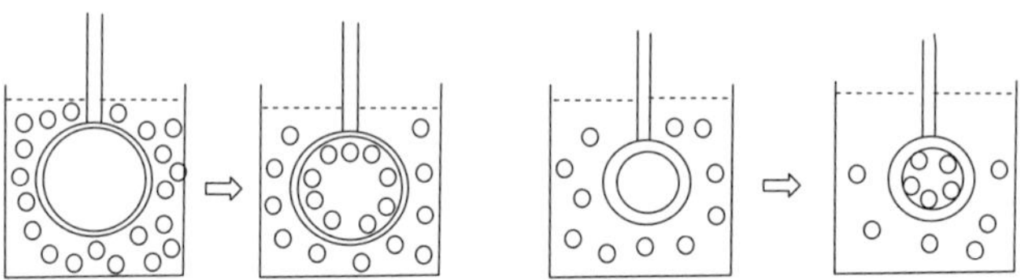

Fig. 20.2 Schematic illustration of the compartmentalization effect: smaller detection units can work in more dilute solutions

mismatch, can actually counteract the primary binding contribution ΔG_A; then one expects even an increased selectivity with decreasing total complexation free energy. The total free energy ΔG_t involved in associations with efficient host compounds such as inhibitors can amount to values of, e.g., K = 10exp9 [M^{-1}] and more (see scheme 20.1); the differences necessary for a relatively good selectivity ratio of, e.g., $K_{Ri}/K_{Rj} = 100$ is then smaller by many orders of magnitude. As a consequence, even percentage-wise minor alterations of ΔG_A at the affinity-dominating site can have the same impact on selectivity as relatively larger changes at the weaker discrimination sites; also for this reason, strictly linear correlations between ΔG_t and ΔΔG are generally not to be expected. The most promising approach to reaching both high selectivity and affinity is obviously to design one binding site that should be as strong as possible, and to provide possibly much weaker or even repulsive interaction sites that serve to discriminate between different analytes.

The Sensitivity of detection of course increases with the affinity in host-analyte complexes, but is not necessarily directly related. First of all, the detection and selectivity limit depends on the actual proportion of complexed to uncomplexed analyte, which, e.g., for high-affinity complexes in more concentrated solution can be so large that only a fully complexed material is present. Although this factor is well known, it is often overlooked that the strength of a detection signal depends also on the size of the detection chamber. Figure 20.2 illustrates that with high-enough affinities, a small chamber obviously needs less

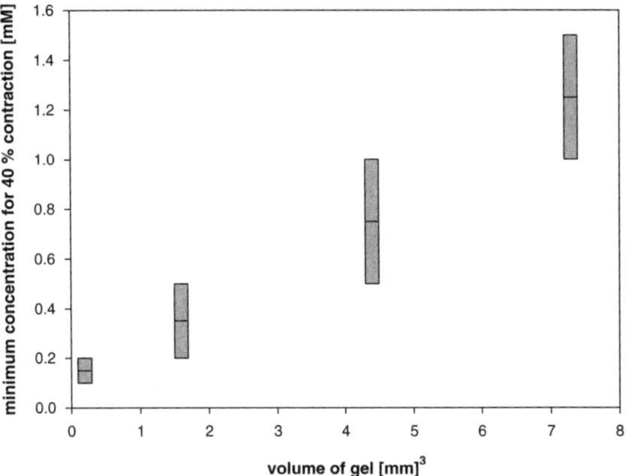

Fig. 20.3 Example for sensitivity enhancement by compartimentalization. Volume change of a polymer gel with binding centers for tartraric acid derivatives; guest concentration necessary for contraction of 40% as function of the gel particle size (With permission of Schneider and Kato (2007), copyright 2008 John Wiley & Sons Inc)

analyte molecules to occupy the available host binding sites, whereas a larger chamber needs more analyte material or higher concentration, in order to reach the same proportional occupation and therefore signal change (Schneider et al. 2004; see also Koblenz et al. 2008). That the sensitivity increases with r^{-3} of such a detection chamber, as has been discussed in the framework of miniature sensors (Clark et al. 1999; Kopelman and Dourado 1996); however, it has been overlooked that this increase is only possible as long as the affinity is large enough to pick up analyte molecules from very dilute solutions. That under those conditions the sensitivity by such a compartmentalization effect is real is shown in Fig. 20.3, with a chemomechanical polymer sensor (Schneider and Kato 2007; Guo et al. 2005).

20.3 Examples of Different Analytes

In the following, we highlight the use of synthetic receptors for a selection of analytes, preferentially those of biological, medicinal, or environmental importance. In view of the vast literature, we discuss only a necessarily arbitrary selection, illustrating the possibilities of such artificial complexes for a selective and sensitive detection for representative analyte classes. Enhanced fluorescence emission by binding of an analyte – such as chelation of a metal ion ("CHEF"-effect, Chelation Enhanced Fluorescence) – is the most preferred detection method and has been discussed in numerous books and reviews (Czarnik 1994; Lakowicz 1994; de Silva et al. 1997; Desvergne and Czarnik 1997; Rettig et al. 1999; Rurack 2001; Prodi 2005). As electrochemical devices such as ion selective electrodes, optical techniques can be well adapted for remote sensing, using waveguides and fiber optics. The CHEF principle usually relies on removal of fluorescence quenching by lone pairs of nitrogen atoms which are placed in the vicinity of a luminophore; if the lone pair is removed by occupation with, e.g., a metal ion, or also by protonation, fluorescence emission appears.

Selectivity is brought about by providing suitable analyte binding units in the vicinity of the nitrogen. As will be illustrated in the following section with some examples, one can design ligands that exhibit a signal only in the presence of two analyte molecules, or, in the more simple case, only at a given pH

20.3.1 Inorganic Cations

The choice of suitable binding units for metal ions follows the known principles of coordination chemistry. Alkali and earth alkali cations are preferentially complexed with oxygen-containing ligands; with macrocyclic systems such as crown ethers, one can achieve selectivity by shape complementarity. Figure 20.4 illustrates that with a closed crown ether cavity, one can reach a selectivity between K and Na ions surpassing that of natural ionophores such as Valinomycin (Ghidini et al. 1990) ; at the same time, one observes the theoretically expected correlation between affinity and selectivity (Schneider and Yatsimirsky 2008). A coumarine derivative with rotatable –C = N double bonds shows a 550-fold fluorescence emission with Mg (II) ions, which this way allows discrimination from other metal ions like Li(I), Ca (II), and Zn(II) or biologically relevant transition-metal ions which all show much weaker emission (Ray and Bharadwaj 2008).

The fluorophore shown in Fig. 20.5 binds potassium ions in basic solutions within ring A, and in acidic solution sodium ions in ring B, due to N-protonation, and represents a logical gate function (Xu et al. 2002). As shown, two PET processes are switched by protonation, leading to distinct fluorescence emission, which amounts to a 7.2-fold enhancement with K-ions, and under acidic conditions,

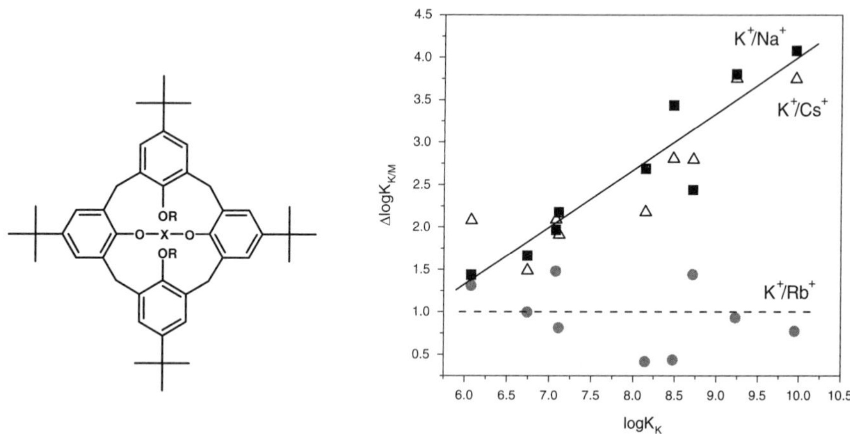

Fig. 20.4 Selectivity ΔlogK with calixarene-crown ether complexes (X = $CH_2CH_2(OCH_2CH_2)_n$ $n = 3$ or 4, R = Me, Et, n-C_3H_7, i-C_3H_7, or $CH_2C_6H_5$), for K^+ *vs.* Na^+ (*squares*), for K^+ *vs.* Cs^+ (*triangles*), and for K^+ *vs.* Rb^+ (*circles*); experimental data: (With permission of Schneider and Yatsimirsky (2008), copyright 2008 the Royal Society of Chemistry)

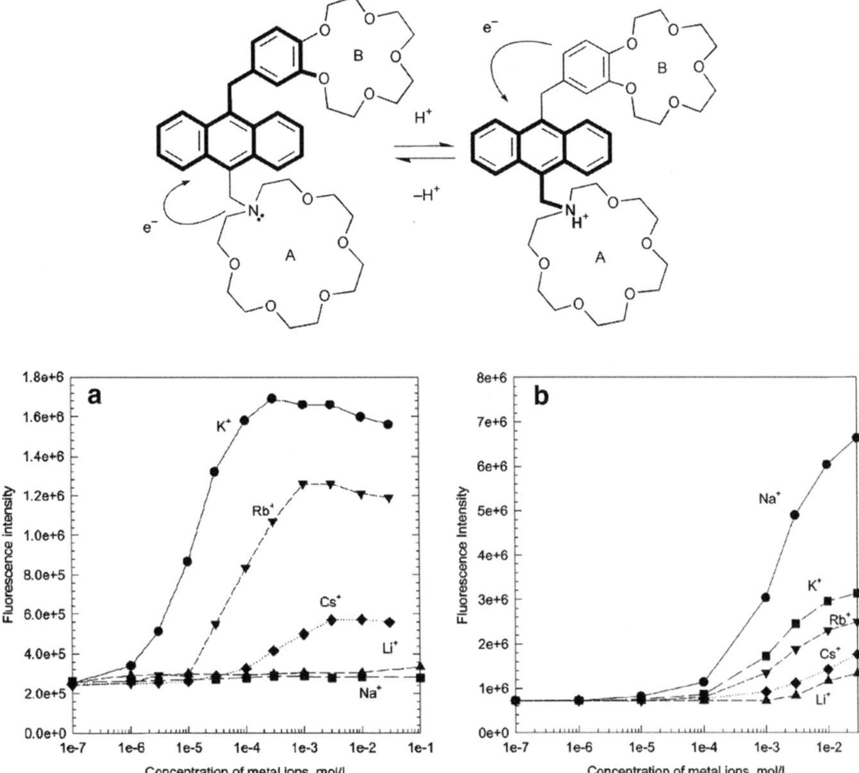

Fig. 20.5 A crown ether derivative for selective alkali ion detection; pH-dependent fluorescence emission, under basic condition (**a**); under acidic condition (**b**), (With permission of Xu et al. (2002), copyright 2008 the Royal Society of Chemistry)

to a 9.4-fold enhancement with Na-ions. Depending on pH, other cations such as Rb^+ also induce a sizeable emission (Fig. 20.5, Xu et al. 2002).

The combination of a cryptand and a macrocycle with an anthracene (Fig. 20.6) leads to a fluorescent signaling system (Bag and Bharadwaj 2005) that gives fluorescence enhancement only in simultaneous presence of alkali complexed by the crown) and transition metal ions (complexed by the cryptand), again by photoinduced electron transfer (PET). Protonation as well as addition of Na or K ions leads to only small emission changes, as the quenching N-lone pair of the cryptand remains undistorted. Only if both transition and alkali metal ions are present, one observes an emission that is $>$ 100-fold enhanced with the Na and Zn(II) or /Fe(II)combination.

Heavy and transition metal ions are preferentially bound to nitrogen-or sulfur-containing ligands such as cryptands, which also can be equipped with an optical signal output (Fabbrizzi et al. 1996; Prodi et al. 2000). With the simple anthracene derivative **1** Zn or Cu cations lead to more than 2,000-fold emission (Kubo, Sakurai

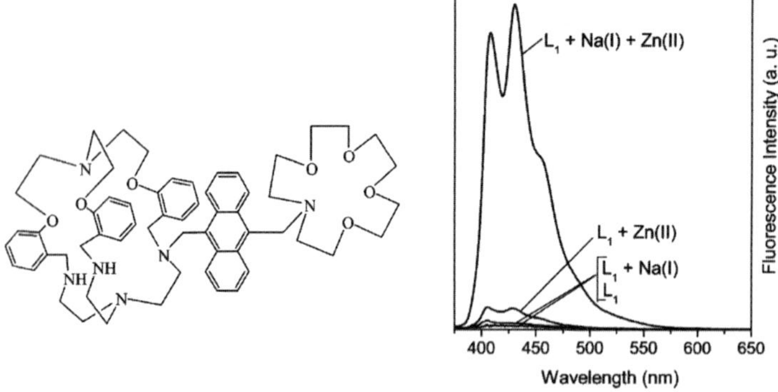

Fig. 20.6 A Cryptand-crownether combination for simultaneous detection of alkali and transition metal ions; maximal fluorescence emission in simultaneous presence of, e.g., Na and Zn ions (With permission of (Bag and Bharadwaj (2005), copyright 2008 the Royal Society of Chemistry)

$(HOC_2H_5)_2N$ $N(C_2H_5OH)_2$

1 **2**

and Mori 1999). Complexation of ferric ions is often achieved with catechol host compounds, reaching affinities close to natural peptidic siderophores (Lützen 2007). With phenol-oxazoline tripods such as **2** fluorescence allows detection limits in the low nanogram per milliliter range methanol/water (80/20) mixtures; the emission is ascribed to an intramolecular proton transfer from an excited state (Kikkeri et al. 2007).

20.3.2 Anion Complexes

Supramolecular anion complexation has been aptly reviewed (Bianchi, et al. 1997; Schmidtchen and Berger 1997; Beer and Gale 2001; Bowman-James 2005; Schmidtchen 2005; Amendola and Fabbrizzi 2006), which allows us to mention

3 4 5

Table 20.1 Association constants (as lgK values) of selected halide receptors, in water

	F	Cl	Br
3, n = 6, pK 5.1	10.5	<2	–
3,n = 3, pK 7.5	3.3	–	–
4 n = 6 , pK 4.4	6.1	5.7	4.4
4 n = 4, pK 6.85	2.75	2.4	1.95

here only some typical complexes, particularly those that lend themselves for sensing application. The distinction of anions such as carboxylates, phosphates, sulphates, etc. in water is achieved by host compounds with complementary and geometrically matching cationic functions, but also hydrogen bond donors.

Many synthetic anion receptors rely on the presence of protonated or peralkylated nitrogen centers in cavities suitable for shape discrimination. Impressive affinities and selectivities have been materialized with protonated polyaza-macrocycles such as with **3** and **4** (Reilly et al. 1995; Dietrich et al. 1996). Table 20.1 illustrates also how the binding depends on the protonation and thus on the pH of the solutions, which obviously limits sometimes application to anions only of strong acids. The data also illustrate again how the selectivity is diminished by a smaller affinity, as theoretically predicted (see Sect. 20.2) The host **5** was designed for complexation of anisotropic anions such as of azide, with lgK >4 (Dietrich et al. 1984). The fully protonated macrocycle **6** binds triphosphates such as ATP with a theoretical value of lgK = 11 (Hosseini, et al. 1988).

Guanidinium systems such as the **7** cation have the advantage of working also at higer pH, as the corresponding pK is 13.5 (Schug and Lindner 2005). Furthermore, they can be fixed within a bicyclic framework such as **7**, which allows to attach substituents securing discrimination of chiral anions (Müller et al. 1988; Echavarren et al. 1989). Host compounds with peralkylated N- or S- centers have the advantage of ion pairing at any pH. The tricyclic receptor **8** has a confined space, allowing with X = $(CH_2)_6$ complexation of iodide, and with X = $(CH_2)_8$ that of larger anions such as phenolate (Schmidtchen and Berger 1997).

Often, one uses the displacement of an indicator, binding not too strongly to a chosen host, which, e.g., quenches the emission of a fluorescence indicator. Addition of an analyte such as an anion displaces the indicator from the receptor's cavity, leading to emission of the detection signal from the indicator (Wiskur et al.

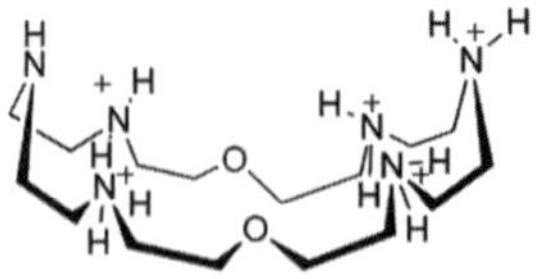

6 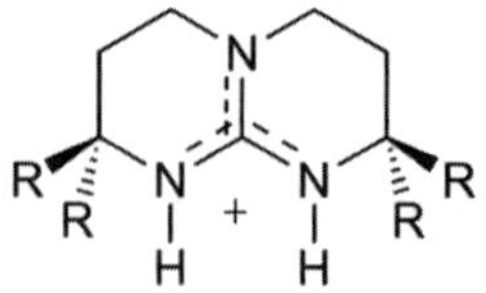 **7** 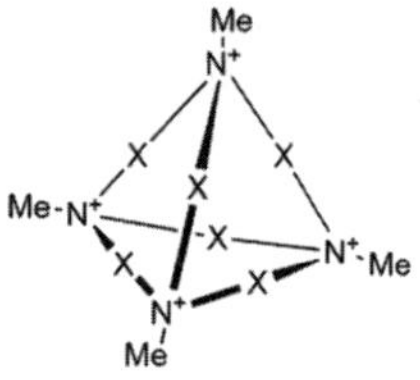 **8**

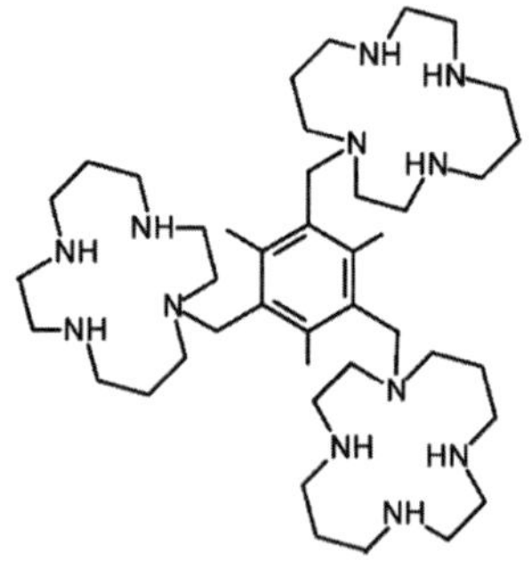

9 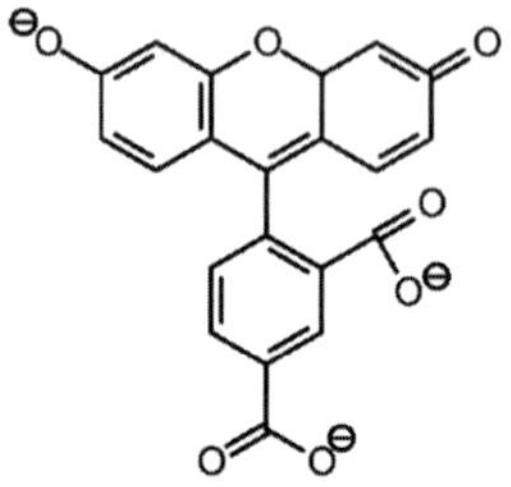 **10**

2001). The trifurcated cyclam host **9** containing three copper(II) ions recognizes tricarboxylates such as citrate , which can be detected through displacement of primarily bound fluorescent indicator **10** (Fabbrizzi et al. 2005). Tricarboxylates such as citrates are bound with lgK $\approx$ 5.5 and dicarboxylates such as maleate or tartrate 10 times less.

Electroneutral host compounds for anion recognition use mostly hydrogen bonding; the strong competition with water requires usually the use of nonaqueous media for measurements; this can also be the lipophilic membrane coating of an ion-sensitive electrode. Simultaneous action of many NH-donors such as in the cyclopeptide **11** leads to an efficient complexation of, e.g., sulfate in 20:80 methanol-water (Kubik and Goddard 2002). Open-chain hosts providing several donors, also in the form of urea derivatives such as **12** complex halides in chloroform with appreciable selectivity as, e.g., $K_{Cl}/K_{Br} = 9$ (Werner and Schneider 2000). If the donor functions are preorganized on a steroid skeletons, higher affinities are reached, with K values of up to 10^{11} M^{-1} for chloride (Clare et al. 2005); the selectivity, however, is similar to that of simple host **12.**

Pyrroles also provide sufficiently acidic donors, as illustrated with the octamethylcalix[4]pyrrole **13**, which binds fluoride with lgK = 4.2 in CD_2Cl_2 (Gale et al. 1996). The macrocycle **14** exhibits not only high binding affinities, e.g., in CD_3CN $K_{Cl} = 2.1 \times 10^6$ M^{-1}, but with $K_F/K_{Cl} = 133$ and $K_{Cl}/K_{Br} = 780$ also a remarkable selectivity (Chang et al. 2005).

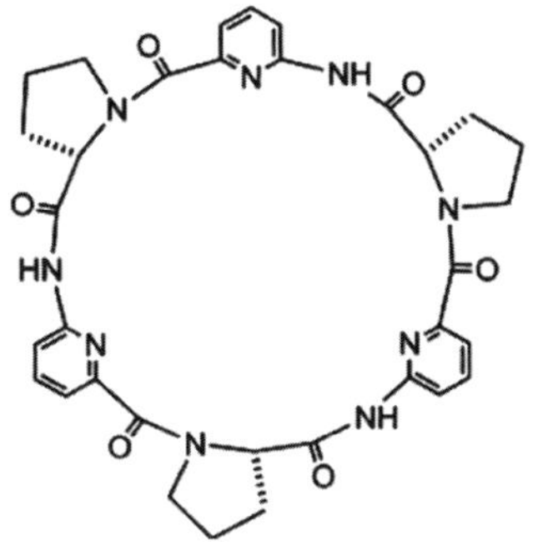

11 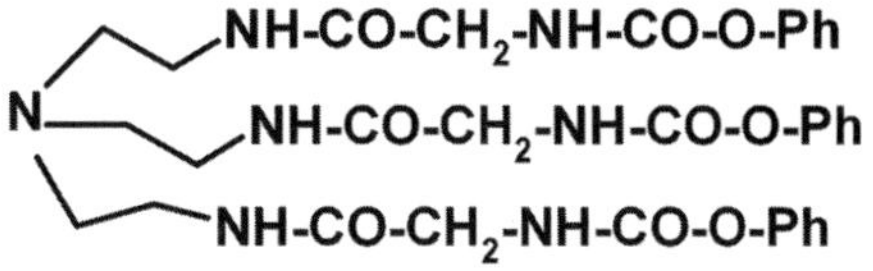 **12**

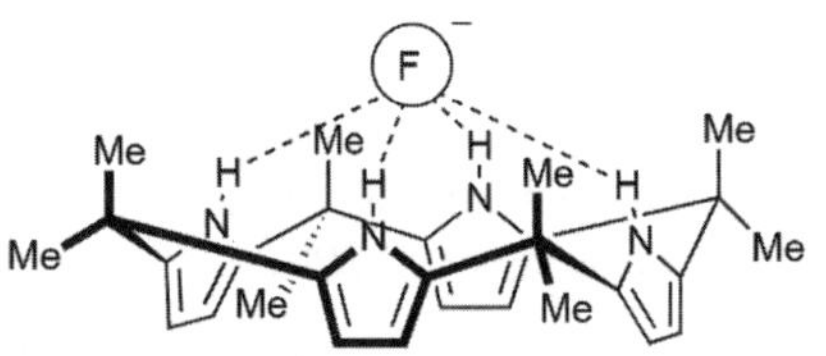

13 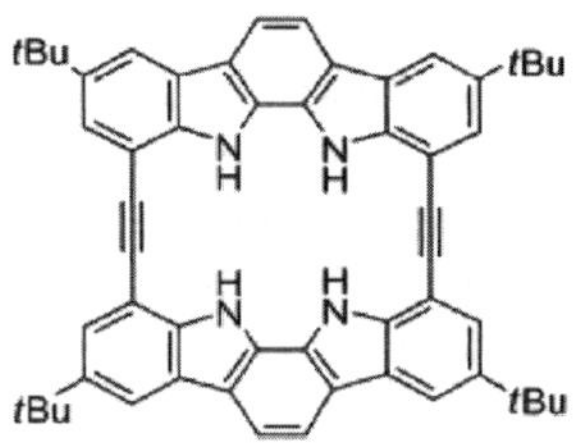 **14**

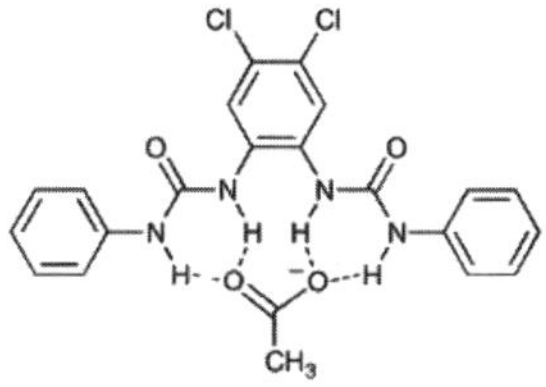

15 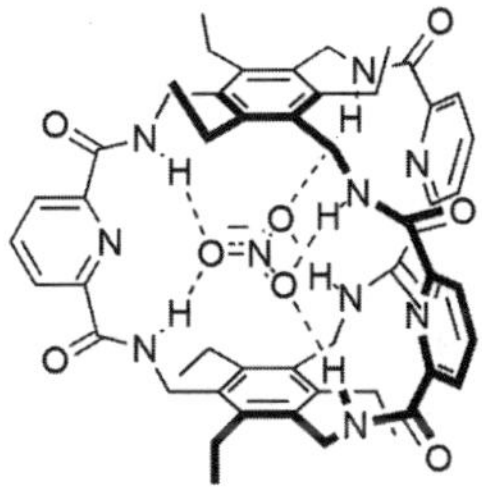 **16** 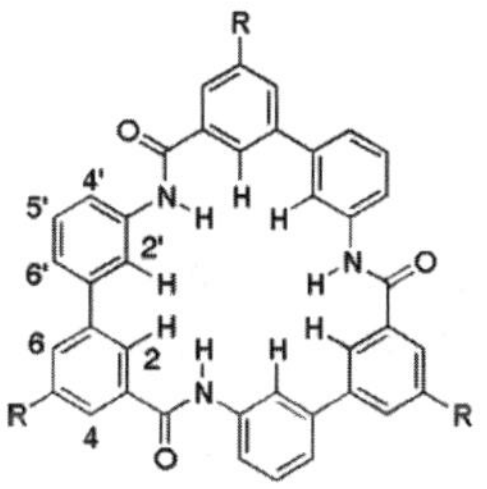 **17**

Hydrogen bonds have the unique feature of a relatively high directionality. This can be used for selective binding of polydentate anions, as illustrated with the complexes **15** and **16.** The first binds in DMSO, e.g., carboxylates or phosphate with lgK $\approx$3, whereas halides show lgK < 2 (Brooks et al. 2006). The convergent hydrogen bonds in **16** secure selective binding of nitrate (Best et al. 2003). The large cavity of host **17** leads to a reversal of the usual binding preference for anions with higher charge density, with $K_I/K_{Cl} = 15$ in $CDCl_3$ + 2% DMSO (Choi and Hamilton 2003).

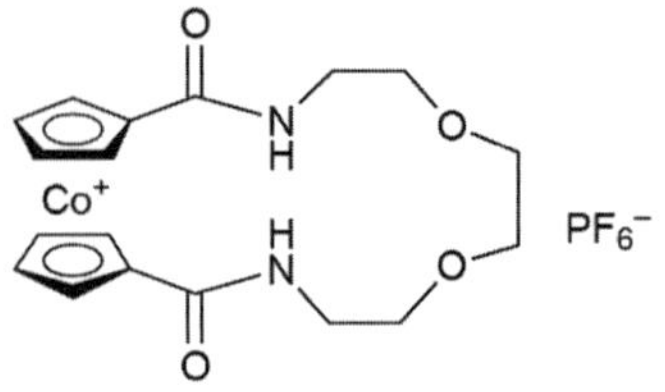

18 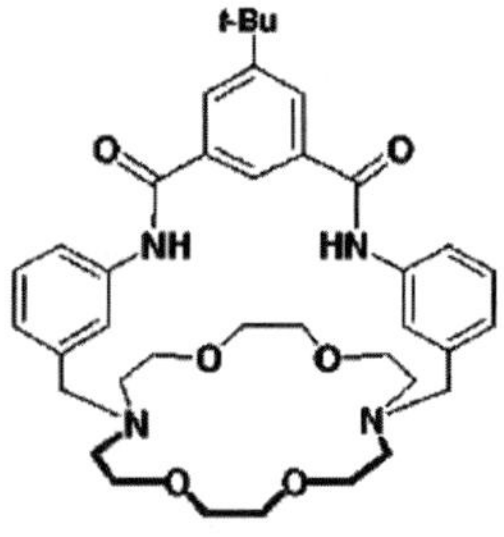 **19**

Electrochemical methods are applicable not only to ion sensitive electrodes with the abovementioned lipophilic host compounds, but also to redoxactive systems. With the receptor **18**, one observes, e.g., upon complexation with phosphates, much higher cathodic shifts in comparison with chloride complexes using simultaneously. Receptors providing binding sites both for a cation and an anion as ion pair, e.g., simultaneously a crown ether and amide functions, allow positive cooperativity as result of the ion pairing between complexed cation and anion; thus, host **19** shows with KCl a much higher affinity than with the K BPh_4 salt, since the tetraphenylborate is too large for the cavity (Mahoney et al. 2001).

20.3.3 Complexation of Aminoacids and Peptides

In many cases, synthetic hosts for aminoacids and peptides (Peczuh and Hamilton 2000) rely in the first place on ion pairing with the charged termini, e.g., between the $–^{+}NH_3$-terminus and the phosphoryl- or sulphonato- group of calixarenes (Arena et al. 1999). The calix **20** complexes in methanol many aminoacids with lgK values above 3, with a stability increasing with the hydrophobicity of the aminoacids (Zielenkiewicz et al. 2006); in particular, aromatic residues are bound within the calixarene cavity (Douteau-Guevel et al. 2002). The complex **21**

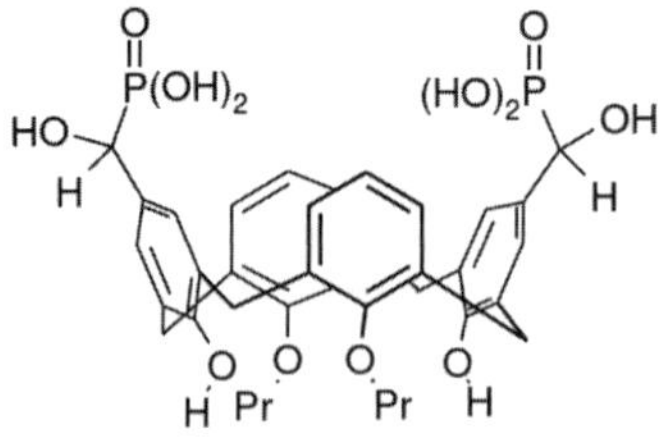

20 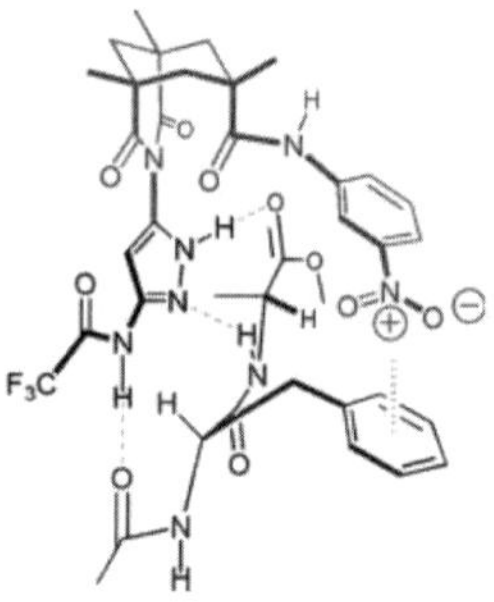 **21**

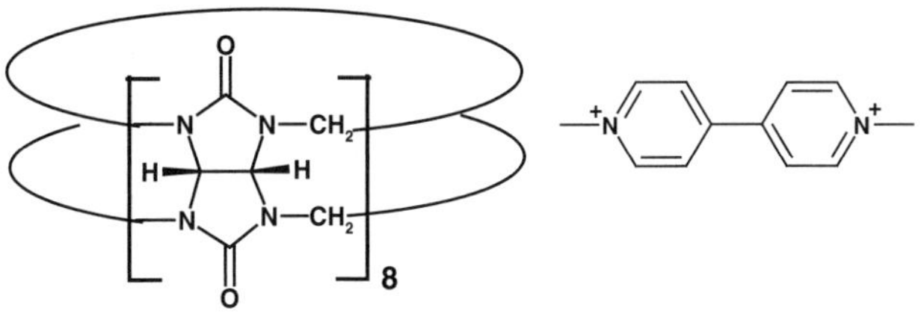

Guest	*K [M-1]*
Trp-Gly-Gly	1.3×105
Gly-Trp-Gly	2.1×104
Gly-Gly-Trp	3.1×103
Trp	4.3×104
Phe	5.3×103
Tyr	2.2×103

Scheme 20.4 Binding constants with the cucurbituril **22** together with methylviologen

illustrates how, besides ion pairing hydrogen bonding and stacking contribute to binding of, e.g., phenylalanine within protected Ac-Phe-Ala-OMe dipeptide by a nitroarene-modified aminopyrazole host, although nonaqueous solvents must be used with such systems (Wehner et al. 2006).

In cucurbiturils like **22** binding with $K = 3 \times 10^7$ $[M^{-1}]$ is observed, e.g., with Phe-Gly, due to strong hydrogen bonding of the $^{+}NH_3$ the polarized carbonyl groups of the host, even in water (Rekharsky et al. 2008). This interaction is absent with the inverted sequence Gly-Phe, which exhibits K = 1,300 $[M^{-1}]$. With the larger methylviologen, MV is bound simultaneously with aminoacids or peptides, which also allows detection by a UV signal (Bush et al. 2005). Scheme 20.4 illustrates the selectivity that can be reached with respect to the aminoacid residues.

Another commonly used cavity for sensing (Shahgaldian and Pieles 2006) is that of *cyclodextrins*, which particularly after substitition with charged groups provides stronger binding (e.g. **23** , R = H, X = $^{+}NH_2Me$ or R = X = $^{+}NH_2Me$). Relatively strong complexation of di- und tripeptides with monoamino-β-CD's is only observed with lipophilic aminoacids, the residues of which are bound within the cavity; with preference for this at the C-terminus (Hacket et al. 2001). Unprotected di- and tripeptides associate with CD bearing one charge at the rim (e.g. **23** R = H,

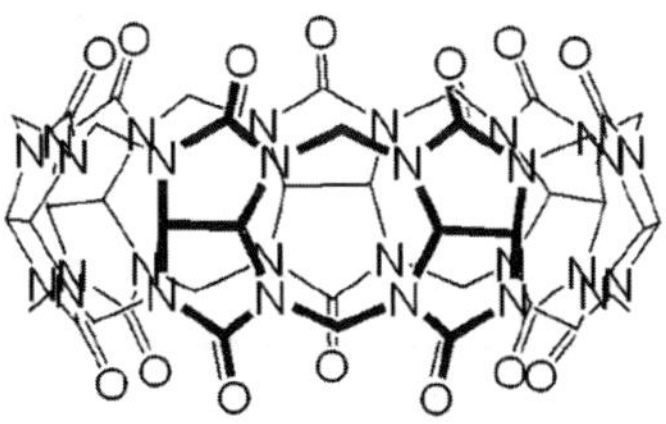

22 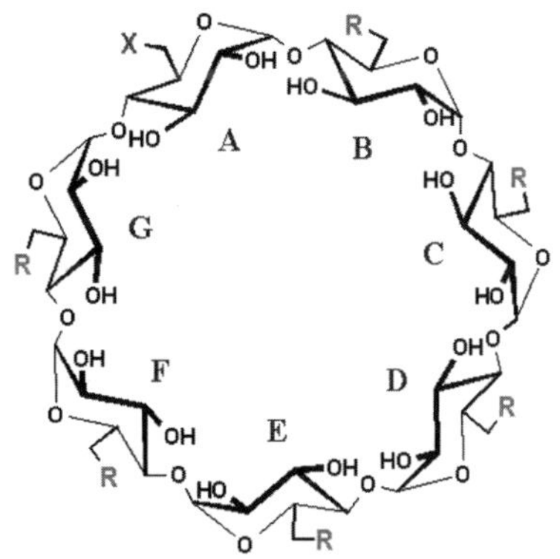 **23**

X = $^{+}NH_2Me$) with up to 200 M^{-1}, after protection at the N terminus with up to 680 M^{-1} (Hacket, Simova and Schneider 2001). A moderate sequence and sequence selectivity is observed monosubstituted CD like **23** R = H, X = $^{+}NH_2Me$), but less with the heptaamino-β-CD such as **23** R = X = $^{+}NH_2Me$, where ion pairing dominates. Substituted cyclodextrins also lead often to improved enantio – and diastereoseletive detection of aminoacids and peptides (Kano 1997). Protected aminoacids are also bound weakly to unsubstituted cyclodextrins, e.g., *N*-acetyl aminoacids with γ-CD bind in water with only K ≈ 50 [M^{-1}]; the *N*-carbobenzyloxy derivative reacts with K ≈100 [M^{-1}], due to insertion of the benzyl residue in the cavity (Rekharsky et al. 2003).

Cyclodextrins with pendant aromatic side chains have been used in particular by Ueno et al. for fluorometric assays of many analytes (Ikeda et al. 1996; Hamasaki et al. 1997; Hamai et al. 1998; Usui et al. 1998; Narita and Hamada 2000). Detailed NMR analyses' (Impellizzeri et al. 2000; Salvatierra et al. 2000; Uccello-Barretta et al. 2000) with such host compounds have already provided clear evidence for self-inclusion of the pendant aromatic side chain within the CD cavity, which can be expelled by addition of stronger binding analytes; alternatively, binding of a smaller analyte together with the fluorophore is possible inside the cavity. In both the cases, large changes of the fluorescence occur.

A general strategy for length- and sequence-selective detection of native peptides in water is based on linear receptors, which bear a permanent charged group for binding to the $–COO^{-}$ -terminus, a crown ether for association to the -$^{+}NH_3$ terminus, and a selector group L for the different aminoacid residues A crown ether instead of a negatively charged end group is necessary in order to avoid self association of the receptor. In addition, the lone pairs of the crown ether oxygens lead to quenching of fluorescence emission of a fluorophor like a dansyl group, which at the same time can serve as selector L by, e.g., stacking with an aminoacid residue. Complexation with the -$^{+}NH_3$ terminus leads to removal of quenching and thus to strong fluorescence emission, which depends on the nature of the aminoacid. Scheme 20.5 illustrates, how peptides are discriminated according to their length and sequence.

A combination of stacking and binding to crown ether can also be used with porphyrin-based receptors such as **24**, which have the advantage of a built-in optical signal in form of the Soret bands, and associate, e.g., with Gly-Gly-Phe with lg K = 4.4 in water (Sirish and Schneider 1999; Sirish et al. 2002). Aminoacids with special functions in the side chain can be recognized by specific reactions. Imidazole in His-containing peptides can bind preferentially with, e.g., Cu(II) ions , as shown in the complex **25** (Kruppa et al. 2005). The large surface of porphyrins and their ability to implement metal ions make such compounds of promising candidates also for complexation of peptides (Ogoshi and Mizutani 1998). With suitable substituents as well as in form of dimeric hosts they can be used for chiral discrimination, e.g., of lysine (Hayashi et al. 2002). Gable-type bisporphyrins promote binding of amines and oligopeptides, with K = 9.4×10^5 [M^{-1}] for the protected peptide H-His-Leu-His-NHBz (Mizutani et al. 2002).

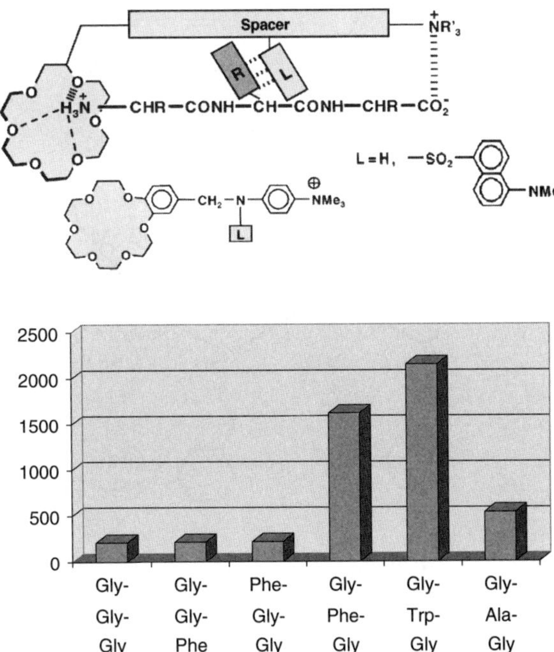

Scheme 20.5 Receptor for length- and sequence selective detection of tripeptides; binding constants with different aminoacids and sequences; di- or tetrapeptides exhibit K values below 50

24 25

20.3.4 *Complexation of Nucleotides and Nucleosides*

This section focuses on the application of noncovalent bonding via fully synthetic organic receptors toward the optical detection of nucleotides and nucleosides in aqueous media.

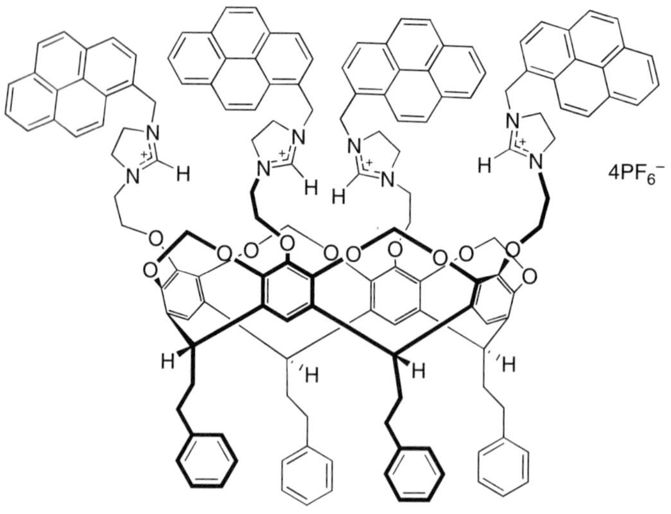

26

Fluorescent cavitand **26** bearing imidazolium and pyrene groups (Kim et al. 2005) shows a moderate selectivity toward GTP over ATP and CTP with binding constants of $7.38 \times 10^4\ M^{-1}$ for GTP, $1.4 \times 10^4\ M^{-1}$ for ATP and $7.7 \times 10^3\ M^{-1}$ for CTP, all in DMSO/water (6/4, 0.02 M HEPES buffer pH 7.4). The stoichiometry of the complexes is 1:1. For related examples see Shi and Schneider 1999; Zadmard and Schrader 2006. In receptor **27**, two porphyrin units are held by an *o*-dioxy-methylphenyl group in the shape of a cleft. Six pyridinium groups attached to the receptor provide solubility in water. This dimeric porphyrin is able to complex nucleotides and nucleosides in water or in aqueous buffer. Moreover, even electro-neutral nucleosides such as cytidine bind reaching K values up to $270{,}000\ M^{-1}$ compared with $430{,}000\ M^{-1}$ for GTP (Table 20.2). The proximity of the values indicates that the predominant interaction is stacking (Schneider and Sirish 2000; Sirish et al. 2002).

Protonated macrocyclic polyamines such as **28** are water-soluble via their positive charge, they can interact with anions such as phosphate (see Sect. 20.3.2). The introduction of acridine moieties such as in **29** and **30** enhances π–π interactions with adenine and nicotinamide groups. While the UV-Vis spectra of **29** and **30** in the presence of one equivalent of ATP (0.5 M, 5 mM Tris acetate, pH 7.6) shows no change in absorbance, the fluorescence intensity increases 150% for **29** and 250% for **30** as a stable 1:1 complex is formed. Table 20.2 displays the binding constants for complexes of **28–30** (Fenniri et al. 1997). For more information on macrocyclic polyamines, see: Kubik et al. 2005.

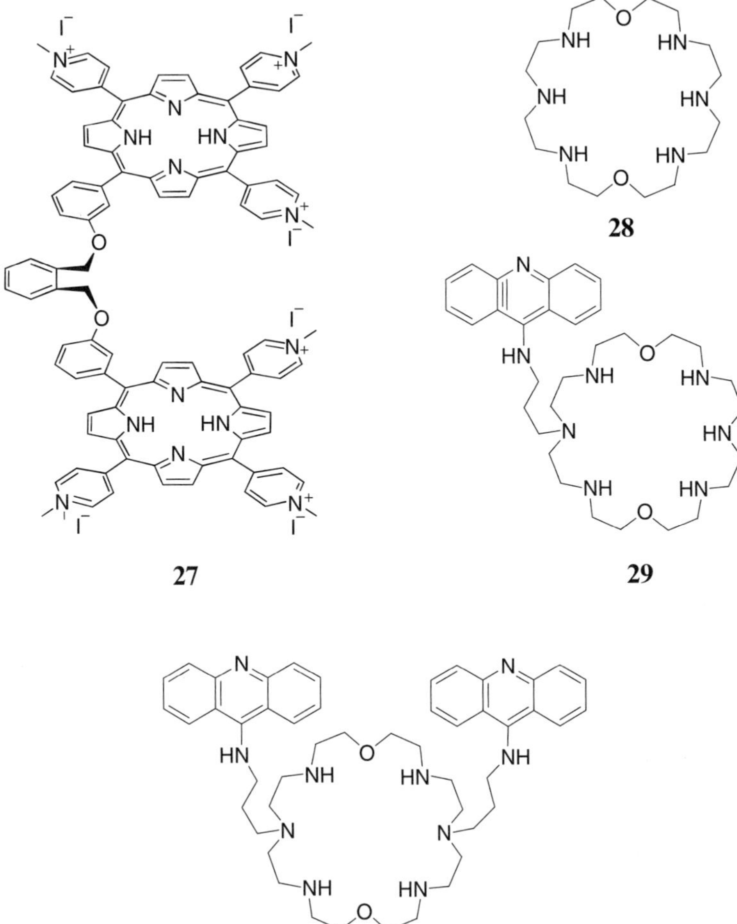

Table 20.2 Binding constants of complexes of **28**, **29** and **30**[a] (With permission of Fenniri et al. (1997), copyright 2008 John Wiley & Sons Inc)

Receptor/Substrate	K $[M^{-1}]$
30/ATP	7×10^7
29/ATP	1×10^7
28/ATP	$\geq 5 \times 10^6$
30/PPP	$\geq 7 \times 10^7$
29/PPP	$\geq 7 \times 10^7$
30/NADP	5×10^5
29/NADP	5×10^5
30/NADPH	$\geq 3 \times 10^8$
29/NADPH	$\geq 1 \times 10^7$

[a]For experimental details see (Fenniri et al. 1997).

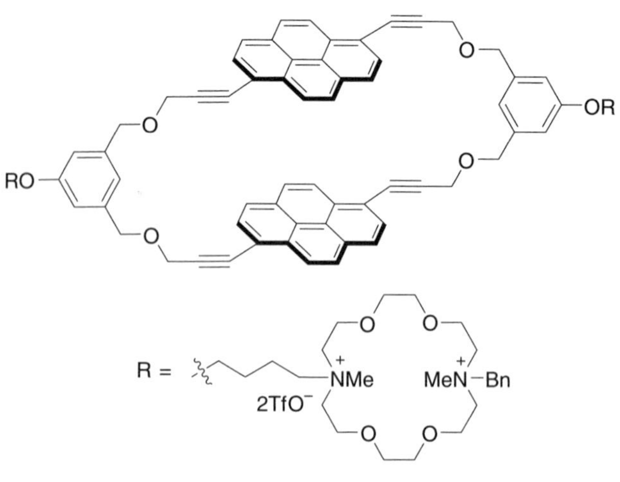

31

Cyclophane **31** consists of pyrene and azoniacrown groups. It exhibits interactions with nucleotides, but modest selectivity with respect to the nucleobase. The binding constants indicate as expected a stronger preference for triphosphates (ATP, GTP, CTP, and UTP) over mono- and di-phosphates (see Table 20.3). A UV-Vis titration study with **31** (1.0×10^{-5}M) and ATP (0–1.4 equiv) in H_2O shows a decrease in absorbance upon ATP addition. The fluorescence titration shows quenching with an excitation wavelength of 370 nm (Abe et al. 2004).

Table 20.3 Association constants and free-energy changes for the complexation of **31** with nucleotides[a] (With permission of Hajime Abe et al. (2004), copyright 2008 American Chemical Society)

Nucleotide	Association constant (M^{-1})	$-\Delta G_{293}$ (kJ mol^{-1})
ATP	1.0×10^6	34
ADP	5.3×10^3	21
AMP	1.9×10^3	18
GTP	1.3×10^6	34
GDP[b]	4.0×10^3	20
GMP	2.7×10^3	19
CTP	2.6×10^5	30
CDP	7.7×10^3	22
CMP[b]	4.0×10^3	20
UTP	7.7×10^5	33
UDP	2.2×10^4	24
UMP	3.0×10^3	20

[a]Calculated by curve-fitting analysis of hypochromism on 390 nm. Conditions: **31** (1.0×10^{-5} M), nucleotide, 20°C in water

[b]The association constant was estimated by the Benesi-Hildebrand method

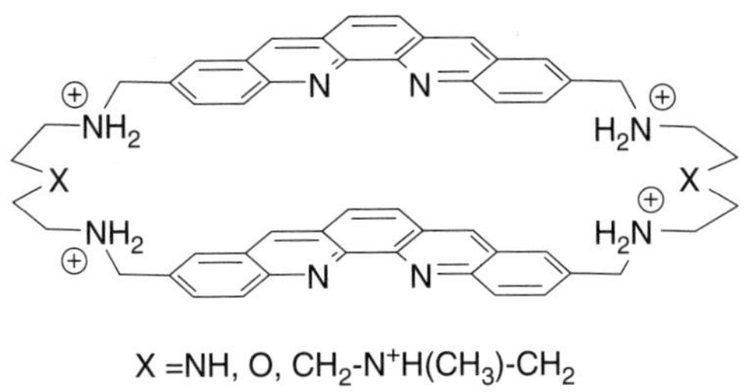

X =NH, O, CH_2-$N^+H(CH_3)$-CH_2

32

Another related example is cyclophane receptor **32,** which consists of two quinacridine units bridged with alkylamino groups (Baudoin et al. 1999). The binding constants for nucleosides (guanosine, adenosine, cytidine, and uridine) are $\leq 1 \times 10^2\ M^{-1}$, the maximum value for nucleotides corresponds to GTP^{2-} $6.31 \times 10^6 M^{-1}$. It is found that all the monophosphates produce 1:2 complexes, while diphosphates and triphosphates produce 1:1 complexes. For related examples see: Moreno-Corral and Lara 2008; Oshovsky et al. 2007.

Peptide-based receptors can mimic interactions between nucleotides and proteins (i.e. enzymes) found in nature. For example, peptide **33** consisting of 12 residues has been designed for ATP recognition (Butterfield et al. 2003). The peptide sequence is: acetyl-Arg-Trp-Val-Lys-Val-Asn-Gly-Orn-Trp-Ile-Lys-Gln-NH_2.

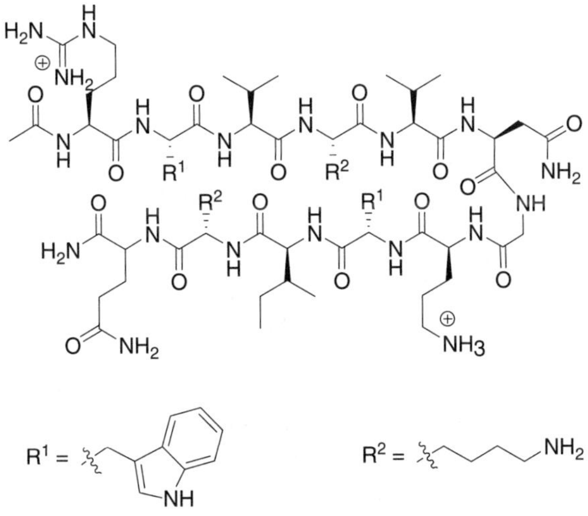

R^1 =

R^2 =

33

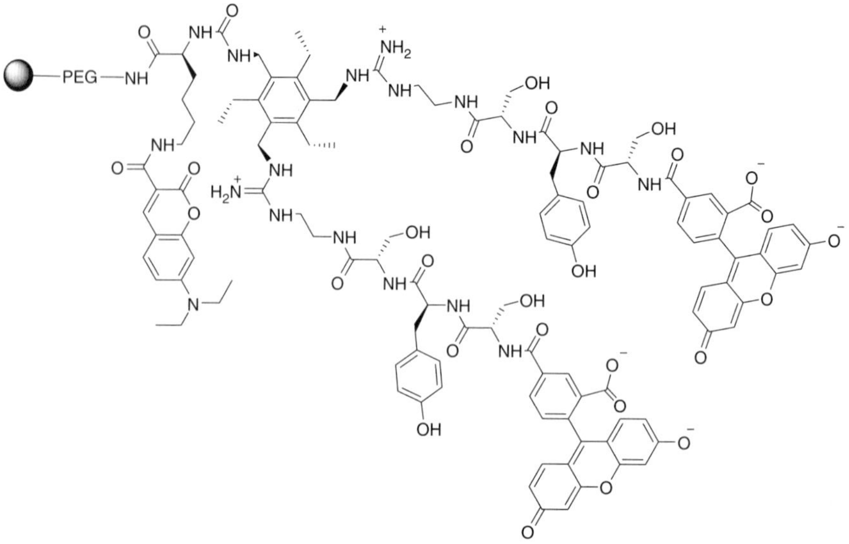

34

The design of peptide host **33** allows nucleobase intercalation. The Trp residues appear to form a cleft. Fluorescence quenching of the tryptophan moieties is observed when the peptide is titrated with ATP. The binding constant of **33** and ATP in H_2O is 5815 M^{-1}. The presence of salt dramatically affects the electrostatic interactions of the complex. For instance, in the presence of NaCl (0.2 M) the binding constant decreases to 70 M^{-1} (see also Butterfield et al. 2005).

Receptor **34** was obtained by combinatorial methods (Schneider, O'Neil, Anslyn, 2000) and developed a strong fluorescence response toward ATP (0.2 M HEPES buffer pH 7.4) with a binding constant for the complex with ATP of 3.4×10^3 M^{-1}. Fluorescence titration studies of **34** with AMP and GTP showed practically no fluorescence changes, indicating that, in comparing ATP and GTP, the peptide sequence Ser-Tyr-Ser is what imparts the nucleobase specificity. This peptide combination promotes π-stacking between the phenolic moiety in tyrosine and the adenine group and hydrogen bonding interactions between the ribose or adenine moieties and the OH of serine. For related work see: Basabe-Desmonts et al. 2007; Anslyn 2007; Diaz-Garcia et al. 2006.

Generally, with positively charged host compounds, one observes the expected binding constant magnitude increase from mono- to di- and –triphosphates. Some receptors in which stacking contributions dominate, such as **27**, show a remarkable independence of guest charge. Similarly, a bis (phenanthridinium) receptor binds nucleotides and nucleosides with similar strength (log K = 5–6, Žinić et al. 1995). If the nucleobase is less exposed to the aqueous environment, as in the complexes with the amino-cyclodextrins **23** R = X = $^{+}NH_2Me$, which binds ATP with K = 3 $\times$ 10^6 [M^{-1}], one finds a slightly increased selectivity, for instance for AMP K = 12.5, in comparison with GMP (4.0), UMP (2.0) and CMP (8.7) (all in 10^4 [M^{-1}] units) (Eliseev and Schneider 1994). Improved discrimination must rely on combination of stacking and hydrogen bonding, which, however, in water is usually too weak.

20.3.5 Carbohydrate Detection

This section describes the sensing of sugar molecules mostly by fully synthetic supramolecular receptors that afford complexation by noncovalent interactions. We exclude here reagents that function via reversible covalent, e.g., boronic acid binding, which have been reviewed extensively elsewhere (for example see Striegler 2003a, b; James 2007; James and Shinkai 2005; James et al. 1996b; James et al. 1996a; Shinkai and Robertson 2001; Yan et al. 2005; Wang et al. 2002).

The design of noncovalent saccharide receptors requires careful approaches to molecular geometry, including three-dimensional binding sites and relatively large frameworks. The advantage of such strategies is the fact that on relatively small "construction sites" numerous binding modes can be incorporated. Various exemplary designs may mimic to a degree the molecular architecture of lectins and facilitate multivalent binding to very specific target saccharides (Schrader and Hamilton 2005; Davis and Wareham 1999; Sears and Wong 1999; James and Shinkai 2002; Penadés and Fuhrhop 2002). Overall, the recognition of natural saccharides in aqueous media remains a fairly challenging task. Existing synthetic receptors can be categorized in several types, varying in their chemical nature and molecular construction. As hydrogen bonding is very weak in aqueous media significant affinity in water is generally feasible with derivatives bearing charged groups for ion pairing with a host of complementary charge. Alternatively one can study selectivity by hydrogen bonding with sugar derivatives soluble in for instance chloroform (see, e.g., Das and Hamilton 1997; Coterón et al 1996; Ishi-i et al. 2003; Vacca et al. 2004; Mazik and Buthe 2008; Mazik and Kuschel 2008). Another possibility is the extraction of carbohydrates into lipophilic solvents which contain receptors binding strongly to the analyte.

In 1988, Aoyama introduced **35**, the first synthetic saccharide receptor for the stereoselective complexation of saccharides and extraction into an organic phase (Aoyama et al. 1988; 1989; Tanaka et al. 1990; Kikuchi et al. 1991, 1992b; Kobayashi et al. 1993; Kurihara et al. 1991; Fujimoto et al. 1998; Hayashida et al. 1999; Goto et al. 2007) via hydrogen bonding, CH-π and hydrophobic interactions. Saccharides such as D-ribose, D-arabinose and L-fucose but not D-xylose or D-lyxose bind next to the surface of macrocycle and can be selectively extracted with **35** from their concentrated water solutions into an organic layer.

Alternative designs based on nonplanar geometries were implemented in the synthesis of polyazamacrocycles and their linear analogues. These nitrogen-

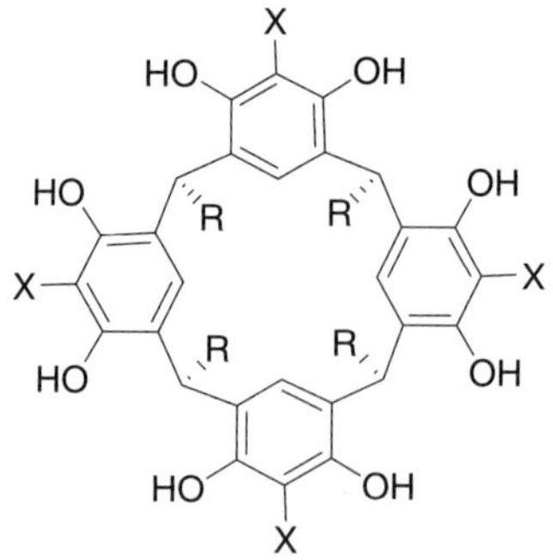

35

R= $(CH_2)_2SO_3Na$; X= H

containing hosts with central or peripheral aminopyridine units are able to form hydrogen bonds to saccharides. Inoye and co-workers proposed a series of polypyridine receptors (Inoye et al. 1995; Inoye et al. 1999a, b) that extracted ribose derivatives into nonpolar media. For instance, macrocycle **36** was able to transfer deoxyribose, arabinose, xylose, lyxose and fructose from water into the organic phase. Glucose, mannose and galactose do not interfere with this process. The selectivity was determined by direction of the OH-groups in saccharide molecules as well as its whole molecular geometry. Extraction was readily observed by tracking of the fluorescence intensity changes in the absence and the presence of the bonded sugar in a 1:1 stoichiometry.

The aforementioned studies by Aoyama and Inoye are included for historical significance. Most current efforts focus on the relatively more challenging binding and detection in aqueous rather than non polar media.

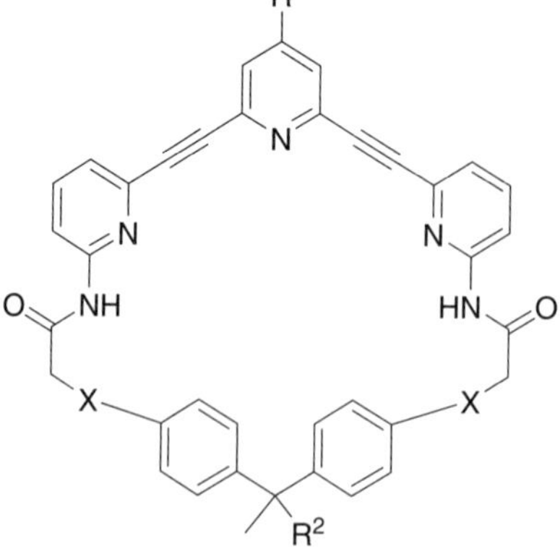

36 **a** . R_2= Me, R_1= H; X= O
b R_2= Me, R_1= H; X= S
c R_2= Me, R_1= H; X= CH_2
d R_2= Bu, R_1= H; X= CH_2
e R_2= Bu, R_1= Bu; X= CH_2

The importance of weak forces in the interactions between saccharides and receptors was illustrated via water-soluble cyclophane **37** (Ferrand et al. 2007). The authors remark that regardless weak host-guest interaction in water, their system was able to bind glucose over galactose (ratio 5:1) with a constant of 9 M^{-1}.

As an alternative to cyclophane cavities, a bulky calixarene **38** with binaphthyl subunits was studied (Rusin and Král 2001). The inclusion of saccharides into the "bowl" in aqueous media was monitored indirectly by the displacement of the dye methyl red from its 1:1 complex with the calixarene. The dye displacement corresponded with distinct color changes and could be observed visually and monitiored by UV-V is spectroscopy. Calculated binding constants for oligosaccharides were in the range of 10^3 M^{-1}.

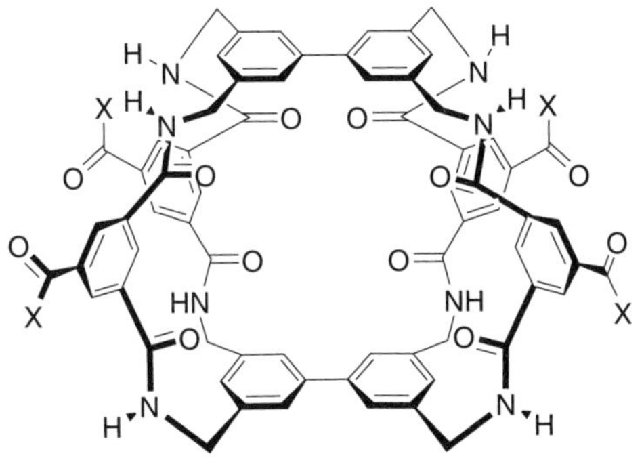

X= $NHC(CH_2OCH_2CH_2CO_2t\text{Bu})_3$

37

Porphyrins and their analogs have unique optical properties. A characteristic sharp and intense absorption peak in the visible region (Soret band) and intense fluorescence make them very attractive substrate materials for artificial receptor design (Biesaga et al. 2000). The introduction of substituents in the *meso*-position of the planar porphyrin core afford access to many possible geometries effective for substrate binding (Vögtle 1995).

Some functionalyzed porphyrins are known for their noncovalent recognition and detection of saccharides in aqueous media (**39**–**45**). Bulky bowl-shaped moieties with hydrophilic substituents, coordinating metals, phosphonate anions on their periphery, spacers between tetrapyrrolic planar rings, or long olygopeptide chains – many types of design elements provide one or more features in the binding and detection of different mono- oligo- and polysaccharides (Král et al. 2000; Král et al. 2001 and 2006; Rusin et al. 2002; Dukh et al. 2003; Králová et al. 2008). The tracking of these binding interactions is readily accomplished by monitoring Soret band shifts and/or fluorescence intensity changes with association constants ranging from 10^2 M^{-1} to 10^4 M^{-1} for different sugars and porphyrin receptors.

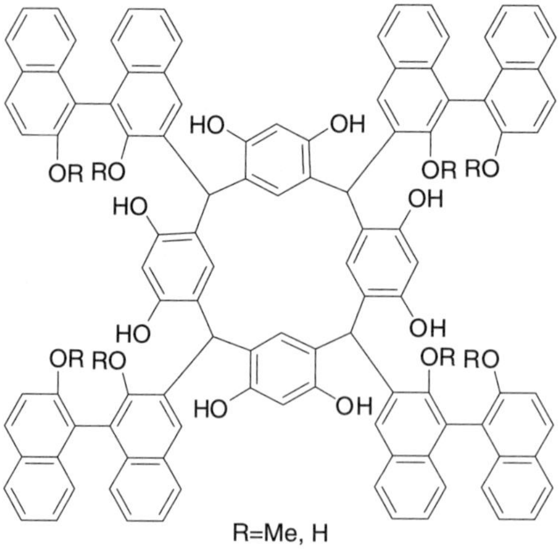

R=Me, H

38

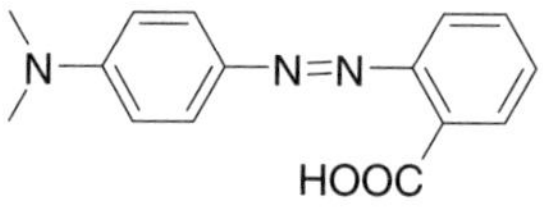

Methyl red

Cyclodextrins (cycloamyloses) can also form complex with saccharides (Schrader and Hamilton 2005; Aoyama et al. 1992; Hirsch et al. 1995; Hacket et al. 1997; de Namor et al. 1994). Optical spectroscopy can be useful via the displacement of dyes from its complexes with cyclodextrins. Certain preferences in the binding for pentoses for example have been reported (for example, see Schrader and Hamilton 2005).

Metal ions incorporated in suitable frameworks can contribute significantly to the binding of sugar moieties. Remarkable results were obtained in the complexation and detection of saccharides via nonmacrocyclic binuclear Cu(II) complexes such as **46** in alkaline aqueous solutions (Striegler 2003a, b, Dittel 2005). This principle was used in the development of new polymeric materials with unique selectivities.

Fluorescent complexes of the simple water-soluble salophene **47** and La(III) – induced binding of neutral saccharides in buffer at pH 7.0 was reported (Alptürk et al. 2006). Different mono- and oligosaccharides showed affinity to the complex with relatively good binding constants ranging from 500 to 2,500 M^{-1}. The authors compared their observations of the binding of neutral sugars with a highly fluorescent tetracycline-Eu(III) complex.

Linear analogs of macrocycles exhibit relatively high flexibility. A number of such structures have been reported effective in the binding and signaling of saccharides, but often only within aprotic environements. However, some studies demonstrate successful approaches in aqueous media. An interesting example of a sugar-binding peptide sequence Trp-Gly-Asp-Glu-Tyr was reported (Schrader

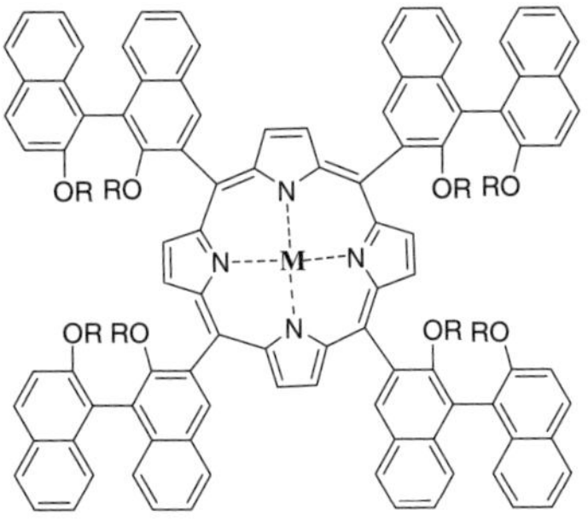

39

R=Me
R=H; M=Fe, Zn

40

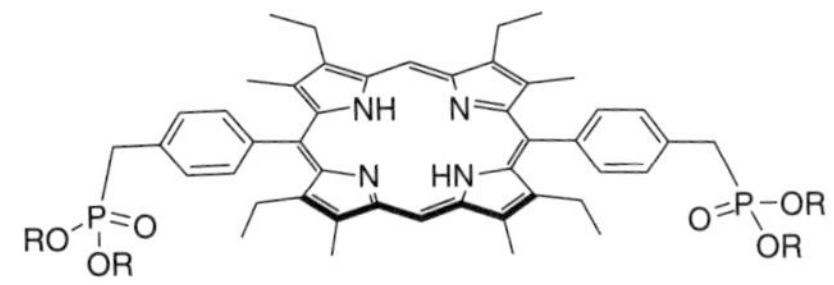

R=C_2H_5, H

41

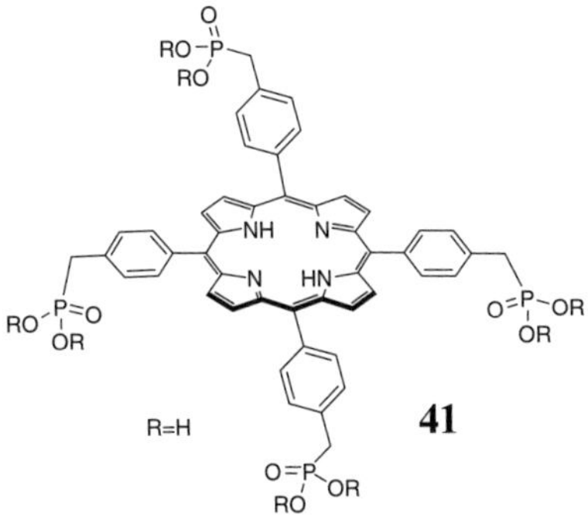

R=H

42

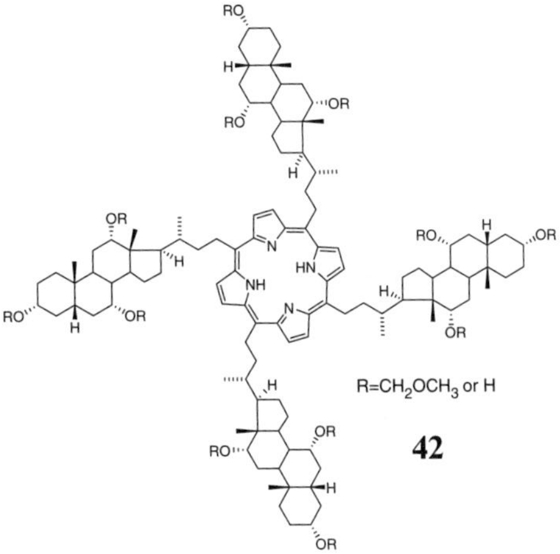

R=CH_2OCH_3 or H

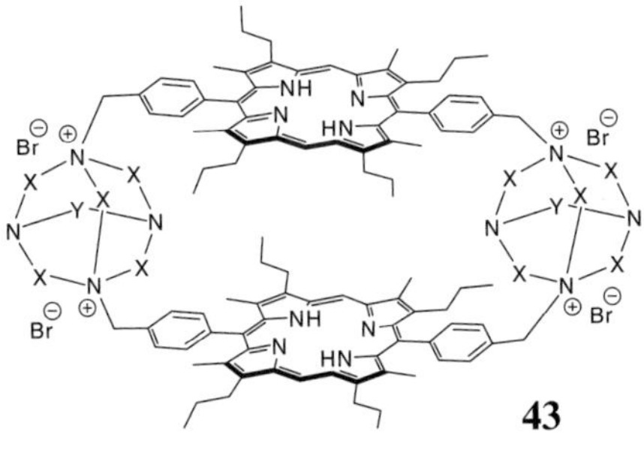

43

X = $(CH_2)_6$, Y = $CO(CH_2)_4CO$

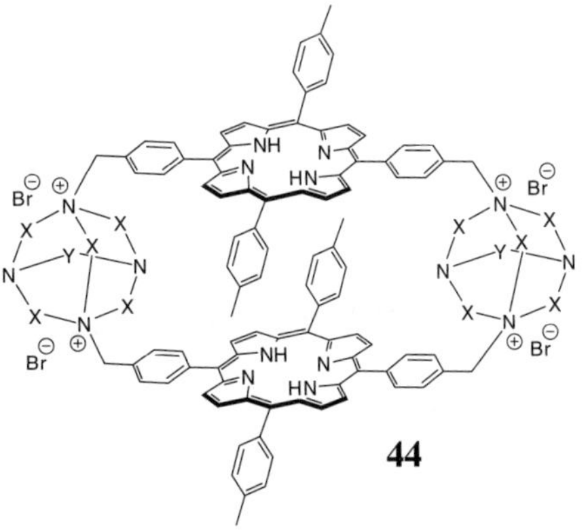

44

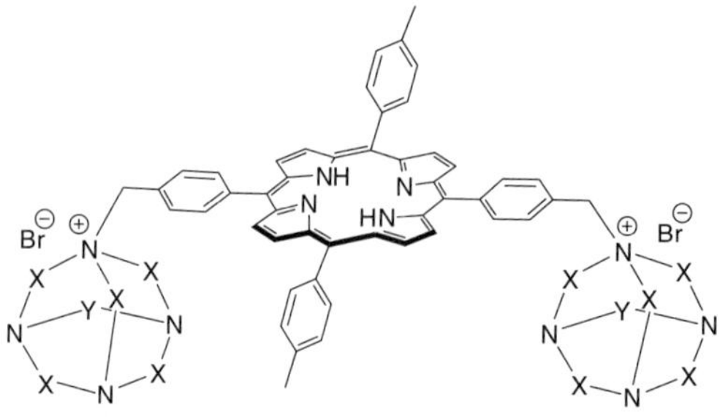

$X = (CH_2)_6$, $Y = CO(CH_2)_4CO$ **45**

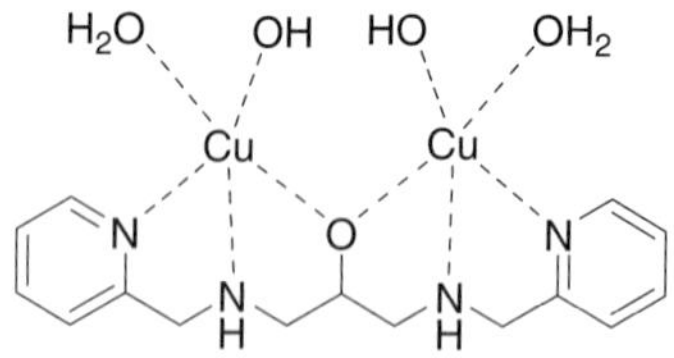

46

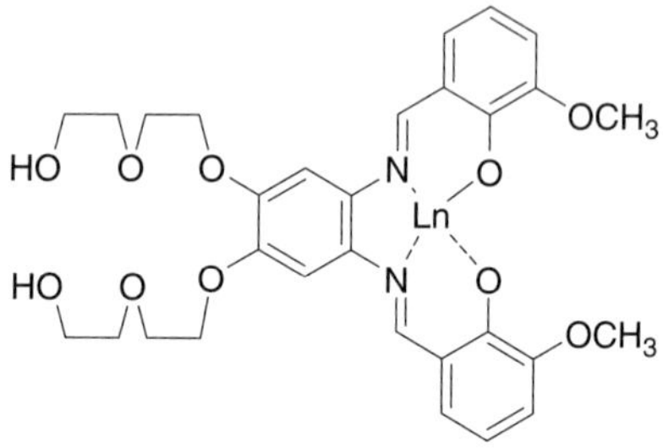

$Ln = La^{3+}$ or $Ln = Eu^{3+}$ **47**

and Hamilton 2005; Sugimoto et al. 2000). This sequence binds erythrose and galactose in neutral buffer with remarkable affinity and the formation of a distinctive yellow color.

Tripodal polypyridine and quinoline ring receptor **48** exhibits large fluorescent responses in the presence of mono alkyl pyranosides in protic organic solvents as reported by Palde et al. 2008. The authors reported remarkable visual fluorescence of the complex with glucopyranoside in methanol at 365 nm.

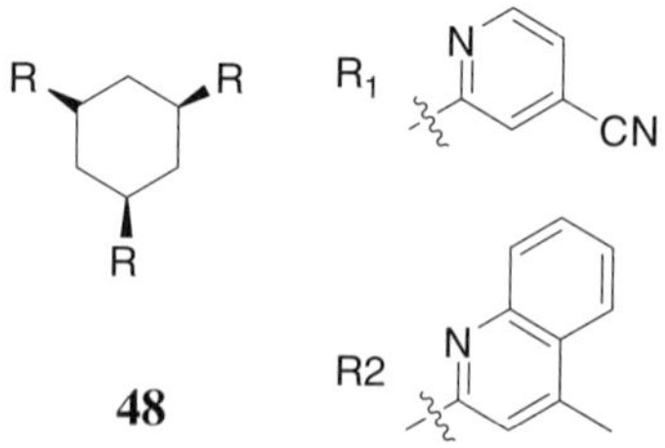

48

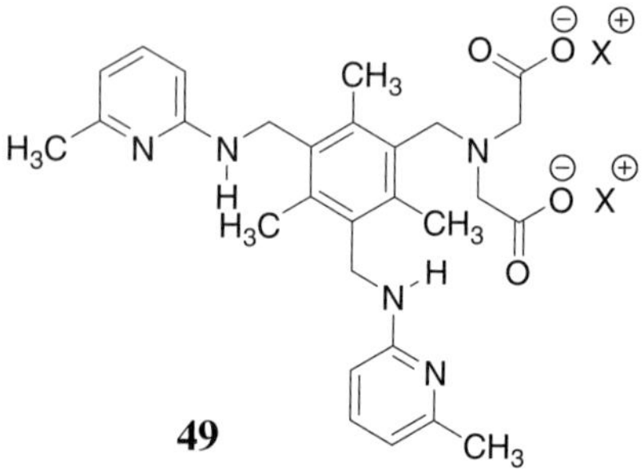

49

The design of the fluorescent receptors prepared by Mazik and Cavga (2007); Mazik and Buthe (2008); Mazik and Kuschel (2008) was based on the principles of saccharide binding by natural ligands. The authors studied binding in $CDCl_3$ and reported significant values for different alkyl glycosides with strong preference to alkyl disaccharides. More importantly, carboxylate/pyridine based receptor **49** was also able to form complexes with nonsubstituted saccharides such as maltose and cellobiose in aqueous media with a distinct preference for cellobiose.

20.3.6 Complexation of Terpenes and Steroids

The complexation of lipophilic natural analytes relies primarily on hydrophobic, dispersive, and stacking interactions which require a close geometric fit between host and guest. This is nicely illustrated by a study of complexation with a nitrophenol-substituted cyclodextrin: the ideal fit is reached for adamantan-1-ol with an equilibrium constant as high as 700.000 $[M^{-1}]$; borneol is less space filling and exhibits $K = 50{,}000$ $[M^{-1}]$, with smaller guest compounds such as nerol one observes $K = 3{,}400$ $[M^{-1}]$, or $K = 2{,}500$ $[M^{-1}]$ with geraniol (Matsushita et al. 1997).

Steroid complexation has been extensively reviewed by Wallimann et al. 1997. As in proteins the binding in aqueous media is favoured by the presence of an nonpolar environment. This can be provided by cyclophanes, as shown as far back as 1989 with host **50** (Scheme 20.6, Kumar and Schneider 1989). Estradiol in the form of the anion is bound 10 times stronger than in the neutral form, indicating contribution of an ion pair; the complexation induced NMR shifts indicate immersion of the A and B ring within the aromatic host cavity. Cyclophanes such as **51** with a more extended aromatic cavity naphthalenes show higher affinities to bile acids, again supported by salt bridges with the steroid carboxylate group. (Wallimann et al. 1997). A macrotricyclic receptor **52** binds cholesterol in water with a very high association constant of $K = 10^6$ $[M^{-1}]$ (Wallimann et al. 1997).

The complexation of steroids in nonpolar media lacks hydrophobic contributions and is usually weaker, but can be quite selective, mostly because of hydrogen bonding. The calixarene **53** is stabilized in the cone conformation by symmetric hydrogen bonds at the upper rim, and shows with cholic acids in chloroform association constants which increase with the number of hydroxygroups in the steroid, from $K = 10$ to $K = 690$ $[M^{-1}]$, indicating the importance of hydrogen bonds (Kikuchi et al. 1992a). The cleft-like resorcin [4]arene cavitand **54** is a selective receptor for steroids, with K values of up to 650 $[M^{-1}]$ in $CDCl_3$, preferring flat A-rings with hydrogen bonding groups (Cacciarini et al. 2005).

The perfectly watersoluble cyclodextrins have been used most often as complexation agent for steroids (Wallimann et al. 1997; Forgacs and Cserhati 2004). Native, unsubstituted CD's bind relatively weakly, e.g., β-CD binds lithocholic acid (**55,** $R_1 = OH$; $R_2 = R_3 = R_4 = R_5 = H$;$R_6 = COOH$) with $K \approx 100$ $[M^{-1}]$. Introduction of a larger aromatic residue such as a dansyl group at the rim of the

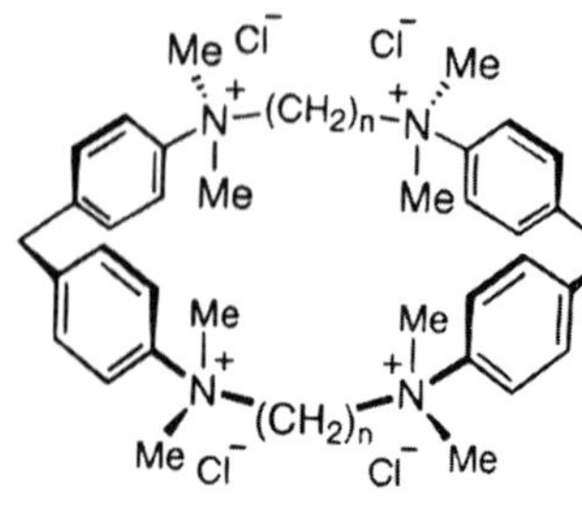

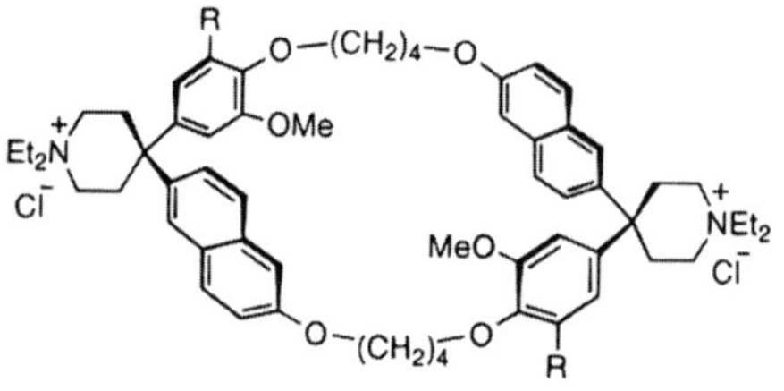

50

51

R = OMe

R = $CH_2NHCOCH_2\overset{+}{N}Me_2Et\ Cl^-$

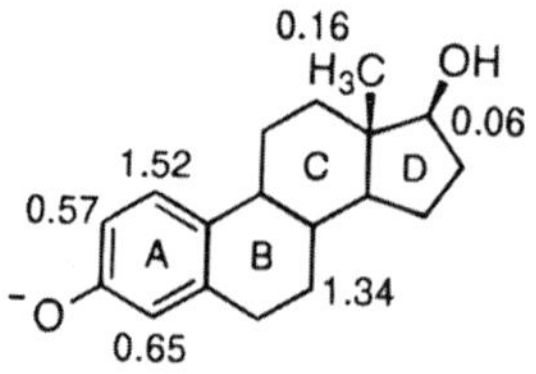

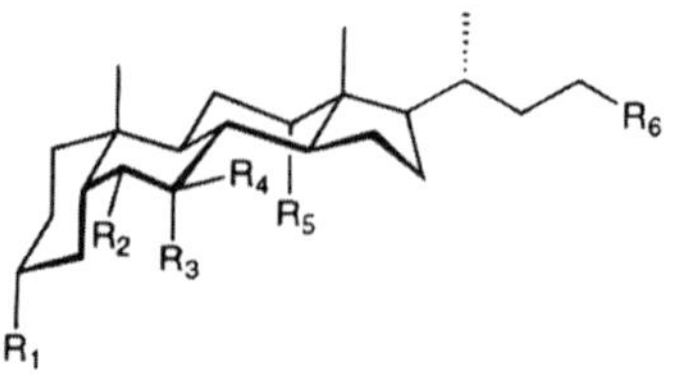

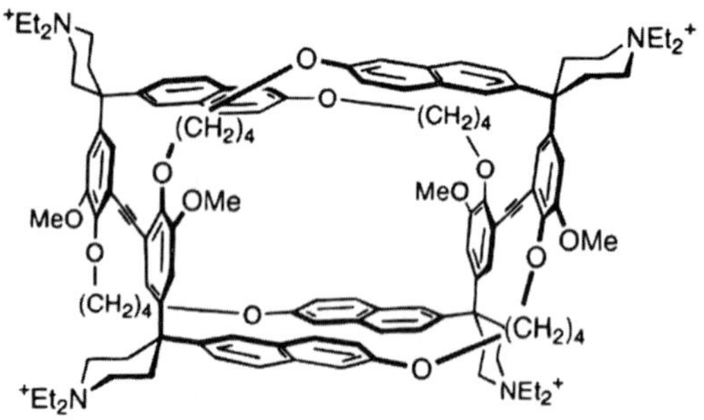

52

Scheme 20.6 Examples for steroid complexation in water

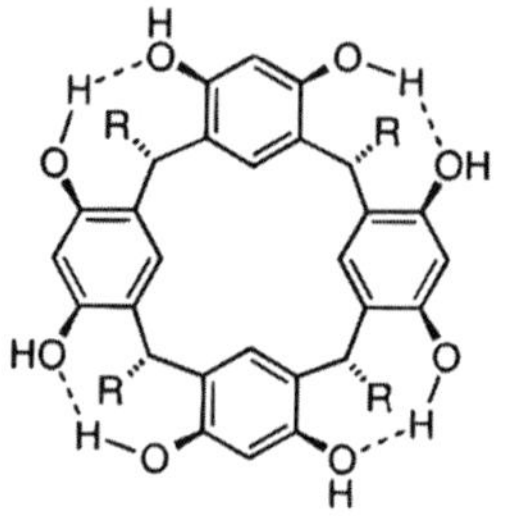

R = $C_{11}H_{23}$

53

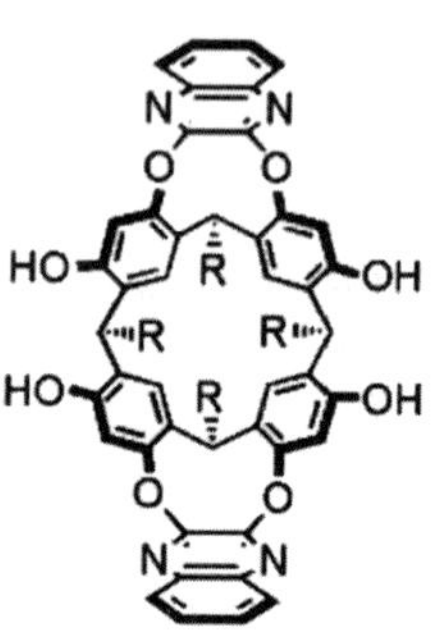

R = $C_{11}H_{23}$

54

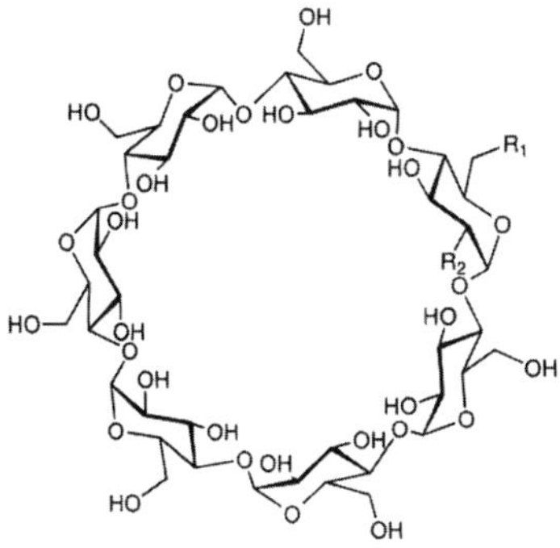

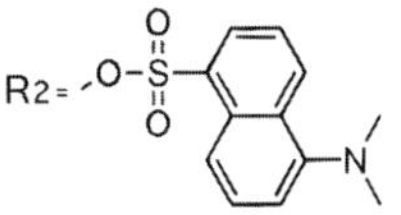

55

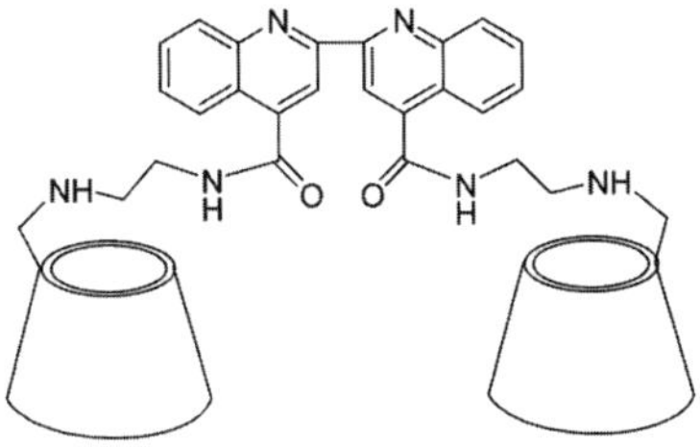

56

CD cavity not only allows for fluorescence measurement but also can lead to dramatic affinity enhancement, e.g., for lithocholic acid to K $\approx 4 \times 10^6$ [M^{-1}] (Wang et al. 1994).

Bridged bis(β-cyclodextrins like **56** bind two cholic acids with the steroid carboxylate groups penetrating into the two hydrophobic CD cavities; the equilibrium constants with cholic acids vary between ≈3,000 and 8,000 K [M^{-1}] , depending on their substitution (Liu et al. 2004; see also Dejong et al. 2000).

20.4 Concluding Remarks

The field of artificial receptors as sensing agents has advanced to the point of enabling the detection of analytes of increasing complexity. For instance, the reader is referred to excellent reviews on the selective sensing of phospholipid membranes (Lambert and Smith 2003; Hanshaw and Smith 2005) and proteins (Schrader and Koch 2007) for recent developments in these exciting areas. The major aim to reach high sensitivity and selectivity can both be approached by implementation of many binding sites in one receptor. The resulting multivalent hosts compounds can be used, e.g., for coating sensitive electrodes for electrochemical detection, or they can be equipped with suitable optical signalling units, which often provide additional binding. In future, synthetic host structures will be used more often in hybrid combination with biological receptors, with the possibility to design micro- and nano-sized systems (Katz and Willner 2004).

References

Abe H, Mawatari Y, Teraoka H, Fujimoto K, Inouye M (2004) Synthesis and molecular recognition of pyrenophanes with polycationic or amphiphilic functionalities: Artificial plate-shaped cavitand incorporating arenas and nucleotides in water. J Org Chem 69:495–504

Abraham MH, Enomoto K, Clarke ED, Sexton G (2002) Hydrogen bond basicity of the chlorogroup; hexachlorocyclohexanes as strong hydrogen bond bases. J Org Chem 67:4782–4786

Albrecht M (2007) Supramolecular chemistry - general principles and selected examples from anion recognition and metallosupramolecular chemistry. Naturwissenschaften 94:951–966

Alptürk O, Rusin O, Fakayode SO, Wang W, Escobedo JO, Warner IM, Crowe WE, Král V, Pruet JM, Strongin RM (2006) Lanthanide complexes as fluorescent indicators for neutral sugars and cancer biomarkers. Proc Natl Acad Sci USA 103:9756–9760

Amendola V, Fabbrizzi L (2006) What anions do to N-H-containing receptors. Acc Chem Res 39:343–353

Anslyn EV (2007) Supramolecular analytical chemistry. J Org Chem 72:687–699

Aoyama Y, Tanaka Y, Toi H, Ogoshi H (1988) Polar host-guest interaction. binding of nonionic polar compounds with resorcinol-aldehyde cyclooligomer as a lipophylic polar host. J Am Chem Soc 110:634–635

Aoyama Y, Tanaka Y, Sugahara S (1989) Molecular Recognition. 5. Molecular recognition of sugars via hydrogen-bonding interaction with a synthetic polyhydroxy macrocycle. J Am Chem Soc 111:5397–5404

Aoyama Y, Nagai Y, Otsuki J, Kobayashi K, Toi H (1992) Selective binding of sugar to β-cyclodextrin: a prototype for sugar-sugar interaction in water. Angew Chem Int Ed Engl 31:745–747

Arduini A, Demuru D, Pochini A, Secchi A (2005) Recognition of quaternary ammonium cations by calix[4]arene derivatives supported on gold nanoparticles. Chem Commun: 645–647

Arena G, Contino A, Gulino FG, Magri A, Sansone F, Sciotto D, Ungaro R (1999) Complexation of native L-alpha-aminoacids by water soluble calix[4]arenes. Tetrahedron Lett 40:1597–1600

Ariga K, Kunitake T (2005) Supramolecular chemistry: fundamentals and applications. Springer, Berlin

Atwood JL, Steed JW (2004) Encyclopedia of supramolecular chemistry. Marcel Dekker, New York

Badjic JD, Nelson A, Cantrill SJ, Turnbull WB, Stoddart JF (2005) Multivalency and cooperativity in supramolecular chemistry. Acc Chem Res 38:723–732

Bag B, Bharadwaj PK (2005) Fluorescence enhancement of a signaling system in the simultaneous presence of transition and alkali metal ions: a potential and logic gate. Chem Commun: 513–515.

Basabe-Desmonts L, Reinhoudt DN, Crego-Calama M (2007) Design of fluorescent materials for chemical sensing. Chem Soc Rev 36:993–1017

Baudoin O, Gonnet F, Teulade-Fichou MP, Vigneron JP, Tabet JC, Lehn JM (1999) Molecular recognition of nucleotide pairs by a cyclo-bis-intercaland-type receptor molecule: a spectrophotometric and electrospray mass spectrometry study. Chem Eur J 5:2762–2771

Beer PD, Gale PA (2001) Anion recognition and sensing: the state of the art and future perspectives. Angew Chem Int Ed 40:486–516

Beer PD, Gale PA, Smith DK (1999) Supramolecular chemistry. Oxford Chemistry Primers Text No. 74, Oxford University Press

Best MD, Tobey SL, Anslyn EV (2003) Abiotic guanidinium containing receptors for anionic species. Coord Chem Rev 240:3–15

Bianchi A, Bowman-James K, García-España E (1997) The supramolecular chemistry of anions. Wiley-VCH, Weinheim

Biesaga M, Pyrzynska K, Trojanowicz M (2000) Porphyrins in analytical chemistry. Talanta 51:209–224

Bowman-James K (2005) Alfred Werner revisited: The coordination chemistry of anions. Acc Chem Res 38:671–678

Brooks SJ, Edwards PR, Gale PA, Light ME (2006) Carboxylate complexation by a family of easy-to-make ortho-phenylenediamine based bis-ureas: studies in solution and the solid state. New J Chem 30:65–70

Buhleier E, Wehner W, Vögtle F (1978) Ligand structure and complexation,.22. 2, 6-bis(aminomethyl)pyridine as a complex ligand and new building block for crown ether synthesis. Liebigs Ann Chem 1978:537–544

Bush ME, Bouley ND, Urbach AR (2005) Charge-mediated recognition of N-terminal tryptophan in aqueous solution by a synthetic host. J Am Chem Soc 127:14511–14517

Butterfield SM, Sweeney MM, Waters ML (2003) A designed beta-hairpin peptide for molecular recognition of ATP in water. J Am Chem Soc 125:9580–9581

Butterfield SA, Sweeney MM, Waters ML (2005) The recognition of nucleotides with model β-hairpin receptors: Investigation of critical contacts and nucleotide selectivity. J Org Chem 70:1105–1114

Cacciarini M, Azov VA, Seiler P, Kunzer H, Diederich F (2005) Selective steroid recognition by a partially bridged resorcin[4]arene cavitand. Chem Comm: 5269–5271

Cao H, Heagy MD (2004) Fluorescent chemosensors for carbohydrates: a decade's worth of bright spies for saccharides in review. J Fluoresc 14:569–584

Chang KJ, Jeong KS, Moon D, Lah MS (2005) Indole-based macrocycles as a class of receptors for anions. Angew Chem Int Ed Engl 44:7926–7929

Choi K, Hamilton AD (2003) Rigid macrocyclic triamides as anion receptors: Anion-dependent binding stoichiometries and H-1 chemical shift changes. J Am Chem Soc 125:10241–10249

Clare JP, Ayling AJ, Joos JB, Sisson AL, Magro G, Perez-Payan MN, Lambert TN, Shukla R, Smith BD, Davis AP (2005) Substrate discrimination by cholapod anion receptors: geometric effects and the "affinity-selectivity principle". J Am Chem Soc 127:10739–10746

Clark HA, Kopelman R, Tjalkens R, Philbert MA (1999) Optical nanosensors for chemical analysis inside single living cells. 2. Sensors for pH and calcium and the intracellular application of PEBBLE sensors. Anal Chem 71:4837–4843

Coterón JM, Hacket F, Schneider H-J (1996) Interactions of hydroxy compounds and sugars with anions. J Org Chem 61:1429–1435

Cragg P (2005) A practical guide to supramolecular chemistry. Wiley, New York

Czarnik AW (1994) Chemical communication in water using fluorescent chemosensors. Acc Chem Res 27:302–308

Das G, Hamilton AD (1997) Carbohydrate recognition: Enantioselective spirobifluorene diphosphonate receptors. Tetrahedron Lett 38:3675–3678

Davis AP, Wareham RS (1999) Carbohydrate recognition through noncovalent interactions: a challenge for biomimetic and supramolecular chemistry. Angew Chem Int Ed 38:2978–2996

de Namor AFD, Blackett PM, Cabaleiro MC, Al Rawi JMA (1994) Cyclodextrin-monosaccharide interactions in water. J Chem Soc Faraday Trans 90:845–847

de Silva AP, Gunaratne HQN, Gunnlaugsson T, Huxley AJM, McCoy CP, Rademacher JT, Rice TE (1997) Signaling recognition events with fluorescent sensors and switches. Chem Rev 97:1515–1566

Dejong MR, Engbersen JFJ, Huskens J, Reinhoudt DN (2000) Cyclodextrin dimers as receptor molecules for steroid sensors. Chem Eur J 6:4034–4040

Desvergne JP, Czarnik AW (1997) Chemosensors of ion and molecule recognition, NATO ASI Ser., Ser. C, vol 492. Kluwer, Dordrecht

Diaz-Garcia ME, Pina-Luis G, Rivero I (2006) Combinatorial solid-phase organic synthesis for developing materials with molecular recognition properties. Trends Anal Chem 25:112–121

Dietrich B, Guilhem J, Lehn JM, Pascard C, Sonveaux E (1984) Molecular recognition in anion coordination chemistry - structure, binding constants and receptor-substrate complementarity of a series of anion cryptates of a macrobicyclic receptor molecule. Helv Chim Acta 67:91–104

Dietrich B, Dilworth B, Lehn JM, Souchez JP, Cesario M, Guilhem J, Pascard C (1996) Anion cryptates: synthesis, crystal structures, and complexation constants of fluoride and chloride inclusion complexes of polyammonium macrobicyclic ligands. Helv Chim Acta 79:569–587

Douteau-Guevel N, Perret F, Coleman AW, Morel JP, Morel-Desrosiers N (2002) Binding of dipeptides and tripeptides containing lysine or arginine by p-sulfonatocalixarenes in water: NMR and microcalorimetric studies. J Chem Soc Perkin Trans 2 3:524–532

Dukh M, Šaman D, Lang K, Pouzar V, Černý I, Drašar P, Král V (2003) Steroid-porphyrin conjugate for saccharide sensing in protic media. Org Biomol Chem 1:3458–3463

Dunitz JD (1995) Win some, lose some - enthalpy-entropy compensation in weak intermolecular interactions. Chem Biol 2:709–712

Echavarren A, Galan A, de Mendoza J, Salmeron A, Lehn JM (1989) Chiral Recognition of aromatic carboxylate anions by an optically-active abiotic receptor containing a rigid guanidinium binding subunit. J Am Chem Soc 111:4994–4995

Eliseev AV, Schneider HJ (1994) Molecular recognition of nucleotides, nucleosides, and sugars by aminocyclodextrins. J Am Chem Soc 116:6081–6088

Fabbrizzi L, Licchelli M, Pallavicini P, Sacchi D, Taglietti A (1996) Sensing of transition metals through fluorescence quenching or enhancement. A review. Analyst 121:1763–1768

Fabbrizzi L, Foti F, Taglietti A (2005) Metal-containing trifurcate receptor that recognizes and senses citrate in water. Org Lett 7:2603–2606

Fenniri H, Hosseini MW, Lehn J-M (1997) Molecular recognition of NADP(H) and ATP by macrocyclic polyamines bearing acrydine groups. Helv Chim Acta 80:786–803

Ferrand Y, Crump MP, Davis AP (2007) A synthetic lectin analog for biomimetic disaccharide recognition. Science 318:619–622

Forgacs E, Cserhati T (2004) Study of the interaction of some steroidal drugs with cyclodextrin derivatives. Analyt Lett 37:1897–1908

Fujimoto T, Shimizu C, Hayashida O, Aoyama Y (1998) Ternary complexation involving protein. molecular transport to saccharide-binding proteins using macrocyclic saccharide cluster as specific transporter. J Am Chem Soc 120:601–602

Gale PA, Sessler JL, Kral V, Lynch V (1996) Calix[4]pyrroles: old yet new anion-binding agents. J Am Chem Soc 118:5140–5141

Ghidini E, Ugozzoli F, Ungaro R, Harkema S, El-Fald AA, Reinhoudt DN (1990) Complexation of alkali-metal cations by conformationally rigid, stereoisomeric calix[4]arene crown ethers: a quantitative-evaluation of preorganization. J Am Chem Soc 112:6979–6985

Goto H, Furusho Y, Yashima E (2007) Double helical oligoresorcinols specifically recognize oligosaccharides via heteroduplex formation through noncovalent interactions in water. J Am Chem Soc 129:9168–9174

Guo CX, Boullanger P, Liu T, Jiang L (2005) Size effect of polydiacetylene vesicles functionalized with glycolipids on their colorimetric detection ability. J Phys Chem B 109:18765–18771

Hacket F, Coterón JM, Schneider H-J, Kazachenko VP (1997) The complexation of glucose and other monosaccharides with cyclodextrins. Can J Chem 75:52–55

Hacket F, Simova S, Schneider H-J (2001) The complexation of peptides by aminocyclodextrins. J Phys Org Chem 14:159–170

Hamai S, Ikeda H, Ueno A (1998) H-1-NMR investigation of the binding of 2-methylnaphthalene to alpha-cyclodextrin in D2O solutions. J Incl Phenom Mol Recogn 31:265–273

Hamasaki K, Usui S, Ikeda H, Ikeda T, Ueno A (1997) Dansyl-modified cyclodextrins as fluorescent chemosensors for molecular recognition. Supramol Chem 8:125–135

Hamilton AD (1996) Supramolecular control of structure and reactivity: perspectives in supramolecular chemistry. Wiley, New York

Hanshaw RG, Smith BD (2005) New reagents for phosphatidylserine recognition and detection of apoptosis. Bioorg Med Chem 13:5035–5042

Hartley JH, James TD, Ward CJ (2000) Synthetic receptors. Perkin Trans 1 19:3155–3184

Hayashi T, Aya T, Nonoguchi M, Mizutani T, Hisaeda Y, Kitagawi S, Ogoshi H (2002) Chiral recognition and chiral sensing using zinc porphyrin dimmers. Tetrahedron 58:2803–2811

Hayashida O, Nishayama K, Aoyama Y (1999) Preparation and host-guest interaction of novel macrocyclic sugar clusters having mono- and oligosaccharides. Tetrahedron Lett 40:3407–3410

Heinz U, Hergetschweiler K, Acklin P, Faller B, Lattmann R, Schnebli HP (1999) 4-[3, 5-Bis (2-hydroxyphenyl)-1, 2, 4-triazol-1-yl]benzoic acid: A novel efficient and selective iron(III) complexing agent. Angew Chem Int Edi 38:2568–2570

Helgeson RC, Koga K, Timko JM, Cram DJ (1973) Complete optical resolution by differential complexation in solution between a chiral cyclic polyether and an alpha-amino-acid. J Am Chem Soc 95:3021–3033

Hirsch W, Muller T, Pizer R, Ricatto PJ (1995) Complexation of glucose by α- and β- cyclodextrins. Can J Chem 73:12–15

Hosseini MW, Blacker AJ, Lehn JM (1988) Multiple molecular recognition and catalysis - nucleotide binding and atp hydrolysis by a receptor molecule bearing an anion binding-site, an intercalator group, and a catalytic site. J Chem Soc Chem Commun 9:596–598

Houk KN, Leach AG, Kim SP, Zhang XY (2003) Binding affinities of host-guest, protein-ligand, and protein-transition-state complexes. Angew Chem Int Ed Engl 42:4872–4897

Hunter CA (2004) Quantifying intermolecular interactions: Guidelines for the molecular recognition toolbox. Angew Chem Int Ed Engl 43:5310–5324

Ikeda H, Nakamura M, Ise N, Oguma N, Nakamura A, Ikeda T, Toda F, Ueno A (1996) Fluorescent cyclodextrins for molecule sensing: Fluorescent properties, NMR characterization, and inclusion phenomena of N-dansylleucine-modified cyclodextrins. J Am Chem Soc 118:10980–10988

Impellizzeri G, Pappalardo G, D'Alessandro F, Rizzarelli E, Savioano M, Iacovino R, Benedetti E, Pedone C (2000) Solid state and solution conformation of 6-{4-[N-tert-butoxycarbonyl-N-(N′-ethyl)propanamide]imidazolyl}-6-deoxycyclomaltoheptaose: evidence of self-inclusion of the Boc group within the beta-cyclodextrin cavity. Eur J Org Chem 6:1065–1076

Inoye M, Miyake T, Furusyo M, Nakazumi H (1995) Molecular recognition of β-ribofuranosides by synthetic polypyridine-macrocyclic receptors. J Am Chem Soc 117:12416–12425

Inoye M, Chiba J, Nakazumi H (1999a) Glucopyranoside recognition by polypyridine-macrocyclic receptors possessing a wide cavity with a flexible lincage. J Org Chem 64:8170–8176

Inoye M, Takahashi K, Nakazumi H (1999b) Remarkably strong, uncharged hydrogen-bonding interactions of polypyridine-macrocyclic receptors for deoxyribofuranosides. J Am Chem Soc 121:341–345

Ishi-i T, Mateos-Timoneda MA, Timmerman P, Crego-Calama M, Reinhoudt DN, Shinkai S (2003) Self-assembled receptors that stereoselectively recognize a saccharide. Angew Chem Int Ed Engl 42:2300–2305

James TD (2007) Saccharide-selective boronic acid based photoinduced electron transfer (PET) fluorescent sensors. In creative chemical sensor systems. Top Curr Chem 277:107–152

James TD, Shinkai S (2002) Artificial receptors as chemosensors for carbohydrates. In Host-Guest Chemistry. Top Curr Chem 218:159–200

James TD, Shinkai S (2005) Fluorescent saccharide sensors. In: Geddes Chris D, Lakowicz Joseph R (eds) Advanced concepts in fluorescence sensing, part B. topics in fluorescence spectroscopy 10:41–67

James TD, Linnane P, Shinkai S (1996a) Fluorescent saccharide receptors: a sweet solution to the design, assembly and evaluation of boronic acid derived PET sensors. Chem Commun 3: 281–288

James TD, Sandanayake KRAS, Shinkai S (1996b) Angew Chem Int Ed 35:1911–1922

Kano K (1997) Mechanisms for chiral recognition by cyclodextrins. J Phys Org Chem 10:286–291

Katz E, Willner I (2004) Integrated nanoparticle-biomolecule hybrid systems: Synthesis, properties, and applications. Angew Chem Int Ed Engl 43:6042–6108

Kikkeri R, Traboulsi H, Humbert N, Gumienna-Kontecka E, Arad-Yellin R, Melman G, Elhabiri M, Albrecht-Gary AM, Shanzer A (2007) Toward iron sensors: Bioinspired tripods based on fluorescent phenol-oxazoline coordination sites. Inorg Chem 46:2485–2497

Kikuchi Y, Kato Y, Tanaka Y, Toi H, Aoyama Y (1991) Molecular recognition and stereoselectivity: geometrical requirements for the multiple hydrogen-bonding interaction of diols with multidentate polyhydroxy macocycle. J Am Chem Soc 113:1349–1354

Kikuchi Y, Kobayashi K, Aoyama Y (1992a) Complexation of chiral glycols, steroidal polyols, and sugars with a multibenzenoid, achiral host as studied by induced circular dichroism spectroscopy: exciton chirality induction in resorcinol-aldehyde cyclotetramer and its use as a supramolecular probe for the assignements of stereochemistry of chiral guests. J Am Chem Soc 114:1351–1358

Kikuchi Y, Tanaka Y, Sutarto S, Kobayashi K, Toi H, Aoyama Y (1992b) Highly cooperative binding of alkyl glucopyranosydes to the resorcinol cyclic tetramer due to intracomplex guest-guest hydrogen bonding: solvophobicity/solvophilicity control by an alkyl group of the geometry, stoichiometry, stereoselectivity, and cooperativity. J Am Chem Soc 114: 10302–10306

Kim SK, Moon B-S, Park JH, Seo YI, Koh HS, Yoon YJ, Lee KD, Yoon J (2005) A fluorescent cavitand for the recognition of GTP. Tetrahedron Lett 46:6617–6620

Kobayashi K, Asakawa Y, Kikuchi Y, Toi H, Aoyama Y (1993) CH-π interactions as an important driving force of host-guest complexation in apolar organic media. binding of monools and acetylated compounds to resorcinol cyclic tetramer as studied by ^{1}h nmr and circular dichroism spectroscopy. J Am Chem Soc 115:2648–2654

Koblenz TS, Wassenaar J, Reek JNH (2008) Reactivity within a confined self-assembled nano-space. Chem Soc Rev 37:247–262

Kopelman R, Dourado S (1996) Is smaller better?–Scaling of characteristics with size of fiber-optic chemical and biochemical sensors. Proc SPIE- Intern Soc Opt Eng 2836:2–11

Král V, Rusin O, Charvátová J, Anzenbacher P Jr, Fogl J (2000) Porphyrin phosphonates: novel anionic receptors for saccharide recognition. Tetrahedron Lett 41:10147–10151

Král V, Rusin O, Schmidtchen FP (2001) Novel porphyrin-cryptand cyclic systems: receptor for saccharide recognition in water. Org Lett 3:873–876

Král V, Králová J, Kaplánek P, Bríza T, Martásek P (2006) Quo vadis porphyrin chemistry? Physiol Res 55(Suppl 2):S3–S26

Králová J, Koivukorpi J, Kejík Z, Poučková P, Sievänen E, Kolehmainen E, Král V (2008) Porphyrin-bile acid conjugates: from saccharide recognition in the solution to the selective cancer cell fluorescence detection. Org Biomol Chem 6:1548–1552

Kruppa M, König B (2006) Reversible coordinative bonds in molecular recognition. Chem Rev 106:3520–3560

Kruppa M, Mandl C, Miltschitzky S, König B (2005) A luminescent receptor with affinity for N-terminal histidine in peptides in aqueous solution. J Am Chem Soc 127:3362–3365

Kubik S, Goddard R (2002) Conformation and anion binding properties of cyclic hexapeptides containing L-4-hydroxyproline and 6-aminopicolinic acid subunits. Proc Nat Acad Sci USA 99:5127–5132

Kubik S, Reyheller C, Stuwe S (2005) Recognition of anions by synthetic receptors in aqueous solution. J Incl Phenom 52:137–187

Kubo K, Sakurai T, Mori A (1999) Complexation and fluorescence behavior of 9, 10-bis[bis(beta-hydroxyethyl)aminomethyl]anthracene. Talanta 50:73–77

Kumar S, Schneider HJ (1989) The Complexation of Estrogens and Tetralins in the Cavities of Azoniacyclophanes. J Chem Soc Perkin Trans 2:245–250

Kurihara K, Ohto K, Tanaka Y, Aoyama Y, Kunitake T (1991) Molecular recognition of sugars by monolayers of resorcinol-dodecanal cyclotetramer. J Am Chem Soc 113:444–450

Lakowicz JR (1994) Probe design and chemical sensing. Plenum, New York

Lambert TN, Smith BD (2003) Synthetic receptors for phospholipid headgroups. Coord Chem Rev 240:129–141

Lehn JM (1995) Supramolecular chemistry: concepts and perspectives. Wiley-VCH, Weinheim

Lehn JM, Atwood JL, Davies JED, MacNicol DD, MacNicol DD, Vögtle F (1996) Comprehensive supramolecular chemistry, vol 1–11. Pergamon. Elsevier, Oxford

Liu Y, Chen Y, Li B, Wada T, Inoue Y (2001) Cooperative multipoint recognition of organic dyes by bis(beta-cyclodextrin)s with 2, 2′-bipyridine-4, 4′-dicarboxy tethers. Chem Eur J 7: 2528–2535

Liu Y, Yang YW, Yang EC, Guan XD (2004) Molecular recognition thermodynamics and structural elucidation of interactions between steroids and bridged bis(beta-cyclodextrin)s. J Org Chem 69:6590–6602

Lützen A (2007) Supramolekulare Chemie - Chemie jenseits des Moleküls. CHEMKON 14:123–130

Mahoney JM, Beatty AM, Smith BD (2001) Selective recognition of an alkali halide contact ion-pair. J Am Chem Soc 123:5847–5848

Mammen M, Choi SK, Whitesides GM (1998) Polyvalent interactions in biological systems: Implications for design and use of multivalent ligands and inhibitors. Angew Chem Int Ed Engl 37:2755–2794

Mandl CP, König B (2005) Luminescent crown ether amino acids: Selective binding to N-terminal lysine in peptides. J Org Chem 70:670–674

Matsushita A, Kuwabara T, Nakamura A, Ikeda H, Ueno A (1997) Guest-induced colour changes and molecule-sensing abilities of p-nitrophenol-modified cyclodextrins. J Chem Soc Perkin Trans 2: 1705–1710.

Mazik M, Buthe AC (2008) Highly effective receptors showing di- vs. monosaccharide preference. Org Biomol Chem 6:1558–1568

Mazik M, Cavga H (2007) An acyclic aminonaphthyridine-based receptor for carbohydrate recognition: binding studies in competitive solvents. Eur J Org Chem: 3633–3638

Mazik M, Kuschel M (2008) Amide, amino, hydroxy and aminopyridine groups as building blocks for carbohydrate receptors. Eur J Org Chem: 1517–1526

Meyer EA, Castellano RK, Diederich F (2003) Interactions with aromatic rings in chemical and biological recognition. Edit Angew Chem-Int 42:1210–1250

Mizutani T, Wada K, Kitagawa S (2002) Hydrophobic environment of gable-type bisporphyrin receptors in water promotes binding of amines and oligopeptides. Chemical Communications: 1626–1627

Moreno-Corral R, Lara KO (2008) Complexation studies of nucleotides by tetrandrine derivatives bearing anthraquinone and acridine groups. Supramol Chem 20:427–435

Mulder A, Huskens J, Reinhoudt DN (2004) Multivalency in supramolecular chemistry and nanofabrication. Org Biomol Chem 2:3409–3424

Müller G, Riede J, Schmidtchen FP (1988) Host-guest bonding of oxoanions to guanidinium anchor groups. Angew Chem Int Ed Engl 27:1516–1518

Narita M, Hamada F (2000) The synthesis of a fluorescent chemo-sensor system based on regioselectively dansyl-tosyl-modified beta- and gamma-cyclodextrins. J Chem Soc Perkin Trans 2:823–832

Nassimbeni LR (2003) Physicochemical aspects of host-guest compounds. Acc Chem Res 36:631–637

O'Neil EJ, Smith BD (2006) Anion recognition using dimetallic coordination complexes. Coord Chem Rev 250:3068–3080

Ogoshi H, Mizutani T (1998) Multifunctional and chiral porphyrins: Model receptors for chiral recognition. Acc Chem Res 31:81–89

Oshovsky GV, Reinhoudt DN, Verboom W (2007) Supramolecular chemistry in water. Angew Chem Int Ed 46:2366–2393

Page D, Zanini D, Roy R (1996) Macromolecular recognition: effect of multivalency in the inhibition of binding of yeast mannan to concanavalin A and pea lectins by mannosylated dendrimers. Bioorg Med Chem 4:1949–1961

Palde PB, Gareiss PC, Miller BL (2008) Selective Recognition of Alkyl Pyranosides in Protic and Aprotic Solvents. J Am Chem Soc 130:9566–9573

Paulini R, Müller K, Diederich F (2005) Orthogonal multipolar interactions in structural chemistry and biology. Edit Angew Chem-Int 44:1788–1805

Peczuh MW, Hamilton AD (2000) Peptide and protein recognition by designed molecules. Chem Rev 100:2479–2493

Penadés S, Fuhrhop J-H (eds) (2002) Host-guest chemistry : mimetic approaches to study carbohydrate recognition (Top. Curr. Chem. ; 218). Springer, Berlin, 241

Prodi L (2005) Luminescent chemosensors: from molecules to nanoparticles. New J Chem 29: 20–31

Prodi L, Bolletta F, Montalti M, Zaccheroni N (2000) Luminescent chemosensors for transition metal ions. Coord Chem Rev 205:59–83

Rao JH, Lahiri J, Isaacs L, Weis RM, Whitesides GM (1998) A trivalent system from vancomycin center dot D-Ala-D-Ala with higher affinity than avidin center dot biotin. Science 280: 708–711

Rao JH, Lahiri J, Isaacs L, Weis RM, Whitesides GM (2000) Design, synthesis, and characterization of a high-affinity trivalent system derived from vancomycin and L-Lys-D-Ala-D-Ala. J Am Chem Soc 122:2698–2710

Ray D, Bharadwaj PK (2008) A coumarin-derived fluorescence probe selective for magnesium. Inorg Chem 47:2252–2254

Raymond KN (1994) Recognition and Transport of Natural and Synthetic Siderophores by Microbes. Pure & Appl Chem 66:773–781

Reilly SD, Khalsa GRK, Ford DK, Brainard JR, Hay BP, Smith PH (1995) Octaazacryptand Complexation of the Fluoride-Ion. Inorg Chem 34:569–575

Rekharsky MV, Yamamura H, Kawai M, Inoue Y (2003) Complexation and chiral recognition thermodynamics of gamma-cyclodextrin with N-acetyl- and N-carbobenzyloxy-dipeptides possessing two aromatic rings. J Org Chem 68:5228–5235

Rekharsky MV, Mori T, Yang C, Ko YH, Selvapalam N, Kim H, Sobransingh D, Kaifer AE, Liu SM, Isaacs L, Chen W, Moghaddam S, Gilson MK, Kim KM, Inoue Y (2007) A synthetic host-guest system achieves avidin-biotin affinity by overcoming enthalpy-entropy compensation. Proc Natl Acad Sci USA 104:20737–20742

Rekharsky MV, Yamamura H, Ko YH, Selvapalam N, Kim K, Inoue Y (2008) Sequence recognition and self-sorting of a dipeptide by cucurbit[6]uril and cucurbit[7]uril. Chem Commun 2236–2238

Rettig W, Strehmel B, Schrader S, Seifert H (1999) Applied fluorescence in chemistry, biology, and medicine. Springer, Berlin, p 179

Rurack K (2001) Flipping the light switch 'ON'–the design of sensor molecules that show cation-induced fluorescence enhancement with heavy and transition metal ions. Spectrochim Acta A Mol Biomol Spectrosc 57:2161–2195

Rusin O, Král V (2001) 1, 1′-Binaphthyl substituted calixresorcinols-methyl red complexes: receptors for saccharide sensing. Tetrahedron Lett. 42:4235–4238

Rusin O, Lang K, Král V (2002) 1, 1′-Binaphthyl-Substituted Macrocycles as Receptor for Saccharide Recognition. Chem Eur J 8:655–663

Salvatierra D, Sanchez-Ruiz X, Garduno R, Cervello E, Jaime C, Vergili A, Sanchez-Ferrando F (2000) Enantiodifferentiation by complexation with beta-cyclodextrin: Experimental (NMR) and theoretical (MD, FEP) studies. Tetrahedron 56:3035–3041

Schalley C (2007) Analytical methods in supramolecular chemistry. Wiley-VCH, Weinheim

Schmidtchen FP (2005) Artificial host molecules for the sensing of anions. Topics in Current Chemistry: Anion Sensing. 255:1–29

Schmidtchen FP, Berger M (1997) Artificial organic host molecules for anions. Chem Rev 97:1609–1646

Schmuck C, Wennemers H (2004) Highlights in bioorganic chemistry: methods and applications. Wiley-VCH, Weinheim

Schneider H-J (1991) Mechanisms of molecular recognition: investigations of organic host guest complexes. Angew Chem Int Ed 30:1417–1436

Schneider H-J (1994) Linear free-energy relationships and pairwise interactions in supramolecular chemistry. Chem Soc Rev 22:227–234

Schneider H-J (2009) Binding mechanisms in supramolecular complexes. Angen Chem Int Ed Engl 48:3294–3977

Schneider H-J, Kato K (2007) Direct translation of chiral recognition into mechanical motion. Angew Chem Int Ed Engl 46:2694–2696

Schneider H-J, Sirish M (2000) Highly efficient complexations of a porphyrin dimer with remarkably small differences between nucleosides and nucleotides/the predominance of stacking interactions for nucleic acid components. J Am Chem Soc 122:5881–5882

Schneider H-J, Yatsimirsky A (2000) Principles and methods in supramolecular chemistry. Wiley, Chichester

Schneider H-J, Yatsimirsky A (2008) Selectivity in supramolecular host-guest complexes. Chem Soc Rev 263–277

Schneider SE, O'Neil SN, Anslyn EV (2000) Coupling rational design with libraries leads to the production of an ATP selective chemosensor. J Am Chem Soc 122:542–543

Schneider H-J, Tianjun L, Lomadze N (2004) Sensitivity increase in molecular recognition by decrease of the sensing particle size and by increase of the receptor binding site - a case with chemomechanical polymers. Chem Commun: 2436–2437.

Schrader T, Hamilton M (2005) Functional synthetic receptors. Wiley-VCH, Weinheim 428 pp

Schrader T, Koch S (2007) Artificial protein sensors. Mol BioSyst 3:241–248

Schug KA, Lindner W (2005) Noncovalent binding between guanidinium and anionic groups: Focus on biological- and synthetic-based arginine/guanidinium interactions with phosph[on]ate and sulf[on]ate residues. Chem Rev 105:67–113

Sears P, Wong C-H (1999) Carbohydrate mimetics: a new strategy for tackling the problem of carbohydrate-mediated biological recognition. Angew Chem Int Ed 38:2300–2324

Shahgaldian P, Pieles U (2006) Cyclodextrin derivatives as chiral supramolecular receptors for enantioselective sensing. Sensors 6:593–615

Shanzer A, Libman J, Lytton SD, Glickstein H, Cabantchik ZI (1991) Reversed siderophores act as antimalarial agents. Proc Natl Acad Sci USA 88:6585–6589

Shi Y, Schneider H-J (1999) Interactions between aminocalixarenes and nucleosides or nucleic acids. J Chem Soc Perkin Trans 2:1797–1803

Shinkai S, Robertson A (2001) The design of molecular artificial sugar sensing systems. In: New trends in fluorescence spectroscopy. Springer Series on Fluorescence 1:173–185

Sirish M, Schneider H-J (1999) Porphyrin derivatives as water-soluble receptors for peptides. Chem Commun: 907–908

Sirish M, Chertkov VA, Schneider H-J (2002) Porphyrin-based peptide receptors: syntheses and NMR analysis. J Chem Eur J 8:1181–1188

Steed JW, Atwood JL (2000) Supramolecular Chemistry. Wiley, Chichester

Steed JW, Turner DR, Wallace KJ (2007) Core concepts in supramolecular chemistry and nanochemistry. Wiley, Chichester

Striegler S (2003a) Carbohydrate recognition in cross-linked sugar-templated poly(acrylates). Macromolecules 36:1310–1317

Striegler S (2003b) Selective carbohydrate recognition by synthetic receptors in aqueous solution. Curr Org Chem 7:81–102

Striegler S, Dittel M (2003) A sugar discriminating binuclear copper(II) complex. J Am Chem Soc 125:11518–11524

Striegler S, Dittel M (2005) A sugar's choice: coordination to a mononuclear or a dinuclear copper (II) complex? Inorg Chemistry 44:2728–2733

Sugimoto N, Miyoshi D, Zou J (2000) Development of small peptides recognizing a monosaccharide by combinatorial chemistry. Chem Commun 2295–2296

Tanaka Y, Khare C, Yonezawa M, Aoyama Y (1990) Highly stereoselective glycosidation of ribose solubilized in apolar organic media via host-guest complexation. Tetrahedron Lett 43:6193–6196

Uccello-Barretta G, Balzano F, Cuzzola A, Menicagli R, Salvadori P (2000) NMR detection of the conformational distortion induced in cyclodextrins by introduction of alkyl or aromatic substituents. Eur J Org Chem 3:449–453

Usui S, Tanabe T, Ikeda H, Ueno A (1998) The structure of mono-(3-amino-3-deoxy)-alpha-cyclodextrin in aqueous solution - Molecular dynamics and NMR studies. J Mol Struct 442:161–168

Vacca A, Nativi C, Cacciarini M, Pergoli R, Roelens S (2004) A new tripodal receptor for molecular recognition of monosaccharides. A paradigm for assessing glycoside binding affinities and selectivities by H-1 NMR spectroscopy. J Am Chem Soc 126:16456–16465

Vögtle F (1993) Supramolecular chemistry: an introduction. Wiley

Vögtle, F (ed) (1995) Comprehensive supramolecular chemistry. vol 2. Pergamon

Wallimann P, Marti T, Furer A, Diederich F (1997) Steroids in molecular recognition. Chem Rev 97:1567–1608

Wang Y, Ikeda T, Ikeda H, Ueno A, Toda F (1994) Dansyl-β-cyclodextrins as fluorescent sensors responsive to organic compounds. Bull Chem Soc Jpn 67:1598–1607

Wang W, Gao X, Wang B (2002) Boronic acid-based sensors. Curr Org Chem 6:1285–1317

Wehner M, Janssen D, Schäfer G, Schrader, T (2006) Sequence-selective peptide recognition with designed modules. Eur J Org Chem: 138–153

Werner F, Schneider HJ (2000) Complexation of anions including nucleotide anions by open-chain host compounds with amide, urea, and aryl functions. Helv Chim Acta 83:465–478

Williams DH, Stephens E, O'Brien DP, Zhou M (2004) Understanding noncovalent interactions: ligand binding energy and catalytic efficiency from ligand-induced reductions in motion within receptors and enzymes. Angew Chem Int Ed 43:6596–6616

Wiskur SL, Ait-Haddou H, Lavigne JJ, Anslyn EV (2001) Teaching old indicators new tricks. Acc Chem Res 34:963–972

Xu H, Xu XH, Dabestani R, Brown GM, Fan L, Patton S, Ji HF (2002) Supramolecular fluorescent probes for the detection of mixed alkali metal ions that mimic the function of integrated logic gates. J Chem Soc Perkin Trans 2:636–643

Yan J, Fang H, Wang B (2005) Boronolectins and fluorescent boronolectins: an examination of the detailed chemistry issues important for the design. Med Res Rev 25:490–520

Zadmard R, Schrader T (2006) DNA recognition with large calixarene dimers. Angew Chem Int Ed 45:2703–2706

Zielenkiewicz W, Marcinowicz A, Cherenok S, Kalchenko VI, Poznanski J (2006) Phosphorylated calixarenes as receptors of L-amino acids and dipeptides: calorimetric determination of Gibbs energy, enthalpy and entropy of complexation. Supramol Chem 18:167–176

Žinić M, Vigneron J-P, Lehn J-M (1995) Binding of nucleotides in water by phenanthridinium bis (intercland) receptor molecules. Chem Commun: 1073–1075

Part IV
Kinetics of Chemo/Biosensors

Chapter 21
Kinetics of Chemo/Biosensors

Ifejesu Eni-Olorunda and Ajit Sadana

Abstract An insight into biosensor kinetics has been presented in this chapter with a focus on analyte–receptor interactions on biosensor surfaces. The vast applications of sensors are rapidly increasing, especially in the areas of medicine, food and drug administration and detection of toxic substances in various fields, and industrial applications. The Surface Plasmon Resonance (SPR) Biosensor is a dynamic equipment for measuring signals in analyte–receptor systems and is the main medium through which all the data in this chapter are obtained. A fractal analysis (single and dual) is presented here in an attempt to model the kinetic data obtained from literature to obtain important kinetic parameters such as the binding and dissociation rate constants, affinity, and fractal dimension values.

Examples are cited for the binding of liquid petroleum gas (LPG) to zinc oxide films prepared by the spray pyrolysis method onto a glass substrate (Sens and Actuator B 120: 551–559, 2007) and the binding of different NH_3 concentrations in air to a sol-gel derived thin film (Sens and Actuator B 110:299–303, 2005). Since the major focus of this chapter is the kinetics, only scant information, if any, is presented on thermodynamics in these system interactions. However, fractal analysis provides a unique perspective of these molecular interactions and could be applied in other systems.

Keywords Biosensors · Kinetics · Binding · Dissociation · Fractals

Abbreviations

SPR	Surface plasmon resonance
LPG	Liquid petroleum gas

A. Sadana (✉)
Chemical Engineering Department, University of Mississippi, University, MS 38677-1848 USA
e-mail: cmsadana@olemiss.edu

M. Zourob (ed.), *Recognition Receptors in Biosensors*,
DOI 10.1007/978-1-4419-0919-0_21,

Sens	Sensors
BST	Barium strontium titanate
λ_{exo}	$\lambda_{exonuclease}$
TYPNK	T4 polynucleotide kinase
FAM-SP-2p	Oligonucleotide sequence (5-3′)
ATP	Adenosine triphosphate
DNA	Deoxyribonucleic acid
ATR	Attenuated total reflection
BIA	Biomolecular interaction analysis
SPs	Surface plasmons
UV	Ultraviolet
IR	Infrared
Ag	Antigen
Ab	Antibody

21.1 Background

Kinetics is the study of rates of change and also concerns itself with the routes of reactions as materials proceed from an initial to a final state. Kinetics has broad applications in our daily living such as rates of growth and changes in biology and medicine, for example, the rate of bacterial population growth or the rate of cell death because of anticancer medication (Wojciechowski 1975). This chapter would focus on the kinetics as it relates to biosensor operations. The aim is to give us a clearer perspective of the properties of biosensors and possible modification to improve their performance in terms of selectivity, sensitivity, and other inherent biosensor characteristics. This would be done through a fractal analysis. At the outset, it is appropriate to indicate that the fractal analysis method is just one way of presenting the analyte–receptor reactions occurring on biosensor surfaces.

21.2 Introduction

Biosensors monitor bimolecular interactions in real time. Simply defined, a biosensor is an analytical device that converts a biological response into an electrical signal. Currently, biosensors represent a rapidly expanding field, with an estimated 60% annual growth rate,, the major impetus coming from the health-care industry (e.g., 6% of the western world are diabetic and would benefit from the availability of a rapid, accurate, and simple biosensor for glucose) but with some pressure from other areas, such as food quality appraisal and environmental monitoring. The estimated world analytical market is about \$17 billion per year,[1] of which 30% is

[1] *Source:* Chaplin (2010) *Adsorption and Phase Behavior in Nanochannels and Nanotubes*

in the health care area. There is clearly a vast market expansion potential, as less than 0.1% of this market is currently using biosensors. Research and development in this field is wide and multidisciplinary, spanning biochemistry, bioreactor science, physical chemistry, electrochemistry, electronics, and software engineering (Chaplin 2010).

A biosensor includes two steps: a recognition step and a transducing step. In the recognition step, the biological element (the immobilized ligand) can recognize the analyte either in solution or in the atmosphere. Immobilized ligands can be antibodies, proteins, receptors, or enzymes. The analyte that binds to the ligands can be antigens, drug molecules, cell and cellular moieties, substrates, and so on. Analyte–receptor binding generates different signals on biosensors that can be measured. The receptor is in close contact with the transducing element and converts the signal into a quantitative optical or electrical signal. The signal can be a (1) change in the resonance units (SPR, Surface Plasmon Resonance biosensors), (2) change in either the UV or IR absorption, (3) change in mass (Piezoelectric biosensors), or (4) change in electrical properties.

The following is an illustration of the basics of the chemistry of biosensor operations using the SPR biosensor. Figure 21.1 shows the mechanism of the SPR biosensor, and Fig. 21.2 shows the various stages of molecule interaction on the biosensor chip.

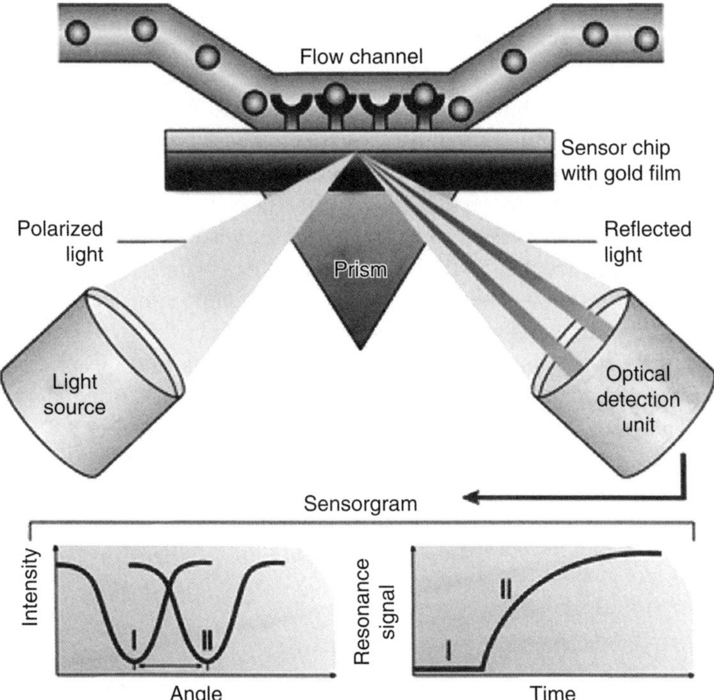

Fig. 21.1 Mechanism of the SPR biosensor

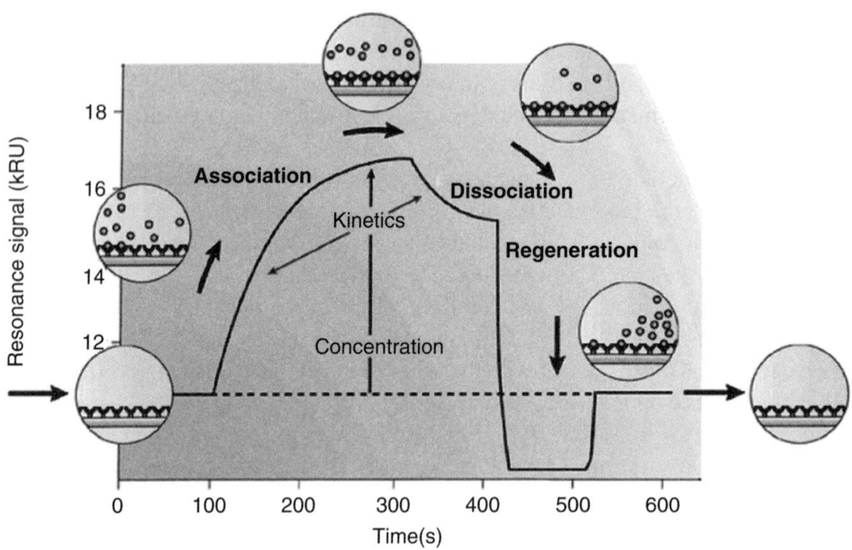

Fig. 21.2 Various stages of molecular interaction on the biosensor chip
Source: Nature Review Drug Discovery, 2009

21.3 Adsorption Models

The analyte in solution binds (adsorbs) to the surface of the biosensor chip and exhibits some kinetic activity that can be modeled based on various characteristic equations. The most important class of adsorption is reversible monolayer adsorption, in which adsorbing molecules reversibly attach themselves to the surface of the solid. Langmuir derived a simple model for an adsorption isotherm originally developed for gases and that is still in use. Researchers in the past have successfully modeled the adsorption behavior of analytes in solution to solid surfaces, using the Langmuir model, although it does not strictly conform to theory. These modifications with their constraints have provided some insights into the physics of the adsorption of absorbates on different surfaces. Langmuir's original derivation assumed that adsorption and desorprtion were elementary processes in which the rate of adsorption (r_{ad}) and that of desorption (r_{d}) are given by the following equations:

$$r_{\mathrm{ad}} = k_{\mathrm{ad}} P_{\mathrm{A}} [S]$$

$$r_{\mathrm{d}} = k_{\mathrm{d}} [A_{\mathrm{ad}}]$$

P_A is the partial pressure of 'A' over the surface, $[S]$ is the concentration of bare sites in number/cm^2, $[A_{\mathrm{ad}}]$ is the surface concentration of 'A' in molecules/cm^2, and k_{ad} and k_{d} are constants. Following a series of derivations, the Langmuir adsorption isotherm for noncompetitive, nondisscociative adsorption is given by θ_A, which represents the fraction of sites covered with 'A' is (Masel 1996);

$$\theta_A = \frac{K_{equ} P_A}{1 + K_{equ} P_A} \quad \text{where } K_{equ} = k_{ad}/k_d$$

Despite the broad applications of the Langmuir approach, it has a major drawback, because it is based on the assumption that the surface of the adsorbent is composed of discrete classes of adsorption sites, which is not a realistic assumption. With the ever increasing use of biosensors as bioanalytical tools, there is a greater need to understand in detail, their mode of operation, which could help to improve their sensitivity, stability and specificity, and other biosensor performance parameters. Mass transfer limitations and heterogeneity of the sensor chip surface complicate the binding assay kinetics. Thus, it is necessary to characterize reactions occurring on the biosensor surface in the presence of diffusion limitations, heterogeneity, and other complications that may be present.

Most literature on macroscopic chemical kinetics treats the rate constant of a reaction as a time-independent study. However, this condition applies to homogenous, well-stirred solutions only. When the reaction occurs in a restricted geometry and depletion zones develop, concentration fluctuations can dominate the kinetics of the process. Under these conditions, the rate constant will have time dependence (Kang and Redner 1985). Considering a bimolecular reaction of the type, A + B → products, one has:

$$-\frac{d[A]}{dt} = k(t)[A][B]$$

where the time-dependent rate constant, $k\,(t)$, has the form; $k = k^{J} t^{-h}$, where k^{J} is a time-independent constant and the exponent h is restricted to values between zero and one (Dewey 1997).

Melikhova et al (1988) has proposed a new method for the analysis of experimental data on ligand–receptor binding at equilibrium. The difference method can also detect heterogeneity of a receptor system in cases where the contribution of the high-affinity site to the total binding is rather small. The method also permits exclusion of experiments for measuring nonspecific binding.

The importance of dimensionality in biological systems has long been recognized. Changes in dimensionality of a system can greatly alter the rates at which a small diffusable molecule binds to a receptor site. A typical example illustrating this is the *lac repressor* (a protein that binds to a specific site on DNA called the *lac operon*). It was discovered that one-dimensional diffusion greatly enhances the rate of the specific binding process, giving bimolecular rate constants that are faster than expected for diffusion in three dimensions.

A number of theoretical works have been considered for the diffusion of a ligand from the bulk solution to the membrane surface, followed by two-dimensional diffusion in the membrane to the receptor. The reduction of dimension at the membrane surface results in a rate enhancement for diffusion-limited reactions. Investigations for steady state (Berg and Purcell 1977) and transient (Zwanzig and Szabo 1991) kinetics of ligand binding have been made and they confirm a ligand binding rate increase with a reduction in dimensionality.

For one, two, and three dimensions, the Einstein relationship $r^2 \sim t$, holds. If a diffusing molecule is visiting sites at a constant rate, then the number of sites visited, N_{visited}, is proportional to time, i.e., $N_{\text{visited}} \sim t$. Within a time t, the diffuser will have access to all the sites within a volume, r^d, where d is the spatial dimension. The total number of sites, N_{total}, available for a visit will be equal to the density times this volume.

Kopelman (1988) was the first to suggest the use of fractals to describe diffusion-limited kinetics coupled with surface heterogeneity. He suggested that surface-diffusion-controlled reactions occurring on clusters or islands exhibit anomalous fractal-like kinetics, which in turn yields anomalous reaction orders and time-dependent (binding and dissociation) rate coefficients. The biosensor surface can be considered complex and "disordered". Therefore, it lends itself to a fractal analysis, which would provide a physical clarification of the reactions occurring on a disordered or heterogeneous biosensor surface.

Fractals are disordered systems described by nonintegral dimensions. A characteristic feature of the fractal structure is its "self-similarity" at different levels of scale. Part of a fractal object resembles the bigger structure of which it is a part (e.g., leaves on a tree). Fractal kinetics has been used to describe other biochemical systems such as ion-channel gating (Liebovitch and Sullivan 1987) and protein dynamics (Dewey and Bann 1992). Fractals are particularly useful in explaining these types of reactions, because the heterogeneity that exists on the surface can be characterized by a single number (lumped parameter), the fractal dimension or D_{f} value. The analysis presented here will be performed on data available from literature. The goal is to promote a better understanding of analyte–receptor reactions occurring on biosensor surfaces.

A fractal analysis via the fractal dimensions and the binding and dissociation rate coefficients provides a convenient and useful lumped parameter(s) analysis of diffusion-limited reactions occurring on surfaces. The analysis makes quantitative the heterogeneity or surface roughness existing on the biosensor surface and requires the input of only very few parameters to obtain the binding and dissociation rate coefficients.

21.4 Theory

Havlin (1989) has reviewed and analyzed the diffusion of reactants toward fractal surfaces. The details of the theory and the equations involved in the binding and the dissociation phases for analyte–receptor binding are available (Sadana 2001). The details are not repeated here, except that just the equations are given to permit an easier reading. These equations have been applied to other biosensor systems (Sadana 2001; Ramakrishnan and Sadana 2001). For most applications, a single- or a dual-fractal analysis is often adequate to describe the binding and the dissociation kinetics. Peculiarities in the values of the binding and the dissociation rate coefficients, as well as in the values of the fractal dimensions with regard to the dilute analyte systems being analyzed, will be carefully noted, if applicable.

21.4.1 Single-Fractal Analysis

21.4.1.1 Binding Rate Coefficient

Havlin (1989) indicates that the diffusion of a particle (analyte [Ag]) from a homogeneous solution to a solid surface (e.g. receptor [Ab]-coated surface) on which it reacts to form a product (analyte–receptor complex; (Ab·Ag)) is given by:

$$(\mathrm{Ab}\cdot\mathrm{Ag}) \approx \begin{cases} t^{(3-D_{\mathrm{f,bind}})/2} = t^{p} & t < t_c \\ t^{1/2} & t > t_c \end{cases} \tag{21.1}$$

Here $D_{\mathrm{f,bind}}$ or D_{f} (used later on in the chapter) is the fractal dimension of the surface during the binding step. t_c is the cross-over value. Havlin (1989) indicates that the cross-over value may be determined by $r_c^2 \sim t_c$. Above the characteristic length, r_c, the self-similarity of the surface is lost and the surface may be considered homogeneous. Above time, t_c the surface may be considered homogeneous, since the self-similarity property disappears, and "regular" diffusion is now present. For a homogeneous surface where D_{f} is equal to two, and when only diffusional limitations are present, $p = ½$ as it should be. Another way of looking at the $p = ½$ case (where $D_{\mathrm{f,bind}}$ is equal to two) is that the analyte in solution views the fractal object, in our case, the receptor–coated biosensor surface, from a "large distance." In essence, in the association process, the diffusion of the analyte from the solution to the receptor surface creates a depletion layer of width $(Đt)^{½}$ where $Đ$ is the diffusion constant. This gives rise to the fractal power law, (Analyte·Receptor) ~ $t^{(3-D_{\mathrm{f,bind}})/2}$. For the present analysis, t_c is arbitrarily chosen and we assume that the value of the t_c is not reached. One may consider the approach as an intermediate "heuristic" approach that may be used in the future to develop an autonomous (and not time-dependent) model for diffusion-controlled kinetics.

21.4.1.2 Dissociation Rate Coefficient

The diffusion of the dissociated particle (receptor [Ab] or analyte [Ag]) from the solid surface (e.g., analyte [Ag]-receptor [Ab]) complex coated surface) into solution may be given, as a first approximation by:

$$(\text{Ab} \cdot \text{Ag}) \approx -t^{(3-D_{\text{f,diss}})/2} = t^{\text{p}}(t > t_{\text{diss}}) \tag{21.2}$$

Here $D_{\text{f,diss}}$ is the fractal dimension of the surface for the dissociation step. This corresponds to the highest concentration of the analyte–receptor complex on the surface. Henceforth, its concentration only decreases. The dissociation kinetics may be analyzed in a manner "similar" to the binding kinetics.

21.4.2 Dual-Fractal Analysis

21.4.2.1 Binding Rate Coefficient

Sometimes, the binding curve exhibits complexities and two parameters (k, D_{f}) are not sufficient to adequately describe the binding kinetics. This is further corroborated by low values of r^2 factor (goodness-of-fit). In that case, one resorts to a dual-fractal analysis (four parameters; k_1, k_2, D_{f1}, and D_{f2}) to adequately describe the binding kinetics. The single-fractal analysis presented earlier is thus extended to include two fractal dimensions. At present, the time ($t = t_1$) at which the "first" fractal dimension "changes" to the "second" fractal dimension is arbitrary and empirical. For the most part, it is dictated by the data analyzed and experience gained by handling a single-fractal analysis. A smoother curve is obtained in the "transition" region, if care is taken to select the correct number of points for the two regions. In this case, the product (antibody-antigen; or analyte–receptor complex, Ab·Ag or analyte·receptor) is given by:

$$(\text{Ab} \cdot \text{Ag}) \approx \begin{cases} t^{(3-D_{\text{f1,bind}})/2} = t^{\text{p1}}(t < t_1) \\ t^{(3-D_{\text{f2,bind}})/2} = t^{\text{p2}}(t_1 < t < t_2) = t_{\text{c}} \\ t^{1/2} \quad (t > t_{\text{c}}) \end{cases} \tag{21.3}$$

In some cases, as mentioned earlier, a triple-fractal analysis with six parameters (k_1, k_2, k_3, D_{f1}, D_{f2}, and D_{f3}) may be required to adequately model the binding kinetics. This is when the binding curve exhibits convolutions and complexities in its shape perhaps because of the very dilute nature of the analyte (in some of the cases to be presented) or for some other reasons. Also, in some cases, a dual-fractal analysis may be required to describe the dissociation kinetics.

21.5 Illustrations

A fractal analysis is presented for (a) the binding of liquid petroleum gas (LPG) to zinc oxide films prepared by the spray pyrolysis method onto a glass substrate (Shinde et al. 2007) and (b) the binding and dissociation of different NH_3 concentrations in air to a sol-gel-derived thin film (Roy et al. 2005).

21.5.1 Illustration 1

Shinde et al. (2007) have recently investigated the use of ZnO thin films prepared by the spray pyroloysis method for the sensing of liquid petroleum gas (LPG). These authors deposited nanocrystalline ZnO films onto glass substrates by the spray pyrolysis of zinc nitrate solution, and used this as a LPG gas sensor. They further indicate that ZnO has replaced the more toxic and expensive materials such as CdS, TIO_2, GaN, and SiO_2 for applications in gas sensors (Rao and Rao 1999). Ismail et al. (2001) have indicated that because of its resistivity control in the range 10^{-3}–10^5 ohm cm, ZnO is particularly suitable for gas sensors. Besides, it exhibits high electrochemical stability, absence of toxicity, and is readily available in nature. Shinde et al. (2007) emphasize that thin films are particularly suited for gas sensors, since the gas-sensing properties of metal oxides (a) may be related to the material surface and (b) gases are readily adsorbed and react with the thin film biosensor surface (Liu et al. 1997). Furthermore, Zhu et al. (1993) and Chai et al. (1995) emphasize that thin film gas-sensing materials have good gas sensitivity and selectivity. Patil (1999) has demonstrated the versatility of using the spray pyrolysis method for the deposition of metal oxides.

Shinde et al. (2007) emphasize that the gas-sensing properties of oxide materials may be related to the surface morphology and are grain size dependent. This surface morphology is what we will attempt to characterize and make quantitative using the fractal analysis method. The fractal dimension (a) provides a quantitative measure of the degree of heterogeneity on the sensing surface, and (b) an increase in the fractal dimension on the sensing surface indicates an increase in the degree of heterogeneity on the sensing surface.

Shinde et al. (2007) used scanning electron micrographs to analyze the surface morphology studies of The ZnO films sprayed on glass substrates. These authors noted that the grain size increased with an increase in the sprayed precursor solution [S_1 (0.1 M), to S_2 (0.15 M), to S_3 (0.25 M) zinc nitrate solution]. This is consistent with the results obtained by Korotcenkov et al. (2001).

Figure 21.3a shows the binding and dissociation of 0.2 volume percent LPG in the gas phase to sample S_1 (0.1 M zinc nitrate solution used in the spray pyrolysis method) (Shinde et al. 2007). A dual-fractal analysis is required to adequately describe the binding and the dissociation kinetics. Table 21.1a and b show (a) the binding rate coefficient, k and the fractal dimension, D_f for a single-fractal analysis, (b) the dissociation rate coefficient, k_d, and the fractal dimension for the dissociation phase for a single-fractal analysis, (c) the binding rate coefficients, k_1 and k_2, and the fractal dimensions, D_{f1} and D_{f2} for a dual-fractal analysis and (d) the dissociation rate coefficients, k_{d1} and k_{d2}, and the fractal dimensions, D_{fd1} and D_{fd2} for the dissociation phase for a dual-fractal analysis.

Figure 21.3b and c show the binding and dissociation of 0.2 volume percent LPG in the gas phase to sample S2 (0.15 M zinc nitrate) and S3 (0.2 M zinc nitrate) used in the spray pyrolysis method, respectively.

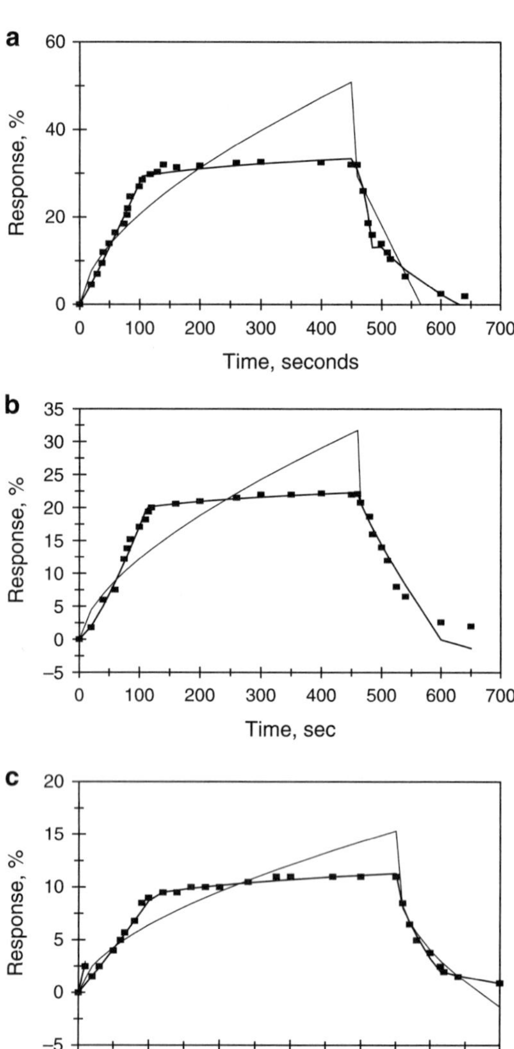

Fig. 21.3 Binding of samples S_1, S_2, and S_3 (LPG; liquid petroleum gas) to zinc oxide (ZnO) films prepared by the spray pyrolysis method onto a glass substrate (Shinde et al. 2007). Influence of zinc oxide concentration (0.1–0.3 M) solution: (**a**) 0.1 M, sample S_1,(**b**) 0.15 M, sample S_2, (**c**) 0.3 M, sample S_3 When only a solid line is used (—) then a single-fractal applies. When both a dotted (. . .) and a solid line (—) is used, then the dotted line is for a single-fractal analysis, and the solid line is for a dual-fractal analysis. In this case, the dual-fractal analysis provides the better fit

The values of the binding and the dissociation rate coefficients and the fractal dimensions for the binding phase presented in Table 21.1a and b were obtained from a regression analysis using Corel Quattro Pro 8.0 (1997) to model the data using (Eq. 21.1–3), wherein (Ab·Ag) $= kt^{(3-D_f)}$ for a single-fractal analysis, and (Ab·Ag) $= k_1 t^{(3-D_{f1})}$ for time, $t < t_1$, and (Ab·Ag) $= k_2 t^{(3-D_{f2})}$ for time, $t = t_1 < t < t_2 = t_c$ for a dual-fractal analysis. The binding and the dissociation rate coefficients presented in Table 21.1a are within 95% limits. For example, for the binding (and dissociation) of LPG to sample S_1 the binding rate coefficient, k_1 is equal

Table 21.1 (**a**) Binding and dissociation rate coefficients for liquid petroleum gas (LPG) on ZnO films prepared by the spray pyrolysis method onto a glass substrate. (**b**) Fractal dimensions for the binding and the dissociation phase for liquid petroleum gas (LPG) on ZnO films prepared by the spray pyrolysis method onto a glass substrate. Influence of zinc nitrate concentration (0.1– 0.25 M) solution (Shinde et al. 2007)

(a) Zinc nitrate concentration, M	k	k_1	k_2	k_d	k_{d1}	k_{d2}
0.1, sample S1	1.284 ± 0.399	0.1825 ± 0.0166	19.334 ± 0.801	0.2685 ± 0.1731	0.00633 ± 0.00124	3.739 ± 0.163
0.15, sample S2	0.6835 ± 0.292	0.0351 ± 0.0041	14.066 ± 0.146	0.3514 ± 0.0615	na	na
0.25, sample S3	0.4449 ± 0.1245	0.06332 ± 0.00393	5.1146 ± 0.1115	0.8868 ± 0.1231	0.6243 ± 0.0590	4.859 ± 0.063
(b) Zinc nitrate concentration, M	D_f	D_{f1}	D_{f2}	D_{fd}	D_{fd1}	D_{fd2}
0.1, sample S1	1.7954 ± 0.1444	0.8196 ± 0.1274	2.8210 ± 0.0480	0.9864 ± 0.3674	0 + 0.3764	2.1732 ± 0.0709
0.15, sample S2	1.7478 ± 0.1969	0.308 ± 0.141	2.849 ± 0.0154	1.3234 ± 0.1073	na	na
0.25, sample S3	1.8404 ± 0.1312	0.8628 ± 0.0731	2.7406 ± 0.0358	1.9492 ± 0.1124	1.7272 ± 1.2726	2.7066 ± 0.0468

to 0.1825 ± 0.0166. The 95 % confidence limit indicates that the k_1 value lies between 0.1659 and 0.1991. This indicates that the values are precise and significant.

An increase in the fractal dimension for a dual-fractal analysis by a factor of 3.44 from a value of D_{f1} equal to 0.8196 to D_{f2} equal to 2.8210 leads to an increase in the binding rate coefficient by factor of 105.93 from a value of k_1 equal to 0.1825 to k_2 equal to 19.334. Note that changes in the fractal dimension or the degree of heterogeneity on the ZnO-glass substrate and in the binding rate coefficient are in the same direction. In other words, an increase in the degree of heterogeneity on the sensing surface leads to an increase in the binding rate coefficient.

Table 21.1a and b show (a) the binding rate coefficient, k and the fractal dimension, D_f for a single-fractal analysis, and (b) the dissociation rate coefficient, k_d and the fractal dimension for dissociation, D_{fd} for a single-fractal analysis, and (c) and the binding rate coefficients, k_1 and k_2 and the fractal dimensions, D_{f1} and D_{f2} for a dual-fractal analysis.

Once again, for the binding phase an increase in the fractal dimension by a factor of 9.25 from a value of D_{f1} equal to 0.8196 to D_{f2} equal to 2.8210 leads to an increase in the binding rate coefficient by a factor of 400.7 from a value of k_1 equal to 0.0351 to k_2 equal to 14.066. An increase in the degree of heterogeneity on the biosensor surface leads, once again, to an increase in the binding rate coefficient.

Figure 21.4a and Table 21.1a and b show for a dual-fractal analysis the increase in the binding rate coefficient, k_2 with an increase in the fractal dimension, D_{f2}. For the data shown in Fig. 21.4a, the binding rate coefficient, k_2 is given by:

$$k_2 = (2.8\text{E} - 13 \pm 1.5\text{E} - 13)D_{f2}^{30.4} \tag{21.4a}$$

The fit is reasonable. Only three data points are available. The availability of more data points would lead to a more reliable fit. The binding rate coefficient, k_2 is extremely sensitive to the fractal dimension, D_{f2} that exists on the biosensor surface as noted by the 30.4 order of dependence exhibited

Figure 21.4b and Table 21.1a and b show for a dual-fractal analysis the increase in the ratio of the binding rate coefficients, k_2/k_1 with an increase in the fractal dimension ratio, D_{f2}/D_{f1}. For the data shown in Fig. 21.4b, the ratio of the binding rate coefficients, k_2/k_1 is given by:

$$k_2/k_1 = (16.43 \pm 1.78)(D_{f2}/D_{f1})^{1.438 \pm 0.122} \tag{21.4b}$$

The fit is very good. Once again, only three data points are available. The availability of more data points would lead to a more reliable fit. The ratio of the binding rate coefficients, k_2/k_1 exhibits close to a one and one-half order of dependence on the ratio of fractal dimensions, D_{f2}/D_{f1}.

Figure 21.4c and Table 21.1a and b show the increase in the ratio, k/k_d, k_1/k_{d1}, and k_2/k_{d2} with an increase in the fractal dimension ratio, D_f/D_{fd}, D_{f1}/D_{fd1}, and D_{f2}/D_{fd2}, respectively. For the data shown in Fig. 21.4c, the ratio of the binding rate coefficients is given by:

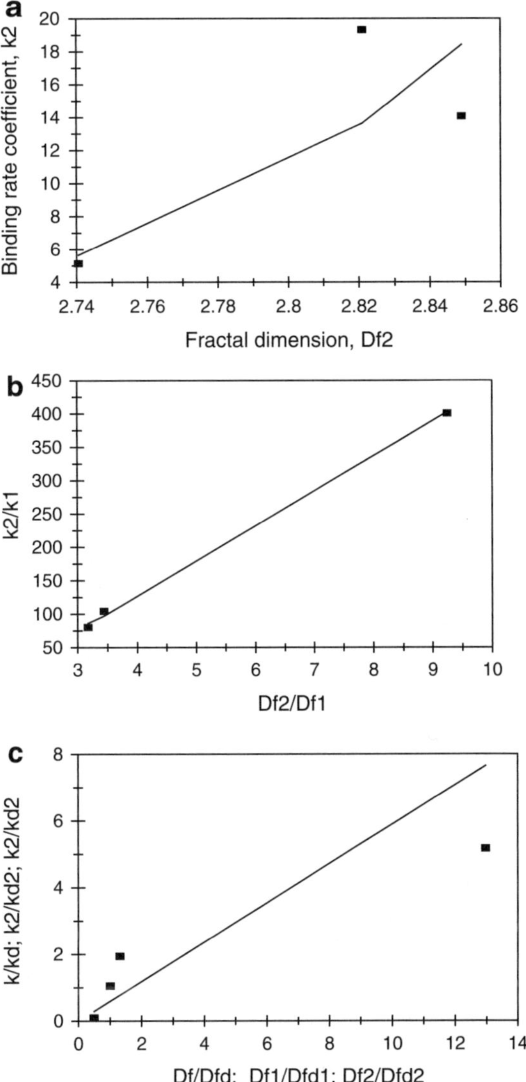

Fig. 21.4 (**a**) Increase in the binding rate coefficient, k_2 with an increase in the fractal dimension, D_{f2} (**b**) Increase in ratio of the binding rate coefficients, k_2/k_1 with an increase in the ratio of fractal dimensions, D_{f2}/D_{f1}. (**c**) Increase in the ratio, k/k_d, k_1/k_{d1}, and k_2/k_{d2} with an increase in the fractal dimension ratio, D_f/D_{fd}, D_{f1}/D_{fd1}, and D_{f2}/D_{fd2}

$$k/k_{\mathrm{d}}, k_1/k_{\mathrm{d1}}, \text{and } k_2/k_{\mathrm{d2}} = (0.5943 + 1.203) \\ (D_{\mathrm{f}}/D_{\mathrm{fd}}, D_{\mathrm{f1}}/D_{\mathrm{fd1}}, \text{and } D_{\mathrm{f2}}/D_{\mathrm{fd2}} D_{\mathrm{f2}}/D_{\mathrm{f1}})^{0.997 \pm 0.453} \tag{21.4c}$$

The fit is reasonable. Only four data points are available. There is scatter in the data, and this is reflected in the error in the ratio of the binding and the dissociation rate coefficient presented. Also, we are presenting the ratio of three different coefficient values, and this too contributes to the error. This is done since the points are very few. In any case, the three different ratios of the binding and their corresponding dissociation rate coefficients presented exhibit very close to a first (equal to 0.997) order of dependence on the ratio of fractal dimensions exhibited in the binding and in the dissociation phases, respectively.

21.5.2 Illustration 2

Roy et al. (2005) have recently developed a sol-gel-derived thin film biosensor to detect ppm concentrations of NH_3 in air. They analyzed the influence of presintering temperature on gas sensitivity. These authors indicate that ammonia gas sensors have been used in chemical plants, food technology, fertilizers, in environmental pollution monitoring, and in food technology. Thin films of polyaniline-insulating matrix polymer blend have been used to detect ammonia near room temperature. Roy et al. (2005) indicate that polymer-based materials may degrade, and often suffer limited cycle of operation. They further indicate that inorganic materials are better for gas sensing.

Roy et al. (2005) have recently used the ammonia-sensing behavior of barium strontium titanate (BST) films to detect ammonia. The BST films were deposited by the sol-gel spin coating technique, and these thin films showed an increase in resistance when exposed to ammonia gas. They further state that the sensitivity variation was from 20 to 60%. The lowest detection limit was around 160 ppm.

Figure 21.5a shows the binding and dissociation of NH_3 in air to the sol-gel-derived thin film sensor where the presintering was performed at 873K (Roy et al. 2005). A single-fractal analysis is adequate to describe the binding and the dissociation kinetics. The values of the rate coefficient and the fractal dimension for the binding and the dissociation phases are given in Table 21.2.

Figure 21.5b shows the binding and dissociation of NH_3 in air to the sol-gel-derived thin film sensor where the presintering was performed at 773K. A single-fractal analysis is adequate to describe the binding and the dissociation kinetics. The values of the rate coefficient and the fractal dimension for the binding and the dissociation phases are given in Table 21.2. In this case, the affinity, $K(=k/k_d)$ value is 4.57.

Figure 21.5c shows the binding and dissociation of NH_3 in air to the sol-gel-derived thin film sensor where the presintering was performed at 673K. A single-fractal analysis is adequate to describe the binding and the dissociation kinetics. The values of the rate coefficient and the fractal dimension for the binding and the dissociation phases are given in Table 21.2. In this case, the affinity, $K(=k/k_d)$ value is 0.420. A decrease in temperature leads to a decrease in the affinity, K value in the range 673–873K temperature.

Figure 21.5d and Table 21.2 show the increase in the binding rate coefficient, k with an increase in the fractal dimension, D_f. For the data shown in Fig. 21.5d, the binding rate coefficient, k is given by:

$$k = (0.1931 \pm 0.0250)D_f^{5.008 \pm 0.3361} \qquad (21.5a)$$

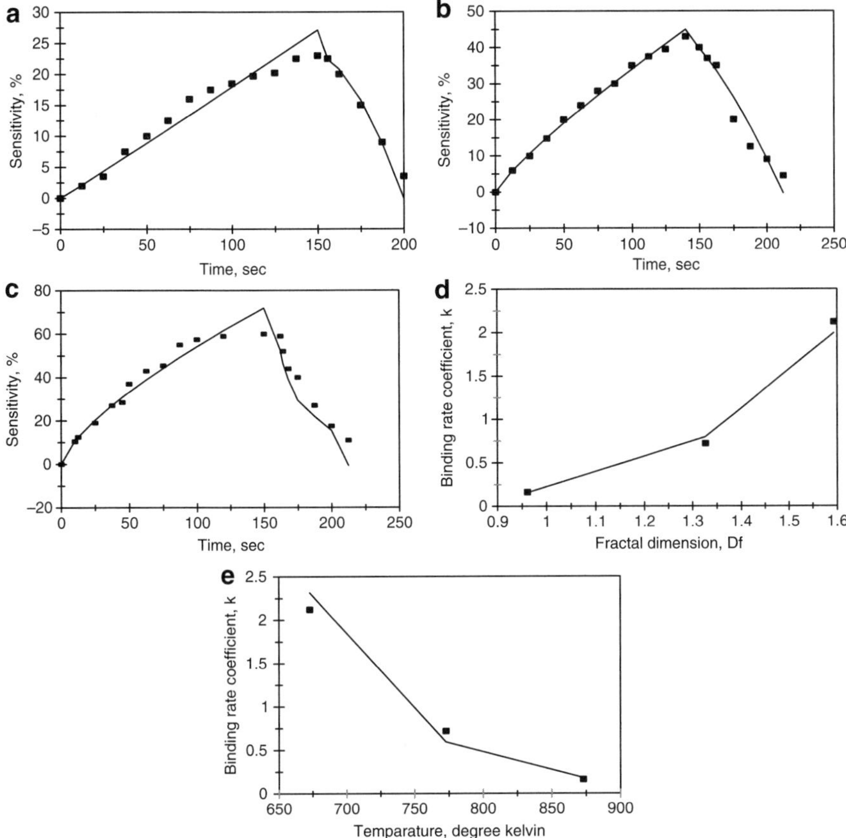

Fig. 21.5 Binding and dissociation of NH_3 concentration in air to a sol-gel derived thin film (Roy et al. 2005). Influence of temperature in K: (**a**) 873 (**b**) 773 (**c**) 673. (**d**): Increase in the binding rate coefficient, k with an increase in the fractal dimension, D_f (**e**): Decrease in the binding rate coefficient, k with an increase in the temperature, T (in K)

The fit is good. Only three data points are available. The availability of more data points would lead to a more reliable fit. The binding rate coefficient, k for a single-fractal analysis is very sensitive to the fractal dimension, D_f or the degree of heterogeneity that exists on the biosensor surface as noted by the order of dependence exhibited which is very slightly higher than five (5.008).

Figure 21.5e and Table 21.2 show the decrease in the binding rate coefficient, k with an increase in the temperature range 673–873K. For the data shown in Fig. 21.5e, the binding rate coefficient, k is given by:

$$k = (1.2\text{E} + 28 \pm 0.3\text{E} + 28)(T, \text{in K})^{-9.80 \pm 1.26} \qquad (21.5b)$$

Table 21.2 Binding and dissociation rate coefficients and fractal dimensions for the binding and the dissociation phases for NH_3 in air to a sol-gel derived thin film biosensor (Roy et al. 2005). Influence of presintering temperature

Analyte in air/receptor on surface	Presintering temperature, K	k	k_d	D_f	D_{fd}	K
ppm NH_3/sol-gel derived thin film biosensor	873	0.1638 ± 0.0245	0.0303 ± 0.0088	$0.9608 \pm \pm 0.111$	$0 + 0.2962$	5.40
ppm NH_3/sol-gel derived thin film biosensor	773	0.7220 ± 0.0305	0.1597 ± 0.0353	1.3270 ± 0.035	0.3832 ± 0.2008	4.5
ppm NH_3/sol-gel derived thin film biosensor	673	2.124 ± 0.206	5.0561 ± 0.0621	1.594 ± 0.649	1.889 ± 0.0654	0.42

The fit is good. Only three data points are available. The availability of more data points would lead to a more reliable fit. The binding rate coefficient, k for a single-fractal analysis is very sensitive to the presintering temperature as noted by the negative 9.8 order of dependence exhibited.

At this point, it is not clear about the influence of presintering temperature on the binding rate coefficient, k in the temperature range 673–873K. Perhaps, there is an Arrhenius-like dependence of the form, $k = k_0 \exp(-A/RT)$, where A and k_0 are the coefficients as in the Arrhenius expression, R is the universal gas constant, and T is the temperature in Kelvin. Thus, the data in Fig. 21.3e are plotted again in Fig. 21.6a with the reciprocal temperature T in K^{-1} as the independent variable. For the data shown in Fig. 21.6a, the binding rate coefficient, k is given by:

$$k = (1.3\text{E} + 28 \pm 0.4\text{E} + 28)(T^{-1},\ \text{in K}^{-1})^{9.81 \pm 1.26} \qquad (21.6a)$$

Note that the coefficients in (Eq. 21.5b) and in (Eq. 21.6a) are very close to each other, as expected. Once again, the fit is good.

Figure 21.6b and Table 21.2 show the increase in the dissociation rate coefficient, k_d with an increase in the reciprocal temperature, T in K^{-1}. For the data shown in Fig. 21.6b, the dissociation rate coefficient, k_d is given by:

$$k_d = (3.8\text{E} + 56 \pm 3.1\text{E} + 56)\left(T^{-1}, \text{in K}^{-1}\right)^{19.79 \pm 3.31} \qquad (21.6b)$$

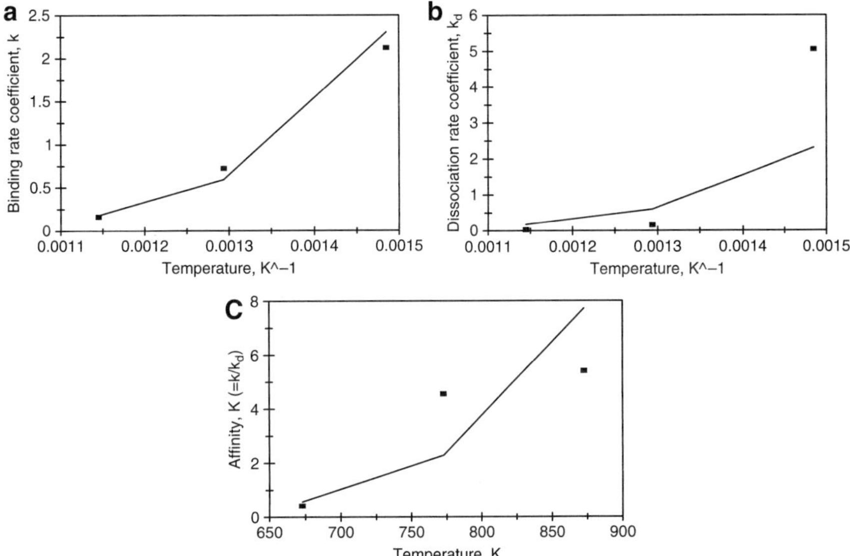

Fig. 21.6 (**a**) Increase in the binding rate coefficient, k with an increase in the reciprocal temperature, T^{-1}(in K^{-1}). (**b**) Increase in the dissociation rate coefficient, k_d with an increase in the reciprocal temperature, T^{-1} (in K^{-1}). (**c**) Increase in the affinity, $K(=k/k_d)$ with an increase in the temperature, T (in K)

There is scatter in the data. This is reflected in the fit and in the error in the binding rate coefficient, k value. The dissociation rate coefficient, k_d is very sensitive to the reciprocal temperature, T^{-1} as reflected in the 19.79 order of dependence exhibited. In other words, both the binding rate coefficient, k and the dissociation rate coefficient, k_d are extremely sensitive to the presintering temperature. Note that the dissociation rate coefficient, k_d exhibits more than twice the order of dependence (equal to 19.79) on temperature, in K than the binding rate coefficient, k in the 673–873K temperature range analyzed.

Figure 21.6c and Table 21.2 show the increase in the affinity, $K(=k/k_d)$ with an increase in the temperature, T in K. For the data shown in Fig. 21.6c, the affinity, K is given by:

$$K(=k/k_{\mathrm{d}}) = (3.3\mathrm{E}-29 \pm 4.3\mathrm{E}-29)(T, \mathrm{in\,K})^{9.98 \pm 4.53} \tag{21.6c}$$

There is scatter in the data. This is reflected in the fit and in the error in the affinity value presented. Only the positive value of the affinity, K is presented since the affinity cannot have a negative value. The affinity, K is extremely sensitive to the temperature, T as reflected in the close to the tenth (equal to 9.98) order of dependence exhibited.

21.6 Illustration 3

The binding of FAM (fluorescein)-SP-2 + 10 units of $\lambda_{\mathrm{exonuclease}}$ in solution to the singly labeled DNA hairpin smart probe was analyzed by Song and Zhao (2009). FAM-SP-2 is the oligonucleotide sequence (5′–3′) GGGCC(AG_{10})GGCCC-FAM, SP: DNA hairpin quenched solely by the DNA base guanine were introduced as smart probes (Knemeyer et al. 2000). A dual-fractal analysis is required to adequately describe the binding kinetics. The values of (a) the binding rate coefficient, k and the fractal dimension, D_f for a single-fractal analysis, and (b) the binding rate coefficients, k_1 and k_2, and the fractal dimensions, D_{f1} and D_{f2} for a dual-fractal analysis are given in Table 21.3.

It is of interest to note that for the dual-fractal analysis, as the fractal dimension increases by a factor of 1.51 from a value of D_{f1} equal to 1.9390 to D_{f2} equal to 2.9296, the binding rate coefficient increase by a factor of 8.34 from a value of k_1 equal to 640.96 to k_2 equal to 5,243.91. Note that changes in the degree of heterogeneity or the fractal dimension on the biosensor surface and in the binding rate coefficient are in the same direction.

The binding of 40 nM FAM-SP-2p + 10 units $\lambda_{\mathrm{exonuclease}}$ + 5.6 nM/s T4 PNK (T4 polynucleotide kinase) in solution to the singly labeled DNA hairpin smart probe was also analyzed by Song and Zhao (2009). In this case, a single-fractal analysis is adequate to describe the binding kinetics. The values of the binding rate coefficient, k and the fractal dimension, D_f for a single-fractal analysis are given in Table 21.3.

Then, these authors analyzed the binding of 40 nM FAM-SP-2 + 5.6 nM/s 4T PNK + 1 mM ATP and 0.5 units of $\lambda_{\text{exonuclease}}$ at pH 8.0 to the singly labeled DNA hairpin smart probe. These authors wanted to analyze the influence of $\lambda_{\text{exonuclease}}$ units on the binding. A dual-fractal analysis is required to adequately describe the binding kinetics. The values of (a) the binding rate coefficient, k and the fractal dimension, D_f for a single-fractal analysis, and (b) the binding rate coefficients, $k_1$and k_2, and the fractal dimensions, D_{f1} and D_{f2} for a dual-fractal analysis are given in Table 21.3.

It is of interest to note that as the fractal dimension increases by a factor of 1.475 from a value of D_{f1} equal to 0.822 to D_{f2} equal to 1.213, the binding rate coefficient increases by a factor of 1.89 from a value of k_1 equal to 58.33 to k_2 equal to 110.31. Note that changes in the binding rate coefficient and in the degree of heterogeneity or the fractal dimension on the biosensor surface are in the same direction.

The binding of 40 nM FAM-SP-2 + 5.6 nM/s 4T PNK + 1 mM ATP and 1.0 units of $\lambda_{\text{exonuclease}}$ at pH 8.0 to the singly labeled DNA hairpin smart probe was also analyzed by Song and Zhao (2009). Once again, a dual-fractal analysis is required to adequately describe the binding kinetics. The values of (a) the binding rate coefficient, k and the fractal dimension, D_f for a single-fractal analysis, and (b) the binding rate coefficients, $k_1$and k_2, and the fractal dimensions, D_{f1} and D_{f2} for a dual-fractal analysis are given in Table 21.3.

It is of interest to note, once again, that as the fractal dimension increases by a factor of 4.375 from a value of D_{f1} equal to 0.458 to D_{f2} equal to 2.004, the binding rate coefficient increases by a factor of 21.86 from a value of k_1 equal to 50.75 to k_2 equal to 1,109.57. Note that changes in the binding rate coefficient and in the degree of heterogeneity or the fractal dimension on the biosensor surface are, once again, in the same direction.

Finally, the binding of 40 nM FAM-SP-2 + 5.6 nM/s 4T PNK + 1 mM ATP and 2.50 units of $\lambda_{\text{exonuclease}}$ at pH 8.0 to the singly labeled DNA hairpin smart probe was analyzed by Song and Zhao (2009). Once again, a dual-fractal analysis is required to adequately describe the binding kinetics. The values of (a) the binding rate coefficient, k and the fractal dimension, D_f for a single-fractal analysis, and (b) the binding rate coefficients, $k_1$and k_2, and the fractal dimensions, D_{f1} and D_{f2} for a dual-fractal analysis are given in Table 21.4.

It is of interest to note, once again, that as the fractal dimension increases by a factor of 3.823 from a value of D_{f1} equal to 0.690 to D_{f2} equal to 2.638, the binding rate coefficient increases by a factor of 48 from a value of k_1 equal to 98.96 to k_2 equal to 4,750.65. Note that changes in the binding rate coefficient and in the degree of heterogeneity or the fractal dimension on the biosensor surface are, once again, in the same direction.

Table 21.4 shows for a dual-fractal analysis the increase in the binding rate coefficient, k_1 with an increase in the λ nuclease units in solution. For the data shown in Table 21.4 the binding rate coefficient, k_1 is given by:

$$k_1 = (64.71 \pm 12.52)\,(\lambda_{\text{exonuclease}}\ \text{units})^{0.3494 \pm 0.2613} \qquad (21.7a)$$

Table 21.3 Binding rate coefficients and fractal dimensions for the binding of FAM-SP-2 + λ_{exo} + T4PNK and FAB-SP-2p + λ_{exo} in solution to the G-quenched singly-labeled DNA-hairpin smart probe coupled with $\lambda_{exonuclease}$ (Song and Zhao 2009)

Analyte in solution/receptor on surface	k	k_1	k_2	D_f	D_{f1}	D_{f2}
40 nM FAM + SP2 + lambda$_{exo}$(10 units)/G quenched smart probe	1972.38 ± 512.55	640.96 ± 124.0	5343.91 ± 63.44	2.5754 ± 0.1088	1.9390 ± 0.2090	2.9296 ± 0.02314
40 nM FAM + SP2 + lambda$_{exo}$(10 units) + 40 nM 4PNK//G quenched smart probe	4.576 ± 0.439	na	na	0.9110 ± 0.0986	na	na

Table 21.4 Binding of 40 nM FAM-SP-2, 5.6 nM/s T4PNK, and 1 mM ATP to the singly-labeled DNA-hairpin probe coupled with $\lambda_{exonuclease}$ at pH 8.0. Influence of $\lambda_{exonuclease}$ units (Song and Zhao 2009)

$\lambda_{exonuclease}$	k	k_1	k_2	D_f	D_{f1}	D_{f2}
0.5	69.96 ± 65.16	58.33 ± 6.93	110.31 ± 7.19	0.906 ± 0.338	0.822 ± 0.114	21.213 ± 0.1578
1.0	118.57 ± 34.29	50.75 ± 9.98	1109.57 ± 29.72	1.0532 ± 0.1124	0.458 ± 0.143	2.004 ± 0.0464
2.5	241.15 ± 87.74	98.96 ± 23.1	4750.65 ± 133.47	1.3224 ± 0.1658	0.690 ± 0.189	2.638 ± 0.0682

The binding rate coefficient, k_1 exhibits only a mild dependence on the $\lambda_{exonuclease}$ units in solution as shown by the close to 0.35 (equal to 0.3494) order of dependence exhibited.

Table 21.4 shows for a dual-fractal analysis the increase in the binding rate coefficient, k_2 with an increase in the $\lambda_{exonuclease}$ units in solution. For the data shown in Table 21.4 the binding rate coefficient, k_2 is given by:

$$k_2 = (1.791 \pm 0.260)(\lambda_{exonuclease}\ \text{units})^{0.475 \pm 00.119} \tag{21.7b}$$

The binding rate coefficient, k_2 exhibits a strong dependence on the $\lambda_{exonuclease}$ units in solution as shown by the greater than second (equal to 2.299) order of dependence exhibited.

21.7 Estimating Kinetic Parameters and Explanation of Fractal Analysis Calculations

21.7.1 Estimating Kinetic Parameters

To give more insights into biosensor operations and kinetics, the Surface Plasmon Resonance (SPR) biosensor is examined to a greater detail. The SPR is popular in the biosensor field and is also the equipment through which data modeled in this chapter is obtained.

The surface plasmons (SPs) are free charge oscillations that occur at the interface between a dielectric medium and a metallic medium. The incidence of light can cause excitation of the SPs. When the momentum of the SPs matches that of the incident light, the so-called surface plasmon resonance (SPR) phenomenon takes place. It is found that this wave-matching condition is very easily disrupted by even very tiny changes in the interface conditions. Hence, if the condition of light excitation is fixed, the SPR technique not only permits the precise measurement of changes in the refractive index or thickness of the medium adjacent to the metal film, but also enables changes in the adsorption layer on the metal surface to be detected. Consequently, various SPR biosensors based on the attenuated total reflection (ATR) method are functionalized by means of an immobilized surface on a metal-sensing surface. These devices provide powerful tools for the real-time investigation of biomolecular interaction analysis (BIA), and have the advantage that prior labeling of the analytes is unnecessary. The SPR biosensors have been widely applied in a diverse range of fields, including molecular recognition, and disease immunoassays, etc. (Liedberg et al. 1983; Kooyman et al. 1988; Owen 1997; Peterlinz et al. 1997; June et al. 1998 and Homola et al. 1999).

Gold and silver are ideal candidates for the metal film in the visible light region. However, due to its stability under chemical reaction, gold is a more suitable choice for SPR biosensing applications. SPR biosensor configurations are generally based

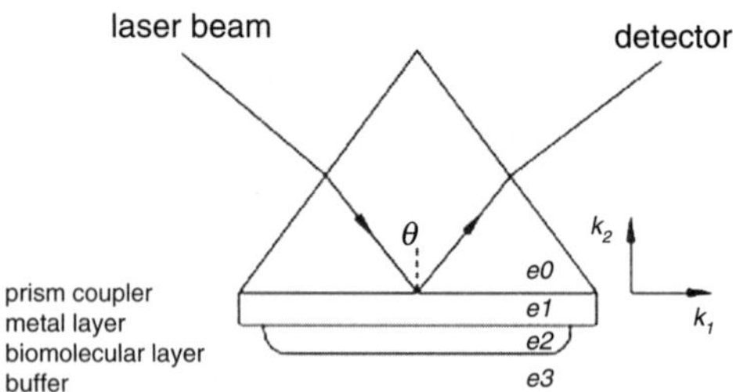

Fig. 21.7 Configuration of the SPR

on a prism coupler ATR system, which provides four basic measurement and interrogation capabilities, namely intensity measurement, angular interrogation, wavelength interrogation, and phase measurement. The following figure shows the configuration of the conventional SPR.

As shown in Fig. 21.7, the Kretschmann configuration of the conventional SPR biosensor comprises four basic layers, namely the prism (ε_0), the metal layer (ε_1, d_1), the biomolecular layer (ε_2, d_2), and the buffer (ε_3), where ε_i and d_i respectively denote the dielectric constant and thickness of the respective layers. Furthermore, $\varepsilon = \varepsilon_r + \varepsilon_j$ expresses the real and imaginary parts of the dielectric constant. If the imaginary part of the dielectric constant is near zero, then the refractive index n is equal to $\surd\varepsilon$.

21.7.2 Explanation of Fractal Analysis Calculations

Modeling of kinetic data is usually done by means of linear regression with the use of a modeling application – *Corel Quattro Pro*. Using the single-fractal model to analyze the binding of analyte–substrate systems, the modified Havlin equation used is given as

$$[\mathrm{Ab} \cdot \mathrm{Ag}] \sim t^b \tag{a}$$

$$[\mathrm{Ab} \cdot \mathrm{Ag}] = k \cdot t^b \tag{b}$$

where [Ab] and [Ag] represent the receptor and analyte concentration respectively. Hence, [Ab·Ag] is the analyte–receptor complex concentration the kinetics of which is being modeled on the biosensor surface. 'k' represents the binding or dissociation rate constant. The value of 'b' in the above equation is given as

$$b = \frac{3 - D_f}{2} \tag{c}$$

D_f in the equation is the fractal dimension of the biosensor surface as mentioned earlier and could be for either the association or dissociation (Dfd) step.

This equation is further simplified by evaluating the natural logarithm of both sides of equation (b) to obtain:

$$\mathrm{Ln}[\mathrm{Ab} \cdot \mathrm{Ag}] = \mathrm{Ln}(k) + b\,\mathrm{Ln}(t) \quad \text{(d)}$$

This follows the form of the equation for a straight line given by;

$$y = mx + c$$

Where m represents the slope and c is the 'y' axis intercept. As such, 'b' would be the slope obtained from the linear regression plot and Ln(k) the intercept on the 'y' axis. The dissociation kinetics may be analyzed in a manner similar to the binding kinetics.

However, sometimes the modeled data do not fit a linear model under the single fractal analysis approach and the kinetic data have to be broken up into convenient portions that satisfy the linear model equations previously discussed. If the number of portions to be modeled is only two, it is called a "dual fractal analysis" and "triple fractal analysis" when there are three independent portions to be modeled.

21.7.3 Example

As an example, an illustration is provided on how to obtain such required results from any given set of kinetic data (Fig. 21.8). Data from a real example of the binding of a certain analyte to the surface of a biosensor is presented as shown:

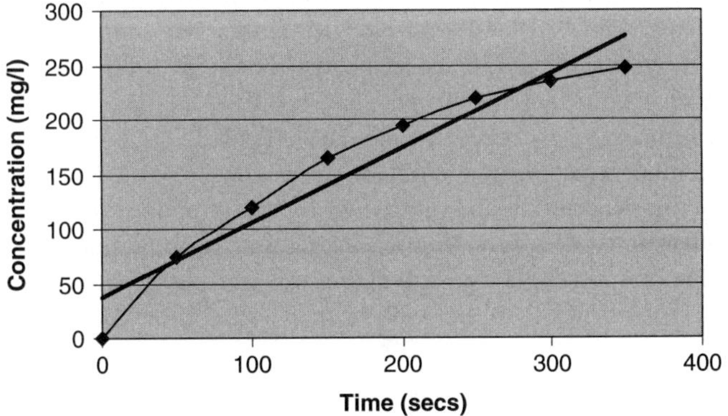

Fig. 21.8 Concentration (in mg/l) versus time (in s) curve

Concentration	Time
0	0
50	75
100	120
150	165
200	195
250	220
300	235
350	248

The solid line in the plot shows the linear regression plot alongside the curve, which reflects the fit of the original data. A fractal analysis is then carried out on the raw kinetic data provided using (Eqs. b and d).

Regression output analysis with *Corel Quattro Pro*:

Constant (k): 1.898766; exp (k) $= 6.677$

Standard error of Y estimate: 0.05023

R squared[2] : 0.988

Number of observations: 7

Degrees of freedom: 5

X Coefficient: 0.628227

Standard Error of Coefficient: 0.0299

Hence, the rate constant is reported as $k = 6.677 \pm 0.344$

$b = 0.62824 \pm 0.0299$, which results in $D_f = 1.7436$

21.7.4 Physical Interpretation of Kinetic Parameters

The binding and dissociation rate constant helps to monitor the progress of the adsorption process, from which other variables can be conveniently inferred. In protein–ligand binding systems, the affinity could be used to describe how tightly a ligand binds to a protein. Hence, the smaller the dissociation constant, the more tightly the bound ligand is. Also, it could be used to investigate the nature of effects or forces existing between the two molecules and/or the complex formed. Such forces could be electrostatic interactions, hydrophobic, and Van der Waals forces. Affinity can also be affected by high concentrations of other macromolecules, which causes macromolecular crowding. The knowledge of affinity can also aid in the preferential selection or identification of a target analyte in a mixture or solution of analytes when binding to certain ligands/proteins. For example, drugs can produce harmful side effects through interactions with proteins for which they were not meant or designed to interact. Therefore, much pharmaceutical research is aimed at designing drugs that bind only to their target proteins with high affinity

[2] The closer the R squared value is to 1, the better the fit of the curve.

(typically 0.1–10 nM) or at improving the affinity between a particular drug and its *in vivo* protein target.

The fractal analysis of the binding and dissociation of analytes in a solution on a biosensor surface has been presented. In addition, the fractal dimensions for the dissociation step, $D_{f,diss}$ and dissociation rate coefficients, k_{diss}, are also presented. This provides a more complete picture of the analyte–receptor reactions occurring on the surface in contrast to an analysis of the binding step alone, as done previously. Besides, the numerical values for the binding and dissociation steps could be used to classify the analyte–receptor biosensor system.

The degree of heterogeneity for both phases is, in general, different for the same reaction. This indicates that the same surface exhibits two degrees of heterogeneity for the binding and the dissociation reaction. Both types of examples are given where either the single or dual fractal analysis is required to describe the binding kinetics. The dual fractal analysis was used only when the single fractal analysis failed to provide an adequate fit. The fractal dimension is not a typical independent variable, such as analyte concentration, that may be manipulated directly. It may be considered as a derived variable.

The predictive relationship developed for the binding rate coefficient as a function of the fractal dimension is of considerable value, because it directly links the binding rate coefficient to the degree of heterogeneity that exists on the surface, and suggests a means by which the binding rate coefficient may be manipulated by changing the degree of heterogeneity that exists on the surface. Generally, it is seen that an increase in the fractal dimension of the surface or the degree of heterogeneity leads to an increase in the binding rate coefficient. One possible explanation for the observed phenomenon could be due to the fact that the fractal surface (roughness) leads to turbulence, which enhances mixing, decreases diffusional limitations, and leads to an increase in the binding rate coefficient (Martin et al. 1991).

References

Berg HC, Purcell EM (1977) Physics of chemoreception. Biophys J 20(2):193–219

Chai CC, Peng J, Yan BP (1995) Preparation and gas-sensing properties of Fe_2O_3 thin films. J Electron Mater 24:82–86

Chaplin MF (2010) Structuring and behaviour of water in nanochannels and confined spaces. In: (ed) Adsorption and phase behavior in nanochannels and nanotubes, Springer, in press

Dewey TG (1997) Fractals in molecular biophysics. Oxford University, New York, NY

Dewey TG, Bann JG (1992) Protein dynamics and 1/f noise. Biophys J 63:594–598

Havlin S (1989) The fractal approach to heterogeneous chemistry: surfaces, colloids, polymers. Wiley, New York, pp 251–269

Homola J, Gauglitz G, Yee SS (1999) Surface plasmon resonance sensors: review. Sens Actuator B 54:3–15

Ismail B, Abaab MA, Rezig B (2001) Structural and electrical properties of ZnO films prepared by screen printing technique. Thin Solid Films 383:92–94

June LS, Campbell CT, Chinowsky TM, Mar MN, Yee SS (1998) Quantitative interpretation of the response of surface plasmon resonance sensors to adsorbed films. Langmuir 12:5636–5648

Kang K, Redner S (1985) Fluctuation-dominated kinetics in diffusion-controlled reactions. Phys Rev A 32:435

Knemeyer JP, Marine N, Sauer M (2000) Probes for detection of specific DNA sequences at the single-molecule level. Anal Chem 72:3717–3724

Kooyman RPH, Kolkman H, Gent JV, Greve J (1988) Surface plasmon resonance immunosensors: sensitivity considerations. Anal Chem Acta 213:35–45

Kopelman R (1988) Fractal reaction kinetics. Science 241:1620–1626

Korotcenkov G, Brinzari V, Schwank J, DiBattista M, Visiliev A (2001) Pecularities of SnO_2 thin film deposition by spray pyrolysis for gas sensor applications. Sens and Actuator B 77:244–252

Liebovitch LS, Sullivan JM (1987) Fractal analysis of a voltage-dependent potassium channel from cultured mouse hippocampal neurons. Biophysics J 52(6):979–988

Liedberg B, Nylander C, Lundström I (1983) Surface plasmon resonance for gas detection and biosensing. Sens and Actuators 4:299–304

Liu XQ, Tao SW, Shen YS (1997) Preparation and characterization of nanocrystalline α-Fe_2O_3 by a sol-gel process. Sens and Actuator B 40:161–165

Martin SJ, Granstaff VE, Frye GC (1991) Effect of surface roughness on the response of thickness-shear mode resonators in liquids. Anal Chem 65:2910–2922

Masel IR (1996) Principles of adsorption and reaction on solid surfaces. John Wiley & Sons, New York, NY

Melikhova EM, Kurochkin IN, Zaitsev SV, Varfolomeev SD (1988) Analysis of ligand-receptor binding by the difference method. Anal Biochem 175(2):507–15

Owen V (1997) Real-time optical immunosensors - a commercial reality. Biosens Bioelectron 12:1–2

Patil PS (1999) Versatility of chemical spray pyrolysis technique. Mater Chem Phys 59:185–198

Peterlinz KA, Georgiadis RM, Herne TM, Tarlov MJ (1997) Observation of hybridization and dehybridization of thiol-tethered DNA using two-color surface plasmon resonance spectroscopy. J Am Chem Soc 119:3401–3402

Ramakrishnan A, Sadana A (2001) A fractal analysis for cellular analyte–receptor binding kinetics; biosensor applications. Automedica, 1–28

Rao GST, Rao DT (1999) Gas sensitivity of ZnO based thick film sensor to NH_3 at room temperature. Sens and Actuator B 55:166–169

Roy SC, Sharma GL, Bhatnagar MC, Samanta SB (2005) Novel ammonia sensing phenomena in sol-gel derived Ba0.5ISr0.5TiO3 thin films. Sens and Actuator B 110:299–303

Sadana A (2001) A kinetic study of analyte–receptor binding and dissociation, and dissociation alone for biosensor application;a fractal analysis. Anal Biochem 291(1):34–47

Shinde VR, Gujar TP, Lokhande CD (2007) LPG sensing properties of ZnO films prepared by spray pyrolysis method. Effect of molarity of precursor solution. Sens and Actuator B 120:551–559

Song C, Zhao M (2009) Real-time monitoring of the activity and kinetics of T4 polynucleotide kinase by a singly-labeled DNA hairpin smart probe coupled with exonuclease cleavage. Anal Chem 81:1383–1388

Wojciechowski BW (1975) Chemical kinetics for chemical engineers. Sterling Swift publishing company, New York, NY

Zhu W, Tan OK, Yan Q, Oh JT (1993) Microstructure and hydrogen gas sensitivity of amorphous BST thin film sensors. Sens and Actuator B 85:205–211

Zwanzig R, Szabo A (1991) Time dependent rate of diffusion-influenced ligand binding to receptors on cell surfaces. Biophys J 60:671–678

Index

B

C

H

N

P

S

Printed in the United States of America